FUSION PROTEIN TECHNOLOGIES FOR BIOPHARMACEUTICALS

FUSION PROTEIN TECHNOLOGIES FOR BIOPHARMACEUTICALS

Applications and Challenges

Edited by

STEFAN R. SCHMIDT

WILEY

Published by John Wiley & Sons, Inc., Hoboken, New Jersey.

Published simultaneously in Canada.

Library of Congress Cataloging-in-Publication Data:

Fusion protein technologies for biopharmaceuticals : applications and challenges / edited by Stefan R. Schmidt.
p. ; cm.
Includes bibliographical references and index.
ISBN 978-0-470-64627-4 (cloth)
I. Schmidt, Stefan R.
[DNLM: 1. Recombinant Fusion Proteins–therapeutic use. 2. Drug Carriers.
3. Immunotoxins–pharmacokinetics. 4. Immunotoxins–therapeutic use. 5. Protein Engineering.
6. Recombinant Fusion Proteins–pharmacokinetics. QU 450]

615.7–dc23

2012039440

10 9 8 7 6 5 4 3 2 1

CONTENTS

PREFACE

This book covers the applications and challenges of fusion proteins; a relatively new class of therapeutic molecules. It is written both for experienced scientists wanting to acquire a deeper understanding of the underlying principles of fusion protein design and for students looking for a comprehensive introduction to this topic. This book shall serve as a source of inspiration by describing a wide range of examples. It gives an overview on the current state of the art and summarizes the huge potential of fusion protein design and functionality. The fascination of fusion proteins is derived from the vast range of possibilities that can be generated by joining individual natural molecules through human creativity. Overall, the rise of fusion proteins is driven by the revolution of genetic engineering, the improved knowledge on protein function, and the unmet medical need of many indications.

One of the many objectives is trying to fulfil the promise of Paul Ehrlich's *Magic Bullet*. This book also picks up the current trend of designing next generation biopharmaceuticals to enable combination therapy by addressing two targets simultaneously. Fusion proteins offer an elegant solution to the challenge of delivering equivalent concentrations of two active molecules to a specific location while keeping manufacturing costs low.

Inspiring the design and therapeutic applications of fusion proteins is the goal of this book. In the first few chapters, the readers are introduced to the concept of fusion proteins while some challenges such as immunogenicity are addressed. The main body contains three large sections on typical paradigms such as strategies to extend the plasma half-life, the design of toxic proteins, and finally utilizing fusion proteins for highly specific targeting. Each of these sections is preceded by an introduction summarizing the respective topic. The last part of the book covers novel concepts including examples of highly relevant multifunctional antibodies. I have tried to maintain a balance between well-established molecules with clinical data and more experimental approaches that nevertheless appear very promising. All chapters are authored by experts in the field who happily share their deep understanding with the readers.

Throughout my professional life I have been studying proteins, starting with proteins whose malfunction causes disease and moving to monoclonal antibodies; always dealing with their isolation and characterization. My particular personal interest in fusion proteins was stimulated years ago when trying to optimize the half-life of an enzyme prototype. By digging deeper into the matter, I came across the many ways how to influence the plasma half-life. However, at this time no textbook about that topic was available, therefore, evaluating the different approaches was quite tedious. In the meantime, some books have been published addressing some isolated aspects that could be of interest to fusion proteins, but none really dedicated to fusion proteins in general. Identifying that gap, I started to collect material to prepare a focussed book and got into contact with several well-known colleagues working on various fusion proteins. Many stimulating discussions shaped the framework of the book. The result of these interactions is this book that covers all aspects of modern therapeutic fusion proteins.

I hope that my readers will enjoy reading this book as much as I enjoyed creating it together with so many talented people.

Finally I would like to thank my wife Gabriele and my son Felix for their encouragement and loving support.

STEFAN R. SCHMIDT

CONTRIBUTORS

Gill Adams, Bone and Joint Research Unit, William Harvey Research Institute, Barts and The London Medical School, London, UK

Ralph F. Alderson, MacroGenics, Inc., Rockville, MD, USA

Wojciech Ardelt, Tamir Biotechnology, Inc., Monmouth Junction, NJ, USA

Peter J. Artymiuk, Department of Molecular Biology and Biotechnology, University of Sheffield, Sheffield, UK

Patrick A. Baeuerle, Amgen Research (Munich) GmbH, Munich, Germany

Bhaswati Barat, MacroGenics, Inc., Rockville, MD, USA

Stefan Barth, Department of Experimental Medicine and Immunotherapy, Institute of Applied Medical Engineering, Helmholtz-Institute for Biomedical Engineering, University Hospital RWTH Aachen, Germany; Department of Pharmaceutical Product Development, Fraunhofer Institute for Molecular Biology and Applied Ecology, Aachen, Germany

Anton Bauer, f-star GmbH, Vienna, Austria

Mark Belsey, WestLB, London, UK

Pedro Berraondo, Division of Hepatology and Gene Therapy, Center for Applied Medical Research, University of Navarra, Pamplona; Liver Unit, University Clinic, CIBER-EHD, Pamplona, Navarra, Spain

Abhijit Bhat, CovX Research, Pfizer Worldwide R & D, San Diego, CA, USA

Hans Kaspar Binz, Molecular Partners AG, Schlieren, Switzerland

Ezio Bonvini, MacroGenics, Inc., Rockville, MD, USA

Chien-Hsing Chang, Immunomedics, Inc., Morris Plains, NJ, USA

Xiaoying Chen, Department of Pharmacokinetics, Dynamics, and Metabolism, Pfizer, San Diego, CA, USA

Yuti Chernajovsky, Bone and Joint Research Unit, William Harvey Research Institute, Barts and The London Medical School, London, UK

Udo Conrad, Leibniz Institute of Plant Genetics and Crop Plant Research (IPK), Gatersleben, Germany

Leslie Cousens, EpiVax Inc., Providence, RI, USA

Leif Dahlberg, Active Biotech AB, Lund, Sweden

Jessica P. Dawson, EMD Serono Research Institute, Billerica, MA, USA

Anne S. De Groot, EpiVax Inc., Providence, RI, USA; Institute for Immunology and Informatics, University of Rhode Island, Providence, RI, USA

Jennifer A. Dumont, Biogen Idec Hemophilia, Weston, MA, USA

Gráinne Dunlevy, Biopharm R & D, Domantis Ltd., Cambridge, UK

Aris N. Economides, Regeneron Pharmaceuticals, Tarrytown, NY, USA

Alan L. Epstein, Department of Pathology, Keck School of Medicine, University of Southern California, Los Angeles, CA, USA

Stefano Evangelista, Menarini Ricerche SpA, Preclinical Development, Florence, Italy

Fuad Fares, Department of Human Biology, Faculty of Natural Sciences, University of Haifa, Mount Carmel, Haifa, Israel

Rewas Fatah, Bone and Joint Research Unit, William Harvey Research Institute, Barts and The London Medical School, London, UK

Eric Ferrandis, IPSEN, Institut Henri Beaufour, Les Ulis, France

Jessica Fioravanti, Division of Hepatology and Gene Therapy, Center for Applied Medical Research, University of Navarra, Pamplona; Liver Unit, University Clinic, CIBER-EHD, Pamplona, Navarra, Spain

Doreen M. Floss, Institute of Biochemistry and Molecular Biology II, Medical Faculty, Heinrich-Heine-University, Düsseldorf, Germany

Göran Forsberg, Active Biotech AB, Lund, Sweden

Arthur Frankel, Cancer Research Institute of Scott and White Memorial Hospital, Texas A&M Health Science Center College of Medicine, Temple, TX, USA

Stefan Gattenlöhner, Institute of Pathology, University Hospital Giessen and Marburg GmbH (UKGM), Giessen, Germany

David M. Goldenberg, Garden State Cancer Center, Center for Molecular Medicine and Immunology, Morris Plains, NJ, USA

David Gould, Bone and Joint Research Unit, William Harvey Research Institute, Barts and The London Medical School, London, UK

Alec W. Gross, EMD Serono Research Institute, Billerica, MA, USA

Namir J. Hassan, Immunocore Ltd, Abingdon, UK

Michael Heartlein, Shire Human Genetic Therapies, Lexington, MA, USA

Gunnar Hedlund, Active Biotech AB, Lund, Sweden

Volkmar Heidecke, Department of Neurosurgery, Klinikum Augsburg, Augsburg, Germany

Henrik Helmfors, Department of Neurochemistry, Stockholm University, Stockholm, Sweden

Gottfried Himmler, f-star GmbH, Vienna, Austria

Björn Hock, Merck Serono, Protein Engineering and Antibody Technologies, Darmstadt, Germany

Peisheng Hu, Department of Pathology, Keck School of Medicine, University of Southern California, Los Angeles, CA, USA

Chichi Huang, Biologics Research, Centocor R&D, A Division of Johnson & Johnson Pharmaceutical Research & Development, Radnor, PA, USA

Bent K. Jakobsen, Immunocore Ltd, Abingdon, UK

Vibha Jawa, Medical Sciences, Amgen, Inc., Thousand Oaks, CA, USA

Syd Johnson, MacroGenics, Inc., Rockville, MD, USA

Serene Josiah, Shire Human Genetic Therapies, Lexington, MA, USA

Kristin Karlsson, Department of Neurochemistry, Stockholm University, Stockholm, Sweden

Dennis Keefe, Shire Human Genetic Therapies, Lexington, MA, USA

Christie Kelton, EMD Serono Research Institute, Billerica, MA, USA

Leslie A. Khawli, Department of Pharmacokinetic and Pharmacodynamic Sciences, Genentech Research and Early Development, Genentech, Inc, San Francisco, CA, USA

Apostolos Koutsokeras, Bone and Joint Research Unit, William Harvey Research Institute, Barts and The London Medical School, London, UK

Snejana Krassova, Biogen Idec Hemophilia, Weston, MA, USA

Ülo Langel, Department of Neurochemistry, Stockholm University, Stockholm, Sweden

Rodney Lappe, CovX Research, Pfizer Worldwide R & D, San Diego, CA, USA

Olivier Laurent, CovX Research, Pfizer Worldwide R & D, San Diego, CA, USA

Hua W. Li, MacroGenics, Inc., Rockville, MD, USA

Nikolai M. Lissin, Immunocore Ltd, Abingdon, UK

Jeffrey A. Loeb, Department of Neurology, Center for Molecular Medicine and Genetics, Wayne State University, Detroit, MI, USA

Susan C. Low, Biogen Idec Hemophilia, Weston, MA, USA

Zhenzhong Ma, Department of Neurology, Center for Molecular Medicine and Genetics, Wayne State University, Detroit, MI, USA

Stefano Manzini, Menarini Ricerche SpA, Preclinical Development, Florence, Italy

Sean D. McKenna, EMD Serono Research Institute, Billerica, MA, USA

Baisong Mei, Biogen Idec Hemophilia, Weston, MA, USA

Hubert J. Metzner, Preclinical Research and Development, CSL Behring GmbH, Marburg, Germany

José Luis Millán, Sanford Children's Health Research Center, Sanford-Burnham Medical Research Institute, La Jolla, CA, USA

Gayatri Mittal, Bone and Joint Research Unit, William Harvey Research Institute, Barts and The London Medical School, London, UK

Paul A. Moore, MacroGenics, Inc., Rockville, MD, USA

Marco Muda, EMD Serono Research Institute, Billerica, MA, USA; Merrimack Pharmaceuticals, Cambridge, MA, USA

Geert C. Mudde, f-star GmbH, Vienna, Austria

Lisa Mullen, Bone and Joint Research Unit, William Harvey Research Institute, Barts and The London Medical School, London, UK

Andrés Muñoz-Alarcón, Department of Neurochemistry, Stockholm University, Stockholm, Sweden

Thore Nederman, Active Biotech AB, Lund, Sweden

Mats Nilsson, Active Biotech AB, Lund, Sweden

Robert T. Peters, Biogen Idec Hemophilia, Weston, MA, USA

Glenn F. Pierce, Biogen Idec Hemophilia, Weston, MA, USA

Sarbendra L. Pradhananga, Department of Human Metabolism, University of Sheffield, Medical School, Sheffield, UK

Jesús Prieto, Division of Hepatology and Gene Therapy, Center for Applied Medical Research, University of Navarra, Pamplona; Liver Unit, University Clinic, CIBER-EHD, Pamplona, Navarra, Spain

Nikolai G. Rainov, Department of Neurosurgery, Klinikum Augsburg, Augsburg, Germany

Tobias Raum, Amgen Research (Munich) GmbH, Munich, Germany

Anne Rigby, Bone and Joint Research Unit, William Harvey Research Institute, Barts and The London Medical School, London, UK

Peter R. Rhode, Altor BioScience Corporation, Miramar, FL, USA

Stefan Rose-John, Institute of Biochemistry, Christian-Albrechts-University, Kiel, Germany

Richard J. Ross, Department of Human Metabolism, University of Sheffield, Medical School, Sheffield, UK

Edmund A. Rossi, IBC Pharmaceuticals, Inc., Morris Plains, NJ, USA

Florian Rüker, Department of Biotechnology, Christian Doppler Laboratory for Antibody Engineering, University of Natural Resources and Life Sciences, Vienna, Austria

Jon R. Sayers, Department of Infection and Immunity, University of Sheffield, Medical School, Sheffield, UK

Jürgen Scheller, Institute of Biochemistry and Molecular Biology II, Medical Faculty, Heinrich-Heine-University, Düsseldorf, Germany

Stefan R. Schmidt, Rentschler Biotechnologie GmbH, Laupheim, Germany

Stefan Schulte, Preclinical Research and Development, CSL Behring GmbH, Marburg, Germany

Michelle Sclanders, Bone and Joint Research Unit, William Harvey Research Institute, Barts and The London Medical School, London, UK

Surinder K. Sharma, Research Department of Oncology, University College London, London, UK

Wei-Chiang Shen, Department of Pharmacology and Pharmaceutical Sciences, University of Southern California School of Pharmacy, Los Angeles, CA, USA

Giles Somers, Datamonitor, London, UK

Neil Stahl, Regeneron Pharmaceuticals, Tarrytown, NY, USA

Anette Sundstedt, Active Biotech AB, Lund, Sweden

Ronald V. Swanson, Biologics Research, Centocore R&D, A Division of Johnson & Johnson Pharmaceutical Research & Development, San Diego, CA, USA

Mikael Tiensuu, Active Biotech AB, Lund, Sweden

Peter Topley, Biopharm R&D, GlaxoSmithKline Medicines Research Centre, Stevenage, UK

Mehmet Kemal Tur, Institute of Pathology, University Hospital Giessen and Marburg GmbH (UKGM), Giessen, Germany

Mark L. Tykocinski, Department of Pathology, Anatomy and Cell Biology, Jefferson Medical College, Thomas Jefferson University, Philadelphia, PA, USA

Sandrine Vessillier, Bone and Joint Research Unit, William Harvey Research Institute, Barts and The London Medical School, London, UK

Adam Walker, Biopharm R & D, Domantis Ltd., Cambridge, UK

Jiajing Wang, Department of Neurology, Center for Molecular Medicine and Genetics, Wayne State University, Detroit, MI, USA

Thomas Weimer, Preclinical Research and Development, CSL Behring GmbH, Marburg, Germany

Ian R. Wilkinson, Department of Human Metabolism, University of Sheffield, Medical School, Sheffield, UK

Maximilian Woisetschläger, f-star GmbH, Vienna, Austria

Jung Hee Woo, Cancer Research Institute of Scott and White Memorial Hospital, Texas A&M Health Science Center College of Medicine, Temple, TX, USA

Gordana Wozniak-Knopp, Department of Biotechnology, Christian Doppler Laboratory for Antibody Engineering, University of Natural Resources and Life Sciences, Vienna, Austria

Jennica Zaro, Department of Pharmacology and Pharmaceutical Sciences, University of Southern California School of Pharmacy, Los Angeles, CA, USA

PART I

INTRODUCTION

1

FUSION PROTEINS: APPLICATIONS AND CHALLENGES

Stefan R. Schmidt

Rentschler Biotechnologie GmbH, Laupheim, Germany

1.1 HISTORY

Proteins as drugs have a long history. In the beginning, natural proteins were extracted from animal, human sources, or in some rare cases even from plants. Large-scale processing of human plasma became a primary source for the isolation of many proteins [1]. For instance, blood factors became available as a therapy against the different forms of hemophilia or the lack of functional (α-1 antitrypsin. The major serum component, albumin, has now been used for more than 50 years as a treatment for shock, trauma, or burns. Immune globulins isolated from human sources are also used successfully in various immunodeficiency diseases. However, despite the great success of plasma products, contaminations with the HIV or hepatitis virus in the 1970–1980s triggered more intensified efforts to prepare virus-free recombinant therapeutic proteins.

Since its identification in the 1920s until the 1980s, insulin from animals was the only treatment for diabetes patients. Particularly, the porcine insulin was widely used since there is only a single amino acid variation from the human form. Chemical processes were developed to obtain the fully human variant from the pig isoform [2]. Finally, the first recombinant human insulin was manufactured by Eli Lilly & Co in partnership with Genentech, approved in 1982 by the FDA and marketed under the name Humulin® [3]. This was also the first therapeutic recombinant protein for human use.

Since that time, the number of recombinant products and approved biopharmaceuticals has increased considerably. Initially, recombinant copies of proteins were made that replaced the natural protein, which until then was harvested from animal or human sources. With the exception of factor VIII against hemophilia, all these proteins such as human growth hormone (hGH) or follicle stimulating hormone (FSH) belonged to the class of hormones. They were soon accompanied by a growing number of first generation therapeutics that could only be obtained recombinantly such as erythropoietin (EPO), interferon (IFN) or tissue plasminogen activator (tPA), just to name a few. After this first enthusiasm and the success with reproducing natural proteins by recombinant DNA technology, researchers started to consider the *de novo* design of therapeutic proteins that do not occur in nature. There is one specific class that can be seen as intermediate link between natural and designed proteins, monoclonal antibodies (mABs).

Antibodies, being a major part of the organism's immune defense, are large proteins that exist in all higher animals. In 1975, a method was developed to generate murine cell lines producing antibody molecules of a single specificity, the so-called monoclonal antibodies (mABs) [4]. The first therapeutic monoclonal antibody, orthoclone OKT3, was of murine origin and approved in 1986. From then, this concept was further refined with the help of modern recombinant DNA technology to obtain the first fully human antibody against tumor necrosis factor-α (TNF-α), marketed under

Fusion Protein Technologies for Biopharmaceuticals: Applications and Challenges, First Edition. Edited by Stefan R. Schmidt.

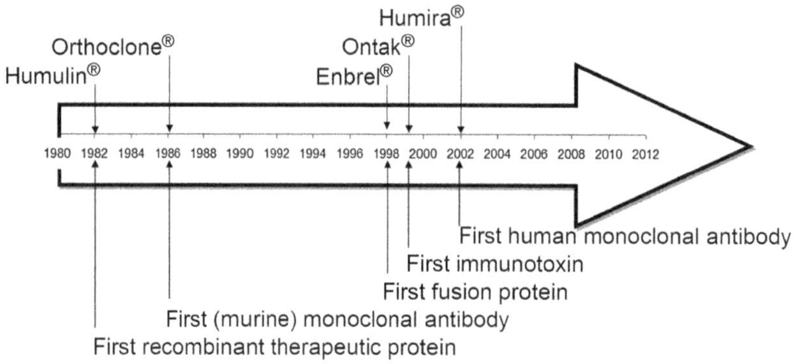

FIGURE 1.1 Milestones of recombinant therapeutic proteins.

the name Humira® in 2002 [5]. The milestones for recombinant therapeutic proteins can be seen in Figure 1.1.

Interestingly, preventing the activity of TNF-α was also the goal for the first fusion protein, Etanercept. It consists of the TNF-α receptor attached to a sequence encoding the Fc portion and hinge region of an IgG1 heavy chain. This drug has been marketed under the name Enbrel® since 1998 and is the best selling fusion protein till date [6]. The two extraordinarily successful drugs Humira and Enbrel can serve as prototype for their molecule classes, exemplifying the different ways to address the same target and present a typical case of competition between antibodies and fusion proteins in the market.

1.2 DEFINITIONS AND CATEGORIES

This book focuses on fusion proteins that are generated by joining two or more genes by genetic engineering that originally code for separate proteins. The result is a single polypeptide with functional properties of both parent proteins. These recombinant proteins are combinations of unrelated domains not occurring in nature. Excluded from the content of this book are multiepitope recombinant vaccines [7], chemical conjugates [8] naturally occurring fusion proteins resulting from chromosomal rearrangements that can be observed in many cancer cells [9] or fusion tags for affinity purification [10].

Bi or multispecific antibodies are special case that do not always represent a single polypeptide chain but usually consist of the combination of heavy and light chains. The Part IIIb of this book discusses some non-natural versions that have more than a single specificity.

The most straightforward classification of these novel proteins can be based on the functions of their incorporated domains. Typically, one part serves molecular recognition or binding, whereas the other part adds certain functionalities such as extending half-life or stability, cytotoxicity, or novel targeting or delivery routes [11].

Recently, a review classified therapeutic proteins according to their pharmacologic activity to (a) replace a deficient or abnormal protein, (b) augment an existing pathway, (c) provide a novel function or activity, (d) interfere with a molecule or organism, or (e) deliver a payload such as a radionuclide, cytotoxic drug, or protein effector [12].

However, this classification is not fully suitable for the scope of this book. Most fusion proteins serve three major purposes that can be summarized under the triple T (T3) paradigm: (a) $t_{1/2}$ (half-life), (b) targeting (or binding), or (c) toxicity (cell killing). Of these three elements, at least two are simultaneously present in fusion proteins (Figure 1.2). Antibodies as natural molecules combine all three aspects in a single molecule. However, antibody derivatives, fragments, or domains have also been used extensively as building blocks for fusion proteins, hence constituting a large part of the portfolio of proteins discussed here, and thus deserving their own category. The main functionality of antibodies, the binding with high affinity and selectivity to a specific epitope, has been reproduced in a number of nonantibody scaffolds that can either be used as single module or by combining two units with different specificity [13]. These molecules together with other bi- or multifunctional therapeutics that do not fit to the T3 categories are classified into the group of novel artificial molecules that is discussed in Part IIIa of this book.

A very practical classification is proposed by the authors of Chapter 3 about "Structural Aspects of Fusion Proteins Determining the Level of Commercial Success" in this book. They suggest sorting fusion proteins based on one of three different functional groups: activity, targeting, or half-life, of which the latter two can be summarized under delivery agents, thus being able to define a two-dimensional landscape of fusion proteins. This is quite similar to a previous scheme using the combination of an effector fragment together with a molecular recognition part as building blocks for fusion proteins [14].

But why should we deal at all with fusion proteins? Several advantages make them very attractive: the combination of two functionalities in a single molecular entity

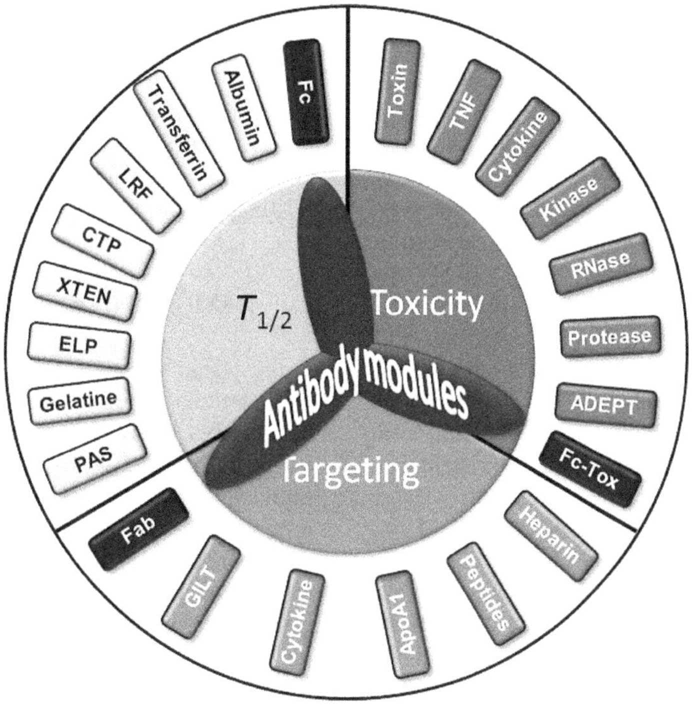

FIGURE 1.2 The triple T paradigm of fusion proteins. Most fusion proteins are composed of mixtures of these modules to combine two functionalities. The multifunctional natural prototype is an antibody combining the constant Fc part that can contribute to half-life extension and toxicity. The Fab part is contained in many fusion proteins for targeting and binding purposes. LRF, ligand receptor fusion; CTP, C-terminal peptide; XTEN, XTENylation; ELP, elastin-like peptide; PAS, PASylation; GILT, glycosylation independent lysosomal targeting; ApoA1, apolipoprotein A1; ADEPT, antibody-directed enzyme prodrug therapy; TNF, tumor necrosis factor family.

simplifies manufacture and drug delivery. Two molecules combined into one will automatically have identical biodistribution profiles instead of two separate molecules that might have a very different distribution. Furthermore, new functionalities can be created that are lacking in natural or separate proteins. This includes the modification of half-life or targeting specificity. Even economic opportunities such as life cycle extension of products with expired patents are possible. This includes also the generation of novel intellectual property for new and non-natural combinations of proteins. Therapeutic benefits derived from reduced side effects or longer dosing intervals and improved activity are strong drivers to promote the generation of fusion proteins.

But besides all these important advantages, there are also a number of challenges. The combination of unrelated proteins might prove difficult to manufacture because in some cases, the fusion partners have noncompatible properties. This can cause aggregation or misfolding of one domain while the conditions might be perfect for the other domain. Despite the fact that some modules of fusion proteins are elements of other well-proven molecules such as antibodies, the established platform processes might not be applicable because other features shield the required property. This can go so far that formulation is not possible due to conflicting stability requirements. Furthermore, it will be difficult to control and tune the relative amounts of each component thus complicating dosing for optimal efficacy and safety. Probably most important challenge is the high potential for immunogenicity due to the formation of novel epitopes at the junction between the fusion partners even if only fully human proteins are connected.

1.3 PATENTING

The first generation biologics that represented a true copy of human proteins used for therapeutic applications have already lost or are about to lose their patent protection [15]. In many cases second generation molecules, for example, with improved half-life, are taking their place. A number of them are fusion proteins that are patented as well.

To be able to file a patent for an invention, three characteristics must be achieved novelty, nonobviousness, and utility or enablement [16]. In the postgenome era, the discovery of novel proteins, at least of human origin, will be difficult. This challenges the first critical parameter on the way to a patent, the novelty. If we focus on the scope of this book, the fusion proteins, novelty still seems to be easy to reach. As described in the paragraph about definitions, "joining two or more genes by genetic engineering that originally code for separate proteins," so the generated fusion protein will be novel if nobody did the same earlier. Therefore, novelty is given on one hand through the composition of matter (the new construct, e.g., long-acting human growth hormone [hGH], consisting of hGH fused to human serum albumin [HSA]) or on the other hand by the use that results from combined features of the new molecule (e.g., to treat dwarfism with longer administration intervals). This is in contrast to a natural polypeptide with multiple functions [17].

Taking the example of antibodies as molecules, their individual characteristics, for example, target or captured epitope, affinity, half-life, or sequence of the variable part should be sufficient to enable patenting based on novelty [18]. Surprisingly, exactly these descriptive features are missing in many antibody patents, which is an effect of patenting human genes that often include hypothetical antibodies to that target in the claims [19]. It might be easier trying to patent antibodies based on unexpected advantages, that is, having lower cross-reactivity since higher affinity could be regarded as obvious [20]. This should be taken into account for patenting antibody derived fusion proteins. Particularly, tumor-specific antibodies have had only limited success so far. This led to new intellectual property of improved molecules by coupling antibodies to toxic proteins, thus combining target specificity with nonspecific cytotoxicity [21].

However, if we continue with the second parameter, the nonobviousness, it starts to get more difficult. In the context of fusion proteins in many cases, the existence or functionality of the potential fusion partners will be well known; thus representing prior art. Interestingly, when combining several "prior arts" into a new concept, this patent will only be rejected as obvious if there existed at the time of invention a known problem for which there was an obvious solution encompassed by the patent's claims [22]. The best argument to get a combination patent is to demonstrate that prior art is not providing a motivation or suggestion to prepare this combination. Even when sufficient suggestions for combinations can be found in prior art, patenting can still be possible if enabling in form of a production method can be claimed additionally. Therefore, when defining patent claims, it can be very beneficial to include methods how to manufacture the protein of interest and what formulations are useful [23].

The third parameter to demonstrate utility is relatively straightforward since the fusion protein was designed with a specific application in mind. Here, both typical aspects of utility, recognition of a benefit and the motivation to make a change to current practice, come into play. Overall it has to be described how the invention can be put into practice, which is ideally done in the form of examples.

In the past, some patent disputes ranked around protected fusion protein technology. For instance, Zymogenetics accused Bristol-Myers Squibb (BMS) to infringe their Fc-fusion technology with Orencia®. Initially, the case was settled by a lump sum payment, but finally BMS acquired Zymogenetics in 2010 together with the rights for Fc fusions [24].

1.4 DESIGN AND ENGINEERING

In the design of a novel fusion protein a number of parameters have to be taken into account. Following questions need to be addressed: Will the proteins be functional on either the N-terminus or the C-terminus? In which orientation will the individual proteins be connected? What linker length and sequence should be used? Is there a need for a specific oligomerization? Are mutations or truncations required to enhance or eliminate certain features? In many cases rational design will guide the generation of innovative protein therapeutics [25]. One of the aims of protein design is certainly to improve the functionality of biological drug (Figure 1.3). A key question in this context is: Will the protein reach the target in a significant quantity? Taking the example of a solid tumor it becomes clear that in order to reach the tumor the protein has to be sufficiently small to penetrate the many cell layers. But on the other hand it should not be too small to be excreted too fast. This requires a delicate balancing of the molecule size and has been demonstrated experimentally with a number of different antibody derivatives [26]. Not only size but also valency of antibodies can be modified. Further details on bi-specific and multifunctional antibodies can be found in Part IIIb of this book.

1.4.1 Orientation of Fusion Proteins

Looking at the largest group, the Fc-fusion proteins, it is well known that naturally the Fc part is positioned at the C-terminus of an antibody. But in artificial fusions it can also form the N-terminal part. Several studies have evaluated cytokine mono- or tandem fusions to full antibodies or Fc parts. Here no dependency on the selection of the respective terminus could be observed [27]. However, comparing N- or C-terminal Fc fusions of peptides blocking angiopoietin-2 (Ang-2), it was found that the N-terminal fusion had shorter half-life and weaker binding but better selectivity [28].

Generally, when combining two proteins, there are two orientations in which the fusion partners can be arranged, either at the amino or at the carboxy terminus of the first protein. In many cases the position is without influence on the functional properties, for instance albumin (HSA) can be fused to either end [29]. But during the manufacture of Albuferon®, a combination of interferon-α2b (IFN-α2b) with albumin, it was observed that due to incomplete

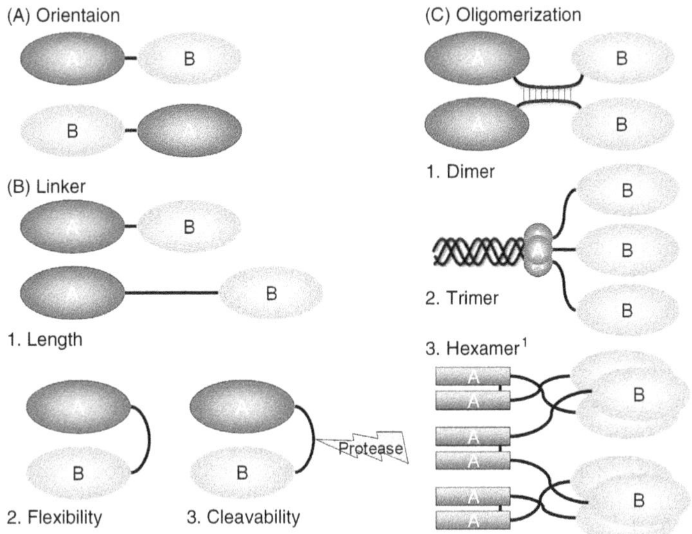

FIGURE 1.3 Fusion protein design. The typical variations of fusion proteins comprise orientation, linkers, and oligomerization. Sometimes proteins require a free N- or C-terminus for their activity. The linker can vary in length, flexibility, and susceptibility to protease cleavage. Dimeric forms of fusion proteins can be generated by including leucine zippers in the linker region. Two examples of higher oligomerization are depicted as well; first, a trimer built from the trimerization domain of human endostatin and second, a hexamer constructed through the combination of three Fc dimers connected via an ILZ coiled coil to two trimers ([1]see Morris et al. [51]).

disulfide bridge formation, aggregates formed that drastically reduced the recovery. Interestingly, this phenomenon could be completely abolished by positioning the IFN-α2b on the N-terminus instead [30]. In another case the order of two angiostatic proteins, human angiostatin (hAS) and endostatin (hES), combined in a fusion protein had to be in a specific orientation to obtain maximum activity. The fusion hES–hAS was 28% more potent than hAS-hES or 7% better than hAS and hES when administered separately [31]. Sometimes a free N-terminus is required. The second partner is then connected at the C-terminus as in the case of receptor traps such as Etanercept. Here, the extracellular receptor domain is combined with an Fc part at the C-terminus to maintain the natural conformation [6]. Another striking example for the positioning effect of fusion partners is the case of elastin-like peptides (ELP). It was observed that C-terminal fusions of ELP resulted in a higher expression level, better yield, and bioactivity. The underlying reason could be increased misfolding, induced by ELPs at the N-terminus; thus reducing the amount of active proteins and increasing their susceptibility to proteolysis [32]. Orientation has a high impact on functionality particularly when fusion proteins contain enzymes that require either a free N- or C-terminus. This has been demonstrated with the Immuno-RNAse consisting of angiogenin (ANG) and a single-chain variable domain (scFv) against CD22. Only constructs in the scFv-ANG orientation did not aggregate and were fully functional [33].

1.4.2 Linker Engineering

Again starting with Fc fusions as example, the hinge region fulfils the function of a linker, allowing some spatial flexibility [34]. Besides this exception where a part of a fusion protein ends with a flexible peptide chain, in most cases specific linkers between protein molecules have to be artificially introduced. The multiple aspects of linker design have recently been reviewed [35]. Many researchers use a simple glycine and serine (G_4S)-containing linkers as proposed by a large study of natural domain separating linkers [36]. Spacer peptides that connect both modules of a fusion protein in a spatial conformation are frequently needed to maintain functionality. For instance, the highest potency could be observed when a spacer was introduced between a single chain variable domain (scFv) and ANG [37]. Since fusion proteins ideally consist only of a single polypeptide chain, a reformatting of Fab fragments to scFv is required. This is done with the help of a linker sequence that frequently consists of repeats of glycine and serine, as for example the popular $(G_4S)_3$ linker. The basis of linker design is the rational engineering of both length and conformation. A controlled distance between domains can be achieved by defined repeats of α-helical peptides $A(EAAAK)_nA$ that maintains the separation of domains in contrast to flexible linkers [38].

When evaluating the optimal linker between IFN-α2b and HSA it could be demonstrated that five amino acids (aa) are already sufficient. However, maximum activity was obtained with a helical 12 aa linker which was 1.7 times or 2.9 times better than a short rigid or the standard five aa long G_4S linker [39]. The knowledge about linkers was continuously expanded [40]. Recently, a database collecting natural linker sequences and their properties could be established.[1] Inter domain linkers have to be differentiated from intradomain loops based on their function, but also because of their amino acid preference. Interestingly proline is the most frequent amino acid in both cases. Proline destroys α-helix and β-sheet structures as well and cannot form hydrogen bonds [41].

But linker peptides can also be engineered to contain additional functionalities. To optimise the pharmacokinetic profile of a granulocyte-colony-stimulating factor (G-CSF) fused to transferrin, a linker with an intramolecular disulfide bridge was introduced. The peptide sequence was thrombin sensitive, which allowed an *in vitro* cleavage by adding this protease, leaving the two domains only connect through the disulfide bridge. This labile bond could then easily be cleaved *in vivo*, releasing free G-CSF that was more active than the fused molecule [42]. Previously, a similar molecule was engineered to improve oral efficacy. Here, a long rigid helical linker composed of 50 aa between G-CSF and transferrin led to a 10 times lower EC_{50} than the initial fusion protein with a two aa spacer [43]. This particular linker also increased the expression level more than 10-fold compared to the direct fusion with only two aa in between. However, this time the construct was in an opposite orientation having transferrin at the N- and G-CSF at the C-terminus. Interestingly an unstructured 50 aa linker inhibited expression totally [44].

The orientation of domains and the impact of spacers were also studied when a scFv against human transferrin receptor was combined with the fungal ribonucleolytic toxin restrictocin. Independent of the orientation in both cases, the introduction of a protease-sensitive linker drastically improved the EC_{50} 2- to 30-fold [45]. The influence of linkers on fusion proteins is discussed in Chapter 4.

1.4.3 Oligomerization of Fusion Proteins

The next level of design involves the correct oligomerization of the protein. In many cases, proteins are dependent on multimerization or have a higher bioactivity as a multimer. One example is the combination of stem cell factor (SCF) and macrophage-colony-stimulating factor (M-CSF) with a 12 aa flexible peptide linker. The fusion protein forms dimers that have a 10- to 20-fold higher potency than the individual monomeric proteins, also benefitting from a synergistic effect [46]. A higher cytoxicity could also be demonstrated for divalent antibody toxin fusions. Combining a Fab fragment with two molecules of a truncated *Pseudomonas* exotoxin A resulted in an almost 40-fold more active fusion protein than with only one toxin molecule [47]. Oligomerization can also help to improve valency of recombinant proteins. Sometimes it is required to induce cross-linking of receptors to execute specific functions such as activation or internalization. It has been demonstrated that by the right choice of linkers between V_H and V_L in scFv, aggregates can spontaneously form having a higher avidity [48].

Using an Fc part automatically delivers dimerization, however there are also different approaches possible to enforce dimers. Frequently leucine zippers are used for that purpose. An early example is the generation of bivalent scFv antibodies with Fos or Jun leucine zippers. A covalent bond could be introduced by positioning cysteine in proximity to the zipper. These molecules spontaneously formed dimers when secreted to *Escherichia coli* periplasm [49]. The leucine zipper GCN4 of *Saccharomyces cerevisiae* has been used to dimerize the soluble insulin receptor. This resulted in an improved binding constant, very similar to the original membrane bound native receptor [50].

As in the case of OX40 ligand (OX40L) even a trimerization can be required for full functionality. Here an active construct in the right conformation could be obtained by fusing OX40L to a GCN4 zipper domain connected to an Fc part. The final hexameric molecule consisted of three Fc-induced disulfide linked dimers that allowed the formation of two trimers [51]. Another molecule having a trimer as active natural conformation is tumor necrosis factor-related apoptosis-inducing ligand (TRAIL). As before, the GCN4 zipper-induced self-assembled trimers of TRAIL based on three-stranded coiled coils [52]. However, TRAIL can also be trimerized as linear fusion in a head to tail configuration interrupted by short flexible peptide loops. To improve targeting, even a scFv could be included at the N-terminus [53]. Trimerization through tandem repeats in a single chain molecule could also be shown with TNF [54]. CD95 ligand (CD95L) as another member of the TNF homology domain (THD) family could be forced into covalent trimerization by fusion to a tenascin-C (TNC) oligomerization domain, again resulting in improved bioactivity [55]. Higher oligomerization of multiple trimers was tested with the CD40 ligand (CD40L). First a di-trimer conformation was achieved by incorporating the N-terminal part of adiponectin (Acrp30). A tetra-trimer construct could be obtained by combining the N-terminal part of mouse surfactant protein-D (SP-D) with the extracellular part of CD40L. Additional leucine zippers were not required because both Acrp30 and SP-D spontaneously self-assemble into the desired multimers. These constructs have been successfully used as adjuvant in DNA vaccines [56]. A similar approach was also initiated with soluble Fas ligand (FasL). The bioactivity of a single FasL trimer was much lower than di-trimer (hexameric) constructs that either relied on the oligomerization effects of

[1] http://www.ibi.vu.nl/programs/linkerdbwww/

Acrp30 or Fc [57]. Other naturally occurring trimer inducing molecules are the carboxyl-terminal noncollagenous domains (NC1 domains) of collagens XV and XVIII coding for endostatin. This was utilized to generate a multi-functional anti-angiogenic compound consisting of a scFv and a NC1 domain. Cleavage of the NC1 domain by tumor-associated proteases released endostatin from the scFv trimers, thus multiplying the anti-angiogenic effect [58]. Collagen trimers were also used to increase the binding strength of scFv 20- to1000-fold though multimerization. It was shown that this collagen-like scaffold $(Gly-Pro-Pro)_{10}$ could either be fused to the N- or C-terminus. Depending on the addition of cysteins flanking the scaffold even a hexameric configuration could be obtained [59]. The trivalent human plasma protein tetranectin builds the core of a novel scaffold, the Atrimer^{TM}. So far Atrimers have not been used as basis for fusion proteins, but their multiple loops have been engineered for diverse binding applications [60].

1.4.4 Immunogenicity

Rational design of fusion proteins must also be used in order to minimize immunogenicity of the new construct. Even if using only human proteins as starting point for a modular assembly, still the region between both molecules represent a novel epitope that could elicit an immune response. Therefore, it is recommended to use at least *in silico* analysis to predict T- and B-cell epitopes. A lot of work on the removal of T-cell epitopes or de-immunization of fusion proteins has been conducted on immune toxins. In one recent example, a truncated version of Pseudomonas exotoxin A (PE38) fused to an anti CD22 scFv was scanned for the presence of B-cell epitopes with a huge set of antibodies. The identified major epitopes were neutralized by specific mutations [61]. Furthermore, two lysosomal protease-sensitive areas were eliminated from PE38. The resulting mutant had a drastically reduced immunogenicity and an even improved cytotoxicity [62].

A very special case is the antibody-directed enzyme prodrug therapy (ADEPT) that utilizes nonhuman enzymes to metabolize inactive prodrugs into highly potent toxins. One of the promising enzymes, a beta-lactamase had to be mutagenized to remove $CD4^{+}$T-cell epitopes. Site-directed mutagenesis replacing individual aa lowered the T-cell response fivefold [63].

In general, Fc-fusion proteins should also display lower immunogenicity. This effect can be based on the presence of inhibitory Fc receptors on B lymphocytes, the FcγRIIb. This hypothesis was proven with the injection of DNA coding for a Exendin-4 Fc-fusion protein that did not result in the generation of neutralizing antibodies, which was the case for Exendin-4 alone [64].

A number of factors determine the immunogenicity potential of proteins in general. For example, protein aggregates might cross-link B-cell receptors, or increase protein internalization of antigen-presenting cells (APC), thus initiating an immune response. However, endocytosis can also happen if the target of a fusion protein is up taken. Furthermore, nonhuman modification (e.g., glcosylation) will also induce an immune reaction. Many of these reactions can be predicted *in silico* or analyzed *in vitro* with a wide range of assays [65]. Since both B- and T-cell epitopes contribute to immunogenicity, it is very advisable to identify and remove these epitopes. Due to the high polymorphism of the major histocompatibility complex (MHC) it can be difficult to remove all T-cell epitopes. B-cell epitopes are not restricted to MHC molecules; therefore, it might be easier to eliminate B-cell epitopes [66]. More details on the immunogenicity of fusion proteins can be seen in Chapter 5 of this book.

1.4.5 Mutagenesis for Molecule Optimization

A lot of work on the optimization of fusion proteins by side-directed mutagenesis is focused toward improving parameters beyond immunogenicity. An important factor is the resistance against proteases, on one hand during manufacture and on the other hand while circulating through the patient's organism. In both cases, it must be distinguished between exo- and endopeptidases. For exopeptidases that can cleave their target from either end, usually a modification of the terminal aa abolishes the degradation. For instance, the extension of a glucagon-like peptide-1 (GLP-1) fusion protein by only a single amino acid at the N-terminal prevented cleavage by dipeptidyl peptidase-IV (DPP-IV) [2]. The circulation half-life of an immunotoxin could be doubled when replacing an endoprotease-sensitive arginine residue by serine or lysine [67]. However, proteolytic processing during the endosome/lysosome trafficking is an important step in the mechanism of action of immunotoxins. Therefore, engineering of cleavage recognition sequences must be done very carefully. Recently, it was demonstrated that a deletion of a protease-sensitive region of PE38 abolishes lysosomal degradation while maintaining efficacy and increasing tolerance of high doses [68]. A combination of increased half-life and improved activity induced by single amino acid changes could also be observed with an interleukin-2 (IL-2) immunocytokine. Here, the linker peptide between the C-terminus of the antibody heavy chain and the N-terminus of IL-2 was modified, primarily with the aim to remove protease cleavage sites [69]. The nonselective toxicity of immunotoxins can be avoided by generating in-frame fusions with ubiquitin that triggers rapid degradation. However, the insertion of a cancer protease-specific cleavage sequence can stabilize the immunotoxin by removal of the ubiquitin moiety. For instance, a saporin-based immunotoxin containing a prostate-specific antigen (PSA) recognition sequence had a

10-fold higher activity in the presence of PSA-producing prostate cancer cells [70].

Other mutagenesis approaches are directed to the presence of disulfide bridges. Cysteins can be added, removed, or repositioned. In many cases, the introduction of disulfide bonds improves stability. Heterodimers between V_H and V_L of antibody fragments are relatively unstable but can be stabilized by adding a cystein in each of the fragments to generate a disulfide bridge. Although there are two possible positions, only one locus retained the full binding activity [71]. Even the rearrangement of existing disulfide bridges is possible as demonstrated with erythropoietin (EPO) where a cystein was moved from position 33 to 88. This modified EPO in fusion to an Fc part exhibited superior dimerization capabilities, better glycosylation stability and improved pharmacokinetic properties [72]. Stabilization by introducing disulfide bridges can also suppress immunogenicity as shown with an immunotoxin that was mutagenized in the domain III of Pseudomonas exotoxin A [73]. However in some cases the presence of an unpaired cystein can cause aggregation problems. A fusion protein between IFN-α2b and HSA (IFN-α2b-HSA) aggregated and caused immunogenicity issues. Only replacing the free cystein by serine abolished the effect, leading to a more stable and less immunogenic protein [74]. Sometimes disulfide-induced dimerization has multiple effects. The Fab-PE38 dimer with PE38 fused to the light chain and linked through a disulfide bridge in the hinge region had a 16-fold higher refolding yield and 2.5-fold better activity than the initial monomer [75].

Not only free cysteins can cause aggregation, much more frequent is the exposure of hydrophobic residues at protein surfaces that leads to aggregation. In a systematic study, these critical positions and their respective ideal aa of scFvs were identified. Two positions on the V_H (82, 85) and also on the V_L (36, 60) were found, whose replacement with more ideal aa resulted in a significant improvement of stability and yield [76]. A similar approach with another scFv revealed three other heavy chain and one light chain residue that contained not conserved aa. Triple V_H mutants resulted in eightfold higher yields that could be increased to 20-fold, when the orientation was reversed. Interestingly, the mutations did not have a negative effect on binding affinity, but improved plasma stability [77]. Aggregation can also occur by the lack of glycosylation when proteins are expressed in *E. coli*. In the case of erythropoietin (EPO) this was counteracted by replacing asparagine residues with lysine. Due to the increased isoelectrical point (*p*I) the protein became positively charged under physiological conditions thus eliminating aggregation [78].

Mutagenesis can also have an effect on the activity profile of proteins. For instance, the replacement of the plasminogen activator inhibitor-1 binding site in Tenecteplase® by exchanging four aa resulted in a significantly longer activity, because the inhibitor could no longer bind [79].

1.5 MANUFACTURING

The first recombinant protein product, insulin, was initially produced as two separate chains that were conjugated chemically. But soon thereafter the commercial process utilized expression in *E. coli* and subsequent enzymatic maturation to generate insulin from proinsulin [80]. Since that time more than 120 therapeutic proteins, including a number of fusion proteins, have been manufactured for human use in bacteria, yeasts, or animal cells [81]. From a manufacturing perspective in most cases it is not necessary to differentiate between fusion proteins or regular singular therapeutic proteins. The production typically covers three complex steps: upstream processing (molecular biology and fermentation), downstream processing (capture and purification), and finally formulation (transforming the protein into a storable and administrable form). One of the many advantages of fusion proteins is the uninterrupted manufacturing process of a single protein molecule with several functions.

1.5.1 Upstream Process

The choice of the expression system depends heavily on the properties of the desired protein product such as glycosylation, disulfide bridges or other post-translational modifications that can only be obtained from eukaryotic cells [82]. Also protein size plays a role; usually, proteins larger than 100 kDa are by default produced in eukaryotic cells, whereas proteins below 30 kDa are expressed in bacteria. A recent analysis revealed that 39% of recombinant proteins were expressed in *E. coli*, 35% by Chinese hamster ovary (CHO) cells, 15% by yeasts, and 10% by other mammalian cells, but only 1% by other systems [83]. Interestingly, 17 of the 58 approved therapeutic proteins between 2006 and 2010 were manufactured in *E. coli* [84]. The following paragraph focuses on the two major host organisms, *E. coli* and CHO cells, while *Pichia pastoris* has been primarily used for albumin and transferrin fusion proteins [85]. Secreted Fc-fusion proteins in *P. pastoris* require individual optimization of upstream conditions [86].

The modern upstream process consists of three elements: the expression construct, the host cell, and the cultivation conditions. The expression construct in microbial expression is mostly a plasmid vector coding for a resistance gene as selection marker, the gene for the protein to be expressed, both accompanied by the respective promoter sequences, and a replication origin to maintain the plasmid during cell division [87]. It is generally recommended to optimize the codon usage according to the selection of the host organism. Usually, microbial cells are transfected and kept as clonal glycerol stock to serve as master cell bank. Alternatively, the expression cassette can also be integrated into the host genome to create a stable transfected cell. This is the preferred option for mammalian cell culture. The integration can occur at multiple sites leading to gene duplication and hence higher expression levels [88]. In microbial cells, a similar effect can be achieved with high copy number plasmids. Molecular biology strategies to optimize recombinant protein expression can target transcriptional and translational regulation. On transcriptional level, promoters and terminators play key roles. For the translation reaction, the binding of ribosomes to mRNA during the initiation and their release at termination are very important as well as the stability of the mRNA in the cytoplasm. Folding of the nascent polypeptide chain can be enhanced by co-expression of chaperones [89]. Further optimization can focus at the selective knock out of proteases that might otherwise destroy the protein product [87].

When working with *E. coli*, we should not forget that post-translational modifications are missing and in some cases the amino terminus might contain an extra methionine residue that is not properly removed during translation [90].

The fermentation of *E. coli* cells depends on media composition (e.g., carbon source, antibiotic selection) and cultivation conditions (temperature and oxygen content). One drawback of high-density culture is the accumulation of acetate that limits proliferation. This problem can be solved by a number of different approaches such as introducing the foreign enzyme acetolactate synthase into *E. coli* that generates a less toxic by-product [91]. Alternatively, glucose uptake can be limited by knocking out the *pts*G gene or the whole phosphotransferase system (PTS) [92].

The attractiveness of *E. coli* is based on the fast, cheap, and simple expression of proteins that can be targeted to the cytoplasm, the periplasm or the cell culture supernatant by secretion. But often intracellular expression of eukaryotic proteins in bacteria results in inclusion bodies (IB) that require refolding approaches during downstream processing. The formation of IBs is mostly caused by an overload of the folding machinery in the bacterial cell that leads to exposure of hydrophobic residues triggering aggregation. In addition, *E. coli* lacks the ability to form disulfide bridges in its reducing environment in the cytoplasm [93].

Particularly, when combining protein toxins and a targeting moiety, the expression level might be significantly lowered due to the translational inhibition by the toxin. One approach to circumvent this difficulty is to express the toxin fusion in form of IB that cannot harm the host cell because of its unfolded, nonactive state [94]. Also secretion to the culture supernatant to keep the toxin away from its target, the ribosomes, has been accomplished [95]. A third variant of producing toxin fusions is the expression in the presence of an inhibitor that is removed during the downstream processing [96]. Another possibility is to express both parts of a fusion protein separately, then fusing them in a later trans-splicing step with the help of split-inteins [97]. Recently, it was shown that an immunotoxin can be successfully manufactured with high density bacterial fermentation under GMP conditions at a yield of 40% and 97% purity [98].

CHO cells came into the focus of manufacturing after Genentech's market approval of Activase®, the recombinant tissue plasminogen activator (tPA), in 1987. Currently, CHO cells are the industry standard for glycosylated complex proteins and benefit from three accomplishments: serum-free production is possible and easy, cell engineering achieves high titers, and high-density large-scale fed-batch cultivation is well established. Cell engineering addresses the parameters expression level, duplication time and stability of the cell, control of proliferation and viability, reduction of toxic metabolites, increasing secretion capacity, and modulation of post-translational modifications [99].

The process starts with expression vector design. Strong viral promoters, the elimination of cryptic splice sites and the increase of G/C content by codon optimization are tools to improve expression levels. A further trick is the insertion of an intron to enhance mRNA export and stability by splicing. The gene silencing effect at the genomic insertion site can be avoided by flanking the coding sequence with DNA elements that can block the formation of heterochromatin [100].

Usually, DNA constructs are integrated randomly into the genome by homologous recombination. Besides the gene of interest, the vector also transfers a selection markers; either dihydrofolate reductase (DHFR) or glutamine synthesis (GS). Cultivating the cells in a medium without the respective metabolites lets only transformed cells survive. Furthermore, genes become amplified when cells are exposed to increasing concentrations of the inhibitor methotrexate (MTX) and methionine sulphoximide (MSX) for DHFR and GS, respectively. High producers with good duplication rates are cloned and a master cell bank is created [101].

The growth and productivity of cells is mainly dependent on the metabolism. Therefore, attempts have been undertaken to reduce waste metabolites, thus making the carbon metabolism more efficient. Other strategies involve extending the cell's lifespan by activating anti-apoptosis genes or engineering cell cycle control. Further improvements were achieved by addressing the secretion apparatus and over expressing chaperones [102]. Mutant CHO cell lines can also be useful for the expression of toxic compounds. Usually, immunotoxins are produced as IB in *E. coli* to prevent cell death. Good secretion levels of a scFv-diptheria toxin fusion could be obtained in an ADP-ribosylation insensitive CHO mutant. However, the glycosylation decreased toxicity. Eliminating the glycosylation sites by mutagenesis improved toxicity 12-fold over the identical molecule prepared from *E. coli* periplasm [103].

In serum-free media nonanimal protein (Sericin) and nonprotein (phosphatidic acid) substitutes enable cultivation. Previously, the most important serum components were albumin, transferrin, and insulin [104]. Proteins can aggregate even when secreted from CHO cells. This should be minimized already in the upstream process to reduce losses in the downstream and to improve formulation. Adding dexamethasone at nanomolar concentrations increases glutathione reductase levels, but not protein disulfide-isomerase. Thus, a balanced redox condition is established in the cells, effectively reducing aggregate formation as demonstrated on the example of an IgG-fusion protein [105].

In addition to cell engineering, the drastic 100-fold increase of productivity within 20 years can mainly be attributed to better understanding of cultivation conditions. This includes media composition, nutrient content, pH, temperature, the addition of histone deacetylase inhibitors such as sodium butyrate or valproic acid, a better dissipation of dissolved oxygen by reactor design, and metabolite triggered feed strategies [106].

1.5.2 Downstream Process

The downstream process starts with harvesting the product of interest. For secreted proteins that are mainly produced by eukaryotic cells (animal cells or yeast), there are two alternative procedures; either centrifugation to sediment the cell mass, or filtration to obtain a cell-free supernatant. Secreted expression is preferable because in state-of-the-art serum-free medium only few contaminants besides the protein of interest are present. A combined unit operation such as expanded bed absorption (EBA) connects cell removal and first product capture in a single step. The protein of interest retained in the cell-free solution is then separated from other contaminants such as other proteins, DNA, or viruses by a series of chromatographic polishing steps. Finally, the isolated protein is pure and can be transferred to the subsequent formulation step [107].

If the protein is not secreted, a product release step, where the microbial cell is homogenized by physical or chemical means, must be included. Most of the employed methods suffer from a rather high unspecific homogenization. This means the product of interest is released together with many contaminants that have to be removed in laborious downstream procedures. A selective product release step minimizes contaminants, increases the adsorption capacity of chromatography and reduces viscosity. Important parameters to consider for the optimal disruption process are particle size and density, recovery rate, viscosity, processing time, and scalability. In general, mechanical (e.g., high pressure, cavitation), chemical (e.g., osmotic shock, buffer conditions), and genetic (e.g., induced cell lysis) methods can be distinguished. Ideally, the product is accumulated in a subcellular compartment or the periplasmatic space that simplifies specific release [108].

A special case for intracellular accumulation is the formation of IB. In many cases, IBs can represent 10–50% of the total cell protein containing up to 95% of a single protein species [109]. Due to their high density IBs can easily be isolated by sedimentation and often allow an enrichment of approximately 90%. Functional proteins from IB require two steps; solubilization under denaturing and reducing conditions, then refolding by removal of denaturants and reformation of disulfide bridges [110].

In some cases, the fusion partner helps by enabling an elegant downstream processing route. One striking example is the utilization of elastin-like peptides (ELP) that facilitate the purification by temperature dependent aggregation that allows simple capturing by sedimentation [111]. A typical antibody-based production, standardized for Fc-based purification, achieves an expression level of 2–5 g/L and an overall yield of 70–80%. Unfortunately, nonantibody Fc fusions do not always achieve such high titers and often they bind less efficiently to Protein A resins [112]. But despite of these deviations, most of the platform solutions derived from antibody processing are also applicable for Fc-fusion proteins. Protein A and its derivatives are the most frequently used capture ligand for Fc-based proteins on chromatography columns. The elution at low pH also inactivates potential viral contaminations if it is below pH 3.7. But an acidic pH might destabilize some Fc-fusion proteins and cause aggregation. This aggregation can be overcome by simultaneously adding a chaotropic agent during elution [113]. Furthermore, ion exchange chromatography step(s) and a final virus filtration complete the standard Fc-dependent purification protocol [114].

If Fab fragments or scFv are fusion partners, also dedicated affinity matrices such as the kappa light chain specific Protein L [115], FabSorbent® [116] or other novel synthetic molecules [117] can be used. Protein L was also successfully applied to purify trispecifc antibodies lacking an Fc part [118]. Even for albumin fusions a platform technology, Albupure™ was created [119]. Especially for unconventional fusion proteins such as Amediplase™ customized affinity matrices have been designed that enable a scalable manufacturing process [120].

One of the challenges of contemporary downstream processes is the constantly increasing protein titer resulting from upstream improvements. Sometimes the column capacity in affinity chromatography is not sufficient to capture these high protein concentrations. Therefore, ion exchange chromatography has been used as alternative. This might also be generally applicable for fusion proteins lacking domains for which platform technologies exist.

As alternative to classical chromatographic procedures, aqueous two phase systems (ATPS) have been used recently for the manufacturing of biopharmaceuticals. The phenomenon of ATPS is based on the incompatibility of two components. The typical pairs are either polymers or salts. By forcing the (soluble) protein of interest into the upper phase, simultaneously aggregates can be removed as well.

Since more parameters (concentration, pH, ionic strength, molecule type, and weight) contribute to the separation, this technique has a high resolution power. ATPS is easily scalable, has a high capacity, and can be used in a continuous operation. ATPS could prove quite useful also for the separation of fusion proteins if predictability and economical sustainability is verified [121].

1.5.3 Formulation

Formulation covers the steps that are necessary to create a drug product out of the drug substance. This means the active pharmaceutical ingredient (API), the therapeutic (fusion) protein, is combined with additives or excipients. The proper selection of excipients is a challenging task because only a limited range of molecules is available [122].

The goal of formulation is to prevent protein degradation during storage and to optimize its delivery. To achieve that, a number of buffer parameters such as pH, ionic strength, and composition must be carefully assessed [123]. This can be very demanding for a fusion protein that could be composed of two proteins with different buffer preferences or different sensitivity to degradation. Proteins can be degraded through physical (denaturation, adsorption, aggregation, precipitation) and chemical (hydrolysis, deamidation, oxidation, isomerization, and disulfide exchange) instability. The protein can also be exposed to harmful conditions during handling such as temperature, physical interfaces, and shear forces [124].

Particularly, non-native aggregation of liquid dosage forms can diminish the concentration of the active drug. To limit the amount of aggregates formed, it is necessary to understand the aggregation routes and to predict its formation rate [125].

From a safety point of view, the drug product has to be sterile, free of aggregates that could cause immunogenicity or reduce the concentration of the active substance. The final product must be free of product or process related impurities within a specified tolerance [126].

It was shown that a change of formulation of a well-tolerated product can cause serious safety issues. For example, a change of formulation and ingredients of epoetin-α caused severe pure red cell aplasia [127]. The final step of formulation, fill to finish, has to be done into vessels that maintain sterility, but also do not contain inacceptable extractables and leachables or cause immunogenicity [128]. Both, leachables and extractables can be set free during the filling procedures or later during storage; therefore these conditions must be defined carefully [129]. Stability can be tested in an accelerated approach, where long-term effects are simulated by elevated stress in form of intense temperature, light exposure, or pH.

Beyond physicochemical parameters also patient compliance must be taken into account. This includes the administration as well. The protein itself can be available either lyophilized (as dry substance) or as liquid formulation, ready for injection. A self-administered therapeutic such as insulin is usually available as prefilled syringe. Noninvasive routes such as oral, pulmonary, nasal, or transdermal delivery would be preferable but suffer from low bioavailability [130]. Recently, a fusion protein approach using transferrin as partner for granulocyte monocyte-colony-stimulating factor (GM-CSF) was successfully applied orally in mice [43]. Pulmonary delivery could be achieved by using monomeric Fc fusions. Here, aerosolized fusion proteins are carried across lung epithelial cells utilizing the neonatal Fc receptor [131]. Part IIc of this book is dedicated to different targeting and delivery approaches with fusion proteins.

Due to the limited stability in the digestive tract therapeutic proteins are usually administered by injection. This can be done through three different ways: intravenous (IV), intramuscular (IM), or subcutaneously (SC). The administration route has an impact on the concentration of the protein. IV injections usually are in the concentration range around 1 mg/mL, IM injections are acceptable up to 100 mg/mL, but only SC injections can have concentrations up to 150 mg/mL. It is very challenging to deliver a soluble, aggregate-free formulation at these very high, viscous concentrations [132]. The concentration and dosage are obviously dependent on the potency and the clearance of the drug. For example, highly active hormone or cytokine drugs are administered at relatively low doses despite their rapid clearance. Antibodies as large molecules on the other side can be required at very high concentrations despite their long half-life. This issue of manufacturability has to be considered from the start [133].

A lot of work has been done to diminish the aggregation. Working with fusion proteins that consist of several unrelated domains that did not evolve together is very challenging. One technique is the selective domain stabilization, where buffer conditions are evaluated that stabilize the least stable domain, because from there aggregation is initiated. Testing an hGH-HSA fusion, best results were obtained when repulsion of protein–protein interaction was increased [134].

1.5.4 Process Economies

As in all other manufacturing processes, cost of goods (COG), processing time, and capital investment are major determinants of fusion protein production. Another important parameter is the usually high dose of therapeutic proteins leading to high demands on capacity. In the last years, a trend to improved processes delivering more output from the same capacity and second generation products with decreased doses and longer half-lives will change biopharmaceutical manufacturing [135].

Many microbial processes generate IBs that need refolding. For instance, the granulocyte macrophage-colony-stimulating factor (GM-CSF) has been produced in *E. coli*. This required a series of washing steps to purify the IBs that are subsequently refolded at low concentration. When combining chemical extraction of IBs with EBA labor costs were fivefold lower and overall costs 50% less [136]. Interestingly, the costs for soluble expression of Heparinise in *E. coli* were twice the expenses of insoluble expression. The difference was primarily dependent on the expression levels that could be achieved in both approaches. Overall it was observed that fermentation related costs represent only a minor fraction, and yield optimization in downstream procedures has the highest impact on cost savings [137].

When comparing the production costs of tissue plasminogen activator stimulating factor (tPA) in *E. coli* and mammalian cells it was found out that refolding concentrations of more than 4 mg and refolding yields above 20% would make the process economically attractive. But it has to be taken into account that the cost of the competing mammalian expression is dominated by the cost for serum in the cultivation medium [138]. This extra cost can nowadays be neglected since most large-scale mammalian cultivations are serum free. Particularly, the absence of animal derived components, minimizes concerns of virus or prion contaminations, and improves the attractiveness of microbial fermentation besides its cost advantage and the short production cycles.

The technological progress of mammalian cell culture with increasing titers for antibodies currently causes bioreactor overcapacity issues. It is anticipated that in 2013 available bioreactor volume will reach 4 million liters compared to 2.4 million liters in 2007. Overall COG projections for antibodies and Fc-fusion proteins currently range from 50 to 100$/g but achieve a median sales price of 8000$/g. Therefore, royalty payments have a huge impact on the total COG. A 10-fold increase in titer would decrease the COG for the drug substance by 85%. For other recombinant (fusion) proteins requiring large doses the product titer is still a major determinant of COG [139]. At an antibody production below 10 kg/year and a titer around 0.1 g/L the ratio between upstream and downstream costs is around 50:50. Increasing the output to 100 kg/year at a titer of 0.5 g/L, upstream costs only represent 20%, indicating that at high levels, savings by the expression system plays only a minor role [140].

The aforementioned overcapacity of mammalian cell culture has been observed for many years and increased from 23.6% in 2003 to 36.7% in 2007. This is a result of constantly increasing product titers and simultaneously expanding the capacity by building new factories [141]. New production installations seem to move away from large single product units to more flexible multiproduct plants that nowadays also include disposable equipment. But disposables can also be used in microbial processes. In a case study, the conventional production of an antibody fragment in *E. coli* was compared to a process with single use equipment. This more than 10-year-old study could demonstrate a 25% advantage of net present value for the conventional manufacture [142]. However, this gap is constantly narrowing, since particularly in downstream processing membrane steps can efficiently replace previous chromatographic column steps. This substitution saves cost in form of 25% less required volume of the stationary phase. Consequently 40% less aqueous waste is generated. Furthermore, labor hours are reduced by 40% because of obsolete cleaning and validation tasks. Overall the antibody purification process based on disposable membranes was 50% faster and 23% cheaper [143]. However, it should be taken into account that smaller and more frequent batches have higher QC and QA costs.

Obviously, the downstream process is a major cost driver suffering from low economy of scale effects. Typical chromatography steps are volume dependent directly impacting space requirements. Further complications are the intermediate cleaning cycles causing downtimes and cost [144].

1.5.5 Glycosylation

Besides production cost also product quality hast to be taken into account when selecting an expression host. One element contributing to product quality is glycosylation, and interestingly about 70% of marketed recombinant proteins are glycoproteins. Huge variations and deviations from a typical human glycan pattern can be observed in economically favorable organisms such as bacteria, yeasts, fungi, insects, plants, or nonhuman mammalian species [145].

Glycosylation has a huge impact on the therapeutic effect by improving pharmacokinetics, pharmacodistribution and the selectivity of binding to receptors. This requires a careful selection of upstream process parameters including the host organism [146]. Even the cultivation conditions can affect the glycosylation pattern. For example, *N*-glycolylneuraminic acid (Neu5Gc) does not exist in humans, but is added when glycoproteins are expressed in Chinese hamster ovary (CHO) cells. The addition of Neu5Gc is dependent on a number if cultivation conditions. Neu5Gc content can be significantly lowered in presence of sodium butyrate, a decrease of temperature after exponential growth, high carbon dioxide concentrations and the utilization of sodium hydroxide for pH control [147]. From a downstream perspective glycosylation makes purification more difficult since often heterogeneity is observed. Since it is not easy to separate glyco-isoforms by preparative chromatography all efforts are taken to suppress heterogeneity already during upstream processes. Nevertheless, to guarantee a reproducible, homogeneous drug substance, high resolution analytical tools must be implemented [148].

Glycosylation is a key parameter of fusion proteins, determining immunogenicity, solubility, and stability. Particularly, the terminal sugars of glycans in the CH_2 domain of antibodies are responsible for influencing antibody-dependent cellular cytotoxicity (ADCC) and complement-dependent cytotoxicity (CDC) activity [149]. ADCC can be significantly enhanced by the lack of core-fucose on the Fc part as demonstrated with scFv-Fc fusion proteins [150]. An interesting example for the importance of glycosylation is the half-life extension technology of Prolor that relies on a carboxy-terminal peptide, originally derived from the chorionic gonadotropin (CG) β-subunit, bearing four serine-linked oligosaccharides. This short peptide can be fused to the N- or C-terminus of other proteins, expanding the hydrodynamic radius by extensive O-glycosylation, thus preventing kidney filtration [151]. Obviously, glycosylation in that case requires secreted expression in mammalian cells to obtain the proper oligosaccharide pattern. Glycoengineering was also used to optimize half-life and potency of an erythropoietin (EPO) variant called Aranesp®. This novel EPO has two additional N-linked carbohydrate chains, resulting in threefold longer serum half-life [152].

Besides half-life extension, glycosylation also contributes to protein stability. The biggest effects are seen in preventing proteolysis by shielding susceptible amino acid sequences, inhibiting aggregation by the hydrophilic nature of oligosaccharides and by protection of the proteins against all kinds of physical denaturation, including freezing [153].

The workhorses of antibody production, CHO cells, deliver a slightly different glycosylation than original human cells. Therefore, CHO cells were engineered to express human glycosyltransferases to improve the ADCC and CDC effects of antibodies [154].

Since cultivation of mammalian cells is rather costly when compared to microbial cells, also attempts were undertaken to obtain human glycoforms in yeasts. As suitable starting point for humanization of yeast the α-1,6-mannose extension must be eliminated. This was followed by the introduction of a α-1,2-mannosidase for mannose trimming. Finally, mannosidase II and N-acetylglucosaminyl transferase II was added to release uniform human-like glycans from yeast [155]. As last step terminal sialysation had to be implemented; this required the introduction of 14 foreign genes into *P. pastoris* [156].

The biggest obstacle of yeast N-glycosylation is the high mannose content that is potentially immunogenic and leads to drastically reduced half lives by the removal of therapeutic glycoproteins through mannose receptors on macrophages, thus reducing the efficacy [157]. But in some cases mannosylation can be beneficial, for example, when an antibody–enzyme fusion protein has to be quickly removed from normal tissue to enable a highly specific antibody-directed enzyme prodrug therapy (ADEPT) approach [158].

1.6 REGULATORY CHALLENGES

Despite the demonstrated success, fusion proteins consisting of two molecules with different functions, still raises concerns. Addressing these issues is of huge interest in a time where biosimilars and biobetters enter clinical development [159]. Several variants of approved drugs can be found within the category of fusion proteins, having improved half-life, potency, stability, or route of administration [11]. In general, three aspects must be clarified satisfactorily: quality, safety, and efficacy. The necessary information is gathered through animal pharmacology and toxicology studies. Here, the pharmacodynamics and pharmacokinetics that cover clearance and absorption, distribution, metabolism, and excretion (ADME), is assessed. The tightly regulated release of proteins in the organisms has to be mimicked by adapting dosing intervals or even activity of the protein [160]. This is much more difficult for non-natural proteins such as fusion proteins.

A separate document contains the chemistry, manufacturing and control (CMC) data. The manufacture must be done under good manufacturing practice (GMP) conditions and demonstrate the ability to consistently and reproducibly supply active batches of the drug. All that must be included in an investigational new drug (IND) application or a clinical trial authorization/exemption (CTA/CTX) for the regulatory authorities in the Unitied States or Europe, respectively.

Since protein therapeutics are produced in living organisms, they could contain intrinsic infectious agents and other process- or product-related impurities. As biological they could have a heterogeneous composition and require extensive analysis. Expected and controllable parameters are size, charge, activity, folding, but on the other hand unexpected and undesired characteristics such as aggregation, amino acid modification or proteolysis could occur as well [161]. Usually, viral clearance with two orthogonal independent steps must be demonstrated. Before release the drug substance's identity, purity, potency, and stability must be verified according to predefined acceptance criteria in a number of validated assays [162].

The key product attributes that determine efficacy, in other words potency and activity, under the conditions of the intended use must be characterized. This includes nonclinical pharmacokinetic, toxicology, and safety studies [163]. Manufacturing conditions have to be chosen that can generate the desired attributes, or product quality, in a reproducible and cost-efficient manner. In the preclinical studies often imperfect replicas of human disease conditions in animal models are used [164]. A further hurdle for human therapeutic (fusion) proteins is that sometimes nonclinical assays must be performed with the respective animal homolog because the human version is nonfunctional in the selected animal species [165].

A fusion protein specific complication is the increased potential of immunogenicity. For instance, immunotoxins contain frequently nonhuman proteins that elicit immune responses and neutralizing antibodies. Also the conjunction site between the two fusion partners forms a novel epitope even when combining only human proteins. In both cases, a detailed study of the potential risk has to be done in form of specialized assays. Chapter 5 of this book explains the current approaches.

Typical challenges for bi-specific molecules are the difficulty to have similar potency/affinity against both targets, the lack of individual dosing or pharmacokinetic and the potential of overlapping and cross-linked toxicity. Dosing and potency can be addressed by combining two molecules with suitable affinities or altering them accordingly. In the case of transferrin fusions, the simultaneous or sequential binding of both fusion partners to their respective cell surface receptors has a huge impact on recycling, degradation, and biological activity [166]. Toxicity issues can be predicted to some degree in simulation models. A number of adverse side effects of antibodies that are primarily dependent on the Fc part have been described in the past [167]. It is important to take the multiple functions of Fc domains into account when designing Fc-fusion proteins. Surprisingly, the majority of Fc fusions contain IgG1 that can trigger ADCC or CDC reactions. Therefore, it is advisable to select the Fc part of IgG2 or 4 as fusion partner. A number of novel scaffolds also serve as building blocks for fusion proteins. It is recommended to adapt the preclinical safety testing to the individual molecules and their mode of action [168].

Recently, some fusion proteins failed in clinical trials for different reasons. In the case of Zalbin™, an interferon-α2b-HSA fusion, the risk benefit ratio was not sufficiently favorable, which let Novartis and Human Genome Science withdraw their biologics license application (BLA) in 2010.[2] The glucagon-like peptide-1 (GLP-1)-Transferrin fusion of Pfizer and BioRexis was discontinued in Phase I because a reversible increase in heart rate was observed.[3] Surprisingly and without disclosing specific reasons, Amevive® (leukocyte function antigen-3 (LFA-3) linked to Fc of IgG1) that was never approved for European markets and was withdrawn from the U.S. market by Astellas Pharma in 2011.[4]

1.7 COMPETITION AND MARKET

Fusion proteins are relative recent entrants to the market of therapeutic proteins, appearing first with Enbrel in 1998. In the years thereafter only a few were approved by the FDA to reach patients, the majority being FC-fusion proteins [169]. The only exception within the eight marketed fusion proteins in 2011 is an immunotoxin, Ontak®. Despite that fact that Ontak does not capture a huge market, many other immunotoxins are in the clinical pipeline, probably expanding the number of approved fusion proteins soon.

In contrast, a total of 25 mABs were accepted by regulatory authorities since 1998. This triple number of products is also reflected in sales. In 2010, the six so far approved fusion proteins collected $8.3 billion in revenues, whereas mABs had sales of $40.8 billion globally.[5] This indicates that the competition of antibodies is really strong, and only Enbrel and Orencia have blockbuster status. On the example of Enbrel, the strong competition and evolution of antibodies in the market segment of TNF-α blockers can be demonstrated (Table 1.1).

However, the big opportunity for fusion proteins can be found in applications where they do not have to face the fierce competition of antibodies. Many arising concepts focus on novel scaffolds that can replace antibodies and represent the binding moiety of fusion proteins [13]. Frequently, these small scaffolds can reach different targets than antibodies, but suffering from pharmacokinetic limitations. Most of them require plasma half-life extension strategies as explained in Part IIa of this book. But due to their small size they are ideal candidates for generating bi-functional molecules by combining two of them.

With regard to bi-specific antibodies, a subtopic of this book, this market is just emerging, with the tri-functional Removab® as the first approved multispecific antibody in 2009 [170]. The rising interest in this class of molecules is also reflected in increased merger and acquisition (M&A) activities such as Amgen buying Micromet for $1.2 billion in 2012.[6]

As described previously, the design of fusion proteins is based on genetic engineering; many of the concepts are patent protected and have been used to build companies. In times when large pharmaceutical corporations are facing expiring patents, these companies became targets for acquisitions to bolster emptying pipelines. In Table 1.2, the M & A deals of the last decade affecting companies with fusion protein technology or drugs in various development stages are collected.

Overall a fast growth for protein therapeutics can be predicted [171]. The top 30 biologicals gathered global sales worth more than $107 billion in 2010, with Enbrel being the most successful drug.[7] Overall the global pharmaceutical market grew by 4.1%, but interestingly sales of

[2] http://www.hgsi.com/latest/human-genome-sciences-announces-withdrawal-of-european-marketing-authorization-application-for-joulferon-zalbin-for-the-treatment-of-chronic-hepati-7.html

[3] http://clinicaltrials.gov/ct2/show/NCT00637338?term=PF-04603629&rank=1

[4] http://www.amevive.com/Patient%20letter.pdf

[5] Datamonitor: PharmaVitae: Monoclonal Antibodies: 2011 – Market Size Update. October 2011.

[6] http://www.amgen.com/media/media_pr_detail.jsp?releaseID=1653062

[7] La Merie S.L., Travesia Balmins, 7, Bajos, 2a, 08870 Sitges, Spain: Top 30 Biologics 2010, 04 March 2011.

TABLE 1.1 Competition Between Antibodies and Fusion Proteins that Capture TNF-α

Drug	Approval	Components	Manufacturing	Dosing	Sales 2009[a]	Sales 2010[a]	Sales 2011[a]
Remicade®	08/1998	Chimeric mAb	Continuous perfusion in CHO	IV 2–4 weekly	5.9	7.3	8.5
Enbrel®	11/1998	Receptor Fc fusion	Fed-batch in CHO	SC 1–2 weekly	6.3	7.4	7.9
Humira®	01/2003	Fully human mAb	Extended fed-batch in CHO	SC bi-weekly	5.5	6.7	7.9
Cimzia®	04/2008	PEGylated humanized Fab	Fed-batch *E. coli*	SC bi-weekly	0.11	0.26	0.31
Simponi®	04/2009	Fully human mAb	Perfusion in Sp2/0	SC monthly	0.11	0.32	0.67

[a]Global sales in billion $.

biological drugs increased by 17% in 2010. Splitting the sales up, it can be seen that the growth rate of proteins was 15%, whereas antibody sales expanded by 21%. This means that of the $15.91 billion growth antibodies contributed 55% and protein 45%. Another important observation is the higher approval success rate of 32% for biological compared to 13% for small molecules between 1993 and 2004 [172]. In Figure 1.4, the annual sales of important therapeutic protein classes in the United States are presented.

Although the growth of fusion protein sales has been relatively modest in comparison to mABs or hormones, there is a huge potential for this molecule class as can be seen in a much deeper analysis in Chapter 2.

1.8 CONCLUSION AND FUTURE PERSPECTIVE

During the last decades, human imagination has assembled a plethora of new protein combinations from the vast array

TABLE 1.2 Mergers and Acquisitions Involving Companies with Fusion Protein Technology

Buyer	Target	Date	Value ($ Million)	Technology	Fusion Protein	Indication	Phase (at Acquis.)
Alexion	Enobia	07/02/2012	610	Bone targeting	ENB-0040	Hypophosphatasia	II
Alexion	Taligen	31/01/2011	111	C3d targeting	TT30	Autoimmune	Preclin.
Insys	Neopharm	29/10/2010	135	Immunotoxins	NK-408	Oncology	III
Bristol-Myers Squibb	Zymogenetics	07/09/2010	885	Fc fusion	Atacicept	Autoimmune	II
Biomarin	Zystor	07/09/2010	22	GILT	ZC-701	Pompe's disease	I
Emergent	Trubion	18/08/2010	97	SMIP and Scorpion	SBI-087 TRU-016	Autoimmune Oncology	II
Teva	CoGenesys	12/08/2010	400	Albumin fusion	Neugranin Cardeva	Febrile neutropenia Congestive heart failure	II II
Pfizer	CovX	22/01/2008	n.a.	CovX body (peptide-mAb fusion)	CVX-045 CVX-060 CVX-096	Oncology Oncology Type 2 diabetes	I II
Topotarget	Apoxis	18/12/2007	19	Mega Ligand	APO010	Oncology	I
Pfizer	Biorexis	11/04/2007	n.a.	Transferrin fusion	BRX-0585	Type 2 diabetes	Preclin.
Biogen Idec	Syntonix	01/02/2007	40	Monomeric Fc fusion	FIX-Fc	Hemophilia B	IND
Amgen	Avidia	29/09/2006	290	Avimer (modular binding domains)	IL-6 inhibitor	Inflammation, autoimmune	I
Medigene	Avidex	31/08/2006	62	TCR technology	EsoDex	Oncology	Preclin.

GILT, glycosylation independent lysosomal targeting; SMIP, small modular immunopharmaceutical; TCR, T-cell receptor.

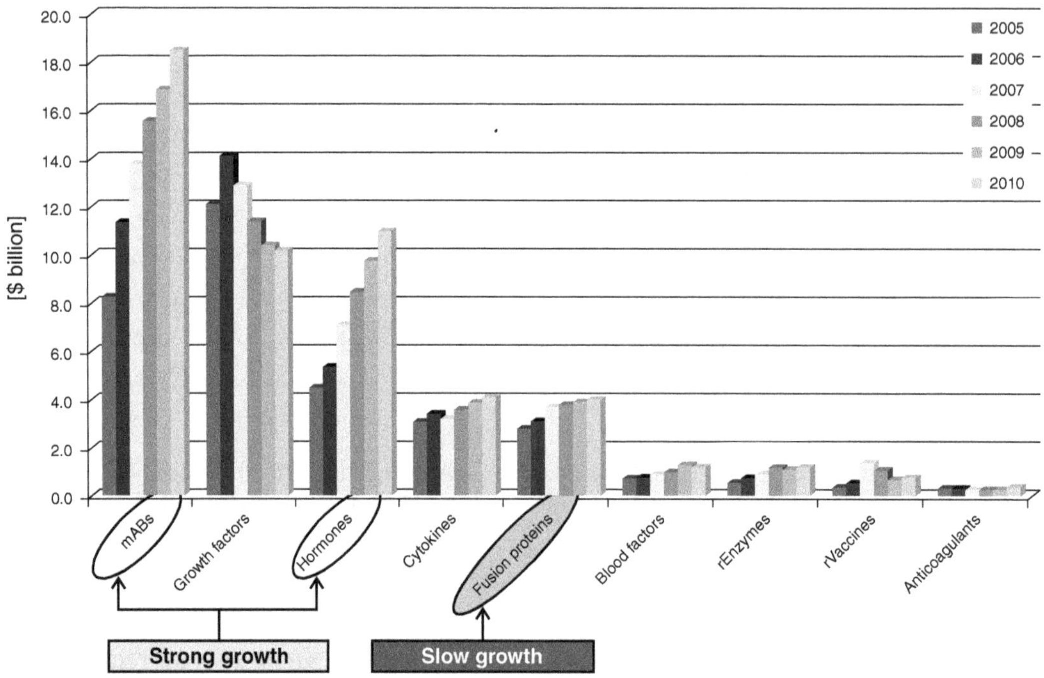

FIGURE 1.4 Annual sales of protein drugs in the United States. The two classes of proteins with a high annual growth rate are monoclonal antibodies (mAB) and hormones. Over the last 4 years, the growth of sales for fusion proteins was minimal, whereas the sales for growth factors even declined.

of nature's building blocks. Many concepts have been heavily influenced by the "magic bullet paradigm" that has been described more than a century ago, but took until now to realize it with artificial recombinations. These novel fusion proteins combine hitherto unrelated functionalities into a single molecule. Several have passed approval from regulatory authorities and many of those currently being in clinical trials will reach the market soon. So far fusion proteins have proven as valuable additions to the arsenal of therapeutic molecules. Still there are many opportunities where innovative fusion proteins can make a significant improvement. For instance, many first generation biopharmaceuticals can benefit from prolonged circulation times. Cancer patients will receive better targeted and more specific drugs with less systemic toxicity based on fusion proteins. So far relapsing tumors can be treated with novel protein drugs that hit two targets simultaneously, so overcoming resistance mechanisms. Novel fusion proteins will make therapies more affordable by lowering manufacture cost and improve quality of life for many patients who benefit from longer administration intervals. However, to fulfill the promises of fusion protein technology still a number of challenges have to be resolved.

A major obstacle is the immunogenicity potential that is always present even in fully human recombination, because at joint between two molecules will always create a new epitope which can provoke immune reactions. Reduction of immunogenicity and understanding the underlying factors is, therefore, a key element to guarantee future success of fusion proteins. Another point to consider is the potential incompatibility between fusion partners that limit manufacturability. Both challenges will require intense efforts of protein engineering.

This book aims to cover the state of the art of fusion proteins. It presents an overview on the multitude of possibilities to design novel protein drugs while balancing between proven concepts and new ideas that have not reached the clinic yet. The book is structured into three larger parts. First general issues and concepts are discussed before in the second part examples on the three categories ($t_{1/2}$, toxicity, and targeting) are presented. Finally novel concepts and the rising class of multispecific antibodies are described. I hope this book will inspire the reader and create enthusiasm for the exciting topic of fusion proteins.

REFERENCES

1. Cohn EJ, Oncley JL, Strong LE, Hughes WL, Armstrong SH. (1944) Chemical, clinical, and immunological studies on the products of human plasma fractionation. I. The characterization

of the protein fractions of human plasma. *J. Clin. Invest.* 23, 417–432.

2. Krauss J, Arndt MAE, Vu BK, Newton DL, Rybak SM. (2005) Targeting malignant B-cell lymphoma with a humanized anti-CD22 scFv-angiogenin immunoenzyme. *Br. J. Haematol.* 128(5), 602–609.
3. Goeddel DV, Kleid DG, Bolivar F, Heyneker HL, Yansura DG, Crea R, et al. (1979) Expression in *Escherichia coli* of chemically synthesized genes for human insulin. *Proc. Natl. Acad. Sci. USA* 76(1), 106–110.
4. Köhler G, Milstein C. (1975) Continuous cultures of fused cells secreting antibody of predefined specificity. *Nature* 256 (5517), 495–497.
5. Kempeni J. (1999) Preliminary results of early clinical trials with the fully human anti-TNFalpha monoclonal antibody D2E7. *Ann. Rheum. Dis.* 58(Suppl. 1), I70–72.
6. Peppel K, Crawford D, Beutler B. (1991) A tumor necrosis factor (TNF) receptor-IgG heavy chain chimeric protein as a bivalent antagonist of TNF activity. *J. Exp. Med.* 174(6), 1483–1489.
7. Cai Q, Peng G, Bu L, Lin Y, Zhang L, Lustigmen S, et al. (2007) Immunogenicity and *in vitro* protective efficacy of a polyepitope plasmodium falciparum candidate vaccine constructed by epitope shuffling. *Vaccine* 25(28), 5155–5165.
8. Teicher BA, Chari RVJ. (2011) Antibody conjugate therapeutics: challenges and potential. *Clin. Cancer Res.* 17(20), 6389–6397.
9. Edwards PAW. (2010) Fusion genes and chromosome translocations in the common epithelial cancers. *J. Pathol.* 220(2), 244–254.
10. Terpe K. (2003) Overview of tag protein fusions: from molecular and biochemical fundamentals to commercial systems. *Appl. Microbiol. Biotechnol.* 60(5), 523–533.
11. Schmidt SR. (2009) Fusion-proteins as biopharmaceuticals: applications and challenges. *Curr. Opin. Drug Discov. Dev.* 12(2), 284–295.
12. Leader B, Baca QJ, Golan DE. (2008) Protein therapeutics: a summary and pharmacological classification. *Nat. Rev. Drug Discov.* 7(1), 21–39.
13. Gebauer M, Skerra A. (2009) Engineered protein scaffolds as next-generation antibody therapeutics. *Curr. Opin. Chem. Biol.* 13(3), 245–255.
14. Schmidt S.R. (2008) *Fusion-Proteins as Biopharmaceuticals: Applications and Challenges*. Protein Expression Europe & Antibodies Europe. Lisbon, Portugal, 20–23 October.
15. Schellekens H. (2004) When biotech proteins go off-patent. *Trends Biotechnol.* 22(8), 406–410.
16. Webber PM. (2007) The patenting of novel biotech products. *IDrugs* 10(7), 459–462.
17. Ilag LL, Ilag LM, Ilag LL. (2002) From patenting genes to proteins: the search for utility via function. *Trends Biotechnol.* 20(5), 197–199.
18. Webber P. (2006) Patenting antibodies. *Nat. Rev. Drug Discov.* 5(2), 97.
19. Hashimoto K, Aida T. (2008) Antibody patenting without antibodies: a global trend. *Nat. Biotechnol.* 26(12), 1341–1343.
20. Holliday L. (2009) Patenting antibodies in Europe. *MAbs* 1 (4), 385–386.
21. Bachran C, Sutherland M, Bachran D, Fuchs H. (2007) Patents on immunotoxins and chimeric toxins for the treatment of cancer. *Recent Pat. Drug Deliv. Formul.* 1(2), 105–115.
22. McCabe KW. (2009) Guardians at the gate: patent protection for therapeutic monoclonal antibodies (Part 1). *MAbs* 1, 382–384.
23. Agris CH. (1998) Patenting protein sequences. *Nat. Biotech.* 16(11), 1075.
24. Dorey E. (2010) Bristol-Myers Squibb reaps biologics in ZymoGenetics windfall. *Nat. Biotech.* 28, 1137–1138.
25. Marshall SA, Lazar GA, Chirino AJ, Desjarlais JR. (2003) Rational design and engineering of therapeutic proteins. *Drug Discov. Today* 8(5), 212–221.
26. Cuesta AM, Sainz-Pastor N, Bonet J, Oliva B, Alvarez-Vallina L. (2010) Multivalent antibodies: when design surpasses evolution. *Trends Biotechnol.* 28(7), 355–362.
27. Ortiz-Sánchez E, Helguera G, Daniels TR, Penichet ML. (2008) Antibody–cytokine fusion proteins: applications in cancer therapy. *Expert Opin. Biol. Ther.* 8, 609–632.
28. Oliner J, Min H, Leal J, Yu D, Rao S, You E, et al. (2004) Suppression of angiogenesis and tumor growth by selective inhibition of angiopoietin-2. *Cancer Cells* 6(5), 507–516.
29. Evans L, Hughes M, Waters J, Cameron J, Dodsworth N, Tooth D, et al. (2010) The production, characterisation and enhanced pharmacokinetics of scFv-albumin fusions expressed in *Saccharomyces cerevisiae*. *Protein Expr. Purif.* 73(2), 113–124.
30. Zhao HL, Xue C, Wang Y, Li XY, Xiong XH, Yao XQ, et al. (2007) Circumventing the heterogeneity and instability of human serum albumin-interferon-alpha2b fusion protein by altering its orientation. *J. Biotechnol.* 131(3), 245–252.
31. Paek S-Y, Kim Y-S, Choi S-G. (2010) The orientation-dependent expression of angiostatin-endostatin hybrid proteins and their characterization for the synergistic effects of antiangiogenesis. *J. Microbiol. Biotechnol.* 20(10), 1430–1435.
32. Christensen T, Amiram M, Dagher S, Trabbic-Carlson K, Shamji MF, Setton LA, et al. (2009) Fusion order controls expression level and activity of elastin-like polypeptide fusion proteins. *Protein Sci.* 18(7), 1377–1387.
33. Arndt MAE, Krauss J, Vu BK, Newton DL, Rybak SM. (2005) A dimeric angiogenin immunofusion protein mediates selective toxicity toward CD22+ tumor cells. *J. Immunother.* 28(3), 245–251.
34. Mariuzza RA, Poljak RJ, Mihaesco C, Mihaesco E. (1983) Crystals of the human heavy chain disease protein Riv and human Fc fragment are isomorphous: further evidence for conformational flexibility in the hinge region of immunoglobulins. *J. Mol. Biol.* 165(3), 559–561.
35. Wriggers W, Chakravarty S, Jennings PA. (2005) Control of protein functional dynamics by peptide linkers. *Biopolymers* 80(6), 736–746.

36. Argos P. (1990) An investigation of oligopeptides linking domains in protein tertiary structures and possible candidates for general gene fusion. *J. Mol. Biol.* 211(4), 943–958.
37. Newton DL, Xue Y, Olson KA, Fett JW, Rybak SM. (1996) Angiogenin single-chain immunofusions: influence of peptide linkers and spacers between fusion protein domains. *Biochemistry* 35(2), 545–553.
38. Arai R, Ueda H, Kitayama A, Kamiya N, Nagamune T. (2001) Design of the linkers which effectively separate domains of a bifunctional fusion protein. *Protein Eng.* 14(8), 529–532.
39. Zhao HL, Yao XQ, Xue C, Wang Y, Xiong XH, Liu ZM. (2008) Increasing the homogeneity, stability and activity of human serum albumin and interferon-alpha2b fusion protein by linker engineering. *Protein Expr. Purif.* 61(1), 73–77.
40. Xue F, Gu Z, Feng J. (2004) LINKER: a web server to generate peptide sequences with extended conformation. *Nucleic Acids Res.* 32(Web Server issue), W562–W565.
41. George RA, Heringa J. (2002) An analysis of protein domain linkers: their classification and role in protein folding. *Protein Eng.* 15(11), 871–879.
42. Chen X, Bai Y, Zaro JL, Shen W-C. (2010) Design of an *in vivo* cleavable disulfide linker in recombinant fusion proteins. *BioTechniques* 49(1), 513–518.
43. Bai Y, Shen W-C. (2006) Improving the oral efficacy of recombinant granulocyte colony-stimulating factor and transferrin fusion protein by spacer optimization. *Pharm. Res.* 23(9), 2116–2121.
44. Amet N, Lee H-F, Shen W-C. (2009) Insertion of the designed helical linker led to increased expression of tf-based fusion proteins. *Pharm. Res.* 26(3), 523–528.
45. Goyal A, Batra JK. (2000) Inclusion of a furin-sensitive spacer enhances the cytotoxicity of ribotoxin restrictocin containing recombinant single-chain immunotoxins. *Biochem. J.* 345(Pt 2), 247–254.
46. Chen T, Yang J, Wang Y, Zhan C, Zang Y, Qin J. (2005) Design of recombinant stem cell factor-macrophage colony stimulating factor fusion proteins and their biological activity *in vitro*. *J. Comput. Aided Mol. Des.* 19(5), 319–328.
47. Won J, Nam P, Lee Y, Choe M. (2009) Higher cytotoxicity of divalent antibody-toxins than monovalent antibody-toxins. *Biochem. Biophys. Res. Commun.* 382(1), 15–20.
48. Liu M, Wang X, Yin C, Zhang Z, Lin Q, Zhen Y, et al. (2008) A novel bivalent single-chain variable fragment (scFV) inhibits the action of tumour necrosis factor α. *Biotechnol. Appl. Biochem.* 50(4), 173.
49. de Kruif J, Logtenberg T. (1996) Leucine zipper dimerized bivalent and bispecific scFv antibodies from a semi-synthetic antibody phage display library. *J. Biol. Chem.* 271(13), 7630–7634.
50. Hoyne PA, Cosgrove LJ, McKern NM, Bentley JD, Ivancic N, Elleman TC, et al. (2000) High affinity insulin binding by soluble insulin receptor extracellular domain fused to a leucine zipper. *FEBS Lett.* 479(1–2), 15–18.
51. Morris NP, Peters C, Montler R, Hu H-M, Curti BD, Urba WJ, et al. (2007) Development and characterization of recombinant human Fc:OX40L fusion protein linked via a coiled-coil trimerization domain. *Mol. Immunol.* 44(12), 3112–3121.
52. Rozanov DV, Savinov AY, Golubkov VS, Rozanova OL, Postnova TI, Sergienko EA, et al. (2009) Engineering a leucine zipper-TRAIL homotrimer with improved cytotoxicity in tumor cells. *Mol. Cancer Ther.* 8(6), 1515–1525.
53. Spitzer D, McDunn JE, Plambeck-Suess S, Goedegebuure PS, Hotchkiss RS, Hawkins WG. (2010) A genetically encoded multifunctional TRAIL trimer facilitates cell-specific targeting and tumor cell killing. *Mol. Cancer Ther.* 9(7), 2142–2151.
54. Krippner-Heidenreich A, Grunwald I, Zimmermann G, Kühnle M, Gerspach J, Sterns T, et al. (2008) Single-chain TNF, a TNF derivative with enhanced stability and antitumoral activity. *J. Immunol.* 180(12), 8176–8183.
55. Berg D, Lehne M, Müller N, Siegmund D, Münkel S, Sebald W, et al. (2007) Enforced covalent trimerization increases the activity of the TNF ligand family members TRAIL and CD95L. *Cell Death Differ.* 14(12), 2021–2034.
56. Stone GW, Barzee S, Snarsky V, Kee K, Spina CA, Yu X-F, et al. (2006) Multimeric soluble CD40 ligand and GITR ligand as adjuvants for human immunodeficiency virus DNA vaccines. *J. Virol.* 80(4), 1762–1772.
57. Holler N, Tardivel A, Kovacsovics-Bankowski M, Hertig S, Gaide O, Martinon F, et al. (2003) Two adjacent trimeric Fas ligands are required for Fas signaling and formation of a death-inducing signaling complex. *Mol. Cell. Biol.* 23(4), 1428–1440.
58. Sánchez-Arévalo Lobo VJ, Cuesta AM, Sanz L, Compte M, García P, Prieto J, et al. (2006) Enhanced antiangiogenic therapy with antibody-collagen XVIII NC1 domain fusion proteins engineered to exploit matrix remodeling events. *Int. J. Cancer* 119(2), 455–462.
59. Fan C-Y, Huang C-C, Chiu W-C, Lai C-C, Liou G-G, Li H-C, et al. (2008) Production of multivalent protein binders using a self-trimerizing collagen-like peptide scaffold. *FASEB J.* 22 (11), 3795–3804.
60. Rohn J. (2010) Newsmaker: anaphore. *Nat. Biotechnol.* 28(11), 1143.
61. Onda M, Beers R, Xiang L, Nagata S, Wang Q-C, Pastan I. (2008) An immunotoxin with greatly reduced immunogenicity by identification and removal of B cell epitopes. *Proc. Natl. Acad. Sci. USA* 105(32), 11311–11316.
62. Pastan I, Onda M, Weldon J, Fitzgerald D, Kreitman R. (2011) Immunotoxins with decreased immunogenicity and improved activity. *Leuk. Lymphoma* 52(Suppl. 2), 87–90.
63. Harding FA, Liu AD, Stickler M, Razo OJ, Chin R, Faravashi N, et al. (2005) A beta-lactamase with reduced immunogenicity for the targeted delivery of chemotherapeutics using antibody-directed enzyme prodrug therapy. *Mol. Cancer Ther.* 4(11), 1791–1800.
64. Liang Y, Qiu H, Glinka Y, Lazarus AH, Ni H, Prud'homme GJ, et al. (2011) Immunity against a therapeutic xenoprotein/Fc construct delivered by gene transfer is reduced

through binding to the inhibitory receptor FcγRIIb. *J. Gene Med.* 13(9), 470–477.

65. De Groot AS, Scott DW. (2007) Immunogenicity of protein therapeutics. *Trends Immunol.* 28(11), 482–490.
66. Onda M. (2009) Reducing the immunogenicity of protein therapeutics. *Curr. Drug Targets* 10(2), 131–139.
67. Brinkmann U, Pai LH, FitzGerald DJ, Pastan I. (1992) Alteration of a protease-sensitive region of Pseudomonas exotoxin prolongs its survival in the circulation of mice. *Proc. Natl. Acad. Sci. USA* 89(7), 3065–3069.
68. Weldon JE, Xiang L, Chertov O, Margulies I, Kreitman RJ, FitzGerald DJ, et al. (2009) A protease-resistant immunotoxin against CD22 with greatly increased activity against CLL and diminished animal toxicity. *Blood* 113(16), 3792–3800.
69. Gillies SD, Lo K-M, Burger C, Lan Y, Dahl T, Wong W-K. (2002) Improved circulating half-life and efficacy of an antibody-interleukin 2 immunocytokine based on reduced intracellular proteolysis. *Clin. Cancer Res.* 8(1), 210–216.
70. Tcherniuk SO, Chroboczek J, Balakirev MY. (2005) Construction of tumor-specific toxins using ubiquitin fusion technique. *Mol. Ther.* 11(2), 196–204.
71. Reiter Y, Brinkmann U, Webber KO, Jung SH, Lee B, Pastan I. (1994) Engineering interchain disulfide bonds into conserved framework regions of Fv fragments: improved biochemical characteristics of recombinant immunotoxins containing disulfide-stabilized Fv. *Protein Eng.* 7(5), 697–704.
72. Way JC, Lauder S, Brunkhorst B, Kong S-M, Qi A, Webster G, et al. (2005) Improvement of Fc-erythropoietin structure and pharmacokinetics by modification at a disulfide bond. *Protein Eng. Des. Sel.* 18(3), 111–118.
73. Liu W, Onda M, Kim C, Xiang L, Weldon JE, Lee B, et al. (2012) A recombinant immunotoxin engineered for increased stability by adding a disulfide bond has decreased immunogenicity. *Protein Eng. Des. Sel.* 25(1), 1–6.
74. Zhao HL, Xue C, Wang Y, Sun B, Yao XQ, Liu ZM. (2009) Elimination of the free sulfhydryl group in the human serum albumin (HSA) moiety of human interferon-alpha2b and HSA fusion protein increases its stability against mechanical and thermal stresses. *Eur. J. Pharm. Biopharm.* 72(2), 405–411.
75. Won J, Choe M. (2008) Disulfide bond bridged divalent antibody-toxin, (Fab-PE38fl)2, with the toxin PE38 fused to the light chain. *J. Microbiol. Biotechnol.* 18(8), 1475–1481.
76. Chowdhury PS, Vasmatzis G, Beers R, Lee B, Pastan I. (1998) Improved stability and yield of a Fv-toxin fusion protein by computer design and protein engineering of the Fv. *J. Mol. Biol.* 281(5), 917–928.
77. McDonagh CF, Beam KS, Wu GJS, Chen JH, Chace DF, Senter PD, et al. (2003) Improved yield and stability of L49-sFv-beta-lactamase, a single-chain antibody fusion protein for anticancer prodrug activation, by protein engineering. *Bioconjug. Chem.* 14(5), 860–869.
78. Narhi LO, Arakawa T, Aoki K, Wen J, Elliott S, Boone T, et al. (2001) Asn to Lys mutations at three sites which are N-glycosylated in the mammalian protein decrease the aggregation of *Escherichia coli*-derived erythropoietin. *Protein Eng.* 14(2), 135–140.
79. Dunn CJ, Goa KL. (2001) Tenecteplase: a review of its pharmacology and therapeutic efficacy in patients with acute myocardial infarction. *Am. J. Cardiovasc. Drugs* 1(1), 51–66.
80. Johnson IS. (1983) Human insulin from recombinant DNA technology. *Science* 219(4585), 632–637.
81. Dingermann T. (2008) Recombinant therapeutic proteins: production platforms and challenges. *Biotechnol. J.* 3(1), 90–97.
82. Demain AL, Vaishnav P. (2009) Production of recombinant proteins by microbes and higher organisms. *Biotechnol. Adv.* 27(3), 297–306.
83. Rader RA. (2008) Expression systems for process and product improvement. *BioProc. Intl.* 6(Suppl. 4), 4–9.
84. Walsh G. (2010) Biopharmaceutical benchmarks 2010. *Nat. Biotech.* 28(9), 917–924.
85. Sheffield WP, McCurdy TR, Bhakta V. (2005) Fusion to albumin as a means to slow the clearance of small therapeutic proteins using the *Pichia pastoris* expression system. *Therapeutic Proteins*, pp. 145–154.
86. Lin H, Kim T, Xiong F, Yang X. (2007) Enhancing the production of Fc fusion protein in fed-batch fermentation of *Pichia pastoris* by design of experiments. *Biotechnol. Prog.* 23(3), 621–625.
87. Sørensen HP, Mortensen KK. (2005) Advanced genetic strategies for recombinant protein expression in *Escherichia coli. J. Biotechnol.* 115(2), 113–128.
88. Kim JY, Kim Y-G, Lee GM. (2012) CHO cells in biotechnology for production of recombinant proteins: current state and further potential. *Appl. Microbiol. Biotechnol.* 93(3), 917–930.
89. Makrides SC. (1996) Strategies for achieving high-level expression of genes in *Escherichia coli. Microbiol. Rev.* 60 (3), 512–538.
90. Kamionka M. (2011) Engineering of therapeutic proteins production in *Escherichia coli. Curr. Pharm. Biotechnol.* 12(2), 268–274.
91. Jana S, Deb JK. (2005) Strategies for efficient production of heterologous proteins in *Escherichia coli. Appl. Microbiol. Biotechnol.* 67(3), 289–298.
92. Swartz JR. (2001) Advances in *Escherichia coli* production of therapeutic proteins. *Curr. Opin. Biotechnol.* 12(2), 195–201.
93. Panda AK. (2003) Bioprocessing of therapeutic proteins from the inclusion bodies of *Escherichia coli. Adv. Biochem. Eng. Biotechnol.* 85, 43–93.
94. Urieto JO, Liu T, Black JH, Cohen KA, Hall PD, Willingham MC, et al. (2004) Expression and purification of the recombinant diphtheria fusion toxin DT388IL3 for phase I clinical trials. *Protein Expr. Purif.* 33(1), 123–133.
95. Woo JH, Liu J-S, Kang SH, Singh R, Park SK, Su Y, et al. (2008) GMP production and characterization of the bivalent anti-human T cell immunotoxin, A-dmDT390-bisFv (UCHT1) for phase I/II clinical trials. *Protein Expr. Purif.* 58(1), 1–11.

96. Su H, Jack M, McIntosh LM, Perdomo L, Choy BSF, Finck BK, et al. (2009) Clinical grade production and characterization of a fusion protein comprised of the chemokine CCL2-ligand genetically fused to a mutated and truncated form of the Shiga A1 subunit. *Protein Expr. Purif.* 66(2), 149–157.
97. Möhlmann S, Bringmann P, Greven S, Harrenga A. (2011) Site-specific modification of ED-B-targeting antibody using intein-fusion technology. *BMC Biotechnol.* 11, 76.
98. Premsukh A, Lavoie JM, Cizeau J, Entwistle J, MacDonald GC. (2011) Development of a GMP Phase III purification process for VB4-845, an immunotoxin expressed in *E. coli* using high cell density fermentation. *Protein Expr. Purif.* 78(1), 27–37.
99. Omasa T, Onitsuka M, Kim W-D. (2010) Cell engineering and cultivation of Chinese hamster ovary (CHO) cells. *Curr. Pharm. Biotechnol.* 11(3), 233–240.
100. Barnes LM, Dickson AJ. (2006) Mammalian cell factories for efficient and stable protein expression. *Curr. Opin. Biotechnol.* 17(4), 381–386.
101. Matasci M, Hacker DL, Baldi L, Wurm FM. (2008) Recombinant therapeutic protein production in cultivated mammalian cells: current status and future prospects. *DDT* 5(2–3), e37–e42.
102. Lim Y, Wong NSC, Lee YY, Ku SCY, Wong DCF, Yap MGS. (2010) Engineering mammalian cells in bioprocessing—current achievements and future perspectives. *Biotechnol. Appl. Biochem.* 55(4), 175–189.
103. Liu YY, Gordienko I, Mathias A, Ma S, Thompson J, Woo JH, et al. (2000) Expression of an anti-CD3 single-chain immunotoxin with a truncated diphtheria toxin in a mutant CHO cell line. *Protein Expr. Purif.* 19(2), 304–311.
104. Grillberger L, Kreil TR, Nasr S, Reiter M. (2009) Emerging trends in plasma-free manufacturing of recombinant protein therapeutics expressed in mammalian cells. *Biotechnol. J.* 4(2), 186–201.
105. Qian Y, Jing Y, Li ZJ. (2010) Glucocorticoid receptor-mediated reduction of IgG-fusion protein aggregation in Chinese hamster ovary cells. *Biotechnol. Prog.* 26(5), 1417–1423.
106. Boado RJ, Hui EK-W, Lu JZ, Pardridge WM. (2011) CHO cell expression, long-term stability, and primate pharmacokinetics and brain uptake of an IgG-paroxonase-1 fusion protein. *Biotechnol. Bioeng.* 108(1), 186–196.
107. Kalyanpur M. (2002) Downstream processing in the biotechnology industry. *Mol. Biotechnol.* 22(1), 87–98.
108. Balasundaram B, Harrison S, Bracewell DG. (2009) Advances in product release strategies and impact on bioprocess design. *Trends Biotechnol.* 27(8), 477–485.
109. Baneyx F, Mujacic M. (2004) Recombinant protein folding and misfolding in *Escherichia coli*. *Nat. Biotechnol.* 22(11), 1399–1408.
110. Fahnert B, Lilie H, Neubauer P. (2004) Inclusion bodies: formation and utilisation. *Adv. Biochem. Eng. Biotechnol.* 89, 93–142.
111. Hassouneh W, Christensen T, Chilkoti A. (2010) Elastin-like polypeptides as a purification tag for recombinant proteins. *Curr. Protoc. Protein Sci.* Chapter 6, Unit 6.11.
112. Ghose S, Hubbard B, Cramer SM. (2007) Binding capacity differences for antibodies and Fc-fusion proteins on protein A chromatographic materials. *Biotechnol. Bioeng.* 96(4), 768–779.
113. Xu X, Didio DM, Leister KJ, Ghose S. (2010) Disaggregation of high-molecular weight species during downstream processing to recover functional monomer. *Biotechnol. Prog.* 26(3), 717–726.
114. Roque ACA, Lowe CR, Taipa MA. (2004) Antibodies and genetically engineered related molecules: production and purification. *Biotechnol. Prog.* 20(3), 639–654.
115. Nilson BH, Solomon A, Björck L, Akerström B. (1992) Protein L from *Peptostreptococcus magnus* binds to the kappa light chain variable domain. *J. Biol. Chem.* 267(4), 2234–2239.
116. Pigeon C, Jackson M, Baines D. (2009) Application of Fabsorbent™ F1P HF, a synthetic ligand adsorbent for capture and purification of a single-domain antibody fragment expressed in *Escherichia coli*. *Bio. Proc. Intl.* 7(7), 90–91.
117. El Khoury G, Rowe LA, Lowe CR. (2012) Biomimetic affinity ligands for immunoglobulins based on the multicomponent Ugi reaction. *Methods Mol. Biol.* 800, 57–74.
118. Willems A, Leoen J, Schoonooghe S, Grooten J, Mertens N. (2003) Optimizing expression and purification from cell culture medium of trispecific recombinant antibody derivatives. *J. Chromatogr. B Analyt. Technol. Biomed. Life Sci.* 786(1–2), 161–176.
119. Lobbezoo B, Dodd R, Morton P, Duell M. (2011) A novel high performance adsorbent as a platform process for the capture & purification of albumin-fusion proteins. PREP 2011—24th International Symposium, 1.
120. (2002) Amediplase: CGP 42935, K2tu-PA, MEN 9036. *BioDrugs* 16(5), 378–379.
121. Rosa PAJ, Azevedo AM, Sommerfeld S, Bäcker W, Aires-Barros MR. (2011) Aqueous two-phase extraction as a platform in the biomanufacturing industry: economical and environmental sustainability. *Biotechnol. Adv.* 29(6), 559–567.
122. Jorgensen L, Hostrup S, Moeller EH, Grohganz H. (2009) Recent trends in stabilising peptides and proteins in pharmaceutical formulation—considerations in the choice of excipients. *Expert Opin. Drug Deliv.* 6(11), 1219–1230.
123. Patro SY, Freund E, Chang BS. (2002) Protein formulation and fill-finish operations. *Biotechnol Annu. Rev.* 8, 55–84.
124. Frokjaer S, Otzen DE. (2005) Protein drug stability: a formulation challenge. *Nat. Rev. Drug Discov.* 4(4), 298–306.
125. Roberts CJ, Das TK, Sahin E. (2011) Predicting solution aggregation rates for therapeutic proteins: approaches and challenges. *Int. J. Pharm.* 418(2), 318–333.
126. Rathore N, Rajan RS. (2008) Current perspectives on stability of protein drug products during formulation, fill and finish operations. *Biotechnol. Prog.* 24(3), 504–514.
127. Haselbeck A. (2003) Epoetins: differences and their relevance to immunogenicity. *Curr. Med. Res. Opin.* 19(5), 430–432.
128. Sharma B. (2007) Immunogenicity of therapeutic proteins. Part 2: impact of container closures. *Biotechnol. Adv.* 25(3), 318–324.

129. Chang JY, Xiao NJ, Zhu M, Zhang J, Hoff E, Russell SJ, et al. (2010) Leachables from saline-containing IV bags can alter therapeutic protein properties. *Pharm. Res.* 27(11), 2402–2413.
130. Brown LR. (2005) Commercial challenges of protein drug delivery. *Expert Opin. Drug Deliv.* 2(1), 29–42.
131. Dumont JA, Low SC, Peters RT, Bitonti AJ. (2006) Monomeric Fc fusions: impact on pharmacokinetic and biological activity of protein therapeutics. *BioDrugs* 20(3), 151–160.
132. Jezek J, Rides M, Derham B, Moore J, Cerasoli E, Simler R, et al. (2011) Viscosity of concentrated therapeutic protein compositions. *Adv. Drug Deliv. Rev.* 63(13), 1107–1117.
133. Shire SJ. (2009) Formulation and manufacturability of biologics. *Curr. Opin. Biotechnol.* 20(6), 708–714.
134. Cordes AA, Platt CW, Carpenter JF, Randolph TW. (2012) Selective domain stabilization as a strategy to reduce fusion protein aggregation. *J. Pharm. Sci.* 101(4), 1400–1409.
135. Werner RG. (2004) Economic aspects of commercial manufacture of biopharmaceuticals. *J. Biotechnol.* 113(1–3), 171–182.
136. Lee GH, Cooney D, Middelberg APJ, Choe WS. (2006) The economics of inclusion body processing. *Bioprocess Biosyst. Eng.* 29(2), 73–90.
137. Ernst S, Garro OA, Winkler S, Venkataraman G, Langer R, Cooney CL, et al. (1997) Process simulation for recombinant protein production: cost estimation and sensitivity analysis for heparinase I expressed in *Escherichia coli. Biotechnol. Bioeng.* 53(6), 575–582.
138. Datar RV, Cartwright T, Rosen CG. (1993) Process economics of animal cell and bacterial fermentations: a case study analysis of tissue plasminogen activator. *Biotechnology (NY)* 11(3), 349–357.
139. Kelley B. (2009) Industrialization of mAb production technology: the bioprocessing industry at a crossroads. *MAbs* 1, 443–452.
140. Farid SS. (2007) Process economics of industrial monoclonal antibody manufacture. *J. Chromatogr. B Analyt. Technol. Biomed. Life Sci* 848(1), 8–18.
141. Langer ES. (2009) Trends in capacity utilization for therapeutic monoclonal antibody production. *MAbs* 1(2), 151–156.
142. Novais JL, Titchener-Hooker NJ, Hoare M. (2001) Economic comparison between conventional and disposables-based technology for the production of biopharmaceuticals. *Biotechnol. Bioeng.* 75(2), 143–153.
143. Varadaraju H, Schneiderman S, Zhang L, Fong H, Menkhaus TJ. (2011) Process and economic evaluation for monoclonal antibody purification using a membrane-only process. *Biotechnol. Prog.* 27(5), 1297–1305.
144. Gottschalk U. (2008) Bioseparation in antibody manufacturing: the good, the bad and the ugly. *Biotechnol. Prog.* 24(3), 496–503.
145. Brooks SA. (2006) Protein glycosylation in diverse cell systems: implications for modification and analysis of recombinant proteins. *Expert Rev. Proteomics* 3(3), 345–359.
146. Werner RG, Kopp K, Schlueter M. (2007) Glycosylation of therapeutic proteins in different production systems. *Acta Paediatr. Suppl.* 96(455), 17–22.
147. Borys MC, Dalal NG, Abu-Absi NR, Khattak SF, Jing Y, Xing Z, et al. (2010) Effects of culture conditions on N-glycolylneuraminic acid (Neu5Gc) content of a recombinant fusion protein produced in CHO cells. *Biotechnol. Bioeng.* 105(6), 1048–1057.
148. Hooker A, James D. (1998) The glycosylation heterogeneity of recombinant human IFN-gamma. *J. Interferon Cytokine Res.* 18(5), 287–295.
149. Raju TS. (2008) Terminal sugars of Fc glycans influence antibody effector functions of IgGs. *Curr. Opin. Immunol.* 20(4), 471–478.
150. Repp R, Kellner C, Muskulus A, Staudinger M, Nodehi SM, Glorius P, et al. (2011) Combined Fc-protein- and Fc-glyco-engineering of scFv-Fc fusion proteins synergistically enhances CD16a binding but does not further enhance NK-cell mediated ADCC. *J. Immunol. Methods* 373(1–2), 67–78.
151. Fares FA, Suganuma N, Nishimori K, LaPolt PS, Hsueh AJ, Boime I. (1992) Design of a long-acting follitropin agonist by fusing the C-terminal sequence of the chorionic gonadotropin beta subunit to the follitropin beta subunit. *Proc. Natl. Acad. Sci. USA* 89(10), 4304–4308.
152. Egrie JC, Browne JK. (2001) Development and characterization of novel erythropoiesis stimulating protein (NESP). *Br. J. Cancer* 84(Suppl. 1), 3–10.
153. Solá RJ, Griebenow K. (2009) Effects of glycosylation on the stability of protein pharmaceuticals. *J. Pharm. Sci.* 98(4), 1223–1245.
154. Umaña P, Jean-Mairet J, Moudry R, Amstutz H, Bailey JE. (1999) Engineered glycoforms of an antineuroblastoma IgG1 with optimized antibody-dependent cellular cytotoxic activity. *Nat. Biotechnol.* 17(2), 176–180.
155. Hamilton SR, Gerngross TU. (2007) Glycosylation engineering in yeast: the advent of fully humanized yeast. *Curr. Opin. Biotechnol.* 18(5), 387–392.
156. Hamilton SR, Davidson RC, Sethuraman N, Nett JH, Jiang Y, Rios S, et al. (2006) Humanization of yeast to produce complex terminally sialylated glycoproteins. *Science* 313 (5792), 1441–1443.
157. Wildt S, Gerngross TU. (2005) The humanization of N-glycosylation pathways in yeast. *Nat. Rev. Microbiol.* 3(2), 119–128.
158. Sharma SK, Pedley RB, Bhatia J, Boxer GM, El-Emir E, Qureshi U, et al. (2005) Sustained tumor regression of human colorectal cancer xenografts using a multifunctional mannosylated fusion protein in antibody-directed enzyme prodrug therapy. *Clin. Cancer Res.* 11(2 Pt 1), 814–825.
159. Schellekens H. (2005) Follow-on biologics: challenges of the "next generation." *Nephrol. Dial. Transplant.* 20(Suppl. 4), iv31–36.
160. Shanafelt AB. (2005) Medicinally useful proteins-enhancing the probability of technical success in the clinic. *Expert Opin. Biol. Ther.* 5(2), 149–151.

161. Rathore AS. (2009) Follow-on protein products: scientific issues, developments and challenges. *Trends Biotechnol.* 27(12), 698–705.

162. Lowe D, Dudgeon K, Rouet R, Schofield P, Jermutus L, Christ D. (2011) Aggregation, stability, and formulation of human antibody therapeutics. In: *Advances in Protein Chemistry and Structural Biology*, Academic Press, 41–61.

163. Baumann A. (2009) Nonclinical development of biopharmaceuticals. *Drug Discov. Today* 14(23–24), 1112–1122.

164. Dixit R, Boelsterli UA. (2007) Healthy animals and animal models of human disease(s) in safety assessment of human pharmaceuticals, including therapeutic antibodies. *Drug Discov. Today* 12(7–8), 336–342.

165. Bussiere JL, Martin P, Horner M, Couch J, Flaherty M, Andrews L, et al. (2009) Alternative strategies for toxicity testing of species-specific biopharmaceuticals. *Int. J. Toxicol.* 28(3), 230–253.

166. Chen X, Lee H-F, Zaro JL, Shen W-C. (2011) Effects of receptor binding on plasma half-life of bifunctional transferrin fusion proteins. *Mol. Pharm.* 8(2), 457–465.

167. Hansel TT, Kropshofer H, Singer T, Mitchell JA, George AJT. (2010) The safety and side effects of monoclonal antibodies. *Nat. Rev. Drug Discov.* 9(4), 325–338.

168. Roberts SA, Woodnutt G, Bradshaw CW, Gad SC. (2010) Strategy considerations for developing the preclinical safety testing programs for protein scaffold therapeutics. In: *Pharmaceutical Sciences Encyclopedia.* John Wiley & Sons, Inc., pp. 1–16.

169. Beck A, Reichert JM. (2011) Therapeutic Fc-fusion proteins and peptides as successful alternatives to antibodies. *MAbs* 3(5), 415–416.

170. Bokemeyer C. (2010) Catumaxomab–trifunctional anti-EpCAM antibody used to treat malignant ascites. *Expert Opin. Biol. Ther.* 10(8), 1259–1269.

171. Hiller A. (2009) Fast growth foreseen for protein therapeutics. *GEN* 29(1), 153–155.

172. Langlands B. (2011) Next-generation protein and peptide therapeutics. *Scrip Bus. Insights* 1–192.

2

ANALYZING AND FORECASTING THE FUSION PROTEIN MARKET AND PIPELINE

MARK BELSEY[1] AND GILES SOMERS[2]

[1]*WestLB, London, UK*
[2]*Datamonitor, London, UK*

2.1 INTRODUCTION

In this chapter, we look at the evolution of the fusion protein (FP) market from the launch of Enbrel® in 1998, through to the end of 2010 where there are seven sales-generating FPs on at least one major western pharmaceutical market. Subsequently, we carry out a market analysis, which indicates that the fusion market could be worth \$10.1 billion in our view by 2016. This chapter also provides an in-depth market analysis of the six drugs that we forecast will account for 93.5% of the estimated \$10.1 billion FP market by 2016 (on a risk-adjusted basis): Enbrel, Orencia®, Nplate®, dulaglutide, albiglutide, and VEGF trap-Eye. Last, we provide an analysis of the FP market pipeline, which has led to the identification of 43 FPs in Phase II development and above. From this analysis, we note that the most common targeted therapy area is oncology, followed by autoimmune diseases: a not uncommon finding for a biotech market. We also find that many Big Pharma/Big Biotech companies have at least one FP in Phase II development and above; however, few have more than two, even though we believe FPs represent attractive licensing targets for these multinational companies. Our analysis leads us to conclude that FPs are unlikely to be viewed as a homologous technology with a single best-in-class technology platform that Big Pharma/Big Biotech companies aim to gain exposure to—instead, FP candidates are selected on a drug-by-drug basis on factors such as their individual efficacy and safety profiles, and disease target. Last, based on our analysis, we think the primary drivers for potential FP in-licensing by multinational companies are (1) to capitalize on already established sales and marketing networks, or (2) to lineup replacement drugs following the loss of exclusivity for marketed drugs, or replace other pipeline drugs they are developing if data are less robust for these other pipeline drugs.

We define FPs as non-native proteins produced from a genetic fusion of sequences encoding two or more protein components (unless specified otherwise, protein component relates to a whole protein, protein fragment, or peptide). As such, all FP components must be combined in gene expression, as opposed to being chemically joined at a later time. A linker or spacer peptide can also be included to aid folding of the proteins and/or biologic activity.

2.2 MARKET SALES DYNAMICS OF THE FP MARKET

In this section, we discuss the evolution of the FP market, what we estimate total FP market sales are currently (as of the end of 2010), and what we think the market will grow to in the future (by 2016), together with discussing the main factors we think will drive or restrict market growth going forward.

Fusion Protein Technologies for Biopharmaceuticals: Applications and Challenges, First Edition. Edited by Stefan R. Schmidt.

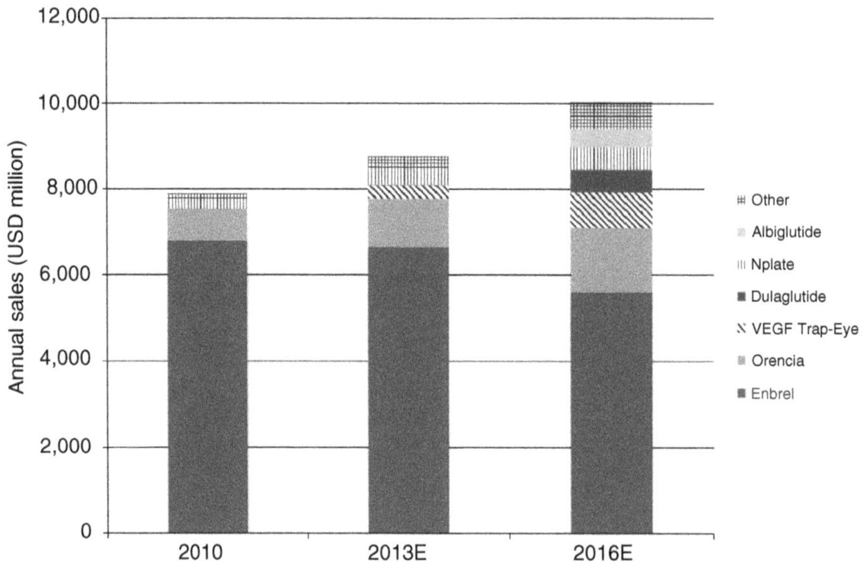

FIGURE 2.1 FP market sales forecasts from 2010-16E for FPs in Phase III clinical trials, in regulatory submission and on the market.

2.2.1 Enbrel Founded the FP Market Following Its Launch in 1998

In 1998, Amgen/Pfizer/Takeda's Enbrel (etanercept) became the first FP with substantial commercial potential to be approved in the United States. It was approved by the FDA for the treatment of moderate to severe rheumatoid arthritis (RA) and was also the first biologic drug to be launched to treat this indication. One year later, Eisai's Ontak®[1] became the first targeted toxin FP to be approved to treat human indications. FPs continued to be launched through the 2000s and as of the end of 2010, we count seven sales-generating FPs on at least one of the major western pharmaceutical markets.[2]

We scanned through resources including Pubmed, ClinicalTrials.gov, FP developer websites and company-related publications, and industry reports, and we have identified a total of 43 FPs in Phase II development or above that meet our definition of an FP. Products known to have been discontinued have been excluded, as are FPs in the Phase I portion of a Phase I/II trial, as these are considered to be in Phase I development. Of these 43 FPs, seven are on the market, 13 are in later-stage clinical development (Phase II/III, Phase III and in registration), and a further 23 FPs are in Phase II. Of these, we forecast six FPs will account for 93.5% of FP market share by 2016: three of these are already marketed, whereas three are in late-stage clinical development. Given the relatively small number of marketed drugs as of 2010, and because only a few FPs generate the vast majority of our 2016 sales forecast for the FP market, we believe this market can be characterized as an early-stage biotech market.

[1] Ontak was developed by Seragen, which was acquired by Ligand Pharmaceuticals, which then sold Ontak in 2006 to Japanese pharmaceutical company Eisai.

[2] In addition to these seven drugs, Bristol-Myers Squibb's Nulojix (belatacept) was approved in the United States in June 2011.

2.2.2 We Forecast the FP Market Will Grow to $10.1 Billion by 2016

We estimate the FP market generated sales of $7.9 billion in 2010 (based on seven marketed drugs) in the major pharmaceutical markets,[3] and we forecast that this market will rise to $10.1 billion on a risk-adjusted basis[4] by 2016 (Figure 2.1), based on forecasts of currently marketed FP drugs, those that have been submitted to regulatory authorities, and those in Phase III clinical trials (these forecasts are detailed in Section 2.3). This 4.1% 2010-16E compound annual growth rate (CAGR) estimate is relatively low compared with other biotech drug classes such as monoclonal antibodies (mAbs; independent market analysts forecast a 2009-15 sales CAGR of 9.5%[5]), although it is stronger than small molecule growth forecasts (independent market analysts forecast a 2009-15

[3] This includes the United States, Europe, and Japan. Unless otherwise specified in this chapter, references to the market refer primarily to major western pharmaceutical markets (the United States and Europe).

[4] Drugs included in the forecast which have not yet been approved by the regulatory agencies are risk adjusted, with the level of risk adjustment correlated to stage, clinical trial data and indication (see also [1] for clinical trial progression and impact of indication on likely success).

[5] Both mAb market forecasts and small molecule forecasts are based on Datamonitor's "Fast growth MAb market to offer great rewards to a select few over 2009-15" from "Monoclonal Antibodies: 2010", November, 1 2010.

sales CAGR of −0.5%). The comparatively low growth of the FP market masks an estimated 3.2% CAGR decline in sales of the key FP drug, Enbrel over this period. Similar to other biotech market (e.g., mAbs or nucleic acid therapeutics), the FP market is dominated by a small number of high-selling drugs. In the FP market, Enbrel is currently the only blockbuster (a blockbuster is a drug generating at least $1 billion annual sales) and is a leading drug in autoimmune diseases (in particular in two major indications: RA and psoriasis). We estimate Enbrel sales in 2010 accounted for 86% of the total FP market; however, we forecast that this will fall to 56% by 2016, based on our assumptions of a ramp up in competition facing Enbrel in the form of branded originators and biosimilars (Figure 2.1).

Stripping out the impact of Enbrel, our 2010-16E FP ex-Enbrel sales CAGR is 26.3% (although we flag that this strong growth is partly the result of having a low base in 2010, together with assumptions that a number of new FP drugs will launch over this period). We forecast that six drugs will account for 93.5% of the estimated $10.1 billion FP market by 2016, including three-marketed drugs: Enbrel, Orencia, and Nplate; and three late-stage pipeline drugs: VEGF trap-Eye, dulaglutide, and albiglutide. We forecast that by 2016, autoimmune[6] will remain the lead therapeutic area in terms of sales, accounting for 78% of total estimated 2016 FP sales. While a full market analysis of all indications targeted by marketed FPs and those in late-stage clinical development is beyond the scope of this chapter, we have provided (1) a brief overview of what we think are factors driving and restricting FP market growth (note: this is a focus for Chapter 3; the section below serves to introduce these topics) and (2) detail on what we consider are key positive and negative factors shaping the sales growth underlying what we think will be the six key FPs (see Section 2.3).

2.2.3 Growth Drivers and Restrictors that Impact the FP Market

Briefly, we believe that the key factors driving and restricting market growth include efficacy, safety, cost, and administration. These are discussed in greater detail in Chapter 3.

2.3 INDIVIDUAL DRUG SALES ANALYSIS

2.3.1 Enbrel

Amgen/Pfizer/Takeda's Enbrel (etanercept), launched in the United States in 1998 for moderate to severe RA, was the first tissue necrosis factor (TNF)-targeting biologic drug to be introduced to treat autoimmune diseases such as RA. It is currently the market leader in both RA [2] and psoriasis [3]. We estimate that it holds one-third of the RA market, and just under one half of the psoriasis market. It is thought by independent market analysts and by key opinion leader physicians to be the most efficacious agent in treating RA [2] and to be in the top-3 psoriasis drugs in terms of overall clinical profile [3]. In 2008, Enbrel was Wyeth's second-highest selling drug [4], and it was the sixth-highest selling pharmaceutical in 2009 globally, according to IMS sales data.[7] We estimate that ~86% of the 2010 FP market came from Enbrel; therefore, the future sales performance of this drug has a substantial impact on the overall FP market growth.

For Enbrel, RA and psoriasis (together with psoriatic arthritis) are the key markets: we estimate that 70% Enbrel current sales are from rheumatology (excluding psoriatic arthritis), whereas ~30% came from psoriasis and psoriatic arthritis [5]. These are sizeable markets with substantial growth potential: independent market analysts have estimated that the RA market generated $8.8 billion sales in 2009 in the seven major markets,[8] and forecast it to rise to $12 billion by 2019 [2], whereas these analysts estimate that the psoriasis market generated $3.4 billion sales in 2009, and forecast it to rise to $6.8 billion by 2019 [6]. A recent rheumatologist survey indicated that while Enbrel remained the most efficacious drug (43%), competition is increasing: Humira® is right behind Enbrel (42%), and a higher percentage of managed care organizations prefer Humira over Enbrel [2]. Enbrel also appears to be facing substantial competition in psoriasis: although we think Enbrel is still one of the top-selling drugs in this market, a recent dermatologist survey indicated that Enbrel is thought to be less efficacious than competitors: 32% voted Humira is the most efficacious, 29% voted for Remicade®, and only 23% selected Enbrel [7]. Furthermore, in both indications, it seems possible that the drug will face competition from biosimilars in the medium term.

Despite this increase in competition, we think specific positive factors will help drive Enbrel sales and prevent a rapid, substantial sales decline in the face of this competition include (1) very strong promotional and marketing support, given the length of time that the drug has spent on the market and the number of patients treated, and because the drug is marketed by Amgen, Pfizer, and Takeda: three leading pharmaceutical companies with substantial experience in autoimmune disease promotion plus extensive key opinion leader networks; (2) a very well-established safety profile, with over 2 million patient years of postmarketing exposure (according to Amgen); (3) the autoinjector pen

[6] We include the following indications in autoimmune: rheumatoid arthritis, psoriasis, psoriatic arthritis, juvenile rheumatoid arthritis, ankylosing spondylitis, idiopathic thrombocytopenic purpura, cryopyrin-associated periodic syndromes and Muckle-wells syndrome.

[7] http://www.imshealth.com/deployedfiles/imshealth/Global/Content/StaticFile/Top_Line_Data/Top%2015%20Global%20Products_2009.pdf

[8] We define the seven major markets as the United States, Japan, France, Germany, Italy, Spain, and the United Kingdom.

device for subcutaneous (SC) administration, which is well established in the market and convenient for patients; (4) favorable formulary status at leading U.S. health insurers, further helped by beneficial formulary tier placing often granted to self-administered biologics [8]; and (5) Enbrel has significantly less dose escalation than competitors such as Abbott/Eisai's mAb Humira (adalimumab) or J&J/Merck/Mitsubishi Tanabe's mAb Remicade (infliximab) [9]. Dose escalation requires increasing dosage sizes to overcome neutralizing antibodies, which boosts treatment costs; therefore, Enbrel's lack of dose escalation is a positive aspect for patient treatment if physicians are concerned about this issue.

However, we think Enbrel sales have reached the mature stage of sales trajectory, with the result that we think sales will be low growth in the near term, before beginning to slowly decline, based on our belief that there are few near-term significant positive catalysts to boost sales growth: (1) market penetration has been substantial but has now slowed as Enbrel reaches peak penetration of its key markets; (2) although Enbrel is in clinical trials for additional indications (e.g., Phase III trials for graft vs. host rejection [10]), we do not see any new indication expansions driving substantially greater sales for two reasons: we think there is already extensive off-label prescription of Enbrel,[9] and we think that the drug already has approval for the indications that generate strong sales; (3) weekly or twice weekly dosing is a disadvantage compared to other biologics for important indications such as RA and psoriasis, where most dosing of competitor biologics is fortnightly, monthly, or less frequent; (4) going forward, newer agents both in late-stage clinical development and those that have been recently launched represent significant threats, due to strong marketing and improved profiles in terms of safety, efficacy, administration (including dosing frequency), and/or novel mechanism of action; and (5) we expect sales decline to increase after 2014/15 when we think the first biosimilars may appear on the market. However, we expect biosimilar erosion of branded Enbrel to be substantially less aggressive (in terms of speed and percentage cannibalization of the branded originator drug) than small molecule generic erosion for three reasons. First, we expect there to be a lower price differential between the branded originator drug and the biosimilar (we expect a 15–30% price difference). Second, we also think there will be lower competition in the biosimilar market, as we think there will be fewer biosimilars developers (given the costs and investment required—particularly in building manufacturing and regulatory functions). Last, we think physicians will remain wary of swapping a branded drug with substantial safety and efficacy data for a slightly cheaper biosimilar with comparatively little safety and efficacy data.

[9] http://www.regencerx.com/docs/physicianRx/biologic-response-modifiers-Humira®-0408.pdf

2.3.2 Orencia

Bristol-Myers Squib's (BMS) Orencia (abatacept) is an FP comprised the extracellular domain of human CTLA4 (Cytotoxic T-Lymphocyte Antigen 4) and the IgG1 Fc region, which is approved for the treatment of adult rheumatoid arthritis (RA) and juvenile idiopathic arthritis (JIA). It was approved in the United States in December 2005 and in Europe in May 2007.

We think the most important positive factors set to drive up Orencia sales are the drug's potential to move into the TNF inhibitor (TNFi) marketspace, and the successful potential launch of a subcutaneous administration. RA biologics are broadly split into those targeting TNF and those with other targets. TNFis such as Enbrel have dominated the autoimmune biologics market since their launch, and the most popular RA drugs—Enbrel and Humira—are both TNFis. Non-TNFi biologics such as Orencia have been held back by U.S. labels that require rheumatologists to treat with TNFis before these drugs can be used. As a result, these therapies have tended to be used in third line [11]. However, in July 2010, Orencia's European label was expanded to include patients who had inadequately responded to DMARD (disease-modifying antirheumatic drug) therapies, which opens up the second-line market for greater penetration.[10] We believe the second major positive driver of Orencia sales growth will be the potential approval of a subcutaneous form, given that most of the drug's competitors are in this easier-to-use form. A self-administered subcutaneous formulation often has more favorable formulary placing [8]. This was submitted to the FDA in October 2010, so we expect an FDA decision on this in H2-2011,[11] and we could see launch soon after if there is a positive FDA response.

Orencia has also been in late-stage clinical trials to treat gastrointestinal (GI) autoimmune diseases. However, the drug failed to meet its primary endpoint in both Crohn's disease [12] and ulcerative colitis [13]. BMS has also terminated earlier-stage clinical trials of the drug in a number of other indications, including sarcoidosis and multiple sclerosis,[12] although there are indications such as uveitis where the drug remains in earlier-stage development. As a result, our forecasts are primarily based on sales from the RA and JIA indications.

[10] http://www.thepharmaletter.com/file/96506/bristol-myers-squibbs-orencia-gets-expanded-indication-in-europe-and-reyataz-cleared-for-pediatric-use.html

[11] The FDA approved the subcutaneous formulation in July 2011.

[12] Terminated clinical trial numbers for these indications are NCT00739960 (sarcoidosis), NCT00035529 (multiple sclerosis).

2.3.3 Nplate

Amgen's Nplate (romiplostim) was approved by the FDA in August 2008 and by the EMEA in February 2009 for the treatment of idiopathic thrombocytopenic purpura (ITP): a autoimmune disease resulting in impaired blood clotting. This bleeding condition is caused by the production of antibodies against platelets, and it has a relatively low incidence of 66 cases per 1 million patients in adults [14]. The treatment of ITP has been revolutionized by the launch of two ITP products over the last few years: Nplate (an injectable peptide thrombopoietin mimetic FP made up of two identical immunoglobulin IgG1 Fc domains, covalently linked to a peptide containing two thrombopoietin receptor-binding domains) and GlaxoSmithKline's (GSK) Promacta®/Revolade® (eltrombopag, a nonpeptide, orally available thrombopoietin mimetic, approved by the FDA in November 2008 and Europe in March 2010). Both drugs are approved for use only after patients have been shown to be refractory to other (often cheaper) therapies (e.g., corticosteroids). Before the approval of these two drugs, the only ITP therapies were to prescribe immune suppressors such as corticosteroids and Rituxan®, administer immunoglobulins, remove the spleen, and/or administer platelet transfusions. The only Nplate Phase III trials are for ITP;[13] therefore, we do not factor in indication expansion as a key sales driver for the drug.

We believe there are two factors that differentiate Nplate positively from Promacta. First, Promacta has a black-box warning (the most serious advisory warning that the FDA provides on labels, which we think restricts sales uptake). From a safety perspective, both drugs may increase the risk for reticulin deposition within the bone marrow, while Promacta may also cause hepatotoxicity: the focus of its black-box warning. As a result, while both drugs require frequent hematologic monitoring, Promacta patients also need frequent hepatic and ocular monitoring.[14] Both may increase the risk for hematological malignancies, particularly in patients with myelodysplastic syndrome, and both have a similar range of adverse reactions. Second, Nplate has a shorter therapeutic time lag.[15]

There are other factors differentiating the two drugs, but we think their impact is minor (e.g., Nplate's once-weekly SC administration profile is likely to be similar in terms of compliance to Promacta's once-daily tablet administration). Both drugs appear to have similar efficacy according to efficacy data on each drug's U.S. label, although direct comparisons of Phase III trial results might be problematic,[16] given differences in trial design. For Nplate, patients were split into splenectomized patients (Study 1) and those who were not (Study 2), and then each group was treated for 24 weeks. For those treated with Nplate, (1) 61% of nonsplenectomized patients and 38% of splenectomized patients had a durable platelet response (vs. 5 and 0% on placebo, respectively); (2) 88% of nonsplenectomized patients and 79% of splenectomized patients had an overall platelet response (vs. 14 and 0% on placebo, respectively); (3) 20% of nonsplenectomized patients and 26% of splenectomized patients required rescue therapy during the trial (vs. 62 and 57% on placebo, respectively); and (4) Nplate patients achieved platelet counts of at least 50×10^9/L on a mean of 12.3 weeks in splenectomized patients and 15.2 weeks in nonsplenectomized patients, compared with 0.2 and 1.3 weeks, respectively, for placebo [15]. Meanwhile, in Studies 1 and 2 for Promacta, patients received 6 weeks of daily drug therapy, then 6 weeks off therapy. The percentage of patients experiencing an increase in platelet count from $<30 \times 10^9$/L to $\geq 50 \times 10^9$/L was 59 and 70% for Promacta patients (vs. 16 and 11% for placebo). In Study 3, patients either received 6 months of Promacta or placebo. In this trial, patients on Promacta for 6 months had 11.3 weeks on average with a platelet count above 50×10^{-9} platelets (vs. 2.4 weeks for placebo), and only 18% required rescue therapy during the trial course (vs. 40% for placebo). Thus, we think both drugs have a similar impact on (1) the number of patients requiring rescue therapy, and (2) the average number of weeks with a platelet count above 50×10^{-9}.

[13] According to clinicaltrials.gov and Amgen's website, as of April 2011.

[14] http://www.oregon.gov/OHA/pharmacy/therapeutics/docs/ps-2009-12-thrombopoeisis.pdf

[15] http://hematology.wustl.edu/conferences/presentations/Sanfilippo 20110318.ppt

[16] The trial results for Promacta and Nplate are taken from the U.S. labels of each drug, unless otherwise stated.

2.3.4 Dulaglutide and Albiglutide

Both Eli Lilly's dulaglutide (LY2189265) and GSK/Human Genome Sciences' albiglutide are GLP-1 (glucagon-like peptide-1) analogs, in development for the treatment of type 2 diabetes (T2D). GLP-1 analogs function by stimulating insulin secretion, inhibiting glucagon secretion and lowering blood glucose. Albiglutide is a FP composed of modified human GLP-1 fused to human albumin. Meanwhile, dulaglutide is a FP comprised GLP-1 (7-37) that has been linked to the Fc section of human IgG4. Both albiglutide's albumin fusion and dulaglutide's IgG4 Fc fragment are designed to extend half-life of the active component, thereby reducing dosing frequency (see Chapter 3).

T2D treatment is shaped by the 2009 update to the 2006 consensus algorithm published by the American Diabetic Association and the European Association for the Study of Diabetes (ADA/EASD) [16], which splits T2D therapy into tiers. Metformin (a generic oral biguanide drug) is generally considered to be the first-line T2D drug, and this forms part of Step 1 therapy in Tier 1 ("well validated") of the algorithm. However, many patients on metformin monotherapy fail to maintain glycemic goals: in a U.K. study, it was shown that 19–86% of patients who had had diabetes for ≥ 4 months achieved glycated hemoglobin (HbA1c) levels of <7% in the first year of metformin treatment, whereas 55 and 70% of these patients reaching these targets had HbA1c measurements above these targets at 2 and 3 years [17]. The primary goal of T2D therapy is to control glucose levels. Measuring hemoglobin A1c (HbA1c) has been used for more than 25 years as a way of determining a patients average blood glucose level over 60–90 days [18]. Step 2 ("additional therapy") in Tier 1 includes sulphonylurea (another generic oral drug often used in second-line therapy), however, this is linked to hypoglycemia and modest weight gain. Tier 2 ("less well validated") therapies then include newer treatments such as thiazolidinediones (TZDs), DPP-IV (dipeptidyl peptidase IV) inhibitors, and GLP-1 (glucagon-like peptide-1) agonists. TDZs and DPP-IV inhibitors tend to be used in second/third line. TDZs are associated with cardiovascular and fracture risks [19], while DPP-IV inhibitors have a favorable safety profile but relatively low efficacy. GLP-1 agonists, because they need to be injected, tend to be used more in third line than in second line [20].

We think the sales performance of GLP-1 analogs such as dulaglutide and albiglutide is dependent on: (1) generating strong efficacy data, in terms of both lowering HbA1c but equally importantly, in weight reduction; (2) demonstrating a favorable safety profile (specifically on cancer and cardiovascular) and a favorable side effect profile (specifically, on emesis); (3) having a formulation that does not negatively impact dosing and administration frequency, given that GLP-1 analogs will likely continue to be administered subcutaneously, we think the most important factor here is low frequency dosing and the scope to use a small needle gauge for delivery; and (4) positioning to market (plus level and effectiveness of marketing support).

We think administration frequency is an important differentiator, and one way of looking at the evolution of the GLP-1 market is to look at improvements in this characteristic. The first GLP-1 analogs on the U.S. market was Lilly/Amylin's Byetta® (exenatide, launched in the United States in 2005), which it has a twice-daily administration profile. Novo Nordisk's Victoza® (liraglutide, approved in the United States in January 2010) is once daily and in the LEAD-6 Phase III trial, patients on liraglutide demonstrated better glycemic control and tolerated the drug better than those on Byetta [21]. However, we think the major growth drivers for the GLP-1 market (provided they are approved) are the longer-acting once-weekly drugs. Of these, Lilly's Bydureon® (exenatide LAR) could be the first once-weekly GLP-1 in the major western markets, after it received a positive CHMP opinion in April 2011.[17] However, it has also received two complete response letters from the FDA, increasing the risk of delays or nonapproval in the United States. We think either albiglutide or dulaglutide will be the second long-acting once weekly GLP-1 agonist to hit major western markets after Bydureon.

As discussed, we think HbA1c and weight loss are the two important efficacy metrics for GLP-1 drugs. ADA's HbA1c target is <7%, while the International Diabetes Federation's target is <6.5% [22]. Of the Tier 1 T2D drugs, both metformin and sulphonylurea reduce HbA1c by 1.0–2.0%, whereas in Tier 2, TZDs reduce HbA1c by 0.5–1.4%, DPP-IV inhibitors reduce HbA1c by 0.5–0.8% and GLP-1 agonists reduce HbA1c by 0.5–1.0% [16]. In addition to glucose level control, weight reduction is increasingly important as part of T2D patient treatment. Metformin and DPP-IV inhibitors are both weight neutral, sulphonylurea and TZDs are both associated with weight gain, whereas GLP-1 agonists are associated with weight loss [16]. A recent meta-analysis looking at randomized controlled trials examining four GLP-1 agonists (Byetta, albiglutide, liraglutide, and taspoglutide) showed (1) all four drugs significantly lowered HbA1c by between 0.62 and 1.16% and (2) mean weight losses with these drugs ranged from 0.7 to 1.3 kg [23]. We believe that provided future GLP-1 agonists can demonstrate late-stage clinical trial HbA1c data that falls within this range, the weight loss efficacy data will be more important in terms of differentiating new GLP-1s from competitors.

Bydureon (likely dulaglutide/liraglutide's key competitor, in our view) has demonstrated relatively strong efficacy in terms of HbA1c reduction in the Phase III DURATION-1 (−1.9% at 30 weeks for Bydureon, vs. −1.5% for Byetta) [21]. However, in a recent head-to-head trial with Victoza (DURATION-6), Bydureon showed slightly less HbA1c reduction (−1.3%) than Victoza (−1.5%).[18] Meanwhile, Phase II data released so far suggests that dulaglutide may have slightly higher efficacy than albiglutide, although we note that it is difficult to compare data between clinical trials. In Phase II trials, HbA1c was 0.9% lower with weekly albiglutide, versus −0.2% for placebo and −0.5% for exenatide [24]. This compares to 1.5% lower HbA1c with dulaglutide over placebo [25]. In terms of weight loss, albiglutide reduced weight by −1.1 to −1.7 kg in this study: numerically better than placebo of −0.7 kg, but less than the −2.4 kg weight loss with exenatide. Meanwhile, dulaglutide also reduced weight by 1.3–2.6 kg (although only overweight/obese patients were included in the trial) [26].

In terms of advantages over Bydureon, GSK has indicated that albiglutide can be administered using a small-gauge needle[19] (other sources have indicated that a 29-gauge needle would be used [27]), likely helping compliance given that Byetta is administered with a 29-, 30-, or 31-gauge

[17] Bydureon received European marketing authorization in June 2011.

[18] https://investor.lilly.com/releasedetail2.cfm?ReleaseID=554248

[19] http://www.gsk.com/investors/reports/q42008/q42008.pdf

needle, according to its U.S. medication guide. The needle used in administering Bydureon is a relatively large 23-gauge needle,[20] potentially negatively impacting compliance. Meanwhile, for dulaglutide, Lilly has indicated that the drug is highly soluble and can be administered using a small-gauge needle[21]. Second, Bydureon is a synthetic drug derived from the Gila lizard, while both albiglutide and dulaglutide include forms of human GLP-1 and GLP-1 analogs, which should carry a lower risk of immunogenicity. A recent (December 2010) publication indicated that there have been no reports of antibodies against dulaglutide [25]. Meanwhile, anti-albiglutide antibodies have been observed, however, these were transient and non-neutralizing, and were not thought to impact efficacy or safety. Furthermore, these antibodies affected much lower percentages of the patient population than exenatide (2.5% for albiglutide [24] vs. 49% [28] for exenatide). Last, it seems that albiglutide pivotal Phase III trials complete in September 2011 to December 2012,[22] while dulaglutide Phase III trials complete between May 2012 and November 2012; therefore, it seems likely that both drugs will receive a regulatory decision (if successfully submitted) at approximately the same time.

The side effect and safety profile are also key for T2D drugs. Longer-acting GLP-1 agonists have been linked to lower nausea and likely better patient compliance [29]. As discussed, T2D treatment failure with monotherapy inevitably leads to combination therapy; however, the side effect profile is less favorable for patients on more than one drug. This is a potential issue, given relatively poor patient compliance in this population: it has been estimated that between 36% and 93% of patients take <80% of the prescribed medications [30]. Thus, safety and side effect profile are key, when looking at new T2D drugs. The main side effects with Byetta are nausea, vomiting, and diarrhea. In the 2-year pooled analysis, there was 39% nausea incidence during weeks 0–10 in the intention to treat population [31]. Similar gastrointestinal side effects were noted in Phase II trials for albiglutide and dulaglutide. In a Phase II trial, patients taking albiglutide most frequently experienced nausea (11.8–54.3%), vomiting (0–41.2%), and headache (5.9–23.5%) [24]. Meanwhile, Phase II data from the EGO trial indicated that, the most common adverse events were nausea (13.8%), diarrhea (13.8%), and abdominal distension (13.8%), at the highest dose of dulaglutide [26]. Phase III data should provide a stronger picture of the safety and side effect profile of these drugs.

In summary, we think both albiglutide and dulaglutide are well placed to carve themselves out part of the GLP-1 market, as they have advantages associated with being composed of human proteins, they have a smaller needle gauge, and although they have generated less data than Bydureon, it seems that they have favorable safety and efficacy profiles. Based on the Phase II efficacy described earlier and because dulaglutide is set to be marketed by Lilly (who have substantial experience with the exentatide franchise), this drug may initially capture a slightly greater share of the market if both drugs are launched at a similar time. However, we caution that other factors emerging between now and a potential market launch could likely play a role in shaping uptake. We note that the performance of Lilly's other long-acting GLP-1 agonist Bydureon will also likely impact dulaglutide's positioning.

2.3.5 VEGFtrap-Eye

Aflibercept is an FP made up from the Ig domain 2 of VEGF receptor-1, fused to the Ig domain 3 of VEGF receptor-2, which is then fused to the IgG1 Fc. In October 2006, a codevelopment/copromotion collaboration between Regeneron and Bayer for the treatment of eye diseases with aflibercept was signed in October 2006. As part of this deal, Bayer has commercialization rights outside the United States and will split profits equally with Regeneron, while Regeneron has exclusive rights to the U.S. market [32]. Aflibercept, termed VEGF trap Eye for ophthalmological indications, was submitted to the FDA as an ophthalmological formulation for the treatment of wet age-related macular degeneration (wet AMD) in February 2011.[23] The drug is also in earlier-stage development for the treatment of central retinal vein occlusion (CRVO) and diabetic macular edema (DME). Aflibercept is also in development in a different formulation as VEGF trap for a range of cancers, as part of a codevelopment/copromotion collaboration with Sanofi–Aventis. However, as aflibercept failed to meet the primary endpoint in Phase III trials for both pancreatic cancer (in 2009) and nonsmall-cell lung cancer (in 2011),[24] and development was stopped on ovarian cancer (in 2009) [33], our risk-adjustments to sales of this drug mean that forecasts fall below the top-6 drug sales forecasts for 2016.[25] Therefore,

[20] See "What's new in diabetes medications", Ackerman, UP Diabetes Outreach, July 2011 (www.diabetesinmichigan.org/PDF/pdfUPDON/Whats%20New%20in%20DM%20Meds.pdf).

[21] http://files.shareholder.com/downloads/LLY/0x0x258459/5acdc123-b0e7-498c-ab5c-92c599f205f2/Eli_Lilly_12-11-08.pdf

[22] Based on Phase III clinical trial completion dates in www.clinicaltrials.gov, as of August 14, 2011.

[23] The drug received a unanimous recommendation from the FDA advisory committee meeting in June 2011. While the FDA does not always follow committee meeting recommendations, we nevertheless view this positively.

[24] According to Regeneron press releases.

[25] We note that after writing this chapter, Regeneron reported that VEGF trap significantly improved survival in previously treated metastatic colorectal cancer patients; see Regeneron press release on June 6, 2011, "ZALTRAP™ (aflibercept) Significantly Improved Survival in Previously Treated Metastatic Colorectal Cancer Patients."

this section focuses on ophtalmological indications of Aflibercept (or VEGF trap-Eye).

Wet AMD is a leading cause of blindness in elderly patients. In the United States, it has been estimated that 1.6 million people over 50 have wet AMD and there are ~200,000 new cases diagnosed each year, while ~500,000 new cases of wet AMD are diagnosed annually worldwide [34]. Wet AMD is caused by abnormal neovascularisation (angiogenesis), leading to the generation of leaky vessels behind the retina growing under the macula. Following approval in 2000, the standard of wet AMD care was Novartis/QLT's Visudyne® (verteporfin/photodynamic therapy), prior to any biologic therapies approved for the indication. Visudyne works by causing localized endothelial cell destruction leading to occlusion of leaking blood vessels. Then, in December 2004, the FDA approved Eyetech Pharmaceuticals' Macugen® (pegaptanib). This nucleic acid therapeutic belongs to the aptamer drug class, and it functioned by inhibiting the interaction of VEGF with VEGF receptors (VEGF is a positive regulator of angiogenesis [35]). In 2003, the year before Macugen was approved, Visudyne generated $357 million, but by 2010, sales were $91 million.

In June 2006, the FDA approved Roche/Novartis's Lucentis® (ranibizumab): a humanized IgG1 kappa mAb antibody fragment that binds to multiple VEGF-A isoforms. The drug was approved in Europe in January 2007, and annual sales by 2010 were $3.1 billion.[26] OSI Pharmaceuticals (Macugen's earlier owner) reported Macugen sales in 2006 of $107 million, but by 2008, annual sales were <$12 million. Roche's Avastin®, an anti-VEGF mAb approved to treat a range of cancers, is closely related to Lucentis and it has been used off-label to treat wet AMD. Avastin use has been driven by the big price differences per treatment, it has been reported that Lucentis costs ~$2000, while Avastin costs ~$50 [36]. A number of U.S. agencies and bodies including the FDA, the NIH, CMS (Centers for Medicare & Medicaid Services), and the Senate Special Committee on Ageing have been involved in discussions over whether Avastin off-label could be an effective substitute for Lucentis, given the substantial cost-savings possible [37]. To help answer this question, the National Eye Institute-sponsored head-to-head CATT (comparison of AMD treatments trials) trial comparing Lucentis with Avastin was initiated in 2008. In April 2011, this trial reported results showing that Avastin and Lucentis are equally effective in treating age-related macular degeneration, in terms of visual acuity improvement and gain or loss in visual function.[27] However, the incidence of serious adverse events was 24% for Avastin patients and 19% for Lucentis patients. Furthermore, there have been reports that patients taking Avastin may be at a greater risk than Lucentis for wet AMD due to (1) the Fc component and corresponding higher protein content, and (2) potential contamination of an Avastin lot [38]. Together, these factors may restrict use of cheaper Avastin, therefore, it seems likely that new wet AMD drugs will continue to target Lucentis rather than Avastin.

Regeneron and Bayer started a Phase III development program for VEGF trap-Eye in wet AMD in August 2007, which involved two Phase III trials ("View 1" and "View 2") comparing VEGF trap-Eye with Lucentis. Data reported in November 2010 [39] for View 1 indicated that vision was maintained or improved in 96% patients receiving monthly aflibercept and 95% of those receiving a dose once or twice a month, versus 94% patients receiving monthly Lucentis. Very similar results were maintained in patients taking each drug after 1 year in the View 2 trial. Both drugs were reported to have similar side effects. Importantly, these data indicate that VEGF trap-Eye could be dosed less frequently than Lucentis [40]. Lower administration frequency is a key driver for wet AMD drugs and is thought to be another factor driving up off-label use of Avastin in wet AMD (as Avastin is thought to remain in the eye longer than Lucentis, potentially meaning lower-frequency administration[28]). However, we flag that pricing is also likely to be an important factor in the uptake of the VEGF trap-Eye, particularly for drugs showing similar efficacy.

In summary, given similar efficacy and safety View 1 and 2 data, we think that Lucentis is VEGF trap-Eye's key competitor, and independent market analysts have indicated that they believe these two drugs will be the leaders of the wet AMD market by 2013 [41] and that there are few differentiators between the drugs. However, we also think that off-label Avastin also represents a competitor, given its low-cost nature in wet AMD, similar efficacy to Lucentis, and the fact that it is thought to stay in the eye longer than Lucentis, potentially allowing for less-frequent dosing. However, the extent of this threat may have been lessened by the safety data from CATT, together potentially with any red flags during the generation of future safety data with greater patient usage.

2.4 PIPELINE DATABASE ANALYSIS

Analysis of our FP database, which includes 43 FPs in Phase II and above clinical development (including FPs on the market) yields two main findings: (1) the most common therapy area target is oncology, followed by autoimmune, and (2) many Big Pharma/Big Biotech companies have at least one FP in development. However, few have more than

[26] Novartis reported 2010 ex-U.S. sales of $1533 million, while Roche reported 2010 U.S. sales of $1565 million (based on exchange rates as of December 31, 2010).

[27] http://www.nih.gov/news/health/apr2011/nei-28.htm

[28] http://www.nei.nih.gov/news/pressreleases/022208.asp

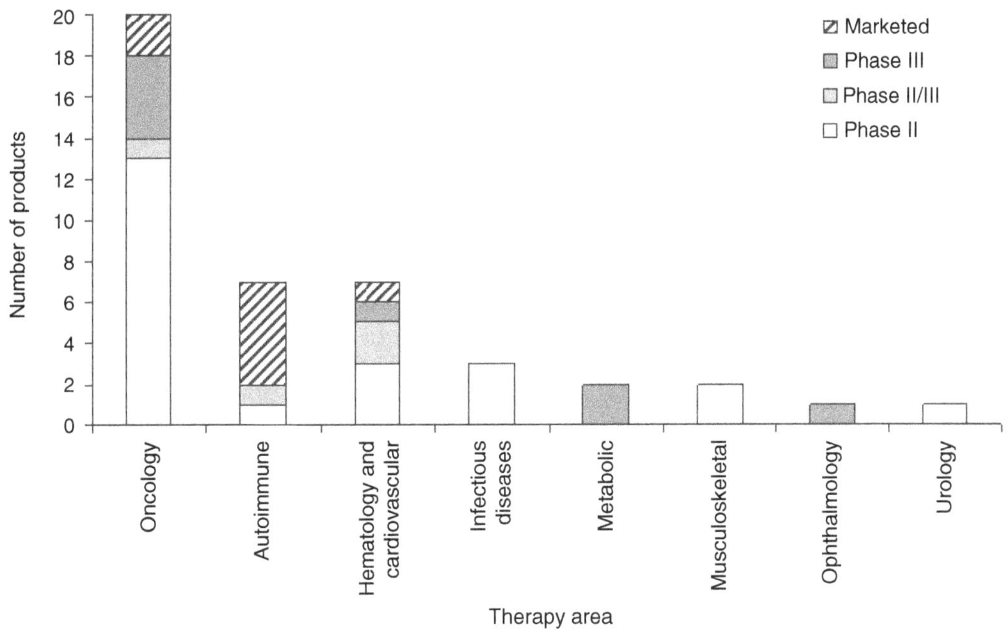

FIGURE 2.2 FP pipeline, split by therapy area and stage of development.

two, even though we believe FPs represent attractive licensing targets for these multinational companies. In this section, we look at the split by stage of development, therapeutic focus, and partnership status and trends in collaboration.

2.4.1 Breakdown by Stage of Development: Half of the FP Pipeline is in Phase II

Out of the 43 FP programs we have identified, 53% are in Phase II (including more advanced Phase I/II trials), 28% are in Phase III (including more advanced Phase II/III trials), and 19% are on the market or have been submitted to regulators (Figure 2.2).

2.4.2 Therapeutic Focus of FPs: Oncology and Autoimmune Dominate

Oncology and autoimmune are the most common focus therapeutic areas for these 43 FP programs, accounting for 63% of programs. Although autoimmune disease leads the FP market in terms of 2016 forecast sales (we estimate it will account for 78% of total 2016 sales), it is a less focused-on therapeutic area for earlier-stage FP development, accounting for only six (14%) of the FP programs identified. While surprising, this can in part be explained by the success of TNFis, meaning it can now be difficult to penetrate high-sales autoimmune disease markets such as RA. Oncology accounts for 47% of all Phase II and above FP development programs, followed by autoimmune, and hematology and cardiovascular; both at 16%. The remaining 21% is made up from infectious disease, metabolic, musculoskeletal, ophthalmology, and urology (in order of weighting).

The focus on autoimmune and oncology is typical of biotech markets [42] (especially those in the earlier stage of development), with a study in 2008 showing that 78% of analyzed ongoing biotech trials were in these therapeutic areas [43]. Autoimmune and oncology are attractive because R&D investment in these areas yields a high number of targets. Also important is the high unmet need in many of these areas (particularly oncology), and the chance for higher reimbursement. Both are popular with developers of drugs that focus on modulating specific targets in the body that are well validated in disease etiology (e.g., Enbrel's targeting of TNF-α) that are not characterized by low bioavailability (in contrast to many indications in therapy areas such as CNS, where bioavailability and well-validated targets are difficult to achieve). This targeted approach is likely to improve clinical progression success rates, which is important as drugs progressing through clinical development for either oncology or autoimmune diseases have had below-average probability of entry into Phase III and into approval, above-average Phase III trial duration and above-average costs associated with clinical development [1]. More recently, a study in 2011 showed that oncology had the lowest success rate in Phase III clinical trials in all therapeutic areas examined, even though it was the most focused-on therapeutic area[29] [44]. The same study also showed that biotech drugs targeting different cancers had different success rates, with those targeting head and neck, renal cell carcinoma, ovarian, and breast cancer having the

[29] See BIO's presentation "BIO/BioMedTracker Clinical Trial Success Rates Study".

greatest success. We note that FPs targeting oncology in our database focus on a wide range of cancers: 20% focus on the four cancer indications with the highest success rates, while no FP in our database focuses on the four least successful cancer indications as a primary target (prostate, colorectal, myelodysplastic syndromes, and nonsmall-cell lung cancer). We note that both glioblastoma and melanoma are also popular target oncology indications for the FPs in our database, together accounting for 35% of the oncology target indications.

In addition to oncology and autoimmune, hematology and cardiovascular also accounted for seven (16%) of the database of FPs in Phase II development and above. Perhaps, the most interesting point to note about these FPs is the diversity of their target indications, ranging across thrombocytopenia, neutropenia, hemophilia, acute myocardial infarction. These are serious medical conditions characterized by a high unmet medical need, but are not typically targeted by biologics such as mAbs, nucleic acid therapeutics or therapeutic vaccines.

2.4.3 FPs Developers and Partners: Many Big Pharma/Big Biotech Companies Have at Least One FP in Phase II Development or Above

A few Big Pharma/Big Biotech companies dominate the two mature biotech sectors: mAbs (e.g., Roche that has mAb blockbusters such as Avastin and Rituxan) and recombinant therapeutic proteins (e.g., recombinant insulin is dominated by Sanofi–Aventis and Novo Nordisk, whereas Amgen and Johnson & Johnson lead the recombinant erythropoietin market). While the FP market is dominated by Enbrel, most Big Pharma/Big Biotech companies (Amgen included) have only a small number of FPs in Phase II and III development, compared to the mAb market where for example Roche has 16 NME (new molecular entity) mAbs in Phase II and III development alone.

Based on our database, many Big Pharma/Big Biotech companies have at least one FP in clinical development, with both the United States and European companies having approximately an equal participation. However, we could find few examples of major drug developers with more than two FPs in Phase II and above active clinical development. Moreover, unlike the mAbs market, there are no FP technology platform companies that we could identify which have been the focus of multiple licensing deals with Big Pharma/Big Biotech companies (e.g., compared to the mAb market, where such deals were signed with "fully human" mAb platform companies such as Medarex [now Bristol-Myers Squibb], Abgenix [now Amgen], and CAT [now AstraZeneca]). Together, this suggests that FPs are not viewed such as mAbs as a technology platform for Big Pharma/Big Biotech to invest in—instead, FP drugs are selected based on factors such as their individual efficacy and safety profiles, administration, and pricing.

The typical evolution for an early-stage biotech market such as therapeutic vaccines and nucleic acid therapeutics is for early-stage developers to build a portfolio of biotech drugs for which they have generated strong preclinical and early-clinical trial data (usually up to Phase II, due to cash constraints) and then search for licensing deals for the most advanced candidates and/or those with the best profile so that potential partners can help carry out large-scale, costly Phase III trials and, where possible, build up a marketing functionality. Looking exclusively at FPs in Phase II clinical development (i.e., at the point that candidates at the same stage from early-stage biotech markets such as nucleic acid therapeutics and therapeutic vaccines would usually start to be licensed), one-third of the programs involve a Big Pharma/Big Biotech partner, while over half are partnered with either a Big Pharma/Big Biotech company or another earlier-stage drug developer. Thus, we think FPs are a more mature market than these earlier-stage biotech markets. One reason for this is potentially the same reason that could explain why Big Pharma/Big Biotech do not have more than two FPs in later-stage clinical development: that FPs are not seen as a homologous technology that Big Pharma/Big Biotech feels they lack access to; instead, each FP appears to be judged on its own characteristics. Supporting this argument is our view that FPs as a class have less of a need for a major Big Pharma/Big Biotech licensing deal to "validate" the technological platform, and our view that there has not been a wave of either FP licensing or major technology platform company acquisitions by Big Pharma/Big Biotech. Another factor likely restricting the acquisition of FP companies to deals done to acquire a specific FP product is that there are not FP technology platform companies with a broad IP estate and a wide range of pipeline candidates.

2.4.4 FP Development: Less Attractive for Earlier-Stage Drug Developers, but More Attractive for Big Pharma Licensees

Compared to other drug classes, we think that FP development could be less attractive for earlier-stage drug developers due to higher operating costs (primarily R&D costs) in the early stage than for other drug classes. To some extent, the FP drug class becomes more attractive for these developers if they are able to retain some of the marketing rights. However, we think that FP licensing is attractive for licensees due to a potentially stronger operating margin resulting from lower marketing spending requirements and often higher prices and reimbursement levels.

From the perspective of a company looking for access to FP drug candidates, the market is attractive for a number of reasons. Developing drugs for indications treated by specialists (i.e., nonprimary care physician indications) keeps sales, general and administrative (S, G, and A) costs down, because there are fewer physicians to target. Meda, a Swedish specialty pharmaceutical company, indicated that "there is a big difference in the cost of marketing and sales efforts to specialists and marketing them to general practitioners in outpatient care."[30] Furthermore, particularly for these markets, there are key physicians that shape the prescription trends of others in the field, meaning that it is cheaper to get a new drug accepted and prescribed. Moreover, we think less direct-to-consumer marketing is required for more serious, life-threatening indications (like the indications being targeted by FP developers, such as oncology), further reducing marketing spending. Targeting these indications is less costly for drug marketers than targeting primary care physician (PCP) market indications such as hypertension, gastroesophageal reflux, and hypercholesterolemia, as there are fewer physicians to market to. Additionally, biologics including FPs tend to be premium priced, with high reimbursement levels (particularly for serious and life-threatening conditions). This makes the operating margin on such drugs even more attractive.

However, from the perspective of the FP developer, FPs can be less attractive than other technologies, because of potentially higher associated operating costs. Notably, R&D expenditure is likely to be higher because of the novel and uncharacterized nature of many FP structures and mechanisms of action. Additionally, setting up manufacturing for novel platforms can be costly, particularly for biologics, particularly if manufacturing protocols are costly and complex. Although cost of goods sold (COGS) spending in early clinical trials is less important for the FP developer until these drugs reach large-scale (usually Phase III) clinical trials, COGS remains a focus for FP developers as licensees could potentially be put off by potential licensing candidates that cost a lot to manufacture. However, the FP drug class becomes more attractive for developers if they are able to retain some of the marketing rights. In addition to making FPs attractive to licensees, a small patient population and/or a population treated by specialist physicians also make it more likely that a small FP developer is able to retain some marketing rights to their drug. FP developers are additionally more likely to retain rights if (1) they already have developed other drugs (particularly if it some of these are already on the market), (2) the new drug is designed to treat the same indication as drugs they already have on the market, and (3) competition from other biotech competitors in the market is low: as the number of these competitors increases, the probability of marketing right retention falls [45].

[30] http://www.meda.se/fileadmin/uploads/MEDA_Corporate/pdf/MEDA_2010_Eng_webb.pdf

2.4.5 FPs Deal Analysis: We Believe FPs are an Attractive Target for Multinational Drug Companies

We think there are two reasons for Big Pharma to buy into FP programs. First, many drug companies are attempting to diversify away from what multinational pharmaceutical company GSK refers to as "white pill, western markets" model, because of low sales growth and substantial generic erosion risk. GSK itself has reduced exposure to this market from 40% in 2007 to 25% in 2010, and many Big Pharma companies have implemented similar strategies by investing in areas such as biotech, consumer health, and vaccines. Biotech, which includes the FP market, is attractive, because of a longer exclusivity period for new biologics in the United States, lower generics competition following patent expiry, potentially higher prices and reimbursement for biologic drugs, and potentially better clinical development success rates. A 2011 survey found that biotech drugs had a higher clinical trial success rate than small molecule drugs (26% vs. 14%) [44].

The second reason why we think Big Pharma companies form deals over FPs is because they are already established in specific therapy areas, and they are looking for next-generation drugs to replace currently marketed drugs and/or ameliorate the risk of depending too heavily on a small number of pipeline drugs they have in development for these therapy areas. For example, GSK had a big type 2 diabetes franchise with Avandia, however, following safety concerns, sales of this drug fell steeply. Therefore, the company's partnership with HGS over albiglutide may give it an opportunity to strengthen its position in this therapy area. Some FP developers/licensees are also developing FPs as direct successors to currently marketed drugs. For example, Lilly is developing dulaglutide as a long-acting GLP-1 agonist, which would be a competitor to both the already marketed GLP-1 analog Byetta and the company's long-acting Byetta formulation, Bydureon. This suggests to us that the primary drivers for these FP deals are (1) to capitalize on already established sales and marketing networks plus knowledge on which parts of the market to target or (2) to lineup replacement drugs following the loss of exclusivity for marketed drugs, or replace other pipeline drugs they are developing if data are less good with these other pipeline drugs. It seems likely that the closer allied a program is to the licensee's therapeutic focus, the less likely a licensed-in FP program will be dropped by the licensee as a result of a strategic rethink, unless it generates weaker data than other pipeline drugs in development.

In terms of deal structure, FPs have been successful in securing relatively generous licensing deals with Big Pharma/Big Biotech companies. Out of 43 FP programs we analyzed, we found information suggesting that significant collaborations had been formed by early-stage FP developers over the development of 11 of these FPs. Of these 11, 9 deals were codevelopment/copromotion deals. Of the seven deals from these nine that provided financial terms of the deal, we found that the up-front payment ranged between \$5 million and \$125 million, with an average of \$50 million, whereas total milestone payments ranged from \$67 million to \$800 million, with an average of \$336 million. Over the past few years, there has been a relatively low level of risk capital both for public and private markets, given the economic climate and the fact that even in areas with well-proven business models (mAbs) private investors have found it tough to exit and public investors have been faced by years of negative profitability. As a result, collaborations have become increasingly important to earlier-stage biotech companies in supporting cashburn and clinical development.

FP programs have not escaped deal attrition. In our deal analysis, we count at least three deal terminations over the past few years, and a further three deal restructurings. Terminations were mainly due to Big Pharma companies changing their strategic direction, while restructurings were mainly due to reorganizations of R&D responsibilities and marketing rights, or because the FP developer was looking for less exposure to development costs.

Meanwhile, we have seen some acquisitions of FP developers (most notably Amgen's \$16 billion acquisition of Immunex to capture Enbrel), but these have been fewer in number than the collaborations. There are four reasons for this: first, it can be cheaper (depending on the level of up-front payments) and easier to form a collaboration than to acquire a company. Second, there is less risk with collaborations—if it goes wrong, or if data is weak, or even if the Big Pharma/Big Biotech company has a strategy change and wants to move out of the area, it is easier for these multinational companies to exit the collaboration. Third, FP developers are likely more incentivized to ensure the FP generates strong data with a milestone structure in place. Last, there is a risk of a brain drain if a Big Pharma acquires an early-stage drug developer, given the change in work environment.

DISCLAIMER

The views expressed are the opinions of the individual authors only and are in no way connected to either WestLB or Datamonitor. No recommendation or opinion concerning any equity security is made in this chapter. The authors received no financial support for this research and declare no competing financial interests.

ACKNOWLEDGMENT

The authors are grateful to Dr Giordano Pula, University of Bath, UK, for his helpful comments on earlier drafts.

REFERENCES

1. Adams CP, Brantner VV. (2006) Estimating the cost of new drug development: is it really 802 million dollars? *Health Aff. (Millwood)* 25(2), 420–428.
2. Comfort C. (2011) Virtually the same percentage of surveyed rheumatologists selected Enbrel® and Humira as the most efficacious agent for the treatment of rheumatoid arthritis. *Decision Resources*, 1 March.
3. Comfort C. (2011) For the treatment of moderate to severe psoriasis, the highest proportion of surveyed dermatologists and MCOS indicate that Humira and Enbrel® have the best overall clinical profile when compared to other currently available agents. *Decision Resources*, 7 February.
4. Singer N. (2009) In Wyeth, Pfizer Sees a Drug Pipeline. *New York Times Magazine*, 26 January.
5. Timmerman L. (2007) Abbott's Humira Threatens Amgen in Psoriasis Market. *Bloomberg*, 7 February.
6. Grogan K. (2010) Psoriasis market worth \$6.8 billion in 2019. *PharmaTimes*, 10 August.
7. Comfort C. (2010) Humira replaces Enbrel® as the therapy perceived by dermatologists to be the most efficacious for the treatment of moderate to severe psoriasis. *Decision Resources*, 23 March.
8. Cohen J, Wilson A. (2009) New challenges to medicare beneficiary access to mAbs. *MAbs* 1(1), 56–66.
9. Moots RJ, Haraoui B, Matucci-Cerinic M, van Riel PLCM, Kekow J, Schaeverbeke T, et al. (2011) Differences in biologic dose-escalation, non-biologic and steroid intensification among three anti-TNF agents: evidence from clinical practice. *Clin. Exp. Rheumatol.* 29(1), 26–34.
10. Choi S. (2011) The use of Etanercept Enbrel® as sole treatment for grade I acute graft versus host disease NCT00726375. *Clinicaltrials*, 30 September.
11. Comfort C. (2011) Bristol-Myers Squibb's Orencia® garners greatest patient share among non-tnf biologics in the treatment of rheumatoid arthritis. *Decision Resources*, 11 July.
12. Luo A. (2009) BMS, Final Clinical Study Report for Study IM101084. 17 December.
13. Luo A. (2010) BMS, Final Clinical Study Report for Study IM101108. 10 April.
14. Silverman M. (2011) Idiopathic thrombocytopenic purpura. *Medscape*, 21 January.
15. Kuter DJ, Bussel JB, Lyons RM, Pullarkat V, Gernsheimer TB, Senecal FM, et al. (2008) Efficacy of romiplostim in patients with chronic immune thrombocytopenic purpura: a double-blind randomised controlled trial. *Lancet* 371(9610), 395–403.
16. Nathan DM, Buse JB, Davidson MB, Ferrannini E, Holman RR, Sherwin R, et al. (2009) Medical management of hyperglycaemia in type 2 diabetes mellitus: a consensus algorithm for the initiation and adjustment of therapy: a consensus statement from the American Diabetes Association and the European Association for the Study of Diabetes. *Diabetologia* 52(1), 17–30.
17. Cook MN, Girman CJ, Stein PP, Alexander CM. (2007) Initial monotherapy with either metformin or sulphonylureas often fails to achieve or maintain current glycaemic goals in patients

with Type 2 diabetes in UK primary care. *Diabet. Med.* 24(4), 350–358.
18. Bozkaya G, Ozgu E, Karaca B. (2010) The association between estimated average glucose levels and fasting plasma glucose levels. *Clinics (Sao Paulo)* 65(11), 1077–1080.
19. Hamnvik O-PR, McMahon GT. (2009) Balancing risk and benefit with oral hypoglycemic drugs. *Mt. Sinai J. Med.* 76(3), 234–243.
20. Adler AI, Shaw EJ, Stokes T, Ruiz F. (2009) Newer agents for blood glucose control in type 2 diabetes: summary of NICE guidance. *BMJ* 338, b1668.
21. Buse JB, Rosenstock J, Sesti G, Schmidt WE, Montanya E, Brett JH, et al. (2009) Liraglutide once a day versus exenatide twice a day for type 2 diabetes: a 26-week randomised, parallel-group, multinational, open-label trial (LEAD-6). *Lancet* 374(9683), 39–47.
22. Chua S, Chan S. (2011) Medication adherence and achievement of glycaemic targets in ambulatory type 2 diabetic patients. *J. Appl. Pharm. Sci.* 1(4), 55–59.
23. Shyangdan DS, Royle PL, Clar C, Sharma P, Waugh NR. (2010) Glucagon-like peptide analogues for type 2 diabetes mellitus: systematic review and meta-analysis. *BMC Endocr. Disord.* 10, 20.
24. Rosenstock J, Reusch J, Bush M, Yang F, Stewart M. (2009) Potential of albiglutide, a long-acting GLP-1 receptor agonist, in type 2 diabetes: a randomized controlled trial exploring weekly, biweekly, and monthly dosing. *Diabetes Care* 32(10), 1880–1886.
25. Jimenez-Solem E, Rasmussen MH, Christensen M, Knop FK. (2010) Dulaglutide, a long-acting GLP-1 analog fused with an Fc antibody fragment for the potential treatment of type 2 diabetes. *Curr. Opin. Mol. Ther.* 12(6), 790–797.
26. Umpierrez GE, Blevins T, Rosenstock J, Cheng C, Anderson JH, Bastyr EJ III. (2011) The effects of LY2189265, a long-acting glucagon-like peptide-1 analogue, in a randomized, placebo-controlled, double-blind study of overweight/obese patients with type 2 diabetes: the EGO study. *Diabetes Obes. Metab.* 13(5), 418–425.
27. Johnson J. (2008) New GLP-1 analogs, mimetics in the pipeline for type 2 diabetes. *Endocrine Today*, December.
28. Kendall DM, Riddle MC, Rosenstock J, Zhuang D, Kim DD, Fineman MS, et al. (2005) Effects of exenatide (exendin-4) on glycemic control over 30 weeks in patients with type 2 diabetes treated with metformin and a sulfonylurea. *Diabetes Care* 28(5), 1083–1091.
29. Freeman JS. (2011) Optimizing outcomes for GLP-1 agonists. *J. Am. Osteopath. Assoc.* 111(2 Suppl. 1), eS15–eS20.
30. Bailey CJ, Kodack M. (2011) Patient adherence to medication requirements for therapy of type 2 diabetes. *Int. J. Clin. Pract.* 65(3), 314–322.
31. Pratley RE, Gilbert M. (2008) Targeting incretins in type 2 diabetes: role of GLP-1 receptor agonists and DPP-4 inhibitors. *Rev. Diabet. Stud.* 5(2), 73–94.
32. Tirrell M. (2010) Regeneron rises as drug matches Roche's in fewer doses. *Bloomberg*, 22 November.
33. Grogan K. (2011) *Sanofi/Regeneron's aflibercept fails in NSCLC trial. PharmaTimes*, 11 March.
34. Melnikova I. (2005) Wet age-related macular degeneration. *Nat. Rev. Drug Discov.* 4(9), 711–712.
35. Brown D. (2005) AMD: underlying causes and treatment: unraveling angiogenesis. *Medscape.*
36. Pollack A. (2011) Cheaper drug to treat eye disease is effective. *New York Times*, 28 April.
37. Baghdadi R. (2007) The Lucentis/Avastin investigation: the story is far from over. *In Vivo*, 13 December.
38. Johnson D, Hollands H, Hollands S, Sharma S. (2010) Incidence and characteristics of acute intraocular inflammation after intravitreal injection of bevacizumab: a retrospective cohort study. *Can. J. Ophthalmol.* 45(3), 239–242.
39. Dennis M. (2010) Studies: Bayer, Regeneron's VEGF Trap-Eye matches Lucentis in patients with wet AMD. *FirstWord*, 22 November.
40. McCallister E. (2010) Wanted: non-inferior pricing, too. *BioCentury*, 6 December.
41. Comfort C. (2010) Genentech/novartis ophthalmics' Lucentis and Regeneron/Bayer's Aflibercept will become decision resources' proprietary clinical gold standard for the treatment of wet age-related macular degeneration in 2013. *Decision Resources*, 16 Februray.
42. Pijpers F, Belsey MJ. (2006) Cancer remains the dominant disease target for biotech through to 2010. *J. Commercial Biotechnol.* 12(4), 294–298.
43. Blazynski C, DeGregorio T. (2008) Boom time for biotech? *Informa*, 31 December.
44. Cooley T. (2011) New study shows the rate of drug approvals lower than previously reported. *Businesswire*, 14 Februray.
45. Levine A. (2009) Licensing and scale economies in the biotechnology pharmaceutical industry. *RWJ Sch. Health Policy Res.* April.

3

STRUCTURAL ASPECTS OF FUSION PROTEINS DETERMINING THE LEVEL OF COMMERCIAL SUCCESS

GILES SOMERS

Datamonitor, London, UK

The previous chapter described the commercial outlook for marketed and upcoming fusion proteins (FPs). In this chapter, we look in more detail at the structural aspects of FPs, drawing out the factors that play a key role in determining the level of commercial success that a product will attain.

3.1 CLASSIFICATION OF FPs

3.1.1 Pharmacokinetic, Targeting, and Activity FP Components

A wide range of FPs exists and in order to better understand their commercial potential, it is useful to investigate their differences in terms of structure and function. In this chapter, FPs have been defined as non-native proteins produced from a genetic fusion of sequences encoding two or more protein components, which unless specified otherwise include whole proteins, protein fragments, and short peptides. The FP components must be combined in gene expression, rather than being chemically joined at a later time. The analysis does not include products consisting only of fused antigens, such as multiepitope vaccines. However, for reference, we identified five such products in Phase II development or beyond. Additionally, FPs produced in the body via administration of a gene or cell therapy are also excluded.

We classified the components of each FP as one of three different functional groupings (Figure 3.1). Allowing for homogenous combinations, a two-component FP could, therefore, have any of six different combinations. The first component type is "activity." To alter the symptoms or processes behind a medical condition, pharmacological treatments must convey some form of activity. Although this activity component can be fused to another, in the vast majority of FPs the partner instead serves to aid delivery of the active moiety to site of action. This can be achieved by either altering the locations to which the protein is distributed (targeting), or by rising the duration for which it remains active (half-life). A particular protein will be predisposed to distribute to certain areas of the body, giving variable levels of activity throughout. We defined targeting moieties as those that have a primary purpose of selectively delivering an activity partner, thereby increasing its concentration and potency at desired sites, while reducing side effects at others. The physical properties of a protein also dictate the rate at which it is broken down or removed from the body. Half-life moieties can be included in an FP to alter the duration for which a partner protein remains active and thereby able to modulate target sites.

The three FP component types (activity, targeting, and half-life) are shown in Figure 3.1. A linker or spacer peptide can also be included to aid folding of the proteins and/or biologic activity. Of note, this chapter is primarily focused on significant secondary and tertiary structural differences or component combinations rather than the fine tuning of specific proteins, such as through sequence modifications.

Fusion Protein Technologies for Biopharmaceuticals: Applications and Challenges, First Edition. Edited by Stefan R. Schmidt.

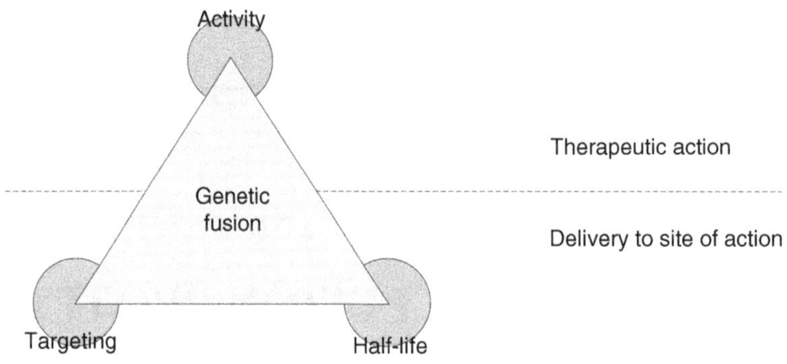

FIGURE 3.1 The three FP component types.

3.1.2 FPs Classified by the Function of Each Component

We identified FPs that are marketed or in development using ClinicalTrials.gov, company websites, PubMed, and internet search engines. A total of 43 FPs are in Phase II development or more advanced. Only those in the Phase II portion of a Phase I/II trial have been considered, while any stated as discontinued have been excluded.

Figure 3.2 shows these FP products classified by the role or function of their activity, targeting, or half-life components. Activity moieties are shown on the *x*-axis, classified according to whether the function is imparted by a receptor, enzyme, toxin, ligand, antigen, antibody (Ab) scFv (binding), or Ab Fc (effector) function. FPs could be classified in any number of ways; however, this approach was chosen since it allows key differences to be drawn out. Identification codes relating to the FP component combinations shown in Figure 3.2 are referenced throughout this chapter.

Targeting moieties have simply been split into those based on antibody-type structures (Ab), and those that use other approaches to bind targets such as receptor domains or ligands (non-Ab). Ab structures are defined as full antibodies, or scFv-type antigen-binding portions. Different types of half-life partner identified have been left grouped but are discussed in detail in this chapter.

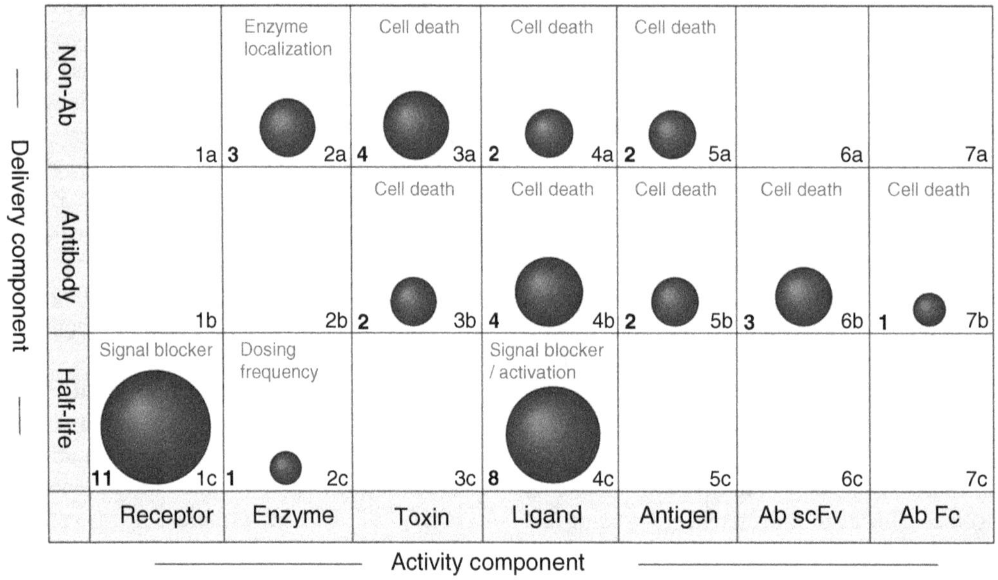

FIGURE 3.2 Classification of FPs in Phase II or above, according to component function. *Note*: The number in the bottom left corner of each segment indicates the number of products, whereas that on the right is the cell identification code.

3.1.3 Half-Life Extension Represents a Dominant Theme (1c–7c)

Of the 43 FPs identified, 20 involve partners that have a primary function of conveying an extended half-life to the activity component. Revolutions in the areas of genetics and biotechnology have given rise to the discovery of proteins with highly specific activity on targets that are either novel, or were not previously druggable with small molecule approaches. However, these biologic structures can often be afflicted by unfavorable pharmacokinetics. Natural metabolic processes result in the activity of a protein diminishing and *in vivo* half-lives can be less than 3 min [1]. Increasing the plasma half-life of a protein can either allow it to become a viable therapeutic option or where extended further, give a commercial advantage over competitors requiring more cumbersome administration regimes. A quarter of the half-life extension FPs relate to next-generation versions of existing marketed therapeutic proteins that are designed to require less-frequent dosing. In addition to dosing benefits, greater efficacy can potentially result from the active component remaining at therapeutically relevant levels for longer, while the avoidance of dosing spikes associated with short-lived molecules can reduce associated side effects.

The half-life of a protein at sites of action in the body is determined by a balance of factors, such as tissue biodistribution, receptor interactions, enzymatic cleavage sites, immune system recognition, susceptibility to processing in the liver, and rate of renal clearance. Available strategies to increase half-life, therefore, seek to address a number of these susceptibilities. The technologies employed can offer different advantages or disadvantages on a case-by-case basis.

3.1.3.1 Fc Fusion is the Half-Life Extension Standard

Aside from sequence modifications to reduce degradation, the FPs identified with dedicated half-life extending partners are designed to exploit a dual strategy. First, addition of a larger partner protein (fusion or conjugated) can reduce clearance in the kidney [2]. Almost all of these FPs use addition of an immunoglobulin (IgG) Fc domain (Fc fusions), which conveys further half-life extension by binding to the neonatal Fc receptor (FcRn), and as a result, protects the partner protein from being broken down. Plasma proteins are ordinarily internalized by endothelial cells and targeted for degradation by the lysosome; however, those bound to FcRn are instead recycled back to the cell surface and protected [3]. Pfizer's BeneFIX (rFactorIX) typically requires 2–3 injections per week to be given for the prophylactic treatment for hemophilia B, due to its half-life of ~18 h. By comparison, the half-life of Biogen Idec's Phase II/III candidate rFIX-Fc is approximately three to four times longer [4]. The immunoglobulins IgG1, IgG2, and IgG4 have serum half-lives of 21 days [5].

While a longer half-life can bring significant commercial advantages, it should be noted that fusion to a large partner, such as an Fc domain, can obscure and impair the FP's activity and this is potentially worsened by the dimerization of most Fc fusions [4]. An FP may also exhibit a shorter half-life than expected due to conformational changes or steric hindrance effects resulting from fusion to the partner component [6]. Strategies undertaken to improve half-life may not therefore yield the benefits hoped for, although reductions in potency can be offset by superior pharmacokinetic properties [4]. Biogen Idec acquired Syntonix in 2007, gaining its monomeric Fc technology that aims to avoid the steric hindrance typical of Fc dimers.

Endogenous IgG Fc domains have effector functions, inducing antibody-dependent cell-mediated cytotoxicity (ADCC) and complement-dependent cytotoxicity (CDC) that can be either beneficial for efficacy or a safety concern. These functions are generally achieved via interaction with Fc receptors on effector cells or via complement binding. Developers of Fc fusions seek to modify sequences and block effector function where resulting reactions such as cell killing, cytokine release, or inflammation are undesirable. Strategies to improve the immunological profile of Fc fusions include humanization, hybridization (e.g., Genexine's HyFc), and removal of sugar chains (deglycosylation).

Despite the challenges, Fc-addition represents the most widely used FP half-life extension approach. Indeed, Fc fusions account for almost half of the FPs in Phase II or more advanced. We consider this to have been driven in part by the experience with Fc domains gained through mAb development. It also supported by the fact that there are six marketed or approved Fc fusions: Pfizer/Amgen's Enbrel® (etanercept), Amgen's Nplate® (romiplostim), Astellas' Amevive® (alefacept), Bristol-Myers Squibb's Orencia® (abatacept), Regeneron's Arcalyst® (rilonacept), and Bristol-Myers Squibb's Nulojix® (belatacept). Enbrel is the most successful FP to date, generating over $7 billion in 2010 to account for the bulk of total FP sales (see Chapter 2). As such, Fc fusions are well validated in both clinically and commercially compared to other novel or less-tested half-life extension technologies.

3.1.3.2 Albumin Fusion is the Most Advanced non-FcFP Half-Life Approach

Albumin is the most abundant human plasma protein, with a physiological role as a carrier of blood substances.[1] Like IgG Fc, it has a long half-life of around 19 days and interacts with FcRn, albeit at distinct and separate sites. However, human albumin does not convey the effector functions associated with unmodified IgG Fc regions. Furthermore, dependent on the fused partner component, albumin fusions can be produced in yeast (as well as

[1] http://www.biospace.com/News/novozymes-biopharma-presents-research-on/206971

fungi or plants). By camparison, identified fusions are all produced in mammalian cells apart from one, Amgen's AMG-386 which is a peptide-Fc fusion (peptibody) and can utilize bacterial fermentation.

The two albumin fusions in late-stage development (both Phase III) are being developed by Human Genome Sciences (HGS) (albiglutide partnered with GlaxoSmithKline) and Teva (albugranin) and Teva. Although the technology has passed through several stages of ownership, HGS gained rights when it acquired Principia in 2000, and Teva after acquiring CoGenesys in 2008. HGS' Zalbin/Joulferon® (albinterferon-α-2b) was previously the most advanced albumin fusion but was discontinued in October 2010. Existing treatment options for hepatitis C virus infection include unmodified interferon (IFN)-α given three times weekly or pegylated IFNs given once-weekly. Pharmacodynamic modeling supported treatment with Zalbin at 2- or 4-week intervals [7]. Although Zalbin met the Phase III primary endpoint of not being statistically inferior to Pegasys (PEG-IFN-α2a), safety and efficacy results were slightly worse [8,9]. This came despite the rationale that the increased half-life of Zalbin would maximize drug exposure and therefore viral suppression, as well as initial studies that suggested lower immunogenicity with Zalbin. Trials of higher dose Zalbin been halted earlier due to the occurrence of serious respiratory adverse events. The FDA took the view that limited clinical benefit did not warrant the increased risk and issued a Complete Response Letter. HGS and partner Novartis subsequently discontinued development. These experiences highlight the low tolerance of regulators to approve more convenient dosing formulations without a suitable risk-benefit ratio.

Other FP extended half-life approaches are available, but currently in earlier stages of development. In 2007, Pfizer acquired BioRexis and its transferrin fusion-based half-life extension technology. Similarly, GlaxoSmithKline acquired Domantis in 2006 to gain access to its domain antibody (dAb) technology. Albumin-binding antibodies (AlbudAbs) are dAbs that are designed to bind to human serum albumin (SA), raising the half-life of a dAb from less than an hour, to several days. By fusing a protein of interest to an AlbudAb, half-life extension can be conveyed without the need for albumin itself to be included in the FP [10]. These smaller FPs potentially allow administration via inhaled, topical, and even oral formulations, depending on the partner protein. The principles behind AlbudAb fusions can also be applied to other antibody fragment forms or binding components. Ablynx is using albumin targeting to extend the half-life of constructs including bispecific nanobodies, while Affibody's Albumod™ technology relates to fusions involving a 5 kDa albumin-binding domain (ABD). Roche (then Genentech), Isogenica, and Philochem have developed further half-life extension structures based on albumin-binding elements. In each case, the active protein can be attached either as part of an FP, or by conjugation.

*3.1.3.3 **Nonfusion Alternatives*** A number of nonfusion half-life extension technologies exist. As with Fc and albumin fusion, these largely relate to increasing the size of the protein in order to reduce degradation and clearance. Attachment of one or more polyethylene glycol (PEG) chains (PEGylation) is the most widely adopted approach and has been validated by several successful marketed products. The large PEG chains can form a shield that protects a protein from proteolytic degradation, while not overly blocking its activity. Renal excretion is also slowed and immunogenicity can be improved. PEG is comparatively low cost to produce and well understood, but it is nonbiodegradable and concerns have been raised regarding potential kidney issues, particularly in patients with impaired renal function [11]. Improvements upon standard PEGylation have been sought, including more consistent manufacture (Polytherics' site-specific conjugation) and altered pharmacokinetics (Enzon's releasable PEGylation).

Alternatives to PEG in development include polysialic acid (Xenetic Biosciences, PolyXen®), hyaluronic acid, and sugars (Fresenius' HESylation®). Approaches that involve peptide chains can potentially also be produced using genetic fusion, such as XL-protein's PASylation® and Amunix's XTEN. Strategies to reduce degradation and extend the half-life of therapeutic proteins do not only include direct attachment of other molecules. Particles or microspheres can be used, potentially protecting proteins to the extent that permits oral dosing [12]. Similarly, Flamel Technologies' Medusa (R) nanogel is a looser entrapment approach (pGluVE or PGA A1).

In an interesting scenario, Eli Lilly was developing two human GLP-1 products—dulaglutide (GLP-1-Fc, LY2189265) and GLP-1-PEG (LY2428757), positioned as alternatives to the GLP-1 analog peptide exenatide, which is available in a twice-daily injectable (Byetta®) and more recently, a once-weekly form (Bydureon™; approved in the EU, but in registration in the United States with a decision date of January 28, 2012). At this time, dulaglutide (GLP-1-Fc) appears to have been preferentially progressed into studies. Several other next-generation GLP-1 candidates are also in development, including HGS/GlaxoSmithKline's albumin fusion, albiglutide. We have forecast dulaglutide and albiglutide to be in the top 6 FP drugs by sales in 2016 (see Chapter 2). Following its acquisition of CovX in 2007, Pfizer is undertaking Phase I studies of PF-04856883/CVX-096, which is based on CovX-Body technology wherein a linker is added to the GLP-1 peptide. This linker is subsequently bound by a specific injected antibody, which serves to increase the half-life of the peptide. ConjuChem was developing an albumin-GLP-1 conjugate CJC-1134-PC, however, development may have ceased following liquidation of the company. The range of approaches highlights the level of competition that can be faced by longer half-life products. Advantages or disadvantages of each can depend on the specific application and may only become clearer in clinical development, if at all. It should also be noted that in terms of facilitated dosing, numerous other competing technologies exist outside of half-life extension (discussed later in this chapter).

3.1.4 Decoy Receptors are the Largest FP Group and Well Validated (1c)

The role of biochemical ligands in nature is to bind to specific target proteins, causing a change in their status, which can for example be linked to initiation of a signaling cascade. Where this is implicated in a disease, pharmacological intervention can take place via either the target receptor or its to inhibit the disease pathway. Decoy receptor therapeutics use a ligand-binding component, such as a receptor fragment, to prevent the ligand from binding with its natural target.

The effectiveness of decoy receptor approaches for a disorder can depend on a multitude of factors, including reaction stoichiometry. Where the ligand is present in high concentrations, receptor blockade may instead represent a more efficient strategy. The specificity and side effect profile may also vary. Decoy receptors can prevent the blocked ligand from interacting with any other receptors that may be unrelated to the disease, while receptor blockade can impact the pathways of other ligands that have affinity with the receptor.

Of the decoy receptor FPs we identified, nine are cytokine traps, whereas the remaining two are T-cell activator traps. These agents target 9 different ligands, with development spanning 10 different lead indications. This highlights the various roles for decoy receptors. In comparison to other Fc fusion subclasses, decoy receptors are the most mature, with four products launched on the market—most notably, Enbrel. Despite its commercial success, Enbrel is the only anti-TNF FP and it has lost a considerable share of autoimmune disorder markets to mAb anti-TNF competitors such as Remicade® and Humira® (see Chapter 2). Primary differences between the FP and mAb products relate to administration method and frequency. Both classes of TNF blocker bind to the target ligand with high affinity, although the mAbs dissociate more slowly and have a longer serum half-life [13,14]. Clinically significant differences have also been revealed in certain indications, allowing the mAbs to command broader labeling [15]. This is thought to relate to subtle differences in the binding of TNF monomers and trimer forms, which most likely resulted from binding epitope differences rather being a function of broader structural differences between the FPs and mAbs [16].

All of the decoy receptor FPs involve partnering with an Fc portion of human IgG1 for half-life. Aside from Bristol-Myers Squibb's Nulojix (belatacept), it is unclear whether the Fc exerts any effector functions. These can cause unintended cytotoxic effects and the labels for anti-TNF mAbs Humira and Remicade warn of the risk to TNF-producing cells [13]. Cimzia® (certolizumab) is a further non-FP TNF blocker, but consists of an Ab-binding domain conjugated to PEG and does not therefore include an Fc component. The anti-TNF biologics have enjoyed significant levels of sales by offering improved efficacy to large populations of autoimmune disorder patients via a mechanism that was not previously possible using small molecule drugs. Although pentoxifylline and bupropion have been shown to have anti-TNF activity, levels of efficacy were only low [17,18]. Other approaches have, however, been proposed, such as by inhibiting TNF transcription, enhancing TNF mRNA degradation, or blocking the formation of TNF receptors via the preligand-binding assembly domain (PLAD) [19,20]. As for all FP classes, developers must stay aware of potential competing technologies that could erode commercial advantages of the FP approach.

3.1.5 Extended Half-Life Ligands (4c)

While decoy receptors involve receptor fragments to bind and block ligands, extended half-life ligand FPs are instead designed to utilize the function of a ligand. In both cases, the ligand or receptor provides activity as well as binding specificity. As such, no targeting partner is required to deliver the active components.

Where the half-life of a receptor fragment or ligand alone is sufficient, the unattached therapeutic proteins can be commercially successful—such as recombinant blood factor proteins or insulins. However, as described in earlier sections, a longer half-life may be required for a product to be efficacious, or to offer benefits over competitors. In this group of FPs, six are partnered with IgG Fc for half-life extension, with the remaining two being albumin fusions. Endogenous ligands are often short peptides as opposed to large proteins. Since the short peptides can be made by synthetic approaches rather than recombinant, decisions of whether to develop an FP or use covalent attachment to a half-life moiety, can be highly dependent on the nature of the ligand involved (discussed further in manufacturing aspects later in this chapter).

The majority of identified extended half-life ligand FPs are designed to bind to specific receptors and activate signaling pathways. In keeping with this, all but one of the lead development indications aim to restore normal function in metabolic or hematological disorders, with the ligand being either a recombinant human protein (such as G-CSF or FVIII-Fc) or a mimetic (GLP-1 analog). Amgen's Nplate is the only marketed product in this group, consisting of a thrombopoietin (TPO) mimetic fused to IgG Fc and approved for the treatment of thrombocytopenia. In a further activation approach, Immutep's ImmuFact acts as an adjuvant by binding to MHC class II receptors, resulting in the production of cytokines required for an immune response. Unlike these activating FPs, two ligand half-life FPs are instead designed to bind to target receptors and thereby block another disease-related ligand from doing so. Biogen Idec/ Astellas' Amevive (R) (alefacept) is the only marketed signal-blocking ligand half-life FP, formed of a ligand fragment fused to IgG Fc that prevents the co-stimulatory molecule interaction with the target receptor. It is approved for the treatment of plaque psoriasis. However, the product has disappointed in the market due to the superior efficacy and safety offered by TNF inhibitors and failure to gain EU approval.

3.1.6 FPs Offer a Wide Range of Cell Death Mechanisms (3a–7b)

Disorders involving malfunctioning cells can potentially be treated in a number of ways. Where associated with disease symptoms or the survival and proliferation of disease-related cells, individual proteins represent potential candidates for manipulation. Alternatively, problematic cells can be killed off. Studies have identified numerous antigens that are associated with cells involved in cancer and other diseases. These provide a rich source of targets that can be employed by antibodies or other molecules that selectively bind to them.

Out of the 43 total FPs, 20 relate to cell death mechanisms. Of these, 16 are in development for cancer indications, with the remainder treating infectious diseases and to a lesser extent, autoimmune and urology (benign prostatic hyperplasia) disorders (see therapy area discussion in Chapter 2). The cell death products span eight different partner combinations with the largest two being ligands (IL-2 or TNF) fused to Abs and microbial toxins fused to non-Abs (Figure 3.2: 4b and 3a). A further approach consists of vaccine FPs that efficiently deliver antigens directly to the immune system (5a and 5b). Similarly, bispecific antibody constructs have been designed to actively recruit the immune response machinery (7b).

3.1.6.1 Targeted Toxins (3a and 3b) Cytotoxic agents historically used for the treatment of cancer, for example, typically have low specificity for tumors, causing problematic side effects. In "magic bullet" approaches, cytotoxics can be targeted to particular areas of the body or cell types, resulting in a more concentrated dose at desired sites. This allows higher overall doses or more potent agents to be used, increasing efficacy, as well as reducing doses in other areas of the body for an improved side effect profile. Of the six targeted toxin FPs, four use non-Ab delivery, including the most advanced, Eisai's Ontak® (denileukin diftitox) which is on the market and NeoPharm's cintredekin besudotox in Phase III. Several of these non-Ab FPs use interleukins to deliver the microbial toxin to locations expressing interleukin receptors, while the two antibody-delivered FPs (recombinant immunotoxin) target cell-surface antigens. In both groups, the FPs are designed to exploit certain targets that are disproportionately associated with cells causing the disease.

Ontak gained FDA approval in 1999 for the treatment of adult patients with recurrent or persistent cutaneous T-cell lymphoma (CTCL) expressing the CD25 component of the IL-2 receptor. It is currently the only marketed targeted toxin. To date, Ontak has failed to achieve annual global sales of over \$100 million—primarily due to a small approved patient population, acute hypersensitivity reactions, and a lack of indication expansion [21].[2,3] CTCL accounts for only 2.2% of total lymphoma cases in the United States.[4] As a technology, such FPs are positioned to compete with conjugated immunotoxins (immunoconjugates/Ab–drug conjugates), such as Pfizer's Mylotarg® (gemtuzumab ozogamicin), which was granted accelerated FDA approval in 2000 for the treatment of patients aged 60 or over with CD33+ recurrent acute myeloid leukemia who were not considered candidates for other chemotherapy. However, the product was withdrawn from the market in June 2010 after further trials showed it to offer no clinical benefit but an increased fatality rate. Seattle Genetics' conjugated Ab, Adcetris™ (brentuximab vedotin) gained FDA approval in August 2011.

Initial targeted toxin development attempts focused on chemo-labeled antibodies, or standard chemotherapy agents conjugated to mAbs. These small molecule toxin approaches offered only limited potency, and so, more efficient agents were employed, including the attachment or fusion to enzymatic peptides that required only one copy to kill a cell. Safety concerns exist regarding the administration of highly potent toxins. Importantly, the toxin and targeting component must contain minimal or no sequences that bind to targets in healthy tissue. The toxicity and fatalities caused by Mylotarg are thought to result from unintended delivery to CD33+ cells in the liver [22]. Aside from delivery to healthy cells expressing the right target, toxicity side effects with targeted toxins can also arise from nonspecific uptake by cells or indirect exposure to the released toxin.

In many cases, the toxin is designed to be internalized via receptor-mediated endocytosis of the linked delivery component. The therapeutic profile of a product will be affected by factors such as the accessibility of targets, target half-life and the internalization half-life of the complex following

[2] http://www.evaluatepharma.com/Universal/View.aspx?type=Entity&entityType=Product&id=35272&lType=modData&componentID=1002

[3] http://www.fda.gov/downloads/Drugs/GuidanceComplianceRegulatoryInformation/EnforcementActivitiesbyFDA/WarningLettersandNoticeofViolationLetterstoPharmaceuticalCompanies/ucm053642.pdf?utm_campaign=Google2&utm_ source=fdaSearch&utm_medium=website&utm_term=ucm053642.pdf&utm_content=1

[4] http://www.fda.gov/ohrms/dockets/ac/05/briefing/2005-4191B1_03_Ligand-Ontak.pdf

binding. Furthermore, once-internalized, additional factors include whether the target receptor is recycled or degraded, as well as how the conditions experienced will affect toxin release or its activity [23]. Levels of active toxin release from target cells can cause bystander effects on surrounding cells. This can aid efficacy, but also increase associated toxicity side effects.

Not all highly specific disease-associated targets will drive internalization of an attached toxin. Once Active Biotech's Phase II/III candidate Anyara (ABR-217620, naptumomab estafenatox) has bound to target cancer cells, the fused toxin recruits T cells that initiate cell death. In a lower specificity approach, Protox's PRX-302 is an FP containing a cleavage site that allows the toxic prodrug to be activated by the enzyme PSA. PSA is produced at high levels in the prostates of patients with prostate cancer but less so in other areas of the body. This mechanism of action is similar to that of antibody-directed enzyme prodrug therapy (ADEPT). Localization of activity can also be achieved by tethering radionucleotides to target cells (non-FP). Spectrum/Bayer's Zevalin® (ibritumomab tiuxetan) launched in 2002, followed by Corixa/Glaxo SmithKline's Bexxar® (tositumomab) in 2003.

The stability of the linker between the toxin and delivery components of an FP or conjugate can have a major impact on its efficacy and safety. The structural chemistry and amino acid sequence of the linker will determine its susceptibility to be broken down in any particular location. Toxin release away from intended sites increases the risk of side effects. Various different linker structures are in development, with an aim of ensuring that toxins remain attached in serum but are appropriately released and active in target cells. Conjugated immunotoxins can potentially employ a wider range of linker designs compared to FPs, since they can utilize synthetic linkers or chemistry not possible with FP protein expression processes. In March 2011, Abbott agreed to license Seattle Genetics' antibody–drug conjugate technology, based on a stable linker and potent cytotoxic agent. As with several other FP classes, conjugation may represent a comparable alternative and the sum of overall advantages and disadvantages of conjugated toxins will vary on a case-by-case basis.

It is notable that the non-Ab delivered toxins do not involve any component specifically added to increase half-life. While this may be due to the targeting partner already conveying sufficient half-life properties to the toxin, a shorter half-life may be desirable for targeted toxins to avoid sensitivity reactions [24]. It can also allow higher maximum tolerated doses, since where off-target release can occur, it is preferable for toxins to be delivered to target cells in a minimal timeframe, or within the window of a short half-life. This is in contrast to several products in other FP groups that require a long or sustained duration of action to be efficacious.

3.1.6.2 Targeted Ligands (4a and 4b) As described earlier, a number of FPs consist of a ligand attached to a half-life moiety to achieve long-acting blockade or activation of signaling. While these products are directed simply by the affinity of the ligand to certain targets, further FP groups involve fused components that provide active delivery. All six cases in Phase II or above are in development for cancer indications, delivering highly potent cytokines, TNF or IL-2, to drive cell death. The majority of these are Ab-delivered "immunocytokines." Targeted cytokines primarily bring about cell death by directing the immune system to target cells. In contrast to most targeted toxins, the targeted ligands typically bind to targets that allow the cytokine to tether and elicit receptor-mediated death-signaling, rather than undergo internalization.

Novartis markets Proleukin®, a recombinant IL-2 for the treatment of various cancers. High-dose IL-2 induces tumor response rates of $\sim$15% in patients with metastatic melanoma, with nearly half of these responses being extremely durable [25]. IL-2 acts by raising the immune system's awareness of cancerous cells. High concentrations of IL-2 through systemic administration are associated with severe side effects. Direct injection can improve the treatment of some solid tumors, but most are not localized or accessible. As such, a number of candidates are in development to provide more targeted delivery of IL-2. Attachment to a partner protein typically also serves to increase half-life—which is less than an hour for IL-2 alone. Against competition from Proleukin and upcoming biosimilars, IL-2-based FPs developed for overlapping indications must offer sufficient increases in efficacy, or reduction in side effects to warrant market share capture.

Of the six targeted ligands, four use Ab delivery. Philogen is developing three of the products. Two are based on the L19 antibody that binds to tumor-associated fibronectin, which is a marker of angiogenesis. Bayer Schering initially licensed rights to three such products in 1999—Darleukin (L19-IL2), L19-TNF, and a non-FP radiolabel radretumab (L19-131I). A further unpartnered FP is in Phase II that consists of IL-2 fused to an alternative antibody. Darleukin is the most advanced, with a Phase III trial for malignant melanoma planned. Bayer had ended its agreement for the product by mid-2009, but in February 2011 also returned rights to L19-TNF and radretumab, citing a change of strategy. As a result, Philogen had to cancel its planned IPO. Philogen has several different products based on the same antibody, allowing some comparison of the approaches to be made. However, head-to-head or same indication trials have not yet been undertaken, which may relate to different approaches being better suited to certain cancers than others. Merck KGaA is developing a further immunocytokine FP, huKS-IL2 (tucotuzumab celmoleukin, EMD 273066) although a Phase II trial for ovarian cancer was terminated by April 2009 due to poor recruitment and a study design that was no longer considered appropriate.[5] The product is no longer shown in the company's pipeline and as such, it has not been included in the FP database.[6]

Other delivery moieties such as receptor fragments can be used to deliver cytokines. Altor BioScience's ALT-801 is in Phase I/II development for metastatic melanoma. It consists of IL-2 fused to a soluble T-cell receptor (TCR) that specifically binds to p53 on cancerous cells, including those that are not readily accessible to larger mAbs. In an alternative approach, MolMed has fused cytokines and other proteins to a tumor-targeting NGR (Asn-Gly-Arg)-containing peptide sequence. The NGR motif is a natural ligand for certain CD13 receptor isoforms that are selectively expressed by endothelial cells in newly formed tumor vessels. NGR-hTNF is in Phase III development for the treatment of mesothelioma. Both NGR-hTNF and ALT-801 are designed to offer convenient dosing of the partnered cytokine with a comparatively low toxicity profile. As well as targeting, NGR binding also has a direct effect on CD13, adding to the antitumoral activity of the attached cytokine [26]. Similarly, studies have found the small molecule drug doxorubicin coupled to an NGR peptide to display greater potency but lower side effects compared to the free drug [27]. FP developers must be careful to ensure that their approaches are not built-upon by others to create commercially stronger competitor products.

It should be noted that targeted ligands in earlier stages of development involve a wider range of ligands than just cytokines. Furthermore, not all targeted cytokines are designed to bring about cell death. Products are also in development for nononcology indications, including F8-IL10 that delivers IL-10, which is an anti-inflammatory cytokine with a potential role in the treatment of rheumatoid arthritis.

3.1.6.3 Antigen-Mediated Cell Death (5a and 5b) Vaccines typically initiate an immune response after processed by that display the antigenic components on their surface via class II MHC molecules. These are then recognized by T cells that cannot otherwise identify foreign antigens. FP immunotherapies in development are designed to be more efficient by interacting directly with the immune system machinery.

Of the 43 FPs in Phase II or above, three act by delivering antigens to APCs. Celldex has two FP APC-targeting candidates in Phase II development for the treatment of cancers. Ab delivery components bind to targets on the surface of APCs that are internalized via receptor-mediated endocytosis. The fused antigen fragments are subsequently processed and presented to T cells, which results in an immune response against cells expressing the antigen. Similarly, VaxInnate uses a non-Ab approach, with an antigen delivered to APCs using a bacterial ligand, flagellin. Toll-like receptors on APCs have evolved to recognize a wide variety of microbial proteins, or pathogen-associated molecular patterns (PAMPs). Using a viral antigen fused to flagellin, VaxInnate is developing an approach for treating influenza.

3.1.7 Antibody Fragment Fusions (6b and 7b)

Advances in protein engineering have facilitated the development of a range of Ab-fragment-based structures offering different properties that are designed to overcome some of the shortcomings of traditional mAbs, or bring about mechanisms of action that were previously not possible. Although mAbs have achieved significant commercial success, with total global sales of ~$40 billion in 2010, their comparatively large size can restrict tissue penetration or access to cellular targets.[7] Full mAbs are comparatively expensive to manufacture, requiring mammalian expression systems. Furthermore, the IP landscape is complex, with many key steps for optimizing designs being covered by several layers of patent protection. These factors have incentivized drug developers to seek alternative structures. FPs in groups 6b and 7b (Figure 3.2) consist of Ab-binding domains fused to other Ab-binding domains, or to an Fc effector.

3.1.7.1 Bispecific Abs: Ab-Binding Domain Fused to Ab-Binding Domain (6b) Three of the FPs identified consist of an Ab-binding domain fused to another binding domain specific for a different target, to achieve bispecific targeting. Trion/Fresenius' marketed product Removab® (catumaxomab) and University Hospital Tuebingen's Phase I/II candidate rM28, also include an Fc effector domain, giving a third function. The structure of Removab resembles that of a traditional mAb, with one binding domain specific to EpCAM that is associated with the surface of certain cancer cells. Through genetic fusion, the second-binding domain is designed to target CD3 on T cells, while the Fc portion accelerates the immune response by recruiting accessory immune system. Similarly, Micromet's bispecific T-cell engager (BiTE®) MT103 (blinatumomab) also uses anti-CD3 binding to bring T cells into the proximity of cancer cells, but lacks an Fc domain.

Bispecific and trifunctional FPs potentially allow for novel mechanisms of action to be employed. However, a range of competing non-FP bispecific scaffold technologies has been developed. Ab-based structures include dimerized single-chain variable fragment (scFv) fragments known as diabodies and F-star's mAb^2 technology which relates to full

[5] http://clinicaltrials.gov/ct2/show/NCT00408967?term=tucotuzumab+celmoleukin&rank=2

[6] http://www.merckserono.net/en/science/pipeline/phase_ii/phase_2.html

[7] http://www.datamonitor.com/store/News/fast_growth_mab_market_to_offer_great_rewards_to_a_select_few_over_2009_15?productid=A31659B5-E3F6-4C49-8C5B-DF524E8EB770

Abs with two additional binding sites engineered into the CH3 domains. Non-Ab bispecific constructs have also been developed. Amgen acquired Avidia in 2006, gaining Avimer technology, formed of two or more linked peptide sequences generated from extracellular receptor domains that convey binding or inhibitory properties. As with other FP segments, the performance of different competing technologies is likely to depend on factors such as the nature of specific targets interacted with and their location. There are though key themes exhibited and smaller structures can be advantaged by greater tissue penetration, but conversely, suffer from shorter half-lives compared to larger alternatives. The importance of these differences will vary from product to product. In terms of in-market pricing and profitability, a number of structures such as Avimers can benefit from the possibility of being produced in bacterial expression systems [28].

3.1.7.2 Miniaturized Abs: Ab-Binding Domain Fused to Fc Effector (7b) One FP uses an scFv fragment for targeting attached to an Fc moiety, giving what is in effect, a scaled down mAb structure. Such fusions benefit from the high specificity targeting capabilities of the Ab-binding domain components, alongside the natural Fc effector function, which is particularly useful for cell-killing activity. Emergent BioSolutions' SBI-087 (PF-05230895) is a Small Modular Immuno Pharmaceutical (SMIP™), partnered with Pfizer and in Phase II trials for rheumatoid arthritis. The product acts by directing an immune response to cells expressing CD20, in a mechanism similar to that of marketed mAb Rituxan® (rituximab). This results in a number of actions including B-cell reduction through ADCC and CDC driven by the Fc component. SBI-087 was found to achieve surprising levels of Fc-mediated cellular toxicity compared to Rituxan. This is thought to be due to it binding primarily to CD20 in soluble regions of the plasma membrane, whereas Rituxan can interact with, or relocate CD20 into insoluble regions [29].

SMIPs consist of scFvs that do not contain the CH1 and CL domains present in conventional mAbs. Like several non-Ab delivery options, the smaller SMIP structures potentially allow improved tissue penetration and biodistribution compared to proteins attached to mAbs [30]. Furthermore, with a shorter distance between binding domains (60–80 A°) compared to Abs (95–140 A°), the differing geometry of SMIPs can result in unique interactions and can be engineered to deliver signaling responses.[8] Both mAbs and SMIPs are produced in mammalian cell expression systems. F-star's Fcabs relate to a further miniaturized Ab technology, with an Ab Fc domain linked to two identical antigen binding sites engineered into the CH3 domains.

[8] http://www.truemergent.com/wp-content/uploads/SMIPS-vs-Mabs-Data-Sheet-11-2010.pdf

3.1.7.3 Other Ab Fragment Structures Other Ab fragment structures are not FPs themselves, but can be fused to other protein components. These scaffolds are often developed as a standalone technology first, before being expanded into FP designs. Early-stage FP developers should therefore consider how technologies that do not compete at present, might later do so. Indeed, an understanding of innovative protein designs can aid FP developers to stake initial patent claims for FP-based extensions of novel protein scaffolds.

Acquired through Domantis, GlaxoSmithKline's Domain Antibodies (dAbs) are 11–15 kDa structures that consist of either an antibody heavy (VH) or light (VL) domain. Bispecific dAbs can be created by fusing one dAb to another. Similarly, Ablynx's Nanobodies® are single-domain structures (sdAb) consisting of a single VH domain. In keeping with the technology overlaps, the two companies have been involved in a number of patent challenges, disputes and settlements. Affibody's Affibodies® are triple helix structures based on the Z domain (the IgG G-binding domain) of protein A. Non-Ab technologies include Archemix's aptamers, which are peptide or nucleotide sequences generated from an *in vitro* selection process that bind to target molecules.

Unlike more complete Ab-based structures, these smaller binding molecules do not exhibit CDC or ADCC activity due to the absence of an Fc region. Their small size facilitates tissue penetration but also leads to a short half-life. Strategies to increase the half-life have been developed, including linking dAbs to ABDs in AlbudAbs (discussed earlier).

3.1.8 Ab- and non-Ab-Targeting Decisions

As shown in Figure 3.2, the direction of FPs to target sites can be driven either by the straight affinity of the active component with its targets (as in simple therapeutic proteins, or FP decoy receptors and ligand half-life products) or by a fused targeting moiety. We have divided targeting partners into those that are Ab-based and those that are not, although in simple terms, the role of these partners is just to bind to target sites in such a way as to alter the distribution of the product's activity in the body. Targeted or localized delivery of a cytotoxic agent, for example, may be desirable to reduce activity in nondiseased tissues and therefore, the side effect profile. Alternatively, targeting may be employed to deliver antigens to the immune system machinery, or in the case of bispecifics, to bring two targets into close proximity.

While half-life extension is a primary purpose of almost half of all FPs in Phase II development or more advanced, the remaining 24 are designed around the concept of more specific (targeted) delivery. These are split equally between Ab targeting and non-Ab targeting. Since non-Ab approaches include various ligand, receptor and enzyme targeting, this highlights the disproportionately high contribution from Ab fusions, which all relate to scFv-type structures. FP developers must choose which approaches to use for the targeted delivery of fused active partners and the factors directing this choice are discussed.

Where other companies hold IP that restricts freedom to operate, choice of targeting mechanism may be limited. Similarly, a company built on a particular targeting platform is likely to focus development efforts on this rather than alternatives. In both cases, drug developers must employ the most effective solutions if they are to raise their chances of success. A product paying royalties on significant sales still offers superior returns to one that fails in clinical trials or cannot compete in the market. However, the best targeting approach for an active partner may only become apparent once several steps along the development pathway have already been undertaken.

Regardless of IP restrictions, choice of an Ab or non-Ab targeting partner can be driven by the availability of known binding components with suitable specificity. There will not always be an endogenous protein that selectively targets a particular antigen, in which case, Abs or alternative targeting moieties must be engineered. Non-Ab FP targeting approaches include AdNexus' fibronectin-based Adnectins®, Altor's STAR™ (soluble T-cell antigen receptor) platform, Molecular Partners' DARPins (ankyrin repeat proteins), Pieris' Anticalin® engineered lipocalins and transferrin for delivery across the blood–brain barrier [31]. Full mAbs are naturally bifunctional and can therefore offer advantages where the effector functions of the Fc domain are wanted. Fusion or conjugation to full Abs can therefore be desirable for therapies aimed at bringing about cell death, such as those developed for oncology indications. Full mAbs offer a well-known safety profile, standardized manufacturing processes, and a long half-life (manufacturing considerations of all FPs are discussed in the following section). However, their large size can dampen activity of the partner or restrict tissue penetration, potentially impacting the product's efficacy. A target may be accessible to FPs based on a ligand or receptor fragment, but not larger full mAbs. Structures based on Ab fragments can potentially circumvent the issue. In both the case of FPs and non-FP Ab fragments, the benefits of smaller targeting components must be weighed up against disadvantages such as more rapid plasma clearance, which can reduce half-life. In each case, the pharmacokinetic properties will depend on further factors such as the tissue type and area of the body targeted, or can be difficult to predict. For example, the clearance rate of Ab-G-CSF FPs from plasma is highly increased compared with the parental Ab [32]. Similarly, the activity of TNF-α or IL-12 FPs can be markedly lower than for the simple recombinant cytokines, although targeting moieties employed in FPs have also been shown to also exert direct action, such as the tumor vascular-targeting NGR sequence used to deliver TNF in MolMed's NGF-hTNF [33].

3.1.9 Targeted Enzymes (2a)

Three targeted enzyme FPs are in development (all Phase II). Two related to structures that localize enzymes. As with targeted toxins, for example, this potentially allows greater potency at the desired site of action while reducing side effects. NexBio's Fludase® consists of the sialidase catalytic domain fused to a respiratory epithelium cell surface-anchoring sequence. It is designed to work by inactivating sialic acid receptors on cells lining the respiratory tract that are otherwise used by influenza viruses to gain entry into cells. The anchoring sequence increases retention time at the sites where the enzyme action is required. Fludase is the only FP in Phase II or above administered by inhalation. Enobia's ENB-0040 (asfotase alfa) also uses a ligand-targeting moiety to localize enzyme activity. A deca-aspartate peptide derived from the bone remodeling protein osteopontin delivers the fused enzyme, tissue nonspecific alkaline phosphatase (TNSALP), to bone.

Localized or targeted enzymes can also be used for activation of an administered prodrug at only desired sites. This approach is employed with mAbs in ADEPT and has been tested with FPs, albeit not any that are currently in Phase II or above [34]. Smaller Ab fragments have also been fused to enzyme domains in FPs to improve the targeting efficiency over that of conjugated ADEPT mAbs, demonstrating very high distribution differentials between tumor cells and normal tissue [35,36]. A key advantage of localized enzymes over nonenzymatic targeted toxins is that as catalysts, they are able to turnover multiple copies of active drug at the target sites. Localized enzymes can result in greater bystander effects on surrounding cells compared to internalized proteins. This wider action on cells can be advantageous for tumor killing for example, paticularly where not all cells exhibit the target toxin, but can also cause side effects in healthy tissue.

A further FP, Menarini's amediplase consists of an enzyme fragment fused to another enzyme fragment. We have classified it within the enzyme localization group since the fusion is inherently designed such that the second enzyme's activity is located near that of the other enzyme component; however, the product is not strictly a targeted enzyme. Amediplase combines a fragment of tissue plasminogen activator (tPA) with a urokinase fragment. Both are human profibrinolytic factors that help to breakdown clots. The fusion is designed to offer the high specificity of one enzyme with the high catalytic activity and resistance to inhibitors of another, as well as increasing half-life. Although amediplase is described as "ready to enter Phase III trials," it is unclear when or if this will occur.[9] Fused or bifunctional enzymes have been shown to hold pharmacokinetic advantages over the separate enzymes [36,37].

[9] http://www.menarini.com/r_d/therapeutic_areas/R_D/amediplase

Furthermore, fused enzyme components can be manufactured more efficiently in a single process.

3.2 FACTORS FOR COMMERCIAL SUCCESS

The purpose of this section is to summaries the characteristics that FPs must possess if they are to stand greater chances of achieving commercial success. The ability of a product to generate profits for a biopharmaceutical company does not depend on the specific protein structure or mechanism of action *per se*, but instead on how it is positioned to meet needs of the marketplace against available alternatives. This positioning can be measured on the basis of four key factors for commercial success: efficacy, safety, administration, and cost (Figure 3.3).

As discussed earlier in this chapter, there will often be a range of FP and non-FP approaches that drug developers can employ to modulate a particular drug target or disease pathway. Where alternative products are available to physicians in the market, commercial success will be dependent on some clinical advantage being offered or comparative cost. Although further factors such as launch timing and promotional activity also have a key bearing on the commercial success of a product, these are not considered in this section, since they are independent from the inherent properties of an FP product. Similarly, there can be wider academic value in scientific discoveries or patients can be given products in a charitable manner and the success of a drug product can also be measured by the improvement to patient health or cost savings to healthcare providers. However, this chapter relates only to the point of view of generating profit as a commercial business.

The development of many FPs has been discontinued in the past, not because of failing to achieve the intended mechanism of action as such, but because the performance against the key factors of commercial success was insufficient versus the offerings of competing products. Positioning

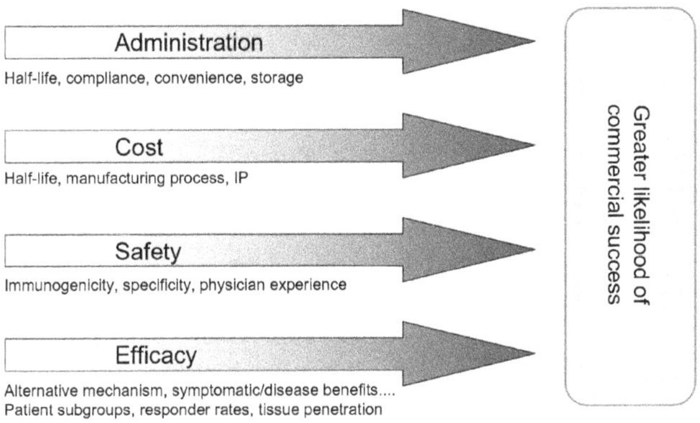

FIGURE 3.3 Key factors for commercial success.

against these key factors is critical for determining the degree to which a product will be prescribed by physicians, and therefore the level of sales or profits it generates. In earlier stages of development, the factors also have a bearing on the ability to attract licensing deals for product candidates or technology platforms.

3.2.1 Administration

Approximately half of the FPs identified are based on half-life extending technologies to allow practical or less-frequent dosing. Of these, 25% are specifically positioned as next-generation versions of existing short half-life therapeutic proteins, each of which is either partnered with or invented by large pharma/biotech companies. Other half-life extending FPs are developed where a novel protein or protein fragment would not be practical without the longer duration of action. However, across the board, drug developers should seek to ensure that the administration of a product is sufficiently robust versus competitors. As well as offering greater convenience, less-frequent dosing can also aid compliance to the treatment regime, improving patient outcomes. It should though be noted that these benefits must be considered in the context of the disorder being addressed and its severity. There may be only limited tolerance from payers to accept higher prices for greater convenience/compliance and this is of particular significance where alternatives are due to become cheaper in generic or biosimilar form. Longer half-life products can though require less physical product to be administered, giving manufacturers a stronger position in terms of pricing and profitability. This is likely to have been taken into account by largest global generics company, Teva, before its February 2008 acquisition of albumin fusion player CoGenesys.

Earlier in this chapter a wide range of half-life extension technologies were highlighted; both FP and non-FP. These typically differ in terms of immunogenicity, level of impact on partner protein activity, freedom to operate restraints and manufacturing cost. A further solution to the inconvenience of frequent dosing is to develop drugs that are less burdensome to administer. Biologics are typically only available as injectable formulations since large hydrophilic peptides are broken down in the gut and do not readily pass across external surfaces of the body. Furthermore, generally being more expensive than small molecule drugs, most biologics are, therefore, only used for comparatively severe indications.

Only one FP in Phase II development or beyond is noninjectable. NexBio's Fludase (DAS181) is a powder formulation administered over 3 days via inhalation for the treatment of influenza. In this exception, inhalation represents a practical option since Fludase is designed to act by binding to the human respiratory tract. However, various emerging protein technologies including novel FP structures, potentially allow inhaled, topical, and even oral alternatives to be developed. Similarly, needle-free injection devices may also reduce the inconvenience of treating disorders that have traditionally been the realm of injectable biologics. These innovations, therefore, threaten to disrupt the market for a number of long half-life FPs, although the burden of a more frequent injection regimen should perhaps not be overstated. Traditional needle and syringes are dependable, low cost, and relatively painless. This became all too apparent to Pfizer with its inhalable insulin product, Exubera®. It was taken off the market in October 2007 after the high price, cumbersome device and dosing concerns resulted in only minimal levels of sales. Similarly, Eli Lilly/Alkermes and Novo Nordisk/Aradigm also terminated development of competing projects. MannKind has though continued with Afrezza®, which may in part relate to the company having few alternative pipeline options compared to larger rivals. The product has struggled with regulators, receiving two Complete Response Letters from the FDA. MannKind is due to commence further clinical studies with a new thumb-sized device that may overcome some of the commercial issues that prevented the 200-mL sized Exubera from achieving success.

Further technologies in development are designed to reduce the frequency of injections by having drug-producing factories set up in the body. Gene therapy approaches use viral or other vectors to insert DNA into a patient's cells, resulting in expression of the desired therapeutic protein. Sanofi/Oxford Biomedica's Retinostat® is a gene therapy in Phase I for age-related macular degeneration (AMD). It is administered via a one-off injection. Long-term follow-up studies have shown gene therapy to be effective for over 10 years without repeat doses [38]. If successful, the product could potentially impact the commercial outlook for AMD treatments such as Roche's Lucentis®, or Regeneron/Bayer's FP, VEGF Trap-Eye, which are required once-monthly in the first year then approximately every 4–12 weeks (see Chapter 2). Similarly, cell therapies involve the administration of whole cells that have been engineered to produce a therapeutic agent and potentially also offer infrequent administration. With Medgenics' Biopump technology, cells are first collected from the patient, which then undergo gene therapy before being re-implanted. The company has reported that a single implantation of erythropoietin (EPO) Biopumps was sufficient to provide sustained treatment of anemia for more than 2 years.[10] These gene and cell therapy approaches can also be used for the expression of FPs to either reduce dosing frequency further, or facilitate novel mechanisms of action. Heat Biologics' HS-110 is a Phase I/II candidate that involves the injection of engineered cells that secrete a gp96-IGg FP for the treatment of cancer—gp96 is a chaperone that binds to tumor antigens and delivers them to APCs. Similarly, Advaxis' ADXS11-001 is a vaccine based on attenuated Listeria that is engineered to produce a truncated virulence factor fused to a tumor-specific antigen, which then generates an anticancer immune response. As stated at the start of this chapter, we have excluded HS-110 and ADXS11-001 from our total of 43 FP products since they are not administered as FPs, but instead are expressed in the body. It is notable that lower administration frequency does not always mean a better profile, and continuous therapeutic exposure can also result in side effects being exacerbated. This was raised as a concern by the FDA regarding Eli Lilly's once-weekly exanatide formulation, Bydureon.[11] Furthermore, many long-acting formulations cannot be removed quickly from the body in the event of adverse reactions occurring.

In this discussion of the factors of commercial success, we flag the importance of focusing on products that offer a sufficiently strong profile against the evolving marketplace. The key patents for several mature therapeutic protein classes, such as erythropoetins, blood factors, IFNs, and insulins, have either expired, or are soon to. As such, there can be greater scope for FP and non-FP platform players to develop next-generation versions of these molecules than for say more recently launched products where there may be less freedom to operate. The downside of this is though that without broad underlying IP, these opportunities will also be available to competitors and it is of no surprise to find a considerable number of next-generation products in these expiring protein classes. Although proven mechanisms of action do represent a lower risk strategy for validating a new platform, promising in-licensed or newly discovered targets should also be considered for in-house development projects once the extended half-life or improved administration technology has been validated. As with many aspects discussed in this chapter, the factors of commercial success should be considered on a case-by-case basis. Furthermore, while we consider it important to properly assess the advantages and disadvantages of competing administration approaches, this should be in the context of wider factors including efficacy, safety, and cost.

3.2.2 Efficacy Profile

A primary factor linked to the commercial success of any FP product is the level of efficacy it can achieve compared to alternatives. Patients are given medicinal agents to treat or prevent a particular disorder. Treatments offering greater efficacy stand out when it comes to making a prescription choice. The stringency of efficacy requirements depends on the risk-reward ratio offered by a product in the indication sought.

[10] http://marketbrief.com/mdgn/8k/events-or-changes-between-quar/2011/5/3/7870770/filing

[11] http://phx.corporate-ir.net/phoenix.zhtml?c=101911&p=irol-newsArticle&ID=1484584&highlight=

For example, a drop in efficacy compared to existing gold standards may be tolerated where a product is associated with fewer side effect concerns in a low severity disease, but a once-weekly cancer treatment is unlikely to be viewed favorably if survival rates are slightly lower.

Comparative efficacy must be taken into account when developing novel or next-generation products designed to be more convenient to administer, cheaper, or simply a me-too in the market. All other aspects being equal, a less-frequently dosed version of an existing therapeutic protein would not require any improvement in efficacy to capture market share. However, factors such as cost and safety are also considered by regulators, physicians, patients, and payers. HGS' albumin fusion, Zalbin/Joulferon (albinterferon-α2b), was discontinued in October 2010 after the FDA declined to approve the product on the basis of its efficacy and safety being slightly worse than for Pegasys®, despite the primary endpoint of noninferiority being met with statistical significance. Combining a partner with an active protein in an FP or conjugate may not always exhibit the desired level of efficacy in clinical trials. Most notably, large partners such as an Fc domain or albumin used for half-life extension can dampen activity of the fusion via steric hindrance [4].

Where a disorder is poorly addressed by existing treatments, the efficacy challenge will be low compared to those that are already covered by several adequate options. For developers exploiting a novel disease pathway to achieve a step change in treatment, sales can be higher and comparatively safe from competing alternatives. Examples of this include the emergence of the anti-TNF class or Genzyme's non-FP enzyme replacement therapy, Cerezyme.® FPs can provide new mechanisms for treating diseases, giving a firm advantage in the market. Compared to larger mAbs, several Ab fragment fusions are designed to offer improved tissue penetration, potentially raising efficacy for the treatment of cancer and other diseases. These benefits will be most applicable where the efficacy of existing approaches is hampered by target accessibility. Similarly, in the CNS therapy area, fusion or conjugation to transferrin has been shown to facilitate the transport of a partner protein across the blood–brain barrier [39,40].

As stated earlier, almost all of the FPs identified are injectable. Furthermore, biologic products are typically priced at a premium to small molecule treatments (see Chapter 2). As such, without providing a sufficient improvement in efficacy or being more commercially attractive in some other aspect, FPs may struggle to capture market share in a number of indications. Established marketed products will have a well-defined efficacy profile and where side effect concerns are not problematic, physicians may feel more comfortable prescribing products that have a longer history of use. It should be noted that new treatments can also be positioned to be used in addition to, rather than instead of existing treatments, potentially reducing competitive pressures. Complementary or synergistic combinations can raise the changes of commercial success. In terms of efficacy, the competitive standing of a development stage product will only be clear after pivotal trials have been undertaken. We consider this as further highlighting the importance of having a strong ranking against several factors for commercial success. One strategy for improving the prospects of a new product against tough competition is to target-specific patient segments, such as those that do not adequately respond to existing options, or a subset in which the new product is more efficacious. Companion diagnostics can aid the identification of such groups. An example outside of the FP space is Roche's Zelboraf (vemurafenib), which is in development for the treatment of melanoma in patients whose cancer is shown as having the mutant BRAF protein. While such stratification reduces the number of applicable patients, using a diagnostic potentially improves the likelihood that stronger efficacy data will be generated. It could be viewed as a lower risk strategy to first ensure approval and launch through a small patient subset, since label expansion can subsequently be sought, with development costs offset by initial sales.

3.2.3 Safety Profile

Medicine regulatory agencies in major markets have a primary objective of ensuring patient safety. Prescribing physicians take a similar view and the safety profile of a product plays a critical role in determining its approval prospects and the level of sales it will achieve. The risk-reward ratio offered by a product will depend on the indication it addresses and side effects will be considered more tolerable where sufficient efficacy benefits are offered, particularly in severe diseases. As highlighted earlier with HGS' Zalbin, falling just slightly short of regulators' safety requirements can prevent even a product with administration benefits and clear efficacy from reaching the market.

Immunogenicity has historically been a key area of concern with biologic products or attached moieties, both in terms of side effects and inactivation due to neutralizing antibodies. Measures such as humanization, use of human-derived proteins, sequence changes, and alternative post-translational modifications have served to reduce the immunogenicity of developed biologics. As such, FP platforms must offer low immunogenicity if they are to compete. FPs specifically designed as therapeutic vaccines to stimulate the immune system are though a clear exception (see Section 3.1.6.3).

As detailed in the half-life extension discussion earlier, a number of FPs aim to improve on the safety of PEGylation. PEG is nonbiodegradable and concerns have been raised regarding potential kidney issues, particularly in patients with impaired renal function [11]. As well as reduced dosing frequency, a longer half-life can produce more consistent plasma levels, avoiding the dose spikes and greater side effects that can be associated with regular dosing of a short half-life product. Similarly, several targeted FPs or comparable conjugates aim to offer improved safety by acting in a more specific and selective manner—such as FPs that use an Ab-binding domain to deliver IL-2 to cancer cells (see Section 3.1.6.2). Tailoring drug delivery to target only certain cell types or certain areas of the body reduces the concentrations experienced elsewhere, potentially reducing side effects or allowing higher doses without safety issues. However, care must be taken to ensure that the target is not also present in other locations, or that the binding domain does not also have affinity for unintended targets. This is highlighted by the case of non-FP immunotoxin Mylotarg, which was withdrawn from sale after further studies showed it to be associated with a higher rate of toxicity and death due to unintended delivery to CD33+ cells in the liver [22]. Possible variation in the genetic makeup or expressed proteome of patients must also be taken into account—a treatment may be safe in the majority of patients but have severe consequences in a subgroup.

As with patient stratification for efficacy purposes, clinical trials and approved indications can also be tailored for only patients who respond to treatment, or trials can be avoided in those who experience side effects. Such approaches can allow FPs or comparable conjugated proteins/mAbs to command a strong commercial advantage over alternatives. It should though be noted that while certain products may be safer in theory, it can take many years before this is clinically demonstrated. Similarly, novel or adjusted approaches can also bring about new safety concerns. The gold standard treatments for osteoporosis, bisphosphonates, are effective but are not broken down in the body and are either excreted via the kidneys or deposited in bone, where they can remain for decades. Long-term exposure and inhibition of bone resorption is thought to cause a number of serious side effects, albeit rare. Amgen's non-FP mAb Prolia® (denosumab) employs a differentiated mechanism of action, but is considerably more expensive and will take many years before it can be shown to offer a superior long-term side effect profile. The Prolia FREEDOM extension study is due to run for 10 years.

3.2.4 Cost Factors

The costs associated with the development and sale of an FP can have a significant bearing on its chances of achieving commercial success. The size and length of clinical trials required for a product will vary depending on the indication sought and the degree to which its efficacy and safety profile must be demonstrated. Novel approaches may offer scope for superior rewards but can be associated with greater uncertainty. The greatest threat to drug developers is that expensive Phase III trials will fail to show the desired results or have to be repeated. Costs and risk will vary on a case-by-case basis; however, FP developers seeking partnership deals should consider at an early stage how the risk-reward balance of a project may be perceived. A product candidate will not only compete for investment funds with others' candidates in the same disease space but also with projects in other business areas.

Aside from clinical trials, a further major cost of drug development can relate to manufacturing facilities. Novel product structures can require new manufacturing facilities to be built and these upfront costs have a disproportionate impact on the financial risk-reward balance. Furthermore, production staff will have less experience of running and optimizing a new process. For these reasons, approaches that are in keeping with existing production capabilities will be viewed more favorably—to a mAb developer, a conjugated Ab may be more likely to be selected over a targeted toxin FP. Further cost factors must also be taken into account.

3.2.4.1 Manufacturing and Cost of Goods Sold We have defined FPs as the genetic fusion of two proteins. Theoretically, most, if not all, FPs could also be produced using chemical attachment. The decision of whether to use an FP or chemical approach relates primarily to manufacturing costs and consistency.

As described in this chapter, there will be a range of FP and non-FP product structures that can potentially be used to target a disease pathway. Only certain manufacturing methods will be practically applicable in each case, affecting the costs of production and therefore the profitability of product. Large human proteins can be complex, with multiple functional domains that undergo extensive post-translational modification or have to be folded in a specific manner. As a result, only certain cell culture media may be capable of appropriate expression of a certain product or structure. Traditionally, mAbs, for example, have been produced only in mammalian cells, in particular Chinese hamster ovary (CHO) cells, which are favored for their ability to assemble complex multisubunit proteins and perform modification such as glycosylation [41].

Out of the 43 FPs identified in Phase II development of above, 29 are produced in mammalian expression systems; 19 of which use CHO cells, while four FPs are produced in HEK-293 cells, which are originally derived from human embryonic kidney cells and can therefore produce modifications more closely aligned with those of natural human proteins. Lower antigenicity is particularly beneficial for products such as recombinant blood factor Fc fusions (rFVIII-Fc and rFIX-Fc) that are intended to be administered over many years [42].

Mammalian cell cultures are challenging to scale-up and comparatively expensive to run. FP or biologic product designs that allow for more straightforward manufacturing can therefore be advantageous. All of the FPs involving an Fc domain use mammalian expression except for two products, which are both peptibodies (Fc-peptide). The most advanced of these, Nplate (romiplostim), is produced using bacterial fermentation and as a result, unlike mammalian cell-expressed Fc fusions, has a nonglycoslated Fc domain. The glycan structure has a key bearing on the ability of an Fc domain to bind to Fc receptors, but more importantly for most mAbs, is critical for effector function [43,44]. Similarly to peptibodies, a number of Ab scFv fragment structures offer scope for bacterial production, although such FPs are currently in earlier stages of development. Approximately 28% of identified FPs in Phase II or above use bacterial fermentation. *Escherichia coli* is used in all cases except for one, Protox's PRX302, which uses *Aeromonas salmonicida*. The applicable manufacturing processes will depend on both components of an FP and certain classes offer more scope for bacterial production than others, such as those based on the short peptides of microbial toxins, antigen fragments, or endogenous ligands (that are frequently peptides rather than larger proteins). Short peptides can also be synthesized by combinatorial chemistry, facilitating conjugation approaches.

Regulators require products to be produced in a sufficiently consistent manner to ensure patient safety. For example, a key challenge of conjugated Abs has been to achieve homogenous product with the same number of partner components attached per molecule. Furthermore, the attached moiety should ideally be bound at the same site in each case, such that its stability or release will be consistent. Severe adverse events could, for example, occur if a patient was to receive an immunotoxin dose with a greater toxin ratio, or where toxin more readily detached and distributed to unintended areas of the body. A key advantage of FPs is that the partner moiety is inherently built in to the sequence of the expressed protein, giving a more consistent position in the final structure. FPs also offer advantages in terms of number of process steps. Formed as a fused molecule, they require only a single fermentation and can be purified in a single stream. In contrast, the two (or more) components of a conjugated protein will usually be produced in separate fermentations (or chemical synthesis stages) and individually purified before being joined. However, having two (or more) components can make purification of FPs more challenging. A single protein may have specific characteristics that can be exploited for it to be preferentially selected over impurities, but these can be altered by addition of a partner component. As such, conjugated products based on a common platform may offer a more regularized process compared to FP alternatives.

These manufacturing differences will alter the cost of producing a product. Although a more complex process is not necessarily a more expensive process to run, having a greater number of processing steps increases variability, potentially raising the risk of product inconsistency. Similarly, in the development stages, a more straightforward manufacturing process facilitates optimization of yield and purity. The manufacturing options available and respective costs for different FP and non-FP structures will vary on a case-by-case basis, but should be assessed in the wider context of other factors for commercial success. That is, in addition to cost considerations, an FP should be administrated in a competitive formulation and offer an adequate efficacy and safety profile. For novel brands, particularly those addressing areas of high unmet need, the in-market price of a product may have less bearing on its ability to capture market share. In contrast, the use of products that are positioned against next-generation or biosimilar versions can be more sensitive to differential pricing.

We were able to identify manufacturers for 37 of the 43 FPs in Phase II development or beyond. These included 30 different companies, of which the vast majority was the innovator or partner, rather than an external contract manufacturer. We consider this an indication of the broad capabilities of both small and large biopharmaceutical companies in terms of development-stage FP manufacturing. It should though be noted that with over half of the FPs currently in Phase II development, we would anticipate that a number will subsequently be transferred to alternative manufacturers, for production of Phase III trial material or commercial supply.

3.2.4.2 Intellectual Property: Freedom to Operate and Protection from Copies Aside from manufacturing costs, the choice of product design against a particular target will be somewhat directed by the indirect costs of any intellectual property (IP) licensing requirements. Development and commercialization costs will be lower where there is freedom to operate and the structure, therapeutic use or manufacturing processes do not infringe others' IP. In keeping with this, a number of FPs and their underlying technology platforms are designed to avoid the patents held by others. This freedom can be exploited either by the inventor for in-house product candidates, or out-licensed to others.

The IP landscape for heavily researched areas such as mAbs can though be a challenge to navigate, particularly where patents are in earlier stages and the full breadth of allowable and granted claims has not yet be issued. Licensure of certain aspects may be unavoidable. We note that while novel, noninfringing approaches can yield greater profit margins, having the right product is critical. It may be better to pay a license fee and/or royalties on a strong product than nothing on one with an inadequate profile due to design compromises made, which fails to launch or gain market share. In June 1999, Immunex (now Pfizer) signed an agreement with Genentech (now Roche) to gain access to its immunoadhesin patent portfolio for Enbrel, which launched in the United States the previous year and is currently the top-selling FP. Infringement cases can take many years to resolve—Abbott Laboratories launched the mAb Humira in 2003, but was sued by Johnson & Johnson for patent infringement in 2007 and subsequently ordered to pay $1.84 billion compensation, royalties, and interest. The verdict was, however, overturned in February 2011 when the Court of Appeals found the underlying patent to be invalid.

Freedom to operate and others' IP estates are of relevance regarding the costs of, or blockades to, developing certain product classes or structures. However, IP and exclusivity will similarly also govern how long an internal product will enjoy a position in the market free from direct copies. To this extent, novel approaches such as those employed by various FPs, can allow additional layers of IP to be secured, potentially delaying the launch of competing products, or generating licensing payments. IP claims must be sufficiently broad to cover potential competition from both alternative FP and non-FP approaches. Developers should stay aware of upcoming competitors bypassing IP that could cut short a market, and look to develop next-generation products offering sufficient advantages with which to retain market share.

As biologic products, FPs are less impacted by cheaper copies after patent expiry than small molecule drugs, which in the United States can typically expect to see the bulk of sales eroded within 1-2 years of generics becoming available. The "Biologics Price Competition and Innovation Act of 2009" (BPCIA) legislation gives originator biologics a 12-year market exclusivity period in the United States and biosimilars can therefore only launch after this has expired. The regulatory approval pathways for biosimilars are still in their infancy in Europe, while in the United States, the FDA has indicated it expects to issue guidance on biosimilar approval requirements in 2011 and has stated that the level of human testing will depend on assessment of how similar the biosimilar is to the originator drug [45]. To date, no biosimilar version of a mAb or FP has been approved in Europe or the United States.

Demonstrating comparability with the originator biologic is likely to be challenging for biosimilar versions in many cases. The BPCIA requires there to be no clinically meaningful differences in safety, purity, and potency between the biosimilar product and the U.S. licensed reference product.[12] These characteristics can be critically affected by differences in the manufacturing process. Biologics are commonly produced using cells in culture or whole organisms, and these are inherently more variable than chemical synthesis.[12] Even when key patents have expired, the production of sufficiently similar product can be delayed by trade secrets. As a result of these challenges, the characteristics in terms of efficacy and safety (including immunogenicity) of an approved biosimilar may still differ somewhat from the originator drug. We take the view that this will dampen the degree to which biosimilars capture market share, as physicians and patients may be apprehensive about switching treatment with an originator biologic that has substantial safety and efficacy data, for a biosimilar with less data and potentially different characteristics. The level of erosion associated with biosimilars is also expected to be distinctly less than for small molecule drugs due to a lower number of entrants. The cost of establishing biosimilar manufacturing capabilities and meeting regulatory demands is comparatively high, acting as a barrier to entry. Furthermore, we expect the price differential between biosimilars and their originator counterparts to be tighter than observed for small molecule branded drugs. Lastly, for certain therapeutic areas such as oncology, the level of patient and physician support provided by the branded drug company may also make it tougher for biosimilars to gain a foothold in the target market without substantial investment. We believe that these factors will together limit the degree to which direct and indirect biosimilar competitors impact the market for FPs.

REFERENCES

1. Bachmair A, Finley D, Varshavsky A. (1986) *In vivo* half-life of a protein is a function of its amino-terminal residue. *Science* 234(4773), 179–186.
2. Knauf MJ, Bell DP, Hirtzer P, Luo ZP, Young JD, Katre NV. (1988) Relationship of effective molecular size to systemic clearance in rats of recombinant interleukin-2 chemically modified with water-soluble polymers. *J. Biol. Chem.* 263(29), 15064–15070.
3. Roopenian DC, Akilesh S. (2007) FcRn: the neonatal Fc receptor comes of age. *Nat. Rev. Immunol.* 7(9), 715–725.
4. Peters RT, Low SC, Kamphaus GD, Dumont JA, Amari JV, Lu Q, et al. (2010) Prolonged activity of factor IX as a monomeric Fc fusion protein. *Blood* 115(10), 2057–2064.
5. Capon DJ, Chamow SM, Mordenti J, Marsters SA, Gregory T, Mitsuya H, et al. (1989) Designing CD4 immunoadhesins for AIDS therapy. *Nature* 337(6207), 525–531.
6. Lobo ED, Hansen RJ, Balthasar JP. (2004) Antibody pharmacokinetics and pharmacodynamics. *J. Pharm. Sci.* 93(11), 2645–2668.

[12] http://www.bio.org/media/press-release/new-study-shows-rate-drug-approvals-lower-previously-reported

7. Subramanian GM, Fiscella M, Lamousé-Smith A, Zeuzem S, McHutchison JG. (2007) Albinterferon alpha-2b: a genetic fusion protein for the treatment of chronic hepatitis C. *Nat. Biotechnol.* 25(12), 1411–1419.

8. Zeuzem S, Sulkowski MS, Lawitz EJ, Rustgi VK, Rodriguez-Torres M, Bacon BR, et al. (2010) Albinterferon Alfa-2b was not inferior to PEGylated interferon-α in a randomized trial of patients with chronic hepatitis C virus genotype 1. *Gastroenterology* 139(4), 1257–1266.

9. Nelson DR, Benhamou Y, Chuang W-L, Lawitz EJ, Rodriguez-Torres M, Flisiak R, et al. (2010) Albinterferon Alfa-2b was not inferior to PEGylated interferon-α in a randomized trial of patients with chronic hepatitis C virus genotype 2 or 3. *Gastroenterology* 139(4), 1267–1276.

10. Holt LJ, Basran A, Jones K, Chorlton J, Jespers LS, Brewis ND, et al. (2008) Anti-serum albumin domain antibodies for extending the half-lives of short lived drugs. *Protein Eng. Des. Sel.* 21(5), 283–288.

11. Bendele A, Seely J, Richey C, Sennello G, Shopp G. (1998) Short communication: renal tubular vacuolation in animals treated with polyethylene-glycol-conjugated proteins. *Toxicol. Sci.* 42(2), 152–157.

12. Malik DK, Baboota S, Ahuja A, Hasan S, Ali J. (2007) Recent advances in protein and peptide drug delivery systems. *Curr. Drug Deliv.* 4(2), 141–151.

13. Scallon B, Cai A, Solowski N, Rosenberg A, Song X-Y, Shealy D, et al. (2002) Binding and functional comparisons of two types of tumor necrosis factor antagonists. *J. Pharmacol. Exp. Ther.* 301(2), 418–426.

14. Nestorov I. (2005) Clinical pharmacokinetics of TNF antagonists: how do they differ? *Semin. Arthritis Rheum.* 34(5 Suppl. 1), 12–18.

15. Gomez-Reino JJ, Carmona L. (2006) Switching TNF antagonists in patients with chronic arthritis: an observational study of 488 patients over a four-year period. *Arthritis Res. Ther.* 8(1), R29.

16. Kohno T, Tam L-TT, Stevens SR, Louie JS. (2007) Binding characteristics of tumor necrosis factor receptor-Fc fusion proteins vs anti-tumor necrosis factor mAbs. *J. Investig. Dermatol. Symp. Proc.* 12(1), 5–8.

17. Kast RE, Altschuler EL. (2004) Bone density loss in Crohn's disease: role of TNF and potential for prevention by bupropion. *Gut* 53(7), 1056.

18. Dubost JJ, Soubrier M, Ristori JM, Beaujon G, Oualid T, Bussière JL, et al. (1997) An open study of the anti-TNF alpha agent pentoxifylline in the treatment of rheumatoid arthritis. *Rev. Rhum. Engl. Ed.* 64(12), 789–793.

19. Qin C, Ding Y, Huang D, Xu J, Ma R, Huang J. (2011) Down-regulation of TNF-alpha by small interfering RNA inhibits particle-induced inflammation *in vitro*. *Artif. Organs* 35(7), 706–714.

20. Deng G-M. (2007) Tumor necrosis factor receptor pre-ligand assembly domain is an important therapeutic target in inflammatory arthritis. *BioDrugs* 21(1), 23–29.

21. Lansigan F, Stearns DM, Foss F. (2010) Role of denileukin diftitox in the treatment of persistent or recurrent cutaneous T-cell lymphoma. *Cancer Manag. Res.* 2, 53–59.

22. Rajvanshi P, Shulman HM, Sievers EL, McDonald GB. (2002) Hepatic sinusoidal obstruction after gemtuzumab ozogamicin (Mylotarg) therapy. *Blood* 99(7), 2310–2314.

23. Erickson HK, Widdison WC, Mayo MF, Whiteman K, Audette C, Wilhelm SD, et al. (2010) Tumor delivery and *in vivo* processing of disulfide-linked and thioether-linked antibody-maytansinoid conjugates. *Bioconjug. Chem.* 21(1), 84–92.

24. Heisler I, Keller J, Tauber R, Sutherland M, Fuchs H. (2003) A cleavable adapter to reduce nonspecific cytotoxicity of recombinant immunotoxins. *Int. J. Cancer* 103(2), 277–282.

25. Atkins MB, Kunkel L, Sznol M, Rosenberg SA. (2000) High-dose recombinant interleukin-2 therapy in patients with metastatic melanoma: long-term survival update. *Cancer J. Sci. Am.* 6(Suppl. 1), S11–14.

26. Rizzardi G, Traversari C, Valentinis B, Di Matteo P, Alberici L, Cota M, et al. (2010) Contribution of NGR and TNF portions of NGR-TNF to tumor vessel-specific homing and apoptosis *in vivo*. *ASCO Meeting Abstracts* 28(15 suppl.), e13554.

27. Arap W, Pasqualini R, Ruoslahti E. (1998) Cancer treatment by targeted drug delivery to tumor vasculature in a mouse model. *Science* 279(5349), 377–380.

28. Silverman J, Liu Q, Lu Q, Bakker A, To W, Duguay A, et al. (2005) Multivalent avimer proteins evolved by exon shuffling of a family of human receptor domains. *Nat. Biotechnol.* 23(12), 1556–1561.

29. Nickerson-Nutter C, Tchistiakova L, Seth NP, Kasaian M, Sibley B, Olland S, et al. (2011) Distinct *in vitro* binding properties of the anti-CD20 small modular immunopharmaceutical 2LM20-4 result in profound and sustained *in vivo* potency in cynomolgus monkeys. *Rheumatology (Oxford)* 50(6), 1033–1044.

30. Jain RK. (1990) Physiological barriers to delivery of monoclonal antibodies and other macromolecules in tumors. *Cancer Res.* 50(3 Suppl.), 814s–819s.

31. Prior CP, Sadeghi H, Turner A. (2011) Oral delivery of modified transferrin fusion proteins. US 2011/0091543 A1. US.

32. Dela Cruz JS, Trinh KR, Morrison SL, Penichet ML. (2000) Recombinant anti-human HER2/neu IgG3-(GM-CSF) fusion protein retains antigen specificity and cytokine function and demonstrates antitumor activity. *J. Immunol.* 165(9), 5112–5121.

33. Corti A, Curnis F, Arap W, Pasqualini R. (2008) The neovasculature homing motif NGR: more than meets the eye. *Blood* 112(7), 2628–2635.

34. Michael NP, Chester KA, Melton RG, Robson L, Nicholas W, Boden JA, et al. (1996) *In vitro* and *in vivo* characterisation of a recombinant carboxypeptidase G2::anti-CEA scFv fusion protein. *Immunotechnology* 2(1), 47–57.

35. Bhatia J, Sharma SK, Chester KA, Pedley RB, Boden RW, Read DA, et al. (2000) Catalytic activity of an *in vivo* tumor targeted anti-CEA scFv::carboxypeptidase G2 fusion protein. *Int. J. Cancer* 85(4), 571–577.

36. Orita I, Sakamoto N, Kato N, Yurimoto H, Sakai Y. (2007) Bifunctional enzyme fusion of 3-hexulose-6-phosphate

synthase and 6-phospho-3-hexuloisomerase. *Appl. Microbiol. Biotechnol.* 76(2), 439–445.

37. Tamada Y, Swanson BA, Arabshahi A, Frey PA. (1994) Preparation and characterization of a bifunctional fusion enzyme composed of UDP-galactose 4-epimerase and galactose-1-P uridylyltransferase. *Bioconjug. Chem.* 5(6), 660–665.
38. Hacein-Bey-Abina S, Hauer J, Lim A, Picard C, Wang GP, Berry CC, et al. (2010) Efficacy of gene therapy for X-linked severe combined immunodeficiency. *N. Engl. J. Med.* 363(4), 355–364.
39. Zhou Q-H, Sumbria R, Hui EK-W, Lu JZ, Boado RJ, Pardridge WM. (2011) Neuroprotection with a brain-penetrating biologic tumor necrosis factor inhibitor. *J. Pharmacol. Exp. Ther.* 339(2), 618–623.
40. Jain A, Chasoo G, Singh SK, Saxena AK, Jain SK. (2011) Transferrin-appended PEGylated nanoparticles for temozolomide delivery to brain: *in vitro* characterisation. *J Microencapsul.* 28(1), 21–28.
41. Chu L, Robinson DK. (2001) Industrial choices for protein production by large-scale cell culture. *Curr. Opin. Biotechnol.* 12(2), 180–187.
42. Glaesner W, Millican J, Vick AM. (2009) GLP-1 analog fusion proteins. US 2009/0074769 A1. US.
43. Forthal DN, Gach JS, Landucci G, Jez J, Strasser R, Kunert R, et al. (2010) Fc-glycosylation influences Fc γ receptor binding and cell-mediated anti-HIV activity of monoclonal antibody 2G12. *J. Immunol.* 185(11), 6876–6882.
44. Ghirlando R, Lund J, Goodall M, Jefferis R. (1999) Glycosylation of human IgG-Fc: influences on structure revealed by differential scanning micro-calorimetry. *Immunol. Lett.* 68(1), 47–52.
45. Beasley D. (2011) *FDA to issue biosimilar guidance in 2011. Reuters* 9. May.

4

FUSION PROTEIN LINKERS: EFFECTS ON PRODUCTION, BIOACTIVITY, AND PHARMACOKINETICS

XIAOYING CHEN,[1] JENNICA ZARO,[2] AND WEI-CHIANG SHEN[2]

[1]*Department of Pharmacokinetics, Dynamics, and Metabolism, Pfizer, San Diego, CA, USA*
[2]*Department of Pharmacology and Pharmaceutical Sciences, University of Southern California School of Pharmacy, Los Angeles, CA, USA*

4.1 INTRODUCTION

With the advancement of recombinant DNA technology in the past few decades, genetic fusion of two or more proteins has created a new class of biological molecules—recombinant fusion proteins. The resultant fusion proteins can obtain bi- or multifunctional properties derived from each protein module. There have been numerous applications of fusion proteins in biomedicine, including development of imaging techniques by fusing a protein of interest with green fluorescent protein (GFP) [1]; improvement of protein expression and purification by fusing with tags [2], and construction of bifunctional proteins such as hybrid enzymes [3]. Recombinant fusion proteins have also become a new category of therapeutic agents [4]. They have been widely applied for drug targeting and delivery, for example, the use of immunotoxins in targeted cancer therapy, and antibody-directed drug delivery across the blood brain barrier (BBB) [5,6]. Additionally, fusing a protein drug with carrier proteins can significantly prolong the plasma half-life, enhance the therapeutic effect, and achieve better pharmacokinetic (PK) and pharmacodynamic (PD) properties than the parent drugs [7–10]. With the rapid development of protein and peptide drugs, it is foreseeable that fusion proteins will have increasing importance in drug development.

Over the years, there has been extensive research about the construction of various fusion proteins by genetically fusing functional domains. The rational choice of protein modules is critical to achieve the desired functions of the resultant fusion protein. However, another crucial component in fusion proteins often overlooked is a suitable linker moiety to join the protein domains. The direct fusion of functional domains without a linker may lead to many undesirable properties, including misfolding of the fusion proteins [11], low yield in protein expression [12], and impaired bioactivity [13,14].

This chapter gives an overview of linkers in recombinant fusion protein technology. First, the properties of linkers in naturally occurring multidomain proteins are discussed. The following section introduces the empirical linkers that have been applied to the successful construction of recombinant fusion proteins. Readers can refer to these examples for their specific applications. Finally, the functionalities of linkers in recombinant fusion proteins are presented to stimulate thoughts about the versatile applications of linkers. By carefully considering the insertion of optimal linkers in the design of recombinant fusion proteins, many improvements, including correct folding [11,12], enhanced stability [15], and increased biological activities [14] can be achieved.

Fusion Protein Technologies for Biopharmaceuticals: Applications and Challenges, First Edition. Edited by Stefan R. Schmidt.

4.2 OVERVIEW OF GENERAL PROPERTIES OF LINKERS DERIVED FROM NATURALLY OCCURRING MULTIDOMAIN PROTEINS

To design a suitable linker for fusion proteins, a direct and reliable way is to find an example in nature. Many multidomain proteins are composed of several functional domains that are joined by linker peptides. Evolution has selected the optimal linker peptides to achieve the desired biological functions of the protein domains. These linkers allow for the correct folding of protein modules, and also keep the necessary distance between domains either to avoid strong steric hindrance or to permit interaction between the protein domains [16].

With the aim to elucidate the general properties of linkers in natural proteins, two studies were performed by George and Heringa [17] and Argos [18]. These groups built databases that contain natural multidomain proteins from which oligopeptide linkers were extracted. By examining these databases, the classification and properties of the linkers were elucidated. Their results suggested that length, composition, hydrophobicity, as well as conformation of the linkers are all important in exerting proper functions such as separating the protein domains and providing structure stability.

TABLE 4.1 Linker Propensities

	George and Heringa [17]	Argos [18]
Pro	*1.299*	*1.35*
Arg	*1.143*	0.84
Phe	*1.119*	0.69
Leu	*1.085*	N/A
Glu	*1.051*	0.87
Gln	*1.047*	*1.13*
Met	*1.032*	0.75
Thr	*1.017*	*1.55*
His	*1.014*	0.55
Tyr	1.000	0.75
Ala	0.964	*1.05*
Val	0.955	1.00
Ser	0.947	*1.46*
Asn	0.944	*1.09*
Lys	0.944	*1.16*
Ile	0.922	0.81
Asp	0.916	*1.25*
Trp	0.895	0.23
Gly	0.835	*1.25*
Cys	0.778	0.35

Calculated from the ratio of single amino acid occurrence in the linker set compared to its occurrence in the full protein set [17,18]. Values greater than 1 (italics) indicate larger than average occurrences in linker sequences compared to the full protein sequences.

Source: Reproduced from Reference [17] by permission of Oxford University Press and reprinted from Reference [18], Copyright (1990), with permission from Academic Press, Elsevier.

4.2.1 Length of Linker Sequence

The average length of oligopeptide linkers in natural multidomain proteins calculated by the Argos study was 6.5 residues [18], while the length calculated by George and Heringa was slightly longer (10.0 residues) [17]. In the latter study, the linkers were further split into small, medium, and large linkers, which have average length of 4.5 ± 0.7, 9.1 ± 2.4, and 21.0 ± 7.6 residues, respectively. The normalized solvent accessibility was then calculated to give insight into the structural environment of each of the groups (i.e., whether the residues in the linkers were buried within the core of the domains giving a value below 20%, or exposed on the surface giving a value above 20%). The average normalized solvent accessibility was calculated to be 14.4% for small linkers, 26.3% for medium sized linkers, and 42.5% for large linkers [17]. This result suggested that longer linkers were more exposed to the solvent compared to the shorter ones. Therefore, when designing a peptide linker for joining functional domains, the length of the linker may affect whether it is buried between the functional domains, or exposed to the solvent.

4.2.2 Preference of Amino Acid Residues in Linker Sequence

The preference of amino acid residues used in the linker sequences in the natural multidomain protein databases was also investigated. George and Heringa [17] suggested proline (Pro), arginine (Arg), phenylalanine (Phe), threonine (Thr), glutamic acid (Glu), and glutamine (Gln) were preferred in the linkers, in the order of descending occurrence. On the other hand, the study by Argos [18] considered Thr, serine (Ser), glycine (Gly), alanine (Ala), as well as Pro, aspartic acid (Asp), lysine (Lys), Gln, and asparagine (Asn) to be desirable linker constituents (Table 4.1). Both studies showed that Pro, Thr, and Gln were the desirable amino acids. Pro was favored probably because it had no amide hydrogen to form a hydrogen bond with other amino acids, and therefore, it structurally prevented the interaction between the linkers and the protein domains. By grouping the linkers into helical and nonhelical linkers, Pro showed a higher preference to constitute nonhelical linkers, with a propensity value of 1.81, compared to 0.8 for helical linkers. Having Pro residues in the nonhelical linkers could increase the stiffness, and therefore, allowed for an effective separation of the protein domains. Small or polar amino acids, such as Thr (both studies) and Ser and Gly (Argos's study), were found to be desirable for the linker sequence, since they provided good flexibility, and also maintained stability of the structure in the solvent through the formation of hydrogen bonds with water or the peptide chain. However, there was a discrepancy between the two research groups about the use of large hydrophobic residues such as Phe, and large basic and acidic residues, such as Arg and Glu, respectively. Whilst George and Heringa [17] showed that these residues were preferred in the linkers, Argos [18] suggested these residues were generally excluded, because their hydrophobicity or bulky size could cause instability of the linker in aqueous solution and affect the structural independency of the linker.

4.2.3 Hydrophobicity of Linkers

The study by George and Heringa [17] also calculated the average hydrophobicity of the linkers using Eisenberg's normalized consensus residue hydrophobicity scale [19], and assigned values ranging from 0 (hydrophilic) to 1 (hydrophobic) for each set of linker lengths. Small linkers displayed a hydrophobicity value of 0.69 ± 0.11, while large linkers showed a lower hydrophobicity of 0.62 ± 0.08. As mentioned earlier, the linkers were more exposed in the solvent with the increase of length. Here the data suggested that longer linkers were more hydrophilic than smaller linkers, and therefore, they could function more independently in the solvent and allow the domains to move freely.

4.2.4 Secondary Structures of Linkers

The linkers from natural proteins could adopt various conformations in secondary structure, such as helical, β-strand, coil or bend, and turns, to provide their functions. Among George and Heringa's database of linkers, 38.3% adopted the α-helical secondary structure, 13.6% were in β-strands, 8.4% were in turns and the rest 37.6% were in coil or bend secondary structures. The conformation preference of linkers also changed with their length. While small linkers mainly adopted strand and coil secondary structures (33.6% and 36.9%, respectively), large linkers were mainly in helical and coil secondary structures (31.4% and 45.4%, respectively). There was a tendency for linkers to adopt more helical structures as their length became longer [17].

On the basis of their conformation, linkers could be classified into two categories: helical and nonhelical. The α-helix was a well-designed and stable structure, with intra-segment hydrogen bonds and a closely packed backbone [20]. Therefore, many natural and empirical linkers (to be discussed later) adopted this conformation. This type of structure might act as a rigid spacer to separate protein domains, facilitate their independent folding, and could reduce unfavorable interactions between domains. On the other hand, without an inherent rigid structure, the nonhelical linkers tended to incorporate a high frequency of Pro to fulfill the requirement for stiffness. As mentioned previously, Pro had a cyclic side chain with no amide hydrogen to form hydrogen bonding with any neighboring amino acids. It also had a cyclic side chain structure and a limited backbone angle for movement [21]. As a result, linkers with a Pro-rich sequence might form relatively rigid structures and avoid unwanted interactions with the protein domains.

In summary, the natural linkers typically exhibited a range of length between 4 and 20 amino acids. Several amino acids (Pro, Thr, and Gln) were found to be favorable linker components. The hydrophobicity of the natural linkers showed a tendency to decrease with the increase of linker length. Various conformations were adopted by the natural linkers, with α-helix and coil/bend secondary structures found to be the most abundant. Longer linkers tended to adopt more helical structures than shorter ones. Nonhelical linkers appeared to incorporate high frequency of Pro to increase the structural rigidity. Taken together, natural linkers exhibited extended conformations, good structural stability, and had no unfavorable interactions with the adjacent protein domains. As a result, these oligopeptide linkers could effectively separate the protein domains, allow independent folding and functions of each domain, and reduce the steric hindrance between them. The properties of natural linkers could serve as a general guideline for the rational design of linkers in fusion proteins.

4.3 EMPIRICAL LINKERS IN RECOMBINANT FUSION PROTEINS

On the basis of the criteria suggested by the natural linker peptides, researchers have designed various empirical linkers for fusion proteins to achieve desirable functions. These linkers can have various lengths, conformations, and compositions to fit into specific applications. Three types of empirical linkers (flexible linkers, rigid linkers, and cleavable linkers) are discussed in this section. Each category of linkers has specific characteristics and usage in the construction of recombinant fusion proteins.

4.3.1 Flexible Linkers

Flexible linkers are generally composed of small amino acids that can be nonpolar, such as Gly, or small, polar amino acids, such as Ser or Thr. These linkers can maintain structural stability in aqueous solutions by forming hydrogen bonds with the water molecules, and also provide good mobility of the connecting functional domains. Flexible linkers can be applied to join two functional domains passively, while either allowing a certain amount of movement of the protein domains, or facilitating the interaction between domains.

4.3.1.1 Flexible Glycine and Serine Linkers The most commonly used flexible linker peptides have sequences consisting primarily of stretches of Gly and Ser residues ("GS" linker). These amino acids are suitable for flexible linker peptides because the polar Ser amino acids allow hydrogen bonding with the solvent, and the small size of Gly and Ser provide the necessary flexibility [18]. The most widely used example of a flexible linker has the sequence of $(\text{Gly-Gly-Gly-Gly-Ser})_n$. By adjusting the copy number "*n*," the length of this GS linker can be fine tuned to achieve suitable separation of the adjacent domains, or to allow for the necessary interaction between domains.

The first notable application of the GS linker is the construction of single-chain variable fragment (scFv), an antigen-binding fusion protein composed of antibody light-chain variable regions (V_L) tethered to heavy chain variable region (V_H) via an oligopeptide linker [22]. By joining via a peptide linker, the V_H and the V_L domains can be synthesized as a single polypeptide chain to obtain equal molar ratios, while the covalent linkage between V_H and V_L facilitates the association of the two domains. An optimal linker for scFv should adopt the proper dimension to bridge the distance between the two domains, and exhibit an independent secondary structure that will not interfere with the protein folding of either domain.

Huston et al. selected $(\text{GGGGS})_3$ to connect the carboxyl terminus of the V_H to the amino terminus of the V_L [23]. This linker was chosen because it has no ordered secondary structure or tendency to interfere with the folding of the protein domains. As a result, the two variable domains could assume a natural orientation and conformation for antigen binding. The length of the linker was also taken into account to allow for the correct conformation of the scFv. The distance between the C terminus of the V_H domain and the N terminus of the V_L domain was about 3.5 nm according to the X-ray data. Three copies of the GGGGS linker were selected since its length, calculated to be about 5.7 nm, is able to bridge the distance between the V_H and V_L domains.

Interestingly, the length of the linkers between the V_H and V_L domains of a scFv could determine the association status of the fusion protein. By shortening the length of the linker to less than 12 residues, the V_H and V_L domains of a single scFv could no longer pair with each other. Instead, two scFv fusion proteins formed a dimer by associating the V_H domain from one fusion protein with the V_L from another. The resultant bivalent scFv dimer was termed *diabody* [24]. Furthermore, when the length of the linker was shortened to less than 3 residues, the distance between V_H and V_L was only enough to allow for the formation of a trimer, termed *triabody* [25].

Besides scFv, the flexible linkers $(\text{GGGGS})_n$ were also widely applied in the construction of many other recombinant fusion proteins. For example, a direct fusion of granulocyte colony-stimulating factor (G-CSF) and transferrin (Tf) was constructed by Bai et al. for oral delivery of G-CSF via Tf receptor-mediated transcytosis [13]. The Tf fusion proteins could bind to the Tf receptors expressed on intestinal epithelium cells, and were efficiently transcytosed across the intestinal absorption barrier. This approach was widely applied for the oral delivery of various protein and peptide drugs and exhibited marked pharmacological effects [26,27]. For the G-CSF-Tf fusion protein, the *in vitro* bioassay revealed a retention of less than 10% bioactivity by the fusion protein compared to the parent free G-CSF. The interference between the two domains most likely decreased the binding of the fusion protein to G-CSF receptors and might account for the suboptimal *in vitro* and *in vivo* activities of the fusion protein. In order to optimize the intrinsic bioactivity of the fusion protein, a flexible GS linker, $(\text{GGGGS})_3$, was inserted to keep the two domains at a distance [14]. The resultant fusion protein displayed a profound increase of bioactivity, with an EC_{50} of 0.20 ng/mL, about 10-fold less than that of the fusion protein without a linker ($EC_{50} = 2.34$ ng/mL). This improvement might be due to the reduced blockage of receptor binding sites on G-CSF with a longer distance to the Tf domain.

In another study, the $(\text{GGGGS})_3$ linker was inserted between two copies of HBsAg preS1 (21–47) peptides to produce a fusion protein for anti-preS1 (21–47) antibody detection [28]. With the linker insertion, the immunoactivity of the fusion protein is much stronger compared to the fusion protein without a linker. This flexible linker allowed more antibodies to bind to the fusion protein probably because of the increased distance between the two domains and their greater freedom of movement.

Another interesting example of the $(\text{GGGGS})_n$ linkers was the construction of self-renaturing enzymes. Enzyme-chaperone chimeric protein was constructed as a new approach to stabilize an enzyme [29]. Three soluble chimeric proteins were constructed, each consisting of a fusion of chaperon domain (rTHS) and the enzyme of interest (model penicillin amidase, PGA) separated by a flexible $(\text{GGGGS})_n$ linker with different copy numbers ($n = 1, 2$, or 4). The bioactivity of the chimeric protein improved as the length of the linker increased. The catalytic activity of the chimeric protein with the longest linker was almost equivalent to that of the parent enzyme PGA. This study indicated that by adjusting the linker length, the unfavorable inter-domain interference could be minimized, and the bioactivity of the fusion protein could achieve optimal conditions.

4.3.1.2 Other Flexible Linkers

Besides the $(\text{GGGGS})_n$ linker, other flexible linkers have been applied in constructing recombinant fusion proteins. These linkers are also rich in G and S residues in order to maintain their flexibility.

Bird et al. [30] selected the flexible linker sequence (EGKSSGSGSESKST) to fuse the V_H and V_L domains of immunoglobulin and constructed a bioactive scFv antibody. Using computational methods and computer graphics, this linker was designed by adding single amino acids or short peptides extending from the carboxyl terminus of the V_L to the amino terminus of the V_H. The linker contained several glycine and serine residues for flexibility, as well as glutamic acid and lysine residues to improve the solubility.

Another application of flexible linkers is in the production of tagged fusion proteins. Various tags, such as a Myc-tag (10 amino acid residues) derived from the c-myc gene product, glutathione *S*-transferase (GST)-tag (220 amino acid residues), and histidine (His)-tag (~6 histidine amino acid residues), can be fused to proteins of interest for improving expression, and for facilitating purification or identification. The direct fusion between a protein and a tag, however, may cause the loss of function because of the interference from the tag sequence. In order to overcome this drawback, Sabourin et al. reported the construction of tagged fusion proteins using a flexible linker, $(Gly)_8$ [31]. When the flexible $(Gly)_8$ linker was inserted between a Myc epitope tag and the protein of interest, Est2p, the functionality of epitope-tagged Est2p was greatly improved compared to the fusion protein without a linker. The positive effects of the flexible protein linkers were likely due to the proper folding of epitope-tagged proteins after linker insertion, as well as the reduced steric hindrance between functional domains.

In another study, a flexible linker, GSAGSAAGSGEF, was described by Waldo et al. [32]. This linker sequence excluded large bulky hydrophobic residues as suggested by Argos [18]. The GSAGSAAGSGEF linker was initially designed to construct GFP-fusion proteins for rapid protein-folding assay. This assay was developed to detect the correct folding of a protein by expressing a protein of interest as an N-terminal fusion with GFP, and has since been applied to numerous other fusion proteins. This linker provided similar performance of the GFP folding reporter as a longer $(GGGS)_4$ linker. It was chosen over the $(GGGS)_4$ linker because it reduced the amount of homologous repeats in the DNA coding sequence, which could result in deletions by homologous recombination during the shuffling protocol for cloning.

In summary, flexible linkers are mainly composed of small or polar amino acids such as Gly and Ser. These linkers can provide distance between functional domains while allowing the necessary interactions or movement of the functional domains. By adjusting the copy number of the linkers, the optimal length of the linkers can be identified to achieve proper folding, good production, or the highest biological activity.

4.3.2 Rigid Linkers

Although the flexible linkers are versatile for constructing a wide variety of chimeric fusion proteins, it has limitations such as ineffectiveness in spatially separating functional domains due to its high flexibility. For instance, Maeda et al. reported that the immunoglobulin binding ability of the protein G domain in a protein G-Vargula luciferase fusion protein is not retained with the insertion of a flexible GGGGS linker [33]. In this case, the insertion of a flexible linker may not be sufficient to separate the functional domains and prevent the unfavorable interactions between the domains. Therefore, for applications where the prevention of interdomain interference is necessary for activity, a rigid linker that has a stiff structure should be applied to effectively separate the domains and allow them to act independently.

4.3.2.1 Alpha Helix-Forming Linker $(EAAAK)_n$ A new type of rigid linkers, the α helix-forming peptides $A(EAAAK)_nA$ ($n = 2$–5) were introduced for the construction of bioactive fusion proteins by Arai et al. [34,35]. These linkers were shown to adopt an α-helix conformation, which was stabilized by a Glu^-–Lys^+ salt bridge within the linker structure [36].

These helical linkers were first characterized by inserting various sequence lengths between enhanced blue fluorescent protein (EBFP) and enhanced green fluorescent protein (EGFP), and measuring the fluorescent resonance energy transfer (FRET) efficiency [34]. FRET is a distance-dependent interaction, where a shorter distance between the two fluorophores (i.e., EBFP and EGFP) will result in a higher FRET efficiency. In this study, the FRET efficiency between the two domains decreased as the length of helical peptides increased, indicating that these linkers can control the distance between domains by changing repetitions of the EAAAK motif. Compared to the flexible linkers with the same number of amino acid residues, the helical linkers induced much less FRET efficiency when inserted into EBFP–EGFP fusion proteins. This result suggested the distance between protein domains did not simply depend on the number of residues in the linker, but could also be affected by the conformation of the linker. The circular dichroism (CD) spectroscopic analysis of the fusion proteins confirmed the α-helical conformation of the linkers, and indicated that the helix content increased as the length of the linker increased [34].

Another study by Werner et al. [37] utilized helical linker technology to develop a virus-protein A system to efficiently manufacture protein A-based immunosorbent nanoparticles for use in industrial purification of monoclonal antibodies. This report showed that a functional fragment of protein A (133 aa) could be displayed on the surface of a tobamovirus as a C-terminal fusion to the viral coat protein via a 15-aa linker $(EAAAK)_3$ [37]. Without a linker, only peptides shorter than 20 aa could be fused to the coat protein of plant rod-shape viruses without preventing the assembly of functional virions [38]. In addition, in the case of tobamovirus, viral particles could form only when excess free coat protein was co-expressed with the fusion protein, leading to the formation of viral particles displaying the protein A fragment on only 5% of the coat protein subunits.

Dramatically, with the help of a 15-aa helical linker between the coat protein and the protein A fragment, the fusion protein could be successfully expressed and assembled into functional virions in the absence of free coat proteins. This example clearly demonstrated the pivotal role of linker insertion to provide proper distance between functional domains and to allow for particle assembly.

Lu et al. used the α-helical peptides $(EAAAK)_n$ ($n = 1$–3) as linkers to construct bifunctional fusions of β-glucanase and xylanase enzymes [3]. The results showed that insertion of helical linkers improved the catalytic efficiencies of both moieties. The β-glucanase thermal stabilities of the fusion proteins also improved with the increased length of the α-helical linkers. More interestingly, all of the fusion proteins with helical linkers were more thermally stable than those with the flexible linkers. Similar to the previous studies, these trends were attributed to the rigid structure of the α-helical linker that might decrease interference between the linked moieties, suggesting that changes in linker structure and length could affect the stability and bioactivity of functional moieties.

4.3.2.2 Rigid $(XP)_n$ Linkers Another type of rigid linkers are composed of a proline rich sequence, $(XP)_n$, where X can be any amino acid, among which the common choices are Ala, Lys, and Glu. Pro is a very unique amino acid, with its side chain cyclized back to the amide on the backbone. As suggested by George and Heringa [17], Pro has no amide hydrogen to form hydrogen bond with other amino acids, and therefore, it structurally prevents the interaction between the linkers and the protein domains. Another characteristic of Pro is that it has a very restricted confirmation, with the backbone angle φ limited to a small range of $-65°$. Having Pro resides in nonhelical linkers can increase the stiffness, and allows for effective separation of the protein domains. As a result, this type of linker usually adopts rigid and extended structure. The $(XP)_n$ linker occurs naturally in the E2 chain of the pyruvate dehydrogenase [39] and in the IIABMan subunit of the *Escherichia coli* mannose transporter [40].

The structure of this type of linker was extensively studied by several groups [39,41,42]. For example, the conformation of the $(\text{Ala-Pro})_7$ dipeptide repeat in the alkali light chain of skeletal muscle (LC1) was studied by Bhandari et al. by using ^{1}H-NMR spectroscopy [41]. They concluded that this peptide linker adopted an extended and rigid conformation because of the conformational restrains imposed by the Pro, which strongly preferred φ backbone angles close to $-65°$ [43]. A similar study of 33-residue peptides containing repeating -Glu-Pro- or -Lys-Pro- by Evans et al. also demonstrated the X-Pro backbone had a relatively stiff elongated conformation [42].

The effect of sequence length in the $(\text{Ala-Pro})_n$ peptide linker (10–34 aa) was evaluated in a study using an interferon (IFN)-γ-gp120 fusion protein [44]. With the short 10-aa linker, the fusion protein possessed a relatively low biological activity of IFN-γ. By gradually increasing the length of the Ala-Pro linker, the bioactivity had been improved. The chimeric protein containing the longest 34-residue linker had the highest activity, measured at ~88% of that of the control protein. In comparison, fusion proteins with cysteine-tryptophan linkers, constructed by inserting the oligonucleotides encoding linker in the reverse orientation, possessed no detectable biological activity, indicating that the improvement of the bioactivity by the $(\text{Ala-Pro})_n$ linker is sequence specific.

4.3.3 *In Vivo* Cleavable Linkers

The linkers in fusion proteins discussed in this chapter so far generally consist of stable peptide sequences, and can provide structure flexibility, improve protein production and stability, or increase biological activity. However, stable peptide linkers do not allow for the separation of the two fusion protein domains *in vivo*, and can have several drawbacks including decreased bioactivity, steric hindrance between functional domains, and altered biodistribution and metabolism of the protein moieties because of interference with each other.

4.3.3.1* In Vivo *Cleavable Disulfide Linker To overcome some of these potential pitfalls in stable peptide linkers, Chen et al. designed an *in vivo* cleavable linker for recombinant fusion proteins utilizing the reversible nature of the disulfide bond [45]. The reversible disulfide linkage has been widely applied in drug delivery by chemical conjugation methods. However, the major pitfall in chemically conjugating protein moieties through disulfide bond formation is the generation of a range of heterogeneous products. This novel recombinant fusion protein approach described by Chen et al. offers the significant advantage in the generation of a precisely constructed, homogeneous product.

The cleavable dithiocyclopeptide linker contains a thrombin-sensitive sequence, as well as an intramolecular disulfide bond formed between two cysteine residues of the linker (Figure 4.1). This linker was inserted between G-CSF and Tf to construct a recombinant fusion protein. Treatment *in vitro* with thrombin results in cleavage of the thrombin-sensitive sequence, while the reversible disulfide linkage between the two domains of the fusion protein remains.

After successful construction of the cleavable linker into G-CSF-Tf fusion protein (designated as G-C-T), the *in vitro* G-CSF biological activity of the fusion protein was determined via cell proliferation assay. The fusion protein was pretreated with thrombin and reduced by DTT to mimic the expected *in vivo* released free G-CSF. With the release of free G-CSF, the fusion protein exhibited a two fold increase

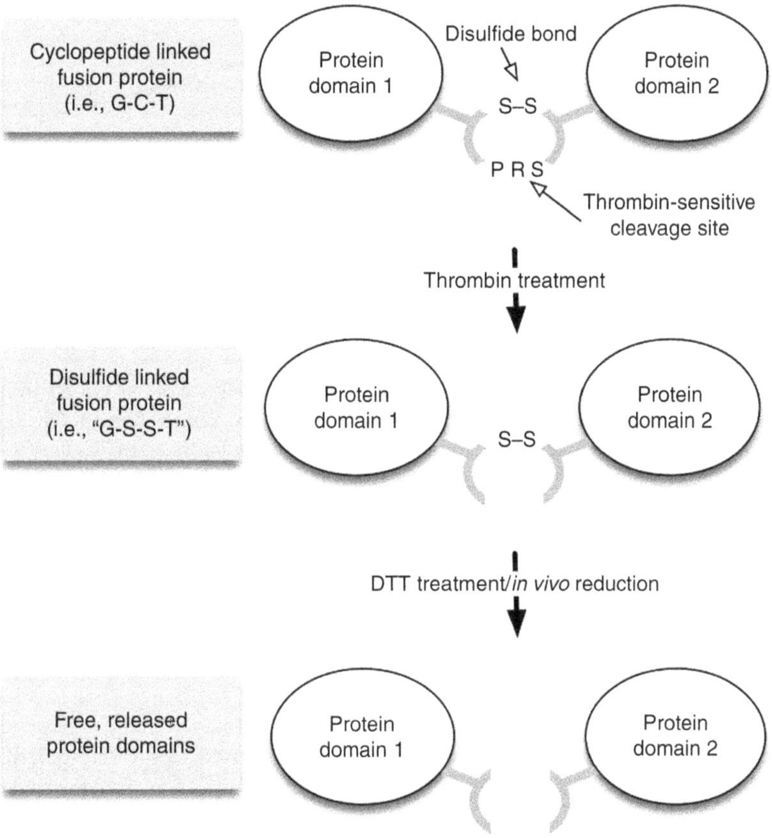

FIGURE 4.1 Design of the dithiocyclopeptide linker. The linker is based on the structure of somatostatin modified to contain a thrombin-specific sequence, PRS. Two cysteinyl-residues on somatostatin naturally form a disulfide bond. *In vitro* thrombin treatment cleaves the thrombin-sensitive sequence while the reversible disulfide bond remains between protein domains. *In vitro* DTT treatment or *in vivo* reduction is able to reduce the disulfide bond, and release free protein domains from the fusion protein. *Source*: Adapted from Reference [45].

of *in vitro* G-CSF bioactivity, with an EC_{50} of 10.12 ng/mL, compared with the noncleaved, nonreduced fusion protein (EC_{50} = 23.69 ng/mL). The maximal response of the cleaved and reduced fusion protein was also much higher than the intact fusion protein (Table 4.2). As previously mentioned, the covalent linkage between two functional domains may cause strong steric hindrance, or block the receptor-binding site of the protein moieties. These reasons may account for the decreased bioactivity of many fusion proteins [13,26]. Therefore, the improved G-CSF bioactivity was probably due to the reduced steric hindrance and decreased blockage effect with the release of free G-CSF from Tf. This result indicated that, owing to the *in vivo* cleavability of the linker, a fusion protein with this disulfide linker might have better *in vivo* biological activity than a comparable fusion protein with a noncleavable linker.

TABLE 4.2 ***In Vitro*** **Biological Activity of G-CSF-Cyclo-Tf Fusion Protein with a Cleavable Linker (G-C-T)**

Fusion Protein	EC_{50} (ng/mL)[a]	E_{max} (A_{570})[b]
G-C-T	23.69	0.8
Thrombin and DTT treated G-C-T	10.12	1.9

[a]EC_{50}, half-maximal effective concentration, as determined by the NSF-60 proliferation assay. The NFS-60 cells proliferate in response to G-CSF treatment.
[b]E_{max}, maximal effect, as determined by the MTT (3-(4,5-dimethylthiazole-2-yl)-2,5-diphenyltetrazolium bromide) assay for NSF-60 cell proliferation. The MTT assay is a colorimetric assay based on the reduction of MTT to purple formazan in the mitochondria of living cells.
Source: Adapted from Reference [45].

The *in vivo* cleavability of the thrombin pretreated, disulfide-linked fusion protein (designated as G-S-S-T) was also investigated in animals. G-S-S-T (with cleavable disulfide linker) or G-C-T (with stable dithiocyclopeptide linker) was administered intravenously to CF1 mice, and the plasma samples were analyzed by nonreducing SDS-PAGE followed by anti-G-CSF Western blot analysis. As shown in

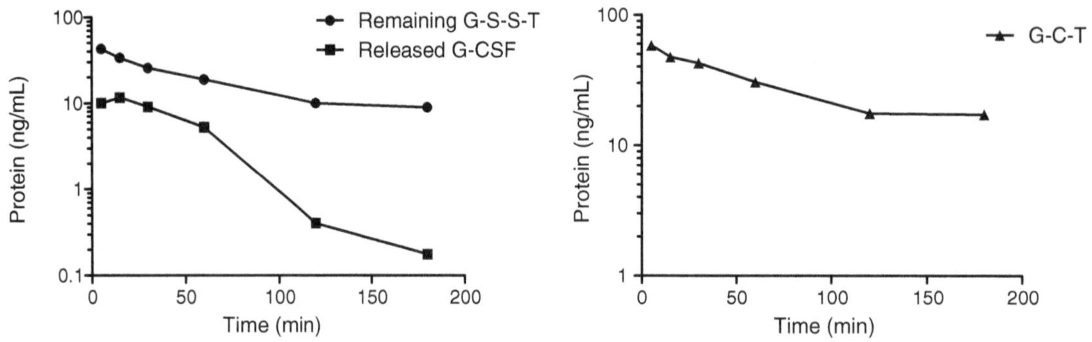

FIGURE 4.2 *In vivo* cleavage of the disulfide linker. (a) *In vivo* release of free G-CSF from G-CSF-S-S-Tf (G-S-S-T) with a cleavable disulfide linker. (b) *In vivo* elimination of G-CSF-*cyclo*-Tf (G-C-T) with a stable dithiocyclopeptide linker. G-S-S-T (thrombin-treated G-C-T) or G-C-T was administered intravenously to CF1 mice via the tail vein at a dose of 4 mg/kg. After administration, the collected blood was analyzed by nonreducing SDS-PAGE followed by anti-G-CSF Western blot analysis to measure the intact fusion protein and free released G-CSF. The *y*-axis represents the concentration of each protein in mouse plasma, quantified using Quantity One software (Bio-Rad Laboratories, Hercules, CA, USA). No free released G-CSF was detected following administration of G-C-T. *Source*: Adapted from Reference [45].

Figure 4.2, free G-CSF could be detected in the blood as early as 5 min after injection of G-S-S-T, with a peak at ~15 min postinjection. A rapid release of G-CSF from the fusion protein and a quick elimination of free G-CSF due to its short *in vivo* half-life were observed. In contrast, no detectable amount of free G-CSF was present in the blood of CF1 mice treated with G-C-T (Figure 4.2). These results demonstrated that the disulfide linkage created by the dithiocyclopeptide linker could indeed be cleaved *in vivo*.

This study demonstrates that the reversible nature of the disulfide bond can be utilized to design an *in vivo* cleavable linker in a recombinant fusion protein to ensure the release of free functional domain into the blood circulation. This *in vivo* cleavable linker can be applied to a wide variety of fusion proteins when the *in vivo* separation of the domains is required for achieving optimal activity, desirable biological functions, or independent actions/metabolism of individual domains.

4.3.3.2 **In Vivo** ***Cleavable Protease-Sensitive Linkers***

Another type of *in vivo* cleavable linkers takes advantage of proteases that are expressed *in vivo* under specific pathological conditions, in specific types of cells or tissues, or are constrained in certain cellular compartments. Linkers in recombinant fusion proteins can be designed to be *in vivo* cleavable specifically where the proteases are present by incorporating a protease sensitive sequence into the linker.

Schulte reported the construction of a recombinant coagulation factor IX (rFIX) and albumin fusion protein for the treatment of hemophilia B [46]. Albumin-fusion technology is widely used to prolong half-life of protein drugs and to improve their efficacy. In the first attempt, a fusion protein of rFIX and albumin with a flexible GS linker exhibited a poor FIX activity. The bulkiness of albumin was thought to be affecting the interaction of FIX with other coagulation factors (e.g., factor VIII and factor FX) and thereby limited its potency. To overcome this limitation, a proteolytically cleavable sequence (VSQTSKLTRAETVFPDV) derived from the N-terminal activation region of FIX was tested as a linker to connect FIX with albumin. The idea was to take advantage of the activation of FIX by either tissue factor/factor VIIa or factor XIa. Both activation sites on FIX and the linker would be simultaneously cleaved. This cleavable linker significantly improved the clotting activity of the cleavable fusion protein by 10- to 30-fold compared to that of the fusion protein with noncleavable linkers. At the meantime, since this linker was only cleavable during clotting, the fusion protein exhibited prolonged half-life compared to rFIX.

In a similar study, Vessillier et al. fused IFN-β with the latency-associated peptide (LAP) of transforming growth factor (TGF) β to create a latent cytokine that was covered by the shell structure provided by the LAP [47,48]. Between the two protein moieties, a linker containing two flexible sequences (GGGGS) flanking a matrix metalloproteinase (MMP) cleavage site was designed. The release of IFN-β would occur only through the cleavage by MMPs that were overexpressed at the disease site during a variety of pathological conditions such as arthritic diseases [49], cancer [50], and inflammation [51,52]. Since the cleavage of the linker was very limited, IFN-β could be masked by LAP and

TABLE 4.3 Cleavable Linker Peptide Sequences Designed in the Fusion Proteins

	Linker No.	Sequence
Type 1 linkers	1	GGGGS-**PLGLWA**-GGGGS
	2	GGGGS-**GPLGLWA**-GGGGS
	3	GGGGS-**GPLGLWAQ**-GGGGS
Type 2 linkers	4	GGGGS-**GVPDVGHFSLFP**-GGGGS
	5	GGGGS-**GVPDVGEFSLFP**-GGGGS
	6	GGGGS-**GVPDVGNFSLFP**-GGGGS
	7	GGGGS-**GVPDVGRFSLFP**-GGGGS
	8	GGGGS-**GVPDVGHYSLFP**-GGGGS
	9	GGGGS-**GVPDVGEYSLFP**-GGGGS
	10	GGGGS-**GVPDVGNYSLFP**-GGGGS
	11	GGGGS-**GVPDVGRYSLFP**-GGGGS
Type 3 linkers	12	GGGGS-*NSG* **PLGLWA** *QSTSAGPTV*-GGGGS
	13	GGGGS-*NS* **GVPDVGHFSLFP** *STSAGPTV*-GGGGS
	14	GGGGS-*NS* **GVPDVGNYSLFP** *STSAGPTV*-GGGGS

Different MMP sites are shown in bold. Underlined amino acids correspond to the positions X_1 and X_2 of the consensus sequence Gly-Val-Pro-Asp-Val-Gly-X_1X_2-Ser-Leu-Phe-Pro based on the MMP prodomain region. Italics amino acids correspond to the hydrophilic loop derived from the flanking sequences of the furin cleavage site from the anthrax protective antigen.

Source: Reproduced from Reference [47] by permission of Oxford University Press.

acted as a latent cytokine. As a result, LAP-IFN-β fusion protein had a long half-life of 55 h *in vivo* (37 times longer than native IFN-β [53]), because the latent cytokine could not interact with its cellular receptors until it was released from the LAP. Conceivably, the side effect of IFN-β could also be minimized because of the restricted exposure to the released cytokine.

To obtain the optimal sensitivity of the linkers to the MMPs, three different types of cleavage sites were introduced into the LAP-IFN-β construct by Vessillier et al. [47] (Table 4.3). The original MMP cleavage site (Pro-Leu-Gly-Leu-Trp-Ala), derived from a fluorogenic substrate for MMPs [54], was extended by addition of one or two amino acids (Gly and Gln), to achieve a higher sensitivity to cleavage by MMPs [55]. The second type of linker was based on cleavage sites derived from the propeptide region of MMPs. A consensus sequence Gly-Val-Pro-Asp-Val-Gly-X1X2-Ser-Leu-Phe-Pro was identified based on the prodomain region of several MMPs. Combinations of X1 (His, Glu, Asn, or Arg) and X2 (Phe or Tyr) were used for the development of new MMP cleavage sites. The last type of linkers added a surface-exposed flexible loop, derived from the amino acids 162–175 around the furin cleavage site of the anthrax toxin protective antigen, to the MMPs cleavage site [56]. The addition of this loop sequence was intended to increase the accessibility of the MMP-sensitive sequence to the MMPs.

These cleavable linkers all exhibited various sensitivities for the MMPs. Among them, the addition of the surface-exposed flexible loop from the furin cleavage site of the anthrax toxin protective antigen increased the sensitivity to MMP-3 to up to 29-fold. The determination of cleavage efficiency and specificity of the cleavage sites to the MMPs could be useful for developing other biological agents that could be activated at the disease sites overexpressing MMPs.

In addition to extracellular spaces, the activation site of *in vivo* cleavable linkers in fusion proteins can also be inside the cells. Another type of *in vivo* cleavable linker takes advantage of furin, a cellular endoprotease that has been implicated in the proteolytic activation of diverse precursor proteins [57]. This endoprotease has the consensus recognition sequence -Arg-X-Arg/Lys-Arg$^{\downarrow}$- ($^{\downarrow}$ identifies the cleavage site). Furin is mainly localized in trans-golgi network (TGN) and also recycles between TGN, early endosomes, and the cell surface.

The furin-sensitive linkers have been applied in the construction of recombinant fusion proteins such as immunotoxins or immunoproapoptotic proteins from which active toxins or protein drugs can be released from scFv after intracellular processing. Inclusion of furin-sensitive linkers in immunotoxins containing ribotoxin, caspase-3, or granzyme B has shown significant improvements in cytotoxicity compared with constructs containing stable linkers [58,59].

Like the disulfide linker, the protease-sensitive linkers can be applied to fusion proteins when the *in vivo* separation of the domains is preferred. Additionally, using protease sensitive linkers also offers a new approach to specifically deliver prodrugs to target sites where the proteases that convert the prodrugs are present.

In conclusion, various types of linkers with different structures or functions have been reported for the construction of fusion proteins. As summarized in Table 4.4, each type of linkers has its specific characteristics, advantages, and disadvantages. The choice of linkers should be based on the properties and applications of the desired fusion proteins.

TABLE 4.4 Summary of Empirical Linkers

	Flexible Linker	Rigid Linker	Cleavable Linker
Characteristics	Rich in small or hydrophilic aa	Rigid structure or rich in Pro	Cleavable by reduction or proteases
Examples	$(GGGGS)_n$, G_8	$(EAAAK)_n$, $(XP)_n$	Disulfide, protease-sensitive sequence
Advantages	Flexibility and mobility	Effective separation	Release of free domain at specific site
Disadvantages	Insufficient separation	Low flexibility	Possible low sensitivity to protease

4.4 FUNCTIONALITY OF LINKERS IN FUSION PROTEINS

As previously discussed, various types of linkers are available for the purpose of protein fusion. They can have different lengths, exhibit various conformations, and even be designed for cleavage at specific sites *in vivo*. These diverse characteristics of linkers provide them a wide spectrum of functions. Besides the properties such as covalently joining domains (e.g., V_H and V_L of antibody), spatially separating protein moieties, linkers can also offer many advantages in improving the biological activities, increasing production, as well as achieving desirable PD and PK properties of the fusion proteins.

4.4.1 Linkers Can Improve Biological Activity of Fusion Proteins

A common issue encountered during the development of fusion protein is the impaired biological activity of the functional domains after protein fusion. One possible reason for this issue is that the two functional domains are brought too close, so that they can interfere with each other and block the accessibility of the protein domains to the respective receptors. To construct bioactive fusion proteins, linkers have been very effective tools to provide enough distance between domains and reduce the steric hindrance between them.

A good example of applying linkers for bioactivity improvement is the development of Tf fusion proteins for oral delivery of protein and peptide drugs [13,14,26]. The Tf-fusion protein can be orally absorbed across intestinal epithelium cells via transcytosis after binding to Tf receptors on the intestinal cell surface [60,61]. A G-CSF and Tf fusion protein with a short dipeptide linker (Leu-Glu, LE) was constructed for oral delivery of G-CSF [13]. With the dipeptide linker, the G-CSF-LE-Tf fusion protein elicited significant pharmacological affect via oral administration at a high dose of 50 mg/kg. However, the intrinsic G-CSF bioactivity, as determined by *in vitro* cell assay, was suboptimal. The fusion protein retained less than 10% of the bioactivity of parent G-CSF. To achieve greater oral efficacy, Bai and Shen reported the construction of G-CSF-Tf fusion proteins by inserting different copies of helical peptide linker $[A(EAAAK)_nA]_m$ ($n=2$–4, $m=1$ or 2), or by inserting a flexible $(GGGGS)_3$ linker between the two functional domains [14]. All of the fusion proteins with either helical or flexible linkers exhibited improved *in vitro* G-CSF biological activities compared to the original fusion protein with the dipeptide linker. Particularly, the fusion protein with the $(H4)_2$ linker, $(A(EAAAK)_4ALEA(EAAAK)_4A)$ exhibited the highest *in vitro* biological activity among all of the linkers tested, with approximately a 10-fold increase in activity compared to the fusion protein with the dipeptide linker. With the $(H4)_2$ linker improved *in vivo* biological activities of G-CSF-Tf were also observed in animal experiments via both subcutaneous and oral administration. The subcutaneously administered G-CSF-$(H4)_2$-Tf protein exhibited much higher efficacy than G-CSF-LE-Tf at three different doses (0.1, 0.5, and 1 mg/kg). At a low dose of 20 mg/kg for oral administration, the G-CSF-$(H4)_2$-Tf fusion proteins elicited a significant pharmacological effect, whereas the original fusion protein with the dipeptide linker only exhibited minimal effect.

The insertion of the helical linker $(H4)_2$ had a similar enhancement effect on the bioactivity of another Tf-based fusion protein—human growth hormone (hGH)-Tf fusion protein [26]. The *in vitro* hGH biological activity in hGH-$(H4)_2$-Tf was improved compared to hGH-Tf which contains a short dipeptide linker LE. The *in vivo* bioactivity was then tested in hypophysectomized rats, which gain body weight in response to hGH treatment. The subcutaneous administration of hGH-$(H4)_2$-Tf promoted higher body weight gain than that of hGH-Tf in a 7-day daily treatment. More remarkably, with the oral administration, only hGH-$(H4)_2$-Tf, but not hGH-Tf, elicited a significant body weight gain [26].

These studies suggest that linker insertion may be a useful method to improve suboptimal intrinsic activity of fusion proteins that are caused by insufficient separation of functional domains. The unwanted interaction and steric hindrance between the protein domains could be greatly decreased, and as a result, the bioactivity may be greatly enhanced. In the study of G-CSF-Tf fusion proteins, the helical peptide linkers appear to be superior to the flexible linker $(GGGGS)_3$, probably because the helical linkers adopt more rigid structures and can separate the protein domains more effectively.

Linker engineering was also applied to the construction of bioactive and stable fusion proteins consisting of human serum albumin (HSA) and interferon-α2b (IFN-α2b). Albumin fusion proteins are commonly used for prolonging the plasma half-life of protein drugs and enhancing *in vivo* efficacy [8,62,63]. Direct fusion of HSA and IFN-α2b can cause interference between domains, and diminish the antiviral activity of IFN-α2b domain [64]. Zhao et al. compared the biological activity of HSA–IFN-α2b fusion proteins with no linker or three types of linkers including flexible linker (GGGGS), rigid linker (PAPAP), and helical linker (AEAAAKEAAAKA) [11]. Previous studies from the same group had indicated that the direct fusion of IFN-α2b to HSA without a linker interfered with the disulfide bond formation between two cysteine residues in the IFN-α2b sequence. Their result from comparing the different types of linkers showed that all the three linkers could separate HSA and IFN-α2b effectively and restore the disulfide formation, as all fusion proteins appeared as a homogenous product on nonreducing SDS-PAGE. Compared to the fusion protein without a linker, all the three linkers demonstrated a significantly improved anti-viral activity of HSA–IFN-α2b by 39% (flexible linker), 68% (rigid linker), and 115% (helical linker). These results suggest that structured linkers such as the rigid and helical linkers may separate HSA and IFN-α2b more effectively than flexible linker, as they increased the anti-viral activity more significantly. The higher activity associated with helical linker could not be simply ascribed to its longer length, as increasing the length of flexible linker from 5 amino acid residues (GGGGS) to 10 amino acid residues (GGGGSGGGGS) could not increase the activity of fusion protein.

These studies provide useful evidences that linker insertion is an effective tool to improve biological activity of fusion proteins. By inserting a flexible or helical peptide linker that adopts a rigid structure, effective separation of the two protein domains can be achieved. The unwanted interaction and steric hindrance between the protein domains could then be diminished, and, as a result, the bioactivity may be greatly enhanced.

4.4.2 Linkers Can Increase Expression Yield of Fusion Proteins

Gene fusion techniques offer the advantage to combine the desired properties from different proteins into a fusion protein. However, direct fusion of two protein moieties may cause folding defects resulting from structural perturbation between different moieties. Many researchers have reported the difficulties in expressing stable and homogenous fusion proteins [11,14]. The instabilities of fusion proteins often result in a low expression yield and a heterogeneous product, which are both unfavorable characteristics for downstream applications. Although the expression level of fusion proteins can sometimes be improved by simply switching the orientation of the two component protein domains [64], the steric hindrance may not be effectively reduced because of the short distance between domains. Many linkers can provide longer distance between domains, and allow for the independent folding of protein moieties. Linker technology has been shown as a promising and practical method to increase the expression yield of recombinant fusion proteins.

The effect of linker insertion on expression efficiency of fusion proteins was observed in Tf-based fusion proteins [12]. Tf was first directly fused with hGH to construct a fusion protein for Tf receptor-mediated oral delivery of hGH. Fusion proteins with two orientations of Tf and hGH (hGH-Tf and Tf-hGH) were constructed to achieve the optimal bioactivity. A helical $(H4)_2$ linker (A(EAAAK)$_4$ALEA(EAAAK)$_4$A) was inserted in the two fusion proteins between hGH and Tf since this linker may improve the bioactivity of the fusion proteins [14]. The resultant fusion proteins were designated as hGH-$(H4)_2$-Tf and Tf-$(H4)_2$-hGH. These fusion proteins were then transiently expressed from HEK293 cells. Surprisingly, the insertion of the helical linker $(H4)_2$ greatly improved the expression level of the Tf-fusion proteins. The insertion of the helical linker in hGH-Tf fusion proteins resulted in a 1.66-fold higher expression in HEK293 cells. Similarly, fusion protein Tf-$(H4)_2$-hGH with the reversed orientation had an expression level 2.39-fold higher than that of Tf-hGH.

A more profound effect of linker insertion on protein production was observed in Tf fusion proteins that were fused to G-CSF. Two fusion proteins with reverse orientation (G-CSF-Tf and Tf-G-CSF) were first constructed. The helical linker $(H4)_2$ was then inserted into G-CSF-Tf to test if it increases the expression level. A moderate enhancement (1.44-fold) was obtained for G-CSF-$(H4)_2$-Tf compared to G-CSF-Tf. Unexpectedly, the Tf-G-CSF fusion protein, which had the reversed orientation, failed to be efficiently expressed from HEK293 cells, with its quantity too low to be detected convincingly by Western blot analysis. Even with the insertion of a flexible linker $(GGGGS)_3$, the expression of the Tf-$(GGGGS)_3$-G-CSF fusion protein was still unsuccessful. Dramatically, when the helical linker $(H4)_2$ was inserted, high level of Tf-$(H4)_2$-G-CSF was expressed from the cells, with an 11.2-fold elevation on production. The effect on the expression level did not simply depend on the length of the linker, because Tf-G-CSF with a linker that was encoded with the reversed coding sequence for the $(H4)_2$ (which produces an identical peptide length without a helical structure) failed to express.

As suggested by these studies, when difficulties are encountered for fusion protein production, a feasible approach is to insert linkers between functional domains. The exact mechanism for the improved expression with helical linkers is still unclear. The rigid, extended nature of the helical linker may effectively separate the two protein domains, and greatly reduce the interference between them.

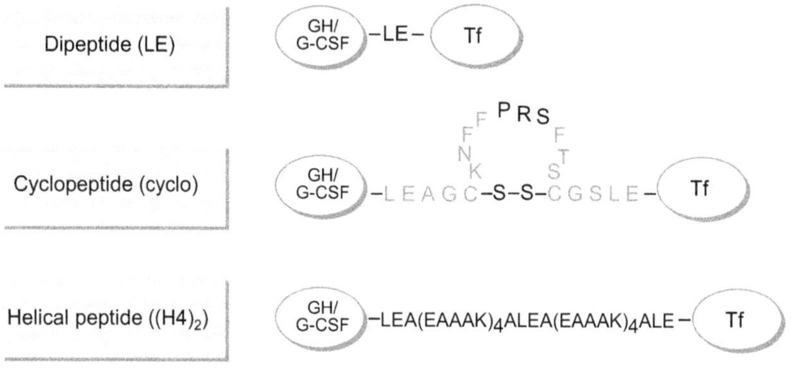

FIGURE 4.3 Illustration of three linkers applied in GH-Tf and G-CSF-Tf fusion proteins for PK study. The dipeptide linker LE is composed of two amino acids—Leu and Glu. The cyclopeptide linker is derived from somatostatin, with a mutated thrombin-sensitive sequence "PRS." Two cysteine residues on the linkers naturally form a disulfide bond. The helical peptide linker adopts an α-helical structure.

The helical linker may provide a relatively large space for the domains to fold independently and properly. A correctly folded fusion protein may have enhanced stability and consequently, an increased expression level from the cells.

4.4.3 Linkers Can Affect the PK of Fusion Proteins

Besides their wide applications in biological and biomedical research, recombinant bifunctional fusion proteins have become an important class of therapeutics for the treatment of human diseases. Several fusion proteins such as Enbrel® (TNF-R/Fc-IgG1), Ontak® (IL-2/diphtheria toxin), Orencia® (CTLA-4/Fc-IgG1), and Amevive® (LFA-3/Fc-IgG1) have been approved by the FDA for clinical usage [65], and many more are currently under development. Fusing two protein domains together may provide many advantages over the parent proteins. By fusing with HSA or the Fc portion of IgG, many protein drugs such as insulin, IFN-α, and TNF have exhibited improved PK and PD properties [62,66,67]. The fusion proteins can also have bifunctional properties where each domain imparts a bioactive property (e.g., human immunoglobulin Fc γ Fc ϵ bifunctional fusion protein) [68], or where a drug targeting effect is combined with biological activity (e.g., immunotoxins) [69].

In contrast to the rapid development of fusion proteins as therapeutics, the PKs of bifunctional fusion proteins are still largely unexplored due to their distinct bifunctional properties. Many bifunctional fusion proteins contain linkers between functional domains to improve their bioactivity, receptor binding affinity, as well as structural stability. However, especially in the case of bifunctional fusion proteins, linker insertion increases the complexity of the PK studies and the impact of linkers on PK is a remaining question. Recently, Chen et al. conducted a study demonstrating that the linkers could affect the PK of fusion proteins, and also established a mechanistic model for PK of fusion proteins.

To elucidate the role of linker insertion on PK of fusion proteins, three linkers with different lengths and conformations (Figure 4.3) were inserted into bifunctional fusion proteins composed of Tf and hGH or G-CSF. The first linker is a short dipeptide, Leu-Glu (LE), resulting from the cloning site *XhoI* between the two protein genes [13]. The second linker is a thrombin-sensitive, disulfide cyclopeptide linker, designated as cyclo. As described in Section 4.3.3, this cyclopeptide linker with a medium length (20 amino acids) and a rigid cyclic structure were originally designed to create a disulfide linker that could be rapidly cleaved *in vivo* by reduction [45]. The third linker is an α-helix forming linker, designated as $(H4)_2$ $(LEA(EAAAK)_4ALEA(EAAAK)_4ALE)$. This linker has the longest length among all the three linkers (50 amino acids) and a rigid, extended structure [14,26]. Previous reports suggest that this long helical linker can spatially separate the functional domains and reduce the steric hindrance between them [34].

After the insertion of three linkers, the receptor binding affinities of hGH-Tf fusion proteins were first compared via competitive receptor binding assays on receptor-bearing cell lines. With the shortest dipeptide linker (LE), fusion protein hGH-LE-Tf exhibited the lowest binding affinities for both hGH receptor (hGHR) and Tf receptor (TfR) (Table 4.5), suggesting that this short linker might generate strong steric hindrance or cause interference between functional domains. On the other hand, the cyclopeptide and helical peptide linkers, which both have longer lengths (20 and 50 amino acids, respectively) and more rigid structures, generated higher receptor binding affinities (Table 4.5). Therefore, by inserting different linkers between the functional domains, the receptor binding affinities of bifunctional fusion proteins could be greatly altered.

TABLE 4.5 Comparison of IC_{50} Values from Competitive Receptor Binding Assays and Plasma-Half Lives of GH-Tf and G-CSF-Tf Fusion Proteins

Protein	IC_{50}, GHR (nM)	IC_{50}, G-CSFR (nM)	IC_{50}, TfR (nM)	$t_{1/2}$ (h)
GH-LE-Tf	17.7	/	21.2	4.97 ± 0.34
GH-*cyclo*-Tf	8.2	/	4.2	1.76 ± 0.27
GH-$(H4)_2$-Tf	7.0	/	8.7	1.87 ± 0.44
G-CSF-LE-Tf	/	38.0	7.5	4.15 ± 0.75
G-CSF-*cyclo*-Tf	/	39.5	0.9	5.69 ± 0.46
G-CSF-$(H4)_2$-Tf	/	31.2	4.5	4.84 ± 1.18

IC_{50} values were determined via competitive receptor binding assay on receptor-bearing cells. Unlabeled fusion proteins were used to compete with radiolabeled free GH, G-CSF, or Tf for the respective receptor binding. Smaller IC_{50} value represents higher receptor binding affinity. Half-life values represent mean $\pm$ SD from three to four mice.

The impact of linker insertion on receptor binding affinities resulted in dramatic differences in the PK profiles of three hGH-Tf fusion proteins. hGH-LE-Tf exhibited an almost threefold longer plasma half-life compared to the others, despite the three fusion proteins having very similar size and sequence (Table 4.5). A strong correlation between hGHR binding affinity (but not TfR binding affinity) and plasma half-life was identified for the hGH-Tf fusion proteins. hGH-LE-Tf, which had the weakest binding for hGHR, exhibited the longest plasma half-life. hGH-cyclo-Tf and hGH-$(H4)_2$-Tf, which exhibited comparable hGHR binding affinities, had similar plasma half-lives.

These results suggested that binding of hGH-Tf fusion proteins to hGHR likely led to endocytosis and lysosomal degradation of the fusion proteins as reported for free hGH [70]. Like many other hormones and cytokines, hGH and hGHR will be endocytosed into endosome following receptor binding, where hGHR is ubiquitinated [71]. The hGH-hGHR complex is then further sorted to lysosome and gets degraded. The receptor binding of hGH domain in hGH-Tf probably leads to similar intracellular processing as free hGH. As a result, hGH-LE-Tf exhibited longer half-life because of its lower binding affinity to hGHR and consequently less degradation via receptor-mediated endocytosis. This result indicated that linker insertion could greatly affect the receptor binding and subsequent intracellular processing, and ultimately alter the plasma half-lives of bifunctional fusion proteins. It also suggested that hGHR binding was the primary binding site for determining the plasma half-lives of the hGH-Tf fusion proteins.

To further test whether linker insertion could affect PK profiles of other bifunctional fusion proteins, the same three linkers were adopted to construct G-CSF-Tf fusion proteins. The impact of linker insertion on receptor binding affinities and PK of fusion proteins was observed again in G-CSF-Tf as in hGH-Tf (Table 4.5). Interestingly, different from hGH-Tf, the G-CSF receptor (G-CSFR) binding affinities of three G-CSF-Tf fusion proteins were quite similar, whereas their Tf receptor (TfR) binding affinities were statistically different. The plasma half-lives of G-CSF-Tf fusion proteins followed the ranking of G-CSF-*cyclo*-Tf (5.69 h) > G-CSF-$(H4)_2$-Tf (4.84 h) > G-CSF-LE-Tf (4.15 h). Stronger TfR receptor binding affinity of G-CSF-Tf fusion protein was correlated with longer plasma half-life.

These results suggested that TfR binding of G-CSF-Tf fusion proteins might follow the classic Tf-TfR recycling pathway [60,71]. For endogenous Tf, its receptor binding to TfR at cell surface usually leads to the endocytosis of the Tf-TfR complex. With the acidification of endosome, Tf releases Fe^{3+} and becomes Apo-Tf, which retains high binding affinity to TfR at acidic pH within endosome, until the complex is recycled back to the cell surface. The neutral extracellular pH promotes the release of Apo-Tf from the TfR. Taking advantage of this property; Tf can be recycled and reused by the cells for intracellular iron delivery. These results suggested that TfR binding of G-CSF-Tf may lead to the recycling of the fusion protein through the Tf-TfR recycling pathway and prolongs the plasma half-life of the fusion proteins. As a result, the stronger TfR binding affinity, the longer plasma half-life was exhibited by the G-CSF-Tf fusion proteins (Table 4.5).

A mechanistic PK model was proposed for hGH/G-CSF-Tf fusion proteins (Figure 4.4). Owing to the competition from high level of endogenous Tf, the fusion proteins likely bind first to the protein drug receptor (e.g., hGHR/G-CSFR) on the plasma membrane of target cells. This binding is considered the *primary binding*, which enriches the fusion proteins onto the target cells for the drug action, and may also increase the accessibility of the Tf domain to TfR. TfR binding, which likely occurs after hGHR/G-CSFR binding either on the plasma membrane or inside the endosomes, is then referred to as *secondary binding*. Following the hGHR/G-CSFR binding, signal transduction is initiated, and the fusion proteins are endocytosed into the early endosome, where TfR is also present [72]. Within the acidic endosome, the fusion proteins may dissociate from hGHR/G-CSFR, and bind tightly to TfR because of the high binding affinity of Tf to TfR at acidic pH [60]. The receptor binding inside the endosomes dictates different fates for the fusion proteins: The hGHR/G-CSFR binding leads to the lysosomal degradation while the TfR binding

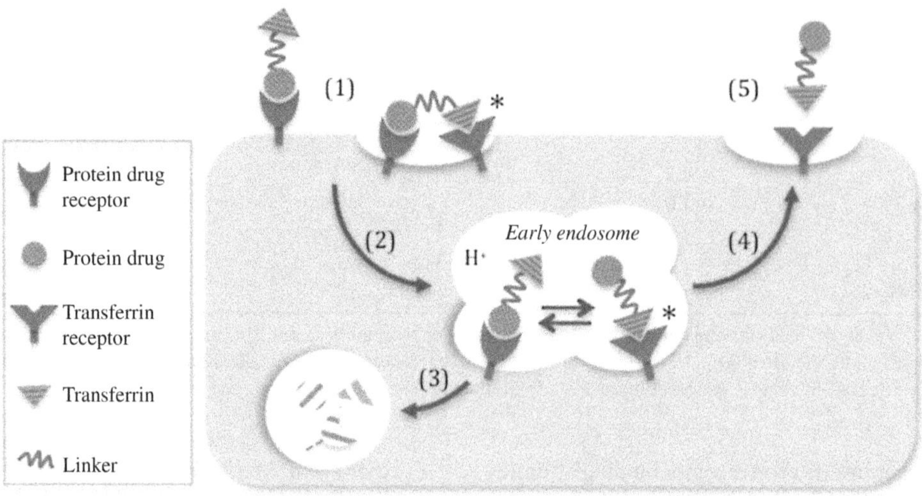

FIGURE 4.4 Endocytic pathway and intracellular metabolism of Tf-fusion proteins. In the presence of abundant endogenous Tf, the Tf-fusion proteins first bind to protein drug receptor (e.g., GHR or G-CSFR) on the target cell membrane via GH domain. This binding is the primary binding, which enriches the fusion proteins onto the target cells. The protein drug receptor binding at the cell surface brings Tf-fusion protein close to the plasma membrane surface, which may lead to bivalent binding of the Tf-domain of the fusion protein to TfRs, which are present in the clathrin-coated pit regions. This binding, indicated as "*" in the figure, is referred to *secondary binding* since it occurs after the protein drug receptor binding [2]. The fusion proteins are endocytosed into the early endosome, where TfR is also present [3]. Fusion proteins that remain bound to protein drug receptor are degraded in the lysosome. With the acidification of endosome, the fusion proteins can dissociate from protein drug receptor, and bind to TfR via their Tf domain, indicated as secondary binding "*." [4]. The binding to TfR allows the fusion protein to be recycled back to the cell surface [5]. The fusion protein is released from TfR into the circulation at cell surface.

promotes the recycling of the fusion protein. The relative receptor binding affinity of the two protein domains inside the endosome will determine the impact of each receptor binding on the plasma half-life of the fusion proteins.

This study highlights the importance of linkers in designing and developing bifunctional fusion proteins as therapeutics. First, the receptor binding affinities of the protein domains could be altered via the insertion of linkers with various lengths and conformations. Second, it is feasible to fine-tune the PK profiles of fusion proteins by linker insertion due to their impact on the receptor binding and subsequent intracellular processing. Third, this study also suggested that the different functional domains in fusion proteins may play distinct roles in determining the PK. Therefore, the impact of linker insertion on each domain's function should be carefully evaluated to achieve the desired PK profiles for the fusion proteins.

4.5 CONCLUSIONS AND FUTURE PERSPECTIVE

During the development of recombinant fusion proteins as therapeutics, linkers have become an indispensable component to achieve success. Linkers deriving from natural multidomain proteins may serve as a good lead for fusing proteins of interest. In addition, many types of empirical linkers such as flexible linkers, rigid linkers, or cleavable linkers have been designed for various applications. An optimal linker can provide many advantages for the fusion proteins, including improving the structural stability, improving bioactivity, increasing expression level, or altering the PK profiles of the fusion proteins.

When linker design is based on intuition, the outcome is less predictable. Many exciting progress have been made toward the rational design of linkers based on the requirements of target fusion proteins. To aid the rational design of linkers, Crasto and Feng presented a program called LINKER which can automatically generate a set of linker sequences that are known to adopt extended conformations as determined by X-ray crystallography and NMR [73]. The inputs include desired linker length, a number of optional parameters such as name of protease, restriction enzyme to eliminate sensitive sequence that may cause instability of the linker. The program output contains a group of linker sequences with a specified length. This program could be a useful tool in facilitating the design of optimal linkers.

The extensive mechanistic studies about the empirical linkers have been an important source for rational linker design. The structures of many linkers such as flexible and helical linkers have been studied by X-ray crystallography and NMR [35,41]. Functions and characteristics are also investigated through various applications. With the development of biotechnology and biomedical research, linkers will play important roles in constructing desired bioactive recombinant fusion proteins.

REFERENCES

1. Yuste R. (2005) Fluorescence microscopy today. *Nat. Meth.* 2, 902–904.
2. Terpe K. (2003) Overview of tag protein fusions: from molecular and biochemical fundamentals to commercial systems. *Appl. Microbiol. Biotechnol.* 60, 523–533.
3. Lu P, Feng MG. (2008) Bifunctional enhancement of a beta-glucanase-xylanase fusion enzyme by optimization of peptide linkers. *Appl. Microbiol. Biotechnol.* 79, 579–587.
4. Schmidt SR. (2009) Fusion-proteins as biopharmaceuticals—applications and challenges. *Curr. Opin. Drug Discov. Dev.* 12, 284–295.
5. Kreitman RJ. (2006) Immunotoxins for targeted cancer therapy. *AAPS J.* 8, E532–E551.
6. Pardridge WM. (2010) Biopharmaceutical drug targeting to the brain. *J. Drug Target.* 18, 157–167.
7. Kim B, Zhou J, Martin B, Carlson O, Maudsley S, Greig N, et al. (2010) Transferrin fusion technology: a novel approach to prolong biological half-life of insulinotropic peptides. *J. Pharmacol. Exp. Ther.* Epub (DOI: 10.1124/jpet.110.166470).
8. Weimer T, Wormsbächer W, Kronthaler U, Lang W, Liebing U, Schulte S. (2008) Prolonged in-vivo half-life of factor VIIa by fusion to albumin. *Thromb. Haemost.* 99, 659–667.
9. Dumont JA, Low SC, Peters RT, Bitonti AJ. (2006) Monomeric Fc fusions: impact on pharmacokinetic and biological activity of protein therapeutics. *BioDrugs* 20, 151–160.
10. Schellenberger V, Wang C, Geething N, Spink B, Campbell A, To W, et al. (2009) A recombinant polypeptide extends the in vivo half-life of peptides and proteins in a tunable manner. *Nat. Biotechnol.* 27, 1186–1190.
11. Zhao H, Yao X, Xue C, Wang Y, Xiong X, Liu Z. (2008) Increasing the homogeneity, stability and activity of human serum albumin and interferon-alpha2b fusion protein by linker engineering. *Protein Expr. Purif.* 61, 73–77.
12. Amet N, Lee HF, Shen WC. (2009) Insertion of the designed helical linker led to increased expression of tf-based fusion proteins. *Pharm. Res.* 26, 523–528.
13. Bai Y, Ann DK, Shen WC. (2005) Recombinant granulocyte colony-stimulating factor-transferrin fusion protein as an oral myelopoietic agent. *Proc. Natl. Acad. Sci. USA* 102, 7292–7296.
14. Bai Y, Shen WC. (2006) Improving the oral efficacy of recombinant granulocyte colony-stimulating factor and transferrin fusion protein by spacer optimization. *Pharm. Res.* 23, 2116–2121.
15. Robinson CR, Sauer RT. (1998) Optimizing the stability of single-chain proteins by linker length and composition mutagenesis. *Proc. Natl. Acad. Sci. USA* 95, 5929–5934.
16. Wriggers W, Chakravarty S, Jennings PA. (2005) Control of protein functional dynamics by peptide linkers. *Biopolymers* 80, 736–746.
17. George R, Heringa J. (2002) An analysis of protein domain linkers: their classification and role in protein folding. *Protein Eng.* 15, 871–879.
18. Argos P. (1990) An investigation of oligopeptides linking domains in protein tertiary structures and possible candidates for general gene fusion. *J. Mol. Biol.* 211, 943–958.
19. Eisenberg D. (1984) Three-dimensional structure of membrane and surface proteins. *Annu. Rev. Biochem.* 53, 595–623.
20. Aurora R, Creamer T, Srinivasan R, Rose G. (1997) Local interactions in protein folding: lessons from the alpha-helix. *J. Biol. Chem.* 272, 1413–1416.
21. Williamson MP. (1994) The structure and function of proline-rich regions in proteins. *Biochem. J.* 297(Pt 2), 249–260.
22. Hagemeyer CE, von Zur Muhlen C, von Elverfeldt D, Peter K. (2009) Single-chain antibodies as diagnostic tools and therapeutic agents. *Thromb. Haemost.* 101, 1012–1019.
23. Huston J, Levinson D, Mudgett-Hunter M, Tai M, Novotný J, Margolies M, et al. (1988) Protein engineering of antibody binding sites: recovery of specific activity in an anti-digoxin single-chain Fv analogue produced in *Escherichia coli*. *Proc. Natl. Acad. Sci. USA* 85, 5879–5883.
24. Holliger P, Prospero T, Winter G. (1993) "Diabodies": small bivalent and bispecific antibody fragments. *Proc. Natl. Acad. Sci. USA* 90, 6444–6448.
25. Iliades P, Kortt A, Hudson P. (1997) Triabodies: single chain Fv fragments without a linker form trivalent trimers. *FEBS Lett.* 409, 437–441.
26. Amet N, Wang W, Shen WC. (2010) Human growth hormone-transferrin fusion protein for oral delivery in hypophysectomized rats. *J. Control. Release* 141, 177–182.
27. Xia CQ, Wang J, Shen WC. (2000) Hypoglycemic effect of insulin-transferrin conjugate in streptozotocin-induced diabetic rats. *J. Pharmacol. Exp. Ther.* 295, 594–600.
28. Hu W, Li F, Yang X, Li Z, Xia H, Li G, et al. (2004) A flexible peptide linker enhances the immunoreactivity of two copies HBsAg preS1 (21-47) fusion protein. *J. Biotechnol.* 107, 83–90.
29. Bergeron LM, Gomez L, Whitehead TA, Clark DS. (2009) Self-renaturing enzymes: design of an enzyme-chaperone chimera as a new approach to enzyme stabilization. *Biotechnol. Bioeng.* 102, 1316–1322.
30. Bird R, Hardman K, Jacobson J, Johnson S, Kaufman B, Lee S, et al. (1988) Single-chain antigen-binding proteins. *Science* 242, 423–426.
31. Sabourin M, Tuzon C, Fisher T, Zakian V. (2007) A flexible protein linker improves the function of epitope-tagged proteins in *Saccharomyces cerevisiae*. *Yeast* 24, 39–45.

32. Waldo G, Standish B, Berendzen J, Terwilliger T. (1999) Rapid protein-folding assay using green fluorescent protein. *Nat. Biotechnol.* 17, 691–695.

33. Maeda Y, Ueda H, Kazami J, Kawano G, Suzuki E, Nagamune T. (1997) Engineering of functional chimeric protein G-Vargula luciferase. *Anal. Biochem.* 249, 147–152.

34. Arai R, Ueda H, Kitayama A, Kamiya N, Nagamune T. (2001) Design of the linkers which effectively separate domains of a bifunctional fusion protein. *Protein Eng.* 14, 529–532.

35. Arai R, Wriggers W, Nishikawa Y, Nagamune T, Fujisawa T. (2004) Conformations of variably linked chimeric proteins evaluated by synchrotron X-ray small-angle scattering. *Proteins* 57, 829–838.

36. Marqusee S, Baldwin RL. (1987) Helix stabilization by Glu-···Lys+ salt bridges in short peptides of de novo design. *Proc. Natl. Acad. Sci. USA* 84, 8898–8902.

37. Werner S, Marillonnet S, Hause G, Klimyuk V, Gleba Y. (2006) Immunoabsorbent nanoparticles based on a tobamovirus displaying protein A. *Proc. Natl. Acad. Sci. USA* 103, 17678–17683.

38. Takamatsu N, Watanabe Y, Yanagi H, Meshi T, Shiba T, Okada Y. (1990) Production of enkephalin in tobacco protoplasts using tobacco mosaic virus RNA vector. *FEBS Lett.* 269, 73–76.

39. Turner S, Russell G, Williamson M, Guest J. (1993) Restructuring an interdomain linker in the dihydrolipoamide acetyltransferase component of the pyruvate dehydrogenase complex of *Escherichia coli*. *Protein Eng.* 6, 101–108.

40. Erni B. (1989) Glucose transport in *Escherichia coli*. *FEMS Microbiol. Rev.* 5, 13–23.

41. Bhandari D, Levine B, Trayer I, Yeadon M. (1986) ^{1}H-NMR study of mobility and conformational constraints within the proline-rich N-terminal of the LC1 alkali light chain of skeletal myosin. Correlation with similar segments in other protein systems. *Eur. J. Biochem.* 160, 349–356.

42. Evans J, Levine B, Trayer I, Dorman C, Higgins C. (1986) Sequence-imposed structural constraints in the TonB protein of *E. coli*. *FEBS Lett.* 208, 211–216.

43. Morris A, MacArthur M, Hutchinson E, Thornton J. (1992) Stereochemical quality of protein structure coordinates. *Proteins* 12, 345–364.

44. McCormick A, Thomas M, Heath A. (2001) Immunization with an interferon-gamma-gp120 fusion protein induces enhanced immune responses to human immunodeficiency virus gp120. *J. Infect. Dis.* 184, 1423–1430.

45. Chen X, Bai Y, Zaro J, Shen WC. (2010) Design of an in vivo cleavable disulfide linker in recombinant fusion proteins. *Biotechniques* 49, 513–518.

46. Schulte S. (2009) Half-life extension through albumin fusion technologies. *Thromb. Res.* 124Suppl. 2, S6–8.

47. Vessillier S, Adams G, Chernajovsky Y. (2004) Latent cytokines: development of novel cleavage sites and kinetic analysis of their differential sensitivity to MMP-1 and MMP-3. *Protein Eng. Des. Sel.* 17, 829–835.

48. Wakefield LM, Winokur TS, Hollands RS, Christopherson K, Levinson AD, Sporn MB. (1990) Recombinant latent transforming growth factor beta 1 has a longer plasma half-life in rats than active transforming growth factor beta 1, and a different tissue distribution. *J. Clin. Invest.* 86, 1976–1984.

49. Martel-Pelletier J, Welsch DJ, Pelletier JP. (2001) Metalloproteases and inhibitors in arthritic diseases. *Best Pract. Res. Clin. Rheumatol.* 15, 805–829.

50. Lampert K, Machein U, Machein MR, Conca W, Peter HH, Volk B. (1998) Expression of matrix metalloproteinases and their tissue inhibitors in human brain tumors. *Am. J. Pathol.* 153, 429–437.

51. Yong VW, Krekoski CA, Forsyth PA, Bell R, Edwards DR. (1998) Matrix metalloproteinases and diseases of the CNS. *Trends Neurosci.* 21, 75–80.

52. Leppert D, Lindberg RL, Kappos L, Leib SL. (2001) Matrix metalloproteinases: multifunctional effectors of inflammation in multiple sclerosis and bacterial meningitis. *Brain Res. Rev.* 36, 249–257.

53. Pepinsky RB, LePage DJ, Gill A, Chakraborty A, Vaidyanathan S, Green M, et al. (2001) Improved pharmacokinetic properties of a polyethylene glycol-modified form of interferon-beta-1a with preserved in vitro bioactivity. *J. Pharm. Exp. Ther.* 297, 1059–1066.

54. Peng KW, Morling FJ, Cosset FL, Murphy G, Russell SJ. (1997) A gene delivery system activatable by disease-associated matrix metalloproteinases. *Hum. Gene Ther.* 8, 729–738.

55. Nagase H, Fields GB. (1996) Human matrix metalloproteinase specificity studies using collagen sequence-based synthetic peptides. *Biopolymers* 40, 399–416.

56. Liu S, Netzel-Arnett S, Birkedal-Hansen H, Leppla SH. (2000) Tumor cell-selective cytotoxicity of matrix metalloproteinase-activated anthrax toxin. *Cancer Res.* 60, 6061–6067.

57. Thomas G. (2002) Furin at the cutting edge: from protein traffic to embryogenesis and disease. *Nat. Rev. Mol. Cell Biol.* 3, 753–766.

58. Goyal A, Batra JK. (2000) Inclusion of a furin-sensitive spacer enhances the cytotoxicity of ribotoxin restrictocin containing recombinant single-chain immunotoxins. *Biochem. J.* 345Pt 2, 247–254.

59. Wang T, Zhao J, Ren JL, Zhang L, Wen WH, Zhang R, et al. (2007) Recombinant immunoproapoptotic proteins with furin site can translocate and kill HER2-positive cancer cells. *Cancer Res.* 67, 11830–11839.

60. Widera A, Norouziyan F, Shen WC. (2003) Mechanisms of TfR-mediated transcytosis and sorting in epithelial cells and applications toward drug delivery. *Adv. Drug Deliv. Rev.* 55, 1439–1466.

61. Shah D, Shen WC. (1994) The establishment of polarity and enhanced transcytosis of transferrin receptors in enterocyte-like Caco-2 cells. *J. Drug Target.* 2, 93–99.

62. Duttaroy A, Kanakaraj P, Osborn B, Schneider H, Pickeral O, Chen C, et al. (2005) Development of a long-acting insulin analog using albumin fusion technology. *Diabetes* 54, 251–258.

63. Halpern W, Riccobene T, Agostini H, Baker K, Stolow D, Gu M, et al. (2002) Albugranin, a recombinant human granulocyte colony stimulating factor (G-CSF) genetically fused to recombinant human albumin induces prolonged myelopoietic effects in mice and monkeys. *Pharm. Res.* 19, 1720–1729.

64. Zhao H, Xue C, Wang Y, Li X, Xiong X, Yao X, et al. (2007) Circumventing the heterogeneity and instability of human serum albumin-interferon-alpha2b fusion protein by altering its orientation. *J. Biotechnol.* 131, 245–252.

65. Leader B, Baca Q, Golan D. (2008) Protein therapeutics: a summary and pharmacological classification. *Nat. Rev. Drug Discov.* 7, 21–39.

66. Osborn B, Olsen H, Nardelli B, Murray J, Zhou J, Garcia A, et al. (2002) Pharmacokinetic and pharmacodynamic studies of a human serum albumin-interferon-alpha fusion protein in cynomolgus monkeys. *J. Pharmacol. Exp. Ther.* 303, 540–548.

67. Müller N, Schneider B, Pfizenmaier K, Wajant H. (2010) Superior serum half life of albumin tagged TNF ligands. *Biochem. Biophys. Res. Commun.* 396, 793–799.

68. Zhu D, Kepley CL, Zhang M, Zhang K, Saxon A. (2002) A novel human immunoglobulin Fc gamma Fc epsilon bifunctional fusion protein inhibits Fc epsilon RI-mediated degranulation. *Nat. Med.* 8, 518–521.

69. Manoukian G, Hagemeister F. (2009) Denileukin diftitox: a novel immunotoxin. *Exp. Opin. Biol. Ther.* 9, 1445–1451.

70. van Kerkhof P, Strous G. (2001) The ubiquitin-proteasome pathway regulates lysosomal degradation of the growth hormone receptor and its ligand. *Biochem. Soc. Trans.* 29, 488–493.

71. Grant B, Donaldson J. (2009) Pathways and mechanisms of endocytic recycling. *Nat. Rev. Mol. Cell Biol.* 10, 597–608.

72. Muller-Eberhard U, Liem H, Grasso J, Giffhorn-Katz S, DeFalco M, Katz N. (1988) Increase in surface expression of transferrin receptors on cultured hepatocytes of adult rats in response to iron deficiency. *J. Biol. Chem.* 263, 14753–14756.

73. Crasto CJ, Feng JA. (2000) LINKER: a program to generate linker sequences for fusion proteins. *Protein Eng.* 13, 309–312.

5

IMMUNOGENICITY OF THERAPEUTIC FUSION PROTEINS: CONTRIBUTORY FACTORS AND CLINICAL EXPERIENCE

VIBHA JAWA,[1] LESLIE COUSENS,[2] AND ANNE S. DE GROOT[2,3]

[1]*Medical Sciences, Amgen Inc., Thousand Oaks, CA, USA*
[2]*EpiVax Inc., Providence, RI, USA*
[3]*Institute for Immunology and Informatics, University of Rhode Island, Providence, RI, USA*

5.1 INTRODUCTION

The past two decades have been remarkable for advances in the field of protein engineering. The impact of these advances on human health is now becoming evident. A wide variety of biologically engineered products are already in clinical use or under preclinical development by the biotechnology industry. Developers are engineering novel therapeutic proteins, monoclonal antibodies (mAbs), fusion proteins, and antibody-like protein scaffolds, intent on expanding the number of clinical applications for these products. The caveat is that virtually all therapeutic proteins elicit some level of immune response that can lead to loss of efficacy or other adverse events, some of them serious; these effects have led to costly failures in clinical trials. Thus, while there is a greater emphasis on biologics as a pathway to developing new drugs as compared to small molecules, drug developers and regulatory agencies are turning their attention to factors that reduce protein immunogenicity as a means of improving clinical success. Fortunately, years of thorough study of the parameters influencing vaccine efficacy allow parallels to be drawn for protein therapeutics. Factors including delivery route, delivery vehicle, dose regimen, aggregation, innate immune system activation, and the ability of the protein to interface with the humoral (B cell) and cellular (T cell) components of the immune system all impact the intended immunogenicity of vaccine immunogens when delivered to humans. The influence of these factors on the unintended immunogenicity of therapeutic proteins is discussed here, with a particular focus on Fc-fusion proteins, together with the use of state-of-the-art technology in predicting and mitigating this risk. Notwithstanding the recent advances discussed here, the current state of science is such that clinical studies are required to evaluate immunogenicity of protein therapeutics.

5.2 BASIS OF THERAPEUTIC PROTEIN IMMUNOGENICITY

Therapeutic protein products encompass diverse proteins such as human cytokines, cellular growth factors, hormones, clotting factors, enzymes, fusion proteins, and mAbs. Therapeutic proteins are attractive drug products, as they are generally considered safe, specific, and nontoxic. However,

Fusion Protein Technologies for Biopharmaceuticals: Applications and Challenges, First Edition. Edited by Stefan R. Schmidt.

their efficacy can be dramatically compromised by the development of anti-therapeutic protein responses [1,2]. Anti-therapeutic antibodies have the potential to neutralize clinical effects of this class of therapeutics. Moreover, anti-therapeutic protein responses can be associated with serious adverse events when they cross-react with autologous protein antigens.

Anti-therapeutic protein responses are not unexpected when the protein is foreign, either as the result of a different species of origin or as a recipient in whom the natural analog of the therapeutic protein is deleted or modified. In these cases, immune responses to foreign proteins are fairly well delineated, owing largely to factors such as B- and T-cell epitopes that have been extensively characterized in the field of vaccinology. However, anti-therapeutic protein responses can also develop against recombinant autologous proteins to which, theoretically, the recipient should be tolerant. Biopharmaceuticals such as antibodies, "humanized" proteins, and fusion proteins fall somewhere between these two poles, as they are homologous with autologous protein sequences, and yet they frequently incorporate point mutations that confer therapeutic activity. Even such small changes may present the risk of introducing new epitopes never before encountered by the host.

For better understanding of immune responses to therapeutic proteins, it is helpful to categorize these responses into one of the two general types: (i) activation of the classical immune system by foreign proteins, such as that elicited against pathogens or vaccines involving T cells, B cells, and the innate immune system and (ii) breach of B- and T-cell tolerance to autologous proteins, a series of immunological events otherwise modulated by T-regulatory cells (T_{reg}) that is yet to be fully understood. In the case of classical immune induction, production of anti-therapeutic protein responses is the end result of a sequence of events that leads to B-cell activation and can be divided into T-cell-independent (T_i) or T-cell-dependent (T_d) antibody production [3,4]. T_i activation of B cells occurs when particular structural patterns, such as polymeric repeats, are able to directly induce stimulation and activation of a B-cell subset. Although T_i activation is often credited as the source of anti-therapeutic protein responses, T_i activation of B cells is easily distinguished, as it generally does not lead to affinity maturation or to the generation of memory B cells [5]. In contrast, T_d activation of B cells is generally characterized by isotype switching, development of B-cell memory, and overall more robust antibody responses. Thus, the induction of IgG-class antibodies (measured by standard immunogenicity assays) through isotype switching indicates that the therapeutic protein contains a T_d antigen. T_d responses, by definition, are contingent upon T-cell recognition of therapeutic protein-derived epitopes through the following basic process. The therapeutic protein is processed by antigen-presenting cells (APCs) and presented on the APC surface by human leukocyte antigen (HLA) major histocompatibility complex (MHC) class II molecules. This complex of epitope–HLA is recognized by antigen-specific helper T cells that facilitate B-cell antibody production and isotype class switching. In the absence of signals provided by the cytokines released downstream of T-cell–APC interactions, naïve B cells do not mature, and activated antigen-specific B cells can be rendered anergic or undergo apoptosis. Therefore, T-cell recognition of peptide epitopes derived from a protein therapeutic is a key determinant of T_d antibody formation and subsequent anti-therapeutic protein responses. Moreover, a number of clinical studies now suggest that serious adverse events are associated with high levels of T-cell-dependent IgG antibodies [6–8].

The measurement of discrete peptide-binding potential to the MHC molecules can be an important and quantitative tool for immunogenicity assessment. Indeed, the greater the affinity of a given peptide for the binding groove of a particular MHC molecule, the greater the opportunity to elicit a T-cell response. However, this piece of the immunogenicity puzzle should be viewed in a broader context. For instance, rather than being randomly distributed throughout a protein, potential T-cell epitopes tend to occur in clusters ranging from 9 to 25 amino acids in length [9,10]. When considered in total, T-cell epitope clusters predict larger sets of binding motifs with affinity across multiple alleles and multiple frames [11]. In addition, it is important to remember that *in vivo*, immunologically relevant epitopes are created by the APC that processes the protein into discrete peptide fragments, assembles the peptides in complexes with MHC molecules, and displays these complexes on its cell surface [12,13]. Thus, while the accurate assessment of peptide-binding potential to MHC molecules is an important tool for honing in on T-cell epitopes and clusters thereof, it is only one facet of the process by which functional T-cell epitopes are generated.

Tolerance to self-proteins is a fundamental feature of the immune system [14,15]. Thus, the development of immune responses to recombinant human or humanized proteins can be considered a breaking of this tolerance. Indeed, although self-reactive B cells are often observed, they are typically not stimulated to produce antibodies in response to natural levels of endogenous proteins [16]. Current efforts are underway to understand the role of T_{reg} in establishing and maintaining tolerance [17,18]. The reduction or the overwhelming of T_{reg} immune responses in the application of therapeutic proteins are potential contributors to anti-therapy immune responses that need to be considered. Moreover, the role of T_{reg} in the development of autoimmune disease may be instructive for better understanding their contributions to the immunogenicity of therapeutic proteins.

In addition to the basic principles of immunology that govern T- and B-cell responses to foreign and self proteins [19], the implications of extrinsic factors that are particular to the therapeutic protein or its administration should also be considered in this context. These include protein modifications, aggregation, and innate immune system activation. Modifications in recombinant proteins that occur in the *in vitro* expression and purification processes, including post-translational modifications and denaturation, could render them sufficiently different from their natural "self" analogs to be immunogenic. Aggregates of a therapeutic protein can be recognized by the immune system in a number of ways. One way that aggregates may trigger immune responses is through the recurrence of B-cell epitopes on the surfaces of the aggregate that could simulate the presence of highly repetitive arrays, such as those seen on the surfaces of bacteria and viruses, facilitating cross-linking of the B-cell receptor to activate proliferation. Alternatively, aggregates are, similar to the virus-like particles (VLPs) that so effectively induce immune responses in the vaccine context, more likely to be taken up by APCs, where degradation and presentation of T-cell epitopes are then able to drive a T_d response [20,21]. Finally, engagement of Toll-like receptor (TLR) family members by products of microbial degradation has been shown to stimulate B-cell responses and antibody production in the absence of T-cell help [22]. Thus, even low-level contaminants or impurities such as endotoxins, microbial agents, or host cell proteins present along with the therapeutic proteins could contribute to their immunogenicity.

5.2.1 Relevance to Fusion Proteins

Full-length Fc-fusion proteins are normally generated by fusing the N-terminus of the protein of interest containing the biologically active site to murine or human Fc (hinge-CH2-CH3). The fusion of proteins with Fc allows for large-scale manufacturing (milligram to gram quantities), easy purification, and an increased *in vivo* half-life, which proves useful in therapeutic applications [23,24]. Taking what we know about general protein immunogenicity, we can envision a number of Fc-fusion protein attributes that should be given special consideration in preclinical development. For example, the joining of two nonimmunogenic or self-proteins together can form junctions never before "seen" by the immune system that could comprise new T- and B-cell epitopes. Likewise, Fc-fusion proteins may be taken up and processed differently than either of their native counterparts, contributing to higher immunogenicity than predicted for either component. Immunogenicity is enhanced by binding to Fc receptors on the surface of APCs thereby targeting the antigen to these cells [25]. If this new combination increases uptake or differentially stimulates APCs, then the host immune system may respond differently. Also, these novel protein sequences could be processed and presented in a way that is sufficiently distinct from the native protein components, resulting in greater immunogenicity. De-immunization of the new immunogenic epitopes generated by fusion at the junctional region can be used to reduce their interaction with a T-cell receptor (TCR) [26]. Owing to these potential concerns of immunogenicity due to the fusion of two or more protein components, a robust immunogenicity evaluation strategy is often required as a part of risk assessment to support the drug development at preclinical and clinical stages [27]. Swanson et al. have reviewed the various assay platforms and their suitability for immunogenicity assessment [28]. From this comparative analysis, surface plasmon resonance-based Biacore assays emerge as ideal for testing the immunogenicity of fusion proteins. With the Biacore approach, each component of the fusion protein, such as the peptide portion, Fc portion, or any linker regions, can be independently assessed for antibody reactivity simultaneously on parallel immobilized surfaces [28].

5.3 TOOLS FOR IMMUNOGENICITY SCREENING

5.3.1 *In Silico* Analysis Tools

As discussed earlier, immunogenicity of a therapeutic protein depends largely on its ability to trigger either a cellular or humoral immune response. In the case of cellular responses, T cells recognize small linear peptides derived from protein antigens upon their uptake and proteolytic processing by an APC. These peptides are then presented to T cells as a complex of epitope in the binding groove of the MHC on the APC surface. One of the critical determinants of T-cell epitope immunogenicity is the affinity of that epitope for the groove of the MHC; the higher the affinity the greater the likelihood that an epitope will be recognized by a T cell [29]. In contrast to T-cell epitope recognition, which is restricted by presentation on MHC molecules, B cells and the antibodies that they produce generally recognize conformational epitopes directly on the surface of native proteins. Compared to MHC–T-cell epitope interactions, those between the B-cell receptor and B-cell epitope are much more difficult to describe mathematically for use in prediction algorithms [30,31]. Thus, the groundbreaking work toward establishing reliable methods for immunogenicity prediction appears somewhat biased toward the definition of T-cell epitopes; accessible, high throughput, and accurate prediction of B-cell epitopes may be beyond the reach of current computational algorithms.

There are a number of factors inherent in the binding between a T-cell epitope and an MHC-binding groove that make this relationship amenable to computer-based prediction algorithms. Because several common HLA-DR types share largely overlapping peptide-binding repertoires, analysis focused on as few as eight MHC molecules can "cover" the genetic backgrounds of most humans worldwide [32]. Likewise, epitopes are generally limited to 9–21 amino acids in length, making their analysis manageable. Databases are available, such as the Immune Epitope Database Analysis Resource (IEDB; www.tools.immunoepitope.org), which manually curate experimentally characterized immune epitopes [33]. Taken together, amino acid preferences for interacting with different parts of the MHC binding groove can be observed and then extrapolated to predict interactions between unknown epitopes and certain MHC molecules.

Many computer-based algorithms created to identify MHC-restricted T-cell epitopes have emerged from the research community over the last 10 or more years (Table 5.1). These *in silico* methods include frequency analysis, support-vector machines, Markov models, and neural networks [34–36]. The common ground between these tools is the ability to quickly screen large datasets, including whole genomes, for prediction of T-cell epitopes. These types of predictions have successfully been applied both to vaccine development [37–41] and to the identification of key epitopes triggering autoimmunity [42].

Additional tools for immunogenicity screening that go beyond these standard prediction methods have been developed by De Groot et al. [9,36,43]. Using their EpiMatrix algorithm, protein sequences are parsed into overlapping 9-mer peptide frames, each of which is then evaluated for binding potential to each of the eight common MHC class II HLA alleles that generally represent the natural breadth of genetic diversity in human HLA [32]. This algorithm uses MHC "pocket profiles" that describe HLA-binding groove coefficients and applies these coefficients to the predict whether a given 9-mer peptide will bind to a given MHC allele. The method initially described by Sturniolo and Hammer [44] has been adapted by De Groot et al. to generate class II prediction tools [45]. One of the unique features of this algorithm is that allele-specific scores are normalized, making it possible to compare scores of any 9-mer across multiple HLA alleles and enabling a more broad assessment of predicted immunogenicity [36]. The predictive value of this algorithm has been extensively tested and is supported by published *in vitro* and *in vivo* studies [40,45–50].

Mapping putative epitopes within a candidate protein therapeutic is a starting point for assessing the potential immunogenicity of a whole protein. It provides information about the number of epitopes in a given protein as well as the tendency of these epitopes to cluster within particular regions of a protein. Subregions of densely packed high-scoring frames or "clusters" of potential

TABLE 5.1 Epitope Prediction Tools

Name	Developer/Institution	Type	Website
EpiScreen	M. Baker and F. Carr Antitope, Ltd, Cambridge, UK	Commercial	www.antitope.co.uk/
Epibase	Lasters and P. Stas Algonomics NV/Lonza, Inc., Gent, Belgium	Commercial	www.lonza.com
EpiMatrix	A.S. De Groot and W.D. Martin EpiVax, Inc., Providence, RI, USA	Collaborative/ commercial	www.epivax.com
IEDB	Vita R, Zarebski L, Greenbaum JA, Emami H, Hoof I, Salimi N, Damle R, Sette A, Peters B. The immune epitope database 2.0. *Nucleic Acids Res.* (2010) January; 38, D854–D862.	Public	www.immuneepitope.com
MHC2PRED	G.P.S. Raghava Bioinformatics Center, Institute of Microbial Technology, Chandigarh, India	Public	www.imtech.res.in/ raghava/mhc2pred/
MHCPRED	D.R. Flower The Jenner Institute, Oxford, UK	Public	www.darrenflower.info/ MHCPRED/
PROPRED/TEPITOPE	G.P.S. Raghava and H. Singh Bioinformatics Center, Institute of Microbial Technology, Chandigarh, India	Public	www.imtech.res.in/ raghava/propred/
RANKPEP	P.A. Reche Harvard Medical School, Cambridge, MA, USA	Public	http://bio.dfci.harvard .edu/RANKPEP/
SVRMHC	P. Donnes, A. Elofsson Division for Simulation of Biological Systems, University of Tubingen, Germany	Public	http://svrmhc.biolead.org/
SYFPEITHI	H.G. Rammensee Department of Immunology, Tubingen, Germany	Public	www.syfpeithi.de/home .htm
SMM-Align/NetMHCII-2.2	M. Nielsen, C.Lundegaard, and O. Lund Center for Biological Sequence Analysis, Department of Systems Biology, Technical University of Denmark	Public	www.cbs.dtu.dk/services/ NetMHCII-2.2/

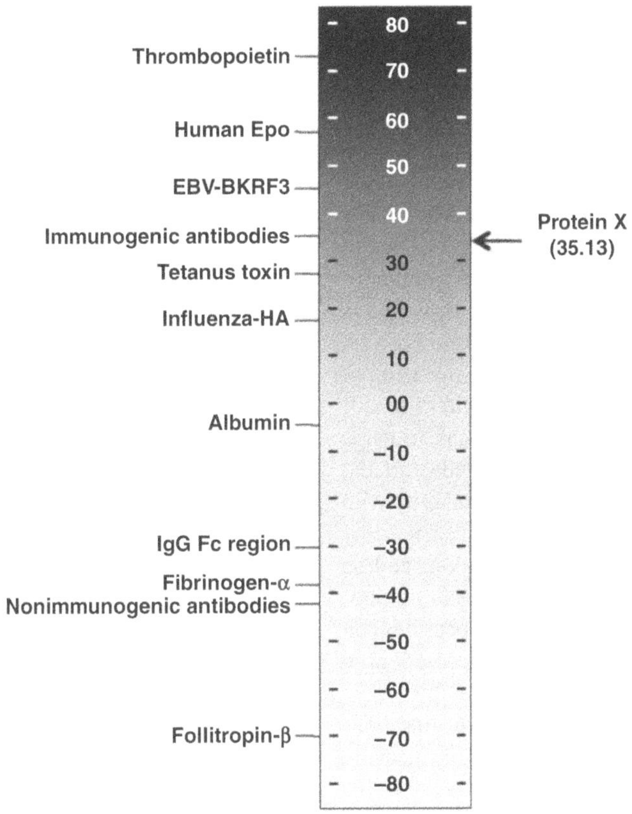

FIGURE 5.1 Protein immunogenicity scale. Protein "X" is mapped onto the protein immunogenicity scale according to its EpiMatrix score. Proteins scoring above +20 are considered to be potentially immunogenic. On the left of the scale are included some well-known proteins for comparison representing a range of immunogenicities, arranged by EpiMatrix score, from highest (black) to lowest (white).

immunogenicity can be identified, and cluster scores calculated and compiled, a process that has been well established by the team of De Groot and Martin [9]. The frequency and density of these T-cell epitope clusters can then be compared to a set of randomly generated pseudoprotein sequences of similar sizes. Proteins containing many, or concentrated clusters of, T-cell epitopes are predicted to be highly immunogenic, while those containing a few sparse epitopes are more likely to be less immunogenic. The relationship between the density of epitope clusters and protein immunogenicity has been validated extensively *in vitro* and *in vivo*. Moreover, these characteristics can be factored into the estimate of a whole protein's overall "immunogenicity score," a useful tool for consideration in the bioengineering of protein therapeutics. The ability to assign this standardized measure of immunogenicity to a protein facilitates informed decisions about the likelihood that a protein will provoke an immune response [43].

To aid in the direct comparison of immunogenicity predictions between proteins, an "immunogenicity scale" that maps immunogenicity predictions of analyzed proteins onto a scale of known proteins likewise analyzed and validated for immunogenicity can be developed (Figure 5.1). Using this approach, the De Groot and Martin group has correctly predicted the clinical immunogenicity of a novel "peptibody," a bioengineered autologous protein, and antibodies in clinical use [9,51,52] among others. Consequently, a number of biotech companies have integrated this approach to immunogenicity screening into their drug development pipeline. Beyond its use in evaluating a single therapeutic candidate, future applications of the immunogenicity scale might include its utility in the early development process by aiding in the selection of a lead low-immunogenicity candidate from among a panel of protein variants with a range of predicted immunogenicities.

5.3.2 *In Vitro* Analysis Tools

For many years, *in vitro* assays have been used in transplantation research to predict the responsiveness of T cells to a target tissue based on HLA haplotype matching. Likewise, given the inherent risks posed by unwanted immunogenicity elicited by therapeutic proteins, there is an urgent need to develop specific and sensitive *in vitro* methods to analyze the potential immunogenicity of new therapeutic proteins. The challenge in developing these types of assays is to provide an appropriate milieu in culture that accurately represents *in vivo* conditions of immune stimulation [52–54]. In this respect, bulk cultures of peripheral blood lymphocytes, either with or without co-stimulatory signals (anti-CD28 antibody, IL-2, IL-7, etc.), have been utilized to assess immunogenicity of therapeutic proteins. A number of biological outcomes for T-cell activation can be measured in these *in vitro* assays, including cytokine secretion (IFN-γ, IL-2, IL-4, etc.), regulation of cell surface markers of activation, signal transduction events, and proliferation. If blood from individuals participating in clinical trials is available, peptides predicted as T-cell epitopes and validated in MHC-binding assays can be tested for their reactivity with T cells *in vitro*. A positive immune response, measured by peptide-specific cytokine production, proliferation, and so on, is a strong indicator that the parent protein is immunogenic. Supported by this type of evidence, these peptides can be considered validated T-cell epitopes with the potential to contribute to an antidrug Ab response.

Enzyme-linked immunosorbent assay (ELISA) and ELISpot are two related methods, both of which measure cytokines secreted by T cells (i.e., IFN-γ, IL-2, and IL-4). ELISAs are a means to measure cytokine levels in culture supernatants generated under conditions of T-cell stimulation. Cytokine levels measured in an ELISA can provide information about the magnitude of the response (how much cytokine is secreted into the supernatant) as well as the quality of the response (which cytokines are or are not secreted into the supernatant). ELISpot assays provide information regarding the numbers of cytokine-producing cells within a cell population stimulated *ex vivo*. ELISpots are sensitive because they are able to identify cells that produce relatively low quantities of the cytokine of interest.

The proliferation of T cells in response to stimulation by a peptide–MHC complex can be measured by (1) the dilution of a fluorescent dye, CFSE, that will decrease in fluorescence intensity by half with each round of cell division, a property measured by flow cytometry and (2) the incorporation of a radioactive nucleotide, tritiated thymidine, into the DNA of dividing but not resting cells. Some of the most precise methodologies available for measuring and defining T-cell responses are measured by flow cytometry. For example, T cells bearing TCR specific for a particular epitope can be directly enumerated using fluorescent-labeled tetrameric complexes of MHC class II molecules loaded with the peptide of interest. Antigen-specific T cells detected in this manner can be further examined for phenotypic and activation markers as well as for intracellular cytokine expression.

HLA-binding assays can be used to measure the affinity of peptides derived from immunoinformatics analyses for binding to soluble MHC class II molecules *in vitro*. Evaluation of MHC binding can be performed as a competition assay, where a "test" peptide added over a range of concentrations is measured for its ability to inhibit the binding of a labeled "reference" peptide with a well-characterized affinity for a particular MHC allele [55]. This standardized method is cell-free and can be adapted for high throughput [56]. The advantage in using *in vitro* HLA binding assays is that detailed information about a peptide's HLA affinity may be obtained prior to performing experiments on live cells. Data acquired from these assays may be indicative of immune responses in later experiments, such as ELISpots or HLA transgenic mouse studies, providing a reliable framework to interpret results. It is important to note, however, that the conditions under which candidate epitopes are tested *in vitro* are different from the cellular environment in ways that make a 1:1 comparison difficult. Artificially synthesized peptides may contain impurities or sequence errors. Solvents used in the experiment may interact negatively with certain amino acids, causing oxidation or forming unwanted disulfide bonds. Quality-controlled peptide manufacturing and storage along with proper reagent selection can minimize the impact of these problems.

Because competition-binding assays yield indirect measures of affinity on the basis of how effectively a peptide inhibits the binding of a labeled reference peptide, only relative measurements proportionate to specific assay parameters can be made. A set of peptides can be "ranked" for binding to a given HLA allele provided they have been tested under the same conditions, but comparisons across alleles are not as straightforward. By fine-tuning the protocol for each allele through repeated experiments, a consensus among data points can be approached.

T-cell restimulation assays can be utilized to identify and measure a recall or memory response in peripheral blood mononuclear cells (PBMCs) derived from subjects who have been exposed to a given biological product within the context of a clinical trial. As the precursor frequency of the antigen-specific cells in a recall response is relatively high, whole PBMC assays are generally both rapid and sensitive with no pretreatment or enrichment required. Epitope mapping of the recall response can be performed using specific peptides from the whole protein at the time of challenge. T-cell responses after restimulation *ex vivo* can include ELISA, ELISpot, proliferation, and flow cytometric analysis of activation markers of intracellular cytokine expression.

Naïve PBMC assays have been used to predict immunogenicity of therapeutic proteins [54,57]. Compared to a recall response, the precursor frequency of antigen-specific cells in a naïve population is quite low. Thus, multiple rounds of antigen stimulation, sometimes over several weeks, are required to expand sufficient numbers of antigen-specific T cells for reliable measurement. By conditioning naïve blood samples *ex vivo* through prolonged and/or repeated exposure to experimental antigen, immune responses can be accurately characterized. A kinetic analysis of responses following stimulation with the protein antigens of interest can be useful for detecting both low and high frequency antigen-specific effector cells.

Both whole proteins and peptides can be used to measure T-cell responses *in vitro*; peptides can be of variable length (9–25 residues). Peptides presented in the context of class I MHC are generally limited to 9 or 10 amino acids in length, although some processing is believed to occur during the T-cell assay and so 15-mers are also used for class I assays. In contrast with class I epitopes, which are short and fit tightly in the MHC molecule to which they bind, class II (T helper) epitopes are displayed on an open-ended groove in the MHC II. As such, longer peptides of 20 or more amino acids in length can shift within this groove, thereby increasing the number of discreet 9-mer epitopes able to interact with the MHC molecule. The only limit on the size of the peptide is its ability to remain in a linear conformation in the groove. Similar to peptides, whole antigens too can be used to measure T-cell responses *in vitro*. The recognition of these antigens requires the presence of an APC that is capable of processing and presenting peptides derived from the antigen. Advantages of whole PBMC assays include the ability to set up several assays and/or assay conditions with a limited blood sample volume and the ease of assay performance, features which lend themselves to high-throughput assay development. However, since PBMC contain several cell types capable of secreting IFN-γ (NK cells, NKT cells, CD4, or CD8 T cells), assays employing PBMC to measure this cytokine may encounter obstacles. As not all IFN-γ-producing cells utilize the HLA class II–TCR activation pathway, the variety of cell types can contribute to a nonspecific signal. This bystander effect can be minimized by use of CD8+ T-cell-depleted PBMCs. The optimized *in vitro* concentration of protein required for challenge may be nonphysiologic, potentially because of a limited antigen presenting population and/or co-stimulation. Moreover, there is a limitation in the number of APCs when either whole PBMCs or enriched PBMCs are used in a long-term assay. The response rate of T cells could be further improved by coculture of optimal APC: T-cell ratios.

While the practice of inducing T-cell responses *in vitro* through MHC-bound peptides on APCs can generate valuable information, the complete nature of the immune response cannot be fully replicated in these experiments.

Future considerations for developing effective *in vitro* assays include the ability to

1. characterize responders by setting statistically derived criteria such as fold increase or stimulation index;
2. distinguish responders from nonresponders;
3. select the optimal set of markers for the identification of activated T cells and distinguishing between T_{reg} and T_{eff} (T-effector cells) subsets;
4. identify parameters that define memory versus naïve T-cell populations; and
5. standardize methods for cell harvest and preparation from whole blood.

Naïve and presensitized blood can differ in the content of antigen-specific T_{eff} and T_{reg}. Hence, immediate handling of whole blood with a stringent freezing process is essential to ensure viability and functionality. On the basis of the nature of response (primed vs stimulated and recalled), stimulation methods and the amount of antigen required for challenge can also differ. The naïve cells will require multiple challenges while antigen-specific recall responses can be elicited even with a single challenge. Similarly, stimulation can be performed with peptides or whole proteins alone, peptides in context of tetramers or APC pulsed with whole proteins.

5.4 APPROACHES FOR RISK ASSESSMENT AND MINIMIZATION

5.4.1 Importance to Product Development

The extent to which one characterizes immunogenicity of a new protein therapeutic is determined in part by the potential risk of immune-mediated sequelae in the human population targeted for its use. Factors to consider include, but are not limited to, inherent potential immunogenicity of the protein, its biological role, indication of use, route of administration, duration of treatment, and health status of the patient (e.g., immune competence).

The immunogenicity of therapeutic protein products is an important issue in product development as it can pose a serious safety risk to the health of patients if not thoroughly addressed during the drug development process. Indeed, serious and unexpected immune responses observed in clinical trial patients can derail a promising therapeutic otherwise on the path to successful licensing and marketing. To preclude these events, strategies for assessing immunological risk of a potential protein therapeutic should be incorporated into the earliest stages of the drug development process. Even before the clinical trial stage of development, immunogenicity should be considered for its impact on the ability to accurately measure pharmacokinetics, pharmacodynamics, bioavailability, and efficacy during preclinical testing. Implementation strategies might include the use of current methods for immunogenicity prediction, the development of *in vitro* techniques to test the immunogenicity of predicted epitopes as well as validated and sensitive *in vivo* assays for detecting immune responses in preclinical animal models and clinical trial patients. Because of the diverse nature of therapeutic proteins, testing strategies should be considered on an individual basis. An aspect of risk-based analysis concerns the design of the molecule, encompassing *in silico* identification of any immunological "hot spots" and their reengineering so as to reduce or eliminate these hot spots. Performing binding, *in vivo*, and cell-based assays to understand the functional role of drug attributes in a physiological context can be an additional component of the risk assessment. Clinical evaluation to confirm acceptable immunogenicity is always required because the *in silico* and *in vitro* methods cannot account for all sources of variability. The types of tools for risk assessment and mitigation described herein can increase the stringency of selecting which therapeutics to move forward into clinical trials, making the process safer and more efficient.

5.4.2 De-immunization by Epitope Modification

Prediction of immunogenic epitopes has led to the strategy of de-immunization by epitope modification to disrupt HLA binding, an underlying requirement for T-cell stimulation. Indeed, rational epitope modification is based on the natural process that occurs when tumor cells and pathogens escape immune pressure by accumulating mutations to reduce binding of their constituent epitopes to host HLA, rendering the host cell unable to alert T cells to the presence of the tumor or pathogen [58,59]. As de-immunized protein therapeutics make their way into the clinic, the initial results appear to support this approach to reducing immunogenicity risk.

5.4.3 Tolerance Induction

Clusters of epitopes with MHC binding potential can stimulate an immune response through T_{eff} cell induction, but the ability of a peptide to bind to MHC does not necessarily imply an effector immune response. Alternatively, a peptide with MHC binding potential may interact with regulatory T cells to modulate an immune response. For example, potential epitopes in autologous proteins may trigger T cells that are absent from the peripheral circulation. While the majority of autoreactive T cells are thought to be deleted during thymic development, it is now well accepted that some T cells specific for autologous proteins escape thymic deletion to become natural regulatory T cells (nT_{reg}). This subset of T cells serves as a suppressor of autoimmune and auto-reactive immune responses [18]. Just as the inadvertent addition of stimulatory T-effector epitopes to proteins may lead to increased immunogenicity, removal or alteration of regulatory T-cell epitopes in the drug development process may alter the natural T_{reg} immune response to recombinant autologous proteins. The link between the HLA-restricted T-cell immune response and the development of autoantibodies is still being defined, but early evidence points to the reduction of T_{reg} immune responses and to the induction of T_{eff} responses as significant contributors in the context of unwanted immunogenicity [16].

During the last several decades, a variety of approaches have been developed for the induction of tolerance in experimental animals. Some of these approaches have progressed to clinical trials. Our goal herein is not to cover this area of research, which has been extensively reviewed elsewhere [15,60]. Rather, the fundamental principles favoring tolerance are outlined, following which we focus on approaches developed in our lab and those of our collaborators.

As noted in the introductory section, the route of immunization with a protein significantly influences its immunogenicity and tolerogenicity. An intravenous injection favors tolerance, whereas a subcutaneous or intramuscular injection leads to direct uptake by dendritic cells. These cells travel to the lymphoid follicles, where antigens are processed, and immune responses follow. In contrast, intravenous antigen is taken up by APCs in the spleen, primarily B cells. Oral administration of antigens can lead to unresponsiveness due to processing of antigens in the gut immune system. This route may also deviate the response toward mucosal immune responses similar to IgA formation [61].

Alternatively, if co-stimulation is blocked (e.g. by CTLA-4-Ig or antibodies to CD80/CD86), anergy may ensue. The strategy is to provide "signal 1" (T-cell epitopes bound to MHC) to the T cell in the absence of "signal 2" (co-stimulation). This approach has been widely employed in transplant models but has not been successful for other applications. Additional methods are based on treatments to block or subvert T-cell signaling using either antibodies to the CD3 co-receptor or drugs such as rapamycin that inhibit downstream signaling pathways. Clinical trials of drugs employing these approaches also are in progress [62,63].

Recently, the use of so-called tolerogenic APCs, which are tolerogenic immature dendritic cells (iDC) pulsed with target antigens, has gained popularity. Such iDC express low amounts of CD80/CD86 and thus provide little of the co-stimulation necessary to generate a robust adaptive response [64]. However, maintaining this immature phenotype *in vivo* can be challenging. In a related approach, B cells as tolerogenic APC in combination with gene therapy are being explored. Similar to iDC, naïve B cells are low in co-stimulatory molecules [65]. Even mature or activated B cells can still be tolerogenic under certain circumstances. This approach has been successful in a variety of mouse and rat models for autoimmune diseases and hemophilia; it is now moving forward to proof of concept in nonhuman primates.

Over two decades ago, B cells with immunoglobulin (Ig) fusion proteins was used as tolerogens. Immunoglobulins, especially IgG subclasses, had long been known to be excellent tolerogenic carriers by virtue of long half-life, favorable pharmacodymamics, and their ability to cross-link inhibitory Fc receptors [66]. The fusion protein *per se* was shown to be tolerogenic [67], a point that we shall discuss further. Animal studies demonstrated that retroviral expression of this Ig fusion in stem cells and B cells induced tolerance [68] and that T_{reg} cells were induced by tolerogenic B-cell presentation of epitopes, an effect that was enhanced by the presence of the IgG heavy chain scaffold [69]. Thus, Fc-fusion proteins are one option to consider for accessing tolerogenic mechanisms and for potentially decreasing the immunogenicity of a biologic drug.

5.4.4 Regulatory T-Cell Epitopes in IgG Framework Regions: Tregitopes

"Tregitopes" are one potential explanation for the tolerogenicity of Fc-fusion proteins as described in 2008 [70]. Tregitopes are highly conserved, tolerogenic peptides discovered during epitope mapping and de-immunization programs with human protein (especially IgG) therapeutics. It was observed that all IgG subclasses contained epitopes in constant regions that were predicted to have strong binding to MHC class II and that were promiscuous and conserved in multiple mammalian species. It was predicted that these epitopes may be recognized by T_{reg} [70]. Further experimental data provided evidence that Tregitopes not only activate FoxP3-expressing T cells, they also suppress effector T-cell reactions *in vitro* and *in vivo* [70]. In their natural state, Tregitopes are integrated into the arms and stem of human antibodies and may serve to balance the inflammatory triggers that are present in the rearranged segments (variable loops) of the antibody arms. Tregitopes are therefore present in intravenous immunoglobulins (IVIGs), a blood-derived product that is used clinically to control autoimmune conditions [71]. Further studies may provide some insight into the mechanisms by which IVIG acts as an immunosuppressive agent in the clinical setting [71–74]. Immunoglobulin therapies have previously been shown to induce expansion of T_{reg} *in vitro* and *in vivo* [72,74]. Consistent with this observation, co-incubation of donor PBMC with Tregitopes derived from IgG has been shown to lead to suppression of immune response to bystander antigens. The corresponding murine epitopes suppress *in vivo* immune response in HLA DR4 transgenic mice [70–72]. De Groot and collaborators have now shown that the Tregitopes are immunosuppressive, dampen inflammation, and induce tolerance in animal models of T1D, gene therapy, and organ transplants [75–81]. These observations have been corroborated by other researchers who have identified similar peptides in IgG but have not recognized them as being T_{reg} epitopes [82–85].

5.5 CASE STUDY AND CLINICAL EXPERIENCE

5.5.1 Observed Versus Predicted Immunogenicity of 21 Antibodies in Clinical Use

The accuracy of *in silico* methods for immunogenicity prediction has been tested on more than 20 antibodies in clinical use, *in vitro* or in a retrospective study. Protein sequence information for the variable regions of heavy and light chains was obtained from GenBank and the United States Patent and Trademark Office. All sequences were analyzed and scored for immunogenicity according to an immunogenicity scale. Observed immunogenicity from multiple studies was obtained, averaged, and regressed against the predicted overall immunogenicity score of the combined heavy and light chains (Figure 5.2). The results of this regression analysis showed that the immunogenicity score was a statistically significant predictor of observed immunogenicity (correlation coefficient = 0.67). Notable in this analysis is the relatively small data set, limited by the numbers of therapeutic antibodies in clinical trials, with only 20–30 participants on average per study. With recent evidence suggesting that observed immunogenicity for some therapeutics has been under-reported due to inadequate assays [86], this retrospective study supports the use of *in silico* immunogenicity assessment as part of a comprehensive preclinical development strategy [9].

5.5.2 Correlation of Clinical Immunogenicity of Fusion Proteins with *In Silico* Risk Estimates

In a prospective study, the sequences from four different peptibody-based fusion proteins were assessed *in silico* for binding to HLA-DR alleles using the EpiMatrix algorithm, which assigns a Z-score to overlapping 9-mer sequences based on their predicted affinity for eight class II MHC alleles [52,54]. All four molecules were also administered in clinical trials, where an independent immunogenicity assessment was conducted. Table 5.2 summarizes the EpiMatrix-generated scores associated with each protein and their respective rates of antibody incidence (binding and neutralizing). An assessment of Tregitope content in each molecule was also performed and the Z-scores adjusted accordingly. FPX1, which

TABLE 5.2 Correlation of Clinical Immunogenicity with *In Silico* Risk Estimates

Protein Therapeutics	FPX 1	FPX 2	FPX 3	FPX 4
EpiMatrix score	21.97	1.76	−0.76	1.63
Tregitope-adjusted EpiMatrix score	21.97	1.62	−1.76	−111.25
Binding antibodies (%)	37	7.8	5.6	4.5
Neutralizing antibodies (%)	40	0.5	n.a.	0

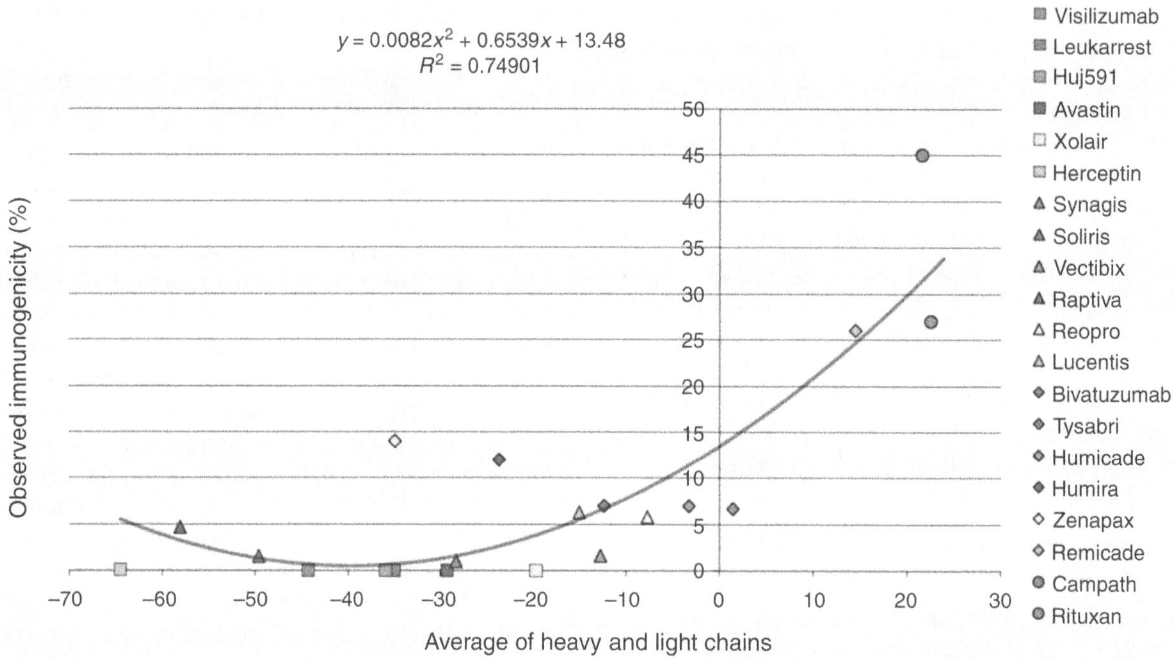

FIGURE 5.2 Predicted versus observed immunogenicity. Twenty human and chimeric antibodies were selected from the literature based on the availability of *in vivo* immunogenicity data and averaged where multiple studies of observed immunogenicity were available. Protein sequence information for the variable regions of heavy and light chains was obtained (GenBank and the United States Patent and Trademark Office). These sequences were analyzed with EpiMatrix and scored according to the EpiMatrix immunogenicity scale. Observed immunogenicity was regressed against predicted immunogenicity score of the combined heavy and light chains.

had a high rate of clinical immunogenicity, was associated with elevated T-cell epitope content reflected by its high Z score. Low Tregitope content was also associated with FPX1. FPX2-4 was associated with a low EpiMatrix score, and Tregitope adjustment reduced the predicted potential for binding further. Accordingly, FPX2-4 showed only minor clinical immunogenicity. Hence, an inverse relationship between Tregitope content and immunogenicity rate could be noted for these fusion proteins.

This retrospective study linking high T-cell epitope content and low Tregitope content with observed clinical immunogenicity of FPX1 was followed by an evaluation of FPX1 immunogenicity *in vitro* by Koren et al. [52]. Whole blood-derived PBMC were obtained from healthy individuals dosed with FPX1. Cells were collected from both antibody-positive and antibody-negative subjects. Peptides, derived from either the N-terminus (predicted as nonimmunogenic by *in silico* algorithm) or the C-terminus (predicted to be immunogenic by *in silico* algorithm) from FPX1, were used for challenge in the recall response. Antigen-specific responses from PBMC were assessed using IFN-γ and IL-4 ELISpots as readouts. A significant induction of IL-4 and IFN-γ spot forming cells was observed in PBMC derived from FPX1 antibody-positive subjects when the *in vitro* cultures were stimulated with the C-terminus peptide as well as the whole protein FPX1 (Table 5.3). Neither IL-4 nor IFN-γ spot forming cells were observed in PBMC derived from antibody-negative subjects and N-terminus peptide stimulation *in vitro*. There was an excellent correlation between the HLA restriction of the peptides, as predicted by EpiMatrix algorithms, and the HLA of the patients who responded to the epitopes [49].

The FPX1 fusion protein was further evaluated in a naïve human PBMC-derived *in vitro* sensitization assay [54]. Whole PBMC were stimulated with FPX1-derived N-terminus and C-terminus peptides multiple times to amplify the frequency of antigen-specific cells. Immune responses were measured by enumerating IFN-γ-secreting cells by ELISpot. A significant induction of IFN-γ-secreting cells was evident for the PBMC challenged with the immunogenic C-terminus of FPX1 peptide. No reactive cells were observed when challenged with nonimmunogenic N-terminus peptide and a clinically proven and EpiMatrix-predicted nonimmunogenic monoclonal antibody therapeutic.

This retrospective study validates the hypothesis that immunogenicity of a biotherapeutic can be predicted through the systematic application of *in silico* and *in vitro* tools such as those described here (Figure 5.3). On the basis

TABLE 5.3 High PBMC Response to FPX Peptides among Donors with High Anti-FPX Titers

Subject ID	HLA-DRB1*	Antibody Concentration (mg/mL)	aa 1–10		aa 11–20		aa 1–24	
			IFN-γ	IL-4	IFN-γ	IL-4	IFN-γ	IL-4
1	0701/1501	20.2	1.8	0.8	26	89	34	92
2	0101/1301	1.5	1.5	1.8	9.9	4.7	26.6	30.1
3	0101/1701	1	0.6	1.4	14.6	6.8	16.9	14.5
4	0405/0101	1.1	1.4	1.2	6	4.7	9.4	7.1

of the *in silico*, *in vitro*, and clinical findings related to FPX1 and their concordance to each other, a similar risk assessment strategy can be proposed for biotherapeutics under early stage development.

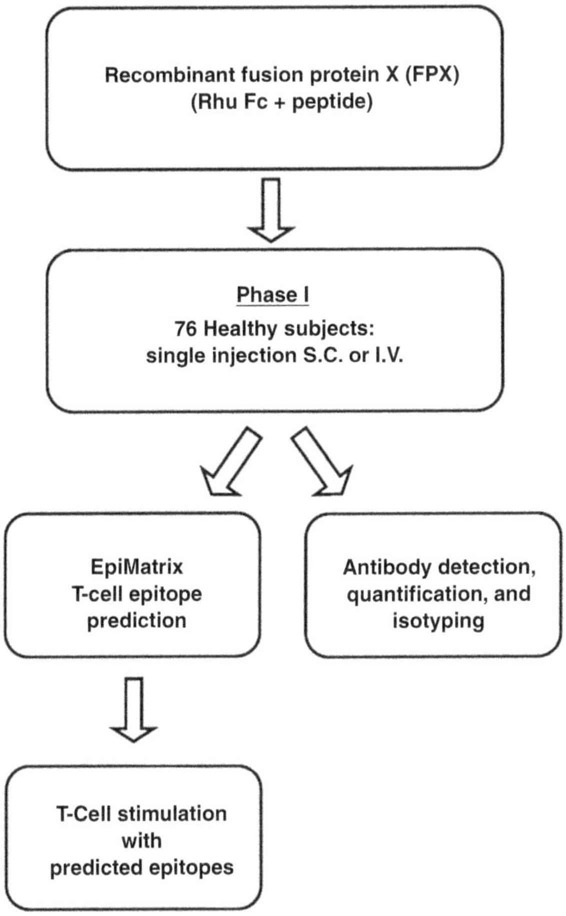

FIGURE 5.3 *In silico* prediction of immunogenicity—proof of concept study in humans with FPX 1. In a prospective study, sequences from four different fusion proteins were assessed *in silico* for binding to HLA-DR alleles using the EpiMatrix algorithm. High T-cell epitope content and low Tregitope content were retrospectively linked with observed clinical immunogenicity of FPX1 and followed by an evaluation of FPX1 immunogenicity *in vitro*. On the basis of the *in silico*, *in vitro*, and clinical findings related to FPX1 and their concordance to each other, a similar risk assessment strategy may be proposed for biotherapeutics under early stage development.

5.6 PRECLINICAL AND CLINICAL IMMUNOGENICITY ASSESSMENT STRATEGY

The assessment of immunogenicity of a biotherapeutic is under considerable scrutiny by regulatory agencies. Hence, as a part of risk minimization and mitigation, drug manufacturers should have a strategy to detect and characterize the potential for immunogenicity. An evaluation of all samples for binding antibodies in a screening assay is one such strategy. The reactive samples are confirmed to contain antibody using a secondary species-specific antibody. Additionally, a drug depletion step confirms the specificity of the sample. The optimal platform for binding antibody assessment is chosen based on the nature and modality of the biotherapeutic. Lastly, a biological assay should be used to test if these antibodies are capable of neutralizing the biological effect of the drug [87,88]. The appropriate assays when developed should be sensitive and specific enough to eliminate false positive results. A fully characterized antibody response to a biotherapeutic enables risk assessment and clinical relevance for the patient [89].

5.6.1 Strategy and Recommendation

A comprehensive approach to preclinical immunogenicity testing could begin with a high throughput *in silico* screening followed by an *in vitro* evaluation and end with testing *in vivo* in transgenic animal models (Figure 5.4). Immunogenicity screening could follow a tiered approach where Tier 1 would entail screening of linear sequences from multiple therapeutic candidates for T-cell epitopes and clusters therein. Candidates could then be ranked on the basis of the quantity and quality of immunogenic epitopes, adjusted for Tregitope content.

Once the field of candidates has been prioritized and/or narrowed, Tier 2 screening would test the immunogenicity of these molecules in one or more *in vitro* assays. At this point, the *in vitro* assays would, in addition to validating *in silico* predictions, bring forward any non-sequence-related immunogenicity concerns such as processing-associated changes, post-translational modifications and alterations owing to misfolding. *In vitro* assays can also be utilized to test therapeutics, which may have target-mediated or agonist effects, and to overcome hurdles during

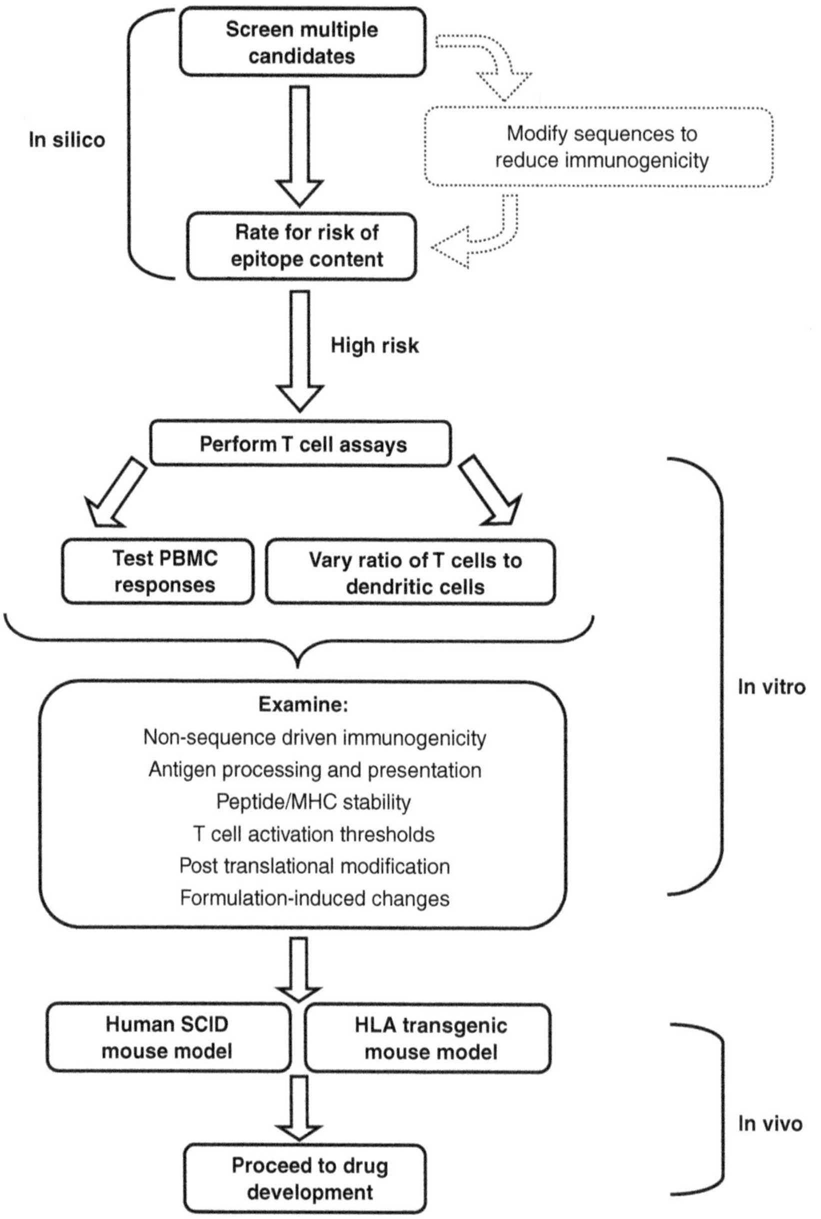

FIGURE 5.4 Immunogenicity prediction strategy. A potential tiered approach to preclinical immunogenicity testing is depicted here. Tier 1 would consist of the *in silico* screening of linear sequences from multiple therapeutic candidates for T-cell epitopes and clusters. At the conclusion of Tier 1, therapeutic candidates would be rated for immunogenic risk. Tier 2 would be an *in vitro* evaluation of immunogenicity in a series of T-cell assays. This phase could include examination of antigen processing and presentation, post-translational modifications, and different formulations. Tier 3 would focus on immunogenicity testing of therapeutic candidates *in vivo*, in established "humanized" animal models such as the Hu-SCID and HLA-transgenic mouse models. The culmination of this immunogenicity prediction strategy would be advancement into the drug development pipeline.

formulation such as interference from aggregates. Such variables can enhance the immunogenicity of the drug product in ways that cannot be captured by *in silico* modeling, which only attends to the amino acid sequence. Even *in vitro* tests cannot be fully indicative of immunogenicity, and hence clinical evaluation of immunogenicity is always required to confirm the development predictions.

5.7 CONCLUSIONS

The ability to analyze and prospectively predict the immunogenicity of a potential protein therapeutic has tremendous benefits at every stage of the drug development process. Fusion-protein-based biotherapeutics can pose an immunogenicity risk due to their structural differences from conventional fully human monoclonal or recombinant protein therapeutics. Reliable *in silico* immunogenicity screening makes it possible to rank lead candidates at the preclinical stage of development and/or reengineer proteins to make them less immunogenic. Moving forward in the preclinical development process, *in vitro* methods and *in vivo* animal models are important for validating the *in silico* findings. The *in vitro* methods can also address any non-sequence post-translational and manufacturing-associated changes such as aggregation or contaminants. Owing to limitations of the *in silico* and *in vitro* approaches, these assessments cannot substitute for clinical studies during drug development. Further downstream, the measurement of immunogenicity in patients enrolled in Phase I/II clinical trials of protein therapeutics should reflect the considerations taken in the preclinical process and provide additional opportunities to refine a lead candidate should anti-drug immunity arise. Finally, rigorous data collection in clinical trials may confer the ability to identify prospectively individuals at a higher risk of developing an antidrug response, such as by HLA typing, where alternative therapies would be indicated.

ACKNOWLEDGMENT

The contributions of Ryan Tassone and Frances Terry, of EpiVax, to the process of illustrating and curating this manuscript, are gratefully acknowledged.

REFERENCES

1. De Groot AS, Scott DW. (2007) Immunogenicity of protein therapeutics. *Trends Immunol.* 28, 482–490.
2. Kuus-Reichel K, Grauer LS, Karavodin LM, Knott C, Krusemeier M, Kay NE. (1994) Will immunogenicity limit the use, efficacy, and future development of therapeutic monoclonal antibodies? *Clin. Diagn. Lab. Immunol.* 1, 365–372.
3. Goodnow CC, Vinuesa CG, Randall KL, Mackay F, Brink R. (2010) Control systems and decision making for antibody production. *Nat. Immunol.* 11, 681–688.
4. Zubler RH. (2001) Naive and memory B cells in T-cell-dependent and T-independent responses. *Springer Semin. Immunopathol.* 23, 405–419.
5. Onda M. (2009) Reducing the immunogenicity of protein therapeutics. *Curr. Drug Targets* 10, 131–139.
6. Casadevall N, Nataf J, Viron B, Kolta A, Kiladjian JJ, Martin-Dupont P, et al. (2002) Pure red-cell aplasia and antierythropoietin antibodies in patients treated with recombinant erythropoietin. *N. Engl. J. Med.* 346, 469–475.
7. Li J, Yang C, Xia Y, Bertino A, Glaspy J, Roberts M, et al. (2001) Thrombocytopenia caused by the development of antibodies to thrombopoietin. *Blood* 98, 3241–3248.
8. Baert F, Noman M, Vermeire S, Van Assche G, G DH, Carbonez A, et al. (2003) Influence of immunogenicity on the long-term efficacy of infliximab in Crohn's disease. *N. Engl. J. Med.* 348, 601–608.
9. De Groot AS, Martin W. (2009) Reducing risk, improving outcomes: bioengineering less immunogenic protein therapeutics. *Clin. Immunol.* 131, 189–201.
10. Valmori D, Levy F, Godefroy E, Scotto L, Souleimanian NE, Karbach J, et al. (2007) Epitope clustering in regions undergoing efficient proteasomal processing defines immunodominant CTL regions of a tumour antigen. *Clin. Immunol.* 122, 163–172.
11. Zhang GL, Khan AM, Srinivasan KN, Heiny A, Lee K, Kwoh CK, et al. (2008) Hotspot hunter: a computational system for large-scale screening and selection of candidate immunological hotspots in pathogen proteomes. *BMC Bioinform.* 9(Suppl. 1), S19.
12. Fooksman DR, Gronvall GK, Tang Q, Edidin M. (2006) Clustering class I MHC modulates sensitivity of T cell recognition. *J. Immunol.* 176, 6673–6680.
13. Rammensee HG. (1996) Antigen presentation—recent developments. *Int. Arch. Allergy. Immunol.* 110, 299–307.
14. Podojil JR, Miller SD. (2009) Molecular mechanisms of T-cell receptor and costimulatory molecule ligation/blockade in autoimmune disease therapy. *Immunol. Rev.* 229, 337–355.
15. Brusko TM, Putnam AL, Bluestone JA. (2008) Human regulatory T cells: role in autoimmune disease and therapeutic opportunities. *Immunol. Rev.* 223, 371–390.
16. Reveille JD. (2006) The genetic basis of autoantibody production. *Autoimmun. Rev.* 5, 389–398.
17. Pulendran B, Tang H, Manicassamy S. (2010) Programming dendritic cells to induce T(H)2 and tolerogenic responses. *Nat. Immunol.* 11, 647–655.
18. Bluestone JA, Abbas AK. (2003) Natural versus adaptive regulatory T cells. *Nat. Rev. Immunol.* 3, 253–257.
19. Lo D, Reilly C, Marconi LA, Ogata L, Wei Q, Prud'homme G, et al. (1995) Regulation of CD4 T cell reactivity to self and non-self. *Int. Rev. Immunol.* 13, 147–160.
20. Nurnberger T, Brunner F, Kemmerling B, Piater L. (2004) Innate immunity in plants and animals: striking similarities and obvious differences. *Immunol. Rev.* 198, 249–266.
21. Rosenberg SA, Sherry RM, Morton KE, Yang JC, Topalian SL, Royal RE, et al. (2006) Altered CD8(+) T-cell responses when immunizing with multiepitope peptide vaccines. *J. Immunother.* 29, 224–231.
22. Xu D, Liu H, Komai-Koma M. (2004) Direct and indirect role of Toll-like receptors in T cell mediated immunity. *Cell Mol. Immunol.* 1, 239–246.

23. Flanagan ML, Arias RS, Hu P, Khawli LA, Epstein AL. (2007) Soluble Fc fusion proteins for biomedical research. *Methods Mol. Biol.* 378, 33–52.
24. Dumont JA, Low SC, Peters RT, Bitonti AJ. (2006) Monomeric Fc fusions: impact on pharmacokinetic and biological activity of protein therapeutics. *BioDrugs* 20, 151–160.
25. Sallusto F, Cella M, Danieli C, Lanzavecchia A. (1995) Dendritic cells use macropinocytosis and the mannose receptor to concentrate macromolecules in the major histocompatibility complex class II compartment: downregulation by cytokines and bacterial products. *J. Exp. Med.* 182, 389–400.
26. Gillies S. (2006) Reducing the immunogenicity of fusion proteins. USPTO, U.P. Office, ed. (US).
27. Wolbink GJ, Aarden LA, Dijkmans BA. (2009) Dealing with immunogenicity of biologicals: assessment and clinical relevance. *Curr. Opin. Rheumatol.* 21, 211–215.
28. Swanson SJ, Ferbas J, Mayeux P, Casadevall N. (2004) Evaluation of methods to detect and characterize antibodies against recombinant human erythropoietin. *Nephron Clin. Pract.* 96, c88–c95.
29. Lazarski CA, Chaves FA, Jenks SA, Wu S, Richards KA, Weaver JM, et al. (2005) The kinetic stability of MHC class II: peptide complexes is a key parameter that dictates immunodominance. *Immunity* 23, 29–40.
30. Van Regenmortel MH. (2009) What is a B-cell epitope? *Methods Mol. Biol.* 524, 3–20.
31. Roggen EL. (2006) Recent developments with B-cell epitope identification for predictive studies. *J. Immunotoxicol.* 3, 137–149.
32. Southwood S, Sidney J, Kondo A, del Guercio MF, Appella E, Hoffman S, et al. (1998) Several common HLA-DR types share largely overlapping peptide binding repertoires. *J. Immunol.* 160, 3363–3373.
33. Kim Y, Sette A, Peters B. (2010) Applications for T-cell epitope queries and tools in the immune epitope database and analysis resource. *J. Immunol. Methods.* 2011; 374:62–69.
34. Sette A, Buus S, Appella E, Smith JA, Chesnut R, Miles C, et al. (1989) Prediction of major histocompatibility complex binding regions of protein antigens by sequence pattern analysis. *Proc. Natl. Acad. Sci. USA* 86, 3296–3300.
35. Zhang GL, Petrovsky N, Kwoh CK, August JT, Brusic V. (2006) PRED(TAP): a system for prediction of peptide binding to the human transporter associated with antigen processing. *Immunome Res.* 2, 3.
36. De Groot AS, Rayner J, Martin W. (2003) Modelling the immunogenicity of therapeutic proteins using T cell epitope mapping. *Dev. Biol. (Basel)* 112, 71–80.
37. Ahlers JD, Belyakov IM, Thomas EK, Berzofsky JA. (2001) High-affinity T helper epitope induces complementary helper and APC polarization, increased CTL, and protection against viral infection. *J. Clin. Invest.* 108, 1677–1685.
38. De Groot AS, Rivera DS, McMurry JA, Buus S, Martin W. (2008) Identification of immunogenic HLA-B7 "Achilles' heel" epitopes within highly conserved regions of HIV. *Vaccine* 26, 3059–3071.
39. De Groot AS, Martin W. (2003) From immunome to vaccine: epitope mapping and vaccine design tools. *Novartis Found. Symp.* 254, 57–72; discussion 72–76, 98–101, 250–252.
40. De Groot AS, Bosma A, Chinai N, Frost J, Jesdale BM, Gonzalez MA, et al. (2001) From genome to vaccine: *in silico* predictions, ex vivo verification. *Vaccine* 19, 4385–4395.
41. De Groot AS, Ardito M, McClaine EM, Moise L, Martin WD. (2009) Immunoinformatic comparison of T-cell epitopes contained in novel swine-origin influenza A (H1N1) virus with epitopes in 2008–2009 conventional influenza vaccine. *Vaccine* 27, 5740–5747.
42. Inaba H, Martin W, De Groot AS, Qin S, De Groot LJ. (2006) Thyrotropin receptor epitopes and their relation to histocompatibility leukocyte antigen-DR molecules in Graves' disease. *J. Clin. Endocrinol. Metab.* 91, 2286–2294.
43. De Groot AS, Moise L. (2007) Prediction of immunogenicity for therapeutic proteins: state of the art. *Curr. Opin. Drug Discov. Devel.* 10, 332–340.
44. Sturniolo T, Bono E, Ding J, Raddrizzani L, Tuereci O, Sahin U, et al. (1999) Generation of tissue-specific and promiscuous HLA ligand databases using DNA microarrays and virtual HLA class II matrices. *Nat. Biotechnol.* 17, 555–561.
45. Koita OA, Dabitao D, Mahamadou I, Tall M, Dao S, Tounkara A, et al. (2006) Confirmation of immunogenic consensus sequence HIV-1 T-cell epitopes in Bamako, Mali and Providence, Rhode Island. *Hum. Vaccin.* 2, 119–128.
46. Bond KB, Sriwanthana B, Hodge TW, De Groot AS, Mastro TD, Young NL, et al. (2001) An HLA-directed molecular and bioinformatics approach identifies new HLA-A11 HIV-1 subtype E cytotoxic T lymphocyte epitopes in HIV-1-infected Thais. *AIDS Res. Hum. Retroviruses* 17, 703–717.
47. Dong Y, Demaria S, Sun X, Santori FR, Jesdale BM, De Groot AS, et al. (2004) HLA-A2-restricted CD8+-cytotoxic-T-cell responses to novel epitopes in Mycobacterium tuberculosis superoxide dismutase, alanine dehydrogenase, and glutamine synthetase. *Infect. Immun.* 72, 2412–2415.
48. McMurry J, Sbai H, Gennaro ML, Carter EJ, Martin W, De Groot AS. (2005) Analyzing Mycobacterium tuberculosis proteomes for candidate vaccine epitopes. *Tuberculosis (Edinb.)* 85, 95–105.
49. Cohen T, Moise L, Ardito M, Martin W, De Groot AS. (2010) A method for individualizing the prediction of immunogenicity of protein vaccines and biologic therapeutics: individualized T cell epitope measure (iTEM). *J. Biomed. Biotechnol.* 2010; 2010: 961752. Published online 2010 July 18. doi: 10.1155/2010/961752.
50. Moise L, Buller RM, Schriewer J, Lee J, Frey SE, Weiner DB, et al. (2010) VennVax, a DNA-prime, peptide-boost multi-T-cell epitope poxvirus vaccine, induces protective immunity against vaccinia infection by T cell response alone. *Vaccine.* 2011; 29:501–511.
51. Tatarewicz SM, Wei X, Gupta S, Masterman D, Swanson SJ, Moxness MS. (2007) Development of a maturing T-cell-mediated immune response in patients with idiopathic Parkinson's disease receiving r-metHuGDNF via continuous intraputaminal infusion. *J. Clin. Immunol.* 27, 620–627.

52. Koren E, De Groot AS, Jawa V, Beck KD, Boone T, Rivera D, et al. (2007) Clinical validation of the "*in silico*" prediction of immunogenicity of a human recombinant therapeutic protein. *Clin. Immunol.* 124, 26–32.

53. Baker MP, Jones TD. (2007) Identification and removal of immunogenicity in therapeutic proteins. *Curr. Opin. Drug Discov. Devel.* 10, 219–227.

54. Wullner D, Zhou L, Bramhall E, Kuck A, Goletz TJ, Swanson S, et al. (2010) Considerations for optimization and validation of an *in vitro* PBMC derived T cell assay for immunogenicity prediction of biotherapeutics. *Clin. Immunol.* 137, 5–14.

55. Steere AC, Klitz W, Drouin EE, Falk BA, Kwok WW, Nepom GT, et al. (2006) Antibiotic-refractory Lyme arthritis is associated with HLA-DR molecules that bind a Borrelia burgdorferi peptide. *J. Exp. Med.* 203, 961–971.

56. McMurry JA, Kimball S, Lee JH, Rivera D, Martin W, Weiner DB, et al. (2007) Epitope-driven TB vaccine development: a streamlined approach using immuno-informatics, ELISpot assays, and HLA transgenic mice. *Curr. Mol. Med.* 7, 351–368.

57. Jaber A, Baker M. (2007) Assessment of the immunogenicity of different interferon beta-1a formulations using ex vivo T-cell assays. *J. Pharm. Biomed. Anal.* 43, 1256–1261.

58. Vossen MT, Westerhout EM, Soderberg-Naucler C, Wiertz EJ. (2002) Viral immune evasion: a masterpiece of evolution. *Immunogenetics* 54, 527–542.

59. De Groot AS, Knopp PM, Martin W. (2005) De-immunization of therapeutic proteins by T-cell epitope modification. *Dev. Biol. (Basel)* 122, 171–194.

60. Podojil JR, Turley DM, Miller SD. (2008) Therapeutic blockade of T-cell antigen receptor signal transduction and costimulation in autoimmune disease. *Adv. Exp. Med. Biol.* 640, 234–251.

61. Tsuji NM, Kosaka A. (2008) Oral tolerance: intestinal homeostasis and antigen-specific regulatory T cells. *Trends Immunol.* 29, 532–540.

62. Chatenoud L, Bluestone JA. (2007) CD3-specific antibodies: a portal to the treatment of autoimmunity. *Nat. Rev. Immunol.* 7, 622–632.

63. Alegre ML, Fallarino F. (2006) Mechanisms of CTLA-4-Ig in tolerance induction. *Curr. Pharm. Des.* 12, 149–160.

64. Pechhold K, Koczwara K. (2008) Immunomodulation of autoimmune diabetes by dendritic cells. *Curr. Diab. Rep.* 8, 107–113.

65. Skupsky J, Su Y, Lei TC, Scott DW. (2007) Tolerance induction by gene transfer to lymphocytes. *Curr. Gene Ther.* 7, 369–380.

66. Borel Y, Borel H. (1988) Oligonucleotide linked to human gammaglobulin specifically diminishes anti-DNA antibody formation in cultured lymphoid cells from patients with systemic lupus erythematosus. *J. Clin. Invest.* 82, 1901–1907.

67. Zambidis ET, Scott DW. (1996) Epitope-specific tolerance induction with an engineered immunoglobulin. *Proc. Natl. Acad. Sci. USA* 93, 5019–5024.

68. Zambidis ET, Kurup A, Scott DW. (1997) Genetically transferred central and peripheral immune tolerance via retroviral-mediated expression of immunogenic epitopes in hematopoietic progenitors or peripheral B lymphocytes. *Mol. Med.* 3, 212–224.

69. Soukhareva N, Jiang Y, Scott DW. (2006) Treatment of diabetes in NOD mice by gene transfer of Ig-fusion proteins into B cells: role of T regulatory cells. *Cell Immunol.* 240, 41–46.

70. De Groot AS, Moise L, McMurry JA, Wambre E, Van Overtvelt L, Moingeon P, et al. (2008) Activation of natural regulatory T cells by IgG Fc-derived peptide "Tregitopes". *Blood* 112, 3303–3311.

71. Maddur MS, Othy S, Hegde P, Vani J, Lacroix-Desmazes S, Bayry J, et al. (2010) Immunomodulation by intravenous immunoglobulin: role of regulatory T cells. *J. Clin. Immunol.* 30(Suppl. 1), S4–S8.

72. Ephrem A, Chamat S, Miquel C, Fisson S, Mouthon L, Caligiuri G, et al. (2008) Expansion of CD4+CD25+ regulatory T cells by intravenous immunoglobulin: a critical factor in controlling experimental autoimmune encephalomyelitis. *Blood* 111, 715–722.

73. Kaveri S, Prasad N, Vassilev T, Hurez V, Pashov A, Lacroix-Desmazes S, et al. (1997) Modulation of autoimmune responses by intravenous immunoglobulin (IVIg). *Mult. Scler.* 3, 121–128.

74. Lopez M, Clarkson MR, Albin M, Sayegh MH, Najafian N. (2006) A novel mechanism of action for anti-thymocyte globulin: induction of CD4+CD25+Foxp3+ regulatory T cells. *J. Am. Soc. Nephrol.* 17, 2844–2853.

75. De Groot AS. (2008) IgG Tregitopes and AITD-ASATI: antigen specific tolerance induction in autoimmune thyroid disease. Tregs and Th17 Cells in Autoimmunity, Washington, DC.

76. De Groot AS, Moise L, Li X, Su Y, Yang W, Desrosiers J, et al. (2009) Effect of Tregitopes on T1D Immune response *in vitro* and on diabetes in NOD mice. 2nd European Congress of Immunology, Berlin, Germany.

77. D'Addio F, Boenisch O, Wattanabe T, De Groot AS, Sayegh MH, Najafian N. (2009) Immuno-modulatory functions of a novel IgG Fc-derived peptides, Tregitopes, in alloimmunity. American Transplant Conference, Boston, MA.

78. Su Y, Moise L, Li X, Rossi R, Skupsky J, Martin WD, et al. (2009) T cell epitopes (Tregitopes) contained in IgG constant domains activate natural Tregs. 1st International Conference on Tolerance, Elsevier, Boston, MA.

79. Mingozzi F, Finn JD, Zhou S, Hui DJ, Pien GC, Basner-Tschakarjan E, et al. (2009) Suppression of CTL responses against AAV-capsid epitopes by peptide-induced regulatory T cells. American Society for Hematology, Oral Session: Gene Therapy and Transfer, New Orleans, LA.

80. Buhlmann JE, Najafian N, Hui DJ, D'Addio F, Mingozzi F, Moise L, et al. (2010) Preclinical design of less immunogenic biologics: tregitopes and tolerance. Annual Meeting of the American Association of Immunologists, Baltimore, MD.

81. De Groot AS, Najafian N, Hui DJ, D'Addio F, Mingozzi F, Moise L, et al. (2010) Preclinical design of less immunogenic biologics: tregitopes and tolerance. AAPS 2010 National Biotechnology Conference, San Francisco, CA.

82. Sharabi A, Lapter S, Mozes E. (2010) Bcl-xL is required for the development of functional regulatory CD4 cells in lupus-

afflicted mice following treatment with a tolerogenic peptide. *J. Autoimmun.* 34, 87–95.

83. Sharabi A, Zinger H, Zborowsky M, Sthoeger ZM, Mozes E. (2006) A peptide based on the complementarity-determining region 1 of an autoantibody ameliorates lupus by up-regulating CD4+CD25+ cells and TGF-beta. *Proc. Natl. Acad. Sci. USA* 103, 8810–8815.
84. Hahn BH, Singh RP, La Cava A, Ebling FM. (2005) Tolerogenic treatment of lupus mice with consensus peptide induces Foxp3-expressing, apoptosis-resistant, TGFbeta-secreting CD8+ T cell suppressors. *J. Immunol.* 175, 7728–7737.
85. Hanvesakul R, Maillere B, Briggs D, Baker R, Larche M, Ball S. (2007) Indirect recognition of T-cell epitopes derived from the alpha 3 and transmembrane domain of HLA-A2. *Am. J. Transplant.* 7, 1148–1157.
86. van Schouwenburg PA, Bartelds GM, Hart MH, Aarden L, Wolbink GJ, Wouters D. (2010) A novel method for the detection of antibodies to adalimumab in the presence of drug reveals "hidden" immunogenicity in rheumatoid arthritis patients. *J. Immunol. Methods.* 362, 82–88.
87. Swanson SJ. (2006) Immunogenicity issues in drug development. *J. Immunotoxicol.* 3, 165–172.
88. Moxness M, Tatarewicz S, Weeraratne D, Murakami N, Wullner D, Mytych D, et al. (2005) Immunogenicity testing by electrochemiluminescent detection for antibodies directed against therapeutic human monoclonal antibodies. *Clin. Chem.* 51, 1983–1985.
89. Koren E. (2002) From characterization of antibodies to prediction of immunogenicity. *Dev. Biol. (Basel)* 109, 87–95.

PART II

THE TRIPLE T PARADIGM: TIME, TOXIN, TARGETING

IIA TIME: FUSION PROTEIN STRATEGIES FOR HALF-LIFE EXTENSION

6

FUSION PROTEINS FOR HALF-LIFE EXTENSION

STEFAN R. SCHMIDT

Rentschler Biotechnologie GmbH, Laupheim, Germany

6.1 INTRODUCTION

Therapeutic proteins are part of our standard medical care for many years. But most of the attractive biological drugs from the classes of hormones, cytokines, coagulation, and growth factors are small and disappear quickly from the blood stream. In the past, this small therapeutic window was compensated by frequent administration, which makes therapy cumbersome for the patients and also creates concentration spikes that might limit dosing.

The following part is dedicated to fusion protein based concepts to extend the half-life of therapeutic proteins. The bioavailability of drugs in general is dependent on factors such as absorption, distribution, metabolism, and excretion (ADME). In Chapter 1 mutagenesis strategies were described to avoid proteolytic degradation. Metabolic removal of proteins is dependent on their cellular uptake through receptor-mediated endocytosis or pinocytosis. As explained later, receptor binding can also serve the purpose of recycling, thus keeping protein longer in circulation and protecting it from lysosomal degradation. A schematic view on the fate of a therapeutic protein can be seen in Figure 6.1.

This chapter introduces the various fusion protein based concepts to modify the excretion of proteins in order to extend their plasma half-life. Neglecting absorption and metabolism, the half-life is primarily dependent on the hydrodynamic radius of the molecule from a strictly biophysical point of view. Globular proteins below 70 kDa are removed quite rapidly by glomerular filtration in the kidney. The other important parameter regulating excretion of macromolecules besides the pore diameter in the glomeruli of 60 Å is the negative charge of the cell surface that repels anionic macromolecules [1]. However, all these designs to increase the hydrodynamic radius must take into account the other side of the metal. Large size limits tissue penetration. This is a major obstacle in the treatment of solid tumors. Particularly, in the case of antibody variants there seems to be an ideal tumor-targeting zone with a size above the renal cutoff and 120 kDa. This area represents a good compromise between penetration and half-life [2]. It has been described that size, charge, valency, and affinity determine the ability of molecule to penetrate solid tumors. Some of these parameters also contribute to the pharmacokinetic profile of the molecule. Therefore, a good balance must be achieved to design the optimal therapeutic protein [3]. It seems that uncharged molecules have a better penetration profile because they do not suffer from charge repulsion.

An important mechanism that prolongs circulation time is the recycling of molecules through receptors. This is the case for some very abundant plasma molecules, albumin, and immunoglobulins and to a lesser extent for transferrin (Tf) as well. The impact of receptor-mediated recycling was

Fusion Protein Technologies for Biopharmaceuticals: Applications and Challenges, First Edition. Edited by Stefan R. Schmidt.

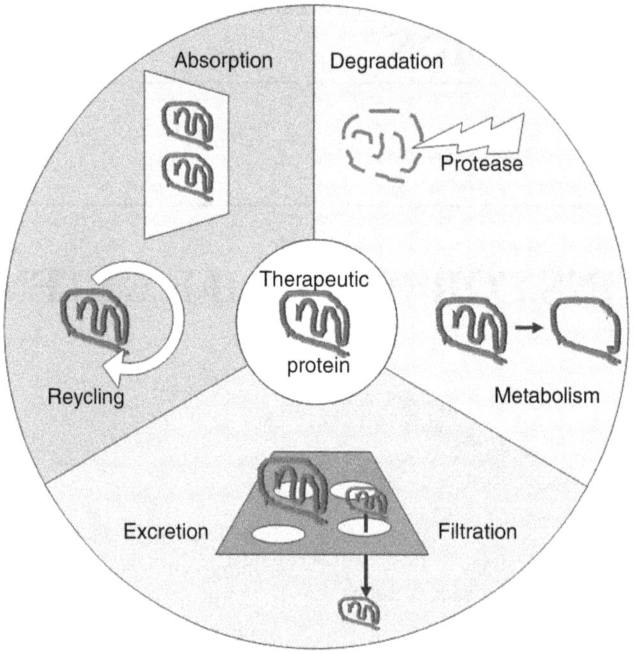

FIGURE 6.1 Factors influencing the bioavailability of a therapeutic protein. Proteins disappear from circulation by proteolysis or metabolism. In the kidney, they are excreted if their size permits passage through the glomeruli. Absorption or transfer into specific organs or compartments can influence the bioavailability as well as receptor-mediated recycling.

demonstrated by a rational design of granulocyte colony-stimulating factor (G-CSF). Optimizing the receptor binding for the different pH conditions in the endosome and on the surface by histidine mutants improved the half-life by 1 order of magnitude [4]. The most important recycling system is the neonatal receptor for the fraction crystallizable (FcRn). This receptor handles both, Fc-containing molecules such as antibodies and serum albumin (HSA), but binding occurs at different positions of the FcRn, hence there is no competition between both molecules.

A good example for the different strategies of half-life prolongation is glucagon-like peptide (GLP-1). The first optimization was mutagenesis to introduce resistance to dipeptidyl peptidase-4 (DPP IV). With GLP-1, a plethora of methods has been tested; therefore, it can serve as benchmarking tool [5]. Some of the currently published data is summarized in Table 6.1. Exenedin-4 acts similar to GLP-1 but has naturally a longer half-life. Nevertheless this peptide has been optimized with fusion partners as well (Table 6.2).

Recently, a global view on half-life extension strategies summarized several nonfusion protein based techniques [6]. In Figure 6.2, molecular mass of many therapeutic and natural proteins versus their half-life in humans is plotted. Most fusion proteins are positioned in the upper right quadrant, indicating that they have been optimized with regard to half-life. The following paragraphs aim to give an overview on the various fusion concepts, but also serve as brief description of strategies that could not be included as full chapters in this book.

6.2 HALF-LIFE EXTENSION THROUGH SIZE AND RECYCLING

6.2.1 Direct Genetic Fusions

6.2.1.1 Albumin Fusions Human serum albumin, the most abundant serum protein with a size of 66 kDa and a diameter of 90 Å, is above the limit of fast kidney clearance. Therefore it remains in circulation with a half-life of up to 19 days [7]. Both, albumin and gamma globulins are recycled through the neonatal Fc receptor. This receptor is present on many epithelia. The primary function of the FcRn during infancy is the absorption of maternal protective antibodies from the milk through the intestinal epithelium. The FcRn in the intestine is downregulated for adults; however, FcRn remains present on vascular cells. The critical regions both on the FcRn and albumin contain several conserved histidines that stabilize flexible loops [8]. Antibodies or albumin are taken up through pinocytosis. When the pH in the endosome is lowered, these

TABLE 6.1 GLP-1 Versions with Extended Half-Life

Name	Company	DPP IV Resistant	Technology	$t_{1/2}$ (h)	Organism	Phase
Liraglutide (Victoza®)	Novo Nordisk	Yes	Albumin binding	12	Human	Market
Albugon	HGSI	No	Albumin fusion	120	Human	III
Dulaglutide	Eli Lilly	Yes	Fc-IgG4 fusion	48	Cynomolgus	III
Albiglutide (Syncria®)	HGSI	Yes	2xGLP-1 albumin fusion	120	Human	III
Taspoglutide	Ipsen	Yes	Bulky side chain	10	Rat *in vitro*	III term
CJC-1131	ConjuChem	Yes	Albumin conjugate	240	Human	II
Semaglutide	Novo Nordisk	No	Bulky side chain	n.a.	n.a.	II
Glymera	PhaseBio	n.a.	ELP fusion	148	Human	IIa
CVX-096	CovX	No	IgG1	96	Rodents	I
CNTO736	Centocor	No	Fc fusion	30	Mouse	Preclin.
n.a.	BioRexis	Yes	Transferrin fusion	44	Cynomolgus	Preclin.

GLP-1 $t_{1/2}$ in human: 7 min.

molecules associate with the membrane-bound receptor, and are transported back to the cell membrane, where they dissociate in a pH-dependent manner from the receptor. Thus, both albumin and immune globulins are protected from proteolytic degradation that would happen in the more acidic lysosome. If the receptors are saturated, excess unbound albumin or antibody are catabolized in the lysosome. Even a transcytosis where albumin or antibody crosses the cell from the apical to the basolateral side is possible [9] (Figure 6.3).

The phenomenon that albumin accumulates in tumors can be utilized for therapeutic approaches in combination to the intrinsic ability to enhance the half-life of proteins fused or attached to it [10]. Another advantage is the tendency of albumin to stabilize proteins despite its own conformational instability. With a G-CSF fusion to HSA it was demonstrated that HSA overcomes the aggregation tendency of G-CSF [11]. However, it has to be taken into account that the presence of multiple domains with different aggregation susceptibility formulation has to be carefully assessed for the whole molecule.

A further advantage is the possibility to attach the fusion partner either at the N- or C-terminus of HSA [12]. This gives sufficient flexibility for the design of novel chimeric molecules. During the manufacture of Albuferon, a combination of interferon-α2b (IFN-α2b) with albumin, it was observed that because of incomplete disulfide bridge formation, aggregates were formed, which drastically reduced the recovery. This phenomenon could be reversed by positioning the IFN-α2b on the N-terminus instead [13]. The half-life extension was sufficient to prepare administration intervals of 2–4 weeks [14]. However, in 2010 the development of the HSA-IFN-α fusion protein was stopped at Phase III because of concerns about the risk benefits of this drug (http://www.novartis.com/newsroom/media-releases/en/2010/1449020.shtml).

Currently, a number of therapeutic proteins have been fused to albumin. For instance, an insulin-HSA variant showed normal activity and drastically prolonged half-life [15]. To the evaluated proteins belong the blood clotting factors FVIIa and FIX with very short half-lives that require frequent dosing. In the case of FVIIa, the albumin fusion had a six- to sevenfold extended half-life [16]. For FIX, not only the circulation time could be improved, but introducing a cleavable linker between albumin and FIX increased specific clotting activity 10- to 30-fold as well. This linker was designed to be cleaved in parallel to FIX activation, thus releasing the small blood factor from its bulky fusion partner

TABLE 6.2 Exenedin-4 Versions with Extended Half-Life

Name	Company	Fusion	$t_{1/2}$ (h)	Organism	Phase
Lixisenatide (Lyxumia®)	Sanofi	Amino acid extension	4	Human	III
Exenatide/NexP™	Alteogen	Alpha 1 antitrypsin	13	Rat	Preclin.
VRS 859	Versatis	XTEN fusion	60	Cynomolgus	II
CJC-1134 PC	ConjuChem	Albumin conjugate	192	Human	II
LAPS-Exenedin	Hanmi	Fc fusion	n.a.	n.a.	II
PF-04603629	Pfizer	Transferrin fusion	n.a.	n.a.	I term

Exenedin-4 $t_{1/2}$ in human: 2 h.

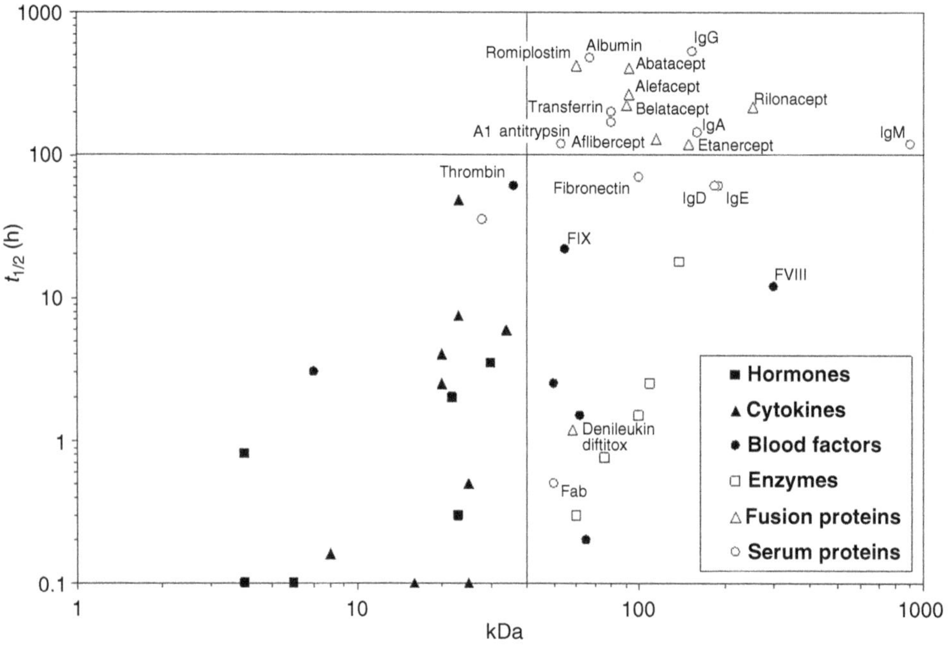

FIGURE 6.2 Molecular mass versus half-life plot of proteins.

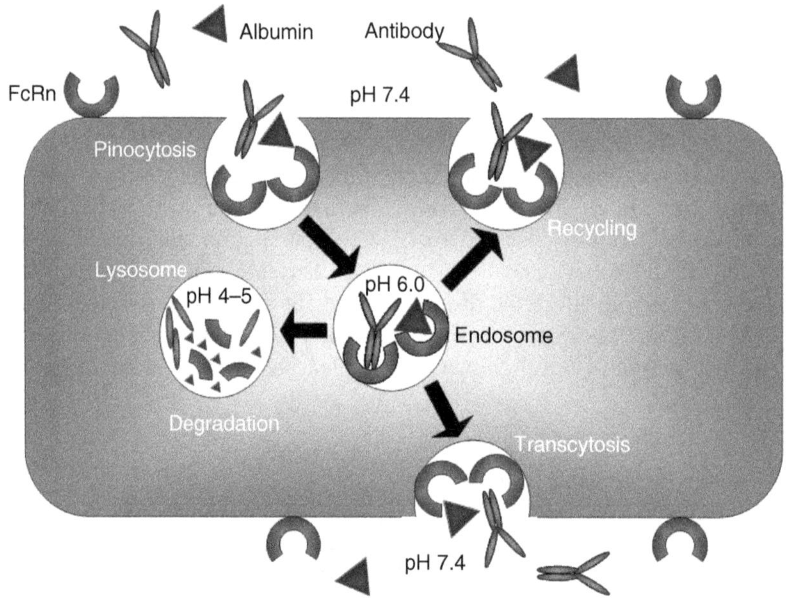

FIGURE 6.3 FcRn-mediated recycling of albumin and IgG. Antibodies or albumin are internalized by pinocytosis. At the lower pH in the endosome, the molecules associate with the FcRn. They are either transported back to the cell membrane, where they dissociate in a pH-dependent manner. Alternatively, transcytosis from the apical to the basolateral side is possible. In the case of saturated receptors, excess albumin and antibodies are degraded in the lysosome.

when blood clotting was required [17]. Interestingly, a similar pharmacodynamic (PD) effect was observed in an IFN-α2b and HSA fusion protein. The insertion of a cleavable linker increased the potency of the molecule. Basic amino acids in the spacers lead to a factor Xa susceptibility. The effect was even tunable by the number of inserted amino acids. Choosing linkers with intermediate release rate had the most balanced pharmacokinetic/-dynamic effect [18].

However, despite the documented success with albumin fusions, in some cases it was found that the bulky partner HSA reduced the activity. One of the examples for this effect is the fusion to soluble human tumor necrosis factor receptor. Independent of an N- or C-terminal fusion, both variants had drastically reduced activity [19]. Albumin fusions are discussed in Chapter 10.

6.2.1.2 Fc Fusions Looking at the currently marketed fusion proteins, it is obvious that Fc fusions dominate the field. The fragment crystallizable (Fc) region of an antibody consists of the second and third constant domain of the heavy chain (CH_2, CH_3), forming a dimer. In the hinge region at the N-terminal end, two disulfide bridges connect both chains. Enbrel®, the first ever approved fusion protein can be found in the group of Fc fusion proteins. This molecule belongs to the family of receptor traps that combine the extracellular domain of a receptor with the Fc part of an antibody [20]. Other members are Amevive® or Orencia®. A complete overview of approved Fc fusion proteins can be seen in Table 6.3. Comparing the first approved fusion protein Enbrel with its antibody competitors it shows that all antibodies have a better half-life than Enbrel. Nevertheless, the fusion protein is still quite successful on the market. Only the immunotoxin Ontak® has a relatively short half-life around 1 h. But this is not necessarily a disadvantage since Ontak is not used for chronic diseases. Chapter 9 deals exclusively with the topic of soluble receptors as traps.

The clinical success of Fc fusion proteins is primarily based on utilizing the high and specific affinity of extracellular receptor domains and the well-known principle of receptor-mediated antibody recycling through the Fc part. Antibodies and albumin share the FcRn, although they bind to different epitopes of this receptor. But the general process is identical; binding at acidic and release at neutral pH. The understanding of this phenomenon paved the way to combine the Fc part with other protein molecules. Of course it should not be neglected that already the addition of a 50-kDa moiety aids escaping fast kidney filtration.

The recycling through FcRn does not only protect Fc-containing molecules from lysosomal degradation, but also strips bound molecules from the antibody at low pH in the endosome, thus freeing the binding domain to become available again after release from the receptor.

The contribution of the FcRn-mediated recycling on half-life was evaluated in a study that correlated the affinity of antibodies and Fc-fusion proteins to the FcRn with their half-life. It was clearly demonstrated, that low affinity corresponds to short half-life and vice versa. To some respect the observed shorter half-lives for Fc-fusions could be attributed to a sterical or conformational effect of the molecules fused to the N-terminus. Some antibodies with exceptionally short half-life despite high affinity are probably eliminated by other antibodies or through internalizing cell-surface antigens [21]. Therefore, engineered Fc domains with higher affinity to FcRn, particularly at low pH should have longer half-lives. This was elegantly proven by optimizing Avastin® to have 11-fold higher affinity, resulting in a threefold longer half-life. An identical effect could be shown by modifying Erbitux® that addresses an internalizing antigen [22].

TABLE 6.3 Half-Life of Fusion Proteins Compared to Antibodies

Type	Drug	Brand	kDa	$t_{1/2}$ range	Average $t_{1/2}$ (h)
Fusion protein	Romiplostim	Nplate®	60	1–34 days	408
	Abatacept	Orencia®	92	8–25 days	400
	Alefacept	Amevive®	92	11 days	264
	Belatacept	Nulojix®	90	8–10 days	220
	Rilonacept	Arcalyst®	251	9 days	216
	Aflibercept	Eylea®	115	5–6 days	130
	Etanercept	Enbrel®	150	3–6 days	120
	Denileukin difitox	Ontak®	58	70–80 min	1.2
Antibody	Infliximab	Remicade®	150	9.5 days	228
	Adalimumab	Humira®	150	10–20 days	360
	Certolizumab pegol	Cimzia®	91	14 days	334
	Golimumab	Simponi®	150	14 days	334

Fc-fusion proteins are preferable for dimeric molecules since the interaction of the antibody heavy chain and the disulfide bride at the hinge region automatically forces the formation of homodimers. However, these large molecules might not always be favorable. Therefore, attempts have been undertaken to establish also Fc-fusion proteins with monomeric effector molecules. This means only one heavy chain is fused to the effector molecule, thus creating a heterodimer that is fully symmetric only for the Fc part. Often Fc-fusion proteins prepared for pulmonary delivery and using the FcRn expressed alveolar epithelial cells have a monomeric conformation. For instance, monomeric erythropoietin (EPO) Fc fusion had a twofold higher affinity to the FcRn and a 30% better half-life than the dimeric variant. Compared to the original EPO, monomeric EPO-Fc reached a much higher systemic concentration when administered at equimolar concentrations [23]. The improved distribution of the monomer is a result of smaller size, better transport, and less steric hindrance. The EPO-Fc dimer achieved 70% bioavailability in the central lung regions during a Phase I clinical study [24]. Recently it was shown with primates that an inhaled interferon-β (IFN-β) monomeric Fc fusion was fourfold more active than the corresponding dimer, but still 2.3-fold less potent than the original IFN-βs However, the presence of the Fc-part extended the half-life threefold [25]. Besides IFN-β, also interferon α (IFN-α) was successfully tested as monomeric Fc-fusion protein for pulmonary delivery [26]. Monomeric Fc-fusion proteins are discussed in detail in Chapter 7.

The attractiveness of monomeric Fc fusion has initiated the search for novel solutions. One approach takes advantage from the ability of antibodies to assemble their CH_3 domains asymmetrically. By alternating the sequences of IgA and IgG complementary Fc domains can be generated. They efficiently form heterodimers, while disfavoring the generation of homodimers. This process is much more efficient than other approaches to produce monomeric Fc-fusions, which suffer from low yields. These findings can establish the basis to generate on one hand bispecific antibodies or on the other hand monomeric Fc-fusions. This technique is called SEED (strand exchange engineered domain) platform and was used to generate a monomeric interleukin 2 (IL-2) Fc-fusion. This molecule exhibited extended serum half-life [27]. In addition, SEED was used to evaluate the ability to form bispecific antibodies [28]. More details can be found in Chapter 37.

Fc-based hybrid molecules can also be built symmetrically. One example is a hybrid Fc domain that combines the properties of IgG4 with those of IgD. This results in a highly flexible hinge region and the lack of secondary toxic functions while maintaining the long half-life of ~20 days. This technology was designed by Genexine in Korea (http://www.genexine.com/20070801_html_e/rnd/rnd_03.asp).

The contribution of recycling to half-life extension is a function of the affinity between the Fc part and FcRn. Therefore, it is worthwhile trying to optimize it. Mutagenesis toward higher affinity to FcRn at pH 6.0 while leaving the binding properties at pH 7.4 unchanged can lead to a 2.5-fold longer half-life in primates. This means less antibody is released from the FcRn during intracellular passage, thus avoiding lysosomal degradation of weakly bound antibody [29]. Mutagenesis at the Fc part always has to take other functionalities into account. This includes effector functions such as antibody-derived cellular cytotoxicity (ADCC) or complement-dependent toxicity (CDC) [30]. The best way to avoid any unwanted activities would be to switch to the Fc part of IgG4, which does not provoke CDC at all and has low ADCC stimulation ability. So far most of the mutagenesis studies were directed to improve ADCC and CDC [31]. However, the Fc region can be engineered for high affinity to FcRn independently of the pH, too. This then blocks the FcRn salvage function for all other antibodies or Fc-fusion proteins and causes much shorter half-life for Fc-containing proteins. This was described as an approach to lower IgG concentration in antibody-mediated diseases or to promote the rapid clearance of immunotoxins or antibody drug conjugates [32].

Despite the well-known behavior of Fc domains, the construction of fusion protein containing additional domains can negatively influence stability. Often conformational stability is pH dependent. In the case of Orencia, consisting of an IgG1 Fc part and the soluble part of T-cell receptor CTLA-4, the instability of the CH_2 domain can cause aggregation [33].

One group of therapeutic molecules that could particularly benefit from half-life extension is peptides. Peptides have important functions in the organism but because of their small size they are quickly removed from circulation by excretion. Their typical half-life is in the range of minutes, therefore, any improvement would have a huge impact. The trusted and proven platform of Fc fusion is certainly one of the first choices. And consequently a number of examples exist. One already approved molecule is Nplate® to treat chronic idiopathic (immune) thrombocytopenic purpura. It consists of thrombopoietin fused to Fc. This leads to an improved half-life of 3.5 days that makes chronic treatment feasible [34]. Another therapeutic peptide utilized in the so-called Mimetibody™ platform is angiopoietin, currently being in clinical Phase III. Beyond the relatively straightforward fusion to Fc further optimization that includes introduction of disulfide bridges or mutagenesis to reduce immunogenicity was performed. The whole Chapter 8 addresses Mimetibodies™ [35].

A bit larger than typical peptides bit still suffering from short circulation times are cytokines. They play a role in regulating immune response and inflammation but vanish from the blood stream within minutes. But not only size-dependent rapid clearance contributes to the short half-life but also degradation and receptor-mediated endocytosis. A wide range of different cytokines has been fused in a dimeric and monomeric form to Fc domains [36].

6.2.1.3 ***Transferrin Fusions*** Transferrin (Tf) is under the 10 most abundant serum proteins having the task of facilitating iron transport from intestine and liver into proliferating cells. Its size of 79 kDa alone prevents fast removal through kidney filtration. But in addition it is capable of receptor-mediated recycling as part of its normal functionality. The observed half-life of unglycosylated Tf is 14–17 days and 7–10 days for the glycosylated form, respectively. The recycling of Tf through the transferrin receptor (TfR) is pH independent. The acidic pH in the endosome dissociates iron form Tf, but not Tf from its receptor [37].

Tf fusion proteins that address a cell surface receptor will bind first to the receptor of the fusion partner in the presence of abundant Tf. Thereafter, a secondary binding can occur between Tf and its receptor. A secondary binding could also happen after endocytosis in the early endosome. If the fusion protein is now bound through the Tf part, it retains binding to the TfR under acidic conditions and can recycle back to the cell surface [38].

The Tf-mediated passage through the endosome can also be utilized to activate protein prodrugs by proteolysis. This was demonstrated with a proinsulin-Tf fusion that was processed to the active insulin-Tf fusion when incubated with hepatocytes [39]. Other peptides to treat diabetes such as GLP-1 were also fused to Tf, resulting in a more than 1000-fold improved half-life [40]. However, the GLP-1-Tf fusion was recently withdrawn from clinical studies, because it increased pulse and blood pressure [41].

Besides the positive effects on half-life of fusion proteins, Tf has another important feature. Iron is essential for neuronal function, but has to be kept at a specific equilibrium to avoid dangerous side effects. The homeostasis is regulated through iron transporters such as the TfR at the blood brain barrier (BBB) between blood and cerebrospinal. Therefore, Tf can be used as vehicle to transfer cargo across the BBB. Early examples comprised the fusion of nerve growth factor (NGF) to Tf [42]. Another approach focused on the treatment of brain tumors with Tf toxin fusion proteins [43]. The unique features of Tf have also been utilized in gene therapy. A genetic construct coding for a Tf α-L-iduronidase (ID) fusion to treat Hurler syndrome was injected into mice. The Tf-ID protein could be detected in the brain of the transfected animals and reduced cerebella glycosaminoglycans storage in MPS I mice [44]. The specific case of Tf fused to the acetylcholine receptor to treat myasthenia gravis is presented in Chapter 12.

The success with abundant serum proteins brought α1 antitrypsin (α1AT) into the focus as fusion partner to increase half-life. A Fab fragment targeting CD28 fused to α1AT extended blood circulation twofold [45]. This technology is patented and called NexP fusion (http://www.alteogen.com/).

6.2.2 Half-Life Extension Through Attachment to Large Proteins

The superior half-life of albumin has triggered a number of approaches to exploit this advantage for therapeutic concepts. Many researchers have designed HSA-binding molecules that can be fused to other proteins.

HSA-binding peptides were identified by phage display and reached up to 40% of the normal albumin half-life. This indicates that the noncovalent binding released the peptide from HSA during the incubation, which is impossible for covalent binding or fusion. This feature is quite useful for preparing slow release formulations of small therapeutic proteins. Since the peptides bound also the albumin from rodents albeit with a lower affinity, they could be used in animal studies without substitution [46]. The same peptide was later used in fusion to human growth hormone (hGH). Besides the typically prolonged half-life, this study also revealed, that the peptide at the N-terminus was more active than at the C-terminus [47]. Other combinations of this peptide included antibody fragments against different cancer cell epitopes. The resulting half-life of the molecules was dependent on the affinity of the peptide to albumin; the stronger the affinity, the longer the half-life [48]. Using antibody fragments fused to the HSA-binding peptide, a much better tumor penetration could be achieved compared to the full-length antibody. This effect is caused by the smaller size of the Fab on one hand, but also by the tumor accumulation of albumin on the other hand [49].

Besides the standard linear peptides also a wide range of affinity scaffolds has been selected to target albumin. The obvious first choice was an antibody Fab fragment binding to albumin. The half-life of a bispecific Fab with albumin binding capability was increased eightfold compared to another bispecific Fab without albumin specificity [50]. Another variation of this theme is a HSA-targeting domain antibody (AlbudAb). In the first example, fusing the interleukin-1 receptor antagonist (IL-1Ra) to AlbudAb, the half-life was extended 100-fold. This was lower than what could be achieved with the AlbudAb alone, and results probably from losses during recycling through the endosome [51]. Another example connecting the same AlbudAb with interferon α2b (IFN-α2b) prolonged the circulation time almost 20-fold [52].

Of course, other novel scaffolds such as Affibodies® have been screened for albumin binding abilities as well. The selected molecule consisting of 46 amino acids only, had a subnanomolar affinity to albumin and the complex between the Affibody® and albumin had a half-life of several days [53]. The same molecule could be further engineered at the side that does not bind to albumin to obtain specificity for tumor necrosis factor-α (TNF-α), thus generating a miniature bispecific molecule. This novel dual binder combines a high affinity to albumin and the resulting half-life with a nanomolar affinity against TNF-α, making it an interesting therapeutic molecule [54]. The minimal albumin binding domain (ABD) derived from streptococcal protein G does not interfere with the FcRn binding of HSA and extends the half-life of a HER2-affibody molecule to values similar to HSA alone [55].

Bispecific molecules incorporating HSA binding can also be designed on the basis of camelid heavy-chain-only antibodies, the so-called nanobodies. As expected also this approach resulted in prolonged circulation. Even more important was the improved tumor accumulation and penetration compared to a full-length antibody with the same specificity against epidermal growth factor receptor (EGFR) [56]. A summary of albumin binding peptides and proteins is given in Chapter 11.

The concept of albumin as carrier should also be transferable to the second most abundant molecule class in human serum, the immune globulins. The simplest case would be a bispecific molecule of the same type, binding to antibodies and having specificity for a second target. Such a diabody had a fivefold longer half-life and could also recruit complement and cellular toxicity through the Fc part of the antibody it bound to. A further advantage is the low manufacturing cost by expression in *Escherichia coli* [57].

Alternatively bacterial immunoglobulin binding proteins can be used for the same purpose as fusion partner. In an experimental study, the half-life prolonging abilities of *Staphylococcus* protein A (SpA), *Streptococcus* protein G (SpG), and *Peptostreptococcus* protein L (PpL) were compared when fused to a single-chain Fv fragment (scFv) or a related bispecific single-chain diabody (scDb). It was found that the C3 domain of SpG had the best effect, which was also supported by the lowest affinity to IgG [58].

A more extreme example combines the strengths of antibodies with the functionality of peptides. The so-called CovX bodies consist of a catalytic antibody with aldolase activity and a small pharmacophore. To this pharmacophore peptides can be chemically grafted. Mixing antibody and the peptide-pharmacophore molecule, creates a covalent bond between a specific lysine residue of the antibody and the azetidinone moiety of the pharmacophore, linking both molecules irreversibly together [59]. This concept is explained in more detail in Chapter 38.

6.3 HALF-LIFE EXTENSION THROUGH INCREASE OF HYDRODYNAMIC RADIUS

6.3.1 Repetitive Peptide Fusions

Neglecting the contribution of receptor-mediated recycling; half-life can be extended by solely increasing the hydrodynamic radius. Large molecules cannot be excreted by kidney filtration (Figure 6.4). Therefore, a number of fusion protein concepts have evolved over the last years utilizing this phenomenon. Most of them rely on the attachment of peptide (repeats) that are attached to one or both of the termini. To achieve the desired effect, the peptide should be hydrophilic, ideally uncharged to avoid aggregation, and without ordered structure. The absence of structure together with Brownian motion supports a cloud-like random conformation of the peptide chain that causes a huge hydrodynamic radius. The size of this "cloud" is only dependent on the number of amino acids if no other intramolecular forces are present.

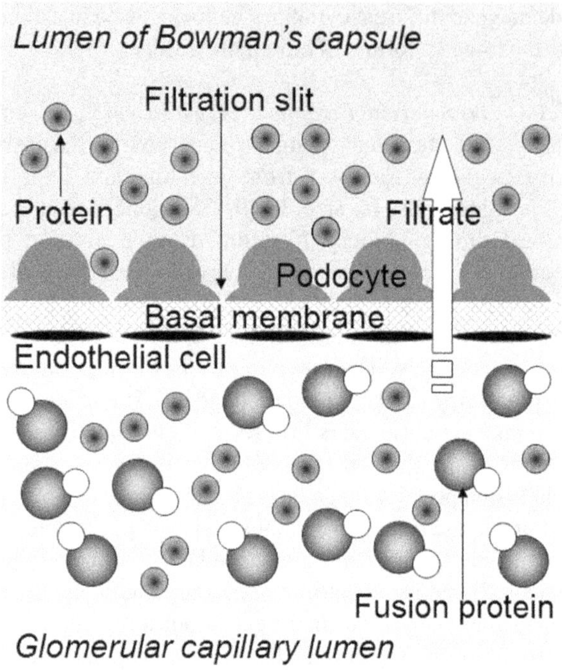

FIGURE 6.4 Kidney filtration. Molecules below a diameter of 60 Å passage from the glomerular lumen to the Bowman capsule through small pores in the endothelium. After crossing the negatively charged network of the basal membrane, they reach the filtration slit between the podocytes. Larger molecules such as some fusion proteins remain in the capillaries.

The phenomenon of a protein cloud can be observed even in nature. Some protozoan parasites seem to utilize this mechanism to prolong the half-life of virulence factors such as *trans*-sialidase. This enzyme is secreted and remains in circulation for up to 35 h because of the presence of a long hydrophilic repetitive carboxy terminal peptide. The same enzyme without this tail has a fivefold decreased half-life. This effect could also be repeated with heterologous proteins and a different natural sequence containing 13 repeats of EPSKA [60]. An engineered less immunogenic variant with the repeat sequence PSTAD still was able to extend the half-life approximately fivefold [61].

The success of natural peptide repeats to extend protein half-lives in plasma has stimulated research to engineer artificial sequences that could even be superior. Initially, single amino acid repeats utilizing polyglycine have been tried because an unstructured conformation could be expected from *in silico* simulations. However, long glycine polymers suffer from a low solubility in water, therefore the hydrophilic amino acid serine was introduced. The typical (glycine$_4$serine)$_n$ repeat has been used in many linker sequences between two domains before. This pentamer was multiplied either 20- or 40-fold to generate a 100 or 200 residue polymer that was fused to the C-terminus of a light chain in a Fab fragment. The Fab maintained its affinity to its target and showed a threefold increased half-life for the 200 residues polymer. With size exclusion chromatography the expansion of its apparent size by 120% could be demonstrated, although the real mass increase in comparison to the original Fab was only 29%. Larger polymers are difficult to generate, first due to some limitations of the genetic construct but mainly because of the tendency to aggregate at higher copy numbers. Nevertheless the threefold half-life increase could be sufficient for *in vivo* imaging [62].

In an attempt to overcome the obstacles of this polymer, it was tested if the addition of proline to the hydrophilic amino acid serine and a small aliphatic one like alanine instead of glycine could suppress the tendency to aggregate. The constant variation of the residues without long stretches of identical amino acids should also reduce the potential to form preferred helical or sheet-like conformations. Besides the higher degree of variability, through a third amino acid, proline also adds the ability for cis/trans isomerization that gives further options for conformational flexibility. Large so-called PAS-sequences result in a drastically increased hydrodynamic radius because of the lack of structural conformation, and should also increase *in vivo* half-life of proteins fused to this PAS-polymer. Interestingly, the PAS polymer does neither contain motifs allowing the cleavage by serum proteases, nor T-cell epitopes. PASylation®, as this technology is called, is currently tested with a number of therapeutic proteins (http://www.xl-protein.com/technology.html).

A similar approach to artificially engineer a peptide polymer with random conformation started with a bigger set of suitable amino acids (alanine, glycine, glutamate, proline, serine, and threonine), excluding larger, hydrophobic, positively charged, or sulfur-containing amino acids. These residues were systematically screened to find combinations that fulfilled the criteria of being unstructured and nonrepetitive. The selected candidate sequence called XTEN® contained 864 amino acids and extended the half-life of Exenatide in mice by a factor of 71. During size exclusion chromatography (SEC) proteins with the attached XTEN sequence eluted much earlier from the column than what could be expected by their nominal molecular mass, thus indicating a significantly increased hydrodynamic radius [63]. The *in vivo* half-life extension is probably also influenced by the high negative net-charge contributed by a high glutamate content that can cause charge repulsion at the glomeruli. This high negative charge lowers the isoelectrical point, pI, which could be beneficial for chromatographic purifications based on ion exchange, but could cause problems when addressing molecules on negatively charged cell surfaces. When testing a glucagon-XTEN fusion protein, it was found that the potency was only 15% of glucagon without XTEN. The reduction of potency was independent of the length of the XTEN polymer. But the polymer increased solubility 60-fold [64]. Table 6.4 summarizes all peptide-based half-life extension strategies.

6.3.2 Glycosylated Peptides

The hydrodynamic radius can also be increased by attaching carbohydrates. An early example is a hyperglycosylated

TABLE 6.4 Peptide Fusion Based Half-Life Extension Strategies

	Name	Amino Acid Composition	References/Websites
Large molecule binding	Albumin-binding peptide	RLIEDICLPRWGCLWEDD	[48]
Soluble peptide polymer	Antigen 13 repeats	$[\text{EPSKA}]_n$	[60]
	Deimmunized antigen 13	$[\text{PSTAD}]_n$	[61]
	Homo-amino-acid polymer	$[\text{G}_4\text{S}]_n$	[62]
	PASylation	P, A, S	http://www.xl-protein.com
	XTENylation	P, A, S, E, T, G	[63,64]
Glycosylation	C-terminal peptide	$[\text{SSSKAPPPSLPSPSRLPGPSDTPILPQ}]_n$	[66]
	Hydroxy proline glycosylation	$[\text{S/AHypS/A/Hyp}]_n$ or $[\text{SHyp4}]_n$	[69,70]
	Genetic polymers	G, N, Q and A, S, T, D, E	http://www.aequusbiopharma.com
Aggregate forming	Elastin-like peptide	$[\text{VPGX}^a\text{G}]_n$	http://www.phasebio.com
	Gelatin-like peptide	$[\text{GX}^b\text{Z}]_n$	[74]

Hyp, hydroxyproline; X^a, any amino acid except P; X^b, E, K, N, P, Q, or S; Z, X^b except S.

erythropoietin (Aranesp®) that had a threefold increased half-life compared to the wild-type molecule [65]. Besides the artificial introduction of glycosylation sites into the protein of interest, a similar effect can be achieved by adding a naturally occurring peptide sequence that contains multiple O-linked glycoslylation sites [66]. The molecular mass of the glycosylated peptide is 8 kDa instead of 2.9 kDa for the peptide without glycans. Although the natural peptide is derived from the carboxyl terminal domain (CTP) of chorionic gonadotropin (CG); it can also be positioned at the amino-terminus. Even larger molecules can be designed that contain repeats of this peptide. Since its first description CTP has been used with several therapeutic proteins. For instance, hGH was tested in a number of configurations with CTP. The best variant of hGH with CTP at both termini was fourfold more potent than the hGH alone and showed a drastically more than 10-fold prolonged half-life [67]. EPO equipped with three copies of CTP had eightfold improved potency and threefold extended half-life. Interestingly, it was even superior to the hyperglycosylated EPO variant Aranesp® [68]. The CTP concept is explained in more detail in Chapter 13.

A more massive glycosylation can be obtained using the hydroxyproline (Hyp)-Glyco technology. Here the recombinant protein is expressed as fusion to the Hyp-rich repetitive peptide (HypRP) tag that directs the formation of proline hydroxylation, and subsequent Hyp-O-glycosylation. This is a plant-specific posttranslational modification. Typically tandem repeats of Ser-Hyp or Ala-Hyp are used since they lead to highly branched heteropolysaccharides with 15–150 residues, similar to the glycosylation that can be found in many plant gums. When testing this method with IFN-α2b, first the secretion from plant cells was drastically improved up to 1500-fold and second the *in vivo* half-life in mice was extended 13-fold [69]. Similar results were obtained with hGH as fusion partner. The 10 Ser-Hyp repeats at the C-terminus of hGH with a corresponding glycosylation of 25 residues in average increased the circulation time more than 6-fold and the secretion from plant cells 500-fold [70]. The downside of this approach is the need to express in plant cells, which is not that widely established and the risk of heterogeneous glycosylation, which might be a regulatory hurdle.

A more versatile glycosylation that works in any eukaryotic organism can be obtained using Genetic Polymers™ that contain the standard N-linked glycosylation motif Asn-Xaa-Ser/Thr with Xaa being any amino acid except proline. The rest of the peptide repeats consists primarily of G, N, Q with minor substitutions of A, S, T, D, E. The repeats can be amino acid triplets up to sextets. Multiple (2–500) peptide repeats together with their corresponding additional glycosylation will obviously increase the hydrodynamic radius of any protein molecule fused to it, thus extending the *in vivo* half-life. Testing this technology with 155 triplet repeats of NNT in conjunction to the carboxy terminus of G-CSF improved the circulation half-life in mice fourfold (http://www.aequusbiopharma.com). It has to be taken into account, that excessive N-linked glycosylation can give rise to high mannose content that might be a target for mannose receptors to eliminate proteins relatively fast from the blood stream.

6.4 AGGREGATE FORMING PEPTIDE FUSIONS

Other naturally occurring peptide repeat sequences can be found in gelatine and elastin. Therefore, it was quite logical to test these proteins also for their ability to extend half-life. The examples using elastin-like peptides (ELP) are quite numerous, and initially this sequence has primarily been used as tool for downstream processing because of its unusual ability to form aggregates at high temperatures and becoming a soluble monomer at lower temperatures. This phenomenon is called reversible thermal phase transition and the critical temperature for the phase transition is dependent on the length of the polymer. The longer the polymer, the higher is the transition temperature [71]. The pentamer repeat motif is VPGXG and typically 90–120 repeats are used. With the exception of proline any amino acid can be incorporated at the X position. Currently, a vasoactive peptide and GLP-1 are in clinical trials (http://www.phasebio.com). Both molecules form depots in the body from which active molecules are slowly dissociating. Therefore, the mechanism that leads to half-life extension is very different to other peptide repeats that increase the hydrodynamic radius. Unfortunately, in the case of interleukin-1 receptor antagonist (IL-1Ra) ELP fusion drastically reduced the bioactivity more than 500-fold [72]. This could be an obstacle for therapeutic applications. But recently an interesting approach combining ELP with cell penetrating peptides (CPP) and hyperthermia was discussed. Here the tumor is artificially heated up above 40°C. At this temperature, ELP can aggregate, thus achieving a high local concentration of therapeutic agents. The combination of CPP, ELP, and anticancer peptides in a single genetic fusion can have a synergistic therapeutic effect [73]. A detailed description of the ELP technology can be found in Chapter 14.

The other natural peptide repeat gelatine has only recently been used as fusion partner for G-CSF. Repeats of the amino acids G, P, E, Q, N, S, and K resulted in hydrophilic 116 residue long gelatine-like protein (GLK). This fusion protein had a more than fivefold improved half-life. Owing to its hydrophilic nature no, aggregates could be found. Interestingly, the mechanism for half-life extension relies not necessarily only on the increase of the hydrodynamic radius, but rather on the additive effect of shielding G-CSF against proteolysis and electrostatic repulsion due to its negative charge [74].

6.5 OTHER CONCEPTS

Totally different approaches were used to improve the half-life of hormones and cytokines. The first example is based on the observation, that growth hormone binds to its receptor even when only the extracellular domain is circulating in plasma. This molecule pair can be generated artificially by connecting the growth hormone to the extracellular receptor domain by a peptide linker. This head-to-tail dimer had a 300-fold improved half-life compared to growth hormone alone. In contrast to other fusion approaches, the hormone retained full activity and stimulated growth for 10 days [75]. Recently, the absence of immunological risk has been demonstrated in animal models [76]. A full description of this concept can be found in Chapter 15.

In a second example, interferon-β (IFN-β) was fused to the latency-associated peptide (LAP) of transforming growth factor-β1 (TGF-β1). On one hand this improved the pharmacokinetics and on the other hand minimized the pleiotropic effects of IFN-β. To enable proper release from LAP at metallo-protease rich inflammation sites, a protease sensitive linker was introduced between both molecules. Besides the impact of the sequence at the cleavage site the orientation of the LAP played a major role. Only LAP at the amino terminus and IFN-β at the carboxy end prevented cleavage-independent activity of IFN-β. The half-life for IFN-β was extended to 55 h [77]. Recently, other small anti-inflammatory peptides that would be cleared very fast were successfully tested [78]. Further examples for this approach are collected in Chapter 16.

6.6 CONCLUSIONS AND FUTURE PERSPECTIVE

In the context of fusion proteins currently the most frequently used technique for half-life extension is certainly Fc fusion. This is emphasized by the fact that with the exception of Ontak only Fc-fusion proteins are approved today. Furthermore, a large number of this protein class can be found in clinical trials. But nevertheless the dominance might soon be broken by alternative technologies that have reached the clinic as well (Figure 6.5). Some as in the case of HSA-fusions rely on a similar mechanism that benefits from FcRn recycling; others focus on the expansion of the hydrodynamic radius to avoid fast kidney clearance. The choice of the appropriate system will be influenced by general parameters such as freedom to orient the fusion partner, immunogenicity, production costs, or the homogeneity of the material. Nonfusion-based half-life extension technologies such as PEGylation suffer from safety concerns due to lacking biodegradability, the generation of neutralizing antibodies, lower potency of the drug, and higher manufacturing costs because of a non-continuous process. However, currently several PEGylated products are on the market.

The limitations of PEGylation will trigger more initiatives with fusion proteins. Novel approaches will enable a fine tuning of half-life in combination with an adaptable size to optimize tissue penetration. However, broad acceptance of novel strategies will only be achievable, if the respective technology has proven efficacious and safe in clinical trials. From an economic point of view, novel concepts will only be successful if both manufacturing and licensing cost are advantageous. The availability of a wide range of new strategies will also break the dominance of some blockbuster

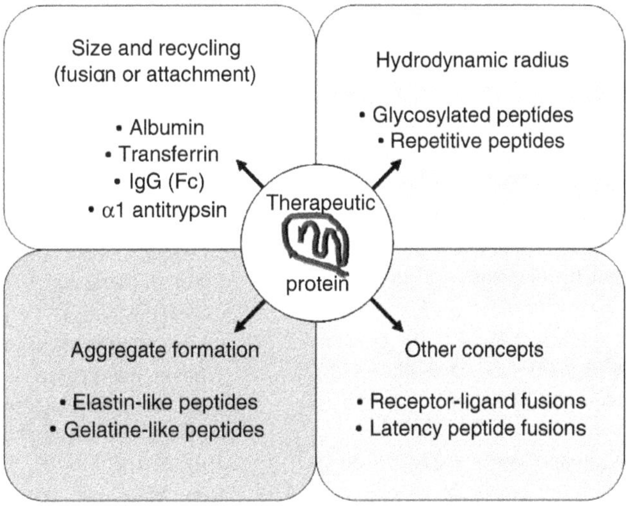

FIGURE 6.5 Fusion protein based approaches for half-life extension.

proteins, by creating second-generation proteins with improved characteristics. This will on one hand enable life-cycle management for some companies with established drugs, but on the other hand enable new market entrants to place novel variants of a proven product.

REFERENCES

1. Tryggvason K, Wartiovaara J. (2005) How does the kidney filter plasma? *Physiology (Bethesda)* 20, 96–101.
2. Cuesta AM, Sainz-Pastor N, Bonet J, Oliva B, Alvarez-Vallina L. (2010) Multivalent antibodies: when design surpasses evolution. *Trends Biotechnol.* 28(7), 355–362.
3. Beckman RA, Weiner LM, Davis HM. (2007) Antibody constructs in cancer therapy: protein engineering strategies to improve exposure in solid tumors. *Cancer* 109(2), 170–179.
4. Sarkar CA, Lowenhaupt K, Horan T, Boone TC, Tidor B, Lauffenburger DA. (2002) Rational cytokine design for increased lifetime and enhanced potency using pH-activated "histidine switching". *Nat. Biotechnol.* 20(9), 908–913.
5. Woodnutt G, Violand B, North M. (2008) Advances in protein therapeutics. *Curr. Opin. Drug Discov. Deve.* 11(6), 754–761.
6. Kontermann RE. (2011) Strategies for extended serum half-life of protein therapeutics. *Curr. Opin. Biotechnol.* 22(6), 868–876.
7. Peters TJr., (1985) Serum albumin. *Adv. Protein Chem.* 37, 161–245.
8. Andersen JT, Dalhus B, Cameron J, Daba MB, Plumridge A, Evans L, et al. (2012) Structure-based mutagenesis reveals the albumin-binding site of the neonatal Fc receptor. *Nat. Commun.* 3, 610.
9. Andersen JT, Sandlie I. (2009) The versatile MHC class I-related FcRn protects IgG and albumin from degradation: implications for development of new diagnostics and therapeutics. *Drug Metab. Pharmacokinet.* 24(4), 318–332.
10. Wunder A, Stehle G, Sinn H, Schrenk H, Hoffbiederbeck D, Bader F, et al. (1997) Enhanced albumin uptake by rat tumors. *Int. J. Oncol.* 11(3), 497–507.
11. Cordes AA, Carpenter JF, Randolph TW. (2012) Selective domain stabilization as a strategy to reduce human serum albumin-human granulocyte colony stimulating factor aggregation rate. *J. Pharm. Sci.* 101(6), 2009–2016.
12. Evans L, Hughes M, Waters J, Cameron J, Dodsworth N, Tooth D, et al. (2010) The production, characterisation and enhanced pharmacokinetics of scFv-albumin fusions expressed in *Saccharomyces cerevisiae. Protein Expr. Purif.* 73(2), 113–124.
13. Zhao HL, Xue C, Wang Y, Li XY, Xiong XH, Yao XQ, et al. (2007) Circumventing the heterogeneity and instability of human serum albumin-interferon-alpha2b fusion protein by altering its orientation. *J. Biotechnol.* 131(3), 245–252.
14. Subramanian GM, Fiscella M, Lamousé-Smith A, Zeuzem S, McHutchison JG. (2007) Albinterferon alpha-2b: a genetic fusion protein for the treatment of chronic hepatitis C. *Nat. Biotechnol.* 25(12), 1411–1419.
15. Duttaroy A, Kanakaraj P, Osborn BL, Schneider H, Pickeral OK, Chen C, et al. (2005) Development of a long-acting insulin analog using albumin fusion technology. *Diabetes* 54(1), 251–258.
16. Weimer T, Wormsbächer W, Kronthaler U, Lang W, Liebing U, Schulte S. (2008) Prolonged *in-vivo* half-life of factor VIIa by fusion to albumin. *Thromb. Haemost.* 99(4), 659–667.
17. Schulte S. (2009) Half-life extension through albumin fusion technologies. *Thromb. Res.* 124(Suppl 2), S6–8.
18. Zhao HL, Xue C, Du JL, Ren M, Xia S, Liu ZM. (2012) Balancing the pharmacokinetics and pharmacodynamics of interferon-α2b and human serum albumin fusion protein by proteolytic or reductive cleavage increases its *in vivo* therapeutic efficacy. *Mol. Pharm.* 9(3), 664–670.
19. Zhan J, Chen Y, Yuan H-Y, Li H, Lu H. (2012) Fusion of HSA influences TNF-α neutralizing activity of shTNFRs. *Biotechnol. Lett.* 34(3), 417–423.
20. Economides AN, Carpenter LR, Rudge JS, Wong V, Koehler-Stec EM, Hartnett C, et al. (2003) Cytokine traps: multi-component, high-affinity blockers of cytokine action. *Nat. Med.* 9(1), 47–52.
21. Suzuki T, Ishii-Watabe A, Tada M, Kobayashi T, Kanayasu-Toyoda T, Kawanishi T, et al. (2010) Importance of neonatal FcR in regulating the serum half-life of therapeutic proteins containing the Fc domain of human IgG1: a comparative study of the affinity of monoclonal antibodies and Fc-fusion proteins to human neonatal FcR. *J. Immunol.* 184(4), 1968–1976.
22. Zalevsky J, Chamberlain AK, Horton HM, Karki S, Leung IWL, Sproule TJ, et al. (2010) Enhanced antibody half-life improves *in vivo* activity. *Nat. Biotechnol.* 28(2), 157–159.
23. Bitonti AJ, Dumont JA, Low SC, Peters RT, Kropp KE, Palombella VJ, et al. (2004) Pulmonary delivery of an erythropoietin Fc fusion protein in non-human primates through an immunoglobulin transport pathway. *Proc. Natl. Acad. Sci. USA* 101(26), 9763–9768.
24. Dumont JA, Bitonti AJ, Clark D, Evans S, Pickford M, Newman SP. (2005) Delivery of an erythropoietin-Fc fusion protein by inhalation in humans through an immunoglobulin transport pathway. *J. Aerosol Med.* 18(3), 294–303.
25. Vallee S, Rakhe S, Reidy T, Walker S, Lu Q, Sakorafas P, et al. (2012) Pulmonary administration of interferon Beta-1a-fc fusion protein in non-human primates using an immunoglobulin transport pathway. *J. Interferon Cytokine Res.* 32(4), 178–184.
26. Dumont JA, Low SC, Peters RT, Bitonti AJ. (2006) Monomeric Fc fusions: impact on pharmacokinetic and biological activity of protein therapeutics. *BioDrugs* 20(3), 151–160.
27. Davis JH, Aperlo C, Li Y, Kurosawa E, Lan Y, Lo K-M, et al. (2010) SEEDbodies: fusion proteins based on strand-exchange engineered domain (SEED) CH3 heterodimers in an Fc analogue platform for asymmetric binders or immunofusions and bispecific antibodies. *Protein Eng. Des. Sel.* 23(4), 195–202.
28. Muda M, Gross AW, Dawson JP, He C, Kurosawa E, Schweickhardt R, et al. (2011) Therapeutic assessment of SEED: a new engineered antibody platform designed to

generate mono- and bispecific antibodies. *Protein Eng. Des. Sel.* 24(5), 447–454.

29. Hinton PR, Xiong JM, Johlfs MG, Tang MT, Keller S, Tsurushita N. (2006) An engineered human IgG1 antibody with longer serum half-life. *J. Immunol.* 176(1), 346–356.
30. Correale P, Cusi MG, Tagliaferri P. (2011) Immunomodulatory properties of anticancer monoclonal antibodies: is the "magic bullet" still a reliable paradigm? *Immunotherapy* 3(1), 1–4.
31. Desjarlais JR, Lazar GA. (2011) Modulation of antibody effector function. *Exp. Cell Res.* 317(9), 1278–1285.
32. Vaccaro C, Zhou J, Ober RJ, Ward ES. (2005) Engineering the Fc region of immunoglobulin G to modulate *in vivo* antibody levels. *Nat. Biotechnol.* 23(10), 1283–1288.
33. Fast JL, Cordes AA, Carpenter JF, Randolph TW. (2009) Physical instability of a therapeutic Fc fusion protein: domain contributions to conformational and colloidal stability. *Biochemistry* 48(49), 11724–11736.
34. Molineux G, Newland A. (2010) Development of romiplostim for the treatment of patients with chronic immune thrombocytopenia: from bench to bedside. *Br. J. Haematol.* 150(1), 9–20.
35. Huang C. (2009) Receptor-Fc fusion therapeutics, traps, and MIMETIBODY technology. *Curr. Opin. Biotechnol.* 20(6), 692–699.
36. Jazayeri JA, Carroll GJ. (2008) Fc-based cytokines: prospects for engineering superior therapeutics. *BioDrugs* 22(1), 11–26.
37. Mayle KM, Le AM, Kamei DT. (2012) The intracellular trafficking pathway of transferrin. *Biochim. Biophys. Acta* 1820(3), 264–281.
38. Chen X, Lee H-F, Zaro JL, Shen W-C. (2011) Effects of receptor binding on plasma half-life of bifunctional transferrin fusion proteins. *MolPharm* 8(2), 457–465.
39. Wang Y, Chen Y-S, Zaro JL, Shen W-C. (2011) Receptor-mediated activation of a proinsulin-transferrin fusion protein in hepatoma cells. *J. Control. Release* 155(3), 386–392.
40. Kim B-J, Zhou J, Martin B, Carlson OD, Maudsley S, Greig NH, et al. (2010) Transferrin fusion technology: a novel approach to prolonging biological half-life of insulinotropic peptides. *J. Pharmacol. Exp. Ther* 334(3), 682–692.
41. Gustavson SM, Chen D, Somayaji V, Hudson K, Baltrukonis DJ, Singh J, et al. (2011) Effects of a long-acting GLP-1 mimetic (PF-04603629) on pulse rate and diastolic blood pressure in patients with type 2 diabetes mellitus. *Diabetes Obes. Metab.* 13(11), 1056–1058.
42. Park E, Starzyk RM, McGrath JP, Lee T, George J, Schutz AJ, et al. (1998) Production and characterization of fusion proteins containing transferrin and nerve growth factor. *J. Drug Target.* 6(1), 53–64.
43. Debinski W. (2002) Local treatment of brain tumors with targeted chimera cytotoxic proteins. *Cancer Invest.* 20(5–6), 801–809.
44. Osborn MJ, McElmurry RT, Peacock B, Tolar J, Blazar BR. (2008) Targeting of the CNS in MPS-IH using a nonviral transferrin-alpha-L-iduronidase fusion gene product. *Mol. Ther.* 16(8), 1459–1466.
45. Vanhove B, Laflamme G, Coulon F, Mougin M, Vusio P, Haspot F, et al. (2003) Selective blockade of CD28 and not CTLA-4 with a single-chain Fv-alpha1-antitrypsin fusion antibody. *Blood* 102(2), 564–570.
46. Dennis MS, Zhang M, Meng YG, Kadkhodayan M, Kirchhofer D, Combs D, et al. (2002) Albumin binding as a general strategy for improving the pharmacokinetics of proteins. *J. Biol. Chem.* 277(38), 35035–35043.
47. Langenheim JF, Chen WY. (2009) Improving the pharmacokinetics/pharmacodynamics of prolactin, GH, and their antagonists by fusion to a synthetic albumin-binding peptide. *J. Endocrinol.* 203(3), 375–387.
48. Nguyen A, Reyes AE2nd, Zhang M, McDonald P, Wong WLT, Damico LA, et al. (2006) The pharmacokinetics of an albumin-binding Fab (AB.Fab) can be modulated as a function of affinity for albumin. *Protein Eng. Des. Sel.* 19(7), 291–297.
49. Dennis MS, Jin H, Dugger D, Yang R, McFarland L, Ogasawara A, et al. (2007) Imaging tumors with an albumin-binding Fab, a novel tumor-targeting agent. *Cancer Res.* 67(1), 254–261.
50. Smith BJ, Popplewell A, Athwal D, Chapman AP, Heywood S, West SM, et al. (2001) Prolonged *in vivo* residence times of antibody fragments associated with albumin. *Bioconjug. Chem.* 12(5), 750–756.
51. Holt LJ, Basran A, Jones K, Chorlton J, Jespers LS, Brewis ND, et al. (2008) Anti-serum albumin domain antibodies for extending the half-lives of short lived drugs. *Protein Eng. Des. Sel.* 21(5), 283–288.
52. Walker A, Dunlevy G, Rycroft D, Topley P, Holt LJ, Herbert T, et al. (2010) Anti-serum albumin domain antibodies in the development of highly potent, efficacious and long-acting interferon. *Protein Eng. Des. Sel.* 23(4), 271–278.
53. Nilvebrant J, Alm T, Hober S, Löfblom J. (2011) Engineering bispecificity into a single albumin-binding domain. *PLoS ONE* 6(10), e25791.
54. Jonsson A, Dogan J, Herne N, Abrahmsén L, Nygren P-A. (2008) Engineering of a femtomolar affinity binding protein to human serum albumin. *Protein Eng. Des. Sel.* 21(8), 515–527.
55. Andersen JT, Pehrson R, Tolmachev V, Daba MB, Abrahmsén L, Ekblad C. (2011) Extending half-life by indirect targeting of the neonatal Fc receptor (FcRn) using a minimal albumin binding domain. *J. Biol. Chem.* 286(7), 5234–5241.
56. Tijink BM, Laeremans T, Budde M, Stigter-van Walsum M, Dreier T, de Haard HJ, et al. (2008) Improved tumor targeting of anti-epidermal growth factor receptor Nanobodies through albumin binding: taking advantage of modular Nanobody technology. *Mol. Cancer Ther.* 7(8), 2288–2297.
57. Holliger P, Wing M, Pound JD, Bohlen H, Winter G. (1997) Retargeting serum immunoglobulin with bispecific diabodies. *Nat. Biotechnol.* 15(7), 632–636.
58. Hutt M, Färber-Schwarz A, Unverdorben F, Richter F, Kontermann RE. (2012) Plasma half-life extension of small recombinant antibodies by fusion to immunoglobulin-binding domains. *J. Biol. Chem.* 287(7), 4462–4469.

59. Doppalapudi VR, Huang J, Liu D, Jin P, Liu B, Li L, et al. (2010) Chemical generation of bispecific antibodies. *Proc. Natl. Acad. Sci. USA* 107(52), 22611–22616.
60. Buscaglia CA, Alfonso J, Campetella O, Frasch AC. (1999) Tandem amino acid repeats from Trypanosoma cruzi shed antigens increase the half-life of proteins in blood. *Blood* 93 (6), 2025–2032.
61. Alvarez P, Buscaglia CA, Campetella O. (2004) Improving protein pharmacokinetics by genetic fusion to simple amino acid sequences. *J. Biol. Chem.* 279(5), 3375–3381.
62. Schlapschy M, Theobald I, Mack H, Schottelius M, Wester H-J, Skerra A. (2007) Fusion of a recombinant antibody fragment with a homo-amino-acid polymer: effects on biophysical properties and prolonged plasma half-life. *Protein Eng. Des. Sel.* 20(6), 273–284.
63. Schellenberger V, Wang C-W, Geething NC, Spink BJ, Campbell A, To W, et al. (2009) A recombinant polypeptide extends the in vivo half-life of peptides and proteins in a tunable manner. *Nat. Biotechnol* 27(12), 1186–1190.
64. Geething NC, To W, Spink BJ, Scholle MD, Wang C, Yin Y, et al. (2010) Gcg-XTEN: an improved glucagon capable of preventing hypoglycemia without increasing baseline blood glucose. *PLoS ONE* 5(4), e10175.
65. Egrie JC, Browne JK. (2001) Development and characterization of novel erythropoiesis stimulating protein (NESP). *Br. J. Cancer* 84(Suppl. 1), 3–10.
66. Fares FA, Suganuma N, Nishimori K, LaPolt PS, Hsueh AJ, Boime I. (1992) Design of a long-acting follitropin agonist by fusing the C-terminal sequence of the chorionic gonadotropin beta subunit to the follitropin beta subunit. *Proc. Natl. Acad. Sci. USA* 89(10), 4304–4308.
67. Fares F, Guy R, Bar-Ilan A, Felikman Y, Fima E. (2010) Designing a long-acting human growth hormone (hGH) by fusing the carboxyl-terminal peptide of human chorionic gonadotropin beta-subunit to the coding sequence of hGH. *Endocrinology* 151(9), 4410–4417.
68. Fares F, Havron A, Fima E. (2011) Designing a long acting erythropoietin by fusing three carboxyl-terminal peptides of human chorionic gonadotropin β subunit to the N-terminal and C-terminal coding sequence. *Int. J. Cell Biol.* 2011, 275063.
69. Xu J, Tan L, Goodrum KJ, Kieliszewski MJ. (2007) High-yields and extended serum half-life of human interferon alpha2b expressed in tobacco cells as arabinogalactan-protein fusions. *Biotechnol. Bioeng.* 97(5), 997–1008.
70. Xu J, Okada S, Tan L, Goodrum KJ, Kopchick JJ, Kieliszewski MJ. (2010) Human growth hormone expressed in tobacco cells as an arabinogalactan-protein fusion glycoprotein has a prolonged serum life. *Transgenic Res.* 19(5), 849–867.
71. Betre H, Liu W, Zalutsky MR, Chilkoti A, Kraus VB, Setton LA. (2006) A thermally responsive biopolymer for intra-articular drug delivery. *J. Control. Release* 115(2), 175–182.
72. Shamji MF, Betre H, Kraus VB, Chen J, Chilkoti A, Pichika R, et al. (2007) Development and characterization of a fusion protein between thermally responsive elastin-like polypeptide and interleukin-1 receptor antagonist: sustained release of a local antiinflammatory therapeutic. *Arthritis Rheum.* 56(11), 3650–3661.
73. Bidwell GL3rd, Raucher D. (2010) Cell penetrating elastin-like polypeptides for therapeutic peptide delivery. *Adv. Drug Deliv. Rev.* 62(15), 1486–1496.
74. Huang Y-S, Wen X-F, Wu Y-L, Wang Y-F, Fan M, Yang Z-Y, et al. (2010) Engineering a pharmacologically superior form of granulocyte-colony-stimulating factor by fusion with gelatin-like-protein polymer. *Eur. J. Pharm. Biopharm.* 74(3), 435–441.
75. Wilkinson IR, Ferrandis E, Artymiuk PJ, Teillot M, Soulard C, Touvay C, et al. (2007) A ligand-receptor fusion of growth hormone forms a dimer and is a potent long-acting agonist. *Nat. Med.* 13(9), 1108–1113.
76. Ferrandis E, Pradhananga SL, Touvay C, Kinoshita C, Wilkinson IR, Stafford K, et al. (2010) Immunogenicity, toxicology, pharmacokinetics and pharmacodynamics of growth hormone ligand-receptor fusions. *Clin. Sci.* 119(11), 483–491.
77. Adams G, Vessillier S, Dreja H, Chernajovsky Y. (2003) Targeting cytokines to inflammation sites. *Nat. Biotechnol.* 21(11), 1314–1320.
78. Vessillier S, Adams G, Montero-Melendez T, Jones R, Seed M, Perretti M, et al. (2012) Molecular engineering of short half-life small peptides (VIP, αMSH and γ_3MSH) fused to latency-associated peptide results in improved anti-inflammatory therapeutics. *Ann. Rheum. Dis.* 71(1), 143–149.

7

MONOMERIC Fc-FUSION PROTEINS

BAISONG MEI, SUSAN C. LOW,[†] SNEJANA KRASSOVA, ROBERT T. PETERS, GLENN F. PIERCE, AND JENNIFER A. DUMONT

Biogen Idec Hemophilia, Weston, MA, USA

7.1 INTRODUCTION

Immunoglobulin G (IgG) is one of the most abundant proteins in the blood, present at ~10–12 mg/mL in the circulation. Moreover, IgG has a uniquely long half-life of ~21 days in humans [1,2]. The persistence of IgG in the circulation is attributed to the neonatal Fc receptor (FcRn), which salvages IgG from intracellular lysosomal degradation and recycles it from the sorting endosome back to the cell surface and back into the circulation.

FcRn is a heterodimer comprising a transmembrane heavy chain with structural homology to the extracellular domain of the major histocompatibility complex (MHC) class I molecules and a soluble light chain consisting of β2-microglobulin (β2 m) [3]. It was originally identified as the receptor that transports IgG from mother's milk to the neonate in rodents during the suckling period and was later found to also mediate IgG homeostasis [4,5]. Evidence that FcRn was responsible for IgG homeostasis came from mice that had genetically deleted β2 m, which resulted in low serum IgG levels [6,7].

[†]Present address: Bind Biosciences, Cambridge, MA, USA

Further proof came when it was observed that short serum IgG half-life, along with low serum IgG levels occurred in mice deficient in the FcRn heavy chain [8]. The only report of a defect in FcRn in humans is for two family members with familial hypercatabolic hypoproteinaemia [9]. The clinical manifestation was that of low serum IgG levels and the genetic defect was a single nucleotide transversion in β2 m [10]. The half-life of IgG was ~3 days in these siblings and although IgG synthesis was determined to be normal, they each showed hypercatabolism of IgG. Therefore, the critical role of FcRn in IgG homeostasis has been demonstrated in what is essentially a naturally occurring knockout for FcRn.

It is now known that FcRn plays various roles and is present in many cell types and tissues [11–18]. The regions of IgG that are critical for the control of IgG catabolism are the CH2–CH3 domains of the Fc fragment [19]. FcRn has been leveraged to prolong the half-life of therapeutic peptides and proteins through fusion of the Fc domain (hinge-CH2–CH3) of IgG with the therapeutic molecule of interest. Because of the disulfide bond formation in the hinge region of Fc, the traditional Fc fusion molecules were dimeric with respect to both the Fc domain and also the therapeutic moiety. Many effector molecules have been coupled to Fc with the aim to extend half-life [20,21]. For example, soluble receptors for cytokines, such as tumor necrosis factor (TNF) and vascular endothelial growth factor (VEGF), have been designed to function as inhibitors of their cognate cytokines as an approach to regulate excessive levels of cytokines. However, these soluble cytokine receptors have a relatively short half-life and are cleared from the body rapidly. As an approach to extend the half-life of these soluble receptors, they were linked to the Fc domain of IgG1. Although fusion

Fusion Protein Technologies for Biopharmaceuticals: Applications and Challenges, First Edition. Edited by Stefan R. Schmidt.

with Fc resulted in marked improvement of the half-life for several therapeutic proteins, it was hypothesized and confirmed that the Fc construct could be improved to further enhance the biological properties of the therapeutic moiety. This improved construct is referred to as a monomeric Fc fusion molecule and is comprised of a single therapeutic protein coupled to dimeric Fc. The applications and challenges of this novel Fc fusion configuration are discussed in the current review.

7.2 FcRn AND MONOMERIC Fc-FUSION PROTEINS

7.2.1 FcRn Expression and Function

Passive immunity is accomplished through the uptake of maternal IgG from ingested colostrum and milk across the epithelial cell layer of the small intestine via FcRn in neonatal rodents during the suckling period. Expression of the receptor is high during the first few weeks of life, but at the time of weaning, it is markedly reduced (~1000-fold) in rodents [22]. FcRn transport of IgG also takes place in the yolk sac of rodents during gestation [23]. Therefore, the importance of FcRn is clear during both the gestational and neonatal time periods in rodents. In contrast, passive immunity in humans occurs through the transport of IgG across the syncytiotrophoblast of the placenta where FcRn mediates the maternal transfer of IgG to the fetus during gestation [17]. Over the years, it has been demonstrated that FcRn expression is not limited to the gestational and neonatal periods in humans, but rather is found in many cell and tissue types throughout life.

The major site of IgG homeostasis is thought to be in the vascular endothelium, primarily in skin and muscle, as well as in liver and adipose tissue [24]. These sites provide a large contact area for FcRn with the blood. Elegant experiments in mice in which FcRn was conditionally deleted in endothelial and hematopoietic cells demonstrated that these tissues are indeed the primary sites for the regulation of IgG homeostasis [25].

FcRn expression remains relatively high throughout life in the epithelial cells lining the intestine, as well as in the lung [12,14,18]. Moreover, the human receptor is functional in adult life in the intestinal epithelium when studied in a humanized mouse model [26]. Further examination of expression in the lung showed that FcRn was most prominent in the central airways, with much less receptor located in the peripheral lung regions [12]. FcRn is also expressed in the kidney where the receptor is thought to contribute to the resorption of IgG [11,15], as well as in hepatocytes in the liver where it may contribute to maintaining circulating IgG levels and protection from catabolism [11,27].

There appears to be limited expression of FcRn at the surface of cells, with expression occurring mostly in intracellular compartments in most cell types in which FcRn is found [24,28]. Initial uptake of IgG is reported to occur by fluid-phase pinocytosis, the process by which other proteins in the blood enter endothelial cells. Binding of IgG to FcRn takes place in the sorting endosomes at acidic pH [29,30]. IgG is then routed along with FcRn back to the cell surface and released at neutral pH in some cell types (e.g., endothelial cell) or transcytosed across the cell in other cell types such as the mucosal epithelial cells lining the intestine. In contrast, proteins that do not bind FcRn are targeted for degradation in lysosomal compartments. It is through this specific interaction with FcRn that IgG or Fc-containing proteins are protected from degradation and conferred a relatively longer half-life.

7.2.2 FcRn for the Delivery of Fc-Fusion Proteins

The FcRn pathway was explored as a means to deliver large therapeutic proteins noninvasively through the lung [12,31–33]. Because FcRn expression was demonstrated in the lungs of adult humans, as was the transcytosis of IgG across human lung epithelial cells, it was hypothesized that this natural pathway could enable delivery of aerosolized Fc-fusion proteins through the lung. The prototype fusion proteins were dimeric with respect to both the therapeutic protein and Fc (Figure 7.1). Erythropoietin (Epo) was selected as the fusion partner because it is a potent cytokine that elicits a readily measured biological response [34]. Proof-of-principle studies were conducted with the EpoFc dimer in cynomolgus monkeys. Monkeys were selected because the expression of FcRn was determined to be similar to human, and also because the characteristics of a monkey lung were reported to be similar to human [12,18]. EpoFc was administered to cynomolgus monkeys targeting the central airways. Absorption of the EpoFc dimer was achieved, while a mutant EpoFc which contained three mutations at Fc domain (I253A, H310A, and H435A) (EpoFc/IHH) that differed from the wild-type EpoFc in three amino acids in the Fc domain known to be necessary for binding to FcRn was taken up very poorly [12]. Pulmonary delivery of EpoFc dimer was also demonstrated in healthy human volunteers and the resulting pharmacokinetic (PK) parameters were consistent with what was expected based on the nonhuman primate studies [35]. While these data were encouraging, it was postulated that the Fc fusion construct could be modified to improve the PK parameters, including bioavailability. EpoFc is ~100 kDa in size, so it is possible that having two effector molecules may pose steric hindrance between the Fc and Epo and on the binding to their respective receptors. In addition, Epo is a highly glycosylated protein. It is known that the charge on the molecule may affect the penetration of mucus that bathes the epithelial cells in the lung, as well as affect endocytosis into cells [36,37]. Therefore, it was hypothesized that removal of one therapeutic moiety may reduce size and charge and possibly relieve some steric hindrance between the

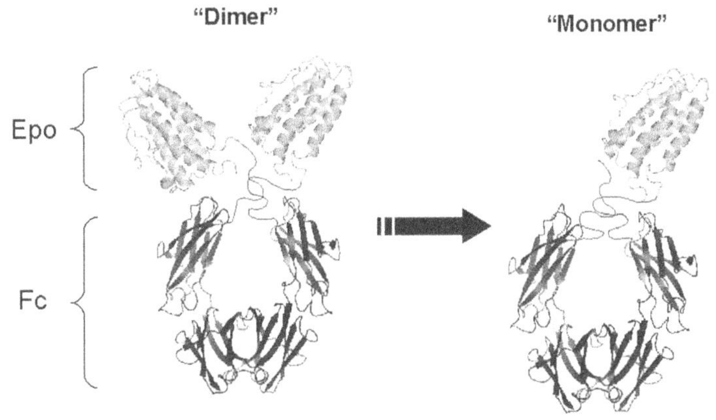

FIGURE 7.1 Schematic diagram for erythropoietin Fc (EpoFc) dimer and monomer. The dimer is composed of two molecules of Epo joined to dimeric Fc and the monomer has a single molecule of Epo joined to dimeric Fc. *Source*: Reproduced from Reference [32] with permission from Adis (© Springer International Publishing AG (2006) All rights reserved).

therapeutic protein and its receptor. Thus, a molecule was created that was comprised of a single molecule of Epo coupled to dimeric Fc, and was described as the EpoFc monomer.

Pulmonary uptake of the EpoFc monomer was compared to the EpoFc dimer after a single dose in cynomolgus monkeys [12]. The systemic exposure of the EpoFc monomer, measured by peak concentration (C_{max}) and area under the curve (AUC), was markedly increased compared to the EpoFc dimer (Figure 7.2). There was a 4- to 5-fold increase in C_{max} and a 6- to 10-fold increase in the AUC. Furthermore, the half-life was improved to 25 h compared to 16 h, for the monomer and dimer, respectively, and bioavailability of the EpoFc monomer was also improved to >30% compared to 5–10% for the dimer. Because of the superior PK parameters of the monomeric EpoFc, this configuration was applied to other therapeutic proteins.

7.3 TYPICAL APPLICATIONS

Several Fc-fusion proteins have been generated which are monomeric with respect to the effector moiety, and in each case improvements relative to the dimeric Fc-fusion proteins have been observed in various characteristics of the molecule including bioactivity and PK parameters [32]. Several examples of monomer Fc-fusion proteins are described below.

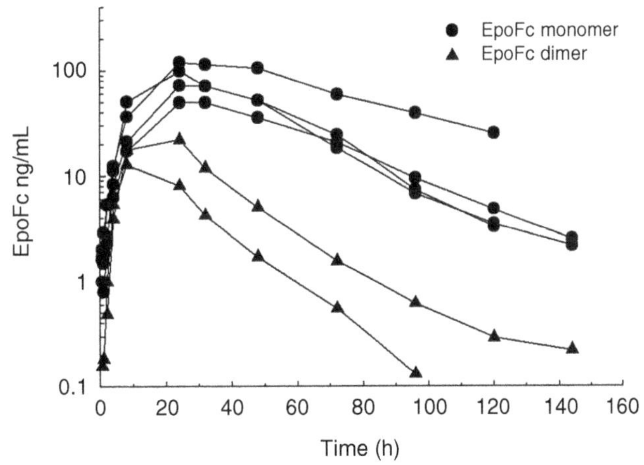

FIGURE 7.2 Pulmonary delivery of EpoFc monomer and dimer in nonhuman primates. Equivalent doses (20 mg/kg) of EpoFc dimer or EpoFc monomer were aerosolized with an Aeroneb Pro® nebulizer and delivered directly into the lungs of intubated cynomolgus monkeys using shallow breathing maneuvers. Serum concentrations of the dimer and monomer were measured with an Epo-specific ELISA. The number of animals per group were $n = 2$ for EpoFc dimer and $n = 4$ for EpoFc monomer. *Source*: This research was originally published in Reference [12].

7.3.1 Interferon Beta-Fc Monomer

Interferon-β (IFN-β) is a 166 amino acid glycoprotein used in the treatment of multiple sclerosis (MS) [38]. The mechanism of action of IFN-β in MS is not clear, but IFN-β is known to protect cells from viral challenge, and an *in vitro* assay was developed in which the protective effect of IFN-β is measured in cells challenged with a cytopathic virus [39]. IFN-β1a is approved for the treatment of MS and is typically administered subcutaneously or intramuscularly several times per week due to its relatively short plasma half-life [40]. Thus, an IFN-βFc fusion monomer that could be administered less frequently and by pulmonary inhalation could be an important advance in the treatment of MS.

The IFN-βFc monomer and dimer were constructed and expressed similarly to that described in Section 7.5. In a cytopathic effect assay, the IFN-βFc monomer was demonstrated to be approximately fourfold more active than IFN-βFc dimer [41] suggesting that having two IFN-β molecules in the fusion protein may hinder the ability of both IFN-β molecules to bind to IFN-β receptor, whereas an Fc-fusion protein with a single IFN-β molecule may have better access to the receptor.

IFN-βFc monomer was administered to cynomolgus monkeys by pulmonary delivery at deposited doses of 1, 3, or 10 μg/kg and the PKs were compared to a subcutaneous dose of 3 μg/kg IFN-β [42]. The PKs of IFN-βFc were dose dependent with C_{max} values of 1.15 ± 0.73, 3.9 ± 0.65, and 17.92 ± 6.13 ng/mL for the 1, 3, and 10 μg/kg doses, respectively [42]. The maximum serum concentration of IFN-β administered as a single 3 μg/kg subcutaneous dose was 3.66 ± 0.92 ng/mL which was similar to that value attained after pulmonary administration of the same dose of 3 μg/mL of IFN-βFc monomer. However, at approximately equal molar IFN-β doses of 0.135 nmol/kg (10 μg/kg) IFN-βFc monomer and 0.125 nmol/kg (3 μg/kg) IFN-β, the C_{max} was approximately 4.6-fold higher for IFN-βFc monomer (17.92 ± 6.13 and 3.66 ± 0.92 ng/mL, respectively). In addition, the AUC of IFN-βFc monomer was approximately 6.8-fold higher and the half-life was approximately threefold longer than for IFN-β [42].

7.3.2 Interferon α-Fc Monomer

Interferon-α (IFN-α) is an 18 kDa glycoprotein used in the treatment of hepatitis C and B and several forms of cancer [43]. The half-life of IFN-α is short (~5 h) [44], resulting in subcutaneous dosing three times per week as typical therapy. Similar to IFN-β, a monomeric Fc-fusion of IFN-α that reduces dosing frequency and provides a more tolerable alternate route of administration could be an important advance in hepatitis therapies.

IFN-αFc monomer and dimer constructs with 15 amino acid linkers of three repeats of the sequence GGGGS inserted between the IFN-α and Fc moieties were generated in a similar manner to that described in Section 7.5 and tested in a cytopathic effect assay [41]. IFN-αFc monomer was demonstrated to be approximately threefold more active than IFN-αFc dimer [41] suggesting that similar to IFN-β, having two IFN-α molecules in the fusion protein causes steric hindrance between the two IFN-α moieties in the dimeric IFN-αFc (Table 7.1). Consistent with this, reducing the linker length between the IFN-α and Fc moieties reduced the activity

TABLE 7.1 Summary of Bioactivity and Pharmacokinetics for Dimeric and Monomeric Fc-Fusion Proteins

		Pharmacokinetics					*In Vitro* Bioactivity	
Fusion Protein	Species and Route	Construct	Dose (μg/kg)	C_{max} (μg/mL)	AUC (μg·h/mL)	$t_{1/2}$ (h)	Assay	Result
EpoFc	Monkey/pulmonary	Dimer	20	0.017	0.557	16	TF-1 proliferation	0.07 nM
		Monomer	20	0.086	5.279	25	assay[a] (EC_{50})	0.09 nM
IFN-β-Fc	Monkey/pulmonary	Dimer	20	0.0038	0.123	11	CPE assay[b]	0.45×10^6 IU/nmol
		Monomer	20	0.022	0.987	27	(specific activity)	1.22×10^6 IU/nmol
IFN-α-Fc[c]	Monkey/pulmonary	Dimer	15	0.107	18.760	107	CPE assay[b]	2.3×10^5 IU/nmol
		Monomer	15	0.102	17.325	87	(specific activity)	5.2×10^5 IU/nmol
FIXFc	Neonatal rat/oral	Dimer	1,370	0.590	9,570	14	Not determined	
		Monomer	1,000	5.830	121,250	19		
	Adult rat/IV	Dimer	5,000	7.500	109	22		
		Monomer	5,000	33.000	509	35		
	Adult FIX-deficient mouse/IV	Dimer	5,000	10.100	167	53		
		Monomer	5,000	33.400	761	46		

AUC, area under the serum or plasma concentration versus time curve; C_{max}, maximum serum or plasma concentration; CPE, cytopathic effect; EC_{50}, concentration of drug that stimulates growth to 50% of maximum; EpoFc, erythropoietin-Fc; FIXFc, factor IX-Fc; IFN, interferon; IU, international units; $t_{1/2}$, terminal half-life in plasma or serum; TF-1, human erythroleukemia cells.

[a]Growth of human erythroleukemia cells were measured for 48 h. Values represent the concentration of drug that stimulated growth by 50% of maximum (EC_{50}).

[b]Antiviral activity was measured in a standard CPE assay using human lung A549 cells and encephalomyocarditis virus.

[c]Constructs with 15 amino acid linkers.

TABLE 7.2 Traditional Dimeric Fc-Fusion Proteins Approved on Market

Drug Name (Generic Name) Manufacturer	Mechanism	Indication(s)	Approval Year	Fc Fragment	Expression System	References
Enbrel® (Etanercept) Amgen/Pfizer	TNF-α in ihibitor	Rheumatoid arthritis and other rheumatologic diseases	1998	IgG1	CHO	[86]
Orencia® (Abatacept) BMS	T-cell modulation	Rheumatoid arthritis and other rheumatologic diseases	2006	IgG1	CHO	[83]
Amevive® (Alefacept) Astellas	T-cell modulation	Psoriasis	2003	IgG1	CHO	[84]
Nplate® (Romiplostim) Amgen	TPO agonist	Chronic idiopathic thrombocytopenic purpura	2008	IgG1	*E. coli*	[85]
Arcalyst® (Rilonacept) Regeneron	IL-1 inhibitor	Cryopyrin-associated periodic syndromes	2008	IgG1	CHO	[87]
Nulojix (Belatacept)	T-Cell Co-Stimulation Blocker	Transplant Rejection	2011	IgG1	CHO	[88]
Eylea (Aflibercept)	Decoy receptor for VEGF-A and PIGF	Age-Related Macular Degeneration	2011	IgG1	CHO	[89]

further, presumably by increasing the steric hindrance between the two IFN-α moieties in the dimeric constructs, and between the IFN and Fc moieties in either construct [41].

Pulmonary administration of IFN-αFc monomer and dimer to the central airways of cynomolgus monkeys resulted in significant and similar absorption of both proteins [41] both of which were significantly greater than those observed following pulmonary uptake of an equivalent dose of IFN-α. Not only was the uptake much greater for both fusion proteins compared to IFN-α, but also the serum half-life was much longer (~100 h compared to 5 h) (Table 7.2). Other PK parameters of pulmonary uptake of IFN-αFc monomer and dimer (i.e., C_{max}, AUC, Vz) were similar after inhalation of the same dose in nonhuman primates.

7.3.3 Factor VIII-Fc Monomer

Hemophilia A is an X chromosome-linked bleeding disorder caused by deficiency or mutations in the coagulation factor VIII (FVIII) gene that affects 1 in 5000 males [45]. As a consequence, the low levels or complete absence of functional FVIII levels result in different degrees of severity of the disease: severe form with levels of functional FVIII at 1% (<1 IU/dL), moderate form at 1–5% and mild at 5–40%. Hemophilia's hallmark is repeated bleeding episodes that can cause hemophilic arthropathy and other sequelae, including death if untreated [46]. The frequency of bleeding episodes in hemophilia varies depending on whether the person has the mild, moderate, or severe form of the disorder. For people with severe hemophilia, bleeding episodes occur more often and spontaneously. Treatment for hemophilia A may be on demand, whereby patients are administered plasma-derived or recombinant FVIII when a bleeding episode occurs. Prophylactic FVIII therapy is a treatment regimen in which the clotting factor is administered at regular intervals in order to maintain FVIII through levels above 1% of normal FVIII plasma levels in order to prevent spontaneous bleeds. Maintenance of a low but consistent level of FVIII activity may potentially prevent life threatening hemorrhages and musculoskeletal bleeds. Both on demand and prophylactic FVIII treatment often requires multiple injections per week to achieve the necessary FVIII plasma levels. This can be attributed, in part, to the relatively short half-life of ~12 h for currently available FVIII products [47,48]. FVIII concentrates are administered by intravenous injection, which requires repeated venous access. This is particularly difficult in the pediatric population and often requires the use of central venous access devices (e.g., ports). Therefore, recombinant FVIII with a longer half-life would be expected to require fewer injections, reducing the need for repeated venous access, and potentially improving the adherence to prophylactic regimens by reducing the frequency of administration.

FVIII is ~280 kDa and is comprised of a heavy (200 kDa) and light chain (80 kDa) with several distinct domains (A1–A2–B–A3–C1–C2). The B-domain which contains 19 of 25 potential glycosylation sites has no impact on the function of FVIII [49]. In fact, B-domain deleted (BDD) FVIII is a normal intermediate during the process of coagulation *in vivo*. FVIII is normally processed intracellularly to generate a 200 kDa heavy chain (A1–A2) and an 80 kDa light chain (A3–C1–C2) that remain associated via metal ion-mediated, noncovalent interactions. Deletion of a large portion of the B-domain (BDD) has no effect on the functional activity of FVIII, but decreases the size of the protein significantly (38% reduction) and increases the FVIII expression levels in eukaryotic cells [49,50]. A recombinant BDD factor VIII-Fc monomer (rFVIIIFc) was created in a similar way to that described in Section 7.5. In the case of FVIII, generation of the rFVIIIFc monomer was enabling to the Fc fusion technology since the homodimer was not well produced, presumably because the large size of the homodimer (~380 kDa) prevented secretion from the cells.

The PKs and clotting activity of the rFVIIIFc monomer (rFVIIIFc) were studied in mouse and dog models of hemophilia A. The ability of rFVIIIFc to clot blood was evaluated in hemophilia A mice by measuring whole blood clotting time (WBCT) after a single 50 IU/kg dose of rFVIIIFc or BDD-rFVIII (ReFacto® [moroctocog alfa]). At selected time points after dosing, blood was collected and visually inspected once per minute for the formation of a clot. Baseline clotting in these mice was >60 min, but after administration of rFVIIIFc or rFVIII, WBCT was achieved in 2–5 min. Blood from mice treated with rFVIII lost the ability to clot after 42 h. In contrast, all of the mice treated with rFVIIIFc (50 IU/kg) maintained effective clotting through 96 h, but lost the ability to clot at 120 h after dosing. These data suggest that there is a threefold improvement in WBCT for rFVIIIFc compared with rFVIII.

The PKs and pharmacodynamics of rFVIIIFc were also evaluated in the in-bred Chapel Hill colony of hemophilia A dogs [51–52]. These dogs have a severe hemophilic phenotype that is comparable to the severe form of human disease with plasma FVIII levels of <1%. A single intravenous dose (125 IU/kg) of rFVIIIFc was administered to four dogs, which immediately corrected blood clotting to normal as measured by WBCT and activated partial thromboplastin time (aPTT) [53]. WBCT remained below 20 min, which is consistent with FVIII levels of >1%, through ~96 h. The concentration of rFVIIIFc in the plasma was measured both by ELISA and by a chromogenic activity assay, which showed that the concentration versus time curves were similar using both methods. The terminal half-life was also similar for both methods [53] at 15.7 ± 1.7 h (ELISA) and 15.4 ± 0.3 h (chromogenic activity).

Two of the dogs also received a single dose of ReFacto, 114 IU/kg for one dog and 120 IU/kg for the other, 72 h prior to receiving rFVIIIFc (125 IU/kg) in a cross over design. A washout period of 3 days was chosen to allow the clotting to return to baseline. Clotting was corrected to normal immediately after dosing with rFVIII (determined by WBCT and clotting activity measured using an aPTT assay) similar to what was observed following dosing with rFVIIIFc. However, the WBCT normalization after rFVIIIFc dosing lasted for approximately twice as long compared to rFVIII. In addition, the half-life of rFVIIIFc determined by ELISA or chromogenic activity was approximately twice as long as that determined for rFVIII (7.0 h and 6.7 h, respectively). No adverse clinical signs were detected with any of the infusions. Therefore, construction of a monomeric Fc fusion of rFVIII produces a molecule with a defined mechanism of action that has an increased half-life and the potential to provide prolonged protection from bleeding.

The safety, tolerability, and PK of a single intravenous dose of rFVIIIFc in previously treated severe hemophilia A patients were evaluated in a two-arm Phase I/IIa study in 16 male patients [54,55]. Patients were administered intravenously with either a single low dose (25 IU/kg) of rFVIII (Advate® [octocog alfa]—rFVIII full-length product) followed by an equal dose of rFVIIIFc, or a single high dose (65 IU/kg) of rFVIII followed by an equal dose of rFVIIIFc. PK assessments were performed following administration of a single dose of rFVIII and rFVIIIFc. rFVIIIFc was well tolerated. The primary PK estimates were derived from FVIII activity data measured by a one-stage (aPTT) clotting assay. rFVIIIFc showed a 1.54- to 1.70-fold longer half-life compared with rFVIII [54,55] and the C_{max} increased proportionally to the dose, but was comparable between equal doses of rFVIII and rFVIIIFc (Figure 7.3a and b) [55]. The total exposure (AUC_{INF}) also increased proportionally to the

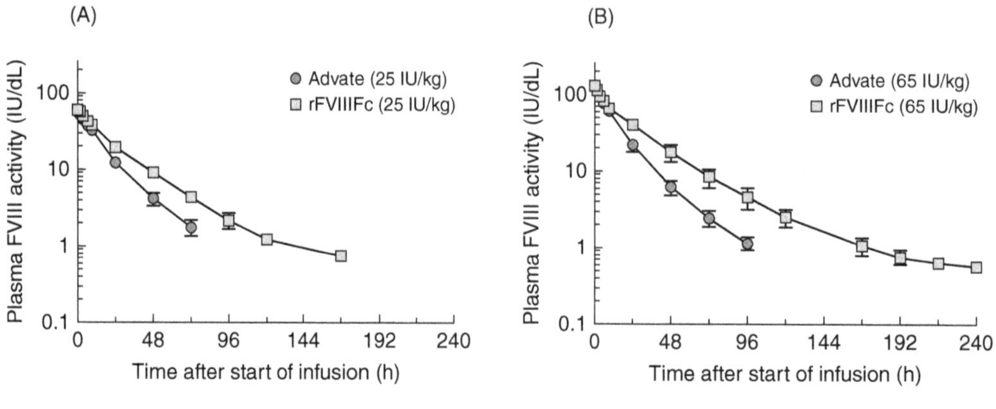

FIGURE 7.3 Group mean plasma FVIII activity pharmacokinetic profiles for low-dose and high-dose cohorts in patients with severe hemophilia A. The plasma FVIII activity (one stage aPTT assay) versus time curve after a single intravenous injection of rFVIIIFc or Advate are shown for (A) 25 IU/kg (low-dose cohort, $n = 6$) and (B) 65 IU/kg (high-dose cohort, $n = 10$ [Advate]; $n = 9$ [rFVIIIFc]). Results presented are group mean ± standard error of mean (SEM) *Source*: This research was originally published in Reference [55].

dose [55]. However, the AUC_{INF} of rFVIIIFc was 1.48 and 1.56-fold greater than that of rFVIII at 25 IU/kg ($p = 0.002$) and 65 IU/kg ($p < 0.001$), respectively [55]. There were no significant differences in incremental recovery between rFVIII and rFVIIIFc. Therefore, within each patient, rFVIIIFc demonstrated an improved PK profile compared with rFVIII. The results from the Phase I/IIa study supported the development of the Phase III study to evaluate the safety, PKs, and efficacy of rFVIIIFc in previously treated patients with severe hemophilia A (www.clinicaltrials.org identifier NCT01181128). In conclusion, rFVIIIFc has been shown to have improved PKs in humans and animal models such as hemophilia A mice and dogs. Treatment with this new long-lasting product has the potential to result in less frequent dosing and improved quality of life for hemophilia A patients.

7.3.4 Factor IX-Fc Monomer

Factor IX (FIX) is a 55 kDa serine protease that is an essential component of the coagulation cascade [56–58]. Deficiency of FIX results from a X-linked inherited disorder known as hemophilia B or Christmas disease [59] that affects 1 in 25,000–30,000 males [45]. In severe form of hemophilia B, levels of functional FIX are at <1% (<1 IU/dL), moderate form at 1–5% and mild at 5–30%. Plasma derived FIX products (AlphaNine® and Mononine®) and the recombinant FIX product (BeneFIX® [nonacog alfa]) are used in the on-demand and prophylactic treatment (2–3 times weekly) of hemophilia B. The coupling of recombinant FIX with recombinant Fc was pursued as an approach to extend the half-life of FIX which could lead to a reduction in the frequency of administration required for the treatment of hemophilia B.

rFIXFc monomer and dimer were constructed and expressed in a way similar to that described in Section 7.5. The PKs of rFIXFc monomer and dimer were evaluated in adult rats and in adult FIX-deficient mice [60] after a single intravenous dose [41]. The systemic exposure, measured by C_{max} and AUC, of rFIXFc monomer was considerably greater (three to fivefold) compared to that of the rFIXFc dimer in rats and mice (Table 7.1). The PKs of the rFIXFc dimer and rFIXFc monomer were also compared in neonatal rats after oral administration where the uptake of the rFIXFc monomer was much greater compared to the rFIXFc dimer after a similar molar dose, with a 10-fold greater C_{max} and greater than 12-fold increase in AUC (Table 7.1).

In rats (neonatal and adult), the terminal half-life of the monomer was also greater compared to the rFIXFc dimer. However, in FIX-deficient mice, the half-lives for rFIXFc monomer and dimer were similar. In both models, the recovery of the monomer appeared to be greater compared to the dimer. Moreover, the half-life for both rFIXFc fusion proteins was markedly longer compared to commercially available rFIX (BeneFIX) (22 and 33 h vs. 5.8 h). Overall, the rFIXFc monomer construct has markedly improved PKs after intravenous administration in rodents in comparison to the dimer configuration.

Additionally, the PKs of rFIXFc monomer clotting activity was compared with recombinant FIX (BeneFIX, rFIX) in FIX-deficient mice. Activity was measured by both a one-stage clotting assay (aPTT) and by WBCT. Hemophilia B mice were administered a single equivalent intravenous dose of rFIXFc monomer or rFIX. The plasma FIX activity declined in a bi-exponential manner following injection of both rFIXFc and rFIX (Figure 7.4a). The systemic exposure was markedly greater for rFIXFc compared to rFIX, with a 27% increase in C_{max} and 3.4-fold increase in AUC, and the half-life was also increased. Thus, the PK parameters are greatly improved for the rFIXFc compared to rFIX.

The functional activity of rFIXFc was also evaluated *ex vivo* by measuring the WBCT. rFIXFc monomer and rFIX were administered as a single intravenous dose to hemophilia B mice and WBCT measured as described [61]. Fifteen minutes after dosing, normal clotting was observed in all mice treated with either rFIXFc or rFIX. WBCT returned to baseline by 72 h in all mice treated with rFIX, whereas blood from rFIXFc mice still clotted normally. At 144 h, all mice treated with rFIXFc returned to baseline. The percentage of animals able to clot blood at each time point is shown in Figure 7.4b.

The PKs of rFIXFc and rFIX have also been studied in FIX-deficient dogs and normal cynomolgus monkeys. In FIX-deficient dogs and cynomolgus monkeys, the half-life for rFIXFc was approximately 3- to 4-fold greater compared to rFIX (Table 7.1). The dog model was one of severe hemophilia B for which the mutation is similar to that found in humans [62]. As a result, the concentration of rFIXFc in the circulation could be readily measured by both activity and antigen levels since there was no endogenous FIX activity in these animals. The half-life for rFIXFc (38.3 h) was markedly longer compared to that reported for rFIX (17–18 h) in FIX-deficient dogs based on activity data [62,63]. In the same way, in cynomolgus monkeys, the half-life for rFIXFc was 47 h compared to 12.7 h for rFIX reported earlier [63].

To prove the mechanism of action for the prolonged half-life, the PK of rFIXFc was assessed in a murine FcRn knockout model. As expected, the elimination half-life of rFIXFc was similar to rFIX in the FcRn knockout mice (16.9 h for rFIXFc and 16.5 h for rFIX). However in mice that lacked murine FcRn (mFcRn −/−, mβ2 m −/−), but were transgenic for hFcRn (hFcRn +/+, hβ2 m +/+), the half-life of FIXFc was 53.0 h while rFIX remained unchanged (14.2 h) compared to the FcRn knockout mice. These data confirm that FcRn mediates the prolonged half-life of rFIXFc [61].

rFIXFc was evaluated in an open-label Phase I/IIa dose escalation study in previously treated patients ($n = 14$) with severe hemophilia B at intravenous doses ranging from 1 to 100 IU/kg [64,65]. The lowest two rFIXFc dose levels

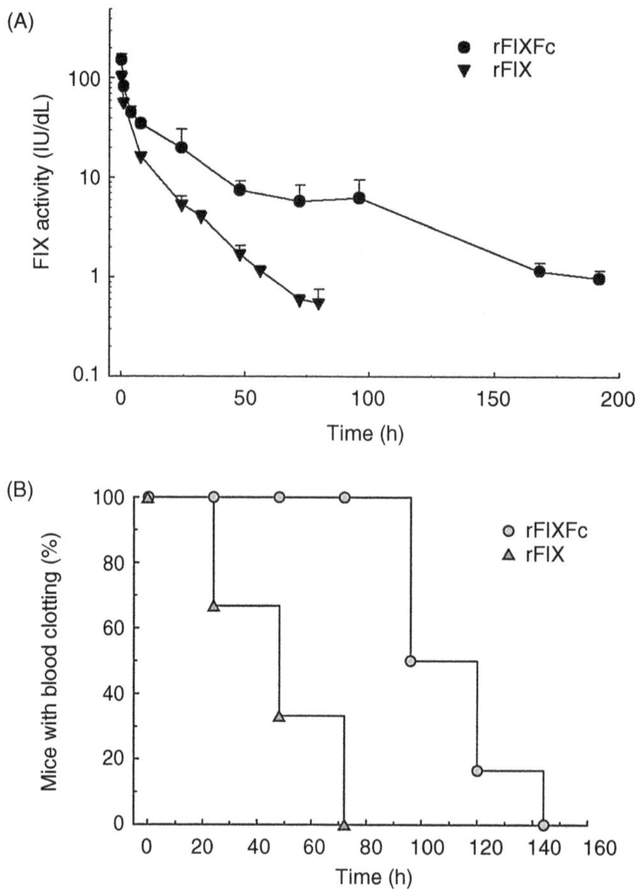

FIGURE 7.4 Functional activities of rFIXFc and rFIX in FIX-deficient mice. (a) Clotting activity measured by modified aPTT assay in FIX-deficient mice. FIX-deficient mice were administered a single intravenous dose of rFIXFc (200 IU/kg, $n = 4$ mice per time point). Blood samples were collected at predetermined time points and analyzed for FIX activity (mean ± SD). (b) FIX-deficient mice (six per group) were dosed intravenously with 50 IU/kg rFIXFc or rFIX. Blood samples were collected before dosing and at various times after dosing. Blood samples were incubated at 37°C and were visually inspected for the presence of a blood clot once per minute. The time needed for a clot to form was recorded, and, once the clotting activity returned to baseline (i.e., no clot formation), no additional samples were obtained (samples were collected 15 min to 144 h for rFIXFc or 15 min to 72 h for rFIX). *Source* (b): This research was originally published in Reference [61].

administered (1 and 5 IU/kg, one subject in each group) were assessed for safety only. In addition to safety, the FIX activity PK parameters were assessed in patients, who received higher rFIXFc dose levels (12.5, 25, 50, and 100 IU/kg). Dose-proportional increase in plasma FIX activity was observed based on C_{max} occurring immediately after infusion and total exposure (AUC_{INF}) (Figure 7.5). Mean activity terminal half-life for rFIXFc was 56.7 h, approximately threefold longer than historical values reported for rFIX [65–67]. The incremental recovery of rFIXFc was 0.93 IU/dL per IU/kg, an improvement compared with rFIX and similar to plasma-derived FIX products [65–67]. These results show that rFIXFc may offer a viable therapeutic approach to achieve prolonged hemostatic protection and less frequent dosing in patients with hemophilia B.

Therefore, the PK parameters determined from preclinical studies in several different animal models compared well with the PKs of rFIXFc determined in humans. The results from the Phase I/IIa rFIXFc study supported development of the Phase III study to evaluate the safety, PKs, and efficacy of rFIXFc in previously treated patients with severe hemophilia B (www.clinicaltrials.gov identifier NCT01027364).

7.4 ALTERNATIVE APPLICATIONS

Fc monomers often have improved bioactivity and PK parameters compared to dimeric counterparts when the effector molecule is naturally a monomer. The concept of

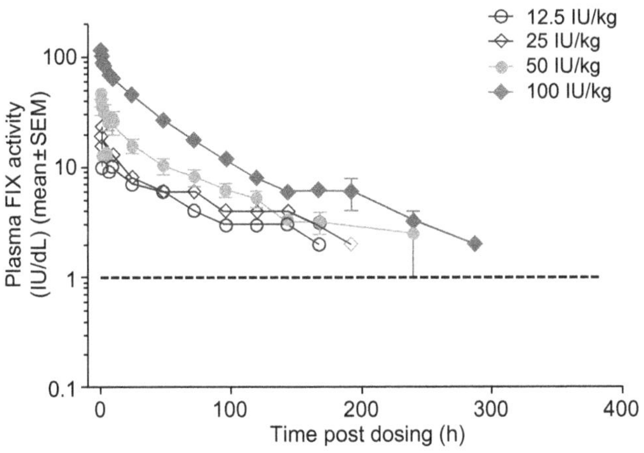

FIGURE 7.5 Dose-dependent pharmacokinetic profiles of rFIXFc after a single intravenous dose of in subjects with severe hemophilia B. Plasma FIX activity levels were measured over time after a single intravenous infusion of 12.5 ($n=1$), 25 ($n=1$), 50 ($n=5$), or 100 IU/kg ($n=5$) rFIXFc. Results presented are mean ± SEM. *Source*: This research was originally published in Reference [65].

monomeric Fc fusion technology is also useful for proteins that require a heterodimeric partner for optimal activity.

7.4.1 Follicle Stimulating Hormone-Fc-Fusion Proteins

Follicle stimulating hormone (FSH) is commonly used in the treatment of infertility in both men and women and is administered daily by subcutaneous or intramuscular injection for several days [68,69]. FSH binds to FSH receptors in the granulosa cells of the ovary resulting in the selection and growth of ovarian follicles. This mechanism of action has commonly been used in rats as a bioactivity assay in which ovarian weight is measured in response to FSH treatment [70]. FSH is a noncovalently linked heterodimer of α and β subunits [71]. Subunit assembly is essential for activity of the heterodimer [72] as well as protection of the β subunit which is unstable when not in complex with the α subunit [73]. Thus, fusion of FSH α and β subunits with Fc in a heterodimeric configuration that maintained full activity of the effector moiety could be useful in stabilizing FSH, reducing the frequency of administration and providing an alternate, less invasive route of administration that could improve the tolerability of current formulations.

FSHFc heterodimers were generated by co-transfecting mammalian cells with FSHαFc and FSHβFc fusion plasmids [33]. Purified FSHFc heterodimer was injected into juvenile female rats subcutaneously resulting in significantly increased ovarian weights compared to vehicle control animals as well as animals administered recombinant FSH [33]. PK studies with FSHFc heterodimer revealed that the protein had a terminal half-life of ~69 h in rats, which is markedly longer than reported for the recombinant protein (11.4 h) [74]. The longer half-life of FSHFc heterodimer likely explains the significantly increased bioactivity of the fusion protein in rats compared to recombinant FSH.

Recombinant FSH and FSHFc heterodimer were also tested in a testis weight gain assay in neonatal rats after oral administration [75]. FcRn expression and IgG transport is relatively high during the first few weeks of life in rodents, while secreted acid in the stomach and the level of digestive enzymes are low. As a result, this model is useful in studying the transport and PKs of Fc-fusion proteins from the epithelial cells in the intestine into the circulation [76–77]. Recombinant FSH had no effect on testis weight compared to vehicle control treated animals while FSHFc heterodimer administration resulted in a significant increase in testis weight [33] indicating the FSHFc heterodimer is more stable than the recombinant protein after oral administration in rats, and that the Fc fusion in combination with FcRn expression has the potential for alternate routes of administration for molecules that are otherwise restricted to subcutaneous and intramuscular injections.

7.4.2 Cytokine-Fc-Fusion Proteins

Cytokines have an impact on many critical biological processes including regulation of gene expression, cell proliferation, and promotion of chronic inflammation. Cytokines are also known to be involved in the initiation and perpetuation of human diseases and these effects can be dampened by the use of soluble receptors that act as competitors with endogenous receptors for cytokine binding. For example, Etanercept is a homodimer of the ectodomain of TNF-α receptor fused to human Fc of IgG1. Etanercept thus inhibits the inflammatory function of TNF-α and is used in the treatment of rheumatoid, juvenile rheumatoid and psoriatic arthritis, plaque psoriasis, and ankylosing spondylitis. In many cases however, cytokines act as multicomponent systems and the use of soluble receptors is limited since cytokine affinity for monomeric receptors is often low suggesting that Fc monomer technology would have limited utility for improving cytokine function. However, it has been demonstrated that heterodimeric Fc-fusion proteins of interleukin (IL)-2/IL-12 and IL-4/granulocyte-macrophage colony stimulating factor results in synergistic biological activity similar to that seen with the combination of individual cytokines [78]. In addition, cytokine traps consisting of the extracellular domains of two distinct cytokine receptors (e.g., IL-6 receptor α and gp-130) each fused to human Fc and expressed in mammalian cells as IL-6 receptor αFc/gp-130Fc heterodimer fusions, have been shown to constitute one of the most potent inhibitors of cytokine action reported [79].

7.5 EXPRESSION AND PURIFICATION OF MONOMERIC Fc-FUSION PROTEINS

The Fc fragment used as fusion partner usually contains the hinge region and CH2 and CH3 domains of immunoglobulin heavy chain. For therapeutic purpose, most Fc-fusion proteins employ the Fc fragment of human IgG1. All seven currently marketed Fc-fusion proteins contain IgG1 Fc (Table 7.2), while Fc fragments of human IgG2 and IgG4 have also been explored as fusion partners in development [80–82]. The Fc domain of human IgG3 is not a preferred fusion partner, presumably in part because of its high complement activation property. Depending on the application and structure-function of the Fc-fusion protein, the Fc fragment is commonly coupled at the C-terminus and sometimes the N-terminus of the effector proteins. In some cases, a small spacer (e.g., a glycine- and serine-rich peptide) between Fc and the effector protein is helpful in preserving the function of the effector protein. In contrast to the expression of traditional dimeric Fc-fusion proteins, which require one expression cassette for the effector protein-Fc fusion, expression of monomeric Fc-fusion proteins usually employs two expression cassettes, one controls the expression of the effector protein coupled with Fc and the other controls expression of Fc alone. The two expression cassettes can be built within one bi-cistronic vector with two expression cassettes, or in two separate vectors.

Several expression systems have been used for Fc-fusion protein expression. For therapeutic purposes, most traditional dimeric Fc-fusion proteins have been produced in mammalian expression systems. Six out of the seven Fc fusion molecules on the market are expressed in CHO cells, while only romiplostim is expressed in *Escherichia coli* [83–89] (Table 7.2).

Expression of Fc-fusion proteins in insect cells (such as Sf9) and yeast systems (such as *Pichia pastoris*) has also been reported [90,91]. The *E. coli* expression system provides high productivity but it cannot carry out post-translational modifications such as glycosylation and may not form disulfide bonds and proper conformation. Refolding of Fc-fusion proteins produced from *E. coli* is often necessary and it can result in scrambled disulfide forms and other misfolded conformations [92,93]. Proteins expressed in yeast cells can form disulfide bonds and are glycosylated, however, glycans made by yeast cells contain high mannose structures, have different oligosaccharide structure from human and are often immunogenic to humans [94]. Over the last decade, however, significant progress has been achieved through genetic engineering of glycosylation pathways in yeast cells for human like glycosylation [94,95]. Expression of Fc-fusion proteins in mammalian systems provides advantages such as correct protein folding, disulfide bond formation, and proper posttranslational modification. All monomeric Fc-fusion proteins reported have been expressed in mammalian cells, namely CHO and HEK293, which are frequently used cell lines for protein expression [12,31,32,61].

Two proteins (effector protein-Fc fusion and Fc alone) are produced using the monomeric Fc-fusion protein expression system described above. After protein folding and the formation of disulfide bonds between the Fc moieties, three forms of products are secreted in the cell culture media, Fc fusion dimer (effector protein-Fc homodimer), Fc fusion monomer (one effector protein-Fc and one Fc), and Fc dimer (Fc homodimer) (Figure 7.6) [12,61]. Because the Fc fusion monomer is the only desired product, it is important that the expression system is tailored to minimize the expression of the other two products (Fc fusion homodimer and Fc homodimer). It can be challenging to separate Fc fusion monomer from Fc fusion homodimer during purification because of the relatively similar biochemical properties of these two molecules, while the Fc dimer can generally be more readily separated. Over-expression of Fc has been shown to increase the ratio of Fc fusion monomer to Fc fusion homodimer, and thereby facilitates purification. Depending on the activity of Fc fusion homodimer, it can contribute significantly to the activity assay of the Fc monomer fusion cell culture harvest when the expression level is analyzed by activity assays, which may pose additional challenges. Analytical size exclusion chromatography HPLC or nonreducing SDS-PAGE can be helpful tools to analyze and quantify the monomer Fc fusion expression levels in the cell culture, and can be coupled to protein A chromatography to facilitate detection.

A flexible polypeptide linker can also be designed to connect the two Fc fragments so that the monomer Fc-fusion protein is expressed as a single protein. This approach eliminates the need to separate the Fc fusion monomer from the Fc fusion homodimer, and can lead to more efficient expression. If desired, intracellular cleavage sites can be added flanking the linker peptide so that the monomeric Fc protein is translated as a single polypeptide and before secretion, processed intracellularly at the cleavable linker to produce Fc-effector protein and Fc heterodimer (the monomer Fc-fusion protein). This may require co-transfection of additional processing enzymes to fully remove the linker from the final proteins.

Similar to traditional dimeric Fc-fusion proteins, the presence of Fc provides an advantage of affinity chromatography purification for Fc monomeric fusion proteins. Monomeric Fc-fusion proteins can usually be purified using Protein A or Protein G affinity chromatography. Both Protein A and Protein G are immunoglobulin binding proteins found in Streptococcal bacteria. While Protein G binds to all four subclasses of human IgG, Protein A binds well to IgG1, IgG2, and IgG4. The binding of IgG to Protein G is often stronger, making elution, and complete recovery of the immunoglobulin more difficult. Additionally, because of the lower cost of Protein A compared to

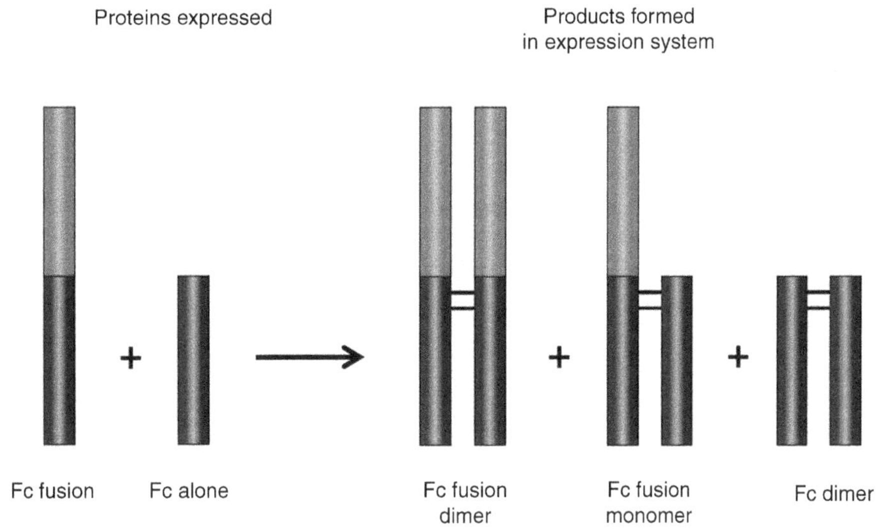

FIGURE 7.6 Expression of monomeric fusion proteins. Two proteins, effector Fc fusion and Fc alone are expressed in the expression system. After disulfide bond formation, three products are produced, including Fc fusion dimer, Fc fusion monomer, and Fc dimer.

Protein G and the fact that Protein A withstands harsher conditions used in cleaning procedures, Protein A is preferred for Fc-fusion protein and antibody purifications. Both Protein A and Protein G chromatography columns employ binding around neutral pH and elution at acidic conditions. When the effector protein is sensitive to the low pH elution, an alternative purification strategy is necessary. In addition to binding to Protein A and Protein G, human IgG and Fc fragment bind to Fc receptors such as neonatal Fc receptor (FcRn) [96] and FcRn affinity chromatography can be a good alternative for purification of Fc-fusion proteins. The Fc fragment binds to FcRn with high affinity at slightly acidic conditions (below pH 6.5) and dissociates at physiological pH or higher (at or above pH 7.4) [97]. This feature can be conveniently employed in FcRn affinity chromatography. The interaction of FcRn and Fc fragment has been found to be similar across species, with different affinity and binding kinetics [98], which provides options to select appropriate FcRn for affinity resin depending on the characteristics of the Fc fusion monomer.

The presence of Fc dimer fusion and Fc dimer during monomer Fc fusion expression requires appropriate chromatography processes to separate them during downstream purification. Protein A or FcRn affinity chromatography are not able to separate monomer from homodimer Fc fusion or Fc dimer. Depending on the characteristics of the effector proteins fused to Fc fragment, Fc monomer fusion and Fc dimer alone can have similar behaviors on the some chromatography columns and cause challenges in purification. Separation of Fc monomer fusion from Fc dimer fusion and Fc dimer can be achieved based on charge, hydrophobicity, and size by using ion exchange chromatography, hydrophobic interaction chromatography, and size exclusion chromatography [12,61]. Immunoaffinity chromatography specific to the effector part of the Fc-fusion protein has also been successfully used for separation of monomer Fc-fusion protein [99]. During the early research phase, including a tag to either the effector protein-Fc fusion or Fc alone can facilitate the separation of monomer from homodimer Fc fusion [12]. The formulation of Fc monomer is usually dependent on the characteristics of effector protein fused to Fc fragment. In general, fusion of Fc fragment could potentially increase the stability of the effector proteins.

The monomeric Fc-fusion protein provides several advantages over traditional dimeric Fc fusion technology. For example, many of the Fc monomer fusion proteins have enhanced biological activity in comparison to Fc dimer fusion proteins, presumably in part due to reduction or elimination of interference between the effector proteins which could occur in dimeric Fc-fusion proteins [12,32]. However, like any other technologies, monomeric Fc fusion has its challenges and limitations. One of them is that Fc monomer fusion proteins are co-expressed with Fc homodimer (Fc dimer fusion protein) and Fc dimer in the expression system. Because a monomeric Fc-fusion protein is a heterodimer of effector protein Fc fusion and Fc alone, two expression cassettes need to be introduced into the expression system during the construction of recombinant cell lines. Appropriate ratio of these two proteins can be important for the productivity of the monomeric Fc-fusion proteins. The ratio of the three products (monomer Fc fusion, dimer Fc fusion, and Fc dimers) in the expression system is affected by multiple factors such as the ratio of effector Fc fusion and Fc alone at a given time, the condition and dynamics of disulfide bond formation inside the cells. Therefore, there can be significant variations in the percentage of monomer Fc fusion among individual cell lines and it is often necessary to establish methods for selecting cell lines that not only express high level of monomer Fc fusions but also higher percentage of monomer Fc-fusion proteins. It should be noted that fusion with Fc fragment often improves the expression level [100] and the monomer Fc fusion expression level is comparable to and often higher than that of effector protein alone. In addition to expression, challenges can also be encountered during purification process in separation of Fc fusion monomer from Fc fusion homodimer and Fc dimer, especially when the effector protein is small in size and has similar charge and hydrophobicity to Fc fragment.

7.6 CONCLUSIONS AND FUTURE PERSPECTIVES

Protein therapeutics are a major source of drug development initiatives in modern medicine. However, one of the most challenging issues can be their short half-life *in vivo*. Significant efforts have been made to develop approaches to prolong the half-life of protein or peptide drugs. Fc fusion technology, by which the Fc fragment of IgG is fused to a protein or peptide drug to extend its *in vivo* half-life by means of the FcRn recycling mechanism, is one of the successful examples. Well-known and validated traditional Fc-fusion drugs, such as etanercept for the treatment of rheumatoid arthritis and the VEGF-trap for the treatment of macular degeneration, consist of two copies of protein or peptide molecules linked to the Fc region to improve PKs, solubility, and production efficiency. This new class of Fc-fusion proteins, Fc monomeric fusion, that links a single copy of effector protein or peptide drug to the Fc region has demonstrated improved biological properties, including further extended half-life and increased bioavailability, compared to traditional Fc fusion dimer. These fusion proteins include EpoFc, IFN-βFc, and IFN-αFc that could be used for the potential treatment of anemia, MS, and hepatitis C, respectively. Additionally, blood coagulation factors-Fc fusions (rFVIIIFc and rFIXFc) have also been generated with this technology to create long-lasting clotting factors for the treatment of hemophilia A and hemophilia B. Both rFVIIIFc and rFIXFc have been shown to have increased systemic half-lives in preclinical models and Phase I clinical trials and are currently being evaluated in global pivotal clinical trials.

In addition to monomeric and dimeric Fc-fusion proteins, heterodimeric Fc-fusion proteins have also been generated and shown to have improved biological properties. For example, FSH which is a heterodimer of α and β subunits has been developed into an Fc heterodimer with FSH-α fused to one arm of Fc and FSH-β fused to the other. FSHFc heterodimer has improved *in vivo* half-life while maintaining FSH bioactivity. Thus the Fc fusion technology has the potential to be applied to a wide range of monomeric as well as heterodimeric peptides and proteins in order to improve biological properties, provide the potential for less frequent dosing and less invasive routes of administration.

REFERENCES

1. Morell A, Terry WD, Waldmann TA. (1970) Metabolic properties of IgG subclasses in man. *J. Clin. Invest.* 49, 673–680.
2. Waldmann TA, Strober W. (1969) Metabolism of immunoglobulins. *Prog. Allergy* 13, 1–110.
3. Simister NE, Mostov KE. (1989) An Fc receptor structurally related to MHC class I antigens. *Nature* 337, 184–187.
4. Brambell F. (1966) The transmission of immunity from mother to young and the catabolism of immunoglobulins. *Lancet* 2, 1087–1093.
5. Rodewald R, Kraehenbuhl JP. (1984) Receptor-mediated transport of IgG. *J. Cell Biol.* 99, 159s–164s.
6. Ghetie V, Hubbard JG, Kim J-K, Tsen M-F, Lee Y, Ward ES. (1996) Abnormally short serum half-lives of IgG in β2-microglobulin-deficient mice. *Eur. J. Immunol.* 26, 690–696.
7. Israel EJ, Wilsker DF, Hayes KC, Schoenfeld D, Simister NE. (1996) Increased clearance of IgG in mice that lack beta 2-microglobulin: possible protective role of FcRn. *Immunology* 89, 573–578.
8. Junghans RP, Anderson CL. (1996) The protection receptor for IgG catabolism is the beta2-microglobulin-containing neonatal intestinal transport receptor. *PNAS USA* 93, 5512–5516.
9. Waldmann TA, Terry WD. (1990) Familial hypercatabolic hypoproteinemia. A disorder of endogenous catabolism of albumin and immunoglobulin. *J. Clin. Invest.* 86, 2093–2098.
10. Wani MA, Haynes LD, Kim J, Bronson CL, Chaudhury C, Mohanty S, et al. (2006) Familial hypercatabolic hypoproteinemia caused by deficiency of the neonatal Fc receptor, FcRn, due to a mutant Î²2-microglobulin gene. *PNAS USA* 103, 5084–5089.
11. Akilesh S, Huber TB, Wu H, Wang G, Hartleben B, Kopp JB, et al. (2008) Podocytes use FcRn to clear IgG from the glomerular basement membrane. *PNAS USA* 105, 967–972.
12. Bitonti AJ, Dumont JA, Low SC, Peters RT, Kropp KE, Palombella VJ, et al. (2004) Pulmonary delivery of an erythropoietin Fc fusion protein in non-human primates through an immunoglobulin transport pathway. *PNAS USA* 101, 9763–9768.
13. Blumberg RS, Koss T, Story CM, Barisani D, Polischuk J, Lipin A, et al. (1995) A major histocompatibility complex class I-related Fc receptor for IgG on rat hepatocytes. *J. Clin. Invest.* 95, 2397–2402.

14. Dickinson BL, Badizadegan K, Wu Z, Ahouse JC, Zhu X, Simister NE, et al. (1999) Bidirectional FcRn-dependent IgG transport in a polarized human intestinal epithelial cell line. *J. Clin. Invest.* 104, 903–911.
15. Haymann J-P, Levraud J-P, Bouet S, Kappes V, Hagege J, Nguyen G, et al. (2000) Characterization and localization of the neonatal Fc receptor in adult human kidney. *J. Am. Soc. Nephrol.* 11, 632–639.
16. Israel EJ, Taylor S, Wu Z, Mizoguchi E, Blumberg RS, Bhan A, et al. (1997) Expression of the neonatal Fc receptor, FcRn, on human intestinal epithelial cells. *Immunology* 92, 69–74.
17. Simister NE, Story CM, Chen H-L, Hunt JS. (1996) An IgG-transporting Fc receptor expressed in the syncytiotrophoblast of human placenta. *Eur. J. Immunol.* 26, 1527–1531.
18. Spiekermann GM, Finn PW, Ward ES, Dumont J, Dickinson BL, Blumberg RS, et al. (2002) Receptor-mediated immunoglobulin G transport across mucosal barriers in adult life. *J. Exp. Med.* 196, 303–310.
19. Kim J-K, Tsen M-F, Ghetie V, Ward ES. (1994) Localization of the site of the murine IgG1 molecule that is involved in binding to the murine intestinal Fc receptor. *Eur. J. Immunol.* 24, 2429–2434.
20. Ashkenazi A, Capon D, Ward R. (1993) Immunoadhesins. *Int. Rev. Immunol.* 10, 219–227.
21. Chamow SM, Ashkenazi A. (1996) Immunoadhesins: principles and applications. *Trends Biotechnol.* 14, 52–60.
22. Martin MG, Wu SV, Walsh JH. (1997) Ontogenic development and distribution of antibody transport and Fc receptor mRNA expression in rat intestine. *Dig. Dis. Sci.* 42, 1062–1069.
23. Roberts DM, Guenthert M, Rodewald R. (1990) Isolation and characterization of the Fc receptor from the fetal yolk sac of the rat. *J. Cell Biol.* 111, 1867–1876.
24. Borvak J, Richardson J, Medesan C, Antohe F, Radu C, Simionescu M, et al. (1998) Functional expression of the MHC class I-related receptor, FcRn, in endothelial cells of mice. *Int. Immunol.* 10, 1289–1298.
25. Montoyo HP, Vaccaro C, Hafner M, Ober RJ, Mueller W, Ward ES. (2009) Conditional deletion of the MHC class I-related receptor FcRn reveals the sites of IgG homeostasis in mice. *PNAS USA* 106, 2788–2793.
26. Yoshida M, Masuda A, Kuo T, Kobayashi K, Claypool S, Takagawa T, et al. (2006) IgG transport across mucosal barriers by neonatal Fc receptor for IgG and mucosal immunity. *Springer Semin. Immunopathol.* 28, 397–403.
27. Telleman P, Junghans RP. (2000) The role of the Brambell receptor (FcRB) in liver: protection of endocytosed immunoglobulin G (IgG) from catabolism in hepatocytes rather than transport of IgG to bile. *Immunology* 100, 245–251.
28. Antohe F, Radulescu L, Gafencu A, Ghetie V, Simionescu M. (2001) Expression of functionally active FcRn and the differentiated bidirectional transport of IgG in human placental endothelial cells. *Hum. Immunol.* 62, 93–105.
29. Ober RJ, Martinez C, Vaccaro C, Zhou J, Ward ES. (2004) Visualizing the site and dynamics of IgG salvage by the MHC class I-related receptor, FcRn. *J. Immunol.* 172, 2021–2029.
30. Ward ES, Ober RJ, Frederick WA. (2009) Chapter 4. Multitasking by exploitation of intracellular transport functions: the many faces of FcRn. In: *Advances in Immunology*. Academic Press, pp. 77–115.
31. Bitonti AJ, Dumont JA. (2006) Pulmonary administration of therapeutic proteins using an immunoglobulin transport pathway. *Adv. Drug Deliv. Rev.* 58, 1106–1118.
32. Dumont JA, Low SC, Peters RT, Bitonti AJ. (2006) Monomeric Fc fusions: impact on pharmacokinetic and biological activity of protein therapeutics. *Biodrugs* 20, 151–160.
33. Low SC, Nunes SL, Bitonti AJ, Dumont JA. (2005) Oral and pulmonary delivery of FSH-Fc fusion proteins via neonatal Fc receptor-mediated transcytosis. *Hum. Reprod.* 20, 1805–1813.
34. Graber SE, Krantz SB. (1978) Erythropoietin and the control of red cell production. *Annu. Rev. Med.* 29, 51–66.
35. Dumont JA, Bitonti AJ, Clark D, Evans S, Pickford M, Newman SP. (2005) Delivery of an erythropoietin-Fc fusion protein by inhalation in humans through an immunoglobulin transport pathway. *J. Aerosol Med.* 18, 294–303.
36. Farquhar MG. (1978) Recovery of surface membrane in anterior pituitary cells. Variations in traffic detected with anionic and cationic ferritin. *J. Cell Biol.* 77, 35–42.
37. Williams MC. (1984) Endocytosis in alveolar type II cells: effect of charge and size of tracers. *Proc. Natl. Acad. Sci. USA* 19, 6054–6058.
38. Noseworthy JH, Lucchinetti C, Rodriguez M, Weinshenker BG. (2000) Multiple sclerosis. *N. Engl. J. Med.* 343, 938–952.
39. Rubinstein S, Familletti PC, Pestka S. (1981) Convenient assay for interferons. *J. Virol.* 37, 755–758.
40. Salmon P, Le Cotonnec J, Galazka A, Abdul-Ahad A, Darragh A. (1996) Pharmacokinetics and pharmacodynamics of recombinant human interferon-beta in healthy male volunteers. *J. Interferon Cytokine Res.* 16, 759–764.
41. Dumont JA, Low SC, Peters RT, Bitonti AJ. (2009) *Therapeutic Proteins from Bench to Clinic*, Chapter 33. John Wiley & Sons, Inc., pp. 779–795.
42. Vallee S, Rakhe S, Reidy T, Walker S, Lu Q, Sakorafas P, et al. (2012) Pulmonary administration of IFN beta1a-Fc fusion protein in non-human primates using an immunoglobulin transport pathway. *J. Interferon Cytokine Res.* 32(4), 178–184.
43. Brassard DL, Grace MJ, Bordens RW. (2002) Interferon alpha as an immunotherapeutic protein. *J. Leukoc. Biol.* 71565–581.
44. Kozlowski A, Milton Harris J. (2001) Improvements in protein PEGylation: pegylated interferons for treatment of hepatitis C. *J. Control. Release* 72, 217–224.
45. Mannucci PM, Tuddenham EGD. (2001) The hemophilias: from royal genes to gene therapy. *N. Engl. J. Med.* 344, 1773–1779.

46. Soucie JM, Nuss R, Evatt B, et al. (2000) Mortality among males with hemophilia relationship with source of medical care: the hemophilia surveillance system project investigators. *Blood,* 96(2), 437–442.
47. Björkman S, Berntorp E. (2001) Pharmacokinetics of coagulation factors: clinical relevance for patients with haemophilia. *Clin. Pharmacokinet.* 40, 815–832.
48. Björkman S, Folkesson A, Jönsson S. (2009) Pharmacokinetics and dose requirements of factor VIII over the age range 3–74 years: a population analysis based on 50 patients with long-term prophylactic treatment for haemophilia A. *Eur. J. Clin. Pharmacol.* June 26, 2009. [Epub ahead of print].
49. Sandberg H, Almstedt A, Brandt J, Gray E, Holmquist L, Oswaldsson U, et al. (2001) Structural and functional characteristics of the B-domain-deleted recombinant factor VIII protein, r-VIII SQ. *Thromb. Haemost.* 85, 93–100.
50. Toole J, Pittman D, Orr E. (1986) A large region (=95 kDa) of human factor VIII is dispensable for *in vitro* procoagulant activity. *Proc. Natl. Acad. Sci. USA* 83, 5939–5942.
51. Graham JB, Buckwalter JA, Hartley LJ, Brinkhous KM. (1949) Canine hemophilia; observations on the course, the clotting anomaly, and the effect of blood transfusions. *J. Exp. Med.* 90, 97–111.
52. Lozier JN, Dutra A, Pak E, Zhou N, Zheng Z, Nichols TC, et al. (2002) The Chapel Hill hemophilia A dog colony exhibits a factor VIII gene inversion. *PNAS USA* 99, 12991–12996.
53. Dumont J, Liu T, Low S, Zhang X, Kamphaus GD, Sakorafas P, Fraley C, Drager D, Reidy T, McCue J, Franck HWG, Merricks EP, Nichols TC, Ashworth T, Franck H, Merricks EP, Nichols TC, Bitonti AJ, Pierce GF, Jiang H. (2012) Prolonged activity of a recombinant factor VIII-Fc fusion protein in hemophilia A mice and dogs. *Blood,* 119(13), 3024–3030.
54. Powell J, Josephson NC, Quon, et al. (2011) Pharmacokinetics and safety of recombinant factor VIII Fc fusion protein (rFVIIIFc) in a Phase 1/2a trial in previously treated patients (PTPs) with severe hemophilia A. *J. Thromb. Haemost.* (s2), O-TU-073, 283.
55. Powell JS, Josephson NC, Quon D, Ragni MV, Cheng G, Li E, et al. (2012) Safety and prolonged activity of recombinant factor VIII Fc fusion protein in hemophilia A patients. *Blood,* 119(13), 3031–3037.
56. Bolton-Maggs PHB, Pasi KJ. (2003) Haemophilias A and B. *Lancet* 361, 1801–1809.
57. Green P, Naylor J, Giannelli F. (1995) The hemophilias. *Adv. Genet.* 32, 99–139.
58. Limentani SA, Roth DA, Furie BC, Furie B. (1993) Recombinant blood clotting proteins for hemophilia therapy. *Semin. Thromb. Hemost.* 19, 62–72.
59. Giangrande P. (2005) Haemophilia B: christmas disease. *Expert Opin. Pharmacother.* 6, 1517–1524.
60. Lin H-F, Maeda N, Smithies O, Straight DL, Stafford DW. (1997) A coagulation factor IX-deficient mouse model for human hemophilia B. *Blood* 90, 3962–3966.
61. Peters RT, Low SC, Kamphaus GD, Dumont JA, Amari JV, Lu Q, et al. (2010) Prolonged activity of factor IX as a monomeric Fc fusion protein. *Blood* 115, 2057–2064.
62. Brinkhous KM, Sigman JL, Read MS, Stewart PF, McCarthy KP, Timony GA, et al. (1996) Recombinant human factor IX: replacement therapy, prophylaxis, and pharmacokinetics in canine hemophilia B. *Blood* 88, 2603–2610.
63. McCarthy K, Stewart P, Sigman J, Read M, Keith JCJr, Brinkhous KM, et al. (2002) Pharmacokinetics of recombinant factor IX after intravenous and subcutaneous administration in dogs and cynomolgus monkeys. *Thromb. Haemost.* 87, 824–830.
64. Shapiro AD, Ragni M, Valentino L, et al. (2010) Safety and prolonged biological activity following a single administration of a recombinant molecular fusion of native human coagulation factor IX and the Fc region of immunoglobulin G (IgG) (rFIXFc) to subjects with hemophilia B. *Haemophilia* 16(Suppl. 4), 30.
65. Shapiro AD, Ragni MV, Valentino LA, Key NS, Josephson NC, Powell JS, et al. (2012) Recombinant factor IX-Fc fusion protein (rFIXFc) demonstrates safety and prolonged activity in a Phase 1/2a study in hemophilia B patients. *Blood* 119(3), 666–672.
66. Roth DA, Kessler CM, Pasi KJ, Rup B, Courter SG, Tubridy KL. (2001) Human recombinant factor IX: safety and efficacy studies in hemophilia B patients previously treated with plasma-derived factor IX concentrates. *Blood* 98, 3600–3606.
67. BeneFIX® (coagulation factor IX Recombinant) Summary of Product Characteristics for BeneFIX, Electronic Medicines Compendium 2010, http://www.medicines.org.uk/emc/medicine/20376/SPC/BeneFIX/#PHARMACODYNAMIC_PROPS. accessed November 6, 2011.
68. Schaison G, Young J, Pholsena M, Nahoul K, Couzinet B. (1993) Failure of combined follicle-stimulating hormone-testosterone administration to initiate and/or maintain spermatogenesis in men with hypogonadotropic hypogonadism. *J. Clin. Endocrinol. Metab.* 77, 1545–1549.
69. Schoot DC, Harlin J, Shoham Z, Mannaerts BMJL, Lahlou N, Bouchard P, et al. (1994) Endocrinology: recombinant human follicle-stimulating hormone and ovarian response in gonadotrophin-deficient women. *Hum. Reprod.* 9, 1237–1242.
70. Steelman SL, Pohley FM. (1953) Assay of the follicle stimulating hormone based on the augmentation with human chorionic gonadotropin. *Endocrinology* 53, 604–616.
71. Pierce JG, Parsons TF. (1981) Glycoprotein hormones: structure and function. *Annu. Rev. Biochem.* 50, 465–495.
72. Jia X-C, Hsueh AJW. (1986) Granulosa cell aromatase bioassay for follicle-stimulating hormone: validation and application of the method. *Endocrinology* 119, 1570–1577.
73. Keene JL, Matzuk MM, Otani T, Fauser BC, Galway AB, Hsueh AJ, et al. (1989) Expression of biologically active human follitropin in Chinese hamster ovary cells. *J. Biol. Chem.* 264, 4769–4775.
74. de Leeuw R, Mulders J, Voortman G, Rombout F, Damm J, Kloosterboer L. (1996) Recombinat hormones: structure–function relationship of recombinant follicle stimulating hormone (Puregon®). *Mol. Hum. Reprod.* 2, 361–369.

75. Meachem SJ, McLachlan RI, de Kretser DM, Robertson DM, Wreford NG. (1996) Neonatal exposure of rats to recombinant follicle stimulating hormone increases adult Sertoli and spermatogenic cell numbers. *Biol. Reprod.* 54, 36–44.
76. Morris B, Morris R. (1974) The effects of cortisone acetate on stomach evacuation and the absorption of ^{125}I-labelled globulins in young rats. *J. Physiol.* 240, 79–89.
77. Simister NE, Rees AR. (1985) Isolation and characterization of an Fc receptor from neonatal rat small intestine. *Eur. J. Immunol.* 15, 733–738.
78. Gillies S, Lan Y, Brunkhorst B, Wong W-K, Li Y, Lo K-M. (2002) Bi-functional cytokine fusion proteins for gene therapy and antibody-targeted treatment of cancer. *Cancer Immunol. Immunother.* 51, 449–460.
79. Economides A, Carpenter L, Rudge J, Wong V, Koehler-Stec E, Hartnett C, et al. (2003) Cytokine traps: multi-component, high-affinity blockers of cytokine action. *Nat. Med.* 9, 47–52.
80. Cox GN, Smith DJ, Carlson SJ, Bendele AM, Chlipala EA, Doherty DH. (2004) Enhanced circulating half-life and hematopoietic properties of a human granulocyte colony-stimulating factor/immunoglobulin fusion protein. *Exp. Hematol.* 32, 441–449.
81. Jazayeri JA, Carroll GJ. (2008) Fc-based cytokines: prospects for engineering superior therapeutics. *BioDrugs* 22, 11–26.
82. Wang Q, Chen K, Liu R, Zhao F, Gupta S, Zhang N, et al. (2010) Novel GLP-1 fusion chimera as potent long acting GLP-1 receptor agonist. *PLoS ONE* 5, e12734.
83. Goldzweig O, Hashkes PJ. (2011) Abatacept in the treatment of polyarticular JIA development clinical utility and place in therapy. *Drug Des. Devel. Ther.* 5, 61–70.
84. Gottlieb AB. (2005) Alefacept for psoriasis and psoriatic arthritis. *Ann. Rheum. Dis.* 64, iv58–iv60.
85. Molineux G. (2011) The development of romiplostim for patients with immune thrombocytopenia. *Ann. N.Y. Acad. Sci.* 1222, 55–63.
86. Moreland LW, Baumgartner SW, Schiff MH, Tindall EA, Fleischmann RM, Weaver AL, et al. (1997) Treatment of rheumatoid arthritis with a recombinant human tumor necrosis factor receptor (p75)-Fc fusion protein. *N. Engl. J. Med.* 337, 141–147.
87. Stahl N, Radin A, Mellis S. (2009) Rilonacept—CAPS and beyond. *Ann. N.Y. Acad. Sci.* 1182, 124–134.
88. Su VCh, Harrison J, Rogers C, Ensom MH. (2012) Belatacept: a new biologic and its role in kidney transplantation. *Ann. Pharmacother.* 46(1), 57–67.
89. Ohr M, Kaiser PK. (2012) Intravitreal aflibercept injection for neovascular (wet) age-related macular degeneration. *Expert Opin. Pharmacother.* 13(4), 585–591.
90. Dwyer MA, Huang AJ, Pan CQ, Lazarus RA. (1999) Expression and characterization of a DNase I-Fc fusion enzyme. *J. Biol. Chem.* 274, 9738–9743.
91. Lee CH, Woo JH, Cho KK, Kang SH, Kang SK, Choi YJ. (2007) Expression and characterization of human growth hormone-Fc fusion proteins for transcytosis induction. *Biotechnol. Appl. Biochem.* 46, 211–217.
92. Hapuarachchi S, Fodor S, Apostol I, Huang G. (2011) Use of capillary electrophoresis-sodium dodecyl sulfate to monitor disulfide scrambled forms of an Fc fusion protein during purification process. *Anal. Biochem.* 414, 187–195.
93. Sahdev S, Khattar S, Saini K. (2008) Production of active eukaryotic proteins through bacterial expression systems: a review of the existing biotechnology strategies. *Mol. Cell. Biochem.* 307, 249–264.
94. De Pourcq K, De Schutter K, Callewaert N. (2010) Engineering of glycosylation in yeast and other fungi: current state and perspectives. *Appl. Microbiol. Biotechnol.* 87, 1617–1631.
95. Jacobs PP, Geysens S, Vervecken W, Contreras R, Callewaert N. (2008) Engineering complex-type N-glycosylation in *Pichia pastoris* using GlycoSwitch technology. *Nat. Protoc.* 4, 58–70.
96. Ghetie V, Ward ES. (2000) Multiple roles for the major histocompatibility complex class I related receptor FcRn. *Annu. Rev. Immunol.* 18, 739–766.
97. Roopenian DC, Akilesh S. (2007) FcRn: the neonatal Fc receptor comes of age. *Nat. Rev. Immunol.* 7, 715–725.
98. Ober RJ, Radu CG, Ghetie V, Ward ES. (2001) Differences in promiscuity for antibody-FcRn interactions across species: implications for therapeutic antibodies. *Int. Immunol.* 13, 1551–1559.
99. McCue JT, Selvitelli K, Walker J. (2009) Application of a novel affinity adsorbent for the capture and purification of recombinant factor VIII compounds. *J. Chromatogr. A* 1216, 7824–7830.
100. Zhang J, Carter J, Siu S, O'Neill JW, Gates AH, Delaney J, et al. (2010) Fusion partners as a tool for the expression of difficult proteins in mammalian cells. *Curr. Pharm. Biotechnol.* 11, 241–245.

8

PEPTIDE-Fc FUSION THERAPEUTICS: APPLICATIONS AND CHALLENGES

CHICHI HUANG[1] AND RONALD V. SWANSON[2]

[1]*Biologics Research, Centocor R&D, A Division of Johnson & Johnson Pharmaceutical Research & Development, Radnor, PA, USA*
[2]*Biologics Research, Centocor R&D, A Division of Johnson & Johnson Pharmaceutical Research & Development, San Diego, CA, USA*

8.1 INTRODUCTION

In 1982, the FDA approved the first recombinant protein drug, Humulin-R, a recombinant form of insulin developed for the treatment of diabetes. This approval is one of the most important milestones in the history of biotechnology and protein therapeutics. Since then, the development and use of biological drugs, including peptides, proteins, recombinant proteins with native or modified sequences, and fusion proteins, have undergone tremendous growth for the treatment of a wide variety of diseases. More than 300 distinct biopharmaceuticals (around 40% of all drug candidates) are in clinical trials for various disease indications including cancer, Alzheimer's disease, and chronic inflammation. In the United States, biopharmaceuticals now account for one in four submissions for FDA approval. Over 130 proteins or peptides have been approved to date for clinical use, and this number is growing rapidly. The top seven major biological drugs, including Remicade®, Enbrel®, Avastin®, Rituxan®, Humira®, Herceptin®, and Lantus®, generated over $40 billion in 2010 [1–3].

Much of the success of biopharmaceuticals comes from recombinant monoclonal antibodies (mAbs), thanks to their favorable biological and pharmacological properties. Four of the top 10 therapeutic drugs of 2010 by sales are mAbs, including Remicade, Avastin, Rituxan, and Humira [3]. It has been predicted that by 2014, five mAbs will be in the top 10 list and that the sales of these mAbs will be over $45 billion [4]. Despite the great success antibody drugs have enjoyed, their utility has mainly been limited to antagonism, in concordance with what antibodies have evolved to do: to neutralize the function and accelerate the clearance of a foreign antigen. Many existing native peptides and receptor ligands are potent agonists, however, and represent a rich source of therapeutic potential. Although there are several successful exceptions such as the GLP-1-analog peptide drug Exanatide, unmodified naturally existing peptides have not traditionally made good drugs.

Several key issues arise when contemplating peptides as a source of drug candidates. Many peptides are very flexible and have lower binding affinity owing to their loss of significant amounts of entropy while binding to targets. Peptides also have a very short half-life owing to rapid renal clearance and proteolytic degradation in the blood, kidneys, or liver. They are typically cleared from the bloodstream

Fusion Protein Technologies for Biopharmaceuticals: Applications and Challenges, First Edition. Edited by Stefan R. Schmidt.

within minutes after administration. As a result, there can be insufficient exposure in the target tissue to have a pharmacodynamic effect [5]. Although transformation of a metabolically labile peptide into a drug remains challenging, considerable efforts and advances have been made to increase the *in vivo* stability and circulating half-life of peptides. Several approaches, have been developed, including conjugation of polyethylene glycol (PEG) to a peptide to increase its hydrodynamic radius and reduce kidney filtration, fusion with albumin-binding peptides, proteins, or lipids to attach to serum albumin, chemical conjugation of peptides to covalently link to human serum albumin (HSA) or other long circulating proteins, including site-specific covalent attachment of peptides to an antibody, genetic fusion of peptides to Fc domains of human IgG, HSA, or transferrin. Among these approaches, Fc fusion is the most advanced and currently the most successful approach. Seven Fc-fusion drugs have been developed and approved for treating patients, and many more targeted at a variety of different disease areas are in the pipeline.

In this chapter, we review the successes and limitations of therapeutic drugs, the technologies developed in the last 20 years for improving peptide stability and serum half-life, and the key for the success of peptide therapeutics, including the Fc-fusion platform. However, the main focus of the discussion is on current approaches in which peptide- Fc-fusion proteins are used as receptor agonist biotherapeutics. Examples of peptide-Fc-fusion proteins that have been on the market or in clinical trials are discussed in detail. In order for an Fc-fusion protein to become a successful therapeutic, it is often necessary to optimize its biological, biophysical, and pharmacological properties, including activity, stability, solubility, heterogeneity, and PK profiles. In this chapter, we discuss in detail the challenges faced in the design and development of Fc-fusion drugs and provide some examples of how we dealt with these challenges in specific cases.

8.2 PEPTIDE DRUGS

One traditional definition of a peptide is a polypeptide chain that is short enough to be chemically synthesized as distinguished from a protein. With the rapid progress of peptide synthesis techniques, especially the development of peptide ligation technology, a polypeptide with hundreds of amino acids long can now be synthesized. Therefore, the size that used to distinguish a peptide from a protein on the basis of synthesis can be ambiguous. An arbitrary definition that a peptide is a chain shorter than 50 amino acids is widely accepted and is used here.

Native endogenous peptides play many very important physiological and biochemical roles in humans. The development of sensitive analytical techniques for isolation, purification, and identification of peptides allowed the many important functional roles of peptides to begin to be elucidated. Some very important peptides in biological process, such as glucagon, corticotrophin, α- and β-melanotropin, oxytocin, vasopressin, and angiotensin, were isolated and their sequences were determined. So far, a large number of naturally existing peptides in our body with their functions in the central and peripheral nervous systems, immunological processes, cardiovascular system, and the intestine were identified [6]. Their potent functions as neurotransmitters, neuromodulators, and hormones involved in a number of physiological processes, including metabolism, pain, reproduction, and immune response give them great potential as sources of therapeutic drugs.

8.2.1 The Limitations of Native Endogenous Peptides in Drug Development

Despite several successful examples, such as the GLP-1-analog peptide drug Exanatide developed by Amylin and Eli Lilly, unmodified peptides have not traditionally made good drugs. Several key issues have prevented peptides from becoming a major source of drug candidates in their native form. Many peptides are very flexible and have lower binding affinity owing to their loss of significant amounts of entropy upon binding to targets. The flexible conformation of peptides renders them easily accessed and cleaved in tissue or cellular compartments by specific endopeptidases and in the systemic circulation by less specific exopeptidases (amino- and carboxy-peptidases). In addition, owing to their very small hydrodynamic radius, peptides are rapidly removed from the circulation by glomerular filtration (renal clearance). Protease susceptibility and rapid clearance are the major reasons that native peptides typically have for their very short serum half-life; these attributes are important physiologically for the inactivation of the peptides after they have exerted their functions. They are usually cleared from the bloodstream within minutes after administration. As a result, there can be insufficient time for them to reach and act in the target tissue to have a pharmacodynamic effect. Although transformation of a metabolically labile natural peptide into a drug remains challenging, considerable advances have been made to increase the *in vivo* stability and circulating half-life of peptides.

8.2.2 Chemical Modification of Peptides to Reduce Proteolytic Degradation

In the last several decades, chemical modifications of peptides have been proven very successful in improving resistance to protease cleavage while preserving the potency and selectivity of the bioactive natural peptides [7]. One commonly used approach is chemically modifying the N- or C-terminus or both ends by N-terminal acetylation or C-terminal amidation. Other terminal modifications, for example, addition of carbohydrate chains such as glycosylation: glucose, xylose, hexose, and so on, have also been used. The modifications of the terminus of a peptide can protect it from exopeptidase cleavage. Another commonly used approach is to identify the protease-sensitive sites by alanine or/and by D-amino acid scanning combined with a variety of analytical tools and then replace the natural amino acids on the sites by nonproteogenic constrained amino acids or unnatural amino acid such as *N*-methyl-α-amino acids, or other nonnative amino acids. Owing to the stereochemical specificity of peptidases, changing some or even all of L-amino acids of a peptide with their corresponding D-amino acids can improve resistance to proteolytic degradation. In addition, D-amino acid peptides have been found to be significantly less immunogenic than their corresponding L-amino acid peptides [7,8].

The chemical modifications mentioned earlier are limited to either the terminus of a peptide or the side chain of residues in the peptide without changing the peptide backbone. In many cases, however, modifications of peptide bonds are necessary. These modifications include replacement of amide bond (amide bond surrogates), attachment of side chains to the amide nitrogen atoms instead of α-carbons (peptoids), and replacement of α-carbons with nitrogen atoms (azapeptides). Peptides with backbone modifications such as these are referred to as pseudopeptides [7].

8.2.3 Constraining Peptides to be More Resistant to Protease Cleavage

Without internal covalent linkages such as disulfide bonding or binding to a larger protein, for example a receptor, peptides alone are flexible and rarely retain a stable tight secondary structure conformation in solution. They take on an ensemble of extended conformations, which are easily accessible and digested by proteases as these enzymes typically bind their substrates in the extended conformation [9]. However, a group of constrained cyclic peptides have been discovered in nature, whose amino and carboxyl termini are linked together with a peptide bond to form a circular peptide. Some naturally existing cyclic peptides such as the amatoxins, amanitin, and phalloidin contain an additional bridge between the side chains of two different residues within the peptide to form a bicyclic structure. The covalent linkages between N- and C-termini or between two side chains of a cyclic or a bicyclic peptide can comprise either normal peptide bonds such as in Cyclosporin, an immunosuppressant drug, or nonpeptide bonds such as in somatostatin, a peptide hormone that is cyclized through a disulfide bond between two cysteines. When constrained by cyclization, naturally existing peptides tend to be extremely resistant to the process of digestion, in some cases enabling them to survive intact in the human digestive tract. The strategies of cyclization of biologically active peptides to stabilize peptides *in vivo* and extend their serum half-life have often been used in drug development [10].

The peptide stapling strategy for stabilizing peptide α-helices is another approach that has demonstrated promise in future drug development. The α-helix features 3.6 residues per complete turn, which places the side chains of residues i, $i+4$, $i+7$, and $i+11$ on the same face of the folded structure. The peptide stapling strategy employs covalent bonds between the i and $i+4$ or i and $i+7$ side chain groups to stabilize the helical conformation. Lactam, disulfide and metal-mediated bridges, and more recently the all-hydrocarbon bridge using a ring-closing metathesis reaction have been used to staple helical peptides. Stapling increases the helical propensity, in particular, stapling with an all-hydrocarbon cross-linker resulted in significant improvements in structure, potency, and protease resistance [11–13].

8.2.4 Peptide Therapeutics Play an Important Role in Drug Development

Besides the advances in techniques, as discussed earlier, which improve *in vivo* stability in peptides, several factors, including the significant progress of industrial-scale synthetic production, formulation and delivery of peptides, widespread acceptance of protein therapeutics by physicians and patients, dramatic increase in the availability of genomic information, and the development of more sensitive analytical techniques, have all contributed to peptides, especially chemically modified peptides, becoming a more important therapeutic drug resource.

Peptide drugs have drawn increasing interest from the pharmaceutical industry and enjoyed steady growth since the 1980s. The average number of peptide drug candidates entering into the clinical trials increased from 1.2 per year in the 1970s to 4.6 per year in the 1980s, 9.7 per year in the 1990s, and 16.9 per year in the first decade of this century [14]. Peptide drugs have been targeted to a wide variety of indications, and the number of therapeutic categories has also expanded significantly during this period. In a number of categories, at least six different compounds progressed into clinical trials; these categories include metabolic disorders, oncology, allergy and immunology, cardiovascular diseases, CNS, infections, pain, and endocrinology [14]. In 2008, global sales of six peptide therapeutic drugs, including Copaxone (random polymer of four amino acids), Lupron®, Zoladex®, Sandostatin®, Forteo® (recombinant parathyroid hormone, PTH), and Byetta® (synthetic version of exendin-4 found in the saliva of Gila monster), reached more than US$750 million [14].

8.2.5 Most Peptide Drugs Act as Receptor Agonists

Receptors and their ligands serve as targets for some of the most successful biological therapeutics on the market, including mAb drugs such as Remicade, Rituxan, and Herceptin, receptor-Fc-fusion protein drug such as Enbrel, and recombinant protein drugs such as Aranesp® and Procrit®. However, unlike mAb drugs, whose mode of action has mainly been limited to antagonism, the majority of the peptide therapeutics in clinical trials or on the market acts as receptor agonists. For example, Lupron (leuprolide), a small nine amino acid peptide analog of Gonadotropin-releasing hormone (GnRH), is an agonist of pituitary GnRH receptors. By stimulating its receptors, Lupron downregulates the secretion of the gonadotropins, luteinizing hormone (LH), and follicle-stimulating hormone (FSH), leading to a dramatic reduction in estradiol levels. The drug is approved for treating prostate and breast cancers [15]. Zoladex (Goserelin acetate), developed by AstraZeneca, is another GbRH analog approved in 1989 for suppressing the production of the sex hormones, testosterone and estrogen, for treating breast and prostate cancers [16]. Sandostatin or Octreo® (Octreotide), developed by Novartis Pharmaceuticals and New Medicon Pharma, is an octapeptide analog of somatostatin which down regulates growth hormone, glucagon, and insulin by stimulating the somatostatin receptor for the treatment of acromegaly [17]. The drug is also used for reducing watery diarrhea and flushing episodes caused by cancerous tumors [18]. Forteo (Teriparatide), a recombinant form of the first 34 amino acids of PTH, is another receptor agonist, developed by Eli Lilly. The drug was approved in 2002 for treating osteoporotic men and postmenopausal women who are at high risk of having a fracture [19]. Byetta (Exenatide, exendin-4), a 39 amino acid peptide isolated from saliva of the Gila monster, is an agonist of the GLP-1 receptor. The drug, developed by Amylin Pharmaceuticals and Eli Lilly, was approved for the treatment of type 2 diabetic patients in 2005.

More than 60% of approved peptide drugs target members of the G-protein coupled receptor (GPCR) family which includes around 800 different receptors predicted from genome sequence analysis (about 4% of the entire protein-coding genome codes for GPCRs). GPCRs activate many cellular responses, such as estrogen secretion, neuropeptide signaling, opiate receptivity and vision pathways (class A GPCRs), calcium and glucose metabolism, growth hormone secretion, and gastrointestinal tract function (class B GPCRs) [14]. Almost 30% of all drugs target GPCRs and most of the peptide drugs targeting them are receptor agonists [20].

The success of peptide drugs acting as receptor agonists highlights several advantages over both small molecules and large protein drugs such as antibodies and receptor-Fc-fusion proteins. The peptide drugs are significantly larger than traditional small molecules, enabling them to more specifically bind targets with higher affinity due to the larger available contact surface; therefore, peptides usually have a more favorable off-target toxicity profile relative to small molecules [21]. Compared to antibodies or soluble receptors, on the other hand, they are much smaller and have the potential to penetrate deeper into tissues and exhibit better bioavailability. They can be generally less immunogenic and can have higher solubility and better stability during the storage and transport. More importantly, many peptide drugs naturally regulate crucial biological processes functioning as agonists, which have proven very difficult for both large recombinant protein therapeutics, such as mAbs and receptor-Fc fusions, and small molecules to achieve.

8.3 TECHNOLOGIES USED FOR REDUCING *IN VIVO* CLEARANCE OF THERAPEUTIC PEPTIDES

Although chemical modification of peptides can significantly improve the pharmacokinetic profiles by increasing the serum half-life from minutes to hours, treatment of many chronic diseases requires single or multiple injections daily. As a comparison, the circulating half-life of many mAb drugs is not hours or days but weeks. Rapid clearance of peptides after administration from the circulation via the kidneys is mainly because of their small size (<5 kDa). A number of techniques have been developed to reduce the renal clearance by fusing or attaching peptides to molecules with larger molecular weight (generally >50 kDa) to retard excretion through the kidneys.

8.3.1 PEGylation

One of the most widely accepted methods to improve pharmacokinetic properties of peptides is to conjugate them to polyethylene glycol (PEG) [22]. Often, a single large PEG (e.g., 40 kDa) or multiple small PEGs (e.g., 5 kDa) molecules are used to increase the peptide size. The PEG can be conjugated to peptides on either terminus or can be conjugated to cysteine or lysine residues at internal positions in the peptide [23]. The conjugation of PEG not only increases the hydrodynamic size of the peptides thus reducing their clearance, but can also protect them from peptidase cleavages *in vivo* by masking sensitive sites [24]. By carefully selecting the conjugation sites and testing different PEG conjugates, a properly conjugated peptide, which retains most of its activities while increasing its serum half-life from minutes to tens of hours, can be identified. PEGylation of peptides can also confer additional positive attributes such as reducing immunogenicity and improving solubility [25–27].

8.3.2 Human Serum Albumin Fusions

Human serum albumin (HSA), comprising about half of the blood serum protein, is the most abundant protein in human blood plasma. It has good solubility and is widely distributed *in vivo*. It has a long circulatory half-life (up to 19 days in humans). It is a monomer with molecular weight of 67 kDa. Similar to Fc-fusion proteins, the idea of improving circulating half-life of peptides or some proteins by fusing them to HSA on either the N- or C-terminus has been developed and proven to be successful in drug development. One major difference between Fc fusions and HSA-fusion protein is that the Fc fusions are homodimers but the HSA fusions are monomers. The most clinically advanced example of peptide-HSA fusion, Syncria® (albiglutide), a glucagon-like peptide-1 (GLP-1) peptide genetically fused to HSA, created originally by Human Genome Sciences and licensed to GlaxoSmithKline, is in Phase III clinical trials for the treatment of type 2 diabetes mellitus [28]. Several other HSA-fusion therapeutics are in different stages of clinical trials, including Zalbin, an interferon (IFN)-α2b fused to HSA which was codeveloped by Human Genome Science and Novartis (discontinued in development), for treatment of chronic hepatitis C; Albugranin, a granulocyte colony-stimulating factor fused to HSA, for neutropenia; Albulin, an insulin fused to HSA for diabetes; and AlbuBNP (also called Cardeva™), the B-type natriuretic peptide fused to HSA for chronic heart failure [29–32].

8.3.3 Other Approaches

Several other approaches, including the use of albumin-binding peptides or proteins to attach to serum albumin, developed by Genentech and Dyax, and the use of chemical conjugation of peptides to covalently link to HSA to extend serum half-life, were reviewed by McGregor [26]. Another interesting approach was developed by CovX (acquired by Pfizer), in which an antibody having aldolase activity catalyzes the site-specific covalent attachment of peptides to itself [33].

8.4 Fc-FUSION PROTEINS IN DRUG DEVELOPMENT

The Fc-fusion technique is the most used and successful technique for extension of serum half-life in both research and drug development. Many of the favorable attributes that have contributed to the successful utilization of antibodies as drugs are embodied in the immunoglobulin constant region or Fc domain. The Fc domain prolongs the serum half-life of proteins, mainly because of the pH-dependent binding of the neonatal Fc receptor (FcRn), which salvages the protein from being degraded in endosomes [34,35] and to limited renal clearance resulting from the increase of mass due to the homodimeric nature of Fc-fusion proteins. Fc-dependent effector functions are important attributes for many indications, particularly for oncology applications. The Fc region in many cases improves the biophysical properties of its fusion partner, such as the solubility and stability of the protein. In addition, high expression, secretion to tissue culture medium, and protein-A affinity purification of Fc-fusion proteins simplify the downstream manufacturing processes by allowing application of the same established procedures for antibody drug manufacturing [36].

8.4.1 The Structure of Fc-Fusion Proteins

In 1989, Daniel Capon at Genetech described the first successful example of using a recombinant soluble receptor to block ligand activity [37]. They fused the ligand-binding domains D1-D2 of the cluster of differentiation 4 (CD4) to the N-terminal of the immunoglobulin G1 (IgG1) Fc domain and demonstrated that the fusion protein can specifically inhibit CD4 activities [38,39]. Since the fusion protein was composed of IgG constant region and ligand-binding domains of CD4, a cell adhesion molecule, it was called an immunoadhesin. These types of proteins are now also commonly referred to as Ig-fusion or Fc-fusion proteins [40]. Figure 8.1A illustrates the overall generic structure of a typical Fc-fusion protein, for example, a receptor-Fc-fusion protein. The C-terminus of the ligand-binding domain or the whole soluble extracellular domain (ECD) of a single receptor chain is genetically fused to the N-terminus of the hinge region, followed by the CH2 and CH3 domains of a human IgG heavy chain. Similar to antibodies or other Fc-containing molecules, the fusion protein forms a homodimer owing to the strong interactions of the CH3 domains. The homodimer is stabilized by interchain disulfide bonds in the hinge regions. This covalently linked homodimer structure provides some of the antibody-like structural properties, such as high affinity binding due to the avidity effect. The hinge regions provide flexible spacers between the two receptor ECDs and the IgG constant regions to allow each part of the molecule the necessary flexibility to function independently.

Several structural variations from the traditional Fc-fusion structure were developed, including the heterodimeric Fc-fusion protein, which contains two distinct fusion partners as shown in Figure 8.1E and monomeric Fc-fusion protein in which only one fusion partner is attached to the dimeric Fc region (Figure 8.1C) [41,42]. The most advanced monomeric Fc fusion is human coagulation Factor IX-Fc for hemophilia B currently in the Phase IIa clinical trials (Table 8.1). The drug is being developed by Biogen Idec (formerly Syntonix) to improve the pulmonary delivery efficiency of Fc-fused proteins by reducing the molecular size. Compared to traditional Fc-fusion proteins, monomeric Fc-fusion proteins have shown higher bioavailability for pulmonary delivery in nonhuman primates, longer circulating half-life, and better biological activities [52,53].

8.4.2 Fc-Fusion Therapeutics

Fc-fusion proteins have been widely used as research tools to study the structure and function of many receptors, identify and isolate novel ligands, and understand better the receptor–ligand interactions [54,55]. More significantly, they

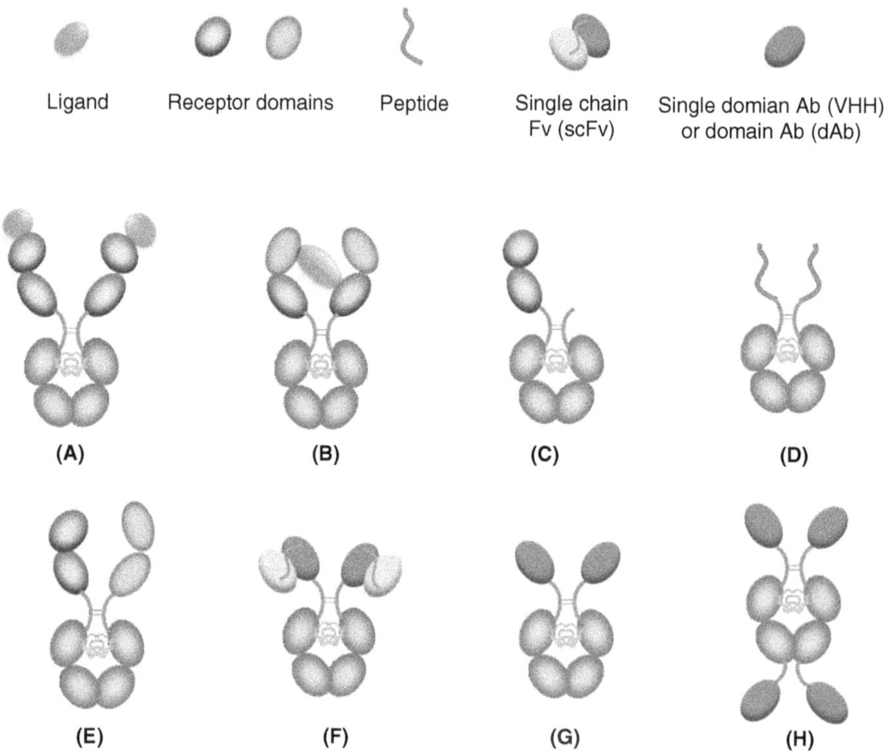

FIGURE 8.1 Schematic representation of selected Fc-fusion proteins. (A) Typical receptor-Fc-fusion protein in which ECD from the same receptor is fused to the Fc domain; (B) multiple ligand binding domains from different receptors are fused to the Fc domain; (C) monomeric-Fc fusion in which only one protein is fused to the Fc domain; (D) MIMETIBODY™ platform; (E) two different Fc-fusion chains form a heterodimer Fc-fusion molecule; (F) single chain Fv (scFv) is fused to the Fc domain; (G) single domain Ab-Fc; (H) single-domain Abs are fused to both N- and C-termini of the Fc domain (Harbour Antibodies). *Source*: Reprinted from *Curr. Opin. Biotechnol.*, **20**(6), Huang C., Receptor-Fc fusion therapeutics, traps, and MIMETIBODY technology. pp. 692–699, Copyright (2009), with permission from Elsevier. (See the color version of this figure in Color Plates section.)

have become an important addition to an ever-growing armamentarium of protein therapeutics. Seven Fc-fusion proteins have been approved by Food and Drug Administration (FDA) as therapeutic drugs and more are in different stages of clinical trials (Tables 8.1 and 8.2).

8.4.2.1 ***Enbrel®*** Enbrel (Etanercept), composed of the 75 kDa soluble ECD of the tumor necrosis factor (TNF)-α receptor II (TNFR II) fused to the Fc domain of human IgG1, is the first successful example of using a soluble receptor as a therapeutic drug. The prototype molecule was designed and was demonstrated to have *in vivo* activity in the early 1990s in laboratory of Bruce Beutler [56,71]. It neutralizes both membrane-bound and soluble forms of TNF-α and thereby indirectly reduces the concentration of serum inflammatory cytokines, serum matrix metalloproteases, and adhesion molecules. The mean circulating half-life of the drug in human is about 68 h [57]. Etanercept was approved by the FDA in 1998 for treatment of rheumatoid arthritis (RA) and other forms of arthritis. Along with other TNF-α antagonist drugs, such as Remicade, an anti-TNF-α chimeric mAb, Etanercept has had a tremendous impact in the treatment and management of RA and enjoyed concomitant commercial success. The sales of this drug reached nearly US$3.6 billion in 2008.

8.4.2.2 ***Amevive®*** Amevive (Alefacept) is the first biological drug approved for the treatment of patients with moderate-to-severe plaque psoriasis. It is composed of the first ECD of lymphocyte function-associated antigen 3 (LFA-3, CD58) fused to the human IgG1 Fc domain [58]. LFA-3, expressed on many cell types including antigen-presenting cells (APCs), is the primary co-receptor for CD2 on T cells and NK cells. The interaction of CD2 on T or NK cells with LFA-3 on APCs provides a costimulatory signal for activating T cells, including the release of inflammatory cytokines, such as IFN-γ. The majority of T cells in psoriatic lesions are of the memory effector phenotype characterized by the presence of the CD45RO

TABLE 8.1 Selected Peptide-Fc and Other Fc Fusion Therapeutics

Drug	Fusion Protein	Target	Mechanism of Action	Indication	Stage	Company	References
Romiplostim (Nplate)	Peptide	Thrombopoietin receptor	Thrombopoietin receptor agonists	Immune thrombocytopenic purpura	Launched	Amgen	[43]
AMG 386	Peptide	Angiopoietin receptor, Tie2	Angiopoietin-1 and -2 inhibitors	Cancers	Phase II	Amgen	[44]
AMG 623	Peptide	Blys	B-cell inhibitors	Systemic lupus erythematosus	Phase I	Amgen Anthera Pharmaceuticals	[45,46]
CNTO 528	EPO mimetic peptide 1	EPO receptor	EPO receptor agonists	Anemia	Phase I	Centocor	[91]
Factor IX-Fc	Factor IX	Factor IX	Factor IX stimulants	Hemophilia B	Phase II	Syntonix Pharmaceuticals and Biovitrum	[53]
SBI 087	Anti-CD20 scFv	CD20	CD20 antigen inhibitors	Rheumatoid arthritis and systemic lupus erythematosus	Phase I	Trubion Pharmaceuticals and Wyeth	[108]
TRU 015	Anti-CD20 scFv	CD20	CD20 antigen inhibitors	Rheumatoid arthritis	Phase II	Trubion Pharmaceuticals and Wyeth	[109,110]
ART 621	Anti-TNF dAb	TNF-α	TNF-α inhibitors	Psoriasis and rheumatoid arthritis	Phase II	Arana Therapeutics and Domantis	[111]

marker. They express high level of cell surface CD2 and therefore are the specific target of Alefacept. By binding to CD2 on T cells, Alefacept can effectively prevent the cross talk of APCs with T cells and therefore inhibits T-cell activation. Alefacept has a significantly better pharmacokinetic profile than Etanercept with a mean circulating half-life of 12 days in humans [72].

8.4.2.3 ***Orencia®*** Orencia (Abatacept) was developed by Bristol-Myers-Squibb (New York, New York) and approved for the treatment of RA. This drug consists of the ECD of human cytotoxic T lymphocyte associated molecule-4 (CTLA-4) fused to a human IgG1 Fc domain. It was designed to block the interactions between CD80 and CD86 on APCs with CD28 on T cells, which provides the secondary costimulatory signal needed for T-cell activation. By binding both CD80 and CD86 with higher affinity than CD28, abatacept effectively prevents the engagement of CD28 on T cells with APCs and therefore blocks T-cell activation and inflammatory cytokine release [73]. To prevent the killing of APCs, both the complement-dependent cytotoxicity (CDC) and the antibody-dependent cellular cytotoxicity (ADCC) activities of abatacept were minimized by the introduction of four single amino acid replacement mutations into the Fc region [74]. The mean terminal half-life of abatacept is 16.7 days in healthy people and 13.1 days in RA patients [69]. Although this drug is currently approved for the treatment of RA, several trials are ongoing for additional indications, including Crohn's disease, systemic sclerodema, ulcerative colitis, psoriatic arthritis, and systemic lupus erythematosus (SLE).

8.4.2.4 ***Arcalyst® (Also Known as IL-1 Trap or Rilonacept)*** Some receptors, including many cytokine receptors, require two or more distinct receptor chains to bind their ligands with high affinity. For example, high affinity binding of IL-1 requires both type I receptors (IL-1RI) and IL-1R accessory proteins (IL-1RacP). For this type of receptor, ligand-binding domains from both receptors have to be incorporated into the Fc-fusion protein to enable high affinity binding. To this end, Regeneron Pharmaceuticals developed the "trap" approach, in which two different ligand-binding domains from different receptor chains were sequentially fused to each other followed by the Fc (Figure 8.1B) [75]. Traps have been shown to be very potent, with IC50s in the single digit picomolar range based on *in vitro* cell-based assays [75]. The high binding affinity is attributed to simultaneous interaction, or "trapping" of a ligand to both chains.

Arcalyst, an example of a cytokine trap, was recently approved by the FDA as an orphan drug for the treatment of patients with a rare autoinflammatory disease known as cryopyrin-associated periodic syndromes (CAPS), which includes familial cold autoinflammatory syndrome, Muckle-Wells syndrome, and neonatal onset multisystem inflammatory disease (Table 8.2). The IL-1 trap consists of the ligand-binding domains of the human IL-1 receptor component (IL-1RI)

TABLE 8.2 Selected Receptor-Fc Fusion Therapeutics

Drug	Fusion Protein	Target	Mechanism of Action	Indication	Stages	Company	References
Etanercept (Enbrel)	p75, TNFaR	TNF-α	TNF-α inhibitors	Juvenile rheumatoid arthritis, plaque psoriasis, psoriatic arthritis, and rheumatoid arthritis	Launched	Amgen	[71,57]
Alefacept (Amevive)	LFA-3	CD2	CD2 antigen inhibitors	Psoriasis and transplant rejection	Launched Phase II	Astellas Pharma	[72]
Abatacept (Orencia)	hCTLA-4	CD80 and CD86	T-cell activation inhibitors	Juvenile rheumatoid arthritis and rheumatoid arthritis	Launched	Bristol-Myers Squibb	[73]
Belatacept (BMS 224818)	Mutated CTLA-4	CD80 andCD86	T-cell activation inhibitors	Renal transplant rejection and liver transplant rejection	Phase III	Bristol-Myers Squibb	[64,65]
Atacicept (TACI-Ig)	TACI receptor	Blys and APRIL	B-cell inhibitors	Lupus nephritis, systemic lupus erythematosus, multiple scelerosis, rheumatoid arthritis	Phase II/III	Merck Serono and ZymoGenetics	[66,67]
Briobacept (BR3-Fc)	BR3	BAFF (Blys)	B-cell inhibitors	Rheumatoid arthritis and Sjogren's syndrome	Phase II	Biogen Idec and Genentech	[68]
YSPSL (PSGL-Ig)	PSGL-1	PSGL-1	Selection inhibitors	Reperfusion injury (associated with liver transplantation)	Phase II	Y's Therapeutics	[48,70]
ACE 011 (ActRIIA-IgG1)	Type II activin receptor IIa	Activin	Osteoblast inhibitors	Osteolysis and postmenopausal osteoporosis	Phase II, Phase I	Acceleron Pharma	[49]
ACE 031 (RAP 011)	Activin receptor IIb	Myostatin	Myostatin inhibitors	Musculoskeletal disorders	Phase I	Acceleron Pharma	[50]
APG101	CD95 receptor	CD95 ligand	Apoptosis inhibitors	Graft-versus-host disease	Phase I	Apogenix	www.apogenix.de
FP 1039	FGFR1	FGFs	FGFR1 antagonists	Solid tumors	Phase I	Five Prime Therapeutics	[51]
Rilonacept (Arcalyst) (IL-1 Trap)	Binding domains of IL-1R and IL-1RacP	IL-1	Interleukin-1 inhibitors	Cryopyrin-associated periodic syndromes (CAPS) and gout	Launched Phase III	Regeneron Pharmaceuticals	[43]
Aflibercept (VEGF Trap(R1R2)	Second domain of VEGFR1 and third domain of VFGFR2	VEGF-A and PIGF	Angiogenesis inhibitors	Age-related macular degeneration and cancers	Phase III	Regeneron Pharmaceuticals	[107]

TNF, tumor necrosis factor; LFA-3, lymphocyte function-associated antigen 3; CTLA-4, cytotoxic T lymphocyte associated molecule-4; TACI, transmembrane activator and CAML interactor; APRIL, a proliferation-inducing ligand; BR3, BAFF receptor 3; PSGL-1, P-selectin glycoprotein ligand-1; ActRIIA, activin type IIa receptor; FGFR1, fibroblast growth factor receptor 1c; IL-1RacP, IL-1R accessory proteins; VEGF, vascular endothelial growth factor; PIGF, placental growth factor.

and the IL-1 receptor accessory protein (IL-1RAcP). In this molecule, the C-terminus of the IL-1RAcP ligand binding region is fused to the N-terminus of the IL-1RI extracellular region followed by the Fc portion of human immunoglobulin G1 (IgG1). It is a very potent IL-1R antagonist with a binding affinity of ~1.5 pM to IL-1β and an IC50 of ~6.5 pM in *in vitro* cell-based assays [75]. One big advantage of Rilonacept is its long circulation half-life (8.6 days) allowing a dosing regimen of once a week. In comparison, another IL-1 antagonist drug, Anakinra (recombinant human IL-1 receptor antagonist), must be administered daily owing to rapid clearance *in vivo* [43].

Other therapeutic traps that have been developed as potential therapeutics include the VEGF, IL-4, IL-6, and IL-4/IL-13 traps [5]. VEGF trap (Aflibercept) was recently approved for wet macular degeneration and is in Phase III clinical trials for multiple oncology indications; however, all other traps listed were ultimately dropped from development.

8.4.2.5 Other Fc-Fusion Protein Therapeutics Several other receptor-Fc-fusion proteins are listed in Table 8.2, including Merck Serono and ZymoGenetics' Atacicept (also called TACI-Ig), Biogen Idec and Genentech's Briobacept (also called BR3-Fc), Y's Therapeutics' YSPSL, Apogenix's APG 101, Acceleron Pharma's ACE-011 and ACE-031, and Five Prime's FP1039. Enzymes, hormones, and ligands fused to Fc domains are all represented in Table 8.2.

The Fc domain has also been used to extend the serum half-life of antibody fragments, such as single chain Fv (scFv) (Figure 8.1F), single-domain antibody (V_{HH}) or domain antibodies (dAbs) (Figure 8.1G and H), and different alternative scaffolds. Several antibody fragment-Fc-fusion therapeutics are currently in clinical trials, including Emergent Biosolutions (formerly Trubion) TRU-015 and SBI-087; these are representatives of Trubion's small modular immunopharmaceutical (SMIP^{TM}) technology. Two anti-CD20 SMIP^{TM} proteins are in clinical trials for the treatment of RA and SLE [109,110]. Arana Therapeutics' anti-TNF-α dAb ART621 is an example of using the Fc domain to extend circulating half-life of dAb [111]. Similarly, Harbour Antibodies has genetically fused V_H-binding domains to both the N- and C-termini of an Fc domain to generate a tetravalent antibody as shown in Figure 8.1H [5].

8.5 PEPTIDE-Fc-FUSION THERAPEUTICS

The peptide-Fc-fusion approach has been used by several pharmaceutical companies, including Amgen, Eli Lilly, and Centocor, to extend the serum half-life of bioactive peptides. A peptide can be genetically fused to either the N- or C-terminus of an IgG Fc domain. Modifications to the linker/hinge and Fc region are customized to fit the different needs for drug development. Some of the peptide-Fc fusions have more than one copy of the peptide fused to one Fc domain. Table 8.1 lists selected peptide-Fc-fusion therapeutics in different stages of development.

8.5.1 Peptibody

Peptibody, developed by Amgen Inc., is a technology in which one functional peptide, or in some cases, two copies of the peptide separated with a flexible linker are fused to either the N-terminus or the C-terminus of the human IgG Fc domain. Unlike other Fc-fusion therapeutics, which are typically produced in mammalian cells, most of Amgen's peptibodies were produced in an *Escherichia coli* expression system. Several specific peptibodies, marketed and in clinical trials, are discussed in detail in the following sections.

8.5.1.1 NPlate® (Romiplostim, AMG531) NPlate (also called AMG 531, Romiplostim), the most advanced peptibody, consists of a human IgG1 Fc domain, genetically fused at the C-terminus with two copies of a 14 amino acid long thrombopoietin (TPO) receptor (also known as c-Mpl)-binding peptide identified by phage display. A five amino acid long flexible glycine linker was placed between the Fc and the first peptide, and another longer flexible glycine linker was inserted between the first and second copy of the mimetic peptide [80]. Similar to TPO, binding of NPlate to c-Mpl activates a number of signaling pathways, including tyrosine phosphorylation of c-Mpl, Janus kinase 2, and signal transducer and activator of transcription 5 (STAT5). Since the active peptide of NPlate shares no amino acid sequence homology to endogenous TPO, it theoretically avoids the risk of eliciting cross-reacting neutralizing antibodies to TPO, and indeed no such antibodies have been observed so far [81]. The serum half-life of NPlate ranged from 1 to 34 (median 3.5) days following subcutaneous administration of 3–15 mg/kg/week [81]. The elimination of the drug in serum is partly dependent on the c-Mpl receptor on platelets. Therefore, for a given dose of the drug, patients with high and low platelet counts are associated with low and high serum drug concentrations, respectively [81].

In Phase III trials, NPlate was able to increase and sustain platelet counts in both splenectomized and nonsplenectomized patients. The drug appears to be generally well tolerated. Eighty-three percent of the NPlate treated patients achieved an overall platelet response compared with 7% of patients receiving placebo. The drug provides a valid treatment option that may bring about a change in the way that patients with immune thrombocytopenia purpura (ITP) are treated in the future. NPlate was approved by the FDA in 2008 for the treatment of thrombocytopaenia in patients with chronic ITP [81–84].

8.5.1.2 AMG 386 (2xCon4(C)) Currently in clinical trials, AMG 386 (previously also called 2xCon4(C)) is an

Amgen peptibody that selectively neutralizes Ang1 and Ang2 interactions with their receptor Tie2 and therefore inhibits endothelial cell proliferation and tumor growth. Similar to AMG 531, AMG 386 uses the same peptibody scaffold with two copies of active peptide fused at the C-terminus of human IgG1 Fc. Flexible glycine linkers are placed between Fc and the peptide as well as between the two copies of peptide [44]. The active peptide was derived by screening phage display peptide libraries against human Ang2. The peptide-Fc-fusion protein was produced in *E. coli*. AMG 386 was shown to neutralize both Ang2 and Ang1 binding to their receptor Tie2 with IC50 values of 23 pM and 900 pM, respectively. The combination of Ang1 and Ang2 inhibition prevented vascular normalization without affecting Ang2-mediated reduction in tumor vascularity and led to greater tumor suppression than by inhibiting Ang1 or Ang2 individually. The total serum clearance was similar in the dose range of 0.3–30 mg/kg suggesting that PK does not vary with the serum concentration of the peptibody. The mean value of terminal phase serum half-life ranged from 3.1 to 6.3 days. Pharmacokinetic profiling supported daily to weekly subcutaneous administration.

In a randomized, double-blind, placebo-controlled Phase II study, AMG 386 combined with weekly Paclitaxel in patients with recurrent ovarian carcinoma extends survival of heavily pretreated patients by almost two thirds (4.6–7.2 months), indicating potential as a first line treatment immediately following surgery [45,46].

AMG 386 has been well tolerated in early clinical trials. The toxicity profile of AMG 386 does not appear to overlap with that of other anti-angiogenesis molecules, including bevacizumab, sorafenib, and sunitinib, in relation to bleeding or thromboembolic events. It is therefore possible that future combinations of AMG 386 with other biologics such as the VEGF inhibitors with or without chemotherapy will allow for a multistrategy approach to the management of advanced ovarian cancer.

8.5.1.3 A-623 (AMG 623) A-623 (previously called AMG 623) is an antagonist peptibody designed to inhibit the B-cell activating factor (BAFF) for the treatment for lupus and other autoimmune diseases. It was originally developed within Amgen Inc. and acquired by Anthera Pharmaceuticals in 2008 for the treatment of lupus and other autoimmune diseases. A-623 consisting of a BAFF-binding peptide fused to the N-terminus of human IgG Fc binds to BAFF and inhibits the interaction of BAFF with its receptors. BAFF, a TNF family member, plays a critical role in the development, maintenance, and survival of B cells. Expressed primarily by macrophages, monocytes, and dendritic cells, BAFF interacts with three different receptors on B cells: BAFF receptor (BAFF-R), B-cell maturation protein (BCMA), and transmembrane activator and cyclophilin ligand interactor (TACI) [85,86]. Phases Ia and Ib clinical trials were completed. Phase Ia was a randomized, placebo-controlled single dose study in which lupus patients were either intravenously dosed with 1, 3, or 6 mg/kg or subcutaneously dosed with 0.1, 0.3, 1, or 3 mg/kg. Phase Ib was a multidose study in which lupus patients were intravenously dosed at 6 mg/kg or subcutaneously dosed at 0.3, 1, or 3 mg/kg. A-623 was safe and well tolerated in both studies. Phase Ib demonstrated a decrease in total B cells as early as 15 days of treatment, and selective but significant reduction in total B cells among lupus patients were achieved. Encouraged by positive results from these clinical studies, Anthera Pharmaceuticals initiated a Phase IIb study of A-623 for the treatment of SLE.

8.5.1.4 Dulaglutide (LY2189265) LY2189265, developed by Eli Lilly (Indianapolis, IN), is a GLP-1 receptor agonist in which a modified GLP-1 peptide is fused to the N-terminus of a modified immunoglobulin G (IgG4) Fc fragment. Several mutations on the peptide are introduced to improve the stability, solubility, and immunogenicity, and these are discussed in more detail in later sections. A relatively long, flexible linker consisting of three repeats of GGGGS is inserted between the peptide and the N-terminus of the IgG hinge to improve its *in vitro* activity. To reduce potential complement-dependent and antibody-dependent cell-mediated cytotoxicity (ADCC), the IgG4 isotype was used for the Fc. To further diminish the ADCC activity by reducing the interactions of the Fc with its high-affinity Fc receptors, both phenylalanine 234 and leucine 235 at the lower hinge region are mutated to alanine. In addition, serine 228 at the core hinge region is mutated to proline to eliminate the potential of forming an intrachain disulfide bond in the core hinge region, which would destabilize the homodimer structure of the Fc. To reduce potential heterogeneity, the C-terminal lysine of the IgG-Fc was deleted from the construct. LY2189265 demonstrated similar *in vitro* and *in vivo* activities to native GLP-1 and significant reduction of weight gain in animal studies [87].

In a Phase I, three-period, crossover, double-blind, placebo-controlled study, healthy volunteers received escalating subcutaneous doses of LY2189265 ranging from 0.1 to 12 mg; the drug was generally well tolerated with some increase in gastrointestinal symptoms with escalating doses. The half-life of LY2189265 was ~90 h, with C_{max} occurring between 24 and 48 h in most subjects. No antidrug antibodies were observed [88].

In a Phase Ib study, type 2 diabetic patients received five once-weekly doses of 0.05, 0.3, 1, 3, 5, or 8 mg. Steady-state concentration levels were obtained after the second weekly dose with an average accumulation ratio (AUCweek 5: AUCweek 1) estimated to be 1.44. The average peak-to-trough concentration ratio ($C_{max} : C_{min}$) was between 2 to 3. The time to maximal plasma drug concentration (t_{max})

ranged from 12 to 72 h and was similar between the first and the fifth dose. The mean plasma half-life determined after the last dose was 95.4 h. The mean steady-state apparent clearance and volume of distribution were 0.157 L/h and 21.5 L, respectively [89]. A flat profile with no burst effect allows once-weekly dosing.

The efficacy and safety of LY2189265 was further evaluated in a placebo-controlled double-blinded study. Two hundred sixty two obese type 2 diabetic patients were randomized to once-weekly subcutaneous injections of either placebo, or 1 of 2 titrated doses of LY2189265 (0.5 mg for 4 weeks followed by 1.0 mg for 12 weeks; or 1.0 mg for 4 weeks followed by 2.0 mg for 12 weeks), or 1.0 mg for 16 weeks [47].

At week 16, glycosylated hemoglobin A1c (HbA1c) (all $p < 0.001$) and both fasting ($p < 0.001$) and postprandial ($p < 0.05$) blood glucose decreased significantly compared to placebo at all LY2189265 doses. Weight loss was dose-dependent and ranged from -1.34 ± 0.39 kg to -2.55 ± 0.40 kg at 16 weeks (all $p < 0.05$ vs. placebo). At the highest LY2189265 dosage, the most common adverse events were nausea (13.8%), diarrhea (13.8%), and abdominal distension (13.8%). Hypoglycemia was uncommon overall ($\leq$0.8 episodes/patient/30 days) but more common with LY2189265 than placebo through the initial 4 weeks ($p < 0.05$). No differences in cardiovascular events or blood pressure were demonstrated between treatments [47]. The positive data from the Phase II study summarized here convinced Lilly to move the Fc-fusion protein to Phase III clinical trials.

8.5.2 MIMETIBODY™ Technology

Centocor's MIMETIBODY™ platform was developed to extend the half-life of different peptides while keeping intrinsic activities similar to that of their parent peptides. Figure 8.1D illustrates a typical MIMETIBODY construct, in which a bioactive peptide sequence is genetically linked to an Fc domain. A variety of bioactive peptides have been successfully engineered into the platform, demonstrating the utility and versatility of the MIMETIBODY technology.

8.5.2.1 CNTO 528/CNTO 530 CNTO 528 is the first successful example using our MIMETIBODY technology. It comprises a human IgG1 Fc and a 20 amino acid erythropoietin mimetic peptide 1 (EMP1), identified through the panning of combinatorial phage peptide libraries; EMP1 can bind and activate the EPO receptor in a variety of cell-based and animal assays [90]. In a first-in-human Phase I clinical study, CNTO 528 was well tolerated and exhibited a dose-dependent half-life, mean $t_{1/2}$ranging from 1.55 days (with 0.03 mg/kg dose) to 7.60 days (with 0.9 mg/kg dose). The maximum effect on the reticulocyte response occurred ~8–9 days after administration. A median increase in hemoglobin to at least 1 g/dL above baseline was achieved 9 to 10 days after administration, with a maximum effect between 19 and 26 days [91].

CNTO 530 is the second generation of the erythropoietin MIMETIBODY protein generated after intensive protein engineering efforts to improve both the biological and biophysical properties of CNTO 528. It addresses heterogeneity and aggregation issues and exhibits a better circulating half-life and significantly better pharmacodynamic activities in animal studies. Single subcutaneous doses of <1 mg/kg in mice caused a long-lived reticulocytosis and an increase in RBC and Hgb that was maintained for at least 29 days [92].

8.5.2.2 CNTO736 CNTO736, another example of MIMETIBODY technology, is a GLP-1 receptor agonist. It incorporates a GLP-1 peptide analog, which has an amino acid substitution rendering it more resistant to dipeptidyl peptidase IV digestion. CNTO736 dose dependently increases cAMP and insulin secretion from islets in a glucose-dependent manner. Studies in rodent models of type 2 diabetes demonstrated that acute dosing with CNTO736 lowers fasting and postprandial blood glucose with a significantly longer duration of action than native GLP-1, and chronic dosing with CNTO736 decreases body weight [93].

8.6 CONSIDERATIONS AND CHALLENGES FOR ENGINEERING PEPTIDE-Fc-FUSION THERAPEUTICS

As a class of biological therapeutics, peptide-Fc-fusion protein therapeutics have some distinct advantages. They have a good circulating half-life (e.g., more than a week for CNTO 528), although in many cases their serum half-lives are shorter than the half-lives of antibodies, probably due to lower serum stability. They can be engineered as both potent antagonists and agonists. Preclinical and clinical studies have reported that Fc therapeutics, in general, are safe, well tolerated, and have low toxicity. In addition, thanks to the Fc-domain, the Fc-fusion proteins share many physical and chemical properties with mAbs. As a result, many advanced technologies developed for mAb drugs can be used for Fc-fusion therapeutics as well, such as engineering, optimization, expression, purification, formulation, and production.

However, unlike antibodies, the physical, chemical, and pharmacological properties of Fc-fusion proteins, such as solubility, stability, and pharmacokinetic properties, are significantly different from one drug to the other. Many of those differences are largely determined by the properties of the region other than Fc domain. Since each individual Fc-fusion partner has its own properties, careful design, engineering, and optimization work on an Fc-fusion protein to improve physical, chemical, and pharmacological properties will be the key to its success in future development, manufacturing, and clinical use.

8.6.1 Improving Activity

In most cases, peptides will lose potency upon fusion to either the N- or C-terminus of an Fc domain. This could be due to any one or a combination of the following factors: the loss of translational/rotational entropy associated with fusion to a much larger partner, steric hindrance, conformational changes of the peptide, loss of interactions with the N- or C-terminus which may be involved in binding, or differences in posttranslational modification (e.g., amidation of the carboxy terminus). In many cases, protein-engineering efforts to restore some of the activity or to minimize the activity losses are necessary. For example, direct fusion of DPP-IV-protected GLP-1 analog (V8-GLP-1) to the Fc-hinge region dramatically reduced *in vitro* activity (by ~95%) compared to that of free V8-GLP-1.

However, by insertion of a linker with optimal length and sequence between the C-terminus of the peptide and the N-terminus of the hinge, the newly engineered peptide-Fc-fusion protein demonstrated approximately fourfold greater *in vitro* potency over that of free peptide [87].

Careful engineering of a peptide-Fc fusion, in some cases, can make the fusion protein significantly more potent than the free peptide as illustrated in the example of EPO MIMETIBODY, CNTO 528. As discussed in Section 8.5.2.1, CNTO 528 contains an EPO mimetic peptide, EMP1, which binds to EPO receptor and exerts effects similar to the native ligand, EPO. However, EMP1 binds to EPO receptor with more than 1000-fold weaker affinity than the native ligand [94]. CNTO 528 showed significantly better binding affinity to the receptor than the free peptide and was 100-fold more active than EMP1 in an *in vitro* UT7 cell proliferation assay (Figure 8.2). By further carefully engineering a linker region between the peptide and the Fc, we were able to improve the activity of CNTO 528 an additional twofold. The new molecule, CNTO 530, demonstrated significantly more activity *in vivo* as measured by its ability to increase hemoglobin levels in normal rats relative to its parent protein [95].

8.6.2 Improving Stability

Many unconstrained peptides are flexible enough to be easily accessed and digested by tissue or serum proteases *in vivo*. Improving their *in vivo* stability and making them more resistant to oxidative stress, changes of temperature, pH, and solution condition are crucial for their use as therapeutic drugs. We routinely evaluate peptide stability by incubating peptides or peptide-Fc-fusion proteins in plasma, tissue homogenates, or cell lysates at 37°C for a period of time and then test their biological activity and molecular integrity. In many cases, we have observed that peptides were significantly better protected from protease cleavage when they are fused to an Fc domain. However, for those more metabolically labile fusions, additional engineering efforts are needed to stabilize the peptide region. In this scenario, several approaches should be considered, including constraining the peptides, modifying the labile sites, or introducing carbohydrate moieties nearby to shield the labile sites [87,93]. For example, one of our MIMETIBODY constructs contained a protease-sensitive peptide. Multiple cleavage products were observed after this protein was purified from tissue culture supernatants. We performed a disulfide scan through the entire peptide region and

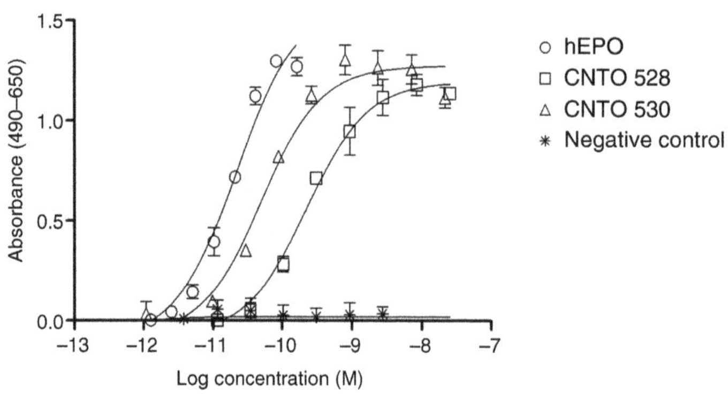

FIGURE 8.2 Supernatants obtained by transient transfection for CNTO528 ($EC_{50} = 2.28 \times 10^{-10}$ M) or CNTO530 ($EC_{50} = 5.00 \times 10^{-11}$ M) were tested for proliferative activity in UT-7 cells. Viable cells monitored by MTS absorbance at 490 nm are plotted as a function EPO receptor agonist concentration in log scale. Human recombinant EPO ($EC_{50} = 2.14 \times 10^{-11}$ M) was used as positive control and cell supernatant from a mock transfection was used as negative control. Data are presented as the mean ± SD ($n = 2$).

identified the best disulfide bond in terms of retention of activity. After introducing the cysteine pair in the peptide region to form the disulfide bridge that constrained the peptide, we were able to eliminate protease cleavage, and consequently, the fusion protein proved to be stable both *in vitro* and *in vivo* while retaining the majority of its potency (data not shown).

8.6.3 Improving Pharmacokinetic Profile

As intended, peptides fused to Fc demonstrate significantly improved PK profiles. For example, we have shown that CNTO 530 has extended clearance kinetics in mice with a terminal $t_{1/2}$ of ~40 h compared to only several minutes for the naked EMP1 peptide [92]. The clearance of CNTO 530 was comparable to the clearance of a humanized IgG4ala-ala in mice [96], suggesting that the mechanism of clearance for CNTO 530 may be similar to that of an antibody and that CNTO 530 may also have extended pharmacokinetics in humans. It suggested that protease cleavage may not play a significant role in the clearance of CNTO 530. The Phase I clinical trial of CNTO 528 showed that the mean circulating half-life reached 7.6 days in humans at high doses (0.9 mg/kg) [92]. Despite the fact that peptide-Fc fusions have a prolonged serum half-life, careful protein engineering, especially at the peptide and Fc junction region, can affect the protein PK profile. This junction region is flexible and probably accessible to proteases. For example, we were able to eliminate the protease cleavage sites by reengineering the junction region between EMP1 and Fc in the CNTO 528 construct. A rat PK study showed improvement of the mean terminal serum half-life of the new construct from 1.54 days to 3.03 days [95]. The improvement of *in vitro* activity and serum half-life results in significantly higher *in vivo* potency in rat [92].

8.6.4 Improving Solubility and Reducing Aggregation

It is highly desirable to formulate protein therapeutics at >100 mg/mL to have a small enough volume for subcutaneous administration. However, some peptides have a tendency to form oligomers and induce aggregation, especially at high concentration, which can cause decreased activity, decreased bioavailability, and increased immunogenicity *in vivo* [97]. Although hydrophobic interactions were shown to be the predominant reason for protein aggregation, there are many other possible mechanisms. Aggregates may be classified in different ways, including soluble/insoluble, covalent/noncovalent, reversible/irreversible, and native/denatured [98]. Understanding the causes of aggregation and then engineering around it is the approach we regularly use. For example, one of our MIMETIBODY constructs contained a metabolically labile peptide, in which a pair of cysteines was introduced to constrain the peptide by a disulfide bond to reduce protease susceptibility. In this example, a substantial portion of the protein-A purified MIMETIBODY protein was aggregated as demonstrated by analytical High Performance Liquid Chromatography (HPLC) (Figure 8.3A). A series of ladder-like high molecular weight bands was observed on an SDS-PAGE gel under nonreducing conditions (Figure 8.3B, left lane), and these bands collapsed to a single band under reducing conditions (Figure 8.3B, right lane), strongly suggesting that the aggregation is disulfide bond mediated. After incubating the aggregates at ~100 μg/mL in a phosphate buffer with the presence of thio-redox reagents (50 mM phosphate buffer, pH 7.2, 1 mM reduced glutathione, 0.1 mM oxidized glutathione) at 4°C for 36 h, more than 50% of aggregate was converted to the monomer (data not shown), further indicating that the aggregation is caused by disulfide bonds. We believed that the misformed intermolecular disulfide bridges cross-linked the proteins together to form oligomers as illustrated in Figure 8.3, which migrated as the ladder-like bands observed on SDS-PAGE (Figure 8.3B). These oligomers can further associate to form larger aggregates as shown in HPLC analysis (Figure 8.3A). We hypothesized that adding carbohydrates to the flanking region of the peptide could prevent the formation of interchain disulfide bonds due to steric hindrance as illustrated in Figure 8.4. To test this hypothesis, several MIMETIBODY constructs were made, in which N-glycosylation sites were introduced near the peptide region. As expected, the aggregation level of some of these constructs was significantly reduced on the basis of size exclusion HPLC analysis (Figure 8.3C) while retaining most of their activity (data not shown).

8.6.5 Reducing Heterogeneity

Protein heterogeneity is another potential challenge for drug development simply because it puts a huge burden on the protein purification process. Many factors, such as protease cleavage, posttranslational modification, and oligomerization, can cause protein heterogeneity. For example, one of our MIMETIBODY proteins migrated on the SDS-PAGE as multiple but very close bands under reducing conditions. Mass spectrum analysis revealed that these bands have an ~900 Da difference in molecular weight, which we hypothesized to be the O-linked carbohydrates populating to varying extents the flexible hinge region. Using site-directed mutagenesis, we eliminated the potential O-linked glycosylation sites, and the multiple bands are reduced to a single band as shown in Figure 8.5.

In another interesting example, we found that sometimes the signal peptidase can potentially act at alternative cleavage sites to generate heterogeneity. Mass spectrometry and amino acid sequencing analysis revealed that a small portion of one of our MIMETIBODY proteins missed three amino acids at the N-terminus of one chain but not the other chain. Interestingly, after a series of experiments, we found that the

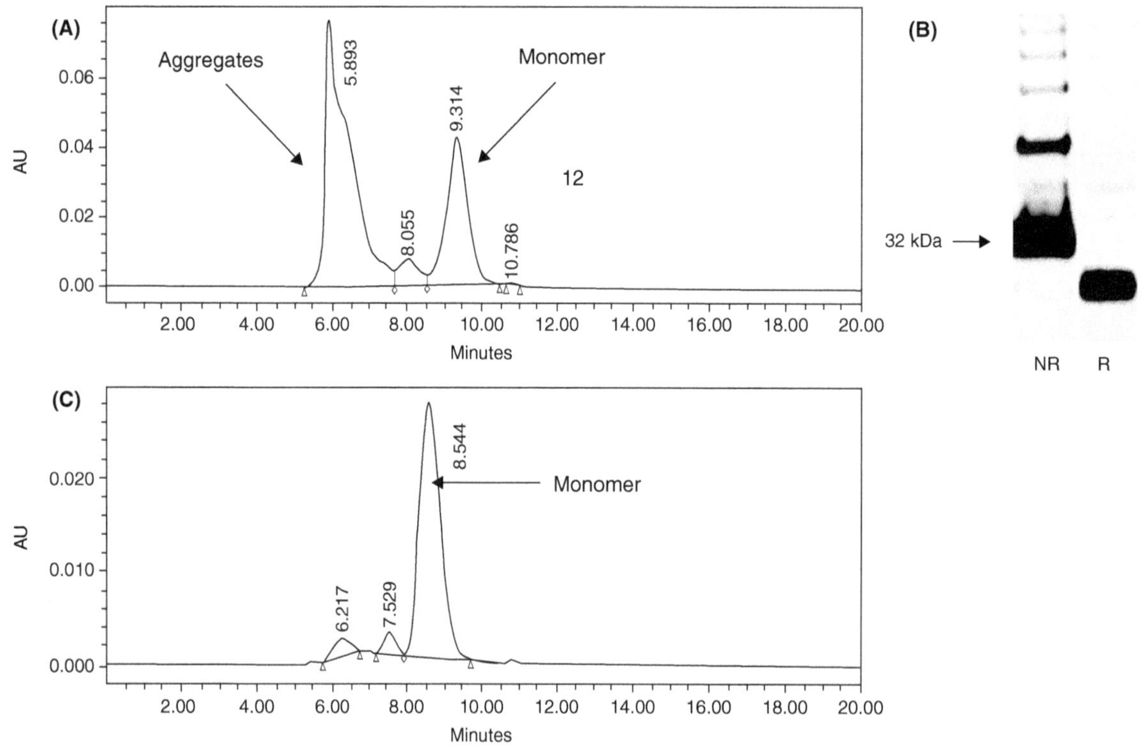

FIGURE 8.3 An example of introduced carbohydrates reducing aggregation. (A) The protein-A affinity column purified MIMETIBODY^{TM} protein was analyzed with size exclusion; HPLC chromatograms show large proportion of aggregates. (B) Western blot analysis of cell supernatants showed that aggregated material when probed with an anti-peptide region antibody. The nonreduced sample was on the left lane and the reduced sample was on the right lane. (C) Size exclusion HPLC analysis of a protein-A purified N-linked variants.

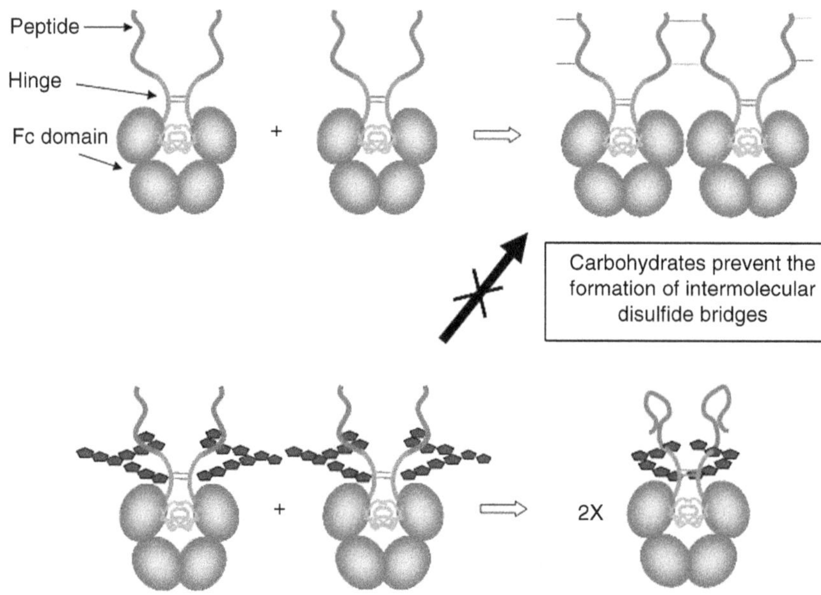

FIGURE 8.4 Schematic representing carbohydrates-mediated steric hindrance of aggregation. The gray line indicates the hinge region. The black line indicates the peptide region; the pentagons indicate carbohydrates; and the balls indicate the Ig constant domain.

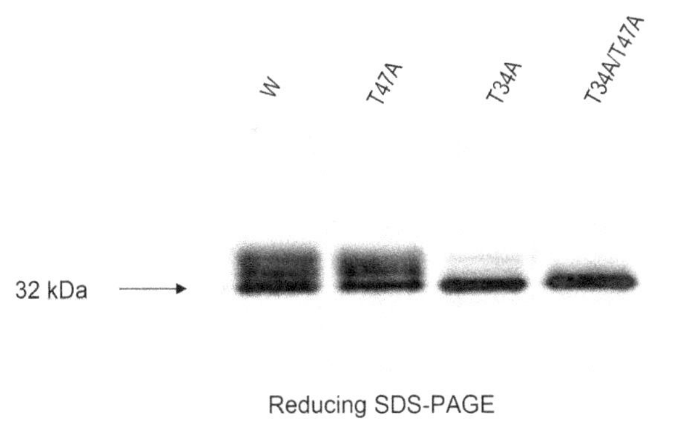

FIGURE 8.5 Western blot analysis. Cell supernatants probed with an anti-peptide region antibody showed that the O-linked glycosylated protein and mutants with removed individual or both O-linked glycosylation sites.

amount of protein missing these three amino acids was not correlated to the protein expression level, harvest time, or cell viability. After ruling out other possibilities, we hypothesized that these three amino acids were removed by a signal peptidase due to a secondary recognition site. The problem was solved by changing to a different signal peptide sequence in this expression construct (data not shown).

8.6.6 Isotype Selection

Fc-dependent activities that recruit or activate components of the innate or adaptive immune system are sometimes very important for the efficacy of antibodies and Fc-fusion proteins. Fc-effector functions include (1) CDC, triggered by the complement cascade at the cell surface, (2) ADCC, and (3) antibody-dependent cellular phagocytosis (ADCP). For the IgG antibodies, both ADCC and ADCP are initiated by the engagement of the lower hinge region of the Fc with Fcγ receptors (FcγRs) on the surface of immune effector cells, leading to the lysis of the targeted cells or phagocytosis.

There are four different isotype classes of human IgG, and each isotype has different levels of effector functions associated with it. IgG4 does not activate the complement cascade and binds weakly to FcγRs except FcγRI [76,77], whereas IgG2 does not bind to FcγRs on mononuclear cells or neutrophils, except the allotypic variants of FcγRIIa [77,78]. Depending on the mechanism(s) of drug action, different isotypes can be selected based on the desired strength of Fc-effector functions. For example, IgG1 is the most common choice for antitumor drugs, which require strong ADCC and CDC activities. In contrast, IgG2 and IgG4 are better choices for agonist drugs, since stimulating certain cellular functions, not killing the target cells, is their main goal. The fact that IgG1can be efficiently purified on a protein-A column also contributes to its popularity. The difference of serum half-life is another important factor to be considered. IgG1, IgG2, and IgG4 all have a circulating half-life of 21 days while IgG3 only has about a week. Other factors for consideration include significant structural differences at the hinge regions between different isotypes, and structural differences in the oligomeric state of IgG, for example, the dimeric structure of IgA and the pentameric structure of IgM, which result in different valencies. The number of allotypic variants of the isotype is also an important factor, since a molecule of one allotype can be potentially immunogenic for individual patients having a different allotype [77,79].

8.6.7 Tailoring Fc-Effector Functions to Fit Individual Needs

Even though the different Fc isotypes have different degrees of Fc-effector functions, further protein engineering efforts to enhance or silence the Fc functions are sometimes required to optimize the particular mode of action of individual drugs.

Efforts to improve Fc-effector functions were fueled by clinical trials of Rituxan, in which patients who expressed a higher binding allotype of FcγRIIIa showed significantly better clinical responses to Rituxan than patients who expressed the lower binding allotype of the receptor [99]. Much effort was focused on engineering the receptor-binding site on the Fc to enhance the FcγRIIIa-binding affinity and as a result, activity was enhanced as much as 100-fold in *in vitro* ADCC assays [100]. XmAb2531, an Fc-engineered anti-CD30 antibody with enhanced ADCC activity from Xencor Inc. (Monrovia, CA), is in Phase 1 clinical trials for the treatment of patients with lymphomas. On the other hand, since Shields et al. demonstrated that antibodies expressed from a fucosylation-deficient Chinese hamster ovary cell line, Lec13 cells, had up to a 50-fold increase of binding activity of FcγRIIIa and a significant increase of ADCC activity [101], efforts lately have been focused on developing production cell lines that produce either low fucosylated or nonfucosylated IgG. A recent structural study demonstrated that Fc with or without core fucose on its CH2 domain have significantly different conformations [102]. A number of antibodies containing no, or significantly reduced, core fucose are currently in clinical trials [103].

For Fc-fusion proteins targeting cell surface receptors, including receptor agonist peptide-Fc therapeutics, there is a need for an Fc domain that does not possess any effector function to destroy cells, yet retains the pharmacokinetics and other favorable properties of IgG. Early efforts to eliminate Fc-effector functions were focused on mutations of the FcγR-binding sites in the lower hinge region of the Fc to reduce FcγR-binding activity [96]. Both IgG1 and IgG4 mutants, in which residues at positions 234 and 235 were changed to Ala, showed at least 100-fold reduction of FcγRI- and FcγRII-binding activities [104]. For example, both teplizumab (MGA031), an anti-CD3 mAb, and LY2189265, a GLP-1-Fc fusion, containing the same Ala^{234} and Ala^{235} mutations, are in late clinical studies for type 1 diabetes [105,106]. Recently, Davis et al. reported that serine substitutions of cysteines at positions 130, 136, and 139 and proline at position 148 in the Fc domain eliminated both CDC and ADCC activities of abatacept, a CTLA4-Ig fusion drug [74]. Other approaches have been developed based on the low level of IgG2 binding to FcγRs [59] and on the inability of IgG4 to activate the complement cascade [60], such as the IgG2/IgG4 hybrid constant region approach used for eculizumab [78] and the recent approach developed by Strohl, in which several IgG4 amino acids on the CH2 domain were introduced into IgG2 at the same position [61].

8.6.8 Reducing Immunogenicity

Although therapeutic proteins are generally safe and nontoxic, the potential of inducing immune responses is always a concern. Antidrug antibody (ADA) responses were observed in patients treated with a variety of human proteins including insulin, factor VIII, IFN, and erythropoietin (Epo) [62,63]. ADAs are of particular concern in that they might decrease the efficacy of the drugs by neutralizing or reducing the circulating half-life of drugs and might also be associated with serious safety issues by cross-neutralizing antibodies to endogenous proteins. Owing to a lack of reliable *in vitro* assays and *in vivo* animal models to predict the immune responses in humans, a few successful examples exist to illustrate successful protein engineering efforts for reducing the immunogenicity of protein therapeutics except for the humanization of murine antibodies. Nevertheless, several factors should be considered in the design and engineering of a therapeutic protein. (1) Since the immune response is generally more severe for nonhuman proteins, efforts should be made to avoid extensive modification of human sequences. (2) Since physical properties such as solubility, stability, and physical state (oligomers or aggregates) of a therapeutic protein as well as the administration dosage, frequency and route of administration can be important factors in the induction of human immune responses, it is well worth the efforts to engineer a therapeutic protein to have better physical, chemical, biological, and pharmacological properties. (3) Since T-cell epitopes are important for T-cell-dependent immune responses, it is worthwhile to remove *in silico* predicted T-cell epitopes from nonhuman protein sequences as long as it does not significantly affect other properties of the protein.

8.7 CONCLUSIONS

Thanks to advances in technologies including chemical modification of peptides to enhance their resistance to protease cleavage and improvement of their physicochemical properties and advances in production which lower costs; peptide-based therapeutics have become one of the fastest growing classes of new drugs in the last few decades. However, peptides currently account for only 2% of drugs on the market and none of them become blockbusters comparable to some recombinant protein therapeutics. The main limitation of peptides is their short half-lives due to rapid clearance *in vivo*. Several techniques, such as PEGylation and HSA fusion, show promise in prolonging the serum half-life for a variety of peptides. However, Fc fusion is still the most advanced strategy in drug development for this purpose.

With multiple drugs on the market and many more exciting new molecules in the pipeline for treating a variety of different diseases, Fc-fusion proteins, which couple the desirable biophysical and pharmacokinetic properties of antibodies with the bioactivity of peptides or proteins fusion partners, have proven to be an important part of the protein therapeutic family. Several peptide-Fc-fusion techniques, including peptibody and MIMETIBODY platforms along with other peptide-Fc-fusion approaches discussed in this chapter, demonstrate the potential for successfully displaying and prolonging the circulating half-lives of bioactive peptides and thus represent valuable approaches for engineering peptide therapeutics. For a peptide-Fc-fusion protein to become a successful therapeutic, it is often necessary to optimize its biological, biophysical, and pharmacological properties, such as activity, stability, solubility, heterogeneity, and PK profiles. Rational protein engineering methods that rely on structural, computational, or biochemical information about the ligand–receptor system of interest are still the most useful approaches for this purpose, in particular for engineering efforts to improve protein agonist activities. The development of combinatorial engineering technologies to identify protein variants with improved biological and biophysical properties, rather than just binding affinity, will be the key to future success of Fc-fusion protein therapeutics.

ACKNOWLEDGMENT

The authors would like to thank Lu Lu and Tracy Spinka-Doms for their technical assistance.

REFERENCES

1. Leader B, Baca QJ, Golan DE. (2008) Protein therapeutics: a summary and pharmacological classification. *Nat. Rev. Drug Discov.* 7, 21–39.
2. Lawrence S. (2007) Pipelines turn to biotech. *Nat. Biotechnol.* 25, 1342–1342.
3. Huggett B, Hodgson J, Lahteenmaki R. (2011) Public biotech 2010–the numbers. *Nature Biotechnology* 29, 585–591.

4. Biotech set to dominate drug industry growth on World Wide Web. 2009 Available at http://www.evaluatepharma.com/Universal/View.aspx?type=Story&id=188700&sectionID=&isEPVantage=yes.
5. Huang, C. (2009). Receptor-Fc fusion therapeutics, traps, and MIMETIBODY technology. *Curr Opin Biotechnol* 20(6), 692-699
6. Kastin AJ. (2006) *Handbook of Biologically Active Peptides*. Academic Press, Amsterdam.
7. Adessi C, Soto C. (2002) Converting a peptide into a drug: strategies to improve stability and bioavailability. *Curr. Med. Chem.* 9, 963–978.
8. Gill TJ, 3rd, Gould HJ, Doty P. (1963) Role of optical isomers in determining the antigenicity of synthetic polypeptides. *Nature* 197, 746–747.
9. Tyndall JD, Nall T, Fairlie DP. (2005) Proteases universally recognize beta strands in their active sites. *Chem. Rev.* 105, 973–999.
10. Li P, Roller PP. (2002) Cyclization strategies in peptide derived drug design. *Curr. Top. Med. Chem.* 2, 325–341.
11. Blackwell HE, Sadowsky JD, Howard RJ, Sampson JN, Chao JA, Steinmetz WE, et al. (2001) Ring-closing metathesis of olefinic peptides: design, synthesis, and structural characterization of macrocyclic helical peptides. *J. Org. Chem.* 66, 5291–5302.
12. Verdine GL, Walensky LD. (2007) The challenge of drugging undruggable targets in cancer: lessons learned from targeting BCL-2 family members. *Clin. Cancer Res.* 13, 7264–7270.
13. Henchey LK, Jochim AL, Arora PS, Henchey LK, Jochim AL, Arora PS. (2008) Contemporary strategies for the stabilization of peptides in the alpha-helical conformation. *Curr. Opin. Chem. Biol.* 12, 692–697.
14. Saladin PM, Zhang BD, Reichert JM, Saladin PM, Zhang BD, Reichert JM. (2009) Current trends in the clinical development of peptide therapeutics. *Idrugs* 12, 779–784.
15. Wilson AC, Meethal SV, Bowen RL, Atwood CS, Wilson AC, Meethal SV, et al. (2007) Leuprolide acetate: a drug of diverse clinical applications. *Expert Opin. Investig. Drugs* 16, 1851–1863.
16. Wingfield M, Healy DL. (1993) Endometriosis: medical therapy. *Baillieres Clin. Obstet. Gynaecol.* 7, 813–838.
17. Lamberts SW, Reubi JC, Krenning EP. (1992) Somatostatin analogs in the treatment of acromegaly. *Endocrinol. Metab. Clin. North Am.* 21, 737–752.
18. Peeters M, Van den Brande J, Francque S. (2010) Diarrhea and the rationale to use Sandostatin. *Acta Gastroenterol. Belg.* 73, 25–36.
19. Blick SK, Dhillon S, Keam SJ, Blick SKA, Dhillon S, Keam SJ. (2008) Teriparatide: a review of its use in osteoporosis. *Drugs* 68, 2709–2737.
20. Overington JP, Al-Lazikani B, Hopkins AL. (2006) How many drug targets are there? *Nat. Rev. Drug Discov.* 5, 4.
21. Cochran AG. (2000) Antagonists of protein-protein interactions. *Chem. Biol.* 7, R85–R94.
22. Bailon P, Won CY, Bailon P, Won CY. (2009) PEG-modified biopharmaceuticals. *Expert Opin. Drug Deliv.* 6, 1–16.
23. Roberts MJ, Bentley MD, Harris JM. (2002) Chemistry for peptide and protein PEGylation. *Adv. Drug Deliv. Rev.* 54, 459–476.
24. Werle M, Bernkop-Schnürch A. (2006) Strategies to improve plasma half life time of peptide and protein drugs. *Amino Acids* 30, 351–367.
25. Stead RB, Lambert J, Wessels D, Iwashita JS, Leuther KK, Woodburn KW, et al. (2006) Evaluation of the safety and pharmacodynamics of Hematide, a novel erythropoietic agent, in a phase 1, double-blind, placebo-controlled, dose-escalation study in healthy volunteers. *Blood* 108, 1830–1834.
26. McGregor DP. (2008) Discovering and improving novel peptide therapeutics. *Curr. Opin. Pharmacol.* 8, 616–619.
27. Lee SH, Lee S, Youn YS, Na DH, Chae SY, Byun Y, et al. (2005) Synthesis, characterization, and pharmacokinetic studies of PEGylated glucagon-like peptide-1. *Bioconjug. Chem.* 16, 377–382.
28. St Onge EL, Miller SA. (2010) Albiglutide: a new GLP-1 analog for the treatment of type 2 diabetes. *Expert Opin. Biol. Ther.* 10, 801–806.
29. Nelson DR, Benhamou Y, Chuang WL, Lawitz EJ, Rodriguez-Torres M, Flisiak R, et al. (2010) Albinterferon Alfa-2b was not inferior to pegylated interferon- in a randomized trial of patients with chronic hepatitis C virus genotype 2 or 3. *Gastroenterology* 139, 1267–1276.
30. Halpern W, Riccobene TA, Agostini H, Baker K, Stolow D, Gu ML, et al. (2002) Albugranin, a recombinant human granulocyte colony stimulating factor (G-CSF) genetically fused to recombinant human albumin induces prolonged myelopoietic effects in mice and monkeys. *Pharm. Res.* 19, 1720–1729.
31. Duttaroy A, Kanakaraj P, Osborn BL, Schneider H, Pickeral OK, Chen C, et al. (2005) Development of a long-acting insulin analog using albumin fusion technology. *Diabetes* 54, 251–258.
32. Wang W, Ou Y, Shi Y. (2004) AlbuBNP, a recombinant B-type natriuretic peptide and human serum albumin fusion hormone, as a long-term therapy of congestive heart failure. *Pharm. Res.* 21, 2105–2111.
33. Abraham S, Guo F, Li LS, Rader C, Liu C, Barbas CF, 3rd, et al. (2007) Synthesis of the next-generation therapeutic antibodies that combine cell targeting and antibody-catalyzed prodrug activation. *Proc. Natl. Acad. Sci. USA* 104, 5584–5589.
34. Lencer WI, Blumberg RS. (2005) A passionate kiss, then run: exocytosis and recycling of IgG by FcRn. *Trends Cell Biol.* 15, 5–9.
35. Roopenian DC, Akilesh S. (2007) FcRn: the neonatal Fc receptor comes of age. *Nat. Rev. Immunol.* 7, 715–725.
36. Goldenberg MM. (1999) Etanercept, a novel drug for the treatment of patients with severe, active rheumatoid arthritis. *Clin. Ther.* 21, 75–87; discussion 71–72.

37. Capon DJ, Chamow SM, Mordenti J, Marsters SA, Gregory T, Mitsuya H, et al. (1989) Designing CD4 immunoadhesins for AIDS therapy. *Nature* 337, 525–531.
38. Byrn RA, Mordenti J, Lucas C, Smith D, Marsters SA, Johnson JS, et al. (1990) Biological properties of a CD4 immunoadhesin. *Nature* 344, 667–670.
39. Chamow SM, Duliege AM, Ammann A, Kahn JO, Allen JD, Eichberg JW, et al. (1992) CD4 immunoadhesins in anti-HIV therapy: new developments. *Int. J. Cancer Suppl.* 7, 69–72.
40. Ashkenazi A, Capon DJ, Ward RH. (1993) Immunoadhesins. *Int. Rev. Immunol.* 10, 219–227.
41. Low SC, Nunes SL, Bitonti AJ, Dumont JA. (2005) Oral and pulmonary delivery of FSH-Fc fusion proteins via neonatal Fc receptor-mediated transcytosis. *Hum. Reprod.* 20, 1805–1813.
42. Bitonti AJ, Dumont JA, Low SC, Peters RT, Kropp KE, Palombella VJ, et al. (2004) Pulmonary delivery of an erythropoietin Fc fusion protein in non-human primates through an immunoglobulin transport pathway. *Proc. Natl. Acad. Sci. USA* 101, 9763–9768.
43. Hoffman HM, Throne ML, Amar NJ, Sebai M, Kivitz AJ, Kavanaugh A, et al. (2008) Efficacy and safety of rilonacept (interleukin-1 Trap) in patients with cryopyrin-associated periodic syndromes: results from two sequential placebo-controlled studies. *Arthritis Rheum.* 58, 2443–2452.
44. Oliner J, Min H, Leal J, Yu D, Rao S, You E, et al. (2004) Suppression of angiogenesis and tumor growth by selective inhibition of angiopoietin-2. *Cancer Cell* 6, 507–516.
45. Karlan BY, Oza AM, Hansen VL, Richardson GE, Provencher DM, Ghatage P, et al. (2010) Randomized, double-blind, placebo-controlled phase II study of AMG 386 combined with weekly paclitaxel in patients (pts) with recurrent ovarian carcinoma [Abstract 5000]. *J. Clin. Oncol.* 28.
46. Herbst RS, Hong D, Chap L, Kurzrock R, Jackson E, Silverman JM, et al. (2009) Safety, pharmacokinetics, and antitumor activity of AMG 386, a selective angiopoietin inhibitor, in adult patients with advanced solid tumors. *J. Clin. Oncol.* 27, 3557–3565.
47. Umpierrez GE, Blevins T, Rosenstock J, Cheng C, Anderson JH, Bastyr EJ, 3rd. (2011) The effects of LY2189265, a long-acting glucagon-like peptide-1 analog, in a randomized, placebo-controlled, double-blind study of overweight/obese patients with type 2 diabetes: the EGO study. *Diabetes Obes. Metab.* 13, 418–425.
48. Amersi F, Farmer DG, Shaw GD, Kato H, Coito AJ, Kaldas F, et al. (2002) P-selectin glycoprotein ligand-1 (rPSGL-Ig)-mediated blockade of CD62 selectin molecules protects rat steatotic liver grafts from ischemia/reperfusion injury. *Am. J. Transplant.* 2, 600–608.
49. Ruckle J, Jacobs M, Kramer W, Pearsall AE, Kumar R, Underwood KW, et al. (2009) Single-dose, randomized, double-blind, placebo-controlled study of ACE-011 (ActRIIA-IgG1) in postmenopausal women. *J. Bone Miner. Res.* 24, 744–752.
50. Lachey JL, Wachtman LM, Mansfield KG, Persall RS, Lane JH, Seehra JS. (2008) Treatment with a soluble activin receptor type IIb results increased muscle mass in marmosets [Abstract]. *New Dir. Biol. Dis. Skelet. Muscle.*
51. Long L, Brennan T, Zanghi J, Palencia S, Cheung R, Aguirre M, et al. (2009) Preclinical antitumor efficacy of FP-1039, a soluble FGF receptor 1:Fc conjugate, as a single agent or in combination with anti-cancer drugs [Abstract]. AACR 100th Annual Meeting.
52. Dumont JA, Low SC, Peters RT, Bitonti AJ. (2006) Monomeric Fc fusions: impact on pharmacokinetic and biological activity of protein therapeutics. *BioDrugs* 20, 151–160.
53. Peters RT, Low SC, Kamphaus GD, Dumont JA, Amari JV, Lu Q, Zarbis-Papastoitsis G, Reidy TJ, Merricks EP, Nichols TC, et al.: Prolonged activity of factor IX as a monomeric Fc fusion protein. *Blood* 2010, 115, 2057–2064.
54. Jazayeri JA, Carroll GJ. (2008) Fc-based cytokines: prospects for engineering superior therapeutics. *BioDrugs* 22, 11–26.
55. Ashkenazi A, Chamow SM. (1997) Immunoadhesins as research tools and therapeutic agents. *Curr. Opin. Immunol.* 9, 195–200.
56. Mohler KM, Torrance DS, Smith CA, Goodwin RG, Stremler KE, Fung VP, et al. (1993) Soluble tumor necrosis factor (TNF) receptors are effective therapeutic agents in lethal endotoxemia and function simultaneously as both TNF carriers and TNF antagonists. *J. Immunol.* 151, 1548–1561.
57. Ducharme E, Weinberg JM. (2008) Etanercept. *Expert Opin. Biol. Ther.* 8, 491–502.
58. Miller GT, Hochman PS, Meier W, Tizard R, Bixler SA, Rosa MD, et al. (1993) Specific interaction of lymphocyte function-associated antigen 3 with CD2 can inhibit T cell responses. *J. Exp. Med.* 178, 211–222.
59. Burton DR, Woof JM. (1992) Human antibody effector function. *Adv. Immunol.* 51, 1–84.
60. Tao MH, Smith RI, Morrison SL. (1993) Structural features of human immunoglobulin G that determine isotype-specific differences in complement activation. *J. Exp. Med.* 178, 661–667.
61. Strohl WR. Non-immunostimulatory antibody and compositions containing the same. US patent application 2007/0148167 A1, 2007.
62. De Groot AS, Scott DW. (2007) Immunogenicity of protein therapeutics. *Trends Immunol.* 28, 482–490.
63. Marshall SA, Lazar GA, Chirino AJ, Desjarlais JR. (2003) Rational design and engineering of therapeutic proteins. *Drug Discov. Today* 8, 212–221.
64. Emamaullee J, Toso C, Merani S, Shapiro AM. (2009) Costimulatory blockade with belatacept in clinical and experimental transplantation - a review. *Expert Opin. Biol. Ther.* 9, 789–796.
65. Larsen CP, Pearson TC, Adams AB, Tso P, Shirasugi N, Strobertm E, et al. (2005) Rational development of LEA29Y (belatacept), a high-affinity variant of CTLA4-Ig with potent immunosuppressive properties. *Am. J. Transplant.* 5, 443–453.

66. Bracewell C, Isaacs JD, Emery P, Ng WF. (2009) Atacicept, a novel B cell-targeting biological therapy for the treatment of rheumatoid arthritis. *Expert Opin. Biol. Ther.* 9, 909–919.
67. Gross JA, Dillon SR, Mudri S, Johnston J, Littau A, Roque R, et al. (2001) TACI-Ig neutralizes molecules critical for B cell development and autoimmune disease impaired B cell maturation in mice lacking BLyS. *Immunity* 15, 289–302.
68. Vugmeyster Y, Seshasayee D, Chang W, Storn A, Howell K, Sa S, et al. (2006) A soluble BAFF antagonist, BR3-Fc, decreases peripheral blood B cells and lymphoid tissue marginal zone and follicular B cells in cynomolgus monkeys. *Am. J. Pathol.* 168, 476–489.
69. Lundquist L. (2007) Abatacept: a novel therapy approved for the treatment of patients with rheumatoid arthritis. *Adv. Ther.* 24, 333–345.
70. Farmer DG, Shen XD, Amersi F, Anselmo D, Ma JP, Ke B, et al. (2005) CD62 blockade with P-Selectin glycoprotein ligand-immunoglobulin fusion protein reduces ischemia-reperfusion injury after rat intestinal transplantation. *Transplantation* 79, 44–51.
71. Peppel K, Crawford D, Beutler B. (1991) A tumor necrosis factor (TNF) receptor-IgG heavy chain chimeric protein as a bivalent antagonist of TNF activity. *J. Exp. Med.* 174, 1483–1489.
72. Strober BE, Menon K. (2007) Alefacept for the treatment of psoriasis and other dermatologic diseases. *Dermatol. Ther.* 20, 270–276.
73. Korhonen R, Moilanen E. (2009) Abatacept, a novel CD80/86-CD28 T cell co-stimulation modulator, in the treatment of rheumatoid arthritis. *Basic Clin. Pharmacol. Toxicol.* 104, 276–284.
74. Davis PM, Abraham R, Xu L, Nadler SG, Suchard SJ. (2007) Abatacept binds to the Fc receptor CD64 but does not mediate complement-dependent cytotoxicity or antibody-dependent cellular cytotoxicity. *J. Rheumatol.* 34, 2204–2210.
75. Economides AN, Carpenter LR, Rudge JS, Wong V, Koehler-Stec EM, Hartnett C, et al. (2003) Cytokine traps: multi-component, high-affinity blockers of cytokine action. *Nat. Med.* 9, 47–52.
76. Bruggemann M, Williams GT, Bindon CI, Clark MR, Walker MR, Jefferis R, et al. (1987) Comparison of the effector functions of human immunoglobulins using a matched set of chimeric antibodies. *J. Exp. Med.* 166, 1351–1361.
77. Jefferis R. (2007) Antibody therapeutics: isotype and glycoform selection. *Expert Opin. Biol. Ther.* 7, 1401–1413.
78. Rother RP, Rollins SA, Mojcik CF, Brodsky RA, Bell L. (2007) Discovery and development of the complement inhibitor eculizumab for the treatment of paroxysmal nocturnal hemoglobinuria. *Nat. Biotechnol.* 25, 1256–1264.
79. Ashkenazi A, Chamow SM. (1995) Immunoadhesins: an alternative to human monoclonal antibodies. *Methods* 8, 104.
80. International Nonproprietary Names for Pharmaceutical Substances (INN). In *WHO Drug Information*. 2009. vol 23.
81. Frampton JE, Lyseng-Williamson KA. (2009) Romiplostim. *Drugs* 69, 307–317.
82. Nichol JL. (2006) AMG 531: an investigational thrombopoiesis-stimulating peptibody. *Pediatr. Blood Cancer* 47, 723–725.
83. Broudy VC, Lin NL, Broudy VC, Lin NL. (2004) AMG531 stimulates megakaryopoiesis *in vitro* by binding to Mpl. *Cytokine* 25, 52–60.
84. Newland A. (2009) Romiplostim: a breakthrough treatment for the management of immune thrombocytopenic purpura. *Eur. J. Haematol. Suppl.* S.71, 20–25.
85. Stohl W, Looney RJ. (2006) B cell depletion therapy in systemic rheumatic diseases: different strokes for different folks? *Clin. Immunol.* 121, 1–12.
86. Kuchimanchi K, Narayanan A, Krzyzanski W, Wild S, Ferbas J, Wang T, et al. Pharmacokinetics and Pharmacodynamics of a Novel Polypeptide Fusion Protein- AMG 623 in: Cynomolgus Monkeys [Abstract]. *AAPS* aaps2007-003196.
87. Glaesner W, Vick AM, Millican R, Ellis B, Tschang SH, Tian Y, et al. (2010) Engineering and characterization of the long-acting glucagon-like peptide-1 analogue LY2189265, an Fc fusion protein. *Diabetes/Metab. Res. Rev.* 26, 287–296.
88. Barrington P, Chien JY, Tibaldi F, Showalter HD, Schneck K, Ellis B. (2011) LY2189265, a long-acting glucagon like-peptide-1 analog, demonstrated dose-dependent insulin secretion in healthy subjects. *Diabetes Obes. Metab.* 13, 434–438.
89. Barrington P, Chien JY, Showalter HD, Schneck K, Cui S, Tibaldi F, et al. (2011) A 5-week study of the pharmacokinetics and pharmacodynamics of LY2189265, a novel, long-acting glucagon-like peptide-1 analog, in patients with type 2 diabetes. *Diabetes Obes. Metab.* 13, 426–433.
90. Wrighton NC, Farrell FX, Chang R, Kashyap AK, Barbone FP, Mulcahy LS, et al. (1996) Small peptides as potent mimetics of the protein hormone erythropoietin. *Science* 273, 458–464.
91. Bouman-Thio E, Franson K, Miller B, Getsy J, Cohen A, Bai SA, et al. (2008) A phase I, single and fractionated, ascending-dose study evaluating the safety, pharmacokinetics, pharmacodynamics, and immunogenicity of an erythropoietin mimetic antibody fusion protein (CNTO 528) in healthy male subjects. *J. Clin. Pharmacol.* 48, 1197–1207.
92. Bugelski PJ, Capocasale RJ, Makropoulos D, Marshall D, Fisher PW, Lu J, et al. (2008) CNTO 530: molecular pharmacology in human UT-7EPO cells and pharmacokinetics and pharmacodynamics in mice. *J. Biotechnol.* 134, 171–180.
93. Picha KM, Cunningham MR, Drucker DJ, Mathur A, Ort T, Scully M, et al. (2008) Protein engineering strategies for sustained glucagon-like peptide-1 receptor-dependent control of glucose homeostasis. *Diabetes* 57, 1926–1934.
94. Johnson DL, Farrell FX, Barbone FP, McMahon FJ, Tullai J, Hoey K, et al. (1998) Identification of a 13 amino acid peptide mimetic of erythropoietin and description of amino acids critical for the mimetic activity of EMP1. *Biochemistry* 37, 3699–3710.

95. Bugelski, P.J., Makropoulos, D. Spinka-Doms, T. Eirikis, E. Volk, A. Jiao, Q. Huang, C. (2011) Differential Effects of Long-lived Erythropoietin Receptor Agonists in *Rats. Pharmaceutica Analytica Acta* 2:7.
96. Hutchins JT, Kull FC, Jr, Bynum J, Knick VC, Thurmond LM, Ray P. (1995) Improved biodistribution, tumor targeting, and reduced immunogenicity in mice with a gamma 4 variant of Campath-1H. *Proc. Natl. Acad. Sci. USA* 92, 11980–11984.
97. Hermeling S, Crommelin DJ, Schellekens H, Jiskoot W. (2004) Structure-immunogenicity relationships of therapeutic proteins. *Pharm. Res.* 21, 897–903.
98. Cromwell ME, Hilario E, Jacobson F. (2006) Protein aggregation and bioprocessing. *AAPS J.* 8, E572–579.
99. Cartron G, Dacheux L, Salles G, Solal-Celigny P, Bardos P, Colombat P, et al. (2002) Therapeutic activity of humanized anti-CD20 monoclonal antibody and polymorphism in IgG Fc receptor FcgammaRIIIa gene. *Blood* 99, 754–758.
100. Lazar GA, Dang W, Karki S, Vafa O, Peng JS, Hyun L, et al. (2006) Engineered antibody Fc variants with enhanced effector function. *Proc. Natl. Acad. Sci. USA* 103, 4005–4010.
101. Shields RL, Lai J, Keck R, O'Connell LY, Hong K, Meng YG, et al. (2002) Lack of fucose on human IgG1 N-linked oligosaccharide improves binding to human Fcgamma RIII and antibody-dependent cellular toxicity. *J. Biol. Chem.* 277, 26733–26740.
102. Crispin M, Bowden TA, Coles CH, Harlos K, Aricescu AR, Harvey DJ, et al. (2009) Carbohydrate and domain architecture of an immature antibody glycoform exhibiting enhanced effector functions. *J. Mol. Biol.* 387, 1061–1066.
103. Raju TS. (2008) Terminal sugars of Fc glycans influence antibody effector functions of IgGs. *Curr. Opin. Immunol.* 20, 471–478.
104. Xu D, Alegre ML, Varga SS, Rothermel AL, Collins AM, Pulito VL, et al. (2000) *In vitro* characterization of five humanized OKT3 effector function variant antibodies. *Cell Immunol.* 200, 16–26.
105. Herold KC, Gitelman S, Greenbaum C, Puck J, Hagopian W, Gottlieb P, et al. (2009) Treatment of patients with new onset type 1 diabetes with a single course of anti-CD3 mAb Teplizumab preserves insulin production for up to 5 years. *Clin. Immunol.* 132, 166–173.
106. Madsbad S, Kielgast U, Asmar M, Deacon C, Torekov SS, Holst JJ. (2011) An overview of once-weekly GLP-1 receptor agonists—available efficacy and safety data and perspectives for the future. *Diabetes Obes. Metab.* 60, 1599–1607.
107. Chu QS. (2009) Aflibercept (AVE0005): an alternative strategy for inhibiting tumour angiogenesis by vascular endothelial growth factors. *Expert Opin. Biol. Ther.* 9, 263–271.
108. Mack GS. (2008) CD20 blockers eye crowded rheumatology market. *Nat. Biotechnol.* 26, 1053–1054.
109. Hayden-Ledbetter MS, Cerveny CG, Espling E, Brady WA, Grosmaire LS, Tan P, et al. (2009) CD20-directed small modular immunopharmaceutical, TRU-015, depletes normal and malignant B cells. *Clin. Cancer Res.* 15, 2739–2746.
110. Burge DJ, Bookbinder SA, Kivitz AJ, Fleischmann RM, Shu C, Bannink J. (2008) Pharmacokinetic and pharmacodynamic properties of TRU-015, a CD20-directed small modular immunopharmaceutical protein therapeutic, in patients with rheumatoid arthritis: a Phase I, open-label, dose-escalation clinical study. *Clin. Ther.* 30, 1806–1816.
111. Saerens D, Ghassabeh GH, Muyldermans S. (2008) Single-domain antibodies as building blocks for novel therapeutics. *Curr. Opin. Pharmacol.* 8, 600–608.

9

RECEPTOR-Fc AND LIGAND TRAPS AS HIGH-AFFINITY BIOLOGICAL BLOCKERS: DEVELOPMENT AND CLINICAL APPLICATIONS

ARIS N. ECONOMIDES AND NEIL STAHL

Regeneron Pharmaceuticals, Tarrytown, NY, USA

9.1 INTRODUCTION

The ability to inhibit the action of growth factors and cytokines via pharmacologic reagents has been of great interest and utility. One important class of blockers is based on extracellular domains (ectodomains) of receptors that function as decoys when engineered as soluble proteins. Proteins composed of receptor ectodomains fused to a stabilizing moiety, usually the fragment crystallizable (Fc) region of IgG, as well as more complex hybrids with Fc such as Traps that incorporate more than one ectodomain, have been used successfully both in research and in the clinic. This review focuses on the various clinically relevant receptor ectodomain-Fc (receptor-Fc) fusion proteins and Traps that target secreted factors and cytokines. Other types of peptide-Fc fusions, such as ligand-Fc fusions, or artificial proteins based on other scaffolds have been reviewed extensively elsewhere [1–9].

The idea that receptor ectodomains can act as modulators of growth factor and cytokine signaling has its origins in the discovery of naturally produced soluble receptors. The presence of these proteins was first demonstrated in cell culture for the epidermal growth factor (EGF) [10] and interleukin-2 receptors (IL-2R) [11], then in rapid succession found to be a property of many cytokine and growth factor receptor systems. These soluble receptors arise either through proteolytic cleavage mediated by "sheddases" or through translation of alternatively spliced transcripts [12]. Furthermore, they are present in bodily fluids, and their production is altered during disease [13–18]. These discoveries led to the realization that soluble receptors may represent endogenous modulators of growth factor and cytokine action, and gave rise to the idea of using soluble receptors as therapeutics in diseases where aberrant expression of a factor or cytokine plays a key role in pathogenesis [19–24].

The ability of soluble receptors to act as blockers or decoys for their cognate ligand was initially demonstrated in the IL-2-R system [25,26], and subsequently was shown to hold true for other cytokines (for reviews, see References [21,27], and [28]). In fact, modulation of cytokine and growth factor signaling by decoy receptors [21], coreceptors [29–31], as well as various other receptor-like modulators [32–39], and secreted blockers [40–45] has been demonstrated for many different types of extracellular signaling

Fusion Protein Technologies for Biopharmaceuticals: Applications and Challenges, First Edition. Edited by Stefan R. Schmidt.

proteins, and likely reflects the need for tight regulation of signaling.

Both naturally occurring and engineered soluble receptors were initially shown to block cytokine action *in vitro*, by effectively acting as decoys. However, *in vivo*, they failed to act efficaciously because of relatively low affinity for their cognate ligand(s) and short *in vivo* half-lives. These limitations were first shown for the monomeric tumor necrosis factor receptor (TNFR) ectodomain [13,17,46]. Hence, it became evident that pharmacologic application of soluble receptors for inhibition of their cognate ligands *in vivo* would either require repeated dosing, or they would need to be engineered to improve their pharmacokinetic properties. A time-tested solution to this problem was provided by assembling recombinant hybrid proteins composed of receptor ectodomains fused to the nonvariable regions of antibodies, as was initially shown for CD4. In this pioneering work, the ectodomain of CD4 was fused either to the constant region of human IgG1 [47] or to the Fc region of mouse IgG2a [48], to generate "immunoadhesins"—that is, hybrid proteins composed of a receptor ectodomain (or part thereof) and the constant or Fc region of an immunoglobulin, now commonly referred to as "receptor-Ig" or "receptor-Fc" fusion proteins. The resulting hybrids were high-affinity blockers of HIV infection in T cells. Although these blockers of HIV did not materialize into therapeutics, they demonstrated several key properties of receptor-Ig or -Fc fusions:

1. Receptor ectodomains can be fused to Ig domains without adversely affecting the function of the resulting hybrid proteins, while improving their expression [47–49].
2. Unlike their soluble receptor counterparts, receptor-Ig or -Fc fusions display long *in vivo* half-lives.
3. Receptor-Ig or -Fc fusions have antibody-like properties that retain at least some of the functions of antibodies, such as binding to Fc receptors including FcRn.
4. Desired effector functions can be potentially engineered into these fusions by choosing the identity of the Ig subclass constant region (along the lines of what is done with recombinant therapeutic antibodies [50]).

Historically, receptor-Ig, receptor-Fc, and other variations of Ig-tagged fusions have been utilized [48,51], but the design that has prevailed is the receptor-Fc version. Although other domains and chemical moieties have been used as "carriers" to engineer proteins with improved pharmacokinetic profiles, it is mainly peptide-Fc fusion proteins that have enjoyed widespread use in research applications. In fact, tagging proteins with Fc is one of the most commonly utilized methods employed for studying growth factor and cytokine function *in vitro* and *in vivo*, and the literature abounds with such examples. Their use has been more limited in the clinic, in part reflecting the time and effort it takes to develop new therapeutics.

9.2 ETANERCEPT AS A PROTOTYPICAL RECEPTOR-Fc-BASED CYTOKINE BLOCKER

The initial work on CD4-Ig fusions was rapidly followed by the generation of TNFR-Fc [46,52], a high-affinity blocker for TNF-α and TNF-β/lymphotoxin (Figure 9.1a). The idea of developing soluble TNFR as a therapeutic arose from a series of preclinical observations that demonstrated a pathologic role of TNF-α in several inflammatory diseases [54]. The initial clinical proof-of-concept that a TNF-α blocker can halt the progression of pathologic inflammatory processes and ameliorate disease was procured using the anti-TNF-α antibody infliximab, and concurrently followed by TNFR-Fc (etanercept). A detailed perspective of the development of etanercept and other TNF blockers can be found in several excellent reviews on the subject, some of which also offer a historical perspective [54–58]. The success of etanercept—it remains one of the most widely used protein-based therapeutics [59]—has been followed by other receptor-Fc fusion proteins as well as Traps with the aim of engineering blockers for their cognate ligands. Some of those that are entering or have completed clinical development or have been approved are listed in Table 9.1.

9.3 HETEROMERIC TRAPS FOR LIGANDS UTILIZING MULTICOMPONENT RECEPTOR SYSTEMS WITH SHARED SUBUNITS

Most of the receptor-Fc-based ligand blockers that have been designed to date have relied on single extracellular domains (Table 9.1, Figure 9.1) dimerized via the Fc portion of immunoglobulins, mainly human and mouse IgG (and the latter mainly for research purposes). However, many receptor systems are composed of more than one receptor subunit. Thus, they do not lend themselves to simple dimerization of a single extracellular domain in order to generate a blocker, at least not one that displays both high affinity and specificity for its cognate ligand(s). For the purposes of this review, the different multicomponent receptor systems can be parsed into two broad categories based on the properties of the subunits comprising the signaling complex:

1. *Heteromeric Receptor Systems that Utilize Two Signaling Receptor Subunits.* Examples include the type I and type II receptors for TGF-β and bone morphogenetic proteins (BMPs) [31,116] or IL-1-RI (IL-1R1) and IL-1-RAcP (IL-1RAP) for IL-1 (Figure 9.2); a similar concept applies to a

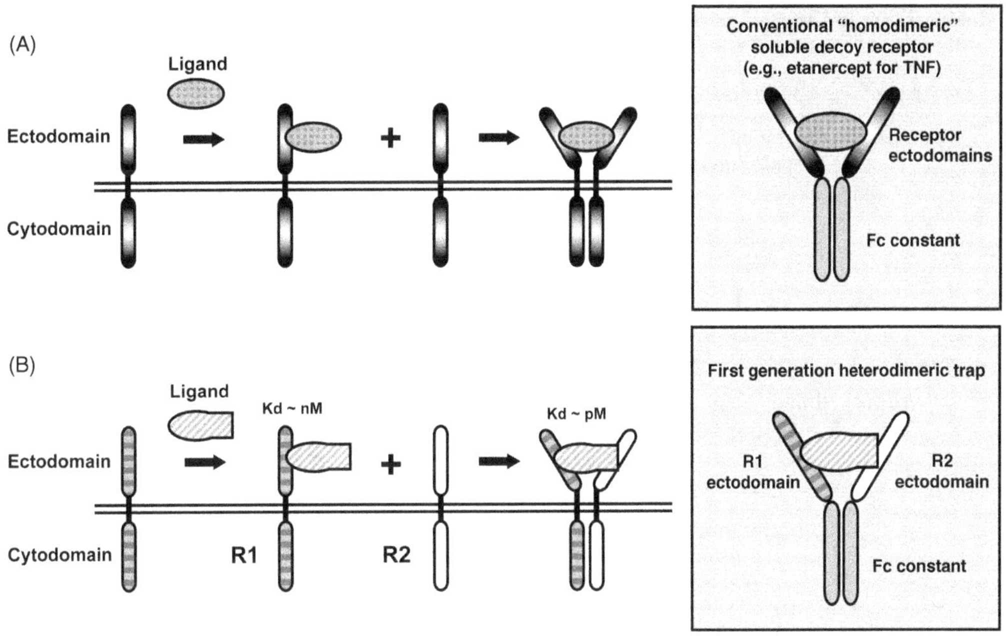

FIGURE 9.1 First-generation Traps for ligands utilizing multicomponent receptor systems. (A) Schematic representation of a ligand-receptor system that utilizes a single receptor. Binding of ligand to a first receptor subunit is followed by dimerization (or multimerization) with another identical subunit and formation of a signaling complex. Although the exact stoichiometry of the number of ligand and receptor monomers that comprise the signaling complex varies depending on the receptor system, simple dimerization of the receptor ectodomains via Fc results in a receptor-Fc-based blocker for the cognate ligand (upper inset). An example of this is etanercept, a receptor-Fc fusion protein composed of TNFR-Fc that acts as a blocker of TNF [46]. (B) Schematic representation of a ligand-receptor system that utilizes a two-component receptor, where the second receptor (R2) may be shared among different members of a family of ligands. Binding of ligand to the first receptor subunit (R1) results in the formation of a low-affinity complex, with K_d in the nanomolar range. This step is required for recognition of the ligand and frequently imparts selectivity, as it is the presence of a specific R1 that determines whether a cell will be responsive to a specific ligand. This first step is followed by engagement of R2 by the R1·ligand complex, resulting in a high affinity, signaling complex composed of R1·ligand·R2. Although the exact stoichiometry of the number of ligand molecules and each of the two different receptor subunits that comprise the signaling complex may vary, artificial heterodimerization of the ectodomains of these two receptor subunits, via the Fc region of IgG, results in a high-affinity ligand trap (lower inset) with low picomolar affinity for its cognate ligands. Examples of such heterodimeric receptor ectodomains-Fc fusion proteins are presented by early versions of Cytokine Traps [53].

multiplicity of other cytokines [117]. These systems may also incorporate additional accessory coreceptors that facilitate binding but are otherwise dispensable, that is, are not obligatory for recognition of ligand or signal transduction [29,30,118–122].

2. *Receptor Systems that Utilize Specificity-Determining Nonsignaling Subunits (Figures 9.1b and 9.3).* These receptor systems employ shared or "common" subunits that are utilized by multiple ligands in any given family, but which neither bind their ligand partners at an appreciable level, nor do they confer specificity. Examples of shared subunits are the common γ chain of the IL-2 family of cytokines (IL-2RG), and gp130 (also known as IL-6ST) for the IL-6 family of cytokines [117]. Consequently, in these systems, binding of ligand is conferred by specificity-determining subunits; this is an obligate step, required to initiate formation of the signaling complex. Examples of such receptor subunits are IL-6R, CNTFR, IL-11RA, GFRA1, and GFRA2 [117,123–125]. These specificity-determining subunits are not capable of signaling either because they are tethered to the membrane by a glycosylphosphatidylinositol (GPI) anchor or because their cytoplasmic domains lack enzymatic activity or binding sites for accessory molecules.

TABLE 9.1 Cytokine Traps in Clinical Development and as Approved Therapeutics

Fusion	Generic Name	Mechanism of Action	Approved Indications	Exploratory Indications
TNFR-Fc [46,55,60]	Etanercept	Inhibition of TNF-α and TNF-β/lymphotoxin [55]	Moderate to severe rheumatoid arthritis (RA) [61] Moderate to severe polyarticular juvenile idiopathic arthritis (JIA) Psoriatic arthritis; Ankylosing spondylitis (AS); Moderate to severe plaque psoriasis	Multiple studies for a variety of different conditions are ongoing
IL1 Trap	Rilonacept	Inhibition of IL1A and IL1β [53,62]	Cryopyrin-associated periodic syndromes (CAPS), specifically familial cold-induced autoinflammatory syndrome (FCAS), and Muckle–Wells syndrome (MWS) [63,64]	Gout (NCT00855920; NCT00856206; NCT00829829; NCT00610363) Familial Mediterranean Fever (NCT00582907); Systemic juvenile idiopathic arthritis (NCT00534495); Diabetes mellitus type 1 (NCT00962026); Schnitzler syndrome (NCT01045772)
VEGFP Trap-Eye	Aflibercept	Inhibition of VEGFA and other VEGF family members and PGF [65]	None	Phase III for diabetes mellitus macular edema (NCT01331681); Phase III for diabetic macular edema (NCT01363440); Phase III for pathologic myopia (NCT01249664)
VEGF Trap	Aflibercept	As above	None	Multiple Phase I trials for oncologic indications (NCT00650923, NCT01148615); multiple Phase II trials in oncologic indications (NCT00357760; NCT00437034; NCT00509249; NCT00828139; NCT01258855)
CD58-IgG1	Alefacept	Exact mechanism remains to be deciphered; mixed agonist/antagonist; decrease in effector memory T cells [66–68]	Chronic moderate to severe plaque psoriasis [69–73]	Pilot study for allogeneic hematopoietic stem cell transplantation (NCT01319851); Phase I for aplastic anemia (NCT01267643); Phase I for de novo kidney transplant (NCT01163799); Phase I/II for resistant chronic graft-versus-host disease (NCT00397332); Phase II for chronic graft-versus-host disease (NCT01226420); Phase II for diabetes mellitus, type 1 (NCT00965458); Phase III for kidney transplantation (NCT00543569); Phase III for graft versus host disease (NCT00361413)
CTLA4-Ig	Abatacept	Inhibition of the CD28-CD80/86-mediated co-stimulatory signal required for T-cell activation [74]	Rheumatoid arthritis [75]; juvenile idiopathic arthritis [76,77]	Phase I for relapsing polychondritis (NCT01272856); Phase II for acute graft versus host Disease (NCT01012492); Phase II for giant cell arteritis and Takayasu's arteritis (NCT00556439); Phase II for lupus nephritis (NCT00774852); Phase II for polymyositis and dermatomyositis (NCT01315938); Phase II for *Alopecia totalis* or *Alopecia universalis* (NCT01314495); Phase II for uveitis (NCT01279954); Phase II for relapsing-remitting multiple sclerosis (NCT01116427); Phase II for asthma (NCT00784459)

TABLE 9.1 (Continued)

Fusion	Generic Name	Mechanism of Action	Approved Indications	Exploratory Indications
CTLA4-Ig, modified (BMS 224818) [78]	Belatacept	Inhibition of the CD28-CD80/86-mediated co-stimulatory signal required for T-cell activation	(Undergoing FDA approval process)	
ACVR2A-Fc (ActRIIA-IgG1; ACE-011)	Sotatercept	Inhibition of activin A (proposed) [79,80]; may also inhibit other BMP family ligands [81,79]	None	Phase I safety, tolerability and pharmacodynamics study in healthy postmenopausal women (NCT00709540) [82]; Phase II for correction of anemia in endstage renal disease (NCT01146574, and Phase II/III for chemotherapy induced anemia (NCT01284348) [83]; Phase II for solid tumors (NCT01190644); Phase II for osteolytic lesions of multiple myeloma (NCT00747123)
ACVR2B-Fc (ActRIIB.hFc; ACE-031) [84]		Inhibition of activin A, myostatin, and GDF11 signaling [85]	None	Phase I for muscular atrophy (NCT00952887); Phase I for muscle loss (NCT00755638); Phase II for Duchenne muscular dystrophy (NCT01239758, suspended based on safety data; NCT01099761 terminated)
ACVRL1-Fc (Alk1-IgG1; ACE-041)		Inhibition of BMP9, BMP10 [86,87]	None	Phase I for advanced solid tumors and multiple myeloma (NCT00996957)
LTBR-Ig (BG9924) [88]	Baminercept	Inhibition of LTa_1b_2 [89] and LIGHT [90] signaling	None	Phase I for chronic HCV hepatitis C (NCT01164384)
TACI-Ig	Atacicept	Inhibition of BLyS and APRIL	None	Phase I for multiple myeloma or active Waldenström's macroglobulinemia [91,92]; Phase I for relapsed and refractory B-cell Non-Hodgkin's lymphoma [93]; Phase II/III for systemic lupus erythematosus (NCT00624338) [94]
gp130-Fc, modified (CR5/18) [95]		Inhibition of "transignaling" via soluble IL-6R [96]	None	Preclinical [97,98]; no active trials listed
IL-17RA-IgG1 or IL-17RA-Fc	–	Inhibition of IL-17A [99–101]	None	Rheumatoid arthritis, multiple sclerosis, psoriasis [102,103]; no active trials listed
IL-17RC-Fc		Inhibition of IL-17A, IL-17F, and IL-17A/F [101,104]	None	Rheumatoid arthritis, multiple sclerosis, psoriasis [102,103]; no active trials listed
BR3-Fc	Briobacept	Inhibition of BLyS [105,106]	None	Completed phase I in rheumatoid arthritis [107]; no active trials listed
CD4-IgG2 (PRO-542) [51]		Prevents attachment of HIV to CD4 [51]		Completed Phase I (NCT00000876) [108] and Phase II (NCT00055185) for HIV infections; no active trials listed
Fc-OPG [109]		Inhibition of RANKL [110,111]	None	Clinical development for bone loss-associated conditions [109,112,113] halted because of superior results and reduced safety concerns with denosumab [114,115]

The fusion protein is indicated using the name under which it has appeared in the literature. Where more than one name has been used, HUGO-standardized nomenclature has been adopted and the names appearing in the literature have been listed in parentheses. Generic names are provided where applicable. The exploratory indications listed are for the most part those still under development; however, indications no longer being pursued are listed in those cases where they illuminate the history of development of the drug.

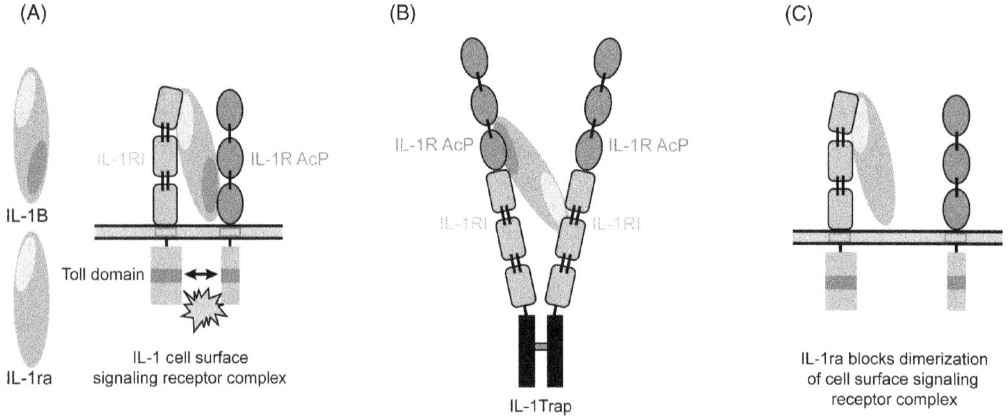

FIGURE 9.2 Rilonacept as an example of a heteromeric ligand trap engineered to mimic the high-affinity binding observed with the natural receptor complex. (A) Interleukin 1 (IL-1) has two receptor binding sites (light and dark gray). The IL-1 receptor type I (IL-1R1, also referred to as IL-1-RI) binds to one site and IL-1 receptor accessory protein (IL-1RAP, also referred to as IL-1-RAcP) to the other, and binding of IL-1 initiates signaling. (B) The IL-1 Trap (rilonacept) is constructed as an "in-line" fusion of the extracellular domains of IL-1RAP and IL-1R1, followed by the Fc domain of human IgG1. The two Fc domains drive dimerization of each trap monomer, and the resulting homodimer is further stabilized and held together via two disulfide bonds. IL-1 binds to rilonacept with high affinity, similar to the binding seen with the receptor complex on the cell surface. (C) The naturally occurring IL-1 receptor antagonist (IL-1RN) binds to only one site and as a result prevents complex formation and receptor activation. In contrast to the rilonacept·IL-1 interaction, which displays a low K_d in the picomolar range, the binding of IL-1RN to IL-1R1 forms a low-affinity complex. *Source*: Reprinted by permission from Macmillan Publishers Ltd Nature Medicine (Dinarello *Nat. Med.* 2003 Jan, 9(1) 20–22), copyright (2003).

These two groupings are artificial, yet they are useful in illuminating the challenges that have to be overcome when attempting to use receptor ectodomains to design specific and high-affinity blockers. For the second type of receptor system, it is usually not possible to utilize the traditional receptor-Fc methodology to create high-affinity blockers for their cognate ligands because simple dimerization of one receptor subunit will either result in a "promiscuous" blocker with relatively low affinity (see, e.g., gp130-Fc, discussed below), or worse yet generate a protein that will act as a carrier for and potentiator of cytokine activity. The latter property is best reflected in the ciliary neurotrophic factor (CNTF) and IL-6 receptor systems, where it has been elegantly demonstrated that the soluble or shed forms of IL-6R [126] and CNTFR [123] enable signaling in cells that do not express them, but which express the shared receptor subunits (e.g., gp130 for IL-6/IL-6R; and gp130 plus LIFRβ for CNTF/CNTFR). This process has been termed "trans-signaling" [23,96] (Figure 9.3). In fact, we have used this property to design cell type-specific potentiators of cytokine activity and achieve exactly the opposite of a blocker, that is, to selectively convert IL-6- or CNTF-unresponsive cells to responsive ones by addressing IL-6R-Fc and CNTFR-Fc to the surface of IL-6R- and CNTFR-negative cells [127].

9.3.1 Proof-of-Concept for High-Affinity Traps for Ligands Utilizing Multicomponent Receptor Systems

As the earlier discussion indicates, the mechanism of receptor assembly for factors and cytokines utilizing multi-component receptor systems imposed unprecedented limitations on the design of receptor ectodomain-based blockers, and therefore required the development of a new kind of technology. A general solution to the problem of generating receptor-based blockers for these types of ligands was provided by the invention of Trap technology [53]. Traps are fusion proteins that are composed of more than one receptor ectodomain (or part thereof) fused to the constant region of immunoglobulins. The first sets of Traps to be developed were for IL-6 and CNTF, but their successful engineering was rapidly followed by development of Traps for multiple other cytokines and growth factors.

The methodology for engineering Traps for ligands utilizing multicomponent receptor systems reflects the mechanism by which certain ligand·receptor signaling complexes form (Figures 9.1b and 9.3). Two key observations informed the design:

1. Signaling receptor complexes form in a sequential, ordered fashion where the ligand binds to a first receptor subunit, usually the specificity-determining

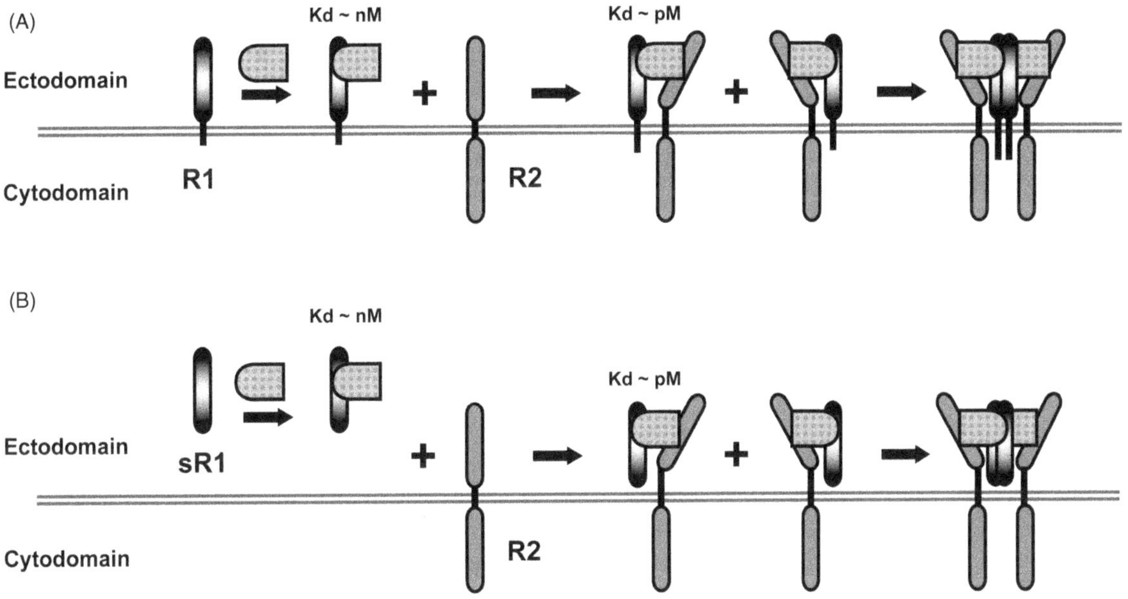

FIGURE 9.3 Trans-signaling. (A) Schematic representation of a receptor system utilizing a specificity-determining (but nonsignaling) receptor (R1). R1 is tethered to the cell membrane either by a transmembrane domain followed by a short cytoplasmic tail as shown here (e.g., IL-6R, IL-11R), or by a GPI anchor (e.g., CNTFR, GFRA1, GFRA2). Mirroring other multicomponent receptor systems (see Figure 9.1), cognate ligands are first recognized by R1. Binding of ligand to R1 results in the formation of a low-affinity R1·ligand complex. The R1·ligand complex engages a second (usually shared or "common") receptor subunit to form a high-affinity R1·ligand·R2. Unlike the multicomponent receptor systems shown on Figure 9.1, this R1·ligand·R2 complex does not transduce signal; a third step involving dimerization of R1·ligand·R2 is required to initiate signaling. (B) Cells that do not express R1 are not responsive to ligand. However, soluble receptor 1 (sR1) can present ligand to these cells and initiate the same cascade of events depicted in the upper panel. This process has been termed trans-signaling.

subunit (e.g., IL-6R for IL-6; CNTFR for CNTF; IL-11RA for IL-11, GDNF family receptor α-1 or -2 (GFRA1 or GFRA2) for glial cell-derived neurotrophic factor, (GDNF)) to generate a receptor 1·ligand complex, followed by binding of a second and sometimes a third receptor subunit, which are usually shared [117,124,125,128].

2. In this stepwise complex formation, the second step is also an "affinity conversion" step, because the resulting complex—receptor 1·ligand·receptor 2—is much more stable than the one resulting from the initial association of the ligand with receptor 1 [117,124].

These observations led to the hypothesis that heteromeric hybrid proteins composed of the ectodomains of receptor 1 (R1) and receptor 2 (R2) should form nonsignaling complexes with their cognate ligands and display the high affinity observed with their natural counterparts. Furthermore, for ligands that share a signaling coreceptor, the specificity of the Trap should be determined by the choice of R1. These hypotheses were then reduced to an engineering problem, that is, how to put together receptor ectodomains in a manner such that resulting hybrid proteins will function as high-affinity ligand traps. In addition, these hybrids should demonstrate high level of expression, easy purification, and desirable pharmacokinetic profiles, which will enable their development into therapeutics.

Generating heterodimers of receptor ectodomains in a way that mirrors their configuration on the cell surface posed a formidable technical problem. It was also not known whether such complexes would act as effective and specific ligand traps. In order to obtain rapid proof-of-concept, heteromeric ligand traps were generated for two members of the IL-6 family of cytokines—CNTF and IL-6—by engineering ectodomain heterodimers between IL-6R and the shared receptor chain gp130 for IL-6 and CNTFR and gp130 for CNTF, respectively (Economides and Stahl, unpublished results). In a first iteration, the ectodomains of IL-6R or CNTFR and gp130 were fused to leucine zippers derived from Fos and Jun, as they assemble into heterodimers [129]. This strategy was successful in generating IL-6R:gp130 and CNTFR:gp130 ectodomain heterodimers that functioned as highly specific and effective blockers of IL-6 and CNTF. Moreover, neither complex cross-inhibited, that is, the IL-6:gp130 heterodimer did not block CNTF, and the CNTF:gp130 heterodimer did not block IL-6, nor did they inhibit signaling initiated by leukemia inhibitory factor (LIF) (AN Economides and N Stahl, unpublished results), another IL-6 cytokine family member whose receptor utilizes gp130 in conjunction with LIFRβ [130]. This first version of cytokine Traps provided proof-of-concept that heterodimeric proteins composed of R1:R2 ectodomains could act as highly specific blockers of their cognate cytokine. Furthermore, this work illustrated that the ectodomain of gp130 is by itself a poor blocker for the different members of the IL-6 family of cytokines (an observation independently verified [131]), but that when it is placed in a preformed complex with a specificity-determining receptor ectodomain, then the resulting hybrid protein acts as an extremely specific blocker for its cognate cytokine. (A notable exception to the low effectiveness of soluble gp130 as a blocker is its ability to selectively inhibit IL-6 trans-signaling [132]; this is discussed in more detail below.) Conversely, this work demonstrated that the properties of the soluble versions of IL-6R, IL-11R, and CNTFR could be altered from potentiators to blockers when engineered as obligate heterodimers with the ectodomain of gp130.

9.3.2 Turning Heterodimeric Soluble Receptor-Based Ligand Traps into Therapeutics

Although this early work provided proof-of-concept, the production of these molecules was inefficient and their purification was difficult. A mixture of homodimers and heterodimers was generated, and the homodimers had to be removed because IL-6R and CNTFR homodimers can potentiate—rather than block—IL-6 or CNTF activity, respectively [123,126] (Figure 9.3). In addition, there was the concern that these artificial heterodimeric hybrid proteins would be immunogenic and display poor pharmacokinetic profiles. The concern for immunogenicity arose from the fact that these were fusions with leucine zipper domains of two transcription factors, that is, proteins that are not naturally found in circulation.

Therefore, taking advantage of the experience gained with immunoadhesins [47,48], and mirroring the experience with TNFR-Fc [46], we engineered the first generation of Fc-based ligand traps. This presented a new engineering problem: how does one bring together the ectodomains of two different receptors and specifically heterodimerize them via the Fc domain of IgG, which will drive dimerization to form homodimers of each ectodomain as well as the desired heterodimers? One solution to this problem was to tag the C-terminus of gp130-Fc with a hexahistidine tag, as this would allow the purification of IL-6R-Fc:gp130-Fc-His6 heterodimers from the corresponding homodimeric proteins—$(\text{IL-6R-Fc})_2$ and $(\text{gp130-Fc-His6})_2$—by nickel affinity chromatography [133]. The technical choice of His6-tagging gp130-Fc was driven by the need to completely remove any traces of IL-6R-Fc from the IL-6R-Fc:gp130-Fc preparation, as IL-6R-Fc would act as a potentiator of signaling. The first-generation Traps utilizing Fc differentially tagged with a hexahistidine tag were generated for IL-6, IL-11, CNTF, and LIF, all using combinations of gp130-Fc-His6 as the common R2 component [53] (Economides and Stahl, unpublished results). All of these Traps were shown to be highly specific and effective blockers for their cognate cytokine. However, their production and purification remained problematic for two reasons: (a) the desired species, R1-Fc·R2-Fc-His6 heterodimer, was only about half of the total Fc-tagged protein, and (b) both the R1-Fc·R2-Fc-His6 heterodimers and the R2-Fc-His6 homodimers have hexahistidine tags. Although, isolation of the desired species (R1·R2 heterodimers) away from the "contaminating" homodimers could be accomplished to generate enough material for laboratory tests, it was deemed difficult and costly for large-scale production (K Bailey, unpublished results).

Therefore, a second generation of cytokine Traps was developed by engineering "in-line" fusion proteins composed of the ectodomain of a specificity-determining subunit (R1) fused to a shared subunit (R2) and tagged with Fc, either in an R1-R2-Fc or an R2-R1-Fc configuration. These were generated for several cytokines and then tested to determine which version worked best. These were the R1-R2-Fc configuration of the IL-6 Trap (IL-6R-gp130-Fc), and the R2-R1-Fc of the IL-4 Trap (IL-2Rγ-IL-4RA-Fc) and the IL-1 Trap (IL-1RAP-IL-1R1-Fc; Figure 9.2). Unlike their first-generation counterparts, these in-line fusion traps were expressed as homogeneous homodimeric proteins that could be easily purified. In addition, just like their first-generation counterparts, these in-line fusions were highly potent and specific blockers and bound their cognate cytokines with high affinity, displaying dissociation constants in the low or subpicomolar range. Furthermore, they were much more effective as blockers than the corresponding homodimeric receptor-Fc fusions [53].

The successful engineering of in-line fusion Cytokine Traps enabled preclinical testing of their pharmacokinetic properties and their ability to block cytokine activity in animal models. This work focused on two Cytokine Traps of therapeutic interest, the IL-1 Trap and the IL-4 Trap. Both of these Traps were shown to persist in circulation for many days after a single subcutaneous or intravenous injection in cynomolgus monkeys. *In vivo*, the IL-1 Trap was shown to block responses to exogenously administered interleukin 1-β (IL-1β), and then in a mouse model of collagen-induced arthritis. Similarly, the IL-4 Trap was shown to be effective in blocking IL-4 responses *in vivo*, both in mice and in cynomolgus monkeys [53].

9.4 DEVELOPMENT AND CLINICAL APPLICATION OF AN INTERLEUKIN 1 TRAP: RILONACEPT

This body of work demonstrated the feasibility of generating therapeutically effective Cytokine Traps composed of in-line receptor ectodomains (or parts thereof) fused to Fc, and opened the way for clinical applications. The IL-1 Trap (rilonacept) was initially tested in rheumatoid arthritis. Although it showed some clinical activity and substantially decreased C-reactive protein (CRP), the study failed to meet its primary endpoints, and development of rilonacept for rheumatoid arthritis was discontinued. Subsequently, rilonacept was demonstrated in clinical trials to be successful in reducing the symptoms of cryopyrin associated periodic syndrome (CAPS), a spectrum of rare autoinflammatory conditions, and became the first approved therapeutic for familial cold autoinflammatory syndrome and Muckle–Wells syndrome [62,134]. In the CAPS clinical program, rilonacept was generally well tolerated, and the most frequent side effects were injection site reactions and upper respiratory tract infections.

Since its FDA approval for CAPS, rilonacept has been tested in other disease settings (Table 9.1), notably for gout. Preclinical research [135] combined with successful application of a mouse IL-1 Trap in a model of gouty arthritis [136] and initial reports of clinical utility of IL-1 blockade [137] opened the way for testing rilonacept in chronic gouty arthritis patients. A first proof-of-concept clinical trial demonstrated that inhibition of IL-1β by rilonacept was well tolerated in patients with chronic active gout and also improved symptoms and reduced flares [138,139]. A Phase II study provided evidence that rilonacept could provide prophylaxis against gout flares induced by the initiation of allopurinol therapy. More recently, Phase III trials were designed to address the safety and efficacy of rilonacept either in patients undergoing an acute gout attack (SURGE trial, NCT00855920), or in a prophylactic setting (PRE-SURGE1, NCT00829829; PRE-SURGE2, NCT00958438; RE-SURGE, NCT00856206). Although the final results of these studies have not been published in a peer-reviewed journal, initial analysis of the SURGE trial showed that rilonacept was not effective in significantly improving ongoing gout pain during an acute gout attack compared to indomethacin, when used either alone or in conjunction with indomethacin. In contrast, when rilonacept was used in a prophylactic mode (PRE-SURGE1, NCT00829829), it met its primary endpoint, reducing the number of gout flares by more than 70%. Rilonacept also met both of its two main secondary endpoints, of reducing the number of patients with one gout flare during the treatment period, and of reducing the number of patients with two or more gout flares during the treatment period. In these studies, rilonacept was generally well tolerated with no reported drug-related serious adverse events. Injection site reaction, generally considered mild, was the most commonly reported adverse event. Although the safety and efficacy of rilonacept in gouty arthritis remain to be evaluated by the US FDA, these results suggest that rilonacept may prove useful in maintenance therapy to prevent gout flares brought about upon initiation of urate-lowering therapy such as allopurinol.

9.5 DEVELOPMENT AND CLINICAL APPLICATION OF A VEGF TRAP

The experience gained in engineering Traps was also utilized during the development of a Trap targeting vascular endothelial growth factor (VEGF) and placental growth factor (PGF). Development of the VEGF Trap presented a new technical challenge: the first versions of VEGFR1-Fc, composed of the first three Ig domains of VEGFR1, were fairly potent blockers of VEGF *in vitro*, but displayed very poor pharmacokinetic properties *in vivo*, with low bioavailability and rapid clearance. The poor pharmacokinetic profile of early versions of the VEGF Trap were traced to a highly basic region present in the third Ig domain of VEGFR1; this basic region was a main contributor to the high-isoelectric point of these traps and conferred interaction with extracellular matrix. This problem was akin to that observed with Noggin protein [140,141]. Following the lessons from Noggin, either chemical modification to negate the positive charges on the third Ig domain of VEGFR1 or deletion of this region resulted in a Trap ("VEGF $Trap_{\Delta B1}$") with an improved pharmacokinetic profile. The properties of VEGF $Trap_{\Delta B1}$ were further improved by deletion of the first Ig domain of VEGFR1, which is also highly basic, and replacement of the third Ig domain with the corresponding domain of VEGFR2 (which does not contain a basic region), to generate the final version, VEGF $Trap_{R1R2}$ (Figure 9.4) that combined a superior pharmacokinetic profile with increased affinity for VEGFA. This new Trap was efficacious in blocking tumor progression in several mouse models [142], and also appeared to be more efficacious than a corresponding mouse monoclonal against human VEGF [143].

The VEGF Trap is in clinical development for both ophthalmologic (VEGF Trap-Eye [144]) and oncologic (aflibercept [145–148]) indications. VEGF Trap has substantially higher affinity for VEGF compared to the antibodies bevacizumab [149] and ranibizumab [32,65,150,151]. This may prove to be an important differentiator in age-related macular degeneration (AMD), where these drugs are administered by intravitreal route, and decreased injection frequency is desirable. A theoretical analysis concluded that effective blockade of VEGF would be maintained for 80 days by VEGF Trap-EYE compared to

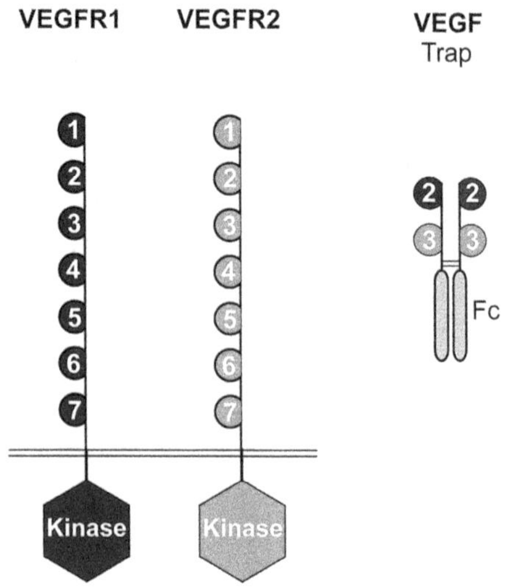

FIGURE 9.4 The VEGF Trap. The VEGF Trap is an in-line fusion protein composed of the second Ig domain of VEGFR1 followed by the third Ig domain of VEGFR2, and fused to the Fc domain of human IgG1, which also drives dimerization of each Trap monomer resulting in 1:1 binding with the VEGF dimer.

30 days for ranibizumab, largely based on their differing affinities for VEGF [152].

Positive results with VEGF Trap-Eye in AMD Phase I and Phase II studies [153,154] led to the design of two parallel Phase III clinical trials called VIEW 1 and VIEW 2. The results of these studies have not yet been presented in peer-reviewed publications, but initial analysis presented in press releases and reviewed in Reference [144] showed that VEGF Trap-Eye met its primary endpoint of noninferiority to the anti-VEGFA antibody fragment ranibizumab in measurements of vision maintenance. Importantly, VEGF Trap-Eye achieved noninferiority with every 8-week administration compared to monthly dosing with ranibizumab, which may result in a decrease in the treatment burden experienced by AMD patients. The safety profile showed no significant differences between VEGF Trap-Eye and ranibizumab. Common adverse events were those typically associated with intravitreal injections or the underlying disease, and included conjunctival hemorrhage, macular degeneration, eye pain, retinal haemorrhage, and vitreous floaters. VEGF Trap-Eye has also demonstrated positive results in a Phase II study for diabetic macular edema (DME), and is currently in additional clinical trials, for example, for macular edema due to central retinal vein occlusion (GALILEO study). In line with adverse events reported in AMD trials, VEGF Trap-Eye was well tolerated in DME; the most frequent adverse events were those associated with the injection procedure, such as conjunctival hemorrhage, eye pain, ocular hyperemia, and increased intraocular pressure [155].

Aflibercept met its primary endpoint of increasing overall survival in a Phase III trial in colorectal cancer in combination with the chemotherapeutic regimen known as FOLFIRI. Recently announced top line results have indicated a positive benefit of the addition of aflibercept to FOLFIRI in terms of overall survival and progression free survival in the second line setting. Adverse events that occurred with more than 2% greater incidence in the aflibercept arm than in the placebo included diarrhoea, fatigue, stomatitis, infections, hypertension, gastrointestinal and abdominal pains, neutropenia, neutropenic complications, and proteinuria. Other studies with aflibercept in different cancer indications are ongoing.

9.6 "TO TRAP OR NOT TO TRAP?" ADVANTAGES AND DISADVANTAGES OF RECEPTOR-Fc FUSIONS AND TRAPS VERSUS ANTIBODIES

In as much as the successful engineering of the Traps opened up new vistas in therapeutic possibilities, their development in the clinic has been limited, reflecting the fact that Traps require careful engineering to achieve good pharmaceutical properties, and robust expression and manufacturing. In addition, the recent technological advancement of VelocImmune® (http://www.regeneron.com/velocimmune), which enables the generation and selection of completely human therapeutic antibodies, has in most cases replaced the preclinical and clinical exploration of new Traps or Trap-like proteins. Therefore, the question arises whether there are any advantages of trap-based therapies over use of therapeutic monoclonal antibodies raised against the same targets.

There are several ways in which both receptor-Fc fusions and Traps differ from monoclonal antibodies, and it is on the basis of these differences that any advantages or disadvantages of these fusion proteins over antibodies should be evaluated.

9.6.1 Multiple Ligand Binding

A key difference between these Fc fusions and monoclonal antibodies is that receptor-Fc fusions and Traps should recapitulate the binding specificities of the receptor component(s) they encompass and therefore should be able to bind multiple ligands. This is clear for the IL-1 Trap (rilonacept), the VEGF Trap (aflibercept), as well as the receptor-Fc fusions TACI-Ig (atacicept), CTLA4-Ig (alefacept and belatacept), gp130-Fc (CR5/18), Fc-OPG, ACVR2A-Fc (ACE-011), ACVR2B-Fc (ACE-031), and ACVRL1-Fc (ACE-041) (Table 9.2). Consequently, the question that arises is whether the ability of a Trap or receptor-Fc fusion to block the activity of multiple ligands is desirable, or whether it can be a complicating factor resulting in undesirable outcomes.

TABLE 9.2 Reported Cognate Ligand Repertoire for Receptor-Fc Fusions and Traps

Trap	Ligands
IL-1 Trap	IL-1A, IL-1β, and to a lesser extent IL-1RN
gp130-Fc	Multiple cytokines of the IL-6 family [117]; *in vivo* repertoire may also be dependent on the presence of soluble receptors of the IL-6 family [123,124]
ACVR2A-Fc (ActRIIA-Fc; ACE-011)	Activin A [79]; multiple other BMPs [81,156]
ACVR2B-Fc (ActRIIB.hFc; ACE-031)	MSTN (GDF8), GDF11, Activin A; BMP2, BMP7 to a lesser extent [85]; multiple other BMPs [81,156]
ACVRL1-Fc (Alk1-IgG1; ACE-041)	BMP9, BMP10 [86,87]
Fc-OPG	RANKL, TRAIL [157]
VEFG Trap	VEGFA isoforms; VEGFB; PGF [65,142,158]

This list of ligands is only those for which there is supporting experimental evidence using the actual receptor-Fc or Trap. It should be noted that for some of these fusion proteins not all the ligands predicted to interact with them have been tested, and also some of these interactions have been tested only *in vitro*.

Therefore, it might be necessary to understand at the organismal level in different disease models, and on a case-by-case basis whether the ability to block multiple ligands is desirable.

There are several examples of ligands for which receptor-Fc blockers or Traps, as well as antibodies have been developed clinically. For TNF-α: etanercept and several anti-TNF-α antibodies (infliximab, adalimumab, goligumab, and certolizumab); for RANKL: Fc-OPG and the anti-RANKL antibody denosumab; for BLyS: atacicept and the anti-BLyS antibody belimumab; for IL-1: rilonacept and the anti-IL-1β antibody canakinumab; for VEGF: aflibercept and the anti-VEGF antibody bevacizumab [149]; and its affinity matured Fab counterpart ranibizumab [150]. It should be noted however that with the exception of the above-referenced Phase III clinical comparison of VEGF Trap-Eye to ranibizumab, there have not been any other side-by-side comparisons in the same clinical setting, and therefore any comparisons regarding clinical efficacy have to rely on the cross-comparison of data from trials performed with individual therapeutics.

Overall, there is currently no clear data from a clinical setting to indicate that the ability of any individual receptor-Fc fusion or Trap to inhibit the action of multiple ligands confers an advantage over monoclonal antibodies that inhibit the action of only single ligands. For example, both rilonacept and canakinumab have been tested in trials of CAPS and gout and both have shown efficacy that is attributed to inhibition of IL-1β [134,139]. Unlike bevacizumab and ranibizumab, which are pure VEGFA blockers, VEGF Trap blocks VEGFA, VEGFB, and PGF [65], and the latter has been hypothesized to also drive pathological neovascularization in ocular and oncology disease settings. However, there has not yet been a clinical finding that conclusively demonstrates a clinical advantage over blockade of VEGFA alone, although further clinical studies are in progress.

For ACVR2A-Fc and ACVR2B-Fc, it is clear that they both interact with multiple BMPs found in mouse and human sera [81], which likely cause numerous physiological effects that may present challenges in clinical development. *In vitro*, they display varying affinities for these BMPs and they can block the majority of them. This contrasts with clinical settings, where an ACVR2A-Fc protein, ACE-011, has been promoted as a fairly specific activin A blocker, yet the possibility that ACE-011 might be blocking other BMPs appears not to have been investigated [79,82]. However, the significance of this in clinical settings is unclear. For example, the reasons for a change in the clinical indication under development for ACE-011—from bone loss [82] to anemia [83]—have not been explicitly attributed to any particular BMP inhibited by ACE-011. In contrast, the preclinical experience with ACE-031 has provided actionable data with respect to which BMPs are important for the regulation of muscle mass, demonstrating that in addition to myostatin, at least one other BMP, activin A, is implicated. However, a Phase II clinical trial of ACE-031 for Duchenne muscular dystrophy (NCT01239758) has been suspended citing preliminary safety concerns, and another, NCT01099761, has been terminated. Still, mirroring ACE-011, it is currently not known whether the ability of ACE-031 to inhibit the action of multiple BMPs [81,156,85] is involved. Finally, it should be noted that given the fact that BMPs interact with multiple extracellular effector proteins, such as their own prodomains [159], multiple receptors [160–162] and coreceptors [29], antagonists [44,163], extracellular matrix, and other binding proteins [116,120,164–166], the question of what are the pharmacologically relevant ligands accessed by ACVR2A-Fc and ACVR2B-Fc in a specific disease setting remains open. Similar comments can be made for ACVRL1-Fc and gp130-Fc.

Fc-OPG is an example where the ability of this decoy receptor-Fc fusion to interact with more than one ligand was deemed to be a drawback. Unlike denosumab, which is exquisitely specific for its target, RANKL, OPG has two known ligands: RANKL and TRAIL [167,168]. Concerns that inhibition of TRAIL by Fc-OPG might enhance tumorigenesis, as well as concerns of immunogenicity and formation of neutralizing antibodies to OPG (see below), combined with the fact that denosumab exhibited a better pharmacokinetic profile, led to the decision to abandon further development of Fc-OPG in favor of denosumab [114,169].

A more complex situation is presented by the comparison of atacicept to belimumab. This is an evolving story; so far, the clinical results obtained with atacicept differ from those obtained with belimumab [170], a monoclonal antibody targeting BLyS [171], but there have been no direct comparisons of the two agents in the same trial. What are the main differences and how might they arise? One key difference in the mechanism of action of these two drugs is that atacicept recognizes two ligands, BLyS or BAFF (also known as TNFSF13B), and APRIL (also known as TNFSF13) [172,173], whereas belimumab inhibits only BLyS. The biological significance of this is illuminated by studies in transgenic and knockout animals as well as studies performed with human cells, all of which have demonstrated that APRIL and BLyS have different roles *in vivo* (reviewed in References [174] and [175]. For example, BLyS signaling through the BAFF receptor (BAFFR; also known as TNFRSF13C) appears to be required for naïve B-cell survival and selection, as well as Ig class switching and plasma cell differentiation in response to infection [176,177]. In contrast, APRIL cannot signal through BAFFR; APRIL signals through TACI and BCMA (also known as TNFRSF17) to maintain mucosal immunity, as well as drive Ig class switching to IgG1 and IgA, but it is not involved in naïve B-cell selection or survival [178]. The differences in biological effects observed between BLyS and APRIL are reflected in the corresponding nonclinical toxicology studies of belimumab [179] and atacicept [180], whereas treatment with either one of these therapeutics results in reversible decreases of circulating mature B cells, only atacicept brings about near complete inhibition—see also a recent review by Reference [170].

The clinical picture is further complicated by the fact that the expression profiles of BLyS and APRIL are differentially regulated in different diseases [181–185]. In addition, BLyS and APRIL form heterotrimers whose levels are elevated in systemic rheumatic diseases (SLE, RA, Reiter's syndrome, psoriatic arthritis, and ankylosing spondylitis) [186,187]. Therefore, it is possible that depending on the target disease it may be desirable to "inhibit" only BLyS or both BLyS and APRIL. It remains to be seen how these differences will affect the clinical application of atacicept (which is still undergoing clinical development) compared to belimumab (which is an approved therapeutic).

Therefore, overall, for the majority of receptor-Fc fusions and Traps that bind multiple ligands it is unclear whether this property is clinically desirable. However, this is still an unfolding story and further laboratory experimentation along with analysis of clinical data will be needed to illuminate this question.

9.6.2 Signaling of Receptor-Fc Fusions

The second potential difference between receptor-Fc fusions or Traps and an antibody is that these Fc fusion proteins might differentially mimic a natural process. One of these processes is reverse signaling—that is, the ability of a transmembrane ligand to act as a receptor and signal transducer. The clinical significance of this process has not been investigated systematically in clinical settings, but has been noted for both anti-BLyS and anti-TNF therapeutics. For example, atacicept and belimumab differ in their ability to induce reverse signaling [22,24,188] in cells expressing the transmembrane form of BLyS. Jeon et al. [189] demonstrated that TACI-Ig induces reverse signaling in cells expressing BLyS, and showed that this can also be accomplished by monoclonal antibodies, albeit with lesser effectiveness. Belimumab was not one of the antibodies tested but given that belimumab binds the soluble but not the transmembrane form of BLyS [171], it is probable that unlike TACI-Ig, belimumab will not induce reverse signaling. The inverse property has been noted for TNF blockers, where etanercept is not as efficient as anti-TNF-α antibodies to induce reverse signaling [22,59,190,191].

Another process that can be selectively blocked by receptor-Fc fusions or Traps is trans-signaling—that is, signaling initiated by a soluble receptor·ligand complex on cells that are otherwise unresponsive to the ligand. The most extensively studied example of trans-signaling is that mediated by IL-6 bound to soluble IL-6R (sIL-6R) and acting on cells that do not express IL-6R, but which do express the shared signal-transducing receptor, gp130 [117,124,192] (Figure 9.3). Trans-signaling by the IL-6/sIL-6R is modulated *in vivo* by endogenous soluble gp130 [131]. In addition, in some animal models of disease, IL-6/sIL-6R trans-signaling appears to be the main driver of pathologic IL-6 signaling, whereas signaling mediated by the membrane-bound form of IL-6R results in desirable anti-inflammatory outcomes [193,194]. These observations have led to the hypothesis that selective inhibition of IL-6/sIL-6R trans-signaling (via gp130-Fc) may be therapeutically desirable over inhibition of total IL-6 signaling [96,132,195,196], and to this effect several versions of gp130-Fc have been engineered [197]. One of them, CR5/18, is currently in preclinical development [23,97,98] (Table 9.1). However, there are no published preclinical studies directly comparing the efficacy of CR5/18 with approved therapeutics targeting IL-6 signaling, such as the anti-IL-6R antibodies tocilizumab [198] or REGN88 (also called sarilumab) [199], or other biologic therapeutics under development [97,98,200]. Therefore, it remains an open question whether CR5/18 will prove superior to other biological therapies targeting IL-6.

9.6.3 Immunogenicity

A third concern that has been raised with receptor-Fc fusions and Traps is the possibility of immunogenicity, particularly, raising autoantibodies that cross-react with their corresponding native ectodomains. The potential for immunogenicity is not limited to these Fc fusions—it has been observed for many protein-based therapeutics [201,202]—and it appears that receptor-Fc fusions and Traps may be less immunogenic than their antibody counterparts [203–205]. There are only few reports of anti-receptor-Fc or anti-Trap antibodies. In one case, for Fc-OPG, the formation of neutralizing antibodies in one patient was listed as one of the reasons for halting further development of this therapeutic [115,157,206]. One well-documented case of antibody formation to a receptor-Fc was observed in clinical trials with TNFR55-IgG1 (lenercept). The antibodies that formed against lenercept were not neutralizing, but they resulted in rapid clearance of the drug that in turn led to suspension of further clinical development [207–209]. Antibodies induced against other Fc fusion proteins have also been nonneutralizing, and their clinical significance remains unclear. For example, anti-etanercept nonneutralizing antibodies have been observed in some clinical settings involving chronic use of this TNF blocker, but the immune response does not appear to decrease the efficacy of the drug, as these antibodies do not dramatically accelerate drug clearance [203,204,210,211]. Similarly, anti-rilonacept nonneutralizing antibodies have been detected in CAPS patients, who have been chronically dosed with rilonacept, but the titers are very low and do not appear to impact pharmacokinetics and efficacy, or have other detrimental effects on the health of the patients [63]. Overall, immunogenicity has not been a greater concern for Fc fusions than other protein-based therapeutics, and therefore it will probably not be the main limiting factor for their clinical application.

9.7 CONCLUSION

This review has summarized how receptor-Fc fusions and Traps, as well as other biologics such as antibodies raised against the same target can all be successfully used in research and in the clinic. The concerns that have been raised for the use of receptor-Fc fusions or Traps as therapeutics have largely not materialized, and in those cases where adverse events have been noted in the clinic, this has happened early on in the development process. The question remains open whether receptor-Fc or Traps might be preferred over other biologics in any specific clinical setting. The main competition to these Fc fusion proteins is presented by monoclonal antibodies—which represent the fastest growing class of protein-based therapeutics [199]. The ability of any given receptor-Fc or Trap to interact with multiple ligands or to elicit additional responses (e.g., reverse signaling), or to selectively block its differentially acting pool of target cytokine (e.g., the trans-signaling pool) remains possible differentiating properties. Therefore, although monoclonal antibodies provide a quicker and perhaps more straightforward path to clinical development, receptor-Fc fusion proteins, and Traps present relevant alternatives. Sometimes, these advantages may be as simple as higher affinity leading to less frequent dosing, but there is still not enough clinical data to allow us to pinpoint all the clinically important differences, advantageous, or otherwise, of each receptor-Fc or Trap in comparison to other drugs that have been developed against the same target. One exception is offered by the receptor-Fc therapeutic, etanercept, which has been widely used in the clinic for a long time. The differences between etanercept and antibody-based TNF-α blockers have been reviewed extensively elsewhere [23,55,191,212]. These reviews as well as overall clinical practice demonstrate that some patients respond better to one type of TNF blocker than another, and this may reflect subtle differences in the mechanism of action of these drugs [213]. The fact that etanercept remains one of the best-selling biologics even in the face of increased competition by antibody-based therapeutics provides credence to the idea that Fc fusions remain a useful and lasting therapeutic modality for a large number of patients.

ACKNOWLEDGMENT

We would like to acknowledge Stephen Jaspers, Scott Mellis, Nicholas Papadopoulos Evangelos Pefanis, John S. Rudge, and Christopher Schoenherr for critical reading of this manuscript.

REFERENCES

1. Huang C. (2009) Receptor-Fc fusion therapeutics, traps, and MIMETIBODY technology. *Curr. Opin. Biotechnol.* 20, 692–699.
2. Andersen JT, Pehrson R, Tolmachev V, Daba MB, Abrahmsen L, Ekblad C. (2011) Extending half-life by indirect targeting of the neonatal Fc receptor (FcRn) using a minimal albumin binding domain. *J. Biol. Chem.* 286, 5234–5241.
3. Gill DS, Damle NK. (2006) Biopharmaceutical drug discovery using novel protein scaffolds. *Curr. Opin. Biotechnol.* 17, 653–658.

4. Kenanova VE, Olafsen T, Salazar FB, Williams LE, Knowles S, Wu AM. (2010) Tuning the serum persistence of human serum albumin domain III: diabody fusion proteins. *Protein Eng. Des. Sel.* 23, 789–798.
5. Kolmar H. (2008) Alternative binding proteins: biological activity and therapeutic potential of cystine-knot miniproteins. *FEBS J.* 275, 2684–2690.
6. McGregor DP. (2008) Discovering and improving novel peptide therapeutics. *Curr. Opin. Pharmacol.* 8, 616–619.
7. Nuttall SD, Walsh RB. (2008) Display scaffolds: protein engineering for novel therapeutics. *Curr. Opin. Pharmacol.* 8, 609–615.
8. Rothe A, Hosse RJ, Power BE. (2006) Ribosome display for improved biotherapeutic molecules. *Expert Opin. Biol. Ther.* 6, 177–187.
9. Schellenberger V, Wang CW, Geething NC, Spink BJ, Campbell A, To W, et al. (2009) A recombinant polypeptide extends the *in vivo* half-life of peptides and proteins in a tunable manner. *Nat. Biotechnol.* 27, 1186–1190.
10. Ullrich A, Coussens L, Hayflick JS, Dull TJ, Gray A, Tam AW, et al. (1984) Human epidermal growth factor receptor cDNA sequence and aberrant expression of the amplified gene in A431 epidermoid carcinoma cells. *Nature* 309, 418–425.
11. Rubin LA, Kurman CC, Fritz ME, Biddison WE, Boutin B, Yarchoan R, et al. (1985) Soluble interleukin 2 receptors are released from activated human lymphoid cells *in vitro*. *J. Immunol.* 135, 3172–3177.
12. Levine SJ. (2008) Molecular mechanisms of soluble cytokine receptor generation. *J. Biol. Chem.* 283, 14177–14181.
13. Engelmann H, Aderka D, Rubinstein M, Rotman D, Wallach D. (1989) A tumor necrosis factor-binding protein purified to homogeneity from human urine protects cells from tumor necrosis factor toxicity. *J. Biol. Chem.* 264, 11974–11980.
14. Marcon L, Fritz ME, Kurman CC, Jensen JC, Nelson DL. (1988) Soluble Tac peptide is present in the urine of normal individuals and at elevated levels in patients with adult T cell leukaemia (ATL). *Clin. Exp. Immunol.* 73, 29–33.
15. Novick D, Engelmann H, Wallach D, Rubinstein M. (1989) Soluble cytokine receptors are present in normal human urine. *J. Exp. Med.* 170, 1409–1414.
16. Schall TJ, Lewis M, Koller KJ, Lee A, Rice GC, Wong GH, et al. (1990) Molecular cloning and expression of a receptor for human tumor necrosis factor. *Cell* 61, 361–370.
17. Seckinger P, Isaaz S, Dayer JM. (1989) Purification and biologic characterization of a specific tumor necrosis factor alpha inhibitor. *J. Biol. Chem.* 264, 11966–11973.
18. Zupan AA, Osborne PA, Smith CE, Siegel NR, Leimgruber RM, Johnson EM, Jr. (1989) Identification, purification, and characterization of truncated forms of the human nerve growth factor receptor. *J. Biol. Chem.* 264, 11714–11720.
19. Fernandez-Botran R. (1999) Soluble cytokine receptors: basic immunology and clinical applications. *Crit. Rev. Clin. Lab. Sci.* 36, 165–224.
20. Fernandez-Botran R. (2000) Soluble cytokine receptors: novel immunotherapeutic agents. *Expert Opin. Investig. Drugs* 9, 497–514.
21. Mantovani A, Bonecchi R, Martinez FO, Galliera E, Perrier P, Allavena P, et al. (2003) Tuning of innate immunity and polarized responses by decoy receptors. *Int. Arch. Allergy Immunol.* 132, 109–115.
22. Eissner G, Kolch W, Scheurich P. (2004) Ligands working as receptors: reverse signaling by members of the TNF superfamily enhance the plasticity of the immune system. *Cytokine Growth Factor Rev.* 15, 353–366.
23. Kopf M, Bachmann MF, Marsland BJ. (2010) Averting inflammation by targeting the cytokine environment. *Nat. Rev. Drug Discov.* 9, 703–718.
24. Shui JW, Steinberg MW, Kronenberg M. (2011) Regulation of inflammation, autoimmunity, and infection immunity by HVEM-BTLA signaling. *J. Leukoc. Biol.* 89, 517–523.
25. Treiger BF, Leonard WJ, Svetlik P, Rubin LA, Nelson DL, Greene WC. (1986) A secreted form of the human interleukin 2 receptor encoded by an "anchor minus" cDNA. *J. Immunol.* 136, 4099–4105.
26. Kondo N, Kondo S, Shimizu A, Honjo T, Hamuro J. (1988) A soluble 'anchorminus' interleukin 2 receptor suppresses *in vitro* interleukin 2-mediated immune responses. *Immunol. Lett.* 19, 299–307.
27. Fernandez-Botran R. (1991) Soluble cytokine receptors: their role in immunoregulation. *FASEB J.* 5, 2567–2574.
28. Gershoni JM. (2008) Molecular decoys: antidotes, therapeutics and immunomodulators. *Curr. Opin. Biotechnol.* 19, 644–651.
29. Corradini E, Babitt JL, Lin HY. (2009) The RGM/DRAGON family of BMP co-receptors. *Cytokine Growth Factor Rev.* 20, 389–398.
30. Bizet AA, Liu K, Tran-Khanh N, Saksena A, Vorstenbosch J, Finnson KW, et al. (2011) The TGF-beta co-receptor, CD109, promotes internalization and degradation of TGF-beta receptors. *Biochim. Biophys. Acta* 1813, 742–753.
31. M. Ehrlich, D. Horbelt, B. Marom, P. Knaus, Y. I. Henis. (2011). Homomeric and heteromeric complexes among TGF-beta and BMP receptors and their roles in signaling. *Cellular signalling*, 23, 1424–1432.
32. Onichtchouk D, Chen YG, Dosch R, Gawantka V, Delius H, Massague J, et al. (1999) Silencing of TGF-beta signalling by the pseudoreceptor BAMBI. *Nature* 401, 480–485.
33. O'Neill LA. (2003) SIGIRR puts the brakes on Toll-like receptors. *Nat. Immunol.* 4, 823–824.
34. Wald D, Qin J, Zhao Z, Qian Y, Naramura M, Tian L, et al. (2003) SIGIRR, a negative regulator of Toll-like receptor-interleukin 1 receptor signaling. *Nat. Immunol.* 4, 920–927.
35. Colotta F, Re F, Muzio M, Bertini R, Polentarutti N, Sironi M, et al. (1993) Interleukin-1 type II receptor: a decoy target for IL-1 that is regulated by IL-4. *Science* 261, 472–475.
36. Sims JE, Smith DE. (2010) The IL-1 family: regulators of immunity. *Nat. Rev. Immunol.* 10, 89–102.

37. Zheng T, Liu W, Oh SY, Zhu Z, Hu B, Homer RJ, et al. (2008) IL-13 receptor alpha2 selectively inhibits IL-13-induced responses in the murine lung. *J. Immunol.* 180, 522–529.

38. Yi KH, Chen L. (2009) Fine tuning the immune response through B7-H3 and B7-H4. *Immunol. Rev.* 229, 145–151.

39. Graham GJ. (2009) D6 and the atypical chemokine receptor family: novel regulators of immune and inflammatory processes. *Eur. J. Immunol.* 39, 342–351.

40. Eisenberg SP, Brewer MT, Verderber E, Heimdal P, Brandhuber BJ, Thompson RC. (1991) Interleukin 1 receptor antagonist is a member of the interleukin 1 gene family: evolution of a cytokine control mechanism. *Proc. Natl. Acad. Sci. USA* 88, 5232–5236.

41. Aizawa Y, Akita K, Taniai M, Torigoe K, Mori T, Nishida Y, et al. (1999) Cloning and expression of interleukin-18 binding protein. *FEBS Lett.* 445, 338–342.

42. Novick D, Kim SH, Fantuzzi G, Reznikov LL, Dinarello CA, Rubinstein M. (1999) Interleukin-18 binding protein: a novel modulator of the Th1 cytokine response. *Immunity* 10, 127–136.

43. Blumberg H, Dinh H, Trueblood ES, Pretorius J, Kugler D, Weng N, et al. (2007) Opposing activities of two novel members of the IL-1 ligand family regulate skin inflammation. *J. Exp. Med.* 204, 2603–2614.

44. Walsh DW, Godson C, Brazil DP, Martin F. (2010) Extracellular BMP-antagonist regulation in development and disease: tied up in knots. *Trends Cell Biol.* 20, 244–256.

45. MacDonald BT, Tamai K, He X. (2009) Wnt/beta-catenin signaling: components, mechanisms, and diseases. *Dev. Cell* 17, 9–26.

46. Peppel K, Crawford D, Beutler B. (1991) A tumor necrosis factor (TNF) receptor-IgG heavy chain chimeric protein as a bivalent antagonist of TNF activity. *J. Exp. Med.* 174, 1483–1489.

47. Capon DJ, Chamow SM, Mordenti J, Marsters SA, Gregory T, Mitsuya H, et al. (1989) Designing CD4 immunoadhesins for AIDS therapy. *Nature* 337, 525–531.

48. Traunecker A, Dolder B, Oliveri F, Karjalainen K. (1989) Solubilizing the T-cell receptor–problems in solution. *Immunol. Today* 10, 29–32.

49. Traunecker A, Schneider J, Kiefer H, Karjalainen K. (1989) Highly efficient neutralization of HIV with recombinant CD4-immunoglobulin molecules. *Nature* 339, 68–70.

50. Carter PJ. (2006) Potent antibody therapeutics by design. *Nat. Rev. Immunol.* 6, 343–357.

51. Allaway GP, Davis-Bruno KL, Beaudry GA, Garcia EB, Wong EL, Ryder AM, et al. (1995) Expression and characterization of CD4-IgG2, a novel heterotetramer that neutralizes primary HIV type 1 isolates. *AIDS Res. Hum. Retroviruses* 11, 533–539.

52. Mohler KM, Torrance DS, Smith CA, Goodwin RG, Stremler KE, Fung VP, et al. (1993) Soluble tumor necrosis factor (TNF) receptors are effective therapeutic agents in lethal endotoxemia and function simultaneously as both TNF carriers and TNF antagonists. *J. Immunol.* 151, 1548–1561.

53. Economides AN, Carpenter LR, Rudge JS, Wong V, Koehler-Stec EM, Hartnett C, et al. (2003) Cytokine Traps: multicomponent, high-affinity blockers of cytokine action. *Nat. Med.* 9, 47–52.

54. Feldmann M, Maini RN. (2003) Lasker Clinical Medical Research Award. TNF defined as a therapeutic target for rheumatoid arthritis and other autoimmune diseases. *Nat. Med.* 9, 1245–1250.

55. Maini RN, Taylor PC. (2000) Anti-cytokine therapy for rheumatoid arthritis. *Annu. Rev. Med.* 51, 207–229.

56. Ducharme E, Weinberg JM. (2008) Etanercept. *Expert Opin. Biol. Ther.* 8, 491–502.

57. Prodanovich S, Ricotti C, Glick BP, Inverardi L, Leonardi CL, Kerdel F. (2010) Etanercept: an evolving role in psoriasis and psoriatic arthritis. *Am. J. Clin. Dermatol.* 11(Suppl. 1), 3–9.

58. Wiens A, Venson R, Correr CJ, Otuki MF, Pontarolo R. (2010) Meta-analysis of the efficacy and safety of adalimumab, etanercept, and infliximab for the treatment of rheumatoid arthritis. *Pharmacotherapy* 30, 339–353.

59. Tansey MG, Szymkowski DE. (2009) The TNF superfamily in 2009: new pathways, new indications, and new drugs. *Drug Discov. Today* 14, 1082–1088.

60. Moreland LW, Baumgartner SW, Schiff MH, Tindall EA, Fleischmann RM, Weaver AL, et al. (1997) Treatment of rheumatoid arthritis with a recombinant human tumor necrosis factor receptor (p75)-Fc fusion protein. *N. Engl. J. Med.* 337, 141–147.

61. Moreland LW. (2001) Tumor necrosis factor inhibitors: new options for treating rheumatoid arthritis. *Isr. Med. Assoc. J.* 3, 686–690.

62. Stahl N, Radin A, Mellis S. (2009) Rilonacept–CAPS and beyond. *Ann. NY Acad. Sci.* 1182, 124–134.

63. Hoffman HM, Throne ML, Amar NJ, Sebai M, Kivitz AJ, Kavanaugh A, et al. (2008) Efficacy and safety of rilonacept (interleukin-1 Trap) in patients with cryopyrin-associated periodic syndromes: results from two sequential placebo-controlled studies. *Arthritis Rheum.* 58, 2443–2452.

64. Hoffman HM. (2009) Rilonacept for the treatment of cryopyrin-associated periodic syndromes (CAPS). *Expert Opin. Biol. Ther.* 9, 519–531.

65. N. Papadopoulos, J. Martin, Q. Ruan, A. Rafique, M. P. Rosconi, E. Shi, E. A. Pyles, G. D. Yancopoulos, N. Stahl, S. J. Wiegand. (2012). Binding and neutralization of vascular endothelial growth factor (VEGF) and related ligands by VEGF Trap, ranibizumab and bevacizumab. *Angiogenesis*, 15, 171–185.

66. Chamian F, Lin SL, Lee E, Kikuchi T, Gilleaudeau P, Sullivan-Whalen M, et al. (2007) Alefacept (anti-CD2) causes a selective reduction in circulating effector memory T cells (Tem) and relative preservation of central memory T cells (Tcm) in psoriasis. *J. Transl. Med.* 5, 27–33.

67. Chamian F, Lowes MA, Lin SL, Lee E, Kikuchi T, Gilleaudeau P, et al. (2005) Alefacept reduces infiltrating T cells, activated dendritic cells, and inflammatory genes in *psoriasis vulgaris*. *Proc. Natl. Acad. Sci. USA* 102, 2075–2080.

68. Haider AS, Lowes MA, Gardner H, Bandaru R, Darabi K, Chamian F, et al. (2007) Novel insight into the agonistic mechanism of alefacept *in vivo*: differentially expressed genes may serve as biomarkers of response in psoriasis patients. *J. Immunol.* 178, 7442–7449.

69. Heffernan MP, Leonardi CL. (2010) Alefacept for psoriasis. *Semin. Cutan. Med. Surg.* 29, 53–55.

70. Krueger GG. (2003) Clinical response to alefacept: results of a phase 3 study of intravenous administration of alefacept in patients with chronic plaque psoriasis. *J. Eur. Acad. Dermatol. Venereol.* 17(Suppl. 2), 17–24.

71. Krueger GG, Gottlieb AB, Sterry W, Korman N, Van De Kerkhof P. (2008) A multicenter, open-label study of repeat courses of intramuscular alefacept in combination with other psoriasis therapies in patients with chronic plaque psoriasis. *J. Dermatolog. Treat.* 19, 146–155.

72. Scheinfeld N. (2007) Alefacept: its safety profile, off-label uses, and potential as part of combination therapies for psoriasis. *J. Dermatolog. Treat.* 18, 197–208.

73. Strober BE, Menon K. (2007) Alefacept for the treatment of psoriasis and other dermatologic diseases. *Dermatol. Ther.* 20, 270–276.

74. Linsley PS, Wallace PM, Johnson J, Gibson MG, Greene JL, Ledbetter JA, et al. (1992) Immunosuppression *in vivo* by a soluble form of the CTLA-4 T cell activation molecule. *Science* 257, 792–795.

75. Korhonen R, Moilanen E. (2009) Abatacept, a novel CD80/86-CD28 T cell co-stimulation modulator, in the treatment of rheumatoid arthritis. *Basic Clin. Pharmacol. Toxicol.* 104, 276–284.

76. Ruperto N, Lovell DJ, Quartier P, Paz E, Rubio-Perez N, Silva CA, et al. (2010) Long-term safety and efficacy of abatacept in children with juvenile idiopathic arthritis. *Arthritis Rheum.* 62, 1792–1802.

77. Ruperto N, Lovell DJ, Quartier P, Paz E, Rubio-Perez N, Silva CA, et al. (2008) Abatacept in children with juvenile idiopathic arthritis: a randomised, double-blind, placebo-controlled withdrawal trial. *Lancet* 372, 383–391.

78. Larsen CP, Pearson TC, Adams AB, Tso P, Shirasugi N, Strobert E, et al. (2005) Rational development of LEA29Y (belatacept), a high-affinity variant of CTLA4-Ig with potent immunosuppressive properties. *Am. J. Transplant.* 5, 443–453.

79. Pearsall RS, Canalis E, Cornwall-Brady M, Underwood KW, Haigis B, Ucran J, et al. (2008) A soluble activin type IIA receptor induces bone formation and improves skeletal integrity. *Proc. Natl. Acad. Sci. USA* 105, 7082–7087.

80. Lotinun S, Pearsall RS, Davies MV, Marvell TH, Monnell TE, Ucran J, et al. (2010) A soluble activin receptor type IIA fusion protein (ACE-011) increases bone mass via a dual anabolic-antiresorptive effect in Cynomolgus monkeys. *Bone* 46, 1082–1088.

81. Souza TA, Chen X, Guo Y, Sava P, Zhang J, Hill JJ, et al. (2008) Proteomic identification and functional validation of activins and bone morphogenetic protein 11 as candidate novel muscle mass regulators. *Mol. Endocrinol.* 22, 2689–2702.

82. Ruckle J, Jacobs M, Kramer W, Pearsall AE, Kumar R, Underwood KW, et al. (2009) Single-dose, randomized, double-blind, placebo-controlled study of ACE-011 (ActRIIA-IgG1) in postmenopausal women. *J. Bone Miner. Res.* 24, 744–752.

83. Raje N, Vallet S. (2011) Sotatercept, a soluble activin receptor type 2A IgG-Fc fusion protein for the treatment of anemia and bone loss. *Curr. Opin. Mol. Ther.* 12, 586–597.

84. Cadena SM, Tomkinson KN, Monnell TE, Spaits MS, Kumar R, Underwood KW, et al. (2010) Administration of a soluble activin type IIB receptor promotes skeletal muscle growth independent of fiber type. *J. Appl. Physiol.* 109, 635–642.

85. Sako D, Grinberg AV, Liu J, Davies MV, Castonguay R, Maniatis S, et al. (2010) Characterization of the ligand binding functionality of the extracellular domain of activin receptor type IIb. *J. Biol. Chem.* 285, 21037–21048.

86. Cunha SI, Pardali E, Thorikay M, Anderberg C, Hawinkels L, Goumans MJ, et al. (2010) Genetic and pharmacological targeting of activin receptor-like kinase 1 impairs tumor growth and angiogenesis. *J. Exp. Med.* 207, 85–100.

87. Mitchell D, Pobre EG, Mulivor AW, Grinberg AV, Castonguay R, Monnell TE, et al. (2010) ALK1-Fc inhibits multiple mediators of angiogenesis and suppresses tumor growth. *Mol. Cancer Ther.* 9, 379–388.

88. Baldassare A, Fiechtner J, Filipowicz-Sosnowska A, Jeka S, O'Gorman J, Weaver M, et al. (2007) Preliminary Safety and Efficacy of LTβR-Ig (BG9924) in the Treatment of Rheumatoid Arthritis (RA), in *American College of Rheumatology*, p. S394.

89. Crowe PD, VanArsdale TL, Walter BN, Ware CF, Hession C, Ehrenfels B, et al. (1994) A lymphotoxin-beta-specific receptor. *Science* 264, 707–710.

90. Mauri DN, Ebner R, Montgomery RI, Kochel KD, Cheung TC, Yu GL, et al. (1998) LIGHT, a new member of the TNF superfamily, and lymphotoxin alpha are ligands for herpesvirus entry mediator. *Immunity* 8, 21–30.

91. Rossi JF. (2011) Phase I Study of atacicept in relapsed/refractory multiple myeloma (MM) and Waldenstrom's macroglobulinemia. *Clin. Lymphoma Myeloma Leuk.* 11, 136–138.

92. Rossi JF, Moreaux J, Hose D, Requirand G, Rose M, Rouille V, et al. (2009) Atacicept in relapsed/refractory multiple myeloma or active Waldenstrom's macroglobulinemia: a phase I study. *Br. J. Cancer* 101, 1051–1058.

93. Ansell SM, Witzig TE, Inwards DJ, Porrata LF, Ythier A, Ferrande L, et al. (2008) Phase I clinical study of atacicept in patients with relapsed and refractory B-cell non-Hodgkin's lymphoma. *Clin. Cancer Res.* 14, 1105–1110.

94. Dall'Era M, Chakravarty E, Wallace D, Genovese M, Weisman M, Kavanaugh A, et al. (2007) Reduced B lymphocyte and immunoglobulin levels after atacicept treatment in patients with systemic lupus erythematosus: results of a multicenter, phase Ib, double-blind, placebo-controlled, dose-escalating trial. *Arthritis Rheum.* 56, 4142–4150.

95. Tenhumberg S, Waetzig GH, Chalaris A, Rabe B, Seegert D, Scheller J, et al. (2008) Structure-guided optimization of the

interleukin-6 trans-signaling antagonist sgp130. *J. Biol. Chem.* 283, 27200–27207.

96. Chalaris A, Garbers C, Rabe B, Rose-John S, Scheller J. (2011) The soluble Interleukin 6 receptor: generation and role in inflammation and cancer. *Eur. J. Cell Biol.* 90, 484–494.
97. Melton L, Coombs A. (2008) Actemra poised to launch IL-6 inhibitors. *Nat. Biotechnol.* 26, 957–959.
98. Ding C, Cicuttini F, Li J, Jones G. (2009) Targeting IL-6 in the treatment of inflammatory and autoimmune diseases. *Expert Opin. Investig. Drugs* 18, 1457–1466.
99. Bush KA, Farmer KM, Walker JS, Kirkham BW. (2002) Reduction of joint inflammation and bone erosion in rat adjuvant arthritis by treatment with interleukin-17 receptor IgG1 Fc fusion protein. *Arthritis Rheum.* 46, 802–805.
100. Lubberts E, Joosten LA, Oppers B, van den Bersselaar L, Coenen-de Roo CJ, Kolls JK, et al. (2001) IL-1-independent role of IL-17 in synovial inflammation and joint destruction during collagen-induced arthritis. *J. Immunol.* 167, 1004–1013.
101. Wright JF, Bennett F, Li B, Brooks J, Luxenberg DP, Whitters MJ, et al. (2008) The human IL-17F/IL-17A heterodimeric cytokine signals through the IL-17RA/IL-17RC receptor complex. *J. Immunol.* 181, 2799–2805.
102. Lubberts E. (2008) IL-17/Th17 targeting: on the road to prevent chronic destructive arthritis? *Cytokine* 41, 84–91.
103. Korn T, Bettelli E, Oukka M, Kuchroo VK. (2009) IL-17 and Th17 Cells. *Annu. Rev. Immunol.* 27, 485–517.
104. Kuestner RE, Taft DW, Haran A, Brandt CS, Brender T, Lum K, et al. (2007) Identification of the IL-17 receptor related molecule IL-17RC as the receptor for IL-17F. *J. Immunol.* 179, 5462–5473.
105. Yan M, Brady JR, Chan B, Lee WP, Hsu B, Harless S, et al. (2001) Identification of a novel receptor for B lymphocyte stimulator that is mutated in a mouse strain with severe B cell deficiency. *Curr. Biol.* 11, 1547–1552.
106. Vugmeyster Y, Seshasayee D, Chang W, Storn A, Howell K, Sa S, et al. (2006) A soluble BAFF antagonist, BR3-Fc, decreases peripheral blood B cells and lymphoid tissue marginal zone and follicular B cells in cynomolgus monkeys. *Am. J. Pathol.* 168, 476–489.
107. Fleischmann R, Wei N, Shaw M, Birbara C, Anand B, Gujrathi S, et al. (2006) BR3-Fc phase i study: safety, pharmacokinetics (PK) and pharmacodynamic (PD) effects of a novel BR3-Fc fusion protein in patients with rheumatoid arthritis. *Arthritis Rheum.* 54, S229–S230.
108. Shearer WT, Israel RJ, Starr S, Fletcher CV, Wara D, Rathore M, et al. (2000) Recombinant CD4-IgG2 in human immunodeficiency virus type 1-infected children: phase 1/2 study. The Pediatric AIDS Clinical Trials Group Protocol 351 Study Team. *J. Infect. Dis.* 182, 1774–1779.
109. Bekker PJ, Holloway D, Nakanishi A, Arrighi M, Leese PT, Dunstan CR. (2001) The effect of a single dose of osteoprotegerin in postmenopausal women. *J. Bone Miner. Res.* 16, 348–360.
110. Simonet WS, Lacey DL, Dunstan CR, Kelley M, Chang MS, Luthy R, et al. (1997) Osteoprotegerin: a novel secreted protein involved in the regulation of bone density. *Cell* 89, 309–319.
111. Yasuda H, Shima N, Nakagawa N, Mochizuki SI, Yano K, Fujise N, et al. (1998) Identity of osteoclastogenesis inhibitory factor (OCIF) and osteoprotegerin (OPG): a mechanism by which OPG/OCIF inhibits osteoclastogenesis *in vitro*. *Endocrinology* 139, 1329–1337.
112. Body JJ, Greipp P, Coleman RE, Facon T, Geurs F, Fermand JP, et al. (2003) A phase I study of AMGN-0007, a recombinant osteoprotegerin construct, in patients with multiple myeloma or breast carcinoma related bone metastases. *Cancer* 97, 887–892.
113. Cundy T, Davidson J, Rutland MD, Stewart C, DePaoli AM. (2005) Recombinant osteoprotegerin for juvenile Paget's disease. *N. Engl. J. Med.* 353, 918–923.
114. Schwarz EM, Ritchlin CT. (2007) Clinical development of anti-RANKL therapy. *Arthritis Res. Ther.* 9(Suppl. 1), S7.
115. Ng KW. (2009) Future developments in osteoporosis therapy. *Endocr Metab Immune Disord Drug Targets* 9, 371–384.
116. Bragdon B, Moseychuk O, Saldanha S, King D, Julian J, Nohe A. (2011) Bone morphogenetic proteins: a critical review. *Cell. Signal.* 23, 609–620.
117. Wang X, Lupardus P, Laporte SL, Garcia KC. (2009) Structural biology of shared cytokine receptors. *Annu. Rev. Immunol.* 27, 29–60.
118. Shen MM, Schier AF. (2000) The EGF-CFC gene family in vertebrate development. *Trends Genet.* 16, 303–309.
119. Lin SJ, Lerch TF, Cook RW, Jardetzky TS, Woodruff TK. (2006) The structural basis of TGF-beta, bone morphogenetic protein, and activin ligand binding. *Reproduction* 132, 179–190.
120. Bier E. (2008) Intriguing extracellular regulation of BMP signaling. *Dev. Cell* 15, 176–177.
121. Angers S, Moon RT. (2009) Proximal events in Wnt signal transduction. *Nat. Rev. Mol. Cell Biol.* 10, 468–477.
122. Glinka Y, Stoilova S, Mohammed N, Prud'homme GJ. (2011) Neuropilin-1 exerts co-receptor function for TGF-beta-1 on the membrane of cancer cells and enhances responses to both latent and active TGF-beta. *Carcinogenesis* 32, 613–621.
123. Davis S, Aldrich TH, Ip NY, Stahl N, Scherer S, Farruggella T, et al. (1993) Released form of CNTF receptor alpha component as a soluble mediator of CNTF responses. *Science* 259, 1736–1739.
124. Stahl N, Yancopoulos GD. (1993) The alphas, betas, and kinases of cytokine receptor complexes. *Cell* 74, 587–590.
125. Ibanez CF. (2010) Beyond the cell surface: new mechanisms of receptor function. *Biochem. Biophys. Res. Commun.* 396, 24–27.
126. Taga T, Hibi M, Hirata Y, Yamasaki K, Yasukawa K, Matsuda T, et al. (1989) Interleukin-6 triggers the association of its receptor with a possible signal transducer, gp130. *Cell* 58, 573–581.
127. Economides AN, Ravetch JV, Yancopoulos GD, Stahl N. (1995) Designer cytokines: targeting actions to cells of choice. *Science* 270, 1351–1353.

128. Le Saux S, Rousseau F, Barbier F, Ravon E, Grimaud L, Danger Y, et al. (2010) Molecular dissection of human interleukin-31-mediated signal transduction through site-directed mutagenesis. *J. Biol. Chem.* 285, 3470–3477.
129. Kouzarides T, Ziff E. (1989) Leucine zippers of fos, jun and GCN4 dictate dimerization specificity and thereby control DNA binding. *Nature* 340, 568–571.
130. Gearing DP, Comeau MR, Friend DJ, Gimpel SD, Thut CJ, McGourty J, et al. (1992) The IL-6 signal transducer, gp130: an oncostatin M receptor and affinity converter for the LIF receptor. *Science* 255, 1434–1437.
131. Narazaki M, Yasukawa K, Saito T, Ohsugi Y, Fukui H, Koishihara Y, et al. (1993) Soluble forms of the interleukin-6 signal-transducing receptor component gp130 in human serum possessing a potential to inhibit signals through membrane-anchored gp130. *Blood* 82, 1120–1126.
132. Jostock T, Mullberg J, Ozbek S, Atreya R, Blinn G, Voltz N, et al. (2001) Soluble gp130 is the natural inhibitor of soluble interleukin-6 receptor transsignaling responses. *Eur. J. Biochem.* 268, 160–167.
133. Crowe J, Dobeli H, Gentz R, Hochuli E, Stuber D, Henco K. (1994) 6xHis-Ni-NTA chromatography as a superior technique in recombinant protein expression/purification. *Methods Mol. Biol.* 31, 371–387.
134. Yu JR, Leslie KS. (2011) Cryopyrin-associated periodic syndrome: an update on diagnosis and treatment response. *Curr. Allergy Asthma Rep.* 11, 12–20.
135. Martinon F, Petrilli V, Mayor A, Tardivel A, Tschopp J. (2006) Gout-associated uric acid crystals activate the NALP3 inflammasome. *Nature* 440, 237–241.
136. Torres R, Macdonald L, Croll SD, Reinhardt J, Dore A, Stevens S, et al. (2009) Hyperalgesia, synovitis and multiple biomarkers of inflammation are suppressed by interleukin 1 inhibition in a novel animal model of gouty arthritis. *Ann. Rheum. Dis.* 68, 1602–1608.
137. So A, De Smedt T, Revaz S, Tschopp J. (2007) A pilot study of IL-1 inhibition by anakinra in acute gout. *Arthritis Res. Ther.* 9, R28.
138. Terkeltaub R, Sundy JS, Schumacher HR, Murphy F, Bookbinder S, Biedermann S, et al. (2009) The interleukin 1 inhibitor rilonacept in treatment of chronic gouty arthritis: results of a placebo-controlled, monosequence crossover, non-randomised, single-blind pilot study. *Ann. Rheum. Dis.* 68, 1613–1617.
139. Burns CM, Wortmann RL. (2011) Gout therapeutics: new drugs for an old disease. *Lancet* 377, 165–177.
140. Glaser DL, Economides AN, Wang L, Liu X, Kimble RD, Fandl JP, et al. (2003) *In vivo* somatic cell gene transfer of an engineered Noggin mutein prevents BMP4-induced heterotopic ossification. *J. Bone Joint Surg. Am.* 85-A, 2332–2342.
141. Paine-Saunders S, Viviano BL, Economides AN, Saunders S. (2002) Heparan sulfate proteoglycans retain Noggin at the cell surface: a potential mechanism for shaping bone morphogenetic protein gradients. *J. Biol. Chem.* 277, 2089–2096.
142. Holash J, Davis S, Papadopoulos N, Croll SD, Ho L, Russell M, et al. (2002) VEGF-Trap: a VEGF blocker with potent antitumor effects. *Proc. Natl. Acad. Sci. USA* 99, 11393–11398.
143. Kim ES, Serur A, Huang J, Manley CA, McCrudden KW, Frischer JS, et al. (2002) Potent VEGF blockade causes regression of coopted vessels in a model of neuroblastoma. *Proc. Natl. Acad. Sci. USA* 99, 11399–11404.
144. Chiang A, Regillo CD. (2011) Preferred therapies for neovascular age-related macular degeneration. *Curr. Opin. Ophthalmol.* 22, 199–204.
145. Teng LS, Jin KT, He KF, Zhang J, Wang HH, Cao J. (2010) Clinical applications of VEGF-trap (aflibercept) in cancer treatment. *J. Chin. Med. Assoc.* 73, 449–456.
146. Twardowski P, Stadler WM, Frankel P, Lara PN, Ruel C, Chatta G, et al. (2010) Phase II study of Aflibercept (VEGF-Trap) in patients with recurrent or metastatic urothelial cancer, a California Cancer Consortium Trial. *Urology* 76, 923–926.
147. Cai J, Han S, Qing R, Liao D, Law B, Boulton ME. (2011) In pursuit of new anti-angiogenic therapies for cancer treatment. *Front. Biosci.* 16, 803–814.
148. J. F. de Groot, K. R. Lamborn, S. M. Chang, M. R. Gilbert, T. F. Cloughesy, K. Aldape, J. Yao, E. F. Jackson, F. Lieberman, H. I. Robins, M. P. Mehta, A. B. Lassman, L. M. Deangelis, W. K. Yung, A. Chen, M. D. Prados, P. Y. Wen. (2011). Phase II study of aflibercept in recurrent malignant glioma: a North American Brain Tumor Consortium study. *Journal of clinical oncology : official journal of the American Society of Clinical Oncology*, 29, 2689–2695.
149. Ferrara N, Hillan KJ, Gerber HP, Novotny W. (2004) Discovery and development of bevacizumab, an anti-VEGF antibody for treating cancer. *Nat. Rev. Drug Discov.* 3, 391–400.
150. Ferrara N, Damico L, Shams N, Lowman H, Kim R. (2006) Development of ranibizumab, an anti-vascular endothelial growth factor antigen binding fragment, as therapy for neovascular age-related macular degeneration. *Retina* 26, 859–870.
151. Muller YA, Chen Y, Christinger HW, Li B, Cunningham BC, Lowman HB, et al. (1998) VEGF and the Fab fragment of a humanized neutralizing antibody: crystal structure of the complex at 2.4 A resolution and mutational analysis of the interface. *Structure* 6, 1153–1167.
152. Stewart MW, Rosenfeld PJ. (2008) Predicted biological activity of intravitreal VEGF Trap. *Br. J. Ophthalmol.* 92, 667–668.
153. Brown DM, Heier JS, Ciulla T, Benz M, Abraham P, Yancopoulos G, et al. (2011) Primary endpoint results of a phase II study of vascular endothelial growth factor trap-eye in wet age-related macular degeneration. *Ophthalmology* 118, 1089–1097.
154. Heier JS, Boyer D, Nguyen QD, Marcus D, Roth DB, Yancopoulos G, et al. (2011) The 1-year results of CLEAR-IT 2, a Phase 2 study of vascular endothelial growth factor trap-eye dosed as-needed after 12-week fixed dosing. *Ophthalmology* 118, 1098–1106.
155. D. V. Do, U. Schmidt-Erfurth, V. H. Gonzalez, C. M. Gordon, M. Tolentino, A. J. Berliner, R. Vitti, R. Ruckert, R. Sandbrink, D. Stein, K. Yang, K. Beckmann, J. S. Heier. (2011). The DA VINCI Study: phase 2 primary results of VEGF Trap-Eye in patients with diabetic macular edema. *Ophthalmology*, 118, 1819–1826.
156. Allendorph GP, Isaacs MJ, Kawakami Y, Izpisua Belmonte JC, Choe S. (2007) BMP-3 and BMP-6 structures illuminate

the nature of binding specificity with receptors. *Biochemistry* 46, 12238–12247.

157. Hamdy NA. (2005) Osteoprotegerin as a potential therapy for osteoporosis. *Curr. Osteoporos. Rep.* 3, 121–125.
158. Koh YJ, Kim HZ, Hwang SI, Lee JE, Oh N, Jung K, et al. (2010) Double antiangiogenic protein, DAAP, targeting VEGF-A and angiopoietins in tumor angiogenesis, metastasis, and vascular leakage. *Cancer Cell* 18, 171–184.
159. Sengle G, Ono RN, Sasaki T, Sakai LY. (2011) Prodomains of transforming growth factor beta (TGFbeta) superfamily members specify different functions: extracellular matrix interactions and growth factor bioavailability. *J. Biol. Chem.* 286, 5087–5099.
160. Mazerbourg S, Hsueh AJ. (2006) Genomic analyses facilitate identification of receptors and signalling pathways for growth differentiation factor 9 and related orphan bone morphogenetic protein/growth differentiation factor ligands. *Hum. Reprod. Update* 12, 373–383.
161. Mazerbourg S, Sangkuhl K, Luo CW, Sudo S, Klein C, Hsueh AJ. (2005) Identification of receptors and signaling pathways for orphan bone morphogenetic protein/growth differentiation factor ligands based on genomic analyses. *J. Biol. Chem.* 280, 32122–32132.
162. Nickel J, Sebald W, Groppe JC, Mueller TD. (2009) Intricacies of BMP receptor assembly. *Cytokine Growth Factor Rev.* 20, 367–377.
163. Canalis E, Economides AN, Gazzerro E. (2003) Bone morphogenetic proteins, their antagonists, and the skeleton. *Endocr. Rev.* 24, 218–235.
164. Zakin L, De Robertis EM. (2009) Extracellular regulation of BMP signaling. *Curr. Biol.* 20, R89–R92.
165. Umulis D, O'Connor MB, Blair SS. (2009) The extracellular regulation of bone morphogenetic protein signaling. *Development* 136, 3715–3728.
166. Rider CC, Mulloy B. (2011) Bone morphogenetic protein and growth differentiation factor cytokine families and their protein antagonists. *Biochem. J.* 429, 1–12.
167. Wiley SR, Schooley K, Smolak PJ, Din WS, Huang CP, Nicholl JK, et al. (1995) Identification and characterization of a new member of the TNF family that induces apoptosis. *Immunity* 3, 673–682.
168. Emery JG, McDonnell P, Burke MB, Deen KC, Lyn S, Silverman C, et al. (1998) Osteoprotegerin is a receptor for the cytotoxic ligand TRAIL. *J. Biol. Chem.* 273, 14363–14367.
169. Kostenuik PJ. (2005) Osteoprotegerin and RANKL regulate bone resorption, density, geometry and strength. *Curr. Opin. Pharmacol.* 5, 618–625.
170. Stohl W, Scholz JL, Cancro MP. (2011) Targeting BLyS in rheumatic disease: the sometimes-bumpy road from bench to bedside. *Curr. Opin. Rheumatol.* 23, 305–310.
171. Baker KP, Edwards BM, Main SH, Choi GH, Wager RE, Halpern WG, et al. (2003) Generation and characterization of LymphoStat-B, a human monoclonal antibody that antagonizes the bioactivities of B lymphocyte stimulator. *Arthritis Rheum.* 48, 3253–3265.
172. Kayagaki N, Yan M, Seshasayee D, Wang H, Lee W, French DM, et al. (2002) BAFF/BLyS receptor 3 binds the B cell survival factor BAFF ligand through a discrete surface loop and promotes processing of NF-kappaB2. *Immunity* 17, 515–524.
173. Bossen C, Cachero TG, Tardivel A, Ingold K, Willen L, Dobles M, et al. (2008) TACI, unlike BAFF-R, is solely activated by oligomeric BAFF and APRIL to support survival of activated B cells and plasmablasts. *Blood* 111, 1004–1012.
174. Mackay F, Schneider P. (2008) TACI, an enigmatic BAFF/APRIL receptor, with new unappreciated biochemical and biological properties. *Cytokine Growth Factor Rev.* 19, 263–276.
175. Mackay F, Schneider P. (2009) Cracking the BAFF code. *Nat. Rev. Immunol.* 9, 491–502.
176. Lesley R, Xu Y, Kalled SL, Hess DM, Schwab SR, Shu HB, et al. (2004) Reduced competitiveness of autoantigen-engaged B cells due to increased dependence on BAFF. *Immunity* 20, 441–453.
177. Thien M, Phan TG, Gardam S, Amesbury M, Basten A, Mackay F, et al. (2004) Excess BAFF rescues self-reactive B cells from peripheral deletion and allows them to enter forbidden follicular and marginal zone niches. *Immunity* 20, 785–798.
178. Castigli E, Scott S, Dedeoglu F, Bryce P, Jabara H, Bhan AK, et al. (2004) Impaired IgA class switching in APRIL-deficient mice. *Proc. Natl. Acad. Sci. USA* 101, 3903–3908.
179. Halpern WG, Lappin P, Zanardi T, Cai W, Corcoran M, Zhong J, et al. (2006) Chronic administration of belimumab, a BLyS antagonist, decreases tissue and peripheral blood B-lymphocyte populations in cynomolgus monkeys: pharmacokinetic, pharmacodynamic, and toxicologic effects. *Toxicol. Sci.* 91, 586–599.
180. Carbonatto M, Yu P, Bertolino M, Vigna E, Steidler S, Fava L, et al. (2008) Nonclinical safety, pharmacokinetics, and pharmacodynamics of atacicept. *Toxicol. Sci.* 105, 200–210.
181. Koyama T, Tsukamoto H, Miyagi Y, Himeji D, Otsuka J, Miyagawa H, et al. (2005) Raised serum APRIL levels in patients with systemic lupus erythematosus. *Ann. Rheum. Dis.* 64, 1065–1067.
182. Vallerskog T, Heimburger M, Gunnarsson I, Zhou W, Wahren-Herlenius M, Trollmo C, et al. (2006) Differential effects on BAFF and APRIL levels in rituximab-treated patients with systemic lupus erythematosus and rheumatoid arthritis. *Arthritis Res. Ther.* 8, R167.
183. Planelles L, Castillo-Gutierrez S, Medema JP, Morales-Luque A, Merle-Beral H, Hahne M. (2007) APRIL but not BLyS serum levels are increased in chronic lymphocytic leukemia: prognostic relevance of APRIL for survival. *Haematologica* 92, 1284–1285.
184. Kimberley FC, Hahne M, Medema JP. (2009) "APRIL hath put a spring of youth in everything": Relevance of APRIL for survival. *J. Cell. Physiol.* 218, 1–8.
185. Morel J, Roubille C, Planelles L, Rocha C, Fernandez L, Lukas C, et al. (2009) Serum levels of tumour necrosis factor family members a proliferation-inducing ligand (APRIL) and B lymphocyte stimulator (BLyS) are inversely correlated in systemic lupus erythematosus. *Ann. Rheum. Dis.* 68, 997–1002.

186. Roschke V, Sosnovtseva S, Ward CD, Hong JS, Smith R, Albert V, et al. (2002) BLyS and APRIL form biologically active heterotrimers that are expressed in patients with systemic immune-based rheumatic diseases. *J. Immunol.* 169, 4314–4321.

187. Dillon SR, Harder B, Lewis KB, Moore MD, Liu H, Bukowski TR, et al. (2010) B-lymphocyte stimulator/a proliferation-inducing ligand heterotrimers are elevated in the sera of patients with autoimmune disease and are neutralized by atacicept and B-cell maturation antigen-immunoglobulin. *Arthritis Res. Ther.* 12, R48.

188. Steinberg MW, Shui JW, Ware CF, Kronenberg M. (2009) Regulating the mucosal immune system: the contrasting roles of LIGHT, HVEM, and their various partners. *Semin. Immunopathol.* 31, 207–221.

189. Jeon ST, Kim WJ, Lee SM, Lee MY, Park SB, Lee SH, et al. (2010) Reverse signaling through BAFF differentially regulates the expression of inflammatory mediators and cytoskeletal movements in THP-1 cells. *Immunol. Cell. Biol.* 88, 148–156.

190. Sun M, Fink PJ. (2007) A new class of reverse signaling costimulators belongs to the TNF family. *J. Immunol.* 179, 4307–4312.

191. Silva LC, Ortigosa LC, Benard G. (2010) Anti-TNF-alpha agents in the treatment of immune-mediated inflammatory diseases: mechanisms of action and pitfalls. *Immunotherapy* 2, 817–833.

192. Taga T, Narazaki M, Yasukawa K, Saito T, Miki D, Hamaguchi M, et al. (1992) Functional inhibition of hematopoietic and neurotrophic cytokines by blocking the interleukin 6 signal transducer gp130. *Proc. Natl. Acad. Sci. USA* 89, 10998–11001.

193. Atreya R, Mudter J, Finotto S, Mullberg J, Jostock T, Wirtz S, et al. (2000) Blockade of interleukin 6 trans signaling suppresses T-cell resistance against apoptosis in chronic intestinal inflammation: evidence in Crohn disease and experimental colitis *in vivo*. *Nat. Med.* 6, 583–588.

194. T. Barkhausen, T. Tschernig, P. Rosenstiel, M. van Griensven, R. P. Vonberg, M. Dorsch, A. Mueller-Heine, A. Chalaris, J. Scheller, S. Rose-John, D. Seegert, C. Krettek, G. H. Waetzig. (2011). Selective blockade of interleukin-6 trans-signaling improves survival in a murine polymicrobial sepsis model. *Critical care medicine*, 39, 1407–1413.

195. Jones SA. (2005) Directing transition from innate to acquired immunity: defining a role for IL-6. *J. Immunol.* 175, 3463–3468.

196. Rose-John S, Schooltink H. (2007) Cytokines are a therapeutic target for the prevention of inflammation-induced cancers. *Recent Results Cancer Res.* 174, 57–66.

197. Silver JS, Hunter CA. (2010) gp130 at the nexus of inflammation, autoimmunity, and cancer. *J. Leukoc. Biol.* 88, 1145–1156.

198. Murakami M, Nishimoto N. (2011) The value of blocking IL-6 outside of rheumatoid arthritis: current perspective. *Curr. Opin. Rheumatol.* 23, 273–277.

199. Reichert JM. (2011) Antibody-based therapeutics to watch in 2011. *MAbs* 3, 76–99.

200. Nishimoto N. (2010) Interleukin-6 as a therapeutic target in candidate inflammatory diseases. *Clin. Pharmacol. Ther.* 87, 483–487.

201. Schellekens H. (2002) Immunogenicity of therapeutic proteins: clinical implications and future prospects. *Clin. Ther.* 24, 1720–1740; discussion 1719.

202. Schellekens H, Casadevall N. (2004) Immunogenicity of recombinant human proteins: causes and consequences. *J. Neurol.* 251(Suppl. 2), II4–II9.

203. Anderson P, Louie J, Lau A, Broder M. (2005) Mechanisms of differential immunogenicity of tumor necrosis factor inhibitors. *Curr. Rheumatol. Rep.* 7, 3–9.

204. Anderson PJ. (2005) Tumor necrosis factor inhibitors: clinical implications of their different immunogenicity profiles. *Semin. Arthritis Rheum.* 34, 19–22.

205. Buttel IC, Chamberlain P, Chowers Y, Ehmann F, Greinacher A, Jefferis R, et al. (2011) Taking immunogenicity assessment of therapeutic proteins to the next level. *Biologicals* 39, 100–109.

206. Hofbauer LC, Schoppet M. (2004) Clinical implications of the osteoprotegerin/RANKL/RANK system for bone and vascular diseases. *JAMA* 292, 490–495.

207. Christen U, Thuerkauf R, Stevens R, Lesslauer W. (1999) Immune response to a recombinant human TNFR55-IgG1 fusion protein: auto-antibodies in rheumatoid arthritis (RA) and multiple sclerosis (MS) patients have neither neutralizing nor agonist activities. *Hum. Immunol.* 60, 774–790.

208. Furst DE, Weisman M, Paulus HE, Bulpitt K, Weinblatt M, Polisson R, et al. (2003) Intravenous human recombinant tumor necrosis factor receptor p55-Fc IgG1 fusion protein, Ro 45-2081 (lenercept): results of a dose-finding study in rheumatoid arthritis. *J. Rheumatol.* 30, 2123–2126.

209. Rau R, Sander O, van Riel P, van de Putte L, Hasler F, Zaug M, et al. (2003) Intravenous human recombinant tumor necrosis factor receptor p55-Fc IgG1 fusion protein Ro 45-2081 (lenercept): a double blind, placebo controlled dose-finding study in rheumatoid arthritis. *J. Rheumatol.* 30, 680–690.

210. Dore RK, Mathews S, Schechtman J, Surbeck W, Mandel D, Patel A, et al. (2007) The immunogenicity, safety, and efficacy of etanercept liquid administered once weekly in patients with rheumatoid arthritis. *Clin. Exp. Rheumatol.* 25, 40–46.

211. Romero-Mate A, Garcia-Donoso C, Cordoba-Guijarro S. (2007) Efficacy and safety of etanercept in psoriasis/psoriatic arthritis: an updated review. *Am. J. Clin. Dermatol.* 8, 143–155.

212. Kaymakcalan Z, Sakorafas P, Bose S, Scesney S, Xiong L, Hanzatian DK, et al. (2009) Comparisons of affinities, avidities, and complement activation of adalimumab, infliximab, and etanercept in binding to soluble and membrane tumor necrosis factor. *Clin. Immunol.* 131, 308–316.

213. Devine EB, Alfonso-Cristancho R, Sullivan SD. (2011) Effectiveness of biologic therapies for rheumatoid arthritis: an indirect comparisons approach. *Pharmacotherapy* 31, 39–51.

10

RECOMBINANT ALBUMIN FUSION PROTEINS

Thomas Weimer, Hubert J. Metzner, and Stefan Schulte

Preclinical Research and Development, CSL Behring GmbH, Marburg, Germany

10.1 CONCEPT

Many peptides and small proteins identified in drug development have failed to produce the desired effect *in vivo* because of a short half-life, often measured in minutes or hours [1]. The relatively small size of these peptides and molecules places them below the threshold for renal clearance. While several therapeutic proteins with a short half-life have made it to the clinic, their application is often limited by the need for frequent administration, which precludes long-term use, hinders compliance, and potentially increases treatment costs and the risk of adverse events. Fusion of these proteins with well-characterized proteins with a longer half-life, such as recombinant albumin, may prolong their time in circulation and thus, increase their efficacy. Recombinant albumin fusion proteins are novel proteins created by fusing the complementary DNA (cDNA) sequence encoding albumin with a sequence encoding a therapeutic protein. The resulting hybrid polypeptide retains the biological activity of the therapeutic protein but has enhanced pharmacokinetic properties due to the inherent long half-life of albumin [2].

Human serum albumin (HSA) is a naturally occurring and widely distributed carrier protein. It is the most abundant plasma protein and, together with immunoglobulin G (IgG), accounts for 80–90% of all plasma proteins, with a serum concentration of 35–50 mg/mL [3,4]. HSA is produced in the liver, is highly stable, and is found mainly in the vascular compartment. Structurally, HSA is composed of 585 amino acids and has a molecular weight of ~67 kDa [5]. It is composed of three structurally homologous domains (DI, DII, and DIII), each of which contains two subdomains (A and B) connected by flexible loops [3]. Functionally, HSA helps to regulate osmotic pressure and pH balance [4,6], and acts as a transporter of many insoluble and hydrophobic ligands, including fatty acids and amino acids [3]. However, HSA has no enzymatic activity and is not involved in immune functioning.

The half-life of albumin in humans is 19–22 days, while most circulating proteins have a half-life, that is measured in a few days or hours [3,6]. The remarkably long half-life of albumin is due in part to pH-dependent recycling mediated by the neonatal Fc receptor (FcRn) [7,8]. FcRn resides in cells and on the surface of cells that have direct contact with circulating blood, such as vascular endothelial cells. Similar to IgG, circulating albumin is pinocytosed by these cells and binds to the FcRn. The complexes IgG-FcRn-albumin, FcRn-IgG, and FcRn-albumin avoid lysosomal degradation and instead are redirected to the cell membrane. Exposure to the physiological pH of circulating blood releases IgG and albumin back into the bloodstream.

In the case of large therapeutic proteins, it is thought that this recycling mechanism contributes to the prolonged half-life of the fusion protein compared to the therapeutic protein by itself. For smaller peptides, however, both the recycling mechanism and the increase in the molecular

Fusion Protein Technologies for Biopharmaceuticals: Applications and Challenges, First Edition. Edited by Stefan R. Schmidt.

weight above the clearance limit of the kidneys may be relevant.

The extraordinarily long half-life of albumin makes it an attractive fusion partner to improve the pharmacokinetics of therapeutic proteins. Given that albumin is a naturally occurring protein, that is widely distributed and has no enzymatic activity, it is unlikely to be toxic or immunogenic [9,10]. The structure of albumin has been well characterized and thus helps to predict how it may interact with other protein moieties [10]. These known properties have allowed recombinant albumin to be fused successfully to both small peptides and large, complex molecules such as cytokines [2] and coagulation factors [11,12]. In addition, recombinant albumin fusion proteins are technically easier and more efficient to produce than other fusion proteins involving the Fc fragment of IgG. Many recombinant albumin fusion proteins can be manufactured efficiently in yeast expression systems on a commercial scale and at typically lower costs than other methods of generating long-acting biopharmaceuticals [13]. Finally, because only monomeric fusions are possible, dosing is simplified [13].

Recombinant albumin fusion proteins can be produced with single-step expression techniques, making production and purification relatively straightforward and economical [14]. This avoids the need for complicated chemical modifications during the manufacturing process [14]. In most cases, fusion protein technology yields a homogenous product, and fusion to naturally occurring and abundant proteins such as albumin can improve the solubility and stability of a therapeutic protein while reducing renal clearance [2] and intracellular degradation. Finally, recombinant fusion protein technology allows genes to be fused in any order, making it possible to construct multiple variants of the same fusion protein; this design flexibility can help optimize the biological activity of the product [2].

Theoretically, fusion to recombinant albumin could allow for less frequent administration of a therapeutic protein while maintaining the same efficacy. Fusion may also lead to more consistent and sustained protein concentrations, thereby improving tolerance and safety by maintaining blood levels within the therapeutic window. The improved pharmacokinetics of the therapeutic protein often means that fewer administrations can be used to achieve the same effect. These features can make therapy more convenient for both clinicians and patients, leading to improvements in treatment compliance. In addition, the reduced dosing frequency and improved safety profile may enable prophylaxis in areas where this was previously unfeasible.

10.2 TECHNOLOGICAL ASPECTS

Several methods have been used to generate recombinant albumin fusion proteins, and the specific methods depend in part on the specific nature of the therapeutic protein in question. In general, cDNA sequences encoding recombinant human albumin (rHA) and the therapeutic protein are isolated using standard laboratory techniques and fused in a plasmid vector. The recombinant albumin fusion technology permits the creation of various orientations that provide design simplicity and flexibility. The albumin molecule has two-preferred sites, particularly the C- and N-termini, where proteins can be attached. Additionally, it is possible to fuse two different proteins together with recombinant albumin to give two different functions (bivalent) at the same time. rHA can be fused with biologically active compounds of varying sizes, from small peptides to large cytokines. Moreover, the two genes can be directly joined together or separated by a linker to maximize the interaction between the therapeutic protein and its target molecule(s). Some linker sequences are cleavable and can therefore separate the two protein moieties under specific conditions.

The cDNA construct is created in a prokaryotic system and then used to transfect eukaryotic cells such as *Pichia pastoris* [2,15]. Although rHA can be expressed in both prokaryotic and eukaryotic systems [6], eukaryotic systems are more efficient in producing the final albumin structure [15]. *P. pastoris* is often used because, compared to other eukaryotic expression systems, it is economical and readily scaled-up for mass production [16]. For proteins with complex posttranslational modifications, however, the use of animal or human cell lines is mandatory. During fermentation, the fusion gene is transcribed and translated as a single polypeptide, that is secreted into the fermentation broth. The broth can then be separated and clarified by centrifugation or filtration, and the protein can be purified using standard separation techniques [15,17–20].

10.3 TYPICAL APPLICATIONS AND INDICATIONS

Recombinant albumin fusion technology has been successfully applied to various proteins and peptides with clinical applications in a broad range of diseases (Table 10.1). Three recombinant albumin fusion proteins are in late-stage clinical development, demonstrating that albumin fusion technology is feasible to improve the pharmacokinetics of therapeutic proteins while retaining their biological activity. Numerous other recombinant albumin fusion proteins are in preclinical or early-stage clinical development; the variety of protein partners fused to recombinant albumin and the many different diseases for which these proteins are being developed are indicative of the broad spectrum of potential applications of recombinant albumin fusion proteins.

TABLE 10.1 **Summary of Selected Recombinant Albumin Fusion Proteins in Preclinical and Clinical Development**

Albumin Fusion Partner	Development Phase	Therapy Area	Extended Half-Life of Albumin Fusion Protein	Reference/Clinical Trial Identifier
G-CSF (Albugranin; Neugranin)	III	Neutropenia	Mean 7.7–13.3 h in monkeys and 5.4–8.3 h in mice: over twofold longer than that of the G-CSF analog filgrastim in mice	[21]; NCT01126190; NCT00837265
GLP-1 (Albiglutide, Albugon)	III	Diabetes mellitus	Mean 3.6–8 days in humans: compared to <2 min for wild-type GLP-1 and 17.9 h for the GLP-1 analog liraglutide	[22–25]; NCT00849017; NCT00839527; NCT01098539; NCT01128894; NCT00976391; NCT00838903; NCT00838916
IFN-α (Albinterferon-α2b [alb-IFN]; Albuferon)	III	Chronic hepatitis C, genotype 1 or 2/3; cancer	Median ∼6 days in humans: two fold longer than that of PEG-IFN-α-2a and fourfold greater than that of PEG-IFN-α-2b	[26,27]; NCT00964665; NCT00759200; NCT00489385; NCT00724776
BNP (AlbuBNP; Cardeva)	I/II	Congestive heart failure	Terminal 12–19 h: compared to 3 min for BNP	[17,28]
FIX (rIX-FP)	I	hemophilia B	Terminal 24–36 h in rats and rabbits: about 4–5-fold longer than that of rFIX in both animal models	[12]; NCT01233440
hGH (Albutropin)	I	Growth hormone deficiency	Terminal 3–6 h in rats and 13–15 h in monkeys: up to 4–6-fold greater than wild-type hGH	[29]
IL-2 (Albuleukin)	I	Cancer	Terminal 6–8 h in mice: compared to 19–57 min for recombinant IL-2	[30,31]
Antibody fragments	Preclinical	Cancer	Mean ∼30–47 h (for all antibody albumin fusions) in rats: 2–10-fold greater than those of the unfused fragments	[19,32]
Barbourin ($BLAH_6$)	Preclinical	Thrombosis	Terminal 3.4 days in rabbits: up to 23-fold longer than that of recombinant barbourin	[33]
BChE (BChE/HSA)	Preclinical	Organophosphorus compound toxicity	Mean ∼32 h in pigs: compared to ∼3 h with BChE	[34]
Cocaine hydrolase (Albu-CocH)	Preclinical	Cocaine toxicity/addiction	Terminal 8 h in rats: compared to 13 min for cocaine hydrolase	[35–37]
Erythropoietin (rHA-EPO)	Preclinical	Anemia	Terminal ∼6 h in rats: up to 2.5-fold longer compared to recombinant human EPO	[38]
FVIIa (rVIIa-FP)	Preclinical	Hemophilia A with anti-FVIII inhibitors	Terminal 4.4 h in rats and 13 h in rabbits: ∼7–8-fold longer than that of wild-type recombinant FVIIa	[11,39]

(continued)

TABLE 10.1 (*Continued*)

Albumin Fusion Partner	Development Phase	Therapy Area	Extended Half-Life of Albumin Fusion Protein	Reference/Clinical Trial Identifier
GLP-1 analog exendin-4 (rEx-4/HSA)	Preclinical	Diabetes mellitus	Terminal 57–77 h in monkeys: compared to 2.4 h for Ex-4 in humans	[15]
Hirudin (HLA, ALH)	Preclinical	Thrombosis	Terminal 4.6 days in rabbits: up to 160-fold longer than that of recombinant hirudin	[40,41]
Infestin-4 (rHA-infestin-4)	Preclinical	Thrombosis	Terminal 4.6 h in mice: over 15-fold longer than that of $(His)_6$-tagged infestin-4	[42]
Insulin (Albulin)	Preclinical	Diabetes mellitus	Terminal 6.7 h in mice: 40-fold longer compared to ~10 min with recombinant human insulin	[43,44]
Thioredoxin (Trx-HSA)	Preclinical	Immunomodulation, anti-inflammatory, antioxidant, apoptosis	Terminal 10.3 h in mice: 10-fold greater than that of Trx	[45]
Thymosin-α1 (rHSA-Tα1)	Preclinical	Hepatitis B/C, cancer	Mean 32–36 h in rats: sixfold greater than that of recombinant human Tα1	[18,46]
TNF ligands (HSA-TNF; HSA-TRAIL; HSA-TWEAK)	Preclinical	Immunomodulation, pro-inflammatory, cancer	Terminal ~15 h in rats: up to 15-fold longer than that of recombinant TNF ligands	[47]

AlbuBNP, BNP albumin fusion protein; alb-IFN, albinterferon-α2b; AlbuCocH, cocaine hydrolase albumin fusion protein; ALH, recombinant hirudin fusion protein linked via its N-terminus to human albumin; BChE, butyrylcholinesterase; BChE/HSA, BChE human serum albumin fusion protein; $BLAH_6$, recombinant barbourin rabbit serum albumin fusion protein with linker; BNP, B-type natriuretic peptide; FIX, factor IX; FVIIa, activated factor VII; FVIII, factor VIII; HLA, recombinant hirudin fusion protein linked via its C-terminus to human albumin; G-CSF, granulocyte colony-stimulating factor; GLP-1, glucagon-like peptide-1; HSA-TNF, TNF human serum albumin fusion protein; HSA-TRAIL, TNF-related apoptosis inducing ligand albumin fusion protein; HSA-TWEAK, TNF-related weak inducer of apoptosis albumin fusion protein; IFN-α, interferon-α; hGH, human growth hormone; IL, interleukin; rIX-FP, recombinant FIX albumin fusion protein; rVIIa-FP, activated recombinant FVII albumin fusion protein; rHA, recombinant human albumin; rHSA-Tα1, recombinant thymosin human serum albumin fusion protein; TNF, tumor necrosis factor; Trx-HSA, thioredoxin human serum albumin fusion protein.

10.3.1 Albumin Fusion Proteins in Late-Stage Development

10.3.1.1 Interferon-α Interferon-α (IFN-α) is a cytokine with antiviral, immunomodulatory, and antiproliferative properties. It is used to treat viral infections, such as hepatitis B and C, and certain malignancies, such as melanoma. The half-life of IFN-α in humans, however, is 2–3 h, so treatment is often given daily or three times weekly [9]. Two versions of IFN-α are available in which polyethylene glycol (PEG) is covalently attached to IFN-α (PEG-IFN-α-2a and PEG-IFN-α-2b). PEGylation improves the pharmacokinetics of IFN-α, resulting in a prolonged half-life and allowing for weekly dosing [48,49]. Nevertheless, ~45% of chronic hepatitis C (CHC) patients do not achieve a sustained virological response (SVR) with these agents [50].

Albinterferon-α-2b (alb-IFN), or Albuferon®, is a recombinant albumin fusion protein that combines the antiviral activity of IFN-α with the prolonged half-life of albumin. The fusion protein has a molecular mass of ~85.7 kDa and is formed by fusing the C-terminus of rHA to the N-terminus of recombinant human IFN-α [51]. It is produced in a *Saccharomyces cerevisiae* expression system [52], and purified using a combination of anion, cation, and semi-affinity chromatography [51]. In monkeys, alb-IFN demonstrated improved pharmacokinetics, including prolonged half-life, reduced clearance, and increased stability, compared to IFN-α [51]. These improvements led to better pharmacodynamics, based on expression levels of 2′,5′-oligoadenylate synthetase (OAS), a biomarker of IFN-α activity, and remained elevated for up to 10 days after a single injection. In serum samples, alb-IFN was shown to maintain antiviral activity for up to 8 days after injection compared to 5 days for PEG-IFN-α-2b and 1 day for IFN-α. The half-life of alb-IFN (93 h) was over twofold greater than the mean half-life of PEG-IFN-α-2b (40 h) and around 18-fold greater than that of IFN-α (5 h). Allometric scaling predicted a half-life in humans of 90–288 h, thereby supporting dosing every 2–4 weeks.

In humans with IFN-α-naive, genotype 1 CHC, the mean absorption half-life of alb-IFN was 23–31 h after subcutaneous administration [26]. The maximum concentration increased linearly with the dose and was reached at a mean of 65–87 h after dosing. Fusion to albumin extended the half-life of IFN-α-2a from 4 to 141 h (~6 days), which was also approximately twofold greater than published results for PEG-IFN-α-2a (72 h) [48,53] and around fourfold greater than that of PEG-IFN-α-2b (40 h) [54]. Similar results were noted in 119 CHC patients (mostly with genotype 1) who had previously received IFN-α [50]. The improved pharmacokinetics compared to conventional IFN-α therapies led to improvements in pharmacodynamic properties: expression of the gene *OAS1* was maintained for more than 28 days after a single injection of alb-IFN.

The safety and efficacy of alb-IFN have been evaluated in two Phase III trials in patients with genotype 1 (ACHIEVE 1) or genotype 2/3 (ACHIEVE 2/3) CHC [27,55,56]. In the ACHIEVE 1 trial, 1331 IFN-α-naive patients were randomized to alb-IFN (900 or 1200 μg every 2 weeks) or PEG-IFN-α-2a (180 μg weekly) [27]. On the basis of an intent-to-treat analysis, alb-IFN met its primary efficacy endpoint of noninferiority to PEG-IFN-α-2a with 48.2% of patients in the 900 μg alb-IFN arm achieving SVR compared to 51.0% of those in the PEG- IFN-α-2a arm. Adverse events were generally similar between the two arms and to those typically reported for IFN-α. However, the 1200 μg dose of alb-IFN was associated with a higher rate of serious pulmonary adverse events and treatment discontinuations. The application for a marketing authorization for the 1200-μg dose of alb-IFN was recently withdrawn by the sponsor because of the regulatory feedback on the relative risk-to-benefit ratio [57]. Consequently, all patients in the ACHIEVE 1 trial receiving alb-IFN 1200 μg were switched to 900 μg given every 2 weeks [27].

10.3.1.2 Glucagon-Like Peptide-1 Glucagon-like peptide 1 (GLP-1) is an incretin hormone secreted in the intestines that stimulates insulin secretion by binding to GLP-1 receptors found on beta cells and various components of the central and peripheral nervous systems [22]. GLP-1 inhibits gastric emptying, regulates appetite and food intake, enhances glucose disposal, and improves endothelial dysfunction. Despite these potent activities, the use of GLP-1 in the treatment of diabetes has been hindered by its rapid degradation in circulation by dipeptidyl peptidase-4. Fusion proteins have been developed that combine albumin with various GLP-1 analogs and mimetics, and several Phase III trials are underway to evaluate these fusion proteins as monotherapy or in combination with other medications for diabetes [22].

Albiglutide (formerly known as albugon) is an albumin fusion protein generated by fusing the gene sequence encoding a dimer of recombinant dipeptidyl-peptidase IV resistant GLP-1 analog with the open reading frame of rHA [24]. Studies conducted *in vivo* indicate that albiglutide activates GLP-1 receptor-dependent central and peripheral pathways that regulate energy intake and glucose homeostasis. These results suggest that the altered molecular configuration that arises when GLP-1 is fused to rHA does not eliminate its ability to recognize and activate GLP-1 receptors. Phase II clinical trial results indicated improved fasting and postprandial plasma glucose levels in diabetes patients and a systemically improved half-life of about 5 days with albiglutide compared to around 5 min for wild-type GLP-1 [24,58].

Another dipeptidyl-peptidase IV resistant GLP-1 fusion protein was developed by attaching a lysine residue at the N-terminus of GLP-1 (termed KGLP-1) with albumin fused to this new N-terminus [14]. The recombinant fusion protein, called KGLP-1/HSA, bound and activated GLP-1 receptors; the activity of which was less potent than that of either KGLP-1 or wild-type GLP-1, possibly due to increased spatial blockade caused by the albumin moiety. Nevertheless, KGLP-1/HSA retained much of the biological activity of GLP-1, and its duration of effect was prolonged; up to eightfold longer for its glucose-lowering effect and up to twofold longer for its insulin-releasing effect [14].

Albumin has also been fused to the C-terminal end of exendin-4, a potent GLP-1 agonist [15]. Exendin-4 is a 39-amino-acid peptide isolated from the salivary gland of the *Heloderma suspectum* lizard [59] and has 53% homology to mammalian GLP-1. Synthetic exendin-4 (exenatide) has greater efficacy compared to GLP-1, but its half-life of 2.4 h necessitates twice-daily injections. Extended-release formulations of exenatide have also been developed that can be given on a weekly basis [60,61]. As an alternative approach, the gene encoding exendin-4 was fused to the rHA N-terminus, linked by a short peptide (GGGGS), to create the novel fusion protein rEx-4/rHA [15]. The rEx-4/rHA fusion protein had improved and prolonged glucose-lowering effects in diabetic mice compared to a GLP-1-albumin conjugated analog (36 vs. 12 h) [62]. In monkeys, the half-life of rEx-4/rHA was 70–80 h, which was 33 times longer than that of exenatide (2.4 h) [63] and much longer than values reported for other GLP-1 modifications that employed fatty acid chain modification (11–15 h) [64] or PEGylation (33.4 min) [65]. These results indicate that albumin fusion prolongs the half-life of exendin-4 while retaining its glucose-lowering effects, potentially allowing for weekly dosing. Indeed, this is currently being evaluated in Phase III clinical trials [61].

10.3.1.3 Granulocyte Colony-Stimulating Factor Recombinant human granulocyte colony-stimulating factor (rhG-CSF; filgrastim) is expressed in bacteria and used in the treatment of neutropenia, but its short circulating half-life necessitates daily or twice-daily injections [21]. Albugranin™ is a fusion protein derived from albumin and rhG-CSF, in which the N-terminus of rhG-CSF is fused to the C-terminus of rHA [21]. In mice and monkeys, Albugranin had improved effects compared to rhG-CSF, including sustained neutrophilic leukocytosis in both species [21]. In mice, clearance was eightfold slower with Albugranin than with rhG-CSF, and the half-life was more than twofold greater with either intravenous or subcutaneous administration. The number of peripheral granulocytes and hematopoietic progenitors was increased for up to 4 days after a single dose, compared to 1 day with rhG-CSF. In monkeys, administration of Albugranin every 4 or 7 days over a 14-day period was well tolerated and increased absolute granulocyte and white blood cell counts. The duration of hematopoietic effects was longer with Albugranin than with rhG-CSF (up to 4 days vs. less than 3 days). Thus, Albugranin may provide similar efficacy as rhG-CSF while allowing for weekly administration, and may therefore improve convenience and, possibly, treatment effects. While a PEGylated form of rhG-CSF (Neulasta®) is available that has extended activity compared to rhG-CSF [66], Albugranin has the added advantages of being simpler to manufacture and easier to produce a uniform product. Albugranin is currently in Phase III clinical trials for chemotherapy-induced neutropenia in breast cancer patients.

10.3.2 Albumin Fusion Proteins in Early-Stage Development

10.3.2.1 Interleukin-2 Albuleukin® is an 81.8-kDa fusion protein derived from albumin and the cytokine interleukin-2 (IL-2) and is produced using an *S. cerevisiae* expression system [30]. A comparison of the biodistribution of radiolabeled albumin, IL-2, and Albuleukin in mice using whole-body autoradiography revealed striking differences. IL-2 was cleared rapidly from the blood: less than 1% of the injected dose (ID) per gram of tissue was detectable 20 min after administration, while levels in the kidney (120% ID/g) suggested rapid uptake that persisted for up to 6 h. In contrast, Albuleukin remained in circulation longer (14% ID/g at 6 h) and had less uptake in the kidneys (less than 6% ID/g) but increased localization in the lymphocyte-rich tissues, such as the liver, spleen, and lymph nodes (maximal uptake ~22% for all three tissues). Albumin also remained in circulation longer than did IL-2, but was found in low levels in the liver, spleen, and lymph nodes, suggesting that uptake of Albuleukin in these tissues is mediated in part by the IL-2 portion of the fusion protein. This targeting of tissues, where lymphocytes reside, indicates that albumin fusion technology may improve the efficacy and reduce the toxicity of IL-2 therapy [30]. Preclinical studies have evaluated the pharmacokinetics of Albuleukin [31]. *In vitro* experiments showed that Albuleukin increased the proliferation of primary human and mouse T cells and B cells as well as primary human natural killer (NK) cells. The fusion protein competed with recombinant IL-2 for binding to IL-2 receptors, and induced the production of IFN-γ in primary human NK cells. In mice, Albuleukin had a prolonged half-life of 6–8 h, compared to only 19–57 min for recombinant IL-2. Administration of Albuleukin inhibited tumor growth in mice with hepatic metastases and reduced the incidence of residual microscopic hepatic tumors, leading to prolonged survival compared to controls and mice treated with recombinant IL-2 [31].

10.3.2.2 Insulin An albumin-insulin fusion protein (albulin) was developed to improve upon the performance of currently available long-acting insulin analogs, such as insulin glargine and insulin detemir, which are used to improve glycemic control in patients with diabetes mellitus. However, these analogs fail to provide stable basal glycemia for more than a few hours and are associated with hyperglycemic and hypoglycemic excursions [43,67]. Albulin was generated by fusing single-chain insulin (containing the B and A chains of mature human insulin linked by a dodecapeptide linker) to the N-terminal end of rHA [43]. The resulting fusion gene was expressed in either yeast or Chinese hamster ovary (CHO) cells. In normoglycemic mice, the elimination half-life of subcutaneous administered albulin was 7 h compared to about 10 min with recombinant insulin. *In vivo* studies in severely hyperglycemic diabetic mice indicated that albulin was capable of maintaining a near-normoglycemic state for long periods of time. A single subcutaneous injection of albulin normalized blood glucose levels in a relatively constant and sustained manner for up to 24 h [43]. In addition, albulin had a lower affinity for the human insulin receptor and insulin growth factor 1 (IGF-1), suggesting that it may be less mitogenic compared to recombinant insulin.

10.3.2.3 Growth Hormone Human growth hormone (hGH) is used to treat adult and pediatric GH deficiency, AIDS-associated wasting in adults, Turner syndrome, and chronic renal insufficiency [29]. hGH is a 191-amino-acid pituitary hormone that stimulates the production and release of IGF-1. The hormone has a very short half-life of less than 20 min *in vivo* and therefore requires frequent administration [68]. Compliance can therefore be an issue, particularly in children. The fusion protein Albutropin® was generated by fusing the N-terminus of recombinant hGH to the C-terminus of albumin. The protein was expressed in *S. cerevisiae* and purified from the supernatants obtained. Following subcutaneous injection in rats, clearance of Albutropin was more than twofold slower than that of hGH, and the volume of distribution was higher [29]. In GH-deficient hypophysectomized rats, body weight increased significantly when Albutropin was given daily, every other day, or every 4 days; these effects on body weight were greater than those of hGH when given at equimolar doses at the same schedule. A marked benefit over hGH was seen when Albutropin was given every 4 days: with this schedule, Albutropin increased body weight and bone growth, whereas hGH had no effect. In monkeys, the half-life of Albutropin was sixfold longer than that of hGH. In addition, its mean residence time was 5- to 10-fold greater, and clearance was eightfold slower. The improved pharmacokinetic properties translated into improved pharmacodynamics: a single subcutaneous dose of Albutropin maintained IGF-1 levels above baseline for 7 days in a manner similar to daily dosing with hGH for 7 days. Thus, Albutropin may provide the same efficacy as does hGH while allowing a reduced dosing frequency [29].

10.3.2.4 Butyrylcholinesterase and Cocaine Hydrolase Albumin has been fused to butyrylcholinesterase (BChE) and its potent analog cocaine hydrolase with the aim of improving the pharmacokinetic properties of these proteins.

BChE is used to prevent nerve toxicity caused by exposure to organophosphorus compounds used in various chemical weapons and pesticides. An albumin-BChE fusion protein was developed to overcome some of the limitations of PEGylation [34]. The sequence encoding rHA was fused to the 3′ end of the human BChE sequence, using a glycine/serine linker (Gly_6Ser). The construct was transfected into baby hamster kidney cells, and transgenic mice and goats were generated. *In vivo*, the fusion protein had a considerably longer half-life in pigs (32 h) than did recombinant human BChE (3 h) and retained much of its biological activity, according to the results of nerve binding and inhibition studies. Notably, the albumin-BChE fusion protein was successfully expressed in goat milk, offering a potential method for large-scale production and stockpiling of BChE for the prevention and treatment of organophosphorus compound toxicity.

A BChE mutant designed to elevate the rate constant for cocaine hydrolysis (called cocaine hydrolase) is a more potent inhibitor of cocaine effects than is BChE, but its effects are also limited by a short half-life. An albumin-cocaine hydrolase fusion protein (Albu-CocH) was therefore developed [35]. After a single injection into rats, the elimination half-life of albumin-cocaine hydrolase was 8 h, suggesting ample stability for short-term applications and the potential for long-term use [36]. The fusion protein was found to block cocaine toxicity and improve relapse behavior in rats exposed to cocaine. Specifically, it prevented seizures in rats that were given typically lethal doses of cocaine, and rescued animals after convulsions had commenced. It also blocked reinstatement behavior in drug-seeking rats that had self-administered cocaine, but it had no effect on amphetamine-primed reinstatement, locomotor behavior, or food-rewarded behavior. These results suggest that albumin-cocaine hydrolase may have applications in not only the acute setting of cocaine overdose but also the chronic setting to help prevent recidivism.

10.3.2.5 B-Type Natriuretic Peptide B-type natriuretic peptide (BNP) is used in the treatment of decompensated congestive heart failure (CHF) but has a short half-life, thereby limiting its use to the acute setting and necessitating continuous intravenous infusion [28]. An albumin-BNP fusion protein (AlbuBNP; Cardeva™) was generated using a mammalian expression system (human embryonic kidney (HEK)-293 cells) [28]. *In vitro*, albumin-BNP has the same maximal bioactivity as BNP in terms of activating cyclic guanosine monophosphate (cGMP). In a hypertensive rat model, a single injection of albumin-BNP reduced systolic and diastolic blood pressure and provided a sustainable reduction in arterial pressure for more than 2 days. When given at a dose of 6 nmol/kg, albumin-BNP increased cGMP levels 5.6-fold compared to baseline. Notably, the elimination half-life was 3 min for wild-type BNP and much longer at 12–19 h for albumin-BNP. Other studies have shown that the renal-enhancing properties of albumin-BNP are prolonged in animal models [17]. Compared to wild-type BNP, the fusion protein provided sustained improvements in markers of BNP activity, natriuresis, diuresis, renal blood flow, and glomerular filtration rate. It also reduced cardiac filling pressure compared to BNP. The development of a long-acting BNP could have important implications in the management of CHF, since it would allow for long-term use in earlier stages of CHF, the postacute follow-up phase, and following myocardial infarction [28].

10.3.2.6 Erythropoietin Recombinant human erythropoietin (rhEpo) is used to treat anemia and has a half-life of 4–13 h when given intravenously. This short half-life necessitates 2–3 doses weekly and produces excessive peak/trough concentrations [38]. Subcutaneous administration is associated with slightly better maintenance of plasma concentrations, but still necessitates dosing at equal frequencies each week. Joung et al. [38] fused recombinant albumin to erythropoietin using a flexible linker $(GGSGG)_4$ to protect against cross-interaction between the two proteins. The fusion protein, rHA-EPO, was expressed in mammalian cells (CHO) and purified from culture supernatant using chromatography. The fusion protein had a similar half-life to that of darbepoetin, a glycosylation variant with increased half-life, when given intravenously in rodents. A cleavage site was found in the linker sequence, and the formulation was modified to include a stabilizer, which prevented 80% of cleavage after 1 month. In terms of biological activity, rHA-EPO was nearly eightfold more potent than was rhEpo in increasing hematocrit in normal mice.

10.3.2.7 Barbourin Barbourin is a 73-amino acid protein that inhibits platelet aggregation by binding to the platelet integrin $\alpha_{IIb}\beta_3$ [33]. It is found in the venom of the southwestern US pigmy rattlesnake *Sistrurus miliarius barbouri* and inspired the rational design of the peptide-based antiplatelet drug eptifibatide. Because barbourin is rapidly cleared from the circulation, a fusion protein of rabbit serum albumin (RSA) and recombinant His_6-tagged barbourin ($BARH_6$) was developed ($BLAH_6$) [33]. The *P. pastoris* expression system was used to produce $BARH_6$, $RSAH_6$, and the fusion protein $BLAH_6$. The fusion protein $BLAH_6$ had inhibitory effects on platelet aggregation *in vitro* comparable to that of $BARH_6$ using both human and rabbit platelets, and both proteins formed a complex with $\alpha_{IIb}\beta_3$. In rabbits, the terminal half-life of $BLAH_6$ approached that of $RSAH_6$ (3.4 and 4.0 days, respectively) and was markedly higher than that of $BARH_6$ (0.15 days). Thus, albumin fusion appeared to slow the clearance of barbourin without interfering with its platelet-aggregating effects [33].

10.3.2.8 Hirudin Hirudin is a small protein (65 amino acids) derived from leeches and is a potent and direct inhibitor of thrombin [69]. Recombinant forms of hirudin, such as desirudin and lepirudin, are used to inhibit thrombus formation. Hirudin increases the risk of hemorrhagic complications, however, and this narrow therapeutic window has limited its clinical use. The small size and compact structure of hirudin leads to rapid clearance from the circulation, and these pharmacokinetic properties present further clinical challenges [70].

Initial attempts to create albumin-hirudin fusion proteins indicated that the biological activity of the fusion protein depended on the way the fusion protein was constructed [40]. Fusion of albumin to the C-terminus of hirudin with a linker separating the two moieties (termed HLA for hirudin linked to albumin) produced a fusion protein that inhibited thrombin, whereas the reverse order, in which albumin was attached to the N-terminus of hirudin (termed ALH for albumin linked to hirudin), did not inhibit thrombin, indicating that a free N-terminus is necessary for hirudin activity. In rabbits, HLA prolonged the half-life by more than twofold compared to hirudin and retained potent anticoagulant and antithrombotic effects [40]. As with its predecessor, however, the HLA fusion protein also increased the risk of bleeding, and the use of lower doses in combination with antiplatelet therapy only moderately reduced this effect [71].

The stable but inactive ALH fusion protein was then reconfigured to include a linker sequence that contained a plasmin cleavage site [69]. With this configuration, hirudin is liberated from albumin in the presence of plasmin, whose activity is tightly restricted to the site of thrombus and/or hemostatic plug formation. When tested *in vivo*, this fusion protein was found to have an elimination profile similar to that of albumin within the 96 h observation period and no hirudin activity until it was liberated by plasmin, after which it inhibited thrombin in solution and bound to fibrin clots. Notably, the fusion protein did not promote bleeding in mice, indicating that this approach may overcome one of the major disadvantages of recombinant hirudin. Compared to other measures to inactivate hirudin by altering its N-terminus, the use of albumin appears to shield hirudin more effectively, thereby reducing bleeding risk [71].

10.3.2.9 Factor VII Hemophilia A is a hereditary bleeding disorder characterized by a lack of coagulation factor VIII (FVIII). Replacement therapy with FVIII products helps to reduce bleeding complications, improve quality of life, and allow hemophilia patients to undergo major surgery [10].

Nevertheless, 10–30% of patients develop inhibitory antibodies against exogenous FVIII [72]. For these patients, the use of recombinant coagulation factors with the ability to bypass FVIII in the coagulation cascade, such as activated recombinant factor VII (rFVIIa, NovoSeven®), can effectively restore homeostasis [73]. The half-life of rFVIIa is ~2.5 h, and frequent injections are therefore necessary and long-term prophylaxis is hardly feasible [74].

A novel rFVIIa fusion protein (rVIIa-FP) was developed using a flexible glycine/serine linker fused between the C-terminus of rFVIIa and recombinant albumin [11]. The length of the linker (31 amino acids) was chosen to minimize potential interactions between the two proteins and to optimize the activity of FVIIa [10,11,76]. The construct was produced in mammalian cell lines (HEK-293 and CHO). A eukaryotic expression system was required to achieve the functionally important posttranslational modifications of the fusion protein, including γ-carboxylation, N- and O-glycosylation, and β-hydroxylation [10].

The pharmacokinetics of rVIIa-FP was assessed in rats [11,39], mice [39], and rabbits [39,75]. In all animal species, the fusion protein demonstrated significant improvements in pharmacokinetic parameters (Table 10.2).

In an *in vitro* evaluation of rVIIa-FP in human FVIII inhibitor whole blood, rVIIa-FP produced a dose-dependent reduction in clot formation time that was similar to that of the wild-type protein [11]. The activity of rVIIa-FP was also evaluated *in vivo* using rats pretreated with phenprocoumon, an anticoagulant agent that blocks the synthesis of FVII over a period of 16 h. When administered 15.75 h after phenprocoumon administration, rVIIa-FP corrected whole blood clotting time measured by thrombelastography to a similar degree as did rFVIIa, confirming the hemostatic activity of the fusion protein. In addition, when rVIIa-FP or rFVIIa was administered immediately after phenprocoumon, the fusion protein effectively corrected clotting time 16 h later, but rFVIIa did not, because of its short half-life. Thus, the fusion protein was shown to be efficacious in an *in vivo* situation and its effects endured much longer than did those of rFVIIa [11]. The results suggest that a single injection of rVIIa-FP may be sufficient per bleeding episode.

TABLE 10.2 Pharmacokinetic Parameters of Activated Recombinant Factor VII Albumin Fusion Protein (rVIIa-FP) and Recombinant Factor IX Albumin Fusion Protein (rIX-FP) Compared with NovoSeven® and BeneFIX®, Respectively, in Different Animal Species [11,12,39]

Ratio Fusion Protein/ Wild-Type Protein	rVIIa-FP			rIX-FP	
Species	Mouse	Rat	Rabbit	Rat	Rabbit
AUC	11.4	14.5	14.2	4.6	7.2
Terminal half-life	4.1	5.8	8.1	4.7	4.0
In vivo recovery	2.0	2.4	2.0	1.7	1.6

Data are expressed as ratios of rVIIa-FP to NovoSeven and rIX-FP to BeneFIX. AUC, area under the curve.

10.3.2.10 Factor IX Hemophilia B is a congenital coagulation disorder characterized by a lack or deficiency of coagulation factor IX (FIX). The management of hemophilia B includes treatment with plasma-derived FIX (e.g., Mononine®) or recombinant FIX (rFIX; BeneFIX). For rFIX, the half-life is 18–34 h [77,78] and therefore, when applied in a prophylactic regimen, requires intravenous administration every 2–3 days to prevent spontaneous bleeding [79]. The need for a long-acting preparation of rFIX that would allow for less frequent dosing prompted the development of recombinant FIX albumin fusion proteins [12,80].

Novel recombinant FIX albumin fusion proteins using linker sequences that contain the same cleavage site as that responsible for proteolytical activation of FIX were developed [12]. Activation of FIX by either FVIIa/tissue factor (TF) or FXIa would therefore also cleave the linker, separating the FIX and rHA moieties of the fusion protein. Thus, FIX would be cleaved from albumin when activated, allowing for enhanced biological activity compared to previous FIX fusion proteins [12]. Because of the complex post-translational modifications of the FIX moiety, the fusion proteins were expressed in mammalian cell lines [12].

Recombinant FIX albumin fusion proteins with cleavable linkers had 10- to 30-fold higher molar specific activity in a one-stage clotting assay compared to those with noncleavable linkers [12]. Fusion proteins with cleavable linkers also showed greater effects in a thrombin generation assay than did those with noncleavable linkers. Despite the improvement in biological activity compared to fusion proteins with noncleavable linkers, the molar specific activity of fusion proteins with cleavable linkers still remained lower than that of rFIX or pdFIX *in vitro*. This suggests that albumin can interfere with the activation process of FIX to some extent, possibly because of steric hindrance [12,81].

Similar to rVIIa-FP, rIX-FP demonstrated improved pharmacokinetics in rats, rabbits, and $FIX^{-/-}$ mice [12,82]. Specifically, the half-life, recovery, and AUC of rIX-FP were superior to those of rFIX (BeneFIX) (Table 10.2) [12].

In a tail-tip bleeding model using FIX-deficient mice, rIX-FP with cleavable linker reduced bleeding time in a dose-dependent manner that was comparable to that of rFIX [12]. The dose calculations were based on a FIX one-stage clotting assay. At a dose of 100 IU per kg body weight, the bleeding time was reduced almost to that of normal mice, which had a mean bleeding time of around 2 min. Moreover, the cleavable rIX-FP minimizes the risk of generating excess FIXa fusion proteins with extended half-life, thereby reducing the potential risk of ensuing prothrombotic effects [12].

10.3.2.11 Infestin-4 Infestin is a Kazal-type serine protease inhibitor found in the midgut of the hematophagus insect *Triatoma infestans*, the main Chagas' disease vector in Brazil [83]. One inhibitory domain of infestin, infestin-4, has been cloned, expressed and purified, showing remarkable inhibitory activity toward the activated human coagulation factor XII [84]. However, infestin-4 has a very short half-life, reducing its efficacy and bioavailability. As an approach to prolong its half-life, infestin-4 has recently been fused to rHA and analyzed *in vitro* [42,85]. The resulting protein, rHA-infestin-4, was created by inserting a flexible $(GGS)_3$ linker between rHA and infestin-4. The fusion protein was expressed in HEK-293 cells. rHA-infestin-4 demonstrated a prolonged half-life compared with that of $(His)_6$-tagged infestin-4 in rats (4.6 h vs. 0.3 h). In addition, the fusion protein inhibited FXIIa and prolonged the activated partial thromboplastin time in human, mouse, and rat plasma. In mouse and rat models of pathological arterial thrombosis, rHA-infestin-4 prevented the formation of arterial thrombosis while leaving hemostasis (tail-bleeding times) fully intact. In addition, rHA-infestin-4 moderately inhibited fibrinolysis, and was highly protective in a mouse model of ischemic stroke [42].

10.3.2.12 Thioredoxin Thioredoxin (Trx) is a small redox-active protein with anti-inflammatory effects [45]. Several reports have indicated protective effects of this protein against various pathological conditions. However, its short half-life of around 1 h necessitates multiple administrations or a constant infusion rate to maintain the required dose. In an attempt to extend its half-life, Ikuta et al. [45] created a thioredoxin recombinant human albumin fusion protein (Trx-rHA) using the *P. pastoris* expression system and examined its characteristics. The half-life of Trx-rHA was prolonged to 10.3 h, a 10-fold increase over the unfused thioredoxin. However, in an insulin-reducing assay, the activity of the fusion protein was only 60% of that of thioredoxin, despite a 10 amino acid linker. Trx-rHA was also highly protective in a mouse model of endotoxic shock; survival rate was twice as high with Trx-rHA compared with thioredoxin.

10.3.2.13 Thymosin-α1 Thymosin-α1 (Tα1) is a naturally occurring thymic peptide with immunomodulatory functions [18,46]. It is indicated for the treatment of certain viral infections, immunodeficiencies, and malignancies. Unfortunately, the very short half-life of Tα1 of ~2 h means that frequent injections are necessary to maintain the therapeutic effect. Recombinant albumin fusion technology has recently been employed to prolong the half-life and efficacy of the peptide [18]. Two fusion proteins, expressed in *P. pastoris*, were created (rHA-Tα1 and rHA-L-Tα1): both were composed of rHA fused at its C-terminus to the N-terminus of recombinant Tα1, but one included a linker peptide (GGGGS) joining rHA and Tα1 (rHA-L-Tα1). While the bioactivity of the two fusion proteins was similar to that of Tα1, rHA-Tα1, and rHA-L-Tα1 demonstrated longer half-lives of 32–36 h following subcutaneous or intravenous administration.

10.3.2.14 Antibody Fragments An increasing number of small recombinant antibodies and their fragments are being developed and investigated for clinical applications, such as for the treatment of cancer or inflammatory diseases [4]. However, as with many proteins with low molecular weight, the half-life of antibody fragments is very short, hampering their therapeutic efficacy. The use of recombinant albumin fusion technology is a strategy, that is proving to be successful in prolonging the half-life of several antibody fragments [4,19,32]. For example, fusion of albumin to the single-chain variable fragment (scFv), tandem scFv molecules (taFv), and single-chain diabodies (scDb) prolonged the half-lives of the fragments by 2–10-fold compared to the unfused antibody fragments [19,32].

10.4 SUCCESSES AND FAILURES IN PRECLINICAL AND CLINICAL RESEARCH

The development of recombinant albumin fusion proteins has produced successes as well as failures that have contributed to our understanding of fusion proteins in general. The recombinant albumin fusion proteins currently in development were often selected from multiple variants because they had the most favorable pharmacokinetics and biological activity. Other areas of potential improvement, such as immunogenicity and stability, have led to further modifications as the optimization of fusion protein technology remains an ongoing process.

10.4.1 Interferon-α

Alb-IFN is in late-stage development for the treatment of CHC, but it tends to dimerize and form covalent aggregations in aqueous solution [86]. The observed heterogeneity and instability is thought to be caused by incomplete disulfide bridges between Cys1 and Cys98 of the IFN moiety. These characteristics have led to a low recovery rate (10%) and necessitated lyophilization. To restore disulfide pairing and improve the stability of alb-IFN, researchers created a unique protein by switching the orientation of the two proteins, placing IFN-α-2b at the N-terminus [86]. Compared to alb-IFN, the new protein (IFN-α-2b-rHA) was relatively more homogenous (i.e., it migrated as a single band in SDS-PAGE) and was easier to purify and more stable (i.e., recovery was increased 2.5-fold). The antiviral activity of IFN-α-2b-rHA *in vitro* was increased compared to IFN-α-2b and alb-IFN.

Unfortunately, the new fusion protein was also susceptible to forming disulfide-linked aggregates in response to mechanical and thermal stress and the formation of these aggregates was associated with increased immunogenicity in mice [87]. In the case of the albumin-hGH fusion protein, the use of nonionic surfactants such as polysorbate 20 and polysorbate 80 was found to minimize aggregation during fermentation, purification, freeze-drying, shipping, and storage [88], and so this approach was applied to IFN-α-2b-rHA. The addition of polysorbate 80 improved the stability of IFN-α-2b-rHA in response to agitation, but not to thermal stress [87]. In fact, the surfactant increased aggregate formation in aqueous solution during inert storage. In an alternative approach to improve stability, the Cys34 residue of albumin was replaced with serine. The stability of the resulting fusion protein (IFN-α-2b-rHA(C34S)) was better than that of IFN-α-2b-rHA. After either agitation for 72 h or incubation at 60°C for 2 h, more than 90% of the protein content remained monomeric. In addition, immunogenicity was reduced after being subjected to these conditions. Pharmacokinetic analysis in rats indicated similar pharmacokinetic properties to that of IFN-α-2b-rHA; both fusion proteins had a half-life of ~50 h.

Another approach to improve the homogeneity, stability, and activity of IFN-α-2b-rHA was to modify the linker [86,87,89]. Variations in linker size and conformation can alter the interaction between the two fused proteins and between the therapeutic protein and its target. Various linker sizes (1, 2, 5, or 10 amino acids) and conformations (flexible glycine/serine linker (GGGGS) or rigid linker (PAPAP) or helix-forming linker (AEAAAKEAAAKA)) were inserted between rHA and IFN-α-2b, and the properties of the resulting fusion proteins were characterized [88]. It was found that a linker length of five residues allowed the fusion proteins to migrate as a single band on SDS-PAGE, facilitating correct disulfide bond formation. Aggregate formation after incubation at 37°C for 10 days was also reduced when these linkers were used. The different linkers varied in terms of susceptibility to hydrolysis; the rigid linker was the least susceptible at pH 6 or 7. Activity assays showed that structured linkers separated rHA and IFN-α-2b more effectively compared to a flexible linker, as they increased the antiviral activity more significantly. A rigid, flexible, or helix-forming linker increased antiviral activity *in vitro* by 39, 68, and 115%, respectively [89].

10.4.2 FVII and FIX

Early attempts to prolong the half-life of FIX using recombinant albumin fusion technology did not or did only moderately improve the pharmacokinetics of FIX [80]. Similar results were expected for rFVIIa, as the two proteins are structurally similar members of the same family of vitamin-K-dependent coagulation factors [11]. A novel rFVIIa fusion protein (rVIIa-FP) was developed using a flexible glycine/serine linker inserted between albumin and the C-terminus of rFVIIa. The length of the linker (31 amino acids) was chosen to minimize potential interactions between the two proteins while retaining a high specific activity of rFVIIa [11,76].

While the use of a flexible glycine/serine linker was successfully used to produce rVIIa-FP, the same approach could not be applied to FIX, as it produced a fusion protein with low FIX-related activity [12,80,90]. A possible explanation for this is that the albumin portion of the protein interferes with the interactions between FIX and its cofactor and substrate, FVIIIa and FX [90]. This led to the development of a FIX-recombinant albumin fusion protein using a linker sequence that was derived from the natural cleavage site of FIX responsible for proteolytic activation [12]. The linker sequence is based on amino acids 137–153 derived from the N-terminus of the activation peptide of FIX. Activation of FIX-recombinant albumin fusion protein by either FXIa or FVIIa/TF cleaves the linker in parallel, separating the FIXa and rHA moieties of the fusion protein. The liberation of FIXa from recombinant albumin at the time of activation allowed for largely enhanced biological activity [12].

10.5 CHALLENGES

One of the inherent challenges of recombinant albumin fusion technology is maintaining specific activity of the therapeutic protein, which in most cases has been affected by the presence of the albumin moiety. While the use of linkers and the flexibility in orientation of the fused genes can help optimize biological activity, the presence of recombinant albumin can interfere with interactions between the therapeutic protein and its target(s). The development of rVIIa-FP and rIX-FP, as discussed earlier, illustrate some of the typical challenges and solutions encountered in recombinant albumin fusion technology, and highlight the fact that, while recombinant albumin fusion technology is relatively simple and easily applied to various settings, each novel protein poses unique conditions and challenges that must be considered. To retain as much specific activity as possible, it may therefore be necessary to develop a specific concept for each fusion protein.

As with all therapeutic proteins, albumin fusion proteins carry a risk of toxicity and immunogenicity [2,13]. However, in a Phase I/II study of alb-IFN in patients with CHC (mostly genotype 1) for whom prior IFN-α therapy had failed, the risk of immunogenicity appeared to be low [50]. In most patients, titers of anti-albumin-IFN-α antibodies remained low throughout the study. Just over 10% of patients had anti-albumin-IFN-α antibodies at baseline, which is not surprising since prior IFN-α-based therapy had not been successful. The immunogenicity of alb-IFN did not correlate with patient demographics, disease status, prior treatment history, or the magnitude or duration of exposure to the fusion protein. Moreover, in the Phase III clinical trial program, the 1200 μg dose of alb-IFN was associated with a higher rate of pulmonary complications and, in some cases, interstitial pulmonary fibrosis [27]. As a result, this treatment arm was eliminated from Phase III trials and all patients receiving 1200 μg alb-IFN were switched to 900 μg every 2 weeks [27].

With regard to the barbourin-albumin fusion protein, *in vivo* studies indicated that the protein was immunogenic in rabbits, but the antibodies generated had no effect on platelet aggregation [91]. Fewer non-neutralizing antibodies were generated when smaller barbourin-derived peptides with antiplatelet activity were fused to albumin [91]. Similarly, the immunogenic risk with the albumin-hirudin fusion protein is not expected to be a major concern, based on previous experiments with recombinant hirudin [92,93] and the experience with albumin-barbourin fusion proteins [91].

Despite the low inherent immunogenicity of rHA, the risk of immunogenicity must be taken into account when developing recombinant albumin fusion proteins. Steps can be taken to minimize the risk of immunogenicity during the design process, such as avoiding peptides that bind to T cell receptors or major histocompatibility complex class II molecules [94]. Adhering to certain production methods can also limit immunogenicity; this includes selecting an appropriate expression system, ensuring sufficient removal of potentially immunogenic aggregates during downstream processing, and minimizing denaturation during formulation [95].

10.6 FUTURE PERSPECTIVES

Among the many fusion protein technologies currently being pursued, recombinant albumin fusion technology represents a promising area of research. Numerous recombinant albumin fusion proteins are in various stages of preclinical and clinical development, demonstrating that this approach is feasible and can extend the half-life of therapeutic proteins while retaining their biological activity. The wide spectrum of current research on recombinant albumin fusion proteins indicates that these fusion proteins have the potential to positively impact treatment in a broad range of diseases. Future applications for recombinant fusion proteins are therefore likely to expand well beyond the established areas of oncology, immunology, hemostasis, and inflammation.

10.7 CONCLUSION

Recombinant albumin fusion technology represents a simple yet flexible platform for the production of proteins with an extended half-life. Albumin is an ideal fusion partner because it is a well characterized, naturally occurring carrier protein with an extraordinarily long half-life. The fact that albumin is distributed widely and has no enzymatic activity suggests that recombinant albumin fusion proteins are unlikely to elicit unwanted immunogenicity or toxicity through the albumin moiety. Recombinant albumin fusion technology allows for high-quality production of a homogenous product while requiring fewer postexpression modifications and purification steps than other technologies such as PEGylation. Compared to other fusion protein approaches, such as fusion proteins using the Fc fragment of IgG, recombinant albumin fusion proteins are more efficient to produce. The use of a well-established yeast expression system facilitates production using robust, reproducible, and scalable processing methods; the same methods can be applied to mammalian expression systems as needed. Albumin fusion has been used to successfully extend the half-life of various therapeutic proteins, including small peptides (GLP-1), cytokines (IFN-α), and complex proteins (FVII and FIX). The improved pharmacokinetic properties have generally translated into improved pharmacodynamic effects. Clinical studies evaluating recombinant albumin fusion proteins are underway that will help further define the potential of recombinant albumin fusion technology to enhance the clinical use of therapeutic proteins.

ACKNOWLEDGMENT

The authors gratefully acknowledge the editorial assistance of Dr. Sandra Cox (Swiss Medical Press GmbH).

REFERENCES

1. Werle M, Bernkop-Schnürch A. (2006) Strategies to improve plasma half life time of peptide and protein drugs. *Amino Acids* 30, 351–367.
2. Subramanian GM, Fiscella M, Lamousé-Smith A, Zeuzem S, McHutchison JG. (2007) Albinterferon α-2b: a genetic fusion protein for the treatment of chronic hepatitis C. *Nat. Biotechnol.* 25, 1411–1419.
3. Andersen JT, Sandlie, I. (2009) The versatile MHC class I-related FcRn protects IgG and albumin from degradation: implications for development of new diagnostics and therapeutics. *Drug Metab. Pharmacokinet.* 24, 318–332.
4. Kontermann RE. (2009) Strategies to extend plasma half-lives of recombinant antibodies. *BioDrugs* 23, 93–109.
5. Meloun B, Morávek L, Kostka V. (1975) Complete amino acid sequence of human serum albumin. *FEBS Lett.* 58, 134–137.
6. Chuang VT, Kragh-Hansen U, Otagiri M. (2002) Pharmaceutical strategies utilizing recombinant human serum albumin. *Pharmaceut. Res.* 19, 569–577.
7. Anderson CL, Chaudhury C, Kim J, Bronson CL, Wani MA, Mohanty S. (2006) Perspective: FcRn transports albumin:

relevance to immunology and medicine. *Trends Immunol.* 27, 343–348.

8. Chaudhury C, Mehnaz S, Robinson JM, Hayton WL, Pearl DK, Roopenian DC, et al. (2003) The major histocompatibility complex-related Fc receptor for IgG (FcRn) binds albumin and prolongs its lifespan. *J. Exp. Med.* 197, 315–322.
9. Chemmanur AT, Wu GY. (2006) Drug evaluation: albuferon-alpha—an antiviral interferon-alpha/albumin fusion protein. *Curr. Opin. Invest. Drugs* 7, 750–758.
10. Schulte, S. (2008) Use of albumin fusion technology to prolong the half-life of recombinant factor VIIa. *Thromb. Res.* 122 (Suppl. 4), S14–S19.
11. Weimer T, Wormsbächer W, Kronthaler U, Lang W, Liebing U, Schulte S. (2008) Prolonged *in-vivo* half-life of factor VIIa by fusion to albumin. *Thromb. Haemost.* 99, 659–667.
12. Metzner HJ, Weimer T, Kronthaler U, Lang W, Schulte S. (2009) Genetic fusion to albumin improves the pharmacokinetic properties of factor IX. *Thromb. Haemost.* 102, 634–644.
13. Schmidt SR. (2009) Fusion-proteins as biopharmaceuticals—applications and challenges. *Curr. Opin. Drug Discov. Devel.* 12, 284–295.
14. Gao Z, Bai G, Chen J, Zhang Q, Pan, P, Bai F, et al. (2009) Development, characterization, and evaluation of a fusion protein of a novel glucagon-like peptide-1 (GLP-1) analog and human serum albumin in *Pichia pastoris. Biosci. Biotechnol. Biochem.* 73, 688–694.
15. Huang YS, Chen Z, Chen YQ, Ma GC, Shan JF, Liu W, et al. (2008) Preparation and characterization of a novel exendin-4 human serum albumin fusion protein expressed in *Pichia pastoris. J. Pept. Sci.* 14, 588–595.
16. Daly R, Hearn MT. (2005) Expression of heterologous proteins in *Pichia pastoris*: a useful experimental tool in protein engineering and production. *J. Mol. Recognit.* 18, 119–138.
17. Chen HH, Martin FL, Cataliotti A, Schirger JA, Burnett JC. (2007) AlbuBNP (Cardeva), a novel recombinant human B-type natriuretic peptide serum albumin fusion protein has prolonged renal enhancing properties when compared to human BNP. *Circulation* 116, II_284 [Abstract 1382].
18. Chen J-H, Zhang X-G, Jiang Y-T, Yan L-Y, Tang LO, Yin Y-W, et al. (2010) Bioactivity and pharmacokinetics of two human serum albumin-thymosin α1-fusion proteins, rHSA-Tα1 and rHSA-L-Tα1, expressed in recombinant *Pichia pastoris. Cancer Immunol. Immunother.* 59, 1335–1345.
19. Müller D, Karle A, Meissburger B, Höfig I, Stork R, Kontermann RE. (2007) Improved pharmacokinetics of recombinant bispecific antibody molecules by fusion to human serum albumin. *J. Biol. Chem.* 282, 12650–12660.
20. Dou W-F, Lei J-Y, Zhang L-F, Xu Z-H, Chen Y, Jin J. (2008) Expression, purification, and characterization of recombinant human serum albumin fusion protein with two human glucagon-like peptide-1 mutants in *Pichia pastoria. Protein Expr. Purif.* 61, 45–49.
21. Halpern W, Riccobene TA, Agostini H, Baker K, Stolow D, Gu ML, et al. (2002) Albugranin, a recombinant human granulocyte colony stimulating factor (G-CSF) genetically fused to recombinant human albumin induces prolonged myelopoietic effects in mice and monkeys. *Pharmaceut. Res.* 19, 1720–1729.
22. Tomkin GH. (2009) Albiglutide, an albumin-based fusion of glucagon-like peptide 1 for the potential treatment of type 2 diabetes. *Curr. Opin. Mol. Therap.* 11, 579–588.
23. Baggio LL, Huang Q, Brown TJ, Drucker DJ. (2004) A recombinant human glucagon-like peptide (GLP)-1-albumin protein (albugon) mimics peptidergic activation of GLP-1 receptor-dependent pathways coupled with satiety, gastrointestinal motility, and glucose homeostasis. *Diabetes* 53, 2492–2500.
24. Matthews JE, Stewart MW, De Boever EH, Dobbins RL, Hodge RJ, Walker SE, et al. (2008) Pharmacodynamics, pharmacokinetics, safety, and tolerability of albiglutide, a long-acting glucagon-like peptide-1 mimetic, in patients with type 2 diabetes. *J. Clin. Endocrinol. Metab.* 93, 4810–4817.
25. Degn KB, Juhl CB, Sturis J, Jakobsen G, Brock B, Chandamouli V, et al. (2004) One week's treatment with the long-acting glucagon-like peptide 1 derivative liraglutide (NN2211) markedly improves 24-h glycemia and α- and β-cell function and reduces endogenous glucose release in patients with type 2 diabetes. *Diabetes* 53, 1187–1194.
26. Bain VG, Kaita KD, Yoshida EM, Swain MG, Heathcote EJ, Neumann AU, et al. (2006) A phase 2 study to evaluate the antiviral activity, safety, and pharmacokinetics of recombinant human albumin-interferon alfa fusion protein in genotype 1 chronic hepatitis C patients. *J. Hepatol.* 44, 671–678.
27. Zeuzem S, Sulkowski MS, Lawitz EJ, Rustgi VK, Rodriguez-Torres M, Bacon BR, et al. (2010) Albinterfeon alfa-2b was not inferior to pegylated interferon-α in a randomized trial of patients with chronic hepatitis C virus genotype 1. *Gastroenterology* 139, 1257–1266.
28. Wang W, Ou Y, Shi, Y. (2004) AlbuBNP, a recombinant B-type natriuretic peptide and human serum albumin fusion hormone, as a long-term therapy of congestive heart failure. *Pharmaceut. Res.* 21, 2105–2111.
29. Osborn BL, Sekut L, Corcoran M, Poortman C, Sturm B, Chen G, et al. (2002) Albutropin®: a growth hormone-albumin fusion with improved pharmacokinetics and pharmacodynamics in rats and monkeys. *Eur. J. Pharmacol.* 456, 149–158.
30. Yao Z, Dai W, Perry J, Brechbiel MW, Sung C. (2004) Effect of albumin fusion on the biodistribution of interleukin-2. *Cancer Immunol. Immunother.* 53, 404–410.
31. Melder RJ, Osborn BL, Riccobene T, Kanakaraj P, Wei P, Chen G, et al. (2005) Pharmacokinetics and *in vitro* and in vivo anti-tumor response of an interleukin-2-human serum albumin fusion protein in mice. *Cancer Immunol. Immunother.* 54, 535–547.
32. Evans L, Hughes M, Waters J, Cameron J, Dodsworth N, Tooth D, et al. (2010) The production, characterization and enhanced pharmacokinetics of scFv-albumin fusions expressed in *Saccharomyces cerevisiae. Protein Expr. Purif.* 73, 113–124.
33. Marques JA, George JK, Smith IJ, Bhakta V, Sheffield WP. (2001) A barbourin-albumin fusion protein that is slowly

cleared *in vivo* retains the ability to inhibit platelet aggregation in vitro. *Thromb. Haemost.* 86, 902–908.

34. Huang YJ, Lundy PM, Lazaris A, Huang Y, Baldassarre H, Wang B, et al. (2008) Substantially improved pharmacokinetics of recombinant human butyrylcholinesterase by fusion to human serum albumin. *BMC Biotechnol.* 8, 50.
35. Brimijoin S, Gao Y, Anker JJ, Gliddon LA, Lafleur D, Shah R, et al. (2008) A cocaine hydrolase engineered from human butyrylcholinesterase selectively blocks cocaine toxicity and reinstatement of drug seeking in rats. *Neuropsychopharmacology* 33, 2715–2725.
36. Gao Y, LaFleur D, Shah R, Zhao Q, Singh M, Brimijoin S. (2008) An albumin-butyrylcholinesterase for cocaine toxicity and addiction: catalytic and pharmacokinetic properties. *Chem. Biol. Interact.* 175, 83–87.
37. Cooper ZD, Narasimhan D, Sunahara RK, Mierzejewski P, Jutkiewicz EM, Larsen NA, et al. (2006) Rapid and robust protection against cocaine-induced lethality in rats by the bacterial cocaine esterase. *Mol. Pharmacol.* 70, 1885–1891.
38. Joung CH, Shin JY, Koo JK, Lim JJ, Wang JS, Lee SJ, et al. (2009) Production and characterization of long-acting recombinant human albumin-EPO fusion protein expressed in CHO cell. *Protein Expr. Purif.* 68, 137–145.
39. Zollner S, Weimer T, Schmidbauer S, Raquet E, Mueller-Cohrs J, Schulte S. (2010) Pharmacokinetics of a recombinant albumin-fused, human coagulation factor VIIa (rVIIa-FP) exhibiting prolonged serum half-life in different animal species. *Haemophilia* 16 (Suppl. 4), 33 [Abstract 07P24].
40. Syed S, Schuyler PD, Kulczycky M, Sheffield WP. (1997) Potent antithrombin activity and delayed clearance from the circulation characterize recombinant hirudin genetically fused to albumin. *Blood* 89, 3243–3252.
41. Hatton MW, Ross-Ouellet B. (1994) Radiolabeled r-hirudin as a measure of thrombin activity at, or within, the rabbit aorta wall *in vitro* and *in vivo*. *Thromb. Haemost.* 71, 499–506.
42. Hagedorn I, Schmidbauer S, Pleines I, Kleinschnitz C, Kronthaler U, Stoll G, et al. (2010) Factor XIIa inhibitor recombinant human albumin Infestin-4 abolishes occlusive arterial thrombus formation without affecting bleeding. *Circulation* 121, 1510–1517.
43. Duttaroy A, Kanakaraj P, Osborn BL, Schneider H, Pickeral OK, Chen C, et al. (2005) Development of a long-acting insulin analog using albumin fusion technology. *Diabetes* 54, 251–258.
44. Cresto JC, Lavine RL, Buchly ML, Pnehos JC, Bhathena SJ, Recant L. (1977) Half-life of injected ^{125}I-insulin in control and ob/ob mice. *Acta Physiol. Latino Ame.* 27, 7–15.
45. Ikuta S, Chuang VTG, Ishima Y, Nakajou K, Furukawa M, Watanabe H, et al. (2010) Albumin fusion of thioredoxin – the production and evaluation of its biological activity for potential therapeutic applications. *J. Control. Rel.* 147, 17–23.
46. Ancell CD, Phipps J, Young L. (2001) Thymosin alpha-1. *Am. J. Health-Syst. Pharm.* 58, 879–885.
47. Müller D, Schneider B, Pfizenmaier K, Wajant H. (2010) Superior serum half life of albumin tagged TNF ligands. *Biochem. Biophys. Res. Commun.* 396, 793–799.
48. Fried MW, Shiffman ML, Reddy KR, Smith C, Marinos G, Gonçales F.L. Jr., et al. (2002) Peginterferon alfa-2a plus ribavirin for chronic hepatitis C virus infection. *N. Engl. J. Med.* 347, 975–982.
49. Manns MP, McHutchison JG, Gordon SC, Rustgi VK, Shiffman M, Reindollar R, et al. (2001) Peginterferon alfa-2b plus ribavirin compared with interferon alfa-2b plus ribavirin for initial treatment of chronic hepatitis C: a randomised trial. *Lancet* 358, 958–965.
50. Balan V, Nelson DR, Sulkowski MS, Everson GT, Lambiase LR, Wiesner RH, et al. (2006) A Phase I/II study evaluating escalating doses of recombinant human albumin-interferon-alpha fusion protein in chronic hepatitis C patients who have failed previous interferon-alpha-based therapy. *Antivir. Ther.* 11, 35–45.
51. Osborn BL, Olsen HS, Nardelli B, Murray JH, Zhou JX, Garcia A, et al. (2002) Pharmacokinetic and pharmacodynamic studies of a human serum albumin-interferon-α fusion protein in cynomolgus monkeys. *J. Pharmacol.Exp. Ther.* 303, 540–548.
52. Yeh P, Landais D, Lemaître M, Maury I, Crenne JY, Becquart J, et al. (1992) Design of yeast-secreted albumin derivatives for human therapy: biological and antiviral properties of a serum albumin-CD4 genetic conjugate. *Proc. Natl. Acad. Sci. USA* 89, 1904–1908.
53. Perry CM, Jarvis B. (2001) Peginterferon-alpha-2a (40 kDa): a review of its use in the management of chronic hepatitis C. *Drugs* 61, 2263–2288.
54. Glue P, Fang JW, Rouzier-Panis R, Raffanel C, Sabo R, Gupta SK, et al. (2000) Pegylated interferon-alpha2b: pharmacokinetics, pharmacodynamics, safety, and preliminary efficacy data. *Clin. Pharmacol. Ther.* 68, 556–567.
55. Nelson D, Benhamou Y, Chuang WL, Lawitz E, Flisiak R, et al. (2009) Efficacy and safety results of albinterferon alfa-2b in combination with ribavirin in interferon-alfa treatment naive patients with genotype 2 or 3 chronic hepatitis C. *J. Hepatol.* 50 (Suppl. 1), S378 [Abstract 1042].
56. Nelson DR, Rustgi V, Balan V, Sulkowski MS, Davis GL, Muir AJ, et al. (2009) Safety and antiviral activity of albinterferon alfa-2b in prior interferon nonresponders with chronic hepatitis C. *Clin. Gastroenterol. Hepatol.* 7, 212–218.
57. European Medicines Agency. Withdrawn Applications: Joulferon. Available at http://www.ema.europa.eu/ema/index.jsp?curl=pages/medicines/human/medicines/002166/wapp/Initial_authorisation/human_wapp_000096.jsp&murl=menus/medicines/medicines.jsp&mid=WC0b01ac058001d128&source=homeMedSearch&category=human (last accessed September 23, 2010).
58. Rosenstock J, Reusch J, Bush M, Yang F, Stewart M. (2009) Albiglutide study group. Potential of albiglutide, a long-acting GLP-1 receptor agonist, in type 2 diabetes. *Diabetes Care* 32, 1880–1886.
59. Eng J, Kleinman WA, Singh L, Singh G, Raufman JP. (1992) Isolation and characterization of exendin-4, an exendin-3 analogue, from Heloderma suspectum venom. Further evidence for an exendin receptor on dispersed acini from guinea pig pancreas. *J. Biol. Chem.* 267, 7402–7405.

60. Kim D, MacConell L, Zhuang D, Kothare PA, Trautmann M, Fineman M, et al. (2007) Effects of once-weekly dosing of a long-acting release formulation of exenatide on glucose control and body weight in subjects with type 2 diabetes. *Diabetes Care* 30, 1487–1493.

61. Jose B, Tahrani AA, Piya MK, Barnet AH. (2010) Exenatide once weekly: clinical outcomes and patient satisfaction. *Patient Prefer Adherence* 4, 313–324.

62. Kim J-G, Baggio LL, Bridon DP, Castaigne J-P, Robitaille MF, Jetté L, et al. (2003) Development and characterization of a glucagon-like peptide 1-albumin conjugate: the ability to activate the glucagon-like peptide 1 receptor in vivo. *Diabetes* 52, 751–759.

63. Byetta prescribing information. Amylin Pharmaceuticals, Inc., September 2010. Available at http://pi.lilly.com/us/byetta-pi.pdf (last accessed October 1, 2010).

64. Elbrønd B, Jakobsen G, Larsen S, Agersø H, Jensen LB, Rolan P, et al. (2002) Pharmacokinetics, pharmacodynamics, safety, and tolerability of a single-dose of NN2211, a long-acting glucagon-like peptide 1 derivative, in healthy male subjects. *Diabetes Care* 25, 1398–1404.

65. Lee SH, Lee S, Youn YS, Na DH, Chae SY, Byun Y, et al. (2005) Synthesis, characterization, and pharmacokinetic studies of PEGylated glucagon-like peptide-1. *Bioconjug. Chem.* 16, 377–382.

66. Tanaka H, Satake-Ishikawa R, Ishikawa M, Matsuki S, Asano K. (1991) Pharmacokinetics of recombinant human granulocyte colony-stimulating factor conjugated to polyethylene glycol in rats. *Cancer Res.* 51, 3710–3714.

67. King AB, Armstrong D. (2003) A comparison of basal insulin delivery: continuous subcutaneous insulin infusion versus glargine. *Diabetes Care* 26, 1322.

68. Haffner D, Schaefer F, Girard J, Ritz E, Mehis O. (1994) Metabolic clearance of recombinant human growth hormone in health and chronic renal failure. *J. Clin. Invest.* 93, 1163–1171.

69. Sheffield WP, Eltringham-Smith LJ, Gataiance S, Bhakta V. (2009) A long-lasting, plasmin-activatable thrombin inhibitor aids clot lysis in vitro and does not promote bleeding *in vivo*. *Thromb. Haemost.* 101, 867–877.

70. Markwardt F, Nowak G, Stürzebecher J. (1991) Clinical pharmacology of recombinant hirudin. *Haemostasis* 21(Suppl. 1), 133–136.

71. Sheffield WP, Gataiance S, Eltringham-Smith LJ. (2007) Combined administration of barbourin–albumin and hirudin – albumin fusion proteins limits fibrin(ogen) deposition on the rabbit balloon-injured aorta. *Thromb. Res.* 119, 195–207.

72. World Federation of Hemophilia (WFH). Guidelines for the Management of Hemophilia. Montreal: World Federation of Hemophilia; 2005. Available at http://www.wfh.org/2/docs/Publications/Diagnosis_and_Treatment/Guidelines_Mng_Hemophilia.pdf (last accessed May 25, 2010).

73. Hedner U, Ingerslev J. (1998) Clinical use of recombinant FVIIa (rFVIIa). *Transfus. Sci.* 19, 163–176.

74. Erhardtsen E. (2000) Pharmacokinetics of recombinant activated factor VII (rFVIIa). *Semin. Thromb. Hemost.* 26, 385–391.

75. Kronthaler U, Schmidbauer S, Liebing U, Lang W, Metzner HJ, Weimer T, et al. (2009) Prolonged half-life of recombinant factor VIIa fusion protein – single dose study in rabbits. XXII International Society on Thrombosis and Haemostasis Congress, Boston, USA, July 11–16, [Abstract P-TH-561]. Available at http://www.isth2009.com/poster_presentations_thursday.html (last accessed May 25, 2010).

76. Horn C, Steuber H, Weimer T, Wormsbächer W, Liebing U, Kuo W-P. (2010) Concept and structure model of factor VIIa albumin fusion proteins. *Haemophilia* 16(Suppl. 4), 37 [Abstract 08P19].

77. White GC 2nd, Beebe A, Nielsen B. (1997) Recombinant factor IX. *Thromb. Haemost.* 78, 261–265.

78. Ewenstein BM, Joist JH, Shapiro AD, Hofstra TC, Leissinger CA, Seremetis SV, et al. (2002) Pharmacokinetic analysis of plasma-derived and recombinant F IX concentrates in previously treated patients with moderate or severe hemophilia B. *Transfusion* 42, 190–197.

79. Björkman S, Shapiro AD, Berntorp E. (2001) Pharmacokinetics of recombinant factor IX in relation to age of the patient: implications for dosing in prophylaxis. *Haemophilia* 7, 133–139.

80. Sheffield WP, Mamdani A, Hortelano G, Gataiance S, Eltringham-Smith L, Begbie ME, et al. (2004) Effects of genetic fusion of factor IX to albumin on *in vivo* clearance in mice and rabbits. *Br. J. Haematol.* 126, 565–573.

81. Horn C, Steuber H, Metzner HJ, Kuo W-P, Weimer T, Schulte S. (2010) Concept and structure model of factor IX albumin fusion proteins. *Haemophilia* 16(Suppl. 4), 36 [Abstract 08P18].

82. Metzner H, Weimer T, Raquet E, Nolte WM, Pragst I, Zollner S. (2010) Prolonged serum half-life of a recombinant, albumin-fused, human coagulation factor IX (rIX-FP) in different animal species. *Haemophilia* 16(Suppl. 4), 40 [Abstract 08P41].

83. Campos IT, Amino R, Sampaio CA, Auerswald EA, Friedrich T, Lemaire HG, et al. (2002) Infestin, a thrombin inhibitor presents in *Triatoma infestans* midgut, a Chagas' disease vector: gene cloning, expression and characterization of the inhibitor. *Insect Biochem. Mol. Biol.* 32, 991–997.

84. Campos IT, Tanaka-Azevedo AM, Tanaka AS. (2004) Identification and characterization of a novel factor XIIa inhibitor in the hematophagous insect, *Triatoma infestans* (Hemiptera: Reduvildae). *FEBS Lett.* 577, 512–516.

85. Schmidbauer S, Nerlich C, Weimer T, Kronthaler U, Metzner H, Schulte S. (2009) Prevention of thrombotic events by Factor XIIa inhibitors. 53rd Annual Meeting Society of Thrombosis and Haemostasis Research, February 4–7, Vienna, Austria, A22 [Abstract FC9-3].

86. Zhao HL, Xue C, Wang Y, Li XY, Xiong XH, Yao XQ, et al. (2007) Circumventing the heterogeneity and instability of human serum albumin-interferon-α2b fusion protein by altering its orientation. *J. Biotechnol.* 131, 245–252.

87. Zhao HL, Xue C, Wang Y, Sun B, Yao XQ, Liu ZM. (2009) Elimination of the free sulfhydryl group in the human serum albumin (HSA) moiety of human interferon-α2b and HSA fusion protein increases its stability against mechanical and thermal stresses. *Eur. J. Pharm. Biopharm.* 72, 405–411.

88. Chou DK, Krishnamurthy R, Randolph, T.W, Carpenter JF, Manning MC. (2005) Effects of Tween 20 and Tween 80 on the stability of Albutropin® during agitation. *J. Pharm. Sci.* 94, 1368–1381.

89. Zhao HL, Yao XQ, Xue C, Wang Y, Xiong XH, Liu ZM. (2008) Increasing the homogeneity, stability and activity of human serum albumin and interferon-α2b fusion protein by linker engineering. *Protein Expr. Purif.* 61, 73–77.

90. Schulte S. (2009) Half-life extension through albumin fusion technologies. *Thromb. Res.* 124(Suppl. 2), S6–S8.

91. Sheffield WP, Wilson B, Eltringham-Smith LJ, Gataiance S, Bhakta V. (2005) Recombinant albumins containing additional peptide sequences smaller than barbourin retain the ability of barbourin-albumin to inhibit platelet aggregation. *Thromb. Haemost.* 93, 914–921.

92. Eichler P, Friesen HJ, Lubenow N, Jaeger B, Greinacher A. (2000) Antihirudin antibodies in patients with heparin-induced thrombocytopenia treated with lepirudin: incidence, effects on aPTT, and clinical relevance. *Blood* 96, 2373–2378.

93. Fischer KG, Liebe V, Hudek R, Piazolo L, Haase KK, Borggrefe M, et al. (2003) Anti-hirudin antibodies alter pharmacokinetics and pharmacodynamics of recombinant hirudin. *Thromb. Haemost.* 89, 973–982.

94. Baker MP, Jones TD. (2007) Identification and removal of immunogenicity in therapeutic proteins. *Curr. Opin. Drug Discov. Dev.* 10, 219–227.

95. De Groot AS, Scott DW. (2007) Immunogenicity of protein therapeutics. *Trends Immunol.* 28, 482–490.

11

ALBUMIN-BINDING FUSION PROTEINS IN THE DEVELOPMENT OF NOVEL LONG-ACTING THERAPEUTICS

ADAM WALKER,[1] GRÁINNE DUNLEVY,[1] AND PETER TOPLEY[2]

[1]*Biopharm R&D, Domantis Ltd., Cambridge, UK*
[2]*Biopharm R&D, GlaxoSmithKline Medicines Research Centre, Stevenage, UK*

11.1 INTRODUCTION

Recombinant proteins and peptides are potentially of huge benefit in the treatment of a wide variety of human diseases. Unfortunately, the majority of such molecules are hampered in their use as therapeutic agents as they are rapidly excreted and degraded following subcutaneous or intravenous administration. In order to maintain a therapeutically active dose, protein- and peptide-based therapeutics must therefore be administered at frequent intervals in many cases, which can have a negative impact on their safety and tolerability profile, clinical efficacy, and associated cost of goods. This has led to the development of a number of technologies aimed at improving the pharmacokinetic profiles of therapeutic molecules, predominantly by reducing the clearance rate by renal filtration and preventing trafficking to degradative pathways, for example, lysosomal breakdown. Rates of excretion and clearance can be reduced by increasing the size of the therapeutic molecule above the threshold for renal clearance, whereas prevention of enzymatic degradation can be achieved by engaging recycling mechanisms to limit proteolytic cleavage following endocytosis of therapeutics which act as ligands for cell surface receptors and would ordinarily be internalized and targeted for destruction in the lysosome.

Covalent attachment of branched or linear polymers of polyethylene glycol (PEG) to recombinant proteins and peptides can result in a significant improvement in their pharmacokinetic profile owing to the increased molecular radius of these PEGylated molecules. Similarly, chemical conjugation or genetic fusion of therapeutic molecules to large carrier proteins such as human serum albumin (HSA) can also be used as a method of half-life extension due to the increased molecular size and may result in additional improvements in pharmacokinetic properties in comparison to PEGylated molecules due to FcRn-mediated recycling of HSA.

Neither of these half-life extension technologies is without significant development challenges with regard to biopharmaceutical production, and to this end, alternative half-life extension technologies such as serum albumin-binding antibodies (AlbudAbs) have been developed with a view to addressing some of the development issues faced by PEGylation and serum albumin fusion technologies. Alternative half-life extension technologies besides AlbudAbs are also

Fusion Protein Technologies for Biopharmaceuticals: Applications and Challenges, First Edition. Edited by Stefan R. Schmidt.

being developed, for example, other albumin-binding moieties, including camelid antibody domains (nanobodies), small molecules and peptides, and addition of other polymers besides PEG, for example, sialic acid (polysialylation), Pro-Ala-Ser containing sequences (PASylation), and non-repetitive sequences containing residues Ala, Glu, Gly, Pro, Ser, and Thr (EXTEN technology).

In this chapter, the relative merits and challenges associated with the use of AlbudAb technology are discussed and compared with clinically validated half-life extension platforms such as PEGylation and serum albumin fusion. The basic concepts behind other half-life extension technologies currently in development are also discussed and their relative advantages and challenges compared with AlbudAb technology highlighted.

11.2 CLINICALLY VALIDATED HALF-LIFE EXTENSION TECHNOLOGIES—AN OVERVIEW

11.2.1 PEGylation

Polyethylene glycol (PEG) is a polyether compound used in a variety of industrial applications. Owing to its high solubility and low toxicity, it can be attached covalently to therapeutic proteins using a range of chemistries. This has the effect of increasing the molecular size of the PEGylated protein, thereby delaying the rate of renal clearance significantly [1]. PEGylated proteins have been used as therapeutic agents in the treatment of a variety of human diseases over the last 30 years, for example, PEGylated interferon-α (IFN-α) (Pegasys® and Pegintron®) in the treatment of chronic hepatitis C virus (HCV) infection [2,3], PEGylated granulocyte colony stimulating factor (Neulasta®) used as an immunostimulator in patients undergoing chemotherapy [4], and PEGylated asparaginase (Oncaspar®) used in the treatment of acute lymphocytic leukemia and non-Hodgkin's lymphoma [5]. PEGylation of human IFN-α2b (Pegintron, Scherring Plough) by attachment of 12 kDa linear PEG moieties results in a preparation consisting of predominantly mono-PEGylated IFN-α2b with multiple positional isomers present. Biological activity is comparable to the non-PEGylated Intron-A® protein, and PEGylation in this manner increases the serum half-life in humans to ~40 h (compared with 7–9 h for non-PEGylated IFN-α2b) and enables once-weekly dosing schedules of PEGylated IFN to replace the once-daily or three times weekly dosing required with the non-PEGylated form. Owing to their improved pharmacokinetics, which translates directly into improved clinical efficacy, PEGylated IFN molecules in combination with the nucleoside analog Ribavirin now form the current standard of care in the treatment of chronic HCV infections. As in the example provided earlier, PEG is most commonly attached to the ε-amino group of surface exposed lysine residues. This approach requires no modification of the recombinant protein prior to PEGylation, and the availability of surface exposed lysine residues means recombinant proteins can be PEGylated with high efficiency using this method. However, because of the presence of multiple residues available for PEGylation, a high degree of heterogeneity is often observed in reaction products following the use of this type of chemistry, which can present further downstream processing challenges. In certain circumstances, addition of PEG chains to randomly distributed lysine residues can also impact negatively on the biological activity of the molecule. Site-specific chemistries, in which thiol-reactive PEG molecules such as maleimide-PEG are attached to reactive cysteine residues, provide an alternative approach and result in much more homogenous reaction products. However, this approach requires the presence of a free cysteine, which is not found in a large number of recombinant proteins. Site-directed mutagenesis can be used to engineer the addition of a reactive cysteine in recombinant proteins [6] although changing the amino acid sequence of the therapeutic protein in this manner may present additional risk in terms of adverse effects on activity and introduction of immunogenicity, which can in turn impact development costs and timelines.

In summary, the applications of PEGylation technologies such as those described earlier have resulted in the development of therapeutics with improved pharmacokinetics for use in the treatment of a variety of diseases; however, the process challenges associated with production of PEGylated proteins have led to the development of alternative half-life extension technologies, which is discussed here.

11.2.2 Serum Albumin Fusion

Human serum albumin (HSA) is the most abundant protein in the serum, functions in the regulation of colloid osmotic pressure in the circulation, and acts as a carrier protein in the transport of fatty acids, hormones, bilirubin, and cations [7]. Synthesized in liver, the mature 585 amino acid polypeptide of ~67 kDa has a circulating half-life of greater than 20 days in man. The exceptionally long half-life of serum albumin is due in part to its size (it is above the ~60 kDa threshold for renal clearance) but also due to protection from intracellular degradation in the lysosome. Serum albumin is pinocytosed nonspecifically by the cell and accumulates in the early endosomal compartment. Owing to the low pH environment found within the early endosome, serum albumin is able to bind to the MHC-related receptor for IgG (FcRn), which facilitates recycling back to the cell surface, releasing albumin into the extracellular space [8]. Serum albumin is thus able to avoid the alternative fate of lysosomal degradation undergone by proteins, which traffic from the endosome to the lysosome in the absence of FcRn binding. This FcRn-mediated "recycling" receptor is also involved in prolonging the half-life of IgG through the same mechanism, although albumin and IgG bind to distinct domains on FcRn.

Owing to its size, wide distribution within the body, relative lack of immunological and enzymatic function, and innate protection from intracellular degradation, conjugation or fusion to HSA presents an attractive mechanism for prolonging the half-life of therapeutic proteins and peptides while addressing some of the development challenges associated with PEGylated molecules. Genetic fusion of therapeutic proteins to HSA results in significant improvements in serum half-life while maintaining biological activity of the therapeutic fusion partner [9]. These HSA fusion proteins can be expressed to high level in yeast systems and purified to homogeneity [10]. Examples of therapeutic proteins and peptides genetically fused to HSA can also be found in the clinic, for example, HSA fused to a glucagon-like peptide 1 (GLP-1) mimetic, currently being developed for the treatment of type 2 diabetes [11]. HSA fused to IFN-α (Albuferon®) has also been developed for potential use in the treatment of chronic HCV infection. Albuferon has similar efficacy and superior pharmacokinetics compared with PEGylated IFN [12,13]. Indeed, because of its increased half-life (7 days in man), once fortnightly or once monthly dosing schedules may be possible with Albuferon, although further clinical development of this molecule has been halted following Phase III clinical trials. However, as with PEGylated molecules, development challenges are also associated with HSA fusion proteins. Owing to its large size and the number of disulfide bonds present in the molecule, HSA expression in bacterial systems such as *Escherichia coli* is challenging; therefore, production of HSA fusion proteins generally requires yeast-based expression systems. A study carried out by Zhao and coworkers also revealed that HSA-IFN fusion proteins expressed in *Pichia pastoris* were unstable during long-term storage and were heterogeneous following purification [14]. In summary, the development of HSA fusion proteins such as those described earlier has resulted in a number of therapeutics in clinical development. HSA fusion proteins can be expressed to high yield and purity and may allow greater improvement in pharmacokinetics compared to those achieved with PEGylation. However, additional process challenges are associated with production of HSA fusion proteins, with expression limited in many cases to yeast-based systems. This has led to the development of alternative half-life extension technologies based on albumin-binding antibody fragments (AlbudAbs), which are described in more detail in Section 2.3.

11.2.3 Anti-Serum Albumin Domain Antibodies in the Development of Novel Half-Life Extension Technologies

Domain antibodies (dAbs) are the smallest binding unit of an immunoglobulin. They are ~12–14 kDa in size and are composed solely of a single human immunoglobulin variable domain (V_H or V_L). dAbs with antigen-binding activity can be expressed in *E. coli* [15], and using phage display methods, dAbs with high affinity and desirable biophysical properties, such as high solubility, resistance to aggregation following thermal denaturation and monomeric state in solution, have been isolated [16,17] although these desirable properties appear to be dependent on certain critical CDR residues affecting the hydrophobicity of the former V_L–V_H interface [18]. Terminal half-lives of these small proteins are typically <25 min in mice; however, dAbs with antigen-binding activity toward serum albumin have been isolated, and these "AlbudAbs" have circulating half-lives approaching that of serum albumin [19]. Many of these AlbudAbs demonstrated a high degree of cross-reactivity with serum albumin from discovery relevant species such as mouse, rat, and cynomolgus monkey meaning they can easily be assayed for *in vivo* activity in preclinical animal models. Owing to their small size, ease of expression in multiple systems (including yeast, bacteria, and mammalian cell culture), desirable biophysical properties, and high degree of species cross-reactivity, the potential for AlbudAbs in the generation of novel therapeutics with improved pharmacokinetics was investigated.

Holt and coworkers have generated AlbudAb fusions to the interleukin-1 receptor antagonist IL1-RA [19], a therapeutic with short *in vivo* half-life approved for use in the treatment of rheumatoid arthritis. The resulting AlbudAb/IL1-RA fusion was expressed in *E. coli* using the pET expression system and the activity of the purified fusion protein determined. Affinity for serum albumin, determined by surface plasmon resonance assays, was largely unaffected by fusion to IL1-RA. Similarly, when the AlbudAb/IL1-RA fusion protein was assayed for biological activity, measured by its ability to neutralize the secretion of IL-8 from MRC-5 cells, the activity was comparable to that of unfused IL1-RA. These experiments demonstrated conclusively that AlbudAbs could potentially be used to generate fusion proteins of potential therapeutic benefit without affecting the activity of either the AlbudAb or its fusion partner. In order to determine whether the *in vitro* activity of the AlbudAb/IL1-RA fusion protein would be likely to translate into clinical efficacy, the activity of the molecule was determined in a collagen-induced mouse model of arthritis in which animals were injected intraperitoneally with IL-1RA or AlbudAb/IL1-RA fusion protein 21 days after collagen injection to induce arthritis in these animals. Using clinical scores for the severity of arthritis, the authors demonstrated that AlbudAb/IL1-RA was significantly more efficacious in lowering the arthritis score compared to IL-1RA when the two were administered at the same dose. Biodistribution of ^{3}H labeled AlbudAb/IL1-RA, IL1-RA, and mouse serum albumin (MSA) was also determined by quantitative whole body autoradioagraphy and the distribution of AlbudAb/IL1-RA shown to be more similar to that of MSA than that of

IL-1RA. These experiments demonstrated conclusively that AlbudAbs could be used to generate therapeutic molecules with improved pharmacokinetics and *in vivo* efficacy due to binding of serum albumin in the circulation.

11.3 INTERFERON-α FUSED TO HUMAN SERUM ALBUMIN OR AlbudAb—A DIRECT COMPARISON OF HSA AND AlbudAb FUSION TECHNOLOGIES

11.3.1 *In Vitro* Potency of HSA and AlbudAb Fusion Proteins

Studies conducted by Holt and coworkers have demonstrated the utility of AlbudAb technology in developing antagonist molecules with improved pharmacokinetics as potential therapeutics. Subsequently, we wished to determine whether AlbudAbs could be used to develop an agonist molecule with improved half-life, while maintaining biological activity, which could potentially be used as a therapeutic in the treatment of chronic HCV infections. To this end, we have developed human IFN-α2b (IFN-α2b) genetically fused to the N-terminus of an AlbudAb composed of a single V_κ domain; clone DOM7h-14, via a short amino acid linker (hereafter referred to as IFN-α2b-DOM7h-14). In order to determine the relative advantages of AlbudAb technology in comparison to HSA fusion technology, we compared the pharmacokinetics, *in vitro* potency and *in vivo* efficacy of IFN-α2b-DOM7h-14 with that of IFN-α2b fused to the C-terminus of HSA with no intervening linker sequence generated in our laboratory (hereafter referred to as HSA-IFN-α2b), which is effectively the same molecule as Albuferon [20]. Owing to the complex nature of these two molecules, we decided to express them in a mammalian cell culture system which uses transiently transfected human embryonic kidney cells (HEK293E) to express proteins of interest from the vector pTT5 into the cell culture supernatant as soluble active protein [21]. This system has successfully been used in the past to express soluble active IFN-α molecules at high yield [22]. Following expression for 5 days post transfection, IFN-α2b-DOM7h-14 and HSA-IFN-α2b were purified from culture supernatants using affinity ligands immobilized on a solid support. IFN-α2b-DOM7h-14 was bound to protein L, an affinity ligand that specifically binds to immunoglobulin light chains [23], and HSA-IFN-α2b was bound to the affinity dye Cibacron Blue F3GA, an affinity ligand with specificity for human IFNs and serum albumin [24,25]. Following extensive washing, the purified fusion proteins could then be eluted and their activity determined using *in vitro* assays.

In order to compare the *in vitro* potency of purified IFN-α2b-DOM7h-14 and HSA-IFN-α2b, activity of the two fusion proteins was determined using the commercially available HEK293 IFN-α/β cell line (a diagrammatic representation of this assay is shown in Figure 11.1), in which HEK293 cells are stably transfected with a secreted alkaline phosphatase reporter gene under the control of an IFN inducible promoter element. Treatment of this cell line with type I IFNs (IFN-α, IFN-β, and IFN-ω) results in induction of reporter gene expression in a manner dependent upon the level of IFN activity, the level of which can be determined by spectrophotometric methods. Both fusion proteins were capable of inducing reporter gene expression in this cell line, although with reduced *in vitro* potency in comparison with a recombinant IFN-α commercial standard (Figure 11.2). In order to determine the likely effect of albumin binding on activity of AlbudAb fusion proteins, the assay was carried out in both the presence and absence of high concentrations of HSA. In the absence of HSA, both IFN-α2b-DOM7h-14 and HSA-IFN-α2b were able to induce reporter gene expression, though IFN-α2b-DOM7h-14 was able to do so with ~55-fold greater potency than HSA-IFN-α2b. This result shows that genetic fusion to the large HSA protein results in a significant loss of IFN activity *in vitro*. In the presence of 100 μM HSA, the

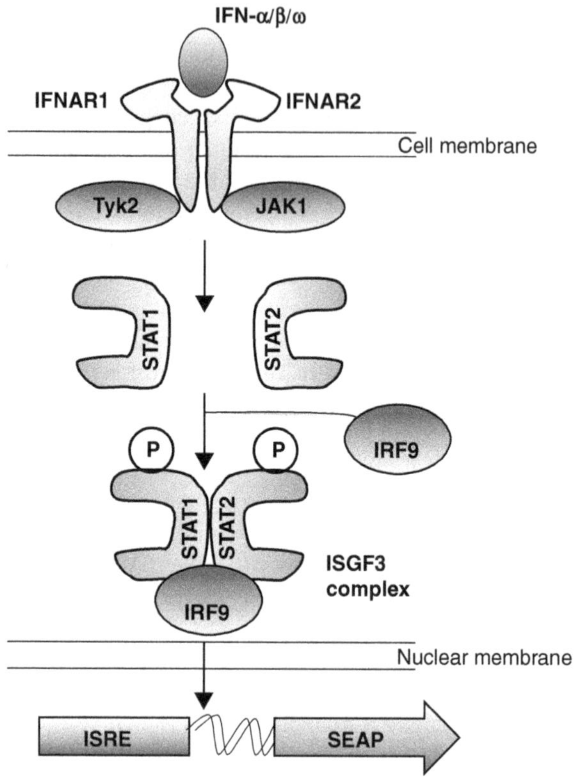

FIGURE 11.1 Schematic diagram of the HEK293 IFN-α/β reporter cell assay. Type I interferon (α, β, or ω) binding to the heterodimeric receptor composed of IFNAR1 and IFNAR2 results in Tyk2 and JAK1 kinase activation. Tyk2 and JAK1 then phosphorylate STAT1 and STAT2, resulting in dimerization and interaction with IRF9, forming a complex named ISGF3. ISGF3 then binds to interferon response elements (ISRE) in the promoters of interferon-stimulated genes, in this case the reporter gene SEAP, regulating their expression.

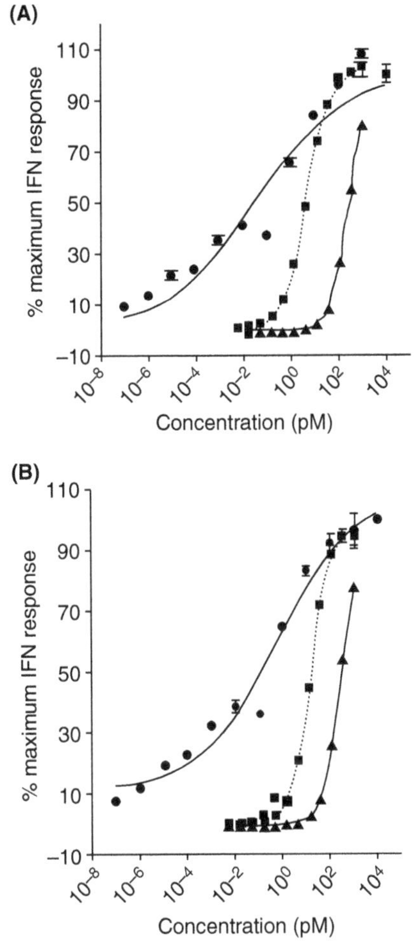

FIGURE 11.2 Activity of interferon fusion proteins in HEK293 IFN-α/β reporter cell assay. To assay for IFN activity *in vitro* IFN-α2b-DOM7h-14 (---■--), HSA-IFN-α2b, (–▲–), and unfused IFN-α (–•–) were tested for their ability to induce SEAP expression in the HEK293 IFN-α/β reporter cell line in the absence (A) or presence (B) of 100 μM HSA. *Source*: Reproduced from Reference [20] by permission of Oxford University Press.

potency of the IFN-α standard and HSA-IFN-α2b was largely unaffected, whereas the potency of IFN-α2b-DOM7h-14 was reduced approximately threefold, presumably due to steric effects following direct binding of the AlbudAb to HSA. However, when comparing the *in vitro* potency of the two fusion proteins directly in the presence of serum albumin, it appears that IFN-α2b-DOM7h-14 is ~18-fold more active than HSA-IFN-α2b. While the concentration of HSA used in this assay is lower than the 600-μM concentration found *in vivo*, no differences were observed in this assay when the two concentrations of HSA were compared directly; therefore, we conclude that even in the presence of physiologically relevant concentrations of HSA, IFN fused to the AlbudAb DOM7h-14 has greater potency than IFN fused directly to HSA.

In order to determine whether potency in the HEK293 IFN-α/β assay correlates directly with *in vitro* antiviral efficacy, we studied the activity of IFN-α2b-DOM7h-14, HSA-IFN-α2b, and unfused IFN-α in a cell-based assay in which the A549 human lung carcinoma cell line is infected with encephalomyocarditis virus, hereafter referred to as the EMCV assay (a diagram of the EMCV assay is shown in Figure 11.3). In this assay, the level of IFN-mediated protection against EMCV infection was determined for each fusion protein and the activity, measured in international units per milligram of protein (IU/mg), calculated by comparison with international reference standard material. Since the *in vitro* potency of IFN-α2b-DOM7h-14 in the HEK293 IFN-α/β assay appears to be slightly reduced in the presence of HSA, we also carried out the EMCV antiviral assay in the absence or presence of 100 μM HSA. When HSA was present in the assay, the activity of both human IFN-α2b fusion proteins and the unfused IFN-α were slightly reduced (Figure 11.4), indicating some nonspecific inhibition of EMCV infection by HSA in this cell line. IFN-α2b-DOM7h-14 and HSA-IFN-α2b activity were calculated at 3.8×10^6 and 2.4×10^5 IU/mg, respectively. In comparison with the unfused IFN-α, this represents a fold reduction in activity of 17.4 for IFN-α2b-DOM7h-14 and 275 for

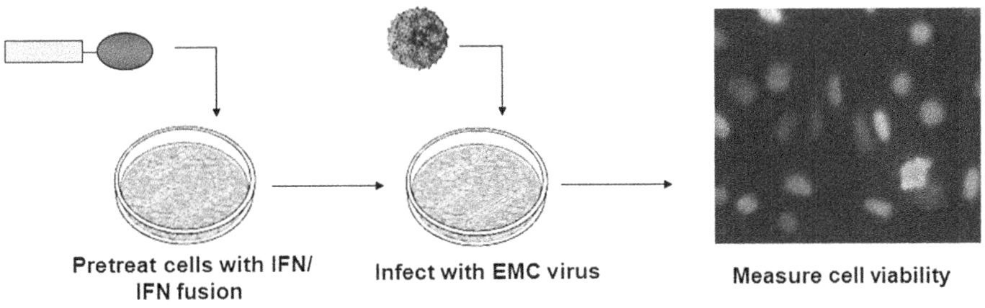

FIGURE 11.3 Schematic diagram of the EMCV assay. A549 cells are preincubated with human IFN-α2b or human IFN-α2b fusion proteins. Following wash off of interferon cells, these cells are then infected with virus, after which protection against virus infection is measured by determining cell viability using dye exclusion methods.

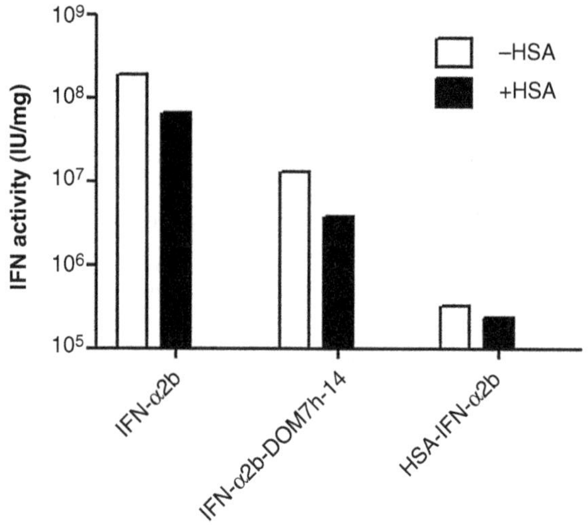

FIGURE 11.4 Antiviral activity of IFN-α2b fusion proteins. The protective effect of IFN-α2b-DOM7h-14 and HSA-IFN-α2b was evaluated in antiviral assays using A549 cells infected with EMCV. Assay was carried out in the presence (black bars) and absence (white bars) of 100 μM human serum albumin. For comparison, data obtained with unfused IFN-α2b are also shown. *Source*: Reproduced from Reference [20] by permission of Oxford University Press.

HSA-IFN-α2b. When compared directly, the antiviral activity of IFN-α2b-DOM7h-14 is 15.8-fold greater than that of HSA-IFN-α2b. On a molar basis, this is equivalent to a 5.8-fold increase in the *in vitro* antiviral efficacy of IFN-α2b-DOM7h-14 in comparison with HSA-IFN-α2b; therefore, even in the presence of HSA, the *in vitro* antiviral efficacy of IFN-α2b-DOM7h-14 is significantly higher than that of HSA-IFN-α2b as would be expected based on the *in vitro* potency of these two molecules observed in the HEK293-Blue IFN-α/β reporter cell assay.

In summary, we have demonstrated using two distinct *in vitro* assays, one of which is a measure of the desired antiviral activity of the therapeutic, that AlbudAb technology can be used to generate fusion proteins with superior potency compared with genetic serum albumin fusions. This superior potency is maintained in the presence of physiologically relevant concentrations of HSA; therefore, it is possible that the improved potency of AlbudAb fusion proteins in comparison to HSA fused molecules will also correlate directly with improved efficacy *in vivo*. Comparison of the pharmacokinetics and *in vivo* efficacy is the subject of Sections 3.2 and 3.3.

11.3.2 Pharmacokinetics of HSA and AlbudAb Fusion Proteins

In order to compare the *in vivo* half-life of IFN-α2b-DOM7h-14 and HSA-IFN-α2b, the pharmacokinetics of these fusion proteins were determined in rodent species. Following intravenous administration in rats, the pharmacokinetics of both IFN-α2b-DOM7h-14 and HSA-IFN-α2b were significantly improved in comparison to unfused IFN-α (Table 11.1). Similar area under the plasma concentration–time curve (AUC) values were obtained for IFN-α2b-DOM7h-14 and HSA-IFN-α2b (737.5 and 689.2 h mg/mL, respectively), which in both cases represents a significant increase over the value of 18.8 h mg/mL observed with unfused IFN-α. The circulating half-life ($T_{1/2}$ elim) of both fusion proteins was also increased when compared to that of unfused IFN-α standard. *In vivo* half-lives of IFN-α2b-DOM7h-14 and HSA-IFN-α2b were calculated at 22.6 and 14.2 h, respectively, which in both cases represent a significant increase in the circulating half-life of both fusion proteins when compared to the 1.2 h half-life calculated for unfused IFN-α. When compared directly, the circulating half-life of IFN-α2b-DOM7h-14 is ~1.5 times longer than that of HSA-IFN-α2b based on the values calculated from these experiments.

As clinical preparations of IFN are patient administered via subcutaneous injection, we also carried out a second study to determine the pharmacokinetics of AlbudAb and HSA-fused IFN-α, compared to unfused IFN-α, following subcutaneous administration in rats. As in the previous study, the *in vivo* half-life of both fusion proteins was increased in comparison to unfused IFN-α (Table 11.2). Both AUC and circulating half-life of IFN-α2b-DOM7h-14 were increased in comparison to values obtained with HSA-IFN-α2b (AUC 342.5 h mg/mL and half-life of 28.3 h for IFN-α2b-DOM7h-14 vs. AUC 137.1 h mg/mL and half-life of 19.7 h for HSA-IFN-α2b), which is in agreement with the results from the study with animals receiving intravenously administered IFNs described earlier. Consistent with the improved *in vivo* half-life of IFN-α2b-DOM7h-14 compared with HSA-IFN-α2b, the AlbudAb fusion protein was also cleared at a lower rate following subcutaneous administration (clearance rate of 5.9 mL/h/kg with IFN-α2b-DOM7h-14 vs. 16.7 mL/h/kg for HSA-IFN-α2b). Bioavailability following subcutaneous administration was also improved with IFN-α2b-DOM7h-14

TABLE 11.1 Pharmacokinetics of Interferon Fusion Proteins Following Intravenous Administration in Rats

	IFN-α2b-DOM7h-14	HSA-IFN-α2b	Unfused IFN-α
$T_{1/2}$ elim (h)	22.6	14.2	1.2
C_{max} (μg/mL)	34.7	60.1	25.3
$AUC_{0-\infty}$	737.5	689.2	18.8
Clearance (mL/h/kg)	2.7	3.1	135.2

Source: Reproduced from Reference [20] by permission of Oxford University Press.

TABLE 11.2 Pharmacokinetics of Interferon Fusion Proteins Following Subcutaneous Administration in Rats

	IFN-α2b-DOM7h-14	HSA-IFN-α2b	Unfused IFN-α
$T_{1/2}$ elim (h)	28.3	19.7	1.5
C_{max} (μg/mL)	5.9	2.9	1.39
$AUC_{0-\infty}$	342.5	137.1	5.5
Clearance (mL/h/kg)	5.9	16.7	357.1
Bioavailability (%)	47.7	19.9	39.9

Source: Reproduced from Reference [20] by permission of Oxford University Press.

compared with HSA-IFN-α2b (46.4% vs. 19.9%) with the reason for this difference currently undetermined.

In summary, we have demonstrated in rodent pharmacokinetic assays that AlbudAb technology can be used to generate fusion proteins with improved pharmacokinetics compared with genetic serum albumin fusions. The reason for this is currently unclear but may be due in part to the difference in interactions with FcRn of the two fusion proteins. It is possible that direct fusion of therapeutic proteins to HSA adversely affects binding of the albumin fusion partner to FcRn, whereas AlbudAb binding to HSA disrupts the albumin–FcRn interaction to a lesser degree, though this remains to be experimentally determined.

11.3.3 *In Vivo* Efficacy of HSA and AlbudAb Fusion Proteins

In order to compare the *in vivo* efficacy of human IFN-α2b AlbudAb and HSA fusion proteins, we used a model in which human tumor cell lines sensitive to type I IFNs were transplanted into mice with a severe combined immunodeficiency due to a recessive mutation in the Prkdc gene (CB17/ICR-Prkdcscid/IcrCrl mice, hereafter referred to as SCID mice). Owing to the lack of a humoral or cellular immune response in these mice, human tumor cell lines could be "xenografted" onto the right or left flank where they form distinct tumor masses [26]. Retardation of tumor growth in response to treatment with human IFN-α2b fusion proteins can be determined as a measure of the *in vivo* efficacy of these molecules, and in these experiments, female SCID mice were xenografted with the 518A2 human melanoma cell line, which previous investigators have shown to be sensitive to human IFN-α following transplantation in this manner [27]. Subcutaneous injections of vehicle control (PBS), or equimolar amounts of either IFN-α2b-DOM7h-14 (8.75 mg/kg) or HSA-IFN-α2b (24 mg/kg) were administered on days 1 and 8 post tumor implantation (Figure 11.5). Both human IFN-α2b fusion proteins were able to significantly reduce the rate of increase in tumor volume as shown by a comparison of the measurements obtained from these two groups with those obtained from the vehicle control group; however, the reduction observed with the IFN-α2b-DOM7h-14 group was greater than that observed in the HSA-IFN-α2b group, consistent with the improved *in vitro* potency and *in vivo* half-life of IFN-α2b-DOM7h-14 compared with that of HSA-IFN-α2b. In order to determine whether these results were statistically significant, slopes of the graphs obtained up to day 34 were compared using random coefficient regression analysis, and the reduction in tumor growth rates in the two treatment groups was shown to be significantly different ($P = 0.001$).

The effect of a 10-fold lower dose (0.875 mg/kg) of IFN-α2b-DOM7h-14 in this model was also investigated and

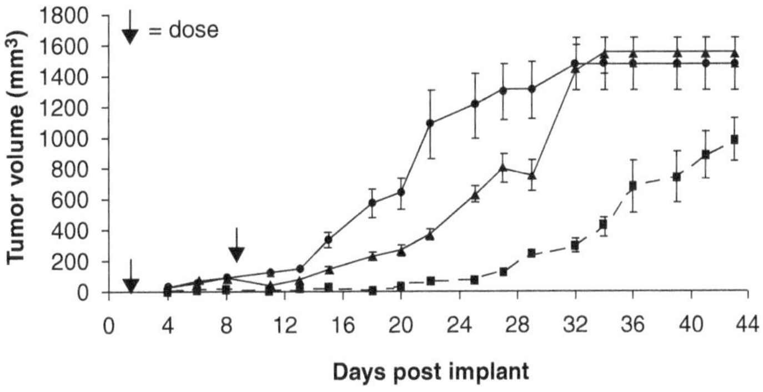

FIGURE 11.5 *In vivo* efficacy of IFN-α2b fusion proteins in A518A2 xenograft model. Tumor volume in A518A2 xenografted mice injected with 8.75 mg/kg IFN-α2b-DOM7h-14 (---■---), 24 mg/kg HSA-IFN-α2b (–▲–), or PBS (–•–) at days 1 and 8 post tumor implant. Tumor volume was measured at the indicated times, with $n = 10$ at each time point (error bars shown represent s.e.m. at each time point). Statistical comparison of the tumor volumes in the IFN-α2b-DOM7h-14 and HSA-IFN-α2b dose groups was carried out using linear regression coefficient analysis and differences found to be significant ($P = 0.0001$). *Source*: Reproduced from Reference [20] by permission of Oxford University Press.

compared with mice receiving the higher 24 mg/kg dose of HSA-IFN-α2b, in this case with dosing on days 6 and 13 post tumor implantation. In both groups, a statistically significant reduction in tumor growth was observed in comparison with the vehicle control group; however, when compared directly, the differences between the two groups were not statistically significant ($P = 0.0715$). These data demonstrate the improved *in vivo* efficacy of IFN-α2b-DOM7h-14 compared with HSA-IFN-α2b in this model. When dosed at molar equivalence, a significantly greater reduction in tumor growth is observed in the IFN-α2b-DOM7h-14 group. When IFN-α2b-DOM7h-14 is administered at a 10-fold lower dose than HSA-IFN-α2b, the level of reduction in tumor growth is indistinguishable between the two groups, demonstrating that even with drastically reduced doses, in comparison to the HSA fusion protein, the AlbudAb fusion protein maintains comparable efficacy in preclinical animal models.

Using IFN as a showcase, we have demonstrated the utility of the AlbudAb platform in creating novel long acting therapeutics with improved *in vitro* potency, *in vivo* half-lives, and preclinical efficacy in animal models, compared to clinically validated half-life extension technologies such as PEGylation and HSA fusion. This novel method of half-life extension potentially addresses issues of heterogeneity associated with PEGylation and can be applied to a wide variety of expression systems, unlike serum albumin fusion proteins which typically require yeast expression systems; however, it should be noted that although AlbudAbs themselves are amenable to expression in bacterial, yeast, and mammalian hosts, the therapeutic molecule to which the dAb is fused may also govern the choice of expression host used to manufacture the AlbudAb fusion protein.

It remains to be seen whether our IFN-AlbudAb fusion proteins exhibit the same heterogeneity and instability during long-term storage and formulation for therapeutic delivery as is seen with some HSA-therapeutic fusions, though data collected to date have shown that these IFN-AlbudAb molecules exhibit the same desirable properties (monomeric solution state, thermal stability, lack of proteolytic cleavage, and heterogeneity following purification) observed with many isolated dAb molecules characterized to date (our unpublished data).

11.4 NANOBODIES IN THE DEVELOPMENT OF ALTERNATIVE HALF-LIFE EXTENSION TECHNOLOGIES BASED ON SINGLE IMMUNOGLOBULIN VARIABLE DOMAINS

Immunoglobulins from the majority of mammalian species consist of two identical heavy and light chain polypeptides. Many species of camelidae, including camels and llamas, have an alternative type of immunoglobulin, which is composed of two identical heavy chain polypeptides in the complete absence of any discernible light chain [28,29]. Referred to as heavy chain antibodies (HcAbs, Figure 11.6), this alternative format of immunoglobulin is found in addition to the classical immunoglobulin (i.e., heavy and light chain containing) format and bears an extensive antigen-binding repertoire because of the presence of a single variable region termed the V_{HH} domain, which retains antigen-binding activity when cloned and expressed as a single domain from *E. coli* [30]. V_{HH} domains bear a number of structural similarities with single variable domains from classical immunoglobulin, and this type of single domain antibody is being developed by Ablynx as nanobody technology for similar clinical and commercial applications as those currently being targeted by dAbs.

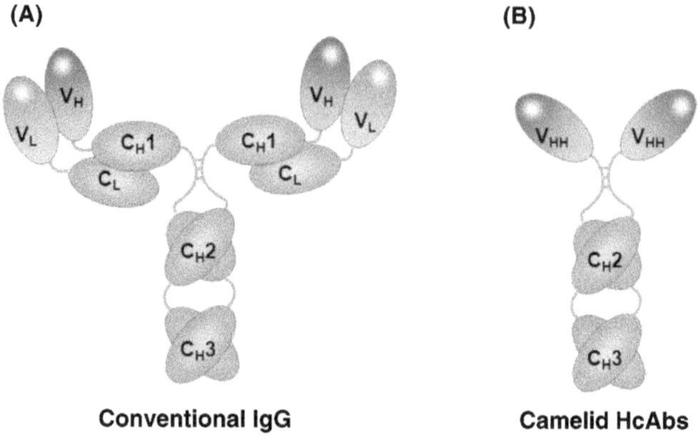

FIGURE 11.6 Structure of conventional immunoglobulins and camelid heavy chain antibodies. Schematic diagram showing domain structure of conventional IgG (A), compared with camelid HcAbs (B). Camelid HcAbs are devoid of any light chain interaction, and possess no C_H1 domain in the heavy chain, with the V_{HH} variable domain fused directly to the C_H2 hinge region.

Although dAbs and nanobodies share many common features, for example, a structure consisting of four constant "framework regions" and three hypervariable "complementarity determining regions" (CDRs numbered CDR1, 2, and 3) responsible for antigen specificity as well as the same immunoglobulin fold composed of two β-sheets [31], the two types of molecule also differ in several minor but important respects. For example, the CDR1 and CDR3 regions of nanobodies are longer than those found in V_H dAbs. In some cases, CDR1 and CDR3 also contain a cysteine residue, which participates in disulfide bond formation and may be involved in maintaining tertiary structure [32]. The framework 2 region of nanobodies and V_H dAbs also exhibits major differences in primary amino acid sequence between the two molecules. In V_H dAbs, this region is composed primarily of residues, which form part of the interface between V_L and V_H domains in the complete immunoglobulin structure. However, framework 2 is composed predominantly of different residues in V_{HH}

domains, consistent with the lack of association with a partner light chain [33]. This may explain the apparent lack of aggregation, high solubility, and desirable biophysical properties of V_{HH} domains in reported cases. Nanobodies with serum albumin binding affinity have been developed and used to increase the circulating half-life of proteins with potential therapeutic value following genetic fusion of the two molecules, for example, EGFR-specific nanobodies fused to albumin specific nanobodies in much the same way as we have described with AlbudAbs [34]. Biodistribution studies with these molecules in SCID mice xenografted with the EGFR expressing tumor cell line A431 show increased deposition of EGFR-specific nanobodies when fused to serum albumin binding nanobodies. These data, together with the increased residence time in blood following serum albumin binding, demonstrate that the bifunctional fusion protein retains binding affinity for both EGFR and albumin *in vivo*.

In summary, nanobodies have the potential for use in the development of half-life extension technologies based on camelid single immunoglobulin variable domains with specificity for serum albumin. In many cases, these V_{HH} domain antibodies require no further engineering to modify their biophysical properties as they contain a number of structural features which allow them to adopt a stable conformation in the absence of light chain binding. However, it remains to be seen whether the presence of camelid sequences in V_{HH} domains results in immunogenicity upon repeat dosing or when adopting extended half-life formats, the removal of which may require further engineering and humanization before nanobodies can be used in a therapeutic capacity.

11.5 NOVEL HALF-LIFE EXTENSION TECHNOLOGIES—ALTERNATIVE APPROACHES TO SINGLE IMMUNOGLOBULIN VARIABLE DOMAINS

11.5.1 Serum Albumin Binding Peptides

Peptides with the core sequence DICLPRWGCLW have been isolated by phage display [35]. These peptides display binding activity to serum albumin from a variety of species including human, mouse, rabbit, and rat and when the pharmacokinetics of these peptides were analyzed in rabbits the circulating half-life of peptides containing this core sequence was found to be ~2.3 h. The half-life of similar peptide sequences with no detectable albumin-binding activity was calculated at <30 min, demonstrating significant half-life extension by virtue of the albumin-binding properties of peptides containing the core sequence DICLPRWGCLW. Further studies have shown utility of these albumin-binding peptides in extending the half-lives of therapeutically relevant proteins to which they are fused, for example, Fab derived from the HER2 specific monoclonal antibody Herceptin® (Trastuzumab) used in the treatment of HER2 positive breast cancer [36]. In these studies, the authors show that Herceptin derived Fab4D5 genetically fused to the peptide QRLME-DICLPRWGCLWEDDF results in a bifunctional molecule which retains both albumin- and HER2-binding activity. When tumor uptake of the peptide fused and unfused Fab4D5 molecules was compared in HER2 overexpressing mammary tumors in mice, both molecules show similar levels of uptake. However, clearance of the albumin-binding peptide fused molecule was significantly delayed in comparison to the unfused Fab, consistent with improved pharmacokinetics because of serum albumin binding *in vivo*. These observations were extended to a direct correlation of albumin-binding affinity of Fab4D5-peptide fusions with increased *in vivo* half-life in pharmacokinetic analyses carried out in rodent species [37].

11.5.2 Serum Albumin Binding Small Molecules

Small molecule albumin-binding moieties have been isolated from a DNA-encoded chemical library [38]. They have been shown to bind both human and mouse serum albumin with reasonable affinity (equilibrium dissociation constants in the ~100 nM range) and significantly increase the circulating half-life of imaging and contrast agents such as fluorescein and Magnevist® (Gd-DTPA) from <5 min to longer than 8 h in mice. These small molecule albumin-binding moieties also can be used to modify the half-lives of recombinant proteins. Trüssel and coworkers have successfully attached maleimide derivatives of albumin binding small molecules to single chain antibodies (ScFv) with an engineered C-terminal cysteine mutation [39]. Modification of ScFv specific to fibronectin, a marker of tumor angiogenesis, with 2-(3-maleimidopropanamido)-6-(4-(4-iodophenyl)butanamido) hexanoate (Albutag) to introduce albumin-binding functionality resulted in a bispecific molecule, which retained both albumin- and fibronectin-binding activity. When injected into mice bearing fibronectin expressing tumors, the Albutag modified ScFv showed an ~10-fold increase in tumor uptake and 35-fold decrease in blood clearance in comparison to the unmodified ScFv, demonstrating the utility of Albutag technology in extending the serum half-life of recombinant proteins.

11.5.3 Hydrophilic Polymers

The use of naturally occurring hydrophilic polymers as an alternative to PEG has been investigated and a study by Constantinou and coworkers has demonstrated the potential for polysialic acid chains in extending the half-life of otherwise short-lived Fab fragments [40]. By chemically coupling polysialic acid chains of 11 and 22 kDa to Fab fragment derived from an antitumor antibody H17E2, the authors were able to produce Fab fragments of increased molecular mass which were stable in serum (indicating that the sialic acid chains would not be cleaved *in vivo*) and had only a moderate (approximately fourfold) decrease in affinity for antigen when compared to the unmodified Fab. Effect of polysialylation on pharmacokinetics was determined in mice and the circulating half-life of H17E2-derived Fab shown to be increased by up to threefold following polysialylation. Although chemical methods were used to couple sialic acid chains in this study, the authors are also investigating the use of engineered glycosylation motifs to allow polysialylation of recombinant proteins during expression, overcoming the requirement for further downstream processing [41].

Other investigators are also studying the use of genetically fused hydrophilic polymers as an alternative to PEGylation. XL-protein has developed PASylation technology, which involves genetic fusion of recombinant proteins to polypeptide sequences containing the amino acids Pro, Ala, and Ser (PAS), leading to an increase in molecular size without affecting the biological activity of the therapeutic protein (Skerra, unpublished work). PAS sequences adopt a conformationally disordered structure under physiological buffer conditions whose spatial dimensions are adjustable by the length of the PAS repeat unit, thereby increasing the apparent molecular size of the therapeutic protein above the pore size of the renal glomeruli and reducing renal clearance rates of PASylated proteins; in fact, these investigators show that this can lead to a reduction of *in vivo* clearance rates of between 10- and 100-fold compared to the unmodified protein.

A similar half-life extension technology has also been developed by Stemmer and colleagues, in which nonrepetitive amino acid sequences containing the residues Ala, Glu, Gly, Pro, Ser, and Thr are genetically fused to proteins and peptides, increasing their hydrodynamic radius and delaying clearance in much the same way as PASylation and PEGylation technologies. Fusion of an 864 amino acid sequence, referred to as EXTEN, to the GLP-1 peptide analog Exenatide increased the circulating half-life in rat from 2.5 to 29 h. The Exenatide-EXTEN fusion protein also displayed significantly improved biological activity in mice compared to the free peptide, consistent with its improved pharmacokinetic profile [42].

Although half-life extension technologies based on albumin-binding molecules other than single domain antibodies are currently being developed, these technologies also have potential limitations in addition to their advantageous properties. Albumin-binding peptides, which can be fused genetically to therapeutic proteins and are much smaller than immunoglobulin variable domains, are not based on naturally occurring epitopes and it therefore remains to be seen whether addition of such synthetic sequences could result in immunogenic responses upon repeat dosing or when using extended half-life formats. Peptide sequences used in this manner may also be susceptible to proteolytic cleavage *in vivo*.

Albutag technology, which is based on addition of chemical groups with affinity for albumin, is unlikely to cause problems with immunogenic responses but the chemical coupling methods used to attach these albumin-binding groups may be subject to the same downstream processing issues associated with the production of PEGylated molecules, in particular, heterogeneity of product or modifications in the amino acid sequence of the protein required to develop site-specific coupling chemistries.

Novel half-life extension technologies, which are not based on albumin-binding functionality, are also currently being developed. The major advantage to such technologies is the reduced requirement for downstream processing compared to PEGylation, for example, glycosylation and PASylation result in addition of hydrophilic polymers during expression via recombinant means rather than chemical coupling of products to PEG following expression and purification. It remains to be seen how the resulting improvements in half-life using these technologies compare with those obtained using serum albumin binding strategies, however.

11.6 CONCLUSIONS

Although PEGylation technologies have been the most widely used method of improving the pharmacokinetics of clinically relevant recombinant proteins over the last 30 years, the limitations associated with this technology, in particular, the associated cost of downstream processing, have led to the development of alternative half-life extension technologies based around the use of serum albumin fusion proteins. Although a number of recombinant HSA fusion proteins are currently in late stage clinical development, HSA fusion technology is also associated with a number of process issues due to the large and complex nature of the serum albumin molecule. This in turn has served as a driver for the development of further half-life extension technologies based on the use of serum albumin binding molecules.

We have developed human immunoglobulin variable domains that specifically bind to serum albumin (AlbudAbs) as a means of extending the half-life of therapeutic molecules to which they are fused. AlbudAbs with desirable biophysical properties have been isolated, and using IFN as an example of a therapeutic molecule with an inherently poor half-life in man, we have used AlbudAb technology to develop a molecule with therapeutic potential, which has an extended half-life, improved *in vitro* potency and greater *in vivo* efficacy in comparison to HSA-fused IFN. AlbudAbs can also be expressed in a variety of bacterial, yeast, and mammalian expression hosts, offering alternatives to the limited choice of expression host required to produce a large and complex molecule such as serum albumin.

Alternative single domain antibodies based on camelid V_{HH} domains (nanobodies) can also be used as a means of half-life extension in much the same way as AlbudAbs. Nanobodies have been shown to possess desirable biophysical properties with little or no further engineering required. However, in contrast to dAbs, which are based on fully human immunglobulin variable domain scaffolds, nanobodies are based on the camelid V_{HH} scaffold, and it therefore remains to be seen whether the presence of nonhuman epitopes will limit their clinical efficacy due to their potential for immunogenic responses in patients.

Serum albumin binding technologies based on alternative formats besides immunoglobulin variable domains are also in development, for example, albumin-binding peptides and small molecules which have both been shown to extend the circulating half-lives of Fab fragments. It is unclear at this stage what benefit these alternative formats of albumin-binding molecule will bring to the development of therapeutics with improved pharmacokinetics. No reports of either technology in comparison to serum albumin fusion, AlbudAb, or nanobody technology have so far been published; therefore, it is difficult to predict how the half-life, potency, or efficacy of therapeutics developed using this technology will differ in comparison to those using alternative half-life extension strategies. Some of the potential disadvantages associated with the use of nonnatural peptides as a means of half-life extension include immunogenicity, whereas small molecule conjugates require further downstream processing in order to couple them to therapeutic proteins, which is also likely to impact negatively on production costs associated with development of such molecules.

Strategies to incorporate hydrophilic polymers into therapeutic proteins by recombinant means during expression, rather than via chemical coupling to purified molecules, have also been described, and this may address such "cost of goods" issues. However, many of these technologies are in early stage development; therefore, the likely improvements in pharmacokinetic profiles of therapeutic proteins developed using such technologies compared to other half-life extension strategies are currently undetermined. In contrast, our studies have demonstrated conclusively that AlbudAbs fused to human IFN-α can be used to develop a molecule which is class leading in terms of pharmacokinetics and efficacy in preclinical models when compared directly with current clinically validated half-life extension strategies.

REFERENCES

1. Fee CJ, Van Altine JM. (2006) PEG-proteins: reaction engineering and separation issues. *Chem. Eng. Sci.* 61, 924–939.
2. Glue P, Fang JWS, Rouzier-Panis R, Raffanel C, Sabo R, Gupta SK, et al. and the Hepatitis C Intervention Therapy Group. (2000) Pegylated interferon-α2b: pharmacokinetics, pharmacodynamics, safety, and preliminary efficacy data. *Clin. Pharmacol. Ther.* 68, 556–567.
3. Wang Y, Youngster S, Grace M, Bausch J, Bordens R, Wyss DF. (2002) Structural and biological characterization of pegylated recombinant interferon alpha-2b and its therapeutic implications. *Adv. Drug Deliv. Rev.* 54, 547–570.
4. Morishita M, Leonard RC. (2008) Pegfilgrastim; a neutrophil mediated granulocyte colony stimulating factor—expanding uses in cancer chemotherapy. *Expert Opin. Biol. Ther.* 8, 993–1001.
5. Rytting M. (2010) Peg-asparaginase for acute lymphoblastic leukemia. *Expert Opin. Biol. Ther.* 10, 833–839.
6. Doherty DH, Rosendahl MS, Smith DJ, Hughes JM, Chlipala EA, Cox GN. (2005) Site-specific PEGylation of engineered cysteine analogs of recombinant human granulocyte-macrophage colony-stimulating factor. *Bioconjug. Chem.* 16, 1291–1298.
7. Peters T. (1977) Serum albumin: recent progress in the understanding of its structure and biosynthesis. *Clin. Chem.* 23, 5–12.
8. Chaudhury C, Brooks CL, Carter DC, Robinson JM, Anderson CL. (2006) Albumin binding to FcRn: Distinct from the FcRn-IgG interaction. *Biochemistry* 45, 4983–4990.
9. Syed S, Schuyler PD, Kulczycky M, Sheffield WP. (1997) Potent antithrombin activity and delayed clearance from the circulation characterize recombinant hirudin genetically fused to albumin. *Blood* 89, 3243–3252.
10. Yeh P, Landais D, Lemaitre, M, Maury, I, Crenne, J, Becquart, J, et al. Design of yeast-secreted albumin derivatives for human therapy: biological and antiviral properties of a serum albumin-CD4 genetic conjugate. *Proc. Natl. Acad. Sci. USA* 89, 1904–1908.
11. Matthews JE, Stewart MW, De Boever EH, Dobbins RL, Hodge RJ, Walker SE, et al. and for the Albiglutide Study Group. (2008) Pharmacodynamics, pharmacokinetics, safety, and tolerability of albiglutide, a long-acting glucagon-like peptide-1 mimetic, in patients with type 2 diabetes. *J. Clin. Endocrinol. Metab.* 93, 4810–4817.
12. Osborn BL, Olsen HS, Nardelli B, Murray JH, Zhou JXH, Garcia A, et al. (2002) Pharmacokinetic and pharmacodynamic studies of a human serum albumin-interferon-α fusion protein in cynomolgus monkeys. *J. Pharmacol. Exp. Ther.* 303, 540–548.
13. Zeuzem S, Yoshida EM, Benhamou Y, Pianko S, Bain VG, Shouval DG, et al. (2008) Albinterferon alfa-2b dosed every two or four weeks in interferon-naïve patients with genotype 1 chronic hepatitis C. *Hepatology* 48, 407–417.
14. Zhao HL, Xue C, Wang Y, Li XY, Xiong XH, Yao XQ, et al. (2007) Circumventing the heterogeneity and instability of human serum albumin-interferon-α2b fusion protein by altering its orientation. *J. Biotechnol.* 131, 245–252.
15. Ward ES, Güssow D, Griffiths AD, Jones PT, Winter G. (1989) Binding activities of a repertoire of single immunoglobulin

variable domains secreted from *Escherichia coli*. *Nature* 341, 544–546.

16. Jespers L, Schon O, Famm K, Winter G. (2004) Aggregation-resistant domain antibodies selected on phage by heat denaturation. *Nat. Biotechnol.* 22, 1161–1165.
17. Jespers L, Schon O, James LC, Veprinstev D, Winter G. Crystal Structure of HEL4, a soluble, refoldable human VH single domain with a germ-line scaffold. *J. Mol. Biol.* 337, 893–903.
18. Ewert S, Cambillau C, Conrath K, Plückthun A. (2002) Biophysical properties of camelid VHH domains compared to those of human VH3 domains. *Biochemistry* 41, 3628–3636.
19. Holt LJ, Basran A, Jones K, Chorlton J, Jespers L, Brewis ND, et al. (2008) Anti-serum albumin domain antibodies for extending the half-lives of short lived drugs. *Protein Eng. Des. Sel.* 21, 283–288.
20. Walker A, Dunlevy G, Rycroft D, Topley P, Holt LJ, Herbert T, et al. (2010) Anti-serum albumin domain antibodies in the development of highly potent, efficacious and long-acting interferon. *Protein Eng. Des. Sel.* 23, 271–278.
21. Durocher Y, Perret S, Kamen A. (2002) High-level and high-throughput recombinant protein production by transient transfection of suspension-growing human 293-EBNA1 cells. *Nucleic Acids Res.* 2, E9.
22. Loignon M, Perret S, Kelly J, Boulais D, Cass B, Bisson L, et al. (2008) Stable high volumetric production of glycosylated human recombinant IFNalpha2b in HEK293 cells. *BMC Biotechnol.* 8, 1–16.
23. Nilson BHK, Solomon A, Björck L, Åkerström B. (1992) Protein L from Peptostreptococcus magnus binds to the κ light chain variable domain. *J. Biol. Chem.* 267, 2234–2239.
24. Jankowski WJ, Von Muenchhausen W, Sulkowski E, Carter WA. (1976) Binding of human interferons to immobilised Cibacron blue F3GA: the nature of molecular interaction. *Biochemistry* 15, 5182–5187.
25. Kassab A, Yavuz H, Odabasi M, Denizli A. (2000) Human serum albumin chromatography by Cibacron Blue F3GA derived microporous polyamide hollow-fiber affinity membranes. *J. Chromatogr. B* 746, 123–132.
26. Ito M, Hiramatsu H, Kobayashi K, Suzue K, Kawahata M, Hioki K, et al. (2002) NOD/SCIDγc^{null}mouse: an excellent recipient mouse model for engraftment of human cells. *Blood* 100, 3175–3182.
27. Krepler C, Certa U, Wachek V, Jansen B, Wolf K, Pehamberger H. (2004) Pegylated and conventional interferon-α induce comparable transcriptional responses and inhibition of tumor growth in a human melanoma SCID mouse xenotransplantation model. *J. Invest. Dermatol.* 123, 664–669.
28. Hamers-Casterman C, Atarhouch T, Muyldermans S, Robinson G, Hamers C, Bajyana Songa E, et al. (1993) Naturally occurring antibodies devoid of light chains. *Nature* 363, 446–448.
29. Muyldermans S, Baral TN, Cortez Retamozzo V, De Baetselier P, De Genst E, Kinne J, et al. (2009) Camelid immunoglobulins and nanobody technology. *Vet. Immunol. Immunopathol.* 128, 178–183.
30. Muyldermans S, Lauwereys M. (1999) Unique single-domain antigen binding fragments derived from naturally occurring camel heavy-chain antibodies. *J. Mol. Recognit.* 12, 131–140.
31. Muyldermans S, Cambillau C, Wyns L. (2001) Recognition of antigens by single-domain antibody fragments: The superfluous luxury of paired domains. *Trends Biochem. Sci.* 26, 230–235.
32. De Genst E, Silence K, Decanniere K, Conrath K, Loris R, Kinne J, et al. (2006) Molecular basis for the preferential cleft recognition by dromedary heavy-chain antibodies. *Proc. Natl. Acad. Sci. USA* 103, 4586–4591.
33. Nguyen VK, Desmyter A, Muyldermans S. (2001) Functional heavy-chain antibodies in camelidae. *Adv. Immunol.* 79, 261–296.
34. Tijink BM, Laeremans T, Budde M, Stigter-van Walsum M, Dreier T, De Haard HJ, et al. (2008) Improved tumor targeting of anti-epidermal growth factor receptor Nanobodies through albumin binding: taking advantage of modular Nanobody technology. *Mol. Cancer Ther.* 7, 2288–2297.
35. Dennis MS, Zhang M, Meng G, Kadkhodayan M, Kirchhofer D, Combs D, et al. (2002) Albumin binding as a general strategy for improving the pharmacokinetics of proteins. *J. Biol. Chem.* 277, 35035–35043.
36. Dennis MS, Jin H, Dugger D, Yang R, McFarland L, Ogasawara A, et al. (2007) Imaging tumors with an albumin-binding Fab, a novel tumor-targeting agent. *Cancer Res.* 67, 255–261.
37. Nguyen A, Reyes AE, Zhang M, McDonald P, Wong WL, Damico LA, et al. (2006) The pharmacokinetics of an albumin-binding Fab (AB.Fab) can be modulated as a function of affinity for albumin. *Protein Eng. Des. Sel.* 19, 291–297.
38. Dumelin CE, Trüssel S, Buller F, Trachsel E, Bootz F, Zhang Y, et al. (2008) A portable albumin binder from a DNA-encoded chemical library. *Angew. Chem. Int. Ed.* 47, 3196–3201.
39. Trüssel S, Dumelin C, Frey K, Villa A, Buller F, Neri D. (2009) New strategy for the extension of the serum half-life of antibody fragments. *Bioconjug. Chem.* 20, 2286–2292.
40. Constantinou A, Epenetos AA, Hrezcuk-Hirst D, Jain S, Deonarain MP. (2008) Modulation of antibody pharmacokinetics by chemical polysialylation. *Bioconjug. Chem.* 19, 643–650.
41. Constantinou A, Chen C, Deonarain MP. (2010) Modulating the pharmacokinetics of therapeutic antibodies. *Biotechnol. Lett.* 32, 609–622.
42. Schellenberger V, Wang C, Geething NC, Spink BJ, Campbell A, To W, et al. (2009) A recombinant polypeptide extends the in vivo half-life of peptides and proteins in a tunable manner. *Nat. Biotechnol.* 27, 1186–1192.

12

TRANSFERRIN FUSION PROTEIN THERAPIES: ACETYLCHOLINE RECEPTOR-TRANSFERRIN FUSION PROTEIN AS A MODEL

Dennis Keefe, Michael Heartlein, and Serene Josiah

Shire Human Genetic Therapies, Lexington, MA, USA

12.1 DISEASE OVERVIEW

Myasthenia gravis (MG) is a rare antibody-mediated autoimmune disease with a prevalence of ~40,000 patients in the United States. The disease pathogenesis is well described and a hallmark of the disease is the development of antibodies targeted to the nicotinic acetylcholine receptor (AChR) located at the surface of neuromuscular junctions (NMJ) (reviewed in Reference [1]). The binding of autoantibodies to the receptor leads to multiple complications leading ultimately to loss of functional NMJ. The antibodies can (1) interfere with the acetylcholine signaling pathway, (2) downregulate AChR expression on the cell surface, and (3) activate the complement cascade promoting formation of the membrane attack complex (MAC). Together, these events lead to necrosis and inflammation [2–4], which results in failure of appropriate neuromuscular synaptic transmission and the subsequent muscle weakness seen in MG patients. In addition to muscle weakness in the arms and legs, other clinical signs include drooping eyelids, ocular motor disturbances, difficulty in chewing and swallowing, breathlessness, and difficulty in speaking. While MG can present at any age, it typically appears in the late 20s (females) to 40s (males), and results in mortality in 3–4% of cases.

Approximately 80% of MG patients develop antibodies to the AChR, while about half of the remaining individuals develop antibodies to muscle-specific tyrosine kinase (MuSK) [1,5] The clinical improvement seen in patients following plasmapheresis supports the assessment that autoantibodies promote disease presentation [6]. Animal studies also provide evidence that the MG phenotype is through autoantibodies against AChR. MG disease presentation can be induced in animal models either by immunization with purified AChR receptor or passive transfer of patient sera or monoclonal antibodies to the main immunogenic region (MIR) of the AChR [7,8]. Thus, it is well established that autoantibodies against AChR are the primary effectors of MG.

The presence of anti-AChR antibodies in serum is diagnostic for MG. Current treatment strategies for MG include thymectomy, steroids, immunosuppressive agents, plasmapheresis, or intravenous immunoglobulin (IVIG) treatment [9]. Clinical improvement associated with a particular therapy or combination of therapies is generally correlated with a notable decrease in circulating autoantibody levels.

The AChR is a transmembrane pentameric protein containing two α-subunits and one each of β, δ, and ε subunits [10]. It is likely that the autoantibodies cross-react with multiple subunits as there is ~40% amino acid sequence homology between the subunits [11]. The antibody response

Fusion Protein Technologies for Biopharmaceuticals: Applications and Challenges, First Edition. Edited by Stefan R. Schmidt.

in patients is varied. For example, MG patients can develop variable antibody responses to multiple subunits, which can range from a polyclonal response to a number of subunit epitopes or a multiple response to a single subunit. Approximately 50% of MG patients develop notable antibody levels to the α-subunit [12]. Anti-α and anti-β subunit antibodies are both able to cross-link and downregulate AChR receptors (antigenic modulation). The anti-α antibodies are approximately four times more effective than anti-β-subunit antibodies at promoting antigenic modulation [13]. Antibodies to the α-subunit are able to induce experimental autoimmune myasthenia gravis (EAMG) in mice, while the effect of the anti-β antibodies is much weaker [8]. These data suggest that anti-α-subunit antibodies play a key role in MG pathogenesis.

Current treatment options for MG patients are invasive and are not targeted therapies for MG. Thus, novel therapeutic approaches, which eliminate the need for surgery and the complications of reoccurring plasmapheresis are desirable. The *in vitro* studies described in this chapter are a positive first step in the characterization of one such approach: specific targeting of α domain specific anti-AChR antibodies and clearing the antibodies from circulation via a fusion protein. Given the broad range of antibodies that can cause MG, it is worthwhile to consider a combinatorial therapeutic strategy. The potential efficacy of a fusion protein targeted to α domain specific anti-AChR antibodies suggests the possibility of developing multiple subunit AChR-transferrin fusion proteins for the treatment of MG. When used in combination with anti-α-subunit therapy, this experimental therapeutic approach could address a range of MG patients.

12.2 FUSION PROTEIN SHG2210 DESIGN

Chimeric proteins containing portions of the α-subunit fused to GST have been shown to effectively bind anti-AChR antibodies and can protect and treat MG-like disease in both chronic and acute EAMG models [14]. There are two key factors that contribute to the design of SHG2210 fusion protein as a targeted therapy for MG; the target molecule implicated in disease is known (α-subunit AChR antibodies) and TFR-mediated cellular uptake can be exploited as a means to clear target molecules from circulation.

Although MG patients have an array of antibodies against various AChR subunits, the majority of patients develop an autoimmune response to the main immunogenic region (MIR) region of the α-subunit of AChR. Thus, to test the concept of a fusion protein, the α-subunit of the AChR containing the MIR domain was chosen to capture disease causing anti-α-subunit AChR antibodies.

TFR-mediated cellular uptake is a complex process that is studied extensively. Transferrin receptors are ubiquitous on the surface of almost all cell types and transferrin recycling mediated via the transferrin receptor occurs very rapidly, on the order of 5–15 min [15–17]. Serum TF, in the iron bound form (holotransferrin) binds to its target cell surface receptor triggering internalization through clathrin-mediated endocytosis into early endosomes. Acidification of the endosome releases iron from transferrin, and this iron-free form (known as apotransferrin) in conjunction with transferrin receptor is returned to the cell surface, via Rab11+ recycling endosomes. Thus, for proof of principle studies to test the concept of a fusion protein, TF was identified as the fusion partner to α-subunit AChR to facilitate rapid and specific receptor-mediated cellular uptake in a wide range of cells.

The targeted mechanism of action (Figure 12.1) is (1) binding of α-subunit AChR antibodies to α-subunit AChR of fusion protein, (2) binding of Ab-bound fusion protein to TFR via TF domain of fusion protein, (3) uptake of fusion protein and antibody complex into cells (serum clearance), (4) separation of fusion protein from cargo (anti-AChR antibody) within the endosome, (5) degradation of cargo in lysosomes or via the proteasome, and (6) ideally, recycling of fusion protein back to the circulation to bind and clear α-subunit AChR antibodies once again.

There are several challenges the α-subunit AChR:TF (SHG2210) fusion protein faces. The fusion protein must be able to bind to anti-α-subunit antibodies specific to native antigen; AChR. The antibody-bound form of the fusion

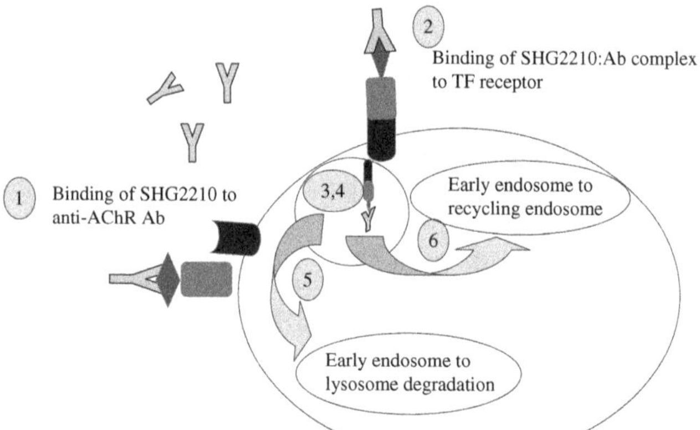

FIGURE 12.1 Schematic representation of proposed mechanism of action of SHG2210. SHG2210 fusion protein binds to anti-AChR α-subunit specific autoantibodies in the circulation through interactions with the AChR α-subunit portion of the fusion protein. Fusion protein:antibody complex is then internalized by cells expressing the transferrin receptor. Internalized complex may then be dissociated to its individual components within the endosome as it acidifies or remain as an associated complex. Associated complex or the individual components of the complex may then follow one of two pathways: default trafficking to the lysosome for degradation or recycling back to the cell surface via fusion with Rab11 + recycling endosomes.

protein must be able to engage the transferrin receptor, and the fusion protein:antibody complex must be efficiently internalized by cells. The fusion protein:antibody complex must either dissociate within the cell, or the entire complex must be degraded within the cell. This latter scenario, in which the fusion protein does not release its cargo and recycle, would impact dosing but, may still be a feasible therapeutic strategy. Further, if the antibody or fusion protein:antibody complex is recycled back into circulation, the fusion protein will lack efficacy or may exacerbate disease. Finally, as MG is caused by antibodies to other AChR domains, SHG2210 would not be able to treat all cases of MG.

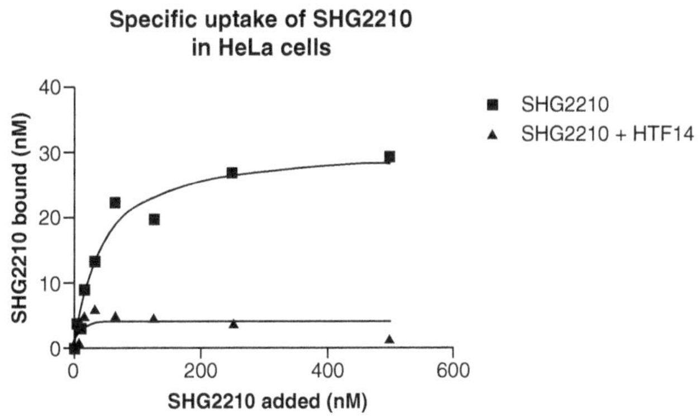

FIGURE 12.2 Uptake of SHG2210 in the presence or absence of antitransferrin neutralizing antibody HTF14 HeLa cells were incubated with increasing doses of SHG2210 or SHG2210 + fivefold molar excess of anti-TF neutralizing antibody HTF14 for 2 h at 37°C. Following trypsin treatment to remove nonincorporated protein, cell lysates were made and the amount of SHG2210 taken into cells quantified by human transferrin ELISA. Values are the mean of triplicate wells extrapolated from a standard curve generated with recombinant SHG2210.

12.3 CHARACTERIZATION OF SHG2210

There are two critical elements to the fusion protein strategy; SHG2210 must bind to anti-α-subunit antibodies within the circulation and SHG2210 must bind transferrin receptor, which facilitates cellular internalization. A corollary to these two prerequisites is that the antibody bound form of SHG2210 must also bind TFR and be internalized by cells. *In vitro* studies were performed to assess the efficacy of SHG2210. Key assays and technology platforms include cell-based binding and uptake assays, molecular imaging and confocal microscopy studies, and antigenic modulation assays. *In vitro* studies show (1) SHG2210 binds to anti-AChR antibodies, (2) SHG2210:antibody complexes are specifically targeted to the transferrin receptor on cells, (3) fusion protein:antibody complex is taken into cells and traffics to the lysosome, and (4) SHG2210 has a protective effect on the antigenic modulation of the AChR induced by serum from select MG patients. These findings suggest that a fusion protein approach may be an effective therapeutic for treating MG.

12.3.1 SHG2210 Binding and Internalization Studies in Cells

To characterize transferrin receptor specific binding and uptake, SHG2210 uptake assays were performed in HeLa cells, which are known to express high levels of transferrin receptor and have been shown to internalize and recycle transferrin [15,16]. Cells were treated with increasing doses of SHG2210 for 2 h at 37°C in the presence or absence of monoclonal antibody HTF14, which specifically binds to human transferrin at an epitope required for recognition by transferrin receptor [18]. Cells were washed to remove surface bound but noninternalized protein, lysed in standard cell lysis buffer, and assayed for fusion protein uptake using a human transferrin enzyme-linked immunosorbent assay (ELISA). As depicted in Figure 12.2, fusion protein uptake occurs primarily through transferrin receptor, with <20% of total cellular uptake occurring through nonspecific (i.e., nontransferrin receptor mediated) mechanisms. ELISA results were further verified with molecular imaging data, which demonstrated that fusion protein is specifically internalized in HeLa cells through transferrin receptor and that fusion protein and transferrin receptor colocalize in the cell cytoplasm [19]. Importantly, neither free fusion protein nor fusion protein:antibody complex were appreciably internalized through AChR receptor itself. Although α-α-subunit interactions are known to occur, SHG2210 is not internalized via AChR. Cross-linking and internalization of native receptor could potentially lead to its trafficking to the lysosome for degradation, which would exacerbate disease rather than have a therapeutic effect.

12.3.2 SHG2210 Anti-AchR Antibody Binding Studies

Binding to α-subunit specific anti-AchR antibodies was first established using the well-characterized anti-AchR monoclonal antibodies mAb198 [20] and mAb 35 [21]. A competition ELISA using immobilized torpedo fish AChR as antigen was used to measure SHG2210 binding to anti-AChR Ab (mAb 198). SHG2210, but not human transferrin, inhibited binding of mAb 198 to AChR, with an $IC_{50} = 0.37 \pm 0.15$ μM [19]. mAb 35 is highly confirmation specific [21] and coimmunoprecipitation studies were used to verify binding of mAb35 by SHG2210. Antibody binding was also verified using serum from patients undergoing treatment for MG disease. SHG2210 bound to anti-AChR antibodies in the sera of patients known to have high anti-α-subunit titers, while antibody binding was weak or absent in patients with low anti-α-subunit titers [19]. Taken together, these data suggest that the α-AChR portion of SHG2210 adopts a native or near native conformation in solution and is capable of binding to a variety of α-subunit specific anti-AChR antibodies. The fusion protein therefore performs as engineered with respect to antibody binding.

12.3.3 SHG2210: Anti-AchR Antibody Complex Binding and Internalization Studies in Cells

SHG2210:antibody complex binding and uptake by HeLa cells was first assayed by ELISA. The ELISA used antihuman transferrin as a capture antibody and peroxidise-conjugated anti-rat IgG as detection antibody. SHG2210: antibody complexes were preformed before adding to cells using mAb35 as model anti-AChR antibody. As with free fusion protein, fusion protein:antibody complex uptake by HeLa cells was transferrin receptor specific, as demonstrated by negligible uptake in the presence of saturating levels of human transferrin as inhibitor [19]. ELISA results were further verified by molecular imaging data. Imaging experiments were performed in the human muscle cell line DB40, a transfected subclone of the TE671 muscle cell line frequently used for *in vitro* studies of human MG disease [22], in order to more closely approximate the cells of interest in actual MG disease. By directly labeling both fusion protein and anti-AChR antibody with fluorescent dyes having distinct emission wavelengths (Alexa fluor 564 and 488, respectively), the fusion protein:antibody complexes bound to transferrin receptor within cells was visualized using three-color laser scanning confocal microscopy (Figure 12.3). Therefore, using a combination of biochemistry and fluorescent imaging, the fusion protein:antibody complex was shown to be taken into cells via TFR.

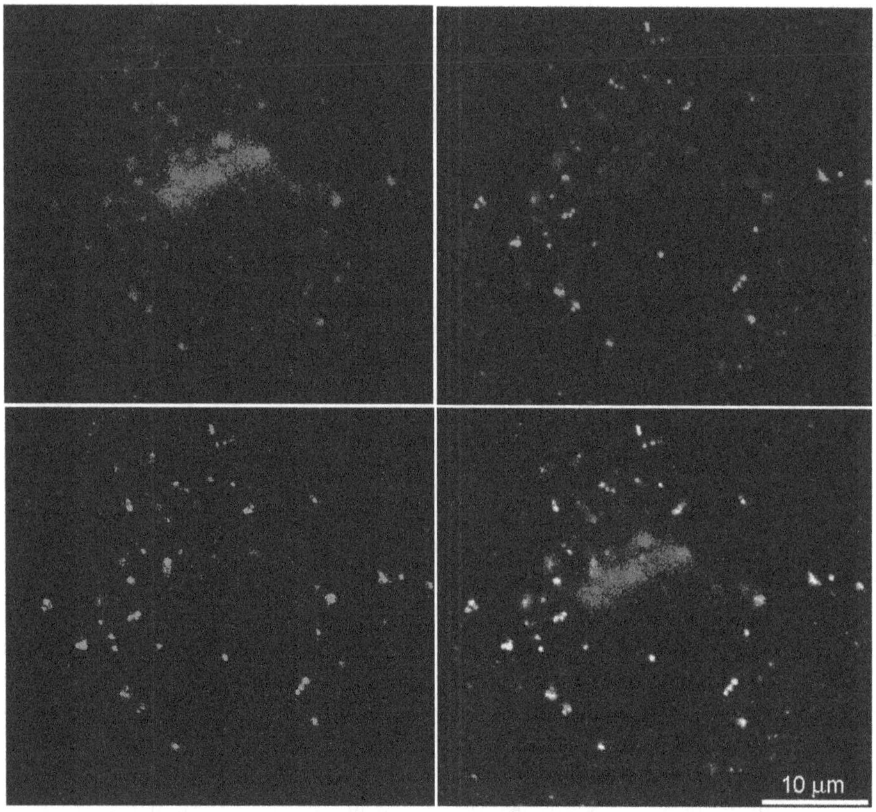

FIGURE 12.3 Molecular imaging of internalized SHG2210:mAb 35 complex in human muscle cells by confocal microscopy SHG2210 fusion protein and mAb 35 were directly labeled with Alexa fluor 546 and 488 dyes, respectively. Fusion protein and antibody were preincubated for 15 min to form complexes and then added to the human muscle cell line DB40 for 2 h on ice. Cells were then washed and switched to 37°C to initiate uptake of the complex. Uptake of complex was allowed to occur for 15 min, at which point cells were fixed and processed for immunofluorescence using antitransferrin receptor monoclonal antibody. Clockwise from top: anti-TF receptor (blue); mAb 35 (green); SHG2210 (red); merged image. (See the color version of this figure in Color Plates section.)

12.3.4 SHG2210 and SHG2210: Anti-AChR Antibody Complex Recycling Studies in Cells

In cells expressing transferrin receptor, iron loaded transferrin (holotransferrin) is recycled back to the cell surface by transferrin receptor after releasing iron within the cell. It was not known if a transferrin fusion protein would also recycle in cells; therefore, fusion protein recycling was assayed with a combination of biochemical and imaging endpoints. Recycling of fusion protein could have both beneficial and deleterious effects on the therapeutic potential of the molecule. For example, if the fusion protein were able to recycle to the cell surface following deposition of its cargo into the cytoplasm, multiple rounds of antibody clearance could be achieved from a single administration of protein, thus prolonging the pharmacokinetic profile of the drug. However, if on the other hand, fusion protein and antibody complex were recycled back to the cell surface and released by cells, pathogenic antibodies could be placed back into circulation.

Unlike native transferrin, very little fusion protein is returned to cell culture media in cells pulsed with SHG2210 (Figure 12.4) suggesting that SHG2210 fusion protein recycles very poorly in cells. In addition, unlike native transferrin, SHG2210 fusion protein does not colocalize with recycling endosomes (unpublished data). Taken together, these data suggest that addition of the AChR α-subunit to transferrin alters its intracellular trafficking properties and prevents efficient recycling back to the cell surface.

The fusion protein and its cargo could dissociate in the endosome or the entire complex could recycle back to the cell surface. *In vitro* biochemical and molecular imaging studies show negligible recycling of fusion protein:antibody complex [19]. Therefore, similar to free fusion protein, the fusion protein:antibody complex does not recycle in a manner analogous to native human transferrin. Rather, the complex appears to traffic directly to the lysosome. As depicted in Figure 12.5, SHG2210:mAb 35 complex colocalizes with the lysosomal antigen Lamp-1 in cellular imaging experiments. In these experiments, lysosomal accumulation of complex required ~30–45 min. Furthermore, molecular imaging data support a temporal progression of SHG2210 to transferrin receptor positive early endosomes at 5–10 min to Lamp-1 positive lysosomes by 30–45 min. In this regard, fusion protein:antibody complex traffic in a manner similar to internalized AChR, which has been shown to traffic to lysosomes both constitutively and following cross-linking by anti-AChR autoantibodies [23–25].

12.3.5 Antigenic Modulation Studies

One of the major mechanisms of disease pathology in MG is antigenic modulation; the binding, internalization, and eventual destruction of AChR at the neuromuscular junction by cross-linking AChR-specific autoantibodies [26–28]. Antigenic modulation ultimately leads to the failure of muscle cells to respond to signals at the NMJ from neurons, leading to muscle weakness and fatigue. For this reason, antigenic modulation of the AChR on cultured cells has become an important diagnostic tool for assessing the pathogenicity of AChR-specific autoantibodies. Thus, the potential therapeutic benefit of SHG2210 was assessed by its ability to prevent antigenic modulation. Antigenic modulation was measured by flow cytometry [29]. Patient sera were used to cross-link receptors. In patients with high titers of α-subunit specific anti-AChR antibodies, the fusion protein had a protective effect. Likewise, in serum samples with low α-subunit composition, protection from antigenic modulation by

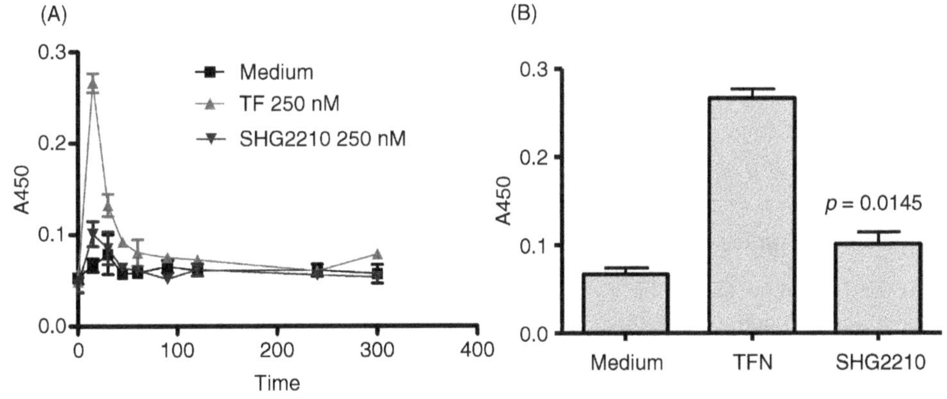

FIGURE 12.4 Recycling of SHG2210 and native human transferrin in DB40 muscle cells. Cells were briefly pulsed with fusion protein or native tranferrin to allow internalization to occur. Following wash steps to remove nonincorporated and cell associated protein, cells were placed at 37°C and aliquots of culture media assayed by human transferrin ELISA at various timepoints for recycled SHG2210 or native transferrin. (A) Recycling over 5 h. (B) Recycling at $t = 15$ min, $p < 0.05$ by one way ANOVA.

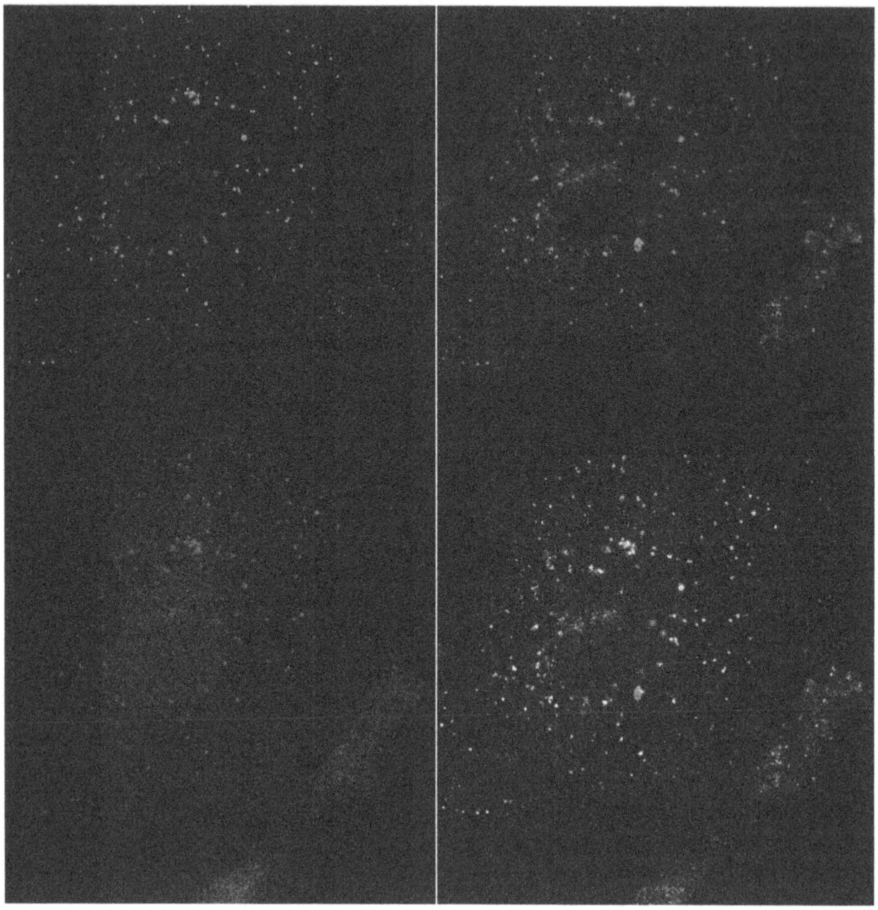

FIGURE 12.5 Molecular imaging of internalized SHG2210:mAb 35 complex in human muscle cells by confocal microscopy SHG2210 fusion protein and mAb 35 were directly labeled with Alexa fluor 546 and 488 dyes, respectively. Fusion protein and antibody were preincubated for 15 min to form complexes and then added to the human muscle cell line DB40 for 2 h on ice. Cells were then washed and switched to 37°C to initiate uptake of the complex. Uptake of complex was allowed to occur for 45 min, at which point cells were fixed and processed for immunofluorescence using anti-Lamp1 monoclonal antibody. Clockwise from top: mAb 35 (green); SHG2210 (red); Lamp-1 (blue); merged image. (See the color version of this figure in Color Plates section.)

SHG2210 was absent [19]. Data from a typical experiment are presented in Figure 12.6.

12.3.6 Summary of *In Vitro* Studies of SHG2210

In vitro studies were performed to assess the efficacy of SHG2210. Key assays and technology platforms include cell-based binding and uptake assays, molecular imaging, and confocal microscopy studies and antigenic modulation assays. *In vitro* studies show (1) SHG2210 binds to anti-AChR antibodies, (2) SHG2210:antibody complexes are specifically targeted to the transferrin receptor on cells, and (3) fusion protein:antibody complex is taken into cells and traffics to the lysosome. The SHG2210 and SHG2210/anti-AchR antibody complex is present in Lamp-1 positive lysosomal compartments after internalization; however, neither SHG2210 nor SHG2210/antibody complex is present in Rab-11 positive recycling endosomes. Thus, SHG2210 and SHG2210/anti-AchR complex is degraded with negligible recycling observed. Significantly, SHG2210 is shown to have a protective effect on antigenic modulation of the AChR induced by serum from selected patients with MG, suggesting that a fusion protein approach may be an effective therapeutic for treating MG.

12.4 APPLICATIONS AND INDICATIONS

SHG2210 is designed as a specific therapeutic for MG. Indeed, it is even specific to a certain subset of MG patients:

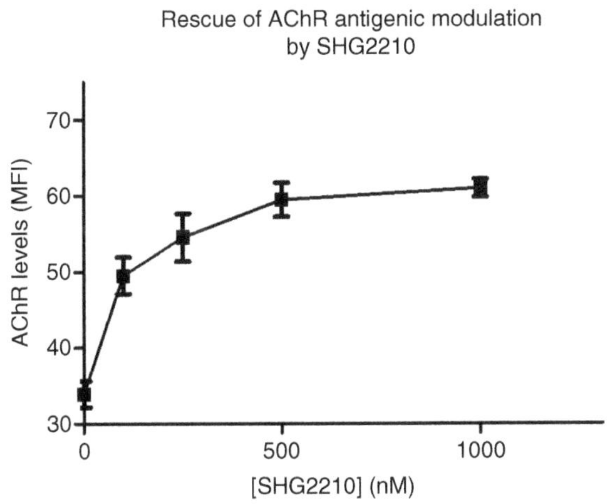

FIGURE 12.6 Prevention of AChR antigenic modulation by SHG2210 fusion protein. Data depict the mean fluorescence intensity (MFI) of AChR staining on DB40 cells by flow cytometry following overnight treatment of cells with myasthenic serum from an MG patient. Data are the mean of triplicate samples, ±SD.

those with predominantly anti-α-subunit responses. The general concept of a neutralizing fusion protein for the binding and sequestration of anti-AChR autoantibodies has been approached with various fusion proteins. Barchan et al. demonstrated that GST fusions of full length or partial α-subunit had a protective effect on antigenic modulation of the AChR both *in vitro* and development of EAMG *in vivo* [14,30]. Recombinant AChR α-subunit fragments have been fused to toxins to produce "smart bombs," which specifically target and eliminate autoreactive B cells expressing anti-AChR antibodies on their cell surface [31]. Finally, a fusion composed of the extracellular domain of the human α1-subunit and the Fc domain of the human IgG1 heavy chain was shown to immunomodulate AChR-specific autoreactive B cells [32].

Apart from myasthenia gravis, fusion proteins that both neutralize an unwanted entity and have an intrinsic uptake or clearance pathway are attractive drug candidates for many indications. For example, an LDL receptor-transferrin fusion protein was shown to reduce levels of plasma LDL cholesterol and has been proposed as a potential therapeutic for the treatment of familial hypercholesterolemia [33]. There is also a wealth of literature describing the use of IgG fusion proteins to scavenge or modulate soluble mediators of inflammation and prevent unwanted interactions between cells and potential pathogens. Several examples are described below and are beyond the scope of this chapter. Tumor necrosis factor receptor/IgG fusion proteins, which neutralize and clear the proinflammatory cytokine TNF-α are well described in the literature and have been commercialized for the treatment of rheumatoid arthritis [34,35], while TrkA-IgG fusions that bind and neutralize intraspinal NGF have been used to help attenuate dysreflexia subsequent to spinal cord injury [36] Recently, a CD4-VH3-IgG fusion protein has been described that partially neutralizes HIV by inhibiting entry of the virus into cells [37], while a TWEAK cytokine scavenging IgG fusion protein has shown promise in the treatment of atherosclerosis in mice [38].

Unlike the IgG fusions outlined above, SHG2210 was engineered to use the TF-TFR. As such, SHG2210 joins a long list of therapeutic candidates and drugs that utilize this ubiquitous ligand/receptor pathway (reviewed in Reference [39]). However, unlike SHG2210, which seeks to use the TFR as a clearance mechanism, most drugs that utilize TF or target the TFR are doing so in an effort to target the therapeutic to a particular cell type or to gain access to difficult drug niches within the body. The most common TFR targeting fusion protein drugs are those used to treat cancer. Examples include TF-doxorubicin conjugates for the treatment of drug-resistant cells [40–42] and transferrin diphtheria toxin subunit fusions for the treatment of solid human gliomas [43]. The TF-TFR axis is also an attractive candidate for the delivery of therapeutics across the blood brain barrier; a host of conjugates based on the anti-TFR monoclonal antibody OX26 have shown promise *in vitro* and in animal models as well [44,45]. Finally, the development of modular drug delivery scaffolds built upon lipoplex and liposome technology have been adapted for TF-TFR specific targeting *in vitro* and *in vivo* [46–48].

12.5 FUTURE PERSPECTIVES

In vitro studies show SHG2210 is able to bind both TF and α-AChR antibodies. However, it is likely that SHG2210 complexed with α-AChR is sterically hindered relative to TF and thus, cellular trafficking is distinct from TF. Although SHG2210 is able to prevent antigenic modulation, its efficacy is limited as not all patient sera responded favorably. The engineering of a novel AChR-TF fusion protein may be needed to improve the efficacy and is faced with several challenges. For example, efficacy may be improved by generating a molecule with multiple AChR subunits, fused to transferrin to enable binding to a broader class of antibodies. Psaridi-Linardaki et al. demonstrated that the extracellular domain of the α-subunit, immobilized on sepharose columns, could remove myasthenic antibodies from the sera of MG patients [49]. In a follow-up to this study, this same group demonstrated that a combinatorial approach, in which α-subunit was combined with β, γ, and ε subunits on the same column, greatly enhanced the efficiency of clearance [11] However, by including the entire extracellular domain of each AChR subunit, it is likely that the resultant fusion protein is simply too large to engage the transferrin receptor. In turn, this would increase the risk of aberrant TFR-mediated uptake and trafficking.

12.6 CONCLUSION

In conclusion, *in vitro* studies are a valuable approach to assessing the mechanism of action and potential efficacy of fusion proteins: (1) SHG2210 is able to bind to α-subunit autoantibodies and the TF receptor (TFR) and (2) SHG2210 and SHG2210-anti-AchR antibody complex are internalized through TFR-mediated endocytosis. SHG2210 is shown to have a protective effect on antigenic modulation of the AChR induced by serum from select patients with MG, suggesting that a fusion protein approach may be an effective therapeutic for treating MG and TF fusion proteins may remove targeted molecules from circulation. The studies outlined in this chapter provide an overview of the characterization of a AChR-TF-specific fusion protein in the context of a potential strategy for the treatment of myasthenia gravis. Key points to consider when characterizing a TF-based fusion protein are (1) binding properties of each portion of the fusion protein, (2) TFR binding and uptake, and (3) cellular trafficking. The use of TF as a fusion partner for tissue targeting and delivery remains a promising area of research.

REFERENCES

1. Conti-Fine BM, et al. (2006) Myasthenia gravis: past, present, and future. *J. Clin. Invest.* 116(11), 2843–2854.
2. Biesecker G, Gomez CM, (1989) Inhibition of acute passive transfer experimental autoimmune myasthenia gravis with Fab antibody to complement C6. *J. Immunol.* 142(8), 2654–2659.
3. Lennon VA, et al. (1978) Role of complement in the pathogenesis of experimental autoimmune myasthenia gravis. *J. Exp. Med.* 147(4), 973–983.
4. Sahashi K, et al. (1980) Ultrastructural localization of the terminal and lytic ninth complement component (C9) at the motor end-plate in myasthenia gravis. *J. Neuropathol. Exp. Neurol.* 39(2), 160–172.
5. Hoch W, et al. (2001) Auto-antibodies to the receptor tyrosine kinase MuSK in patients with myasthenia gravis without acetylcholine receptor antibodies. *Nat. Med.* 7(3), 365–368.
6. Newsom-Davis J, et al. (1978) Function of circulating antibody to acetylcholine receptor in myasthenia gravis: investigation by plasma exchange. *Neurology* 28(3), 266–272.
7. Christadoss P, et al. (2000) Animal models of myasthenia gravis. *Clin. Immunol.* 94(2), 75–87.
8. Tzartos S, et al. (1987) Passive transfer of experimental autoimmune myasthenia gravis by monoclonal antibodies to the main immunogenic region of the acetylcholine receptor. *J. Neuroimmunol.* 15(2), 185–194.
9. Garcia-Carrasco M, et al. (2007) Therapeutic options in autoimmune myasthenia gravis. *Autoimmun. Rev.* 6(6), 373–378.
10. Kalamida D, et al. (2007) Muscle and neuronal nicotinic acetylcholine receptors. Structure, function and pathogenicity. *FEBS J.* 274(15), 3799–3845.
11. Kostelidou K, et al. (2007) Extracellular domains of the beta, gamma and epsilon subunits of the human acetylcholine receptor as immunoadsorbents for myasthenic autoantibodies: a combination of immunoadsorbents results in increased efficiency. *J. Neuroimmunol.* 190(1,2), 44–52.
12. Loutrari H, et al. (1997) Expression of human-Torpedo hybrid acetylcholine receptor (AChR) for analysing the subunit specificity of antibodies in sera from patients with myasthenia gravis (MG). *Clin. Exp. Immunol.* 109(3), 538–546.
13. Sideris S, et al. (2007) Isolation and functional characterization of anti-acetylcholine receptor subunit-specific autoantibodies from myasthenic patients: receptor loss in cell culture. *J. Neuroimmunol.* 189(1,2), 111–117.
14. Barchan D, et al. (1998) Modulation of the anti-acetylcholine receptor response and experimental autoimmune myasthenia gravis by recombinant fragments of the acetylcholine receptor. *Eur. J. Immunol.* 28(2), 616–624.
15. Bleil JD, Bretscher MS, (1982) Transferrin receptor and its recycling in HeLa cells. *EMBO J.* 1(3), 351–355.
16. Lamb JE, et al. (1983) Internalization and subcellular localization of transferrin and transferrin receptors in HeLa cells. *J. Biol. Chem.* 258(14), 8751–8758.
17. Sager PR, et al. (1984) Analysis of transferrin recycling in mitotic and interphase HeLa cells by quantitative fluorescence microscopy. *Cell* 39(2 Pt 1), 275–282.
18. Hradilek A, Neuwirt J, (1986) Iron uptake by MOLT 3 cells from transferrin/monoclonal antitransferrin antibody complexes. *Br. J. Haematol.* 62(1), 21–30.
19. Keefe D, et al. *In vitro* characterization of an acetylcholine receptor-transferrin fusion protein for the treatment of myasthenia gravis. *Autoimmunity* 43(8), 628–639.
20. Tzartos SJ, et al. (1988) Localization of the main immunogenic region of human muscle acetylcholine receptor to residues 67–76 of the alpha subunit. *Proc. Natl. Acad. Sci. USA* 85(9), 2899–2903.
21. Tsouloufis T, et al. (2000) Reconstitution of conformationally dependent epitopes on the N-terminal extracellular domain of the human muscle acetylcholine receptor alpha subunit expressed in *Escherichia coli*: implications for myasthenia gravis therapeutic approaches. *Int. Immunol.* 12(9), 1255–1265.
22. Beeson D, et al. (1996) A transfected human muscle cell line expressing the adult subtype of the human muscle acetylcholine receptor for diagnostic assays in myasthenia gravis. *Neurology* 47(6), 1552–1555.
23. Bruneau E, et al. (2005) Identification of nicotinic acetylcholine receptor recycling and its role in maintaining receptor density at the neuromuscular junction *in vivo*. *J. Neurosci.* 25(43), 9949–9959.
24. Bruneau EG, Akaaboune M, (2006) The dynamics of recycled acetylcholine receptors at the neuromuscular junction *in vivo*. *Development* 133(22), 4485–4493.
25. Kumari S, et al. (2008) Nicotinic acetylcholine receptor is internalized via a Rac-dependent, dynamin-independent endocytic pathway. *J. Cell. Biol.* 181(7), 1179–1193.

26. Heinemann S, et al. (1977) Modulation of acetylcholine receptor by antibody against the receptor. *Proc. Natl. Acad. Sci. USA* 74(7), 3090–3094.
27. Lindstrom J, Einarson B. (1979) Antigenic modulation and receptor loss in experimental autoimmune myasthenia gravis. *Muscle Nerve* 2(3), 173–179.
28. Reiness CG, et al. (1978) Antibody to acetylcholine receptor increases degradation of junctional and extrajunctional receptors in adult muscle. *Nature* 274(5666), 68–70.
29. Keefe D, et al. (2009) A rapid, fluorescence-based assay for detecting antigenic modulation of the acetylcholine receptor on human cell lines. *Cytometry B. Clin. Cytom.* 76(3), 206–212.
30. Barchan D, et al. (1999) Antigen-specific modulation of experimental myasthenia gravis: nasal tolerization with recombinant fragments of the human acetylcholine receptor alpha-subunit. *Proc. Natl. Acad. Sci. USA* 96(14), 8086–8091.
31. Hossann M, et al. (2006) Novel immunotoxin: a fusion protein consisting of gelonin and an acetylcholine receptor fragment as a potential immunotherapeutic agent for the treatment of Myasthenia gravis. *Protein Expr. Purif.* 46(1), 73–84.
32. Chang T, et al. Selective recognition and elimination of nicotinic acetylcholine receptor-reactive B cells by a recombinant fusion protein AChR-Fc in myasthenia gravis in vitro. *J. Neuroimmunol.* 227(1,2), 35–43.
33. Parise F, et al. (1999) Construction and in vitro functional evaluation of a low-density lipoprotein receptor/transferrin fusion protein as a therapeutic tool for familial hypercholesterolemia. *Hum. Gene. Ther.* 10(7), 1219–1228.
34. Moreland LW, et al. (1997) Treatment of rheumatoid arthritis with a recombinant human tumor necrosis factor receptor (p75)-Fc fusion protein. *N. Engl. J. Med.* 337(3), 141–147.
35. Moreland LW, et al. (1997) Biologic agents for treating rheumatoid arthritis. Concepts and progress. *Arthritis Rheum.* 40(3), 397–409.
36. Marsh DR, et al. (2002) Neutralizing intraspinal nerve growth factor with a trkA-IgG fusion protein blocks the development of autonomic dysreflexia in a clip-compression model of spinal cord injury. *J. Neurotrauma.* 19(12), 1531–1541.
37. Meyuhas R, et al. (2009) Enhanced HIV-1 neutralization by a CD4-VH3-IgG1 fusion protein. *Biochem. Biophys. Res. Commun.* 386(2), 402–406.
38. Schapira K, et al. (2009) Fn14-Fc fusion protein regulates atherosclerosis in ApoE-/- mice and inhibits macrophage lipid uptake in vitro. *Arterioscler. Thromb. Vasc. Biol.* 29(12), 2021–2027.
39. Qian ZM, et al. (2002) Targeted drug delivery via the transferrin receptor-mediated endocytosis pathway. *Pharmacol. Rev.* 54(4), 561–587.
40. Fritzer M, et al. (1996) Cytotoxic effects of a doxorubicin-transferrin conjugate in multidrug-resistant KB cells. *Biochem. Pharmacol.* 51(4), 489–493.
41. Lai BT, et al. (1998) Mechanism of action and spectrum of cell lines sensitive to a doxorubicin-transferrin conjugate. *Cancer Chemother. Pharmacol.* 41(2), 155–160.
42. Singh M, et al. (1998) Transferrin directed delivery of adriamycin to human cells. *Anticancer Res.* 18(3A) 1423–1427.
43. Laske DW, et al. (1994) Efficacy of direct intratumoral therapy with targeted protein toxins for solid human gliomas in nude mice. *J. Neurosurg.* 80(3), 520–526.
44. Bickel U, et al. (1993) Pharmacologic effects in vivo in brain by vector-mediated peptide drug delivery. *Proc. Natl. Acad. Sci. USA* 90(7), 2618–2622.
45. Pardridge WM, et al. (1998) Combined use of carboxyl-directed protein pegylation and vector-mediated blood-brain barrier drug delivery system optimizes brain uptake of brain-derived neurotrophic factor following intravenous administration. *Pharm. Res.* 15(4), 576–582.
46. Cheng PW, (1996) Receptor ligand-facilitated gene transfer: enhancement of liposome-mediated gene transfer and expression by transferrin. *Hum. Gene Ther.* 7(3), 275–282.
47. Simoes S, et al. (1999) Cationic liposomes as gene transfer vectors: barriers to successful application in gene therapy. *Curr. Opin. Mol. Ther.* 1(2), 147–157.
48. Simoes S, et al. (1998) Gene delivery by negatively charged ternary complexes of DNA, cationic liposomes and transferrin or fusigenic peptides. *Gene Ther.* 5(7), 955–964.
49. Psaridi-Linardaki L, et al. (2005) Specific immunoadsorption of the autoantibodies from myasthenic patients using the extracellular domain of the human muscle acetylcholine receptor alpha-subunit. Development of an antigen-specific therapeutic strategy. *J. Neuroimmunol.* 159(1,2), 183–191.

13

HALF-LIFE EXTENSION THROUGH O-GLYCOSYLATION

Fuad Fares

Department of Human Biology, Faculty of Natural Sciences, University of Haifa, Mount Carmel, Haifa, Israel

13.1 INTRODUCTION

Recombinant DNA technology has been used to develop long-acting therapeutic proteins. One strategy is to add O-linked or N-linked oligosaccharide chains to the backbone of the protein.

The interest in carbohydrate (COH) chains of glycoproteins has grown because of accumulating data concerning the importance of these chains in a wide array of biological processes. The COH chains of glycoprotein hormones can be classified into two groups: one contains an *N*-acetyl-galactosamine (GalNac) residue that linked to the hydroxyl group of either a serine (Ser) or a threonine (Thr) residue of a polypeptide and is called O-linked oligosaccharides. The recognition signal of O-linked oligosaccharides is unknown. O-linked glycosylation occurs at a later stage during protein processing, probably in the Golgi apparatus. This is the addition of *N*-acetyl-galactosamine to serine or threonine residues by the enzyme *UDP*-N-*acetyl-*D*-galactosamine: polypeptide* N-*acetylgalactosaminyltransferase* followed by other COHs (such as galactose and sialic acid). This process is important for certain types of proteins such as proteoglycans, which involves the addition of glycosaminoglycan chains to an initially unglycosylated "proteoglycan core protein."

The other group, which called N-linked oligosaccharides, contains an *N*-acetylglucosamine (GlcNac) residue at its reducing terminal and is linked to an amide group of an asparagine (Asn) residue of a polypeptide. There are three major classes of N-linked saccharides resulting from this core: high mannose oligosaccharides, complex oligosaccharides, and hybrid oligosaccharides. High mannose contains two *N*-acetylglucosamines with many mannose residues. Complex oligosaccharides contain almost any number of the other types of saccharides including more than the original two *N*-acetylglucosamines. Hybrid-type oligosaccharides possesses structural features characteristic of both types. The addition of N-linked glycans starts in the lumen of the endoplasmic reticulum (ER), with the transfer of oligosaccharides to the asparagine residues in the sequence Asn-X-Ser/Thr (NXS/T) in the nascent polypeptide chain where X could be any amino acid except proline. The N-linked glycosylation process occurs in eukaryotes and widely in archaea but very rarely in bacteria. In eukaryotes, most N-linked oligosaccharides begin with addition of a 14-sugar precursor to the asparagine in the polypeptide chain of the target protein. A complex set of reactions attaches this branched chain to a carrier molecule called dolichol, and then it is transferred to the appropriate point on the polypeptide chain as it is translocated into the ER lumen.

The structures of N- and O-linked oligosaccharides are different where different sugar residues are found in each type. O-linked oligosaccharides are generally short and containing one to four sugar residues. In contrast, N-linked oligosaccharides usually have several branches each

Fusion Protein Technologies for Biopharmaceuticals: Applications and Challenges, First Edition. Edited by Stefan R. Schmidt.

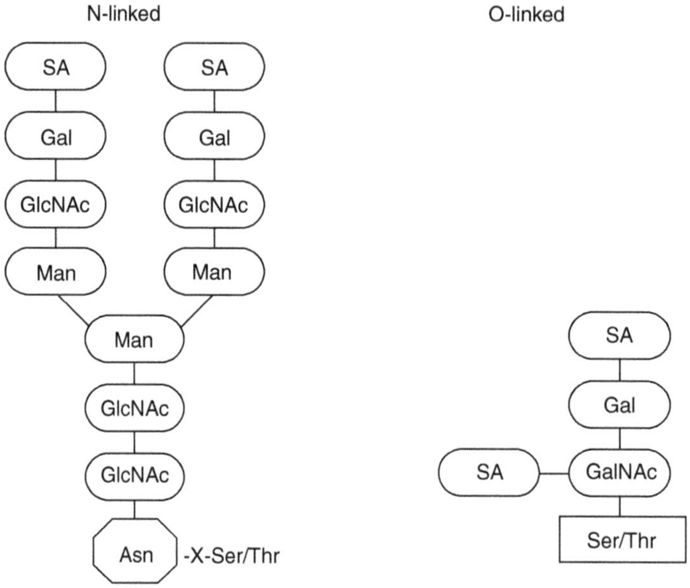

FIGURE 13.1 Structure of typical N- and O-linked oligosaccharide chains.

terminating with negatively charged sialic acid residue (Figure 13.1).

Protein glycosylation is believed to be the most complicated post-translational event. COHs are added to proteins in two organelles: the ER and the Golgi apparatus. COHs addition to proteins occurs both co- and post-translationally. The RNA coding the protein sequence enters the cytoplasm where it binds to ribosomes, which are the site for protein synthesis. Ribosomes bind to the ER and the nascent protein chain enters the lumen and a core oligosaccharide is added to the protein. Further additions of monosaccharides are preformed in the lumen until a final core mannose structure has been added. The ER lumen contains high concentrations of molecular chaperones to assist protein folding. Additional COH modifications (post-translational) are made as the protein moves from the lumen of the ER to the Golgi apparatus. Here, terminal COH modification is completed. The Golgi does not contain molecular chaperons since protein folding is complete when the proteins arrive. Rather they have high concentrations of membrane bound enzymes, including glycosidases and glycosyltransferases.

13.2 THE ROLE OF O-LINKED OLIGOSACCHARIDE CHAINS IN GLYCOPROTEIN FUNCTION

Many reports suggest that O-GlcNAc might be implicated in pathological conditions such as cancer, cardiovascular, or neurodegenerative disorders. Several reports closely link O-GlcNAc to glucose toxicity and insulin resistance. This tight connection between O-GlcNAc and the development of insulin resistance involves the hexosamine biosynthesis pathway (HBP), HBP being the metabolic pathway leading to synthesis of UDP-GlcNAc, the direct donor substrate for O-GlcNAc. The role of O-GlcNAc in striated muscle has to be considered in a larger extent than its implication in insulin resistance. Indeed, the activity of O-GlcNAc transferase is twofold to fourfold higher in skeletal muscle and heart than in liver, therefore suggesting an important role of this post-translational modification in muscle "normal" physiology [1–3]. More generally, different experiments have showed that blocking or reducing O-GlcNAc increased the sensitivity of cells to stress, causing a decrease in cell survival, whereas an increase in O-GlcNAc level protected cells against stress [4].

Other two main functions of O-GlcNAc were described. One function involves secretion to form components of the extracellular matrix, adhering one cell to another by interactions between the large sugar complexes of proteoglycans. The other main function is to act as a component of mucosal secretions, and it is the high concentration of COHs that tends to give mucus its "slimy" feel.

One apparently universal consequence of O-linked oligosaccharide chains is relatively resistance to proteases of O-glycosylated regions in glycoproteins. The most likely explanation for protease resistance is simply that the attached COH blocks access to the peptide core since these same sequences are quite susceptible to proteases in the absence of attached COH. The second consequence of O-glycosylation is the induction of a specific conformation [5].

The human chorionic gonadotropin (hCG) β-subunit is distinguished from the other human β-subunits of the

N-Terminal—Ser Ser Ser **Ser** Lys Ala Pro Pro Pro **Ser**

Leu Prol Ser Pro **Ser** Arg Leu Pro Gly Pro

Ser Asn Thr Prol Ile Leu Pro Gln-*C—Terminal*

FIGURE 13.2 Amino acid sequence of hCG carboxyl-terminal peptide. The positions of O-linked oligosaccharide chains are indicated. O-linked oligosaccharides are indicated as .

glycoprotein hormone family (follitropin [FSH], thyrotropin [TSH], luetropin [LH]) in that it contains a unique 28 amino acid carboxyl-terminal peptide bearing four O-linked oligosaccharide chains (Figure 13.2). Glycoprotein hormones are composed of two subunits: the α-subunit that is common to FSH, TSH, LH, and hCG and a specific β-subunit, which confers the hormone bioactivity. Subunits assembly is the rate-limiting step for secretion, receptor binding and bioactivity of the hormone.

It has been suggested that the O-linked oligosaccharide chains play an important role in the secretion of intact hCG from the cell, enhanced bioactivity *in vivo* and prolonged its circulating half-life [6]. Deletion of the CTP including the O-linked oligosaccharide chains from hCG using site-directed mutagenesis, did not affect assembly of the subunits or secretion of the dimer from the cell. On the other hand, it was shown that truncated hCG without the CTP is three times less potent than intact hCG *in vivo* [7]. On the other hand, the O-linked oligosaccharide chains play a minor role in receptor binding and signal transduction. These findings indicate that the CTP of hCG β and the associated O-linked oligosaccharides are not important for receptor binding or *in vitro* signal transduction but are critical for *in vivo* bioactivity and half-life [6]. It was reported that the kidney is the main site of clearance for glycoprotein hormones. On the other hand, much less hCG, which contains the CTP associated with the four O-linked oligosaccharide chains, is cleared by the kidney [8,9]. Other studies indicated that sialic acids attached to the end of oligosaccharide chains, play an important role in the survival of glycoproteins in the circulation [10]. It has been suggested that more negatively charged forms of glycoprotein hormones have longer half-lives, which may be related to a decrease glomerular filtration [11]. Thus, it was hypothesized that the presence of CTP with its sialylated O-linked oligosaccharides may prolong the circulating half-life of the hormone secondary to a decrease in renal clearance.

Previous studies have shown the important role of sialylation in the pharmacokinetics and distribution of glycoproteins. It was demonstrated that asialoceruloplasmin was cleared rapidly in rabbits, on the other hand, sialic acid capping of the glycan chains of erythropoietin reduces the clearance *in vivo*. Similar observations have been made with other glycoproteins including orosomucoid and hCG [5,12,13]. The primary reason for the rapid clearance of the asialoglycoproteins in previous studies seems to be asialo (exposed) Galactose (Gal) or GalNAc that lead to uptake by lectin receptors in the liver [14]. Circulating glycoproteins usually do not contain O-linked oligosaccharides, and previous data focused on glycoproteins bearing terminal N-linked oligosaccharides. Asialoglycoproteins bearing terminal N-linked Gal or GalNAc are internalized by the asialoglycoprotein (ASGP) receptor via the clathrin-coated pit pathway and are delivered to lysosomes for degradation [15]. The ASGP receptor has been reported to be on the plasma membrane of parenchymal cells/ hepatocytes.

13.3 DESIGNING LONG-ACTING AGONISTS OF GLYCOPROTEIN HORMONES

One major issue regarding the clinical use of glycoprotein hormones is their relatively short half-life *in vivo*. To address this issue, the CTP sequence of hCG β containing four recognition sites of O-linked oligosaccharides was fused to the carboxyl-terminal of hFSH β, hTSH β, and erythropoietin (EPO) (Figure 13.3). Crystallographic studies indicated that both sides of human growth hormone (hGH): N-terminal and C-terminal are not involved in receptor binding and they are accessible (Figure 13.4). Therefore, CTP was ligated to the N-terminal and to the C-terminal of hGH coding sequence (Figure 13.3).

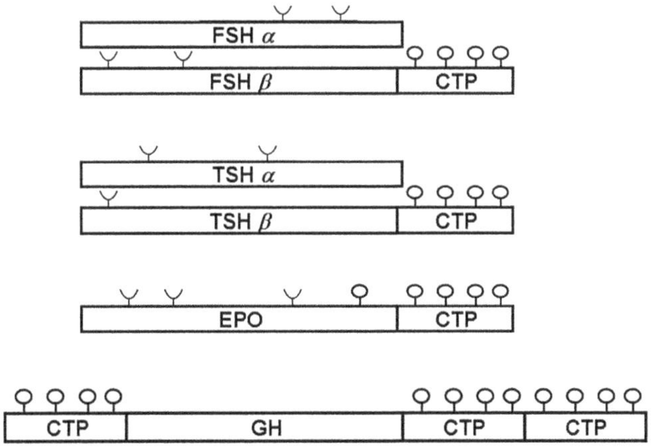

FIGURE 13.3 Construction of glycoprotein hormone chimeras containing the translated sequence of FSH β-subunit, TSH β-subunit, EPO, GH, and hCG β-carboxyl-terminal peptide. O-linked oligosaccharides were indicates as , and N-linked oligosaccharides were indicated as .

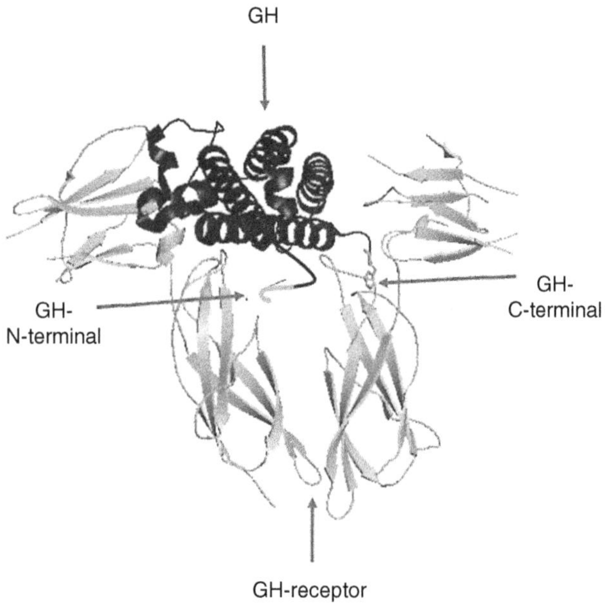

FIGURE 13.4 Crystallographic diagram of GH and its receptor complex. The N-terminal and C-terminal ends of GH are not involved in the binding site of the hormone.

13.3.1 Construction of Chimeric Genes and Expression Vectors

Overlapping mutagenesis PCR technique was used to ligate the CTP to the coding sequence of the hormone. Using this technique, the cassette gene containing the recognition sites of the O-linked oligosaccharides on CTP was ligated in tandem to the coding sequence of the hormone as a single peptide chain. It was hypothesized that hormone containing the CTP would have a prolonged half-life and higher bioactivity *in vivo*.

The FSHβ wild-type (WT), FSH β-CTP, and FSH α genes were inserted into the eukaryotic expression vector, pM^2-HA, that contains the ampicillin (Amp^R) and the neomycin (Neo^R) resistance genes and a strong promoter of the HaMuSV virus, LTR. EPO-WT and EPO-CTP casset genes were inserted into the cloning site of the eukaryotic expression vector, pGT, contains the ampicillin (Amp^R), zeocin resistance genes and a strong promoter of the CMV virus. hGH variant genes, WT and Chimera, were inserted into the pCI-DHFR plasmid, an eukaryotic expression vector. TSH-WT and chimeric subunits were ligated into the expression vector pLBCMV and transfected into COS-7 and 293 cells.

The PCR generated constructs were completely sequenced to ensure that no errors were introduced during the PCR.

13.3.2 Expression of Chimeric Genes

Expression vectors containing wild-type or CTP-chimeric genes were transfected into Chinese hamster ovary (CHO) cells. Stable clones expressing hormones were selected using antibiotics (neomycin or zeocin) and Western blot analysis. Assembly of hFSH β-CTP with α-subunit was examined by transfecting stable clones expressing the α-subunit with hFSH β-CTP. Immunoprecipitation studies prove that dimerization has occurred. It seems that α- and β-subunits of FSH-WT migrate to the same level, meaning that the molecular weight of both subunits is similar (~22 kDa). However, FSH-CTP migrates slower in the gel and exhibits higher molecular weight (~30 kDa) comparing to α-subunit (Figure 13.5). Similarly, The EPO-WT migrated faster than EPO-CTP. EPO-CTP exhibited high molecular weight (~48 kDa) comparing to EPO-WT (~36 kDa) (Figure 13.5). The hGH-WT migrated faster than GH-CTPs and exhibited molecular weight of ~22 kDa. GH variants contain three CTPs exhibited higher molecular weight of ~47.5 kDa. The increase of molecular weight of the variants containing CTP is due to the addition

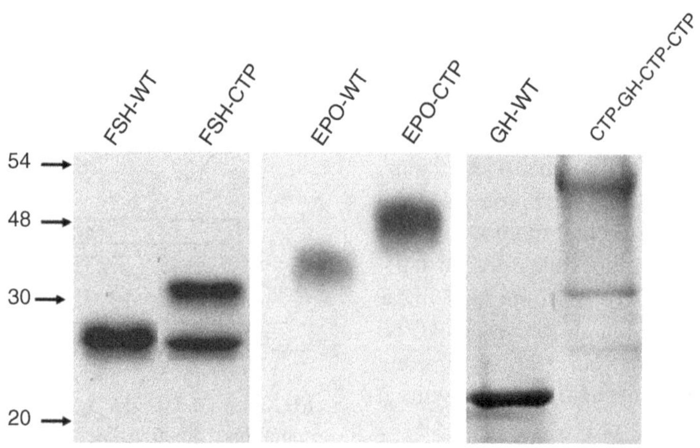

FIGURE 13.5 Expression of FSH, EPO, and GH chimeras from transfected CHO cells. Condition media from transfected cells were prepared for SDS-PAGE and proteins were detected by Western blot analysis using specific antisera.

of 28 amino acids and the O-linked oligosaccharides. These data may indicate that the O-linked glycosylation recognition site of the C-terminal region is preserved even though the sequence is fused to different proteins.

13.3.3 Bioactivity of Designed Long-Acting Glycoproteins

13.3.3.1 ***Follicle-Stimulating Hormone*** (follitropin, FSH) is a pituitary glycoprotein hormone essential for maintenance of ovarian follicle and testicular tubule development. Because FSH specifically induces aromatase enzyme in granulosa cells, *in vitro* signal transduction of the modified FSH dimers was assessed in the granulosa cell aromatase bioassay by measuring hormone-stimulated estrogen production. The detected steroidogenic activity *in vitro* of FSH-CTP chimeras was comparable to that of wild-type FSH. Thus, ligation of CTP to the coding sequence of FSH subunit, did not affect receptor binding affinity or *in vitro* bioactivity. The *in vivo* biopotencies of wild-type FSH and the chimeras were examined by determining ovarian weight augmentation and granulosa cell aromatase induction. It was found that ovarian weight increased significantly between animals treated with wild-type FSH and the FSH chimeras. In addition, estrogen production by granulosa cells from chimera-treated rats increased threefold to fivefold over that seen in rats treated with wild-type FSH.

Because the increased biopotency of the chimeras may reflect a change in their *in vivo* longevity, the circulatory half-life of the hormones was detected. The clearance patterns of the wild-type and chimera reveal multiple phases. It seems that the clearance of the chimera is much slower than that of wild-type FSH, presumably. RIA determinations show that a high level of the chimera is still detectable in serum after 24 h and yet injected wild-type hFSH reaches basal level between 8 and 24 h [16,17].

The safety, pharmacokinetics, and pharmacodynamics of FSH-CTP were studied in hypogonadotrophic hypogonadal male subjects as a Phase I multicenter study. The results indicated that FSH-CTP use is safe and does not lead to detectable formation of antibodies. Furthermore, pharmacokinetic and dynamic profile of FSH-CTP seemed to be promising. Compared with recombinant FSH-WT (Puregon®), the half-life of FSH-CTP was increased two to three times [18].

Further studies in *in vitro* fertilization (IVF) patients, indicated that a single dose of FSH-CTP is able to induce multifollicular growth compared to daily injection of FSH-WT for 7 days [19,20]. According to the promising results described earlier in clinical trials, on January 28, 2010 the European Commission (EC) gave Merck & Co. marketing approval with unified labeling valid in all European Union Member States for FSH-CTP, now branded as Elonva®.

13.3.3.2 ***Thyrotropin (TSH).*** Thyrotropin is a member of the glycoprotein hormone family, which includes lutropin (LH), follitropin (FSH), and chorionic gonadotropin (hCG). These hormones are heterodimeric proteins that share a common α-subunit but differ in their hormone-β specific subunit, which is unique and confers biological specificity. These hormones activate the target cells via G-protein coupled receptors that, in the case of TSH, lead to increased production of the thyroid hormones, triiodothyronine (T_3), and thyroxine (T_4). Thyroid hormones act on organs throughout the body to regulate growth and various metabolic processes.

Recombinant TSH is important in clinical studies of thyroid cancer. Because of its therapeutic potential, a longer acting analog of TSH was constructed by fusing the carboxy-terminal extension peptide (CTP) of hCG β onto the coding sequence of TSH β-subunit. When co-expressed either with α-subunit complementary DNA or alpha minigene in African green monkey (COS-7) or in human embryonic kidney (HEK293) cells, the chimera was fully bioactive *in vitro* and exhibited enhanced *in vivo* potency associated with a prolonged plasma half-life. Ligation of the CTP to the C-terminal of TSH β-subunit did not affect the assembly and secretion of chimeric TSH. Wild-type and chimeric TSH secreted by COS-7 and 293 cells displayed wide differences in their plasma half-lives, presumably due to the presence of terminal sialic acid and SO_4 on their oligosaccharide chains, respectively. Chimeric and WT-TSH secreted by both cell lines demonstrated similar bioactivity in cAMP production. Chimeric TSH appears to be more effective in COS-7 cells than in 293 cells, as judged by growth assay. COS-7-produced chimeric TSH showed the maximum increase in half-life, indicating the importance of sialic acid in prolonging half-life and *in vivo* potency. Sulfation of both subunits, predominantly beta and to a lesser extent alpha, appears to be responsible at least in part for the increased metabolic clearance of WT and chimeric TSH secreted by HEK293 cells. Apart from its therapeutic potential, chimeric TSH produced in various cell lines can be used as a tool to delineate the roles of sulfate and sialic acid in the *in vivo* clearance and, thereby, the *in vivo* bioactivity [21].

Assembly of the glycoprotein hormone subunits is the rate-limiting step in the production of functional heterodimers. To bypass the problem of dimerization of deglycosylated subunits, hTSH β- and α-subunits were genetically fused in a single chain hormone with or without the CTP as a linker between the subunits. Single chains of hTSH retained a biologically active conformation similar to that of the wild-type heterodimer [22,23].

13.3.3.3 ***Erythropoietin (EPO).*** Erythropoietin is a glycoprotein hormone produced primarily by cells of the peritubular capillary endothelium of the kidney. EPO production is stimulated by reduced oxygen content in the renal arterial circulation. Circulating EPO binds to EPO receptors on the surface of erythroid progenitors resulting in replication and maturation to functional erythrocytes by an incompletely understood mechanism. Recombinant human (rhu) EPO has been used successfully in a variety of clinical situations to increase production of red blood cells. Currently, this agent is licensed for use in the treatment of the anemia of renal failure, the anemia associated with HIV infection in zidovudine (AZT) treated patients, and anomie associated with cancer chemotherapy. Administration of rhuEPO has become routine in the treatment of anemia secondary to renal insufficiency, where doses of 50–75 U/kg given three times per week are used to gradually restore hematocrit and eliminate transfusion dependency.

One major issue regarding the clinical use of EPO is its relatively short half-life *in vivo* due to its rapid clearance (~5 h) from the circulation when it is injected intravenously. Thus, the therapeutic protocol used in the treatment of patients, required frequent injections of EPO. The recommended therapy with rhuEPO is two to –three times per week by subcutaneous or intravenous injections. Therefore, it was anticipated that enhancing the *in vivo* half-life of EPO will reduce the number of injections per week.

The CTP was ligated to the coding sequence at the C-terminal end of EPO. The *in vitro* biological activity of EPO analogs was demonstrated by measuring their ability to stimulate proliferation of erythroid burst forming colonies (BFU-E) from human peripheral blood. BFU-E colonies were grown from blood of healthy donors using a microwell modification of the methylcellulose technique. The optimal formation of BFU-E colonies achieved by EPO-CTP was similar to that achieved by EPO-WT and rhuEPO using 1 U/mL of the protein. Receptor-binding assay indicated that ligation of the CTP to the carboxyl-terminal of EPO has no significant effect on the affinity of the hormone to the receptor.

For further pharmacological evaluation of EPO-CTP, comparative pharmacodynamic studies of EPO-CTP and commercial rhuEPO were performed in male C57BL mice ($n = 7$/group) using different frequencies and wide dose range. The *in vivo* efficacy was obtained by measuring the mean values of hematocrit percentage in the blood. The results indicated that EPO-CTP is significantly ($P < 0.05$) more efficient than EPO-WT when administered *IV* once a week with a dose of 5 μg/kg. EPO-CTP can successfully increase the hematocrit when administered once a week with a dose of 15 μg/kg. Once weekly dosing with the same concentration of commercial rhuEPO or EPO-WT was significantly ($P < 0.001$) less efficient than once weekly dosing of EPO-CTP.

An interesting observation from the present study was the ability of a single injection once a week of EPO-CTP (15 μg/kg) to increase the levels of hematocrit, whereas the same effect was achieved by administration of the same total dose of rhuEPO administered three times a week as 5 μg/kg per injection. These results indicated the importance of sustained blood levels, rather than total dose of EPO. These findings are consistent with the hypothesis that the ability of a single injection of EPO-CTP to increase hematocrit, results from its increased stability in the circulation. The increased biopotency of the chimera may reflect a change in their metabolic clearance *in vivo*. Detecting the half-lives of EPO analogs in mice indicated that a higher level of the chimera is still detectable in serum after 24 h. The half-life of EPO-CTP is increased twofold to threefold comparing to EPO-WT. These data suggest that the mechanism of EPO metabolic clearance is affected by the presence of CTP [24].

Other variant of long-acting EPO was design by introduction of additional N-linked glycosylation into the protein backbone. Using site-directed mutagenesis, the DNA sequence was changed to design hyperglycosylation recombinant EPO analog, (novel erythropoisis-stimulating protein, NESP) which contain five N-linked oligosaccharide chains (two more than EPO-WT) [25,26]. It was found that NESP distinct from rHuEPO having an increased molecular weight and greater negative charge. Compared with rHuEPO, it has an approximate threefold longer serum half-life, greater *in vivo* potency, and can be administered less frequently to obtain the same biological response. On the other hand, removal of the sialic acid from either native EPO or rHuEPO resulted in molecules having an increased activity *in vitro*, but very low activity *in vivo*, presumably due to removal from circulation by the asialoglycoprotein receptor in the liver. Similarly, it was shown that EPO molecules, which have been deglycosylated to remove COH (or produced in *Escherichia coli* to allow expression of only the EPO polypeptide), are active *in vitro*, but have very low *in vivo* activity [27]. RHuEPO and NESP were tested for EPO receptor-binding activity in the radioreceptor assay. Consistent with the results obtained in this assay for the isolated isoforms and the hypothesis, the relative EPO receptor-binding affinity was inversely correlated with the COH content. The relative affinity of NESP for the EPO receptor was 4.3-fold lower than that of rHuEPO [25].

13.3.3.4 Human Growth Hormone (GH). Growth hormone is secreted by the somatotrophs of the anterior pituitary gland and regulates a wide variety of physiological processes including growth and differentiation of muscle, bone, and cartilage cells.

The use of hGH for the treatment of children with impaired linear growth has been accepted as an important therapeutic modality for many years. In addition, hGH substitution in adults increases muscle mass by 5–10%, but part of the effect is attributed to rehydration rather than protein accretion [28–31]. One major issue regarding the clinical use of hGH is its short half-life due to its rapid clearance (~12 min) from the circulation. Previous studies were carried out to stabilize hGH and extend its half-life. It was shown that complexation of hGH with heparin did not cause a major distribution in the tertiary structure of hGH but decreased the hydrophilic environment and stabilize the hormone. Other study demonstrated that crystals of hGH coated with positively charged poly (arginine) allowed delivery of hGH over a period of several days. Other studies designed a long-acting hGH by fusing the CTP of hCG β-subunit that contains four O-linked oligosaccharide recognition sites to the coding sequence of the hormone. Crystallographic studies indicated that N-terminal and C-terminal of hGH are not important for binding of the hormone to its receptor. Therefore, CTP was ligated to the N-terminal and C-terminal of hGH. The results indicate that ligation of CTP to the coding sequence of GH did not affect secretion of the chimeric protein into the medium. *In vivo* studies in hypophysectomized rats indicated that, bioactivity and pharmacokinetic parameters, area under the curve (AUC) and mean residence time (MRT), T_{max}, C_{max}, and half-life, of hGH bearing the CTP were dramatically enhanced.

Efficacy of hGH-CTPs was assessed by weight gain in a hypophysectomized rat model. Weight gain was measured in all animals before treatment, 24 h after first injection and then twice weekly until the end of the study on day 13. Subcutaneous injections of GH-WT, or CTP-GH-CTP-CTP (50 μg/rat) once a week for 2 weeks or with Biotropin (10 μg/rat) daily for 2 weeks, resulted a dramatic increase ($P < 0.001$) in weight gain. These results indicated that accumulation of CTP-hGH-CTP-CTP variant induces an incremental weight gain of 16.5 and 16.8 g, respectively. Similar results were obtained when Biotropin was injected daily.

Injection of GH-CTPs (0.4, 0.8, or 4 mg/kg) once every 4 days significantly ($P < 0.001$) increased the weight gain compared to controls. A similar effect was achieved by daily administration of 0.1 mg/kg Biotropin. Injection of higher amounts (1.05 mg/kg) of CTP-hGH-CTP-CTP once every 4 days dramatically increased the weight gain. These results indicated the importance of sustained blood levels, rather than total dose of hGH for higher bioactivity *in vivo*. These findings are consistent with the hypothesis that the ability of a single injection of hGH-CTPs to increase weight gain, results from its increased stability in the circulation. This may support the strategy of hGH administration of at least once a week.

Pharmacokinetic studies indicated the importance of sustained blood levels, rather than total dose of hGH. These findings are consistent with the hypothesis that the ability of a single injection of hGH-CTP to increase weight gain, results from its increased stability in the circulation. The increased biopotency of the chimera may reflect a change in their metabolic clearance *in vivo*. Therefore, the circulatory half-lives of the hormones were determined. hGH-WT or CTP-hGH-CTP-CTP chimera were injected IV into hypophysectomized male mice. At selected intervals after injection of 50 μg/kg, blood samples were collected and the concentrations of hGH were determined. The results indicated that a higher level of the chimera is still detectable in serum after 50 h where the level of Biotropin after 24 h was undetectable. The estimated half-life of CTP-hGH-CTP-CTP is increased by fourfold tofivefold comparing to Biotropin. These data suggest that the mechanism of hGH metabolic clearance is affected by the presence of CTP [32]. Preliminary studies of hGH-CTPs clinical trials indicated that this peptide is safe for use and clinical trials Phase II is in the way.

13.4 CONCLUSIONS

In this chapter, we have described the role of O-linked oligosaccharide chains on glycoprotein function. Moreover, the design of long-acting glycoprotein hormone (FSH, TSH, EPO, and hGH) agonists was explained. This was achieved by ligation of the sequence of hCG β-carboxyl-terminal peptide, containing the recognition signal of four O-linked oligosaccharide chains, to the coding sequence of the hormone. Ligation of the CTP to different proteins indicated that the O-linked glycosylation recognition sites of the CTP are preserved. Moreover, this ligation is not involved in assembly of the subunits, secretion of the dimer, receptor-binding affinity and *in vitro* bioactivity. CTP can readily be attached to a wide array of existing therapeutic proteins, stabilizing the therapeutic protein in the bloodstream and greatly extending its life span without additional toxicity or loss of desired biological activity. Both the *in vivo* bioactivity and half-life in circulation of hormones bearing the CTP were significantly enhanced. CTP-modified proteins can be manufactured using established recombinant DNA techniques in widely used mammalian protein expression systems. Clinical trials of proteins containing the CTP indicated that these chimeras are not immunogenic and safe for use. This strategy may have wide applications for enhancing the *in vivo* bioactivity and half-life of diverse proteins.

REFERENCES

1. Copeland RJ, Bullen JW, Hart GW. (2008) Cross-talk between GlcNAcylation and phosphorylation: roles in insulin resistance and glucose toxicity. *Am. J. Physiol. Endocrinol. Metab.* 295(1), E17–E28.
2. Wopereis S, Lefeber DJ, Morava E, Wevers RA. (2006) Mechanisms in protein O-glycan biosynthesis and clinical and molecular aspects of protein O-glycan biosynthesis defects: a review. *Clin. Chem.* 52(4), 574–600.

3. Yki-Järvinen H, Vogt C, Iozzo P, Pipek R, Daniels MC, Virkamäki A, et al. (1997) UDP-*N*-acetylglucosamine transferase and glutamine: fructose 6-phosphate amidotransferase activities in insulin-sensitive tissues. *Diabetologia* 40(1), 76–81.
4. Zachara NE, Hart GW. (2004) O-GlcNAc a sensor of cellular state: the role of nucleocytoplasmic glycosylation in modulating cellular function in response to nutrition and stress. *Biochim. Biophys. Acta* 1673(1,2), 13–28.
5. Fares F. (2006) The role of O-linked and N-linked oligosaccharides on the structure-function of glycoprotein hormones: development of agonists and antagonists. *Biochim. Biophys. Acta* 1760(4), 560–567.
6. Matzuk MM, Hsueh AJ, Lapolt P, Tsafriri A, Keene JL, Boime I. (1990) The biological role of the carboxyl-terminal extension of human chorionic gonadotropin [corrected] beta-subunit. *Endocrinology* 126(1), 376–383.
7. Matzuk MM, Krieger M, Corless CL, Boime I. (1987) Effects of preventing O-glycosylation on the secretion of human chorionic gonadotropin in Chinese hamster ovary cells. *Proc. Natl. Acad. Sci. USA* 84(18), 6354–6358.
8. Kalyan NK, Bahl OP. (1983) Role of carbohydrate in human chorionic gonadotropin. Effect of deglycosylation on the subunit interaction and on its in vitro and in vivo biological properties. *J. Biol. Chem.* 258(1), 67–74.
9. Sowers JR, Pekary AE, Hershman JM, Kanter M, DiStefano JJIII. (1979) Metabolism of exogenous human chorionic gonadotrophin in men. *J. Endocrinol.* 80(1), 83–89.
10. Van Den Hamer CJ, Morell AG, Scheinberg IH, Hickman J, Ashwell G. (1970) Physical and chemical studies on ceruloplasmin. IX. The role of galactosyl residues in the clearance of ceruloplasmin from the circulation. *J. Biol. Chem.* 245(17), 4397–4402.
11. Wide L. (1986) The regulation of metabolic clearance rate of human FSH in mice by variation of the molecular structure of the hormone. *Acta Endocrinol.* 112(3), 336–344.
12. Morell AG, Gregoriadis G, Scheinberg IH, Hickman J, Ashwell G. (1971) The role of sialic acid in determining the survival of glycoproteins in the circulation. *J. Biol. Chem.* 246(5), 1461–1467.
13. Briggs DW, Fisher JW, George WJ. (1974) Hepatic clearance of intact and desialylated erythropoietin. *Am. J. Physiol.* 227(6), 1385–1388.
14. Ashwell G, Kawasaki T. (1978) A protein from mammalian liver that specifically binds galactose-terminated glycoproteins. *Meth. Enzymol.* 50, 287–288.
15. Yik JHN, Saxena A, Weigel JA, Weigel PH. (2002) Non-palmitoylated human asialoglycoprotein receptors recycle constitutively but are defective in coated pit-mediated endocytosis, dissociation, and delivery of ligand to lysosomes. *J. Biol. Chem.* 277(43), 40844–40852.
16. Fares FA, Suganuma N, Nishimori K, LaPolt PS, Hsueh AJ, Boime I. (1992) Design of a long-acting follitropin agonist by fusing the C-terminal sequence of the chorionic gonadotropin beta subunit to the follitropin beta subunit. *Proc. Natl. Acad. Sci. USA* 89(10), 4304–4308.
17. LaPolt PS, Nishimori K, Fares FA, Perlas E, Boime I, Hsueh AJ. (1992) Enhanced stimulation of follicle maturation and ovulatory potential by long acting follicle-stimulating hormone agonists with extended carboxyl-terminal peptides. *Endocrinology* 131(6), 2514–2520.
18. Bouloux PM, Handelsman DJ, Jockenhövel F, Nieschlag E, Rabinovici J, Frasa WL, et al. (2001) First human exposure to FSH-CTP in hypogonadotrophic hypogonadal males. *Hum. Reprod.* 16(8), 1592–1597.
19. Devroey P, Fauser BC, Platteau P, Beckers NG, Dhont M, Mannaerts BM. (2004) Induction of multiple follicular development by a single dose of long-acting recombinant follicle-stimulating hormone (FSH-CTP, corifollitropin alfa) for controlled ovarian stimulation before in vitro fertilization. *J. Clin. Endocrinol. Metab.* 89(5), 2062–2070.
20. Abyholm T, Andersen A, Balen A, Braat D, Devroey P, D'Hooghe T, et al. (2008) A randomized dose-response trial of a single injection of corifollitropin alfa to sustain multi-follicular growth during controlled ovarian stimulation. *Hum. Reprod.* 23(11), 2484–2492.
21. Joshi L, Murata Y, Wondisford FE, Szkudlinski MW, Desai R, Weintraub BD. (1995) Recombinant thyrotropin containing a beta-subunit chimera with the human chorionic gonadotropin-beta carboxy-terminus is biologically active, with a prolonged plasma half-life: role of carbohydrate in bioactivity and metabolic clearance. *Endocrinology* 136(9), 3839–3848.
22. Fares FA, Yamabe S, Ben-Menahem D, Pixley M, Hsueh AJ, Boime I. (1998) Conversion of thyrotropin heterodimer to a biologically active single-chain. *Endocrinology* 139(5), 2459–2464.
23. Grossmann M, Wong R, Szkudlinski MW, Weintraub BD. (1997) Human thyroid-stimulating hormone (hTSH) subunit gene fusion produces hTSH with increased stability and serum half-life and compensates for mutagenesis-induced defects in subunit association. *J. Biol. Chem.* 272(34), 21312–21316.
24. Fares F, Ganem S, Hajouj T, Agai E. (2007) Development of a long-acting erythropoietin by fusing the carboxyl-terminal peptide of human chorionic gonadotropin beta-subunit to the coding sequence of human erythropoietin. *Endocrinology* 148(10), 5081–5087.
25. Egrie JC, Browne JK. (2001) Development and characterization of novel erythropoiesis stimulating protein (NESP). *Br. J. Cancer* 84(Suppl. 1), 3–10.
26. Nissenson AR. (2001) Novel erythropoiesis stimulating protein for managing the anemia of chronic kidney disease. *Am. J. Kidney Dis.* 38(6), 1390–1397.
27. Higuchi M, Oh-eda M, Kuboniwa H, Tomonoh K, Shimonaka Y, Ochi N. (1992) Role of sugar chains in the expression of the biological activity of human erythropoietin. *J. Biol. Chem.* 267(11), 7703–7709.
28. Raben MS. (1958) Treatment of a pituitary dwarf with human growth hormone. *J. Clin. Endocrinol. Metab.* 18(8), 901–903.

29. de Boer H, Blok GJ, Van der Veen EA. (1995) Clinical aspects of growth hormone deficiency in adults. *Endocr. Rev.* 16(1), 63–86.

30. Carroll PV, Christ ER, Bengtsson BA, Carlsson L, Christiansen JS, Clemmons D, et al. (1998) Growth hormone deficiency in adulthood and the effects of growth hormone replacement: a review. Growth Hormone Research Society Scientific Committee. *J. Clin. Endocrinol. Metab.* 83(2), 382–395.

31. Drake WM, Howell SJ, Monson JP, Shalet SM. (2001) Optimizing GH therapy in adults and children. *Endocr. Rev.* 22(4), 425–450.

32. Fares F, Guy R, Bar-Ilan A, Felikman Y, Fima E. (2010) Designing a long-acting human growth hormone (hGH) by fusing the carboxyl-terminal peptide of human chorionic gonadotropin beta-subunit to the coding sequence of hGH. *Endocrinology* 151(9), 4410–4417.

14

ELP-FUSION TECHNOLOGY FOR BIOPHARMACEUTICALS

DOREEN M. FLOSS,[1] UDO CONRAD,[2] STEFAN ROSE-JOHN,[3] AND JÜRGEN SCHELLER[1]

[1]*Institute of Biochemistry and Molecular Biology II, Medical Faculty, Heinrich-Heine-University, Düsseldorf, Germany*
[2]*Leibniz Institute of Plant Genetics and Crop Plant Research (IPK), Gatersleben, Germany*
[3]*Institute of Biochemistry, Christian-Albrechts-University, Kiel, Germany*

14.1 INTRODUCTION

Fusion proteins are successfully used as biopharmaceuticals and a variety of such therapeutics are investigated in clinical trials or already reached the market [1]. Beyond the fusion of effector molecules to Fc domains, albumin or transferrin to increase the plasma half-life of the resulting therapeutic, cytotoxicity effects are improved by fusion to toxins, enzymes, or cytokines. Over the past years, a new fusion technology based on artificial repetitive polypeptides originated from mammalian elastin came into the interest of scientists.

In connective tissues of the large arterial blood vessels, lung parenchyma, elastic ligaments, and skin, which are subjected to repetitive and reversible deformation, the major protein within the extracellular matrix (ECM) is elastin, an extremely insoluble protein. Furthermore, this structural protein is also present in some cartilaginous tissues [2]. Elastin is found in all vertebrate species, including cartilaginous fish, but not in invertebrates and it is an important component of elastic fibers [3]. The protein is synthesized as a soluble monomer, tropoelastin, which is subsequently assembled into a polymeric structure in the ECM and stabilized by cross-links at Lys residues through the action of lysyl oxidase [2]. This is actually the property that confers the conspicuous physical features of the natural elastin protein [4]. Human, chicken, bovine, and rat tropoelastin have a significant homology at both DNA and amino acid levels [5].

Elastic fibers are composed of two entities: amorphous elastin and a fibrillar component, the microfibrils. During the formation of the elastic fiber, microfibrils, which are made up of at least five distinct proteins [6], act as a scaffold, on which elastin is deposited [4]. Elastin from higher vertebrates contains over 30% Gly and ~75% of the entire sequence consists of just four hydrophobic amino acids (Gly, Val, Ala, and Pro) characteristically present as tetra- (Val-Pro-Gly-Gly), penta- (Val-Pro-Gly-Val-Gly) or hexa-repeats (Ala-Pro-Gly-Val-Gly-Val) and forming the hydrophobic domain of tropoelastin, which are principally responsible for the elasticity of the protein [7]. In contrast, the hydrophilic domains of tropoelastin rich in Lys and Ala are involved in the cross-linking process. These domains often consist of Lys stretches separated by Ala residues such as Ala-Ala-Ala-Lys-Ala-Ala-Lys-Ala-Ala [5]. Derivatives of artificial repetitive polypeptides derived from mammalian elastin, for example, poly(Val-Pro-Gly-Xaa-Gly), where the guest residue Xaa is any natural amino acid except proline [8], have been designated as elastin-like polypeptides (ELPs) and the process of complementing target proteins by their C- or N-terminal fusion to an ELP has been termed

Fusion Protein Technologies for Biopharmaceuticals: Applications and Challenges, First Edition. Edited by Stefan R. Schmidt.

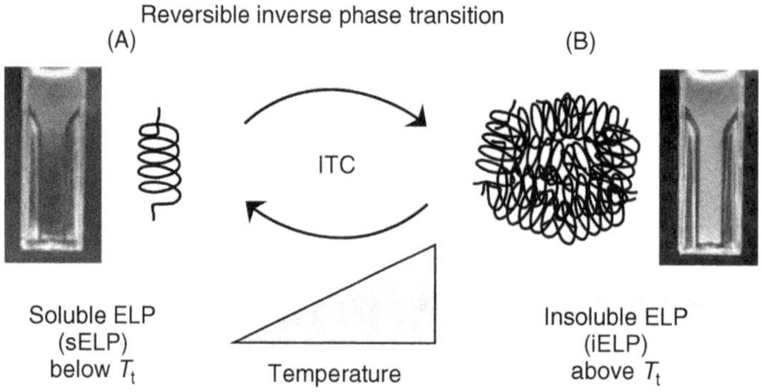

FIGURE 14.1 Protein purification of ELPs by inverse transition cycling (ITC) ELPs are soluble (soluble ELPs, sELPs) at a temperature below their transition temperature (T_t) and become insoluble when the temperature is increased above T_t. The insoluble ELP-aggregates (iELPs) are precipitated by centrifugation and thereby separated from contaminating proteins, such as cellular proteins. The precipitate containing the iELPs is dissolved in buffer with a temperature below T_t. Increased turbidity of ELPylated mini-sgp130 [113] was visualized using standard cuvettes below (A, sELPs) and above (B, iELPs) T_t. *Source*: Reprinted from Reference [9]. Copyright (2010), with permission from Elsevier.

ELPylation [9]. Initially, chemical methods were used for the synthesis of ELPs [10]. Nowadays, applications of ELP-fusion technology exclusively use genetically encoded ELPs because the incorporation of a gene encoding the desired ELP sequence into an expression vector allows the biosynthesis of a monodisperse polymer with the exact molecular weight and the correct stereochemistry along with a high yield [11]. Therefore, a synthetic oligonucleotide (<100 bp) designed to encode the desired ELP sequence is oligomerized by recombinant DNA methods to create the synthetic ELP gene. ELPs exhibit a thermally responsive phase transition; they are monomeric and soluble at temperatures below the transition temperature (T_t), and aggregate and become insoluble at temperatures above [8]. This process, termed inverse phase transition, is reversible, occurs within a 2–3°C range and can be monitored spectrophotometrically by measuring the turbidity of the protein solution (Figure 14.1). This phenomenon can be explained by the development of closer associations between single β-spiral strands of ELP and the subsequent formation of interstrand β-sheet structures that mediate close association or aggregation of different ELP chains [12]. Temperature-triggered insolubility and aggregation is reversed completely when lowering the temperature below T_t, which itself is defined by the combination of buffer composition, salt concentration, ELP concentration, ELP chain length, the degree of ionization of any functional side chains and the guest residue Xaa, and the polarity arrangement along the molecule [13–16]. These tuneable properties are transferred to ELP-fusion proteins making ELPylation useful for biologically inspired polymers that can respond to thermal and other environmental signals. The exploitation of ELPs in medicine and the suitability for *in vivo* applications is due to the high level of biocompatibility of ELP. There exists a chemical similarity of ELP to native elastin, which is immunological inert and therefore attractive for biomedical applications. Poly(Gly-Val-Gly-Val-Pro) has been used in a number of tissues, tissue-fluid and blood tests as approved by the relevant regulatory body of the United States [17]. Furthermore, *in vitro* and *in vivo* studies using poly(Val-Pro-Ala-Val-Gly) did not reveal any evidence of cytotoxicity, nonspecific depression of cellular respiration in macrophages, or general inflammatory responses [18]. Recently, the effect of purified poly(Val-Pro-Gly-Xaa-Gly) on mouse and bovine immune cells was investigated and no immunomodulatory role was detected [19]. The design of ELPs allows the manipulation of their stimulus-responsive properties and other physical or functional characteristics and has led to biological applications of ELPs as structural tools for recombinant protein purification and drug delivery, and as material for tissue engineering. The following sections highlight the use of ELPylation as a new tool to efficiently purify recombinant proteins and summarize the application of ELPs in medicine and biotechnology.

14.2 ELP-BASED PROTEIN PURIFICATION

Heterologous recombinant protein production in living prokaryotic and eukaryotic cells by biotechnological methods provides the basis for therapeutic application of biologics in autoimmune and chronic inflammatory diseases and cancer. Biologics such as antagonistic monoclonal antibodies or soluble cytokine receptors fused to the constant Fc part of an IgG antibody were developed to inhibit molecules including the inflammatory cytokines tumor necrosis factor or interleukin-6 [20,21]. A cost-intensive issue of recombinant protein expression methodology is the development of efficient downstream protein purification procedures. Traditional protein purification strategies such as size exclusion, ion exchange, or affinity chromatography are rather cost intensive. The commonly applied affinity chromatography to purify recombinant proteins relies on a reversible highly specific immobilization of the target protein. Antibodies might be purified via their constant Fc parts using *Staphylococcus aureus* Protein A affinity chromatography. Nonimmunoglobulin target proteins might be fused to either a carrier protein, such as Fc region of an immunoglobulin, maltose-binding protein, glutathione S-transferase, or thioredoxin, to carrier peptides such as oligohistidine or the FLAG-peptide (DYKDDDDK) allowing the purification of the target protein almost independently of its physiochemical properties. For therapeutic applications, these tags are usually removed to obtain the therapeutic protein.

The thermally stimulated, reversible aggregation/precipitation behavior of ELPs as carrier fused to a target protein provides an alternative mode of protein purification and avoids the rather cost-intensive approaches of other purification strategies such as affinity chromatography. ELP-based protein purification bypasses the need for affinity chromatography; the complete process becomes simpler, less expensive and is readily upgradable to a preparative mode, which is discussed in the following section. Several methods that allow the cleavage of the carrier ELP tag from the target protein are also introduced.

The thermally stimulated purification of heterologous expressed ELPs was first demonstrated for *Escherichia coli* by McPherson and colleagues in 1996 [22]. Later on, the thermally stimulated purification of recombinant carrier ELP-fusion proteins by inverse transition cycling (ITC) was achieved from bacterial cells [23] (Figure 14.2A). Here, aggregation of ELPylated target proteins was achieved by raising the temperature above T_t. ELPylated target proteins were pelleted by subsequent centrifugation. The resulting supernatant, containing the undesirable host proteins, was removed and the remaining recombinant ELP-target protein-containing pellet was dissolved in low-salt buffer at a temperature below T_t. This strategy leads to an almost complete purification of the ELP-target proteins. In this example from Meyer and Chilkoti [23], the synthetic ELP component of the fusion protein was an oligomerized series of Val-Pro-Gly-Xaa-Gly pentapeptides, where Xaa possessed a 5:2:3 ratio of Val:Ala:Gly, engineered to produce a specific T_t. It became evident that the ELP-fusion partner also influences T_t. The hydrophobic tendamistat decreased T_t, whereas the hydrophilic thioredoxin slightly increased T_t. The authors speculated that interactions between the ELP chain and solvent-exposed hydrophobic regions in tendamistat lead to the observed decrease in T_t. However, biologic activity of both target proteins as ELP fusion remained unaffected [23]. Even though average yields of ELP-fusion proteins were in the range of 10–100 mg/L of *E. coli* culture volume, optimization of culture conditions and medium, by addition of frequent ELP-containing amino acids, boosted the yield about 36-fold to the level of 1.6 g/L for GFP-ELP [24]. Also, high cell density *E. coli* fermentation was suitable for high-level production of ELP-fusion proteins [25]. Recently, it has been shown that positioning the ELP component at the C-terminus but not at the N-terminus of the target protein can enhance expression levels, leading to an improved yield of purified protein [26]. In the case of very low ELP-fusion protein expression (defined as $<100\,\mu g$ soluble expressed protein per liter of culture medium), the addition of excess free ELP mediates efficient phase transition of ELP-fusion protein as the co-aggregation between the ELP and the ELPylated target is highly specific (Figure 14.2B) [27].

The ELP moiety can be cleaved from the target protein by treatment with site-specific proteases such as factor X or thrombin. The addition of proteases might interfere with the structure of the target protein, such as minimal unspecific cleavage of the protease within the ELP-fusion partner or could raise cost issues for large-scale protein production. An attractive source of proteases was recently developed by Lan and co-workers. Here, the target protein and a protease were ELPylated and purified from *E. coli*. After purification, the ELP-target and the ELP-protease fusion proteins were mixed, resulting in the cleavage of the target protein from the ELP component. Finally, the target protein was separated from ELP as well as the ELP-tagged protease by ITC (Figure 14.2C) [28].

The intein technology, designed to induce protein self-cleavage, was combined with ELP technology to avoid cost issues of postpurification protein cleavage [29]. Intein-ELP fusions were designed to purify a number of native recombinant target proteins from *E. coli* [30,31]. This method requires the introduction of an intein coding sequence between the sequence encoding the target protein and the ELP carrier protein resulting in the triple fusion "target protein-intein-ELP". First of all, the triple fusion is purified by inverse transition cycling (ITC), and subsequently intein self-cleavage is induced by pH/temperature shift or DTT treatment to release the target protein from the intein-ELP fragment. The intein-ELP carrier fragment could be then efficiently removed by a second round of ITC because the temperature-based precipitation step specifically affects the ELP carrier protein (Figure 14.2D) [30,31]. For scale-up, separation of aggregated ELP-fusion proteins could include continuous centrifugation or tangential-flow microfiltration [32]. The use of microfiltration including efficient prefiltration steps and the fine-tuned temperature and salt conditions depending on the properties of the purified protein could result in significant improvement of yield and purity [33]. High concentrations of sodium chloride, which are commonly used decreasing the T_t are potentially corrosive, thereby limits their use in industrial processes. Fong and

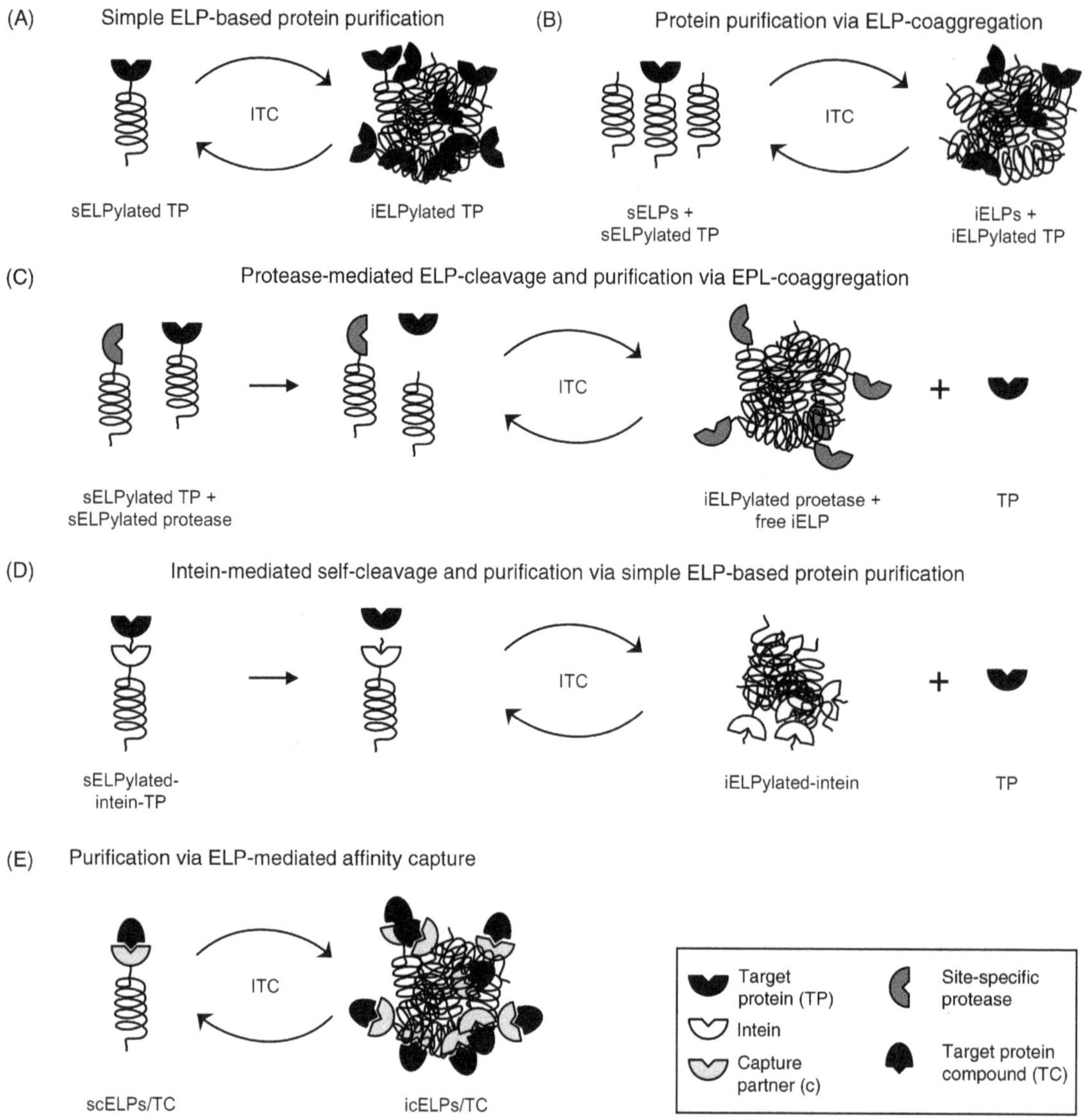

FIGURE 14.2 Schematic illustration of ELP-based protein purification strategies by inverse transition cycling (ITC). (A) Soluble ELPylated target proteins (sELPylated TP) became reversibly insoluble above T_t (iELPylated TP). (B) Excess of free sELPs facilitated co-aggregation of sparse amounts of sELPylated TP above T_t to support purification via ITC. (C) Target protein and a protease are ELPylated. After cleavage of the ELP-target protein, both ELPylated protease and the free ELPs are precipitated by ITC. (D) Intein-mediated cleavage of ELPylated target protein. The intein-ELP-fusion protein was precipitated by ITC after cleavage. (E) ELP-mediated affinity capture (EMAC) for the purification of target compounds without the need for direct ELP fusion. ELPs were fused to a capture protein (cELPs) that binds specifically and reversibly to the target compound (e.g., protein, plasmid-DNA, glycopeptides, or heavy metals). Soluble cELPs complexes with the target compound (sELPs/TC) can be specifically aggregated above T_t to form insoluble cELPs/TC complexes (icELPs/TC). *Source*: (A), (B), and (E) are reprinted from Reference [9]. Copyright (2010), with permission from Elsevier.

co-workers found that ammonium sulfate could substitute sodium chloride at a fourfold lower concentration to achieve equivalent ELP precipitation. Moreover, the precipitation step could be performed at room temperature instead of the commonly used 37°C, allowing a cheaper, gentler purification procedure [34]. Although ammonium sulfate triggered the phase transition at a lower molar concentration, it should be noted that ammonium sulfate precipitated more contaminant proteins than sodium chloride. Therefore, more rounds of ITC were necessary to obtain pure ELP-fusion proteins when ammonium sulfate was used in the precipitation procedure [34]. Furthermore, ELP-mediated affinity capture (EMAC) procedures, which avoid the need of ELPylation, have been developed (Figure 14.2E). Here, ELP is genetically or chemically fused to a caption partner that binds specifically but reversibly to the target compound, which could be proteins, plasmid DNA, glycopeptides, or heavy metals [35–45]. This procedure also eliminates the need for postpurification enzymatic or chemical cleavage of the ELP, and allows a reuse of the ELP capture molecule.

The introduction of ELPs in biotechnology has not just added one more way for the purification of recombinant proteins but has revealed a generally new principle for protein purification. ITC-based purification could make conventional affinity chromatography-based protein purification methods almost dispensable. The following section is about the present status of how ELPs could aid drug application in medicine or gave rise to new applications in nanobiotechnology.

14.3 ELPYLATED PROTEINS IN MEDICINE AND NANOBIOTECHNOLOGY

Besides the aforementioned property of ELPs, namely, the inverse temperature transition, it is the extraordinary biocompatibility that makes this type of polymer usable for medical applications. This was initially demonstrated for poly(Gly-Val-Gly-Val-Pro) by generic biological tests for materials and devices recommended by the American Society for Testing and Materials (ASTM) in contact with tissues, tissue fluids, and blood [17]. *In vitro* and *in vivo* studies of poly(Val-Pro-Ala-Val-Gly) revealed no induction of cytotoxicity, no nonspecific depression of cellular respiration on macrophages and no inflammatory response caused by ELPs [18]. This unmatched biocompatibility could be based on the similarity of ELPs to native elastin, making them undistinguishable for the immune system. Due to the reported biocompatibility and the fact that the simple polypeptide nature of ELPs prevented the generation of monoclonal antibodies, ELP-containing proteins have already found their way into numerous biomedical applications [11,46,47].

One application area of ELPs is to form scaffolds for the growth of tissue, which can be used for tissue repair or *de novo* tissue engineering. *In vitro*, genetically engineered ELPs, resembling the ocular surface ECM, were useful as a substratum to culture cells from the ocular surface [48]. Furthermore, ELP-containing polymers were used to construct an ECM for thermally reversible cell sheet recovery of cultured cells [49], and an ELP-hydrogel was used as a growing matrix for human adipose-derived adult stem (hADAS) cells, which showed signs of chondrogenesis without addition of differentiation supplements [50].

In addition, ELPs represent homing devices for pharmaceuticals *in vivo* and carrier for thermally targeted delivery of therapeutics [51]. The effectiveness of pharmaceutical treatments as well as the severity of the corresponding side effects is largely dependent on the precise targeting of the applied medicament. Especially. the application of toxic substances, for example, in chemotherapy, can cause severe ailments in nonafflicted tissues. By fusion to ELPs, the active compound was precipitated in the targeted tissue with a locally induced hyperthermia [52,53]. Micro-sized ELP aggregates adhere to tumor vasculature only when tumors are locally heated. Upon return to normothermia and dissolving of the vascular ELP particles into the plasma, the local vascular concentration was increased and drove more ELPs across the tumor blood vessel thus enhancing its extravascular accumulation [54]. Due to the hydrophobic folding and the assembly into nano- and/or microparticles to form a phase separated state above T_t (Figure 14.3A) [55]. Microspheres are small spherical particles, with diameters in the nano- to micrometer range, that might serve as drug depot delivery devices in inflammation and tumor therapy. Self-assembled particles of poly(Val-Pro-Ala-Val-Gly) as vehicles for controlled release of dexamethasone phosphate (DMP) were generated and once formed; they remain stable either at ambient or body temperature. Significant amounts of the model drug were encapsulated after self-assembly and sustained DMP release for about 30 days was reported [56]. The temperature-dependent release of trapped particles from cross-linked ELP-made microspheres was also demonstrated for prednisone acetate, an immunosuppressant drug. Pore sizes increased above T_t, leading to an increased particle release from ELP microspheres [57]. Some of the recent studies focused on controlling size and homogeneity of ELP microspheres, which are critical factors in improving therapeutic efficacy [58–62]. To generate nanomeso scale bioresponsive peptide-based particles of defined morphology, which can encapsulate drugs, electrospraying was used for preparing ELP particles containing doxorubicin [63]. Functionalized ELP-fusion proteins might benefit from thermally induced local multivalent ELP biopolymer formation by triggering avidity resulting in simultaneous interaction of multiple ELP-conjugated ligands with multiple receptors. The strategy to use ELPs for the assembly of low-affinity into high-avidity states was named dynamic affinity modulation (DAM). Since the switch is thermally and locally inducible, avidity effects can be targeted to the desired site of action, thereby limiting unwanted interaction in nontarget sites [64].

Since ELPs are biopolymers capable of *in vivo* drug depot formation through thermally stimulated complexes at physiological temperatures, fusion of such polymers with soluble tumor necrosis factor receptors (sTNFR) have been designed and used as *in situ* anti-TNF drug depots. TNF is a major inflammatory cytokine and blocking of TNF with monoclonal antibodies or sTNFR-fragments could lead to remission

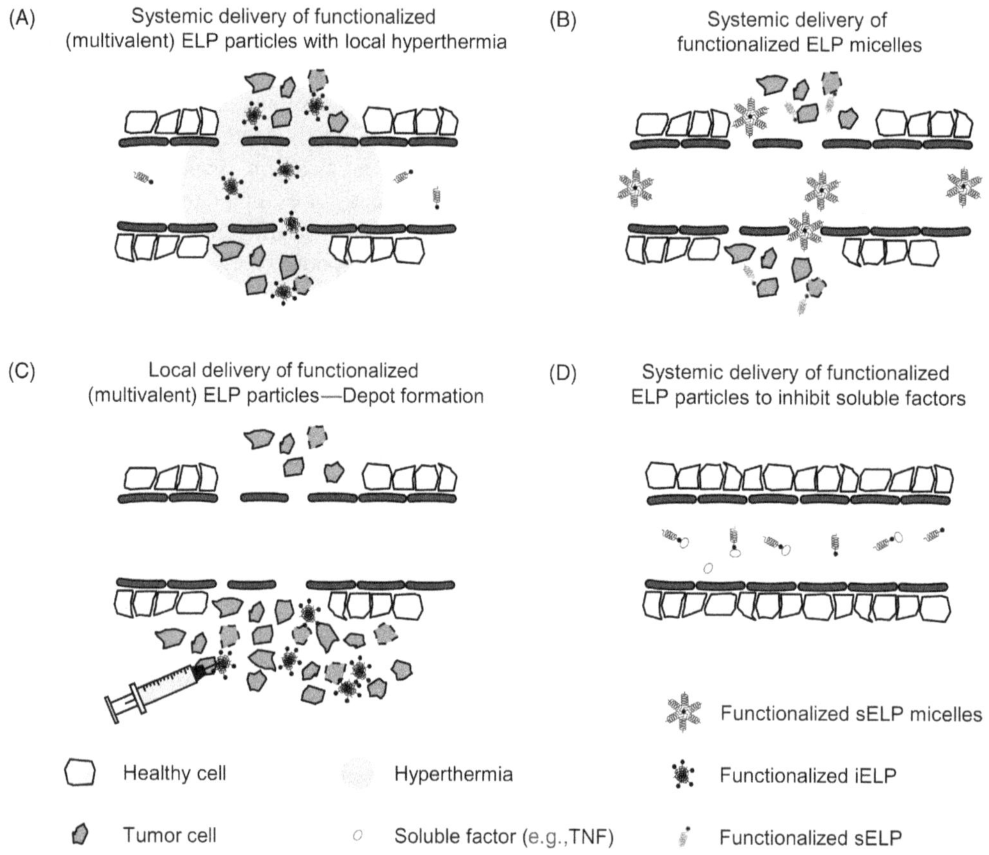

FIGURE 14.3 Delivery strategies and working principles of functionalized ELPs. (A) Targeting of a functionalized ELP by local hyperthermia to a tumor by the phase transition triggered aggregation of the ELP in tumor vasculature and passive passing through the leaky tumor vasculature, functionalized ELP might have multivalent properties. After return to normothermia, the aggregates might dissolve, generating a large concentration gradient that drives the ELP that dissolves from the aggregates into the tumor. (B) Functionalization of ELP with a C-terminal hydrophobic drug such as Paclitaxel, trigger the self-assembly of ELP micelles, which are passing through the leaky tumor vasculature. (C) Depot formation of functionalized ELP-aggregates (T_t below body temperature) by local delivery upon intratumoral injection. (D) Systemic delivery of functionalized ELPs (monomeric or multivalent) can bind and neutralize soluble factors within the circulation or inflamed tissues (e.g., anti-TNF, anti-IL-6). *Source*: Modified from Reference [124].

of autoimmune and chronic inflammatory diseases [65]. The association of ELP-sTNFRII fusion proteins caused a significantly reduced *in vitro* bioactivity [66]. However, TNF-mediated effects in cultured astrocytes and microglia or dorsal root ganglion explants were attenuated by ELP-sTNFRII-fusions. This is consistent with the anti-TNF effects of unfused sTNFRII [67]. The same group analyzed the aggregation kinetics of radioactively labeled ELP. Aggregated ELPs demonstrated a sevenfold longer perineural half-life as compared with the soluble ELPs. They concluded that thermally stimulated aggregation of ELP fused therapeutics might provide a perineural drug depot for longevity drug delivery to an inflamed nerve [68].

ELPs were fused to a cell penetrating peptide and a cell-cycle inhibitory peptide derived from the p21 protein, which caused a portion of the polypeptide to reach the nucleus [69]. The engineered protein aggregated at temperatures above 39°C and showed enhanced proliferation inhibitory effects in this stage, thus rendering the protein a thermally responsive carrier that might be targeted to solid tumors by application of focused hyperthermia [70]. Furthermore, Massodi and colleagues constructed a fusion protein of the cell penetrating Tat peptide with ELP, which was shown to inhibit cell adhesion, spreading, invasion, and migration of ovarian cancer cells in culture. *In vivo*, the same protein exhibited an anti-metastatic potential in an ovarian cancer metastasis model suggesting a novel role of the fusion protein as a therapeutic intervention in cancer metastasis [71]. A lytic peptide (L12) that has been shown to exhibit anticancer activity was further fused to the C-terminus of Tat-ELP and the resulting fusion protein possessed cytotoxic properties to cancer cells *in vitro* indicating that Tat-ELP-L12 has the potential to present an effective vehicle to thermally target tumors [72]. A similar strategy makes use of the cell penetrating peptide Bac-7. Bac-7 was N-terminally fused to ELP and the cell-cycle inhibitory protein p21 was C-terminally fused. The fusion protein Bac-ELP-p21 revealed antiproliferative effects on SKOV-3 cells specifically after hyperthermia-induced biopolymer formation [70]. Paclitaxel is a hydrophobic antitumor drug interfering with the cell cycle by inhibiting the destruction of microtubules. Paclitaxel was fused to ELP with an acid-labile linker and increased the overall solubility of Paclitaxel by forming of ELP micelles (Figure 14.3B). Hyperthermia induced ELP biopolymer formation increased the local concentration of ELP-Paclitaxel and forced the uptake into MCF-7 cells. The acid-labile linker resulted in Paclitaxel release from ELP within the lysosome and induction of MCF-7 cell apoptosis [73]. Furthermore, ELP was conjugated to a therapeutic radionuclide and injected into tumors in mice (Figure 14.3C). Intratumoral drug delivery might be an attractive alternative to systemic drug delivery, which is, however, hampered by rapid clearance of the drug. ELPs might serve as general framework for local drug depot formation. Here, a thermally sensitive ELP-conjugate was compared to a thermally insensitive ELP-conjugate, with only the first resulting in depot formation, leading to longer residence time in the tumor and improved antitumor efficacy of iodine-131. Interestingly, ELP-polymer formation protected the radionucleotide from cleavage by dehalogenation [74].

The self-assembly of ELPs into basic forms such as capsules and aggregates might be utilized to bridge the gap from biomaterials to nanodevices [75]. ELPs were cross-linked to form matrices with mechanical properties comparable to those of native elastin [76]. With a better understanding of the 3D architecture and structural transformations of ELP aggregation, it will be possible to produce tailor-made ELPs for biotechnological applications. A first example for this development is soluble ELP (sELP), which combines physical and mechanical features of silkworm silk and ELPs in one polypeptide. sELPs formed temperature and pH responsive hydrogels [77] and were spun into fibers [78,79] with a high tensile strength and high deformability [80]. A second example is the formation of an ELP layer with pH-controllable nanopore formation [81]. Considering the responsiveness of ELPs to temperature, it should be possible to improve this feature to literally build nanostructures with switch-mechanisms, for example, to open or close nanopores at will.

14.4 MOLECULAR PHARMING: A NEW APPLICATION FOR ELPYLATION

For centuries, plant products have been used with different intention, as part of foods and powders to cure and prevent diseases or, as plant raw materials. The twenty-first century brought plants a new function, namely, the synthesis of heterologous proteins. Traditional production systems for recombinant proteins, microbial fermentation, insect or mammalian cell cultures, and transgenic animals, hold drawbacks in terms of production costs, scalability, and product safety. These limitations can be addressed through the use of plants as production hosts. In developing countries with a low level of technological infrastructure, where the demand for pharmaceuticals is greatest, but where the financial resources to pay for them are the most limited the production of recombinant proteins in transgenic plants is probably more realistic than the current alternatives of bacterial, yeast, insect, and mammalian cells [82–84]. Neutralizing and infection-preventing antibodies against HIV-1 as components of microbicides [85] or recombinant vaccines against avian influenza [86] are promising candidate proteins and currently under investigation. The term molecular pharming refers to the large-scale production and purification of pharmaceutical proteins in transgenic plants or plant-based expression systems. Four systems for plant-based expression of recombinant proteins have been developed so far; plant cell cultures (including duck weed and mosses), transient expression systems using *Agrobacterium tumefaciens* or plant viruses for introducing the expression cassette into plant leaves, expression in transplastomic plants, and expression in stably transformed transgenic plants [87]. Several plant species are used as expression systems [88]. Since the successful expression of assembled immunoglobulins in transgenic tobacco plants [89] and the first report of plant-based vaccine production [90], a large number of different vaccines, antibodies, as well as antibody fragments have been produced in plants for both medical and veterinary purposes and have entered clinical trials [91]. However, only two plant-produced vaccine-related products have gone all the way through the production and regulatory hurdles, and only one, a plant made single-chain variable fragment (scFv) is used in the production of a recombinant hepatitis B virus (HBV) vaccine in Cuba [92]. A major step was made at the beginning of 2006 by Dow AgroSciences (DAS, Indianapolis, USA). Their plant cell expressed veterinary vaccine against the Newcastle disease virus (NDV), produced in a suspension-cultured tobacco cell line, has gained regulatory approval by the U.S. Department of Agriculture (USDA) Center for Veterinary Biologics—the final authority for veterinary vaccines in the United States [93]. However, this vaccine has not been introduced to the market yet. Dow AgroSciences apparently wished to demonstrate that their Concert™ Plant-Cell-Produced system was useful for the production of safe and effective vaccines, fulfilling the approval requirements of the regulatory system [94]. Recently, Pfizer has acquired the exclusive worldwide licensing rights to commercialize Protalix BioTherapeutics Inc.'s taliglucerase alfa, a plant cell expressed form of glucocerebrosidase for the treatment of Gaucher's disease (www.protalix.com). This drug is produced in a contained disposable bioreactor system with suspension-cultured carrot cells and is optimistically predicted to receive FDA approval, making it the first plant-made pharmaceutical [95]. A recent press release announced that Bayer started a clinical Phase I study with personalized vaccines from tobacco plants, produced with the magnICON® system [96], for treatment of non-Hodgkin's lymphoma.[1]

The combination of plant-based expression and the ELP-fusion technology represents an auspicious chance to solve aforementioned challenges of traditional protein production systems. Already 14 years ago, first attempts were undertaken to use plant-based expression systems for the production of the bioelastic polymer with 121 repeats of the amino acid sequence Gly-Val-Gly-Val-Pro either in cultured cells of tobacco [97] or in stable nuclear transgenic tobacco plants [98]. Despite a marginal accumulation of polymer protein (0.01–0.05% of total soluble protein, TSP), (Gly-Val-Gly-Val-Pro)$_{121}$ was purified using the temperature-dependent aggregation, but only a portion of polymer proteins (~0.003–0.03% of TSP) was extractable resulting in a yield between 0.5 and 5 μg of polymer protein from 1 g of fresh weight of leaf tissue [98]. The unexpected low level of production was explained on the one side with the prokaryotic-preferred polymer and on the other side with the availability of the amino acids glycine, valine, or proline [98]. To overcome this problem, (Gly-Val-Gly-Val-Pro)$_{121}$ was expressed in chloroplasts due to the presence of certain amino acid pools. Despite of the presence of higher polymer transcript levels in transplastomic plants, the protein expression was quite low, which was justified with glycine as the limiting factor for production in chloroplasts [99].

In 2004, the traditional ELP-fusion strategy from Meyer and Chilkoti was transferred to the plant biotechnology field to produce spider silk elastin in the endoplasmic reticulum (ER) of transgenic tobacco and potato plants, and furthermore to purify the recombinant protein by a simple method using heat treatment and ITC (Figure 14.4A) [100]. Generated transgenic plants accumulated recombinant synthetic spider silk protein SO1 [101] fused with the elastic biopolymer 100xELP from 0.5% up to 4% of TSP. Unexpectedly, fusion of SO1 with 100xELP resulted in a twofold increase in the accumulation level compared to the naked SO1 protein (0.5–2% [101]. This was the first report on the successful purification of recombinant ELP-fusion proteins from crude leaf extracts and confirmed the portability of ITC from bacterial production systems to eukaryotic expression systems. Furthermore, biocompatibility of spider silk elastin was demonstrated by growth of anchorage-dependent mammalian cells (CHO K1, human chondrocytes) on SO1-100xELP coated culture plates compared to conventional coatings [100]. ELPylated SO1 was used for the preparation of layers and foils, which were more elastic and harder than thermoplastic polymers currently used for industrial applications [102,103]. A second spider silk protein, the native major ampullate spidroin protein 2 (MaSp2) from *Nephila clavipes* was also fused to an ELP part consisting of 27 repeats of Val-Pro-Gly-Xaa-Gly and expressed in tobacco. Despite a reported 60-fold higher level of recombinant protein in transgenic plants, the actual accumulation was about 0.0125% TSP for MaSp2 and 0.75% TSP for MaSp2-ELP, respectively [104].

The unexpected enhancement in recombinant protein yield due to the C-terminal ELP part was further confirmed for seed-produced single-chain Fv (scFv) ELP fusions. Here, the ELP-fusion strategy resulted in a 40-fold increase in scFv accumulation with levels approaching 25% of total soluble seed protein [105]. Plant seeds are particularly attractive for the production of recombinant antibodies because they remain stable and functional even if mature seeds are stored at ambient temperature for several years [106,107]. The scFv-ELP fusions from tobacco seeds exhibit similar binding behavior to the corresponding unfused antibodies indicating that ELP does not abrogate antigen binding [105]. A similar approach was designed by Joensuu and colleagues to produce scFvs against the foot-and-mouth disease virus (FMDV) in *N. tabacum* as ELP-fusion proteins with 28 repeats of the pentapeptide Val-Pro-Gly-Val-Gly [108]. Furthermore, a tobacco etch virus (TEV) cleavage side was included between the scFv and the ELP part for the removal of the ELP tag. The highest scFv accumulation level was 0.08% TSP (transient expression) and 0.8% TSP for stable expression, respectively. The ITC process in combination with StrepII affinity chromatography was used to purify 1.5 mg recombinant scFv-ELP fusions from 660 mg leaf material. After overnight cleavage the ELP part and the protease were removed by sepharose capture using the His-tag on the TEV protease and the StrepII-tag on the ELP. The binding of purified scFv proteins prior to and after ELP removal to FMDV was investigated and a higher binding of scFv without ELP was observed [108].

The reputation of ELP tags have been increased due to their salt- and temperature-dependent phase transition, which can be used for nonchromatographic separation of recombinant proteins, a significantly increase in the accumulation of the fusion protein and their biocompatibility. Therefore, the ELP-fusion strategy is a promising approach for the *in planta* production of various pharmaceutical proteins. Among the plant-based production of ELPylated antibodies or antibody derivatives, vaccines and other proteins

[1] http://www.icongenetics.com/html/5954.htm

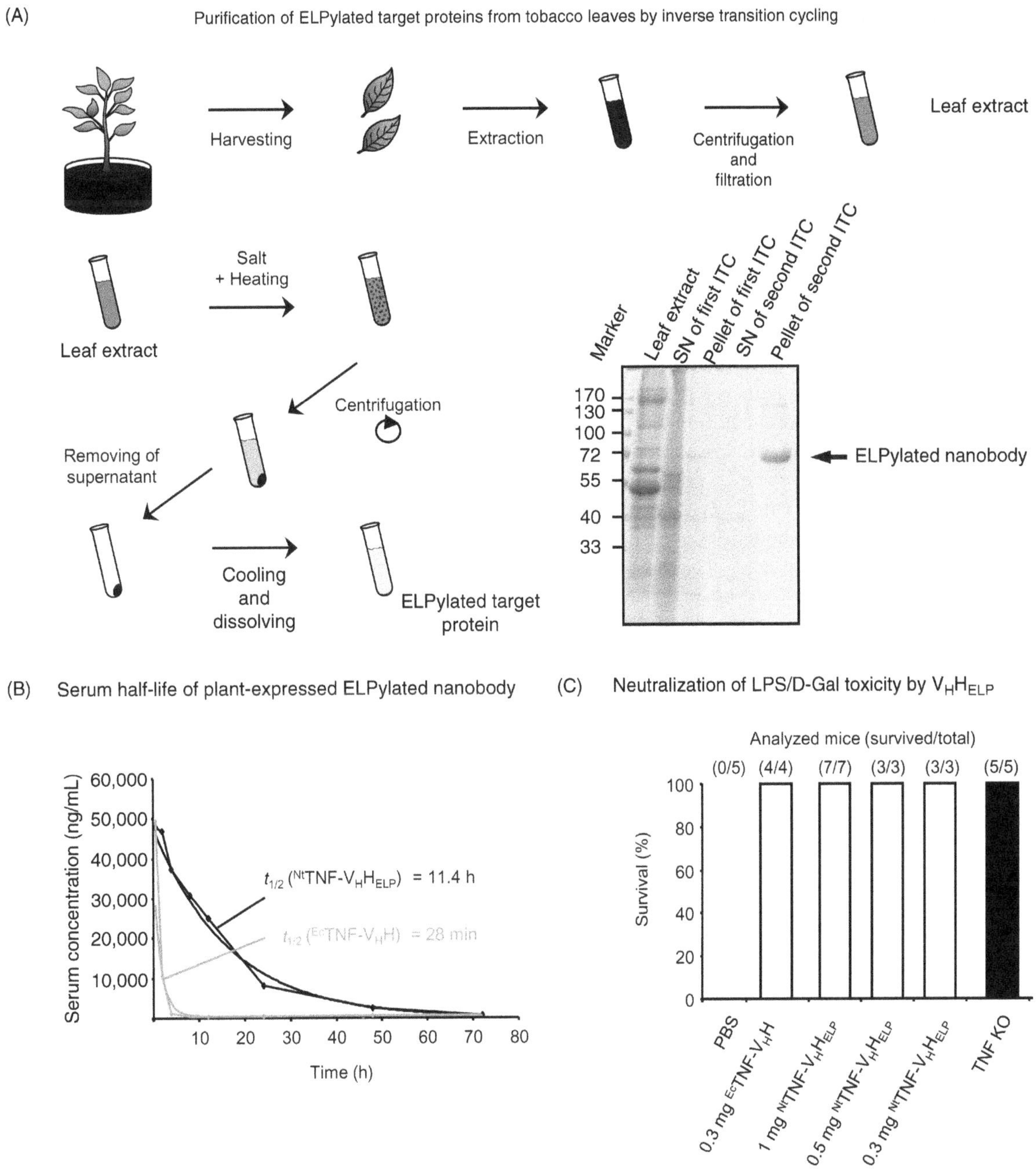

FIGURE 14.4 Pharmaceutical ELPylated target proteins from transgenic tobacco plants. (A) Leaves of transgenic tobacco plants were harvested and grinded. Leaf extract was prepared by solving grounded leaves in buffer followed centrifugation and filtration. The resulting extract was supplement with sodium chloride at a final concentration of 2 M and the temperature was increased above T_t to initiate the aggregation of ELPylated target proteins. Aggregated ELP fusions were precipitated by centrifugation. After removing the supernatant the ELPylated target protein, present in the resulting pellet, was dissolved in buffer with a temperature below T_t. To concentrate the ELPylated target proteins the ITC process might be repeated. The ITC-based purification of ^{Nt}TNF-V_HH_{ELP} is illustrated. Lane 1: molecular mass markers (kDa); lane 2: leaf extract after centrifugation and filtration; lane 3: supernatant (SN) after first round ITC; lane 4: soluble proteins after first round ITC; lane 5: supernatant after second round ITC; lane 6: soluble proteins after second round ITC. Proteins were separated by 10% SDS-PAGE and stained with coomassie brilliant blue. (B) Serum half-life of a plant-produced and ITC-purified ELPylated nanobody against

for pharmaceutical purpose have been expressed as ELP fusions in plants. First attempts were done with human and murine interleukins. An increase in the accumulation of hIL-10 and mIL-4 was reported but biological activity of the plant-derived cytokines was not determined [104]. The first ELPylated, biologically active recombinant protein from transgenic plants was a fragment of naturally occurring soluble glycoprotein 130 (sgp130), which is the natural inhibitor of IL-6 trans-signaling [109,110] and has been shown to be a potent candidate for therapy of rheumatoid arthritis and colon cancer [111,112]. The mini-sgp130-ELP was expressed in transgenic tobacco plants and purified via ITC followed by size-exclusion chromatography. This leads to a yield of 141 μg purified protein per gram fresh leaf material. The plant-derived recombinant mini-sgp130-ELP inhibited IL-6 trans-signaling but not classical IL-6 signaling [113].

Two antibodies, mAb 2F5 and mAb 2G12, with neutralizing capacity of the human immunodeficiency virus 1 (HIV-1) were expressed in tobacco leaves and seeds as ELP-fusion proteins with ELP on either the light or heavy chains, on both or on neither. Floss and co-workers confirmed the increased accumulation of independent antibody chains due to C-terminal fusion with 100 repeats of the pentapeptide Val-Pro-Gly-Xaa-Gly for both 2F5 and 2G12 immunoglobulin chains. In addition, they verify a higher accumulation of the full-length IgGs with ELP either on the light or heavy chains or on both [114,115]. Detailed characterization of the purified antibodies revealed that ELP fusion to the light chain, heavy chain or both chains had no adverse effects on protein quality, and ELPylation did not interfere with folding, assembly, trafficking in the secretory pathway or post-translational modification, but enhance stability while at the same time simplifying recovery [114]. In seeds, however, N-glycans on the naked antibody were extensively trimmed compared to those on the ELP-fusion formats, and were localized to a different subcellular compartment [115]. Surface plasmon resonance (SPR) spectroscopy showed that the kinetic binding parameters of the antibody formats were identical to the immunoglobulins lacking ELP produced in Chinese hamster ovary (CHO) cells. Purification of antibody ELP fusions by ITC was demonstrated and importantly, protein purification from seeds by ITC did not affect the binding kinetics [114,115]. The *in vitro* HIV-neutralization properties of the tobacco leaf-derived 2G12 and ELP fusions were equivalent to or better than those of the CHO counterpart [115].

Recently, the ELP-fusion technology has been used for the ELPylation of an anti-human TNF Camelidae heavy chain monoclonal antibody (anti-TNF-V_HH) [116]. This single-domain antibody (anti-TNF-V_HH_{ELP}) was produced in transgenic tobacco plants (Figure. 14.4A) [117]. A cell-based cytotoxicity assay revealed that the inhibitory properties of the plant-derived anti-TNF-V_HH_{ELP} were indistinguishable from the *E. coli* produced anti-TNF-V_HH without ELP. This demonstrates that neither the *in planta* production, purification via ITC nor the fusion of 100 repeats of Val-Pro-Gly-Xaa-Gly (Xaa: Ala, Val or Gly) to the antibody format compromised the biological activity. In contrast, the serum half-life of the plant-produced V_HH was dramatically increased compared to the traditional *E. coli* produced anti-TNF-V_HH [117]. Furthermore, the authors could show for the first time that an antibody ELP-fusion protein was biologically active *in vivo* as demonstrated in the LPS/D-gal septic shock model applied to humanized TNF mice (Figure 14.4B).

The production of recombinant human erythropoietin (rhEPO) for therapeutically applications is currently done in mammalian cell culture, which is technically complex and expensive. The fusion of ELP (28xVal-Pro-Gly-Val-Gly) to EPO had a negligible effect on many of the reported expression constructs, although the presence of an ELP-fusion partner significantly increased the accumulation of specific EPO constructs. The authors argued that his limited effect of ELP fusion on EPO accumulation may be due to the potential toxicity of EPO-ELP to the plant [118]. However, the presence of an ELP-fusion partner appears to have no effect on the receptor binding of prEPO, demonstrating that ELP has no effect on the biological activity of recombinant proteins [118].

Fusion of $(\text{Val-Pro-Gly-Xaa-Gly})_{100}$ to a mycobacterial antigen and the expression of the ELPylated protein in

FIGURE 14.4 ***(Continued)*** human TNF. Mice were intravenously injected with 100 μg plant derived ^{Nt}TNF-V_HH_{ELP} or with 100 μg ^{Ec}TNF-V_HH from *E. coli* as control. Serum samples were prepared and levels of the respective nanobodies were quantified by ELISA. The *in vivo* persistence of the ELPylated nanobody from tobacco plants was ~24-fold longer than that of non-ELPylated TNF-V_HH produced in bacteria. (C) The therapeutic application of the tobacco produced and ITC purified TNF-V_HH-ELP-fusion protein was tested in humanized TNF mice using a LPS/D-Gal sepsis model. Mice treated with LPS/D-Gal died during the first 24 h (PBS), but mice deficient for human TNF (TNF KO) survived. All mice treated with either bacterial produced TNF-V_HH (^{Ec}TNF-V_HH) or plant-expressed TNF-V_HH_{ELP} (^{Nt}TNF-V_HH_{ELP}) survived, demonstrating the *in vivo* ability of the ELPylated nanobody from transgenic plants to block the activity of human TNF. *Source*: (A), (B) and (C) are reprinted from Reference [117]. Copyright (2011), with permission from Wiley.

transgenic tobacco plants was initiated to obtain an efficient vaccine against *Mycobacterium tuberculosis*, which is the main tuberculosis (TB) causing microorganism in humans. Two mycobacterial antigens, Ag85B and ESAT-6, which are common in *M. tuberculosis* and *M. bovis* and might therefore be effective for both human and bovine TB, have been connected to the fusion antigen TBAg [119]. Effective vaccination of mice and guinea pigs was achieved with this fusion protein [120–122]. ELPylated TBAg was produced in transgenic tobacco plants and immunogenicity of the recombinant protein was assessed in mice, as a reference laboratory model, and piglets, as a preclinical model of human vaccination based on similarities between swine and human immune responses [19]. Both mice and piglets, injected with the ITC-purified ELPylated TBAg or crude leaf extract, produced antibodies against the plant-derived antigen and showed no side effects. Furthermore, a T-cell immune response able to recognize the native antigens, Ag85B and ESAT-6 contained in CFP (culture filtrate proteins) and PPD (purified protein derivatives) was detected in mice [19].

Avian flu hemagglutinin H5 (HA1) and neuraminidase (NA) N1 fused to $(\text{Val-Pro-Gly-Xaa-Gly})_{100}$ have been produced in the ER of transgenic tobacco plants and purified using the salt- and temperature-dependent phase transition of ELP. Here, an effective membrane-based ITC method based on Ge et al. [32] was developed to overcome the proteolytic cleavage and/or partial denaturation of the ELPylated protein. This procedure resulted in a large improvement of the protein purity of recombinant ELPylated NA. Enzymatic activity of plant-derived NA-ELP, assessed on the basis of its ability to cleave 2′-(4-methylumbelliferyl)-α-D-*N*-acetylneuraminic acid, was enhanced compared to the classical centrifugation-based ITC purification method. In addition, the efficiency of the purification was significantly improved as shown for hemagglutinin, neuraminidase, and anti-TNF-V_HH_{ELP} [33].

Recently, a detailed study of the ELP-fusion technology demonstrated the effect of ELP size, orientation, and amino acid sequence composition on the accumulation of different recombinant proteins (GFP, IL-10, EPO, scFv) as well as the effects of ELP size on the efficiency of ITC purification from plant extracts and the biological activity of EPO were investigated [123]. The C-terminal ELP_{28} (28xVal-Pro-Gly-Val-Gly) significantly increased the concentration of IL-10 and EPO in the transient system. A random rearrangement of ELP amino acids (rELP tag) showed a significant reduction of protein yield suggesting that the proper protein sequence and structure of ELP is important for its beneficial effects on the accumulation of recombinant proteins [123]. The concentration of all four recombinant proteins tested as well as nonfused ELP was reduced when utilizing an ELP tag comprising 80 repeats of Val-Pro-Gly-Val-Gly. In contrast, a 4.25-fold enhancement of the recombinant proteins in leaves was demonstrated when comparing the ELP_{80} to the ELP_{20} fusion protein levels. Furthermore, the size of ELP has a significant impact on the purity of the target proteins. The highest concentration of purified recombinant proteins was obtained when using ELP_{20}, ELP_{30}, or ELP_{40} as fusion partner but larger ELP tags (ELP_{80}, ELP_{120}, ELP_{240}) resulted in the highest recovery rates indicating that ELP size is more important than ELP concentration for effective ITC purification [123]. For the investigation of effects of ELP tags on the biological activity of recombinant proteins different EPO expression constructs were evaluated. The resulting recombinant EPO-ELP variants were tested for their ability to bind the human EPO receptor (EPOR). The presence of ELP tags appeared to have no adverse effect on biological activity of recombinant proteins as demonstrated by the receptor binding capacity of recombinant EPO derived from plants [123].

14.5 CHALLENGES AND FUTURE PERSPECTIVES

The development of ELPylation in the 90s of the last century has opened a novel way of laboratory-scale and presumably industrial-scale protein expression and purification. The attractiveness of ELPylation is based on the thermally stimulated inverse phase transition, which can be used for non-chromatographic purification of recombinant proteins. However, the combination of ELPylation and intein-technology and the ELP-mediated affinity capture (EMAC) allows the production of ELP-free target proteins. The good biocompatibility of ELPs and their controllable aggregation *in vitro* and *in vivo* will enable novel medical applications, such as hyperthermia-induced locally extravascular accumulation, ELP-based drug depot formation or materials for tissue repair or engineering. In the near future, we will see systematic studies of relations between ELP size and steric influences on antigenicity (vaccine), functionality (recognition by antibodies), and enzymatic activity. Moreover, more sophisticated mouse studies with ELPylated proteins are needed to evaluate the potential of ELPylation for therapeutic application.

Plant-made industrial and pharmaceutical proteins are still in their early days, with no protein products being in the clinic. However, current clinical studies are on the way to demonstrate the general feasibility of application of plant made pharmaceuticals with current state-of-the-art production systems. The adaptation of ELPylation to plant biotechnology has revealed a general increase in accumulation rate, with good ITC-based recovery and high bioactivity of the final target protein. Future work is needed to fully explore the potential of the ELPylation technology on plant biotechnology for the *in planta* production of industrial and pharmaceutical proteins. Scale-up experiments for purification and processing have to be performed and mass production systems have to be developed.

14.6 CONCLUSION

Recombinant protein expression is an important process but cost-intensive process for the production of therapeutically active biologics. ELPs derived from mammalian elastin are extraordinarily biocompatible and they exhibit a thermally responsive reversible soluble to insoluble phase transition. These properties make ELPs attractive for drug delivery, as materials for tissue repair or engineering, or to facilitate the purification of recombinant ELP-fusion proteins by a process termed inverse phase transition. Recently, plant-based expression systems have been evaluated for the production of ELP-fusion proteins. The simplified purification and the great potential of transgenic plants as bioreactors open new possibilities for the large-scale production of recombinant proteins such as biopharmaceuticals and industrial enzymes.

REFERENCES

1. Schmidt SR. (2009) Fusion-proteins as biopharmaceuticals—applications and challenges. *Curr. Opin. Drug Discov. Dev.* 12, 284–295.
2. Keeley FW, Bellingham CM, Woodhouse KA. (2002) Elastin as a self-organizing biomaterial: use of recombinantly expressed human elastin polypeptides as a model for investigations of structure and self-assembly of elastin. *Philos. Trans. R. Soc. Lond. B Biol. Sci.* 357, 185–189.
3. Rosenbloom J, Abrams WR, Mecham R. (1993) Extracellular matrix 4: the elastic fiber. *FASEB J.* 7, 1208–1218.
4. Daamen WF, Hafmans T, Veerkamp JH, Van Kuppevelt TH. (2001) Comparison of five procedures for the purification of insoluble elastin. *Biomaterials* 22, 1997–2005.
5. Vrhovski B, Weiss AS. (1998) Biochemistry of tropoelastin. *Eur. J. Biochem.* 258, 1–18.
6. Gibson MA, Kumaratilake JS, Cleary EG. (1989) The protein components of the 12-nanometer microfibrils of elastic and nonelastic tissues. *J. Biol. Chem.* 264, 4590–4598.
7. Urry DW, Long MM. (1976) Conformations of the repeat peptides of elastin in solution: an application of proton and carbon-13 magnetic resonance to the determination of polypeptide secondary structure. *CRC Crit. Rev. Biochem.* 4, 1–45.
8. Urry DW. (1992) Free energy transduction in polypeptides and proteins based on inverse temperature transitions. *Prog. Biophys. Mol. Biol.* 57, 23–57.
9. Floss DM, Schallau K, Rose-John S, Conrad U, Scheller J. (2010) Elastin-like polypeptides revolutionize recombinant protein expression and their biomedical application. *Trends Biotechnol.* 28, 37–45.
10. Urry DW, McPherson DT, Xu J, Daniell H, Guda C, Gowda DC, et al. (1996) Protein based polymeric materials: synthesis and properties. In: Salamone JC (Ed), *The Polymeric Materials Ebcyclopedia: Sythesis, Properties, and Applications*. Boca Raton, CRC Press, pp. 2645–2699.
11. Simnick AJ, Lim DW, Chow D, Chilkoti A. (2007) Biomedical and biotechnological applications of elastin-like polypeptides. *J. Macromol. Sci. C Poly. Rev.* 47, 121–154.
12. Serrano V, Liu W, Franzen S. (2007) An infrared spectroscopic study of the conformational transition of elastin-like polypeptides. *Biophys. J.* 93, 2429–2435.
13. Girotti A, Reguera J, Arias FJ, Alonso M, Testera AM, Rodriguez-Cabello JC. (2004) Influence of the molecular weight on the inverse temperature transition of a model genetically engineered elastin-like pH-responsive polymer. *Macromolecules* 37, 3396–3400.
14. Meyer DE, Chilkoti A. (2004) Quantification of the effects of chain length and concentration on the thermal behavior of elastin-like polypeptides. *Biomacromolecules* 5, 846–851.
15. Ribeiro A, Arias FJ, Reguera J, Alonso M, Rodriguez-Cabello JC. (2009) Influence of the amino-acid sequence on the inverse temperature transition of elastin-like polymers. *Biophys. J.* 97, 312–320.
16. Urry DW, Luan CH, Parker TM, Gowda DC, Prasad KU, Reid MC, et al. (1991) Temperature of polypeptide inverse temperature transition depends on mean residue hydrophobicity. *J. Am. Chem. Soc.* 113, 4346–4348.
17. Urry DW, Parker TM, Reid MC, Gowda DC. (1991) Biocompatibility of the bioelastic materials, poly(GVGVP) and its gamma irradiation cross-linked matrix—summary of generic biological test results. *J. Bioactive Compatible Polym.* 6, 263–282.
18. Rincon AC, Molina-Martinez IT, de Las Heras B, Alonso M, Bailez C, Rodriguez-Cabello JC, et al. (2006) Biocompatibility of elastin-like polymer poly(VPAVG) microparticles: in vitro and in vivo studies. *J. Biomed. Mater. Res. A.* 78A, 343–351.
19. Floss DM, Mockey M, Zanello G, Brosson D, Diogon M, Frutos R, et al. (2010) Expression and immunogenicity of the mycobacterial Ag85B/ESAT-6 antigens produced in transgenic plants by elastin-like peptide fusion strategy. *J. Biomed. Biotechnol.* 274346.
20. Jones DS, Silverman AP, Cochran JR. (2008) Developing therapeutic proteins by engineering ligand-receptor interactions. *Trends Biotechnol.* 26, 498–505.
21. Jones S, Scheller J, Rose-John S. (2011) Therapeutic strategies for the clinical blockade of IL-6/gp130 signaling. *J. Clin. Invest.* 121, 3375–3383.
22. McPherson DT, Xu J, Urry DW. (1996) Product purification by reversible phase transition following *Escherichia coli* expression of genes encoding up to 251 repeats of the elastomeric pentapeptide GVGVP. *Protein. Expr. Purif.* 7, 51–57.
23. Meyer DE, Chilkoti A. (1999) Purification of recombinant proteins by fusion with thermally-responsive polypeptides. *Nat. Biotechnol.* 17, 1112–1115.
24. Chow DC, Dreher MR, Trabbic-Carlson K, Chilkoti A. (2006) Ultra-high expression of a thermally responsive recombinant fusion protein in *E. coli*. *Biotechnol. Prog.* 22, 638–646.

25. Fong BA, Wood DW. (2010) Expression and purification of ELP-intein-tagged target proteins in high cell density *E. coli* fermentation. *Microb. Cell Factories* 9, 77.
26. Christensen T, Amiram M, Dagher S, Trabbic-Carlson K, Shamji MF, Setton LA, et al. (2009) Fusion order controls expression level and activity of elastin-like polypeptide fusion proteins. *Protein Sci.* 18, 1377–1387.
27. Christensen T, Trabbic-Carlson K, Liu W, Chilkoti A. (2007) Purification of recombinant proteins from *Escherichia coli* at low expression levels by inverse transition cycling. *Anal. Biochem.* 360, 166–168.
28. Lan D, Huang G, Shao H, Zhang L, Ma L, Chen S, et al. (2011) An improved nonchromatographic method for the purification of recombinant proteins using elastin-like polypeptide-tagged proteases. *Anal. Biochem.* 415, 200–202.
29. Fong BA, Wu WY, Wood DW. (2010) The potential role of self-cleaving purification tags in commercial-scale processes. *Trends Biotechnol.* 28, 272–279.
30. Banki MR, Feng L, Wood DW. (2005) Simple bioseparations using self-cleaving elastin-like polypeptide tags. *Nat. Methods* 2, 659–661.
31. Ge X, Yang DS, Trabbic-Carlson K, Kim B, Chilkoti A, Filipe CD. (2005) Self-cleavable stimulus responsive tags for protein purification without chromatography. *J. Am. Chem. Soc.* 127, 11228–11229.
32. Ge X, Trabbic-Carlson K, Chilkoti A, Filipe CDM. (2006) Purification of an elastin-like fusion protein by microfiltration. *Biotechnol. Bioeng.* 95, 424–432.
33. Phan HT, Conrad U. (2011) Membrane-based inverse transition cycling: an improved means for purifying plant-derived recombinant protein-elastin-like polypeptide fusions. *Int. J. Mol. Sci.* 12, 2808–2821.
34. Fong BA, Wu WY, Wood DW. (2009) Optimization of ELP-intein mediated protein purification by salt substitution. *Protein. Expr. Purif.* 66, 198–202.
35. Fong BA, Gillies AR, Ghazi I, LeRoy G, Lee KC, Westblade LF, et al. (2010) Purification of *Escherichia coli* RNA polymerase using a self-cleaving elastin-like polypeptide tag. *Protein Sci.* 19, 1243–1252.
36. Ge X, Filipe CD. (2006) Simultaneous phase transition of ELP tagged molecules and free ELP: an efficient and reversible capture system. *Biomacromolecules* 7, 2475–2478.
37. Kim JY, Mulchandani A, Chen W. (2005) Temperature-triggered purification of antibodies. *Biotechnol. Bioeng.* 90, 373–379.
38. Kim JY, O'Malley S, Mulchandani A, Chen W. (2005) Genetically engineered elastin-protein A fusion as a universal platform for homogeneous, phase-separation immunoassay. *Anal. Chem.* 77, 2318–2322.
39. Kostal J., Mulchandani A., Chen W., (2001) Tunable biopolymers for heavy metal removal. *Macromolecules* 34, 2257–2261.
40. Kostal J, Mulchandani A, Gropp KE, Chen W. (2003) A temperature responsive biopolymer for mercury remediation. *Environ. Sci. Technol.* 37, 4457–4462.
41. Kostal J, Yang R, Wu CH, Mulchandani A, Chen W. (2004) Enhanced arsenic accumulation in engineered bacterial cells expressing ArsR. *Appl. Environ. Microbiol.* 70, 4582–4587.
42. Kostal J, Mulchandani A, Chen W. (2004) Affinity purification of plasmid DNA by temperature-triggered precipitation. *Biotechnol. Bioeng.* 85, 293–297.
43. Lao UL, Kostal J, Mulchandani A, Chen W. (2007) Affinity purification of plasmid DNA by temperature-triggered precipitation. *Nat. Protoc.* 2, 1263–1268.
44. Prabhukumar G, Matsumoto M, Mulchandani A, Chen W. (2004) Cadmium removal from contaminated soil by tunable biopolymers. *Environ. Sc. Technol.* 38, 3148–3152.
45. Stiborova H, Kostal J, Mulchandani A, Chen W. (2003) One-step metal-affinity purification of histidine-tagged proteins by temperature-triggered precipitation. *Biotechnol. Bioeng.* 82, 605–611.
46. Chilkoti A, Christensen T, MacKay JA. (2006) Stimulus responsive elastin biopolymers: applications in medicine and biotechnology. *Curr. Opin. Chem. Biol.* 10, 652–657.
47. Chow D, Nunalee ML, Lim DW, Simnick AJ, Chilkoti A. (2008) Peptide-based biopolymers in biomedicine and biotechnology. *Mater. Sci. Eng. R Rep.Rev. J.* 62, 125–155.
48. Martinez-Osorio H, Juarez-Campo M, Diebold Y, Girotti A, Alonso M, Arias FJ, et al. (2009) Genetically engineered elastin-like polymer as a substratum to culture cells from the ocular surface. *Curr. Eye Res.* 34, 48–56.
49. Mie M, Mizushima Y, Kobatake E. (2008) Novel extracellular matrix for cell sheet recovery using genetically engineered elastin-like protein. *J. Biomed. Mater. Res.B Appl. Biomater.* 86, 283–290.
50. Betre H, Ong SR, Guilak F, Chilkoti A, Fermor B, et al. (2006) Chondrocytic differentiation of human adipose-derived adult stem cells in elastin-like polypeptide. *Biomaterials* 27, 91–99.
51. Raucher D, Massodi I, Bidwell GL. (2008) Thermally targeted delivery of chemotherapeutics and anti-cancer peptides by elastin-like polypeptide. *Expert Opin. Drug Deliv.* 5, 353–369.
52. Meyer DE, Kong GA, Dewhirst MW, Zalutsky MR, Chilkoti A. (2001) Targeting a genetically engineered elastin-like polypeptide to solid tumors by local hyperthermia. *Cancer Res.* 61, 1548–1554.
53. Meyer DE, Trabbic-Carlson K, Chilkoti A. (2001) Protein purification by fusion with an environmentally responsive elastin-like polypeptide: effect of polypeptide length on the purification of thioredoxin. *Biotechnol. Prog.* 17, 720–728.
54. Dreher MR, Liu W, Michelich CR, Dewhirst MW, Chilkoti A. (2007) Thermal cycling enhances the accumulation of a temperature-sensitive biopolymer in solid tumors. *Cancer Res.* 67, 4418–4424.
55. Rodriguez-Cabello JC, Reguera J, Alonso M, Parker TM, McPherson DT, Urry DW. (2004) Endothermic and exothermic components of an inverse temperature transition for hydrophobic association by TMDSC. *Chem. Phys. Lett.* 388, 127–131.

56. Herrero-Vanrell R, Rincon AC, Alonso M, Reboto V, Molina-Martinez IT, Rodriguez-Cabello JC. (2005) Self-assembled particles of an elastin-like polymer as vehicles for controlled drug release. *J. Control. Release.* 102, 113–122.

57. Na K, Jung J, Lee J, Hyun J. (2010) Thermoresponsive pore structure of biopolymer microspheres for a smart drug carrier. *Langmuir* 26, 11165–11169.

58. Dash BC, Mahor S, Carroll O, Mathew A, Wang W, Woodhouse KA, et al. (2011) Tunable elastin-like polypeptide hollow sphere as a high payload and controlled delivery gene depot. *J. Control. Release* 152, 382–392.

59. Kim W, Chaikof EL. (2010) Recombinant elastin-mimetic biomaterials: emerging applications in medicine. *Adv. Drug Deliv. Rev.* 62, 1468–1478.

60. Osborne JL, Farmer R, Woodhouse KA. (2008) Self-assembled elastin-like polypeptide particles. *Acta Biomater.* 4, 49–57.

61. MacKay JA, Chen M, McDaniel JR, Liu W, Simnick AJ, Chilkoti A. (2009) Self-assembling chimeric polypeptide-doxorubicin conjugate nanoparticles that abolish tumours after a single injection. *Nat. Mater.* 8, 993–999.

62. Fujita Y, Mie M, Kobatake E. (2009) Construction of nanoscale protein particle using temperature-sensitive elastin-like peptide and polyaspartic acid chain. *Biomaterials* 30, 3450–3457.

63. Wu YQ, MacKay JA, McDaniel JR, Chilkoti A, Clark RL. (2009) Fabrication of elastin-like polypeptide nanoparticles for drug delivery by electrospraying. *Biomacromolecules* 10, 19–24.

64. Simnick AJ, Valencia CA, Liu R, Chilkoti A. (2010) Morphing low-affinity ligands into high-avidity nanoparticles by thermally triggered self-assembly of a genetically encoded polymer. *ACS Nano* 4, 2217–2227.

65. Silva LC, Ortigosa LC, Benard G. (2010) Anti-TNF-alpha agents in the treatment of immune-mediated inflammatory diseases: mechanisms of action and pitfalls. *Immunotherapy* 2, 817–833.

66. Shamji MF, Chen J, Friedman AH, Richardson WJ, Chilkoti A, Setton LA. (2008) Synthesis and characterization of a thermally-responsive tumor necrosis factor antagonist. *J. Control. Release* 129, 179–186.

67. Shamji MF, Jing L, Chen J, Hwang P, Ghodsizadeh O, Friedman AH, et al. (2008) Treatment of neuroinflammation by soluble tumor necrosis factor receptor type II fused to a thermally responsive carrier. *J. Neurosurg. Spine* 9, 221–228.

68. Shamji MF, Whitlatch L, Friedman AH, Richardson WJ, Chilkoti A, Setton LA. (2008) An injectable and in situ-gelling biopolymer for sustained drug release following perineural administration. *Spine (Phila Pa 1976)* 33, 748–754.

69. Bidwell GL, Davis AN, Raucher D. (2009) Targeting a c-Myc inhibitory polypeptide to specific intracellular compartments using cell penetrating peptides. *J. Control. Release* 135, 2–10.

70. Massodi I, Moktan S, Rawat A, Bidwell GL, III, Raucher D. (2010) Inhibition of ovarian cancer cell proliferation by a cell cycle inhibitory peptide fused to a thermally responsive polypeptide carrier. *Int. J. Cancer* 126, 533–544.

71. Massodi I, Bidwell GL, III Davis A, Tausend A, Credit K, Flessner M, et al. (2009) Inhibition of ovarian cancer cell metastasis by a fusion polypeptide Tat-ELP. *Clin. Exp. Metastasis* 26, 251–260.

72. Massodi I, Thomas E, Raucher D. (2009) Application of thermally responsive elastin-like polypeptide fused to a lactoferrin-derived peptide for treatment of pancreatic cancer. *Molecules* 14, 1999–2015.

73. Moktan S, Ryppa C, Kratz F, Raucher D. (2012) A thermally responsive biopolymer conjugated to an acid-sensitive derivative of Paclitaxel stabilizes microtubules, arrests cell cycle, and induces apoptosis. *Invest. New. Drugs* 30, 236–248.

74. Liu W, MacKay JA, Dreher MR, Chen M, McDaniel JR, Simnick AJ, et al. (2010) Injectable intratumoral depot of thermally responsive polypeptide-radionuclide conjugates delays tumor progression in a mouse model. *J. Control. Release* 144, 2–9.

75. Rodriguez-Cabello JC, Prieto S, Reguera J, Arias FJ, Ribeiro A. (2007) Biofunctional design of elastin-like polymers for advanced applications in nanobiotechnology. *J. Biomater. Sci. Polym. Ed.* 18, 269–286.

76. Di Zio K, Tirrell DA. (2003) Mechanical properties of artificial protein matrices engineered for control of cell and tissue behavior. *Macromolecules* 36, 1553–1558.

77. Nagarsekar A, Crissman J, Crissman M, Ferrari F, Cappello J, Ghandehari H. (2003) Genetic engineering of stimuli-sensitive silkelastin-like protein block copolymers. *Biomacromolecules* 4, 602–607.

78. Martin DC, Jiang T, Buchko CJ. (1997) In: McGrath KP, Kaplan DL (Eds), *Protein-Based Materials*. Birkhäuser, Boston, pp. 339–370.

79. Cappello J, McGrath KP. (1994) In: Kaplan DL, Adams WW, Farmer B, Viney C (Eds), *Silk Polymers: Materials Science and Biotechnology*. ACS Symposium Series, American Chemical Society, Washington, pp. 311- 327.

80. Qiu W, Teng W, Cappello J, Wu X. (2009) Wet-spinning of recombinant silk-elastin-like protein polymer fibers with high tensile strength and high deformability. *Biomacromolecules* 10, 602–608.

81. Reguera J, Fahmi A, Moriarty P, Girotti A, Rodriguez-Cabello JC. (2004) Nanopore formation by self-assembly of the model genetically engineered elastin-like polymer [(VPGVG)(2)(VPGEG)(VPGVG)(2)](15). *J. Am. Chem. Soc.* 126, 13212–13213.

82. Boehm R. (2007) Bioproduction of therapeutic proteins in the 21st century and the role of plants and plant cells as production platforms. In: Lal S (Ed), *Biology of Emerging Viruses: Sars, Avian and Human Influenza, Metapneumovirus, Nipah, West Nile, and Ross River Virus*. Blackwell Publishing, Uxford, UK, pp. 121–134.

83. Sparrow PAC, Irwin JA, Dale PJ, Twyman RM, Ma JKC. (2007) Pharma-Planta: road testing the developing regulatory guidelines for plant-made pharmaceuticals. *Transgenic Res.* 16, 147–161.

84. Twyman RM, Schillberg S, Fischer R. (2005) Transgenic plants in the biopharmaceutical market. *Expert Opin. Emerg. Drugs* 10, 185–218.

85. Whaley KJ, Hanes J, Shattock R, Cone RA, Friend DR. (2010) Novel approaches to vaginal delivery and safety of microbicides: biopharmaceuticals, nanoparticles, and vaccines. *Antiviral. Res.* 88, S55–S66.

86. Singh N, Pandey A, Mittal SK. (2011) Avian influenza pandemic preparedness: developing prepandemic and pandemic vaccines against a moving target. *Expert Rev. Mol. Med.* 12, e14.

87. Yusibov V, Rabindran S. (2008) Recent progress in the development of plant-derived vaccines. *Expert Rev. Vaccines* 7, 1173–1183.

88. Rybicki EP. (2010) Plant-made vaccines for humans and animals. *Plant Biotechnol. J.* 8, 620–637.

89. Hiatt A, Cafferkey R, Bowdish K. (1989) Production of antibodies in transgenic plants. *Nature* 342, 76–78.

90. Mason HS, Lam DMK, Arntzen CJ. (1992) Expression of hepatitis B surface antigen in transgenic plants. *Proc. Natl. Acad. Sci. U. S. A.* 89, 11745–11749.

91. Yusibov V, Streatfield SJ, Kushnir N. (2011) Clinical development of plant-produced recombinant pharmaceuticals: vaccines, antibodies, and beyond. *Human Vaccines* 7, 1–9.

92. Pujol M, Ramirez NI, Ayala M, Gavilondo JV, Valdes R, Rodriguez M, et al. (2005) An integral approach towards a practical application for a plant-made monoclonal antibody in vaccine purification. *Vaccine* 23, 1833–1837.

93. Santi L. (2009) Plant derived veterinary vaccines. *Vet. Res. Commun.* 33, S61–S66.

94. Rybicki EP. (2009) Plant-produced vaccines: promise and reality. *Drug Discov. Today* 14, 16–24.

95. Ahmad A, Pereira EO, Conley AJ, Richman AS, Menassa R. (2010) Green biofactories: recombinant protein production in plants. *Recent Patents Biotechnol.* 4, 242–259.

96. Gleba Y, Klimyuk V, Marillonnet S. (2005) Magnifection—a new platform for expressing recombinant vaccines in plants. *Vaccine* 23, 2042–2048.

97. Zhang X, Guda C, Datta R, Dute R, Urry DW, Daniell H. (1995) Nuclear expression of an environmentally friendly synthetic protein based polymer gene in tobacco cells. *Biotechnol. Lett.* 17, 1279–1284.

98. Zhang XR, Urry DW, Daniell H. (1996) Expression of an environmentally friendly synthetic protein-based polymer gene in transgenic tobacco plants. *Plant Cell Rep.* 16, 174–179.

99. Guda C, Zhang X, Mcpherson DT, Xu J, Cherry JH, Urry DW, et al. (1995) Hyper expression of an environmentally friendly synthetic-polymer gene. *Biotechnol. Lett.* 17, 745–750.

100. Scheller J, Henggeler D, Viviani A, Conrad U. (2004) Purification of spider silk-elastin from transgenic plants and application for human chondrocyte proliferation. *Transgenic Res.* 13, 51–57.

101. Scheller J, Guhrs KH, Grosse F, Conrad U. (2001) Production of spider silk proteins in tobacco and potato. *Nat. Biotechnol.* 19, 573–577.

102. Junghans F, Morawietz M, Conrad U, Scheibel T, Heilmann A, Spohn U. (2006) Preparation and mechanical properties of layers made of recombinant spider silk proteins and silk from silk worm. *Appl. Phys. A Mater. Sci. Process.* 82, 253–260.

103. Junghans F, Conrad U, Heilmann A, Spohn U. (2008) Micromechanical investigations on films made of recombinant spider silk proteins and silk fibroin. *Mater. Res. Soc. Symp. Proc Mater. Res. Soc.* 1062.

104. Patel J, Zhu H, Menassa R, Gyenis L, Richman A, Brandle J. (2007) Elastin-like polypeptide fusions enhance the accumulation of recombinant proteins in tobacco leaves. *Transgenic Res.* 16, 239–249.

105. Scheller J, Leps M, Conrad U. (2006) Forcing single-chain variable fragment production in tobacco seeds by fusion to elastin-like polypeptides. *Plant Biotechnol. J.* 4, 243–249.

106. Fiedler U, Conrad U. (1995) High-level production and long-term storage of engineered antibodies in transgenic tobacco seeds. *Biotechnology (N Y)* 13, 1090–1093.

107. Stoger E, Ma JKC, Fischer R, Christou P. (2005) Sowing the seeds of success: pharmaceutical proteins from plants. *Curr. Opin. Biotechnol.* 16, 167–173.

108. Joensuu JJ, Brown KD, Conley AJ, Clavijo A, Menassa R, Brandle JE. (2009) Expression and purification of an anti-foot-and-mouth disease virus single chain variable antibody fragment in tobacco plants. *Transgenic Res.* 18, 685–696.

109. Jostock T, Mullberg J, Ozbek S, Atreya R, Blinn G, Voltz N, et al. (2001) Soluble gp130 is the natural inhibitor of soluble interleukin-6 receptor transsignaling responses. *Eur. J. Biochem.* 268, 160–167.

110. Rose-John S, Waetzig GH, Scheller J, Grotzinger J, Seegert D. (2007) The IL-6/sIL-6R complex as a novel target for therapeutic approaches. *Expert Opin. Ther. Targets* 11, 613–624.

111. Nowell MA, Richards PJ, Horiuchi S, Yamamoto N, Rose-John S, Topley N, et al. (2003) Soluble IL-6 receptor governs IL-6 activity in experimental arthritis: blockade of arthritis severity by soluble glycoprotein 130. *J. Immunol.* 171, 3202–3209.

112. Becker C, Fantini MC, Wirtz S, Nikolaev A, Lehr HA, Galle PR, et al. (2005) IL-6 signaling promotes tumor growth in colorectal cancer. *Cell Cycle* 4, 217–220.

113. Lin M, Rose-John S, Grotzinger J, Conrad U, Scheller J. (2006) Functional expression of a biologically active fragment of soluble gp130 as an ELP-fusion protein in transgenic plants: purification via inverse transition cycling. *Biochem. J.* 398, 577–583.

114. Floss DM, Sack M, Stadlmann J, Rademacher T, Scheller J, Stoger E, et al. (2008) Biochemical and functional characterization of anti-HIV antibody-ELP fusion proteins from transgenic plants. *Plant Biotechnol. J.* 6, 379–391.

115. Floss DM, Sack M, Arcalis E, Stadlmann J, Quendler H, Rademacher T, et al. (2009) Influence of ELP fusions on the quantity and quality of a tobacco-derived HIV-neutralizing antibody. *Plant Biotechnol. J.* 7, 899–913.

116. Plagmann I, Chalaris A, Kruglov AA, Nedospasov SA, Rosenstiel P, Rose-John S, et al. (2009) Transglutaminase-catalyzed covalent multimerization of camelidae anti-human TNF single domain antibodies improves neutralizing activity. *J. Biotechnol.* 142, 170–178.

117. Conrad U, Plagmann I, Malchow S, Sack M, Floss DM, Kruglov AA, et al. (2011) ELPylated anti-human TNF therapeutic single-domain antibodies for prevention of lethal septic shock. *Plant Biotechnol. J.* 9, 22–31.

118. Conley AJ, Mohib K, Jevnikar AM, Brandle JE. (2009) Plant recombinant erythropoietin attenuates inflammatory kidney cell injury. *Plant Biotechnol. J.* 7, 183–199.

119. Frutos R, Denise H, Vivares C, Neuhaus JM, Vitale S, Pedrazzini E, et al. (2008) Pharmaceutical proteins in plants. A strategic genetic engineering approach for the production of tuberculosis antigens. *Ann. N. Y. Acad. Sci.* 1149, 275–280.

120. Olsen AW, van Pinxteren LAH, Okkels LM, Rasmussen PB, Andersen P. (2001) Protection of mice with a tuberculosis subunit vaccine based on a fusion protein of antigen 85B and ESAT-6. *Infect. Immun.* 69, 2773–2778.

121. Olsen AW, Williams A, Okkels LM, Hatch G, Andersen P. (2004) Protective effect of a tuberculosis subunit vaccine based on a fusion of antigen 85B and ESAT-6 in the aerosol guinea pig model. *Infect. Immun.* 72, 6148–6150.

122. Derrick SC, Yang AL, Morris SL. (2004) A polyvalent DNA vaccine expressing an ESAT6-Ag85B fusion protein protects mice against a primary infection with *Mycobacterium tuberculosis* and boosts BCG-induced protective immunity. *Vaccine* 23, 780–788.

123. Conley AJ, Joensuu JJ, Jevnikar AM, Menassa R, Brandle JE. (2009) Optimization of elastin-like polypeptide fusions for expression and purification of recombinant proteins in plants. *Biotechnol. Bioeng.* 103, 562–573.

124. McDaniel JR, Callahan DJ, Chilkoti A. (2010) Drug delivery to solid tumors by elastin-like polypeptides. *Adv. Drug Deliv. Rev.* 62, 1456–1467.

15

LIGAND-RECEPTOR FUSION DIMERS

SARBENDRA L. PRADHANANGA,[1] IAN R. WILKINSON,[1] ERIC FERRANDIS,[2] PETER J. ARTYMIUK,[3] JON R. SAYERS,[4] AND RICHARD J. ROSS[1]

[1]*Department of Human Metabolism, University of Sheffield, Medical School, Sheffield, UK*
[2]*IPSEN, Institut Henri Beaufour, Les Ulis, France*
[3]*Department of Molecular Biology and Biotechnology, University of Sheffield, Sheffield, UK*
[4]*Department of Infection & Immunity, University of Sheffield, Medical School, Sheffield, UK*

15.1 INTRODUCTION

Growth hormone (GH) is a cytokine hormone that regulates linear growth in childhood and normal body composition in adults [1]. GH therapy is used to treat GH deficiency; in children, treatment has obvious benefits with the deficient child showing an increase in growth rate within a few months of starting treatment, other benefits may also be noticed such as increased strength and reduction of body fat. In adults, the effects of GH treatment are less obvious but still measurable and include improved bone density, increased muscle mass, decreased adipose tissue, a strengthened immune system, improved circulatory system, and blood lipid levels. GH has also been used to treat wasting due to HIV, and short stature in Turner's and Prader–Willi syndrome and chronic renal failure.

The current therapeutic regimen for GH therapy is once-daily subcutaneous injections, which are expensive and inconvenient to the patient. A number of approaches have been taken to extend the clearance half-life of GH, thereby decreasing the frequency of the GH injections and so providing greater convenience to the patient.

PEGylation of the GH molecule (where poly(ethylene glycol) molecules are covalently bound to the GH molecule) increases the size of the therapeutic molecule hence reducing its clearance rate via the kidneys, decreasing the hormone's immunogenic profile and also protecting the GH molecule from proteases [2]. However, PEGylation greatly reduces the affinity of the hormone for its receptor [2] and has been associated with lipoatrophy [3] and hepatotoxic effects [4,5]. It is thought that the lack of renal clearance may allow the levels of the PEGylated proteins to accumulate to toxic levels [6].

A sustained-release or depot preparation, where GH is encapsulated in microspheres, is another means of increasing the intervals of injections during GH treatment [7–9]. However, this method also has some disadvantages; namely, bruising and pain at the injection site [10], the formation of small erythematous nodules at the injection site [8,10] and initial supra-physiological GH levels followed by a slow decline over the dosing period [11].

PEGylation and depot preparations of GH are both relatively expensive to manufacture due to the complex processing required to produce these therapies. A therapeutic that does not require the complex processing required by the aforementioned products, while maintaining their prolonged activity would be highly desirable to the manufacturer, the clinician, and patient.

Fusion Protein Technologies for Biopharmaceuticals: Applications and Challenges, First Edition. Edited by Stefan R. Schmidt.

Molecule size has a profound effect on the clearance rate from the vascular system via the kidneys as demonstrated by the PEGylation of GH. Another means of increasing the molecule size of GH is to create a fusion of GH with another "inert" molecule; the obvious advantage of a fusion molecule is that the increase in size, and hence the reduced clearance rate, could be achieved without costly processing of the expressed molecule. Fusion of GH with albumin has been shown to have delayed clearance [12] as has the approach discussed in this chapter, the fusion of GH with its receptor [13,14].

GH has two binding sites for its receptor and acts by forming a complex with a GHR dimer at the cell surface; this initiates a conformational change in the complex and triggers intracellular signaling (Figure 15.1A). The extracellular domain of GHR is proteolytically cleaved and circulates as GH-binding protein (GHBP). Circulating GH binds to GHBP as a 1:1 molar complex and this biologically inactive complex is protected from degradation and has a reduced rate of clearance [15,16]. A covalently cross-linked complex of GH-GHR has delayed clearance but is biologically inactive [17], and an augmentation of GH activity is observed when separately purified GH and GHBP is administered at a ratio of 1:1 [18]. Thus, the clearance rate of GH is reduced due to the circulating reservoir of GH-GHBP complex, which is in equilibrium with active free GH [19]. A similar system is utilized in many other hormonal systems to increase the circulating half-life of the hormones.

This concept of protecting GH from degradation and clearance by linking it to the extracellular domain of GHR (GHRec) was the basis of the design of the ligand-receptor (LR) fusions. Fusion proteins of GH covalently linked end-to-end with GHRec were designed; based on our observations that a truncated variant of GHR at the cell surface acts a dominant negative inhibitor of GHR [20,21] we initially hypothesized that the LR-fusion would be an antagonist (Figure 15.1B). Unexpectedly, the GH LR-fusions generated were potent long-acting agonists (Figure 15.1C).

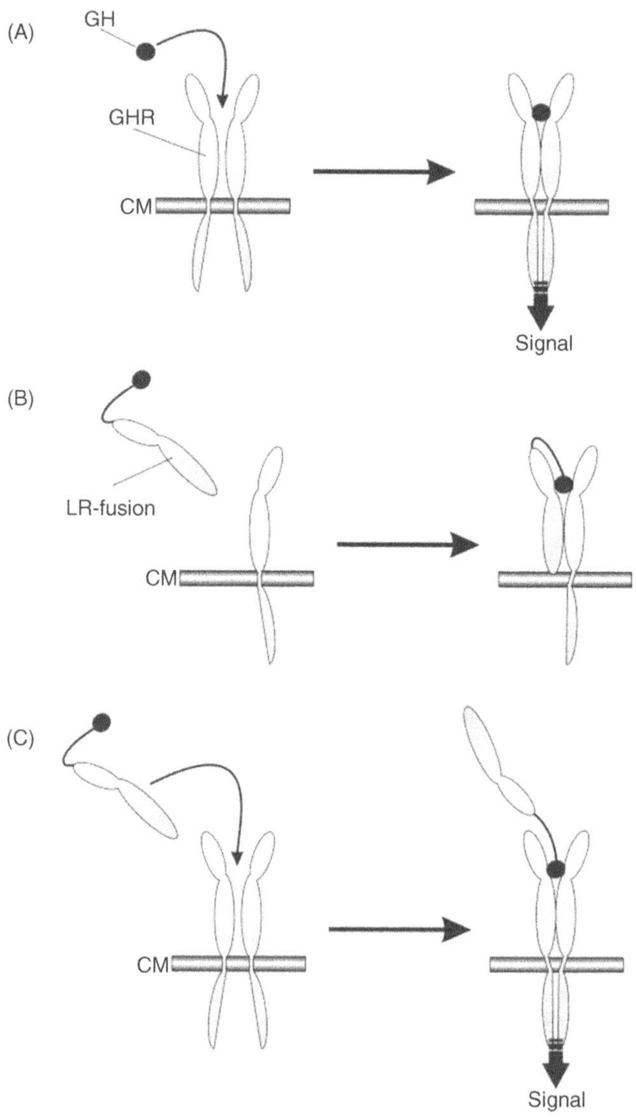

FIGURE 15.1 (A) The signaling mechanism of GH, a GH molecule binds to a GHR dimer at the cell membrane (CM) this causes a conformational change that initiates signaling. (B) A diagram of how the LR-fusions may act as an antagonist by forming an inactive complex with a single GHR molecule at the cell surface. (C) A diagram of how the LR-fusion may act as an agonist by mimicking the action of GH.

15.2 THE GHLR-FUSIONS

The GH LR-fusions were generated at the genetic level by cloning the different components of the fusion protein sequentially into the expression vector. The LR-fusions were then generated by expression of the protein from these expression vectors.

Initial testing of the LR-fusions was performed on different combinations of GH, the extracellular domains of GHR and length of linker between the two constituents of the fusion protein. The lead LR-fusion chosen for this study consisted of GH (growth hormone, 1-191) linked to GHRec (growth hormone extracellular domain, 1-238) by a flexible $(Gly_4Ser)_4$ linker (Figure 15.2). This prototype molecule, variant 0 (GH-LRv0), consisted of extraneous sequence as an artifact of the cloning processes required to generate the LR-fusion gene, for example, restriction sites.

To reduce the immunogenic profile of the LR-fusion, all nonimportant sequences were removed from the fusion protein. Extraneous sequence at the N-terminus and the C-terminus of the gene was removed, to give variant 1 (GH-LRv1). The restriction sites at either end of the linker region were then removed, to produce variant 2 (GH-LRv2). The N-terminal region of GHR is disordered in the crystal

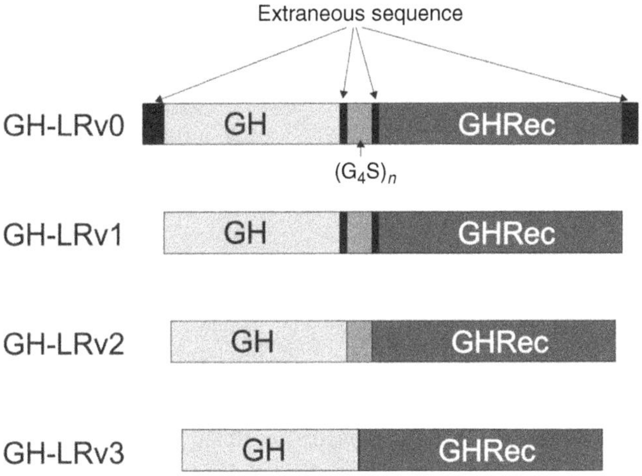

FIGURE 15.2 A schematic diagram of the expressed GH LR-fusions; GH-LRv0 contains extraneous sequence that occurred as a consequence of the cloning strategy used to synthesize the gene for expression. The extraneous sequences at the N- and C-termini were removed to produce GH-LRv1. The restriction sites used to insert the linker into the molecule were removed to generate GH-LRv2. GH-LRv3 was then synthesized by removing the linker region.

structure [22] and so we questioned whether there was a need for the $(Gly_4Ser)_4$ linker. We, therefore, linked the C-terminus of GH directly to the N-terminus of GHRec, to generate variant 3 (GH-LRv3).

15.3 EXPRESSION AND PURIFICATION

The genes encoding the LR-fusions were cloned into an expression vector under the control of a CMV promoter and the GH signal sequence. These vectors were then transfected into Chinese hamster ovary (CHO) cells and stable clones generated. These clones were adapted to suspension culture and grown in protein-free media; the expressed LR-fusion was secreted into the growth media.

The LR-fusions were purified from the growth media by affinity chromatography using columns containing immobilized anti-GH antibodies.

15.4 ANALYSIS OF THE LR-FUSIONS

15.4.1 Characterization

The purified GH-LRv0 was screened by ELISA using 16 conformationally sensitive anti-GHBP monoclonal antibodies; the affinities of these antibodies were similar those obtained against serum-derived GHBP.

Coomassie staining of a SDS-PAGE gel showed the LR-fusion protein separated as a doublet at a mass of ~75 kDa (Figure 15.3A). Immunoblotting using anti-GH antibody as a probe confirmed this band-pattern. The formula mass of the GH-LRv0 is ~55 kDa; it was assumed that differential glycosylation of the 55 kDa protein would account for the two bands observed at 75 kDa.

Native-PAGE gels of the LR-fusion also showed two bands (Figure 15.3B); however, it was thought that the differences in sizes between the two bands were unlikely to be simply due differential glycosylation. The "fast" and "slow" bands from the native gel were excised and reanalyzed by SDS-PAGE under reducing conditions; both the reanalyzed "fast" and "slow" bands showed similar band-patterning to that of the original SDS-PAGE, that is, a doublet of bands at 75 kDa separated by ~5 kDa (Figure 15.3C); this result suggested that

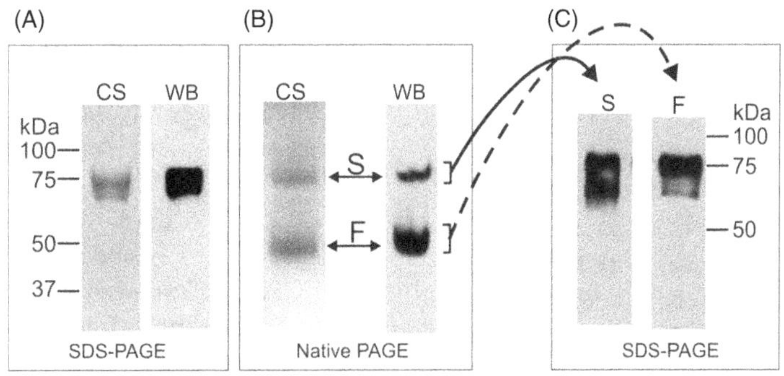

FIGURE 15.3 (A) SDS-PAGE of GH-LRv0, followed by coomassie staining (CS) or Western blotting (WB) using an anti-GH antibody. GH-LRv0 is ~75 kDa and resolves into two bands, ~5 kDa apart. (B) Native gel of GH-LRv0 showing that there are two protein forms; fast (F) and slow (S). (C) The individual F and S bands were excised from the native gel and subjected to SDS-PAGE under reducing conditions, followed by Western blotting using an anti-GH antibody. When rerun on a SDS-PAGE gel the two bands, F and S, run at ~75 kDa and separate as a doublet, suggesting that they both comprise the 75 kDa LR-fusion, which may exist in an equilibrium of monomers and dimers under native conditions. *Source*: This research was originally published in Nature Medicine, Reference [13]. © Nature Publication Group.

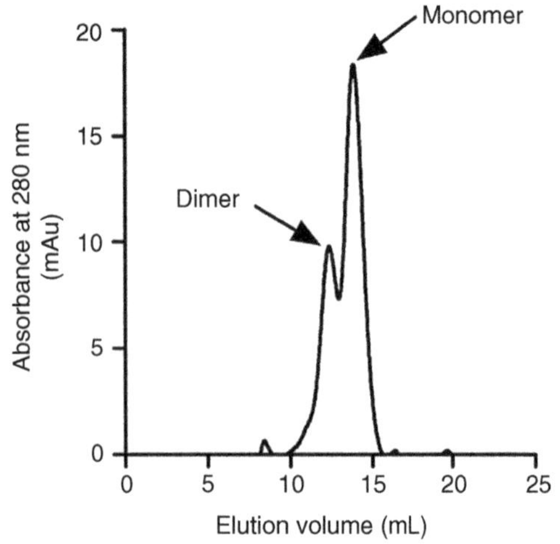

FIGURE 15.4 Analytical gel filtration shows that GH-LRv0 is present in two forms, monomeric and dimeric. *Source*: This research was originally published in Nature Medicine, Reference [13]. © Nature Publication Group.

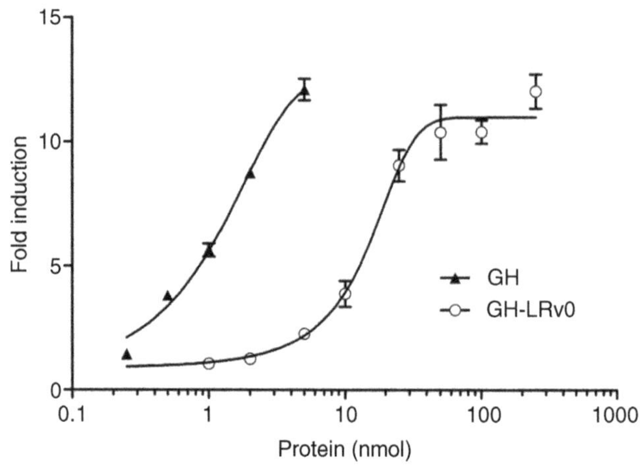

FIGURE 15.5 The *in vitro* biological activity of hGH and GH-LRv0. The LR-fusions all have similar activities and are ~10 times less active than hGH. *Source*: This research was originally published in Nature Medicine, Reference [13]. © Nature Publication Group.

the LR-fusion was in two distinct forms. Analytical gel filtration confirmed the presence of two distinct forms of LR-fusion (Figure 15.4) and this was further corroborated by analytical ultracentrifugation. These analyses suggested that the LR-fusion existed in solution in monomeric and dimeric forms, with the monomer being ~75 kDa in size.

15.4.2 Biological Activity

Measurement of biological activity was carried out using a dual luciferase assay. Briefly, human embryonic kidney cells, which stably expressed GHR at a high level (HEK 293 Hi), were transfected with an inducible firefly luciferase gene and a constitutive *Renilla* luciferase gene. Firefly luciferase expression was stimulated by GH activity through the Lactogenic hormone response element (LHRE) sequence; the *Renilla* luciferase expression was used to correct for differences in transfection efficiency and cell number between the various samples.

GH-LRv0 had agonistic activity; however, the activity was ~10-fold lower than the ligand alone (Figure 15.5), GH-LRv2 and GH-LRv3 showed similar activity to GH-LRv0. The decreased *in vitro* bioactivity of the LR-fusions may be a consequence of the dimer formation of these molecules; the LR-fusions dimers maybe inactive but in dynamic equilibrium with the active monomeric form, hence the reduced *in vitro* bioactivity.

15.4.3 Pharmacokinetics

Seven-week-old Sprague-Dawley rats were injected with vehicle, GH and the LR-fusions. The rats ($n = 4$–6 per group) were injected subcutaneously or intravenously (penile vein) with 0.1 mg/kg (body weight) of GH or GH-LRv0. Blood samples were collected periodically from the retro-orbital plexus under isoflurane anesthesia, the serum samples were stored at −70°C. The serum concentrations of GH and GH-LRv0 in these samples were measured using an ELISA against GH. The pharmacokinetic profiles were estimated by fitting the values of hormone concentration versus time to compartmental models using nonlinear least-square regression analysis. The values were normalized for animal weight and the clearance rates per animal weight and terminal half-lives were calculated using coefficient and exponents obtained from an intravenous bolus model.

GH-LRv0 had delayed clearance when administered intravenously and subcutaneously, delayed absorption was also observed for the subcutaneously administered LR-fusion (Figure 15.6; Table 15.1). When administered by an intravenous route, GH-LRv0 had a clearance rate of 3.3 ± 0.9 mL/h/kg and a terminal half-life of 21 ± 2 h; this is 300 times slower than that of GH [2,18]. After a single subcutaneous dose, the GH-LRv0 had a delayed peak at ~30 h, compared to 1 h for GH. GH-LRv0 was also still detectable 8 days after subcutaneous administration, whereas GH was undetectable after 0.25 days. GH-LRv2 and GH-LRv3 showed similar pharmacokinetic profiles to GH-LRv0 with clearance rates of 2.8 ± 0.2 mL/h/kg and 2.7 ± 0.2 mL/h/kg, respectively (Figure 15.6A; Table 15.1).

To investigate if the pharmacokinetic profiles of the LR-fusions were a consequence of size alone, the pharmacokinetic profiles of two other LR-fusions were also evaluated. One LR-fusion consisted of GH linked to only domain B of GHRec (55 kDa) the other consisted of two GH molecules linked to GHRec, that is, GH-GH-GHRec, (100 kDa); identical linkers were used between the domains in these LR-fusions. Both the 55 and the 100 kDa proteins showed greater clearance rates compared to the 75 kDa GH-LRv0, the terminal half-life of both were less than 4 h after

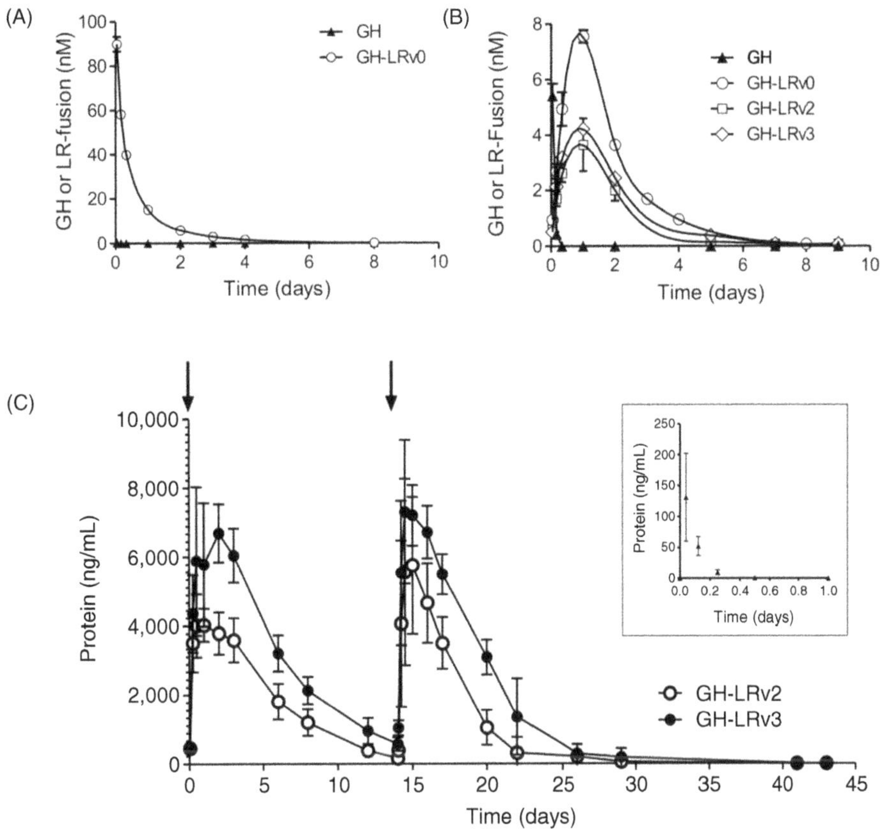

FIGURE 15.6 (A) Pharmacokinetic profiles of hGH and GH-LRv0, intravenously injected into rat, the LR-fusion shows a much reduced rate of clearance. (B) Pharmacokinetic profiles of hGH, GH-LRv0, GH-LRv2, and GH-LRv3, subcutaneously injected into rats, all the LR-fusions show a much reduced rate of clearance and delayed absorption—the peak concentration of the LR-fusions is observed at ~30 h, and the fusion molecules are still detectable at 8 days. (C) Pharmacokinetic profiles of GH-LRv2 and GH-LRv3, subcutaneously injected into cynomolgous monkeys on days 0 and 14 (arrows)—the LR-fusions show a much reduced rate of clearance compared to hGH (inset). *Source*: (C) This research was originally published in Clinical Science, Reference [14].

intravenous administration (precise determination was not possible since the sampling protocol that was used was used expected a longer half-life). Therefore, molecular weight alone is not the cause of the exceptional pharmacokinetic profiles of the LR-fusions.

The pharmacokinetics of GH-LRv2 and GH-LRv3 was also evaluated in cynomolgous monkeys (*Macaca fascicularis*) as part of the immunogenicity study [14] discussed later in this chapter. Male 2- to 3-year-old cynomolgous monkeys weighing ~2.5 kg were used for this study; the monkeys ($n = 3$–4 per group) were treated with subcutaneous injections of vehicle (on days 0 and 14), GH-LRv2 (1 mg/kg of body weight on days 0 and 14), GH-LRv3 (1 mg/kg of body weight on days 0 and 14), and rhGH (0.1 mg/kg of body weight, daily for 27 days). The animals were retained for a total of 6 weeks (42 days) after the initial injection and periodic blood samples were taken. The serum levels of GH or LR-fusion were measured using an ELISA against GH.

A greatly delayed clearance of the LR-fusions was observed in the monkeys (Figure 15.6C; Table 15.1); the

TABLE 15.1 Pharmacokinetic Parameters for hGH and the LR-Fusions for Rats and Monkeys

	Rat		Monkey	
Molecule	Clearance s.c. (mL/h/kg)	$t_{1/2}$ s.c. (h)	Clearance s.c. (mL/h/kg)	$t_{1/2}$ s.c. (h)
hGH [12]	820 ± 94	0.66 ± 0.1	154 ± 4.4	1.7 ± 0.4
GH-LRv0	–	21 ± 2	–	–
GH-LRv2	2.8 ± 0.2	25 ± 6.4	1.7 ± 0.2	54 ± 6.8
GH-LRv3	2.7 ± 0.2	26 ± 6.8	0.96 ± 0.2	76 ± 4.9

clearance was even more delayed than that observed in rats, the clearance rates for GH-LRv2 and GH-LRv3 were, respectively, 91 and 160 times less than that of GH [12]. Both of the LR-fusions evaluated in this pharmacokinetic study did not show any appreciable accumulation or reduction in systemic exposure ($AUC_{0\text{-}\tau}$) over the dosing period, and the average of systemic exposure of GH-LRv3 was ~1.7 times greater than that of GH-LRv2.

Previous studies of allometric scaling have shown that the clearance of human GH (hGH) is fourfold less in monkeys compared to rats and twofold less in humans compared to monkeys [23]. A similar ratio was observed for the clearance rates of the LR-fusions between rat and monkey, there was a 1.6- to 2.8-fold reduction in the clearance rate in monkeys compared to rats. If the difference in clearance rats observed for hGH can be extrapolated to the LR-fusions in humans then the fusion proteins could provide a 2–3 weekly GH replacement therapy [14].

15.4.4 Pharmacodynamics

The pharmacodynamic effects of the LR-fusions were studied using Sprague-Dawley rats, which had been hypophysectomized (made GH deficient) at 4 weeks of age under isoflurane anesthesia. The effects of the LR-fusions on the rats were studied over a 10-day period. The rats were injected with vehicle, hGH or GH-LRv0. Dosing of hGH was daily (10 doses), every 2 days (five doses), every 5 days (two doses), or only on day 1 (one dose). The LR-fusions and vehicle were injected every 2 days (five doses), every 5 days (two doses), or only on day 1 (one dose). The injection solutions of vehicle, hGH or LR-fusion never exceeded 2 mL/kg and an equimolar total dose, of ~10 nmol, of hGH and LR-fusion was given over the 10-day period of the study. The rats were weighed daily. Terminal bleeds were collected and assayed using an ELISA against GH and an ELISA against IGF-1 (insulin-like growth factor 1) (Octeia Rat/Mouse IGF-1 Assay Kit, iDS).

The daily dose of hGH induced a body weight increase of 15–20 g over the 10-day period of the study, reduction in the frequency of treatment caused a reduction in the increase in body weight over the 10-day period of the study (Figure 1.7A). GH-LRv0 produced a similar increase in body weight in all the treatment regimens studied, this increase was comparable to that produced by the daily dose of hGH (Figure 1.7B). Closer investigation of the growth profiles shows that hGH produced a weight gain on the day it was injected but did not have an effect after

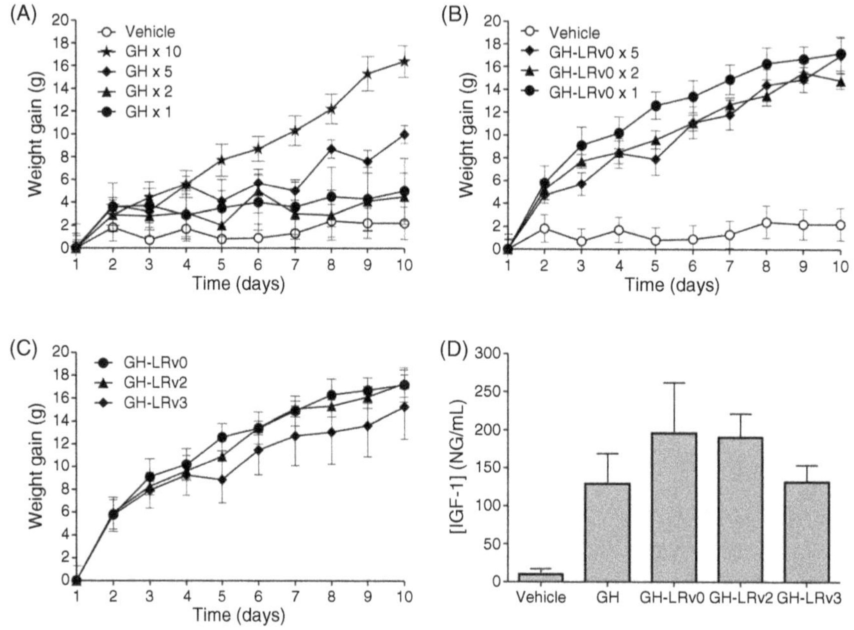

FIGURE 15.7 (A) The effects of hGH on hypophysectomised rats over a 10-day period using different dose regimens; daily (×10), every 2 days (×5), every 5 days (×2) and once on day 1 (×1). The increase in body weight decreases as the frequency of injections decreases. (B) The effects of GH-LRv0 on hypophysectomized rats over a 10-day period using different dose regimens. There is little difference in the weight gain after 10 days for all the dosing regimens; the increase in weight is comparable to the daily dose of GH. (C) A comparison of the pharmacodynamics of all the LR-fusions. (D) Blood serum IGF-1 levels in the terminal bleeds of the rats that underwent the pharmacodynamic study. The graph shows the measurements for vehicle, rats given a daily dose (×10) of hGH, and the rats given a single dose (×1) of the LR-fusions.

this; while the GH-LRv0 maintained its effect for 10 days postinjection. GH-LRv2 and GH-LRv3 had similar effects to GH-LRv0 (Figure 15.7C).

Analysis of the 10-day terminal bleeds from the rats for the GH-dependent biomarker IGF-1 showed elevated levels of IGF-1 in the animals given the LR-fusions; the IGF-1 levels were similar to that of the IGF-1 levels in the animals treated with daily doses of hGH (Figure 1.7D).

15.4.5 Immunogenicity

The immunogenicity of the LR-fusions was evaluated in cynomolgous monkeys, male 2- to 3-year-old cynomolgous monkeys weighing ~2.5 kg were used for this study; the monkeys ($n = 3$–4 per group) were treated with subcutaneous injections of vehicle (on days 0 and 14), GH-LRv2 (1 mg/kg of body weight on days 0 and 14), GH-LRv3 (1 mg/kg of body weight on days 0 and 14), and hGH (0.1 mg/kg of body weight, daily for 27 days). The animals were retained for a total of 6 weeks (42 days) after the initial injection and periodic blood samples were taken.

An ELISA-based assay was used to analyze the blood serum samples for the presence of antibodies against the respective injected molecule. The cutoff for a positive result was the absorbance measured spectrophotometrically at 450 nm for the negative control plus 3 standard deviations (SD). Serum samples giving a positive reading were further analyzed using an inhibition assay to determine if the result was a false positive or a true positive.

None of the animals in the study showed anti-test molecule antibodies (Table 15.2) with the exception of anti-GH-LRv2 antibodies detected in one animal on day 44. One animal treated with GH-LRv3 had a positive reading before drug administration and this remained positive throughout the study, this animal was excluded from the study.

TABLE 15.2 Immunogenicity of hGH and the LR-Fusions in Cynomolgous Monkeys

	Vehicle			1 mg/kg per Dose GH-LRv2			
Animal No.	101	103	105	201	203	205	207
Time							
Day 0 (predose)	0.060	0.066	0.052	0.067	0.053	0.046	0.132
Day 6 (144 h)	0.047	0.064	0.068	0.071	0.072	0.051	0.119
Day 14 (predose)	0.060	0.046	0.050	0.064	0.077	0.080	0.106
Day 20 (144 h)	0.058	0.077	0.065	0.059	0.050	0.077	0.129
Day 26 (288 h)	0.067	0.054	0.055	0.049	0.078	0.068	0.118
Day 43 (696 h)	0.068	0.071	0.055	0.071	0.134	**0.210**	0.078
	Vehicle			1 mg/kg per Dose GH-LRv3			
Animal No.	101	103	105	301	303	305	307
Time							
Day 0 (predose)	0.060	0.066	0.052	0.056	**(0.371)**	0.069	0.085
Day 6 (144 h)	0.047	0.064	0.068	0.060	**(0.196)**	0.064	0.082
Day 14 (predose)	0.060	0.046	0.050	0.065	**(0.305)**	0.064	0.086
Day 20 (144 h)	0.058	0.077	0.065	0.073	**(0.403)**	0.068	0.083
Day 26 (288 h)	0.067	0.054	0.055	0.056	**(0.371)**	0.069	0.085
Day 43 (696 h)	0.068	0.071	0.055	0.060	**(0.196)**	0.064	0.082
	Vehicle			0.1 mg/kg per Dose hGH			
Animal No.	101	103	105	401	403	405	407
Time							
Day 0 (predose)	0.062	0.071	0.079	0.059	0.066	0.049	0.060
Day 6 (144 h)	0.063	0.074	0.070	0.070	0.067	0.059	0.058
Day 14 (predose)	0.062	0.075	0.069	0.066	0.180	0.064	0.083
Day 20 (144 h)	0.064	0.069	0.081	0.132	0.185	0.081	0.109
Day 26 (288 h)	0.063	0.083	0.062	0.167	0.060	0.109	0.085
Day 43 (696 h)	0.062	0.096	0.065	0.147	0.136	0.076	0.074

Table shows the absorbencies at 450 nm for the sera from cynomolgous monkeys given hGH, GH-LRv2 or GH-LRv3. Positive immunogenic results are shown in bold and false positives are shown in brackets. The cutoffs for positive reactions were ≥0.157 for GH-LRv2, ≥0.156 for GH-LRv3, and ≥0.210 for hGH, these cutoff values were determined during the qualification of the respective assays.

While not definitive proof that antibodies will not be generated against the LR-fusions in humans, the fact that the fusion proteins, given at 10 times the human physiological dose, do not initiate an immune response in these animals gives strong reassurance of their nonimmunogenicity when used as a biopharmaceutical in humans.

15.4.6 Toxicology

The animals used in the immunology study were also used to investigate the toxicological effects of the LR-fusions. Blood and urine samples were taken for analysis prior to the start of the treatment and then during week six. After necropsy, external and internal examinations of the monkeys were performed; this included an examination of the major organs and the injection site.

No treatment-related effects on clinical signs, body weights, hematology, blood chemistry, urine analysis, organ weight, or necropsy were observed during the study. There was no evidence of lipoatrophy or reaction at the injection site.

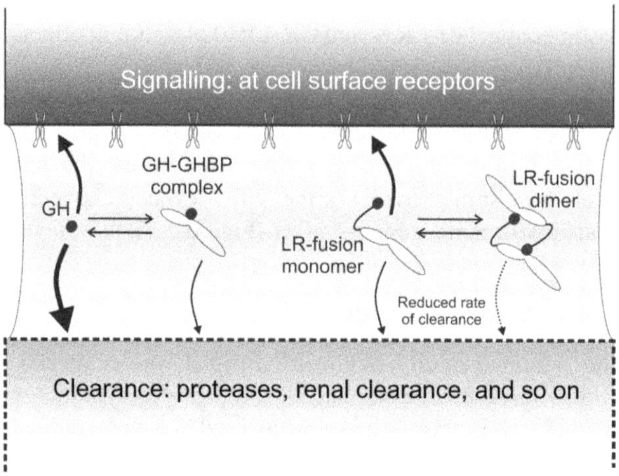

FIGURE 15.8 The rate of clearance of GH is reduced by the formation a GH-GHBP complex. The GH LR-fusions have a reduced rate of clearance and this is thought to be due to the formation of an inactive LR-fusion dimer.

15.5 LR-FUSIONS: THE NEXT GENERATION IN HORMONE TREATMENT

The GH LR-fusions demonstrate a potential long-acting GH therapy. The use of GH LR-fusion proteins as a therapeutic in lieu of hGH is attractive for a number of reasons. The main advantage of the GH LR-fusions over hGH is their exceptional pharmacokinetic and pharmacodynamic effects. Which could enable 3–4 weekly injections of the GH LR-fusion rather than daily injections of hGH. This would increase convenience to the patient and so increase his/her "quality of life."

They are also produced as a single amino acid chain and do not require further modification after purification, and so both the heterogeneity and cost of production of the final product are greatly reduced in comparison to PEGylated GH and depot preparations. The GH LR-fusions show low immunogenicity and no toxicological effects of the GH LR-fusions were observed in the studies in monkeys and there were no reactions at the injection sites or evidence of lipoatrophy.

These studies demonstrate that the reduced rate of clearance shown by the GH LR-fusion is due to the dimerization of the GH-GHRec molecules. We propose a hypothesis in which the LR-fusion circulates as an inactive dimer "reservoir" which is in equilibrium with the active monomer form. The fusion monomers are active and have a reduced rate of clearance compared to GH while the dimers are inactive and are even less readily cleared from the vascular system (Figure 15.8).

The LR-fusion technology can be used in other ligand-receptor systems and may provide the basis of a range of next-generation hormone replacement therapeutics.

15.6 CONCLUSION

Fusions of growth hormone (GH) and its receptor (GHR) have been generated to produce molecules that mimic the effects of GH with the added advantage of a much reduced rate of clearance from the vascular system. These ligand-receptor (LR) fusions were observed to form head to tail reciprocal dimers; this characteristic of the GH LR-fusions is thought to further extend the circulating half-life of these molecules. LR fusions have the potential to generate potent long-acting biologicals.

REFERENCES

1. Woodhouse LJ, Mukherjee A, Shalet SM, Ezzat S. (2006) The influence of growth hormone status on physical impairments, functional limitations, and health-related quality of life in adults. *Endocr. Rev.* 27, 287–317.
2. Clark R, Olson K, Fuh G, Marian M, Mortensen D, Teshima G, (1996) Long acting growth hormones produced by conjugation with polyethylene glycol. *J. Biol. Chem.* 271, 21969–21977.
3. Touraine P, D'Souza GA, Kourides I, Abs R, Barclay P, et al. (2009) Lipoatrophy in GH deficient patients treated with a long-acting PEGylated GH. *Eur. J. Endocrinol.* 161, 533–540.
4. Hodish I, Barkan A. (2008) Long-term effects of pegvisomant in patients with acromegaly. *Nat. Clin. Pract.* 4, 324–332.
5. Bernabeu I, Cameselle-Teijeiro J, Casanueva FF, Marazuela M. (2009) Pegvisomant-induced cholestatic hepatitis with jaundice in a patient with Gilbert's syndrome. *Eur. J. Endocrinol.* 160, 869–872.
6. Lou K-J. (2009) Growing a PEG-like alternative. *SciBX* 2, 1–2.

7. Reiter EO, Attie KM, Moshang TJ, Silverman BL, Kemp SF, Neuwirth RB, et al. (2001) A multicenter study of the efficacy and safety of sustained release GH in the treatment of naive pediatric patients with GH deficiency. *J. Clin. Endocrinol. Metab.* 86, 4700–4706.

8. Cook DM, Biller BMK, Vance ML, Hoffman AR, Phillips LS, Ford KM, et al. (2002) The pharmacokinetic and pharmacodynamic characteristics of a long-acting growth hormone (GH) preparation (Nutropin Depot) in GH-deficient adults. *J. Clin. Endocr. Metab.* 87, 4508–4514.

9. Jostel A, Mukherjee A, Alenfall J, Smethurst L, Shalet SM. (2005) A new sustained release preparation of human growth hormone and its pharmacokinetic, pharmacodynamic and safety profile. *Clin. Endocrinol.* 62, 623–627.

10. Silverman BL, Blethen SL, Reiter EO, Attie KM, Neuwirth RB, Ford KM. (2002) A long-acting growth hormone (Nutropin Depot): efficiency and safety following two years of treatment in children with growth hormone deficiency. *J. Pediatr. Endocrinol. Metab.* 15, 715–722.

11. Kemp SK, Fielder PJ, Attie KM, Blethen SL, Reiter EO, Ford KM, et al. (2004) Pharmacokinetic and pharmacodynamic characteristics of long-acting growth hormone (GH) preparation (Nutropin Depot) in GH deficient children. *J. Clin. Endocr. Metab.* 89, 3234–3240.

12. Osborn BL, Sekut L, Corcoran M, Poortman C, Sturm B, Chen G, et al. (2002) Albutropin: a growth hormone-albumin fusion with improved pharmacokinetics and pharmacodynamics in rats and monkeys. *Eur. J. Pharmacol.* 456, 149–158.

13. Wilkinson IR, Ferrandis E, Artymiuk PJ, Teillot M, Soulard C, Touvay C, et al. (2007) A ligand-receptor fusion of growth hormone forms a dimer and is a potent long-acting agonist. *Nat. Med.* 13, 1108–1113.

14. Ferrandis E, Pradhananga SL, Touvay C, Kinoshita C, Wilkinson IR, Stafford K, et al. (2010) Immunogenicity, toxicology, pharmacokinetics and pharmacodynamics of growth hormone ligand–receptor fusions. *Clin. Sci.* 119, 483–491.

15. Baumann G, Amburn KD, Buchanan TA. (1987) The effect of circulating growth hormone-binding protein on metabolic clearance, distribution, and degradation of human growth hormone. *J. Clin. Endocrinol. Metab.* 64, 657–660.

16. Baumann G. (1991) Growth hormone heterogeneity: genes, isohormones, variants, and binding proteins. *Endocr. Rev.* 12, 424–449.

17. Baumann G, Shaw MA, Buchanan TA. (1989) *In vivo* kinetics of a covalent growth hormone-binding protein complex. *Metabolism* 38, 330–333.

18. Clark RG, Mortensen DL, Carlsson LM, Spencer SA, McKay P, Mulkerrin M, et al. (1996) Recombinant human growth hormone (GH)-binding protein enhances the growth-promoting activity of human GH in the rat. *Endocrinology* 137, 4308–4315.

19. Baumann G. (1995) Growth hormone binding protein-errant receptor or active player? *Endocrinology* 136, 377–378.

20. Ayling RM, Ross R, Towner P, Von Laue S, Finidori J, Moutoussamy S, (1997) A dominant-negative mutation of the growth hormone receptor causes familial short stature. *Nat. Genet.* 16, 13–14.

21. Ross RJ, Esposito N, Shen XY, Von Laue S, Chew SL, Dobson PR, et al. (1997) A short isoform of the human growth hormone receptor functions as a dominant negative inhibitor of the full-length receptor and generates large amounts of binding protein. *Mol. Endocrinol.* 11, 265–273.

22. de Vos AM, Ultsch M, Kossiakoff AA. (1992) Human growth hormone and extracellular domain of its receptor: crystal structure of the complex. *Science* 255, 306–312.

23. Mordenti J, Chen SA, Moore JA, Ferraiolo BL, Green JD. (1991) Interspecies scaling of clearance and volume of distribution data for five therapeutic proteins. *Pharm. Res.* 8, 1351–1359.

16

DEVELOPMENT OF LATENT CYTOKINE FUSION PROTEINS

LISA MULLEN, GILL ADAMS, REWAS FATAH, DAVID GOULD, ANNE RIGBY, MICHELLE SCLANDERS, APOSTOLOS KOUTSOKERAS, GAYATRI MITTAL, SANDRINE VESSILLIER, AND YUTI CHERNAJOVSKY

Bone and Joint Research Unit, William Harvey Research Institute, Barts and The London Medical School, London, UK

16.1 INTRODUCTION

Cytokines are mediators of cell–cell communication. The majority of cytokines are low molecular-weight glycoproteins that play key roles in a wide range of biological processes, including the regulation of immunity, inflammation, and hematopoiesis [1]. These soluble hormone-like proteins are a key component of the immune system, regulating lymphocyte turnover, differentiation, and activation. Hence, it is unsurprising that in chronic pathological conditions, the cytokine network is deregulated.

Therefore, given the importance and central role of cytokines in health and disease, it has long been recognized that these proteins may be developed as therapeutics. To develop cytokine-based therapies, it must first be established whether the cytokine of interest has a beneficial or harmful effect in the pathophysiological situation under scrutiny. The therapeutic intervention can then consist of either the cytokine itself or cytokine antagonists.

However, cytokines have a remarkably complex biology, and there are a number of characteristics of these molecules that must be considered prior to any attempt to use them as therapeutic agents. First, cytokines are pleiotropic, meaning that they have the ability to interact with not one, but a variety of cellular targets. In practice, this means that cytokines have different effects on a wide variety of cells. Second, cytokines share an attribute known as redundancy, which describes the fact that the same biological function can be exerted by several distinct cytokines. The third characteristic of cytokines that is abundantly clear is that these potent signaling molecules do not function in isolation. Indeed, these proteins are known to function as networks or cascades of interacting factors, which are able to affect the synthesis of other cytokines.

So, how do these characteristics of cytokines affect their potential use as therapeutics? The fact that these molecules are not only secreted by, but also recognized by a wide variety of both immune and nonimmune cells means that systemic administration can result in a number of undesirable side-effects. The redundant nature of cytokines also means that blocking of a particular cytokine may not necessarily have the desired affect on a pathological condition. There is also the possibility that blocking or augmenting a specific cytokine may result in unpredictable outcomes due to unforeseen effects on other cytokine pathways.

A further complication of the use of cytokines as therapeutics lies in the paracrine and autocrine manner in which these molecules function. Cytokines are local mediators that are designed to exert biological effects in their local

Fusion Protein Technologies for Biopharmaceuticals: Applications and Challenges, First Edition. Edited by Stefan R. Schmidt.

microenvironment. Hence, they tend to work in pharmacological doses, and once secreted by the producing cell, have very short half-lives, often in the order of a few minutes. The pleiotropic nature of these molecules means that they are exceptionally potent; it is therefore necessary for their secretion to be carefully regulated to avoid cytokine-mediated tissue damage. One of the best-characterized examples of deregulation of cytokines leading to tissue destruction is in rheumatoid arthritis (RA), where up-regulation of tumor necrosis factor-alpha (TNF-α), and the interleukins IL-1 [2], IL-6 [3], and IL-15 [4] have been demonstrated.

Despite the challenges that cytokine biology presents as discussed above, insights into the biological functions of individual cytokines have emerged through molecular, biochemical, and immunological analyses so that sufficient information pertaining to the structure and function of these molecules has enabled the development of a number of therapeutic cytokines and cytokine antagonists. One of the best-known and most successful groups of therapeutic cytokines to date is the interferons (IFNs). The IFNs are a group of naturally occurring cytokines that are induced in response to viral infections. IFN-β was one of the first cytokines to be purified and produced in recombinant form for clinical application [5] and is now used in recombinant form to treat a number of conditions including relapsing forms of multiple sclerosis (MS), chronic hepatitis C (Hep C), and some forms of cancer [6].

However, research continues to focus on ways to increase the serum stability and half-life of recombinant cytokines as the short serum half-life makes frequent administration necessary and increases the possibility of side-effects, thus limiting their clinical uses. Strategies such as modification with polyethylene glycols (PEG) or by fusing the cytokine to an antibody Fc fragment to improve the pharmacokinetics have been in use for some time.

16.2 DESCRIPTION OF CONCEPT

The recent development of "latent" or "shelled" cytokines [7,8] aims to rectify not only the short serum half-life of recombinant cytokines, but also dramatically reduce the risk of undesirable side-effects. The idea underpinning this technology is to produce cytokines that are biologically inactive (i.e., latent) until they arrive at the site of disease, at which point they are released so that they can exert their biological effects. These latent cytokines are engineered by using a naturally occurring "shell" structure from the cytokine transforming growth factor-β (TGF-β). The release of cytokines *in vivo* is usually tightly regulated at the transcription and translation stages to prevent cytokine-related tissue damage. However, in the case of TGF-β, a different and unique strategy is employed. TGF-β is secreted from cells in a latent form; this is achieved by "packaging" of the active cytokine inside a disulfide-linked shell-like structure formed by dimerization of its long precursor amino terminal end, known as the latency-associated peptide (LAP). There are a number of isoforms of TGF-β, but all are produced as large precursor proteins that are 390–412 amino acids in size. This precursor protein is then processed intracellularly prior to secretion from the producing cell. A critical processing step is the proteolytic digestion of the precursor protein by the endopeptidase furin. This proteolysis cleaves the TGF-β precursor protein to yield two products that then form a complex. The N-terminal portion is the LAP, while the C-terminal 25 kDa dimer constitutes the mature TGF-β. Crucially, after cleavage of the precursor protein, the mature TGF-β remains noncovalently attached to the LAP and it is this association that renders the secreted TGF-β latent, due to the fact that mature TGF-β cannot interact with its cellular receptors while associated with LAP in this manner.

It is this LAP portion of TGF-β that we harnessed to make latent cytokines. Therapeutic cytokines are engineered with the LAP at the N-terminus (the same arrangement as that found in native TGF-β) so that the expressed protein consists of the recombinant therapeutic cytokine, surrounded by the LAP (Figure 16.1), which shields the cytokine from interacting with its cellular receptors, thus rendering it biologically inactive. The presence of the LAP also increases the molecular weight of the cytokine and improves serum stability and half-life. A further aim of the design and engineering of these latent cytokines was to address the side-effects that are all too common features of treatment with these pleiotropic molecules. Hence, the latent cytokines are designed to be released from their protective LAP shell only at the sites of disease. This was achieved by cloning in a specific cleavage site for matrix metalloproteinase (MMP) between the LAP and the active cytokine. Therefore, at sites where the latent cytokine is exposed to MMP activity, the LAP is cleaved from the cytokine, thereby allowing the therapeutic cytokine to interact with its cellular receptors and exert its biological effects (Figure 16.2).

MMPs are a family of zinc-dependent endopeptidases that are responsible for the timely degradation of the extracellular matrix (ECM) [9,10,11]. This degradation is essential for tissue repair and remodeling, morphogenesis, and wound healing. Indeed, the levels of MMPs in normal steady-state tissues are extremely low [12], with expression under transcriptional control by pro-inflammatory cytokines, cell–matrix interactions, and growth factors. The activity of MMPs is also subject to further regulation via the activation of precursor zymogens and by the presence of endogenous inhibitors, known as tissue inhibitors of MMP (TIMPs).

There are 23 MMPs in humans, most of which can be grouped according to substrate specificity [13] (Table 16.1). MMPs are extracellular proteins, although at least three (MMP-1, MMP-2, and MMP-11; [14]) have been shown to be present intracellularly and may act on intracellular proteins. However, these enzymes are also implicated in a number of pathological processes, including the destruction of arthritic joints, atherosclerosis, spread of metastatic

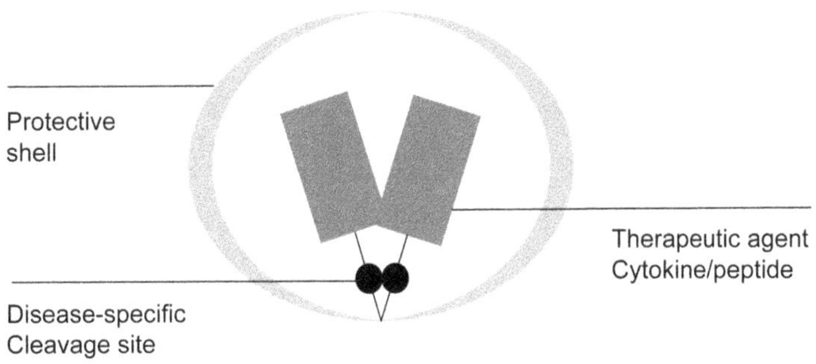

FIGURE 16.1 Latent cytokines. The latency-associated peptide (LAP) from TGF-β forms a shell around the therapeutic cytokine, rendering it biologically inactive. A cleavage site specific for matrix metalloproteinase (MMP) is engineered in between the LAP and the cytokine, enabling the release of the cytokine at the sites of disease. *Source:* Reprinted from Reference [8]. Copyright (2004), with permission from Elsevier.

cancer cells, pulmonary fibrosis, and neuroinflammation [15]. The role of MMPs in many of these conditions is concerned with tissue destruction caused by elevated levels of these enzymes. It has long been noted that the pathologies of autoimmune diseases and cancer share many similar facets [16], namely, aberrant cell proliferation, ECM remodelling, and alteration of the cell microenvironment. It is now abundantly clear that MMP activity contributes significantly to the progression of these diseases by remodeling the ECM and altering the microenvironment. There are now many reports implicating a number of different MMPs in cancer (MMPs-2, -7, and -11) and in RA (MMPs-1, -2, -3, -9, and -13).

MMPs are likely to play a pathological role in any condition where tissue destruction forms part of the pathology, for example, periodontal disease, brain injury [17], and neuroinflammatory diseases [18]. In addition to their effects on tissue destruction, MMPs have also been implicated in fibrosis, most notably in atherosclerosis, which is associated with alterations in the arterial ECM [19]. There are a number of reports detailing the role of a range of MMPs in many of the processes that occur in atherosclerosis [20].

Therefore, there are many pathological conditions where the concentration of one, or more, MMP(s) is elevated at the site of disease. The latent cytokine technology takes advantage of this increase in local MMP concentration to provide a targeted release of therapeutic cytokine only at these sites. In addition, the release of the cytokine at the site of disease will also be dependent on the degree of MMP activity, thereby

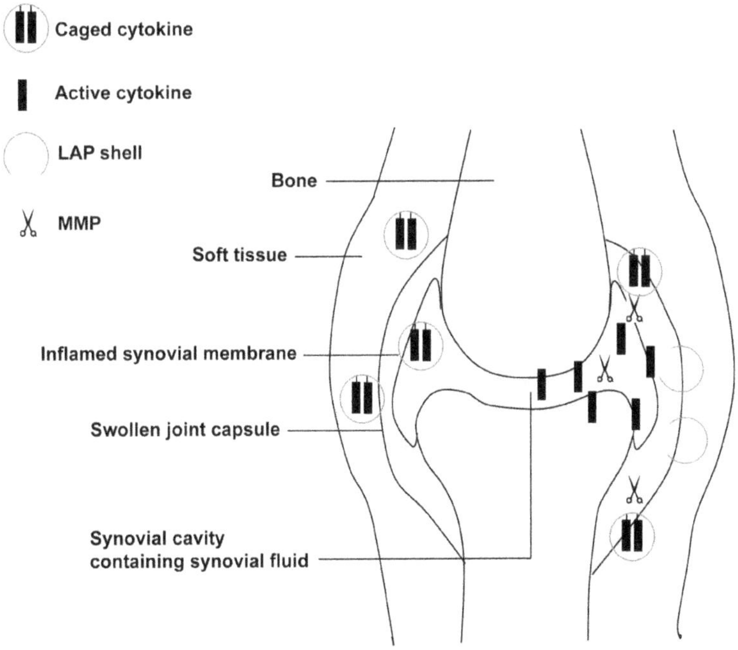

FIGURE 16.2 Illustration of the use of the latent cytokine technology to treat rheumatoid arthritis. The latent cytokine remains biologically inactive until cleaved by MMP at the inflamed joint.

TABLE 16.1 Human Matrix Metalloproteinases (MMPs)

Name	MMP	Location
Collagenases		
Collagenase 1	MMP-1	Secreted
Collagenase 2	MMP-8	Secreted
Collagenase 3	MMP-13	Secreted
Gelatinases		
Gelatinase A	MMP-2	Secreted
Gelatinase B	MMP-9	Secreted
Stromelysins		
Stromelysin 1	MMP-3	Secreted
Stromelysin 2	MMP-10	Secreted
Matrilysins		
Matrilysin 1	MMP-7	Secreted
Matrilysin 2	MMP-26	Secreted
Stromelysin 3	MMP-11	Secreted
Membrane-type MMPs		
MT1-MMP	MMP-14	Membrane associated
MT2-MMP	MMP-15	Membrane associated
MT3-MMP	MMP-16	Membrane associated
MT4-MMP	MMP-17	Membrane associated
MT5-MMP	MMP-24	Membrane associated
MT6-MMP	MMP-25	Membrane associated
Others		
Macrophage elastase	MMP-12	Secreted
	MMP-19	Secreted
Enamelysin	MMP-20	Secreted
	MMP-21	Secreted
CA-MMP	MMP-23	Secreted
	MMP-27	Secreted
Epilysin	MMP-28	Secreted

providing a means by which a more severe disease phenotype will result in the release of more of the therapeutic cytokine from its latent form.

The effectiveness of latent cytokines in treating disease has been demonstrated by Adams et al. [7] using latent IFN-β to treat collagen-induced arthritis (CIA) in mice, a well-known mouse model of RA. It had previously been shown that IFN-β, delivered as gene therapy had therapeutic effects in CIA [21] so a plasmid encoding the latent IFN-β (including the MMP cleavage site) was delivered via a single intramuscular injection into mice with CIA. The latent construct resulted in inhibition of paw swelling and reduced clinical score [7].

This study also demonstrated three very important aspects of the latent cytokine technology. First, the latent IFN-β was biologically inactive until such time as the molecule was treated with MMP to remove the LAP part of the molecule and release the IFN-β. This result shows that true latency can be achieved, even when the cytokine inside the LAP shell is bigger in size than the native TGF-β. Second, synovial fluid from arthritic joints, or cerebral spinal fluid from meningitis or MS patients, was effective in activating the latent IFN-β, as opposed to sera from these patients, which did not result in release of active IFN-β from the latent complex. This indicates that the nature and quantity of MMP in biological fluids in disease conditions is sufficient to cleave the latent molecule and release the active cytokine. Finally, comparison of the half-lives of the latent IFN-β and free IFN-β showed a highly significant (37-fold) increase in the half-life of the latent cytokine compared with the free cytokine [7]. The half-life of the latent IFN-β is also greater than that of pegylated IFN-β [22], which is a commonly used method of extending the half-life of therapeutic proteins.

One of the main strengths of this technology is that targeting of the therapeutic cytokine to the sites of disease is achieved without having to identify specific disease antigens, as is the case when using antibody-cytokine fusions. Another important aspect of the latent approach is that because the LAP is a native protein, the immunogenicity of the latent complex should be virtually nonexistent, and there are fewer problems with interactions with immune cells that are characteristic of antibody Fc fusions. Neither of these types of fusion proteins with immunoglobulin fragments or pegylation confers latency. Hence, the latent cytokine technology is applicable to a much wider range of pathological conditions and causes fewer side effects.

16.3 LIMITATIONS OF THE LATENT CYTOKINE TECHNOLOGY

The latent cytokine technology was initially designed to deliver therapeutic cytokines to the sites of disease. However, we have also used this technology to produce cytokine antagonists using receptors for pro-inflammatory cytokines inside the LAP shell as anti-inflammatory therapeutics. These efforts have focused in particular on the engineering of a dimeric TNF receptor 2 (p75) (dTNFR) [23] inside the LAP shell, as the central role of TNF-α in initiating and maintaining inflammation is well established. It was hypothesized that this LAP-MMP-dTNFR protein could be used to block TNF-α activity in a manner similar to that achieved by the anti-TNF-α antibodies used in the treatment of RA patients [6], with the possible added advantage of activity only at sites of disease. To test this hypothesis, LAP-MMP-dTNFR was expressed in 293 T cells. Cell supernatants containing the recombinant protein were treated with MMP-1, followed by SDS–PAGE and Western blotting using both anti-LAP and anti-TNFR antibodies. These experiments showed the presence of recombinant protein of the expected size, and after treatment with MMP-1, the release of individual components of TNFR and LAP (Figure 16.3a).

Subsequently, experiments to examine the biological activity of LAP-MMP-dTNFR were performed, using HeLa cells that respond to hTNF stimulation with luciferase expression

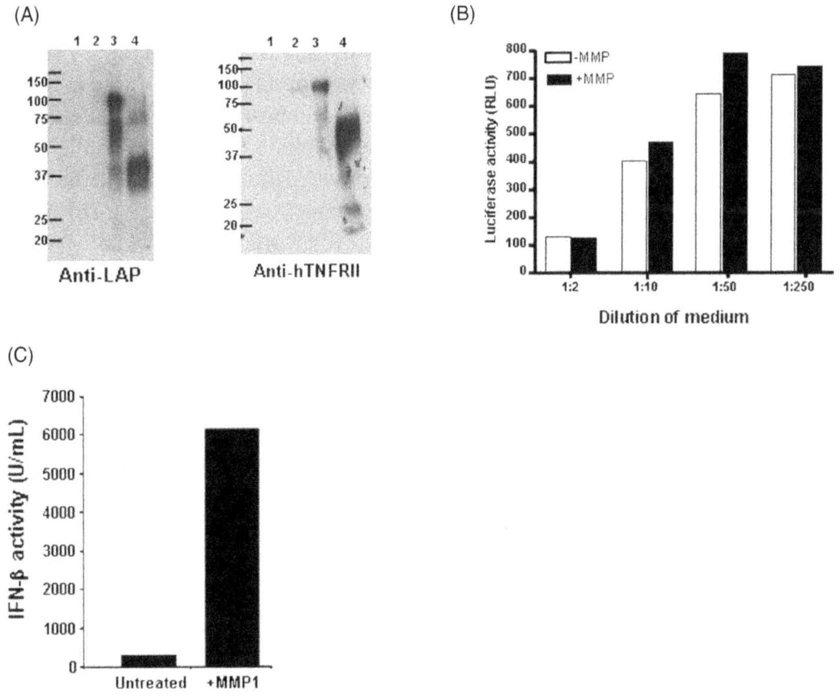

FIGURE 16.3 The LAP technology cannot be used to make latent molecules that bind soluble ligands. (A) LAP-dTNFR detected on Western blot after transient expression from 293 T cells. Lanes 1 and 2 are media from nontransfected cells; 3 and 4 are from cells expressing LAP-dTNFR. Media in 2 and 4 was treated with MMP-1 for 18 h. After digestion with MMP-1 the individual components LAP and dTNFR are clearly released. (B) Nonlatent activity of LAP-dTNFR. 57A HeLa cells respond to hTNF-α stimulation with luciferase expression from an NF-κB responsive promoter. Supernatants from 293 T cells containing LAP-MMP-TNFR2 were either treated with MMP-1 for 18 h or kept untreated prior to addition to 57A HeLa cells along with TNF-α (10 ng/mL). Both MMP treated and untreated LAP-dTNFR equally inhibit the activity of TNF-α. (C) Latency of LAP-IFN-b. Measurement of IFN-b-mediated inhibition of the cytopathic effect of EMC virus on L929 cells before and after treatment with MMP.

from an NF-κβ-responsive promoter. Supernatants from 293T cells containing LAP-MMP-dTNFR, with or in the absence of treatment with MMP-1, were added to HeLa cells along with TNF (10 ng/mL). The results showed that both MMP-treated and untreated LAP-dTNFR equally inhibited the activity of TNF (Figure 16.3b), therefore indicating that, contrary to expectations, LAP-dTNFR was not in fact latent. We believe that the failure of the LAP to provide latency in this instance is because the ligand for the molecule inside the LAP is soluble. LAP confers latency to cytokines by blocking access to cellular receptors, rather than completely enclosing the cytokine within the LAP shell. Hence, in the case of LAP-dTNFR, TNF-α in solution is able to access the dTNFR inside the LAP, nullifying the latent properties of the presence of LAP. We believe that these observations indicate a limitation of the latent cytokine technology, in that it is not suitable for molecules with soluble receptors or ligands.

Hence, we were puzzled by the report of a "latent" human soluble TNF receptor 1 (hsTNFR1) engineered by fusing LAP to the hsTNFR1 via an MMP site to treat endometriosis [24]. The initial phase of endometriosis, involving the ectopic implantation of endometrial tissue, is an invasive event, requiring breakdown of the ECM [25]. In a similar way to cancers, MMP activity is increased in endometriosis, allowing endometriotic cells to invade the ECM. These increases in MMP activity are stimulated by proinflammatory cytokines such as TNF-α and IL-1 [26], so blocking the biological effects of TNF-α could be an effective therapy for this condition.

Cleavage of the LAP-MMP-hsTNFR1 by peritoneal fluid from endometriosis patients is reported [24], indicating that there is sufficient MMP activity to release the hsTNFR1 from the LAP, and is similar to our results with biological fluids from a number of inflammatory conditions [7]. However, it is also claimed that the LAP-MMP-hsTNFR1 is fully latent on the basis of *in vitro* MTT assays with L929 cells. On the basis of our experience of LAP-MMP-dTNFR (and the data shown in Figure 16.3), it is not possible that LAP-MMP-hsTNFR1 is latent. The conclusions drawn by Xiong et al. [24] are based on data from cytotoxicity protection of L929 cells treated with TNF-α. However, the differences reported between the cells treated with intact LAP-MMP-hsTNFR1 and MMP-treated LAP-MMP-hsTNFR1 are minimal. A much greater difference in biological activity between the intact and cleaved LAP-fusion protein is expected if the LAP-fusion protein is truly latent. An example of the effect of MMP treatment on latent LAP-MMP-IFN-β is shown in Figure 16.3c. Prior to MMP-1 treatment, the biological activity of the IFN-β in protecting cells from viral infection is virtually absent. Treatment of LAP-MMP-IFN-β with MMP-1 results in a dramatic increase in biological activity (Figure 16.3c; also see [7]). In addition, a monomeric soluble TNF-R1 has to be in great excess to free TNF in order to properly antagonize. Hence, we have designed the dTNFR that binds TNF in a stable one-to-one molar interaction [27]. The molecular design described by Xiong et al. [24] is not a viable TNF inhibitor.

16.4 GENERATION OF LATENT CYTOKINES

16.4.1 Molecular Design

The cloning of recombinant latent cytokines is relatively straightforward, involving cloning the LAP of TGF-β, followed by a MMP cleavage site, and then the therapeutic cytokine of choice into a suitable expression vector. One important point to note is that the LAP must be at the N-terminus of the molecule, that is, N-terminal with regard to the cytokine, as Adams et al. [7] have clearly shown that cloning of the LAP at the C-terminus of the fusion protein fails to confer complete latency. This is most likely due to the fact that when dimerization of the molecules occur; the cytokine is not enclosed within the LAP shell, and so retains the ability to interact with its cellular receptors.

It is necessary to make two changes to the native LAP sequence when cloning LAP for use in constructing latent cytokines. The first is to remove the furin cleavage site at position 278/279 of the LAP sequence, to prevent processing of the latent cytokine by furin and consequent release of the cytokine from the protective LAP shell. Hence, amino acids Met^1-Ser^{273} of the LAP sequence are cloned. It is also necessary to mutate the cysteine residue at position 33 in the LAP sequence (Figure 16.4). This residue is involved in binding of native LAP-TGF-β to a 125–210 kDa protein known as latent TGF-β binding protein (LTBP), the function of which is to ensure efficient secretion from the cell and to target and tether latent TGF-β to the ECM, allowing subsequent activation of TGF-β [28,29,30]. Complexing of LAP

(A)

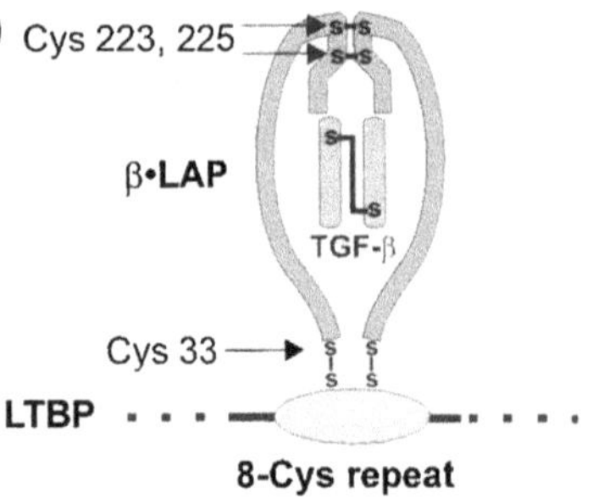

(B)

MPPSGLRLLPLLLPLLWLLVLTPGPPAAGLST**C**KTIDMELV**KRKR**IEAIRGQILSK
(Signal peptide: MPPSGLRLLPLLLPLLWLLVLTPG; 33; Heparin-binding)

LRLASPPSQGEVPPGPLPEAVLALYNSTRDRVAGESAEPEPEPEADYYAKEVT

RVLMVETHNEIYDKFKQSTHSIYMFFNTSELREAVPEPVLLSRAELRLLR**RLKLK**
(Heparin-binding)

VEQHVELYQKYSNNSWRYLSNRLLAPSDSPEWLSFDVTGVVRQWLSRGGEIE

GFRLSAH**C**S**C**DSRDNTLQVDINGFTTGR**RGD**LATIHGMNRPFLLLMATPLERAQ
(223, 225)

HLQSEF*GGGGSGPLGLWAGGGGSA*
(Linker + MMP cleavage site)

FIGURE 16.4 Structure of the latency-associated peptide (LAP) from TGF-β. (A) Schematic representation of the native LAP structure. Printed from [30] with permission from the publisher. Formation of disulfide bonds at $Cys^{223/225}$ results in dimerization of the LAP molecules to form a shell surrounding the active cytokine. The native LAP/cytokine complex interacts with LTBP through Cys^{33}. (B) Amino acid sequence of human LAP indicating the location of the cysteine residues that are important in binding of LAP to LTBP (Cys33) and in dimer formation (Cys^{223} and Cys^{225}) in bold, as well as the location of the RGD site. The locations of the heparin-binding sites are also shown, as well as the glycine/serine linker and the MMP cleavage site. *Source*: (a) This research was originally published in Reference [30].

with LTBP is not required for latency [31] and it is obviously not desirable for the latent therapeutic cytokine to be tethered to the ECM. It has previously been demonstrated that endogenous LTBP from Chinese hamster ovary cells does interact with recombinant LAP expressed in these cells [7]. To prevent this interaction, the cysteine at position 33 in the LAP is mutated to a serine residue to ensure that the latent cytokine does not interact with LTBP during expression. This mutation also prevents the formation of spurious disulfide bonds between this cysteine and any cysteines that may be present in the therapeutic cytokine of interest, or indeed, between LAP molecules. The MMP cleavage site is flanked at either side by a flexible glycine/serine linker to provide a hydrophilic area that will facilitate access to enzymatic attack.

16.4.2 Expression of LAP-Fusion Proteins

Expression of the recombinant fusion protein should be performed in mammalian cells because of their ability to correctly fold, assemble, glycosylate, and secrete the protein. This is particularly important for the production of these latent cytokines, given that the structure of the LAP used to confer latency. The LAP is a homodimer formed by two disulfide bonds formed between two cysteine residues (Cys^{223} and Cys^{225}; Figure 16.4) on each monomer. The formation of both of these disulfide bonds is necessary for the dimerization of the LAP monomers and consequent latency. Substitution of one or both of these cysteine residues with serine results in loss of latency [32]. If dimerization does not occur, the LAP shell is not formed; the cytokine is not surrounded by the LAP, and is biologically active. Hence, the correct formation of these disulfide bonds, and consequent dimerization of the LAP-cytokine molecules is absolutely critical for the production of latent, rather than active, cytokine.

The nature and extent of glycosylation of the recombinant protein are also essential for the production of these latent cytokines and is one of the main reasons why expression of these proteins in nonmammalian systems such as *Escherichia coli* or insect cell lines is not advisable. Under-glycosylation is known to affect the folding and assembly of many proteins [33,34] and can affect the yield of correctly folded protein quite dramatically [35]. Indeed, the carbohydrate structures of LAP have been shown to be important in the secretion of latent TGF-β from cells [36,37,38]. However, a more pressing reason for ensuring the correct levels of glycosylation of the recombinant protein is that this has profound implications for the half-life of the protein *in vivo*. Recombinant proteins that are incorrectly or under-glycosylated are cleared from the circulation much more quickly than correctly glycosylated proteins [39,40]. LAP contains three N-linked glycosylation sites, all three of which are glycosylated [41] and two of which contain mannose-6-phosphate residues [42]. The presence of sialic acid residues at these sites has also been demonstrated by treatment of recombinant LAP-TGF-β with neuraminidase [41]. The degree of sialylation of native latent TGF-β has a critical bearing on the pharmacokinetics of this molecule. Recombinant LAP-TGF-β that is under-sialylated has a half-life over 100-times shorter than fully sialylated protein [43]. The nature and extent of glycosylation also determines the tissue distribution of recombinant LAP-TGF-β, with most under-sialylated material found in the liver, as opposed to the sialylated protein which is found predominantly in the circulation [43]. It was initially believed that glycosylation of the LAP may be important in TGF-β latency, at least *in vitro* [44], as enzymatic removal of the carbohydrate structures from the LAP-TGF-β complex produced active TGF-β from the latent complex. However, subsequent studies have shown conclusively that unglycosylated recombinant LAP produced in inclusion bodies in *E. coli* and re-folded *in vitro*, can associate with mature TFG-β and render it latent [45]. The role of these carbohydrate moieties of LAP in maintaining latency *in vivo* has not been investigated.

As one of the advantages of the latent cytokine technology is that the half-life of the latent form is significantly longer than that of the active cytokine, producing poorly glycosylated latent cytokines actually defeats the purpose of this approach. Hence, the choice of expression system is of the utmost importance. In our experience, Chinese hamster ovary (CHO) cells, grown either as adherent cells or in suspension, are more adept at producing recombinant proteins that are sufficiently sialylated than other mammalian cell lines such as human embryonic kidney (HEK) cells [46].

16.4.3 Purification of LAP-Fusion Proteins

As is the case for the purification of any recombinant protein from the supernatant of cell cultures, there are a number of approaches that may be taken for the purification of recombinant cytokines. Latent cytokines can be purified from cell culture supernatants using conventional chromatography techniques, all of which are based on using the LAP part of the molecule, rather than the cytokine part of the molecule. Although other chromatographic methods may be available depending on the cytokine of choice, for example, IFN-β can be purified using blue sepharose, any such method would depend on the ability of the cytokine to access the chromatography medium in the presence of the LAP. To date, methods of purification based on the cytokine part of the molecule have not been tested.

There are a number of methods that can be used to purify latent cytokines based on the LAP part of the molecule. Affinity chromatography using a commercially available antibody against human LAP immobilized on sepharose beads was used very effectively by Adams et al. [7]. However, this protocol is costly and is not very easily scaled up to deal with large volumes of supernatants.

The alternative protocol that has been developed for the purification of latent cytokines consists of three chromatography steps. Cell supernatants from transfected cells are dialyzed and applied to a heparin-sepharose column, taking advantage of the heparin-binding sites in the LAP part of the molecule (Figure 16.4). The binding of the latent cytokine to the heparin column is of relatively low affinity and can be eluted with an increasing salt gradient. The LAP fusion proteins elute at the beginning of the gradient. This first affinity chromatography step serves not only to partially purify the latent cytokine, but also to concentrate the protein greatly (although further concentration is generally necessary) for application to a gel filtration column.

The gel filtration step is particularly useful as any LAP protein aggregates are separated from the latent LAP dimers at this stage. Aggregation is much less problematic as a result of the Cys^{33} mutation (as discussed above). This step also removes any endogenous LAP that may be present from the cultured cells. This material can also be purified from the heparin column, but the difference in size means that it is effectively removed by the gel filtration column. A final anion exchange chromatography step is then used to purify the latent cytokines to homogeneity.

The purified latent cytokines that have been produced thus far are soluble in aqueous media and are stable when stored at $-80°C$ for extended periods (>6 months). All of the experiments using latent cytokines to date have been conducted in murine models of disease (e.g., the CIA model of RA, experimental allergic encephalitis (EAE) and the ApoE knock-out/high-fat diet model of atherosclerosis) where latent cytokines using either human LAP or mouse LAP have been well tolerated ([7]; unpublished data).

16.5 APPLICATIONS AND POTENTIAL CLINICAL INDICATIONS

There are many potential clinical applications for latent cytokines, but the two indications that show particular promise are inflammatory conditions and cancer. A common denominator between these two groups of disease is the local increases in the expression and activity of MMPs at the sites of disease. As discussed above, this family of proteinases are essential for normal development, but are also implicated in a wide variety of pathological processes, where tissue remodeling takes place, for example, osteoarthritis, autoimmunity, neuroinflammation, spreading of metastatic cancer cells, and atherosclerosis. It is for this reason that the latent cytokines described here are engineered with a MMP cleavage site, situated between the LAP and the therapeutic cytokine. This approach means that these latent molecules can potentially be used to deliver therapeutic cytokines to the sites of disease in a range of inflammatory conditions including cancer.

There are a whole host of pathological conditions in which inflammation is a feature and may occur in response to an exogenous stimulus, for example, in response to an infectious agent, or as part of an autoimmune response. Treatment of inflammation with latent cytokines is most likely to be appropriate in those inflammatory conditions that are chronic in nature, for example, atherosclerosis, or in auto-immune diseases that are characterized by chronic inflammation, such as RA. The central role of cytokines in regulating inflammatory processes means that these molecules are of potentially great therapeutic use, notwithstanding the issues surrounding their short half-lives and pleiotropic effects, as previously discussed.

The onset of inflammation is characterized by the release of large amounts of proinflammatory cytokines such as TNF-α, IL-6, and IL-1 from leukocytes such as neutrophils and macrophages that have been recruited to the site of inflammation. Of course, the roles of cytokines in inflammatory processes are not confined to the actions of the proinflammatory cytokines and it is the cytokines with anti-inflammatory properties (e.g., IL-4, IL-10, TGF-β, and IFN-β) that are the focus of the latent cytokine technology for the treatment of inflammatory conditions.

There are already a small number of therapeutic cytokines that have gained approval for clinical use. Examples include IFN-β for the treatment of MS, IFNs for a variety of conditions including hepatitis B and C, hairy cell leukemia and AIDS-related Kaposi's sarcoma. Unsurprisingly, given the potency of these molecules, there are many more cytokine-based therapeutics in development, most of which are based on the "classic" anti-inflammatory cytokines IL-4, IL-10, and IFN-β. However, the pleiotropic effects and short half-lives of these cytokines continue to hamper these efforts in many cases. The development of latent versions of these anti-inflammatory cytokines could potentially allow the anti-inflammatory potential of these molecules to be fully harnessed for the treatment of disease.

16.5.1 Multiple Sclerosis

16.5.1.1 Interferon β As mentioned above, the use of IFN-β as a therapeutic is already in use in the treatment of MS, a neurodegenerative autoimmune disease. The neurodegeneration in MS is caused by an inflammatory response (the cause of which has remained elusive) that strips myelin from neurons in the white matter of the CNS [47]. This demyelination results in a range of neurological deficits. As this condition is an inflammatory disease, there is obvious interest in attempting to treat this disease with anti-inflammatory cytokines. Indeed, IFN-β is now the treatment of choice for the relapsing/remitting form of MS and is effective in ~50% of patients, in whom relapse rates are reduced by half [48]. Although the mechanism of action of IFN-β in MS has not yet been fully elucidated, there is convincing evidence from *in vitro* studies that IFN-β exerts its effects by modulating the functions of immune cells, with consequent changes in circulating cytokines. There is also evidence that IFN-β down-regulates the T_H1 pathway via its effects on the expression of IFN-γ [48], so that the ratio of T_H1 to T_H2 cells is reduced [49]. IFN-β treatment also results in a transient increase in the level of IL-10 and IL-1ra in the plasma.

16.5.1.2 Interleukin 10 IL-10 has a number of anti-inflammatory actions, including the inhibition of the production of proinflammatory cytokines such as IL-1 and TNF-α and the proliferation for T cells *in vitro*. In addition, it down-regulates the expression of both major histocompatibility complex (MHC) Class II molecules and co-stimulatory molecules, which also contribute to the anti-inflammatory effects. There are a number of studies that have investigated the role of IL-10 both in the mouse model of MS, EAE, and in MS patients. It has been demonstrated that IL-10 down-regulates the immune response in EAE [50,51] and in addition may also have neuroprotective effects in a stroke model [52] and in a spinal cord injury model [53].

16.5.1.3 Erythropoietin Another cytokine that shows promise as a candidate for therapeutic use in MS is erythropoietin (EPO). Recent research has identified potent neuroprotective properties of EPO [54,55,56], and many groups have now established beneficial effects of EPO in EAE including improvements in clinical symptoms, expression of pro-inflammatory cytokines, and increased integrity of the blood–brain-barrier [57,58,59].

Therefore, there are at least two other cytokines that could be developed as treatments for MS: IL-10 as an anti-inflammatory agent to prevent further demyelination of neurons and EPO as a neuroprotective agent. Given the fact that the common therapeutic option for MS, IFN-β, is effective in only 50% of patients, and with an efficacy of ~30%, there is obviously a need to continue the search for other, better treatments. Production of these cytokines as latent molecules could be used to reduce the risk of side-effects and increase the serum half-life of the cytokines. The up-regulation of MMPs in MS [60,61,62], particularly MMP-9, represents an effective means by which the active cytokine can be released from the latent form at the blood–brain-barrier (BBB), the integrity of which is compromised by MMP-9 in MS.

The delivery of EPO to the CNS using the latent cytokine technology is particularly attractive, as this would allow much higher doses of EPO to be used. Ehrenreich et al. [63] have shown that high doses of recombinant EPO resulted in a significant clinical improvement in chronic progressive MS as opposed to lower doses. The authors suggest that the failure of the lower dose of EPO used may be due to insufficient concentrations of the cytokine in the CNS, that is, at the site of disease. Work is now underway to establish whether latent EPO, that is, EPO protected by LAP, can penetrate the BBB in EAE mice.

16.5.1.4 Transforming Growth Factor-β The other cytokine of interest in the treatment of MS is TGF-β, which has been shown to inhibit EAE in mice with this condition [64]. Furthermore, neutralization of TGF-β resulted in enhancement of the disease [65]. However, the pleiotropic nature of TGF-β is well established and administration of TGF-β for therapeutic reasons is extremely risky [66]. Delivery of TGF-β as a latent complex may allow this cytokine to be safely used in a therapeutic setting. In contrast to native TGF-β, which is secreted from cells in a latent from, the engineered molecules would also contain the MMP cleavage site to allow release of the active TGF-β at the site of disease. The engineered molecules would also be devoid of the LTBP that is a part of the native TGF-β complex. This ensures that the therapeutic TGF-β is not tethered to the ECM, but rather, is free to migrate to the site of disease.

16.5.2 Rheumatoid Arthritis

Another auto-immune inflammatory disease that could benefit from treatment with latent cytokines is RA. This is a chronic inflammatory disease affecting the synovial membrane, bone, and cartilage. Cytokines are directly involved in many of the immune processes that occur in the rheumatoid joint. Currently available biological treatments are based on blocking of the pro-inflammatory cytokine, TNF-α, and while these treatments are very successful in some patients, there are a proportion of patients (i.e., 30%) who do not respond to these treatments. Use of the anti-inflammatory IFN-β has been investigated in the murine model of RA, collagen-induced arthritis (CIA).

16.5.2.1 Interferon β Amelioration of CIA by IFN-β delivered via gene therapy was first demonstrated over a decade ago by Triantaphyllopoulos et al. [21]. IFN-β gene therapy was able to prevent the onset of CIA and ameliorate existing disease, with concurrent decreases in IFN-γ, IL-12, and TNF-α. More recently, recombinant IFN-β, injected daily, was shown to be effective in reducing inflammation and cartilage destruction in CIA [67]. These effects appear to be mediated by increased IL-10 expression, with concomitant decreases in TNF-α, IL-6, and IL-8 expression. These data suggest that IFN-β could potentially be an effective therapy for joint inflammation and destruction in RA. If the latent form was used, the daily injections used previously [67] could perhaps be avoided due to greater serum stability and higher local concentrations of active cytokine at the site of disease.

The therapeutic potential of latent cytokines may also be applied to chronic inflammatory diseases such as Crohn's disease (CD) and atherosclerosis. The chronic inflammation, that is, a hallmark of these diseases results in increased expression of MMPs at the sites of disease. Indeed, up-regulation of the expression of a whole range of MMPs has been demonstrated for both CD (particularly MMP-1, -3, and -9; [68,69]) and for atherosclerosis (especially MMPs -8 and -13; [70]). There is long-standing interest in the use of IL-10 to treat both of these diseases due to the anti-inflammatory properties of this cytokine, and there are many tantalizing hints that IL-10 could indeed be effective in treating these conditions.

16.5.2.2 Interleukine 10 IL-10 knockout mice suffer from accelerated atherosclerosis, in contrast to IL-10 transgenic mice that are relatively protected [71]. Patients with unstable angina have decreased serum levels of IL-10 compared with patients with chronic stable angina [72]. Similarly, there are encouraging results from treatment of animal models of colitis with IL-10 [73], and *in vitro* studies have shown that IL-10 can down-regulate the pro-inflammatory cytokine release from cells isolated from the lamina propria isolated from patients with CD. However, despite these results, early clinical studies using recombinant IL-10 to treat CD have been unsuccessful. This is likely due to the complex biology of IL-10. In addition to the down-regulation of immune responses, IL-10 is also known to exert immunostimulatory effects via up-regulation of MHC Class II molecules on B lymphocytes and the induction of cytotoxic T cell differentiation. Certainly, the picture that is now emerging from these *in vivo* studies on the use of IL-10 in the treatment of CD is that IL-10 may have both anti- and pro-inflammatory effects, depending on the local concentrations achieved, and that high systemic doses of IL-10 can alter the balance between immunoregulatory and immunostimulatory effects [74]. It is also possible that systemic administration of IL-10 does not allow delivery of sufficiently high concentrations of the cytokine to the site of disease, given that the half-life is in the region of 1.1–2.6 h. Production of latent IL-10 could solve this problem and prevent the side-effects of high systemic concentrations of IL-10 found in some studies [75,76].

16.5.2.3 Interleukine 4 The other cytokine of interest in terms of a potential treatment for atherosclerosis is IL-4. This is a classic TH2 cytokine, and also has immunosuppressive effects on macrophages, including suppressing the secretion of pro-inflammatory cytokines [77]. IL-4 deficient mice are protected from susceptibility to diet-induced atherosclerosis [78] and mice deficient in Stat6 (one of the transcription factors activated by IL-4) develop larger atherosclerotic lesions than their wild-type counterparts [79]. However, despite these anti-atherogenic effects, the highly pleiotropic nature of IL-4 means that this cytokine also has a number of pro-atherogenic affects such as up-regulation of P-selectin [80] and cholesterol esterification in macrophages [81] and the net effect of IL-4 may also vary with the stage of the disease. Use of latent IL-4 could be used to overcome the pleiotropic effects of this cytokine and it has recently been demonstrated that latent IL-4 is more effective in the treatment of fat-diet-induced atherosclerosis in Apo-E −/− mice than free IL-4 ([82] submitted). These effects on atherosclerotic plaque reduction are thought to be most likely due to the increased serum half-life of the latent molecule.

The uses of therapeutic cytokines in treating inflammatory conditions as discussed above, is obviously focused on the potential of cytokines with anti-inflammatory effects, the most obvious candidates being IFN-β, IL-10, and IL-4. Developing the latent cytokine technology as a treatment for cancer may involve a quite different pool of therapeutic cytokines, from agents that enhance NK cell and T cell function (e.g., IL-2, IL-7, IL-6, and IL-4) to molecules that enhance tumor antigen presentation (e.g., IFN-β, IFN-γ, and IL-13). One of the most effective cytokines in the treatment of cancer to date is IFN-α, which has anti-tumor effects in several hematological malignancies.

Given the evidence that certain cytokines are potentially extremely effective in the treatment of tumors, but that have poor pharmacokinetic and/or toxicity profiles, there is a need to develop novel strategies that will enable these cytokines to be used effectively in the treatment of these diseases and this is one area where the LAP technology may prove invaluable. The up-regulation of MMPs at tumor sites is well documented [83,84] and their role is to promote tumor growth and metastatic spread [85]. This MMP activity at the tumor site will activate any latent cytokine in the vicinity, but ensures that the potent actions of the cytokines are exerted only at the sites of disease.

16.6 ALTERNATIVES/VARIANTS OF APPROACH

The latent cytokine technology, as it currently stands, consists of a therapeutic cytokine enclosed within the LAP shell, which can be removed by MMP activity due to the MMP cleavage site located between the LAP and cytokine. As described previously, the LAP is situated at the N-terminal end of the fusion protein to ensure full latency. It has already been demonstrated that production of the fusion protein with the LAP at the C-terminus of the fusion protein results in a cytokine that retains most, although not all, of its biological activity, and thus, is not fully latent [7]. Therefore, any variant of the technology should not involve any change in the positioning of the LAP relative to the cytokine.

There is, however, much scope for producing fusion proteins that have different cleavage sites and/or different cytokines/signaling molecules. It is clear that the LAP shell can accommodate cytokines that are larger in size than the

TABLE 16.2 Latent Cytokines Engineered to Date

Molecule	Therapeutic Application	Reference
LAP-MMP–IFN-β	RA	[7]
LAP-agg-IFN-β	RA	Unpublished
LAP-MMP-VIP	RA	[82]
LAP-MMP-αMSH	RA	[82]
LAP-MMP-γMSH	RA	[82]
LAP-MMP-IL-4	Atherosclerosis	Unpublished
LAP-MMP-IL-10	Atherosclerosis	Unpublished
LAP-MMP-TGF-β	MS	Unpublished
LAP-MMP-EPO	MS	Unpublished
LAP-BMP-7	OA	Unpublished
LAP-IGF-1	MS	Unpublished

native TGF-β, as shown by the production of latent IFN-β [7], which is a significantly larger molecule. A number of LAP fusion proteins have now been produced and are listed in Table 16.2, along with the potential applications of these molecules. It is of note that this technology works just as well for the production of much smaller molecules than native TGF-β, such as vasoactive intestinal peptide (VIP), α-melanocortin (α-MSH), and γ-melanocortin (γ-MSH), and that the LAP technology has been extended to molecules outside of the classical cytokines to include other anti-inflammatory agents such as those aforementioned.

VIP is a neuropeptide/hormone that was first identified as a vasodilator and hypotensive peptide [86,87], but over the last decade has been shown to have potent anti-inflammatory effects on both innate and adaptive immune responses [88]. A potential problem with the use of VIP as a therapeutic agent is that its neuromodulatory functions are widespread, including heart, lung, thyroid gland, urinary tract, genitals, kidney, and immune system [89] and so increases the risk of causing unwanted side-effects. Both α- and γ-MSH are parts of the proopiomelanocortin (POMC) system, which consists of the POMC preprohormone and its numerous products. These peptides have anti-inflammatory effects [90] as well as effects on cardiovascular function such as decreasing blood pressure that may be useful in treating cardiovascular disease. The production of these much smaller signaling molecules (VIP has a molecular mass of about 3.6 kDa, while the MSH peptides are ~1 kDa) as LAP fusion proteins is instructive in demonstrating that full latency, and subsequent activation following cleavage of the LAP can be achieved with a wide range of peptides and proteins, regardless of size.

A further variation to the basic latent cytokine technology is to change the MMP cleavage site so that the resultant latent molecule is more sensitive to particular MMPs. The original MMP cleavage site used was derived from a collagen substrate of MMPs [91–93], but this cleavage site could be made more specific for individual MMPs by site-specific mutagenesis (see [94]). This could make the latent cytokines produced even more specific to a particular disease or disease stage, given that the overexpression of certain MMPs is disease-specific. Examples include the overexpression of MMP-1 and -3 in RA, MMP-1, -8, and -13 in atherosclerosis and MMP-2, -9, and -14 in tumors.

There is also the possibility of exchanging the MMP cleavage site for a site that is cleaved by other enzymes that are overexpressed at the site of disease. Work is in progress on an alternative latent cytokine that is produced with an aggrecanase cleavage site, rather than an MMP cleavage site. Aggrecanases are a family of proteinases that degrade a large proteoglycan called aggrecan [95,96]. Aggrecan is the most abundant proteoglycan in the soft tissues of the joint, where it is found interspersed within the collagen network of the cartilage. The function of aggrecan is to endow the cartilage with the ability to resist compression on load-bearing. One of the main manifestations of arthritis is the loss of type II collagen and aggrecan, causing the cartilage to become thin and weak [97]. It is now well-established that degradation of aggrecan in articular cartilage is due to the action of aggrecanases [98]. Cleavage sites for aggrecanases were identified via phage display by Hills et al. [99] and it is one of these sequences that has been inserted into a latent IFN-β molecule in between the LAP and the cytokine. Although the efficacy of LAP-MMP-IFN-β administered to mice with CIA has been demonstrated [7], it is hypothesized that use of the aggrecanase cleavage site will prove just as effective in the treatment of disease, but will also make this therapy even more "arthritis specific". Preliminary experiments have shown that these new latent proteins are efficiently cleaved by synovial fluid from arthritic joints (unpublished data).

16.7 CHALLENGES (PRODUCTION AND DEVELOPMENT)

It is clear that the latent cytokine technology provides a novel and efficient method for targeting therapeutic cytokines to the sites of disease (see Reference [7]). The next step in the development of this technology is to be able to treat disease by administering purified latent cytokines, rather than having to rely on a gene therapy approach. The production of sufficient quantities of recombinant latent cytokines is likely to present the greatest challenge to the use of this technology in a therapeutic setting.

The LAP fusion proteins are not easy to purify using conventional protein purification techniques [7] and there appear to be two reasons for this. The first is that LAP demonstrates a propensity to aggregate in standard cell culture conditions and, although this tendency can be abrogated to a large extent by the mutation of Cys^{33} to Ser, some slight aggregation does still occur. The second is the presence of LAP monomers, that is, LAP-cytokine fusions that have not dimerized to form the LAP shell around the cytokine and are therefore biologically active. As discussed previously, correct dimerization of the LAP molecules is an absolute requirement to achieve full latency.

The appearance of LAP monomers is likely the result of the overexpression of the recombinant protein which overtaxes the cell glycosylation and folding machinery, so that only a proportion of the expressed protein undergoes dimerization, and the rest is secreted as monomers. Attempts to supplement the relevant cellular enzymes (e.g., protein disulfide isomerase) to increase the rate of formation of the disulfide bonds required for dimerization of the LAP molecules have not, thus far, been successful. These are not insurmountable problems in terms of purification because chromatography using a gel filtration column effectively separates the three different species: aggregates, dimers, and monomers (unpublished data). However, these issues do present a problem when considering the total amount of LAP protein that is "lost" in aggregates and monomers. In practical terms, these species do in fact represent lost material as these forms of the protein are not suitable for therapeutic use.

The nature and extent of the glycosylation present on the recombinant proteins produced is a critical factor for ensuring an extended half-life *in vivo*, but is also important for expression and purification. There are three N-glycosylation sites in the LAP molecule, the sialylation of which are necessary to prevent clearance of the protein from the circulation [43]. Expression of the protein in CHO cells appears to result in the production of LAP molecules that have a range of levels of glycosylation. This is evident from lectin chromatography using immobilized wheat germ agglutinin (WGA). The elution profile of LAP fusion proteins from WGA is indicative of LAP molecules with differing levels of sugar moieties, as evidenced by the higher concentration of *N*-acetyl-glucosamine required to elute some of the LAP protein (unpublished data). However, although this makes the use of lectin chromatography unsuitable for the purification of latent cytokines, the glycosylation levels achieved in CHO cells appear to be satisfactory in terms of the half-life and tissue distribution of the latent cytokine achieved *in vivo* as demonstrated by Adams et al. [7].

Once the latent cytokine is cleaved by MMP, there is release of the LAP part of the molecule that could conceivably have biological effects independent of those of the cytokine. In fact, LAP has been shown to associate with mature TGF-β in mice, thereby inhibiting the cellular effects of the cytokine [100,101]. However, the expectation is that latent cytokine cleavage, and subsequent release of LAP will occur at sites of inflammation where there are high levels of free radicals also abound. The capacity of LAP to bind to TGF-β is abrogated by nitrosylation [102] and thus it is highly unlikely that the LAP released by MMP cleavage will antagonize TGF-β function. Despite early reports that LAP does not have biological activity independent of TGF-β [103] a recent study has suggested that LAP could indeed have immune functions such as the promotion of monocyte chemotaxis and the blocking of inflammation *in vivo* [104]. However, in our experience, we have not seen any anti-inflammatory effects by expressing free LAP by gene delivery in several models of inflammatory disease. No adverse effects of the LAP released as part of treatment with latent cytokines has ever been observed, but it is perhaps worth re-investigating this issue in particular experimental settings to show conclusively that LAP has no biological effects when used as part of a latent molecule to treat disease.

16.8 CONCLUSIONS AND FUTURE PERSPECTIVES

The latent cytokine technology is undergoing continual development and refinement. Current work is focused on repeating the therapeutic effects of latent cytokines on disease pathology as demonstrated by Adams et al. [7], this time using recombinant protein, rather than the gene therapy approach used previously. This work also aims to demonstrate cleavage of the latent fusion proteins at the sites of disease *in vivo*. Release of active cytokine from the latent complex by MMPs present in the biological fluids of patients with inflammatory disease has already been demonstrated by Adams et al. [7], and to show that this process also occurs efficiently *in vivo* is an important step in the evolution of this technology.

The latent cytokine technology described here was developed as a targeting mechanism for use with therapeutic cytokines. The biological potency and pleiotropy of these molecules dictate that the risk of unwanted side-effects are extremely high and in many cases presented insurmountable barriers to the use of cytokines as therapeutics. This technology has obvious advantages when applied to the use of cytokines as therapeutics. However, the use of this technology is by no means limited to use for the delivery of cytokines to sites of disease. In fact, the LAP technology could be used to "package" and deliver any therapeutic molecule to sites of disease where MMP activity is up-regulated.

There are already some applications of the latent cytokine technology to other therapeutic molecules. Holle et al. [105,106] have used this technology to develop a fusion protein to deliver the pore-forming toxin from honey bees, melittin, to tumor cells. As lytic peptides are fast-acting and have short systemic half-lives [107] as well as demonstrating high potency, the advantages of producing melittin as a LAP-fusion protein are very similar to those for therapeutic cytokines. The latent melittin produced is different from the latent cytokines described by Adams et al. [7] in that LAP is positioned at the C-terminus of the melittin. The LAP-melittin fusion protein was constructed in this way out of necessity, as fusing the melittin at its N-terminal removed all lytic activity. *In vitro* studies have shown that melittin can be released by cleavage with MMP-2, while *in vivo* studies have shown that treatment with melittin-LAP can decrease tumor volume in mice without significant systemic toxicity [105]. It is not clear at this time why this melittin-LAP fusion is fully latent, as it has been clearly demonstrated by Adams et al. [7] that construction of IFN-β with LAP located at the C-terminus of the fusion protein is not latent.

These examples illustrate the fact that the latent cytokine technology is by no means limited to cytokines as the therapeutic molecule of choice. In fact, any therapeutic molecule can theoretically be made latent and targeted to the site of disease using this system. The latent cytokine technology is proving to be an elegant, effective, and versatile method of targeting therapeutic molecules to sites of disease. There is certainly scope for using this technology to make ever more specific treatments for disease by harnessing disease-specific changes that occur locally at the sites of disease, including, but not limited to the up-regulation of MMPs. Similarly, the range of therapeutic molecules that can be produced in a latent form is not limited to cytokines, but can also be used for other therapeutic peptides. This versatility means that there are numerous potential applications for this technology.

ACKNOWLEDGMENTS

Funding for this work was obtained from the MRC, EU, Wellcome Trust, Arthritis Research UK, and the British Heart Foundation.

REFERENCES

1. Oppenheim JJ. (2001) Cytokines: past, present, and future. *Int. J. Hematol.* 74, 3–8.
2. Buchan G, Barrett K, Turner M, Chantry D, Maini RN, Feldmann M. (1988) Interleukin-1 and tumour necrosis factor mRNA expression in rheumatoid arthritis: prolonged production of IL-1 alpha. *Clin. Exp. Immunol.* 73, 449–455.
3. Wood NC, Symons JA, Dickens E, Duff GW. (1992) In situ hybridization of IL-6 in rheumatoid arthritis. *Clin. Exp. Immunol.* 87, 183–189.
4. McInnes IB, Leung BP, Sturrock RD, Field M, Liew FY. (1997) Interleukin-15 mediates T cell-dependent regulation of tumor necrosis factor-alpha production in rheumatoid arthritis. *Nat. Med.* 3, 189–195.
5. Tayal V, Kalra BS. (2008) Cytokines and anti-cytokines as therapeutics—an update. *Eur. J. Pharmacol.* 579, 1–12.
6. Vilcek J, Feldmann M. (2004) Historical review: cytokines as therapeutics and targets of therapeutics. *Trends Pharmacol. Sci.* 25, 201–209.
7. Adams G, Vessillier S, Dreja H, Chernajovsky Y. (2003) Targeting cytokines to inflammation sites. *Nat. Biotechnol.* 21, 1314–1320.
8. Nissim A, Gofur Y, Vessillier S, Adams G, Chernajovsky Y. (2004) Methods for targeting biologicals to specific disease sites. *Trends Mol. Med.* 10, 269–274.
9. Brinckerhoff CE, Matrisian LM. (2002) Matrix metalloproteinases: a tail of a frog that became a prince. *Nat. Rev. Mol. Cell Biol.* 3, 207–214.
10. Burrage PS, Brinckerhoff CE. (2007) Molecular targets in osteoarthritis: metalloproteinases and their inhibitors. *Curr. Drug Targets* 8, 293–303.
11. Burrage PS, Mix KS, Brinckerhoff CE. (2006) Matrix metalloproteinases: role in arthritis. *Front Biosci.* 11, 529–543.
12. Visse R, Nagase H. (2003) Matrix metalloproteinases and tissue inhibitors of metalloproteinases: structure, function, and biochemistry. *Circ. Res.* 92, 827–839.
13. Nagase H, Visse R, Murphy G. (2006) Structure and function of matrix metalloproteinases and TIMPs. *Cardiovasc. Res.* 69, 562–573.
14. Nagase H, Woessner JF, Jr. (1999) Matrix metalloproteinases. *J. Biol. Chem.* 274, 21491–21494.
15. Mandal M, Mandal A, Das S, Chakraborti T, Sajal C. (2003) Clinical implications of matrix metalloproteinases. *Mol. Cell Biochem.* 252, 305–329.
16. Sporn MB, Harris ED, Jr. (1981) Proliferative diseases. *Am. J. Med.* 70, 1231–1235.
17. Rosenberg GA. (1995) Matrix metalloproteinases in brain injury. *J. Neurotrauma* 12, 833–842.
18. Rosenberg GA. (2002) Matrix metalloproteinases in neuroinflammation. *Glia* 39, 279–291.
19. Shah PK, Falk E, Badimon JJ, Fernandez-Ortiz A, Mailhac A, Villareal-Levy G, et al. (1995) Human monocyte-derived macrophages induce collagen breakdown in fibrous caps of atherosclerotic plaques. Potential role of matrix-degrading metalloproteinases and implications for plaque rupture. *Circulation* 92, 1565–1569.
20. Galis ZS, Khatri JJ. (2002) Matrix metalloproteinases in vascular remodeling and atherogenesis: the good, the bad, and the ugly. *Circ. Res.* 90, 251–262.
21. Triantaphyllopoulos KA, Williams RO, Tailor H, Chernajovsky Y. (1999) Amelioration of collagen-induced arthritis and suppression of interferon-gamma, interleukin-12, and tumor necrosis factor alpha production by interferon-beta gene therapy. *Arthritis Rheum.* 42, 90–99.
22. Pepinsky RB, LePage DJ, Gill A, Chakraborty A, Vaidyanathan S, Green M, et al. (2001) Improved pharmacokinetic properties of a polyethylene glycol-modified form of interferon-beta-1a with preserved in vitro bioactivity. *J. Pharmacol. Exp. Ther.* 297, 1059–1066.
23. Gould DJ, Bright C, Chernajovsky Y. (2004) Inhibition of established collagen-induced arthritis with a tumour necrosis

factor-alpha inhibitor expressed from a self-contained doxycycline regulated plasmid. *Arthritis Res. Ther.* 6, R103–113.

24. Xiong ZF, Hu S, Wang ZH. (2010) Role of a latent soluble TNF receptor type I (hsTNFRI) fusion protein in targeted treatment of endometriosis. *Zhonghua Yi Xue Za Zhi* 90, 768–771.
25. Spuijbroek MD, Dunselman GA, Menheere PP, Evers JL. (1992) Early endometriosis invades the extracellular matrix. *Fertil. Steril.* 58, 929–933.
26. Sillem M, Prifti S, Koch A, Neher M, Jauckus J, Runnebaum B. (2001) Regulation of matrix metalloproteinases and their inhibitors in uterine endometrial cells of patients with and without endometriosis. *Eur. J. Obstet. Gynecol. Reprod. Biol.* 95, 167–174.
27. Neve R, Kissonerghis M, Clark J, Feldmann M, Chernajovsky Y. (1996) Expression of an efficient small molecular weight tumour necrosis factor/lymphotoxin antagonist. *Cytokine* 8, 365–370.
28. Moren A, Olofsson A, Stenman G, Sahlin P, Kanzaki T, Claesson-Welsh L, et al. (1994) Identification and characterization of LTBP-2, a novel latent transforming growth factor-beta-binding protein. *J. Biol. Chem.* 269, 32469–32478.
29. Olofsson A, Ichijo H, Moren A, ten Dijke P, Miyazono K, Heldin CH. (1995) Efficient association of an amino-terminally extended form of human latent transforming growth factor-beta binding protein with the extracellular matrix. *J. Biol. Chem.* 270, 31294–31297.
30. Saharinen J, Keski-Oja J. (2000) Specific sequence motif of 8-Cys repeats of TGF-beta binding proteins, LTBPs, creates a hydrophobic interaction surface for binding of small latent TGF-beta. *Mol. Biol. Cell* 11, 2691–2704.
31. Wakefield LM, Smith DM, Broz S, Jackson M, Levinson AD, Sporn MB. (1989) Recombinant TGF-beta 1 is synthesized as a two-component latent complex that shares some structural features with the native platelet latent TGF-beta 1 complex. *Growth Factors* 1, 203–218.
32. Brunner AM, Marquardt H, Malacko AR, Lioubin MN, Purchio AF. (1989) Site-directed mutagenesis of cysteine residues in the pro region of the transforming growth factor beta 1 precursor. Expression and characterization of mutant proteins. *J. Biol. Chem.* 264, 13660–13664.
33. Leavitt R, Schlesinger S, Kornfeld S. (1977) Impaired intracellular migration and altered solubility of nonglycosylated glycoproteins of vesicular stomatitis virus and Sindbis virus. *J. Biol. Chem.* 252, 9018–9023.
34. Machamer CE, Rose JK. (1988) Influence of new glycosylation sites on expression of the vesicular stomatitis virus G protein at the plasma membrane. *J. Biol. Chem.* 263, 5948–5954.
35. Dorner AJ, Bole DG, Kaufman RJ. (1987) The relationship of N-linked glycosylation and heavy chain-binding protein association with the secretion of glycoproteins. *J. Cell Biol.* 105, 2665–2674.
36. Sha X, Brunner AM, Purchio AF, Gentry LE. (1989) Transforming growth factor beta 1: importance of glycosylation and acidic proteases for processing and secretion. *Mol. Endocrinol.* 3, 1090–1098.
37. Lopez AR, Cook J, Deininger PL, Derynck R. (1992) Dominant negative mutants of transforming growth factor-beta 1 inhibit the secretion of different transforming growth factor-beta isoforms. *Mol. Cell Biol.* 12, 1674–1679.
38. Brunner AM, Lioubin MN, Marquardt H, Malacko AR, Wang WC, Shapiro RA, et al. (1992) Site-directed mutagenesis of glycosylation sites in the transforming growth factor-beta 1 (TGF beta 1) and TGF beta 2 (414) precursors and of cysteine residues within mature TGF beta 1: effects on secretion and bioactivity. *Mol. Endocrinol.* 6, 1691–1700.
39. Sinclair AM, Elliott S. (2005) Glycoengineering: the effect of glycosylation on the properties of therapeutic proteins. *J. Pharm. Sci.* 94, 1626–1635.
40. Werner RG, Kopp K, Schlueter M. (2007) Glycosylation of therapeutic proteins in different production systems. *Acta Paediatr. Suppl.* 96, 17–22.
41. Brunner AM, Gentry LE, Cooper JA, Purchio AF. (1988) Recombinant type 1 transforming growth factor beta precursor produced in Chinese hamster ovary cells is glycosylated and phosphorylated. *Mol. Cell Biol.* 8, 2229–2232.
42. Purchio AF, Cooper JA, Brunner AM, Lioubin MN, Gentry LE, Kovacina KS, (1988) Identification of mannose 6-phosphate in two asparagine-linked sugar chains of recombinant transforming growth factor-beta 1 precursor. *J. Biol. Chem.* 263, 14211–14215.
43. Wakefield LM, Winokur TS, Hollands RS, Christopherson K, Levinson AD, Sporn MB. (1990) Recombinant latent transforming growth factor beta 1 has a longer plasma half-life in rats than active transforming growth factor beta 1, and a different tissue distribution. *J. Clin. Invest.* 86, 1976–1984.
44. Miyazono K, Heldin CH. (1989) Role for carbohydrate structures in TGF-beta 1 latency. *Nature* 338, 158–160.
45. Yang Y, Dignam JD, Gentry LE. (1997) Role of carbohydrate structures in the binding of beta1-latency-associated peptide to ligands. *Biochemistry* 36, 11923–11932.
46. Chitlaru T, Kronman C, Zeevi M, Kam M, Harel A, Ordentlich A, et al. (1998) Modulation of circulatory residence of recombinant acetylcholinesterase through biochemical or genetic manipulation of sialylation levels. *Biochem. J.* 336, 647–658.
47. Poser CM, Brinar VV. (2004) The nature of multiple sclerosis. *Clin. Neurol. Neurosurg.* 106, 159–171.
48. Arnason BG. (1999) Immunologic therapy of multiple sclerosis. *Annu. Rev. Med.* 50, 291–302.
49. Hohlfeld R, Wekerle H. (2001) Immunological update on multiple sclerosis. *Curr. Opin. Neurol.* 14, 299–304.
50. Rott O, Fleischer B, Cash E. (1994) Interleukin-10 prevents experimental allergic encephalomyelitis in rats. *Eur. J. Immunol.* 24, 1434–1440.
51. Bettelli E, Das MP, Howard ED, Weiner HL, Sobel RA, Kuchroo VK. (1998) IL-10 is critical in the regulation of autoimmune encephalomyelitis as demonstrated by studies of IL-10- and IL-4-deficient and transgenic mice. *J. Immunol.* 161, 3299–3306.

52. Spera PA, Ellison JA, Feuerstein GZ, Barone FC. (1998) IL-10 reduces rat brain injury following focal stroke. *Neurosci. Lett.* 251, 189–192.
53. Plunkett JA, Yu CG, Easton JM, Bethea JR, Yezierski RP. (2001) Effects of interleukin-10 (IL-10) on pain behavior and gene expression following excitotoxic spinal cord injury in the rat. *Exp. Neurol.* 168, 144–154.
54. Ehrenreich H, Aust C, Krampe H, Jahn H, Jacob S, Herrmann M, (2004) Erythropoietin: novel approaches to neuroprotection in human brain disease. *Metab. Brain Dis.* 19, 195–206.
55. Juul S. (2004) Recombinant erythropoietin as a neuroprotective treatment: *in vitro* and *in vivo* models. *Clin. Perinatol.* 31, 129–142.
56. Brines M, Cerami A. (2005) Emerging biological roles for erythropoietin in the nervous system. *Nat. Rev. Neurosci.* 6, 484–494.
57. Agnello D, Bigini P, Villa P, Mennini T, Cerami A, Brines ML, et al.(2002) Erythropoietin exerts an anti-inflammatory effect on the CNS in a model of experimental autoimmune encephalomyelitis. *Brain Res.* 952, 128–134.
58. Li W, Maeda Y, Yuan RR, Elkabes S, Cook S, Dowling P. (2004) Beneficial effect of erythropoietin on experimental allergic encephalomyelitis. *Ann. Neurol.* 56, 767–777.
59. Zhang J, Li Y, Cui Y, Chen J, Lu M, Elias SB, et al. (2005) Erythropoietin treatment improves neurological functional recovery in EAE mice. *Brain Res.* 1034, 34–39.
60. Rosenberg GA. (2002) Matrix metalloproteinases and neuroinflammation in multiple sclerosis. *Neuroscientist* 8, 586–595.
61. Boz C, Ozmenoglu M, Velioglu S, Kilinc K, Orem A, Alioglu Z, et al. (2006) Matrix metalloproteinase-9 (MMP-9) and tissue inhibitor of matrix metalloproteinase (TIMP-1) in patients with relapsing-remitting multiple sclerosis treated with interferon beta. *Clin. Neurol. Neurosurg.* 108, 124–128.
62. Yong VW, Zabad RK, Agrawal S, Goncalves Dasilva A, Metz LM. (2007) Elevation of matrix metalloproteinases (MMPs) in multiple sclerosis and impact of immunomodulators. *J. Neurol. Sci.* 259, 79–84.
63. Ehrenreich H, Fischer B, Norra C, Schellenberger F, Stender N, Stiefel M, et al. (2007) Exploring recombinant human erythropoietin in chronic progressive multiple sclerosis. *Brain* 130, 2577–2588.
64. Schluesener HJ, Lider O. (1989) Transforming growth factors beta 1 and beta 2: cytokines with identical immunosuppressive effects and a potential role in the regulation of autoimmune T cell function. *J. Neuroimmunol.* 24, 249–258.
65. Johns LD, Flanders KC, Ranges GE, Sriram S. (1991) Successful treatment of experimental allergic encephalomyelitis with transforming growth factor-beta 1. *J. Immunol.* 147, 1792–1796.
66. McCartney-Francis NL, Frazier-Jessen M, Wahl SM. (1998) TGF-beta: a balancing act. *Int. Rev. Immunol.* 16, 553–580.
67. van Holten J, Reedquist K, Sattonet-Roche P, Smeets TJ, Plater-Zyberk C, Vervoordeldonk MJ. (2004) Treatment with recombinant interferon-beta reduces inflammation and slows cartilage destruction in the collagen-induced arthritis model of rheumatoid arthritis. *Arthritis Res. Ther.* 6, R239–249.
68. Schuppan D, Freitag T. (2004) Fistulising Crohn's disease: MMPs gone awry. *Gut* 53, 622–624.
69. von Lampe B, Barthel B, Coupland SE, Riecken EO, Rosewicz S. (2000) Differential expression of matrix metalloproteinases and their tissue inhibitors in colon mucosa of patients with inflammatory bowel disease. *Gut* 47, 63–73.
70. Shah PK, Galis ZS. (2001) Matrix metalloproteinase hypothesis of plaque rupture: players keep piling up but questions remain. *Circulation* 104, 1878–1880.
71. Pinderski Oslund LJ, Hedrick CC, Olvera T, Hagenbaugh A, Territo M, Berliner JA, et al. (1999) Interleukin-10 blocks atherosclerotic events in vitro and *in vivo*. *Arterioscler. Thromb. Vasc. Biol.* 19, 2847–2853.
72. Smith DA, Irving SD, Sheldon J, Cole D, Kaski JC. (2001) Serum levels of the antiinflammatory cytokine interleukin-10 are decreased in patients with unstable angina. *Circulation* 104, 746–749.
73. Rennick DM, Fort MM. (2000) Lessons from genetically engineered animal models. XII. IL-10-deficient (IL-10(-/-) mice and intestinal inflammation. *Am. J. Physiol. Gastrointest. Liver Physiol.* 278, G829–833.
74. Li MC, He SH. (2004) IL-10 and its related cytokines for treatment of inflammatory bowel disease. *World J. Gastroenterol.* 10, 620–625.
75. Tilg H, Ulmer H, Kaser A, Weiss G. (2002) Role of IL-10 for induction of anemia during inflammation. *J. Immunol.* 169, 2204–2209.
76. Tilg H, van Montfrans C, van den Ende A, Kaser A, van Deventer SJ, Schreiber S, et al. (2002) Treatment of Crohns disease with recombinant human interleukin 10 induces the proinflammatory cytokine interferon gamma. *Gut* 50, 191–195.
77. Paul WE. (1991) Interleukin-4: a prototypic immunoregulatory lymphokine. *Blood* 77, 1859–1870.
78. George J, Mulkins M, Shaish A, Casey S, Schatzman R, Sigal E, et al. (2000) Interleukin (IL)-4 deficiency does not influence fatty streak formation in C57BL/6 mice. *Atherosclerosis* 153, 403–411.
79. Huber SA, Sakkinen P, David C, Newell MK, Tracy RP. (2001) T helper-cell phenotype regulates atherosclerosis in mice under conditions of mild hypercholesterolemia. *Circulation* 103, 2610–2616.
80. Khew-Goodall Y, Wadham C, Stein BN, Gamble JR, Vadas MA. (1999) Stat6 activation is essential for interleukin-4 induction of P-selectin transcription in human umbilical vein endothelial cells. *Arterioscler. Thromb. Vasc. Biol.* 19, 1421–1429.
81. Cornicelli JA, Butteiger D, Rateri DL, Welch K, Daugherty A. (2000) Interleukin-4 augments acetylated LDL-induced cholesterol esterification in macrophages. *J. Lipid Res.* 41, 376–383.
82. Vessillier S, Adams G, Montero-Melendez T, Jones R, Seed M, Perretti M, Chernajovsky Y. (2012) Molecular engineering of short half-life small peptides (VIP,

alpha-MSH and gamma3-MSH) fused to latency-associated peptides results in improved anti-inflammatory therapeutics. Ann Rheum Dis 17, 143–149.

83. Fingleton B. (2006) Matrix metalloproteinases: roles in cancer and metastasis. *Front Biosci.* 11, 479–491.
84. Ii M, Yamamoto H, Adachi Y, Maruyama Y, Shinomura Y. (2006) Role of matrix metalloproteinase-7 (matrilysin) in human cancer invasion, apoptosis, growth, and angiogenesis. *Exp. Biol. Med. (Maywood)* 231, 20–27.
85. Amalinei C, Caruntu ID, Giusca SE, Balan RA. (2010) Matrix metalloproteinases involvement in pathologic conditions. *Rom. J. Morphol. Embryol.* 51, 215–228.
86. Piper PJ, Said SI, Vane JR. (1970) Effects on smooth muscle preparations of unidentified vasoactive peptides from intestine and lung. *Nature* 225, 1144–1146.
87. Said SI, Mutt V. (1970) Polypeptide with broad biological activity: isolation from small intestine. *Science* 169, 1217–1218.
88. Ganea D, Delgado M. (2002) Vasoactive intestinal peptide (VIP) and pituitary adenylate cyclase-activating polypeptide (PACAP) as modulators of both innate and adaptive immunity. *Crit. Rev. Oral Biol. Med.* 13, 229–237.
89. Henning RJ, Sawmiller DR. (2001) Vasoactive intestinal peptide: cardiovascular effects. *Cardiovasc. Res.* 49, 27–37.
90. Lipton JM, Ceriani G, Macaluso A, McCoy D, Carnes K, Biltz J, et al. (1994) Antiinflammatory effects of the neuropeptide alpha-MSH in acute, chronic, and systemic inflammation. *Ann. N. Y. Acad. Sci.* 741, 137–148.
91. Stack MS, Gray RD. (1989) Comparison of vertebrate collagenase and gelatinase using a new fluorogenic substrate peptide. *J. Biol. Chem.* 264, 4277–4281.
92. Ye QZ, Johnson LL, Yu AE, Hupe D. (1995) Reconstructed 19 kDa catalytic domain of gelatinase A is an active proteinase. *Biochemistry* 34, 4702–4708.
93. Peng KW, Morling FJ, Cosset FL, Murphy G, Russell SJ. (1997) A gene delivery system activatable by disease-associated matrix metalloproteinases. *Hum. Gene. Ther.* 8, 729–738.
94. Vessillier S, Adams G, Chernajovsky Y. (2004) Latent cytokines: development of novel cleavage sites and kinetic analysis of their differential sensitivity to MMP-1 and MMP-3. *Protein Eng. Des. Sel.* 17, 829–835.
95. Tortorella MD, Burn TC, Pratta MA, Abbaszade I, Hollis JM, Liu R, et al. (1999) Purification and cloning of aggrecanase-1: a member of the ADAMTS family of proteins. *Science* 284, 1664–1666.
96. Sandy JD, Verscharen C. (2001) Analysis of aggrecan in human knee cartilage and synovial fluid indicates that aggrecanase (ADAMTS) activity is responsible for the catabolic turnover and loss of whole aggrecan whereas other protease activity is required for C-terminal processing *in vivo*. *Biochem. J.* 358, 615–626.
97. Fosang AJ, Rogerson FM, East CJ, Stanton H. (2008) ADAMTS-5: the story so far. *Eur. Cell Mater.* 15, 11–26.
98. Sandy JD. (2006) A contentious issue finds some clarity: on the independent and complementary roles of aggrecanase activity and MMP activity in human joint aggrecanolysis. *Osteoarthritis Cartilage* 14, 95–100.
99. Hills R, Mazzarella R, Fok K, Liu M, Nemirovskiy O, Leone J, et al. (2007) Identification of an ADAMTS-4 cleavage motif using phage display leads to the development of fluorogenic peptide substrates and reveals matrilin-3 as a novel substrate. *J. Biol. Chem.* 282, 11101–11109.
100. Gentry LE, Nash BW. (1990) The pro domain of pre-pro-transforming growth factor beta 1 when independently expressed is a functional binding protein for the mature growth factor. *Biochemistry* 29, 6851–6857.
101. Bottinger EP, Factor VM, Tsang ML, Weatherbee JA, Kopp JB, Qian SW, et al. (1996) The recombinant proregion of transforming growth factor beta1 (latency-associated peptide) inhibits active transforming growth factor beta1 in transgenic mice. *Proc. Natl. Acad. Sci. U S A* 93, 5877–5882.
102. Vodovotz Y, Chesler L, Chong H, Kim SJ, Simpson JT, DeGraff W, et al. (1999) Regulation of transforming growth factor beta1 by nitric oxide. *Cancer Res.* 59, 2142–2149.
103. McMahon GA, Dignam JD, Gentry LE. (1996) Structural characterization of the latent complex between transforming growth factor beta 1 and beta 1-latency-associated peptide. *Biochem. J.* 313, 343–351.
104. Ali NA, Gaughan AA, Orosz CG, Baran CP, McMaken S, Wang Y, et al. (2008) Latency associated peptide has in vitro and *in vivo* immune effects independent of TGF-beta1. *PLoS One* 3, e1914.
105. Holle L, Song W, Holle E, Wei Y, Li J, Wagner TE, et al. (2009) In vitro- and *in vivo*-targeted tumor lysis by an MMP2 cleavable melittin-LAP fusion protein. *Int. J. Oncol.* 35, 829–835.
106. Holle L, Song W, Holle E, Wei Y, Wagner T, Yu X. (2003) A matrix metalloproteinase 2 cleavable melittin/avidin conjugate specifically targets tumor cells in vitro and *in vivo*. *Int. J. Oncol.* 22, 93–98.
107. Leuschner C, Hansel W. (2004) Membrane disrupting lytic peptides for cancer treatments. *Curr. Pharm. Des.* 10, 2299–2310.

IIB TOXIN: CYTOTOXIC FUSION PROTEINS

17

FUSION PROTEINS WITH TOXIC ACTIVITY

STEFAN R. SCHMIDT

Rentschler Biotechnologie GmbH, Laupheim, Germany

17.1 INTRODUCTION

Targeted therapy was first defined in the second half of the nineteenth century when Paul Ehrlich described his concept of a "magic bullet" that selectively targets a disease-causing organism and delivers a toxin to eliminate the pathogen. This paradigm is still an ideal many scientists strive after, particularly to treat cancer without affecting normal tissue [1]. Initially, the targeting moiety and the toxin were chemically conjugated, but this was causing manufacturing problems. Other difficulties were the often-unacceptable high off-target effects or immunogenicity due to toxins of nonhuman origin. The existence of only a single approved immunotoxin, Ontak® just underlines the many hurdles that must be overcome. Since its market introduction in 1999 no other toxin-based fusion protein was accepted by regulatory authorities, despite the many clinical trials that have been ongoing during the last 20 years. Currently, selective administration, for instance, into bladder by catheters, direct injection into brain, or application under positive pressure, is tried to minimize systemic toxicity. In addition, genetic engineering of immunotoxins addresses these issues.

A number of mechanisms are available to destroy cells in a targeted therapy. It can be distinguished between molecules that have to be internalized and those that work directly at the cell surface or through interactions with other cells or extracellular molecules (Figure 17.1). Functional screening with an antibody fragment library can be applied to identify internalizing surface antigens [2].

Two parameters are important for the binding molecule: the affinity and the ability to cross-link, since cross-linking triggers receptor-mediated internalization. Therefore, scFv might not be the best choice if internalization of a monomeric toxin is required. Often a high affinity is not desirable because a binding site barrier effect may arise. This means high affinity ligands bind extremely tight to the first binding sites they reach, thus creating a zone with blocked access that limits further penetration. Affinities above 10^{-7} M are required for sufficient tumor uptake. Saturation can be observed around 10^{-9} M, but at 10^{-11} M, a model single-chain (scFv) antibody traversed only 2–3 cell diameters [3].

Overall, the ideal cytotoxic fusion protein has to be optimized for half-life and tissue penetration ability, parameters that can often be addressed through fine-tuning the hydrodynamic radius of the molecule (see Chapter 6). Usually, a relatively short half-life is desirable to minimize off-target effects that can increase by prolonged exposure. In general, three strategies have been employed: improving affinity, decreasing size, or increasing doses. Since most of the natural toxins have intracellular activity, an efficient passage from the surface to the cytosol must be established as well. Frequently, endosomal escape strategies are required [4].

Besides the direct elimination of malignant cells through toxic molecules, a similar effect can be achieved indirectly through the attraction and stimulation of immune cells. This requires refined strategies since many tumor mechanisms evolved to escape the immune response of the host. Much

Fusion Protein Technologies for Biopharmaceuticals: Applications and Challenges, First Edition. Edited by Stefan R. Schmidt.

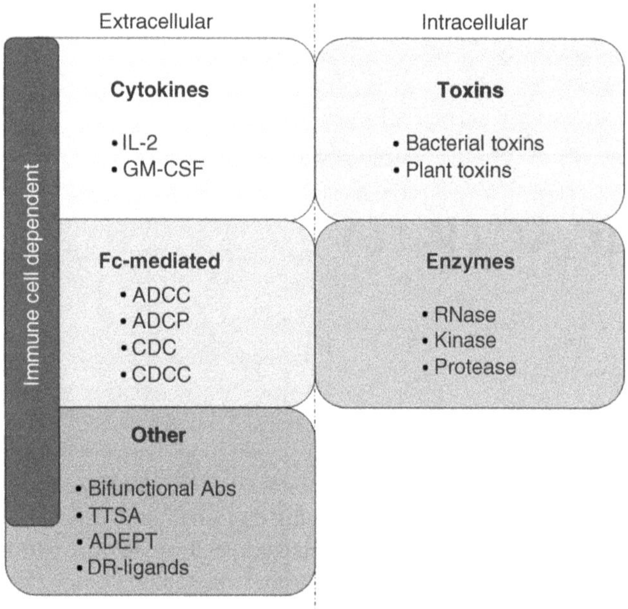

FIGURE 17.1 Toxic activities of fusion proteins. Fusion proteins that do not require internalization frequently involve the activity of immune cells. They either attract them by cytokines or engage them through Fc receptors. In the case of bifunctional antibodies, instead of an Fc-receptor leucocyte specific surface molecules such as CD3 are addressed (see part IIIb of this book). Fc receptors can cause antibody-derived cellular cytotoxicity (ADCC), antibody-dependent cell-mediated phagocytosis (ADCP), complement-dependent toxcitiy (CDC), and complement-dependent cell-mediated cytotoxicity (CDCC). Another case involving immune cells is tumor-targeted superantigens (TTSA). Special applications such as ADEPT (antibody-directed enzyme prodrug therapy) generate extracellularly toxic molecules that can enter the cell. Death receptor (DR) ligands trigger apoptosis. The two other classes of toxic fusion proteins exert their activities only within the cell. They contain either toxins from bacteria or plants, or human enzymes that destroy cellular functions.

progress has been achieved in engaging cytotoxic T-cells with bifunctional antibodies. Part IIIb of this book is dedicated to elucidate the concept of multifunctional antibodies.

Ideally, the stimulation of the immune system is restricted to the tumor site to avoid any unwanted autoimmune effects when distributed systemically. However, on the other hand, immunotherapy is most efficient if a long lasting memory effect can be obtained [5].

Owing to the immunogenic limitations of natural toxins, immunotoxins based on human proteins have been established. This group consists of enzymes that interfere with cellular signaling and protein translation [6], cause proteolytic degradation, or trigger apoptosis [7].

The following paragraphs introduce the different concepts of targeted cell killing therapy based on fusion proteins. They describe the different designs and improvements researchers have employed to specifically eliminate malignant cells while leaving normal cells unaffected [8].

17.2 TOXINS

A general mechanism of microbial or plant toxins is their ability to escape lysosomal degradation. Besides the many variables such as antigen expression, trafficking kinetics, extracellular immunotoxin concentration, and exposure time, the real rate-limiting step determining potency seems to be the endosomal escape [9]. Most of the naturally occurring toxins contain a binding domain that enables attachment to cell surface molecules. This binding domain can easily be replaced by another domain that binds specifically to markers on malignant cells. Nowadays, a multitude of these extracellular markers are known.

On the basis of their mechanism of action, toxins can be classified into three categories: (a) disruption of membrane integrity, (b) interruption of electrical signaling, and (c) interference with enzymatic processes. Interestingly, so far only members of the last group (c) have been used for therapeutic approaches in the form of fusion proteins. Since the early 1970s with chemically conjugated complex toxins, much progress has been achieved in the last 40 years. With the rise of genetic engineering, toxin variants could be generated, which solved many problems of first generation targeted toxins, such as heterogeneity and low production yields, poor tumor penetration due to size limitations, and unspecific toxicity. In general, immunotoxins have been primarily used to treat hematological cancers because of better accessibility of the malignant cells and the immunocompromised status of the patients [10]. Table 17.1 summarizes the receptors and activities of natural toxins. An overview on the current status of immunotoxins is presented in Chapter 18.

17.2.1 Microbial Toxins

Bacterial toxins typically consist of several domains and are able to separate the binding domain form the catalytic domain. This is their key advantage when compared with plant toxins. Most research has been done with diphtheria toxin (DT) and Pseudomonas exotoxin A (PE) [11]. Their mode of action is illustrated in Figure 17.2. Bacterial toxins are usually highly immunogenic; therefore, mutagenesis or truncation was performed to optimize the protein sequence [12].

(DT) is the prototype for ADP-ribosylating toxins. It is secreted by the Gram-positive, aerobic *Corynebacterium diphtheria*. The full-length toxin comprises three domains and stretches over 535 amino acid residues. The toxic activity is contained in the N-terminal fragment (A). The C-terminal domain is divided into two parts, the translocation domain (T) and the receptor-binding domain (B) at the carboxyl end. Upon binding to its native receptor

TABLE 17.1 Natural Receptors and Enzymatic Activity of Toxins

Toxin	Receptor	Enzymatic Activity	Cellular Target
Diphtheria toxin	Heparin binding epidermal growth factor precursor	ADP-ribosyl transferase	Elongation factor 2
Pseudomonas exotoxin A	Low-density lipoprotein receptor-related protein 1	ADP-ribosyl transferase	Elongation factor 2
Antrax protective antigen	Integrin-like cell surface receptor (TEM8 and CMG2)	None	None
Antrax endema factor	None	Adenlylate cyclase	cAMP modulated proteins
Antrax lethal factor	None	Zinc endoprotease	MAPKK
Shiga toxin	Glycosphingolipid globotriaosylceramide (Gb3)	*N*-glycosylase	28S rRNA
Botulinum neurotoxin	Synaptotagmin (Syt) and synaptic vessel protein (SV2)	Zinc endoprotease	SNARE proteins
Ricin	Galactose or *N*-acetylgalactosamine residues	*N*-glycosylase	28S rRNA
Gelonin, bouganin	None	*N*-glycosylase	28S rRNA

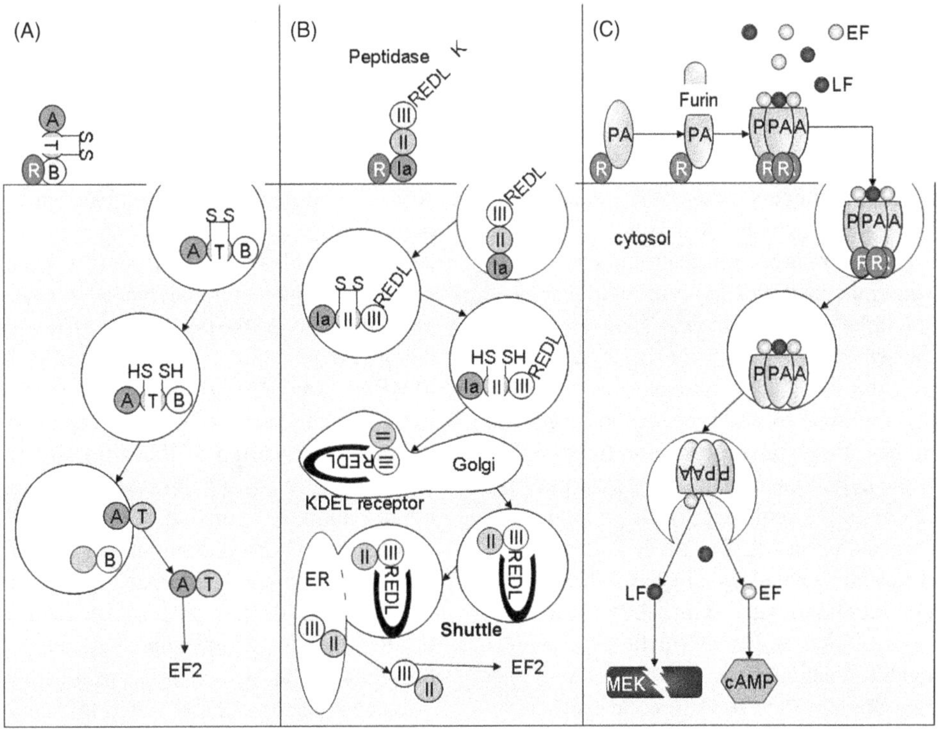

FIGURE 17.2 Mechanism of some frequently used bacterial toxins. (A) After binding to its receptor (R) and cleavage, Diphtheria toxin is internalized. The reduction of the remaining disulfide bond separates the binding domain (B) from the catalytic (A) and translocation (T) domain. A pH induced conformational change enables the transfer of A and T to the cytosol, where A blocks the elongation factor 2 (EF2). (B) The C-terminal lysine (K) of Pseudomonas exotoxin is cleaved off by a serum peptidase before internalization through coated pits. Reduction of its disulfide bridge and furin cleavage separates binding domain (Ia) from the catalytic domain (III) that is still bound to the translocation domain (II). In the Golgi apparatus, the exposed REDL sequence is captured by the KDEL receptor and the toxin is shuttled to the endoplasmic reticulum (ER). There the translocation domain mediates the escape to the cytosol where it inactivates EF2. (C) The PA domain of the anthrax toxin binds to its receptor and is truncated by furin. PA assembles into homo-heptamer that associates with three molecules of LF and/or EF. The lowering of the pH in the endosome leads to a conformational change of the pore-like PA structure into a real pore through which LF and/or EF translocate to the cytoplasm, where LF degrades MEK kinases while EF increases cAMP concentration.

on the cell surface, the heparin binding epidermal growth factor precursor, it is cleaved by a furin-like protease localized on the surface and internalized through clathrin-coated pits. After cleavage, the two fragments are still joined through a single disulfide bridge. The lowering of the pH in the endosome induces a conformational change of the translocation domain, exposing hydrophobic residues that enable membrane insertion and finally escape from the endosome. In the cytosol, the catalytic domain transfers adenosine diphosphate-ribose (ADP-ribose) to the eukaryotic translation elongation factors 2 (EF2), thus inhibiting protein synthesis and ultimately causing cell death. Since the toxin has catalytic activity, a single molecule is sufficient to kill a cell [13]. Various truncations were evaluated to identify the optimal construct with regard to immunogenicity, unspecific binding, and potency.

Ontak, also called denileukin diftitox, the second ever approved fusion protein, is a recombinant immunotoxin consisting of diphtheria toxin whose binding domain is replaced by interleukin 2 (IL-2). The molecule binds to cells expressing the high affinity IL-2 receptor that can be frequently found on T-cell lymphoma [14]. After the success with Ontak, several other DT-containing immunotoxins were evaluated in preclinical and clinical studies, but today no study is active.

The exotoxin A of the Gram-negative, aerobic *Pseudomonas aeruginosa* bacterium contains 613 amino acid residues and is composed of three domains. In principle, the same components are present as in DT, however at different positions. This time, the recognition domain is located at the N-terminus (Ia), followed by the translocation domain (II). At the C-terminus, the catalytic domain (III) can be found. Removal of the carboxy-terminal lysine by a plasma carboxypeptidase exposes the sequence REDL, which will be later used for intracellular sorting. The toxin binds to low-density lipoprotein receptor-related protein 1 (CD91) and is internalized through clathrin-coated pits. After low pH induced receptor dissociation and conformational change, furin cleaves the toxin. Reduction of the only disulfide bridge separates the binding domain from the still connected catalytic and translocation domain that continues with traveling to the trans-Golgi network. Binding of the exposed REDL to the KDEL receptor sorts the fragment to the endoplasmic reticulum. There the translocation domain mediates the escape and brings the catalytic domain to the cytosol where it inactivates EF2 in a similar manner as DT [15].

The majority of immunotoxins is nowadays based on PE. Currently, most advanced in clinical Phase III is cintredekin besudotox. This immunotoxin is a combination of interleukin 13 (IL-13) with a de-immunized PE variant. It targets cells expressing receptors for IL-13 in glioblastoma [16].

A different toxic mechanism is executed by anthrax toxin (AnTx), which is secreted by the Gram-positive, spore-forming bacterium *Bacillus anthrachis*. AnTx is composed of three large separately encoded proteins: protective antigen (PA), lethal factor (LF), and edema factor (EF). After binding of PA to its integrin-like cell surface receptor, a 20-kDa fragment is cleaved off by furin, co-localized at the surface. The still bound large fragment of PA assembles into a pore-like homo-heptamer that associates with three molecules of LF and/or EF. This complex reaches the early endosome after clathrin-mediated endocytosis. The lowering of the pH leads to a conformational change of the pore-like structure into a real pore through which LF and/or EF translocate to the cytoplasm. Both translocated proteins have toxic activity: LF has a protease activity that degrades MEK kinases while EF as adenylate cyclase increases cAMP concentration, ultimately leading to cell death [17].

Anthrax toxin can be used to design novel binary toxins. The first component comprises a modified protective antigen in which the original furin cleavage motif is replaced by another protease motif, for instance that of urokinase plasminogen activator (uPA). This motif was selected because uPA is heavily expressed on the surface of lung cancer cells and mostly absent on normal cells. The second part consists of the ADP-ribosylation domain of Pseudomonas exotoxin A fused to truncated LF to enable translocation through the pore. A higher specificity is obtained by including a cancer cell specific cleavage motif and by separating binding and catalytic component to become active only when interacting properly inside the cell [18]. This concept has been employed in combination to other protease motifs (e.g., MMP2 and MPP9 [19]) or with other bacterial toxins (Shiga toxin or diphtheria toxin [20]) fused to the truncated LF domain. Engineering of the furin cleavage site contributes also to an up to 20-fold improved circulating half-life. So far, binary anthrax toxins have been very potent against a number of tumor types in animal models.

A totally different approach is taken to apply botulinum neurotoxin (BT) from the anaerobic, Gram-positive bacteria of the genus *Clostridium*, specifically of strains of *botulinum*. The toxin consists of a heavy and a light chain. The heavy chain includes two domains: one to bind surface molecules on neurons, another to enable translocation into the cytosol. The light chain contains the catalytic domain, which encodes for an endopeptidase that cleaves soluble *N*-ethylmaleimide-sensitive factor attachment protein receptor (SNARE) proteins, thus inhibiting vesicular secretion and ultimately neurotransmission. Different BT serotypes cleave different SNARE proteins. Heavy and light chains are connected via noncovalent interactions and a covalent disulfide bridge. The mechanism of BT activity starts with binding of the heavy chain to polysialogangliosides at the presynaptic terminal. After internalization and acidification of the endosome, the translocation domain forms a pore through which the light chain escapes into the cytosol. There the endopeptidases specifically cleave SNARE proteins, preventing SNARE complex assembly and exocytosis.

It has to be noted that this approach does not lead to cell killing, but prevention of secretion [21].

Therapeutic variants of BT replace the binding domain by a specific targeting domain, for example, a peptide ligand. Furthermore, the light chain can be directly linked to the catalytic domain by a protease-sensitive spacer to allow release to the cytosol. These engineered proteins have a huge therapeutic potential to treat diseases that involve secretion processes. Currently, there is one fusion protein termed AGN-214868 in Phase II trials [22].

Another bacterial toxin tested for therapeutic potential is Shiga toxin from *Shigella dysenteriae*. It is composed of the catalytic A subunit associated to a homopentamer of the small B subunit. The B subunits bind to a cell surface glycolipid, and the toxin complex enters the cell by endocytosis. In the endosome, the A domain is proteolytically processed into an A1 and A2 fragment that is still linked by a disulfide bridge. After retrograde transport through Golgi and endoplasmic reticulum, the bond is reduced and the catalytically active *N*-glycosidase of A1 is released into the cytosol, where it modifies an adenine of the 28S ribosomal ribonucleic acid (rRNA) of the ribososme which disrupts the interaction with EF2 and protein synthesis [23].

Fusion proteins combining a chemokine as targeting moiety and the A1 chain of Shiga toxin are called leukocyte population modulators since they can deplete certain unwanted leukocyte subtypes. A molecule consisting of human CCL2 (monocyte chemoattractant protein-1) chemokine fused to a truncated form of the enzymatically active A1 domain of *S. dysenteriae* holotoxin (SA1) was prepared for clinical trials of glomerulonephritis [24].

Cells infected by viruses often contain a high concentration of proteases. The presence of these proteases can be utilized to increase the specificity of therapeutic toxins that can eliminate infected cells. This concept was successfully tested by designing the so-called Zymoxins consisting of a targeting, translocation, and a toxin domain that was kept inactive by an inhibitory protein. This inhibitor can only be removed in cells expressing a certain virus protease. This then activates the toxin, which kills the cell. In an *in vitro* model, the viral NS3 protease in hepatitis C virus (HCV) infected cells cleaved the linker between diphtheria toxin and a protective defensin or the spacer connecting ricin catalytic domain with a stalk peptide [25].

A special case utilizing toxins is the generation of tumor-targeted superantigens (TTSA). Antibody linked superantigens target cytotoxic T-cells to tumors. Besides the direct killing, secondary effects are induced by the release of cytokines. This secondary effect is very useful for eliminating cancer cells that do not present the specific antigen. In general, superantigens secreted from bacteria help pathogens evade the immune system. They bind to major histocompatibility complex (MHC) class II molecules on antigen-presenting cells (APC) and particular T-cell receptors activating the T cells in an antigen independent manner, which leads to the elimination of APCs [26].

The first generation of TTSA consisted of staphylococcal enterotoxin A (SEA) fused to a Fab fragment directed against the 5T4 oncofetal antigen that can be found on many solid tumors but not in normal adult tissues. However, despite its reduced binding to APC, this molecule, called anatumomab mafenatox, caused systemic T-cell activation and required individual dosing in Phase II clinical trial. Furthermore, neutralizing antibodies could be found in patients [27]. Therefore, it was replaced by naptumomab estafenatox consisting of an improved Fab against the same target and a mutagenized SEA with a 100-fold reduced systemic toxicity fused to the heavy chain. This molecule has demonstrated a beneficial safety and efficacy profile in the clinic [28]. Further details can be seen in Chapter 24.

17.2.2 Plant Toxins

Plant toxins are derived from the family of ribosome inactivating proteins (RIP). They are smaller than bacterial toxins and can be dissected into three groups. Type I RIP contain only an enzymatic domain of $\sim$30 kDa and cannot enter cells without conjugation to a targeting moiety nor translocate to the cytosol easily. Members of this group are saporin from *Saponaria officinalis*, bouganin from *Bougainvillea spectabilis*, and gelonin from *Gelonium multiforum*. Ricin from *Ricinus communis* belongs to the group of type II RIP. They are heterodimers with a catalytic domain (A chain) and a lectin-like domain (B chain). Both domains have a size of $\sim$30 kDa and are connected by a disulfide bridge. The last group is type III RIP that have no lectin domain and require proteolytic processing to become active [29].

One advantage of class I RIP is their low systemic toxicity, which is a result of their lack of binding to cells. This renders them particularly suitable for coupling to specific targeting molecules. The small size also limits the number of potential antigenic epitopes and minimizes mutagenesis efforts to reduce immunogenicity. The least toxic type I RIP is bouganin of *B. spectabilis*. The numerous evaluations of plant-derived toxins have now yielded a variant, which is in early stage clinical trials. This toxin, called citatuzumab bogatox (VB6-845), is a de-immunized variant of bouganin combined with a Fab fragment against epithelial cell adhesion molecule (EpCAM). Between the targeting Fab and the toxin is a furin-sensitive linker to enable specific uncoupling of bouganin to execute its toxic function. This immunotoxin demonstrated nanomolar potency and high selectivity toward several cancer cell lines [30].

Another member of the type I RIP is saporin (S6) from *S. officinalis*. The toxin has primarily been used as conjugate to antibodies or fragments [31]. Only recently attempts were successful to produce it as fusion protein secreted from *Pichia pastoris* [32].

The lectin domain (B chain) of ricin binds to galactose or *N*-acetylgalactosamine residues that can be found on the surface of many eukaryotic cells. Since ricin is heavily glycosylated, even attachment to mannose receptors on endothelial cells or macrophages might occur; therefore, therapeutic variants are preferably deglycosylated. Endocytosis by clathrin-coated pits internalizes the surface bound ricin. From the early endosome ricin travels in a retrograde transport via the Golgi to the endoplasmic reticulum. There the disulfide bond is reduced by protein disulfide isomerase, and the catalytic domain (A chain) is translocated to the cytoplasm. The *N*-glycosidase activity of the A chain modifies the 28S rRNA. Therefore, the ribosome cannot interact with EF2 anymore, which irreversibly interrupts protein synthesis [33].

The most recent application of ricin is its conjugation to two antibodies targeting CD22 and CD19. This combination therapy, called Combotox, is now in clinical Phase I [34]. Currently, no genetically fused version of ricin is under development. Particularly when using ricin, a severe complication, the vascular leak syndrome, is observed. Endothelial cells are unspecifically affected by high toxin doses and become leaky, releasing fluid into the tissues [35].

Overall plant toxins have been less successful than microbial toxins, probably because of difficulties regarding the association of toxin and ligand during internalization and translocation [6].

17.3 IMMUNOCYTOKINES

The immune stimulatory activities of several cytokines are well understood; therefore, it was logical to use them as fusion partners. In contrast to directly acting toxins, cytokines attract and activate immune cells that then eliminate malignant cells through various mechanisms. Combining the cytokines with a targeting motif also increases the size of the resulting molecule, thus extending the usually short half-life of cytokines. This approach is discussed in Chapter 19. Initially a lot of work was done fusing cytokines to full antibodies. Using the several N- or C-termini that are available for coupling to an antibody, more than a single cytokine molecule can be coupled. Alternatively, more than one type of cytokine can be bound to an antibody thus allowing synergistic activity [36]. Immunocytokines should not be confused with targeted toxins where cytokines as fusion partner take the role of targeting the toxin to cells overexpressing the respective cytokine receptor [37].

17.3.1 Interleukin 2 Fusion Proteins

The activity of interleukin 2 (IL-2) is transferred through its binding to the respective receptor (IL-2R) and initiates intracellular signaling. The high affinity heterotrimeric receptor consists of three subunits: α (CD25), β (CD122), and γ (CD132). However, an intermediate affinity variant assembled only from βγ subunits triggers a different pathway. When designing an antibody-based immunocytokine directed against the necrotic core of tumors, the IL-2 was mutagenized to exhibit lower affinity to the βγ receptor to reduce unwanted toxicity. This fusion protein had full antitumor activity against neuroblastoma and nonsmall cell lung cancer metastases in mice, but was better tolerated [38]. HER2 positive tumors can be addressed with the dimer of an anti-HER2 scFv linked to an Fc domain that was connected to an IL-2 molecule. Owing to the presence of a functional Fc part, this fusion protein is capable to induce antibody-derived cellular cytotoxicity (ADCC) and IL-2-dependent immune stimulation, leading to high antitumor activity [39]. In principle, IL-2 can be linked to any known antibody that targets a tumor antigen, thus increasing the potency of ADCC by additionally activating natural killer (NK) cells. Furthermore, a fusion protein is more efficacious than administering IL-2 and the antibody, since now a colocalization of the antibody and the cytokine is achieved, increasing the local concentration of the stimulating agent. This has been demonstrated with the anti-GD2 disialoganglioside antibody hu14.18, fused to IL-2 [40].

17.3.2 GM-CSF Fusion Proteins

The known properties of granulocyte-macrophage colony-stimulating factor (GM-CSF) to enhance antigen presentation and co-stimulation lead to the concept of dual cytokines. Fusion of GM-CSF to IL-2 resulted in highly active NK cells that released more interferon-γ (IFN-γ) than expected. But this molecule, called GIFT2, did not enhance proliferation of immune cells. Replacing IL-2 by IL-21 to generate GIFT21 differentiated dendritic cells and inhibited tumor growth *in vivo*. A third tumor killing variant contains CCL2 as fusion partner to GM-CSF. This chimera induced the upregulation of Bax and ultimately apoptosis of CCR2 positive cells that contribute to inflammation in an arthritis model. The utilization of GM-CSF fused to different cytokines and the sometimes-unexpected results show one dilemma of fusion proteins. It is difficult to predict effects or synergy when combining molecules of the same type. It is therefore more advisable to combine proteins for targeting and toxicity [41].

17.3.3 Other Leukocyte Attracting Cytokines and Chemokines

Another important target that has been evaluated with a wide range of different immunocytokines is single stranded DNA, which becomes exposed in degenerating cells. For this purpose, the so-called tumor necrosis therapy (TNT) monoclonal antibody was generated and linked to IL-2, IFN-γ,

GM-CSF, or tumor necrosis factor α (TNF-α) at the carboxy terminal end of the heavy chain. Surprisingly, all variants had a much shorter half-life than the parental antibody, which is useful since this reduces unwanted effects through rapid clearance, but retains the fusion protein in the tumor because of the high specificity of the antibody. All protein variants achieved only transient tumor regression because no memory cells were generated. A promising candidate to design immunocytokines is the human chemokine LEC (liver-expression chemokine) that strongly attracts a number of different lymphocytes and monozytes. Chemokines require a free amino terminus; therefore, fusions to antibodies have to be arranged at the variable domain. Similarly, to other immunocytokines, fusions with LEC result in a high tumor to organ ratio, minimizing unwanted side effects. An alternative to the classical cytokines to attract and stimulate immune cells is the utilization of co-stimulatory molecules such as B7 to induce T-cell proliferation. B7 and LEC fusion proteins were able to achieve complete regression in a number of tumor models.

The next class of molecules suitable to induce indirect killing of malignant cells through immune cells are members of tumor necrosis factor superfamily (TNFSF) such as OX40L, CD137L, and GITRL. These proteins can be found on the surface of many activated immune cells and trigger a sustained T-cell response after initial activation, thus creating an immunological memory. Tumor-targeted fusion proteins containing these ligands are highly efficient to eradicate tumors and establish long-term protection [5].

Overall, three effects of immunocytokines have to be orchestrated to obtain a long lasting cure: first, tumor infiltration of immune cells must be achieved by chemokine attraction and antibody-mediated targeting; second, immune cells must be activated, ideally in a co-stimulatory way to obtain a memory effect; third, tumor-mediated immune tolerance must be abolished. All these can be achieved by a careful design of immune modulating fusion proteins [42]. Further details on targeted immunotherapy can be found in Chapter 19.

17.4 HUMAN ENZYMES

Enzymes as proteins with a wide range of functions can also be utilized to kill aberrant cells. As before, it can be distinguished between internal and external activity. Internal targets for enzymes in principle are all functions that are either essential for survival or trigger cell death through apoptosis signals. Many of the toxins address the protein generating process, blocking the ribosome or transcription factors. The essential element for the translation is ribonucleic acid (RNA). The sensitive nature of this molecule makes it an ideal target for enzymatic damage through RNA degrading enzymes, the RNAses. Other intracellular targets are the multiple signal transduction cascades that are regulated through phosphorylation. Aberrant phosphorylation often causes uncontrolled proliferation. The obvious countermeasure to trigger self-destruction or to halt proliferation would be to introduce kinases. Enzymes not requiring cellular uptake for their toxic activity have been elegantly designed as antibody-dependent enzyme prodrug therapy (ADEPT). Novel therapeutic molecules to fight cancer are the various antibody-enzyme fusion proteins [43].

17.4.1 Kinases

The utilization of kinases to force tumors into cell death is a relatively new strategy. In order to achieve specificity and receptor-mediated uptake, the kinase has to be fused to a targeting molecule that identifies a marker or surface antigen that can be found only on malignant cells. The kinase of choice should restore cellular protection mechanisms such as tumor suppressors. A good example is the inactivation of the tumor suppressor death associated protein kinase (DAPK2) in Hodgkin lymphoma. DAPK2 itself is a kinase. Introducing a constitutively active version of DAPK2 fused to a ligand of a surface molecule such as CD30 induces apoptosis [44]. Since both molecules are of human origin, no immunogenicity should be expected. A further advantage is the lack of a toxic effect as long as DAPK2 is not introduced into a cancer cell. The kinase itself is not harmful when present in the bloodstream. This example is illustrated in more detail in Chapter 21.

17.4.2 RNAses

The process to transcribe a gene and translate it into a protein is a vital step for all organisms. The central molecule linking transcription with translation is the ribonucleic acid (RNA) in form of the messenger RNA (mRNA) or as amino acid carrying transfer RNA (tRNA). Interrupting translation or transcription ultimately kills cells. Therefore, RNA was selected as an important cytotoxic target. RNA is a labile molecule that can be destroyed enzymatically by RNAses. One complication limiting the therapeutic access could be the presence of RNAse inhibitor (RI) in the cytosol. However, it was demonstrated that an RNAse fusion protein neutralizes RI and causes apoptosis [45]. Alternatively, RNAses of nonhuman origin being insensitive to RI could be utilized. Most frequently, human pancreatic RNAse, angiogenin (ANG) or amphibian RNAses were used. All RNAse fusion proteins are notoriously difficult to produce, forming inclusion bodies or killing the production host. Here, the same production strategies can be applied as mentioned in Chapter 1. Of course, the enzymes can be linked to any targeting molecule, either in the form of antibody fragments or the cancer receptor specific ligands such as growth factors for instance. However, RNAses typically require both a free N- or C-terminus for activity and a flexible linker connecting the enzyme to the targeting domain. This has to be taken into account when designing the fusion protein [46]. Potency could be enhanced 5–40-fold by combining two RNAse molecules with a single antibody moiety [47]. The inefficient production and, sometimes, low potency could delay the introduction of immunoRNAses [48]. A complete summary of the current state of the art of ImmunoRNAses is given in Chapter 22.

17.4.3 Proteases

Other mechanisms to disturb cellular functions rely on proteolytic activities. Some proteases are key enzymes that mediate apoptosis, such as the members of the caspase family. In particular, a constitutively active caspase 3 has been fused to various targeting moieties such as a single-chain anti-HER2 antibody for instance. To enable the escape from the endosome, this molecule is also equipped with a translocation domain based on Pseudomonas exotoxin A. Jurkat cells expressing and secreting this construct remained healthy, whereas co-cultivated HER2-positive cells such as SKBR-3 or SKOV-3 were selectively killed [49]. Another construct containing caspase 3 was directed against immune cells expressing the interleukin-2 receptor (IL-2R). The IL-2-caspase 3 fusion protein eliminated regulatory T-cells responsible for autoimmunity in a diabetes mouse model [50]. A second example was the successful treatment of experimental inflammatory colitis in mouse with the same construct [51].

The lamin A cleaving, apoptotic function of caspase 6 was utilized to treat HER2 positive osteosarcoma lung metastasis in mice. As before, domain II of Pseudomonas exotoxin A was included again, and the therapy was performed by genetic transduction [52]. The DNA construct coding for constitutively active caspase 6 fused to a single-chain anti-HER2 antibody was injected into BALB/c athymic mice bearing human breast SK-BR-3 tumors. The survival of all treated animals was significantly prolonged [53].

Another gene therapeutic example for targeted cell killing is the granzyme B (GrB) fusion protein directed against HER2. This protease can initiate apoptosis through multiple mechanisms. As described before, the same single-chain antibody and translocation domain was used. After transfection, the expression of this DNA construct selectively destroyed HER2-positive tumor cells in nude mice. Continuous secretion of the fusion protein is responsible for a substantially prolonged survival of these animals [54]. GrB serves also as the cell-killing moiety for a single-chain anti-melanoma antibody. Melanoma cells were selectively killed through apoptosis when exposed to the fusion protein [55]. Owing to the low immunogenicity of human enzymes, Gr B and other proteases have now been combined with many cell-targeting motifs to form a novel class of human immunotoxins [56].

17.4.4 Antibody-Directed Enzyme Prodrug Therapy

For decades, small molecules have been used in chemotherapy. However, these highly potent drugs cause massive, sometimes dose limiting, side effects and tumor cells develop resistance. Therefore, strategies are desired that limit the distribution of the drug but not the toxicity. Specificity both for the drug and the targeted cell is needed. This goal can be fulfilled by combining the specific activity of an enzyme to activate a nontoxic small molecule with the high target specificity of an antibody. So, a new approach was born [57], and called ADEPT [58]. This approach consists of two elements: an antibody-enzyme fusion and a prodrug. ADEPT uncouples binding from killing to increase specificity. Both the large fusion protein and the small prodrug alone are nontoxic, but they generate cytotoxicity when co-localized. Nowadays no longer full antibodies are used; a single scFv or a target specific ligand is sufficient to enable binding to a surface receptor. Monovalency is even preferred, since cross-linking of receptors by a bivalent binder could lead to internalization, thus removing the required enzymatic activity from the surface.

The whole procedure comprises three steps. First, the antibody-coupled enzyme is injected. After completion of administration, unbound enzyme fusion proteins have to be removed from the organism to focus enzymatic activity only on targeted cells. This can be achieved either with small constructs with a short half-life or through phagocytosis if the fusion protein is mannosylated [59]. The clearance should be fast to have almost maximum concentration of the tumor bound enzyme. After removal of unbound enzyme finally the prodrug is added. This prodrug is metabolized by the enzyme, thus generating a high local concentration of the active drug molecule in proximity of malignant cells. ADEPT is the only internalization independent enzyme strategy [60].

As additional level of safety, preferably nonhuman enzymes such as cytosine deaminase and β-lactamase are used to avoid unspecific drug activation by endogenous enzymatic activity. But these foreign proteins can sometimes cause an immune response. In the case of β-lactamase, mutagenesis reduced immunogenicity fivefold [61]. If enzymes of human origin such as β-glucuronidase are utilized, a prodrug formulation must be selected that cannot reach the intracellular enzyme. Alternatively, the substrate specificity of a human enzyme can be modified by mutagenesis to introduce prodrugs that cannot be metabolized by the native enzyme. Besides the short half-life of unbound enzyme fusions, the generated drug should also have a short circulation time to avoid diffusion to healthy tissue. The diffusion radius defines the bystander effect that kills even tumor cells not expressing the targeted antigen, as long as the cells are in proximity to the released activated drug [62]. In some cases, even catalytic antibodies have been used for prodrug therapies. Taking into account this recent progress, bispecific antibodies combining targeting with enzymatic activity could become feasible [63]. A more detailed overview of ADEPT is given in Chapter 23.

17.5 APOPTOSIS INDUCTION

Another possibility to destroy malignant cells is to trigger apoptosis through death receptors (DR) that belong to the TNF receptor superfamily. The receptors contain a cytoplasmic death domain (DD) and can transduce an extracellular signal that results from binding of a suitable ligand into intracellular protease-mediated cell death. Homotrimeric ligands such as the tumor-necrosis factor-related apoptosis-inducing ligand (TRAIL) induce receptor oligomerization and the activation of a proteolytic cascade, which involves maturation of caspases. The adapter protein FADD (Fas-associated protein with death domain) generates a death-inducing signaling complex (DISC) by recruiting and autocatalytically activating caspase 8 and 10 that in turn activate a series of other caspases that ultimately perform cell killing. Owing to the involvement of extracellular ligands, this pathway is called extrinsic pathway. Addressing this pathway through death receptor ligands is an attractive therapeutic approach since here apoptosis is induced independently of the tumor suppressor p53, which is frequently nonfunctional in human cancer cells [64]. Obstacles that might prevent the utilization of the soluble TNF such as ligands are their relative short half-life, their tendency to form high molecular aggregates, and their systemic toxicity. However, most of the problems can be addressed by an intelligent design of fusion proteins. A schematic overview on apoptosis pathways is presented in Figure 17.3.

17.5.1 TNF Fusion Proteins

Directly targeting death receptors with their respective ligands seems to be a promising solution for targeted therapy. As a first candidate, TNF itself was selected. The primary function of TNF is not necessarily direct killing of cancer cells but rather the destruction of vasculature.

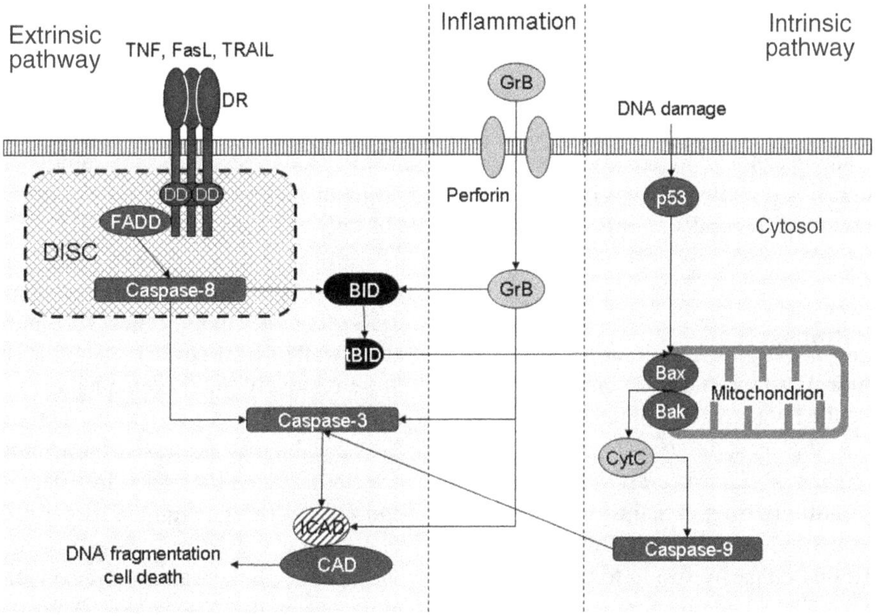

FIGURE 17.3 Apoptosis induction. Binding of death receptor (DR) ligands such as tumor necrosis factor (TNF), TRAIL, or FasL, triggers the formation of a death-inducing signaling complex (DISC) at their death domains (DD) that interact with Fas-associated death-domain (FADD) and activates caspase-8. This protease truncates BID to form tBID, which connects the extrinsic with the intrinsic pathway. tBID or the p53-induced oligomerization of Bax and Bak cause cytochrome C (CytC) release which triggers ultimately the activation of caspase-9. This protease links the intrinsic with the extrinsic pathway by activating caspase-3 that degrades ICAD to release CAD which in turn causes cell death by DNA fragmentation. Several important molecules such as BID, caspase-3, and ICAD can also be activated by granzyme B (GrB), which can enter the cell during inflammation through perforin pores.

Despite the impressive preclinical results, in the clinic systemic administration caused severe side effects. Initially, TNF alone was used. Since the noncovalent trimer showed low stability, trimerization through tandem repeats in a single-chain molecule was chosen. This molecule had also a better antitumoral activity [65].

First generation TNF fusion proteins received a targeting domain to specifically address only selected cells exposing a typical antigen and to increase the local concentration while keeping the systemic concentration low. A number of classical TNF fusion proteins containing only a targeting domain such as a scFv or a peptide are currently in development [66].

Consequently, second generation TNF fusion proteins were designed, which included a targeting domain and a second domain that could shield TNF. This shielding or neutralizing domain is equipped with a protease-sensitive linker that can be cleaved by tumor proteases such as MMP-2 or uPA, thus releasing the active TNF. As TNF-neutralizing agent a fragment of the extracellular domain of the TNF receptor 1 was selected. In order to maintain the natural trimeric conformation of TNF, the trimerization domain of tenascin-C (TNC) was included as well. Upon administration of this complex multidomain protein, elimination of target antigen negative cells in close proximity could be also observed which might be beneficial for heterogeneous tumors. Targeted TNF fusion proteins create a similar conformation as the original membrane bound TNF, which is more potent since it not only activates TNF receptor 1 but also the more important receptor 2 as well. Overall, this novel class of prodrug fusion proteins contains two levels of safety, first the targeting and second the target-dependent activation [67].

17.5.2 FasL Fusion Proteins

Another death receptor ligand is CD95L binding to its receptor, CD95, also called apoptosis stimulating fragment (Fas). This receptor is the most intensely studied death receptor. Initially its ligand, FasL, was evaluated and compared with a series of multimerizing domains. It was found that soluble FasL does not induce cell death. But hexameric variants of FasL generated either by fusion to the collagen domain of adiponectin ACRP30 or fusion to an Fc domain were very cytotoxic and mediated apoptosis [68]. The fusion protein is also known as Mega-Fas-ligand or APO010 [69].

Of course, targeting FasL with antibody fragments was also tested. In one example, CD20 binding variable domain of Rituximab was converted into a scFv and linked to FasL. This fusion protein had a high cytotoxic activity in a wide range of malignant B-cell lines. Cell killing by this chimeric molecule was more efficient than co-treatment with Rituximab and FasL. Interestingly, this targeted FasL did not affect normal B-cells or show any systemic toxicity [70].

On the basis of the experiences with shielded TNF, molecules were designed that shielded the activity of FasL while in circulation. Alternative formats were even equipped with a targeting moiety, in this case a scFv against the tumor marker fibroblast activation protein (FAP). In order to allow desired activation of the FasL domain in tumor proximity, the protective domain was equipped with a protease-sensitive linker. Experimentally, it was observed that FasL requires a free C-terminus for optimal activity, and that the shielding domain consisting of the extracellular part of Fas only inhibits FasL activity when trimerized. Therefore, the most potent fusion protein had FasL positioned at the N-terminus, followed by the trimerization domain of tenascin-C and the protease-sensitive linker. Between linker and FasL that formed the C-terminus, the anti-FAP-scFv was inserted. On the basis of size exclusion chromatography, this multidomain protein forms noncovalent head to tail hexamers. This engineered protein had a highly specific potency with no negative systemic effects in a mouse tumor model [71].

Besides antibodies targeting peptides such as the integrin-binding RGD motif were also used in combination with FasL. The RGD-FasL fusion protein could suppress tumor growth of a murine hepatocellular carcinoma more efficiently than the untargeted FasL [72].

17.5.3 TRAIL Fusion Proteins

The TNF-related apoptosis-inducing ligand (TRAIL) exists as soluble and membrane bound form similarly as the other members of the TNF superfamily. TRAIL binds to four different death receptors (DR4, DR5, DCR1, and DCR2), but only the interaction with DR4 or 5 triggers apoptosis. The process follows the aforementioned extrinsic pathway. Soluble TRAIL was selected as therapeutic molecule owing to its preference to kill cancer cells but not normal cells, thus being well tolerated even in systemic administration. Despite these relatively good prerequisites of TRAIL, efforts were undertaken to improve specific activity and bioavailability. ErbB2, which is expressed on a multitude of tumors, was selected as target for a scFv. In order to simplify manufacture and to stabilize the required trimer of TRAIL, three TRAIL molecules were linearly fused together separated only by short peptide linkers and equipped with one anti-ErbB2-scFv molecule at the N-terminus. This fusion protein induced apoptosis in a xenotransplant model more efficiently and had a longer half-life than the nontargeted TRAIL trimer lacking the scFv component [73].

Even without the forced trimerization, TRAIL fusion proteins showed a good dose-dependent activity. One example is the addition of an integrin targeting peptide (RGD) to the amino terminus of TRAIL. This molecule formed monomers, dimers, and trimers and suppressed tumor growth in mice bearing COLO-205 tumors more efficiently than untargeted TRAIL [74].

Apoptosis induction through TRAIL can even be accompanied by a secondary inhibition of proliferation if the targeting molecule contains this functionality. This was demonstrated with an epidermal growth factor receptor (EGFR) blocking scFv fused to TRAIL. The synergistic effect of downregulating EGFR signaling in combination with apoptosis induction eliminated EGFR positive tumor cells very efficiently. The binding of the fusion protein to the EGFR on cell surfaces converted TRAIL from the soluble into a membrane bound form that cross-links agonistic TRAIL receptors to create an even stronger apoptosis effect. This works also through a bystander effect on EGFR negative cells [75].

17.5.4 Intrinsic Apoptosis Inducing Fusion Proteins

Besides the extrinsic pathway, apoptosis can also be mediated through the intrinsic pathway that is primarily stimulated by DNA damage caused by irradiation or cell permeable drugs. A key step of intrinsic apoptosis is the release of cytochrome C from the mitochondria. The two molecules Bak (Bcl-2 homologous antagonist/killer) and Bax (Bcl-2-associated X protein) act on mitochondrial permeability by forming pores after oligomerization. Therefore, attempts were made to introduce these proteins into malignant cells to trigger apoptosis. In any case, proteins that initiate intrinsic cell death must be internalized to execute their function.

In a conceptual study, IL-2 receptor positive cells were targeted by a chimeric protein containing IL-2 and Bax. After internalization, the IL-2-Bax molecule is found first in the nucleus. Then it stimulates the expression of endogenous Bax while suppressing apoptosis inhibitors such as Bcl-2. Human T-cell lymphoma (HUT-102) cells incubated with IL-2-Bax were killed in a time-dependent manner [76]. Other proapoptotic members of the Bcl2 family such as Bik (Bcl-2-interacting killer) or Bak were combined to gonadotropin releasing hormone (GnRH) since its respective receptor is expressed in many tumors of hormone-dependent tissue such as the prostate. Exposing GnRH receptor positive adenocarcinoma cells with either GnRH-Bik, -Bak, or -Bax fusion protein killed the cells in a time- and dose-dependent manner through intrinsic apoptosis [77].

Both extrinsic and intrinsic apoptosis pathways are connected and the protein BH3-interacting domain death agonist (BID) is positioned at the interface between them. Therefore, a novel BID containing molecule was designed. It included an antihuman epidermal growth factor receptor-2 (HER2) scFv connected via a Pseudomonas exotoxin A derived translocation domain. Since BID is cleaved by caspase 8 to obtain the active truncated BID, the construct was designed to contain only the shortened carboxyterminal domain. Cells transfected with this DNA construct secreted the fusion protein and selectively killed HER2 positive cells. Mice obtaining the DNA by intramuscular injection suppressed HER2 positive tumor growth and showed a prolonged survival [78]. Replacing the translocation domain by a short furin-sensitive sequence of diphtheria toxin reduced potential immunogenicity and maintained full activity of the fusion protein [79].

A key molecule upstream in the apoptosis pathway is the tumor suppressor p53. Many tumors lack functional p53, so the introduction of active p53 could restore apoptosis. Combination of full-length p53 with a nucleus targeting scFv fragment reached in the majority of tested cancer cells a killing efficiency of 90% [80]. The same fusion protein significantly reduced liver metastases in a colon cancer model in mice *in vivo* [81].

17.6 Fc-BASED TOXICITY

To some respect, antibodies already fulfill the promise of a "magic bullet" by being able to selectively address a molecule through the variable domain and induce ADCCor complement-dependent cytotoxcitiy (CDC) [82]. Dissecting the antibody into these two parts, molecular targeting/binding and Fc receptor mediated effector functions, a new recombination with other molecules becomes possible. On one hand, the binding arms in form of a Fab fragment could be connected to a powerful toxin as described in the many examples earlier. On the other hand, the ADCC/CDC conferring Fc domain can be connected to a nonantibody derived binding moiety. However, it has to be taken into account that IgG1 and IgG3 immunoglobulins are the most potent mediators of ADCC or CDC, whereas IgG4 cannot stimulate CDC at all. This lack of cytotoxic activity in IgG4 variants could be demonstrated with OX40-Fc fusion proteins and swapping IgG1 against IgG4 regions [83]. The importance of the Fc domain triggered approaches to optimize effector functions by amino acid sequence adaptation or specific glycosylation [84]. In Figure 17.4, all different Fc mediated cell killing mechanisms are summarized.

17.6.1 Antibody-Derived Cellular Cytotoxicity (ADCC)

Antibody-derived cellular cytotoxicity is particularly useful to deplete cells with certain surface antigens. This mechanism acts in three steps. First, the antibody binds to the respective antigen on the target cell. Then NK cells with their CD16 Fc receptor (FcγRIIIa) bind to the CH3 domain of antibodies on the cell surface. In the last step, the cross-linking of the CD16 receptors leads to degranulation at a so-called lytic synapse. The release of GrB and perforin ultimately causes apoptosis of target cells [85].

Recently, a Fc domain double mutant S239D/I332E was designed, which had a 40-fold higher affinity to

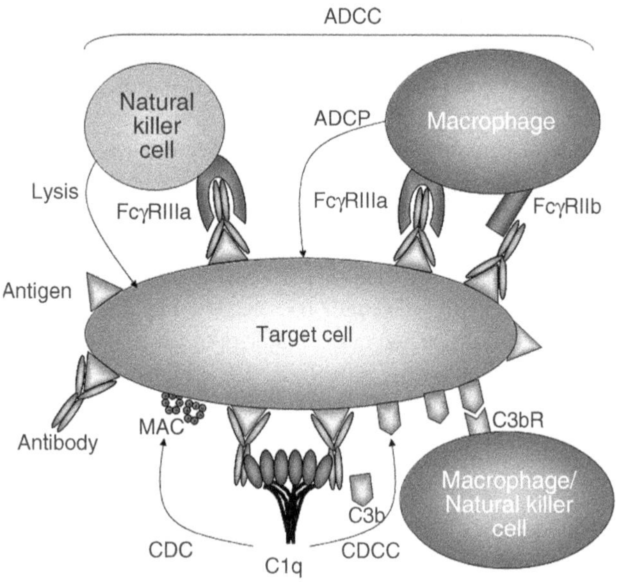

FIGURE 17.4 Fc-mediated cell killing. An antibody bound to its respective antigen on the surface of a target cell can induce antibody-dependent cellular cytotoxicity (ADCC) through the FcγRIIIa on natural killer cells or antibody-dependent cell mediated phagocytosis (ADCP) by macrophages that contain the FcγRIIa as well. Complement-dependent cytotoxicity (CDC) is caused by the interaction of C1q complement factor with the CH2 constant region of the antibody, which induces the formation of a membrane-attack complex (MAC) that lyses the cell. Alternatively, C3b that is generated during this cascade facilitates phagocytosis and cytolysis by labeling the cell as opsonin and interacting with the C3b receptor (C3bR) on a macrophage or natural killer cell. This reaction is called complement-dependent cell-mediated cytotoxicity (CDCC).

FcγRIIIa [86]. Alternatively, binding to FcγRIIIa can be enhanced by a lack of fucosylation at amino acid 297. It could be shown that the affinity to FcγRIIIa beyond a certain threshold irrespective of the optimization method will not further improve ADCC effects [87]. This was also verified for scFv-Fc fusion proteins undergoing Fc mutagenesis or glycoengineering, or both [88].

Fc fusion proteins could benefit from improved effector functions as well. Duplications of the Fc region were used to increase ADCC of best-understood candidate, Enbrel®. These tandem Fc repeats increased ADCC and CDC 10-fold and threefold [89]. A nonfucosylated variant of Amevive® had 100-fold improved ADCC activity [90].

17.6.2 Antibody-Dependent Cell-Mediated Phagocytosis (ADCP)

Since the FcγRIIIa is found on macrophages too, antibody decorated cells can also be eliminated by phagocytosis. This process, also called antibody-dependent cell-mediated phagocytosis (ADCP), can be blocked through FcγRIIb, which are also present on macrophages, thus allowing a fine-tuning of ADCP/ADCC reactions. ADCP is primarily dependent on FcγRIIa, but FcγRIIIa and FcγRI seem to be involved as well [91]. Recently, it could be demonstrated that three tandem repeats of Fc domains induced ADCP five times more efficiently than the original antibody [92]. The aforementioned Fc domain double mutant S239D/I332E also had a 10-fold better binding to FcγRIIa and consequently a much better ADCP activity [93].

17.6.3 Complement-Dependent Cytotoxcitiy (CDC)

Complement-dependent cytotoxcitiy is mediated through the interaction of the CH2 constant region of antibodies with the C1q complement factor, which consists of six heterotrimeric subunits. This triggers a proteolytic cascade which induces the generation of a membrane-attack complex (MAC), which represents a transmembrane channel composed of a complex of C5b, C6, C7, C8, and C9 proteins [94]. Since 10 times more surface bound antibody is required to trigger the classical complement pathway than for ADCC, only in rare cases a CDC reaction can be observed. Originally, CDC serves as pathogen defense mechanism because it can be regulated by complement regulatory proteins that are only present on human cells. In some cases, multiple mutations in the Fc region could dramatically improve CDC activity up to a factor of 600 [95].

Enbrel, a Fc fusion protein of the receptor-trap type, binds both soluble and transmembrane tumor necrosis factor α (TNF-α). Although it contains a normal CH2 constant region, only a very low CDC activity was found. This might be caused by the absence of the CH1 region in Enbrel, which is required for C3 binding and activation of subsequent complement components [96].

17.6.4 Complement-Dependent Cell-Mediated Cytotoxicity (CDCC)

Alternative to the generation of MAC, decorating the cell surface of the target cell with C3b as opsonin can cause cell killing. This is mediated by the interaction of C3b with its corresponding receptor (C3bR) on macrophages or NK cells that facilitates cytolysis and phagocytosis. This process is called complement-dependent cell-mediated cytotoxicity (CDCC). Besides C3, the chemo-attractant C5a is also involved in destroying labeled cells by facilitating an inflammatory response [97].

17.7 PEPTIDE-BASED TOXICITY

Many of the aforementioned strategies rely on a combination of large targeting molecules such as antibodies with other large molecules exerting the toxic activity. Despite the many positive examples, this approach suffers from manufacturing difficulties and cost because often, the toxic component reduces expression levels. Furthermore, toxic proteins of nonhuman origin will cause an immune response that over time will eliminate the fusion proteins through neutralizing antibodies. All these disadvantages can be circumvented when instead of large molecules small peptides are used. For this purpose, toxic peptides are needed, which can be identified from nature or artificially designed and synthetically produced. Let us focus on natural peptides found in insects, arthropods, spiders, reptiles, or amphibians.

Recently, insect toxins such as melittin became exploited as anti-proliferative agents. Melittin is the major component of bee venom. It is expressed as a 70 amino acid long promelittin, which is maturated through proteolytic removal of the amino terminus, resulting in a 26 residue long amphiphilic active peptide. This small size is one of the advantages of melittin. The peptide has a strong anti-proliferative effect in many cancer cell lines, which is probably based on the inhibition of calmodulin, the Akt pathway, and the stimulation of phospholipase A2. Furthermore, melittin can also disrupt cell membranes. The ability of proteolytic activation has been utilized to generate a number of targeted fusion proteins. One construct targeting tumor cells expressing integrin $\alpha_v\beta_3$ contained uPA cleavable linker and successfully inhibited proliferation of uPA expressing cells [98]. In a different approach, melittin was connected through a protease-sensitive linker with a latency peptide of TGF-β that shielded the cytotoxic activity. The proteolytic cleavage of the linker by tumor-associated proteases released free melittin that killed the respective cancer cells [99].

An example for an artificial peptide is a lytic peptide consisting of lysine and leucine repeats that selectively destroys cancer cells by disrupting the cell membrane. Targeting was enabled by fusion to a second peptide that binds to the transferrin receptor (TfR). Overexpression of TfR is typical for a number of cancer cell types. Exposure of 12 cancer cell lines to this synthetic 32 AA fusion peptide showed a clear dose-dependent response with an IC_{50} of 4–9 μM. Cell killing was a function of TfR expression level. Normal cell lines did not show cell death up to a concentration of 40 μM. Membrane disruption was fast with more than 80% dead cells after 60 min [100]. The same toxic peptide was used in a similar approach, now fused to an EGFR targeting peptide. This fusion peptide did also kill cancer cells that were resistant to tyrosine kinase inhibitors. The observed IC_{50} values ranged from 6 to 12 μM. In these experiments, it could be verified that the mechanism of cell killing is apoptosis. *In vivo* administration of this fusion peptide drastically reduced tumor size in three different xenograft models while having no negative influence on animal health, in general. The synthetic peptide could not be degraded by serum proteases owing to its unique diastereometric conformation [101].

17.8 CONCLUSIONS AND FUTURE PERSPECTIVES

Specific and long-lasting eradication of malignant cells, which is the holy grail of targeted therapy, has not been achieved yet. But in the last four decades, since the successful introduction of antibodies and their derivates, many new weapons have been added to the arsenal. Fusion proteins can contribute heavily in generating novel combinations of targeting moieties and toxic molecules. From a targeting perspective, the choice is between well-established specific natural ligands and novel designer molecules based on engineered scaffolds. Similarly, the eliminating function can be derived from natural toxic proteins or factors that trigger cell death either through their endogenous function or by attracting cells from the immune system specialized in eradicating malignant cells. Fusion proteins give new options by allowing novel combinations of well-proven molecules or integrating novel activities.

Despite the enormous progress and the huge body of knowledge, a number of issues still remain to be solved. The highly potent targeted therapies based on natural toxins that frequently require only a single molecule to reach the cytosol to efficiently kill a cell still need improvement with regard to immunogenicity to allow several therapeutic cycles without eliciting neutralizing antibodies. On the other hand, replacing foreign proteins with human enzymes is limited by the required escape from the endosomes. Very promising seems the approach to attract immune cells by fusion proteins containing leukocyte specific ligands or Fc domains interacting with their specific receptors. Immune therapy might also provide long-term protection from recurring and relapsing cancers or metastases by creating memory cells. This concept is one of the driving forces that led to the invention of bi- or multi-specific antibodies that are explained later in this book in part IIIb. Good results have been obtained by addressing hematological diseases because single cells are much easier to reach by large molecules than the hidden cells in the center of a solid tumor. Probably, combination therapies mixing small molecule drugs with engineered therapeutic fusion proteins could be a solution in the near future. Alternatively, smaller variants with tailored tissue penetration capability and optimized half-life can be engineered.

REFERENCES

1. Strebhardt K, Ullrich A. (2008) Paul Ehrlich's magic bullet concept: 100 years of progress. *Nat. Rev. Cancer* 8(6), 473–480.
2. Cizeau J, Torres MGP, Cowling SG, Stibbard S, Premsukh A, Entwistle J, et al. (2011) Fusogenics: a recombinant immunotoxin-based screening platform to select internalizing

tumor-specific antibody fragments. *J. Biomol. Screen.* 16(1), 90–100.

3. Adams GP, Schier R, McCall AM, Simmons HH, Horak EM, Alpaugh RK, et al. (2001) High affinity restricts the localization and tumor penetration of single-chain fv antibody molecules. *Cancer Res.* 61(12), 4750–4755.
4. Lu Y, Yang J, Sega E. (2006) Issues related to targeted delivery of proteins and peptides. *AAPS J.* 8(3), E466–E478.
5. Khawli LA, Hu P, Epstein AL. (2008) Cytokine, chemokine, and co-stimulatory fusion proteins for the immunotherapy of solid tumors. *Handb. Exp. Pharmacol.* 181, 291–328.
6. Lorberboum-Galski H. (2011) Human toxin-based recombinant immunotoxins/chimeric proteins as a drug delivery system for targeted treatment of human diseases. *Expert Opin. Drug Deliv.* 8(5), 605–621.
7. Mathew M, Verma RS. (2009) Humanized immunotoxins: a new generation of immunotoxins for targeted cancer therapy. *Cancer Sci.* 100(8), 1359–1365.
8. Hetzel C, Bachran C, Tur MK, Fuchs H, Stöcker M. (2009) Improved immunotoxins with novel functional elements. *Curr. Pharm. Des.* 15(23), 2700–2711.
9. Pirie CM, Hackel BJ, Rosenblum MG, Wittrup KD. (2011) Convergent potency of internalized gelonin immunotoxins across varied cell lines, antigens, and targeting moieties. *J. Biol. Chem.* 286(6), 4165–4172.
10. Shapira A, Benhar I. (2010) Toxin-based therapeutic approaches. *Toxins (Basel)* 2(11), 2519–2583.
11. Pastan I, Hassan R, Fitzgerald DJ, Kreitman RJ. (2006) Immunotoxin therapy of cancer. *Nat. Rev. Cancer* 6(7), 559–565.
12. Kreitman RJ. (2009) Recombinant immunotoxins containing truncated bacterial toxins for the treatment of hematologic malignancies. *BioDrugs* 23(1), 1–13.
13. Murphy JR. (2011) Mechanism of diphtheria toxin catalytic domain delivery to the eukaryotic cell cytosol and the cellular factors that directly participate in the process. *Toxins (Basel)* 3(3), 294–308.
14. Frankel AE, Rossi P, Kuzel TM, Foss F. (2002) Diphtheria fusion protein therapy of chemoresistant malignancies. *Curr. Cancer Drug Targets* 2(1), 19–36.
15. Weldon JE, Pastan I. (2011) A guide to taming a toxin–recombinant immunotoxins constructed from Pseudomonas exotoxin A for the treatment of cancer. *FEBS J.* 278(23), 4683–4700.
16. Mut M, Sherman JH, Shaffrey ME, Schiff D. (2008) Cintredekin besudotox in treatment of malignant glioma. *Expert Opin. Biol. Ther.* 8(6), 805–812.
17. Moayeri M, Leppla SH. (2009) Cellular and systemic effects of anthrax lethal toxin and edema toxin. *Mol. Aspects Med.* 30(6), 439–455.
18. Liu S, Aaronson H, Mitola DJ, Leppla SH, Bugge TH. (2003) Potent antitumor activity of a urokinase-activated engineered anthrax toxin. *Proc. Natl. Acad. Sci. U.S.A.* 100(2), 657–662.
19. Liu S, Netzel-Arnett S, Birkedal-Hansen H, Leppla SH. (2000) Tumor cell-selective cytotoxicity of matrix metalloproteinase-activated anthrax toxin. *Cancer Res.* 60(21), 6061–6067.
20. Arora N, Leppla SH. (1994) Fusions of anthrax toxin lethal factor with shiga toxin and diphtheria toxin enzymatic domains are toxic to mammalian cells. *Infect. Immun.* 62(11), 4955–4961.
21. Brunger AT, Rummel A. (2009) Receptor and substrate interactions of clostridial neurotoxins. *Toxicon* 54(5), 550–560.
22. Foster K, Chaddock J. (2010) Targeted secretion inhibitors—innovative protein therapeutics. *Toxins (Basel)* 2(12), 2795–2815.
23. Engedal N, Skotland T, Torgersen ML, Sandvig K. (2011) Shiga toxin and its use in targeted cancer therapy and imaging. *Microbiol. Biotechnol.* 4(1), 32–46.
24. McIntosh LM, Barnes JL, Barnes VL, McDonald JR. (2009) Selective CCR2-targeted macrophage depletion ameliorates experimental mesangioproliferative glomerulonephritis. *Clin. Exp. Immunol.* 155(2), 295–303.
25. Shapira A, Gal-Tanamy M, Nahary L, Litvak-Greenfeld D, Zemel R, Tur-Kaspa R, et al. (2011) Engineered toxins "zymoxins" are activated by the HCV NS3 protease by removal of an inhibitory protein domain. *PLoS ONE* 6(1), e15916.
26. Petersson K, Forsberg G, Walse B. (2004) Interplay between superantigens and immunoreceptors. *Scand. J. Immunol.* 59(4), 345–355.
27. Shaw DM, Connolly NB, Patel PM, Kilany S, Hedlund G, Nordle O, et al. (2007) A phase II study of a 5T4 oncofoetal antigen tumour-targeted superantigen (ABR-214936) therapy in patients with advanced renal cell carcinoma. *Br. J. Cancer* 96(4), 567–574.
28. Forsberg G, Skartved N-J, Wallén-Ohman M, Nyhlén HC, Behm K, Hedlund G, et al. (2010) Naptumomab estafenatox, an engineered antibody-superantigen fusion protein with low toxicity and reduced antigenicity. *J. Immunother.* 33 (5), 492–499.
29. Stirpe F, Battelli MG. (2006) Ribosome-inactivating proteins: progress and problems. *Cell. Mol. Life Sci.* 63(16), 1850–1866.
30. Cizeau J, Grenkow DM, Brown JG, Entwistle J, MacDonald GC. (2009) Engineering and biological characterization of VB6-845, an anti-EpCAM immunotoxin containing a T-cell epitope-depleted variant of the plant toxin bouganin. *J. Immunother.* 32(6), 574–584.
31. Polito L, Bortolotti M, Pedrazzi M, Bolognesi A. (2011) Immunotoxins and other conjugates containing saporin-s6 for cancer therapy. *Toxins (Basel)* 3(6), 697–720.
32. Lombardi A, Bursomanno S, Lopardo T, Traini R, Colombatti M, Ippoliti R, et al. (2010) Pichia pastoris as a host for secretion of toxic saporin chimeras. *FASEB J.* 24(1), 253–265.
33. Endo Y. (1988) Mechanism of action of ricin and related toxins on the inactivation of eukaryotic ribosomes. *Cancer Treat. Res.* 37, 75–89.
34. Schindler J, Gajavelli S, Ravandi F, Shen Y, Parekh S, Braunchweig I, et al. (2011) A phase I study of a combination

of anti-CD19 and anti-CD22 immunotoxins (Combotox) in adult patients with refractory B-lineage acute lymphoblastic leukaemia. *Br. J. Haematol.* 154(4), 471–476.

35. Vitetta ES. (2000) Immunotoxins and vascular leak syndrome. *Cancer J.* 6(Suppl. 3), S218–S224.
36. Ortiz-Sánchez E, Helguera G, Daniels TR, Penichet ML. (2008) Antibody–cytokine fusion proteins: applications in cancer therapy. *Expert Opin. Biol. Ther.* 8, 609–632.
37. Rainov NG, Söling A. (2006) Clinical studies with targeted toxins in malignant glioma. *Rev. Recent Clin. Trials* 1(2), 119–131.
38. Gillies SD, Lan Y, Hettmann T, Brunkhorst B, Sun Y, Mueller SO, et al. (2011) A low-toxicity IL-2-based immunocytokine retains antitumor activity despite its high degree of IL-2 receptor selectivity. *Clin. Cancer Res.* 17(11), 3673–3685.
39. Shi M, Zhang L, Gu H-T, Jiang F-Q, Qian L, Yu M, et al. (2007) Efficient growth inhibition of ErbB2-overexpressing tumor cells by anti-ErbB2 ScFv-Fc-IL-2 fusion protein *in vitro* and *in vivo*. *Acta Pharmacol. Sin.* 28(10), 1611–1620.
40. Shusterman S, London WB, Gillies SD, Hank JA, Voss SD, Seeger RC, et al. (2010) Antitumor activity of hu14.18-IL2 in patients with relapsed/refractory neuroblastoma: a children's oncology group (COG) phase II study. *J. Clin. Oncol.* 28(33), 4969–4975.
41. Williams P, Galipeau J. (2011) GM-CSF-based fusion cytokines as ligands for immune modulation. *J. Immunol.* 186(10), 5527–5532.
42. Lechner MG, Russell SM, Bass RS, Epstein AL. (2011) Chemokines, costimulatory molecules and fusion proteins for the immunotherapy of solid tumors. *Immunotherapy* 3(11), 1317–1340.
43. Andrady C, Sharma SK, Chester KA. (2011) Antibody-enzyme fusion proteins for cancer therapy. *Immunotherapy* 3(2), 193–211.
44. Tur MK, Neef I, Jäger G, Teubner A, Stöcker M, Melmer G, et al. (2009) Immunokinases, a novel class of immunotherapeutics for targeted cancer therapy. *Curr. Pharm. Des.* 15(23), 2693–2699.
45. De Lorenzo C, Di Malta C, Calì G, Troise F, Nitsch L, D'Alessio G. (2007) Intracellular route and mechanism of action of ERB-hRNase, a human anti-ErbB2 anticancer immunoagent. *FEBS Lett.* 581(2), 296–300.
46. Krauss J, Arndt MAE, Vu BK, Newton DL, Rybak SM. (2005) Targeting malignant B-cell lymphoma with a humanized anti-CD22 scFv-angiogenin immunoenzyme. *Br. J. Haematol.* 128(5), 602–609.
47. Krauss J, Arndt MAE, Vu BK, Newton DL, Seeber S, Rybak SM. (2005) Efficient killing of CD22+ tumor cells by a humanized diabody-RNase fusion protein. *Biochem. Biophys. Res. Commun.* 331(2), 595–602.
48. Rybak SM, Arndt MAE, Schirrmann T, Dübel S, Krauss J. (2009) Ribonucleases and immunoRNases as anticancer drugs. *Curr. Pharm. Des.* 15(23), 2665–2675.
49. Jia L-T, Zhang L-H, Yu C-J, Zhao J, Xu Y-M, Gui J-H, et al. (2003) Specific tumoricidal activity of a secreted proapoptotic protein consisting of HER2 antibody and constitutively active caspase-3. *Cancer Res.* 63(12), 3257–3262.
50. Yarkoni S, Kaminitz A, Sagiv Y, Askenasy N. (2010) Targeting of IL-2 receptor with a caspase fusion protein disrupts autoimmunity in prediabetic and diabetic NOD mice. *Diabetologia* 53(2), 356–368.
51. Sagiv Y, Kaminitz A, Lorberboum-Galski H, Askenasy N, Yarkoni S. (2009) A fusion protein composed of IL-2 and caspase-3 ameliorates the outcome of experimental inflammatory colitis. *Ann. N. Y. Acad. Sci.* 1173, 791–797.
52. Wang L-F, Zhou Y, Xu Y-M, Qiu X-C, Zhou B-G, Wang F, et al. (2009) A caspase-6 and anti-HER2 antibody chimeric tumor-targeted proapoptotic molecule decreased metastasis of human osteosarcoma. *Cancer Invest.* 27(7), 774–780.
53. Xu Y-M, Wang L-F, Jia L-T, Qiu X-C, Zhao J, Yu C-J, et al. (2004) A caspase-6 and anti-human epidermal growth factor receptor-2 (HER2) antibody chimeric molecule suppresses the growth of HER2-overexpressing tumors. *J. Immunol.* 173(1), 61–67.
54. Zhao J, Zhang L-H, Jia L-T, Zhang L, Xu Y-M, Wang Z, et al. (2004) Secreted antibody/granzyme B fusion protein stimulates selective killing of HER2-overexpressing tumor cells. *J. Biol. Chem.* 279(20), 21343–21348.
55. Liu Y, Cheung LH, Hittelman WN, Rosenblum MG. (2003) Targeted delivery of human pro-apoptotic enzymes to tumor cells: *in vitro* studies describing a novel class of recombinant highly cytotoxic agents. *Mol. Cancer Ther.* 2(12), 1341–1350.
56. Rosenblum MG, Barth S. (2009) Development of novel, highly cytotoxic fusion constructs containing granzyme B: unique mechanisms and functions. *Curr. Pharm. Des.* 15(23), 2676–2692.
57. Philpott GW, Bower RJ, Parker KL, Shearer WT, Parker CW. (1974) Affinity cytotoxicity of tumor cells with antibody-glucose oxidase conjugates, peroxidase, and arsphenamine. *Cancer Res.* 34(9), 2159–2164.
58. Bagshawe KD. (1987) Antibody directed enzymes revive anti-cancer prodrugs concept. *Br. J. Cancer* 56(5), 531–532.
59. Kogelberg H, Tolner B, Sharma SK, Lowdell MW, Qureshi U, Robson M, et al. (2007) Clearance mechanism of a mannosylated antibody-enzyme fusion protein used in experimental cancer therapy. *Glycobiology* 17(1), 36–45.
60. Bagshawe KD, Sharma SK, Begent RHJ. (2004) Antibody-directed enzyme prodrug therapy (ADEPT) for cancer. *Expert Opin. Biol. Ther.* 4(11), 1777–1789.
61. Harding FA, Liu AD, Stickler M, Razo OJ, Chin R, Faravashi N, et al. (2005) A beta-lactamase with reduced immunogenicity for the targeted delivery of chemotherapeutics using antibody-directed enzyme prodrug therapy. *Mol. Cancer Ther.* 4(11), 1791–1800.
62. Schellmann N, Deckert PM, Bachran D, Fuchs H, Bachran C. (2010) Targeted enzyme prodrug therapies. *Mini Rev. Med. Chem.* 10(10), 887–904.
63. Wójcik T, Kieć-Kononowicz K. (2008) Catalytic activity of certain antibodies as a potential tool for drug synthesis and

for directed prodrug therapies. *Curr. Med. Chem.* 15(16), 1606–1615.

64. Guicciardi ME, Gores GJ. (2009) Life and death by death receptors. *FASEB J.* 23(6), 1625–1637.
65. Krippner-Heidenreich A, Grunwald I, Zimmermann G, Kühnle M, Gerspach J, Sterns T, et al. (2008) Single-chain TNF, a TNF derivative with enhanced stability and antitumoral activity. *J. Immunol.* 180(12), 8176–8183.
66. Bauer S, Adrian N, Fischer E, Kleber S, Stenner F, Wadle A, et al. (2006) Structure-activity profiles of Ab-derived TNF fusion proteins. *J. Immunol.* 177(4), 2423–2430.
67. Gerspach J, Pfizenmaier K, Wajant H. (2009) Improving TNF as a cancer therapeutic: tailor-made TNF fusion proteins with conserved antitumor activity and reduced systemic side effects. *Biofactors* 35(4), 364–372.
68. Holler N, Tardivel A, Kovacsovics-Bankowski M, Hertig S, Gaide O, Martinon F, et al. (2003) Two adjacent trimeric Fas ligands are required for Fas signaling and formation of a death-inducing signaling complex. *Mol. Cell. Biol.* 23(4), 1428–1440.
69. Eisele G, Roth P, Hasenbach K, Aulwurm S, Wolpert F, Tabatabai G, et al. (2011) APO010, a synthetic hexameric CD95 ligand, induces human glioma cell death *in vitro* and *in vivo*. *Neuro-oncol.* 13(2), 155–164.
70. Bremer E, ten Cate B, Samplonius DF, Mueller N, Wajant H, Stel AJ, et al. (2008) Superior activity of fusion protein scFvRit: sFasL over cotreatment with rituximab and Fas agonists. *Cancer Res.* 68(2), 597–604.
71. Watermann I, Gerspach J, Lehne M, Seufert J, Schneider B, Pfizenmaier K, et al. (2007) Activation of CD95L fusion protein prodrugs by tumor-associated proteases. *Cell Death Differ.* 14(4), 765–774.
72. Liu Z, Wang J, Yin P, Qiu J, Liu R, Li W, et al. (2009) RGD-FasL induces apoptosis in hepatocellular carcinoma. *Cell. Mol. Immunol.* 6(4), 285–293.
73. Schneider B, Münkel S, Krippner-Heidenreich A, Grunwald I, Wels WS, Wajant H, et al. (2010) Potent antitumoral activity of TRAIL through generation of tumor-targeted single-chain fusion proteins. *Cell Death Dis.* 1, e68.
74. Cao L, Du P, Jiang S-H, Jin G-H, Huang Q-L, Hua Z-C. (2008) Enhancement of antitumor properties of TRAIL by targeted delivery to the tumor neovasculature. *Mol. Cancer Ther.* 7(4), 851–861.
75. Bremer E, Samplonius DF, van Genne L, Dijkstra MH, Kroesen BJ, de Leij LFMH, et al. (2005) Simultaneous inhibition of epidermal growth factor receptor (EGFR) signaling and enhanced activation of tumor necrosis factor-related apoptosis-inducing ligand (TRAIL) receptor-mediated apoptosis induction by an scFv:sTRAIL fusion protein with specificity for human EGFR. *J. Biol. Chem.* 280(11), 10025–10033.
76. Aqeilan R, Kedar R, Ben-Yehudah A, Lorberboum-Galski H. (2003) Mechanism of action of interleukin-2 (IL-2)-Bax, an apoptosis-inducing chimaeric protein targeted against cells expressing the IL-2 receptor. *Biochem. J.* 370(Pt 1), 129–140.
77. Azar Y, Lorberboum-Galski H. (2000) GnRH-Bik/Bax/Bak chimeric proteins target and kill adenocarcinoma cells; the general use of pro-apoptotic proteins of the Bcl-2 family as novel killing components of targeting chimeric proteins. *Apoptosis* 5(6), 531–542.
78. Qiu X-C, Xu Y-M, Wang F, Fan Q-Y, Wang L-F, Ma B-A, et al. (2008) Single-chain antibody/activated BID chimeric protein effectively suppresses HER2-positive tumor growth. *Mol. Cancer Ther.* 7(7), 1890–1899.
79. Shan L-Q, Ma S, Qiu X-C, Wang T, Yu S-B, Ma B-A, et al. (2011) A novel recombinant immuno-tBid with a furin site effectively suppresses the growth of HER2-positive osteosarcoma cells *in vitro*. *Oncol. Rep.* 25(2), 325–331.
80. Weisbart RH, Hansen JE, Chan G, Wakelin R, Chang SS, Heinze E, et al. (2004) Antibody-mediated transduction of p53 selectively kills cancer cells. *Int. J. Oncol.* 25(6), 1867–1873.
81. Hansen JE, Fischer LK, Chan G, Chang SS, Baldwin SW, Aragon RJ, et al. (2007) Antibody-mediated p53 protein therapy prevents liver metastasis *in vivo*. *Cancer Res.* 67(4), 1769–1774.
82. Correale P, Cusi MG, Tagliaferri P. (2011) Immunomodulatory properties of anticancer monoclonal antibodies: is the "magic bullet" still a reliable paradigm? *Immunotherapy* 3(1), 1–4.
83. Taylor L, Bachler M, Duncan I, Keen S, Fallon R, Mair C, et al. (2002) *In vitro* and *in vivo* activities of OX40 (CD134)-IgG fusion protein isoforms with different levels of immune-effector functions. *J. Leukoc. Biol.* 72(3), 522–529.
84. Desjarlais JR, Lazar GA. (2011) Modulation of antibody effector function. *Exp. Cell Res.* 317(9), 1278–1285.
85. Topham NJ, Hewitt EW. (2009) Natural killer cell cytotoxicity: how do they pull the trigger? *Immunology* 128(1), 7–15.
86. Lazar GA, Dang W, Karki S, Vafa O, Peng JS, Hyun L, et al. (2006) Engineered antibody Fc variants with enhanced effector function. *Proc. Natl. Acad. Sci. U S A* 103(11), 4005–4010.
87. Masuda K, Kubota T, Kaneko E, Iida S, Wakitani M, Kobayashi-Natsume Y, et al. (2007) Enhanced binding affinity for FcgammaRIIIa of fucose-negative antibody is sufficient to induce maximal antibody-dependent cellular cytotoxicity. *Mol. Immunol.* 44(12), 3122–3131.
88. Repp R, Kellner C, Muskulus A, Staudinger M, Nodehi SM, Glorius P, et al. (2011) Combined Fc-protein- and Fc-glyco-engineering of scFv-Fc fusion proteins synergistically enhances CD16a binding but does not further enhance NK-cell mediated ADCC. *J. Immunol. Methods* 373(1–2), 67–78.
89. Nagashima H, Kaneko K, Yamanoi A, Motoi S, Konakahara S, Kohroki J, et al. (2011) TNF receptor II fusion protein with tandemly repeated Fc domains. *J. Biochem.* 149(3), 337–346.
90. Shoji-Hosaka E, Kobayashi Y, Wakitani M, Uchida K, Niwa R, Nakamura K, et al. (2006) Enhanced Fc-dependent cellular cytotoxicity of Fc fusion proteins derived from TNF receptor II and LFA-3 by fucose removal from Asn-linked oligosaccharides. *J. Biochem.* 140(6), 777–783.

91. Nimmerjahn F, Ravetch JV. (2011) FcγRs in health and disease. *Curr. Top. Microbiol. Immunol.* 350, 105–125.

92. Nagashima H, Ootsubo M, Fukazawa M, Motoi S, Konakahara S, Masuho Y. (2011) Enhanced antibody-dependent cellular phagocytosis by chimeric monoclonal antibodies with tandemly repeated Fc domains. *J. Biosci. Bioeng.* 111(4), 391–396.

93. Richards JO, Karki S, Lazar GA, Chen H, Dang W, Desjarlais JR. (2008) Optimization of antibody binding to FcgammaRIIa enhances macrophage phagocytosis of tumor cells. *Mol. Cancer Ther.* 7(8), 2517–2527.

94. Gelderman KA, Tomlinson S, Ross GD, Gorter A. (2004) Complement function in mAb-mediated cancer immunotherapy. *Trends Immunol.* 25(3), 158–164.

95. Moore GL, Chen H, Karki S, Lazar GA. (2010) Engineered Fc variant antibodies with enhanced ability to recruit complement and mediate effector functions. *MAbs* 2(2), 181–189.

96. Mitoma H, Horiuchi T, Tsukamoto H, Tamimoto Y, Kimoto Y, Uchino A, et al. (2008) Mechanisms for cytotoxic effects of anti-tumor necrosis factor agents on transmembrane tumor necrosis factor alpha-expressing cells: comparison among infliximab, etanercept, and adalimumab. *Arthritis Rheum.* 58(5), 1248–1257.

97. van Lookeren Campagne M, Wiesmann C, Brown EJ. (2007) Macrophage complement receptors and pathogen clearance. *Cell. Microbiol.* 9(9), 2095–2102.

98. Zhu W, Sun M, Wang Y, Sun de J, Zhang S. (2011) Expression and functional characterization of a recombinant targeted toxin with an uPA cleavable linker in *Pichia pastoris. Protein Expr. Purif.* 76(2), 184–189.

99. Holle L, Song W, Holle E, Wei Y, Li J, Wagner TE, et al. (2009) *In vitro-* and *in vivo*-targeted tumor lysis by an MMP2 cleavable melittin-LAP fusion protein. *Int. J. Oncol.* 35(4), 829–835.

100. Kawamoto M, Horibe T, Kohno M, Kawakami K. (2011) A novel transferrin receptor-targeted hybrid peptide disintegrates cancer cell membrane to induce rapid killing of cancer cells. *BMC Cancer* 11, 359.

101. Kohno M, Horibe T, Haramoto M, Yano Y, Ohara K, Nakajima O, et al. (2011) A novel hybrid peptide targeting EGFR-expressing cancers. *Eur. J. Cancer* 47(5), 773–783.

18

CLASSIC IMMUNOTOXINS WITH PLANT OR MICROBIAL TOXINS

Jung Hee Woo and Arthur Frankel

Cancer Research Institute of Scott and White Memorial Hospital, Texas A&M Health Science Center College of Medicine, Temple, TX, USA

18.1 INTRODUCTION

Immunotoxins are protein molecules consisting of one or two cell surface-directed ligands covalently linked to a peptide toxin either from plants or microorganisms. The ligand moiety of immunotoxins recognizes cell surface proteins on target cells and the toxin moiety kills targeted cells according to the mechanism of action of its toxin. Ligands include monoclonal antibodies and antibody fragments, adhesion molecules, growth factors, and cytokines. There are two types of peptide toxins that have been used in immunotoxin construction (Table 18.1). Type I toxins are proteins whose enzymatic action takes place in the cytosol. They enzymatically inhibit or modify critical intracellular functions, mostly protein synthesis. Most of plant and microbial toxins belongs to type I. Type II toxins are membrane-acting peptides, which cause changes or leaks in the plasma membrane, resulting in cell death. No internalization of toxins is required for cell intoxication but nonspecific cell toxicity is a major drawback. Most of the immunotoxins have been made with type I toxins. Toxins are either holotoxins or hemitoxins. Holotoxins (ricin, diphtheria toxin [DT], Pseudomonas exotoxin [PE]) consist of a catalytic moiety and a binding moiety in either a single chain protein or two proteins. Hemitoxins (saporin, gelonin, pokeweed antiviral protein [PAP]) have only a catalytic chain. Type I toxins must be internalized and then translocated into the cytosol.

The major application is cancer treatment due to the targeted cell killing effect of immunotoxins. Some immunotoxins have been tested in patients with autoimmune diseases. Cancer treatments include surgery, radiation therapy, and chemotherapy. Chemotherapy is the last treatment option for widespread metastatic cancer patients. However, the success of chemotherapy in many cancers is hindered by acquired chemotherapy resistance. Most chemotherapy drugs are small molecules that target DNA, mitotic mechanisms, or transcription factors. These small molecules can be exported by various pump mechanisms that can be enhanced by selection processes leading to cancer chemotherapy resistance. Alternative approaches to cancer treatment are needed. Immunotoxins can selectively target malignant cells and induce cell death by a different mechanism than conventional chemotherapy drugs. Therefore, targeted therapy with immunotoxins is a potential treatment option for relapsed/refractory cancers that have resistance to chemotherapy drugs. Although immunotoxins have a unique mechanism of action toward malignant cells, their clinical use and repeated dosing are limited by nonspecific cell toxicity and immunogenicity. Nonspecific cell toxicity causes transient elevation of liver enzymes and endothelial damage. The latter often results in vascular leak syndrome. Classic immunotoxins are immunogenic. Immunogenicity of immunotoxins

Fusion Protein Technologies for Biopharmaceuticals: Applications and Challenges, First Edition. Edited by Stefan R. Schmidt.

TABLE 18.1 Toxins Used in Immunotoxin Preparation

Type	Function	Toxins
Cytosol-acting toxins	Protein synthesis inhibition by ADP-ribosylation of EF-2	Holotoxins: diphtheria toxin [1], Pseudomonas extotoxin [2]
	Protein synthesis inhibition by cleavage of ribosomal RNA	Holotoxins: ricin [3], abrin [4], viscumin [5], nigrin b [6] Hemitoxins: gelonin [7], saporin [8], pokeweed antiviral protein [9], luffin [10], bouganin [11], trichokirin [12], trichosanthin [13], barley ribosome inactivating protein [14], ebulin [15], bryodin [16], momordin [17], momorchochin [18], moschatin [19], dianthin 30 [20], ocymoidine [21], pyramidatine [21], colocin 1 [22], α-sarcin [23], mitogillin [24], clavin [25], restrictocin [26], barnase [27]
	Modification of signaling pathways	Holotoxins: Anthrax toxin [28]
	Inhibition of exocytosis (secretion)	Holotoxins: Botulinum neurotoxins [29]
Membrane-acting toxins	Disintegration of vital membrane function	Hemitoxins: *Clostridia perfringens* phospholipase C [30], Pyrularia thionin [31]
	Membrane pore formation	Hemitoxins: proaerolysin [32], Cyt1Aa toxin [33]

induces anti-immunotoxin antibodies that neutralize biological activities of immunotoxins. In particular, most individuals have anti-diphtheria toxin antibody titers due to childhood and adult vaccination programs of DPT (diphtheria, pertussis, and tetanus). The preexisting anti-diphtheria antibodies may reduce biological activities of diphtheria immunotoxins. In this chapter, overview of toxins used in immunotoxin preparation, recent progress in immunotoxin design and synthesis, clinical updates of immunotoxin trials, and challenges and perspectives in the immunotoxin field are discussed.

18.2 TOXINS USED IN IMMUNOTOXIN PREPARATION

Many toxins have been used in immunotoxin preparation (Table 18.1). DT, PE, ricin, saporin, gelonin, and PAP immunotoxins have been tested in clinical trials. DT and PE immunotoxins currently in clinical trials are mostly recombinant proteins because recombinant DT and PE immunotoxins can be easily produced in *Escherichia coli* and/or *Pichia pastoris*. DT and PE immunotoxins are more potent against target cells and have less side effects in trials as compared to plant toxins including ricin, saporin, gelonin, and PAP. Ricin, saporin, gelonin, and PAP immunotoxins tested in humans are chemically conjugated proteins.

18.2.1 Diptheria Toxin

DT is a single chain protein of 535 amino acid residues produced by *Corynebacterium diphtheriae*. It consists of two subunits, an A chain (catalytic domain) and a B chain, connected by a furin-cleavable linker that is an arginine-rich loop formed by one disulfide bond [34]. The cleavage of the furin-cleavable linker is a prerequisite for DT intoxication of cells [35,36]. The B chain contains a translocation domain and a receptor-binding domain [34]. In recombinant DT immunotoxins, the receptor-binding domain is replaced with tumor-selective ligands. Binding induces receptor-mediated endocytosis followed by localization to the endosomes. Because different receptors have different preferences for intracellular routing some receptors achieve a limited penetration to the late acidic endosome. These DT-ligand combinations exhibit suboptimal target toxicity [37]. In the acidic environment of the endosomes, DT unfolds partially. This exposes hydrophobic regions on the translocation domain and triggers membrane insertion to form a channel for translocation of the catalytic domain [38–40]. Unfolding of the catalytic domain is also required for cytosolic entry through the translocation channel [41,42]. Chaperonic action of cytosolic Hsp90 (heat shock protein 90) and the membrane-inserted translocation domain translocates the unfolded catalytic domain to the cytosol [43,44]. For the release of the catalytic domain into the cytosol, cytosolic TrR-1 (thioredoxin reductase) reduces the disulfide bond between the catalytic and translocation domains [44]. The translocated catalytic domain refolds and catalytically inactivates cellular protein synthesis by ADP-ribosylating the diphthamide residue in domain IV of elongation factor-2 (EF-2) [45]. One single molecule of catalytic domain in the cytosol is sufficient to block protein synthesis and lead to death of the target tumor cell [46]. Three forms of DT have been used in immunotoxin preparation (Figure 18.1). CRM107 and CRM9 are full-length DT mutant forms lacking binding function. DAB_{486}, DAB_{389}, DT_{388}, and DT390 are truncated forms of DT. The first 388 amino acid residues of mature DT contain the catalytic domain and translocation domain essential for cell intoxication.

18.2.2 Pseudomonas Exotoxin

PE is a single chain protein of 613 amino acid residues produced by *Pseudomonas aeruginosa*. It consists of three

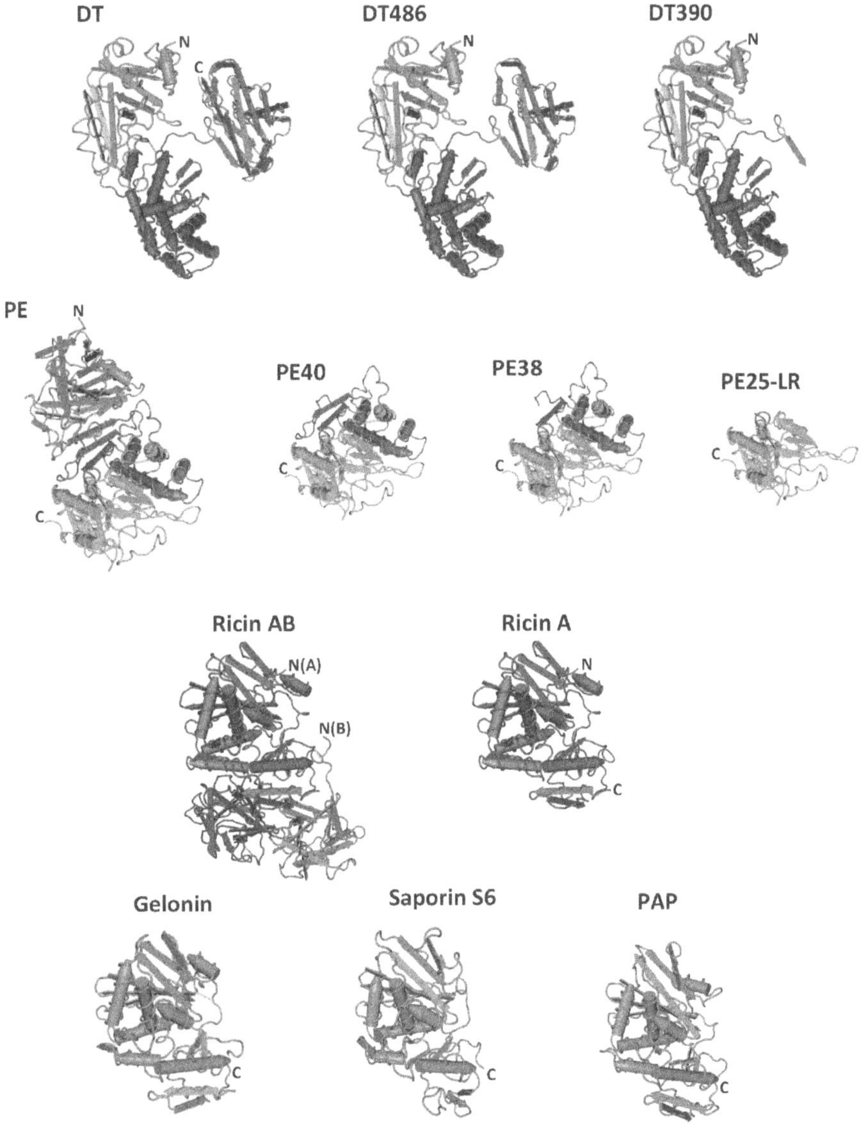

FIGURE 18.1 Backbone diagram of X-ray crystal structures of toxins used in immunotoxin preparation. N and C indicate N and C termini of proteins, respectively. PAP: pokeweed antiviral protein. *Diphtheria toxin*: catalytic domain (magenta), translocation domain (blue), binding domain (brown), DT (residues 1–535), DT486 (residues 1–486), DT390 (residues 1–390). *Pseudomonas exotoxin*: domain 1a (residues 1–252; magenta), domain 1b (residues 365–404; blue), domain II (residues 253–364; brown), domain III (405–613; green), PE (residues 1–613), PE40 (residues 253–613), PE38 (residues 253–364 and residues 381–613), and PE25-LR (residues 274–284 and residues 395–613). *Ricin*: chain A (brown) and chain B (magenta and blue). Ricin, gelonin, saporin, and PAP have a similar structure. (See the color version of this figure in Color Plates section.)

distinct structural regions: domains Ia (residues 1–252) and Ib (residues 365–404), domain II (residues 253–364) and domain III (residues 405–613) [47]. The full-length PE binds the α_2-macroglobulin receptor/low-density lipoprotein receptor-related protein 1 (LRP1) or LRP1b on the surface of mammalian cells [48,49]. The PE and receptor complex is internalized by receptor-mediated endocytosis. The internalized PE in endocytic vesicles undergoes furin cleavage between Arg-279 and Gly-280, a pH-dependent conformation change followed by reduction of the disulfide bond between Cys-265 and Cys-287. This creates a 28-kDa N-terminal fragment and a 37-kDa C-terminal fragment. The 37 kDa fragment is further routed to the endoplasmic reticulum. In the ER, the 37 kDa fragment is translocated to the cytosol. The cytosolic C-terminal fragment then catalyzes the ADP-ribosylation of EF2, resulting in cell death similar to DT.

PE-based immunotoxins have been made with various truncated forms of PE and full-length PE by either chemical conjugation or protein engineering (Figure 18.1). The full-length PE is conjugated to the OVB3 monoclonal antibody that has affinity to ovarian cancer cell lines and colon cancer cell lines. PE40 is a 40-kDa truncated form of PE in which residues 1–250 of domain Ia have been removed. Recombinant NLysPE40, a modified form of PE40 having an N-terminal lysine residue for chemical conjugation, is conjugated to a monoclonal antibody. The truncated PE40 is used to synthesize TP-40 consisting of transforming growth factor α (TGFα) as the ligand. PE38 is a 38-kDa truncated form of PE in which residues 1–250 of domain Ia and residues 365–384 of domain Ib have been removed. PE38KDEL, PE38QQR, and PE38-8X are modified forms of PE38. PE38KDEL has C-terminal KDEL residues as a endoplasmic reticulum (ER) retention signal that enhances internal routing to ER and subsequently improves potency. To minimize conjugation of heterobifunctional cross-linking reagents to the C terminal catalytic domain of PE38, lysines 590, 606, and 613 of PE38 are replaced with glutamines and arginine (PE38QQR) [50]. PE38-8X is a deimmunized version of PE38 with eight point mutations on seven major B-cell epitopes of PE38. The deimmunized version is less immunogenic in animals [51]. PE25-LR is lysosomal protease-resistant and the smallest truncated form of PE containing residues 274–284 and residues 395–613 [52].

18.2.3 Ricin

Ricin is a two-chain toxin extracted from castor beans (*Ricinus communis*). The A chain of 267 residues (RTA) is linked to the B chain of 262 residues (RTB) by a disulfide bond (Figure 18.1). RTA has N-glycosidase activity that specifically depurinates an invariant adenine residue of the 28S ribosomal subunit. RTB has three lectin binding sites, which bind galactosyl residues on mammalian cell surface glycoproteins [53,54]. After cell binding, ricin internalizes into endosomes using both clathrin-dependent and clathrin-independent pathways [55] and reaches the trans-reticular Golgi compartment [56]. In the Golgi, the toxin is retrograde transported to the ER via COP-I-independent, Rab6-regulated vesicles [57,58]. In the ER, the toxin is reduced, the A chain unfolds with the help of chaperones, and the A chain translocates to the cytosol using the Sec61p transposon [59]. The A chain depurinates the adenine base of the 28S ribosomal RNA, resulting in irreversible blocking of protein elongation [60].

A blocked ricin, a ricin A chain and a deglycosylated ricin A chain have been used in preparation of ricin immunotoxin. The blocked ricin has the A and B chains. The B chain is blocked by covalent attachment of a glycopeptide ligand with a sulfhydryl group, which provides a reactive group for chemical conjugation with an antibody [61]. The deglycosylated ricin A has less nonspecific toxicity than the blocked ricin and the ricin A chain. Deglycosylation of ricin A chain is required to reduce nonspecific normal tissue binding because mannose receptors on liver cells recognize terminal oligosaccharides of ricin A chain [62,63]. Deglycosylated ricin A chain is obtained by chemical treatment with a mixture of sodium metaperiodate and sodium cyanoborohydride [64] or expression in *Escherichia coli* [65].

18.2.4 Gelonin and Saporin

Gelonin, PAP, and saporin are potent single-chain ribosome-inactivating proteins (28–30 kDa) from plants of *Gelonium multiflorum, Phytolacca americana,* and *Saponaria officinalis,* respectively. These toxins and ricin A chain have similar protein structure (Figure 18.1). They are all N-glycosidases which remove the same adenine base from the 28S ribosomal RNA of the mammalian ribosomes [66]. Unlike ricin, these toxins lack any cell-binding or cytoplasmic delivery domains. For immunotoxins, these toxins need to escape from the endosomes following internalization by ligand-directed endocytosis in order to intoxicate a cell. However, the endosomal escape is an inefficient rate-limiting step [67]. A conserved C-terminal hydrophobic motif of ribosome-inactivating proteins [68] may play an important role in the endosomeal escape.

18.3 IMMUNOTOXIN DESIGN AND SYNTHESIS

In the past three decades, over 1000 immunotoxins have been engineered. The first generation immunotoxins are chemical conjugates of toxins and tumor-selective ligands. Large amounts of toxin and ligand molecules are needed to synthesize chemical conjugates because of low conjugation yield. Chemical conjugation yields a heterogeneous product due to the random reaction of derivatizing reagents with multiple amino acid residues in the ligand and/or toxin. Serum half-lives of chemical conjugates are relatively longer (9–24 h), compared to recombinant immunotoxins (1–5 h). Many chemical conjugates have reached clinical testing. The second-generation immunotoxins are recombinant fusion proteins. Toxin structure information has been used in immunotoxin design to minimize unwanted toxin effect and maximize specific cytotoxicity to target cells. They have only one tumor-selective ligand and a truncated toxin in a single chain protein. Owing to a relatively simple protein structure, these molecules are produced in *E. coli* with reasonable yields. One of good examples is the first approved immunotoxin, $DAB_{389}IL2$ (Ontak®, denileukin diftiftox) for the treatment of cutaneous T-cell lymphoma (CTCL) in 1999 by the FDA. Most recombinant immunotoxins belong to this class.

The third-generation immunotoxins have either a modified toxin moiety to reduce immunogenicity and lysosomal protein degradation or a more complicated structure to increase binding affinity for targeted antigens. For less immunogenic and protease-resistant molecules, systemic studies on PE immunotoxins have been performed [51,52]. In this study, seven major B-cell epitopes have been indentified in PE38 and eliminated by introduction of eight point mutations (R313A, Q332S, R432G, R467A, R490A, R513A, E548S, and K590S). The new mutant, PE38-8X, is significantly less immunogenic in three strains of mice, yet retains full cytotoxic and antitumor activities [51]. In the other study, a lysosomal protease-resistant PE immunotoxin has been developed by serial deletion of PE38 domain II. The immunotoxin with PE25LR containing a small portion of domain II (residues 274–284) and full-length domain III, retains cytotoxic activity and is less toxic to mice [52]. Combination of these point mutations and major deletion of domain II would generate a new PE immunotoxin with less immunogenicity, less toxicity, and protease-resistance.

The molecules with complicated protein structure to increase efficacy are designed by addition of an identical ligand or a different ligand. The third-generation DT immunotoxins, A-dmDT390-bisFv(UCHT1) and DT2219ARL, are being investigated in clinical trials. For molecules with two identical ligands (normally sFv), there are two formats, bivalent and fold-back formats. In the bivalent format (DT-sFv-$(G_4S)_3$-sFv), the additional sFv (V_L-$(G_4S)_3$-V_H) is fused to C-terminus of monovalent DT immunotoxin through the $(G_4S)_3$ linker. The bivalency provides much higher binding to targeted cells (20- to 30-fold increase) [69]. In the fold-back format (DT-V_L-G_4S-V_H-$(G_4S)_3$-V_L-G_4S-V_H), the G_4S linkers between V_L and V_H are shorter than the $(G_4S)_3$ linker between the first sFv and second sFv. The shorter linkers prevent adjacent V_L and V_H domains from forming a sFv. The $(G_4S)_3$ linker is long enough so that the second V_L and V_H domains can fold back to pair with the first V_H and V_L respectively, forming two sFv antigen-binding sites. This provides a more rigid structure compared to the bivalent, resulting in increased potency by fivefold to sevenfold [70].

A good example of molecules with two different ligands is DT2219ARL, which has the first 389 amino acid residues of DT, and two different sFv moieties recognizing different antigens (CD22 and CD19). The V_L and V_H domains in each sFv moiety are connected by a flexible ARL linker derived from human muscle aldolase (the first 20 amino acid residues). Two sFv molecules are linked by the $(G_4S)_3$ linker. This flexible ARL linker increased potency and product yield in *E. coli* [71]. This bispecific molecule was a 1000-fold more potent than DT22 alone, or an equal mixture of DT22 and DT19. Three other bispecific DT immunotoxins (DTEGF13, DTEpCAM23, and DTAT13) have been reported [72–74]. More recently a deimmunized bispecific PE immunotoxin (2219KDEL7mut) has shown *in vivo* efficacy in a xenograft B-cell metastases model. This molecule, an upgraded version of DT2219ARL, has two different sFv moieties recognizing CD22 and CD19, and the less immunogenic PE38-8X with KDEL [75]. Since this molecule has much less immunogenicity, multiple drug treatment may be given in future clinical trials.

For chemical conjugation, the conjugation reaction must not severely alter the ligand affinity or impair the ability of the toxin moiety to translocate to the cytosol and enzymatically inactivate cell functions. Typically large amounts of toxin and ligand molecules are needed to synthesize chemically conjugated immunotoxins because of low conjugation yield. Chemical conjugations yield a heterogeneous product because of the random reaction of derivatizing reagents with multiple amino acid residues in the ligand and/or toxin. Bifunctional conjugation reagents used include *N*-succinimidyl-oxycarbonyl-α-methyl-(2-pyridyl-dithio) toluene (SMPT), *N*-succinimidyl 3-(2-pyridyl-dithio)-propionate (SPDP), *m*-maleimidobenzoyl-*N*-hydrosuccinimide ester (MBS), and *N*-succinimidyl 4-(*N*-maleimidomethyl) cyclohexane-1-carboxylate (SMCC). Amine-reactive *N*-hydroxy-succinimide (NHS) ester that reacts with amine group of lysine residues forms a stable amide bond. The pyridyl disulfide group and the maleimide group of bifunctional conjugation reagents react with thiol group of cysteine residues to form a reversible disulfide bond and a stable thioether bond, respectively. DTNB (5,5′-dithiobis (2-nitobenzoic acid) is used as a thiol-protecting agent for synthesis of Fab′ and deglycosylated ricin A chain conjugate. Traut agent (2-iminothiolane) is often used to add a thiol group to a ligand for conjugation. Synthesis method and formulation of chemical conjugates tested in clinical trials are summarized in Table 18.2. Typically, a whole IgG or Fab2′ fragment is derivatized with a bifunctional conjugation reagent. A free thiol group is introduced into a toxin if it does not have the free thiol group. The derivatized ligand and toxin are conjugated through a disulfide bond or thioether bond. Then unconjugated molecules are removed from conjugates by gel filtration, anion exchange, and/or affinity chromatography.

Genetic fusions by recombinant DNA technology are well defined and controlled, and generally have less effect on toxin functions than chemical conjugation. *E. coli* is a typical expression system for production of recombinant immunotoxins. The trc and T7 promoters are popularly used to express recombinant immunotoxins in *E. coli*. Since most of recombinant immunotoxins are expressed in insoluble inclusion bodies, an *in vitro* refolding step is required to obtain functional immunotoxins. Immunotoxins in the inclusion body preparation are washed extensively with detergent to remove endotoxin, solubilized, denatured, and reduced. The denatured and reduced recombinant immunotoxins are then renatured in a refolding buffer containing L-arginine and glutathione. The renatured proteins can be purified by

TABLE 18.2 Synthesis Method and Formulations of Chemically Conjugated Immunotoxins Tested in Clinical Trials

Chemically Conjugated Immunotoxin	Target Antigen	Conjugation Reagents	Conjugate Purification	Formulation, Concentration, and Storage	References
Anti-CD3 IgG (SPV-T3a)-dgA	CD3	SMPT	Blue affinity/S-300 GF/ultrafiltration	PBS, pH 7.4, 0.05% Tween 20; 1 mg/mL; −20°C	[76]
Anti-CD5 IgG (H65)-RTA	CD5	SPDP	Protein A affinity/ ultrafiltration	PBS, pH 7.4; 1 mg/mL; 4°C	[77,78]
Anti-CD7 IgG (WT1)-dgA	CD7	SMPT	Blue affinity/S-300 GF/ultrafiltration	PBS, pH 7.4, 0.05% Tween 20; 1 mg/mL; −20°C	[76]
Anti-CD22 Fab′ (RFB4)-dgA	CD22	DTNB	Blue affinity/S-200 GF	0.145 M NaCl, pH 6.5; 1 mg/mL; −70°C	[79,80]
Anti-CD22 IgG (RFB4)-dgA	CD22	SMPT	Blue affinity/S-200 GF/ultrafiltration	0.15 M NaCl/5 mM lysine, pH 9; 0.5 mg/mL; −70°C	[81]
Anti-CD19 IgG (HD37)-dgA	CD19	SMPT	S-200 GF/blue affinity	0.15 M NaCl, pH 6.5; 0.5 mg/mL; −70°C	[81]
Anti-CD19 IgG (anti-B4)-bRicin	CD19	SMCC	S cation/anti-bRicin affinity/ S-300 GF	PBS, 1 mg/mL HSA, pH 7.3; 0.1 mg/mL; 4°C	[61,82]
Anti-CD19 IgG (B43)-PAP	CD19	SPDP and Traut	TSK-3000-SW GF/CM cation/Affi-Prep polymyxin	40 mM sodium phosphate, 150 mM NaCl, pH 7.5; 1 mg/mL; −70°C	[83]
Anti-CD19 IgG (BU12)-saporin	CD19	SPDP	CM cation/dialysis	PBS, pH 7.2; −80°C	[84]
Anti-CD25 IgG (RFT5)-dgA	CD25	SMPT	S-200 GF/blue affinity	0.85% NaCl, 5 mM lysine; 0.5 mg/mL; −70°C	[85,86]
Antimelanoma IgG (XMMME-001)-RTA	Melanoma	SPDP	S-200 GF	0.9% PBS, pH 7.0; 1.0 mg/mL; 4°C	[87–89]
Anti-breast carcinoma IgG (260F9)-rRA	55 kDa antigen	SPDP	TSK-300 GF	0.1 M sodium sulfate, 0.02 M sodium phosphate, pH 6.8	[90,91]
Anti-CD30 IgG (Ber-H2)-saporin	CD30	Traut	S-200 GF	PBS, pH 7.4; 0.21 mg/mL; −150°C	[92,93]
Anti-CD33 IgG (HuM195)-rGel	CD33	SPDP and Traut	S-300 GF/blue affinity	PBS, pH 7.4	[94,95]
B3(IgG)-NLysPE38 or LMB-1	LeY	Traut and SMCC	Mono Q anion/TSK 250 GF	PBS; 1 mg/mL	[50,96]
Transferrin-CRM107	TfR	Traut and MBS	TSK3000 GF	Unpublished	[97]

dgA, deglycosylated ricin A chain; bRicin, blocked ricin; RTA, ricin A chain; rRA, recombinant ricin A chain; PAP, pokeweed antiviral protein; rGel, recombinant gelonin; TfR, transferrin receptor; HSA, human serum albumin; DTNB, 5,5′-dithiobis(2-nitorbenzoic acid); SPDP, *N*-succinimidyl 3-(2-pyridyldithio)-propionate; SMCC, *N*-succinimidyl 4-(*N*-maleimidomethyl) cyclohexane carboxylate; SMPT, *N*-succinimidyl-oxycarbonyl-α-methyl-(2-pyridyldithio) toluene; GF, gel filtration; Traut, 2-iminothiolane; MBS, *m*-maleimidobenzoyl-*N*-hydrosuccinimide ester.

ion exchange and gel filtration chromatography according to protein characteristics such as isoelectric point (pI) and molecular weight. For $DAB_{389}IL2$, $DAB_{486}IL2$, and $DAB_{389}EGF$, these recombinant immunotoxins are expressed in periplasm, lysed and purified by affinity, anion exchange, and size-exclusion chromatography. As discussed earlier, the second-generation immunotoxins have only one ligand and relatively simple structure so that these molecules can be produced in *E. coli* with reasonable yields. However, production of the third-generation immunotoxins with two ligands by *in vitro* refolding from *E. coli* is very inefficient because of the multidomain structure [98]. Good refolding could be achieved in a toxin-resistant CHO cell line, but production capacity is limited to 5 μg/mL [99]. *P. pastoris* is an ideal host strain to express recombinant immunotoxins via the secretory route, because secretory expression minimizes exposure of immunotoxins to the cytosol where the catalytic domain inactivates protein synthesis [98,100]. Further, a DT-resistant *P. pastoris* strain has been developed by a point mutation (G710R) of elongation factor 2 [101]. The DT-resistant strain can produce an anti-CD3 DT immunotoxin at three times higher concentrations (120 mg/L) than wild-type strain (37 mg/L) after optimization of fermentation conditions [102,103]. Production method and formulation of recombinant immunotoxins are summarized in Table 18.3.

TABLE 18.3 Production and Formulation of Recombinant Immunotoxins Tested in Clinical Trials

Recombinant Immunotoxin	Target Antigen	Expression System	Purification	Formulation, Concentration, and Storage	References
$DAB_{389}IL2$	IL2 receptor	*E. coli*; trc promoter	Anti-IL2 affinity/G4000PW GF	20 mM citrate, 0.05 mM EDTA, 1% polysorbate 20, pH 6.9–7.2; 0.15 mg/mL; −20°C	[104,105]
$DAB_{486}IL2$	IL2 receptor	*E. coli*; trc promoter	Anti-IL2 affinity/G4000PW GF	Tris-buffered saline; −20°C	[104,105]
A-dmDT390-bisFv (UCHT1)	CD3	*P. pastoris*; AOX1 promoter; secretion	Butyl HIC/borate Poros HQ anion/Poros HQ anion	5 mM Tris, 1 mM EDTA, 150 mM NaCl, 5% glycerol, pH 8; 0.8 mg/mL; −80°C	[106,107]
$DT_{388}IL3$	IL3 receptor	*E. coli*; T7; refolding	Q Anion/S-200 GF/polymixin B	5 mM Tris, 250 mM NaCl, pH 8; 1.8–1.9 mg/mL; −80°C	[108]
DT2219ARL	CD22 and CD19	*E. coli*; T7; refolding	Q Anion	10 mM sodium phosphate, 150 mM NaCl, 0.5% polysorbate 80, pH 7.4; 1 mg/mL; −80°C	[71]
$DAB_{389}EGF$	EGF receptor	*E. coli*; trc promoter	anti-DT affinity/DEAE anion/G4000PW GF	PBS, 1% mannitol, 0.05 mM EDTA, pH 7.2; 0.25 mg/mL; −20°C after lyophilization	[109]
$DT_{388}GMCSF$	GM-CSF receptor	*E. coli*; T7; refolding	Q Anion/S-200 GF/polymixin B	PBS, pH 7.4; 1.5 mg/mL; −80°C	[110]
SS1(dsFv)-PE38 or SS1P	Mesothelin	*E. coli*; T7; refolding	Mono Q anion/TSK250 GF	PBS; −70°C	[111,112]
e23(dsFv)-PE38 or erb-38	erbB2/Her2	*E. coli*; T7; refolding	Q anion/Mono Q anion/S-200 GF	PBS; 0.5 mg/mL; −80°C	[113,114]
IL13-PE38QQR	IL13R	*E. coli*; T7; refolding	Q anion/Mono Q anion/Sephacryl S-100 GF	PBS; −70°C	[115]
Anti-Tac(Fv)-PE38 or LMB2	CD25	*E. coli*; T7; refolding	Q anion/Mono Q anion/Superdex GF	PBS; −70°C	[116,117]
TP-40 or TGFα-PE40	EGFR	*E. coli*; T7; refolding	DEAE anion/S-300 GF/Q anion/ultrafiltration	PBS, pH 7.4; 0.5 mg/mL; −70°C	[118]
RFB4(dsFv)-PE38 or BL22	CD22	*E. coli*; T7; refolding	Q anion/Mono Q anion/TSK G3000SW GF	PBS; 763 g/mL; −70°C	[119]
HA22	CD22	*E. coli*; T7; refolding	Q anion/Mono Q anion/TSK G3000SW GF	PBS; −70°C	[119,120]
IL4(38-37)-PE38KDEL	IL4R	*E. coli*; T7; refolding	Q anion/Mono Q anion/TSK250 GF	0.1 M NaH_2PO_4, 1 mM EDTA, pH 7; −80°C	[121–123]
scFv(FRP5)-ETA	erbB2/Her2	*E. coli*; Tac; refolding	His-tag affinity/dialysis	PBS; 0.3 mg/mL; −70°C	[124,125]
B3(Fv)-PE38 or LMB-7	Lewis Y	*E. coli*; T7; refolding	Mono Q anion/TSK250 GF	PBS; 0.25 mg/mL; −70°C	[112,126]
B3(dsFv)-PE38 or LMB-9	Lewis Y	*E. coli*; T7; refolding	Mono Q anion/TSK250 GF	PBS; −70°C	[112]
BR96(sFv)-PE40	Lewis Y	*E. coli*; T7; refolding	Q anion/TSK 3000 GF	25 mM sodium phosphate, 50 mM NaCl, 0.05% Tween 80, pH 7.5; 1 mg/mL; −70°C	[127,128]

18.4 CLINICAL UPDATE OF IMMUNOTOXIN TRIALS

Among over a thousand immmunotoxins, only about 40 immunotoxins have been tested in clinical trials (Table 18.4). Most of immunotoxins have failed to reach clinical testing because of poor *in vivo* efficacy. Immunotoxins with high response rates in clinical studies are highly active in cell proliferation inhibition assays using target cell lines. Typical half maximal inhibitory concentrations (IC_{50}) of immunotoxins in the assays are in the picomolar range. In contrast, many immunotoxins discontinued at the preclinical stages have poor cytotoxic activity in cell proliferation inhibition assays (the nanomolar range of IC_{50}) and poor antitumor activity in animal tumor models. Tumor regression is typically observed at high doses >1 mg/kg/day. These dose levels are at least twofold higher than the highest MTD (0.5 mg/kg/day) observed in immunotoxin clinical trials.

High response rates have been observed in clinical trials of immunotoxins for hematologic malignancies. Target antigens for hematologic malignancies are CD3, CD5, CD19, CD22, CD25, and IL2R. Expression of these antigens is restricted on hematologic cells. This allows immunotoxins to treat patients at sufficient doses producing clinical responses without severe toxicities. However, some target antigens such as EGF receptor, Her2, Lewis Y, transferrin receptor (TfR), and a 55 kDa antigen expressed on breast carcinomas causes dose-limiting toxicities because of the presence of target antigens on normal tissues.

$DAB_{389}IL2$ (Ontak®, denileukin diftitox) is the first FDA-approved immunotoxin for the treatment of persistent or recurrent CTCL. In the first Phase III trial, $DAB_{389}IL2$ was administered at 9 or 18 μg/kg/day by intravenous infusion over 30–60 min for five consecutive days every 3 weeks for eight cycles. Thirty percentage of the 71 patients with CTCL experienced an objective response (10% complete remission [CR] and 20% partial remission [PR]) [129]. In the other Phase III placebo-controlled trial of $DAB_{389}IL2$, 10% CR and 34% PR were observed in CTCL patients [130]. In other trials of $DAB_{386}IL2$ for expansion of indications, $DAB_{386}IL2$ produced transient T-cell depletion in melanoma patients, causing regression of melanoma metastases in 4 out of 16 patients, suggesting that transient T-cell depletion may disrupt the homeostatic control of cognate immunity and allow for the expansion of effector T cells with specificity against neoplastic cells during homeostatic T-cell proliferation [131]. In a Phase II trial for T-cell NHL (peripheral T-cell lymphoma, anaplastic large cell lymphoma, angioimmunoblastic T-cell lymphoma, Sezary syndrome, and NK/T-cell lymphoma), six CRs and seven PRs were observed in 27 patients (overall response rate 48.1%). Interestingly objective responses were achieved in 8 of 13 patients (61.5%) with CD25+ tumors (4 CRs and 4 PRs) [132]. $DAB_{389}IL2$ is currently being evaluated in clinical trials, either alone or in combination with other treatment for treatment of various cancers including B-cell NHL, T-cell NHL, melanoma, advanced breast cancer, pancreatic cancer, ovarian cancer, and AML (acute myeloid leukemia).

BL22 (RFB4(dsFv)-PE38) consists of a disulfide-stabilized anti-CD22 Fv fused to PE38. When given at intravenous doses of 3–50 μg/kg every other day for three doses every month for up to 14 cycles to patients with chemotherapy-resistant B-cell malignancies, there were 77% overall response rate (61% CR and 16% PR) in hairy cell leukemia (HCL) [133]. In a Phase II for HCL, BL22 treatment at doses of 40 μg/kg every other day for three doses every month produced 72% overall response rate (47% CR and 25% PR) [134]. HA22 is an affinity optimized, more active variant of BL22 and is currently in a clinical trial for refractory B-cell malignancies. The affinity of HA22 to CD22 was improved 14-fold by phage display affinity maturation [135]. In a Phase I clinical trial for HCL, HA22 was administered at intravenous doses of 5–50 μg/kg every other day for three doses every month up to 10 cycles. The overall response rate was 79% including 12 CR (43%) of 28 HCL patients. No dose-limiting toxicity was observed after completion of expansion at the highest dose (50 μg/kg × 3) cohort (12 patients) [136].

LMB2 (anti-Tac(sFv)-PE38) is composed of an anti-CD25 sFv fused to PE38. LMB2 was given to 35 patients with lymphomas at intravenous doses of 2–63 μg/kg every other day for three doses. LMB2 therapy produced one durable CR and 7 PR (overall response rate: 23%) [117]. LMB2 was administered with MART-1/gp-100 vaccine to patients with metastatic melanoma in an effort to augment antimelanoma immunity via T_{reg} depletion. LMB2 administration at dose of 30 μg/kg every other day for three doses transiently reduced 79.1% of CD25 positive T cell [137].

A-dmDT390-bisFv(UCHT1) contains diphtheria toxin (DT), preceded by an alanine residue (A), truncated at amino acid residue 390, doubly mutated (dm) to remove glycosylation sites within the toxin that reduce bioactivity when expressed in *P. pastoris* and eukaryotic cells [99], and two tandem sFv molecules derived from UCHT1 parental antibody. A-dmDT390-bisFv(UCHT1) is being tested in a Phase I clinical trial [138]. After eight infusions of 2.5 or 5.0 μg/kg over a 4-day period blood T cells are depleted by three logs. CD4 and CD8 T cell numbers return to pretreatment levels by day 20 but the naïve subsets are markedly depressed for several months. One course of A-dmDT390-bisFv(UCHT1) treatment produced durable remissions lasting over two years. Of eight eligible patients with CTCL, two patients had CR (25%) and three patients had PR (37.5%). Overall response rate was 62.5%.

Transferrin-CRM107 (TransMID) is a chemical conjugate of transferrin and a mutant DT (CRM7) lacking receptor-binding function [1] for the treatment of malignant brain tumors. In a Phase I trial, 18 patients with recurrent

TABLE 18.4 Clinical Results of Classic Immunotoxins

Target Antigen	Drug	Phase	Indications	Dose Levels: Doses/Course	Response Rate (CR)	Toxicities: MTD	References
IL2R	$DAB_{389}IL2$	I	CTCL NHL HD	3–31 μg/kg: 5	13(5)/35 3(1)/17 0(0)/21	Acute infusion events, VLS: 27 μg/kg × 5 (135 μg/kg/cycle)	[165]
IL2R	$DAB_{389}IL2$	III	CTCL	9 or 18 μg/kg: 5	21(7)/71	Flu-like symptoms, acute infusion events, VLS	[129]
IL2R	$DAB_{389}IL2$	III	CTCL	Placebo 9 or 18 μg/kg: 5	7(1)/44 44(10)/100	Flu-like symptoms, acute infusion events, VLS	[130]
IL2R	$DAB_{389}IL2$+steroid premedication	II	CTCL	9 or 18 μg/kg: 5	9(1)/15	Acute infusion events, VLS	[166]
IL2R	$DAB_{389}IL2$	I	GVHD	9 μg/kg: 2 9 μg/kg: 6 9 μg/kg: 7	5(1)/7 9(6)/13 17(8)/24	Transaminasemia: 9 μg/kg × 6 (54 μg/kg/cycle)	[167]
IL2R	$DAB_{389}IL2$	I	GVHD	4.5 or 9 μg/kg: 9	9(9)/22	VLS, opportunistic infection: 4.5 μg/kg × 9 (40.5 μg/kg/cycle)	[168]
IL2R	$DAB_{389}IL2$	II	CLL	9 or 18 μg/kg: 5 18 μg/kg: 5 18 μg/kg: 5	2(0)/12 2(0)/7 6(1)/22	Transaminasemia, VLS Transaminasemia, VLS, anorexia Transaminasemia, VLS, opportunistic infection	[169–171]
IL2R	$DAB_{389}IL2$	II	Melanoma	12 μg/kg: 4	4(0)/16	Mild dermatitis and arthritis	[131]
IL2R	$DAB_{389}IL2$	II	Psoriasis	Placebo 5, 10, or 15 μg/kg: 3	2(0)/12 7(0)/29	Flu-like symptoms, rash, VLS	[172]
IL2R	$DAB_{389}IL2$+ bexarotene	II	CTCL	18 μg/kg: 3	8(4)/14	Lymphopenia, leukopenia	[173]
IL2R	$DAB_{389}IL2$	II	T-cell NHL	18 μg/kg: 5	13(6)/27	VLS, transaminasemia, skin reaction	[132]
IL2R	$DAB_{389}IL2$	II	B-cell NHL	18 μg/kg: 5	11(3)/45	Transaminasemia, leucopenia, thrombocytopenia, VLS	[174]
IL2R	$DAB_{389}IL2$+ rituximab	II	B-cell NHL	18 μg/kg: 5	12(6)/38	Transaminasemia, fatigue, VLS	[175]
IL2R	$DAB_{389}IL2$	II	Indolent NHL	18 μg/kg: 5	3(0)/29	Fatigue, transminasemia, neutropenia, VLS	[176]
IL2R	$DAB_{389}IL2$	II	NSCLC	18 μg/kg: 5	0(0)/41	VLS, constitutional symptoms	[177]
IL2R	$DAB_{486}IL2$ (bolus infusion)	I	NHL, HD, CLL, CTCL	0.7–200 μg/kg: 10–11	3(1)/18	Transaminasemia, rash: 100 g/kg × 10 (1000 μg/kg/cycle)	[178]
IL2R	$DAB_{486}IL2$ (bolus infusion)	I	HD	75–200 μg/kg: 5	1(1)/15	Transaminasemia, hypoalbuminemia: 200 μg/kg × 5 (1000 μg/kg/cycle)	[179]
IL2R	$DAB_{486}IL2$ (bolus infusion)	I	CTCL	75 or 100 μg/kg: 5	3(1)/5	Transaminasemia: 100 μg/kg × 5 (500 μg/kg/cycle)	[180]
IL2R	$DAB_{486}IL2$ (90-min infusion)	I	NHL, CLL, CTCL, HD, Kaposi's sarcoma	150–400 μg/kg: 5	2(0)/23	Renal insufficiency, hemolysis, thrombocytopenia, transaminasemia: 300 μg/kg × 5 (1500 μg/kg/cycle)	[181]
IL2R	$DAB_{486}IL2$ (6-h infusion)	I	NHL, CLL, CTCL, PLL, AML/MDS	100, 200, or 300 μg/kg: 6	1(0)/17		[182]

(continued)

TABLE 18.4 (*Continued*)

Target Antigen	Drug	Phase	Indications	Dose Levels: Doses/Course	Response Rate (CR)	Toxicities: MTD	References
						Transaminasemia, hypersensitivity, anemia, fever, thrombocytopenia: >300 μg/kg × 6 (>1800 μg/kg/cycle)	
IL2R	$DAB_{486}IL2$ (90-min infusion)	II	CTCL	200 μg/kg: 5	1(0)/14	Transaminasemia, hypoalbuminemia	[183]
IL2R	$DAB_{486}IL2$ (bolus infusion)	I	Rheumatoid arthritis	40–100 μg/kg: 6 or 7	6/19	Transaminasemia, arthralgias, IV site phlebitis, hypersensitivity: 100 μg/kg × 7 (700 μg/kg/cycle)	[184]
IL2R	$DAB_{486}IL2$ (bolus infusion)	II	Rheumatoid arthritis	75 μg/kg: 5	4/22: first cycle 11/36: second cycle 11/33: third cycle	Transaminasemia, arthralgias, fever	[185]
IL3R	$DT_{388}IL3$	I	AML, MDS	4–12.5 μg/kg: 6	3(1)/45	Flu-like symptoms, VLS: >12.5 μg/kg × 6 (75 μg/kg/cycle)	[186]
IL3R	$DT_{388}IL3$	I	AML, MDS	7.07 μg/kg: 5	Ongoing		
IL4R	IL4(38-37)-PE38KDEL or NBI-3001	I	Glioma	6–720 μg: intratumoral injection)	1(1)/9	Tumor necrosis-related edema, seizures: >720 μg/cycle	[121]
IL4R	IL4(38-37)-PE38KDEL or NBI-3001	I	Solid tumors	8, 16, or 27 μg/m^2: 5	0(0)/14	Liver toxicity: 16 μg/m^2 × 5 (80 μg/m^2 total)	[187]
IL13R	IL13-PE38QQR[a]	I	Glioma, glioblastoma	72 mL of 0.25–1.0 g/mL: 1	45.9 weeks: 51	Necrosis and inflammation: 72 mL of 0.5 μg/mL (36 μg/cycle)	[141]
IL13R	IL13-PE38QQR[a]	III	Glioblastoma	Gliadel wafer 72 mL of 0.5 μg/mL: 1	35.3 weeks: 93 36.4 weeks: 183	Similar safety profiles	[142]
Breast cancer	260F9-rRA	I	Breast cancer	10 or 50 μg/kg: 6–8	0(0)/4	VLS, neuropathy: not determined	[91]
Breast cancer	260F9-rRA	I	Breast cancer	50 or 100 μg/kg: 6–8 (CI)	0(0)/5	VLS, neuropathy: not determined	[158]
CEA	Anti-CEA IgG-bRicin	I	Hepatic metastases	2–35 μg/m^2: 2–4 (intralesional infusion)	0(0)/27	Pain, anaphylaxis: >35 μg/m^2 × 4 (140 μg/m^2/cycle)	[188]
Colon cancer	XMMCO-791-RTA_{30} (Xomazyme-791)	I	Metastatic colon cancer	20–200 μg/kg: 5	2(0)/17	VLS, fever, mental status change: 150 μg/kg × 5 (750 μg/kg/cycle)	[145]
Colon cancer	XMMCO-791-RTA_{33} (Xomazyme-791)	I	Metastatic colon cancer	20, 30, or 40 μg/kg: 5	0(0)/12	VLS: 40 μg/kg × 5 (200 μg/kg/cycle)	[189]
EGFR	$DAB_{389}EGF$	1	Metastatic solid tumors	0.3–15 μg/kg: 5 6~9 μg/kg: 3	1(0)/72	Liver and renal toxicity: unpublished	[161]
EGFR	TP-40 (TGFα-PE40)	I	Bladder cancer	0.15–9.6 mg: 6 (intravesical instillation)	8(1)/43	Mild urinary tract irritation: >9.6 mg × 6 (57.6 mg/cycle)	[146]

erbB2/Her2	e23(dsFv)-PE38 or erb-38	I	Breast cancer, esophageal cancer	1 or 2 μg/kg: 3	0(0)/6	Liver toxicity (transaminasemia): not determined	[113]
erbB2/Her2	scFv(FRP5)-ETA (PE40)	I	Breast cancer	60–900 μg: 10 (intratumoral injection)	6(2)/11	Inflammation at injection sites: >600–900 μg × 7 (>5100 μg/cycle)	[147]
erbB2/Her2	scFv(FRP5)-ETA (PE40)	I	Breast cancer	2, 4, 10, 12.5, or 20 μg/kg: 10	0(0)/18	Liver toxicity: 12.5 μg/kg × 10 (125 μg/kg/cycle)	[124]
GM-CSFR	DT_{388}GMCSF	I	AML	1–5 μg/kg: 5	3(1)/31	Liver and renal toxicity, VLS: 4 μg/kg × 5 (20 μg/kg/cycle)	[162]
Lewis Y	B3(IgG)-NLysPE38 or LMB-1	I	Adenocarcinoma	10–100 μg/kg: 3	2(1)/38	VLS: 75 μg/kg × 3 (225 μg/kg total)	[190]
Lewis Y	B3(Fv)-PE38 or LMB-7	I	Adenocarcinoma	Unpublished	No response	Renal and GI toxicity: 24 μg/kg × 3 (72 μg/kg/cycle)	[191]
Lewis Y	B3(dsFv)-PE38 or LMB-9	I	Adenocarcinoma	4, 8, or 12 μg/kg: 10	No response	Renal toxicity: unpublished	[191,192]
Lewis Y	BR96(sFv)-PE40 or SGN-10	I	Adenocarcinoma	24–962 μg/m^2: 4	0(0)/46	Hypoalbuminemia, GI toxicity: 641 μg/m^2 × 4 (2564 μg/m^2/cycle)	[127]
Mela-noma	XMMME-001-RTA	I	Melanoma	0.01–1 mg/kg: 5 or 4	5(1)/22	VLS: 0.5 mg/kg × 5 (2 mg/kg/cycle)	[88]
Mela-noma	XMMME-001-RTA + cyclophosphamide	II	Melanoma	0.4 mg/kg: 1	4(0)/20	Hypoalbuminemia, fatigue	[87]
Meso-thelin	SS1(dsFv)-PE38 or SS1P	I	Mesothelioma, ovarian, pancreatic cancers	8–60 μg/kg: 6 or 3	0(0)/34	Uticaria, VLS: 18 μg/kg × 6 (108 μg/kg/cycle) 45 μg/kg × 3 (135 μg/kg/cycle)	[111]
Meso-thelin	SS1(dsFv)-PE38 or SS1P	I	Mesothelioma, ovarian, pancreatic cancers	2–25 μg/kg: 10 (CI)	1(0)/24	VLS, transaminasemia: 25 μg/kg × 10 (250 μg/kg/cycle)	[193]
Ovary cancer antigen	OVB3-PE	I	Ovarian cancer	1, 2, 5, 10 μg/kg: 2–4 (intraperitoneal infusion)	0(0)/23	CNS toxicity, transaminasemia, gastrointestinal toxicity: 5 μg/kg × 3 (15 μg/kg/cycle)	[159]
TfR	Transferrin-CRM107	I	Glioma	5–180 mL of 0.1–1.0 g/mL: 1	9(2)/15	Cerebral edema, peritumoral brain injury: 40 mL of 0.67 μg/mL/cycle	[139]
TfR	Transferrin-CRM107	II	Glioma	40 mL of 0.67 g/mL: 1	12(5)/44	Cerebral edema, seizures	[194]
TfR	IgG (454A12)-rRA	I	Peritoneal cavity related cancers	5–50 μg/kg: 5 (intraperitoneal infusion)	No response	Hyperalbuminemia, myalgias: unpublished	[160]
CD3	A-dmDT390-bisFv (UCHT1)	I	CTCL, PTCL	2.5 μg/kg: 8 5.0 μg/kg: 8	3(2)/6 2(0)/2	Flu-like symptoms, VLS, opportunistic infection: ongoing	[138]
CD3/CD7	SPV-T3a-dgA/WT1-dgA	I	GVHD	2 or 4 mg/m^2: 3 or 4	2(0)/4	Opportunistic infection, VLS: not determined	[148]
CD5	Anti-CD5 IgG (H65)-RTA	I/II	GVHD	50–330 μg/kg: 14	16(9)/32	VLS, tremors, fever, hematuria: 100 μg/kg × 14 (1400 μg/kg/cycle)	[143]
CD5	Anti-CD5 IgG (H65)-RTA + glucocorticoids	III	GVHD	Placebo 100 μg/kg: 14	78(43)/114 91(54)/129	VLS, allergic reaction (no long term clinical benefit)	[144]

(*continued*)

TABLE 18.4 (*Continued*)

Target Antigen	Drug	Phase	Indications	Dose Levels: Doses/Course	Response Rate (CR)	Toxicities: MTD	References
CD5	Anti-CD5 IgG (H65)-RTA	I	CTCL	200–330 μg/kg: 10	4(0)/14	VLS: 330 μg/kg × 10 (3.3 mg/kg/cycle)	[78]
CD19	Anti-B4-bR	I	B-cell malignancy	1–60 μg/kg: 5	3(1)/25	Transaminasemia, hypoalbuminemia: 50 μg/kg × 5 (250 μg/kg/cycle)	[149]
CD19	Anti-B4-bR	I	NHL, CLL, ALL	10–70 μg/kg: 7 (CI)	5(2)/34	Transaminasemia, VLS, myalgias: 50 μg/kg × 7 (350 μg/kg/cycle)	[82]
CD19	Anti-B4-bR	II	NHL	40 μg/kg: 5	0(0)/16	Thrombocytopenia, neutropenia, opportunistic infection	[195]
CD19	Anti-B4-bR	II	Multiple myeloma	40 μg/kg: 7 (CI)	0(0)/5	Transaminasemia, myalgias, thrombocytopenia, neurotoxicity	[196]
CD19	Anti-B4-bR	I	NHL after ABMT	20, 40, or 50 μg/kg: 7	11(11)/12[b]	Thrombocytopenia, transaminasemia, VLS: 40 μg/kg × 7 (280 μg/kg/cycle)	[197]
CD19	Anti-B4-bR	II	NHL after ABMT	30 μg/kg: 7	26(26)/49[b]	Thrombocytopenia, HUS, myalgias, transaminasemia	[163]
CD19	Anti-B4-bR + mBACOD chemo	II	AIDS-related NHL	20 μg/kg: 7	25(12)/44[c]	Transaminasemia, flu-like symptoms	[198]
CD19	IgG-HD37-dgA	I	B-cell lymphoma	0.5–6 mg/m^2: 4 9.6 or 19.2 mg/m^2/8 days	2(1)/23 1(0)/9	VLS, aphasia, rhabdomyolysis, acrocyanosis: 16 mg/m^2/cycle	[150]
CD19	B43-PAP + chemotherapy	I	pediatric B-ALL	1 or 1.5 mg/m^2: 10	12(10)/24	Myalgias, transaminasemia: 1 mg/m^2 × 10 (10 mg/m^2/cycle)	[151]
CD19	Anti-CD19 IgG (BU12)-saporin	I	B-cell malignancy	Unpublished	Unpublished	VLS: unpublished	[199]
CD19/CD22	IgG-HD37-dgA IgG-RFB4-dgA	I	B-cell lymphoma	10, 20, or 30 mg/m^2/192h: 1	2(0)/22	VLS, microangiopathic hemolysis: 30 mg/m^2/192 h	[164]
CD19/CD22	IgG-HD37-dgA IgG-RFB4-dgA	I	Pediatric B-ALL	2, 4, 5, or 6 mg/m^2: 4	3(3)/17	Pancreatitis, GVHD, anaphylaxis: 5 mg/m^2 × 4 (20 mg/m^2/cycle)	[152]
CD19/CD22	DT2219ARL	I	B-cell malignancy		Ongoing		
CD22	IgG (RFB4)-dgA	I	B-cell lymphoma	9.6, 19.2, or 28.8 mg/m^2/192 h: 1	4(0)/18	VLS: 19.2 mg/m^2/192 h	[153]
CD22	IgG (RFB4)-dgA	I	B-cell lymphoma	2.5–13.9 mg/m^2: 2–12	6(1)/26	VLS, myalgias: 8 mg/m^2 × 4 (32 mg/m^2/cycle)	[154]
CD22	Fab′(RFB4)-dgA	I	B-cell lymphoma	12.5–100 mg/m^2/cycle: 2–6	5(0)/13	VLS: 75 mg/m^2/cycle	[79]
CD22	RFB4(dsFv)-PE38 (BL22)	I	NHL CLL Hairy cell leukemia	3–50 μg/kg: 3	0(0)/4 1(0)/11 24(19)/31	HUS, VLS, transaminasemia: 40 μg/kg × 3 (120 μg/kg/cycle)	[133]
CD22	RFB4(dsFv)-PE38 (BL22)	II	Hairy cell leukemia	40.0 μg/kg: 3	26(17)/36	Reversible hemolytic uremic syndrome (HUS)	[134]

CD22	RFB4(dsFv)-PE38 (BL22)	I	Pediatric CD22+ malignancy	10–40 μg/kg: 3–6	0(0)/23	Transaminasemia: >40 μg/kg × 6 (>240 μg/kg/cycle)	[200]
CD22	HA22 (a higher affinity variant of BL22)	I	Hairy cell leukemia	5–50 μg/kg: 3	22(12)/28	HUS: >50 μg/kg × 3 (>150 μg/kg/cycle)	[136]
CD25	RFT5-SMPT-dgA	I	HD	1.25, 2.5, 3.75, or 5 mg/m^2: 4	2(0)/15	VLS, myalgia, tachycardia: 3.75 mg/m^2 × 4 (15 mg/m^2/cycle)	[86]
CD25	Anti-Tac(Fv)-PE38 or LMB2	I	CD25+ hematologic malignancy	2–63 μg/kg: 3	8(1)/35	Transaminasemia, fever: 40 μg/kg × 3 (120 μg/kg/cycle)	[117]
CD30	Ki-4.dgA	I	CD30+ HD, NHL	1.25, 1.88, or 2.5 mg/m^2: 4	2(0)/15	VLS: 1.25 mg/m^2 × 4 (5 mg/m^2/cycle)	[155]
CD33	HuM195-rGel	I	AML, MDS, CMML	2.5–10 mg/m^2: 4	0(0)/28	Allergic reaction: 7 mg/m^2 × 4 (28 mg/m^2/cycle)	[94]
CD38	OKT10-saporin	I	Multiple myeloma	Unpublished	Unpublished	Unpublished	
CD56	N901-bRicin	I	SCLC	5–40 μg/kg: 7	1(0)/21	VLS: 30 μg/kg × 7 (350 μg/kg/cycle)	[201]
CD56	N901-bRicin	II	SCLC	30 μg/kg: 7	1(1)/6	VLS	[156]

CTCL, cutaneous T-cell lymphoma; NHL, non-Hodgkin lymphoma; HD, Hodgkin's disease; GVHD, graft-versus-host disease; CLL, chronic lymphocytic leukemia; ALL, acute lymphoblastic leukemia; NSCLC, non-small cell lung cancer; SCLC, small cell lung cancer; PTCL, peripheral T-cell lymphoma; AML, acute myeloid leukemia; MDS, myelodysplasia; PLL, prolymphocytic leukemia; CI, continuous infusion; CR, complete remission; VLS, vascular leak syndrome; HUS, hemolytic uremic syndrome; ABMT, autologous bone marrow transplantation; c, cycle.

[a]Endpoint is overall survival. Gliadel wafer is a FDA approved local treatment.

[b]Assessment of efficacy is difficult because patients in CR are enrolled in this trial and there is no control group.

[c]Of the 44 patients, 26 received anti-B4-bR therapy. Assessment of efficacy includes untreated patients with anti-B4-bR therapy.

malignant glioma were intratumorally treated with high-flow interstitial microinfusion of transferrin-CRM107 by convection-enhanced delivery (CED), which permitted direct drug delivery to the brain tumor interstitum. Of 15 eligible patients, 9 had objective responses (2 CR and 7 PR) [139]. In a Phase II trial, 44 patients received 40 mL of 0.67 μg/mL over 4–5 days by CED. Of 34 patients evaluated for efficacy, there were 5 CR and 7 PR (overall response rate: 35.3%). Owing to lack of objective efficacy in a Phase III trial for comparison with best standard treatment, further clinical development of transferrin-CRM107 was discontinued [140]. Detailed clinical data have not been published.

IL13-PE38QQR is composed of interleukin-13 fused to PE38QQR. IL13-PE38QQR was administered by CED to recurrent glioblastoma multiforme tumors for up to 6 days. Nine patients (17.6%) and seven patients (13.7%) had a prolonged progression-free survival beyond one and two years, respectively [141]. IL13-PE38QQR treatment showed similar overall survival and safety profiles as compared to Gliadel wafer treatment in a Phase III trial for glioblastoma patients [142].

Anti-CD5 IgG (H65)-RTA is a chemical conjugate of anti-CD5 antibody (H65) and ricin A chain. In a Phase I for CTCL, patients received intravenous doses of 200–330 μg/kg daily for 10 days. There were 4 PRs out of 14 patients. No CR was observed in this trial [78]. In a Phase I/II for patients with GVHD in the liver, gut, or skin, anti-CD5 IgG (H65)-RTA was administered 14 daily intravenous infusions of 50–330 μg/kg. There were 9 CR and 7 PR out of 32 evaluated patients. Anti-CD5 IgG (H65)-RTA treatment depleted 80% of CD5+ blood lymphocytes [143]. Anti-CD5 IgG (H65)-RTA was further evaluated in a Phase III clinical trial. Combination therapy of anti-CD5 IgG (H65)-RTA with glucocorticoids showed no long-term clinical benefits in GVHD patients [144].

Overall response rates >10% were observed in trials of NBI-3001 [121], Xomazyme-791 [145], TGFα-PE40 [146], scFv(FRP5)-ETA [147], XMMME-001-RTA [88], SPV-T3a-dgA/WT1-dgA [148], anti-B4-bR [82,149], IgG-HD37-dgA [150], B43-PAP [151], Combotox [152], IgG-FRB4-dgA [79,153,154], RFT5-SMPT-dgA [86], Ki-4.dgA [155], and N901-bRicin [156].

Common toxicities observed with different immunotoxin trials are (1) transient vascular leak syndrome (VLS) consisting of edema, hypoalbuminemia, weight gain, and dyspnea; (2) transient hepatotoxicity with elevated transaminases; and (3) infusion-related hypersensitivity with fever, chills, nausea, vomiting, myalgias, arthralgias, and asthenia. These common toxicities may be due to binding of the toxin moieties to normal tissues. Reaction of immunotoxins with macrophages, lymphocytes, or endothelium may lead to cytokine release with consequent infusion-related hypersensitivity.

The presence of target antigens on normal tissues can lead to significant toxicities. LMB-1, LMB-7, LMB-9, and BR96 (sFv)-PE40 are PE immunotoxins that bind the Lewis Y antigen on many carcinomas. The Lewis Y antigen is also expressed on normal gastrointestinal (GI) epithelium and within the pancreas [157]. Dose-limiting toxicity was GI toxicity in these trials except for a Phase I trial of LMB-1. 260F9-rRA is a ricin immunotoxin that binds a 55-kDa antigen expressed on breast carcinomas and normal Schwann cells [158]. Two Phase I clinical trials of 250F9-rRA have been stopped without clinical responses after the occurrence of reversible but prolonged peripheral neutropathies [91,158]. Clinical trials with OVB3-PE (anti-ovarian cancer antigen PE immunotoxin) and 454A12-rA (antitransferrin receptor ricin immunotoxin) were discontinued after CNS toxicity [159,160]. OVB3-PE and 454A12-rA cross-react with a normal antigen on the pons and transferrin receptors on normal brain capillaries, respectively. $DAB_{389}EGF$ and erb-38 react with EGFR and Her2 receptors present on normal hepatocytes, resulting in dose-limiting liver injury [113,124,161]. $DT_{388}GMCSF$ reacts with normal liver macrophages, Kupffer cells. Subsequent cytokine release triggers dose-limiting hepatocyte damage [162]. Transferrin-CRM107 produces cerebral edema because of cross-reactivity with trasferrin receptors on normal brain capillaries and subsequent damage [139]. HUS (hemolytic uremic syndrome) seems to be a dose-limiting toxicity restricted to CD22 and CD19 targeting immunotoxins [133,134,163,164], because other immunotoxins do not cause HUS.

18.5 CHALLENGES AND PERSPECTIVE OF CLASSIC IMMUNOTOXINS

18.5.1 Adverse Events

As discussed above, three major adverse events (vascular leak syndrome (VLS), infusion-related hypersensitivity, and liver toxicity) in clinical trials were commonly observed in all classic immunotoxin trials. Acute infusion-related hypersensitivity is characterized by fever, chills, nausea, vomiting, myalgias, arthralgias, hypotension, dyspnea, chest pain, and back pain whose symptoms occur within 24 h of drug infusion. Acute infusion-related hypersensitivity occurred in up to 80% of patients receiving monoclonal antibody or other protein-based therapies [166]. Acute hypersensitivity events are common in most of protein-based therapies. Liver toxicity may be related to nonspecific uptake of imunotoxins by Kupffer cells. The major dose-limiting toxicity of classic immunotoxin therapies is VLS. VLS is characterized by hypotension, peripheral edema, and hypoalbuminemia. VLS in severe form causes pulmonary and cardiovascular failure. Symptoms are variable among patients and the causes remain unclear. The pathogenesis of endothelial cell (EC) damage is complex and includes activation or damage to ECs and leukocytes, release of cytokines and of inflammatory mediators, and change in cell–cell and cell–matrix adhesion. VLS is a syndrome observed with IL-2, monoclonal antibodies, some chemotherapy drugs such as taxotere, and many immunotoxins. Higher incidence of VLS is found in immunotoxin trials than in other protein-based therapy trials. This may be related to the presence of potential VLS motifs in classic immunotoxins. VLS motifs, $(x)D(y)$ where $x = L, I, G$, or V and $y = V, L$, or S, are commonly found in toxins, ribosome-inactivating proteins, and IL-2 [202]. VLS restricts the doses of classic immunotoxins and necessitates the cessation of therapy in severe cases. Owing to VLS, congestive heart failure (CHF) is a contraindication to the treatment with classic immunotoxins. A total of 65% of CHF patients have clinically unrecognized fluid overload [203]. Under clinical setting, fluid infusion for adequate hydration and immunotoxin treatment could induce fluid overload and VLS, aggravating CHF. It is also difficult to distinguish between VLS and CHF because symptoms are overlapping. One way to reduce the VLS incidence is to use prophylactic steroids. Premedication of steroids improved tolerability of $DAB_{386}IL2$ without compromising the clinical response [166]. In one trial, the VLS incidence was significantly reduced from 27% to 13% as compared to the Phase III studies [166]. In addition to steroid, theophylline, terbutaline, thalidomide, and immunoglobulins are used to treat systemic capillary leak syndrome (SCLS) [204–207]. Prophylactic use of theophylline, terbutaline, and thalidomide may enhance tolerability of immunotoxins. Since interactions of the toxin moiety with endothelial cells and/or cell matrix appear to be a major cause of VLS, integrin-binding motif peptides of fibronectin and vitronectin as antagonists may also reduce the VLS incidence. Clinical experience with VLS can also minimize the risk of the VLS complications due to early intervention.

18.5.2 Immunogenicity

Immunogenicity of immunotoxins hinders repeated dosing so that a small number of immunotoxins have produced objective clinical efficacy in clinical trials. Immunotoxins have two parts, ligands and toxins. Except for monoclonal antibodies and antibody fragments, all ligands are derived from human. Monoclonal antibodies and antibody fragments can be humanized. The immunogenicity problem of the immunotoxins is attributed to the toxin parts. Most individuals are also immunized with diphtheria toxoid as children. Some patients have subclinical Pseudomonas infection. And patients exposed to castor oil may have anti-ricin antibodies. These circulating antibodies reduce the half-life, $C_{\max}$ (maximum plasma concentration of drug) and AUC (area under the curve) for immunotoxins. Further, treatment of immunotoxins boosts antitoxin antibody production in most patients. Interestingly, in seven out of eight patients treated with A-dmDT390-bisFv(UCHT1), antidrug antibody titers declined below an acceptable level for retreatment within a year (unpublished data). In this trial, a patient with an antidrug antibody titer of 61.1 μg/mL has a durable PR lasting 22+ months. These results are encouraging because patients can be retreated yearly with DT immunotoxins. In addition, 67.6% of children and adolescents have anti-DT antibody titers <2 μg/mL, which did not block *in vivo* efficacy of DT_{388}-GMCSF [208].

There are three ways to overcome the blocking effect of the preexisting antitoxin antibodies in human sera. One is to modify the toxin structure for deimmunization, another is to neutralize the antibodies with nontoxic toxin mutants, and the other is to use two or three different immuntoxins with different toxin moieties but the same ligand. Many efforts have been made for humanization or deimmunization of the toxins. In particular, PE and DT have been modified to minimize immune responses. The deimmunized DT by extensive mutagenesis on T-cell epitopes of DT [209] and the deimmunized PE by eight point mutations on B-cell epitopes of PE38 [51] will be tested in clinical trials in near future. Time to patient retreatment with deimmunized DT and PE immunotoxins may be shorter as compared to immunogenic DT and PE immunotoxins. The other method is to neutralize existing antitoxin antibodies by infusion of nontoxic toxin mutants such as CRM197 (DT mutant). This hypothesis has been tested in rhesus monkeys using FN18-CRM9 chemical conjugate (a monkey equivalent of anti-CD3 DT immunotoxin) [210]. 100-fold excess CRM197 (20 mg/kg) before FN18-CRM9 (0.2 mg/kg) is infused in two rhesus monkeys with preexisting anti-DT antibodies. In one monkey, this treatment produces a similar T-cell depletion effect of FN18-CRM9 conjugate as observed in rhesus monkeys with no anti-DT antibodies. But the other monkey died because of multiple kidney infarcts. This could be related to either formation of immune complex precipitates or weak toxicity of CRM197 or both. Although CRM197 is known as a nontoxic DT, further studies demonstrated that CRM197 has about 10^{-6} the toxicity of wild-type DT because of weak ADP-ribosylating activity. In addition, a large dose of CRM197 (50 mg/kg) caused death of DT-sensitive animals [211]. Therefore, it would be worth testing safety and neutralizing efficacy of different nontoxic DT mutants in animals. The last approach is to use two or three different immuntoxins with different toxin moieties but the same ligand. For example, the first course of immunotoxin therapy for CTCL patients would use an anti-CD3 DT immunotoxin, the second course would use an anti-CD3 PE immuntoxin, and the third course of immunotoxin therapy would use an anti-CD3 saporin immunotoxin.

18.6 CONCLUSIONS

Cytotoxic therapy with classic immunotoxins is a potential treatment option for chemo-resistant relapsed/refractory cancers because classic immunotoxins can selectively kill target malignant cells by a different mechanism than conventional chemotherapy. However, their clinical uses are limited by immunogenicity and unwanted toxicity such as VLS. To overcome these problems, studies on (1) development of a standard clinical protocol for premedication and antagonists for VLS motif binding, (2) improvement of classic immunotoxin potencies, and (3) a new strategy to avoid immunogenicity such as the use of deimmunized toxin are needed. The success of these studies would warrant safety and efficacy of classic immunotoxins in clinic.

REFERENCES

1. Greenfield L, Johnson VG, and Youle RJ. (1987) Mutations in diphtheria toxin separate binding from entry and amplify immunotoxin selectivity. *Science* 238, 536–539.
2. Pirker R, FitzGerald DJ, Hamilton TC, Ozols RF, Willingham MC, Pastan I. (1985) Anti-transferrin receptor antibody linked to Pseudomonas exotoxin as a model immunotoxin in human ovarian carcinoma cell lines. *Cancer Res.* 45, 751–757.
3. Vitetta ES, Krolick KA, Miyama-Inaba M, Cushley W, Uhr JW. (1983) Immunotoxins: a new approach to cancer therapy. *Science* 219, 644–650.
4. Godal A, Fodstad O, Pihl A. (1987) Studies on the mechanism of action of abrin-9.2.27 immunotoxin in human melanoma cell lines. *Cancer Res.* 47, 6243–6247.
5. Wiedlocha A, Sandvig K, Walzel H, Radzikowsky C, Olsnes S. (1991) Internalization and action of an immunotoxin containing mistletoe lectin A-chain. *Cancer Res.* 51, 916–920.
6. Munoz R, Arias Y, Ferreras JM, Rojo MA, Gayoso MJ, Nocito M. (2007) Targeting a marker of the tumour neovasculature using a novel anti-human CD105-immunotoxin containing the non-toxic type 2 ribosome-inactivating protein nigrin b. *Cancer Lett.* 256, 73–80.
7. Lambert JM, Senter PD, Yau-Young A, Blattler WA, Goldmacher VS. (1985) Purified immunotoxins that are reactive with human lymphoid cells. Monoclonal antibodies conjugated to the ribosome-inactivating proteins gelonin and the pokeweed antiviral proteins. *J. Biol. Chem.* 260, 12035–12041.
8. Siena S, Lappi DA, Bregni M, Formosa A, Villa S, Soria M. (1988) Synthesis and characterization of an antihuman T-lymphocyte saporin immunotoxin (OKT1-SAP) with *in vivo* stability into nonhuman primates. *Blood* 72, 756–765.
9. Ramakrishnan S, Houston LL. (1984) Prevention of growth of leukemia cells in mice by monoclonal antibodies directed against Thy 1.1 antigen disulfide linked to two ribosomal inhibitors:pokeweed antiviral protein or ricin A chain. *Cancer Res.* 44, 1398–1404.
10. Ramakrishnan S, Enghlid JJ, Bryant HL. Jr., Xu FJ. (1989) Characterization of a translation inhibitory protein from *Luffa aegyptiaca. Biochem. Biophys. Res. Commun.* 160, 509–516.
11. Bolognesi A, Polito L, Tazzari PL, Lemoli RM, Lubelli C, Fogli M. (2000) *In vitro* anti-tumor activity of anti-CD80 and anti-CD86 immunotoxins containing type 1 ribosome-inactivating proteins. *Br. J. Haematol.* 110, 351–361.
12. Casellas P, Dussossoy D, Falasca AI, Barbieri L, Guillemot JC, Ferrara P. (1988) Trichokirin, a ribosome-inactivating protein from the seeds of Trichosanthes kirilowii Maximowicz. Purification, partial characterization and use for preparation of immunotoxins. *Eur. J. Biochem.* 176, 581–588.
13. Wang QC, Ying WB, Xie H, Zhang ZC, Yang ZH, Ling LQ. (1991) Trichosanthin-monoclonal antibody conjugate specifically cytotoxic to human hepatoma cells *in vitro*. *Cancer Res.* 51, 3353–3355.
14. Ovadia M, Hager CC, Oeltmann TN. (1990) An antimelanoma-barley ribosome inactivating protein conjugate is cytotoxic to melanoma cells *in vitro*. *Anticancer Res.* 10, 671–675.
15. Citores L, Ferreras JM, Munoz R, Benitez J, Jimenez P, Girbes T. (2002) Targeting cancer cells with transferrin conjugates containing the non-toxic type 2 ribosome-inactivating proteins nigrin b or ebulin l. *Cancer Lett.* 184, 29–35.
16. Francisco JA, Gawlak SL, Siegall CB. (1997) Construction, expression, and characterization of BD1-G28-5 sFv, a single-chain anti-CD40 immunotoxin containing the ribosome-inactivating protein bryodin 1. *J. Biol. Chem.* 272, 24165–24169.
17. Porro G, Bolognesi A, Caretto P, Gromo G, Lento P, Mistza G, et al. (1993) *In vitro* and *in vivo* properties of an anti-CD5-momordin immunotoxin on normal and neoplastic T lymphocytes. *Cancer Immunol. Immunother.* 36, 346–350.
18. Bolognesi A, Tazzari PL, Tassi C, Gromo G, Gobbi M, Stirpe F. (1992) A comparison of anti-lymphocyte immunotoxins containing different ribosome-inactivating proteins and antibodies. *Clin. Exp. Immunol.* 89, 341–346.
19. Xia HC, Li F, Li Z, Zhang ZC. (2003) Purification and characterization of Moschatin, a novel type I ribosome-inactivating protein from the mature seeds of pumpkin (*Cucurbita moschata*), and preparation of its immunotoxin against human melanoma cells. *Cell Res.* 13, 369–374.
20. Bolognesi A, Tazzari PL, Legname G, Olivieri F, Modena D, Conte R. (1995) Anti-CD30 immunotoxins with native and recombinant dianthin 30. *Cancer Immunol. Immunother.* 40, 109–114.
21. Di Massimo AM, Di Loreto M, Pacilli A, Raucci G, D'Alatri L, Mele A. (1997) Immunoconjugates made of an anti-EGF receptor monoclonal antibody and type 1 ribosome-inactivating proteins from *Saponaria ocymoides* or *Vaccaria pyramidata. Br. J. Cancer* 75, 822–828.
22. Dosio F, Brusa P, Crosasso P, Fruttero C, Cattel L, Bolognesi A. (1996) Synthesis of different immunotoxins composed by

ribosome inactivating proteins non-covalently bound to monoclonal antibody. *Farmaco* 51, 477–482.

23. Rathore D, Nayak SK, Batra JK. (1997) Overproduction of fungal ribotoxin alpha-sarcin in *Escherichia coli*: generation of an active immunotoxin. *Gene* 190, 31–35.
24. Better M, Bernhard SL, Lei SP, Fishwild DM, Carroll SF. (1992) Activity of recombinant mitogillin and mitogillin immunoconjugates. *J. Biol. Chem.* 267, 16712–16718.
25. Parente D, Raucci G, Celano B, Pacilli A, Zanoni L, Canevari S, et al. (1996) Clavin, a type-1 ribosome-inactivating protein from *Aspergillus clavatus* IFO 8605. cDNA isolation, heterologous expression, biochemical and biological characterization of the recombinant protein. *Eur. J. Biochem.* 239, 272–280.
26. Goyal A, Batra JK. (2000) Inclusion of a furin-sensitive spacer enhances the cytotoxicity of ribotoxin restrictocin containing recombinant single-chain immunotoxins. *Biochem. J.* 345(Pt 2), 247–254.
27. Martsev SP, Tsybovsky YI, Stremovskiy OA, Odintsov SG, Balandin TG, Arosio P. (2004) Fusion of the antiferritin antibody VL domain to barnase results in enhanced solubility and altered pH stability. *Protein Eng. Des. Sel.* 17, 85–93.
28. Liu S, Bugge TH, Leppla SH. (2001) Targeting of tumor cells by cell surface urokinase plasminogen activator-dependent anthrax toxin. *J. Biol. Chem.* 276, 17976–17984.
29. Foster KA, Adams EJ, Durose L, Cruttwell CJ, Marks E, Shone CC, et al. (2006) Re-engineering the target specificity of Clostridial neurotoxins—a route to novel therapeutics. *Neurotox. Res.* 9, 101–107.
30. Chovnick A, Schneider WP, Tso JY, Queen C, Chang CN. (1991) A recombinant, membrane-acting immunotoxin. *Cancer Res.* 51, 465–467.
31. Gasanov SE, Rael ED, Gasanov NE, Vernon LP. (1995) *In vitro* evaluation of Pyrularia thionin-anti-CD5 immunotoxin. *Cancer Immunol. Immunother.* 41, 122–128.
32. Osusky M, Teschke L, Wang X, Wong K, Buckley JT. (2008) A chimera of interleukin 2 and a binding variant of aerolysin is selectively toxic to cells displaying the interleukin 2 receptor. *J. Biol. Chem.* 283, 1572–1579.
33. Cohen S, Cahan R, Ben-Dov E, Nisnevitch M, Zaritsky A, Firer MA. (2007) Specific targeting to murine myeloma cells of Cyt1Aa toxin from *Bacillus thuringiensis* subspecies israelensis. *J. Biol. Chem.* 282, 28301–28308.
34. Choe S, Bennett MJ, Fujii G, Curmi PM, Kantardjieff KA, Collier RJ. (1992) The crystal structure of diphtheria toxin. *Nature* 357, 216–222.
35. Tsuneoka M, Nakayama K, Hatsuzawa K, Komada M, Kitamura N, Mekada E. (1993) Evidence for involvement of furin in cleavage and activation of diphtheria toxin. *J. Biol. Chem.* 268, 26461–26465.
36. Gordon VM, Klimpel KR, Arora N, Henderson MA, Leppla SH. (1995) Proteolytic activation of bacterial toxins by eukaryotic cells is performed by furin and by additional cellular proteases. *Infect. Immun.* 63, 82–87.
37. Neville DM, Scharff J, Srinivasachar K. (1993) Anti-T cell immunotoxins: a look at post-endocytotic receptor-mediated routing. *J. Control. Release* 24, 133–144.
38. London E. (1992) Diphtheria toxin: membrane interaction and membrane translocation. *Biochim. Biophys. Acta* 1113, 25–51.
39. Donovan JJ, Simon MI, Draper RK, Montal M. (1981) Diphtheria toxin forms transmembrane channels in planar lipid bilayers. *Proc. Natl. Acad. Sci. USA* 78, 172–176.
40. Kagan BL, Finkelstein A, Colombini M. (1981) Diphtheria toxin fragment forms large pores in phospholipid bilayer membranes. *Proc. Natl. Acad. Sci. USA* 78, 4950–4954.
41. Wiedlocha A, Madshus IH, Mach H, Middaugh CR, Olsnes S. (1992) Tight folding of acidic fibroblast growth factor prevents its translocation to the cytosol with diphtheria toxin as vector. *Embo J.* 11, 4835–4842.
42. Falnes PO, Choe S, Madshus IH, Wilson BA, Olsnes S. (1994) Inhibition of membrane translocation of diphtheria toxin A-fragment by internal disulfide bridges. *J. Biol. Chem.* 269, 8402–8407.
43. Ren J, Kachel K, Kim H, Malenbaum SE, Collier RJ, London E. (1999) Interaction of diphtheria toxin T domain with molten globule-like proteins and its implications for translocation. *Science* 284, 955–957.
44. Ratts R, Zeng H, Berg EA, Blue C, McComb ME, Costello CE, et al. (2003) The cytosolic entry of diphtheria toxin catalytic domain requires a host cell cytosolic translocation factor complex. *J. Cell Biol.* 160, 1139–1150.
45. Van Ness BG, Howard JB, Bodley JW. (1980) ADP-ribosylation of elongation factor 2 by diphtheria toxin. Isolation and properties of the novel ribosyl-amino acid and its hydrolysis products. *J. Biol. Chem.* 255, 10717–10720.
46. Yamaizumi M, Mekada E, Uchida T, Okada Y. (1978) One molecule of diphtheria toxin fragment A introduced into a cell can kill the cell. *Cell* 15, 245–250.
47. Allured VS, Collier RJ, Carroll SF, McKay DB. (1986) Structure of exotoxin A of *Pseudomonas aeruginosa* at 3.0-Angstrom resolution. *Proc. Natl. Acad. Sci. USA* 83, 1320–1324.
48. Kounnas MZ, Morris RE, Thompson MR, FitzGerald DJ, Strickland DK, Saelinger CB. (1992) The alpha 2-macroglobulin receptor/low density lipoprotein receptor-related protein binds and internalizes Pseudomonas exotoxin A. *J. Biol. Chem.* 267, 12420–12423.
49. Pastrana DV, Hanson AJ, Knisely J, Bu G, Fitzgerald DJ. (2005) LRP 1 B functions as a receptor for Pseudomonas exotoxin. *Biochim. Biophys. Acta* 1741, 234–239.
50. Debinski W, Pastan I. (1994) An immunotoxin with increased activity and homogeneity produced by reducing the number of lysine residues in recombinant Pseudomonas exotoxin. *Bioconjug. Chem.* 5, 40–46.
51. Onda M, Beers R, Xiang L, Nagata S, Wang QC, Pastan I. (2008) An immunotoxin with greatly reduced immunogenicity by identification and removal of B cell epitopes. *Proc. Natl. Acad. Sci. USA* 105, 11311–11316.
52. Weldon JE, Xiang L, Chertov O, Margulies I, Kreitman RJ, FitzGerald DJ, et al. (2009) A protease-resistant immunotoxin against CD22 with greatly increased activity against CLL and diminished animal toxicity. *Blood* 113, 3792–3800.

53. Frankel AE, Burbage C, Fu T, Tagge E, Chandler J, Willingham MC. (1996) Ricin toxin contains at least three galactose-binding sites located in B chain subdomains 1 alpha, 1 beta, and 2 gamma. *Biochemistry* 35, 14749–14756.
54. Steeves RM, Denton ME, Barnard FC, Henry A, Lambert JM. (1999) Identification of three oligosaccharide binding sites in ricin. *Biochemistry* 38, 11677–11685.
55. Llorente A, Rapak A, Schmid SL, van Deurs B, Sandvig K. (1998) Expression of mutant dynamin inhibits toxicity and transport of endocytosed ricin to the Golgi apparatus. *J. Cell. Biol.* 140, 553–563.
56. Grimmer S, Iversen TG, van Deurs B, Sandvig K. (2000) Endosome to Golgi transport of ricin is regulated by cholesterol. *Mol. Biol. Cell* 11, 4205–4216.
57. Girod A, Storrie B, Simpson JC, Johannes L, Goud B, Roberts LM, et al. (1999) Evidence for a COP-I-independent transport route from the Golgi complex to the endoplasmic reticulum. *Nat. Cell Biol.* 1, 423–430.
58. Sandvig K, van Deurs B. (1999) Endocytosis and intracellular transport of ricin: recent discoveries. *FEBS Lett.* 452, 67–70.
59. Wesche J, Rapak A, Olsnes S. (1999) Dependence of ricin toxicity on translocation of the toxin A-chain from the endoplasmic reticulum to the cytosol. *J. Biol. Chem.* 274, 34443–34449.
60. Endo Y, Mitsui K, Motizuki M, Tsurugi K. (1987) The mechanism of action of ricin and related toxic lectins on eukaryotic ribosomes. The site and the characteristics of the modification in 28 S ribosomal RNA caused by the toxins. *J. Biol. Chem.* 262, 5908–5912.
61. Lambert JM, Goldmacher VS, Collinson AR, Nadler LM, Blattler WA. (1991) An immunotoxin prepared with blocked ricin: a natural plant toxin adapted for therapeutic use. *Cancer Res.* 51, 6236–6242.
62. Blakey DC, Skilleter DN, Price RJ, Watson GJ, Hart LI, Newell DR. (1988) Comparison of the pharmacokinetics and hepatotoxic effects of saporin and ricin A-chain immunotoxins on murine liver parenchymal cells. *Cancer Res.* 48, 7072–7078.
63. Thorpe PE, Wallace PM, Knowles PP, Relf MG, Brown AN, Watson GJ. (1988) Improved antitumor effects of immunotoxins prepared with deglycosylated ricin A-chain and hindered disulfide linkages. *Cancer Res.* 48, 6396–6403.
64. Blakey DC, Thorpe PE. (1986) Effect of chemical deglycosylation on the *in vivo* fate of ricin A-chain. *Cancer Drug Deliv.* 3, 189–196.
65. O'Hare M, Roberts LM, Thorpe PE, Watson GJ, Prior B, Lord JM. (1987) Expression of ricin A chain in *Escherichia coli*. *FEBS Lett.* 216, 73–78.
66. Stirpe F, Bailey S, Miller SP, Bodley JW. (1988) Modification of ribosomal RNA by ribosome-inactivating proteins from plants. *Nucleic Acids Res.* 16, 1349–1357.
67. Pirie CM, Hackel BJ, Rosenblum MG, Wittrup KD. (2011) Convergent potency of internalized gelonin immunotoxins across varied cell lines, antigens, and targeting moieties. *J. Biol. Chem.* 286, 4165–4172.
68. Baykal U, Tumer NE. (2007) The C-terminus of pokeweed antiviral protein has distinct roles in transport to the cytosol, ribosome depurination and cytotoxicity. *Plant J.* 49, 995–1007.
69. Thompson J, Stavrou S, Weetall M, Hexham JM, Digan ME, Wang Z, et al. (2001) Improved binding of a bivalent single-chain immunotoxin results in increased efficacy for *in vivo* T-cell depletion. *Protein Eng.* 14, 1035–1041.
70. Kim GB, Wang Z, Liu YY, Stavrou S, Mathias A, Goodwin KJ. (2007) A fold-back single-chain diabody format enhances the bioactivity of an anti-monkey CD3 recombinant diphtheria toxin-based immunotoxin. *Protein Eng. Des. Sel.* 20, 425–432.
71. Vallera DA, Chen H, Sicheneder AR, Panoskaltsis-Mortari A, Taras EP. (2009) Genetic alteration of a bispecific ligand-directed toxin targeting human CD19 and CD22 receptors resulting in improved efficacy against systemic B cell malignancy. *Leuk. Res.* 33, 1233–1242.
72. Oh S, Ohlfest JR, Todhunter DA, Vallera VD, Hall WA, Chen H. (2009) Intracranial elimination of human glioblastoma brain tumors in nude rats using the bispecific ligand-directed toxin, DTEGF13 and convection enhanced delivery. *J. Neurooncol.* 95, 331–342.
73. Stish BJ, Chen H, Shu Y, Panoskaltsis-Mortari A, Vallera DA. (2007) Increasing anticarcinoma activity of an anti-erbB2 recombinant immunotoxin by the addition of an anti-EpCAM sFv. *Clin. Cancer Res.* 13, 3058–3067.
74. Todhunter DA, Hall WA, Rustamzadeh E, Shu Y, Doumbia SO, Vallera DA. (2004) A bispecific immunotoxin (DTAT13) targeting human IL-13 receptor (IL-13R) and urokinase-type plasminogen activator receptor (uPAR) in a mouse xenograft model. *Protein Eng. Des. Sel.* 17, 157–164.
75. Vallera DA, Oh S, Chen H, Shu Y, Frankel AE. (2010) Bioengineering a unique deimmunized bispecific targeted toxin that simultaneously recognizes human CD22 and CD19 receptors in a mouse model of B-cell metastases. *Mol. Cancer Ther.* 9, 1872–1883.
76. van Oosterhout YV, van Emst JL, Bakker HH, Preijers FW, Schattenberg AV, Ruiter DJ. (2001) Production of anti-CD3 and anti-CD7 ricin A-immunotoxins for a clinical pilot study. *Int. J. Pharm.* 221, 175–186.
77. Kernan NA, Knowles RW, Burns MJ, Broxmeyer HE, Lu L, Lee HM. (1984) Specific inhibition of *in vitro* lymphocyte transformation by an anti-pan T cell (gp67) ricin A chain immunotoxin. *J. Immunol.* 133, 137–146.
78. LeMaistre CF, Rosen S, Frankel A, Kornfeld S, Saria E, Meneghetti C. (1991) Phase I trial of H65-RTA immunoconjugate in patients with cutaneous T-cell lymphoma. *Blood* 78, 1173–1182.
79. Vitetta ES, Stone M, Amlot P, Fay J, May R, Till M, et al. (1991) Phase I immunotoxin trial in patients with B-cell lymphoma. *Cancer Res.* 51, 4052–4058.
80. Ghetie V, Ghetie MA, Uhr JW, Vitetta ES. (1988) Large scale preparation of immunotoxins constructed with the Fab' fragment of IgG1 murine monoclonal antibodies and chemically deglycosylated ricin A chain. *J. Immunol. Methods* 112, 267–277.

81. Ghetie V, Thorpe P, Ghetie MA, Knowles P, Uhr JW, Vitetta ES. (1991) The GLP large scale preparation of immunotoxins containing deglycosylated ricin A chain and a hindered disulfide bond. *J. Immunol. Methods* 142, 223–230.
82. Grossbard ML, Lambert JM, Goldmacher VS, Spector NL, Kinsella J, Eliseo L, et al. (1993) Anti-B4-blocked ricin: a Phase I trial of 7-day continuous infusion in patients with B-cell neoplasms. *J. Clin. Oncol.* 11, 726–737.
83. Myers DE, Irvin JD, Smith RS, Kuebelbeck VM, Uckun FM. (1991) Production of a pokeweed antiviral protein (PAP)-containing immunotoxin, B43-PAP, directed against the CD19 human B lineage lymphoid differentiation antigen in highly purified form for human clinical trials. *J. Immunol. Methods* 136, 221–237.
84. Flavell DJ, Flavell SU, Boehm DA, Emery L, Noss A, Ling NR. (1995) Preclinical studies with the anti-CD19-saporin immunotoxin BU12-SAPORIN for the treatment of human-B-cell tumours. *Br. J. Cancer* 72, 1373–1379.
85. Winkler U, Gottstein C, Schon G, Kapp U, Wolf J, Hansmann ML. (1994) Successful treatment of disseminated human Hodgkin's disease in SCID mice with deglycosylated ricin A-chain immunotoxins. *Blood* 83, 466–475.
86. Engert A, Diehl V, Schnell R, Radszuhn A, Hatwig MT, Drillich S, et al. (1997) A Phase-I study of an anti-CD25 ricin A-chain immunotoxin (RFT5-SMPT-dgA) in patients with refractory Hodgkin's lymphoma. *Blood* 89, 403–410.
87. Oratz R, Speyer JL, Wernz JC, Hochster H, Meyers M, Mischak R. (1990) Antimelanoma monoclonal antibody-ricin A chain immunoconjugate (XMMME-001-RTA) plus cyclophosphamide in the treatment of metastatic malignant melanoma: results of a Phase II trial. *J. Biol. Response Mod.* 9, 345–354.
88. Spitler LE, del Rio M, Khentigan A, Wedel NI, Brophy NA, Miller LL, et al. (1987) Therapy of patients with malignant melanoma using a monoclonal antimelanoma antibody-ricin A chain immunotoxin. *Cancer Res.* 47, 1717–1723.
89. Scannon PJ, Spitler LE, Lee HM, Kawahata RT, Mischak PR. (1986) Human melanoma specific immunotoxins. *United States Patent*, 4590071.
90. Bjorn MJ, Ring D, Frankel A. (1985) Evaluation of monoclonal antibodies for the development of breast cancer immunotoxins. *Cancer Res.* 45, 1214–1221.
91. Weiner LM, O'Dwyer J, Kitson J, Comis RL, Frankel AE, Bauer RJ. (1989) Phase I evaluation of an anti-breast carcinoma monoclonal antibody 260F9-recombinant ricin A chain immunoconjugate. *Cancer Res.* 49, 4062–4067.
92. Falini B, Bolognesi A, Flenghi L, Tazzari PL, Broe MK, Stein H, et al. (1992) Response of refractory Hodgkin's disease to monoclonal anti-CD30 immunotoxin. *Lancet* 339, 1195–1196.
93. Tazzari PL, Bolognesi A, de Totero D, Falini B, Lemoli RM, Soria MR, et al. (1992) Ber-H2 (anti-CD30)-saporin immunotoxin: a new tool for the treatment of Hodgkin's disease and CD30+ lymphoma: *in vitro* evaluation. *Br. J. Haematol.* 81, 203–211.
94. Dean A, Talpaz M, Kantarjian H, Faderl S, Jabbour E, FKashani FR. (2010) Phase I clinical trial of the anti-CD33 immunotoxin HuM195/rgel in patients (pts) with advanced myeloid malignancies. *J. Clin. Oncol.* 28, 6549.
95. Pagliaro LC, Liu B, Munker R, Andreeff M, Freireich EJ, Scheinberg DA. (1998) Humanized M195 monoclonal antibody conjugated to recombinant gelonin: an anti-CD33 immunotoxin with antileukemic activity. *Clin. Cancer Res.* 4, 1971–1976.
96. Batra JK, Jinno Y, Chaudhary VK, Kondo T, Willingham MC, FitzGerald DJ. (1989) Antitumor activity in mice of an immunotoxin made with anti-transferrin receptor and a recombinant form of Pseudomonas exotoxin. *Proc. Natl. Acad. Sci. USA* 86, 8545–8549.
97. Johnson VG, Wilson D, Greenfield L, Youle RJ. (1988) The role of the diphtheria toxin receptor in cytosol translocation. *J. Biol. Chem.* 263, 1295–1300.
98. Woo JH, Liu YY, Mathias A, Stavrou S, Wang Z, Thompson J, (2002) Gene optimization is necessary to express a bivalent anti-human anti-T cell immunotoxin in *Pichia pastoris*. *Protein Expr. Purif.* 25, 270–282.
99. Liu YY, Gordienko I, Mathias A, Ma S, Thompson J, Woo JH, et al. (2000) Expression of an anti-CD3 single-chain immunotoxin with a truncated diphtheria toxin in a mutant CHO cell line. *Protein Expr. Purif.* 19, 304–311.
100. Lombardi A, Bursomanno S, Lopardo T, Traini R, Colombatti M, Ippoliti R. (2010) *Pichia pastoris* as a host for secretion of toxic saporin chimeras. *FASEB J.* 24, 253–265.
101. Liu YY, Woo JH, Neville DM. (2003) Targeted introduction of a diphtheria toxin resistant mutation into the chromosomal EF-2 locus of *Pichia pastoris* and expression of immunotoxin in the EF-2 mutants. *Protein Expr. Purif.* 30, 262–274.
102. Woo JH, Liu YY, Stavrou S, Neville DM. Jr. (2004) Increasing secretion of a bivalent anti-T-cell immunotoxin by *Pichia pastoris*. *Appl. Environ. Microbiol.* 70, 3370–3376.
103. Woo JH, Liu YY, Neville DM. Jr. (2006) Minimization of aggregation of secreted bivalent anti-human T cell immunotoxin in *Pichia pastoris* bioreactor culture by optimizing culture conditions for protein secretion. *J. Biotechnol.* 121, 75–85.
104. Williams DP, Snider CE, Strom TB, Murphy JR. (1990) Structure/function analysis of interleukin-2-toxin (DAB486-IL-2). Fragment B sequences required for the delivery of fragment A to the cytosol of target cells. *J. Biol. Chem.* 265, 11885–11889.
105. Williams DP, Parker K, Bacha P, Bishai W, Borowski M, Genbauffe F, et al. (1987) Diphtheria toxin receptor binding domain substitution with interleukin-2: genetic construction and properties of a diphtheria toxin-related interleukin-2 fusion protein. *Protein Eng.* 1, 493–498.
106. Woo JH, Liu JS, Kang SH, Singh R, Park SK, Su Y. (2008) GMP production and characterization of the bivalent anti-human T cell immunotoxin, A-dmDT390-bisFv (UCHT1) for Phase I/II clinical trials. *Protein Expr. Purif.* 58, 1–11.
107. Woo JH, Neville DM.Jr. (2003) Separation of bivalent anti-T cell immunotoxin from *Pichia pastoris* glycoproteins by borate anion exchange. *Biotechniques* 35, 392–398.

108. Urieto JO, Liu T, Black JH, Cohen KA, Hall PD, Willingham MC, (2004) Expression and purification of the recombinant diphtheria fusion toxin DT388IL3 for Phase I clinical trials. *Protein Expr. Purif.* 33, 123–133.

109. Shaw JP, Akiyoshi DE, Arrigo DA, Rhoad AE, Sullivan B, Thomas J, (1991) Cytotoxic properties of DAB486EGF and DAB389EGF, epidermal growth factor (EGF) receptor-targeted fusion toxins. *J. Biol. Chem.* 266, 21118–21124.

110. Frankel AE, Ramage J, Latimer A, Feely T, Delatte S, Hall P, (1999) High-level expression and purification of the recombinant diphtheria fusion toxin DTGM for Phase I clinical trials. *Protein Expr. Purif.* 16, 190–201.

111. Hassan R, Bullock S, Premkumar A, Kreitman RJ, Kindler H, Willingham MC, (2007) Phase I study of SS1P, a recombinant anti-mesothelin immunotoxin given as a bolus I.V. infusion to patients with mesothelin-expressing mesothelioma, ovarian, and pancreatic cancers. *Clin. Cancer Res.* 13, 5144–5149.

112. Batra JK, FitzGerald D, Gately M, Chaudhary VK, Pastan I. (1990) Anti-Tac(Fv)-PE40, a single chain antibody Pseudomonas fusion protein directed at interleukin 2 receptor bearing cells. *J. Biol. Chem.* 265, 15198–15202.

113. Pai-Scherf LH, Villa J, Pearson D, Watson T, Liu E, Willingham MC, (1999) Hepatotoxicity in cancer patients receiving erb-38, a recombinant immunotoxin that targets the erbB2 receptor. *Clin. Cancer Res.* 5, 2311–2315.

114. Reiter Y, Brinkmann U, Kreitman RJ, Jung SH, Lee B, Pastan I. (1994) Stabilization of the Fv fragments in recombinant immunotoxins by disulfide bonds engineered into conserved framework regions. *Biochemistry* 33, 5451–5459.

115. Joshi BH, Puri RK. (2005) Optimization of expression and purification of two biologically active chimeric fusion proteins that consist of human interleukin-13 and Pseudomonas exotoxin in *Escherichia coli*. *Protein Expr. Purif.* 39, 189–198.

116. Kreitman RJ, Pastan I. (1993) Purification and characterization of IL6-PE4E, a recombinant fusion of interleukin 6 with Pseudomonas exotoxin. *Bioconjug. Chem.* 4, 581–585.

117. Kreitman RJ, Wilson WH, White JD, Stetler-Stevenson M, Jaffe ES, Giardina S, (2000) Phase I trial of recombinant immunotoxin anti-Tac(Fv)-PE38 (LMB-2) in patients with hematologic malignancies. *J. Clin. Oncol.* 18, 1622–1636.

118. Edwards GM, DeFeo-Jones D, Tai JY, Vuocolo GA, Patrick DR, Heimbrook DC, (1989) Epidermal growth factor receptor binding is affected by structural determinants in the toxin domain of transforming growth factor-alpha-Pseudomonas exotoxin fusion proteins. *Mol. Cell. Biol.* 9, 2860–2867.

119. Mansfield E, Amlot P, Pastan I, FitzGerald DJ. (1997) Recombinant RFB4 immunotoxins exhibit potent cytotoxic activity for CD22-bearing cells and tumors. *Blood* 90, 2020–2026.

120. Alderson RF, Kreitman RJ, Chen T, Yeung P, Herbst R, Fox JA, et al. (2009) CAT-8015: a second-generation pseudomonas exotoxin A-based immunotherapy targeting CD22-expressing hematologic malignancies. *Clin. Cancer Res.* 15, 832–839.

121. Rand RW, Kreitman RJ, Patronas N, Varricchio F, Pastan I, Puri RK. (2000) Intratumoral administration of recombinant circularly permuted interleukin-4-Pseudomonas exotoxin in patients with high-grade glioma. *Clin. Cancer Res.* 6, 2157–2165.

122. Debinski W, Puri RK, Kreitman RJ, Pastan I. (1993) A wide range of human cancers express interleukin 4 (IL4) receptors that can be targeted with chimeric toxin composed of IL4 and Pseudomonas exotoxin. *J. Biol. Chem.* 268, 14065–14070.

123. Debinski W, Karlsson B, Lindholm L, Siegall CB, Willingham MC, FitzGerald D, et al. (1992) Monoclonal antibody C242-Pseudomonas exotoxin A. A specific and potent immunotoxin with antitumor activity on a human colon cancer xenograft in nude mice. *J. Clin. Invest.* 90, 405–411.

124. von Minckwitz G, Harder S, Hovelmann S, Jager E, Al-Batran SE, Loibl S, et al. (2005) Phase I clinical study of the recombinant antibody toxin scFv(FRP5)-ETA specific for the ErbB2/HER2 receptor in patients with advanced solid malignomas. *Breast Cancer Res.* 7, R617–626.

125. Altenschmidt U, Schmidt M, Groner B, Wels W. (1997) Targeted therapy of schwannoma cells in immunocompetent rats with an erbB2-specific antibody-toxin. *Int. J. Cancer* 73, 117–124.

126. Pastan IH, Archer GE, McLendon RE, Friedman HS, Fuchs HE, Wang QC, et al. (1995) Intrathecal administration of single-chain immunotoxin, LMB-7 [B3(Fv)-PE38], produces cures of carcinomatous meningitis in a rat model. *Proc. Natl. Acad. Sci. USA* 92, 2765–2769.

127. Posey JA, Khazaeli MB, Bookman MA, Nowrouzi A, Grizzle WE, Thornton J, et al. (2002) A Phase I trial of the single-chain immunotoxin SGN-10 (BR96 sFv-PE40) in patients with advanced solid tumors. *Clin. Cancer Res.* 8, 3092–3099.

128. Friedman PN, McAndrew SJ, Gawlak SL, Chace D, Trail PA, Brown JP, et al. (1993) BR96 sFv-PE40, a potent single-chain immunotoxin that selectively kills carcinoma cells. *Cancer Res.* 53, 334–339.

129. Olsen E, Duvic M, Frankel A, Kim Y, Martin A, Vonderheid E, et al. (2001) Pivotal Phase III trial of two dose levels of denileukin diftitox for the treatment of cutaneous T-cell lymphoma. *J. Clin. Oncol.* 19, 376–388.

130. Prince HM, Duvic M, Martin A, Sterry W, Assaf C, Sun Y, et al. (2010) Phase III placebo-controlled trial of denileukin diftitox for patients with cutaneous T-cell lymphoma. *J. Clin. Oncol.* 28, 1870–1877.

131. Rasku MA, Clem AL, Telang S, Taft B, Gettings K, Gragg H, et al. (2008) Transient T cell depletion causes regression of melanoma metastases. *J. Transl. Med.* 6, 12.

132. Dang NH, Pro B, Hagemeister FB, Samaniego F, Jones D, Samuels BI, et al. (2007) Phase II trial of denileukin diftitox for relapsed/refractory T-cell non-Hodgkin lymphoma. *Br. J. Haematol.* 136, 439–447.

133. Kreitman RJ, Squires DR, Stetler-Stevenson M, Noel P, FitzGerald DJ, Wilson WH, et al. (2005) Phase I trial of recombinant immunotoxin RFB4(dsFv)-PE38 (BL22) in patients with B-cell malignancies. *J. Clin. Oncol.* 23, 6719–6729.

134. Kreitman RJ, Stetler-Stevenson M, Margulies I, Noel P, Fitzgerald DJ, Wilson WH, et al. (2009) Phase II trial of recombinant immunotoxin RFB4(dsFv)-PE38 (BL22) in patients with hairy cell leukemia. *J. Clin. Oncol.* 27, 2983–2990.

135. Salvatore G, Beers R, Margulies I, Kreitman RJ, Pastan I. (2002) Improved cytotoxic activity toward cell lines and fresh leukemia cells of a mutant anti-CD22 immunotoxin obtained by antibody phage display. *Clin. Cancer Res.* 8, 995–1002.

136. Kreitman RJ, Tallman MS, Coutre SE, Robak T, Wilson WH, Stetler-Stevenson M, et al. (2010) Phase I trial of recombinant immunotoxin CAT-8015 (HA22) in multiply relapsed hairy cell leukemia. *J. Clin. Oncol.* 28, 6523.

137. Powell DJ. Jr., Felipe-Silva A, Merino MJ, Ahmadzadeh M, Allen T, Levy C, et al. (2007) Administration of a CD25-directed immunotoxin, LMB-2, to patients with metastatic melanoma induces a selective partial reduction in regulatory T cells *in vivo*. *J. Immunol.* 179, 4919–4928.

138. Frankel AE, Zuckero SL, Mankin AA, Grable M, Mitchell K, Lee YJ, et al. (2009) Anti-CD3 recombinant diphtheria immunotoxin therapy of cutaneous T cell lymphoma. *Curr. Drug Targets* 10, 104–109.

139. Laske DW, Youle RJ, Oldfield EH. (1997) Tumor regression with regional distribution of the targeted toxin TF-CRM107 in patients with malignant brain tumors. *Nat. Med.* 3, 1362–1368.

140. Noonan M. (2007) Celtic Pharma terminates Transmid trial KSB311R/CIII/001 (Feb 7. 2007). Drugscom.

141. Kunwar S, Prados MD, Chang SM, Berger MS, Lang FF, Piepmeier JM, et al. (2007) Direct intracerebral delivery of cintredekin besudotox (IL13-PE38QQR) in recurrent malignant glioma: a report by the Cintredekin Besudotox Intraparenchymal Study Group. *J. Clin. Oncol.* 25, 837–844.

142. Kunwar S, Chang S, Westphal M, Vogelbaum M, Sampson J, Barnett G, et al. (2010) Phase III randomized trial of CED of IL13-PE38QQR vs Gliadel wafers for recurrent glioblastoma. *Neuro. Oncol.* 12, 871–881.

143. Byers VS, Henslee PJ, Kernan NA, Blazar BR, Gingrich R, Phillips GL, et al. (1990) Use of an anti-pan T-lymphocyte ricin a chain immunotoxin in steroid-resistant acute graft-versus-host disease. *Blood* 75, 1426–1432.

144. Martin PJ, Nelson BJ, Appelbaum FR, Anasetti C, Deeg HJ, Hansen JA, et al. (1996) Evaluation of a CD5-specific immunotoxin for treatment of acute graft-versus-host disease after allogeneic marrow transplantation. *Blood* 88, 824–830.

145. Byers VS, Rodvien R, Grant K, Durrant LG, Hudson KH, Baldwin RW, et al. (1989) Phase I study of monoclonal antibody-ricin A chain immunotoxin XomaZyme-791 in patients with metastatic colon cancer. *Cancer Res.* 49, 6153–6160.

146. Goldberg MR, Heimbrook DC, Russo P, Sarosdy MF, Greenberg RE, Giantonio BJ, et al. (1995) Phase I clinical study of the recombinant oncotoxin TP40 in superficial bladder cancer. *Clin. Cancer Res.* 1, 57–61.

147. Azemar M, Djahansouzi S, Jager E, Solbach C, Schmidt M, Maurer AB, et al. (2003) Regression of cutaneous tumor lesions in patients intratumorally injected with a recombinant single-chain antibody-toxin targeted to ErbB2/HER2. *Breast Cancer Res. Treat.* 82, 155–164.

148. van Oosterhout YV, van Emst L, Schattenberg AV, Tax WJ, Ruiter DJ, Spits H, et al. (2000) A combination of anti-CD3 and anti-CD7 ricin A-immunotoxins for the *in vivo* treatment of acute graft versus host disease. *Blood* 95, 3693–3701.

149. Grossbard ML, Freedman AS, Ritz J, Coral F, Goldmacher VS, Eliseo L, et al. (1992) Serotherapy of B-cell neoplasms with anti-B4-blocked ricin: a Phase I trial of daily bolus infusion. *Blood* 79, 576–585.

150. Stone MJ, Sausville EA, Fay JW, Headlee D, Collins RH, Figg WD, et al. (1996) A Phase I study of bolus versus continuous infusion of the anti-CD19 immunotoxin, IgG-HD37-dgA, in patients with B-cell lymphoma. *Blood* 88, 1188–1197.

151. Seibel NL, Krailo M, O'Neill K, Franklin J, Uckun F, Reaman GH. (1999) Phase I study of B43-PAP immunotoxin in combination with standard 4-drug induction for patients with CD19+ acute lymphoblastic leukemia (ALL) in relapse, a children's cancer group study. *Pediatr. Res.* 45, 775.

152. Herrera L, Bostrom B, Gore L, Sandler E, Lew G, Schlegel PG, et al. (2009) A Phase 1 study of Combotox in pediatric patients with refractory B-lineage acute lymphoblastic leukemia. *J Pediatr Hematol Oncol* 31, 936–941.

153. Sausville EA, Headlee D, Stetler-Stevenson M, Jaffe ES, Solomon D, Figg WD, et al. (1995) Continuous infusion of the anti-CD22 immunotoxin IgG-RFB4-SMPT-dgA in patients with B-cell lymphoma: a Phase I study. *Blood* 85, 3457–3465.

154. Amlot PL, Stone MJ, Cunningham D, Fay J, Newman J, Collins R, et al. (1993) A Phase I study of an anti-CD22-deglycosylated ricin A chain immunotoxin in the treatment of B-cell lymphomas resistant to conventional therapy. *Blood* 82, 2624–2633.

155. Schnell R, Staak O, Borchmann P, Schwartz C, Matthey B, Hansen H, et al. (2002) A Phase I study with an anti-CD30 ricin A-chain immunotoxin (Ki-4.dgA) in patients with refractory CD30+ Hodgkin's and non-Hodgkin's lymphoma. *Clin. Cancer Res.* 8, 1779–1786.

156. Fidias P, Grossbard M, Lynch TJ. Jr. (2002) A Phase II study of the immunotoxin N901-blocked ricin in small-cell lung cancer. *Clin. Lung Cancer* 3, 219–222.

157. Hellstrom I, Garrigues HJ, Garrigues U, Hellstrom KE. (1990) Highly tumor-reactive, internalizing, mouse monoclonal antibodies to Le(y)-related cell surface antigens. *Cancer Res.* 50, 2183–2190.

158. Gould BJ, Borowitz MJ, Groves ES, Carter PW, Anthony D, Weiner LM, et al. (1989) Phase I study of an anti-breast cancer immunotoxin by continuous infusion: report of a targeted toxic effect not predicted by animal studies. *J. Natl. Cancer Inst.* 81, 775–781.

159. Pai LH, Bookman MA, Ozols RF, Young RC, Smith JW, 2nd, Longo DL, et al. (1991) Clinical evaluation of intraperitoneal Pseudomonas exotoxin immunoconjugate OVB3-PE in patients with ovarian cancer. *J. Clin. Oncol.* 9, 2095–2103.

160. Bookman MA, Godfrey S, Padavic K, Griffin T, Corda JP, Hamilton TC, et al. (1990) Anti-transferrin receptor immunotoxin (IT) therapy: Phase I intraperitoneal (i.p.) trial. *Proc. ASCO* 9, 187.

161. Cohen KA, Liu T, Bissonette R, Puri RK, Frankel AE. (2003) DAB389EGF fusion protein therapy of refractory glioblastoma multiforme. *Curr. Pharm. Biotechnol.* 4, 39–49.

162. Frankel AE, Powell BL, Hall PD, Case LD, Kreitman RJ. (2002) Phase I trial of a novel diphtheria toxin/granulocyte macrophage colony-stimulating factor fusion protein (DT388GMCSF) for refractory or relapsed acute myeloid leukemia. *Clin. Cancer Res.* 8, 1004–1013.

163. Grossbard ML, Multani PS, Freedman AS, O'Day S, Gribben JG, Rhuda C, et al. (1999) A Phase II study of adjuvant therapy with anti-B4-blocked ricin after autologous bone marrow transplantation for patients with relapsed B-cell non-Hodgkin's lymphoma. *Clin. Cancer Res.* 5, 2392–2398.

164. Messmann RA, Vitetta ES, Headlee D, Senderowicz AM, Figg WD, Schindler J, et al. (2000) A Phase I study of combination therapy with immunotoxins IgG-HD37-deglycosylated ricin A chain (dgA) and IgG-RFB4-dgA (Combotox) in patients with refractory CD19(+), CD22(+) B cell lymphoma. *Clin. Cancer Res.* 6, 1302–1313.

165. LeMaistre CF, Saleh MN, Kuzel TM, Foss F, Platanias LC, Schwartz G, et al. (1998) Phase I trial of a ligand fusion-protein (DAB389IL-2) in lymphomas expressing the receptor for interleukin-2. *Blood* 91, 399–405.

166. Foss FM, Bacha P, Osann KE, Demierre MF, Bell T, Kuzel T. (2001) Biological correlates of acute hypersensitivity events with DAB(389)IL-2 (denileukin diftitox, Ontak®) in cutaneous T-cell lymphoma: decreased frequency and severity with steroid premedication. *Clin. Lymphoma* 1, 298–302.

167. Ho VT, Zahrieh D, Hochberg E, Micale E, Levin J, Reynolds C, et al. (2004) Safety and efficacy of denileukin diftitox in patients with steroid-refractory acute graft-versus-host disease after allogeneic hematopoietic stem cell transplantation. *Blood* 104, 1224–1226.

168. Shaughnessy PJ, Bachier C, Grimley M, Freytes CO, Callander NS, Essell JH, et al. (2005) Denileukin diftitox for the treatment of steroid-resistant acute graft-versus-host disease. *Biol. Blood Marrow Transplant.* 11, 188–193.

169. Frankel AE, Fleming DR, Hall PD, Powell BL, Black JH, Leftwich C, et al. (2003) A Phase II study of DT fusion protein denileukin diftitox in patients with fludarabine-refractory chronic lymphocytic leukemia. *Clin. Cancer Res.* 9, 3555–3561.

170. Morgan SJ, Seymour JF, Prince HM, Westerman DA, Wolf MM. (2004) Confirmation of the activity of the interleukin-2 fusion toxin denileukin diftitox against chemorefractory chronic lymphocytic leukemia, including cases with chromosome 17p deletions and without detectable CD25 expression. *Clin. Cancer Res.* 10, 3572–3575.

171. Frankel AE, Surendranathan A, Black JH, White A, Ganjoo K, Cripe LD. (2006) Phase II clinical studies of denileukin diftitox diphtheria toxin fusion protein in patients with previously treated chronic lymphocytic leukemia. *Cancer* 106, 2158–2164.

172. Bagel J, Garland WT, Breneman D, Holick M, Littlejohn TW, Crosby D, et al. (1998) Administration of DAB389IL-2 to patients with recalcitrant psoriasis: a double-blind, Phase II multicenter trial. *J. Am. Acad. Dermatol.* 38, 938–944.

173. Foss F, Demierre MF, DiVenuti G. (2005) A Phase-1 trial of bexarotene and denileukin diftitox in patients with relapsed or refractory cutaneous T-cell lymphoma. *Blood* 106, 454–457.

174. Dang NH, Hagemeister FB, Pro B, McLaughlin P, Romaguera JE, Jones D, et al. (2004) Phase II study of denileukin diftitox for relapsed/refractory B-Cell non-Hodgkin's lymphoma. *J. Clin. Oncol.* 22, 4095–4102.

175. Dang NH, Fayad L, McLaughlin P, Romaguara JE, Hagemeister F, Goy A, et al. (2007) Phase II trial of the combination of denileukin diftitox and rituximab for relapsed/refractory B-cell non-Hodgkin lymphoma. *Br. J. Haematol.* 138, 502–505.

176. Kuzel TM, Li S, Eklund J, Foss F, Gascoyne R, Abramson N, et al. (2007) Phase II study of denileukin diftitox for previously treated indolent non-Hodgkin lymphoma: final results of E1497. *Leuk. Lymphoma* 48, 2397–2402.

177. Gerena-Lewis M, Crawford J, Bonomi P, Maddox AM, Hainsworth J, McCune DE, et al. (2009) A Phase II trial of denileukin diftitox in patients with previously treated advanced non-small cell lung cancer. *Am. J. Clin. Oncol.* 32, 269–273.

178. LeMaistre CF, Meneghetti C, Rosenblum M, Reuben J, Parker K, Shaw J, et al. (1992) Phase I trial of an interleukin-2 (IL-2) fusion toxin (DAB486IL-2) in hematologic malignancies expressing the IL-2 receptor. *Blood* 79, 2547–2554.

179. Tepler I, Schwartz G, Parker K, Charette J, Kadin ME, Woodworth TG, et al. (1994) Phase I trial of an interleukin-2 fusion toxin (DAB486IL-2) in hematologic malignancies: complete response in a patient with Hodgkin's disease refractory to chemotherapy. *Cancer* 73, 1276–1285.

180. Hesketh P, Caguioa P, Koh H, Dewey H, Facada A, McCaffrey R, et al. (1993) Clinical activity of a cytotoxic fusion protein in the treatment of cutaneous T-cell lymphoma. *J. Clin. Oncol.* 11, 1682–1690.

181. LeMaistre CF, Craig FE, Meneghetti C, McMullin B, Parker K, Reuben J, et al. (1993) Phase I trial of a 90-minute infusion of the fusion toxin DAB486IL-2 in hematological cancers. *Cancer Res.* 53, 3930–3934.

182. Platanias LC, Ratain MJ, O'Brien S, Larson RA, Vardiman JW, Shaw JP, et al. (1994) Phase I trial of a genetically engineered interleukin-2 fusion toxin (DAB486IL-2) as a 6 hour intravenous infusion in patients with hematologic malignancies. *Leuk. Lymphoma* 14, 257–262.

183. Foss FM, Borkowski TA, Gilliom M, Stetler-Stevenson M, Jaffe ES, Figg WD, et al. (1994) Chimeric fusion protein toxin DAB486IL-2 in advanced mycosis fungoides and the Sezary syndrome: correlation of activity and interleukin-2 receptor expression in a Phase II study. *Blood* 84, 1765–1774.

184. Sewell KL, Parker KC, Woodworth TG, Reuben J, Swartz W, Trentham DE. (1993) DAB486IL-2 fusion toxin in refractory rheumatoid arthritis. *Arthritis Rheum.* 36, 1223–1233.

185. Moreland LW, Sewell KL, Trentham DE, Bucy RP, Sullivan WF, Schrohenloher RE, et al. (1995) Interleukin-2 diphtheria fusion protein (DAB486IL-2) in refractory rheumatoid arthritis. A double-blind, placebo-controlled trial with open-label extension. *Arthritis Rheum.* 38, 1177–1186.

186. Frankel A, Liu JS, Rizzieri D, Hogge D. (2008) Phase I clinical study of diphtheria toxin-interleukin 3 fusion protein in patients with acute myeloid leukemia and myelodysplasia. *Leuk. Lymphoma* 49, 543–553.

187. Garland L, Gitlitz B, Ebbinghaus S, Pan H, de Haan H, Puri RK, et al. (2005) Phase I trial of intravenous IL-4 pseudomonas exotoxin protein (NBI-3001) in patients with advanced solid tumors that express the IL-4 receptor. *J. Immunother.* 28, 376–381.

188. Zalcberg JR, Pietersz G, Toohey B, Laird J, Huggins R, Zimet AS, et al. (1994) A Phase I/II study of the intralesional injection of ricin-monoclonal antibody conjugates in patients with hepatic metastases. *Eur. J. Cancer* 30A, 1227–1231.

189. LoRusso PM, Lomen PL, Redman BG, Poplin E, Bander JJ, Valdivieso M. (1995) Phase I study of monoclonal antibody-ricin A chain immunoconjugate Xomazyme-791 in patients with metastatic colon cancer. *Am. J. Clin. Oncol.* 18, 307–312.

190. Pai LH, Wittes R, Setser A, Willingham MC, Pastan I. (1996) Treatment of advanced solid tumors with immunotoxin LMB-1: an antibody linked to Pseudomonas exotoxin. *Nat. Med.* 2, 350–353.

191. Pastan I, Hassan R, FitzGerald DJ, Kreitman RJ. (2007) Immunotoxin treatment of cancer. *Annu. Rev. Med.* 58, 221–237.

192. Hausner PF, Karp JE, Edelman MJ, Kennedy NL, Tathineni N, Willingham M, et al. (2003) Phase I study of LMB-9 (B3 (dsFv)-PE38), a recombinant disulfide stabilized anti-Lewis Y immunotoxin administered by coninuous infusion. *Proc. Am. Soc. Clin. Oncol.* 22, 736.

193. Kreitman RJ, Hassan R, Fitzgerald DJ, Pastan I. (2009) Phase I trial of continuous infusion anti-mesothelin recombinant immunotoxin SS1P. *Clin. Cancer Res.* 15, 5274–5279.

194. Weaver M, Laske DW. (2003) Transferrin receptor ligand-targeted toxin conjugate (Tf-CRM107) for therapy of malignant gliomas. *J. Neurooncol.* 65, 3–13.

195. Multani PS, O'Day S, Nadler LM, Grossbard ML. (1998) Phase II clinical trial of bolus infusion anti-B4 blocked ricin immunoconjugate in patients with relapsed B-cell non-Hodgkin's lymphoma. *Clin. Cancer Res.* 4, 2599–2604.

196. Grossbard ML, Fidias P, Kinsella J, O'Toole J, Lambert JM, Blattler WA, et al. (1998) Anti-B4-blocked ricin: a Phase II trial of 7 day continuous infusion in patients with multiple myeloma. *Br. J. Haematol.* 102, 509–515.

197. Grossbard ML, Gribben JG, Freedman AS, Lambert JM, Kinsella J, Rabinowe SN, et al. (1993) Adjuvant immunotoxin therapy with anti-B4-blocked ricin after autologous bone marrow transplantation for patients with B-cell non-Hodgkin's lymphoma. *Blood* 81, 2263–2271.

198. Scadden DT, Schenkein DP, Bernstein Z, Luskey B, Doweiko J, Tulpule A, et al. (1998) Immunotoxin combined with chemotherapy for patients with AIDS-related non-Hodgkin's lymphoma. *Cancer* 83, 2580–2587.

199. Ortin M. (2005) Immunotherapy of hematological malignancies: What is new? *Ann. Oncol.* 16(Suppl 2), ii53–62.

200. Wayne AS, Kreitman RJ, Findley HW, Lew G, Delbrook C, Steinberg SM, et al. (2010) Anti-CD22 immunotoxin RFB4 (dsFv)-PE38 (BL22) for CD22-positive hematologic malignancies of childhood: preclinical studies and Phase I clinical trial. *Clin. Cancer Res.* 16, 1894–1903.

201. Lynch TJ. Jr., Lambert JM, Coral F, Shefner J, Wen P, Blattler WA, et al. (1997) Immunotoxin therapy of small-cell lung cancer: a Phase I study of N901-blocked ricin. *J. Clin. Oncol.* 15, 723–734.

202. Baluna R, Rizo J, Gordon BE, Ghetie V, Vitetta ES. (1999) Evidence for a structural motif in toxins and interleukin-2 that may be responsible for binding to endothelial cells and initiating vascular leak syndrome. *Proc. Natl. Acad. Sci. USA* 96, 3957–3962.

203. Androne AS, Hryniewicz K, Hudaihed A, Mancini D, Lamanca J, Katz SD. (2004) Relation of unrecognized hypervolemia in chronic heart failure to clinical status, hemodynamics, and patient outcomes. *Am. J. Cardiol.* 93, 1254–1259.

204. Kinoshita Y, Kasaoka S, Fujita M, Oshima C, Kawamura Y, Tsuruta R, et al. (2010) Synchronized changes in serum vascular endothelial growth factor during the clinical course of chronic systemic capillary leak syndrome. *Intern. Med.* 49, 791–794.

205. Droder RM, Kyle RA, Greipp PR. (1992) Control of systemic capillary leak syndrome with aminophylline and terbutaline. *Am. J. Med.* 92, 523–526.

206. Staak JO, Glossmann JP, Esser JM, Diehl V, Mietz H, Josting A. (2003) Thalidomide for systemic capillary leak syndrome. *Am. J. Med.* 115, 332–334.

207. Lambert M, Launay D, Hachulla E, Morell-Dubois S, Soland V, Queyrel V, et al. (2008) High-dose intravenous immunoglobulins dramatically reverse systemic capillary leak syndrome. *Crit. Care. Med.* 36, 2184–2187.

208. Hall PD, Razzouk BI, Willoughby TE, McLean TW, Frankel AE. (2002) The majority of children and adolescents with acute myeloid leukemia have detectable anti-DT388-GMCSF IgG concentrations, but at concentrations that should not preclude *in vivo* activity. *J. Pediatr. Hematol. Oncol.* 24, 521–526.

209. Davis CG, Datta D. (2008) Modified toxins. *PTC Patent Application*, PCT/US08/86858.

210. Thompson J, Hu H, Scharff J, Neville DM. Jr. (1995) An anti-CD3 single-chain immunotoxin with a truncated diphtheria toxin avoids inhibition by pre-existing antibodies in human blood. *J. Biol. Chem.* 270, 28037–28041.

211. Kageyama T, Ohishi M, Miyamoto S, Mizushima H, Iwamoto R, Mekada E. (2007) Diphtheria toxin mutant CRM197 possesses weak EF2-ADP-ribosyl activity that potentiates its anti-tumorigenic activity. *J. Biochem.* 142, 95–104.

19

TARGETED AND UNTARGETED FUSION PROTEINS: CURRENT APPROACHES TO CANCER IMMUNOTHERAPY

LESLIE A. KHAWLI,[1] PEISHENG HU,[2] AND ALAN L. EPSTEIN[2]

[1]*Department of Pharmacokinetic and Pharmacodynamic Sciences, Genentech Research and Early Development, Genentech, Inc, San Francisco, CA, USA*
[2]*Department of Pathology, Keck School of Medicine, University of Southern California, Los Angeles, CA, USA*

19.1 INTRODUCTION

The generation of monoclonal antibodies (mAbs) has resulted in major therapeutic advances across a spectrum of diseases, including cancer, inflammatory, and autoimmune diseases. Antibodies or immunoglobulins (Ig) are generally multifunctional glycoproteins that are produced and secreted by plasma cells and are capable of binding with high affinity and specificity to foreign substances known as antigens. In humans, there are five major classes of antibodies: IgG, IgA, IgD, IgE, and IgM that differ from each other based on their structure and immune function. The IgG class, whose molecular weight is ~150 kDa, is the principle Ig that participates in the adaptive immune response and comprises ~85% of the Ig in normal human serum. The detailed knowledge of antibody structure (e.g., size, charge, valency, amino acid composition, and carbohydrate content) has been considerably extended and therefore helped the intelligent design to improve their pharmacological properties [1]. The full IgG molecule is a tetramer composed of two light chains (LCs) and two heavy chains (HCs) assembled in the "Y" shape motif. The LCs contain an N-terminal variable domain and a C-terminal constant domain, while the HCs consist of a variable N-terminal domain and three to four C-terminal constant domains. The Fab domain (antigen binding fragment) is composed of the LC and the first two N-terminal domains of the HC. The Fc region (crystallizing fragment) is a dimer of the two or more C-terminal domains of the HC that are not part of the Fab. The Fc regulates the biological functions of the antibody by binding to Fc receptors and eliciting immune effector functions. Of particular importance for therapeutic mAbs, the Fc portion of IgG binds pH-dependently to the neonatal Fc receptor (FcRn), which protects mAb from degradation and partially accounts for their long serum half-lives [2]. The unique complementarity-determining regions (CDRs) usually define the antigen specificity and reside in the variable fragment (Fv) portion of the Fab. There are four human IgG subclasses (IgG1, IgG2, IgG3, and IgG4), which are distinguished from one another by differences in the constant region of the HC sequence. Different IgG subclasses bind to different Fc receptors with varying affinities. Thus, IgGs of one subclass may be more or less efficient at eliciting certain effector functions than IgGs of another subclass.

As advances in antibody engineering, including hybridoma and phage-display technologies, have enabled the

Fusion Protein Technologies for Biopharmaceuticals: Applications and Challenges, First Edition. Edited by Stefan R. Schmidt.

development of murine, chimeric, humanized, and fully human mAbs, several novel therapeutic mAb currently approved for cancer immunotherapy have been mass produced [3]. Moreover, it is possible to generate a variety of smaller fragments of the IgG molecule that retain or have improved binding affinities. One of the most common forms is the single chain variable fragment (scFv), produced by combination of V_H and V_L joined by a flexible peptide linker. These mAbs as a class of oncology therapeutics are best used in patients with tumors driven by numerous antigens that are overexpressed on certain types of cancer cells compared with normal tissues. Therefore, immunotherapy is a powerful therapeutic modality that is marked by excitement since it can be employed for the treatment of cancer patients [4]. In addition, many investigators have proposed that if the proper immunoregulatory ligands, toxins, or enzymes can be targeted to the tumor, an effective antitumor response can be mounted to treat patients with both high-risk primary tumors and distant metastatic lesions. Another approach to delivery of these large proteins to the tumor is to create untargeted Fc fusion proteins consisting of a single protein molecule conjugated to a dimeric Fc fragment of an IgG. This FcRn-mediated approach offers evidence that it can exhibit enhanced pharmacokinetic and pharmacodynamic properties [5]. Thus, the present review demonstrates the current use of multiple antibody-targeted and untargeted soluble Fc fusion proteins with emphasis on *in vitro* and *in vivo* properties including the identification of relevant cellular processes associated with effective solid tumor immunotherapy. Those cellular and molecular components of the immune system that have the capacity to treat tumor need to be harnessed and properly presented in order to generate effective cancer therapy.

19.2 IMMUNOTHERAPEUTIC STRATEGY FOR CANCER: FUSION PROTEINS

Harnessing the immune system to treat tumor-bearing hosts is a major goal of immunotherapy, which has been used for over a century [6]. Effective cancer immunotherapy requires the establishment of an immune response to reject cancer in patients. However, impediments to this goal include host failure to identify tumor antigens by the generation of tolerance to self and negative immunoregulatory mechanisms. As part of the malignant phenotype, tumors have evolved to overcome innate and adaptive immune responses of the host in order to grow and metastasize [6]. Historically, investigators have been of the opinion that the immune response against tumors is inherently weak and has insufficient T effector cells to destroy large, established lesions. As evidenced by recent research, however, cancer immunotherapy may have sufficient therapeutic potential to treat and cure deep-seated malignancies if properly invoked [7]. For this to happen, however, two key events must occur, namely, the reversal of immune tolerance and the activation of effective immunity with memory. One of the most powerful methods nature has evolved to control the immune system is the generation of T regulatory (T_{reg}) cells, the same cells now believed to be one of the major mechanisms responsible for the protection of the fetus in the mother's womb [8]. As confirmed previously, histopathologic, and flow cytometric studies, respectively, have shown an increase in T_{reg} cells in the parenchyma of solid tumors or circulating in the blood or cavity fluids of cancer patients [9]. Experimentally, methods to suppress T_{reg} cells in tumor-bearing animals using anti-cytotoxic T lymphocyte-associated antigen 4 (CTLA4) [10], anti-CD4 [11], or anti-CD25 [12,13] antibodies, have provided improved immunotherapy when used alone or with tumor vaccines. The second event that is required for active immunotherapy is the use of potent immunostimulatory agents to evoke an effective immune response against tumor antigens. It has previously been confirmed that combination therapy with an anti-T_{reg} cell antibody and several different fusion proteins is an effective method of reactivating the immune system to tumor [14–17]. By comparison, when these reagents are used alone, they produce only partial responses compared to untreated controls [18,19]. For example, B7-CD28 co-stimulation (second signal) is an important and vital step in the activation of T cells [20]. Interestingly, it has been also demonstrated that B7-CD28 co-stimulation can activate both the Th1 and Th2 differentiation pathways and that the latter is dependent on the induction of interleukin 4 (IL-4). Used systemically, however, this approach can induce unwanted autoimmune side effects such as those seen with anti-CTLA-4 antibody treatment, which affects immunologically privileged sites in the host [21]. These toxic side effects may be substantially decreased if co-stimulation is targeted to the tumor site and not distributed systemically [22].

Despite numerous *in vitro* and *in vivo* studies demonstrating the stimulatory effects of various cytokines to induce active leukocyte responses against tumor cells, favorable and consistent responses have remained elusive in the actual treatment of human disease. The limitation in clinical efficacy is partly due to the toxic effects observed with traditional administration methods as well as the inability to recreate the microenvironmental conditions required for the generation of specific antitumor immune responses using these reagents [23]. Systemic infusion produces elevated cytokine levels that lead to a nonspecific activation of immune cells. Since cytokines are meant to be released at local sites of infection or tissue perturbation, systemic infusion often leads to dysfunctional immune activation with significant dose-limiting and potentially fatal toxicities [24]. Additionally, treatment protocols which fail to localize these bioactive reagents may not only be unsuccessful in the generation of specific immune activity, but may predispose the patient to immunodeficiency states or tolerance [25]. Most recently, it has become recognized that in addition to the induction of an effective immunologic response to tumor, it is also important to produce a lasting response by the development of immunologic memory [26]. Placed in the correct context, the immune system is capable of massive cell destruction as witnessed by graft versus host responses and certain autoimmune states. On the basis of these observations, it is highly feasible that once harnessed, the immune system will be capable of destroying deep-seated primary tumors and metastatic lesions in cancer patients.

In general, there is a variety of antibody fusion proteins targeting various tumor antigens in which different immunoregulatory ligands, toxins, or enzymes have been fused to full-length antibodies or their derivatives. Immune activators, including cytokines, chemokines, and costimulatory fusion proteins, can serve as one class of these potent modulators and have been shown to generate antitumor responses in animal models. A second class of fusion proteins has been developed by combining antibody molecules with truncated tissue factor (tTF) that target the thrombogenic capacity of the tumor vasculature, thereby enabling the initiation of coagulation and occlusion of blood flow within the tumor [27,28]. In addition, attempts have been made to engineer a third class fusion molecules combining antibody and human enzyme together as a single fusion protein [29–31]. This approach is called antibody directed enzyme prodrug therapy (ADEPT), and specifically aims at causing bystander effects by targeting enzymes to the tumor cell and delivering a prodrug that is converted to a chemotherapeutic by the targeted enzyme [32]. Overall, these multiple fusion protein constructs have demonstrated significant antitumor activity as direct therapeutics in preclinical studies, paving the way for their clinical evaluation.

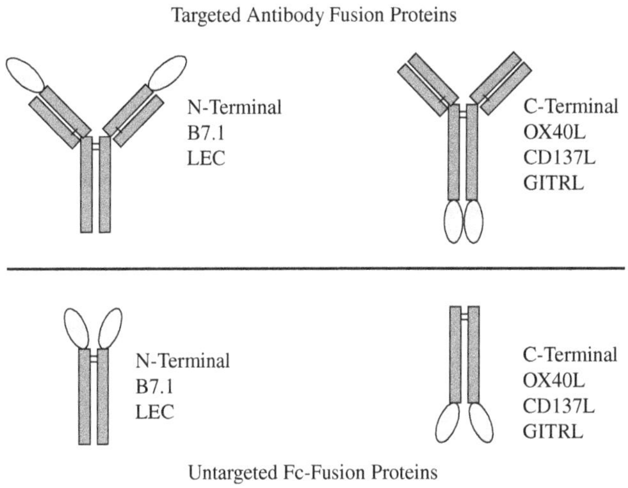

FIGURE 19.1 Schematic of targeted and untargeted fusion proteins.

19.3 IMMUNOTHERAPEUTIC APPLICATIONS OF ANTIBODY-TARGETED AND UNTARGETED Fc FUSION PROTEINS

The development and applications of molecular engineering technology led to the design of several new fusion proteins targeting various tumors antigens in which different immune activators/ligands have been fused to full-length antibodies, single chain Fv (scFv), Fab, $F(ab')_2$, or their Fc derivatives (Figure 19.1). A broad variety of these molecules resulted in improved functions to enhance the tumoricidal activity of the antibodies and/or activate a secondary antitumor immune response; all with promising results, particularly from both academic institutions and biotechnology companies [33]. Each immune activator or ligand can be fused at the amino- (N-terminal) or carboxy-terminus (C-terminal) of the antibody or Fc derivatives (Figure 19.1) to conserve its biological activity. For instance, in an attempt to identify reagents that can provide an enhanced immune stimulation and treatment of cancer, investigators [34–36] have developed a novel mAb targeting approach, designated tumor necrosis therapy (TNT), which targets single stranded DNA exposed in degenerating cells. Because of these attributes, TNT has been used to deliver radionuclides [37,38] to treat experimental and human tumors [39,40].

To ensure the clinical potential of this approach, a variety of ways have been investigated to enhance the antitumor response of the host at the tumor site, thereby minimizing systemic toxicity. To do so, TNT mAb linked to immune effector or ligands molecules was produced by genetic engineering as fusion proteins. These fusion proteins have diverse effects including ability to target solid tumors in immunocompetent mice, overcome peripheral tolerance, and incite an effective antitumor immune response that brings about tumor regression in solid tumor models. As a first approach, a number of TNT fusion proteins were successfully generated with cytokines IL-2 and IL-12, granulocyte macrophage-colony stimulating factor (GM-CSF), tumor necrosis factor-α (TNF-α), and interferon-γ (IFN-γ), which elicited significant immune responses in tumor-bearing hosts. Secondarily, fusion proteins consisting of the cysteine-cysteine class of chemokines (CC) liver-expression chemokine (LEC, CCL16) were generated to chemoattract polymorphonuclear leukocytes (PMNs), T cells, macrophages, B-cells, and dendritic cells into the tumor microenvironment [41]. These studies confirmed the ability of LEC/chTNT-3 to induce a multiarmed immune response against solid tumors by attracting antigen-presenting cells (APC) and effector cells into the tumor to induce significant tumor regression [14]. More recently, it was found that tumor regression can be achieved through the use of fusions proteins consisting of co-stimulatory molecules such as B7.1

[16], the APC molecule that activates T cells by delivering a second signal via binding to CD28. While investigating which T-cell subpopulations are required for LEC and B7.1 activity, it was found that depletion of CD4+ or CD4+CD25+ T_{regs} greatly enhanced LEC and B7.1 activity, resulting in complete and lasting remissions of established solid tumors in BALB/c mice [15,16]. Furthermore, our work provided compelling evidence that fusion proteins consisting of glucocorticoid-induced tumor necrosis factor receptor (TNFR)-related gene ligand (GITRL), CD137L, and OX40L co-stimulatory proteins which are members of the tumor necrosis factor superfamily (TNFSF) can also enhance immunity and are very effective in suppressing tumor growth in syngeneic tumor models of the mouse.

Other applications have been employed including the fusion of soluble co-stimulatory proteins linked to the N-terminus of the Fc portion of antibodies [42,43]. One example, which uses the B7.2 as the Fc-fusion molecule, produced complete tumor rejection in some tumor models including the poorly immunogenic B16/F10 tumor [18]. This kind of therapy used as an adjuvant in cancer vaccines was mediated by type 2 CD8+ T cells as well as dependent on IL-4 induction, therefore inducing tumor rejection [44]. Generally, soluble Fc fusion proteins represent a "nontargeting" approach to providing co-stimulation or immune activation systemically. Because they do not target tumor *in vivo*, however, they could produce autoimmune side effects, which may limit their clinical utility. Conversely, they may be very useful if immune modulation in tumor draining lymph nodes or peripheral blood is important to reversing tolerance in the tumor-bearing host. To determine which reagent(s) are most promising for translation to man, several targeting and nontargeting fusion proteins as shown schematically in Figure 19.1 have been generated and compared in representative murine tumor models. Therefore, the following sections of this review intend to show the effects of different fusion proteins and compare their activity as targeted and untargeted reagents.

19.3.1 Cytokine Fusion Proteins

Cytokines are extracellular protein messenger molecules produced by cells involved in inflammation, immunity, differentiation, cell division, fibrosis, and repair [45]. A distinctive feature of cytokines is that they are not constitutively produced, but are generated in response to stimulation locally. The usual mechanism of action of cytokines is on neighboring cells and because of their short half-life, which is in the order of minutes, only small amounts ever enter the systemic circulation during inflammation. When administered systemically, cytokines are in general very toxic causing multiple symptoms including fever, hypotension, headache, weakness, and capillary leak syndrome thereby making it difficult to achieve clinically relevant dosages. Importantly, cytokines can act in networks and cascades influencing different immune cells once secreted into an inflamed tissue. Because of their potency, a number of mechanisms exist to regulate their effects in tissues.

The first cytokine to obtain approval for cancer therapy was IL-2 [45]. IL-2 is a potent cytokine involved in both the cellular and humoral arms of the immune system. Its primary role is to stimulate the growth and proliferation of T lymphocytes. It has also been shown, however, to have stimulatory effects on a variety of immune cells, including natural killer (NK) cells, lymphokine-activated killer (LAK) cells, and tumor infiltrating lymphocytes (TIL). While it has shown some promise in the treatment of renal cancer and melanoma [46], its potential is offset by the occurrence of serious adverse reactions with its use, including damage to the blood vessels of the body (capillary leak syndrome). Direct applications in which IL-2 is infused intratumorally, has been somewhat more effective and has resulted in a higher level of therapeutic response, including the control of clinical complications, prevention of the growth of established tumors, and reduction in the size of the tumor mass [47]. Unfortunately, this technique is not feasible for disseminated disease.

As mentioned previously, one promising approach to using these potent cytokines is to link them to tumor targeting mAbs. In these early investigations, cytokines were covalently conjugated to various targeting mAbs and demonstrated their potential to cause localized stimulation and activation of immune effector cells in multiple tumor models [48,49]. Toward this end, a series of antibody-cytokine fusion proteins consisting of mAb and IL-2 [50–52], murine IFN-γ [53], human IFN-γ, TNF-α, and muGM-CSF [54] have been genetically linked to the C-terminal portion of a TNT mAb. The biological activity of each of the cytokine moieties of the four fusion proteins was determined by *in vitro* assays, thus measuring their ability to support the proliferation of cytokine dependent cell lines. Clearance and biodistribution studies were also performed to determine their pharmacokinetic clearance half-life and tumor uptake. The results of these studies are shown in Table 19.1, which show marked differences in clearance times between all fusion proteins. The parental antibodies (i.e., chTNT-3 and muTNT-3) were found to have an unusually long half-life in these studies. For all the cytokine fusion proteins, tumor uptake is significantly lower than that seen for parental antibodies. The tumor-to-organ ratios, however, which reflect normal organ uptake and provide an indication of possible toxicity, are either comparable or even slightly higher for all the fusion proteins.

It is important to stress that the above characterization studies demonstrate that cytokine fusion proteins are able to maintain their binding affinity to antigen as well as their direct cytotoxic effect and immunomodulatory functions.

TABLE 19.1 Summary of Pharmacokinetic Clearance and Tumor Uptake Studies of Antibody/Cytokine Fusion Proteins

Antibody/Cytokine Fusion Protein	Half-Life (h)	Tumor Model	% Injected Dose Per Gram (%ID/g) of Tumor at 3 Days
chTNT-3/muIFN-γ	46 (134)	MAD 109	2.4 (13.5)
chTNT-3/huIFN-γ	19.1 (134)	LS 174T	1.2 (7.1)
chTNT-3/huIL-2	12 (134)	LS 174 T	1.7 (7.1)
chTNT-3/huTNF-α	8 (134)	LS 174 T	1.3 (7.1)
muTNT-3/muGM-CSF	15 (150)	COLON 26	1.27 (5.3)

Values in parentheses refer to the representative naked antibody.

In vivo, the fusion proteins were found to have a substantially shorter whole body half-life than parental antibodies, yet were able to target tumor as shown by biodistribution analyses. Because of their retention in tumor and rapid clearance from normal tissues, the fusion proteins were found to have equivalent or higher normal tissue/tumor ratios than parental antibody. In particular, different treatment studies have broadly explored the potential of using cytokine fusion proteins combinations. These observations are presented in Section 4.1 and open the door to superior tumor therapeutic efficacy for these reagents.

19.3.2 Chemokine Fusion Proteins

Chemokines are small (7–15 kDa), secreted, and structurally related soluble proteins that are involved in leukocyte and dendritic cell chemotaxis, PMN degranulation, and angiogenesis [55]. Because of their ability to recruit leukocytes into tumors, alter tumor vasculature structure, and stimulate host anti-cancer immune responses, chemokines are promising reagents in cancer therapy. The human chemokine LEC (CCL16, also known as NCC-4, LMC, and HCC-4) was originally found in an expression sequence tag library and was also found to chemoattract monocytes, lymphocytes, and PMNs on binding to chemokine receptors [56]. LEC is unique because unlike any other chemokine, it is the first chemokine whose mRNA expression is strongly increased and stabilized by the presence of IL-10. The potential therapeutic applications of LEC were first studied by Giovarelli et al. [41], who showed that mammary carcinoma TSA cells engineered to express LEC inhibit the metastatic spread of tumor and induce tumor rejection due to an impressive infiltration of macrophages, dendritic cells, T cells, and PMNs, and the production of IFN-γ and IL-12. Furthermore, LEC is a potent chemotactic factor for both human monocytes and dendritic cells (APC cells) [57]. The ability of LEC to improve markedly the recognition of poorly immunogenic tumor cells by promoting APC-T cell interaction establishes it as a prime candidate for targeted immunotherapy.

Li et al. [14] first described the generation and testing of an antibody/chemokine fusion protein consisting of chTNT-3 and LEC that targets central necrotic areas of tumor to chemoattract and amplify responding lymphoid and dendritic cells capable of inducing an effective antitumor immune response. Since the N-terminus of chemokines is essential for bioactivity, we fused the C-terminus of the LEC gene to the N-terminus of the chTNT-3 HC gene with a five amino acid universal linker (Gly4Ser). The biological activity of the LEC fusion protein was demonstrated by measuring the migration of human leukemia TPH-1-0 cells in a microchemotaxis chamber in a dose-dependent manner or by binding to receptor positive TPH-1-0 cells by flow cytometry. Biodistribution studies demonstrated that ^{25}I-labeled LEC/chTNT-3 specifically bound to tumor with excellent retention at the tumor site (2.4% injected dose per gram at both 12 and 24 h postinjection). The immunotherapeutic potential of LEC/chTNT-3 was also demonstrated by treating different experimental solid tumors of the mouse. As shown in Figure 19.2, LEC/chTNT-3 treatment produced a 37% reduction in the MAD109 lung tumor model, as compared to untreated controls. These observations fit with other solid tumors that produced a 55% tumor growth reduction in the COLON 26 tumor model and a 42% reduction in the RENCA tumor model. Recently, a nontargeted LEC fusion protein consisting of human LEC and soluble Fc (LEC-Fc) was constructed and showed similar tumor regression as the targeted fusion protein (Figure 19.2). This indicates that sufficient amount of LEC reached and bound to tumor to reverse tolerance.

19.3.3 Co-Stimulatory Fusion Proteins

Most tumors express antigens that can be recognized to a variable extent by the host immune system, but in many cases, an inadequate immune response is elicited because of the ineffective activation of effector T cells via the immune synapse. Studies showed that the weak immunogenicity of tumor antigens might be due to inappropriate or absent expression of co-stimulatory molecules on tumor cells [58]. For most T cells, proliferation and IL-2 production will not occur unless a co-stimulatory signal is also provided. In the absence of co-stimulatory signals during T cell receptor (TCR) engagement, T cells enter a functionally unresponsive state, referred to as clonal anergy. One of the major signaling pathways involved in delivering the co-stimulatory signal is mediated by interaction between CD28 on T cells and the B7 family members including B7.1 (CD80) and B7.2 (CD86) on APC [59]. T cell-mediated rejection of tumors is achieved by presenting co-stimulatory signals directly on the tumor cells surface. Transfection of

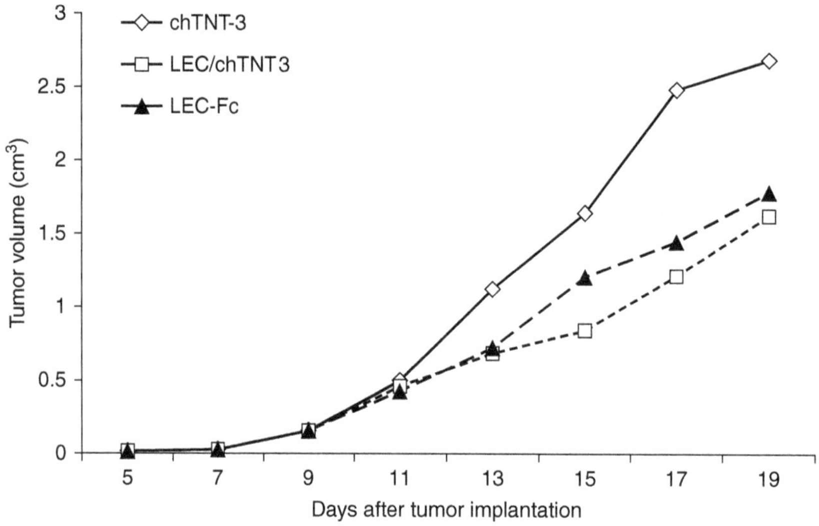

FIGURE 19.2 Immunotherapy in MAD 109-bearing BALB/c mice treated daily for 5 days with intravenous injection of 0.1 mL LEC/chTNT-3 (20 μg), Fc-LEC (20 μg), or control chTNT-3 (20 μg).

several murine tumor cells with B7.1 or B7.2 genes induced T cell-dependent rejection of B7-expressing tumors in mice and protected against tumor challenge with parental tumor cells. These data suggest that the presentation of B7 molecules on the tumor cell surface *in vivo* might be a promising novel approach for cancer immunotherapy. Other forms of B7.1 and B7.2, namely soluble B7-Igs (now termed B7-Fc), were studied and found to be very effective for the immunotherapy of solid tumors [18], but little work with these reagents were done past these initial studies.

Co-stimulatory molecules of the TNFSF are also protein-based conjugates that have been investigated as fusion protein partners. Members of the TNFSF are type II cell surface glycoproteins that are normally expressed on APCs. Their receptors are type I transmembrane proteins, characterized by cysteine-rich motifs in their extracellular domains, and are transiently expressed on T cells after initial activation. Several members of the TNFSF family including CD137L, GITRL, and OX40L, on binding to their receptors, provide critical signals for T cells to sustain their response after initial activation (immunological memory). In addition, CD137 (previously called 4-1BB) is another important co-stimulatory molecule that has been recently described [60]. Generally, CD137 is found on activated T cells, and its natural ligand CD137L is found on activated B cells, activated macrophages and differentiated dendritic cells. CD137L has been shown to co-stimulate T cell responses independently of signals through the CD28 molecule [61] and can stimulate both primary [62] and secondary [63] responses of both CD4+ and CD8+ T cells. Several experiments have shown that enhanced CD137/CD137L interactions amplify T cell-mediated antitumor immunity in several mouse models and that systemic administration of the agonistic anti-CD137 2A antibody eradicates established subcutaneous tumors in mice [64]. It is not surprising therefore, that a human 2A antibody is in development for the treatment of cancer.

19.3.3.1 B7.1 Fusion Proteins B7.1 is a prototypic co-stimulatory molecule and has the ability to provide T cells with the "second signal" that ensures T cell activation supersedes T cell anergy. To target B7.1, Sharifi et al. [65] linked the molecule to NHS76, a human TNT mAb generated by phage display that is capable of binding intracellular antigens accessible and abundant in necrotic regions of tumors. These intracellular antigens show preferential localization in malignant tumors because of the presence of abnormally permeable, degenerating cells not found in normal tissues. Because the N-terminus of B7.1 is critical for interaction with its counter-receptors, the C-terminus of B7.1 was linked to the N-terminus of NHS76. Studies confirmed that the B7.1/NHS76 fusion protein retained both the co-stimulatory activity of B7.1 and the tumor-targeting ability of NHS76 antibody [17]. For comparison, Liu et al. [16] also constructed a B7.1 fusion protein consisting of human B7.1 and the Fc portion of human IgG_1, referred to as B7.1-Fc. Relative tumor uptake of the fusion protein was determined by tissue biodistribution studies in COLON 26 tumor-bearing BALB/c mice using radioiodinated B7.1/NHS76. Despite the rapid clearance of ^{125}I-B7.1/NHS76, these data demonstrate that B7.1/NHS76 targets the tumor with good retention as compared to normal organs and blood. Since human B7.1 can interact functionally with murine B7.1 counter-receptors, the immunotherapeutic potential of this fusion protein was tested in different mouse tumor models. In testing the

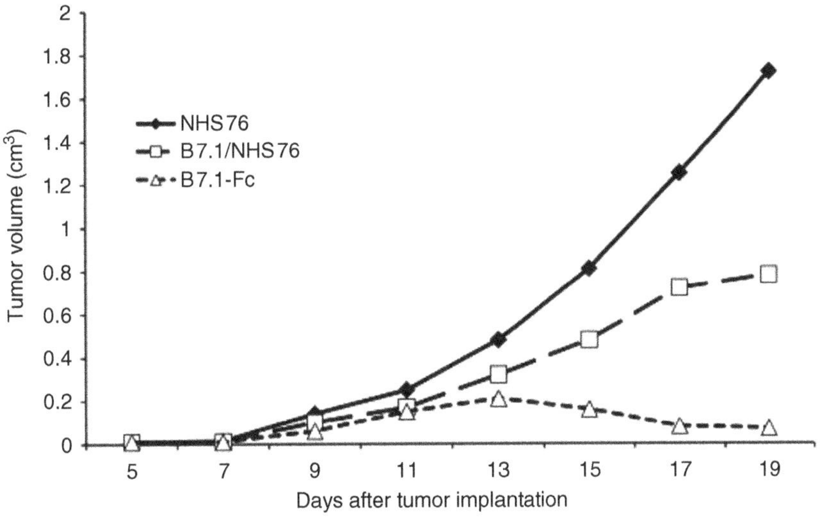

FIGURE 19.3 Targeted human B7.1/NHS76 and untargeted B7.1-Fc immunotherapy in COLON 26-bearing BALB/c mice. Control studies showed that pretreatment with NHS76 antibody alone does not show any effect.

targeted molecule, mice treated with B7.1/NHS76 alone showed a 35% reduction in the COLON 26 tumor volume (Figure 19.3). These observations fit with other solid tumors that produced a 55% tumor growth reduction in the MAD109 and RENCA tumor models [16]. Identical treatment studies performed with the soluble B7.1-Fc fusion protein showed complete regression of COLON 26 tumors after a 5-day treatment regimen (Figure 19.3). This was true even in mice with established MAD109 tumors, which are known to be poorly immunogenic.

19.3.3.2 CD137L Fusion Proteins Besides its relationship with conventional T cells, CD137 signaling on CD4+CD25+ T_{reg} cells, although presumed, has been controversial. For example, Zheng et al. [66] have verified the expression of CD137 on T_{reg} cells isolated from mice, and demonstrated that CD137L ligation augmented the proliferation of functional T_{reg} cells. Choi et al. [67], however, using the agonist 2A antibody, showed that CD137 signaling resulted in negligible enhancement of proliferation of T_{reg} cells and actually inhibited T_{reg} cell function *in vitro* and *in vivo*.

Taking into account the importance of the extracellular C-terminus of CD137L for its bioactivity, the N-terminus of the extracellular CD137L gene was fused to the C-terminus of the mTNT-3 HC gene to construct both TNT-3/CD137L and Fc-CD137L fusion genes [68]. Whole-body clearance studies were performed in healthy BALB/c mice to establish the *in vivo* half-life of TNT-3/CD137L and Fc-CD137L, which was found to be 18 and 24 h, respectively. Tissue biodistribution studies in COLON 26 tumor-bearing BALB/c mice were performed to determine the relative tumor uptake of each fusion protein. The uptake of radioiodinated TNT-3/CD137L per gram of tumor was significantly higher than the uptake in normal organs at 24 h (3.1% ID/g) and showed even higher retention at 48 h postinjection (4.5% ID/g). As expected, Fc-CD137L demonstrated low tumor retention over time (0.6% ID/g at 48 h postinjection). Examination of the tumor draining lymph nodes, an important site for tumor immunotherapy, revealed that TNT-3/CD137L (1.2%ID/g) had slightly better uptake than Fc-CD137L (0.5% ID/g) at 48 h postinjection. However, the average tumor draining lymph node retention of TNT-3/CD137L was still much lower than that measured in tumor.

A dosing study was also performed in COLON 26-bearing BALB/c mice with different doses ranging from 1 to 100 ug/dose of TNT-3/CD137L, Fc-CD137L, and the anti-CD137 agonist antibody 2A (Figure 19.4). Higher doses of TNT-3/CD137L and Fc-CD137L treatment resulted in 30–40% tumor reduction. At the highest doses, however, the TNT-3/CD137L group showed 70% tumor reduction, whereas the Fc-CD137L-treated mice achieved only 30% tumor reduction. This difference indicates that at this particular dosage, localization of CD137L in tumor is critical and therefore more effective. All 2A treated groups showed 95% tumor reduction; eventually became tumor free and were free of signs of toxicity throughout the 140-day observation period. Additional studies are underway to improve the potency of antibody/CD137L fusion proteins since biological activity of CD137 depends on functional homotrimer structures.

19.3.3.3 GITRL Fusion Proteins Recently, there have been a number of studies that have attempted to characterize the cell surface of T_{reg} cells to evaluate whether individual proteins play a role in the activation or inhibition of

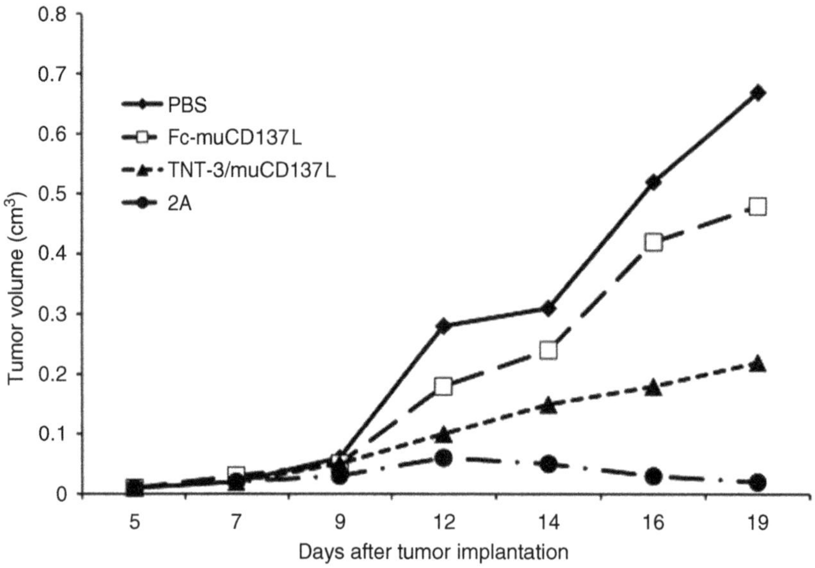

FIGURE 19.4 Dose response of mTNT-3/CD137L, Fc-CD137L, and agonist CD137 antibody 2A in COLON 26-bearing BALB/c mice.

T_{reg}-induced suppression. One such cell surface marker is a member of TNFSF designated GITR, which is constitutively expressed on CD4+CD25+ (T_{reg}) cells and is up-regulated in CD4+ and CD8+ cells [69]. GITR has a high homology in the cytoplasmic region with other TNFSF members such as CD137, OX40, and CD40L. The ligand for GITR (GITRL) is a type II transmembrane protein that belongs to TNFSF and is normally expressed on APCs such as macrophages, dendritic cells, and B cells. More importantly, GITR activation by GITRL leads to the proliferation of $CD4+CD25^-$ and CD8+ effector T cells and a reduction of tumor volume *in vivo*. Also, an interesting characteristic of GITR activation on T_{reg} cells results in their inability to suppress responding $CD4+CD25^-$ T cells [66]. Therefore, GITR engagement by GITRL results in the inhibition of T_{reg} cells and the activation of responding T cells. This dual function of GITR activation may be unique for co-stimulatory interactions and could be responsible for its highly potent anti-tumor activity *in vivo*.

Both C- and N-terminal Fc fusion proteins containing the extracellular domain of murine GITRL were also constructed [70]. In order to determine the therapeutic efficacy of the GITRL fusion proteins, immunotherapy studies were performed with tumor-bearing BALB/c mice using a COLON 26 tumor model. The activity of Fc-GITRL (C-terminal) and GITRL-Fc (N-terminal) was first examined at different doses in order to analyze its potency *in vivo*. For comparison, Fc fusion proteins consisting of the lymphocyte function-associated antigen (LFA) were used in several aspects of this work. The group that received 11 μg of Fc-GITRL fusion protein or the maximum dosage of 90 μg both resulted in a 96% tumor reduction. The group that received the maximum dose (90 μg) of Fc-GITRL protein displayed the longest survival out of all of the treatment groups. In addition, the dosing study of GITRL-Fc demonstrated an 87% reduction in tumor volume at a dose of 11 μg compared to the control LFA-Fc treatment group. Therefore, these data show that treatment using as little as 11 μg of either Fc-GITRL or GITRL-Fc can activate an effective anti-tumor response *in vivo*.

19.3.3.4 OX40L Fusion Proteins OX40 (CD134), a membrane-associated glycoprotein, is transiently expressed on the surface of T cells after TCR ligation [71]. Its natural ligand, OX40L (TNFSF4, CD134L), is found primarily on APCs such as activated B cells, activated endothelial cells, dendritic cells, and macrophages [72,73]. The OX40L delivers a potent co-stimulatory signal to OX40+ T cells leading to optimal T cell function. Studies of the primary T cell responses revealed that OX40L is a potent co-stimulatory molecule for sustaining CD4+ T cell response [74]. OX40 signaling strikingly prolongs T cell division initially induced by CD28, and enhances the survival of CD4+ cells during the initial response by promoting Bcl-XL and Bcl-2 expression. OX40−/− T cells fail to maintain high levels of Bcl-XL and Bcl-2 4–8 h after activation, and undergo apoptosis [75]. *In vivo*, OX40 signaling would augment tumor-specific priming through stimulating and expanding the natural repertoire of the host's tumor-specific CD4+ T cells [73]. Therefore, it is believed that OX40-OX40L interactions are crucial for the survival of effector T cells and the generation of memory T cells. Although the effects of OX40 co-stimulation on CD4+ T cells were initially studied, more recent work has shown that OX40 signaling also

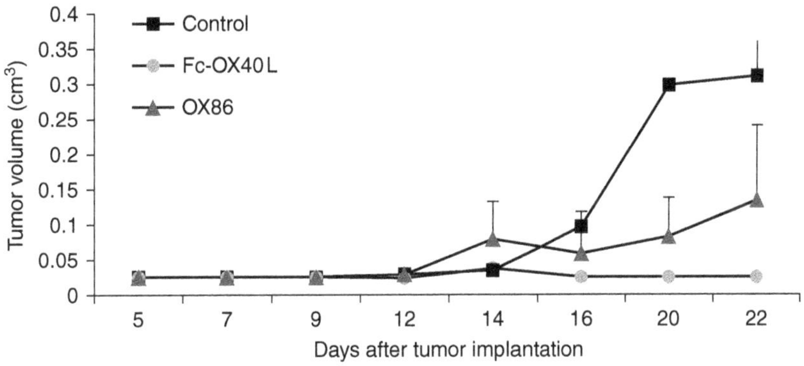

FIGURE 19.5 Immunotherapy response with Fc-OX40L and agonist anti-OX40 antibody (OX86) in COLON 26-bearing BALB/c mice.

co-stimulates CD8+ T cells [76]. Profound CD8+ T cell clonal expansion and a massive burst of CD8+ T cell effector function were induced by dual co-stimulation of OX40 and CD137 [77]. This synergistic response of the specific CD8+ T cells has been shown to last for several weeks, and is sufficient to treat established tumors even under immunocompromised conditions.

Interestingly, OX40 is also found to be expressed on both naïve and activated CD4+CD25+ T_{reg} cells [79]. OX40 signals may abrogate T_{reg}-mediated suppression when they are delivered directly to antigen-engaged naïve T cells [80]. Using the agonist antibody, OX86, OX40 signaling on T_{reg} cells inhibited their capacity to suppress, and restored effector T cell proliferation and cytokine production [81]. It has been confirmed that OX40 abrogation of T_{reg} cells occurs *in vivo* using a graft-versus-host disease model [81].

Since the C-terminus of OX40L is extracellular and essential for its bioactivity, the N-terminus of the extracellular OX40L gene was fused to the C-terminus of the immunoglobulin HC gene [82]. A dosing study was performed on COLON 26-bearing BALB/c mice with Fc-OX40L and the anti-OX40 agonist antibody OX86 (Figure 19.5). Treatment groups receiving high doses of Fc-OX40L showed 80–90% tumor reduction, whereas the OX86-treated mice achieved less tumor reduction. This difference potentially indicates that at this particular dosage, the Fc-OX40L is more effective. Identical treatment studies performed with the soluble fusion protein showed similar regression of RENCA tumors. All treated groups eventually became tumor free and were free of signs of toxicity throughout the 120-day observation period.

19.3.4 Enzyme Fusion Proteins

Mammalian enzymes, such as carboxypeptidase A and β-glucuronidase (βG) that are unlikely to be found in high basal concentrations in the body, have been also used resulting in ADEPT technology for the treatment of cancer [83]. This ADEPT strategy is a two-step approach which seeks to generate a potent cytotoxic agent selectively at a tumor site [31]. In the first step, a tumor selective antibody linked to an enzyme (fusion) is administered and allowed to localize at the tumor site. This is followed, in the second step, by the administration of a relatively nontoxic prodrug that will be enzymatically converted to active drug only at the tumor site (optimum at low pH) to minimize the usual nonspecific toxicity. Potentially, one of the major advantages of ADEPT is the ability to deliver, to the tumor, a highly potent cytotoxic drug in concentrations that cannot be achieved by conventional chemotherapy. In principle, to maximize the benefit of ADEPT therapy it is beneficial to (a) have high tumor to normal tissue ratio of the conjugate, (b) ensure the prodrug to drug conversion is maximal at the tumor site, and (c) generate a potent drug that will act locally and not diffuse into normal tissues.

To test this hypothesis, fusion proteins consisting of single chain Fv (scFv), Fab, or $F(ab')_2$ fragments of chTNT-3 and the human βG enzyme were constructed for ADEPT [84]. Each of these reagents was tested to assess specificity and avidity of antigen binding as compared to the parental antibody. In addition, studies were performed to demonstrate enzymatic function of the fusion proteins and retention of catalytic activity in circulating blood, specific tissues, and tumor after *in vivo* administration. The results of these experiments show that all three constructs retained their antigen binding capability and demonstrated active enzymatic function against substrate *in vitro*. Pharmacokinetic and biodistribution studies of radiolabeled fusion proteins in MAD109 murine lung carcinoma tumor model transplanted into BALB/c mice confirmed specific localization of the fusion proteins and rapid clearance from blood and normal tissues over time. For example, tumor uptake (% injected dose per gram) of the fusion proteins at 1 day postinjection varied from 2.35% for scFv/βG to 2.52% for Fab/βG, and to 2.91% for $F(ab')_2$/βG. Despite this rather low uptake, the rapid clearance of the fusion proteins

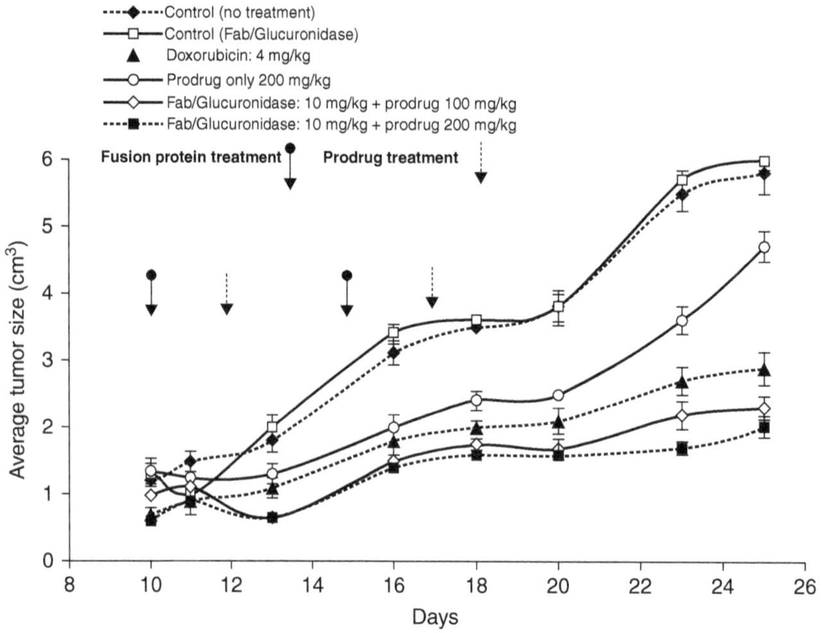

FIGURE 19.6 Treatment of MAD109 in BALB/c mice using combination of the Fab/human β-glucuronidase fusion protein and a prodrug of doxorubicin.

produced higher tumor/normal organ ratios (>1) for many normal tissues, illustrating the specificity of tumor targeting with the fusion proteins and their rapid elimination from the blood and normal organs. Finally, one of the constructs (chTNT-3 Fab/βG was used in a pilot treatment study with a glucuronide prodrug of doxorubicin to demonstrate the anti-tumor activity of ADEPT using the chemoresistant MAD109-bearing BALB/c mice. As shown in Figure 19.6, fusion protein pretreatment was very effective in slowing the rate of tumor growth. Although doxorubicin-treated mice showed a slow rate of growth, the prodrug-treated groups demonstrated more profound effects and had outstanding anti-tumor responses compared to control groups. Importantly, no toxicity or adverse effects were seen. In addition, the group receiving the combination fusion protein and higher dose of prodrug (200 mg/kg) had a higher survival rate (62%) compared to the group treated with doxorubicin alone (43%). These studies show that the use mAbs to target the enzyme to the tumor is a significant advance in ADEPT and that further studies are warranted to test this novel therapeutic approach in the treatment of solid tumors.

19.3.5 Vascular Targeting Fusion Proteins

The primary objective of this approach is to explore the use of mAbs to deliver potent cytotoxic agents capable of inducing sustained, effective therapy that selectively block the blood flow to tumors by targeting different antigens. A potential mediator of this event is the vascular cell membrane receptor protein tissue factor (TF) that is the initiator of the extrinsic pathway of the blood coagulation cascade and normally released from damaged tissues [85]. By substituting the attachment site with a tumor delivery agent, this potent thrombogenic protein in its tTF form can be targeted to the tumor where it can initiate clotting, thereby occluding the tumor's blood supply and causing rapid tumor destruction. Several advantages of this approach over conventional antitumor therapies have been suggested [27] including (a) the target molecules are directly accessible to antigen, permitting rapid localization of a high percentage of the injected dose; (b) cellular degeneration caused by the occlusion of tumor vessels is microregional, amplifying the effects of therapy; (c) microvascular endothelial cells are a normal, genetically stable cell population, so target antigens remain relatively the same regardless of selective pressures exerted by cytotoxic therapies; and (d) the same target drug can be used for a variety of solid tumors because tumor vessels share common morphological, immunological, and biochemical properties.

To test the therapeutic potential of this vascular targeting approach, three fusion proteins, chTNT-3/tTF, chTV-1/tTF, and RGD/tTF, which target DNA exposed in degenerative areas of tumors, fibronectin on the tumor vascular basement membrane, and αvβ3 on the luminal side of tumor vessels, respectively, were developed and tested for their anti-tumor effects [28]. Antigen binding and clotting assays demonstrated that each of the fusion proteins retained their antigen binding and thrombogenic activities. *In vivo* studies in COLON 26-bearing mice revealed that all three reagents induced histological evidence of microregional thrombosis

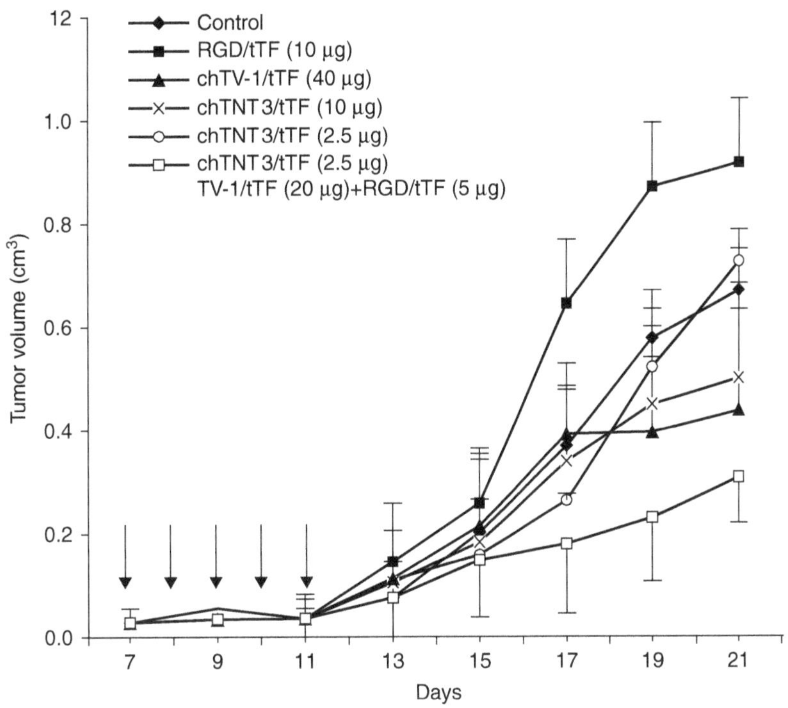

FIGURE 19.7 Treatment of COLON 26-bearing BALB/c mice using different tTF fusion proteins. Mice treated with RGD/tTF and chTV-1/tTF were injected daily × 5. Mice treated with chTNT-3/tTF were injected at three-day intervals × 3.

and massive cell necrosis. Of interest, the chTV-1/tTF and the RGD/tTF fusion proteins induced thrombosis in small- and medium-sized tumor vessels while the chTNT-3/tTF induced clotting in relatively larger vessels. Treatment studies also showed that chTNT-3/tTF and chTV-1/tTF but not RGD/tTF had a significant inhibition of tumor growth (Figure 19.7). To attain a significant antitumor effect, however, these thrombogenic agents had to occlude medium and large vessels within the tumor. Further studies are warranted to identify maximal conditions for inducing therapeutic vascular coagulation as a new and potent method of cancer therapy.

19.4 COMBINATION FUSION PROTEINS THERAPY

In recent years, several groups have begun to study the potential synergistic activity of the combination of fusion proteins fused to the same antibody molecule [86,87]. For instance, a large body of data has been generated showing the advantage in using combinations of cytokine and chemokine fusion proteins, compared with their single use, for enhancing anticancer immune response [88–90]. In particular, different studies have broadly explored the potential of combination cancer therapy using cytokine, chemokine, and co-stimulatory fusion proteins. Other approaches, which explore the use of fusion proteins in combination with immunoregulatory T-cells (T_{reg}) depletion, are also presented in this section and promise to afford superior tumor therapeutic efficacy through normal biological processes and the induction of a memory response.

19.4.1 Cytokine Fusion Protein Combinations

Owing to the growing number of cytokine fusion proteins that are being generated, it has been demonstrated that it is possible to improve antibody-cytokine fusion protein mediated therapy by using them in combination with other cancer therapies [33]. For example, chTNT-3/cytokine fusion proteins have been evaluated in multiple solid tumor models of the BALB/c mouse [91]. Analysis of the cytokine activity of these fusion proteins using standard *in vitro* tests shows that they have between 50% and 85% of the activity of the comparable free cytokine tested in parallel. Despite these somewhat lower activity levels, these reagents produced good tumor suppression in the three tumor models during and directly after treatment, but 5–7 days after therapy, the tumors continued to progress in size (Table 19.2). In two of these models, those mice receiving the combination of chTNT-3/IL-2, chTNT-3/TNF-α, and chTNT-3/IFN-γ had the best tumor regression estimated to be about 80% of control groups at day 17. Although individual fusion proteins had varying degrees of effectiveness in these tumor

TABLE 19.2 Combination Immunotherapy of TNT-3/IL-2 (20 ug) in RENCA-Bearing BALB/C Mice at 17 Days Post-Tumor Implantation

Antibody Fusion Protein	Dose (ug)	Tumor Reduction (%)
Antibody alone control	20	0
chTNT-3/IL2	20	40
chTNT-3/muIFN-γ	20	50
chTNT-3/TNF-α	20	25
chTNT-3/muGM-CSF	5	40
chTNT-3/muIFN-γ + chTNT-3/TNF-α	20 + 20	80
chTNT-3/muIFN-γ + chTNT-3/muGM-CSF	20 + 5	65

n-treated controls.

models, this combination or the one in which muTNT-3/muGM-CSF was substituted for the chTNT-3/IL-2, was most effective. Since the tumors were found to continue growing at about the same pace as control treated mice, the therapeutic effect of these treatments was found to be transient in nature. This incomplete immune response may have been due to the nongeneration of immune memory cells.

19.4.2 Chemokine and Cytokine Synergy

In order to enhance the therapeutic effectiveness of the LEC/chTNT-3, Khawli et al. [91] focused their attention on combining this reagent with other fusion proteins that others have found to be critical components of cancer immunotherapy. In this new combination therapy, groups of MAD109-bearing BALB/c mice were injected with 20 μg of LEC/chTNT-3 alone or with comparable doses of chTNT-3/IFN-γ, chTNT-3/TNF-α, chTNT-3/GM-CSF, or chTNT-3/IL-2. After tumors reached 0.5 cm in diameter, all groups were treated daily for 5 days and tumor growth was monitored every other day by caliper measurement in three dimensions. As shown in Table 19.3, combination therapy with LEC/chTNT-3 and each of the four chTNT-3/cytokine fusion proteins produced only minimal improvement but the combination containing the chTNT-3/IL-2 did show flattening of the growth curve by day 17. Importantly, LEC/chTNT-3 immunotherapy was found to be entirely nontoxic (even at 100 μg), in marked contrast to the use of chTNT-3/IL-2, chTNT-3/TNF-α, and TNT-3/mGM-CSF fusion proteins, which demonstrated varying degrees of toxicity when increased above 20 μg per dose. Treatment with these three fusion proteins at doses exceeding 20 μg, made the mice lethargic, induced a ruffled fur appearance, produced a loss of appetite, and lowered activity levels of treated mice.

TABLE 19.3 Combination Immunotherapy of LEC/chTNT-3 (20 ug) in MAD109-Bearing BALB/c Mice at 17 Days Post-Tumor Implantation

Antibody Fusion Protein	Dose (ug)	Tumor Reduction (%)
Antibody alone control	20	0
LEC/chTNT-3	20	40
chTNT-3/muGM-CSF	20	55
chTNT-3/TNF-α	20	55
chTNT-3/muIFN-γ	20	55
chTNT-3/IL-2	20	70

19.4.3 Co-Stimulatory Fusion Protein Combinations

Effective against multiple solid tumor types, TNFSF ligand fusion proteins serve to provide second signal activation of T cells and work well with T_{reg} depletion. The co-stimulatory abilities of these ligands may prove to be a powerful combination with, for example, the chemoattraction of the LEC or the co-stimulation of the B7.1 and may possess an unrivalled ability to attract and activate the immune cells necessary for tumor regression and immunologic memory. As described above, our laboratory had established the effectiveness of B7.1-Fc N-terminal fusion protein *in vivo* in order to determine whether it could be used in combination with Fc-GITRL. As shown in Figure 19.8A, B7.1-Fc, in combination with Fc-GITRL, did indeed produce ~40–50% reduction in tumor volume (COLON-26) in comparison to using B7.1-Fc or Fc-GITRL alone.

In order to assess whether signaling through GITR and CD137 receptors would lead to a synergistic antitumor effect, a combination treatment was performed using Fc-GITRL and Fc-CD137L. The data shown in Figure 19.8B demonstrate that in comparison to their use alone, the combination of these two C-terminal fusion proteins led to ~44–56% reduction in tumor volume (COLON 26) using very low doses of each. In general, these data provide evidence in mice that combination fusion proteins consisting of these immune activators may indeed be especially effective for the immunotherapy of cancer.

19.5 MECHANISM OF ACTION: IMMUNOREGULATORY T-CELL (T_{REG}) DEPLETION AND FUSION PROTEIN COMBINATION THERAPY

T-cell anergy is one of the major impediments to the ability of the immune system to eradicate established tumors. To elicit an effective T-cell response, the engagement of the T-cell receptor by peptide-major histocompatibility complex needs to occur in the setting of a second signal provided by co-stimulation. In order to create more effective

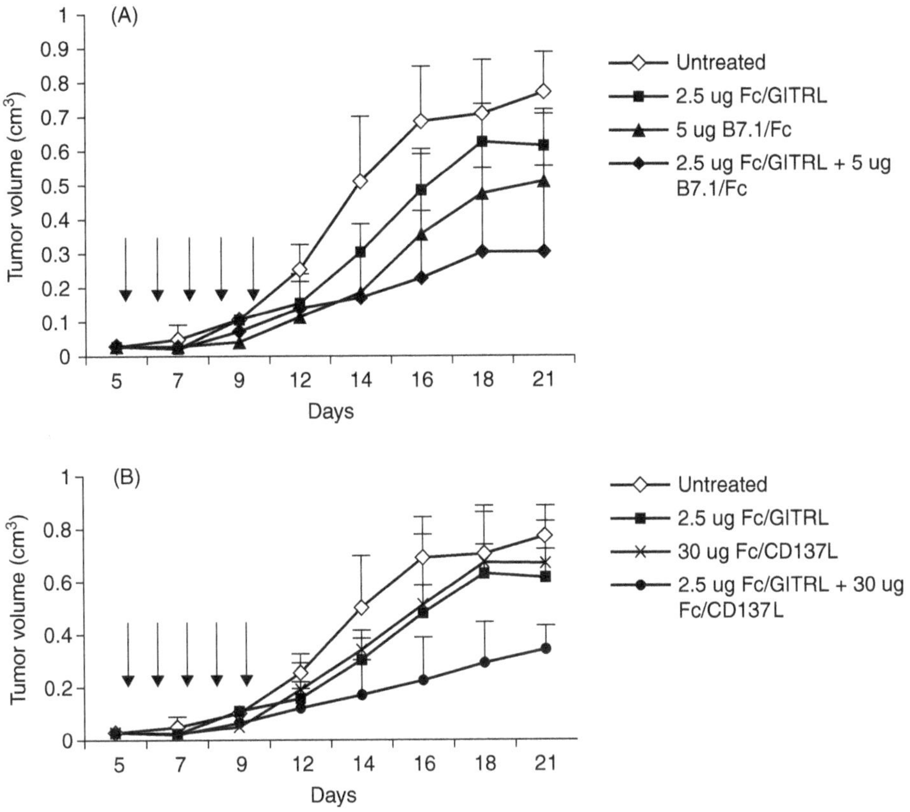

FIGURE 19.8 (A) Fc-GITRL + B7.1-Fc and (B) Fc-GITRL + Fc-CD137L combination therapy in COLON 26-bearing BALB/c mice showing enhanced tumor regression compared to treatment with either reagent alone.

immunotherapy for the treatment of cancer, it is important to have a comprehensive understanding of how the immune system is regulated. There is now abundant evidence from a number of diverse experimental systems that a subpopulation of CD4+ T cells, collectively termed T_{reg} cells, is present in normal mice and is essential for both homeostasis and the maintenance of tolerance to tissue-specific antigens [92–95]. It is now generally agreed that these cells display the CD4+CD25+ phenotype and the FOX3p genotype [96] and that they exert their activity either by cell–cell contact [21] or by secretion of inhibitory IL-10 or TGF-β cytokines [97,98]. Another discovery with the immunosuppressive effects of CD4+CD25+ T_{reg} cells is that immunocompetent mice bearing syngeneic tumors become tolerant to their tumors by day 9 after transplantation when CD4+CD25+ appeared in the peripheral circulation [99,100]. Depleting T_{reg} cells before or at the time of implantation with a rat mAb PC61 will produce either incomplete tumor reduction or a delay in tumor growth. Moreover, tumor suppression could also be produced by low dose whole-body irradiation or cyclophosphamide chemotherapy, which has been found to be cytotoxic to T_{reg} cells [101]. Although CD4+CD25+ cells were first discovered in mice [102], a population with similar phenotypic and functional properties has been defined in humans [103–107]. Since clinical trials to reduce or delete T_{reg} cells in the setting of cancer immunotherapy have not been performed, the potential of this procedure in humans is currently unknown. Evidence however, is accumulating that T_{reg} cells are more abundant in the peripheral blood and tumor microenvironment of cancer patients and may be responsible for the observed tolerance displayed in patients [108].

19.5.1 LEC Fusion Proteins in Combination with T_{reg} Depletion

To evaluate the subpopulation(s) of T cells responsible for the observed tumor destruction, depletion studies were performed in conjunction with the above therapy studies [15]. One day after COLON 26 tumor implantation, mice received cytotoxic antisera specific for CD4+, CD8+, or NK cells. Each of these antisera was administered every 5 days, reducing the appropriate cell subpopulation to <2%

TABLE 19.4 Combination Cytokine or Chemokine Fusion Protein Immunotherapy and CD4+ T-cell Depletion in the Treatment of COLON 26 Carcinoma

Immunotherapy[a]	T-cell Subset Depletion[b]	Tumor Reduction[c] (%)
chTNT-3 (control)	CD4+ depletion	33 (0)
LEC/chTNT-3	CD4+ depletion	100 (60)
chTNT-3/IL-2	CD4+ depletion	64 (38)
chTNT-3/IFN-γ	CD4+ depletion	33 (32)
chTNT-3/TNF-α	CD4+ depletion	33 (10)

Values in parentheses refer to group of animals which are not treated with CD4+ depletion.
[a]Antibodies and fusion proteins (20 μg/dose) were injected i.v. for five consecutive days after tumors reached 0.5 cm in diameter.
[b]CD4+ depletion (0.5 mg/dose of GK1.5) was performed i.p. 1 day after tumor implantation and repeated every 5 days.
[c]Percentage tumor reduction on day 19 after treatment.

in the peripheral blood. As expected, depletion of CD8+ and NK cells negated the antitumor activity of LEC/chTNT-3 treatment, indicating that these lymphocyte subpopulations are critical to the function of the fusion protein (Table 19.4). By contrast, mice receiving CD4+ depletion regardless of the time of administration in combination with LEC/chTNT-3 therapy demonstrated 100% regression of tumor. These unexpected results were highly significant and turned our attention to the necessity of deleting or suppressing immunoregulatory cells and mechanisms.

To probe this observation further, only the CD4+CD25+ T_{reg} subpopulation was depleted in repeat experiments. Since T_{reg} represents ~10% of the CD4+ population in both mice and humans, depletion of this small subpopulation using the rat antimouse CD25+ mAb PC61 would be highly instructive. Depletion of T_{reg} cells in combination with LEC/chTNT-3 did indeed produce complete remissions in COLON 26 tumor-bearing BALB/c mice, suggesting that it is this small population of CD25+ cells that is responsible minimizing the effects of LEC/chTNT-3. Treating with PC61 alone or in combination with control chTNT-3, also produced significant tumor regressions, but never a complete response [15].

Rechallenge experiments demonstrated that combining LEC/chTNT-3 treatment with CD25+ depletion produces long-acting memory cells. Tumor-free mice from the experiments described above were again injected subcutaneously with COLON 26 and MAD109 tumor cells in their contralateral flanks, alongside naïve mice injected with each tumor cell line. Two weeks after tumor implantation, all naïve mice had both tumors growing on their flanks, while the tumor-remission mice (now 2 months after initial tumor implantation) had only one tumor (MAD109) growing. Interestingly, a second series of rechallenge experiments performed 6 months after the initiation of therapy, however, showed no difference between naïve and remission mice. To determine if regenerated CD4+CD25+ T cells in tumor-draining lymph nodes correlated with the ability of tumor to take in these mice, the percent of T_{reg} cells was measured in each of these groups of mice. The data showed that CD4+CD25+ T cells in naïve mice constituted 9.7% of total T cells, while in 2–3 month tumor-regressed mice they constituted only 5% of total T cells and in 5–6 months tumor-regressed T_{reg} cells constituted 13% of total T cells. This increase in T_{reg} cells may explain the failure to reject tumor seen at 6 months. Additionally, these experiments suggest that combination LEC/chTNT-3 treatment and CD25+ depletion does in fact produce long-acting memory cells capable of preventing re-engraftment of the same tumor, so long as T_{reg} regeneration does not suppress the memory response.

To show that targeted LEC is critical to the above results, similar combination studies were performed with chTNT-3/cytokine fusion proteins consisting of human IL-2, murine IFN-γ, and murine TNF-α using identical treatment regimens. The results of these studies presented in Table 19.4, showed only modest improvement indicating that combination therapy with anti-CD25+ antisera requires LEC not cytokine localization to tumor in order to produce complete regression.

19.5.2 CD137L and GITRL Fusion Proteins in Combination with T_{reg} Depletion

As described above (Figure 19.3), mice treated with B7.1/NHS76 alone showed a 35–55% reduction in tumor volume. To determine if T_{reg} deletion could enhance the effects of B7.1 fusion proteins, additional tumor-bearing mice were treated with combination therapy using PC61 and B7.1/NHS76 [17]. Indeed, combination therapy produced complete regression of established lung tumors (MAD109) and was associated with increased effector T cell tumor infiltration (Table 19.5). When soluble B7.1-Fc was used in combination with T_{reg} depletion PC61 in MAD109-bearing mice, all mice also went into complete remission and remained so for the observation period of 6 months. Rechallenge experiments performed with these mice showed that they had immunological memory and were able to reject subsequent tumor implants.

Using knockout mice, it was determined that the antitumor activity of B7.1-Fc immunotherapy was dependent both on IFN-γ and perforin, but not on IL-4, as demonstrated by the failure of IFN-γ and perforin knockout mice to reject tumors after treatment (Table 19.5). In these studies, anti-IL-4 was found to have no tumor suppressive activity. By these experiments, it appears that perforin may not be the exclusive method of tumor cell killing since the activity of the B7.1-Fc was not entirely eliminated as seen in the IFN-γ knock-out experiments. In summary, these data show that

TABLE 19.5 Summary of T_{reg} and Fusion Protein Combination Therapy in COLON 26-Bearing BALB/C Mice at 19 Days Post-tumor Implantation

Antibody Fusion Protein	Dose (ug)	Tumor Reduction (%) CD25 Depleted	CD8 Depleted	CD4 Depleted	Nondepleted
B7.1/NHS76	30	100	<5	100	40
B7.1-Fc	10	70	NA	NA	45
chTNT-3/muCD137L	30	NA	0[a]	80	80
Fc-muCD137L	30	NA	0[a]	85	70
Fc-muGITRL	90	NA	45	100	90
LEC/chTNT-3	30	100	<5	100	55

[a]Tumors grew faster than nontreated controls.

B7.1-Fc immunotherapy is dependent on the induction of IFN-γ, activated CD8+ T cells, and partially by the use of perforin for tumor cell killing. When used in combination with T_{reg} cell depletion, memory cells are generated to assure the complete destruction of tumor in the host.

In order to assess the role of CD4+ and CD8+ T cells in CD137L mediated immunotherapy, T-subset depletion studies were performed with cytotoxic antibodies against CD4+ and CD8+ cells. Anti-CD4 antibody (Clone GK1.5) and anti-CD8 antibody (Clone H35) were injected i.p. on the 5th day after tumor implantation. Importantly, CD8+ T cell depletion completely abrogated the antitumor effects of TNT-3/CD137L (Figure 19.1) and Fc-CD137L (Table 19.5), indicating that tumor suppression was dependent on this subpopulation of T cells [68]. Depletion studies were also performed in mice treated with Fc-GITRL. Groups that received soluble Fc-GITRL or DTA-1 (agonist rat anti-murine GITR) treatment alone resulted in a 90% reduction in tumor volume at day 19 posttumor implantation compared to the control LFA-Fc group, respectively (Table 19.5). The anti-CD4+ alone treatment group resulted in a 96% decrease in tumor volume compared with LFA-Fc, while the anti-CD8+ alone treatment group resulted in no significant difference in tumor volume compared to control LFA-Fc treated mice. The group treated with anti-CD4+ in combination with Fc-GITRL resulted in the same percent tumor reduction as the group treated with anti-CD4+ alone. In contrast, the group that received Fc-GITRL in combination with anti-CD8+ resulted in an 89% *increase* in tumor volume compared to Fc-GITRL treatment alone. These results demonstrate that CD8+ T cells are critical for tumor destruction by Fc-GITRL as with the other fusion proteins described above. Since the Fc-GITRL was by itself so potent, it is difficult to show the contribution of anti-T_{reg} deletion in these mice. Additionally, Fc-GITRL may by itself suppress T_{reg} function thereby contributing to its outstanding therapeutic effects *in vivo*.

19.6 FUTURE DIRECTIONS

The overall objective of this chapter was to evaluate the efficacy of genetically engineered fusion proteins developed in our laboratory for the immunotherapy of solid tumors. The use of chTNT-3, which targets necrotic areas of solid tumors regardless of the species or origin of the tumors, enabled us to compare by the same targeting antibody, the relative therapeutic potential of selected cytokines, chemokines, and co-stimulatory molecules in tumor-bearing immunocompetent mice. This point cannot be overemphasized since there are few antibodies available that target murine tumor models. By our studies, it has become clear that the targeting of specific but not all immune modulators to the tumor microenvironment can cause effective destruction of established tumors. As controls, soluble nontargeting Fc fusion proteins were also prepared for most of these immunoregulatory molecules, and to our surprise, many of these appear to be as effective as the corresponding antibody fusion proteins that target tumor. It appears that when administered in high enough concentrations, soluble Fc fusion proteins can also reverse immune tolerance to tumor. Various cytokines, chemokines, and co-stimulatory molecules are highly potent proteins, requiring a much lower dose than antibodies [31]. This is seen clearly in dosing studies in which a threshold amount of soluble Fc is required for the induction of tumor suppression. Threshold amounts were less often seen in those mice receiving antibody fusion proteins. Although very potent in their antitumor action, it remains to be seen if these soluble Fc fusion proteins are more toxic in humans than corresponding antibody fusion proteins or if they are effective against larger, well-established tumors. The other major finding of these studies, namely the role of T_{reg} cells in tumor growth, illustrates the importance of understanding existing mechanisms of immune tolerance in cancer patients. For instance, do all cases of a given tumor type evoke a similar mechanism of tumor escape, and do metastatic lesions display the same mechanism as primary lesions in the same patient? These types of questions should be answered prior to proceeding with the testing of specific immunotherapeutic reagents to maximize the therapeutic potential of a given reagent. We should also be aware that histopathologic diagnosis may not predict which immunosuppressive mechanisms are at play in individual patients. Instead, a comprehensive analysis of the immune phenotype in patients may be necessary to identify correctly which immunotherapeutic approach will be effective [109].

19.7 CONCLUSION

In conclusion, in the last 10 years, we have generated a panel of cytokine, chemokine, and co-stimulatory fusion proteins [110] and tested them similarly in multiple solid tumor models of immunocompetent mice. Cell deletion experiments have identified the importance of NK+ and CD8+ effector T cells for tumor immunotherapy and the negative influence of CD25+ T_{reg} cells. Both nontargeted soluble Fc and targeted antibody fusion proteins have been found to be effective agents for the delivery of immunoregulatory molecules that can induce significant tumor regression or cure. In particular, the chemokine LEC and several different co-stimulatory molecules including B7.1, OX40L, GITRL, and CD137L were found to be especially effective in the induction of effective cancer treatment. These co-stimulatory molecules are not the only protein-based conjugates that have been investigated as fusion proteins. We have also demonstrated that treatment with H60/TNT-3, a new immunotherapeutic fusion protein drug, could dramatically enhance antitumor effect of experimental murine tumors [111]. Interestingly, these results proved that injection of H60/TNT-3 could stimulate innate antitumor immunity via NK cell activation and thereby suppress tumor growth. Additional antibody conjugates such as those with CpG and other toll-like receptor agonists may also prove to be highly effective in stimulating innate immunity in the tumor microenvironment. In general, our work demonstrates that several fusion proteins may have clinical potential for the immunotherapy of cancer. Although several challenges in optimizing the development of fusion proteins remain, advances in protein engineering should provide ready solutions to enable these promising immunomodulatory reagents to reach the clinic.

ACKNOWLEDGMENTS

We would like to thank all our past and present students for their dedication and commitment during the long hours of experimental studies. We also would like to express appreciation to the many colleagues and peers who have contributed to expanding our knowledge of fusion proteins and regret the absence herein of many valuable reference citations due to space limitations.

REFERENCES

1. Beck A, Wurch T, Bailly C, Corvaia N. (2010) Strategies and challenges for the next generation of therapeutic antibodies. *Nat. Rev. Immunol.* 10, 345–352.
2. Roopenian DC, Akilesh S. (2007) FcRn: the neonatal Fc receptor comes of age. *Nat. Rev. Immunol.* 7, 715–725.
3. Sidhu SS. (2007) Full-length antibodies on display. *Nat. Biotechnol.* 25, 537–538.
4. Mocellin S, Rossi CR, Lise M, Marincola FM. (2002) Adjuvant immunotherapy for solid tumors: from promise to clinical application. *Cancer Immunol. Immunother.* 5, 1583–1595.
5. Bitonti AJ, Dumont JA, Low SC, Peters RT, Kropp KE, Palombella VJ, et al. (2004) Pulmonary delivery of an erythropoietin Fc fusion protein in non-human primates through an immunoglobulin transport pathway. *Proc. Natl. Acad. Sci. U S A* 10, 9763–9768.
6. Waldmann TA. (2003) Immunotherapy: past, present, and future. *Nat. Med.* 9, 269–277.
7. Blattman JN, Greenberg PD. (2004) Cancer immunotherapy: a treatment for the masses. *Science* 305, 200–205.
8. Aluvaihare VR, Kallikourdis M, Betz AG. (2004) Regulatory T-cells mediate maternal tolerance to the fetus. *Nat. Immunol.* 5, 266–271.
9. Marshall NA, Christie LE, Munro LR, Culligan DJ, Johnston PW, Barker RN, et al. (2004) Immunosuppressive regulatory T-cells are abundant in the reactive lymphocytes of Hodgkin's lymphoma. *Blood* 103, 1755–1762.
10. Leach DR, Krummel MF, Allison JP. (1996) Enhancement of antitumor immunity by CTLA-4 blockade. *Science* 271, 1734–1736.
11. North RJ, Awwad M. (1990) Elimination of cycling $CD4^+$ suppressor T cells with an anti-mitotic drug releases non-cycling $CD8^+$ T cells to cause regression of an advanced lymphoma. *Immunology* 71, 90–95.
12. Tanaka H, Tanaka J, Kjaergaard J, Shu S. (2002) Depletion of $CD4^+CD25^+$ regulatory cells augments the generation of specific immune T cells in tumor-draining lymph nodes. *J. Immunother.* 25, 207–217.
13. Pardoll D. (2003) Does the immune system see tumors as foreign or self? *Annu. Rev. Immunol.* 21, 807–839.
14. Li J, Hu P, Khawli LA, Epstein AL. (2003) LEC/chTNT-3 fusion protein for the immunotherapy of experimental solid tumors. *J. Immunother.* 26, 320–331.
15. Li J, Hu P, Khawli LA, Epstein AL. (2003) Complete regression of experimental solid tumors by combination LEC/chTNT-3 immunotherapy and $CD25^+$ T-cell depletion. *Cancer Res.* 63, 8384–8392.
16. Liu A, Hu P, Khawli LA, Epstein AL. (2005) B7-Fc fusion protein treatment induces tumor regressions with long-term memory and is enhanced by Treg cell depletion. *Clin. Cancer Res.* 11, 8492–8502.
17. Liu A, Hu P, Khawli LA, Epstein AL. (2006) B7.1/NHS76: A new co-stimulator fusion protein for the immunotherapy of solid tumors. *J. Immunother.* 29, 425–435.
18. Sturmhoefel K, Lee K, Gray GS, Jenifer Thomas J, Zollner R, O'Toole M, et al. (1999) Potent activity of soluble B7-IgG fusion proteins in therapy of established tumors and as vaccine adjuvant. *Cancer Res.* 59, 4964–4972.
19. Runyon K, Lee K, Zuberek K, Collins M, Leonard JP, Dunussi-Joannopoulos K. (2001) The combination of chemotherapy and systemic immunotherapy with soluble B7-immunoglobulin G leads to cure of murine leukemia and lymphoma and demonstration of tumor-specific memory responses. *Blood* 97, 2420–2426.
20. Chen L. (2004) Co-Inhibitory molecules of the B7-CD28 family in the control of T-cell immunity. *Nat. Rev. Immunol.* 4, 336–347.

21. Sutmuller RP, van Duivenvoorde LM, van Elsas A, Schumacher TNM, Manon E. Wildenberg ME, et al. (2001) Synergism of cytotoxic T lymphocyte-associated antigen-4 blockade and depletion of CD25$^+$ regulatory T cells in antitumor therapy reveals alternative pathways for suppression of autoreactive cytotoxic T lymphocyte responses. *J. Exp. Med.* 194, 823–832.
22. Munn DH, Mellor AL. (2004) IDO and tolerance to tumors. *Trends Mol. Med.* 10, 15–18.
23. Parmiani G, Rivoltini L, Andreoda G, Carrabba M. (2000) Cytokines in cancer therapy. *Immunol. Lett.* 74, 41–44.
24. Ben-Efraim S. (1999) One hundred years of cancer immunotherapy: a critical appraisal. *Tumor Biol.* 20, 1–24.
25. Bubenik J, Otter W, Huland E. (2000) Local cytokine therapy of cancer: interleukin-2, interferon, and related cytokines. *Cancer Immunol. Immunother.* 49, 116–122.
26. Di Cailo E, Forni G, Lollini P, Colombo MP, Modesti A, Musiani P. (2001) The intriguing role of polymorphonuclear neutrophils in anti-tumor reaction. *Blood* 97, 339–345.
27. Ran S, Gao B, Duffy S, Watkins L, Rote N, Thorpe PE. (1998) Infarction of solid Hodgkin's tumors in mice by antibody-directed targeting of tissue factor to tumor vasculature. *Cancer Res.* 58, 4646–4653.
28. Hu P, Yan J, Sharifi J, Bai T, Khawli LA, Epstein AL. (2003) Comparison of three different targeted tissue factor fusion proteins for inducing tumor vessel thrombosis. *Cancer Res.* 63, 5046–5053.
29. Dowell RI, Springer CJ, Davies DH, Hadley EM, Philip J, Burke PJ, et al. (1996) New mustard prodrugs for antibody-directed enzyme prodrug therapy: alternatives to the amide link. *J. Med. Chem.* 39, 1100–1105.
30. Haisma HJ, Brakenhoff RH, van der Muelen-Muileman IH, Boven E. (1998) Construction and characterization of a fusion protein of single chain anti-carcinoma antibody 323/A3 and human β-glucuronidase. *Cancer Immunol. Immunother.* 45, 266–272.
31. Schmidt SR. (2009) Fusion proteins as biopharmaceuticals—applications and challenges. *Curr. Opin. Drug Discov. Devel.* 12, 284–295.
32. Schrama D, Reisfeld RA, Becker JC. (2006) Antibody targeted drugs as cancer therapeutics. *Nat. Rev. Drug Discov.* 5, 147–159.
33. Ortiz-Sánchez E, Helguera G, Daniels TR, Penichet ML. (2008) Antibody–cytokine fusion proteins: applications in cancer therapy. *Expert Opin. Biol. Ther.* 8, 609–632.
34. Epstein AL, Chen FM, Taylor CR. (1988) A novel method for the detection of necrotic lesions in human cancers. *Cancer Res.* 48, 5842–5848.
35. Chen FM, Hansen EB, Taylor CR, Epstein AL. (1991) Diffusion and binding of monoclonal antibody TNT-1 in multicellular tumor spheroids. *J. Natl. Cancer Inst.* 83, 200–204.
36. Miller GK, Naeve GS, Gaffar SA, Epstein AL. (1993) Immunologic and biochemical analysis of TNT-1 and TNT-2 monoclonal antibody binding to histones. *Hybridoma* 12, 689–698.
37. Epstein AL, Chen D, Ansari A, Najafi A, Siegel M, Lee K, et al. (1991) Radioimmunodetection of necrotic lesions in human tumors using I-131 labeled TNT-1 F(ab')$_2$ monoclonal antibody. *Antibody Immunoconj. Radiopharm.* 4, 151–161.
38. Hornick JL, Hu P, Khawli LA, Hu P, Biela BH, Mizokami MM, et al. (1998) A new chemically modified chimeric TNT-3 monoclonal antibody directed against DNA for the radioimmunotherapy of solid tumors. *Cancer Biother. Radiopharm.* 13, 255–268.
39. Chen S, Yu L, Jiang C, Zhao Y, Sun D, Li S, et al. (2005) Pivotal study of iodine-131-labeled chimeric tumor necrosis treatment radioimmunotherapy in patients with advanced lung cancer. *J. Clin. Oncol.* 23, 1538–1547.
40. Yu L, Ju D, Chen W, Li T, Xu Z, Jiang C, et al. (2006) I-131 chTNT radioimmunotherapy of 43 patients with advanced lung cancer. *Cancer Biother. Radiopharm.* 21, 5–14.
41. Giovarelli M, Cappello P, Forni G, Salcedo T, Moore PA, LeFleur DW, et al. (2000) Tumor rejection and immune memory elicited by locally released LEC chemokine are associated with and impressive recruitment of APCs, lymphocytes, and granulocytes. *J. Immunol.* 164, 3200–3206.
42. Moro M, Gasparri AM, Pagano S, Bellone M, Tornaghi P, Veglia F, et al. (1999) Induction of therapeutic T-cell immunity by tumor targeting with soluble recombinant B7-immunoglobulin costimulatory molecules. *Cancer Res.* 59, 2650–2656.
43. Zheng G, Chen A, Sterner RE, Zhang PJ, Pan T, Kiyatkin N, et al. (2001) Induction of autitumor immunity via intratumoral tetra-costimulator protein transfer. *Cancer Res.* 61, 8127–8134.
44. Yamaguchi N, Hiraoka SI, Mukai T, Takeuchi N, Zhou XY, Ono S, et al. (2004) Induction of tumor regression by administration of B7-Ig fusion proteins: mediation by type 2 CD8$^+$ T-cells and dependence on IL-4 production. *J. Immunol.* 172, 1347–1354.
45. Smith KA. (1993) Lowest dose interleukin-2 immunotherapy. *Blood* 81: 1414–1423.
46. Silagi S, Shaefer A. (1986) Successful immunotherapy of mouse melanoma and sarcoma with rIL-2 and cyclophosphamide. *J. Biol. Response Modif.* 5, 411–422.
47. Sone S, Ogura T. (1994) Local interleukin-2 therapy for cancer, and its effect induction mechanisms. *Oncology* 51, 170–176.
48. LeBerthon B, Khawli LA, Alauddin M, Miller GK, Bishan S. Charak BS, et al. (1991) Enhanced tumor uptake of macromolecules induced by a novel vasoactive interleukin-2 immunoconjugate. *Cancer Res.* 51, 2694–2698.
49. Lode HN, Xiang R, Becker JC, Gillies SD, Reisfeld RA. (1998) Immunocytokines: a promising approach to cancer immunotherapy. *Pharmacol. Ther.* 80, 277–292.
50. Khawli LA, Miller GK, Epstein AL. (1994) Effect of seven new vasoactive immunoconjugates on the enhancement of monoclonal antibody uptake in tumors. *Cancer* S73, 824–831.
51. Khawli LA, Hornick JL, Sharifi J, Epstein AL. (1997) Improving the chemotherapeutic index of IUdR using a vasoactive immunoconjugate. *Radiochim. Acta* 79, 83–86.

52. Hornick JL, Khawli LA, Hu P, Sharifi J, Khanna C, Epstein AL. (1999) Pretreatment with a monoclonal antibody/interleukin 2 fusion protein directed against DNA enhances the delivery of therapeutic molecules to solid tumors. *Clin. Cancer Res.* 5, 51–60.
53. Mizokami MM, Hu P, Khawli LA, Jaili L, Epstein AL. (2003) Chimeric TNT-3/murine interferon-(fusion protein for the immunotherapy of solid malignancies. *Hybrid. Hybridomics* 22, 197–207.
54. Sharifi J, Khawli LA, Hu P, Li J, Epstein AL. (2002) Generation of human interferon gamma and tumor necrosis alpha chimeric TNT-3 fusion proteins. *Hybrid. Hybridomics* 21, 421–432.
55. Mackay CR. (2001) Chemokines: immunology's high impact factors. *Nat. Immunol.* 2, 95–101.
56. Howard OMZ, Dong HF, Shirakawa AK, Oppenheim JJ. (2000) LEC induces chemotaxis and adhesion by interacting with CCR1 and CCR8. *Blood* 96, 840–845.
57. Thelen M. (2001) Dancing to the tune of chemokines. *Nat. Immunol.* 2, 129–134.
58. Hombach A, Abken H. (2007) Costimulation tunes tumor-specific activation of redirected T cells in adoptive immunotherapy. *Cancer Immunol. Immunother.* 56, 731–737.
59. Hathcock KS, Laszlo G, Pucillo C, Linsley P, Hodes RJ. (1994) Comparative analysis of B7-1 and B7-2 co-stimulatory ligands: expression and function. *J. Exp. Med.* 180, 631–640.
60. Futagawa T, Akiba HK, Kodama T, Takeda K, Hosoda Y, Yagita H, et al. (2002) Expression and function of 4-1BB and 4-1BB ligand on murine dendritic cells. *Int. Immunol.* 14, 275–286.
61. Vinay DS, Kwon BS. (1998) Role of 4-1BB in immune responses. *Semin. Immunol.* 10, 481–489.
62. Gramaglia I, Cooper D, Miner K, Kwon BS, Croft M. (2000) Co-stimulation of antigen-specific CD4 T cells by 4-1BB ligand. *Eur. J. Immunol.* 30, 392–402.
63. DeBenedette MA, Chu NR, Pollok KE, Hurtado J, Wade WF, Kwon BS, et al. (1995) Role of 4-1BB ligand in costimulation of T lymphocyte growth and its upregulation on M12 B lymphomas by cAMP. *J. Exp. Med.* 181, 985–992.
64. Melero I, Shuford WW, Newby SA, Aruffo A, Ledbetter JA, Hellström KE, et al. (1997) Monoclonal antibodies against the 4-1BB T-cell activation molecule eradicate established tumors. *Nat. Med.* 3, 682–685.
65. Sharifi J, Khawli LA, Hu P, King S, Epstein AL. (2001) Characterization of a phage display-derived human monoclonal antibody (NHS76) counterpart to chimeric TNT-1 directed against necrotic regions of solid tumors. *Hybrid. Hybridomics* 20, 305–312.
66. Zheng G, Wang B, Chen A. (2004) The 4-1BB costimulation augments the proliferation of $CD4^+CD25^+$ regulatory T cells. *J. Immunol.* 173, 2428–2434.
67. Choi BK, Bae JS, Choi EM, Kang WJ, Sakaguchi S, Vinay DS, et al. (2004) 4-1BB-dependent inhibition of immunosuppression by activated $CD4^+CD25^+$ T cells. *J. Leukoc. Biol.* 75, 785–791.
68. Zhang N, Sadun RE, Arias RS, Flanagan ML, Sachsman SM, Nien YC, et al. (2007) Targeted and untargeted CD137L fusion proteins for the immunotherapy of experimental solid tumors. *Clin. Cancer Res.* 13, 2758–2767.
69. Nocentini G, Riccardi C. (2005) GITR: A multifaceted regulator of immunity belonging to the tumor necrosis factor receptor superfamily. *Eur. J. Immunol.* 35, 1016–1022.
70. Hu P, Arias RS, Sadun RE, Nien YC, Zhang N, Sabzevari H, et al. (2008) Construction and preclinical characterization of Fc-mGITRL for the immunotherapy of cancer. *Clin. Cancer Res.* 14, 579–588.
71. Mallett SS, Fossum S, Barclay AN. (1990) Characterization of the MRC OX40 antigen of activated CD4 positive T lymphocytes—a molecule related to nerve growth factor receptor. *Eur. Mol. Biol. Organ. J.* 9, 1063–1068.
72. Weinberg AD, Wegmann KW, Funatake C, Whitham RH. (1999) Blocking OX-40/OX-40 ligand interaction *in vitro* and *in vivo* leads to decreased T cell function and amelioration of experimental allergic encephalomyelitis. *J. Immunol.* 162, 1818–1826.
73. Pippig SD, Pena-Rossi C, Long J, Godfrey WR, Fowell DJ, Reiner SL, et al. (1999) Robust B cell immunity but impaired T cell proliferation in the absence of CD134 (OX40). *J. Immunol.* 163, 6520–6529.
74. Maxwell JR, Weinberg A, Prell RA, Vella AT. (2000) Danger and OX40 receptor signaling synergize to enhance memory T cell survival by inhibiting peripheral deletion. *J. Immunol.* 164, 107–112.
75. Rogers PR, Song J, Gramaglia I, Killeen N, Croft M. (2001) OX40 promotes Bcl-xL and Bcl-2 expression and is essential for long-term survival of CD4 T cells. *Immunity* 15, 445–455.
76. Wang HC, Klein JR. (2001) Multiple levels of activation of murine $CD8^+$ intraepithelial lymphocytes defined by OX40 (CD134) expression: effects on cell-mediated cytotoxicity, IFN-gamma, and IL-10 regulation. *J. Immunol.* 167, 6717–6723.
77. Lee HW, Park SJ, Choi BK, Kim HH, Nam KO, Kwon BS. (2002) 4-1BB promotes the survival of $CD8^+$ T lymphocytes by increasing expression of Bcl-xL and Bfl-1. *J. Immunol.* 169, 4882–4888.
78. Lee SJ, Myers L, Muralimohan G, Dai J, Qiao Y, Li Z, et al. (2004) 4-1BB and OX40 dual costimulation synergistically stimulate primary specific CD8 T cells for robust effector function. *J. Immunol.* 17, 3002–3012.
79. Gavin MA, Clarke SR, Negrou E, Gallegos A, Rudensky A. (2002) Homeostasis and anergy of $CD4^+CD25^+$ suppressor T cells *in vivo*. *Nat. Immunol.* 3, 33–41.
80. Takeda I, Ine S, Killeen N, Ndhlovu LC, Murata K, Satomi S, et al. (2004) Distinct roles for the OX40-OX40 ligand interaction in regulatory and nonregulatory T cells. *J. Immunol.* 172, 3580–3589.

81. Valzasina B, Guiducci C, Dislich H, Killeen N, Weinberg AD, Colombo MP. (2005) Triggering of OX40 (CD134) on $CD4^+CD25^+$ T cells blocks their inhibitory activity: a novel regulatory role for OX40 and its comparison with GITR. *Blood* 105, 2845–2851.

82. Sadun RE, Hsu WE, Zhang N, Nien YC, Bergfeld SA, Sabzevari H, et al. (2008) Fc-mOX40L fusion protein produces complete remission and enhanced survival in 2 murine tumor models. *J. Immunother.* 31, 235–245.

83. McCarron PA, Olwill SA, Marouf WM, Buick RJ, Walker B, Scott CJ. (2005) Molecular conjugates and therapeutic strategies. *Mol. Intervent.* 5, 368–380.

84. Biela BH, Khawli LA, Hu P, Epstein AL. (2003) Chimeric TNT-3/human β-glucuronidase fusion proteins for antibody-directed enzyme prodrug therapy (ADEPT). *Cancer Biother. Radiopharm.* 18, 339–353.

85. Stone MJ, Ruf W, Miles DJ, Edgington TS, Wright PE. (1995) Recombinant soluble human tissue factor secreted by *Saccharomyces cerevisiae* and refolded from *Escherichia coli* inclusion bodies: glycosylation of mutants, activity and physical characterization. *Biochem. J.* 310, 605–614.

86. Wigginton JM, Komschlies KL, Back TC, Franco JL, Brunda MJ, Wiltrout RH. (1996) Administration of interleukin 12 with pulse interleukin 2 and the rapid and complete eradication of murine renal carcinoma. *J. Natl. Cancer Inst.* 88, 38–43.

87. Ohe Y, Podack ER, Olsen KJ, Miyahara Y, Ohira T, Miura K, et al. (1993) Combination effect of vaccination with IL2 and IL4 cDNA transfected cells on the induction of a therapeutic immune response against Lewis lung carcinoma cells. *Int. J. Cancer* 53, 432–437.

88. Wakimoto H, Abe J, Tsunoda R, Aoyagi M, Hirakawa K, Hamada H. (1996) Intensified antitumor immunity by a cancer vaccine that produces granulocytemacrophage colony-stimulating factor plus interleukin 4. *Cancer Res.* 56, 1828–1833.

89. Hill HC, Conway TF, Sabel MS, Jong YS, Mathiowitz E, Bankert RB, et al. (2002) Cancer immunotherapy with interleukin 12 and granulocyte-macrophage colony-stimulating factor-encapsulated microspheres: coinduction of innate and adaptive antitumor immunity and cure of disseminated disease. *Cancer Res.* 62, 7254–7263.

90. Gillies SD, Lan Y, Brunkhorst B, Wong WK, Li Y, Lo KM. (2002) Bi-functional cytokine fusion proteins for gene therapy and antibody-targeted treatment of cancer. *Cancer Immunol. Immunother.* 51, 449–460.

91. Khawli LA, Hu P, Epstein AL. (2008) *Therapeutic Antibodies. Handbook of Experimental Pharmacology*. Springer-Verlag, Berlin Heidelberg, 291–328.

92. Hill JO, Awwad M, North RJ. (1989) Elimination of $CD4^+$ suppressor T cells from susceptible BALB/c mice releases $CD8^+$ T lymphocytes to mediate protective immunity against Leishmania. *J. Exp. Med.* 169, 1819–1827.

93. Rakhmilevich AL, North RJ. (1993) Presence of $CD4^+$ T suppressor cells in mice rendered unresponsive to tumor antigens by intravenous injection of irradiated tumor cells. *Int. J. Cancer* 55, 338–343.

94. Rakhmilevich AL, North RJ. (1994) Elimination of $CD4^+$ T cells in mice bearing an advanced sarcoma augments the antitumor action of interleukin-2. *Cancer Immunol. Immunother.* 38, 107–112.

95. Shevach EM. (2002) $CD4^+CD25^+$ suppressor T cells: more questions than answers. *Nat. Rev. Immunol.* 2, 389–400.

96. Javia LR, Rosenberg SA. (2003) $CD4^+CD25^+$ Suppressor lymphocytes in the circulation of patients immunized against melanoma antigens. *J. Immunother.* 26, 85–93.

97. Hara M, Kingsley C. (2001) IL-10 is required for regulatory T cell to mediate tolerance to alloantigens *in vivo*. *J. Immunother.* 166, 3789–3796.

98. Somasundaram R, Jacob L. (2002) Inhibition of cytolytic T lymphocyte proliferation by autologous $CD4^+CD25^+$ regulatory T cells in a colorectal carcinoma patient is mediated by transforming growth factor-β. *Cancer Res.* 62, 5267–5272.

99. Onizuka S, Tawara I, Shimizu J, Sakaguchi S, Fujita T, Nakayama E. (1999) Tumor rejection by *in vivo* administration of anti-CD25 (interleukin-2 receptor alpha) monoclonal antibody. *Cancer Res.* 59, 3128–3133.

100. Shimizu J, Yamazaki S, Sakaguchi S. (1999) Induction of tumor immunity by removing $CD25^+CD4^+$ T cells: a common basis between tumor immunity and autoimmunity. *J. Immunol.* 163, 5211–5218.

101. Awwad M, North RJ. (1990) Radiosensitive barrier to T-cell-mediated adoptive immunotherapy of established tumors. *Cancer Res.* 50, 2228–2233.

102. Levings MK, Sangregorio R, Roncarolo MG. (2001) Human $CD25^+CD4^+$ T cells suppress naïve and memory T-cell proliferation and can be expanded *in vitro* without loss of suppressor function. *J. Exp. Med.* 193, 1295–1302.

103. Dieckmann D, Plottner H, Berchtold S, Berger T, Schuler G. (2001) *Ex vivo* isolation and characterization of $CD4^+CD25^+$ T cells with regulatory properties from human blood. *J. Exp. Med.* 193, 1303–1310.

104. Jonuleit H. (2001) Identification and functional characterization of human $CD4^+CD25^+$ T cells with regulatory properties isolated from peripheral blood. *J. Exp. Med.* 193, 1285–1294.

105. Stephens LA, Mottet C, Mason D, Powrie F. (2001) Human of $CD4^+CD25^+$ thymocytes and peripheral T cells have immune suppressive activity. *Eur. J. Immunol.* 31, 1247–1254.

106. Ng WF. (2001) Human of $CD4^+CD25^+$ cells: a naturally occurring population of regulatory T cells. *Blood* 98, 2736–2744.

107. Baecher-Allen C, Brown JA, Freeman GJ, Haefler DA. (2001) $CD4^+CD25^+$ regulatory cells in human peripheral blood. *J. Immunol.* 167, 1245–1253.

108. Liyanage UK, Moore TT, Joo HG, Tanaka Y, Herrmann V, Doherty G, et al. (2002) Prevalence of regulatory T cells is increased in peripheral blood and tumor microenvironment of patients with pancreas or breast adenocarcinoma. *J. Immunol.* 169, 2756–2761.

109. Sadun RE, Sachsman SM, Chen X, Christenson KW, Morris WZ, Hu P, et al. (2007) Immune signatures of murine and human cancers reveal unique mechanisms of tumor escape and new targets for cancer immunotherapy. *Clin. Cancer Res.* 13, 4016–4025.

110. Flanagan ML, Arias RS, Hu P, Khawli LA, Epstein AL. (2007) *Methods in Molecular Biology: Monoclonal Antibodies: Methods in Protocol.* Humana Press, Totowa, NJ, pp. 33–52.

111. Flanagan ML, Khawli LA, Hu P, Epstein AL. (2006) H60/TNT-3 Fusion protein activates NK cells *in vitro* and improves immunotherapeutic outcome in murine syngeneic tumor models. *J. Immunother.* 29, 274–283.

20

DEVELOPMENT OF EXPERIMENTAL TARGETED TOXIN THERAPIES FOR MALIGNANT GLIOMA

Nikolai G. Rainov and Volkmar Heidecke

Department of Neurosurgery, Klinikum Augsburg, Augsburg, Germany

20.1 INTRODUCTION

Glia-derived neoplasms of the brain (gliomas) of the histological grades 3 (anaplastic astrocytoma, anaplastic oligoastrocytoma, and anaplastic oligodendroglioma) and 4 (*glioblastoma multiforme*, GBM) according to the morphology-based classification of the World Health Organization (WHO) [1] represent different stages of the same fatal neoplastic disease of the central nervous system (CNS). WHO grade 3 and 4 gliomas are also known as malignant gliomas and are the most frequent intrinsic brain tumors in adults [2].

Malignant gliomas typically arise in the lobar white matter or in the deep gray matter of the brain and are characterized by diffusely infiltrating growth. These gliomas consist of highly proliferative and exceptionally migratory tumor cells with a variety of acquired genetic alterations, making them extremely resistant to antiproliferative therapies [3–5]. Neuroradiological imaging such as computed tomography (CT) or magnetic resonance imaging (MRI) can only visualize areas with high percentage of tumor cells and strong neoangiogenesis, but not such with low density of invading glioma cells and a few neoplastic capillaries [6,7]. Gliomas expand both by mitotic activity of tumor cells and by migration of these cells away from the initial tumor mass and into the surrounding brain [8]. It is now generally accepted that widespread migration of glioma cells is a very early feature of these tumors, and by the time of clinical manifestation and histological diagnosis cells have spread throughout the CNS and are located beyond the extent of any reasonable surgical resection [9].

Supporting the notion that malignancy declines with distance from the main tumor, the fact is well known that most malignant gliomas will recur within a few centimeters of the walls of the resection cavity or the residual tumor [8,10,11]. Simple diffusion of cells could potentially explain this observation. On the other hand, recent insights in glioma cell biology suggest that more malignant and actively proliferating cells might be less migratory and less invasive, while less malignant cells may migrate far and tend to invade more remote areas of the brain which surrounds the tumor mass [12].

Survival of patients with malignant glioma has remained essentially the same in the last three decades, despite a quantum leap in medical and surgical technology and major advancements in understanding the natural history and biology of the disease and the molecular genetics of glioma cells [3,9,13,14]. There are no curative treatments for malignant glioma at present and no treatment is able to prevent eventual tumor progression or recurrence. Even the most radical surgical resection is unable to cure a malignant glioma although it may significantly prolong survival and improve quality of life [10,11,15,16].

Current standard therapy of newly diagnosed malignant gliomas consists of surgery, preferably a gross total resection

Fusion Protein Technologies for Biopharmaceuticals: Applications and Challenges, First Edition. Edited by Stefan R. Schmidt.

of contrast-enhancing areas, subsequent fractionated external radiotherapy up to a total dose of 60 Gy, and concomitant and adjuvant chemotherapy in WHO grade 4 tumors, with various combinations of these three approaches in grade 3 tumors [5,10,14]. The use of chemotherapy in addition to surgery and radiation was considered somewhat controversial until recent studies demonstrated significantly improved outcome for patients with previously untreated GBM receiving radiotherapy plus concomitant and adjuvant chemotherapy after surgery [17]. However, in all clinical studies, significant improvements in overall survival are demonstrated mostly in selected subpopulations of patients with somewhat better prognostic factors, such as higher Karnofsky Performance Status (KPS) and younger age, or with specific genetic alterations [9,15,16,18].

Perhaps the most prominent feature which renders malignant glioma a preferred target for tumor selective drugs and local treatments is the unique combination of high mitotic activity of tumor cells on the background of the mostly post-mitotic environment of the adult brain [6,8]. In addition, malignant gliomas do not metastasize systemically, but are extremely aggressive in their local environment within the closed compartment of the CNS. Distinct biological features common in malignant gliomas, but unusual in normal brain tissue, have been the subject of many research studies. Such therapeutically useful features are specific genetic alterations in glioma cells. Expression of specific proteins common for malignant glioma cells but uncommon in normal brain cells, and aberrant activation of signal transduction cascades, seems to provide the best opportunity for selective molecular targeting of therapeutic modalities to tumor cells [8]. All these features make gliomas a unique model disease for the design and testing of novel treatments with tumor-selective toxicity and intracerebral rather than systemic mode of administration.

This review summarizes the most recent clinically relevant experimental approaches to intracerebral immunotoxin therapy of malignant gliomas. All of these were employed in an adjuvant setting in addition to the aforementioned and well-established standard modalities of surgery and fractionated external radiotherapy with chemotherapy.

20.2 TARGETED TOXINS—GENERAL CONSIDERATIONS

Targeted toxins (immunotoxins) represent a new class of anticancer agents providing high specificity for tumor cells selectively overexpressing some surface proteins [18,19]. Clinically used toxins are recombinant polypeptide molecules consisting of a tumor-selective ligand coupled to a highly potent peptide toxin, which is truncated to abolish native toxicity. The most frequently used and best researched ligands bind to tumor-associated molecules with receptor signaling properties, such as epidermal growth factor receptor (EGFR), transferrin receptor (TfR), interleukin-13 receptor (IL-13R), or interleukin-4 receptor (IL-4R). The toxin part of the molecule in all clinically used toxins is a polypeptide derived from bacteria (*Corynebacterium diphtheriae, Pseudomonas aeruginosa*), which has been modified by deleting the native targeting and internalization domains of the polypeptide and replacing them with one of the above ligands (Figure 20.1) [18–20].

The mechanism of action of targeted toxins may have important advantages over that of radiation and classic chemotherapeutic agents. Toxins are effective against radiation-resistant hypoxic tumor cells and far more potent than any chemotherapy drug—one single molecule of toxin is sufficient to cause tumor cell death independent of any malignancy-associated genetic alterations and/or mutations (Figure 20.2) [21]. Multidrug resistance and apoptosis resistance are therefore not an issue with toxins—after receptor binding and internalization, no tumor cell is able to survive the toxin part of the molecule [22–24]. A few targeted toxins have advanced from the laboratory to the stage of clinical studies, with two of these reaching Phase III trials.

20.3 DELIVERY MODE AND PHARMACOKINETICS OF TARGETED TOXINS IN THE BRAIN

The therapeutic success of a targeted toxin will depend not only on its ability to selectively target tumor cells but to a high degree also on the delivery mode and distribution throughout the tumor and the surrounding normal brain tissue [25,26]. Diffusion of peptide molecules such as recombinant toxins in tissue is a rather inefficient way of distribution and will depend not only on the concentration of the compound but also on its molecular weight, polarity, and its avidity for the target receptor [27]. The time required for a molecule to diffuse a distance L can be approximated by the formula $L^2/4D$, where D is the diffusion coefficient of that particular molecule [28]. For example, the time required for an IgG antibody to diffuse a distance of 1 mm will be ~2 days. To circumvent this limitation, distribution of toxins by convection-enhanced delivery (CED), a much more efficient and fast mode for interstitial delivery by high-flow infusion, has been studied in several animal models [28].

Morrison et al. [29] published a paper on mathematical modeling of the pharmacodynamics of high-flow (0.5–6.0 μL/h) microinfusion of macromolecules in the brain. For a representative protein of 180-kDa molecular weight infused into homogenous brain tissue at a constant flow rate of 3 μL/min, delivery of the molecule within 12 h would occur to a sphere with 1.5 cm radius from the infusion tip. Tissue doses were described as rather uniform, with the

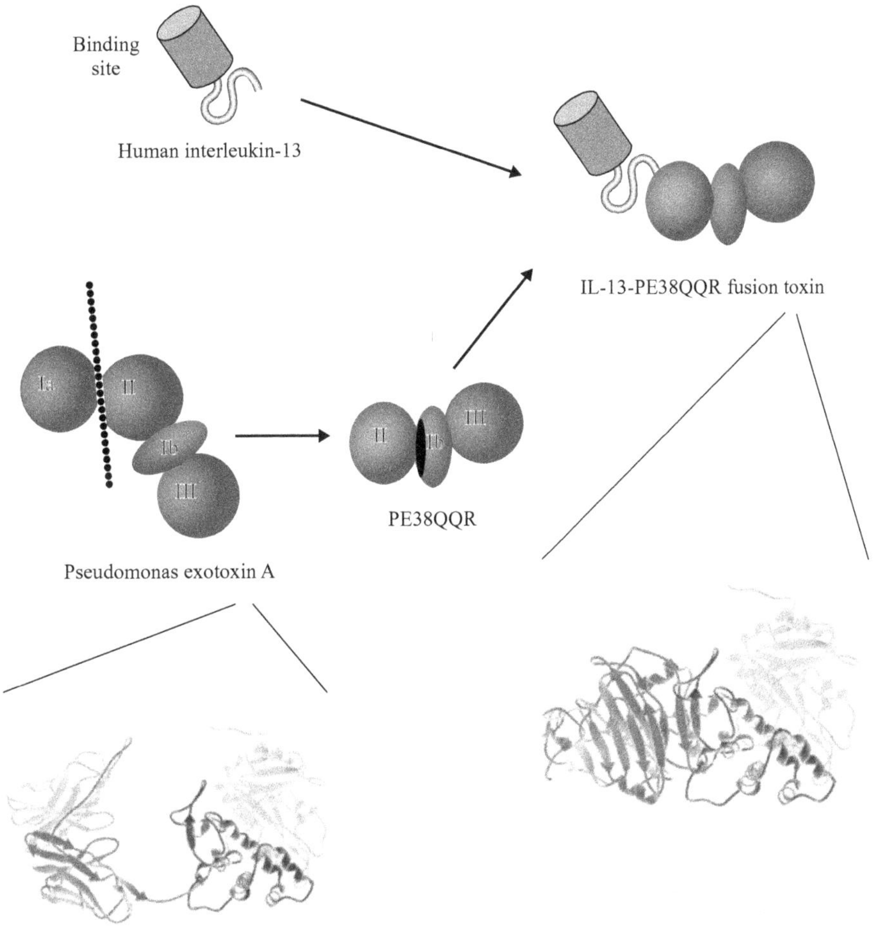

FIGURE 20.1 Schematic and 3D-ribbon representations of the structure of the recombinant fusion toxin IL-13-PE38 (IL-13-PE38QQR). A derivative of the native Pseudomonas exotoxin A (PE), PE38QQR, in which domain Ia (amino acids 1–252) has been completely deleted and domain Ib (amino acids 365–380) has been partially deleted, lysine residues at positions 590 and 606 have been replaced by glutamine, and at position 613 by arginine, has been fused to the carboxy terminus of the human IL-13 to produce the fusion toxin IL-13-PE38QQR (IL-13-PE38).

benefit of control over undesired toxicity that may occur with alternative delivery methods depending on large concentration gradients for tissue transport. There was a significant penetration advantage of high-flow (convective) over low-flow (diffusive) microinfusion. Using the same example as described earlier, a 12-h high-flow microinfusion of a 180-kDa protein would provide 5–10-fold increases in volume over low-flow infusion, with total treatment volumes $>10\,cm^3$. Degradation rates of the protein would play a role in the final treatment volume, and more rapid degradation rates would reduce the volume [29].

Tf labeled with ^{111}In (80 kDa) was intracerebrally infused in cats and compared with the distribution of a 0.4-kDa ^{14}C-sucrose molecule [30]. Systemic concentrations of ^{111}In-Tf were determined in blood, liver, kidneys, and muscle. A pressure gradient was maintained during infusion so as to generate bulk flow in the interstitial space of the brain. The volume of distribution was shown to increase linearly with the infusion volume for both compounds. The systemic concentrations of ^{111} In-Tf were less than 1% of the brain concentrations for all infused volumes. These data indicated that by using convection (bulk flow) instead of diffusion, increased distribution of large and small molecules could be achieved in the brain with minimal peripheral losses of infusate [30].

A subsequent study in primates used single-photon emission computerized tomography (SPECT) to measure brain distribution of ^{111}In-DTPA-Tf (81 kDa) by CED [31]. ^{111}In-DTPA-Tf was infused at a rate of 114 μL/h for 87 h, and CT and SPECT scans were taken in regular intervals. Distribution of the protein ranged between 2 and 3 cm from the needle tip and infusion volumes were 3.9–6.7 cm^3. These data for time of infusion and area of distribution compared favorably with the calculated diffusion of an IgG molecule,

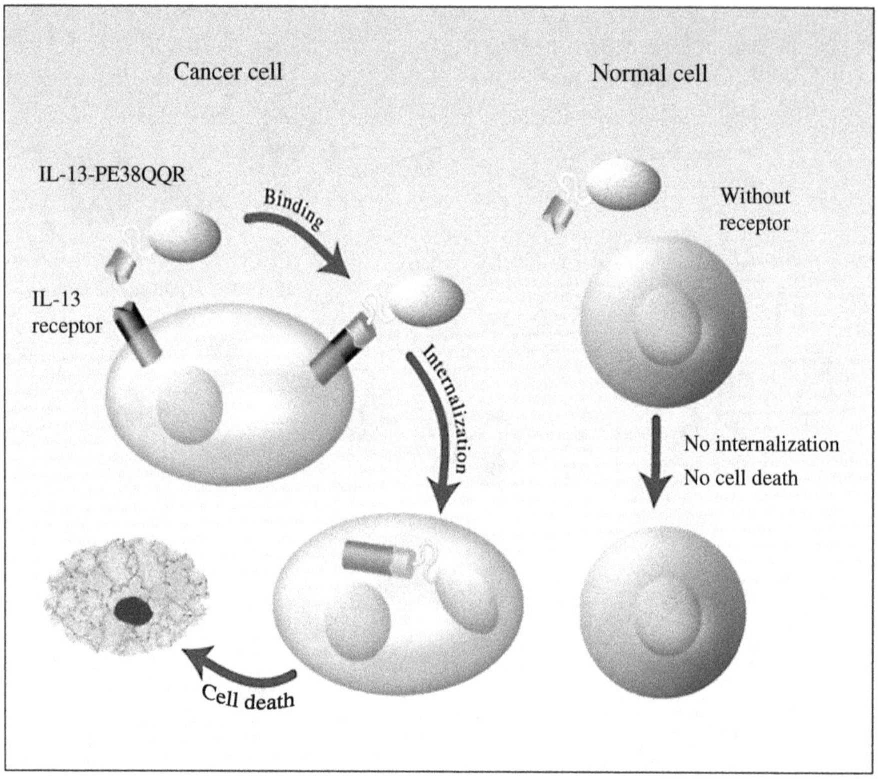

FIGURE 20.2 Schematic representation of the mode of action of the recombinant fusion toxin IL-13-PE38 (IL-13-PE38QQR). IL-13 receptor (IL-13R-α2 subunit) is a tumor-specific protein in human malignant glioma cells. Immune cells, endothelial cells and normal glia and neurons express none or very low amounts of IL-13R. IL-13-PE38 is highly selective and potent in killing human GBM cells by binding to IL-13R, but does not kill normal cells, which are IL-13R-negative.

which would require ~8 months to diffuse 1 cm. Blood perfusion was measured in the infused hemisphere by ^{99}Tc-HMPAO-SPECT and showed no significant reduction (<5%) compared to the control hemisphere. Histology demonstrated only mild gliosis immediately surrounding the needle tract. This study showed that CED could distribute a high molecular weight protein over centimeter distances in nonhuman primate brain and could do so reproducibly and safely [31].

In nude mice with intracerebral glioma receiving an intratumoral bolus injection of IL-13-PE38, drug distribution was maximal at 1 h after injection and the volume of distribution was 19.3 ± 5.8 mm [32]. Intratumoral infusion of IL-13-PE38 via CED in the same glioma model resulted in better distribution than intratumoral bolus administration, with the drug remaining in the brain for 6 h after the treatment [33].

Because human GBM is a neoplastic disease not metastasizing outside the CNS, local delivery of therapeutic agents such as toxins via CED seems to be the best approach to circumvent the limitations of the blood–brain barrier (BBB) and to increase therapeutic efficacy by high local concentrations of drug. All clinical trials with targeted toxins have adopted CED as the delivery mode of choice.

20.4 PRECLINICAL AND CLINICAL STUDIES WITH TARGETED TOXINS

20.4.1 IL4-PE (NBI-3001)

The chimeric recombinant fusion protein IL-4-PE (proprietary designation of IL-4-PE, originally owned by Neurocrine Inc., S. Diego, CA) is composed of circularly permuted interleukin-4 (IL-4) and a truncated form of *P. aeruginosa* exotoxin (PE) A [18,19,23]. PE is a 66-kDa protein with three domains: Ia/Ib, II, and III. The N-terminal domain Ia binds to the α_2-macroglobulin receptor (α_2-macroglobulin receptor/low-density lipoprotein receptor-related protein, LRP), and the ligand-receptor complex undergoes receptor-mediated internalization and processing of the toxin. When the domain Ia is removed, the resulting molecule (termed PE-40) retains its translocation function and elongation factor 2 (EF-2) inhibition properties, but is unable to bind to and kill human cells [18]. Domain II is the site of proteolytic cleavage and is responsible for catalyzing translocation of the toxin into the cytosol. Domain III, located at the C-terminus, possesses ADP-ribosylation activity, which in turn leads to inactivation of EF-2, eventually leading to cell death. In the genetically engineered PE molecule PE38KDEL, amino acids (AA) 253–364 were linked to AA 381–608, which in turn were fused to KDEL (an endoplasmic reticulum retaining sequence) at position 609–612. To improve the binding of IL-4 toxin to the IL-4R, a circularly permuted form of IL-4 was fused to the toxin. This new agent was termed IL-4(38-37)-PE38KDEL or IL-4-PE [34,35]. IL-4-PE was found to have a 16-fold higher affinity for binding to GBM cell lines than the native PE toxin, and was 3–30-fold more toxic to GBM cells [36,37].

IL-4 is a pleiotropic cytokine composed of four α-helices, which is primarily produced by Th_2-type T lymphocytes [38,39]. It has been suggested that residues on the A and C helices bind to the IL-4R and are important for binding affinity, whereas residues near the carboxy terminus bind to a different subunit, presumably the common γ chain of the receptor, and are important for signal transduction [40,41]. However, the amino and carboxy termini are close together, suggesting that attachment of a toxin moiety to the carboxy terminus of IL-4 can disrupt receptor binding [18,34]. IL-4R is a transmembrane protein composed of two subunits with different functions, a 140-kDa chain and a 70-kDa γ chain [40]. IL-4R is also a component of the IL-13 receptor (IL-13R) [41].

While normal CNS cells such as endothelial cells, microglia, or astrocytes express low levels of IL-4R, malignant glioma cells express high levels of this receptor on their surface [23]. In one series, 16 of 21 surgical samples were determined to express high numbers of IL-4R [19]. In another series, all of the 25 tested surgical specimens of high-grade gliomas expressed IL-4R. In addition, primary cell cultures of glioma also expressed IL-4R, as determined by RT-PCR and immunohistochemical analysis [42]. Another recent study demonstrated that 15 of 18 GBM and 12 other brain tumor samples were moderately to intensely positive for IL-4R. In contrast, no detectable IL-4R was expressed in normal brain as determined by immunohistochemistry [43].

Cell culture studies and animal models confirmed that IL-4-PE is highly cytotoxic to immortalized GBM cell lines and to primary glioma cell cultures [39,44]. *In vivo* efficacy was evaluated in a nude mouse model with subcutaneously injected human GBM cells [35]. Antitumor activity was observed in 80% and 100% of animals treated with IL-4-PE at 50 or 100 μg/kg for 3 days. To define the safety profile of the toxin, IL-4-PE was administered intravenously (i.v.), directly into the forebrain, and into the *cisterna magna* in mice, rats, and nonhuman primates, respectively. In the mouse, the approximate LD_{50} following i.v. administration was 475 μg/kg. At high doses (>400 μg/kg) hepatic necrosis was seen. In the rat, direct injection into the forebrain of 275 ng/kg toxin caused no side effects, and in primates, the maximum tolerated dose (MTD) following injection into the *cisterna magna* was 6 μg/kg [35].

Weber et al. [45] carried out an open-label, dose-escalation trial of intratumoral administration of IL-4-PE in patients with recurrent malignant glioma. A total of 31 adult patients were enrolled, 25 with GBM and 6 with anaplastic astrocytoma. Patients were assigned to one of four groups receiving different doses of toxin in different volumes of infusate. Toxin was administered by CED via an external pump connected to stereotactically placed catheters. The overall median survival for the whole group of WHO grade 3 and 4 tumors was 8.2 months, with a median survival of 5.8 months for GBM patients. Six months survival was 52% and 48%, respectively. Drug-related grade 3 or 4 CNS toxicity was seen in all groups in a total of 39% of all patients, and in 22% of patients at the MTD of 6 μg/mL in 40 mL. Treatment-related adverse events (AE) were limited to the CNS. No deaths were attributable to treatment. No drug-related systemic toxicity was apparent in any patient. Gadolinium-enhanced MRI of the brain in some of the treated patients showed areas of decreased signal intensity within the tumor consistent with toxin-mediated tumor necrosis [45].

A multicenter, randomized, open-label Phase II study with IL-4-PE was carried out in patients with recurrent GBM to investigate continuous intratumoral infusion of the toxin followed by surgical resection of tumor. The study was designed to evaluate efficacy of IL-4-PE, with a secondary objective to evaluate safety and tolerability of the toxin [46]. In the toxin group, patients received intratumoral infusion of toxin at total doses of up to 90 μg and underwent surgical resection of tumor between 2 and 7 days after the end of toxin infusion. Patients in the control group underwent tumor resection without prior toxin treatment. A total of 30 adult patients with unilateral, unifocal tumors with a volume <100 mL and a KPS ≤60 have been enrolled. Recruitment was completed in 2003, but no final published results of the study are available yet [46]. There are currently no Phase III protocols using IL-4-PE.

20.4.2 TP-38

TP-38 (Teva Pharmaceuticals, Miami, FL, formerly IVAX Inc., Miami) is a recombinant chimeric protein molecule composed of transforming growth factor α (TGF-α), an EGFR ligand, and the genetically engineered form of PE described earlier (PE38KDEL).

The 170–180-kDa transmembrane glycoprotein, epidermal growth factor (EGF) receptor, is one of four members of the ErbB family of receptor tyrosine kinases, which consist of an extracellular domain that can bind ligands, a transmembrane domain, and an intracellular tyrosine kinase domain. Binding of a ligand (EGF or TGF-α) to EGFR causes receptor dimerization leading to tyrosine kinase activation. The resultant receptor autophosphorylation initiates signal-transduction cascades involved in cell proliferation and survival. After ligand binding, the whole receptor–ligand complex undergoes endocytosis and is translocated to the lysosomes, where the ligand dissociates from the receptor [47]. The tyrosine kinase part of the receptor is required for rapid internalization, transportation through the cell, and amplification of gene synthesis. Studies of the amplification of EGFR have shown that the transmembrane and tyrosine kinase domains are invariably amplified, while the ligand-binding domain and the entire gene are amplified intermittently [48,49].

Human malignant gliomas and many malignant tumors that metastasize to the brain express EGFR, and this is commonly associated with amplification and/or mutation of the EGFR gene during neoplastic transformation [50]. Amplification and high expression of EGFR in gliomas may drive to a significant degree tumor growth and proliferation [51]. By contrast, EGFR is expressed at very low levels or is undetectable on normal human glial cells and neurons. The ratio of EGFR expression in glioma versus normal control brain specimens has been shown to be as high as 300-fold [48].

EGFR-specific binding of TP-38 toxin has been investigated in cell culture. TP-38 may inhibit the binding of ^{125}I-EGF to EGFR-expressing nonsmall cell lung carcinoma at an IC_{50} of 100–2000 ng/mL, depending on the cell line tested [52]. Concentration of TP-38 required to inhibit protein synthesis in malignant tumor cells may range from 5–6 ng/mL (in EGFR-expressing lung cancer cells) to 0.005–0.5 ng/mL in GBM [53].

Animal studies with TP-38 have been carried out in rodent and primate models. In tumor-bearing mice, a total dose of 0.1 μg TP-38 at a concentration of 5 μg/mL was found to be safe and efficacious in prolonging survival. The MTD of intracerebral TP-38 in normal rats was 0.666 μg at a toxin concentration of 33.3 μg/mL. In nonhuman primates, it was demonstrated that an intracerebral bolus dose of 2 μg TP-38 at a concentration of 10 μg/mL is safe [44].

Sampson et al. [54] investigated TP-38 in a Phase I clinical trial. The primary objective of the study was to define MTD and dose-limiting toxicity of TP-38 delivered by CED in patients with recurrent malignant glioma. A secondary objective was to detect efficacy of the toxin. A total of 20 patients were enrolled in the study and doses were escalated from 25 to 100 ng/mL. TP-38 was infused by two stereotactically placed catheters at a flow rate of 0.4 mL/h for each catheter. A total volume of 40 mL was infused. TP-38 was tolerated well, and a MTD was not established. Nonspecific toxicity was not found at any of the dose levels. Toxicity encountered was solely neurologic and mostly related to infusion volume, recurrent tumor, or stereotactic catheter placement, but not directly to TP-38. A total of 15 of the patients in this study died from progressive disease. When the study closed, four further patients did not have recurrence of tumor and were 55, 56, 69, and 116 weeks from the time of TP-38 therapy. Overall median survival after TP-38 for all patients was 23 weeks, whereas for those without radiographic evidence of residual disease at the time of therapy, the median survival was 31.9 weeks. Overall, 3 of 15 patients with residual disease at the time of therapy have demonstrated radiographic responses [54].

A multicenter randomized open-label Phase II study was conducted in adult patients with recurrent GBM [55]. Patients were randomized to two dose levels of TP-38, 50 or 100 ng/mL, and toxin was administered by CED via stereotactically placed catheters. Patients did not have to undergo tumor resection prior to treatment. End points of the study were time to progression, progression-free survival, and overall survival. Three catheters were stereotactically placed in investigator-determined locations within the enhancing tumor area. The infusion rate was 200 μL/h per catheter. Each catheter delivered 13.4 mL over 67 h. The total volume infused was ~40 mL, and the total dose of TP-38 infused was 2 μg (50 ng/mL) or 4 μg (100 ng/mL). One patient had a complete response 48 weeks after infusion, and another patient showed a partial response stable over 60 weeks. A total of 24 patients remained stable. Post-infusion MRI scans in most patients showed unspecific treatment-related changes such as halo contrast enhancement around the infusion sites, which made assessment of response rather difficult. These changes usually resolved by 20 weeks post treatment. There were no grade 3 and 4 toxicities related to TP-38, and all adverse effects of the treatment were reversible [55]. Currently, there is no Phase III clinical protocol employing TP-38.

20.4.3 IL13-PE38

IL-13-PE38 (also known as cintredekin besudotox, NeoPharm Inc., Lake Bluff, IL) is a recombinant chimeric toxin consisting of human IL-13 fused to a mutated form of PE [18–20]. Since domain Ia of PE binds to almost all types of eukaryotic cells, several mutated forms of PE were generated in order to avoid binding to its original ubiquitous receptors. One of the derivatives of PE was PE38QQR, in which domain Ia (amino acids 1–252) and domain Ib (amino acids 365–380) were deleted, and lysine (K) residues at positions 590 and 606 were replaced by glutamine (Q) and at position 613 by arginine (R). PE38QQR was fused to the carboxy terminus of the human IL-13 (hIL-13) to produce IL-13-PE38QQR (IL-13-PE38) [56–59].

The T-helper cell 2-derived immunoregulatory cytokine interleukin-13 (IL-13) inhibits the production of inflammatory cytokines in monocytes. The cell surface receptor for

IL-13 (IL-13R) is expressed on many human cancer cells, including GBM, AIDS-associated Kaposi's sarcoma, ovarian carcinoma, renal cell carcinoma, and fibroblast cell lines [42]. The IL-13R structure varies according to cell type, existing in three different forms known as type I, type II, and type III [59,60].

Subunit structure studies of the IL-13R complex in primary brain tumor cells established the IL-13Rα2 subunit as a tumor-specific protein [42]. Immune cells, endothelial cells, and normal glia and neurons generally express none or very low amounts of IL-13R [20].

IL-13-PE38 was found to be highly selective and potent in killing human GBM cells in culture. The activity of IL-13-PE38 is mediated via IL-13R, as its cytoxicity is blocked by exposing cell lines to 10–100-fold excess of human IL-13 [41]. Furthermore, IL-13R-negative cell lines or cell lines expressing low numbers of IL-13R, including human bone marrow-derived cells and PHA-activated T-cells, are not susceptible to IL-13-PE38-mediated cytotoxicity [58].

Significant antitumor activity of IL-13-PE38 was observed in animal models of human glioma, with intratumoral injection proving more efficacious than i.v. or intraperitoneal (i.p.) administration [20]. In nude mice with intracerebral glioma, the MTD of IL-13-PE38 was 4 μg when delivered by bolus injection, and 10 μg when delivered by CED. Intracerebral injections of IL-13-PE38 into the right frontal lobe of normal rats were carried out at total doses of 0.001, 0.01, 0.1, and 1.0 μg in a constant volume of 10 μL. No morphological changes were evident in the brain tissue removed after 4 days and assessed by light microscopy. There was no evidence of neuroglial cell toxicity with total toxin doses of up to 1.0 μg. In a similar intracerebral toxicity study, IL-13-PE38 at 100, 500, 1000, and 2000 μg/mL was injected into the right frontal lobe of rats. There were no changes detected in weight, behavior, or hematologic or chemistry values after 4 days. No residual drug was detected in serum. However, IL-13-PE38 at high dose levels (500 μg/mL and above) caused severe brain necrosis and edema around the injection site. Inflammatory reaction was comparatively minimal [20]. Interstitial infusion of IL-13-PE38 into the rat brain stem was well tolerated. IL-13-PE38 (10 μg/mL) delivered as a 24-h infusion of 200 μL total volume did not cause any neurological deficits. Post-mortem examination of the brain up to 2 weeks after treatment did not reveal any histological changes beyond the site of the cannula [61].

Several Phase I and II studies have been initiated to investigate IL-13-PE38 as an antitumor agent for the treatment of patients with recurrent malignant glioma. In a Phase I/II study, IL-13-PE38 was infused intratumorally in adult patients with recurrent malignant glioma (including anaplastic oligoastrocytoma) to determine the MDT of the drug [62]. Patients received two IL-13-PE38 infusions at 8-week intervals via two stereotactically placed intratumoral catheters at a constant infusion rate. One aim of the study was to determine MTD and toxicity of IL-13-PE38 administered by CED via intratumoral catheters at 200 μL/h per catheter for 96 h (total 38.4 mL), in two courses 8 weeks apart. Escalation was planned through nine levels ranging from 0.125 to 12.0 μg/mL (total dose 4.8 to 460.8 μg) in cohorts of three patients per level. Cohorts of three patients received 0.5, 1.0, 2.0, and 4.0 μg/mL. Concentrations of up to 2.0 μg/mL were safe and well tolerated. Adverse events reported across all dose ranges were mild and mainly neurological. The incidence of the most frequent drug-related AE was 24%, 24%, 14%, 10%, and 10%, respectively, for headache, hemiparesis, brain edema, aphasia, and ataxia. In the Phase 1 portion of the study, two histopathological and two radiographic responses were observed, with progression-free survival ranging from 3 to 88 weeks and overall survival of 147 weeks. No results have been published yet for the Phase II portion of this study [62].

Another Phase I study investigated IL-13-PE38 in adult supratentorial malignant glioma. In this four-stage study, the primary objectives were to determine effective dose of IL-13-PE38 either prior to or post-resection of the tumor [63,64]. In stage 1, after biopsy and intratumoral catheter placement on day 1, IL-13-PE38 was administered for 48 h at 400 μL/h on days 2 to 4 at escalating doses ranging from 0.25 to 2 μg/mL. Tumor was resected on day 8 and tissue was evaluated for necrosis adjacent to the catheter. Following resection, 2 or 3 catheters were placed into the brain adjacent to the tumor resection cavity, and on days 10 to 14, IL-13-PE38 was given at 750 μL/h for 96 h at a fixed dose of 0.25 μg/mL (total dose 18 μg). In stage 2, intratumoral infusion was omitted and patients underwent resection of tumor followed by a 96-h peritumoral infusion of IL-13-PE38 at 0.5 and 1.0 μg/mL. In stage 3, the duration of peritumoral infusion was increased from 5 to 7 days at a fixed dose of 0.5 μg/mL. Tumor necrosis up to 2.5 cm from the catheter tip was demonstrated radiologically in at least five patients with preoperative intratumoral toxin infusion. Tumor specimens from at least two patients after intratumoral injection of 0.5 μg/mL IL-13-PE38 toxin revealed regional necrosis in an ovoid zone extending 1–2 cm from the catheter tip. Peritumoral post-resection MTD was defined as 0.5 μg/mL. Similar adverse events occurred across all cohorts, most being neurological. Incidence of the most frequent AE was 53%, 27%, 20%, 17%, 13%, 13%, 13%, 10%, 10%, and 10%, respectively, for headache, hemiparesis, sensory disturbance, fatigue, aphasia, facial paresis, abnormal gait, convulsions, hypesthesia, and lymphopenia. Prolonged individual patient survival has been observed after peritumoral therapy at concentrations of 0.25 μg/mL (total dose 18 μg) and above [20,65,66].

A further multicenter Phase I/II study investigated IL-13-PE38 in adult recurrent supratentorial malignant glioma with confirmed radiographic evidence of recurrent or progressive tumor. The primary objective of the Phase I portion of the study was to determine the optimum duration of infusion and drug concentration of IL-13-PE38 delivered by CED via 1 or 2 intratumoral catheters prior to surgical resection of tumor [63,67]. The Phase 2 objective was to determine the proportion of patients surviving at 6 months. In Phase 1, cohorts of 3–6 patients received a fixed drug concentration (0.5 μg/mL) with escalating duration of infusion (4–7 days) to determine MTD based on duration. After determining maximum duration, drug concentration was escalated from 1.0 to 4.0 μg/mL in cohorts of 3–6 patients to determine a MTD based on concentration. Tumor necrosis was assessed 6–13 days after the end of infusion, at the time of tumor resection. There were no severe drug-related AE in the 10 treated patients. Five patients experienced AE grade 2-3. Most frequent AE were headache (70%), hemiparesis (40%), seizures (40%), edema (40%), and aphasia (30%). Progression-free survival ranged from 6 to 30 weeks and overall survival from 10 to 30 weeks [20,63].

In all of these studies, intratumoral infusion of IL-13-PE38 with or without tumor resection in patients with recurrent or progressing malignant glioma seemed to be well tolerated and did not result in any grade 3 and 4 AE, or in any AE in peripheral organs such as liver, kidney, and lung. Neurological AE during and after toxin infusion were encountered in a significant proportion of the treated patients. These included brain edema, meningitis, seizures, headache, and symptoms of increased intracranial pressure. All these AE were temporary and mostly controllable by administration of steroids. In general, IL-13-PE38 doses (0.5–2 μg/mL) showing biological activity (e.g., necrosis on MRI scans) in tumors were below the threshold of widespread neurotoxicity (4–12 μg/mL). Selection of patients without incipient mass effect in the brain (due to tumor size and/or peritumoral edema) seemed particularly important in avoiding serious unwanted side effects. Some prolonged survival was observed in this selected population of patients [20,65,66].

A multicenter, Phase III, randomized, open label, active control, parallel assignment efficacy study (PRECISE) was carried out in order to determine whether overall survival duration, safety, and quality of life were improved for patients treated with IL-13-PE38 compared to patients treated with Gliadel® wafers following surgical tumor removal in the treatment of first recurrence of GBM after initial surgery and external beam radiation therapy [68,69]. Patients were randomized 2:1 to receive IL-13-PE38 or Gliadel. IL-13-PE38 toxin (0.5 μg/mL, total flow rate 0.75 mL/h) was administered over 96 h via 2–4 intraparenchymal catheters placed around the surgical cavity after tumor resection. The primary endpoint was overall survival from the time of randomization. Secondary and tertiary endpoints were safety and health-related quality-of-life assessments.

From March 2004 to December 2005, 296 patients were enrolled at 52 centers. Median survival was 36.4 weeks (9.1 months) for the toxin group and 35.3 weeks (8.8 months) for the Gliadel group ($p = 0.476$). For the efficacy evaluable population, the median survival was 45.3 weeks (11.3 months) for toxin and 39.8 weeks (10 months) for Gliadel ($p = 0.310$). The adverse-events profile was similar in both arms, except that pulmonary embolism was higher in the toxin arm (8% vs 1%, $p = 0.014$).

This was the first randomized Phase III evaluation of an agent administered via CED and the first with an active control group in GBM patients. There was no survival difference between IL-13-PE38 toxin administered via CED and Gliadel wafers placed at the time of surgery. Drug distribution was not assessed but may be crucial for evaluating future CED-based therapeutics [68,69].

20.4.4 TransMID-107 (Tf-CRM107)

TransMID-107 (known as TransMID™, KS Biomedix Holdings plc, UK; now Xenova Group plc, UK/Celtic Pharmaceutical Holdings PL, Bermuda), also known as Tf-CRM107 of KSB-311, is a thioether conjugate of human transferrin (Tf) with a truncated natural mutant form of diphtheria toxin (DT) known as CRM-107, which lacks receptor binding [26]. Native DT is produced by the bacterium *C. diphtheriae* and is composed of two disulfide-linked subunits, both of which are involved in its mechanism of toxicity. The A subunit catalyzes ADP-ribosylation of EF-2, thereby stopping protein synthesis and killing the cell. The B subunit has two functions: binding to cell-surface receptors and translocation of the A subunit into the cytosol [70]. It has been shown previously that the binding function of DT can be separated from the translocation function. Laird and Groman (1976) published data on the isolation of nontoxic DT mutants belonging to four major classes, I to IV [71]. Mutant proteins serologically related to DT were designated as cross-reacting material (CRM). In the class I CRM-107 mutant, two point mutations (phenylalanine at positions 390 and 525 of the DT sequence) reduced the binding 8000-fold and its toxicity 10,000-fold, compared to native DT [71,72]. However, conjugation of CRM-107 by thioether linkage to various new binding moieties was able to reconstitute full toxicity of the mutant [70]. DT entry via different receptor molecules was examined and results showed that the DT receptor was not required for translocation of DT to the cytosol, and that it apparently had no other function than to mediate rapid internalization to low pH compartments. The B chain of DT appeared to contain all the membrane translocation activity, and this activity was distinct and separable from the receptor-binding site on the toxin B subunit [70]. By linking CRM-107 with Tf, a receptor-ligand complex was created that is rapidly internalized to an environment that facilitates toxin translocation.

Translocation out of the endosome to the cytosol appears to be carried out by the B chain and conjugates without it are nontoxic [27,70].

TransMID™ targets tumor cells via the Tf receptor (TfR). TfR is a transmembrane glycoprotein that mediates cellular uptake of iron. TfR are overexpressed on rapidly dividing cells, such as hematopoietic and neoplastic cells, including GBM cells [73]. This fact is thought to reflect an increased requirement for iron in rapidly dividing cells [74]. TfR in normal brain are sparse and their expression is largely restricted to the luminal surface of brain capillaries [75]. It is this relative difference in the density of TfR that TransMID uses to generate differential toxicity to highly TfR expressing neoplastic cells, while sparing low expressing normal brain cells [76]. Low-density background expression of the respective target receptor in normal tissue remains a major concern, however, with TransMID (and with all other targeted toxin-based therapies).

TransMID has been thoroughly evaluated in cell culture. Hall et al. [77] showed that human Tf conjugated to either the *Abrus* plant toxin, abrin, or the *P. aeruginosa* exotoxin A had comparable cytotoxic potency in glioma and melanoma cell lines. All toxins had an ID_{50} in the low picomolar range. Martell et al. [78] compared the efficacy and cytotoxic properties of TransMID to 454A12-rRA, a conjugate of a monoclonal antibody to the human TfR receptor (454A12) with the recombinant *Ricinus* plant toxin, ricin A chain (rRA). Cell lines were sensitive to toxin doses of 10^{-10} M and less. Highly malignant tumor cells such as GBM or medulloblastoma had lower IC_{50} values (10^{-12} M), indicating high sensitivity of such tumors to the targeted toxin. Expression of TfR on the cell lines and resection specimens correlated with sensitivity to the toxin in cell culture [78]. Chignola et al. [79] treated monolayer and multicellular spheroid cultures of human breast cancer, human GBM, and rat gliosarcoma with a toxin composed of TfR and ricin A chain (TfR-RTA) and with TransMID™, which yielded similar patterns of toxicity. The results indicated that the efficacy of targeted toxins against three-dimensional (3D) tumors might be heavily influenced by the number of target molecules expressed by the tumor cells, as well as by the affinity of toxin-cell interaction.

In vivo efficacy of locally administered TransMID against human glioma in nude mice was demonstrated by Laske et al. [80]. These authors studied TransMID compared to 454A12-rRA and administered toxins intratumorally in a subcutaneous model. Repeated intratumoral injections of 10 μg TransMID or 10 μg 454A12-rRA were compared to equimolar doses of the untargeted native toxin CRM-107, 454A12 antibody, rRA, and phosphate-buffered saline. By day 14, TransMID administration resulted in almost complete tumor regression (>95%), without evidence of recurrence by day 30. Treatment with 454A12-rRA caused a less significant decrease (30%) in tumor volume. Unconjugated toxin components (CRM-107, 454A12, or rRA) caused significant, but less potent tumor growth inhibition than the conjugated toxins [80]. In a more recent study, Engebraaten et al. [81] compared the efficacy of TransMID with that of PE conjugated with the 425.3 antibodies directed against EGFR. Mice with subcutaneous gliomas or rats with intracranial gliomas received different doses (1–10 μg) of TransMID or 425.3-PE injected intratumorally in established tumors. Both toxins showed significant antitumor effects in subcutaneous tumors, but only 425.3-PE was effective in intracranial gliomas in rats. Intracerebral TransMID was toxic in doses above 10 ng, while intracerebral 425.3-PE was tolerated up to a dose of 4 μg per animal. The authors concluded that both toxins have promising efficacy in brain tumor models, but that 425.3-PE is better tolerated and has a more specific activity at higher doses.

Laske et al. [82] treated 18 patients with recurrent malignant glioma with intratumoral high flow interstitial microinfusion of TransMID in a dose-escalating single arm Phase I clinical trial. The drug was infused at a maximum flow rate of 4–10 μL/min with toxin concentration of 0.1 μg/mL. On the basis of preclinical data, the rate of infusion was limited to a maximum of 10 μL/min per catheter to avoid leakage via the catheter track. Drug concentration at a given infusion volume was escalated by half log increments in every 3–4 patients, unless severe toxicity was detected in one of the first 2–3 patients. Nine of the 15 evaluable patients responded to treatment by at least 50% reduction in tumor volume on MRI, including two complete responses. Reduction in tumor volume occurred no earlier than 1 month after completion of the first toxin infusion. In four patients, response was not maximal until 6–14 months after the first treatment. Tumor response appeared to be concentration and dose dependent. Only two of the eight patients in the first two treatment groups (0.1–0.32 μg/mL) had partial responses. In contrast, two of four patients treated with >1.0 μg/mL had complete responses and the other two patients had partial responses. Those patients treated with a total dose of 60 μg or higher all had partial responses; however, toxicity was noted 1–4 weeks after treatment. At intermediate toxin concentrations, responses correlated more with total dose than with concentration of drug. Although pretreatment tumor volume seemed to be related to the likelihood of response, it did not reach statistical significance. Median survival after treatment in the responder group was 74 weeks, with three of the responders still alive at 102–142 weeks after the first treatment. Nonresponders survived a median of 36 weeks. Intratumoral infusions of TransMID with total volumes of 5–180 mL were well tolerated. There were no treatment-related deaths or life-threatening or irreversible toxicity. Transient worsening of a pre-existing neurological deficit during infusion occurred 3 times in 44 treatments. No symptomatic systemic toxicity occurred [82].

In this first clinical trial, the relationship between dose, drug concentration, and sustained neurotoxicity established a MTD of 26.8 μg (40 mL at 0.67 μg/mL). The trial results indicated that therapy with TransMID can reduce the size of malignant brain tumors refractory to conventional therapy without producing severe neurologic or systemic toxicity [82].

An open-label, single-arm, multicenter Phase II study investigated intratumoral CED infusion of TransMID in recurrent or progressive malignant glioma in adults [83]. The primary study objective was evaluation of efficacy and safety of TransMID, and the primary endpoint was 50% reduction in tumor volume (measured by MRI) within 12 months after the second treatment. Inclusion criteria called for unifocal, unilateral, and supratentorial tumors <3.5 cm in diameter on contrast-enhanced MRI. Patients received TransMID (0.67 μg/mL) at an escalating rate up to 200 μL/h per catheter for 4–5 days, until a total volume of 40 mL was delivered. Four to 10 weeks after the initial infusion patients were due to receive a second round of toxin treatment with the same parameters as the initial treatment. A total of 44 patients were enrolled in total and all received at least one TransMID infusion. A total of 31 out of the 44 patients (70%) completed two treatment cycles. Of the 34 patients evaluable for analysis, there were a total of five complete responders and seven partial responders (total of 35% response), which was a statistically significant result. Median survival time for all 44 patients was 37 weeks. Infusions of TransMID within this clinical protocol resulted in symptomatic progressive brain edema in 8 of 44 patients (14%). Seizures were seen in three patients, but all of them responded to anticonvulsant therapy. Patients that developed increased cerebral edema and clinical neurotoxicity demonstrated some manifest changes on MRI, such as cortical high signal on nonenhanced T1 sequences. These MRI changes were suggestive of venous thrombosis, which may correlate with the higher expression of TfR on endothelial cells in normal blood vessels. There were no grade 3 and 4 AE due to failure of peripheral organs such as liver, kidney, and lung. The results of this Phase II clinical trial have confirmed the safety and tumor response data of the Phase I trial. The Phase II study indicated that tumor response can be obtained in a significant percentage of recurrent or progressing malignant glioma (35% of evaluable patients) and that tumor response as seen on MRI appears to correlate with a significantly prolonged survival [83].

A multicentric randomized, open label, active control, parallel assignment, safety, and efficacy Phase III study (KSB311R/CIII/001 trial) was launched in 2004 in order to compare TransMID with the best standard treatment available for patients with progressive and/or recurrent nonresectable GBM [84,85]. Best standard treatment involved a chemotherapeutic regimen considered to be best standard of care at the respective institution, and consisting of nitrosoureas, platinum compounds, temozolomide, procarbazine, or PCV (procarbazine, CCNU [lomustine], vincristine). The primary endpoint of the study was overall survival time. Adult patients eligible for enrolment into this trial had to be diagnosed with histologically confirmed GBM and must have had undergone conventional treatment, including surgery (biopsy or debulking) and/or radiation therapy and/or chemotherapy. Patients were required to have a recurrent and/or progressive tumor ≥1.0 and ≤4.0 cm in diameter [84,85]. The trial was however prematurely terminated in 2007 after an interim conditional power analysis showed that it was extremely unlikely that TransMID would meet the trial criteria for efficacy [85]. No data from this clinical trial have been published yet.

20.5 CONCLUSIONS AND FUTURE DEVELOPMENTS OF TARGETED TOXINS

A number of targeted toxins for tumor specific treatment of solid and hematological malignancies have been advanced to the stage of clinical trials in the last decade. Ligands such as IL-4, IL-13, EGF, TGF-α, and Tf have been used in conjunction with plant or bacterial toxins. All targeted toxins have a common principle: the ligand portion defines the tumor specificity of the construct and directs the toxic portion to the cell surface; the peptide toxin is then internalized into the cell, translocates to the cytosol, and catalytically inactivates protein synthesis leading to cell death. Targeted toxins are highly potent and able to kill tumor cells independent of any malignancy-associated genetic alterations and/or mutations. Multidrug resistance and apoptosis resistance is therefore not an issue with toxins—after receptor binding and internalization, no tumor cell is able to survive the toxin part of the molecule [24,86,87]. Apart from being tumor selective in cell culture, targeted toxins retain this selectivity *in vivo* if used at an appropriate concentration. The MTD differs with every type of toxin and with delivery mode and delivery volume. However, there is currently a strong need for the development of improved delivery methods *in vivo* and for non-invasive imaging of the intracerebral and intratumoral distribution of toxin in patients [25]. In animal studies, SPECT scanning and gamma camera imaging have been successfully used to study the distribution of TransMID [80]; however, these studies have not been replicated in patients.

Targeted toxins have shown considerable promise in Phase I and II clinical trials with recurrent malignant gliomas. There are however at least two major obstacles that need to be overcome before targeted toxins may enter the mainstream of brain tumor therapies. These are their nonuniform and poorly predictable distribution in the brain, and nonspecific associated toxicity to glial and neuronal cells [20,25]. With the introduction of MRI-based functional imaging and 3D volumetric techniques in routine clinical practice, it will be possible to determine quantitatively brain tumor volumes and volume of infusate in tumor and normal brain [88,89]. Diffusion tensor imaging (DTI) is a new MRI technique sensitive to directional movements of water molecules induced by tissue barriers [90]. CED makes use of the extracellular space of the brain as a natural pathway for the widespread distribution of agents infused in an aqueous solution, and therefore, DTI could be used for imaging of CED delivery without the need for contrast agents. Further key developments in MRI and MR-related computer software as well as infusion catheter design and computerized simulation and delivery modeling within the brain may provide much needed new resources to be combined with the biologically targeted toxins [25].

To avoid unspecific toxicity to neural cells, future clinical trials using targeted toxins may involve combinations of agents with different targets expressed either on tumor cells or on neovasculature, or on both. This should allow reduction of dose-based toxicity of individual toxins without decrease in overall efficacy. Weaver and Laske [26] reported the effects of an EGF targeted diphtheria toxin, DAB389EGF, on brain tumor growth in an animal model. When DAB389EGF was combined with TransMID at only inhibitory doses of both toxins, a marked decrease in tumor growth was seen, which was greater than the growth inhibition of each single drug. This preliminary study illustrates the potential for combining at least two targeted toxins to achieve a higher antitumor effect, while reducing toxicity encountered with each individual treatment.

Clinical protocols have explored several points of presumed significance, such as the use of different numbers of catheters, positioning of catheters, surgical resection of tumor before or after toxin infusion, and single versus repeated infusion, but there is no clear answer to any of these questions. It remains unknown whether there are benefits from combining targeted toxins with chemotherapy and/or with fractionated external radiation. Protocols have investigated patients with recurrent or progressing GBM; the role of targeted toxin infusion in newly diagnosed malignant glioma remains to be determined. Evidence of prolonged survival and late local recurrences in treated glioma patients has raised the question of periodically repeated application of toxin for extended control of local recurrence and progression. The two Phase III randomized clinical trials were able to provide some answers as to the efficacy of the respective studied toxins and clinical protocol; however, more clinical research is needed to address the aforementioned global issues.

DISCLOSURE

NGR has received research grants from Neurocrine Inc. (San Diego, CA) and IVAX Inc. (Miami, FL). The authors have no financial interests in any of the biotechnology companies mentioned in the review.

REFERENCES

1. Kleihues P, Cavenee WK. (2000) *Pathology and Genetics of Tumors of the Nervous System*. IARC Press, Lyon.
2. McCarthy BJ, Kruchko C. (2005) Central brain tumor registry of the United States: consensus conference on cancer registration of brain and central nervous system tumors. *Neurooncology* 7, 196–201.
3. Ohgaki H, Kleihues P. (2005) Population-based studies on incidence, survival rates, and genetic alterations in astrocytic and oligodendroglial gliomas. *J. Neuropathol. Exp. Neurol.* 64, 479–489.
4. Janus TJ, Kyritsis AP, Forman AD, Levin VA. (1992) Biology and treatment of gliomas. *Ann. Oncol.* 3, 423–433.
5. Salcman M. (1990) Malignant glioma management. *Neurosurg. Clin. N. Am.* 1, 49–63.
6. Schiffer D, Cavalla P, Dutto A, Borsotti L. (1997) Cell proliferation and invasion in malignant gliomas. *Anticancer Res.* 17, 61–69.
7. Swanson KR, Alvord, EC, Jr, Murray JD. (2002) Virtual brain tumors (gliomas) enhance the reality of medical imaging and highlight inadequacies of current therapy. *Br. J. Cancer* 86, 14–18.
8. Guha A, Mukherjee J. (2004) Advances in the biology of astrocytomas. *Curr. Opin. Neurol.* 17, 655–662.
9. Mitchell P, Ellison DW, Mendelow AD. (2005) Surgery for malignant gliomas: mechanistic reasoning and slippery statistics. *Lancet Neurol.* 4, 413–422.
10. Burton EC, Prados MD. (2000) Malignant gliomas. *Curr. Treat. Options Oncol.* 1, 459–468.
11. Hess CF, Schaaf JC, Kortmann RD, Schabet M, Bamberg M. (1994) Malignant glioma: patterns of failure following individually tailored limited volume irradiation. *Radiother. Oncol.* 30, 146–149.
12. Giese A, Bjerkvig R, Berens ME, Westphal M. (2003) Cost of migration: invasion of malignant gliomas and implications for treatment. *J. Clin. Oncol.* 21, 1624–1636.
13. Davis FG, Freels S, Grutsch J, Barlas S, Brem S. (1998) Survival rates in patients with primary malignant brain tumors stratified by patient age and tumor histological type: an analysis based on surveillance, epidemiology, and end results (SEER) data, 1973–1991. *J. Neurosurg.* 88, 1–10.
14. DeAngelis LM. (2001) Brain tumors. *N. Engl. J. Med.* 344, 114–123.
15. Schiff D, Shaffrey ME. (2003) Role of resection for newly diagnosed malignant gliomas. *Expert Rev. Anticancer Ther.* 3, 621–630.

16. Whittle IR. (2002) Surgery for gliomas. *Curr. Opin. Neurol.* 5, 663–669.
17. Stupp R, Mason WP, van den Bent MJ, Weller M, Fisher B, Taphoorn MJ, et al. (2005) Radiotherapy plus concomitant and adjuvant temozolomide for glioblastoma. *N. Engl. J. Med.* 352, 987–996.
18. Pastan I, Chaudhary VK, Fitzgerald DJ. (1992) Recombinant toxins as novel therapeutic agents. *Annu. Rev. Biochem.* 61, 331–354.
19. Pastan I, Fitzgerald D. (1992) Recombinant toxins for cancer treatment. *Science* 254, 1173–1177.
20. Husain SR, Puri RK. (2003) Interleukin-13 receptor-directed cytotoxin for malignant glioma therapy: from bench to bedside. *J. Neurooncol.* 65, 37–48.
21. Hall WA, Fodstad O. (1992) Immunotoxins and central nervous system neoplasia. *J. Neurosurg.* 76, 1–12.
22. Gottstein C, Winkler U, Bohlen H, Diehl V, Engert A. (1994) Immunotoxins: is there a clinical value? *Ann. Oncol.* 5(Suppl. 1), 97–103.
23. Puri RK, Siegel JP. (1993) Interleukin-4 and cancer therapy. *Cancer Invest.* 11, 473–486.
24. Reiter Y. (2001) Recombinant immunotoxins in targeted cancer cell therapy. *Adv. Cancer Res.* 81, 93–124.
25. Chiocca EA, Broaddus WC, Gillies GT, Visted T, Lamfers ML. (2004) Neurosurgical delivery of chemotherapeutics, targeted toxins, genetic and viral therapies in neuro-oncology. *J. Neurooncol.* 69, 101–117.
26. Weaver M, Laske DW. (2003) Transferrin receptor ligand-targeted toxin conjugate (Tf-CRM107) for therapy of malignant gliomas. *J. Neurooncol.* 65, 3–13.
27. Rustamzadeh E, Low WC, Vallera DA, Hall WA. (2003) Immunotoxin therapy for CNS tumor. *J. Neurooncol.* 64, 101–116.
28. Baxter LT, Jain RK. (1989) Transport of fluid and macromolecules in tumors I. Role of interstitial pressure and convection. *Microvasc. Res.* 37, 77–102.
29. Morrison PF, Laske DW, Bobo H, Oldfield EH, Dedrick RL. (1994) High-flow microinfusion: tissue penetration and pharmacodynamics. *Am. J. Physiol.* 266, R292–305.
30. Bobo RH, Laske DW, Akbasak A, Morrison PF, Dedrick RL, Oldfield EH. (1994) Convection-enhanced delivery of macromolecules in the brain. *Proc. Natl. Acad. Sci. USA* 91, 2076–2080.
31. Laske DW, Morrison PF, Lieberman DM, Corthesy ME, Reynolds JC, Stewart-Henney PA, et al. (1997) Chronic interstitial infusion of protein to primate brain: determination of drug distribution and clearance with single-photon emission computerized tomography imaging. *J. Neurosurg.* 87, 586–594.
32. Baxter LT, Jain RK. (1989) Transport of fluid and macromolecules in tumors I. Role of interstitial pressure and convection. *Microvasc. Res.* 37, 77–102.
33. Kawakami K, Kawakami M, Kioi M, Husain SR, Puri RK. (2004) Distribution of bolus or convection-enhanced delivery of IL-13 receptor-directed cytotoxin to intracranial brain tumors in a mouse model [Abstract]. *Proc. Am. Assoc. Cancer Res.* 45, 5402.
34. Kreitman RJ, Puri PK, Pastan I. (1994) A circularly permuted recombinant interleukin 4 toxin with increased activity. *Proc. Natl. Acad. Sci. USA* 91, 6889–6893.
35. Kreitman RJ, Puri RK, Pastan I. (1995) Increased antitumor activity of a circularly permuted IL-4-toxin in mice with IL-4 receptor bearing human carcinoma. *Cancer Res.* 55, 3357–3363.
36. Frankel AE, Kreitman RJ, Sausville EA. (2000) Targeted toxins. *Clin. Cancer Res.* 6, 326–334.
37. Puri RK, Hoon DS, Leland P, Snoy P, Rand RW, Pastan I, et al. (1996) Preclinical development of a recombinant toxin containing circularly permuted interleukin 4 and truncated Pseudomonas exotoxin for therapy of malignant astrocytoma. *Cancer Res.* 56, 5631–5637.
38. Nelms K, Keegan AD, Zamorano J, Ryan JJ, Paul WE. (1999) The IL-4 receptor: signaling mechanisms and biologic functions. *Ann. Rev. Immunol.* 17, 701–738.
39. Kruse N, Shen B, Arnold S, Tony HP, Muller T, Sebald W. (1993) Two distinct functional sites of human interleukin 4 are identified by variants impaired in either receptor binding or receptor activation. *EMBO J.* 12, 5121–5129.
40. Muller T, Dieckmann T, Sebald W, Oschkinat H. (1994) Aspects of receptor binding and signalling of interleukin-4 investigated by site-directed mutagenesis and NMR spectroscopy. *J. Mol. Biol.* 237, 423–436.
41. Debinski W, Obiri NI, Powers SK, Pastan I, Puri RK. (1995) Human glioma cells overexpress receptors for interleukin 13 and are extremely sensitive to a novel chimeric protein composed of interleukin 13 and Pseudomonas exotoxin. *Clin. Cancer Res.* 1, 1253–1258.
42. Joshi BH, Plautz, GE, Puri RK. (2000) Interleukin-13 receptor alpha chain: a novel tumor-associated transmembrane protein in primary explants of human malignant gliomas. *Cancer Res.* 60, 1168–1172.
43. Joshi BH, Leland P, Asher A, Prayson RA, Varricchio F, Puri RK. (2001) In situ expression of Interleukin-4 receptors in human brain tumors and cytotoxicity of a recombinant interleukin-4 (IL-4) cytotoxin in primary glioblastoma cell cultures. *Cancer Res.* 61, 8058–8061.
44. Husain SR, Behari N, Kreitman RJ, Pastan I, Puri RK. (1998) Complete regression of established human glioblastoma tumor xenograft by interleukin-4 toxin therapy. *Cancer Res.* 58, 3649–3653.
45. Weber F, Asher A, Bucholz R, Berger M, Prados M, Chang S, et al. (2003) Safety, tolerability, and tumor response of IL4-Pseudomonas exotoxin (NBI-3001) in patients with recurrent malignant glioma. *J. Neurooncol.* 64, 125–137.
46. http://clinicaltrials.gov, study identifier NCT00014677. Last accessed on Oct. 28th, 2012.
47. Hunter T. (1984) The epidermal growth factor receptor gene and its product. *Nature* 311, 414–416.
48. Liberman TA, Nusbaum HR, Razon N, Kris R, Lax I, Soreq H, et al. (1985) Amplification, enhanced expression, and possible rearrangement of EGF receptor gene in primary human brain tumors of glial origin. *Nature* 313, 144–147.

49. Malden LT, Novak U, Kaye AH, Burgess AW. (1988) Selective amplification of the cytoplasmic domain of the epidermal growth factor receptor gene in glioblastoma multiforme. *Cancer Res.* 48, 2711–2714.
50. Wong AJ, Bigner SH, Bigner DD, Kinzler KW, Hamilton SR, Vogelstein B. (1987) Increased expression of the epidermal growth factor receptor gene in malignant gliomas is invariably associated with gene amplification. *Proc. Natl. Acad. Sci. USA* 84, 6899–6903.
51. Humphrey PA, Wong AJ, Vogelstein B, Friedman HS, Werner MH, Bigner DD, et al. (1988) Amplification and expression of the epidermal growth factor receptor gene in human glioma xenografts. *Cancer Res.* 48, 2231–2238.
52. Draoui M, Siegall CB, FitzGerald D, Pastan I, Moody TW. (1994) TGF alpha-PE40 inhibits non-small cell lung cancer growth. *Life Sci.* 54, 445–453.
53. Mesri EA, Kreitman RJ, Fu YM, Epstein SE, Pastan I. (1993) Heparin-binding transforming growth factor alpha-Pseudomonas exotoxin A. A heparin sulfate-modulated recombinant toxin cytotoxic to cancer cells and proliferating smooth muscle cells. *J. Biol. Chem.* 268, 4853–4862.
54. Sampson JH, Akabani G, Archer GE, Bigner DD, Berger MS, Friedman AH, et al. (2003) Progress report of a Phase I study of the intracerebral microinfusion of a recombinant chimeric protein composed of transforming growth factor (TGF)-alpha and a mutated form of the Pseudomonas exotoxin termed PE-38 (TP-38) for the treatment of malignant brain tumors. *J. Neurooncol.* 65, 27–35.
55. Weber, F. Preliminary results of a Phase II study of TP-38 immunotoxin delivered via convection-enhanced delivery to patients with recurrent glioblastoma multiforme: report of the TP-38 study group. Ann. Meeting German Soc. Neurosurg., http://www.egms.de/static/en/meetings/dgnc2005/05dgnc0094.shtml, last accessed on Oct. 28th, 2012.
56. Debinski W, Pastan I. (1994) An immunotoxin with increased activity and homogeneity produced by reducing the number of lysine residues in recombinant Pseudomonas exotoxin. *Bioconjug. Chem.* 5, 40–46.
57. Debinski W, Obiri NI, Pastan, I, Puri RK. (1995) A novel chimeric protein composed of interleukin 13 and Pseudomonas exotoxin is highly cytotoxic to human carcinoma cells expressing receptors for interleukin 13 and interleukin 4. *J. Biol. Chem.* 270, 16775–16780.
58. Puri RK, Leland P, Obiri NI, Husain SR, Kreitman RJ, Haas GP, et al. (1996) Targeting of interleukin-13 receptor on human renal cell carcinoma cells by a recombinant chimeric protein composed of interleukin-13 and a truncated form of Pseudomonas exotoxin A (PE38QQR). *Blood* 87, 4333–4339.
59. Joshi BH, Kawakami K, Leland P, Puri RK. (2002) Heterogeneity in interleukin-13 receptor expression and subunit structure in squamous cell carcinoma of head and neck: differential sensitivity to chimeric fusion proteins comprised of interleukin-13 and a mutated form of Pseudomonas exotoxin. *Clin. Cancer Res.* 8, 1948–1956.
60. Kawakami K, Takeshita F, Puri RK. (2001) Identification of distinct roles for a dileucine and a tyrosine internalization motif in the interleukin (IL)-13 binding component IL-13 receptor alpha 2 chain. *J. Biol. Chem.* 276, 25114–25120.
61. Souweidane MM, Occhiogrosso G, Mark EB, Edgar MA. (2004) Interstitial infusion of IL13-PE38QQR in the rat brain stem. *J. Neurooncol.* 76, 287–293.
62. Weingart J, Strauss LC, Grossman SA. (2002) Phase I/II study: intratumoral infusion of IL13-PE38QQR cytotoxin for recurrent supratentorial malignant glioma. *Neuro-oncology* 4, 379.
63. Kunwar S. (2003) Convection enhanced delivery of IL13-PE38QQR for treatment of recurrent malignant glioma: presentation of interim findings from ongoing Phase 1 studies. *Acta Neurochir. Suppl.* 88, 105–111.
64. Kunwar S, Prados M, Chang S, Berger MS, Lang FF, Piepmeier JM, et al. (2003) Peritumoral convection-enhanced delivery of IL13-PE38QQR in patients with recurrent malignant glioma—Phase I interim results. *Neuro-oncology* 5, 4.
65. Chang SM. (2003) Neuroradiographic changes following convection-enhanced delivery (CED) of the conjugated toxin IL13-PE38QQR following resection of recurrent malignant glioma. *Neuro-oncology* 5, 4.
66. Prados MD, Lang FF, Stauss LC, Fleming CK, Aldape K, Kunwar S, et al. Intratumoral and intracerebral microinfusion of IL13-PE38QQR cytotoxin: Phase I/II study of pre- and post-resection infusions in recurrent resectable maliganant glioma. *Proc. Am. Soc. Clin. Oncol.* 21, 2087.
67. Ram Z, Barnett G, Vogelbaum M, Constantini S, Lillehei K, Sherman J, et al. (2003) Pre-operative infusion of IL13-PE38QQR cytotoxin by convection-enhanced delivery (CED) in recurrent malignant glioma: a Phase I/II study. *Proc. Am. Soc. Clin. Oncol.* 22, 403.
68. http://clinicaltrials.gov, study identifier NCT00076986, s. also www.precisetrial.com. Last accessed on Oct. 28th, 2012.
69. Kunwar S, Chang S, Westphal M, Vogelbaum M, Sampson J, Barnett G, et al. (2010) Phase III randomized trial of CED of IL13-PE38QQR vs. Gliadel wafers for recurrent glioblastoma. *Neuro-oncol.* 12, 871–881.
70. Johnson VG, Wilson D, Greenfield L, Youle RJ. (1988) The role of the diphtheria toxin receptor in cytosol translocation. *J. Biol. Chem.* 263, 1295–1300.
71. Laird W, Groman N. (1976) Isolation and characterization of tox mutants of corynebacteriophage beta. *J. Virol.* 19, 220–227.
72. Greenfield L, Johnson VG, Youle RJ. (1987) Mutations in diphtheria toxin separate binding from entry and amplify immunotoxin selectivity. *Science* 238, 536–539.
73. Recht L, Torres CO, Smith TW, Raso V, Griffin TW. (1990) Transferrin receptor in normal and neoplastic brain tissue: implications for brain tumor immunotherapy. *J. Neurosurg.* 72, 941–945.
74. Gatter KC, Brown G, Trowbridge IS, Woolston RE, Mason DY. (1983) Transferrin receptors in human tissues: their distribution and possible clinical relevance. *J. Clin. Pathol.* 36, 539–545.
75. Jefferies WA, Brandon MR, Hunt SV, Williams AF, Gatter KC, Mason DY. (1984) Transferrin receptor on endothelium of brain capillaries. *Nature* 312, 162–163.

76. Johnson VG, Wrobel C, Wilson D, Zovickian J, Greenfield L, Oldfield EH, et al. (1989) Improved tumor-specific immunotoxins in the treatment of CNS and leptomeningeal neoplasia. *J. Neurosurg.* 70, 240–248.
77. Hall WA, Godal A, Juell S, Fodstad O. (1992) In vitro efficacy of transferrin-toxin conjugates against glioblastoma multiforme. *J. Neurosurg.* 76, 838–844.
78. Martell LA, Agrawal A, Ross DA, Muraszko KM. (1993) Efficacy of transferring receptor-targeted immunotoxins in brain tumor cell lines and pediatric brain tumors. *Cancer Res.* 53, 1348–1353.
79. Chignola R, Foroni R, Candiani C, Franceschi A, Pasti M, Stevanoni G, et al. (1994) Cytoreductive effects of anti-transferrin receptor immunotoxin in a multicellular tumor spheroid model. *Int. J. Cancer* 57, 268–274.
80. Laske DW, Ilercil O, Akbasak A, Youle RJ, Oldfield EH. (1994) Efficacy of direct intratumoral therapy with targeted protein toxins for solid human gliomas in nude mice. *J. Neurosurg.* 80, 520–526.
81. Engebraaten O, Hjortland GO, Juell S, Hirschberg H, Fodstad O. (2002) Intratumoral immunotoxin treatment of human malignant brain tumors in immunodeficient animals. *Int. J. Cancer* 97, 846–852.
82. Laske, DW, Youle RJ, Oldfield EH. (1997) Tumor regression with regional distribution of the targeted toxin TF-CRM107 in patients with malignant brain tumors. *Nat. Med.* 3, 1362–1368.
83. Weaver M, Laske DW. (2003) Transferrin receptor ligand-targeted toxin conjugate (Tf-CRM107) for therapy of malignant gliomas. *J. Neurooncol.* 65, 3–13.
84. http://clinicaltrials.gov, study identifier NCT00083447, s. also NCT00088400 and NCT00087230. Last accessed on Oct. 28th, 2012.
85. http://www.drugs.com/news/celtic-pharma-terminates-trans-mid-trial-ksb311r-ciii-001-5246.html. Last accessed on Oct. 28th, 2012.
86. Puri RK. (2003) Introduction to special issue: targeted toxins for glioma therapy. *J. Neurooncol.* 65, 1–2.
87. Pastan I, Kreitman RJ. (1998) Immunotoxins for targeted cancer therapy. *Adv. Drug Deliv. Rev.* 31, 53–88.
88. Briellmann RS, Pell GS, Wellard RM, Mitchell LA, Abbott DF, Jackson GD. (2003) MR imaging of epilepsy: state of the art at 1.5 T and potential of 3 T. *Epileptic Disord.* 5, 3–20.
89. Tummala RP, Chu RM, Liu H, Truwit CL, Hall WA. (2003) Application of diffusion tensor imaging to magnetic-resonance-guided brain tumor resection. *Pediatr. Neurosurg.* 39, 39–43.
90. Cruz Junior LC, Sorensen AG. (2005) Diffusion tensor magnetic resonance imaging of brain tumors. *Neurosurg. Clin. N. Am.* 16, 115–134.

21

IMMUNOKINASES

STEFAN BARTH,[1,2] STEFAN GATTENLÖHNER,[3] AND MEHMET KEMAL TUR[3]

[1]*Department of Experimental Medicine and Immunotherapy, Institute of Applied Medical Engineering, Helmholtz-Institute for Biomedical Engineering, University Hospital RWTH Aachen, Aachen, Germany*
[2]*Department of Pharmaceutical Product Development, Fraunhofer Institute for Molecular Biology and Applied Ecology, Aachen, Germany*
[3]*Institute of Pathology, University Hospital Giessen and Marburg GmbH (UKGM), Giessen, Germany*

21.1 INTRODUCTION

Specific pathophysiological properties of tumor cells, particularly the aberrant expression of surface and intracellular markers, make it possible to explore rational strategies for immunological targeting of cancer cells without harming the surrounding healthy tissue. Therefore, successful immunotherapy depends on the identification of reliable tumor-associated molecular targets. Therapeutic fusion proteins can be rationally engineered to take advantage of these significant changes. They consist of a major component that binds specifically to a disease-specific cell-surface target molecule and another that confers selective cytotoxicity in terms of induction of apoptosis by substantial inhibition of protein biosynthesis (immunotoxins) [1], degradation of RNA molecules (immunoRNases) [2,3], proteolytic cleavage of multiple apoptosis-related cytoplasmic protein substrates (Granzyme B fusion proteins) [4,5], or functional reconstitution of down-regulated tumor suppressor functions (immunokinases) [6,7].

Many forms of cancer are caused by the inactivation of tumor suppressor genes, whose normal function is to impose limits on the extent of cell division and proliferation (often by inducing apoptosis) [8], where the missing function provided by the tumor suppressor gene is a pro-apoptotic protein kinase, it should be possible to restore normal apoptotic function to tumor cells by importing the appropriate kinase activity, leading to cell death [9]. Such is the reasoning behind the development of immunokinases, that is, targeted fusion proteins comprising a tumor-specific targeting moiety and an apoptosis-inducing protein kinase [6]. The targeting moiety may be an antibody or antibody fragment (hence *immuno*kinase) but the same term is used when the targeting moiety is a ligand specific for a receptor preferentially displayed on the surface of tumor cells. Targeting is necessary because the unregulated delivery of a pro-apoptotic kinase would result in indiscriminate cell death; also, the target must be an internalizing receptor to ensure the kinase is delivered into the cytoplasm, allowing it to interact with the apoptotic machinery. Downregulation of pro-apoptotic DAP kinase family members such as DAPK1 and DAPK2 have been described for different tumors in the process of oncogenic transformation [10,11]. Therefore, to increase specificity of targeted drugs, we developed a new dual-target concept based on the targeted restoration of a missing or downregulated tumor suppressor protein in tumor surface antigen-positive cells [6,7]. In this chapter, we

Fusion Protein Technologies for Biopharmaceuticals: Applications and Challenges, First Edition. Edited by Stefan R. Schmidt.

describe the principles underlying the activity of immunokinases and recent work to develop an immunokinase against Hodgkin lymphoma. The targeting moiety in this case is the CD30 ligand (CD30L), which binds to the internalizing receptor CD30 preferentially expressed on Hodgkin lymphoma cells, and the kinase moiety is the pro-apoptotic serine/threonine kinase DAPK2 [7].

21.2 PROTEIN KINASES, APOPTOSIS, AND CANCER

21.2.1 General Properties of Protein Kinases

Protein kinases are enzymes that add phosphate groups to specific target proteins. Phosphate groups have a strong negative charge, and their presence therefore has a profound effect on the net charge of the modified protein and its interactions with other molecules in the cell. Phosphorylation by protein kinases is a useful mechanism to control protein activity, and it is thought that more than 30% of all proteins in mammalian cells are regulated in this manner [12]. Protein phosphorylation is usually reversible, with specific protein phosphatases responsible for removing the phosphate groups. The three basic components of the phosphorylation system are therefore phosphoproteins whose properties are altered by phosphorylation and dephosphorylation, protein kinases that transfer phosphate groups from a donor substrate such as ATP or GTP to target phosphoproteins, and protein phosphatases that dephoshorylate phosphorylated proteins, thereby restoring the protein phosphorylation system to its nonphosphorylated state [13].

Because protein kinases are proteins that can themselves be regulated by phosphorylation, kinase cascades are often used to transduce signals within the cell, and protein phosphorylation is one of the most common mechanisms used to convert extracellular signals into intracellular responses [12]. Such signals are used to control every conceivable cellular process, including cell growth, proliferation, migration, differentiation, and (most relevant in the context of this chapter) apoptosis. Signaling must be tightly controlled to avoid these fundamental processes going awry [14]. Both mutations and epigenetic modifications can disrupt kinase-mediated signaling by interfering with the expression and/or activity of individual protein kinases, and this can have profound effects on the core cellular processes listed above, including the prevention of apoptosis in cells that would normally be programmed to die. Such effects often contribute to the progression of cancer [8,15] and many kinases and phosphatases have been identified as the products of oncogenes and tumor suppressor genes [16]. The human genome encodes 518 protein kinases, representing ~2% of all genes, and many of these have been linked with cancer and other diseases [17].

21.2.2 Death-Associated Protein Kinases (DAPKs)

The eukaryotic protein kinases are divided into eight broad classes based on catalytic domain structure, substrate specificity, and regulation mechanism [18,19] (Figure 21.1). The calcium/calmodulin-dependent kinases (CAMKs) include the death-associated protein kinases (DAPKs), which are soluble, cytosolic kinases that play an important role in the promotion of apoptosis in tumor cells [20–22]. DAPK1 was identified using a functional gene cloning approach based on the transfection of mammalian cells with an antisense cDNA library and the subsequent isolation of antisense cDNA fragments that protected HeLa cells from apoptosis induced by interferon-γ [23]. This led to the discovery of four additional members—DAPK2 (also known as DRP-1), DAPK3 (also known as ZIP), DRAK-1, and DRAK-2—all of which share significant sequence and functional homology including the ability to induce apoptosis when overexpressed [24]. Phylogenetic analysis based on the alignment of catalytic domain sequences has shown that DAPK1, DAPK2, and DAPK3 can be grouped into a distinct clade with high bootstrap probabilities, whereas DRAK1 and DRAK2 form another clade sharing a putative common ancestor with other DAPK-related proteins. Although the catalytic domains of the five DAPKs are

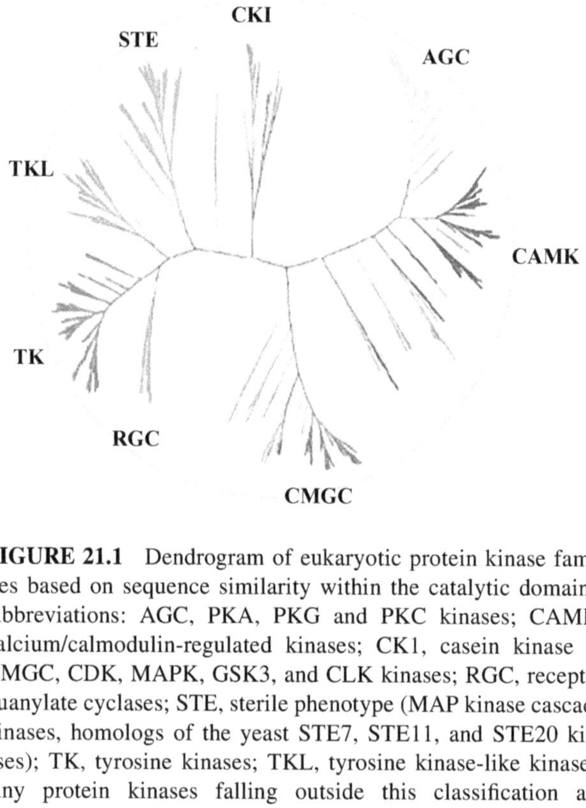

FIGURE 21.1 Dendrogram of eukaryotic protein kinase families based on sequence similarity within the catalytic domains. Abbreviations: AGC, PKA, PKG and PKC kinases; CAMK, calcium/calmodulin-regulated kinases; CK1, casein kinase 1; CMGC, CDK, MAPK, GSK3, and CLK kinases; RGC, receptor guanylate cyclases; STE, sterile phenotype (MAP kinase cascade kinases, homologs of the yeast STE7, STE11, and STE20 kinases); TK, tyrosine kinases; TKL, tyrosine kinase-like kinases. Any protein kinases falling outside this classification are described as atypical kinases. *Source*: From Reference 17. Reprinted with permission from AAAS.

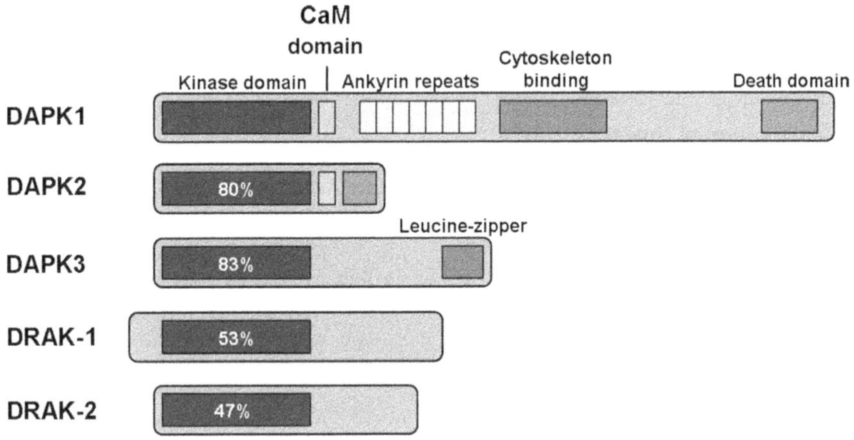

FIGURE 21.2 The DAPK family of proteins shares a highly conserved catalytic domain (numbers show similarity relative to DAPK1). Only DAPK1 and DAPK2 share the conserved regulatory CaM domain. *Source*: Reproduced from Reference 6.

conserved, the extracatalytic sequences diverge significantly and show little similarity (Figure 21.2).

DAPK1 and DAPK2 share a highly conserved N-terminal kinase domain and a calcium/calmodulin regulatory domain that controls the catalytic activity by sensing intracellular calcium levels. The unique regulatory mechanism is based on autophosphorylation, which is relieved when calcium-activated calmodulin binds to the calcium/CaM domain [25]. Therefore, the deletion of this domain in mutants such as DAPK1ΔCaM and DAPK2Δ73 generates a constitutively active kinase that enhances apoptosis via a range of pro-apoptotic and/or autophagic signals [26–28]. Both the pro-apoptotic and tumor suppression functions of DAPK1 and DAPK2 depend strictly on their serine/threonine kinase activity [23,29]. Potential physiological substrates for DAPK1 include the regulatory light chain of myosin II (MLC), syntaxin-1A, ZIP-kinase, CaMKK, protein kinase D, ribosomal protein L5, and Mcm3 [30]. The pro-apoptotic tumor-suppressing activity of DAPK1 is mediated through the activation of p53 in a p19ARF-dependent fashion [31]. The signaling mechanisms upstream of DAPK1 and DAPK2 remain unclear, but both are localized exclusively in the cytoplasm. In contrast, DAPK3, DRAK1, and DRAK2 are localized in the nucleus and do not require Ca^{2+}/CaM for activation.

21.2.3 The Role of DAPK1 and DAPK2 in Cancer

DAPK2 mRNA is present at high levels in normal heart, lung, skeletal muscle, colon, breast and spleen tissue, as well as leukocytes, but the gene appears to be silent in various leukemic myeloid cell lines [22,32] and the Hodgkin lymphoma cell lines L540 and L1236 [7]. In L540 cells, the main mechanism is likely to be epigenetic inactivation since the *DAPK2* promoter region is hypermethylated and inactive, whereas a methylation-independent mechanism is likely in L1236 cells line where DAPK2 responds to an alternate regulatory pathway [7]. Similarly, DAPK1 appears to be silent in several human primary tumor cells and tumor cell lines, and its expression is inversely correlated with recurrence and metastasis in lung carcinoma [33–36]. The inactivation of tumor suppressor genes by aberrant promoter hypermethylation is an important epigenetic mechanism in the induction and progression of human cancers, resulting in transcriptional repression and loss of the relevant protein activity [37]. The link between *DAPK1* and *DAPK2* silencing and epigenetic phenomena such as promoter hypermethylation in certain cell lines explains why treatment with DNA demethylating agents such as 5-aza-2′-deoxycytidine (Decitabine, DAC) can prevent tumor proliferation and reactivate both *DAPK1* [38] and *DAPK2* [7].

21.3 THERAPEUTIC STRATEGIES TO RESTORE MISSING KINASE EXPRESSION

21.3.1 Systemic Delivery of Demethylating Agents

The systemic administration of demethylating agents such as DAC has met with some success in the treatment of myelodysplastic syndromes [39,40], but can have unpredictable effects due to the lack of specificity and the reactivation of multiple epigenetically silenced genes [41]. Furthermore, even if the target gene is reactivated, full protein activity may not be restored. This is demonstrated by the recent discovery that the gene encoding C/EBPα (which is hypermethylated and silenced in several solid tumors and in acute myelogenous leukemia) can be reactivated by DAC, but protein expression does not recover to normal levels because of the concomitant disruption of micro-RNAs, resulting in the suppression of protein synthesis [42].

21.3.2 Nontargeted Reactivation of Specific Protein Kinases

Because demethylating drugs lack specificity and can reactivate dormant oncogenes, strategies have been developed to target individual tumor suppressor genes, including those encoding protein kinases [43]. In the case of DAPK1, the expression of *DAPK1* cDNA in stably transfected Lewis carcinoma cells prevented metastasis in a murine model of lung carcinoma by promoting apoptosis [12] and the overexpression of DAPK1 and DAPK2, but not of their catalytically inactive mutants K42A and K52A, induced apoptosis in several transfected malignant cell lines through a range of apoptotic and/or autophagic signals [23,26,27]. Conceptually, these strategies are similar to those applied to reactivate the retinoblastoma protein [44] and p53 [45,46]. In all cases, however, the expression construct was delivered using a plasmid or viral vector, which is unsuitable for specific tumor targeting *in vivo*.

21.3.3 Targeted Reactivation of Specific Protein Kinases Using DAPK2-Based Immunokinases

The disadvantages of systemic delivery can be overcome through the development of therapeutic modalities that are targeted specifically to tumor cells, thus avoiding off-target effects on normal cells. Certain drugs aspire to this objective by targeting proteins that are solely expressed in cancer cells, meaning that all cells receive the drug but only cancer cells should be affected. An example is imatinib, which blocks an aberrantly expressed serine/threonine kinase activity in chronic myelogenous leukemia cells [47]. Other drugs are universally cytotoxic but achieve specificity by targeting receptors that are only displayed on the surface of tumor cells (immunotoxins) [1]. This can be achieved using specific targeting moieties such as monoclonal antibodies, as in the case of gemtuzumab ozogamicin (Mylotarg®) for the treatment acute myeloid leukemia [48]. Unfortunately, these approaches are less effective in practice due to the emergence of clinical resistance, as well as significant off-target effects resulting from, inter alia, the expression of the tumor-associated antigen on nonmalignant cells, albeit at lower levels [49]. Mylotarg was withdrawn voluntarily from the market in 2010 for this reason.

The immunokinase fusion protein strategy combines the advantages of the targeted delivery and targeted activity approaches discussed above, therefore reducing the potential for off-target effects to very low levels. The concept has been discussed for DAPK1, which is downregulated in chronic lymphocytic leukemia (CLL) cells, leading to speculation that the restoration of DAPK1 activity in CLL cells might increase their sensitivity to apoptosis [9,35]. In our proof-of-principle study with DAPK2, the same goal is achieved by targeting Hodgkin lymphoma cells with a functional tumor suppressor protein that is only downregulated in Hodgkin lymphoma cells (i.e., its delivery to other cells would have no impact) [7] (Figure 21.3). As discussed above, we determined that *DAPK2* expression is epigenetically silenced in Hodgkin lymphoma L540 cell lines, so we set out to determine whether the selective reconstitution of DAPK2 catalytic activity in these cells could induce apoptosis. We therefore created immunokinase fusion proteins comprising the catalytic domain of human DAPK2 joined by a linker to the extracellular domain of human CD30L [3] or the murine anti-CD30 single chain variable fragment antibody Ki-4 [50]. CD30 is an attractive candidate target for immunotherapy because it is internalized efficiently

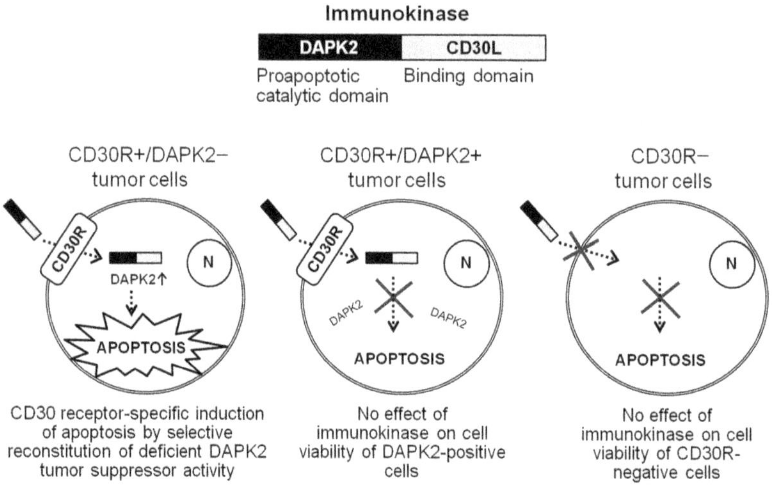

FIGURE 21.3 Schematic diagrams showing principle of operation of immunokinase fusion proteins for dual-targeted cancer treatment.

following ligand binding, and although its expression is normally restricted to subpopulations of activated T- and B-lymphocytes, it is also expressed strongly in hematopoietic malignancies such as Hodgkin lymphoma and anaplastic large cell lymphoma [51,52]. Several anti-CD30 immunotherapeutics are currently undergoing clinical trials [53].

21.4 ANALYSIS OF IMMUNOKINASE EFFICACY

21.4.1 *In Vitro* Activity of Recombinant Immunokinase DAPK2′-CD30L

The recombinant immunokinase DAPK2′-CD30L comprises a truncated DAPK2 protein retaining full catalytic activity but lacking the CaM domain joined by a 11-aa peptide linker to CD30L [7]. We initially tested the specific activity of the immunokinase in proliferation and apoptosis experiments. Specific binding to several CD30+ target cells was confirmed by flow cytometry, and the dissociation constant (KD) values were in the 7–17 nM range. Potent and cell-specific elimination of CD30+/DAPK2$^-$ target cells was confirmed using proliferation and apoptosis assays, revealing that the immunokinase had an IC_{50} value of ~20 nM against CD30+/DAPK2-negative L540 cells and ~63 nM against L1236 cells, respectively. There was no cytotoxic effect against CD30+/DAPK2$^+$ L428 cells (Table 21.1).

TABLE 21.1 DAPK2 Expression and DAPK2′-CD30L Activity in Three Hodgkin Lymphoma-Derived Cell Lines

	L540	L1236	L428
Methylation	+	−	−
DAPK2 RT-PCR	−	+	+
DAPK2 RT-PCR after Decitabine treatment	+	+	+
Western blot analysis of DAPK2	−	−	+
Western blot after Decitabine treatment	+	−	+
Toxicity after targeted DAPK2 restoration	+	+	−

Source: Reproduced from Reference 6.

21.4.2 *In Vivo* Activity of Recombinant Immunokinase DAPK2′-CD30L

In SCID mice, the intravenous injection of L540 cells inevitably leads to the disseminated growth of tumor cells within 5–6 weeks and ultimately to death from cancer [54]. To investigate the influence of DAPK2′-CD30L on overall survival, we started anti-tumor treatment 1 day after tumor challenge by administrating DAPK2′-CD30L to one group of mice, a nonspecific immunokinase called DAPK2′-H22 (scFv), which targets the irrelevant antigen CD64, to a second group, and saline solution (PBS) to a third. The recombinant fusion proteins were administrated at the maximum tolerable dose of 70 μg per mouse. The overall survival rate of the PBS control and nonspecific immunokinase groups was significantly lower than that of the DAPK2′-CD30L group ($p = 0.028$). The mean survival times for the PBS-treated and nonspecific immunokinase control groups ($n = 10$) were 55 and 60 days, respectively, whereas, 8 out of 10 mice in the DAPK2′-CD30L treatment group were long-term survivors (>175 days) and showed no signs of disease. Two animals died (days 60 and 100) without developing any macroscopic signs of disease, suggesting that cancer was not the cause of death (Figure 21.4).

Histological examination of the PBS-treated SCID mice revealed the presence of disseminated tumors in the liver, lung, spleen, and brain, as previously reported [54].

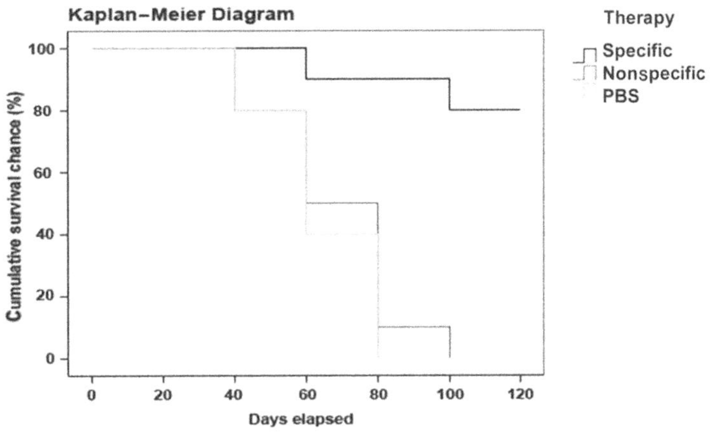

FIGURE 21.4 Effect of DAPK2′-CD30L treatment on disseminated Hodgkin lymphoma-derived tumors in SCID mice. Mice were injected intravenously with L540 tumor cells and were left untreated (PBS control) (none, light grey line), nonspecifically treated (nonspecific, grey line), or treated with maximum tolerable doses of DAPK2′-CD30L (specific, black line). Survival was plotted using the Kaplan–Meier method. *Source*: Reproduced from Reference 6.

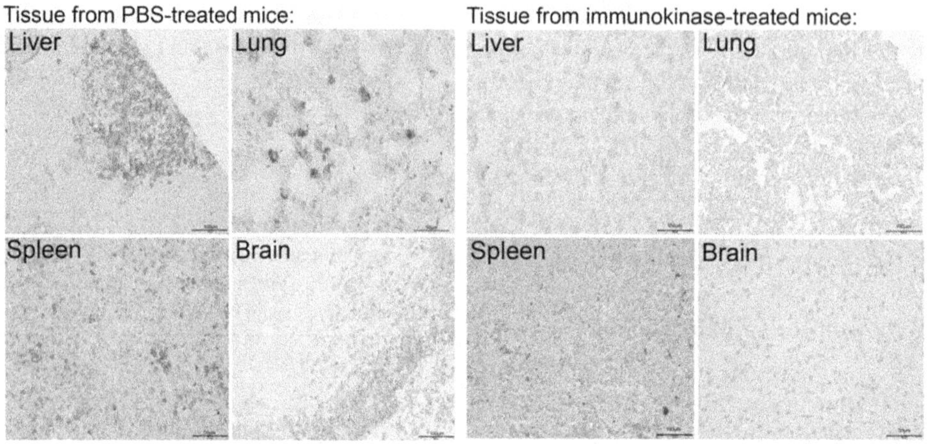

FIGURE 21.5 Immunohistochemical analysis of selected organ sections after intravenously injection of SCID mice with Hodgkin lymphoma-derived L540 cells, magnification x 4000. (a) PBS control group shows tumor cell infiltration as analyzed by immunological detection of CD30+ cells. (b) DAPK2′-CD30L immunokinase group shows no evidence of tumor cell infiltration. *Source*: Reproduced from Reference 6.

Immunohistochemical analysis with a mouse anti-human CD30 monoclonal antibody revealed intensive staining around the tumors, confirming the presence of the CD30 antigen (Figure 21.5). No tumor cells were detected in mice treated with DAPK2′-CD30L.

21.5 OUTLOOK

Protein kinases are often deregulated in cancer, some being more active and others less active than usual. DAPK2, a pro-apoptotic serine/threonine kinase, falls into the latter category, which means that it is a tumor suppressor protein and that targeted restoration of its activity could help to induce apoptosis in certain tumors. To avoid nonspecific effects on normal cells, the immunokinase strategy was devised to take advantage of both specific targeting and specific activity. By targeting CD30, a receptor that is predominantly expressed on Hodgkin lymphoma-derived tumor cells, it was possible to deliver DAPK2 directly into tumor cells lacking kinase activity, resulting in selective restoration in precisely those cells where the activity was required. We were able to show that the purified recombinant immunokinase fusion protein prevented tumor development in a disseminated Hodgkin lymphoma model in SCID mice, and that treated animals remained disease free more than 100 days after single injection of DAPK2′-CD30L. Our experiments provide an important proof of principle for the immunokinase concept, and there is no reason why the same strategy should not be applicable to other kinases that are inactivated in tumor cells.

REFERENCES

1. Barth S, (2009) Recombinant immunotoxins—the next generation. *Curr. Pharm. Des.* 15, 2650–2651.
2. Schirrmann T, Krauss J, Arndt MA, Rybak SM, Dubel S, (2009) Targeted therapeutic RNases (ImmunoRNases). *Expert Opin. Biol. Ther.* 9, 79–95.
3. Huhn M, Sasse S, Tur MK, Matthey B, Schinkothe T, Rybak SM, et al. (2001) Human angiogenin fused to human CD30 ligand (Ang-CD30L) exhibits specific cytotoxicity against CD30-positive lymphoma. *Cancer Res.* 61, 8737–8742.
4. Rosenblum MG, Barth S, (2009) Development of novel, highly cytotoxic fusion constructs containing granzyme B: unique mechanisms and functions. *Curr. Pharm. Des.* 15, 2676–2692.
5. Stahnke B, Thepen T, Stocker M, Rosinke R, Jost E, Fischer R, et al. (2008) Granzyme B-H22(scFv), a human immunotoxin targeting CD64 in acute myeloid leukemia of monocytic subtypes. *Mol. Cancer Ther.* 7, 2924–2932.
6. Tur MK, Neef I, Jager G, Teubner A, Stocker M, Melmer G, (2009) Immunokinases, a novel class of immunotherapeutics for targeted cancer therapy. *Curr. Pharm. Des.* 15, 2693–2699.
7. Tur MK, Neef I, Jost E, Galm O, Jager G, Stocker M, et al. (2009) Targeted restoration of down-regulated DAPK2 tumor suppressor activity induces apoptosis in Hodgkin lymphoma cells. *J. Immunother.* 32, 431–441.
8. Egan SE, Wright JA, Greenberg AH. (1991) Molecular determinants of metastatic transformation. *Environ. Health Perspect.* 93, 91–95.
9. Debatin KM. (2007) Chronic lymphocytic leukemia: keeping cell death at bay. *Cell* 129, 853–855.

10. Gozuacik D, Kimchi A. (2006) DAPk protein family and cancer. *Autophagy* 2, 74–79.
11. Michie AM, McCaig AM, Nakagawa R, Vukovic M. (2010) Death-associated protein kinase (DAPK) and signal transduction: regulation in cancer. *Febs. J.* 277, 74–80.
12. Maller JL. (1993) On the importance of protein phosphorylation in cell cycle control. *Mol. Cell Biochem.* 128, 267–281.
13. Barik S. (1996) Protein phosphorylation and signal transduction. *Subcell. Biochem.* 26, 115–164.
14. Graves JD, Krebs EG. (1999) Protein phosphorylation and signal transduction. *Pharmacol. Ther.* 82, 111–121.
15. Hanahan D, Weinberg RA. (2000) The hallmarks of cancer. *Cell* 100, 57–70.
16. Cross TG, Scheel-Toellner D, Henriquez NV, Deacon E, Salmon M, Lord JM. (2000) Serine/threonine protein kinases and apoptosis. *Exp. Cell. Res.* 256, 34–41.
17. Manning G, Whyte DB, Martinez R, Hunter T, Sudarsanam S. (2002) The protein kinase complement of the human genome. *Science* 298, 1912–1934.
18. Hanks SK, Quinn AM, Hunter T. (1988) The protein kinase family: conserved features and deduced phylogeny of the catalytic domains. *Science* 241, 42–52.
19. Hanks SK, Quinn AM. (1991) Protein kinase catalytic domain sequence database: identification of conserved features of primary structure and classification of family members. *Methods Enzymol.* 200, 38–62.
20. Deiss LP, Feinstein E, Berissi H, Cohen O, Kimchi A. (1995) Identification of a novel serine/threonine kinase and a novel 15-kD protein as potential mediators of the gamma interferon-induced cell death. *Genes Dev.* 9, 15–30.
21. Inbal B, Cohen O, Polak-Charcon S, Kopolovic J, Vadai E, Eisenbach L, Kimchi A. (1997) DAP kinase links the control of apoptosis to metastasis. *Nature* 390, 180–184.
22. Inbal B, Shani G, Cohen O, Kissil JL, Kimchi A. (2000) Death-associated protein kinase-related protein 1, a novel serine/-threonine kinase involved in apoptosis. *Mol. Cell. Biol.* 20, 1044–1054.
23. Cohen O, Feinstein E, Kimchi A. (1997) DAP-kinase is a Ca2+/calmodulin-dependent, cytoskeletal-associated protein kinase, with cell death-inducing functions that depend on its catalytic activity. *Embo J.* 16, 998–1008.
24. Bialik S, Kimchi A. (2006) The death-associated protein kinases: structure, function, and beyond. *Annu. Rev. Biochem.* 75, 189–210.
25. Shani G, Henis-Korenblit S, Jona G, Gileadi O, Eisenstein M, Ziv T, Admon A, Kimchi A. (2001) Autophosphorylation restrains the apoptotic activity of DRP-1 kinase by controlling dimerization and calmodulin binding. *Embo J.* 20, 1099–1113.
26. Gozuacik D, Kimchi A. (2004) Autophagy as a cell death and tumor suppressor mechanism. *Oncogene* 23, 2891–2906.
27. Inbal B, Bialik S, Sabanay I, Shani G, Kimchi A. (2002) DAP kinase and DRP-1 mediate membrane blebbing and the formation of autophagic vesicles during programmed cell death. *J. Cell. Biol.* 157, 455–468.
28. Shohat G, Spivak-Kroizman T, Cohen O, Bialik S, Shani G, Berrisi H, Eisenstein M, Kimchi A. (2001) The pro-apoptotic function of death-associated protein kinase is controlled by a unique inhibitory autophosphorylation-based mechanism. *J. Biol. Chem.* 276, 47460–47467.
29. Kawai T, Nomura F, Hoshino K, Copeland NG, Gilbert DJ, Jenkins NA, Akira S. (1999) Death-associated protein kinase 2 is a new calcium/calmodulin-dependent protein kinase that signals apoptosis through its catalytic activity. *Oncogene* 18, 3471–3480.
30. Bialik S, Berissi H, Kimchi A. (2008) A high throughput proteomics screen identifies novel substrates of death-associated protein kinase. *Mol. Cell. Proteomics* 7, 1089–1098.
31. Raveh T, Droguett G, Horwitz MS, DePinho RA, Kimchi A. (2001) DAP kinase activates a p19ARF/p53-mediated apoptotic checkpoint to suppress oncogenic transformation. *Nat. Cell. Biol.* 3, 1–7.
32. Rizzi M, Tschan MP, Britschgi C, Britschgi A, Hugli B, Grob TJ, et al. (2007) The death-associated protein kinase 2 is up-regulated during normal myeloid differentiation and enhances neutrophil maturation in myeloid leukemic cells. *J. Leukoc. Biol.* 81, 1599–1608.
33. Raveh T, Kimchi A. (2001) DAP kinase-a proapoptotic gene that functions as a tumor suppressor. *Exp. Cell. Res.* 264, 185–192.
34. Katzenellenbogen RA, Baylin SB, Herman JG. (1999) Hypermethylation of the DAP-kinase CpG island is a common alteration in B-cell malignancies. *Blood* 93, 4347–4353.
35. Raval A, Tanner SM, Byrd JC, Angerman EB, Perko JD, Chen SS, et al. (2007) Downregulation of death-associated protein kinase 1 (DAPK1) in chronic lymphocytic leukemia. *Cell* 129, 879–890.
36. Bai T, Tanaka T, Yukawa K, Maeda M, Umesaki N. (2004) Reduced expression of death-associated protein kinase in human uterine and ovarian carcinoma cells. *Oncol. Rep.* 11, 661–665.
37. Herman JG, Baylin SB. (2003) Gene silencing in cancer in association with promoter hypermethylation. *N. Engl. J. Med.* 349, 2042–2054.
38. Kissil JL, Feinstein E, Cohen O, Jones PA, Tsai YC, Knowles MA, et al. (1997) DAP-kinase loss of expression in various carcinoma and B-cell lymphoma cell lines: possible implications for role as tumor suppressor gene. *Oncogene* 15, 403–407.
39. Daskalakis M, Nguyen TT, Nguyen C, Guldberg P, Kohler G, Wijermans P, et al. (2002) Demethylation of a hypermethylated P15/INK4B gene in patients with myelodysplastic syndrome by 5-Aza-2′-deoxycytidine (decitabine) treatment. *Blood* 100, 2957–2964.
40. Silverman LR, Mufti GJ. (2005) Methylation inhibitor therapy in the treatment of myelodysplastic syndrome. *Nat. Clin. Pract. Oncol.* 2 Suppl 1, S12–23.
41. Esteller M. (2005) DNA methylation and cancer therapy: new developments and expectations. *Curr. Opin. Oncol.* 17, 55–60.
42. Hackanson B, Bennett KL, Brena RM, Jiang J, Claus R, Chen SS, et al. (2008) Epigenetic modification of CCAAT/enhancer

binding protein alpha expression in acute myeloid leukemia. *Cancer Res.* 68, 3142–3151.

43. Friedmann T. (1992) Gene therapy of cancer through restoration of tumor-suppressor functions? *Cancer* 70, 1810–1817.
44. Bowen C, Spiegel S, Gelmann EP. (1998) Radiation-induced apoptosis mediated by retinoblastoma protein. *Cancer Res.* 58, 3275–3281.
45. Levine AJ, Wu MC, Chang A, Silver A, Attiyeh EF, Lin J, et al. (1995) The spectrum of mutations at the p53 locus. Evidence for tissue-specific mutagenesis, selection of mutant alleles, and a "gain of function" phenotype. *Ann. N Y Acad. Sci.* 768, 111–128.
46. Ventura A, Kirsch DG, McLaughlin ME, Tuveson DA, Grimm J, Lintault L, et al. (2007) Restoration of p53 function leads to tumor regression *in vivo*. *Nature* 445, 661–665.
47. Capdeville R, Buchdunger E, Zimmermann J, Matter A. (2002) Glivec (STI571, imatinib), a rationally developed, targeted anticancer drug. *Nat. Rev. Drug. Discov.* 1, 493–502.
48. Pagano L, Fianchi L, Caira M, Rutella S, Leone G. (2007) The role of Gemtuzumab Ozogamicin in the treatment of acute myeloid leukemia patients. *Oncogene* 26, 3679–3690.
49. Kreitman RJ. (2004) Confirmation and prevention of targeted toxicity by a recombinant fusion toxin. *Mol. Cancer Ther.* 3, 1691–1692.
50. Barth S, Tur MK, Stöcker M, Fischer R. (2005) Immunokinases. EP 1704228 (A2).
51. Smith CA, Gruss HJ, Davis T, Anderson D, Farrah T, Baker E, et al. (1993) CD30 antigen, a marker for Hodgkin's lymphoma, is a receptor whose ligand defines an emerging family of cytokines with homology to TNF. *Cell* 73, 1349–1360.
52. Herszfeld D, Wolvetang E, Langton-Bunker E, Chung TL, Filipczyk AA, Houssami S, et al. (2006) CD30 is a survival factor and a biomarker for transformed human pluripotent stem cells. *Nat. Biotechnol.* 24, 351–357.
53. Klimm B, Schnell R, Diehl V, Engert A. (2005) Current treatment and immunotherapy of Hodgkin's lymphoma. *Haematologica* 90, 1680–1692.
54. Barth S, Huhn M, Matthey B, Tawadros S, Schnell R, Schinkothe T, et al. (2000) Ki-4(scFv)-ETA', a new recombinant anti-CD30 immunotoxin with highly specific cytotoxic activity against disseminated Hodgkin tumors in SCID mice. *Blood* 95, 3909–3914.

22

IMMUNORNASE FUSIONS

WOJCIECH ARDELT

Tamir Biotechnology, Inc. Monmouth Junction, NJ, USA

22.1 INTRODUCTION

More than 100 years ago, Paul Ehrlich predicted that it would be possible to develop "magic bullets" that would selectively kill cancers or pathogenic microorganisms [1–3]. He reasoned that compounds specifically targeting cancer cells or bacteria could be designed and toxins could be delivered along with those targeting agents.

22.1.1 Immunotoxins

Many years later, in 1980, monoclonal antibodies were used first time to deliver a plant toxin to neoplastic cells [4,5]. The antibodies (targeting, binding, or immune moieties) were covalently linked to ricin (effector or cell-killing moiety). Such constructs were later termed immunotoxins [6]. Immunotoxins are covered in another chapter of this book; they have also been discussed in numerous review articles [3,7–11].

Initially, entire molecules of monoclonal antibodies were chemically conjugated to toxins. Later, various antibody fragments retaining the ability to bind specific antigens on the surfaces of targeted cancer cells were developed and fused to the effector moieties at the genetic level by recombinant DNA technology (recombinant immunotoxins). Antibody fragments and other "alternative antibody formats" have recently been reviewed [12].

Effector moieties are usually toxic proteins from plants (like ricin, saporin, gelonin, and viscumin) or from bacteria (Pseudomonas exotoxin and diphtheria toxin). They have also been extensively modified in the course of immunotoxins development, for example, their natural binding domains were frequently removed to facilitate specific immune targeting. Immunotoxins bind to cells as a result of specific interaction between the targeting moieties and cell surface antigens. Next, they are internalized and translocated to cell cytosol where their effector moieties interfere with the translation step of protein biosynthesis. This leads to cell death.

Immunotoxins have been used in the clinic mainly to treat cancer (see Reference 7 for recent review). Main problems with this therapy are usually large molecular size of those drugs (that limits their penetrability to solid tumors) and the immunogenicity of the bacterial or plant proteins constituting their effector moieties. The latter usually precludes multiuse of these drugs because neutralizing antibodies are formed in patients. Immunotoxins may also damage endothelial cells and this frequently leads to a severe condition known as vascular leakage syndrome [7,13]. Liver and kidney toxicity are other complications. The adverse side effects of immunotoxins inspired a search for alternative effector moieties, which would be less or not immunogenic and better tolerated by patients.

22.1.2 Cytotoxic Activity of Ribonucleases

Ribonucleases (RNases) are potentially cytotoxic by virtue of their ability to catalyze cleavages of phosphodiester bonds in RNA. Intracellular degradation of various RNA species

Fusion Protein Technologies for Biopharmaceuticals: Applications and Challenges, First Edition. Edited by Stefan R. Schmidt.

engaged in the machinery of protein biosynthesis would eventually inhibit translation and cause cell death. However, most known RNases do not exhibit innate cytotoxic or cytostatic functions although some were found cytotoxic when microinjected to *Xenopus* oocytes [14,15]. Therefore, it seems that the cell entry is a prerequisite for RNase cytotoxicity [16]. A number of RNases demonstrate inherent cytostatic and cytotoxic activities and have been considered as potential drugs for anticancer therapy. Their structures and functions, as well as mechanisms of toxicity and therapeutic potentials, were exhaustively reviewed [17–28]. Cytotoxic RNases are small basic proteins of 10-28 kDa molecular mass. They were found among members of both, RNase A (reviewed in References 29 and 30) and T1 (reviewed in References 31) superfamilies. They bind to cancer cell membranes by ionic interactions, enter cells by endocytosis, avoid degradation by proteinases inside the endosomes [19] (that is largely due to their high conformational stability), then translocate to the cytosol, where they evade mammalian protein ribonuclease inhibitor [32] and degrade various species of RNA that leads to apoptotic cell death.

The most studied enzyme of this group is Onconase® (ranpirnase or, formerly, P30-protein) discovered by Alfacell Corporation (presently Tamir Biotechnology, Monmouth Junction, NJ, USA) in oocytes of a frog *Rana pipiens* [33]. Onconase® (recently reviewed [17,18]) is the smallest known member of pancreatic RNase A superfamily [33]. It consists of 104 amino acid residues and has a molecular weight 11,820. Onconase® is an unusually stable protein and is remarkably resistant to proteolysis. Its very high isoelectric point (9.7) facilitates binding to highly negative [34] cancer cell surfaces. The enzyme is then internalized by energy-dependent endocytosis [35] based on clathrin adaptors and the GTPase dynamin (clathrin/Ap-2 mediated endocytic pathway) [36,37]. From endosomes, Onconase® is delivered to the cell cytosol through the receptor recycling compartment, apparently bypassing the Golgi and endoplasmic reticulum [36,37] (intracellular routing of other cytotoxic RNases is discussed in Reference 37). In the cytosol, the enzyme resists proteolysis and does not interact with mammalian protein RNase inhibitor [38–40]. This allows for the degradation of RNA, preferentially tRNA. This in turn, leads to inhibition of protein synthesis, induction of G_1 arrest and apoptosis (see References 17 and 18 for more discussion). Inhibition of protein synthesis did not, however, explain certain effects of Onconase®. For example, it was found that the enzyme was able to suppress the expression of certain proteins and simultaneously upregulate other proteins [41] (similar results were recently published [42]). Therefore it was postulated that Onconase®, in addition to tRNAs [43,44], may also target other noncoding RNAs–microRNAs involved in the regulation of gene expression through RNA interference (RNAi) [45]. Indeed, it was recently found that the silencing of the gene encoding glyceraldehyde-3-phosphate dehydrogenase in human lung adenocarcinoma A549 cells by a siRNA was effectively prevented by this RNase [46].

Onconase® is the only RNase that has advanced to clinical trials and has been tested in the United States and Europe. It has been tested as a single anticancer agent or in combination with other drugs [47–51]. Even though Onconase® is not a human protein, no serious immunogenicity complications were noticed in the clinic. Approximately, a total of 1000 of patients have been treated to date and there have been few problems with repeated administering. This may be due to a high degree similarity between the 3-D structure of Onconase® and human RNases. The lack of complications may also manifest from the enzyme's ability to specifically enhance activation-induced apoptosis of peripheral blood lymphocytes [52]. Thus, Onconase® has been well tolerated by cancer patients; the main side effect, renal toxicity at high doses was found reversible and less pronounced or prevented by patient hydration.

Onconase® and other naturally cytotoxic RNases demonstrate substantial specificity to rapidly proliferating cells, for example, cancer cells. Endocytic internalization and translocation to the cell's cytosol are limiting factors for the cytotoxicity of these enzymes as well as of RNases with engineered cytotoxicity [16,37]. Therefore, linking of RNases by chemical conjugation or recombinant fusion to cell binding, internalizing molecules has been used to improve the specificity and cellular uptake of these enzymes.

22.1.3 Ribonuclease Targeting to Cancer Cells

The first evidence that RNases could be specifically targeted to cancer cells was generated nearly two decades ago. Bovine pancreatic RNase A [53,54] and Onconase® [55] were chemically conjugated to transferrin or to antibodies binding transferrin receptor. All these early conjugates were found highly cytotoxic to cancer cells and demonstrated IC_{50} values below 1 μM. RNases were also chemically conjugated to a number of other targeting molecules like growth factors, cytokines, monoclonal antibodies, or their fragments (see References 8,56, and 57 for reviews). Chemical immunoconjugates of RNases are discussed later in this chapter.

Two groups of authors provided excellent proofs of concept for immune targeting of RNases to cancer cells. Rybak et al. [58,59] described a chemical conjugate of LL2 anti-CD22 antibody with Onconase®. Both, immune and effector molecules were modified prior to conjugation. Sulfhydryl groups were introduced to the antibody in the reaction with 2-iminothiolane and a heterobifunctional crosslinker was incorporated to Onconase® through its amino-reactive arm. Then, the sulfhydryl-reactive arm of the incorporated crosslinker formed a disulfide bond linking the two moieties. This conjugate was several orders of magnitude more cytotoxic to Daudi B lymphoma cells overexpressing CD22 (a sugar binding transmembrane protein) as compared to unbound Onconase®. It was also very active *in vivo*, able to extend the life of mice with disseminated Daudi lymphoma up to 200%. As a result, this conjugate has been considered to be potential therapy for non-Hodkin lymphoma.

D' Alessio and coworkers [60] employed a single-chain variable region (scFv) of the murine antibody to ErbB2 transmembrane receptor in order to target human pancreatic RNase to breast, ovary and lung carcinomas. The targeting moiety was genetically fused to the N-terminus of human pancreatic RNase (effector moiety) through a six-residue spacer. Like LL2-Onconase® conjugate, this fusion protein named by the authors "immunoRNase," was highly specific and very toxic to target cells, whereas human pancreatic RNase itself was inactive.

As it is discussed in Section 4, production of chemical conjugates of RNases to antibodies (e.g., LL2-Onconase®), especially their large-scale manufacturing, poses numerous technical and regulatory challenges. Therefore, more attention has been recently paid to immunoRNase fusion proteins, like the mentioned anti-ErbB2 targeted enzyme. The following sections of this chapter are devoted to such constructs.

22.2 DEVELOPMENT OF IMMUNORNASE FUSION PROTEINS AS BIOPHARMACEUTICALS

A fusion protein may be defined as a genetic combination of originally separate proteins [61] expressed from a single synthetic gene. Most immunoRNase fusions are single chain engineered proteins comprising an antibody fragment able to bind the target antigen (targeting domain) and an entire ribonucleolytic enzyme (effector domain) combined by recombinant DNA technology. The domains are usually separated by a short "spacer" amino acid sequence that allows for independent folding of the construct components afterexpression of the synthetic gene in a host cell. Targeting domain may be composed of two "subdomains" like V_H and V_L fragments of the antibody connected by a "linker" amino acid sequence segment of various lengths. Other additional short sequence segments frequently built in RNase fusion proteins are "tag" sequences that facilitate detection or isolation and purification of these constructs.

So far, immunofusion proteins of seven RNases were designed and characterized as potential biopharmaceuticals. The enzymes studied were: angiogenin, eosinophil-derived neurotoxin, bovine seminal plasma RNase, human pancreatic RNase, Onconase®, *Rana pipiens* liver RNase, and barnase. These fusion proteins were in part discussed in previous review articles [3,8,27,56,57,62–68].

Table 22.1 shows immunoRNase fusions designed and characterized to date. They are listed in chronologic order illustrating the progress made in the field since 1992.

22.2.1 Immuno-Fusion Proteins of Eosinophil-Derived Endotoxin or Angiogenin

Eosinophil-derived neurotoxin (Edn, RNase 2) [69,70] recently reviewed [71] and angiogenin (Ang) [72] are members of the superfamily of RNase A [71,73]. Both enzymes are not inherently cytotoxic but they acquire cytotoxic activity when fused to antibodies or other ligands. Then, they interact with internalizing antigens that are overexpressed on cancer cell surface.

The early papers by Rybak, Youle, and coworkers [74,75] had a great influence on the field of RNase targeting. The authors constructed a chimeric gene encoding a heavy chain of mouse/human antibody to the human transferrin receptor (CD71) and angiogenin. Expression of the synthetic gene in E12B5 cells producing an antitransferrin receptor light chain [76] resulted in a very low yield of secreted CH2-Ang fusion protein. The product was very active (Table 22.1) against K562 human erythroleukemia cells and activity was blocked by the parent murine E6 antitransferrin receptor antibody. The individual components of the fusion protein, the antibody or angiogenin, did not affect K562 cells. Therefore, it was clear that toxicity was acquired as a result of their fusion. The unusually high activity of this fusion protein was later ascribed to its ability to form dimers [63,64].

The same group of authors followed up with a fusion of the single chain (scFv) of the same E6 antibody and another RNase, Edn [77]. This time the antibody moiety was fused to the carboxy terminus of the RNase. The enzyme and antibody domains were separated by a 13 amino acid residues spacer (48–60 residues segment of the staphylococcal protein A) of the following sequence: AKKLNDAQAPKSD [78]. The V_L and V_H fragments of the scFv were connected by a 14 amino acid residue linker. A six histidine affinity tag was attached at the carboxyl end of the V_H fragment. The entire construct was expressed in *Escherichia coli* BL21 (DE3) cells using the T7 system. Most of the fusion protein was found in inclusion bodies. After denaturation and renaturation it was purified by metal chelate chromatography. The fusion protein was active against cell lines expressing transferrin receptor, K562 and A431 human epidermoid carcinoma (Table 22.1) but not toward a cell line devoid of the antigen. Another Edn-scFv (E6) fusion protein constructed in a similar way, except for a different V_L–V_H linker, was toxic to human cell lines derived from melanoma, renal, and breast carcinomas with IC_{50} values of 8, 1.2, and 3 nM, respectively [79] (Table 22.1).

Angiogenin was also fused to E6 derived scFv [80]. The authors used two different peptides linking the V_L and V_H fragments, EGKSSGSGSESKEF or $(GGGGS)_3$. Ang-scFv

TABLE 22.1 Anti-Cancer ImmunoRNase Fusion Proteins

Design	Parent Antibody	Target Antigen	IC_{50} (nM)	Potential Indication	References
CH2-Ang	E6	CD71	0.05	Carcinomas	[56,74,75]
Edn-scFv	E6	CD71	0.2–1	Carcinomas	[77]
scFv-bsR	H17	hPLAP	0.3–4	Carcinomas	[94,95]
Ang-sFv	E6	CD71	4– 10	Carcinomas	[80]
hpR-scFv	E6	CD71	5–10	Carcinomas	[79]
Edn-scFv	E6	CD71	1– 8	Carcinomas	[79]
scFv-hpR	MgR6	ErbB2	50, 120	Carcinomas	[60]
Ang-scFv	Ki4	CD30	0.5–0.6	Lymphomas	[84]
scFv-Ang	Ki4	CD30	0.7–0.9	Lymphomas	[84]
scFv-Bar	4D5	ErbB2	NA	Carcinomas	[146]
scFv-hpR	PhL	ErbB2	12.5–60	Carcinomas	[103]
rpLR-scFv	HMFG1	MUC1	80	Carcinomas	[127]
V_L-Bar	F11	hFER	NA	Carcinomas	[151]
rpLR-scFv	RFB4	CD22	132–185	Lymphomas	[129]
$(rpLR\text{-}scFv)_2$	RFB4	CD22	3–20	Lymphomas	[129]
sFv-Ang	MLT-7	CD22	360–380	Lymphomas	[86]
$(sFv\text{-}Ang)_2$	MLT-7	CD22	74–118	Lymphomas	[86]
Ang-scFv	RFB4	CD22	57	Lymphomas	[87,88]
$(Onc)_2$-IgG	LL1	CD74	0.6–1.3	Lymphomas	[133]
scFv-hpR	BerH2	CD30	100	Lymphomas	[64,115]
$(scFv\text{-}hpR)_2$	PhL	ErbB2	3.0	Carcinomas	[107]
scFv-Fc-hpR	4E3, 2A1	CD30	3.3	Lymphomas	[116]
scFv-Ang	h22	CD64	0.2	Leukemias	[89]
scFv-AD-Ang	h22	CD64	0.03	Leukemias	[89]
scFv-diBar	4D5	ErbB2	1.8–4.1	Carcinomas	[152–154]
$(scFv)_2$-Fc-hpR	PhL	ErbB2	NA	Carcinomas	[111]
$(Onc)_2$-IgG	RS7	Trop-2	1.3–8.5	Lymphomas	[135]

Ang, angiogenin; Edn, eosinophil-derived neurotoxin; bsR, bovine seminal plasma RNase, hpR, human pancreatic RNase; Bar, barnase; rpLR, *Rana pipiens* liver RNase; NA, not available; PhL, phage Library (fully human scFv isolated by phage display); Onc, onconase; diBar, dibarnase; AD, "adapter" (membrane transfer peptide followed by a cytosolic cleavable peptide).

fusion protein with the latter linker was more active when tested against renal carcinoma, colon carcinoma, and human breast cancer cell lines. The respective IC_{50} values are given in Table 22.1. Elimination of the FB spacer between the enzyme and antibody moieties resulted in distinctly lower activity.

CD71, transferrin receptor used as a target surface antigen in the construction of early immunoRNase fusions discussed above is overexpressed in many cancers and rapidly internalizes after antibody binding. However, it was found to be able to cross the blood brain barrier together with the toxic moieties associated with those constructs [63,66]. Therefore, other target antigens were searched for. A cell transmembrane type 1 glycoprotein CD30 (Ki-1), a 120 kDa member of the tumor necrosis factor receptor family, is expressed in activated lymphocytes and related tumors as well as in some carcinomas and sarcomas [81–83]. Stöker et al. [84] were first to fuse the anti-CD30 single-chain variable fragment of Ki4 monoclonal antibody either to the C- or N-terminus of recombinant angiogenin. For the first time, these fusion proteins were for the first time expressed in eukaryotic (human embryonic kidney HEK293T) cells and secreted to the culture supernatant. They were highly cytotoxic to Hodgkin-derived CD30 positive L540 cells (Table 22.1).

CD22, a sugar binding membrane glycophosphoprotein, is highly overexpressed in non-Hodgkin's lymphoma [85] and most other B-cell lymphomas. It is therefore considered to be a potential target in the treatment of hematologic malignancies like non-Hodgkin's lymphoma, Burkitt's lymphoma, multiple myeloma, chronic lymphocytic leukemia, hairy cell leukemia, and prolymphocytic leukemia. An anti-CD22 antibody was previously used for targeting Onconase® [58,59] and, as discussed later in this chapter, scFv of a similar antibody was fused to a close variant of this RNase. Angiogenin was at first fused to a C-terminus of MLT-7 scFv via $(G_4S_3)_2$ spacer [86]. The product expressed in *E. coli* periplasmic space was poorly active. Elimination of the V_L–V_H linker from this fusion protein resulted in its noncovalent dimerization through the scFv moieties and substantially improved activity (Table 22.1). Kraus et al. [87] generated several highly stable humanized single-chain fragments of murine anti-CD22 RFB4 antibody. One of the fragments was fused to the carboxyl terminus of angiogenin through the $(G4S3)_3$ spacer. The construct was expressed in transiently transfected Chinese hamster ovary (CHO) cells. It was highly active against $CD22^+$ Daudi cells (Table 22.1) but not toward Jurkat cells which do not overexpress this antigen. A high-level concentration (5–10 mg/L) of this fusion protein was recently attained from supernatants of stably transfected myeloma (NSO) cells from the same laboratory [88].

Hetzel et al. [89] constructed a fusion protein comprising scFv of h22 humanized antibody and angiogenin. The antibody binds FcγR (CD64) receptor, which is frequently overexpressed by acute myeloid leukemia cells [90]. The fusion protein was produced in *E. coli* BL21 Star (DE3) cells using a periplasmic space secretion system under osmotic shock [91]. Interestingly, insertion of the membrane transfer peptide followed by the cytosolic-cleavable peptide [92,93] between h22 scFv and angiogenin [89] substantially increased fusion protein activity against $CD64^+$ U937 cells (Table 22.1). The use of "cleavable adaptors" for the construction of immunoRNases is discussed later in this chapter (Section 3.3).

22.2.2 Immuno-Fusion Proteins of Bovine Seminal Plasma RNase or Human Pancreatic RNase

Bovine seminal plasma RNase (BsRNase) is another member of pancreatic ribonuclease A superfamily [29,30]. This RNase was recently reviewed along with other inherently cytotoxic RNases [17]. BsRNase is the only enzyme in the superfamily with quaternary structure. It is a natural homodimer consisting of two identical 124 residue subunits linked by two disulfide bonds and noncovalent interactions. The subunits are 80% identical to bovine pancreatic RNase A. Deonarain and Epenetos [94,95] fused a monomer subunit of BsRNase to the C-terminus of single-chain fragment of murine H17E2 antibody. It was specific to the human placental-like alkaline phosphatase antigen (PLAP) that is expressed in ovarian, testicular, bladder, and head and neck cancers [96,97]. V_H and V_L of this scFv were connected by the $(G_4S)_3$ linker. In the original construct, the scFv was directly fused to the N-terminus of the BsRNase subunit. Insertion of the APAASPADA spacer and/or a diphtheria toxin disulfide loop between the immune and the effector moieties increased its activity. Also, C-terminal extension of the constructs with the KDEL sequence segment (endoplasmic reticulum retention signal) lowered the IC_{50} values against PLAP positive human oral epidermal carcinoma KB cells. The most active fusion protein comprised the scFv, the disulfide loop followed by the spacer, and the RNase extended with KDEL (Table 22.1). The fusion proteins were $\sim 2 \times 10^4$ fold more active than native BsRNase. All variants were expressed in *E. coli*, renatured and purified by metal chelate chromatography through the N-terminally located His_6 tag.

Along with angiogenin and RNase 2, human pancreatic RNase (RNase 1) [98] was an attractive candidate for immune targeting to cancer cells. This is attributed to its high catalytic activity and expected low immunogenic potential as a human protein. Zewe et al. [79] designed a fusion protein of this enzyme and a single-chain antibody to transferrin receptor (CD71). It was expressed in *E. coli* and recovered from inclusion bodies by denaturation and renaturation. This fusion protein was only slightly less active against three different human cancer cell lines compared to the analogous RNase 2 construct described in the same paper and discussed in Section 2.1 (Table 22.1). As mentioned earlier, the anti-CD71 antibody is no longer considered a candidate for construction of fusion proteins because it is at risk of crossing the blood brain barrier.

ErbB2, a transmembrane tyrosine kinase receptor (also known as HER2 or CD340) emerged as a promising target antigen because of its preferential expression in tumors [99]. It is highly expressed in breast, ovary, and lung carcinomas [100,101] and only marginally in normal tissues [102]. As mentioned in Section 1.3, Lorenzo et al. [60] were first to employ an anti ErbB2 scFv for RNase targeting. This fusion protein comprised a murine scFv with $(G_4S)_3$ linker, a GSPEFM spacer, and RNase 1. It was expressed in *E. coli*, isolated from cell lysate by metal chelate chromatography (His_6 tag located at the N-terminus) and re-natured. The product (immunoRNase) was toxic to $ErbB2^+$ SKBR3 and MDA-MB453 breast carcinoma cells (Table 22.1) but not to ErbB2 negative cells. Two years later, the same group reported the construction of a similar fusion protein with a human instead of a murine immune moiety [103]. In their earlier work [104], the authors successfully isolated a novel, fully human ErbB2-specific scFv (Erbicin) from a phage display library [105]. Erbicin was then fused to RNase 1 to obtain the first fully human immunoRNase [103]. The novel fusion protein contained a modified V_H–V_L linker, SS $(G_4S_2)_3$GGS as well as a longer spacer between the immune and effector moieties (AAASGGPEGGS). The His_6 tag located at the C-terminal position. It was expressed in *E. coli* as a soluble protein (molecular mass 46 kDa) in the periplasmic space. The imunoRNase was cytotoxic to four different ErbB2-positive cell lines with IC_{50}s 12.5, 47, 52, and 60 nM, respectively (Table 22.1). It also strongly inhibited the growth of tumor in mice, which were inoculated with ErbB2-positive TUBO cells. Studies of the mechanism of action of this novel fusion protein [106] revealed expected cytotoxicity dependence on the fusion protein's catalytic activity. It was shown that inactivation of the RNase moiety by alkylation abolished cell killing activity. The authors also demonstrated that the construct translocates directly from endosomes to the cytosol, where it interacts with ribonuclease inhibitor protein. However, because of high cellular uptake, the immunoRNase neutralizes the inhibitor and elicits apoptotic cell death.

A dimeric immunoRNase was then designed [107] based on Erbicin and a previously engineered cytotoxic dimer of

RNase 1 [108,109]. The latter was constructed by several amino acid replacements. Two of them introduced Cys residues in positions analogous to those in naturally dimeric BsRNase. They formed two intermolecular disulfide bonds. Other replacements either favored dimerization by creating a hydrophobic interaction or enhanced cytotoxicity. The novel immunoRNase was obtained by fusing the scFv (with a modified linker) to the engineered enzyme through a previously developed spacer [107]. This single-chain construct dimerized by forming intermolecular disulfide bonds in its enzyme moiety (molecular mass 92-kDa). The dimeric form did not interact with the ribonuclease inhibitor and was 2.5-fold more cytotoxic than the monomer [107].

Another Erbicin-derived immunoagent was a compact, fully functional anti-ErbB2 antibody [110]. In this construct, Erbicin was fused to the Fc domain of human IgG. The resulting antibody-like molecule (100 kDa) showed anticancer activity *in vitro* and *in vivo* (see Reference 66 for further discussion). Logical consequence of this project was very recent fusion of RNase 1 to this compact antibody [111]. The novel immunoRNase (140 kDa) was endowed with RNase- and complement-dependent toxicities as well as antibody-dependent cytostatic activity against ErbB2 positive tumor cells. Its catalytic activity was slightly lower than that of the monomeric human immunoRNase developed earlier by the authors [103]; however, the mammalian RNase inhibitor affected the novel construct to a significantly lesser degree. Both effects might be due to steric hindrances within the larger molecule. The novel immunoRNase was also more potent, both *in vitro* (toward ErbB2 positive SKBR3 cells) and *in vivo* (mice TUBO tumors) than the parental monomeric compound.

The Erbicin-based immunoRNases described in this section were found to interact with different ErbB2 epitope than those targeted by Herceptin (trastuzumab) [112]. Trastuzumab is the only ErbB2 specific monoclonal antibody approved for cancer treatment. Therefore, these Erbicin-based immunoRNases are active against Herceptin-resistant tumors but, unlike Herceptin, do not elicit cardiotoxicity [112–114].

Human pancreatic RNase was also targeted to the CD30 antigen overexpressed in lymphomas like Hodgkin- and anaplastic large cell lymphomas, but not in resting hematolymphoid cells [81]. An anti-CD30 scFv of murine Ber-H2 monoclonal antibody was fused to the N-terminus of RNase 1; the fusion protein (calculated mass 46.3 kDa) was expressed in Drosophila S2 cells and purified from the cell culture supernatant by cationic exchange chromatography [115]. It was highly toxic to Karpas 299 and two other CD30-positive cell lines and only slightly active against cells that do not express this antigen. It also significantly reduced the growth of CD30 positive tumors in the BALB/c mice model. Interestingly, this construct was insensitive to placental RNase inhibitor (interaction of immunoRNases with RNase inhibitor is discussed later in this chapter).

More recently, Menzel et al. [116] designed the first entirely human immunoRNase targeting CD30-positive lymphomas. Human anti-CD30 scFv (4E4) generated by a phage display methodology was fused to the Fc fragment of human IG1 and then to HpRNase through the AAASSG spacer. The V_H and V_L parts of the scFv were connected by the $(G_4S)_3$ linker. The fusion protein (75 kDa) was expressed in human embryonic kidney cells (HEK 293T) with a good yield of 4 mg/L. Purification from the supernatant by protein A column chromatography predominantly yielded a homodimer of about 150 kDa molecular mass. The product was stable in human serum and strongly inhibited the growth of CD30-positive Karpas-299 cells (Table 22.1).

22.2.3 Immuno-Fusion Proteins of Onconase® or Rana Pipiens Liver RNase

Onconase® (ranpirnase), already discussed in Section 1.2, was isolated from extracts of *Rana pipiens* (Leopard frog) oocytes and early embryos; it was the major cytotoxic component. Onconase® was identified as an RNase only after amino acid sequencing [33]. *Rana pipiens* Liver RNase (*Rap*LR) is a close variant of Onconase® that was identified by cloning from genomic DNA [117]. The two proteins differ in four (of 104) amino acid residues. In *Rap*LR, Ile11 of Onconase® is replaced by Leu, Asp20 by Asn, Lys 85 by Thr, and Ser103 by His in *Rap*LR. Since three other natural variants of Onconase® were found in *R. pipiens* oocytes [18], it seems that the frog genome contains at least five genes encoding Onconase® variants with replacements at positions 11, 20, 25, 85, and 103.

N-terminal pyroglutamate of Onconase® and other known frog RNases, with the exception of amphinase [18,118], is an integral part of the active sites of these enzymes and is critical for their catalytic and antiproliferative/cytotoxic activities [119–122]. Unexpectedly, *Rap*LR was found to be fully active without the N-terminal pyroglutamic acid residue [123]. This finding is difficult to explain, as the 3-D structure of this enzyme has not been refined. The lack of N-terminal pyroglutamic acid has an important practical implication. Unlike Onconase®, this RNase may be fused to targeting molecules through its N-terminus. However, fusion proteins of *Rap*LR obtained to date have been designed in the RNase-scFv format. The enzyme was targeted to MUC1 (DF3) antigen, which a high molecular weight, transmembrane glycoprotein, overexpressed in adenocarcinomas and hematological malignances [124–126]. Single-chain Fv derived from humanized anti-MUC1 HMFG1 antibody was engineered to a stable form by an Arg → Ala replacement at position 71 in V_H fragment [127]. The product was fused to the C-terminus of *Rap*LRI through the G_4S spacer, the construct was expressed in

E. coli, recovered from inclusion bodies by denaturation and re-naturation and finally purified by cation exchange chromatography. The novel fusion protein was cytotoxic to the MCF7 breast cancer cell line (Table 22.1).

The same group of authors designed a series of humanized scFv fragments [128] derived from murine RFB4 antibody, which recognizes the CD22 antigen (discussed in Section 2.1). Some of the fragments were later employed for the construction of monomeric and dimeric fusion proteins with *Rap*LR [129]. Shortening of the V_L–V_H $(G_4S)_3$ linker to the five residue form (G_4S) based on the work of other authors [130] allowed for generation of a diabody and, consequently a dimeric fusion protein with *Rap*LR [129]. The scFvs to be fused to the enzyme were further engineered by replacing the V_L Leu 36 with Tyr in order to improve their solubility and stability. Both mono- and dimeric fusion proteins were expressed in *E. coli* (strain TG1) periplasmic space and purified by immobilized metal chromatography and size exclusion. Dimerization caused substantial increase of cytotoxicity to CD22 positive cell lines, Daudi, Raji, and CA46 cells (Table 22.1).

Onconase® was immunotargeted to CD74, a type II transmembrane glycoprotein that is highly expressed in hematologic malignances like non-Hodgkin lymphoma and multiple myeloma [131,132]. Humanized murine LL1 anti-CD74 internalizing antibody was first engineered [134] by changing its constant region from $\gamma 1$ to $\gamma 4$ and introducing an S228 to P replacement in the hinge region (to prevent half molecule formation [134]). Then, its' (V_L) N-terminus was genetically fused to the C-terminus of Onconase® through the $(G_4S)_3$ spacer [133]. The construct, 2L-Onc-hLL1-V4P, was expressed in transfected Sp2/0-Ag14 myeloma cells and purified by protein A affinity chromatography. The molecular weight of the product corresponded to one IgG fused to two enzyme molecules. This first recombinant RNase-IgG fusion protein was highly cytotoxic to CD74 positive non-Hodgkin lymphoma (Daudi and Rai) or multiple myeloma (MC/Car) cells with IC_{50} values 0.63, 1.3, and 0.63 nM, respectively. CD74 negative, small cell lung cancer cell line was insensitive.

The same group recently succeeded in the construction of a similar immunoRNase [135] targeting Onconase® to the Trop-2 (EGP-1or TACSTD2) antigen, a tumor-associated calcium signal transducer [136] highly expressed by human carcinomas but not by most normal tissues ([135] and references therein). The humanized murine RS7 anti-Trop-2 antibody [137,138] was used for generation of a fusion protein with Onconase®. The single putative N-glycosylation site in the enzyme was disrupted (by the replacement of Asn69 with Gln) to prevent glycan heterogeneity of the fusion protein when expressed in mammalian cells. Two molecules of this Onconase® variant were then recombinantly fused to the antibody [135]. Each was connected to the N-terminus of one L chain in a similar way that used in the previous paper [133]. The fusion protein was expressed in stably transfected Sp2/0 myeloma cells and purified from the supernatant on a Protein A column. The product was apparently homogenous based on size exclusion high performance liquid chromatography and demonstrated two bands in reducing SDS–PAGE. The bands were of molecular mass ~50 and ~37 kDa and, respectively, corresponded to the heavy chain of the antibody and the light chain fused to Onconase®. It was very active against cultured breast cancer cells MDA-MB-468 and T-47D (IC_{50}s 3.8 and 2.0 nM, respectively). It was also active toward nonsmall cell lung cancer Calu-3 (IC50 8.5 nM) and cervical cancer ME-180 (IC50 1.5 nM) cells. The immunoRNase also significantly inhibited the growth of Calu-3 xenographs on athymic *nu/nu* mice.

Very recently, a development of Onconase® fusion proteins with humanized anti-CD22 antibody was announced [139].

22.2.4 Barnase Immuno-Fusion Proteins

So far, barnase is the only microbial RNase targeted to cancer cells. It is a small (110 amino acid residues, 12.4 kDa) extracellular enzyme secreted by *Bacillus amyloliquefaciens*. It was first described in 1959 [140] and has been well characterized since then [141–143]. Unlike most known RNases it contains no disulfide bonds. Barnase and its specific protein inhibitor barstar are reviewed [144,145].

Deyev et al. [146] designed fusion proteins composed of a single-chain variable antibody fragment specific to the ErbB2 (HER2) surface receptor (discussed in Section 2.2) and barnase. The scFv derived from a humanized 4D5 antibody [147] was fused to the N-terminus of barnase through a peptide spacer (EFPKPSTPPGSSGGAP). The spacer was derived from murine IgG3 hinge region. The entire construct, scFv-barnase, had molecular mass of 42 kDa. Another immunoRNase contained the same scFv and the "dibarnase" component as an effector moiety. The latter was a barnase molecule directly fused to a variant of the same enzyme with five N-terminal residues removed (des1-5-barnase). The entire construct was scFv-dibarbase (54 kDa). Both fusion proteins had "FLAG" and pentahistidine tags at the N- and C-termini, respectively. They were expressed in *E. coli* SB536 cells [146]. Since barnase is lethal to bacterial cells, the constructs were co-expressed with barstar [145,148,149]. Notably, an elegant method was developed to separate the enzymes from the inhibitor and purify the fusion proteins from cell lysates. This was accomplished with immobilized ion chromatography followed by protein A-affinity chromatography. The 4D5 scFv-barnase-His5 fusion protein was also successfully expressed in human embryonic kidney (HEK) 293T cells and purified from cell lysates [150].

In the earlier paper, the same group described the construction and characterization of a fusion protein that contained the antiferritin antibody (F11) light chain variable domain and barnase [151]. The enzyme was fused to the C-terminus of the V_L through a 13 amino acid spacer (ALGTVDSASAGIL); the hexahistidine tag was placed at the C-terminus of barnase. The construct was expressed in *E. coli*, mostly as a soluble protein in a dimeric form (~53 kDa). It was able to specifically bind ferritin and remained active as an RNase.

The only barnase fusion protein actually tested for anticancer activity is the 4D5 scFv-dibarnase. Edelweiss et al. [152] found that this construct as described above [146] was 2800-fold more active against SKOV-3 ovarian carcinoma cell line overexpressing ErbB2 antigen (IC_{50} 1.8 nM) than did barnase alone (Table 22.1). This fusion protein was almost as active toward another highly ErbB2 positive breast carcinoma BT-474 cell line. On the other hand, embryonal kidney HEK293 cells, with normal level of this antigen, were over 270-fold less sensitive. ErbB2 negative human peripheral blood mononuclear cells were unaffected [152]. The sensitive cells displayed typical features of apoptosis, including membrane blebbing, DNA fragmentation, and caspase activation. Apoptotic activity was inhibited by either 4D5 scFv or barstar. This confirmed that both, binding and effector moieties of this immunoRNase contributed to cytotoxicity. In the next papers [153,154], the authors reported high activity against ErbB2-overexpressing human breast cancer cell lines, SKBR-3 or BT-474 (IC_{50} 4.1 and 2.4 nM, respectively) (Table 22.1) as well as the *in vivo* anticancer activity of 4D5 scFv-dibarnase in mice bearing human breast cancer xenografts. Treatment of BALB/c nude mice bearing SKBR-3 breast cancer with a tolerated dose of the immunoRNase reduced the tumor growth by 77%.

22.3 ASPECTS OF IMMUNORNASE DESIGN AND PRODUCTION

22.3.1 Molecular Size and Valency

Single-chain Fv fragments are the smallest constructs comprising the entire antigen binding sites of parent antibodies. They are composed of variable heavy and light chain domains linked with flexible peptide segments. Their molecular masses range from 27 to 30 kDa. When compared with larger antibody formats like entire IgG (150 kDa), $F(ab')_2$ (110 kDa), or Fab' (55 kDa) [155], tumor xenograft studies demonstrated superior penetration of a radiolabeled scFv fragment from the vascular system into a solid tumor. While most of the entire IgG penetrated tumor regions directly adjacent to the vasculature, the scFv was more evenly (homogeneously) distributed in the tumor mass. The remaining forms penetrated the tumor in a size-related manner. On the other hand, tumor retention of scFvs is low due to their rapid clearance from the circulation and monovalent binding [156–161]. Shortening the flexible peptide linker connecting V_H and V_L domains to less than 12 (usually five) residues prevents the scFv to assemble in a normal monovalent orientation and, as a result, two scFvs form a bivalent dimer (diabody, 60 kDa) [131,162,163, and references therein]. Similar diabodies were also designed by engineering a C-terminal cysteine residue to scFvs to form a disulfide bond [158]. Reducing the linker to less than three residues leads to the formation of trimers or tetramers with molecular masses of 90 or 120 kDa, respectively [162]. Tumor retention improves with multimerization. This is a result of an increase in size and/or valency [158,162,163]. On the basis of these and similar observations, an optimal molecular size range of 60–120 kDa for therapeutic molecules targeting solid tumors has been proposed [161] and widely accepted [65,66,163] as reasonably balancing tumor penetration, retention, and clearance from the circulation. Monovalent immunoRNases comprising scFvs and monomeric enzymes (42–46 kDa) fall below the lower limit of this range but some were found quite effective *in vitro* (Table 22.1) as well as *in vivo*. Single Fv-dibarnase [152,153,155] with molecular mass of 54 kDa is slightly closer to the optimal range but dimeric/bivalent immunoRNases (82–92 kDa) fit in perfectly and, indeed, are profoundly more cytotoxic to cancer cells that their monomeric/monovalent counterparts [63–65,86,107,129]. ImmunnoRNases comprising scFvs and Fc fragments of compact [111] or full size [116] antibodies (140 and 150 kDa, respectively) are outside of the optimal molecular mass range and may be more suitable for the treatment hematological malignances (if properly targeted) rather than solid tumors. The largest fusion protein of Onconase® and the anti-Trop-2 antibody (180 kDa) targets lymphomas [135].

22.3.2 Interaction with Mammalian RNase Inhibitor

RNase inhibitor is an ~50 kDa, leucine and cysteine rich protein present in the cytosol of mammalian cells (reviewed in References 164–166). It forms extremely stable complexes with most mammalian RNases (K_d values 10^{-16}–10^{-14} M) and thus, abolishes their enzymatic activities. RNases with innate cytotoxic activity toward cancer cells are not inhibited by this protein [17]. As discussed earlier, some of them (Onconase®, *Rana pipiens* liver RNase or barnase) were used as effector moieties of ImmunoRNases and these constructs are certainly able to evade the inhibitor when translocated to cytosol. Mammalian RNases, angiogenin, RNase 1, RNase 2, or a single-chain subunit of bovine seminal plasma RNase used in the construction of immunoRNases strongly interact with the inhibitor. Therefore, the effectiveness of these fusion proteins could be potentially affected by this interaction. As discussed by Rutkoski and Raines [32], fusion of various binding moieties to RNases may or may not influence their affinity for the inhibitor. For example, the previously mentioned fusion protein, which comprises an anti-CD30 single-chain Fv fragment of murine Ber-H2 antibody and RNase 1, evaded the inhibitor [115]. It was apparently due to a steric hindrance generated by the scFv, preventing the enzyme–inhibitor interaction. On the other hand, another fusion protein created with the same enzyme, specific to ErbB2-positive solid tumors, strongly interacted with the inhibitor but, simultaneously effectively killed target cells [103]. Evidently, the immunoRNase reached the cytosol in large enough amount to overwhelm the inhibitor (usually present in cytosol in roughly 4 μM concentration [32]) and elicit apoptosis in tumor cells [103]. It seems, therefore, that lack of cytotoxic activity of some other RNase 1 immunofusions [167] could result from their inadequate internalization and/or intracellular routing and therefore, low uptake to the cytosol. Consequently, a molar excess of RNase inhibitor over the enzyme is created. Most immunofusions with mammalian RNases as effector moieties were active against target cells. It seems, however, that some fusion proteins could have been more effective if RNase variants with engineered resistance to the inhibitor [32] were used instead. Generally though, the inhibitor cannot compromise the effects of immunoRNases with high cytosol uptake.

22.3.3 Cleavable ImmunoRNases

In chemical conjugates of Onconase® or RNase A with antibodies, the functional components are connected either by disulfide or thioether linkages [58,59,168–170]. The former bond is cleavable in the reducing environment of the cell cytosol while the latter remains intact. The cleavable, disulfide conjugates were found more active against target cells than stable thioether conjugates [167–170]. Similar observations were also made in our laboratory (Saxena SK, Ardelt W, unpublished). Most probably, separation of the reducible conjugates in cytosol augmented the enzymatic activity of the released RNases. It is possible that large antibody molecules attached to the enzymes could have obstructed their interaction with intracellular substrates (steric hindrance). In fusion proteins, stable peptide segments (spacers) [171] do not allow for separation of the functional components inside the cell. In this respect, therefore, fusion proteins are similar to the thioether-linked chemical conjugates. Component inseparability creates even more limitations in fusion proteins than in chemical conjugates. Steric hindrance may not only directly affect catalytic activity but also impair the process of protein folding. This was observed in an antibody-barnase fusion protein [151]. Thus, development of cleavable immunoRNases may be in some cases advantageous. Hetzel et al. [89] first worked in this direction. "Cleavable adapters" were initially inserted between the binding and toxic moieties of classic immunotoxins in an attempt to reduce nonspecific toxicity [93,172]. They were peptide segments containing sequences cleavable by proteolytic enzymes present either in cytosol or endosomes. The "cytosolic cleavable unit," YVHDEVDRGP, contained cleavable sites for caspases 1 and 3. They were -DE- and -DR-, respectively. The 13 amino acid "endosomal cleavable unit" had two sites available for furin (-RG- and -RS-) [93]. Insertion of both cleavable units between the anti-CD64 scFv and angiogenin enhanced cytotoxicity of this immunoRNase by a factor of 20 but simultaneously significantly reduced stability of this construct in serum [89]. This could be due to degradation of the endosomal cleavable unit by proteases with trypsin specificity present in the circulatory system. The fusion protein with cytosolic cleavable unit alone was stable and retained its acquired increased activity against target cells [89]. Chen et al. [173] recently presented a novel approach toward cleavable fusion proteins. The authors designed a dithiocyclopeptide based on the amino acid sequence of somatotropin. It was used as a cleavable (reducible) linker/spacer between functional components of a fusion protein that contained colony stimulating factor and transferrin. The amino acid sequence stretch of somatotropin, AGCKNFFWKTFTSC (C3–C14 disulfide bond), was modified by replacing the WKT segment with PRS (the thrombin specific site). The peptide was inserted between the C-terminus of colony stimulating factor and the N-terminus of transferrin through short Leu-Glu linkers. The fusion protein expressed in HEK293 cells, contained a circular linker/spacer with the original disulfide bond of somatotropin. *In vitro* treatment with thrombin cleaved the circular spacer but the components of the fusion protein remained connected by the disulfide bond. Pretreatment with thrombin improved activity of this fusion protein but the disulfide bond was not very stable in the mouse circulatory system. This may be a limiting factor of this method when used for the construction of other fusion proteins such as immunoRNases.

22.3.4 High-Level Production of ImmunoRNases

Numerous classic immunotoxins and immunoRNases were expressed in *E. coli* where they accumulated in inclusion bodies as miss-folded, nonfunctional aggregates [60,77,79,80,94,95,127]. Their recovery required *in vitro* unfolding by denaturation and reduction of miss-formed disulfide bonds followed by refolding at native conformation and purification. Buchner et al. [174] designed the original protocol for this process. It is a tedious, complicated, and costly procedure with usually low product yield. It is therefore, not suitable for large-scale manufacturing and another methodology clearly needed to be developed.

Periplasmic space expression in *E. coli*, with the use of the pET22b(+) expression vector leads to soluble and properly folded proteins. This approach was used for the preparation of immunoRNases [86,103,107,129,152,155] but the product yield was usually very low. For example, in the case of dimeric fusion proteins, yield from culture was as low as ~20 μg/L [63,86,129]. In an attempt to improve periplasmic expression yield, Barth et al. [91] introduced a method entailed the growth of bacteria under osmotic stress (4% sodium chloride, 0.5 M sorbitol) in the presence of natural osmolytes (glycine betaine and hydroxyectoine) protecting proteins at high salt concentration. Expression of angiogenin fused to an anti-CD64 single-chain Fv under those conditions increased yield to 1 mg/L [89]. However, this method although improved, was still not robust enough for large-scale production. In addition to an inadequate yield, bacterial systems had other limitations (discussed in Reference 84). For example, it is very difficult to produce larger proteins of the optimal 60–120 kDa range in bacteria. Therefore, other approaches had to be sought which led to the exploration of eukaryotic host systems.

Stöker et al. [84] expressed previously mentioned anti-CD30 scFv fusion proteins with angiogenin in transiently transfected human embryonic kidney cells, HEK 293T. The constructs were not toxic to host cells and were equipped with leader sequences that mediated secretion of fusion products into the culture supernatant. The yield, however, did not exceed 0.5 mg/L. CHO cells were also tested as potential host cells [87]. Unfortunately, in an effort to produce a humanized anti-CD22 scFv-angiogenin in CHO cells, a disappointing yield of 0.13 mg/L was obtained. Further studies with HEK 293 T led to significantly higher yields of anti-CD30 scFv-RNase 1 (4 mg/L) [116]. Recently, the same group reported that production of anti-CD30 scFv-Fc-RNase 1 in transiently transfected HEK 293T cells resulted in the yield of 2–6 mg/L. Notably, when a derivative cell line, HEK293-EBNA was used; the authors were able to recover 100 mg/L of the immunoRNase [175].

Mouse myeloma cells were also employed to produce immunoRNases. Transiently, transfected NSO or Sp2/0-Ag14 cells were used as hosts to express the construct that was composed of two Onconase® molecules fused to humanized hLL1 anti-CD74 antibody [133]. Quite recently, two groups succeeded in the stable integration of immunoRNase genes into the genome of myeloma cells, NSO [88] or Sp2/0 [135]. The expressed fusion proteins were not toxic to the host cells and they were secreted into culture supernatants. The former paper reported that the yield of previously designed [87] anti-CD22 scFv-angiogenin immunoRNase produced in stably transfected NSO cells was 5–10 mg/L. Anti-CD22 specific Onconase® fusion proteins were produced in the same cells with the yield of 20 mg/L [139].

The results from human embryonic and mouse myeloma cell lines are very promising and it seems that one of the major obstacles in the clinical development of immuno-RNases, the lack of efficient production systems, may be resolved with the use of mammalian hosts. Production of immunoRNases using this expression system in industrial scale bioreactors should be possible in the near future.

Large-scale purification of secreted fusion proteins should not cause major problems. Oligo-histidine tags have been engineered to most immunoRNases to use immobilized metal affinity chromatography (IMAC) for their isolation; this technique may be used on large scale. Furthermore, it may be possible to purify larger constructs based on fully functional antibodies [133,135], on scaled-up protein A columns.

22.4 ALTERNATIVES

The concept of immunoRNase fusions encompasses both, classic immunotoxins and chemical conjugates of RNases to antibodies, which are specific to cancer antigens. Both approaches are still under development. Some immunotoxins have been effective in the clinical trials. However, immunotoxins with effector domains that are plant are usually highly immunogenic. This is a significant problem because the patient's immune system produces antibodies to the toxins and the drugs desired effects are neutralized. In addition, in the clinic immunotoxins have induced serious hepato- and nefro-toxicities as well as a vascular leakage syndrome. Immmunogenicity and general toxicity usually preclude repeated administration of these drugs. Intensive research toward the reduction of these adverse effects by modification or replacing the effector moieties is under way and recent advances were reviewed [176,177].

Rybak [170] recently reviewed Onconase® and antibody chemical conjugates. The enzyme covalently linked with anti-CD22 antibodies, LL2 [58] and RFB4 [170,178] demonstrated unusually high activity against the human lymphoma Daudi cell line. IC_{50} values of 20–70 pM were observed. Both conjugates were also very active *in vivo* in a murine model of human lymphoma—significantly increased the survival of tumor bearing mice. These are very encouraging results but the drawback of this approach is the lack a reliable, large-scale production method. The recently published method of preparation [179] is essentially the same as the original method [58] and requires chemical modification of both, the enzyme and the antibody. Introduction of free sulfhydryl groups to the latter by the interaction of the antibody (Ab) primary amino groups with 2-iminothiolate (IT) leads to heterogeneity of the intermediate. This is because typical IgG contains ~50 Lys residues bearing ε-amino groups reactive with IT. One molecule of Onconase® contains 12 lysine residues (including two necessary for enzymatic activity). These residues are potentially available for the amino-reactive arm of a hetero-bifunctional (amino- and sulfhydryl reactive) crosslinker used for conjugation of both moieties. Consequently, the conjugates are also heterogeneous. It is possible to minimize the heterogeneity by careful optimization of the reaction conditions. This however, negatively affects the yield and does not guarantee exactly the same Onc to Ab ratio in each preparation (Ardelt W, unpublished observations). The "classic" synthesis of the Ab-Onc conjugates proceeds according to the following scheme:

$$\text{Ab-NH}_2 + \text{IT} \rightarrow \text{Ab-SH}$$

$$\text{Onc} + \text{Crosslinker} \rightarrow \text{Onc-Crosslinker}$$

$$\text{Ab-SH} + \text{Onc-Crosslinker} \rightarrow \text{Ab-S-S-Crosslinker-Onc Conjugate.}$$

An alternative strategy has been explored in our laboratory (Saxena SK and Ardelt W, unpublished). Saxena [180] engineered a recombinant variant of Onconase® containing a single, free sulfhydryl group by Ser_{72}/Cys replacement (C72-Onc) that serves as an attachment point for antibodies or other targeting molecules such as, growth factors, other proteins, peptides, and low molecular weight compounds. That approach is illustrated as follows:

$$\text{Ab-NH}_2 + \text{Crosslinker} \rightarrow \text{Ab-Crosslinker}$$

$$\text{C72-Onc} + \text{Ab-Crosslinker} \rightarrow \text{Ab-Crosslinker-S-S-Onc Conjugate.}$$

This approach allows for elimination of (a) the heterogeneity introduced by Onconase® derivatization and (b) inactivation of a percent of enzyme molecules by random derivatization of the critical for activity Lys-9 or Lys-31 residues. However, the products are not fully homogenous and it is not easy to scale-up production. Thus, antibody-RNase conjugate chemistry still needs improvement.

22.5 CONCLUSIONS AND FUTURE PERSPECTIVES

With the exception of barnase, RNases used for the construction of fusion proteins discussed in this chapter are members of the super-family of bovine pancreatic RNase A. Three of them: angiogenin, RNase 1 and RNase 2 are human proteins. They are therefore unlikely to induce immune response or mediate nonspecific toxicity in patients. *Rana pipiens* Onconase®, the only RNase that advanced to clinical trials as a single agent, did not show appreciable immunogenicity or severe toxicity when repeatedly administered to over 1000 patients. *Rana pipiens* liver RNase is a close variant of Onconase® and is expected to exhibit similar ant-cancer activity with comparable lack of immunogenicity. As members of the same superfamily, these two frog RNases as well as bovine seminal RNase have 3-D structures that are very similar to those of human RNases. This may explain their low immunogenic potential. In addition, the propensity of Onconase® to increase activation-induced apoptosis of peripheral blood lymphocytes [52] may be a factor contributing to its low immunogenicity. Barnase, a member of microbial family of RNase T1, structurally differs from the enzymes discussed earlier. However, On the basis of the results of two close barnase variants, it is expected that this enzyme will not be substantially immunogenic in mammals [153]. In conclusion, it seems that RNases are promising alternatives for highly immunogenic and toxic plant or bacterial proteins used as effector moieties in classic immunotoxins.

Murine origin of binding moieties of some immunoRNases also poses a risk of immunogenicity. However, enormous progress in the field of antibody engineering allows for isolation of fully human scFvs and other antibody formats from phage display libraries [105,181–185]. Several immunoRNases with human immune and effector moieties (entirely human) have already been designed [103,107,111,116] and this trend will certainly continue in the future. Thus, it seems that imunogenicity is not longer an essential challenge for clinical development of immunoRNases.

Phage display technology was also used to improve stability of immune domains [128] and has a great potential in optimization of immunoRNase selectivity to particular cancer types. Substantial progress has been also achieved in the development of effective production methods suitable for scaling up to the industrial level. Therefore, it seems probable that some immunoRNases may advance to clinical trials in a reasonable time and possibly join an armament of anticancer drugs in the future.

REFERENCES

1. Bosch F, Rosich L. (2008) The contributions of Paul Ehrlich to pharmacology: a tribute on the occasion of the centenary of his Nobel Prize. *Pharmacology* 82, 171–179.
2. Gensini GF, Conti AA, Lippi D. (2007) The contributions of Paul Ehrlich to infectious disease. *J. Infect.* 54, 221–224.
3. De Lorenzo C, D'Alessio G. (2008) From immunotoxins to immunoRNases. *Curr. Pharm. Biotechnol.* 9, 210–214.
4. Krolick KA, Villemez C, Isakson P, Uhr JW, Vitetta ES. (1980) Selective killing of normal or neoplastic B cells by antibodies coupled to the A chain of ricin. *Proc. Natl. Acad. Sci. U S A* 77, 5419–5423.
5. Youle RJ, Neville DM, Jr. (1980) Anti-Thy 1.2 monoclonal antibody linked to ricin is a potent cell-type-specific toxin. *Proc. Natl. Acad. Sci. USA* 77, 5483–5486.
6. Vitetta ES, Krolick KA, Miyama-Inaba M, Cushley W, Uhr JW. (1983) Immunotoxins: a new approach to cancer therapy. *Science* 219, 644–650.
7. Pastan I, Hassan R, FitzGerald DJ, Kreitman RJ. (2007) Immunotoxin treatment of cancer. *Annu. Rev. Med.* 58, 221–237.

8. Rybak SM, Arndt MA, Schirrmann T, Dübel S, Krauss J. (2009) Ribonucleases and immunoRNases as anticancer drugs. *Curr. Pharm. Des.* 15, 2665–2675.
9. Kreitman RJ. (2006) Immunotoxins for targeted cancer therapy. *AAPS J.* 18, E532–551.
10. Pastan I, Hassan R, Fitzgerald DJ, Kreitman RJ. (2006) Immunotoxin therapy of cancer. *Nat. Rev. Cancer* 6, 559–565.
11. Frankel AE, Kreitman RJ, Sausville EA. (2000) Targeted toxins. *Clin. Cancer Res.* 6, 326–334.
12. Kontermann RE. (2010) Alternative antibody formats. *Curr. Opin. Mol. Ther.* 12, 176–183.
13. Vitetta ES. (2000) Immunotoxins and vascular leak syndrome. *Cancer J.* 6, 218–224.
14. Saxena SK, Rybak SM, Winkler G, Meade HM, McGray P, Youle RJ, et al. (1991) Comparison of RNases and toxins upon injection into Xenopus oocytes. *J. Biol. Chem.* 266, 21208–21214.
15. Rybak SM, Saxena SK, Ackerman EJ, Youle RJ. (1991) Cytotoxic potential of ribonuclease and ribonuclease hybrid proteins. *J. Biol. Chem.* 266, 21202–21207.
16. Leich F, Stöhr N, Rietz A, Ulbrich-Hofmann R, Arnold U. (2007) Endocytotic internalization as a crucial factor for the cytotoxicity of ribonucleases. *J. Biol. Chem.* 282, 27640–27646.
17. Ardelt W, Ardelt B, Darzynkiewicz Z. (2009) Ribonucleases as potential modalities in anticancer therapy. *Eur. J. Pharmacol.* 625, 181–189.
18. Ardelt W, Shogen K, Darzynkiewicz Z. (2008) Onconase® and Amphinase, the antitumor ribonucleases from *Rana pipiens* oocytes. *Curr. Pharm. Biotechnol.* 9, 215–225.
19. Arnold U. (2008) Aspects of the cytotoxic action of ribonucleases. *Curr. Pharmacol. Biotechnol.* 9, 161–168.
20. Lee JE, Raines RT. (2008) Ribonucleases as novel chemotherapeutics: the ranpirnase example. *Biodrugs* 22, 53–58.
21. Ramos-Nino ME. (2007) Cytotoxic ribonuclease-based cancer therapies. *Drugs Fut.* 32, 1–10.
22. Benito A, Ribó M, Vilanova M. (2005) On the track of antitumor ribonucleases. *Mol. Biosyst.* 1, 294–302.
23. Saxena SK, Shogen K, Ardelt W. (2003) Onconase® and its therapeutic potential. *Lab. Med.* 34, 380–387.
24. Makarov AA, Ilinskaya ON. (2003) Cytotoxic ribonucleases: molecular weapons and their targets. *FEBS Lett.* 540, 15–20.
25. Leland PA, Raines RT. (2001) Cancer chemotherapy: ribonucleases to the rescue. *Chem. Biol.* 8, 405–413.
26. Matoušek J. (2001) Ribonucleases and their antitumor activity. *Comp. Biochem. Physiol.* C 129, 175–191.
27. Rybak SM, Newton DL. (1999) Natural and engineered cytotoxic ribonucleases: therapeutic potential. *Exp. Cell Res.* 253, 325–335.
28. Youle RJ, D'Alessio G. (1997) Antitumor RNases. In: D'Alessio G, Riordan JF. (Eds.), *Ribonucleases: Structures and Functions*. Academic Press, San Diego, CA 491–514.
29. Beintema JJ, Schuller C, Irie M, Carsana A. (1988) Molecular evolution of the ribonuclease superfamily. *Prog. Biophys. Mol. Biol.* 151, 165–192.
30. Dyer KD, Rosenberg HF. (2006) The RNase A superfamily: generation of diversity and innate host defense. *Mol. Divers.* 10, 585–597.
31. Irie M. (1997) RNase T_1/RNase T_2 family RNases. In: D'Alessio G., Riordan JF. (Eds.), *Ribonucleases: Structures and Functions*. Academic Press, San Diego, CA, pp. 101–130.
32. Rutkoski TJ, Raines RT. (2008) Evasion of ribonuclease inhibitor as a determinant of ribonuclease cytotoxicity. *Curr. Pharm. Biotechnol.* 9, 185–189.
33. Ardelt W, Mikulski SM, Shogen K. (1991) Amino acid sequence of an anti-tumor protein from *Rana pipiens* oocytes and early embryos. Homology to pancreatic ribonucleases. *J. Biol. Chem.* 26, 245–251.
34. Márquez M, Nilsson S, Lennartsson L, Liu Z, Tammela T, Raitanen M, et al. (2004) Charge-dependent targeting: results in six tumor cell lines. *Anticancer Res.* 24, 1347–1351.
35. Haigis MC, Raines RT. (2003) Secretory ribonucleases are internalized by a dynamin-independent endocytic pathway. *J. Cell Sci.* 116(Pt 2), 313–324.
36. Rodriguez M, Torrent G, Bosch M, Dubremetz J-F, Ribo M, Benito A, et al. (2007) Intracellular pathway of Onconase® that enables its delivery to the cytosol. *J. Cell Sci.* 120, 1405–1411.
37. Benito A, Vilanova M, Ribó M. (2008) Intracellular routing of cytotoxic pancreatic-type ribonucleases. *Curr. Pharm. Biotechnol.* 9, 169–179.
38. Boix E, Wu Y-N, Vasandani VM, Saxena SK, Ardelt W, Ladner J, et al. (1996) Role of the N-terminus in RNase A homoloques: differences in catalytic activity, ribonuclease inhibitor interaction and cytotoxicity. *J. Mol. Biol.* 257, 992–1007.
39. Kelemen BR, Klink TA, Behlke MA, Eubanks SR, Leland PA, Raines RT. (1999) Hypersensitive substrates for ribonucleases. *Nucleic Acids Res.* 27, 3696–3701.
40. Turcotte RF, Raines RT. (2008) Interaction of Onconase® with the human ribonuclease inhibitor protein. *Biochem. Biophys. Res. Commun.* 377, 512–514.
41. Juan G, Ardelt B, Mikulski SM, Shogen K, Ardelt W, Mittelman A, et al. (1998) G_1 arrest of U-937 cells by Onconase® is associated with suppression of cyclin D3 expression, induction of p16^{INK4A,} p21$^{WAF1/CIP1}$ and p27^{KIP} and decreased pRb phosphorylation. *Leukemia* 12, 1241–1248.
42. Altomare DA, Rybak SM, Pei J, Maizel JV, Cheung M, Testa JR, et al. (2010) Onconase® responsive genes in human mesothelioma cells: implications for an RNA damaging therapeutic agent. *BMC Cancer* 10, 34.
43. Saxena SK, Sirdeshmukh R, Ardelt W, Mikulski SM, Shogen K, Youle RJ. (2002) Entry into cells and selective degradation of tRNAs by a cytotoxic member of the RNase A family. *J. Biol. Chem.* 277, 15142–15146.
44. Suhasini AN, Sirdeshmukh R. (2006) Transfer RNA cleavages by Onconase® reveal unusual cleavage sites. *J. Biol. Chem.* 281, 12201–12209.
45. Ardelt B, Ardelt W, Darzynkiewicz Z. (2003) Cytotoxic ribonucleases and RNA interference. *Cell Cycle* 2, 22–24.

46. Zhao H, Ardelt B, Ardelt W, Shogen K, Darzynkiewicz Z. (2008) The cytotoxic ribonuclease Onconase® targets RNA interference (siRNA). *Cell Cycle* 7, 3258–3261.
47. Mikulski SM, Costanzi JJ, Vogelzang NJ, McCachren S, Taub RN, Chun H, et al. (2002) Phase II trial of a single weekly intravenous dose of ranpirnase in patients with unresectable malignant mesothelioma. *J. Clin. Oncol.* 20, 274–281.
48. Costanzi J, Sidransky D, Navon A, Goldsweig H. (2005) Ribonucleases as a novel pro-apoptotic anticancer strategy: review of the preclinical and clinical data for ranpirnase. *Cancer Invest.* 23, 643–650.
49. Vogelzang NJ, Porta C, Mutti L. (2005) New agents in the management of advanced mesothelioma. *Semin. Oncol.* 32, 336–350.
50. Pavlakis N, Vogelzang NJ. (2006) Ranpirnase—an antitumor ribonuclease: its potential role in malignant mesothelioma. *Expert Opin. Biol. Ther.* 6, 391–399.
51. Beck AK, Pass HI, Carbone M, Yang H, (2008) Ranpirnase as a potential antitumor ribonuclease treatment for mesothelioma and other malignancies. *Future Oncol.* 4, 341–349.
52. Halicka DH, Pozarowski P, Ita M, Ardelt W, Mikulski SM, Shogen K, et al. (2002) Enhancement of activation-induced apoptosis of lymphocytes by the cytotoxic ribonuclease Onconase® (Ranpirnase). *Int. J. Oncol.* 21, 1245–1250.
53. Rybak SM, Saxena SK, Ackerman EJ, Youle RJ. (1991) Cytotoxic potential of ribonuclease and ribonuclease hybrid proteins. *J. Biol. Chem.* 266, 21202–21207.
54. Newton DL, Ilercil O, Laske DW, Oldfield E, Rybak SM, Youle RJ. (1992) Cytotoxic ribonuclease chimeras: targeted tumoricidal activity *in vitro* and *in vivo*. *J. Biol. Chem.* 267, 19572–19578.
55. Rybak SM, Newton DL, Mikulski SM, Viera A, Youle RJ. (1993) Cytotoxic Onconase® and ribonuclease A chimeras: comparison and *in vitro* characterization. *Drug Deliv.* 1, 3–10.
56. Rybak SM, Newton DL. (1999) Natural and engineered cytotoxic ribonucleases: therapeutic potential. *Exp. Cell Res.* 253, 325–335.
57. Rybak SM. (2008) Antibody-Onconase® conjugates: cytotoxicity and intracellular routing. *Curr. Pharm. Biotech.* 9, 226–230.
58. Newton DL, Hansen HJ, Mikulski SM, Goldenberg DM, Rybak SM. (2001) Potent and specific antitumor effects of an anti-CD22-targeted cytotoxic ribonuclease: potential for the treatment of non-Hodgkin lymphoma. *Blood* 97, 528–535.
59. Newton DL, Hansen HJ, Liu H, Ruby D, Iordanov MS, Magun BE, et al. (2001) Specifically targeting the CD22 receptor of human B-cell lymphomas with RNA damaging agents. *Crit. Rev. Oncol. Hematol.* 39, 79–86.
60. De Lorenzo C, Nigro A, Piccoli R, D'Alessio G. (2002) A new RNase-based immunoconjugate selectively cytotoxic for ErbB2-overexpressing cells. *FEBS Lett.* 516, 208–212.
61. Schmidt SR. (2009) Fusion proteins as biopharmaceuticals—applications and challenges. *Curr. Opin. Drug. Discov. Devel.* 12, 284–295.
62. Rybak SM, Newton DL, Xue Y. (1995) RNase and RNase immunofusions for cancer therapy. *Tumor Target.* 1, 141–147.
63. Krauss J, Arndt MA, Dübel S, Rybak SM. (2008) Antibody-targeted RNase fusion proteins (immunoRNases) for cancer therapy. *Curr. Pharm. Biotechnol.* 9, 231–234.
64. Schirrmann T, Krauss J, Arndt MA, Rybak SM, Dübel S. (2009) Targeted therapeutic RNases (ImmunoRNases). *Expert Opin. Biol. Ther.* 9, 79–95.
65. Rybak SM, Arndt MA, Schirrmann T, Dübel S, Krauss J. (2009) Ribonucleases and immunoRNases as anticancer drugs. *Curr. Pharm. Des.* 15, 2665–2675.
66. De Lorenzo C, D'Alessio G. (2009) Human anti-ErbB2 immunoagents—immunoRNases and compact antibodies. *FEBS J.* 276, 1527–1535.
67. Fang EF, Ng TB. (2010) Ribonucleases of different origins with a wide spectrum of medicinal applications. *Biochim. Biophys. Acta* 1815, 65–74.
68. Andrady C, Shrama SK, Chester KA. (2011) Antibody-enzyme fusion proteins for cancer therapy. *Immunotherapy* 3, 193–211.
69. Durack DT, Ackerman SJ, Loegering DA, Gleich GJ. (1981) Purification of human eosinophil-derived neurotoxin. *Proc. Natl. Acad. Sci. U S A* 785165–5169.
70. Peterson CG, Venge P. (1983) Purification and characterization of a new cationic protein--eosinophil protein-X (EPX)—from granules of human eosinophils. *Immunology* 50, 19–26.
71. Rosenberg HF. (2008) Eosinophil-derived neurotoxin/RNase 2: connecting the past, the present and the future. *Curr. Pharm. Biotechnol.* 9, 135–140.
72. Fett JW, Strydom DJ, Lobb RR, Alderman EM, Bethune JL, Riordan JF, et al. (1985) Isolation and characterization of angiogenin, an angiogenic protein from human carcinoma cells. *Biochemistry* 24, 5480–5486.
73. Rosenberg HF. (2008) RNase A ribonucleases and host defense: an evolving story. *J. Leukoc. Biol.* 83, 1079–1087.
74. Rybak SM, Hoogenboom HR, Meade HM, Raus JC, Schwartz D, Youle RJ. (1992) Humanization of immunotoxins. *Proc. Natl. Acad. Sci. U S A* 89, 3165–3169.
75. Rybak SM, Hoogenboom HR, Newton DL, Raus JC, Youle RJ. (1992) Rational immunotherapy with ribonuclease chimeras: an approach toward humanizing immunotoxins. *Cell Biophys.* 21, 121–138.
76. Hoogenboom HR, Raus JC, Volckaert G. (1990) Cloning and expression of a chimeric antibody directed against the human transferrin receptor. *J. Immunol.* 144, 3211–3217.
77. Newton DL, Nicholls PJ, Rybak SM, Youle RJ. (1994) Expression and characterization of recombinant human eosinophil-derived neurotoxin and eosinophil-derived neurotoxin-anti-transferrin receptor sFv. *J. Biol. Chem.* 269, 26739–26745.
78. Tai MS, Mudgett-Hunter M, Levinson D, Wu GM, Haber E, Oppermann H, et al. (1990) A bifunction0al fusion protein containing Fc-binding fragment B of staphylococcal protein A amino terminal to antidigoxin single-chain Fv. *Biochemistry* 29, 8024–8030.
79. Zewe M, Rybak SM, Dubel S, Coy JF, Welschof M, Newton DL, et al. (1997) Cloning and cytotoxicity of a human pancreatic RNase immunofusion. *Immunotechnology* 3127–136.

80. Newton DL, Xue Y, Olson KA, Fett JW, Rybak SM. (1996) Angiogenin single-chain immunofusions: influence of peptide linkers and spacers between fusion protein domains. *Biochemistry* 35, 545–553.

81. Stein H, Mason DY, Gerdes J, O'Connor N, Wainscoat J, Pallesen G, et al. (1985) The expression of the Hodgkin's disease associated antigen Ki-1 in reactive and neoplastic lymphoid tissue: evidence that Reed-Sternberg cells and histiocytic malignancies are derived from activated lymphoid cells. *Blood* 66, 848–858.

82. Mechtersheimer G, Möller P. (1990) Expression of Ki-1 antigen (CD30) in mesenchymal tumors. *Cancer* 66, 1732–1737.

83. Dürkop H, Latza U, Hummel M, Eitelbach F, Seed B, Stein H. (1992) Molecular cloning and expression of a new member of the nerve growth factor receptor family that is characteristic for Hodgkin's disease. *Cell* 68, 421–427.

84. Stöcker M, Tur MK, Sasse S, Krüssmann A, Barth S, Engert A. (2003) Secretion of functional anti-CD30-angiogenin immunotoxins into the supernatant of transfected 293T-cells. *Protein Expr. Purif.* 28, 211–219.

85. Pawlak-Byczkowska EJ, Hansen HJ, Dion AS, Goldenberg DM. (1989) Two new monoclonal antibodies, EPB-1 and EPB-2, reactive with human lymphoma. *Cancer Res.* 49, 4568–4577.

86. Arndt MA, Krauss J, Vu BK, Newton DL, Rybak SM. (2005) A dimeric angiogenin immunofusion protein mediates selective toxicity toward CD22+ tumor cells. *J. Immunother.* 28, 245–251.

87. Krauss J, Arndt MA, Vu BK, Newton DL, Rybak SM. (2005) Targeting malignant B-cell lymphoma with a humanized anti-CD22 scFv-angiogenin immunoenzyme. *Br. J. Haematol.* 128, 602–609.

88. Krauss J, Exner E, Mavratzas A, Seeber S, Arndt MA. (2009) High-level production of a humanized immunoRNase fusion protein from stably transfected myeloma cells. *Meth. Mol. Biol.* 525, 471–490.

89. Hetzel C, Bachran C, Fischer R, Fuchs H, Barth S, Stöcker M. (2008) Small cleavable adapters enhance the specific cytotoxicity of a humanized immunotoxin directed against CD64-positive cells. *J. Immunother.* 31, 370–376.

90. Ball ED, McDermott J, Griffin JD, Davey FR, Davis R, Bloomfield CD. (1989) Expression of the three myeloid cell-associated immunoglobulin G Fc receptors defined by murine monoclonal antibodies on normal bone marrow and acute leukemia cells. *Blood* 73, 1951–1956.

91. Barth S, Huhn M, Matthey B, Klimka A, Galinski EA, Engert A. (2000) Compatible-solute-supported periplasmic expression of functional recombinant proteins under stress conditions. *Appl. Environ. Microbiol.* 66, 1572–1579.

92. Ogata M, Fryling CM, Pastan I, FitzGerald DJ. (1992) Cell-mediated cleavage of Pseudomonas exotoxin between Arg279 and Gly280 generates the enzymatically active fragment which translocates to the cytosol. *J. Biol. Chem.* 267, 25396–25401.

93. Heisler I, Keller J, Tauber R, Sutherland M, Fuchs H. (2003) A cleavable adapter to reduce nonspecific cytotoxicity of recombinant immunotoxins. *Int. J. Cancer* 103, 77–82.

94. Deonarain MP, Epenetos AA. (1995) Construction, refolding and cytotoxicity of a scFv-seminal RNase fusion protein. *Tumor Target.* 1, 177–182.

95. Deonarain MP, Epenetos AA. (1998) Design, characterization and antitumor cytotoxicity of a panel of recombinant, mammalian ribonuclease-based immunotoxins. *Br. J. Cancer* 77, 537–546.

96. Epenetos AA, Travers P, Gatter KC, Oliver RD, Mason DY, Bodmer WF. (1984) An immunohistological study of testicular germ cell tumors using two different monoclonal antibodies against placental alkaline phosphatase. *Br. J. Cancer* 49, 11–15.

97. Iles RK, Ind TE, Chard T. (1994) Production of placental alkaline phosphatase (PLAP) and PLAP-like material by epithelial germ cell and non-germ cell tumors *in vitro*. *Br. J. Cancer* 69, 274–278.

98. Russo N, de Nigris M, Ciardiello A, Di Donato A, D'Alessio G. (1993) Expression in mammalian cells, purification and characterization of recombinant human pancreatic ribonuclease. *FEBS Lett.* 333, 233–237.

99. Ishida T, Tsujisaki M, Hanzawa Y, Hirakawa T, Hinoda Y, Imai K, et al. (1994) Significance of erbB-2 gene product as a target molecule for cancer therapy. *Scand. J. Immunol.* 39, 459–466.

100. Slamon DJ, Godolphin W, Jones LA, Holt JA, Wong SG, Keith DE, et al. (1989) Studies of the HER-2/neu proto-oncogene in human breast and ovarian cancer. *Science* 244, 707–712.

101. Klapper LN, Kirschbaum MH, Sela M, Yarden Y. (2000) Biochemical and clinical implications of the ErbB/HER signaling network of growth factor receptors. *Adv. Cancer Res.* 77, 25–79.

102. Press MF, Cordon-Cardo C, Slamon DJ. (1990) Expression of the HER-2/neu proto-oncogene in normal human adult and fetal tissues. *Oncogene* 5, 953–962.

103. De Lorenzo C, Arciello A, Cozzolino R, Palmer DB, Laccetti P, Piccoli R, et al. (2004) A fully human antitumor immunoRNase selective for ErbB-2-positive carcinomas. *Cancer Res.* 64, 4870–4874.

104. De Lorenzo C, Palmer DB, Piccoli R, Ritter MA, D'Alessio G. (2002) A new human antitumor immunoreagent specific for ErbB2. *Clin. Cancer Res.* 8, 1710–1719.

105. Griffiths AD, Williams SC, Hartley O, Tomlinson IM, Waterhouse P, Crosby WL, et al. (1994) Isolation of high affinity human antibodies directly from large synthetic repertoires. *EMBO J.* 13, 3245–3260.

106. De Lorenzo C, Di Malta C, Calì G, Troise F, Nitsch L, D'Alessio G. (2007) Intracellular route and mechanism of action of ERB-hRNase, a human anti-ErbB2 anticancer immunoagent. *FEBS Lett.* 581, 296–300.

107. Riccio G, Borriello M, D'Alessio G, De Lorenzo C. (2008) A novel human antitumor dimeric immunoRNase. *J. Immunother.* 31, 440–445.

108. Piccoli R, Di Gaetano S, De Lorenzo C, Grauso M, Monaco C, Spalletti-Cernia D, et al. (1999) A dimeric mutant of

human pancreatic ribonuclease with selective cytotoxicity toward malignant cells. *Proc. Natl. Acad. Sci. U S A* 96, 7768–7773.

109. Di Gaetano S, D'alessio G, Piccoli. R. (2001) Second generation antitumor human RNase: significance of its structural and functional features for the mechanism of antitumor action. *Biochem. J.* 358, 241–247.

110. De Lorenzo C, Tedesco A, Terrazzano G, Cozzolino R, Laccetti P, Piccoli R, et al. (2004) A human, compact, fully functional anti-ErbB2 antibody as a novel antitumor agent. *Br. J. Cancer* 91, 1200–1204.

111. Borriello M, Laccetti P, Terrazzano G, D'Alessio G, De Lorenzo C. (2011) A novel fully human antitumor immunoRNase targeting ErbB2-positive tumors. *Br. J. Cancer* 104, 1716–1723.

112. De Lorenzo C, Troise F, Cafaro V, D'Alessio G. (2007) Combinatorial experimental protocols for Erbicin-derived immunoagents and herceptin. *Br. J. Cancer* 97, 1354–1360.

113. Gelardi T, Damiano V, Rosa R, Bianco R, Cozzolino R, Tortora G, et al. (2010) Two novel human anti-ErbB2 immunoagents are active on trastuzumab-resistant tumors. *Br. J. Cancer* 102, 513–519.

114. De Lorenzo C, Fedele C, Malara AE, Troise F, Riccio G, Laccetti P, et al. Novel human anti-ErbB2 immunoagents. Proceedings of the 8th RNase Congress; 2010 October 20–22; Naples, Italy. 25.

115. Braschoss S, Hirsch B, Dübel S, Stein H, Dürkop H. (2007) New anti-CD30 human pancreatic ribonuclease-based immunotoxin reveals strong and specific cytotoxicity *in vivo*. *Leuk. Lymphoma* 48, 1179–1186.

116. Menzel C, Schirrmann T, Konthur Z, Jostock T, Dübel S. (2008) Human antibody RNase fusion protein targeting $CD30^+$ lymphomas. *Blood* 111, 3830–3837.

117. Chen S-I, Le S-Y, Newton DL, Maizel JV, Rybak SM. (2000) A gender specific mRNA encoding a cytotoxic ribonuclease contains a 3′ UTR of unusual length and structure. *Nucleic Acids Res.* 28, 2375–2382.

118. Singh UP, Ardelt W, Saxena SK, Holloway DE, Vidunas E, Lee H-S, et al. (2007) Enzymatic and structural characterisation of amphinase, a novel cytotoxic ribonuclease from *Rana pipiens* oocytes. *J. Mol. Biol.* 371, 93–111.

119. Mosimann SC, Ardelt W, James MNG. (1994) Refined 1.7⊠ X-ray crystallographic structure of P-30 protein, an amphibian ribonuclease with anti-tumor activity. *J. Mol. Biol.* 236, 1141–1153.

120. Boix E, Wu Y-N, Vasandani VM, Saxena SK, Ardelt W, Ladner J, et al. (1996) Role of the N-terminus in RNase A homologues: differences in catalytic activity, ribonuclease inhibitor interaction and cytotoxicity. *J. Mol. Biol.* 257, 992–1007.

121. Notomista E, Cafaro V, Fusiello R, Bracale A, D'Alessio G, Di Donato A. (1999) Effective expression and purification of recombinant Onconase®, an antitumor protein. *FEBS Lett.* 463, 211–215.

122. Liao YD, Wang SC, Leu YJ, Wang CF, Chang ST, Hong YT, et al. (2003) The structural integrity exerted by N-terminal pyroglutamate is crucial for the cytotoxicity of frog ribonuclease from *Rana pipiens*. *Nucleic Acids Res.* 31, 5247–5255.

123. Hursey M, Newton DL, Hansen HJ, Ruby D, Goldenberg DM, Rybak SM. (2002) Specifically targeting the CD22 receptor of human B-cell lymphomas with RNA damaging agents: a new generation of therapeutics. *Leuk. Lymphoma* 43, 953–959.

124. Taylor-Papadimitriou J, Burchell J, Miles DW, Dalziel M. (1999) MUC1 and cancer. *Biochim. Biophys. Acta* 1455, 301–313.

125. Brossart P, Schneider A, Dill P, Schammann T, Grünebach F, Wirths S, et al. (2001) The epithelial tumor antigen MUC1 is expressed in hematological malignancies and is recognized by MUC1-specific cytotoxic T-lymphocytes. *Cancer Res.* 61, 6846–6850.

126. Beatson RE, Taylor-Papadimitriou J, Burchell JM. (2010) MUC1 immunotherapy. *Immunotherapy* 2, 305–327.

127. Krauss J, Arndt MA, Zhu Z, Newton DL, Vu BK, Choudhry V, et al. (2004) Impact of antibody framework residue VH-71 on the stability of a humanised anti-MUC1 scFv and derived immunoenzyme. *Br. J. Cancer* 90, 1863–1870.

128. Krauss J, Arndt MA, Martin AC, Liu H, Rybak SM. (2003) Specificity grafting of human antibody frameworks selected from a phage display library: generation of a highly stable humanized anti-CD22 single-chain Fv fragment. *Protein Eng.* 16, 753–759.

129. Krauss J, Arndt MA, Vu BK, Newton DL, Seeber S, Rybak SM. (2005) Efficient killing of CD22+ tumor cells by a humanized diabody-RNase fusion protein. *Biochem. Biophys. Res. Commun.* 331, 595–602.

130. Holliger P, Prospero T, Winter G. (1993) "Diabodies": small bivalent and bispecific antibody fragments. *Proc. Natl. Acad. Sci. U S A* 90, 6444–6448.

131. Stein R, Qu Z, Cardillo TM, Chen S, Rosario A, Horak ID, et al. (2004) Antiproliferative activity of a humanized anti-CD74 monoclonal antibody, hLL1, on B-cell malignancies. *Blood* 104, 3705–3711.

132. Burton JD, Ely S, Reddy PK, Stein R, Gold DV, Cardillo TM, et al. (2004) CD74 is expressed by multiple myeloma and is a promising target for therapy. *Clin. Cancer Res.* 10, 6606–6611.

133. Chang CH, Sapra P, Vanama SS, Hansen HJ, Horak ID, Goldenberg DM. (2005) Effective therapy of human lymphoma xenografts with a novel recombinant ribonuclease-/anti-CD74 humanized IgG4 antibody immunotoxin. *Blood* 106, 4308–4314.

134. Aalberse RC, Schuurman J. (2002) IgG4 breaking the rules. *Immunology* 105, 9–19.

135. Chang CH, Gupta P, Michel R, Loo M, Wang Y, Cardillo TM, et al. (2010) Ranpirnase (frog RNase) targeted with a humanized, internalizing, anti-Trop-2 antibody has potent cytotoxicity against diverse epithelial cancer cells. *Mol. Cancer Ther.* 9, 2276–2286.

136. Ripani E, Sacchetti A, Corda D, Alberti S. (1998) Human Trop-2 is a tumor-associated calcium signal transducer. *Int. J. Cancer* 76, 671–676.

137. Stein R, Basu A, Chen S, Shih LB, Goldenberg DM. (1993) Specificity and properties of MAb RS7-3G11 and the antigen defined by this pancarcinoma monoclonal antibody. *Int. J. Cancer* 55, 938–946.

138. Qu Z, Griffiths GL, Wegener WA, Chang CH, Govindan SV, Horak ID, et al. (2005) Development of humanized antibodies as cancer therapeutics. *Methods* 36, 84–95.

139. Arndt M, Mavratzas A, DiRosa ML, Kontermann R, Rybak SM, Shogen K, et al. (2010) Onconase®-based immunoRNases. Proceedings of the 8th RNase Congress; 2010 October 20–22; Naples, Italy. 24.

140. Nishimura S, Nomura M, (1959) Ribonuclease of *Bacillus subtilis*. *J. Biochem. (Tokyo)* 46, 161–167.

141. Bycroft M, Ludvigsen S, Fersht AR, Poulsen FM. (1991) Determination of the three-dimensional solution structure of barnase using nuclear magnetic resonance spectroscopy. *Biochemistry* 130, 8697–8701.

142. Martínez JC, el Harrous M, Filimonov VV, Mateo PL, Fersht AR. (1994) A calorimetric study of the thermal stability of barnase and its interaction with 3′GMP. *Biochemistry* 33, 3919–3926.

143. Martin C, Richard V, Salem M, Hartley R, Mauguen Y. (1999) Refinement and structural analysis of barnase at 1.5 A resolution. *Acta Crystallogr. D Biol. Crystallogr.* 55, 386–398.

144. Hartley RW. (1997) Barnase and Barstar In: D'Alessio G, Riordan JF. (Eds.), *Ribonucleases: Structures and Functions*. Academic Press, San Diego, CA, pp. 51–100.

145. Hartley RW. (2001) Barnase-barstar interaction. *Methods Enzymol.* 341, 599–611.

146. Deyev SM, Waibel R, Lebedenko EN, Schubiger AP, Plückthun A. (2003) Design of multivalent complexes using the barnase∗barstar module. *Nat. Biotechnol.* 21, 1486–1492.

147. Eigenbrot C, Randal M, Presta L, Carter P. Kossiakoff AA. (1993) X-ray crystal structures of the antigen-binding domains from three variants of humanized antip185HER2 antibody 4D5 and comparison with molecular modeling. *J. Mol. Biol.* 229, 969–995.

148. Deyev SM, Yazynin SA, Kuznetsov DA, Jukovich M, Hartley RW. (1998) Ribonuclease-charged vector for facile direct cloning with positive selection. *Mol. Gen.* 259, 379–382.

149. Prior TI, Kunwar S, Pastan I. (1996) Studies on the activity of barnase toxins *in vitro* and *in vivo*. *Bioconjug. Chem.* 723–29.

150. Glinka EM, Edelweiss EF, Sapozhnikov AM, Deyev SM. (2006) A new vector for controllable expression of an anti-HER2/neu mini-antibody-barnase fusion protein in HEK 293T cells. *Gene* 366, 97–103.

151. Martsev SP, Tsybovsky YI, Stremovskiy OA, Odintsov SG, Balandin TG, Arosio P, et al. (2004) Fusion of the antiferritin antibody VL domain to barnase results in enhanced solubility and altered pH stability. *Protein Eng. Des. Sel.* 17, 85–93.

152. Edelweiss E, Balandin TG, Ivanova JL, Lutsenko GV, Leonova OG, Popenko VI, et al. (2008) Barnase as a new therapeutic agent triggering apoptosis in human cancer cells. *PLoS ONE* 3, e2434.

153. Balandin TG, Edelweiss E, Andronova NV, Treshalina EM, Sapozhnikov AM, Deyev SM. (2009) Antitumor activity and toxicity of anti-HER2 immunoRNase scFv 4D5-dibarnase in mice bearing human breast cancer xenografts. *Invest. New Drugs* 29, 22–32.

154. Yokota T, Milenic DE, Whitlow M, Schlom J. (1992) Rapid tumor penetration of a single-chain Fv and comparison with other immunoglobulin forms. *Cancer Res.* 52, 3402–3408.

155. Balandin TG, Edelweiss E, Andronowa NV, Treshalina EM, Sapozhnikov AM, Deyev SM. (2011) Antitumor activity and toxicity of anti-HER2 immunoRNase scFv 4D5-dibarnase in mice bearing human breast cancer xenografts. *Invest. New Drugs* 29, 22–32.

156. Colcher D, Bird R, Roselli M, Hardman KD, Johnson S, Pope S, et al. (1990) In vivo tumor targeting of a recombinant single-chain antigen-binding protein. *J. Natl. Cancer Inst.* 82, 1191–1197.

157. Milenic DE, Yokota T, Filpula DR, Finkelman MA, Dodd SW, Wood JF, et al. (1991) Construction, binding properties, metabolism, and tumor targeting of a single-chain Fv derived from the pancarcinoma monoclonal antibody CC49. *Cancer Res.* 51, 6363–6371.

158. Adams GP, McCartney JE, Tai MS, Oppermann H, Huston JS, Stafford WF., 3rd, et al. (1993) Highly specific *in vivo* tumor targeting by monovalent and divalent forms of 741F8 anti-c-erbB-2 single-chain Fv. *Cancer Res.* 53, 4026–4034.

159. Adams GP. (1998) Improving the tumor specificity and retention of antibody-based molecules. *In Vivo* 12, 11–21.

160. Colcher D, Pavlinkova G, Beresford G, Booth BJ, Choudhury A, Batra SK. (1998) Pharmacokinetics and biodistribution of genetically-engineer antibodies. *Q. J. Nucl. Med.* 42, 225–241.

161. Nielsen UB, Adams GP, Weiner LM, Marks JD. (2000) Targeting of bivalent anti-ErbB2 diabody antibody fragments to tumor cells is independent of the intrinsic antibody affinity. *Cancer Res.* 60, 6434–6440.

162. Todorovska A, Roovers RC, Dolezal O, Kortt AA, Hoogenboom HR, Hudson PJ. (2001) Design and application of diabodies, triabodies and tetrabodies for cancer targeting. *J. Immunol. Methods* 248, 47–66.

163. Hudson PJ, Souriau C. (2003) Engineered antibodies. *Nat. Med.* 9, 129–134.

164. Hofsteenge J. (1997) Ribonuclease Inhibitor. In: D'Alessio G, Riordan JF. (Eds.), *Ribonucleases: Structures and Functions*. Academic Press, San Diego, CA, pp. 621–658.

165. Shapiro R. (2001) Cytoplasmic ribonuclease inhibitor. *Methods Enzymol.* 341, 611–628.

166. Dickson KA, Haigis MC, Raines RT. (2005) Ribonuclease inhibitor: structure and function. *Prog. Nucleic Acid Res. Mol. Biol.* 80, 349–374.

167. Erickson HA, Jund MD, Pennell CA. (2006) Cytotoxicity of human RNase-based immunotoxins requires cytosolic access and resistance to ribonuclease inhibition. *Protein Eng. Des. Sel.* 19, 37–45.

168. Rybak SM, Newton DL, Mikulski SM, Viera A, Youle RJ. (1993) Cytotoxic Onconase® and ribonuclease A chimeras: comparison and *in vitro* characterisation. *Drug Deliv.* 1, 3–10.

169. Newton DL, Pearson JW, Xue Y, Smith MR, Fogler WE, Mikulski SM, et al. (1996) Anti-tumor ribonuclease, combined with or conjugated to monoclonal antibody MRK16, overcomes multidrug resistance to vincrinstine *in vitro* and *in vivo*. *Int. J. Oncol.* 8, 1095–1104.

170. Rybak SM. (2008) Antibody-Onconase® conjugates: cytotoxicity and intracellular routing. *Curr. Pharm. Biotechnol.* 9, 226–230.

171. Arai R, Ueda H, Kitayama A, Kamiya N, Nagamune T. (2001) Design of the linkers which effectively separate domains of a bifunctional fusion protein. *Protein Eng.* 14, 529–532.

172. Keller J, Heisler I, Tauber R, Fuchs H. (2001) Development of a novel molecular adapter for the optimization of immunotoxins. *Control Release* 74, 259–261.

173. Chen X, Bai Y, Zaro JL, Shen WC. (2010) Design of an *in vivo* cleavable disulfide linker in recombinant fusion proteins. *Biotechniques* 49, 513–518.

174. Buchner J, Pastan I, Brinkmann U. (1992) A method for increasing the yield of properly folded recombinant fusion proteins: single-chain immunotoxins from renaturation of bacterial inclusion bodies. *Anal. Biochem.* 205, 263–270.

175. Schirmann T, Wezler X, Dübel S. Human antibody RNase fusion proteins targeting CD30+ lymphoma. Proceedings of the 8th RNase Congress; 2010 October 20–22; Naples, Italy. 23.

176. Kreitman RJ. (2009) Recombinant immunotoxins for the treatment of chemoresistant hematologic malignancies. *Curr. Pharm. Des.* 15, 2652–2664.

177. Hetzel C, Bachran C, Tur MK, Fuchs H, Stöcker M. (2009) Improved immunotoxins with novel functional elements. *Curr. Pharm. Des.* 15, 2700–2711.

178. Rybak SM, Newton DL. (2007) Immunotoxins and beyond: targeted RNases. In: Dübel S. (Ed), *Handbook of Therapeutic Antibodies*. Wiley-VCH, Weinheim, pp. 379–410.

179. Newton DL, Stockwin LH, Rybak SM. (2009) Anti-CD22 Onconase®: preparation and characterization. *Methods Biol.* 525, 425–443.

180. Saxena SK. inventor; Alfacell Corporation, assignee. US patent 6,175,003B1.(2001) January 16.

181. Marks JD, Hoogenboom HR, Bonnert TP, McCafferty J, Griffiths AD, Winter G. (1991) By-passing immunization. Human antibodies from V-gene libraries displayed on phage. *J. Mol. Biol.* 222, 581–597.

182. Vaughan TJ, Williams AJ, Pritchard K, Osbourn JK, Pope AR, Earnshaw JC, et al. (1996) Human antibodies with subnanomolar affinities isolated from a large non-immunized phage display library. *Nat. Biotechnol.* 14, 309–314.

183. de Haard HJ, van Neer N, Reurs A, Hufton SE, Roovers RC, Henderikx P, et al. (1999) A large non-immunized human Fab fragment phage library that permits rapid isolation and kinetic analysis of high affinity antibodies. *J. Biol. Chem.* 274, 18218–18230.

184. Poul MA, Becerril B, Nielsen UB, Morisson P, Marks JD. (2000) Selection of tumor-specific internalizing human antibodies from phage libraries. *J. Mol. Biol.* 301, 1149–1161.

185. O'Connell D, Becerril B, Roy-Burman A, Daws M, Marks JD. (2002) Phage versus phagemid libraries for generation of human monoclonal antibodies. *J. Mol. Biol.* 321, 49–56.

23

ANTIBODY-DIRECTED ENZYME PRODRUG THERAPY (ADEPT)

SURINDER K. SHARMA

Research Department of Oncology, University College London, London, UK

23.1 INTRODUCTION

Antibody–enzyme conjugates were first reported in the early 1970s [1] but it was not until 1987 that it was suggested [2] that these could be used in combination with prodrugs. This led to the development of what is known as antibody-directed enzyme prodrug therapy or ADEPT. ADEPT (Figure 23.1) is a selective form of chemotherapy where an antibody directed at a tumor-associated antigen is used to deliver a unique enzyme to tumor sites [3,4]. After localization of enzyme in the tumor and clearance from blood, a nontoxic prodrug is given with the expectation that it would be converted, by the enzyme, to a potent cytotoxic drug at tumor sites. The enzyme provides an amplification effect as one enzyme can convert many prodrug molecules generating higher concentration of the drug within tumors. The drug is generated extracellularly and being a small molecule can diffuse throughout the tumor killing antigen positive as well as the antigen negative cell, thus providing a by-stander effect [5]. Many preclinical studies [reviewed in References 6–8] have confirmed feasibility of this approach but only one system has progressed to clinical trials. Although preclinical studies provide valuable information, it is only through clinical studies that the real challenges can be identified. These need to be addressed when developing new ADEPT systems for cancer.

The aim of ADEPT is to restrict the cytotoxic action of the drug to tumors only. Although the idea may appear simple, its successful application requires multidisciplinary approach and an understanding of all the parameters involved.

23.2 THE COMPONENTS

23.2.1 The Target

Ideally, the target molecule should be unique to the surface membrane of cancer cells. Such targets are available for only certain types of cancer at present. In so far as a potential target has some normal tissue expression, it should not be on a vital or irreplaceable organ. It is also desirable that the targeted molecule should not internalize the antibody–enzyme conjugate since this would obviously result in inactivation of the enzyme, whereas one enzyme molecule remaining on the cell surface may convert many prodrug molecules into cytotoxic agents. The target molecule should be expressed on a high proportion of the cancer cells but as is well-known marker antigens are not invariably expressed on all the cells of epithelial cancers. It may be noted that one of the virtues of the ADEPT approach is that by generating small toxic drugs unmarked cells are as vulnerable as those expressing the marker antigen. It may also be supposed that secretion of the target antigen into extracellular space and into the blood would prevent an antibody–enzyme conjugate from localizing at tumor sites. However, this does not appear

Fusion Protein Technologies for Biopharmaceuticals: Applications and Challenges, First Edition. Edited by Stefan R. Schmidt.

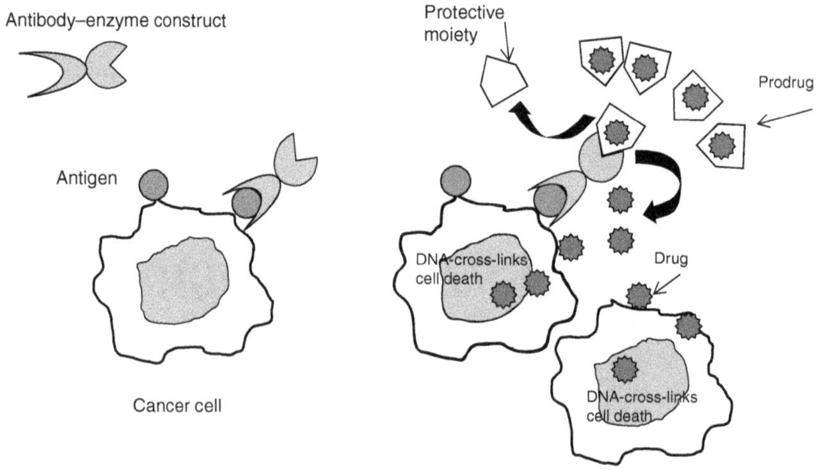

FIGURE 23.1 Antibody-directed enzyme prodrug therapy (ADEPT); the basic principle. First the antibody-enzyme construct binds to the antigen. Then the prodrug is added. The enzyme converts the prodrug into a cytotoxic drug by removing the protective moiety. The active drug enters the cell and causes cell death.

to be the case as it was shown that antibody–enzyme conjugates localized in human chorionic gonadotrophin (hCG) producing tumors even in the presence of high concentrations of hCG in blood [9].

Our studies have used carcino embryonic antigen (CEA) as the target molecule. CEA is virtually universally expressed by colorectal carcinomas and perhaps less regularly by several other epithelial cancers. It also has the advantage that its site of expression in normal intestinal tissue is not accessible to intravenously administered anti-CEA antibodies.

23.2.2 The Antibody

The most important characteristics of the antibody component are specificity and the ability to retain the enzyme at tumor sites for prolonged periods of up to several days. This would allow prodrug to be administered over longer periods. Antibody or the antibody–enzyme conjugate will accumulate in tumors when the concentration in blood exceeds that in the tumor. When the blood concentration falls below that of tumor, there may be loss of conjugate from the tumor. Thus, the binding affinity of the antibody is important for the retention of antibody–enzyme conjugate in the tumor such that a situation is attained where there is an effective concentration of enzyme at tumor sites and an absence of enzyme in blood.

Our early studies were influenced by experience with radioimmunolocalization of cancers using radiolabeled antibodies, in which it was found that $F(ab')_2$ fragments were preferred because these cleared from blood more rapidly resulting in a better ratio of tumor to background than intact IgG [10]. This resulted in our using $F(ab')_2$ fragments conjugated to enzyme in early ADEPT studies in preference to intact antibodies that might have proved to be better delivery agents. Antibody fragments tend to be cleared from blood more rapidly than intact antibodies but binding affinity and, maybe, other factors come into the equation. However, prolonged retention of enzyme in tumor remains an important objective whatever antibody format is employed.

The argument that antibody fragments will penetrate tumors more effectively does not carry much weight in ADEPT partly because of the additional size of the enzyme and also because tumor penetration is dependent on the small drug molecules that are generated.

Most of the antibodies used in clinical studies reported thus far have been murine in origin although this should not remain a problem in the future.

23.2.3 The Enzyme

The use of native human enzymes in ADEPT is excluded by their normal tissue distribution. Nonhuman mammalian enzymes are excluded by their analogous human forms and by their inherent immunogenicity. This restricts the choice to a limited repertoire of bacterial enzymes that have the advantage of high efficiency and the disadvantage of potent immunogenicity that could limit their repeated use. The issue of immunogenicity is discussed later.

The possibility of using mutated human enzymes such that the prodrug is activated only by the mutated enzyme and not by the native form is a challenge that awaits completion.

The enzymes used in ADEPT systems have included:

Mammalian/human enzymes which included placental alkaline phosphatise [4], human β-glucuronidase [11], human carboxypeptidase A [12], human carboxylesterase [13], and a heat-stabilized human propyl endopeptidase [14].

Nonmammalian enzymes include the widely used β-lactamase [15], cytosine deaminase [16], and carboxypeptidase G2 (CPG2) [3].

23.2.4 Abzymes or Catalytic Antibodies

It was proposed early on in ADEPT [17] that catalytic antibodies could be used to convert prodrugs to drugs with the main advantage that these could be humanized thus avoiding immunogenicity issues. Although early studies met with limited success [18], recent progress shows more promise [19,20].

23.2.5 Prodrugs

The fundamental requirement for the prodrug is that it should be a substrate for the selected enzyme only. Other critical factors include that the prodrug itself should be nontoxic but should generate a drug that would be highly toxic but with a short half-life. A series of prodrugs synthesized by Tietze et al. [21] are 4000–6000-fold less toxic than the generated drugs although there is no indication of the half-life of these drugs.

The potential advantage of ADEPT is to deliver a much higher concentration of drug to cancer sites than can be delivered by conventional chemotherapy. It follows that the cytotoxic action of drug should be concentration dependent that is the more that can be delivered, the more cell killing results. This led to the use of alkylating agents, the cytotoxicity of which appears to be dose dependent over a wide range of concentrations [22]. It was pointed out in the early publications that if a high concentration of drug is generated at cancer sites then some of it will leak back into the blood and access normal cell renewal systems. To minimize toxicity, it is necessary for the generated drug to have a very short half-life such that it has enough time to diffuse and kill tumor cells but decays by the time it enters the blood. This factor has been ignored in many ADEPT studies as prodrugs have been made from already licensed drugs with considerable half-lives [23,24].

23.3 ADEPT SYSTEMS WITH CARBOXYPEPTIDASE G2 (CPG2)

23.3.1 System 1: Antibody–Enzyme Conjugates

Antibody–enzyme conjugates were initially constructed using bifunctional agents, where CPG2 was chemically conjugated to $F(ab')_2$ fragments of monoclonal antibodies directed at hCG and CEA. CPG2 cleaves the terminal glutamate from folate-type molecules. A series of benzoic acid mustard drugs, which could be inactivated by a terminal glutamate, were synthesized and a 4-((chloroethyl)(2-methoxy))ethyl)amino benzoyl-L-glutamic acid commonly known as CMDA, was selected for further study [25].

23.3.1.1 Preclinical Studies The first ADEPT system was tested in a drug-resistant choriocarcinoma xenograft model known as CC3 established in nude mice. CC3 expresses hCG in the tumor. hCG is also secreted in blood of mice bearing CC tumors with typical values of >40 IU/mL at the time of anti-hCG-CPG2 conjugate administration resulting in the clearance of enzyme activity from blood such that the prodrug could be given safely between 56 and 72 h after the conjugate injection. This regimen resulted in the eradication of established xenografts [26].

The same protocol was applied to a study in the nude mice bearing the human colon carcinoma xenograft, LS174T. An anti-CEA antibody-CPG2 conjugate was injected followed by the prodrug at 56–72 h. This resulted in toxicity with no effect on the tumor growth. It was concluded that the difference between the two models was that the high hCG level in the blood in the CC3 case resulted in accelerated clearance of the enzyme from blood because of immune complex formation. However, with the LS174T xenograft, although CEA was expressed in the tumor, it is not detectable in circulation. This would result in higher residual enzyme activity in blood when the prodrug was given causing toxicity. When prodrug injection was delayed by 1 week after antibody–enzyme conjugate was given, there was no toxicity and no growth delay [17]. These studies confirmed the need for a clearance system to remove enzyme activity from blood.

23.3.1.2 Removal of Enzyme Activity from Blood It is clearly essential for this residual conjugate to be removed from blood before giving prodrug. It is also essential that the clearing agent should not access tumor sites and inactivate enzyme there. Monoclonal antibodies to CPG2 had been produced in anticipation applying accelerated clearance of the enzyme. One of the monoclonal anti-CPG2 antibodies (SB43) inactivated the enzyme *in vitro* and rapidly reduced enzyme activity in blood (Figure 23.2) [27]. It was necessary to ensure that this enzyme-inactivating antibody cleared quickly from blood to prevent its penetration of tumor sites. This was achieved by glycosylation of the antibody. Administration of the galactosylated antibody SB43gal by intravenous infusion reduced enzyme to undetectable levels within hours without affecting the tumor enzyme levels. When the LS174T studies were repeated using the three-phase ADEPT system comprising administration of the antibody–enzyme conjugate followed by SB43gal at 19 h, the prodrug could be given safely at 24 h, resulting in a growth delay of the tumor. Similar results were obtained in a drug resistant, CEA expressing ovarian carcinoma model [28,29].

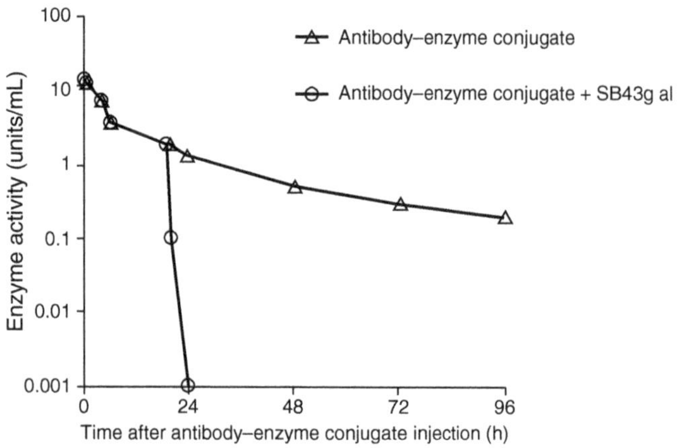

FIGURE 23.2 Enzyme inactivation/clearance by SB43gal.

23.3.1.3 First in Man ADEPT Study Following successful ADEPT studies in various xenograft models, it was important to test the system in man to assess its feasibility. This obviously required a scale-up production of antibody–enzyme conjugate [30] and all other components. The anti-CEA antibody, A5B7 had already been used in humans in the context of radioimmunolocalization of CEA-positive tumors. The enzyme, CPG2 had also been used to treat methotrexate toxicity in patients. The CMDA prodrug had not been used in man before. Therefore, a dose-escalation study of the prodrug alone was carried out to assess its safety and to determine the maximum tolerated dose that could be used in the three-phase ADEPT system. The dose range of 0.2–2.4 g/m^2 was studied. The CMDA was well tolerated in patients with a minimal evidence of toxicity. This was probably because the CMDA was only soluble in dimethyl sulfoxide (DMSO).

In the main part of the ADEPT study, 18 patients aged 28–75 years were studied [31,32]. All had advanced carcinoma of colon, rectum, or appendix with multisite metastatic disease that expressed CEA on histological examination, had undergone initial surgery and chemotherapy but now had progressive drug-resistant disease. Expected survival times were <2 months.

As this was the first ever clinical study of ADEPT, multiple variables in terms of the components dose and interval had to be explored. The dose of antibody–enzyme conjugate [A5B7-F(ab$'$)$_2$-CPG2 commonly referred to as A5CP] ranged between 20,000 and 30,000 units/m^2. The dose range for SB43gal was 90–240 mg/m^2. The total dose of the CMDA prodrug given ranged between 1.4 and 10 g/m^2.

Several variations in the time interval between AEC and SB43 and the duration of SB43 were explored. The CMDA was initially given as a bolus but later as an infusion. A small component of the A5CP conjugate was radiolabeled with 131-iodine to allow SPECT imaging. In an attempt to delay the immune response to the murine antibody and the bacterial enzyme, six patients received cyclosporine that allowed up to three cycles of ADEPT to be given over 3 weeks.

23.3.1.4 Results of the First in Man ADEPT Study Eight patients received the highest total doses of the prodrug (>3.0 g/m^2) and four of these achieved partial remission (>50% reduction in volume of all visible tumors). The fifth patient had a mixed response in which all but one metastasis shrank by more than 50%. Seven patients survived 6 months or more including 3 who died 18, 25, and 36 months. These were remarkable results in patients with advanced and resistant cancer. However, it was surprising to find that most patients had grade 3 or 4 reversible myelotoxicity [33]. Subsequent analysis of blood samples from these patients showed the presence of prodrug and drug in plasma, even though there was no detectable circulating enzyme [34]. The half-life of the generated drug was found to be about 30 min. It was speculated that the drug generated at cancer sites had leaked back into circulation and because of its long half-life, caused myelosuppression.

23.3.1.5 Conclusions from the First in Man ADEPT Study Conclusions from the first in man ADEPT study are as follows:

- The A5CP conjugate and SB43gal were well tolerated.
- SB43gal was effective in eliminating the enzyme from blood.
- The amount of SB43gal given needed to be proportionate to the A5CP dose for the first few hours after administration but could then be substantially reduced such that a low dose infusion would inactivate any enzyme leaking back from tumor sites.
- Cyclosporine delayed immune response to the enzyme but added to toxicity [35].
- The prodrug, CMDA, was nontoxic but was soluble only in DMSO, which was a substantial disadvantage.
- The generated drug had a long half-life, which was a probable cause of myelotoxicity.
- It was concluded that a prodrug/drug system in which the drug had a very short half-life (<1 min) was required.

Given the limitations of the reagents available, the results were most encouraging.

23.3.2 The Second Clinical Study of ADEPT

The second three-phase ADEPT clinical study included 10 patients with unresectable or metastatic but CEA-positive colorectal carcinoma [36]. The study used the same reagents as before but with lower dose of the conjugate (10,000 units/m^2). The CMDA was given as a bolus to allow prodrug and drug pharmacokinetics to be carried out. Again, a small proportion of the conjugate was radiolabeled to allow γ-camera images as well as to estimate enzyme concentrations in the tumor over time. Quantitative SPECT γ-camera images confirmed localization of the conjugate in tumors. SB43gal effectively removed enzyme from blood. In addition, tumor biopsies obtained after completion of SB43gal and at the time of prodrug administration showed median enzyme activity of 0.47 units per gram tumor (range 0.32–0.62), whereas normal liver obtained in biopsies from two patients showed no detectable enzyme activity. These findings confirmed the selectivity of the three-phase system in that SB43gal eliminated enzyme from blood without affecting tumor enzyme such that tumor to plasma enzyme ratio of >10,000:1 was observed at the time of prodrug administration. Efficacy was demonstrated in some patients. One patient with drug-resistant tumor growing through the abdominal wall at a colostomy site had complete resolution of that lesion and a partial response lasing 4 months. Six patients had stable disease.

Pharmacokinetic studies demonstrated that the generated drug had a median half-life of 46 min (range 5–27 min) [37], which was sufficient to allow leak back from tumors causing myelosuppression.

23.3.2.1 Conclusions from the Second Clinical Study of ADEPT Conclusions from the second clinical study of ADEPT are as follows:

- SB43gal eliminated enzyme from blood without affecting concentration of tumor-located enzyme.
- Tumor to plasma ratios of enzyme activity were >10,000:1 [36].
- Tumor to liver ratios in two patients biopsied had ratio >10,000:1.
- Enzyme measurements in tumor biopsies correlated well with the γ-camera estimates from radioactivity.
- Drug half-life in plasma was longer than the prodrug half-life [37] resulting in diffusion of drug into the circulation.
- There was evidence of antitumor response in patients after only one ADEPT cycle.

These studies confirmed the need for a drug with a short half-life and a program of prodrug development was initiated. Several new prodrugs were synthesized and a new prodrug, 4-[*N*,*N*-bis(2-iodoethyl)amino]phenoxycarbonyl L-glutamic acid (short name BIP prodrug) was identified for potential clinical studies [38]. BIP prodrug generated a drug that was highly potent and with a shorter chemical and biological half-life than the previous benzoic acid mustard drug. Preclinical studies in the xenograft systems with the conjugate followed by and the new BIP prodrug at 72 h after the conjugate demonstrated efficacy in a two-component system [39]. Based on these studies, a clinical trial of two-phase system where the enzyme-inactivating antibody SB43gal was omitted was initiated.

23.3.3 The Third Clinical Study of ADEPT

The trial was conducted to study the localization and clearance of the monoclonal anti-CEA antibody A5B7-F(ab′)$_2$ fragment conjugated to CPG2 without the use of the clearing antibody, SB43gal. The study also investigated safety and tolerability of the new BIP prodrug in a two-phase system [40].

As no clearing agent was applied, it was decided that the prodrug would be given when the enzyme activity in serum was below a predefined value. Twenty-seven patients were entered in this study. They were given a fixed dose of A5B7-F(ab′)$_2$-CPG2 conjugate, similar to the one used in the previous clinical trials, followed by escalating dose of the BIP prodrug when serum enzyme levels were below 0.05 units/mL. The time at which prodrug was administered to patients ranged between 2 and 9 days. The maximum tolerated dose of prodrug was reached at 15.5 mg/m^2 when serum enzyme level was <0.05 units/mL. The BIP prodrug at the MTD level showed elimination half-life of 14.5 min. The half-life of the drug was too short to be measured.

23.3.3.1 Conclusions from the Third Clinical Study of ADEPT Conclusions from the third clinical study of ADEPT are as follows:

- The tumor localization of the conjugate was poor.
- The tumor to serum enzyme ratios were <1.
- There were no objective responses.
- However, the study demonstrated that the new BIP prodrug was safe in patients and provided valuable information for future use.

23.4 FUSION PROTEINS

Our ADEPT studies indicated that it is desirable to use a clearance system with the antibody–enzyme conjugates or that a better targeting molecule needs to be developed. Advances in the recombinant technology made it possible to construct antibody-enzyme fusion proteins [41]. One of the earliest recombinant fusion protein used in ADEPT was an anti-CEA Fab fused to human β-glucuronidase [42]. Recombinant fusion proteins have advantages in terms of reproducibility of the product as well as stability. We have utilized a recombinant fusion protein comprising an anti-CEA scFv antibody derived from an immunized mouse phage library fused to CPG2 [43]. The construct was first expressed in *Escherichia coli*, which gave good tumor localization *in vivo* but the yield of the product was low [44]. However, expression in the yeast *Pichia pastoris* gave higher yields but the fusion protein was glycosylated. Preclinical studies showed it to localize in CEA-expressing tumors and a rapid clearance from blood and other normal tissues. This mannosylated fusion protein was termed MFECP. Studies with MFECP showed that prodrug could be given safely at 6 h after the fusion protein resulting in a growth delay of the tumors. Furthermore, ADEPT cycles could be repeated without toxicity to give sustained regressions in the xenograft models [45].

MFECP was produced for clinical studies under the UK compliance for an investigational medicinal product [46] and termed MFECP1.

23.4.1 The Fourth Clinical Study of ADEPT

This is the first ADEPT clinical study using the recombinant antibody-enzyme fusion protein MFECP1. The purpose of this study was to investigate the safety and performance of the fusion protein in combination with the BIP prodrug and to assess if the design criteria for selective ADEPT were met. A total of 31 patients were entered in the study. All had advanced locally recurrent or metastatic colorectal or other CEA expressing cancer, which was nonresponsive to standard chemotherapy [47].

MFECP1 was given at 5000 units/m^2 as a starting dose, diluted in 500 mL 0.9% sodium chloride and infused over 90–120 min via a Hickman line. The BIP prodrug was given as a bolus in three doses, over 2 h with a starting dose of 12.42 mg/m^2 $\times$ 3 as established in the previous clinical trial. The dose of the prodrug was escalated until maximum tolerated dose had reached. In addition, a small dose of MFECP1 was trace labeled with either 131-iodine or 123-iodine and given to some of the patients for single-photon emission computed tomography (SPECT) imaging to determine localization and clearance. Serum samples were taken to measure CPG2 activity.

MFECP1 was safe and well tolerated in all patients. γ-Camera imaging showed localization in tumors. A number of tumor biopsies were taken and selective localization of enzyme was confirmed in tumors using immunohistochemistry with anti-CPG2 antibody. Prodrug activation was confirmed in one tumor biopsy where DNA cross-links were observed by comet assay. No cross-links were present in lymphocytes of this patient. Prodrug dose was escalated up to 1.075 mg/m^2 $\times$ 3 where dose limiting toxicity was observed. The results of this study are published by Myers et al. [47].

Of the 31 patients entering the trial, 28 were evaluable for response at 8 weeks after treatment. Eleven of these had stable disease at 8 weeks but 17 progressed. A 10% reduction by CT was observed in one patient. The study provided design information for repeated ADEPT. A number of patients were given 2 or 3 ADEPT cycles over 2–10 days. Efficacy was demonstrated by ^{18}F FDG-PET and four of nine patients showed PR at total prodrug dose of $\geq$900 mg/m^2 and stable disease by RECIST criteria in 69% of these patients [48]. These results obtained in the patients, who were heavily treated with conventional chemotherapy are probably better than any other available treatment.

23.5 IMMUNOGENICITY

Immune response to the murine anti-CEA antibody conjugated to the bacterial enzyme was anticipated and confirmed in the clinical studies when all patients given the antibody–enzyme conjugate developed antibodies to both components after only a single administration of the conjugate [49]. However, in the first ADEPT study, cyclosporine was given to some patients, which allowed three ADEPT cycles to be given [35] within 3 weeks. However, cyclosporine added to toxicity and better immunosuppressive agents need to be explored. The immunogenicity of the antibody can be controlled by use of human antibody but the problem of CPG2 immunogenicity is more complex. In the case of CPG2, immunodominant epitopes were identified on the C-terminus and a hexa-His-tag was engineered in the fusion protein, MFECP1, to allow purification as well as hypothesize that the His-tag would mask the epitope [50]. This was confirmed in the clinical study where for a single ADEPT cycle, only 30% of the patients developed antibodies to CPG2 compared to 100% for the conjugate. In addition for CPG2, potentially immunogenic T-cell epitopes have been identified, which could be modified or eliminated using new technologies [51]. Interestingly, T-cell modifications were reported for β-lactamase enzyme [52] and promising results demonstrated in mouse and human T-cell proliferation assays but there are no reports of this being used in clinical studies as yet. It was also proposed in the early ADEPT work [17] that catalytic antibodies may be used for prodrug conversion with the advantage that these could be humanized. There were some attempts to generate a catalytic antibody for a nitrogen mustard carbamate prodrug [18] but with limited success. The field has progressed and recent reports show feasibility of this approach with the further promise of selecting catalytic human antibodies from combinatorial libraries [53,54].

Humanized ADEPT system comprising an engineered human purine nucleoside phosphorylase fused to an anti-HER-2/neu peptide mimetic has been reported to convert 2-fluoro-2′-deoxyadenosine prodrug to a cytotoxic agent [55,56]. More recently, Chen et al. [57] have reported promising therapeutic effects in xenograft models with an ADEPT system comprising a humanized immunoenzyme, hcc49-S2 in combination with a glucuronidase prodrug. The real potential of these systems can only be explored in clinical studies.

23.6 CONCLUSIONS AND FUTURE OUTLOOK

ADEPT has shown feasibility as a treatment for cancer and opportunities for developing it for diverse tumors and targets with many different enzyme and prodrug systems. It has the potential to be a most selective form of therapy but a number of critical points need to be addressed some of which tend to be ignored.

A successful ADEPT system requires the following:

- Sustained high concentrations of the enzyme in tumors
- Elimination of enzyme from blood and other normal tissue (may require enzyme inactivation/clearance step) without affecting enzyme concentration in the tumor
- Nontoxic prodrug
- Highly toxic drug with short half-life and nonrepairable tumor cell DNA damage
- Nonimmunogenic system components for repeated treatments to enhance efficacy.

These goals are achievable and we are currently developing an enhanced ADEPT for cancer, utilizing the original idea of a three-phase system [17].

ACKNOWLEDGMENTS

I am grateful to Professor KD Bagshawe for helpful advice and critical comments.

REFERENCES

1. Philpott GW, Bower RJ, Parker CW. (1973) Selective iodination and cytotoxicity of tumor cells with an antibody-enzyme conjugate. *Surgery* 74, 51–58.
2. Bagshawe KD. (1987) Antibody directed enzymes revive anticancer prodrugs concept. *Br. J. Cancer* 56, 531–532.
3. Bagshawe KD, Springer CJ, Searle F, Antoniw P, Sharma SK, Melton RG, et al. (1988) A cytotoxic agent can be generated selectively at cancer sites. *Br. J. Cancer* 58, 700–703.
4. Senter PD, Saulnier MG, Schreiber GJ, Hirschberg DL, Brown JP, Hellstrom I, et al. (1988) Anti-tumor effects of antibody-alkaline phosphatase conjugates in combination with etoposide phosphate. *Proc. Natl. Acad. Sci. USA* 85, 4842–4846.
5. Cheng TL, Wei SL, Chen BM, Chern JW, Wu MF, Liu PW, et al. (1999) Bystander killing of tumor cells by antibody-targeted enzymatic activation of a glucuronide prodrug. *Br. J. Cancer* 79, 1378–1385.
6. Bagshawe KD. (2006) Antibody-directed enzyme prodrug therapy (ADEPT) for cancer. *Expert Rev. Anticancer Ther.* 6, 1421–1431.
7. Senter PD, Springer CJ. (2001) Selective activation of anticancer prodrugs by monoclonal antibody-enzyme conjugates. *Adv. Drug Deliv. Rev.* 53, 247–264.
8. Schellmann N, Deckert PM, Bachran D, Fuchs H, Bachran C. Targeted enzyme prodrug therapies. *Mini Rev. Med. Chem.* 10, 887–904.
9. Begent RH, Bagshawe KD, Green AJ, Searle F. (1987) The clinical value of imaging with antibody to human chorionic gonadotrophin in the detection of residual choriocarcinoma. *Br. J. Cancer* 55, 657–660.
10. Lane DM, Eagle KF, Begent RH, Hope-Stone LD, Green AJ, Casey JL, et al. (1994) Radioimmunotherapy of metastatic colorectal tumors with iodine-131-labelled antibody to carcinoembryonic antigen: phase I/II study with comparative biodistribution of intact and F(ab′)2 antibodies. *Br. J. Cancer* 70, 521–525.
11. Bosslet K, Czech J, Lorenz P, Sedlacek HH, Schuermann M, Seemann G. (1992) Molecular and functional characterisation of a fusion protein suited for tumor specific prodrug activation. *Br. J. Cancer* 65, 234–238.
12. Haenseler E, Esswein A, Vitols KS, Montejano Y, Mueller BM, Reisfeld RA, et al. (1992) Activation of methotrexate-alpha-alanine by carboxypeptidase A-monoclonal antibody conjugate. *Biochemistry* 31, 891–897.
13. Senter PD, Beam KS, Mixan B, Wahl AF. (2001) Identification and activities of human carboxylesterases for the activation of CPT-11, a clinically approved anticancer drug. *Bioconjug. Chem.* 12, 1074–1080.
14. Heinis C, Alessi P, Neri D. (2004) Engineering a thermostable human prolyl endopeptidase for antibody-directed enzyme prodrug therapy. *Biochemistry* 43, 6293–6303.
15. Svensson HP, Kadow JF, Vrudhula VM, Wallace PM, Senter PD. (1992) Monoclonal antibody-beta-lactamase conjugates for the activation of a cephalosporin mustard prodrug. *Bioconjug. Chem.* 3, 176–181.
16. Wallace PM, MacMaster JF, Smith VF, Kerr DE, Senter PD, Cosand WL. (1994) Intratumoral generation of 5-fluorouracil mediated by an antibody-cytosine deaminase conjugate in combination with 5-fluorocytosine. *Cancer Res.* 54, 2719–2723.
17. Bagshawe KD. (1989) The First Bagshawe lecture. Towards generating cytotoxic agents at cancer sites. *Br. J. Cancer* 60, 275–281.
18. Wentworth P, Datta A, Blakey D, Boyle T, Partridge LJ, Blackburn GM. (1996) Toward antibody-directed "abzyme" prodrug therapy, ADAPT: carbamate prodrug activation by a catalytic antibody and its *in vitro* application to human tumor cell killing. *Proc. Natl. Acad. Sci. USA* 93, 799–803.
19. Kakinuma H, Fujii I, Nishi Y. (2002) Selective chemotherapeutic strategies using catalytic antibodies: a common pro-moiety for antibody-directed abzyme prodrug therapy. *J. Immunol. Methods* 269, 269–281.
20. Abraham S, Guo F, Li LS, Rader C, Liu C, Barbas CF, 3rd, et al. (2007) Synthesis of the next-generation therapeutic

antibodies that combine cell targeting and antibody-catalyzed prodrug activation. *Proc. Natl. Acad. Sci. USA* 104, 5584–5589.

21. Tietze LF, Schmuck K, Schuster HJ, Muller M, Schuberth, I. (2000) Synthesis and biological evaluation of prodrugs based on the natural antibiotic duocarmycin for use in ADEPT and PMT. *Chemistry* 17, 1922–1929.
22. Frei E, 3rd, Teicher BA, Holden SA, Cathcart KN, Wang YY. (1988) Preclinical studies and clinical correlation of the effect of alkylating dose. *Cancer Res.* 48, 6417–6423.
23. Kuefner U, Lohrmann U, Montejano YD, Vitols KS, Huennekens FM. (1989) Carboxypeptidase-mediated release of methotrexate from methotrexate alpha-peptides. *Biochemistry* 28, 2288–2297.
24. Kerr DE, Li Z, Siemers NO, Senter PD, Vrudhula VM. (1998) Development and activities of a new melphalan prodrug designed for tumor-selective activation. *Bioconjug. Chem.* 9, 255–259.
25. Springer CJ, Antoniw P, Bagshawe KD, Searle F, Bisset GM, Jarman M. (1990) Novel prodrugs which are activated to cytotoxic alkylating agents by carboxypeptidase G2. *J. Med. Chem.* 33, 677–681.
26. Springer CJ, Bagshawe KD, Sharma SK, Searle F, Boden JA, Antoniw P, et al. (1991) Ablation of human choriocarcinoma xenografts in nude mice by antibody-directed enzyme prodrug therapy (ADEPT) with three novel compounds. *Eur. J. Cancer* 27, 1361–1366.
27. Sharma SK, Bagshawe KD, Burke PJ, Boden RW, Rogers GT. (1990) Inactivation and clearance of an anti-CEA carboxypeptidase G2 conjugate in blood after localisation in a xenograft model. *Br. J. Cancer* 61, 659–662.
28. Sharma SK, Bagshawe KD, Springer CJ, Burke PJ, Rogers GT, Boden JA, et al. (1991) Antibody directed enzyme prodrug therapy (ADEPT): a three phase system. *Dis. Markers* 9, 225–231.
29. Sharma SK, Boden JA, Springer CJ, Burke PJ, Bagshawe KD. (1994) Antibody-directed enzyme prodrug therapy (ADEPT). A three-phase study in ovarian tumor xenografts. *Cell Biophys.* 24–25, 219–228.
30. Melton RG, Boyle JM, Rogers GT, Burke P, Bagshawe KD, Sherwood RF. (1993) Optimisation of small-scale coupling of A5B7 monoclonal antibody to carboxypeptidase G2. *J. Immunol. Methods* 158, 49–56.
31. Bagshawe KD, Sharma SK, Springer CJ, Antoniw P, Boden JA, Rogers GT, et al. (1991) Antibody directed enzyme prodrug therapy (ADEPT): clinical report. *Dis. Markers* 9, 233–238.
32. Bagshawe K.D, Begent R.H.J. (1996). First Clinical experience with ADEPT. *Adv. Drug Deliv. Rev.* 22, 365–367.
33. Bagshawe KD, Sharma SK, Springer CJ, Antoniw P. (1995) Antibody-directed enzyme prodrug therapy: A pilot scale clinical trial. *Tumor Targeting* 1, 17–29.
34. Springer CJ, Poon GK, Sharma SK, Bagshawe KD. (1993) Identification of prodrug, active drug, and metabolites in an ADEPT clinical study. *Cell Biophys.* 22, 9–26.
35. Bagshawe KD, Sharma SK. (1996) Cyclosporine delays host immune response to antibody enzyme conjugate in ADEPT. *Transplant Proc.* 28, 3156–3158.
36. Napier MP, Sharma SK, Springer CJ, Bagshawe KD, Green AJ, Martin J, et al. (2000) Antibody-directed enzyme prodrug therapy: efficacy and mechanism of action in colorectal carcinoma. *Clin. Cancer Res.* 6, 765–772.
37. Martin J, Stribbling SM, Poon GK, Begent RH, Napier M, Sharma SK, et al. (1997) Antibody-directed enzyme prodrug therapy: pharmacokinetics and plasma levels of prodrug and drug in a phase I clinical trial. *Cancer Chemother. Pharmacol.* 40, 189–201.
38. Springer CJ, Dowell R, Burke PJ, Hadley E, Davis DH, Blakey DC, et al. (1995) Optimization of alkylating agent prodrugs derived from phenol and aniline mustards: a new clinical candidate prodrug (ZD2767) for antibody-directed enzyme prodrug therapy (ADEPT). *J. Med. Chem.* 38, 5051–5065.
39. Blakey DC, Burke PJ, Davies DH, Dowell RI, East SJ, Eckersley KP, et al. (1996) ZD2767, an improved system for antibody-directed enzyme prodrug therapy that results in tumor regressions in colorectal tumor xenografts. *Cancer Res.* 56, 3287–3292.
40. Francis RJ, Sharma SK, Springer C, Green AJ, Hope-Stone LD, Sena L, et al. (2002) A phase I trial of antibody directed enzyme prodrug therapy (ADEPT) in patients with advanced colorectal carcinoma or other CEA producing tumors. *Br. J. Cancer* 87, 600–607.
41. Neuberger MS, Williams GT, Fox RO. (1984) Recombinant antibodies possessing novel effector functions. *Nature* 312, 604–608.
42. Bosslet K, Czech J, Hoffmann D. (1994) Tumor-selective prodrug activation by fusion protein-mediated catalysis. *Cancer Res.* 54, 2151–2159.
43. Chester K, Pedley B, Tolner B, Violet J, Mayer A, Sharma S, et al. (2004) Engineering antibodies for clinical applications in cancer. *Tumor Biol.* 25, 91–98.
44. Bhatia J, Sharma SK, Chester KA, Pedley RB, Boden RW, Read DA, et al. (2000) Catalytic activity of an *in vivo* tumor targeted anti-CEA scFv::carboxypeptidase G2 fusion protein. *Int. J. Cancer* 85, 571–577.
45. Sharma SK, Pedley RB, Bhatia J, Boxer GM, El-Emir E, Qureshi U, et al. (2005) Sustained tumor regression of human colorectal cancer xenografts using a multifunctional mannosylated fusion protein in antibody-directed enzyme prodrug therapy. *Clin. Cancer Res.* 11, 814–825.
46. Tolner B, Smith L, Begent RH, Chester KA. (2006) Production of recombinant protein in *Pichia pastoris* by fermentation. *Nat. Protoc.* 1, 1006–1021.
47. Mayer A, Francis RJ, Sharma SK, Tolner B, Springer CJ, Martin J, et al. (2006) A phase I study of single administration of antibody-directed enzyme prodrug therapy with the recombinant anti-carcinoembryonic antigen antibody-enzyme fusion protein MFECP1 and a bis-iodo phenol mustard prodrug. *Clin. Cancer Res.* 12, 6509–6516.
48. Wilkins D, Mayer A, Sharma S, Tolner B, Francis R, Springer C, et al. (2008) Evidence of efficacy of antibody directed enzyme prodrug therapy (ADEPT) in a Phase I trial in patients with advanced carcinoma. AACR Meeting Abstracts, 2008, LB-200.

49. Sharma SK, Bagshawe KD, Melton RG, Sherwood RF. (1992) Human immune response to monoclonal antibody-enzyme conjugates in ADEPT pilot clinical trial. *Cell Biophys.* 21, 109–120.
50. Mayer A, Sharma SK, Tolner B, Minton NP, Purdy D, Amlot P, et al. (2004) Modifying an immunogenic epitope on a therapeutic protein: a step towards an improved system for antibody-directed enzyme prodrug therapy (ADEPT). *Br. J. Cancer* 90, 2402–2410.
51. Cantor JR, Yoo TH, Dixit A, Iverson BL, Forsthuber TG, Georgiou G. (2000) Therapeutic enzyme deimmunization by combinatorial T-cell epitope removal using neutral drift. *Proc. Natl. Acad. Sci. USA* 108, 1272–1277.
52. Harding FA, Liu AD, Stickler M, Razo OJ, Chin R, Faravashi N, et al. (2005) A beta-lactamase with reduced immunogenicity for the targeted delivery of chemotherapeutics using antibody-directed enzyme prodrug therapy. *Mol. Cancer Ther.* 4, 1791–1800.
53. Cesaro-Tadic S, Lagos D, Honegger A, Rickard JH, Partridge LJ, Blackburn GM, et al. (2003) Turnover-based *in vitro* selection and evolution of biocatalysts from a fully synthetic antibody library. *Nat. Biotechnol.* 21, 679–685.
54. Golynskiy MV, Seelig B. De novo enzymes: from computational design to mRNA display. *Trends Biotechnol.* 28, 340–345.
55. Afshar S, Olafsen T, Wu AM, Morrison SL. Characterisation of an engineered human purine nucleoside phosphorylase fused to an anti-her2/neu single chain Fv for use in ADEPT. *J. Exp. Clin. Cancer Res.* 2009, 28, 147.
56. Afshar S, Asai T, Morrison SL. (2009) Humanized ADEPT comprised of an engineered human purine nucleoside phosphorylase and a tumor targeting peptide for treatment of cancer. *Mol. Cancer Ther.* 8, 185–193.
57. Chen KC, Wu SY, Leu YL, Prijovich ZM, Chen BM, Wang HE, et al. A humanized immunoenzyme with enhanced activity for glucuronide prodrug activation in the tumor microenvironment. *Bioconjug. Chem.* 22, 938–948.

24

TUMOR-TARGETED SUPERANTIGENS

GUNNAR HEDLUND, GÖRAN FORSBERG, THORE NEDERMAN, ANETTE SUNDSTEDT, LEIF DAHLBERG, MIKAEL TIENSUU, AND MATS NILSSON

Active Biotech AB, Lund, Sweden

24.1 INTRODUCTION: TUMOR-TARGETED SUPERANTIGENS—A UNIQUE CONCEPT OF CANCER TREATMENT

The tumor-targeted superantigen (TTS) concept uses a unique approach to target large numbers of T cells to tumors [1–3]. T lymphocytes are very effective in eliminating virus-infected cells, allogeneic transplants, and some experimental tumors. However, in humans, tumor-specific T cells occur in low numbers and are insufficient to interfere with an established growing tumor. In TTS, bacterial superantigens (SAgs), the most potent activators of human T cells known today, are used to activate and target large numbers of T cells to the tumor. SAgs bind to major histocompatibility complex (MHC) class II molecules and activate large numbers of cytotoxic and helper T lymphocytes. The fundamental knowledge regarding T-cell activation by the SAg staphylococcal enterotoxin A (SEA) was lined out at the Wallenberg laboratory in Lund, Sweden, where it was shown that SEA binds to MHC class II expressing accessory cells and activates both $CD4^+$ and $CD8^+$ T cells [4,5]. Subsequently, the binding sites between SEA (Figure 24.1) and MHC class II and the T-cell receptor (TCR) containing certain Vβ sequences were identified [6–10]. The capacity of SEA to associate with MHC class II and activate T cells was further elucidated, and it was demonstrated that MHC class II expressing cells were killed by cytotoxic T lymphocytes (CTLs) when targeted by SEA [11–14].

A tumor-specific SAg can be created by fusing the SAg to the Fab moiety of a tumor reactive monoclonal antibody (mAb). Fab-SAg fusion proteins have been successfully used to cure mice with established experimental tumors (for references see 24.3). The therapeutic effect was associated with a massive infiltration throughout the tumor of T cells that actively produced tumoricidal cytokines such as tumor necrosis factor (TNF)-α and interferon (IFN)-γ. The therapeutic efficacy is most likely related to the dual mechanism of tumor cell killing: direct lysis of tumor cells expressing the antigen recognized by the antibody moiety of the fusion protein and secretion of cytokines eliminating antigen-negative tumor cell variants (Figure 24.2). The induction of multiple antitumor mechanisms is considered clinically important because it addresses the problem of tumor cell heterogeneity inevitably found in all human tumors. The inflammatory response induced in the tumor also facilitates uptake of the fusion proteins, thereby overcoming the poor penetration of macromolecules in a tumor. The TTS therapeutic proteins are typically used in cycles of 4–5 daily intravenous injections. In the first phase of a cycle,

Fusion Protein Technologies for Biopharmaceuticals: Applications and Challenges, First Edition. Edited by Stefan R. Schmidt.

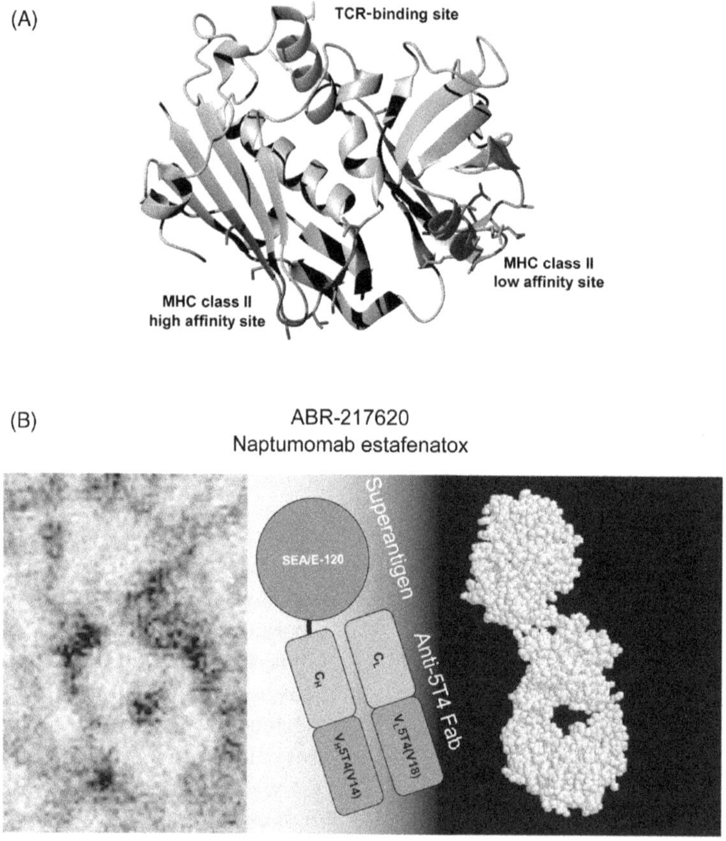

FIGURE 24.1 (A) Model of staphylococcal enterotoxin A (SEA) showing TCR (yellow), MHC class II high affinity (purple), and low affinity (green and red) binding sites. (B) Schematic picture of naptumomab estafenatox (middle), a three-dimensional model (right), and a scanning transmission electron microscopy (STEM) image of the 4×12 nm structure at $1{,}000{,}000 \times$ magnification (left). The STEM image is shown with the courtesy from Dr Tudor Arvinte. (See the color version of this figure in Color Plates section.)

the T lymphocytes are activated and differentiate into effector cells which later in the cycle localize to the tumor and mediate their antitumor functions (Figure 24.2). This treatment schedule can be repeated and is easily combined with other anticancer drug modalities.

24.2 STRUCTURE AND PRODUCTION OF TUMOR-TARGETED SUPERANTIGENS

To generate a TTS molecule, the SAg can be fused to any tumor-targeting moiety. There are several potential targeting molecules that have relatively high specificity for tumor cells; however, during our development, there has been a focus on the use of tumor-specific mAbs. The initial experiments to challenge the TTS concept were done with protein conjugates. The TTS conjugates were prepared by introducing thiol groups on the SAg and iodoacetyl or maleimide groups on the mAb forming a stable thioether linkage between the SAg and the mAb [15]. A hydrophilic spacer, composed of repeated ethylene oxide units, was constructed to increase the distance between the SAg and the mAb, preserving biological activity of both proteins. After having demonstrated proof of principle with TTS chemical conjugates, different types of recombinant fusion proteins were produced including Fab and single chain Fv variants [16]. Whole mAb conjugates as well as Fab and single-chain Fv recombinant fusion proteins showed excellent *in vitro* antitumor activity. Two major motives have guided the design further: (1) the possibility to express the TTS molecules at high levels in *Escherichia coli* and (2) obtaining products with good and stable properties that can be purified in a robust way. Fusion proteins between Fab moieties and SAg moieties fused to the C-terminus of the heavy chain of the Fabs have generally fulfilled these design criteria. The products have been expressed using secretion in a bicistronic construct. Interestingly, using appropriate signal peptides and *E. coli* strains, product levels well above 100 mg/L can

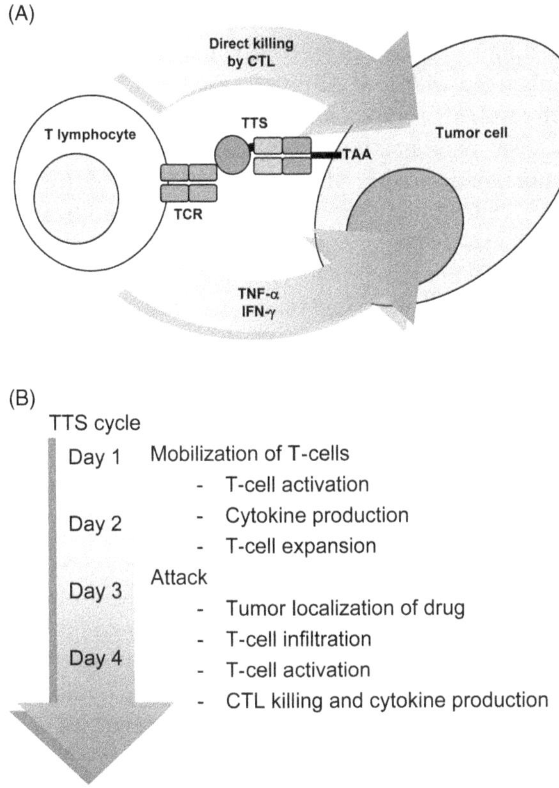

FIGURE 24.2 (A) Mechanism of action for tumor-targeted superantigens. CTL, cytotoxic T lymphocyte; IFN, interferon; TAA, tumor-associated antigen; TCR, T-cell receptor; TNF, tumor necrosis factor; TTS, tumor-targeted superantigen. (B) Schematic representation of immune/drug functions during a TTS cycle.

be obtained in the growth medium [17]. A useful feature of these fusion proteins is that they bind protein G and can therefore be purified using standard affinity chromatography.

As an example, the production of naptumomab estafenatox [18] is outlined here. The murine antibody 5T4 is directed against a tumor antigen found in many carcinomas (see 24.3). When fusing the mAb 5T4 with the SAg and expressing the fusion protein in *E. coli*, it was found that the 5T4 light chain was limiting high expression levels. Therefore, the 5T4 Fab moiety was engineered to improve production [17]. The resulting Fv with seven substitutions in the variable light chain and four substitutions in the variable heavy chain was designated variant 18 (V18) of 5T4Fv. This variant displays higher affinity for the 5T4 antigen than the wild-type Fab. The 5T4 Fv genes were fused to the sequences coding for the constant regions of the murine IgG1/κ antibody C242 in the expression vector. The 5T4FabV18 coding regions were fused to the coding regions for a SAg variant extensively engineered to be optimal for therapy in humans (see 24.4), the SEA/E-120 [19].

The two moieties in the 5T4FabV18-SEA/E-120 molecule (naptumomab estafenatox) also contain a tripeptide between the C-terminus of heavy chain of the antibody fragment and the N-terminus of the SAg to allow for structural flexibility. The two Fab chains are held together by noncovalent forces (Figure 24.1), as the disulfide bond between the Fab heavy and light chain has been removed. The protein contains 672 amino acid residues, consisting of two protein chains of 458 and 214 amino acids.

During production, the expression of naptumomab estafenatox is under the control of the lacUV5 promoter, which is induced by the addition of isopropyl-β-D-thiogalactoside (IPTG) to the culture broth. As the product is found in the growth medium, standard technology such as centrifugation, filtration, or expanded bed adsorption is used to remove the cells. The purification of naptumomab estafenatox consists of four chromatographic steps. (1) The capture step, performed on a SP Sepharose FF matrix, concentrates the protein and reduces the levels of process related impurities. (2) Main purification is performed by affinity chromatography on a protein G Sepharose, which utilizes the specific interaction between protein G and the Fab part of naptumomab estafenatox. This removes virtually all non-Fab-containing proteins, as well as further reducing the levels of DNA and endotoxins. (3) The polishing step, by ion-exchange chromatography, is performed on a Source 15S matrix, which separates some of the isomers of the fusion protein and reduces impurities. (4) The final step is a gel permeation chromatography using a Sephadex G-25 Fine matrix. At this step, a buffer exchange to the formulation buffer is performed.

The process has been scaled up to 1000 L by Richter-Helm BioLogics GmbH & Co KG (former Strathmann Biotec AG), Germany, and material was manufactured for clinical studies. After process optimization, a total of nearly 40 g has been achieved from the 1000-L process.

24.3 TUMOR-TARGETED SUPERANTIGENS ARE POWERFUL TARGETED IMMUNE ACTIVATORS AND USEFUL FOR ALL TYPES OF MALIGNANCIES

The TTS prototype fusion molecule (C215Fab-SEA) was built to target the human tumor-associated antigen EpCAM/GA733-2/CD326 [20] recognized by the C215 mAb. In this case, the Fab of the C215 mAb was used to target the SAg-induced T-cell activity against tumor cells by genetically fusing it with the SAg SEA. A series of novel TTS molecules have been developed from the prototype (see 24.4), and as a consequence of the engineering process to optimize the SAg part of the TTS fusion proteins for human use, many of the new compounds do not activate murine T cells and are therefore not therapeutically active in nonhumanized experimental mouse models. Thus, we have used the C215Fab-SEA fusion protein, which adequately activates murine T cells, for different pharmacological studies in mice. In an attempt to establish a mouse model with an intact immune system being activated with the new SAgs, a double transgenic mouse, expressing human TCR Vβ7.9 and HLA-DR4, was constructed (to be published). T lymphocytes from these mice have recently shown to be activated by a SAg that stimulates human T lymphocytes but is inert with wild-type mouse T cells, demonstrating that these human molecules are central for the engineered SAg function.

The effectiveness of TTS therapy has been demonstrated both in syngeneic and xenogeneic tumor models in mice [2,21–30]. The poorly immunogenic B16 melanoma was transfected with the human carcinoma antigen EpCAM [2] and used to evaluate the antitumor effects of the C215Fab-SEA fusion protein in syngeneic tumor models. Treatment with repeated injections of C215Fab-SEA eradicated >90% of lung tumors in mice carrying established B16-EpCAM melanoma tumors [2,27] (Figure 24.3). Syngeneic mice with EpCAM expressing tumors growing as lung metastases were treated with C215Fab-SEA at various doses and intervals to determine optimal therapy schedules. Therapeutic efficiency and induction of effector functions such as CTLs, IL-2, IFN-γ and IL-10 were evaluated. Significant levels of IL-2 and IFN-γ were recorded during the first three daily injections, but reduced levels were recorded in response to the fourth injection [27]. CTL activity was induced 1–2 days after the first injection of C215Fab-SEA and declined to background levels after 4 days of treatment. Repeated intermittent injections (every fourth day) resulted in low levels of IL-2 and IFN-γ, reduced CTL activity, and increased levels of IL-10. IL-10 production increased with prolonged treatment intervals between injections in the treatment cycle [31].

The selective tumor targeting and uptake of the TTS have been shown to be dependent on targeting via the antibody moiety using radiolabeled compounds (Figure 24.3). It was clearly shown that C215Fab-SEA was accumulated in B16-EpCAM tumors and that the control C242Fab-SEA only was detected in MHC class II expressing tissues such as the spleen. Significant therapeutic activity was only recorded if the SAg moiety was targeted to the tumor (Figure 24.3) [2]. Maximum therapeutic effect was obtained with four daily injections of C215Fab-SEA, which resulted in a marked reduction in the number of lung metastases in treated mice compared to untreated controls or mice treated with SEA conjugated to an irrelevant control Fab moiety [23,27]. C215Fab-SEA treatment induced massive tumor infiltration of $CD4^+$ and $CD8^+$ T cells (Figure 24.3), whereas only scattered T cells were observed in untreated tumors [3]. In addition, immunohistochemical analysis revealed substantial accumulation of TNF-α and IFN-γ locally at the site of the tumor. Depletion of $CD4^+$ and $CD8^+$ T cells *in vivo*, revealed the $CD8^+$ T cells to be the most important effector cells for tumor remission. For optimal tumor eradication, however, both $CD4^+$ and $CD8^+$ T cells were necessary. Taken together, this suggests that the therapeutic mechanisms of Fab-SEA proteins involve targeting of cytotoxic and cytokine-producing T cells to the tumor, which results in a significant reduction of tumor burden (Figure 24.3).

The TTS concept for targeted tumor immunotherapy facilitates targeting of T cells toward any type of malignant cell or tissue, including hematological malignancies, sarcomas, and carcinomas. The choice of the targeting moiety, for example, the mAb, is dependent of cancer indication. A series of different antibodies has been tested as targeting moieties in TTSs (Table 24.1). TTS fusion proteins containing mAbs recognizing carcinomas, malignant melanoma, hematological malignancies, and endothelial cells are all effective. So far, however, the clinical development has focused on relevant TTSs for different carcinomas. The TTS prototype fusion molecule C215Fab-SEA recognizes EpCAM, which is expressed on human carcinomas but unfortunately also on many normal cells. Therefore, we selected mAbs with higher tumor specificity for clinical development and produced TTS fusion proteins with the Fab part from the C242 mAb for the early clinical trials [32,33] and later with the Fab from the 5T4 mAb for the optimized therapeutic fusion proteins.

The 5T4 antigen is a 72-kDa oncofetal trophoblast protein originally isolated from placenta and has been

TABLE 24.1 The TTS Concept Has Been Shown Effective with Many Different Antitumor mAbs

mAb/Antigen	Tumor/Tissue Type	Reference
C215/EpCAM, CD326	Carcinomas	[1,2,15,16,20,34]
C242/CA242	Carcinomas	[1,25,32,35–37]
5T4/5T4, TPBG	Carcinomas	[18,24,38,39]
K305	Malignant melanoma	[40]
A3 and F1	Carcinomas of the colon and the pancreas	[41]
K594	Endothelial cells	[42]
B-B2, B-B4, and Mi15/CD138	Multiple myeloma	[43]
HuM195/CD33	Leukemia	[44]
CLB-B4-1/CD19	B-CLL	[45]

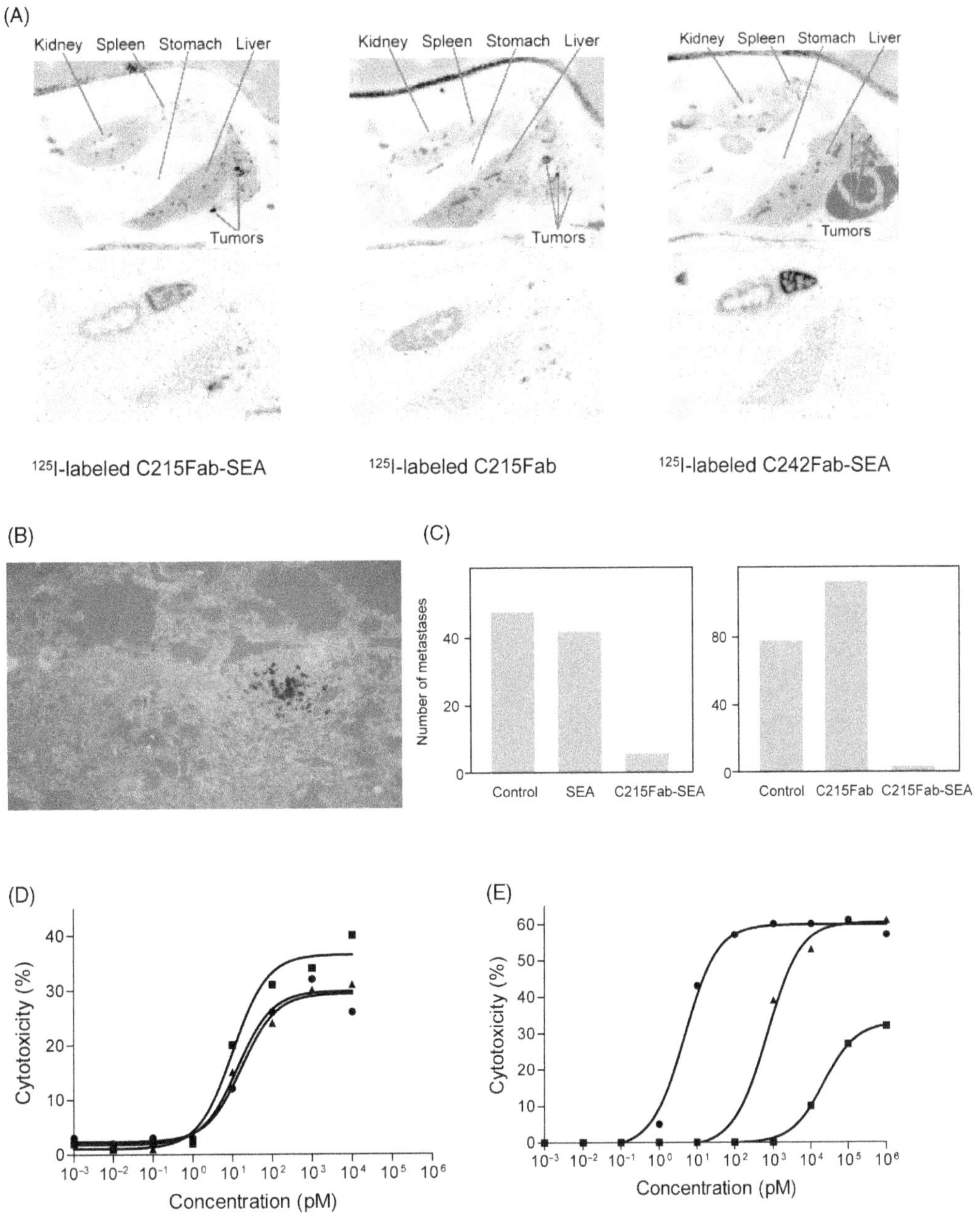

FIGURE 24.3 (A) Radioluminogram showing the localization of TTS/mAb molecules to the targeted tumors (B16 melanoma transfected with EpCAM antigen). The corresponding tissue section is shown above. (B) Tumor-selective recruitment of $CD8^+$ T cell (stained red) by C215Fab-SEA to a B16-EpCAM melanoma (black) in a syngeneic mouse lung. (C) Inhibition of growth of B16-EpCAM melanoma lung metastases after treatment with different TTS/mAb/SAg molecules. (D) SADCC against 5T4-expressing Caki-2 cells mediated by 4-h incubation with 5T4Fab-SEA (circle), anatumomab mafenatox (triangle), or naptumomab estafenatox (square) at the indicated concentrations. (E) SDCC against MHC class II expressing Raji cells mediated by 4-h incubation with 5T4Fab-SEA (circle), anatumomab mafenatox (triangle), or naptumomab estafenatox (square) at the indicated concentrations. *Source*: Permission from Wolters Kluwer Health. (See the color version of this figure in Color Plates section.)

TABLE 24.2 5T4 Reactivity in a Variety of Tumor Types

Cancer	Stage/Type	Reactivity	Reference	Conclusions
Breast	Mixed	5/5	[50]	85–95% of patients express the 5T4 antigen
		51/58	Active Biotech unpublished data	
Cervix	Mixed	5/5	[50]	85–90% of patients express the 5T4 antigen
	I	22/25	[56]	
	II	22/26	[56]	
	III	9/10	[56]	
	IV	5/6	[56]	
Colorectal	Mixed	4/13	[50]	40–50% of patients express the 5T4 antigen. In Dukes stage D patients, more than 70% overexpress the antigen
		7/9	Active Biotech unpublished data	
	A	2/8	[53]	
	B	7/34	[53]	
	C	13/21	[53]	
	D	7/9	[53]	
Gastric	Mixed	6/7	[50]	Approximately 50% of patients express the 5T4 antigen on tumor cells. Most samples with 5T4 negative tumor cells have 5T4 positive tumor stroma [54]
		35/86	[54]	
	I	1/2	[53]	
	II	1/4	[53]	
	III	1/1	[53]	
	IV	12/20	[53]	
NSCLC	Mixed	10/10	[50]	99% of patients express the 5T4 antigen
		249/251	Active Biotech unpublished data	
Ovarian	Mixed	4/7	[50]	More than 70% of patients express the 5T4 antigen. For stage IV patients, 90–95% of the patients are 5T4 positive
	I	2/10	[55]	
	II	4/57	[55]	
	III	21/29	[55]	
	IV	24/26	[55]	
Pancreatic	Mixed	3/3	[50]	100% of patients express the 5T4 antigen
		20/20	Active Biotech unpublished data	
Prostate	Mixed	26/26	Active Biotech unpublished data	100% of patients express the 5T4 antigen
Renal	Clear cell	1/1	[50]	>95% of patients express the 5T4 antigen
		214/221	Active Biotech unpublished data	
	Papillary	18/18	Active Biotech unpublished data	
	Mixed	16/17	Active Biotech unpublished data	
		20/20	[57]	

characterized by the use of the murine IgG1 mAb 5T4 [46,47]. 5T4 antigen expression has been shown to influence adhesion, cytoskeletal organization, and motility, properties that might account for its association with poorer clinical outcome in some cancers. Recently it was shown that the 5T4 glycoprotein is associated with functional cell surface expression of the chemokine receptor CXCR4- and CXCL12-mediated chemotaxis [48]. A general screen by immunohistochemistry on frozen sections showed that many different carcinomas express the 5T4 antigen [49–56] (Table 24.2). Studies have shown tumor reactivity, which in most cases is moderate to strong, in more than 95% of nonsmall cell lung cancer (NSCLC), pancreatic cancer (PC), and renal cell cancer (RCC). In many cases, tumor stromal reactivity has been observed in addition to the tumor cell reactivity. Outside the placenta, 5T4 normal tissue reactivity is considered to be most limited, and there is no evidence of circulating 5T4 antigen.

The TTS therapeutic proteins are applicable in all malignant diseases where tumor selective mAbs can be produced. The 5T4 mAb, selected for clinical development primarily in the indications of NSCLC and RCC, targets the TTS fusion proteins to a tumor antigen broadly expressed on tumor cells, with very low expression on normal tissue and without interference by soluble antigen.

24.4 INCREASING THE THERAPEUTIC WINDOW AND EXPOSURE BY THE CREATION OF A NOVEL TTS FUSION PROTEIN WITH MINIMAL MHC CLASS II AFFINITY; NAPTUMOMAB ESTAFENATOX

The early clinical trials with TTS fusion proteins were performed using wild-type SEA as the T-cell activator [32,33]. However, this fusion protein (nacolomab tafenatox) was associated with a most potent T-cell activation, and therefore, it was necessary to give doses as low as a few nanogram per kilogram to avoid unacceptable toxicity. In addition, the dose had to be individualized based on preformed anti-SAg (anti-SEA) antibodies. *Primum movens* for TTS function is TCR binding and T-cell activation. Wild-type SAgs bind to MHC class II cell membrane proteins, and the complexes subsequently bind to TCRs containing certain Vβ sequences and activate these T cells. After proliferation and differentiation, the T lymphocytes are capable to infiltrate the tumor tissue presenting the TTS fusion proteins specifically bound to the tumor cells. These T cells are activated in the tumor due to TCR-binding to the SAgs of the TTS protein, produce cytokines, and perform direct perforin-dependent tumor killing. To achieve maximal targeted antitumor effects, a balanced relationship between binding affinities of the three functional binding sites in the TTS is required. The objective to mimic the T-effector cell (e.g., the CTL) interaction to a tumor target cell, the TCR binding to the MHC/tumor peptide target, drove the development to the TTS naptumomab estafenatox showing distinct characteristics regarding binding to the tumor antigen (5T4), the TCR, and to the MHC class II proteins [18] (Hedlund et al., to be published). The optimal TTS should have very high affinity for the tumor antigen, intermediate/low affinity for TCR, and very low affinity for MHC class II products.

A fusion protein consisting of the 5T4Fab moiety fused to SEA with a point mutation decreasing MHC class II protein binding (associated to decreased toxicity), anatumomab mafenatox [24], has been investigated in clinical Phase I and Phase II trials with very encouraging results [38,58] (see 24.5.1). However, this fusion protein was also associated with some of the drawbacks seen with nacolomab tafenatox, necessity to give low doses as well as individualized dosing on the basis of preformed anti-SAg (anti-SEA) antibodies.

Affinity of distinct SAgs to MHC class II proteins is correlated to T-cell activation capacity, toxicity, and capacity to direct CTLs in SAg-dependent cell-mediated cytotoxicity (SDCC, killing of MHC class II protein expressing target cells incubated with SAg). The targeting of CTLs by a TTS fusion protein to tumor cells can be measured in assays for SAg-antibody-dependent cell-mediated cytotoxicity (SADCC; killing of target cells incubated with TTS fusion protein specifically bound to target with the mAb part). TTS fusion proteins stimulate human CTLs to kill tumor cells in a dose-dependent manner. Naptumomab estafenatox displayed an increased affinity (by more than a factor 1.5) to the 5T4 antigen as compared to the preceding compound anatumomab mafenatox [18]. The affintiy for naptumomab estafenatox was in the order of 1.6 nM. Naptumomab estafenatox was ~50% more efficient as compared with anatumomab mafenatox, in killing human 5T4-positive Caki-2 tumor cells as measured by SADCC (Figure 24.3). In the killing of MHC class II positive Raji cells by SDCC, naptumomab estafenatox displayed 100-fold and 10,000-fold decreased activity when compared with anatumomab mafenatox and 5T4Fab-SEA (SAg similar to nacolomab tafenatox), respectively, owing to decreased MHC class II binding. Naptumomab estafenatox induced an MHC class II and dose dependent proliferation in human lymphocytes, however, 50-fold and >1000-fold higher concentrations were needed compared to anatumomab mafenatox and to a fusion protein with wild-type SEA, respectively [18]. Naptumomab estafenatox mediated a 1000-fold more potent 5T4 targeted cytotoxicity (EC_{50} ~10^{-11} M) compared with the MHC class II directed cytotoxicity (EC_{50} ~10^{-8} M) [18] (Figure 24.3). Furthermore, preformed human anti-SAg antibodies do not recognize the functional regions in the SAg part of naptumomab estafenatox to the same extent as in the predecessor construct [18]. A 10-fold higher concentration of anti-SAg (anti-SEA) antibodies was needed to neutralize the bioactivity of naptumomab estafenatox compared with anatumomab mafenatox measured as inhibition of SADCC activity. These findings show that the activity of naptumomab estafenatox is significantly less inhibited by human anti-SAg than that of anatumomab mafenatox. Using various *in vitro* assays, it is estimated that only 10–15% of human anti-SEA antibodies can bind naptumomab estafenatox [19].

Thus, even though the novel TTS naptumomab estafenatox has most potent antitumor properties, it contains a SAg moiety without classic SAg activity. The pivotal rescued functional part of the two hybridized and mutated Staphylococcal enterotoxins (A and E) is the TCR-binding region. The MHC class II binding capacity is minimal and the T-cell proliferation induced by naptumomab estafenatox has an EC_{50} >1000 times higher than conventional wild-type SAgs resulting in dramatically reduced toxicity. Furthermore, the target epitopes for naturally occurring anti-SAg antibodies were reduced and dosing of naptumomab estafenatox was therefore considered to be independent of baseline anti-SAg antibodies.

24.5 CLINICAL EXPERIENCE WITH TTS THERAPEUTIC FUSION PROTEINS

Three TTS therapeutic fusion proteins have been investigated in the clinic: nacolomab tafenatox, anatumomab mafenatox, and naptumomab estafenatox. The latter two contain Fab moieties recognizing the 5T4 antigen and engineered SAgs. The 5T4 antigen is expressed by essentially all NSCLC, PC, and RCC tumors (Table 24.2), and the clinical trials have so far been focused on patients having these tumor types.

24.5.1 Anatumomab Mafenatox

Two Phase I studies (USA study and Europe study) to evaluate anatumomab mafenatox in patients with advanced NSCLC were conducted in parallel. The primary objective was to determine the MTD of anatumomab mafenatox as a function of pretreatment anti-SEA antibody levels [58]. The patients received a daily, 3 h infusion of anatumomab mafenatox on four consecutive days (one treatment cycle). Each patient was treated with an individual dose based on the pretreatment anti-SEA antibody concentration and adjusted for body weight. Patients were followed for at least 28 days, and additional cycles were considered in patients with responding or stable disease. Altogether 78 NSCLC patients received one cycle, and 26 of these patients received a second cycle. The toxicity profile was predictable, transient and reversible fever and hypotension, and easily manageable. A total of eight dose-limiting toxicities (DLTs) were recorded, consisting of hypotension in four cases, hypotension and dyspnea in two cases, and hypertension in two cases. A dosing algorithm was constructed that would allow patients in subsequent clinical investigations to be treated with a dose of anatumomab mafenatox that is tailored to their specific tolerance for the agent, as reflected by their pretreatment anti-SEA concentrations. Blood levels of TNF-α were very low, IL-2 levels increased in some patients and IL-10 levels increased following each infusion, but rapidly declined after termination of therapy. The anti-SEA antibody levels increased rapidly during the days after a treatment cycle. After the first infusion, the maximum values were reached by day 28. Thereafter, the anti-SEA antibody concentration declined slowly. Low levels of human anti-mouse antibodies (HAMAs) were observed in some patients after therapy. The results from antitumor efficacy evaluation using CT scans of anatumomab mafenatox treated patients with advanced NSCLC suggested a treatment effect (disease stabilization or tumor regression), and a retrospective analysis of survival was performed. The median overall survival (OS) was 8.3 months, with a 43% 1-year survival and 14% 2-year survival. Interestingly, there was a significant relationship between exposure and survival regardless of performance status of the patients.

A noncontrolled Phase IIa study in advanced RCC was conducted with anatumomab mafenatox in the UK between 2001 and 2003 [38]. A total of 43 RCC patients were treated and evaluated for antitumor effects. One patient showed a sustained partial response (PR). After 2 and 4 months of anatumomab mafenatox treatment, 68% and 40% of the patients showed stable disease (SD), respectively. Of the total number of RCC patients included in the study, 72% had previously received treatment with IL-2, IFN-α or chemotherapy but had continued to progress before initiating treatment with anatumomab mafenatox. The high frequency of patients with SD during treatment in the study indicated its efficacy, and the safety profile showed that anatumomab mafenatox was well tolerated. Survival data were collected for all 43 treated patients and each patient was classified as high, intermediate, or low risk using the Memorial Sloan-Kettering Cancer Center (MSKCC) prognostic criteria [59,60] for untreated or previously treated patients. Median OS for the patients was 19.7 months as compared to simulated control patients having an expectation of 13.7 months (Figure 24.4). Two-year survival for the anatumomab mafenatox treated patients was 42% versus 27% for the simulated control patients. The 43 patients were divided into two groups depending on drug exposure (dose/anti-SEA), that is, high, above or low, below median exposure groups (Figure 24.4). Patients receiving higher, above median exposure showed a survival advantage compared to the low, below median exposure group, 26.6 months versus 12.1 months. Patients who received a high exposure and showed high-sustained IL-2 response were more likely to have disease control (SD or PR) and prolonged survival (Figure 24.4). In conclusion, the recorded prolonged survival suggested treatment benefit, on the basis of two independent comparisons. First, the patients had a median survival of 19.7 months, considerably longer than the 13.7 months survival in matched controls predicted from the previous extensive analysis [59,60]. Second, there was a relationship between drug exposure and a reduction in the rate of tumor growth. Furthermore, a relationship was shown between drug exposure and survival, where high exposure patients lived twice as long as low exposure patients. In addition, sustained IL-2 seemed to be beneficial for antitumor activity induced by anatumomab mafenatox.

24.5.2 Naptumomab Estafenatox

Naptumomab estafenatox was developed in parallel with the initial clinical studies with anatumomab mafenatox and has been evaluated in Phase I trials with NSCLC, PC, and RCC patients [39,61]. Altogether 52 patients were investigated in the monotherapy (MONO) trial and the combination therapy (COMBO) trial with docetaxel. The primary aim was to establish the MTD of naptumomab estafenatox. The secondary objectives were to determine safety profile, pharmacokinetic parameters, immunological response, and effects on tumor disease. The MONO study was carried out in Norway, the United Kingdom, and the United States. Main inclusion criteria were histologically or cytologically confirmed disease, which was refractory to available standard therapies. Patients should have an Eastern Cooperative Oncology Group (ECOG) performance status of 0 or 1 and life expectancy greater than 3 months. Three different indications were investigated, consisting of 20 cases of NSCLC, 11 cases of RCC, and 8 cases of PC. The starting dose was 500 ng/kg/day administrated as a 5-min bolus injection, and naptumomab estafenatox was given for five consecutive days. Six DLTs were recorded and the highest dose investigated was 27.4 μg/kg/day. Patients with SD or response were considered for treatment with a second five-day cycle, unless they had experienced a DLT. The most notable side effects were fever, nausea, and rigors, and these were easily managed. It was concluded that the side effects were not dependant on the

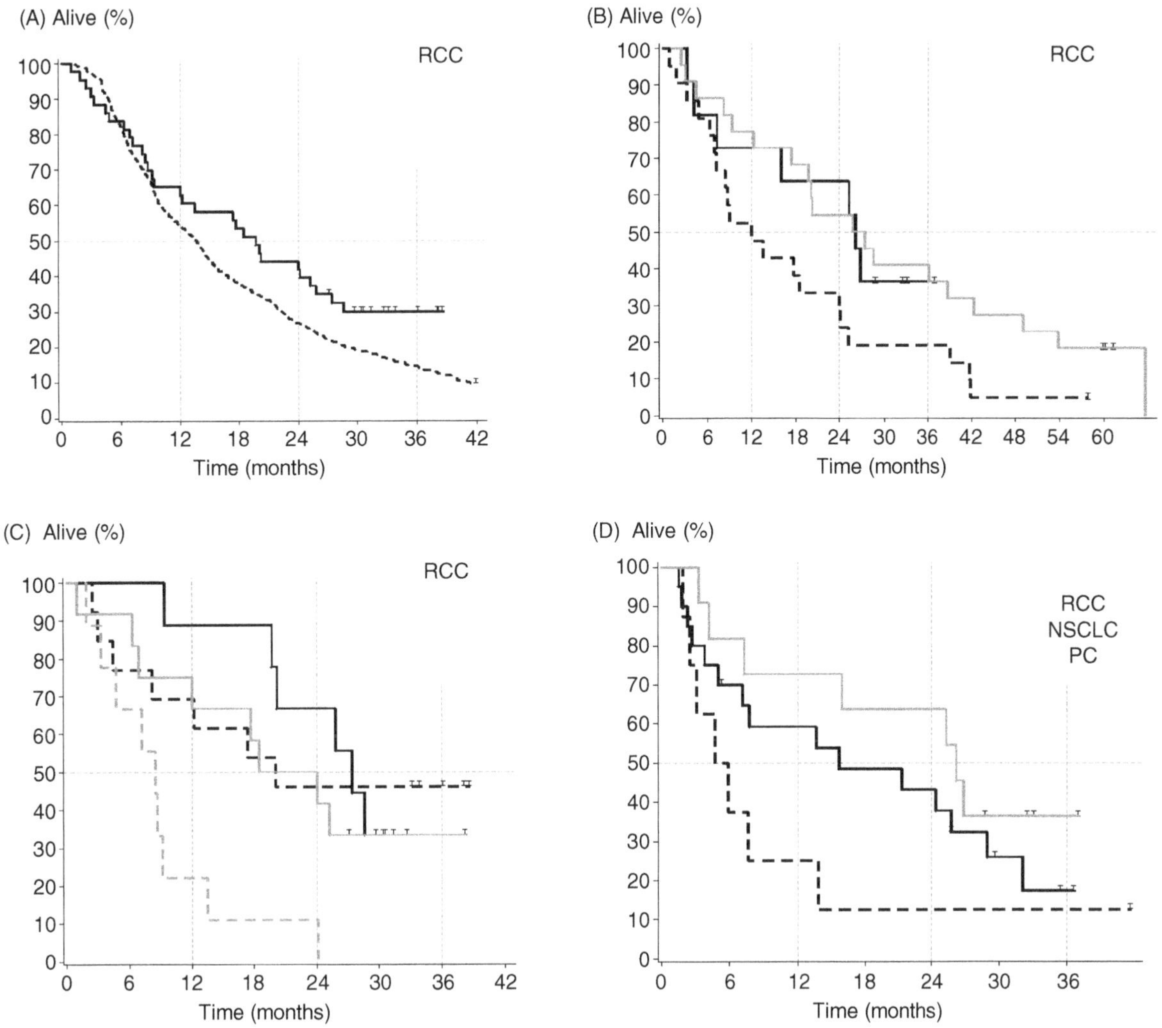

FIGURE 24.4 (A) Kaplan–Meier plot showing survival of the RCC patients treated with anatumomab mafenatox in a Phase IIa study ($n = 43$). Matched simulated control population with equivalent MSKCC prognostic criteria characteristics is represented by the dotted line. (B) Kaplan–Meier plot showing survival of the RCC patients treated with anatumomab mafenatox in a Phase IIa study (solid gray and dashed black) or naptumomab estafenatox in a Phase I study (solid black; $n = 11$). Subgroups of patients receiving high (solid gray) versus low (dashed black) exposure of anatumomab mafenatox are depicted where the high exposure group received a total dose/anti-SEA of >15.3 during the 4 days of the first cycle. (C) Kaplan–Meier plot showing survival of the RCC patients treated with anatumomab mafenatox in a Phase IIa study. Subgroups of patients receiving high (black) versus low (gray) exposure of anatumomab mafenatox and responding with high (solid) versus low (dashed) IL-2 are depicted. (D) Kaplan–Meier plot showing survival of the PC (dashed black; $n = 8$), RCC (solid gray; $n = 11$), and NSCLC (solid black; $n = 20$) patients in the Phase I MONO study with naptumomab estafenatox. *Source*: Reprinted by permission from Macmillan Publishers Ltd on behalf of Cancer research UK, Reference 38.

anti-SAg (anti-SEA/E-120) antibody titers. The recommended Phase II dose was set to 15 μg/kg/day for patients with RCC and 22 μg/kg/day for patients with NSCLC and PC.

As part of the immune activation, increased systemic levels of IL-2, IFN-γ, and IL-10 were observed 3 h after TTS treatment, and there was a correlation between dose and cytokine induction. It can be assumed that the cytokine responses may be used as parameters for beneficial pharmacologic effect (see Figure 24.4 and [38]), and we have therefore evaluated the cytokine responses in all patients from the MONO and COMBO studies. Plasma IL-2 levels were dose-dependently increased 3 h after injection of naptumomab estafenatox in the first treatment cycle (Figure 24.5). RCC patients had larger increases as compared to NSCLC and PC patients. The IL-2 response was declining with proceeding treatment cycle showing marginal IL-2 levels on day 4 and 5. Plasma IFN-γ was also dose-dependently increased 3 h after TTS injection (Figure 24.5). The IFN-γ response was slightly declining as the treatment cycle proceeded, but in opposite to IL-2, high IFN-γ levels were also recorded on day 4 and 5. Starting with induced low levels on day 1, plasma IL-10 was dose-dependently increased 3 h after injection of naptumomab estafenatox (Figure 24.5). In opposite to IL-2, the IL-10 response was increasing as the treatment cycle proceeded. It may be anticipated that a powerful immune activation has to be controlled by inhibitory actions such as IL-10 production in an immune wave similar to a vigorous antivirus response that is naturally downregulated.

A mechanistically important finding was that the SAg reactive population of T cells expressing TCR-Vβ7.9 was selectively expanded after start of treatment and was back to normal levels at day 28. The anti-SAg (anti-SEA/E-120) levels were stable or only moderately increased in ~50% of the patients after one cycle, and HAMA levels were very low.

A major objective in the development of the TTS fusion proteins has been to give them high tumor-targeting capability and a short general exposure to gain maximal effects in the tumor without unacceptable toxicity. The pharmacokinetics of naptumomab estafenatox showed a small volume of distribution (about 0.14 L/kg) and a low clearance (about 0.13 L/h/kg). The mean terminal half-life was determined to be 1.4 h. No time dependency was seen in the pharmacokinetics in treatment cycle 1 with similar plasma concentrations on day 1 and day 5. Furthermore, no dose-dependency in the pharmacokinetics was indicated at doses ≥10 μg/kg and no dependency in the pharmacokinetic parameters of naptumomab estafenatox was seen due to sex or the indications RCC, PC, and NSCLC. During treatment cycle 1, no significant dependency in the pharmacokinetic parameters was seen due to anti-SAg (anti-SEA/E-120) antibody levels.

Efficacy was analyzed with CT scan, and several patients in the MONO study had tumor shrinkage, but not big enough to be formal responses. The median OS times were 15.8 months for the NSCLC patients, 26.2 months for the RCC patients, and 5.2 months for the PC patients (Figure 24.4). The median OS for the NSCLC patients was longer than expected as was the OS for the RCC patients. The RCC patients treated with naptumomab estafenatox showed almost identical median OS as compared to the high exposure RCC patients from the Phase IIa study with anatumomab mafenatox (Figure 24.4).

An open label dose escalation study of naptumomab estafenatox in combination with docetaxel (COMBO) was carried out in Denmark, Russia, and the United States [39]. In this COMBO study, 13 patients with advanced NSCLC were treated. The primary endpoint was to describe the side effect profile of the combination treatment and establish

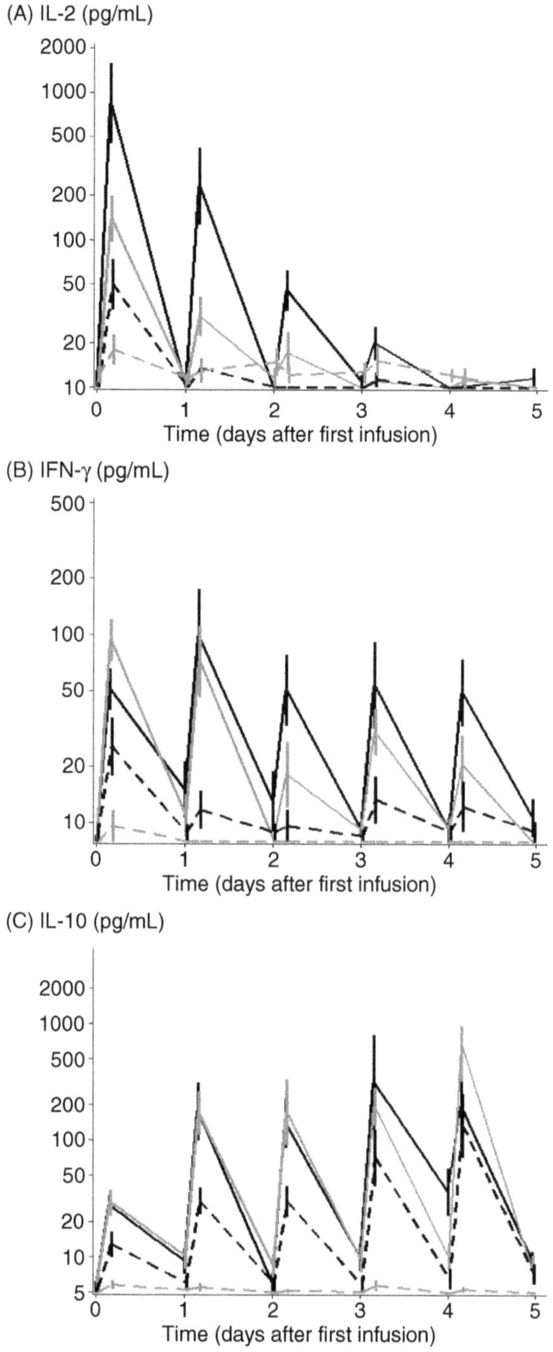

FIGURE 24.5 Plasma cytokines IL-2 (A), IFN-γ (B), and IL-10 (C) (geomean ± SEM) predose and 3 h after naptumomab estafenatox injections during five consecutive days in treatment cycle 1 in patients from the MONO and the COMBO studies. Doses given were <2.5 (dashed gray), 2.5–15 (dashed black), PC and NSCLC >15 (solid gray), RCC >15 μg/kg/day (solid black).

MTD for naptumomab estafenatox in combination with docetaxel. Secondary endpoints included pharmacokinetics, immunological responses, changes in anti-SAg (anti-SEA/E-120) levels, and effects on tumor disease. Patients were treated with a 5-min bolus injection of naptumomab estafenatox for four consecutive days followed by 75 mg/m^2 docetaxel on day 5. Three weeks later, this procedure was repeated (Figure 24.6). Three patients received

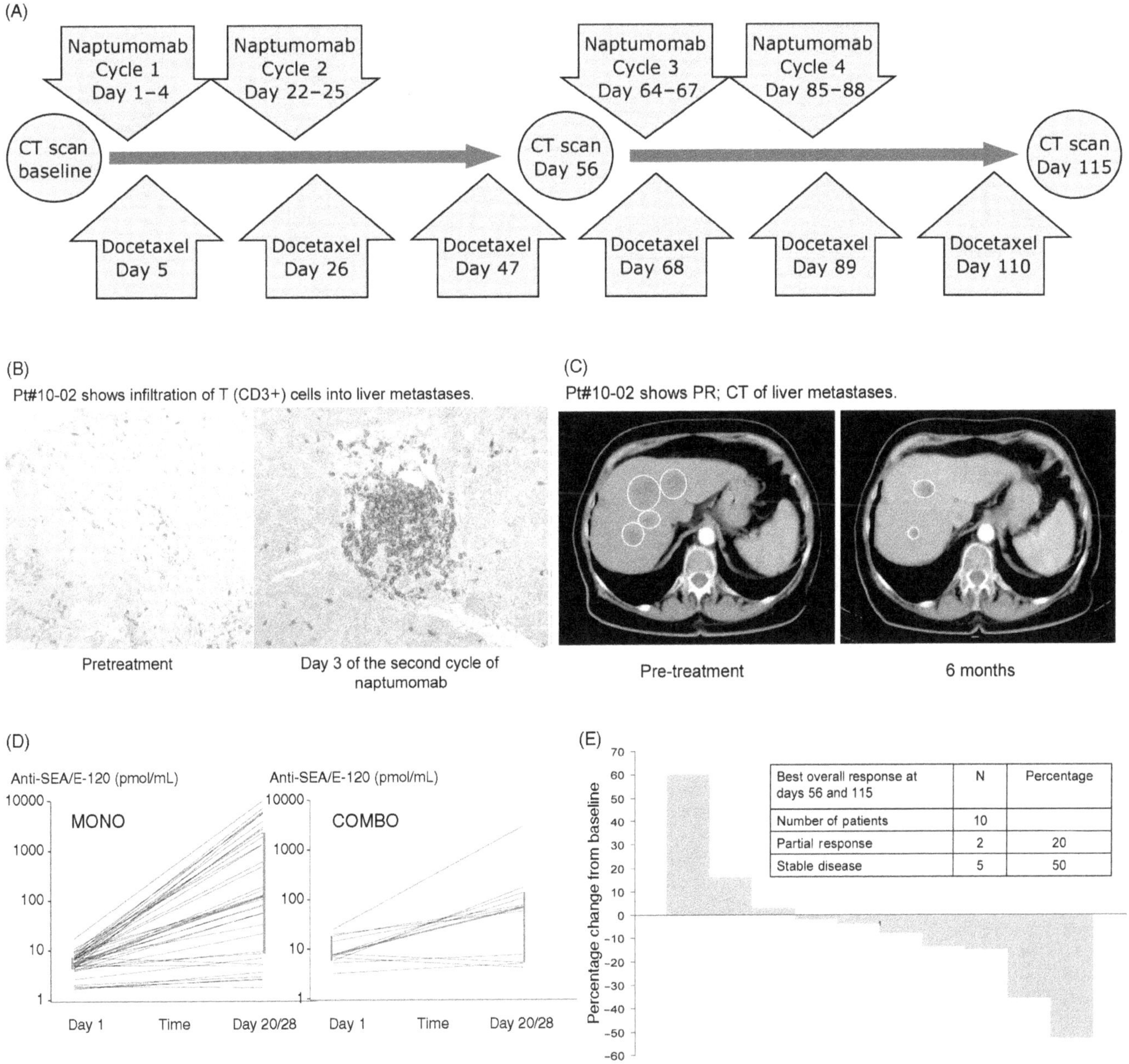

FIGURE 24.6 Naptumomab estafenatox was investigated in combination with docetaxel in a Phase I study with 13 NSCLC patients (COMBO). (A) Study treatment scheme. (B) Immunohistochemical staining (brown) for T lymphocytes (anti-CD3) of paraffin sections of liver metastasis from samples taken pretreatment and day 3 of the second cycle of 22 μg/kg naptumomab estafenatox treatment from a patient with a PR (pt#10-02). This patient had previously received three treatment regimes (including docetaxel), but had a PR after COMBO treatment. (C) CT scans pretreatment and 6 months after naptumomab estafenatox treatment start of pt#10-02. (D) Anti-SEA/E-120 antibody concentrations of individual patients at baseline (day 1) and after 3–4 weeks (Day 20/28); left: population MONO $n = 35$; right: population COMBO $n = 11$; red lines depict median ± quartiles. (E) Best overall tumor response according to RECIST criteria at days 56 and 115 of evaluable patients depicted as a waterfall plot. *Source*: Reprinted from Reference 39 with permission from American Society of Clinical Oncology. (See the color version of this figure in Color Plates section.)

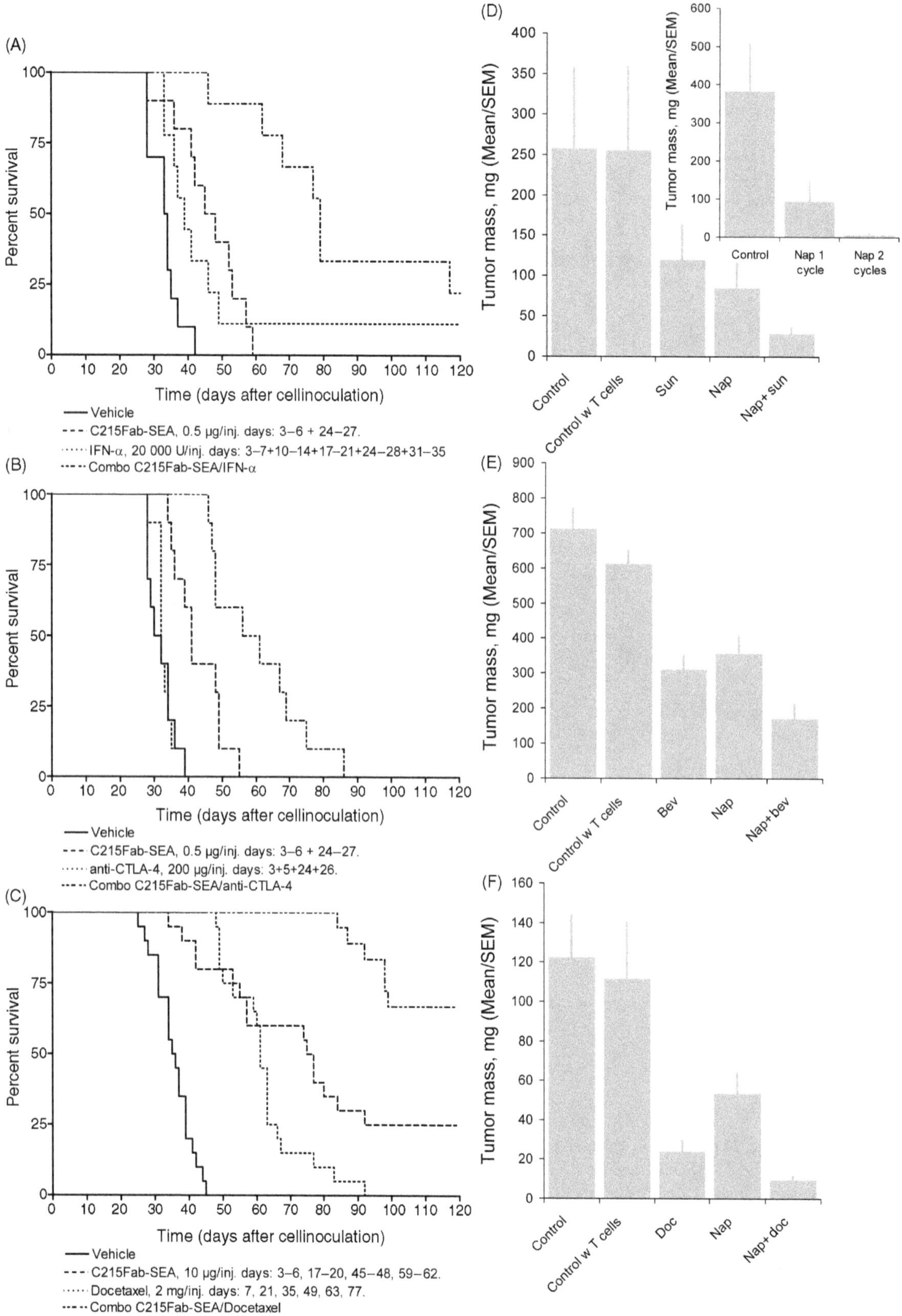
(A)
Percent survival
Time (days after cellinoculation)
Vehicle
C215Fab-SEA, 0.5 μg/inj. days: 3–6 + 24–27.
IFN-α, 20 000 U/inj. days: 3–7+10–14+17–21+24–28+31–35
Combo C215Fab-SEA/IFN-α
(B)
Percent survival
Time (days after cellinoculation)
Vehicle
C215Fab-SEA, 0.5 μg/inj. days: 3–6 + 24–27.
anti-CTLA-4, 200 μg/inj. days: 3+5+24+26.
Combo C215Fab-SEA/anti-CTLA-4
(C)
Percent survival
Time (days after cellinoculation)
Vehicle
C215Fab-SEA, 10 μg/inj. days: 3–6, 17–20, 45–48, 59–62.
Docetaxel, 2 mg/inj. days: 7, 21, 35, 49, 63, 77.
Combo C215Fab-SEA/Docetaxel
(D)
Tumor mass, mg (Mean/SEM)
Control
Control w T cells
Sun
Nap
Nap+sun
Tumor mass, mg (Mean/SEM)
Control
Nap 1 cycle
Nap 2 cycles
(E)
Tumor mass, mg (Mean/SEM)
Control
Control w T cells
Bev
Nap
Nap+bev
(F)
Tumor mass, mg (Mean/SEM)
Control
Control w T cells
Doc
Nap
Nap+doc

10 μg/kg/day, three patients received 16.5 μg/kg/day, and seven patients got 22 μg/kg/day of naptumomab estafenatox. No synergistic toxicities were observed. Best overall responses at 2 or 4 months were two patients with partial remissions (PR), five with SD, and three with progressive disease (Figure 24.6) while three patients were not evaluable. Notably, 11/13 patients had stage IV NSCLC. Biopsies were taken from two patients at day 3 of the second cycle. Both patients showed confirmation of tumor 5T4 expression and increased T-cell infiltration in association to the metastases (Figure 24.6). One of these patients had PR (Figure 24.6). The median survival for the patients was 9.3 months. Antibody formation against SEA/E-120 was low in these patients. The hypothesis included reduced increase of anti-SAg (anti-SEA/E-120) antibodies by the combination of TTS with docetaxel. We compared anti-SEA/E-120 levels after the first treatment cycle in the COMBO study with the MONO study and showed that docetaxel reduced anti-SAg antibody production when it was given one day after the last injection in a treatment cycle with naptumomab estafenatox (Figure 24.6). Since the avoidance of high titer neutralizing antibodies would be advantageous for multicycle treatment with naptumomab estafenatox, combinations with docetaxel or other inhibitors of anti-SAg antibody production would probably be beneficial. In addition, docetaxel interferes with immune suppression resulting in enhanced or sustained CTL activity [62], a feature also considered to be important for an optimal antitumor effect with naptumomab estafenatox treatment.

Promising clinical results were accomplished using naptumomab estafenatox and its predecessor in Phases I and IIa monotherapy studies with RCC patients. In preclinical models, IFN-α was shown to enhance and sustain TTS induced CTL activity when given in combination [21]. In early 2007, patient recruitment started in a pivotal add-on to IFN-α combination trial with naptumomab estafenatox in advanced RCC. This trial is conducted as a multicenter (involving ~50 sites in Bulgaria, Romania, Russia, the United Kingdom, and Ukraine), randomized, open-label, parallel-group, Phase II/III trial in patients with confirmed metastatic or inoperable locally advanced RCC eligible for standard therapy with IFN-α. Recruitment was completed in May 2009 with 527 patients randomized for treatment with IFN-α with or without naptumomab estafenatox. OS (primary endpoint) results are anticipated in due time.

24.6 COMBINING TTS WITH CYTOSTATIC AND IMMUNOMODULATING ANTICANCER DRUGS

Many established cancer treatments utilize multiple mechanisms of action to add therapeutic effect by combining two or more compounds. Therefore, establishment of combinations with other antitumor treatments was integrated within the development of the TTS fusion proteins. Apart from gaining therapeutic effect by combining with other antitumor mechanisms, we also searched for combination schedules interfering with anti-SAg antibody production and hindrance of tumor-induced immune suppression or anergy. In addition, combination with certain anticancer drugs with distinct mechanisms of action would potentially result in not only additive but even synergistic tumor disease inhibition.

We have elaborated a series of combinations between TTS fusion proteins and other anticancer or immune modulating drugs: the cytostatic drugs docetaxel [63] and gemcitabine [64]; the anti-angiogenic drugs sunitinib [65] and bevacizumab [66]; the cytokines IL-2 [67,68] and IFN-α [67,68]; and the immune modulatory mAbs anti-CD25 [69], anti-CTLA-4 [67,70], and anti-Gr-1 [71].

Preclinical studies of C215Fab-SEA in the B16-EpCAM model showed enhanced antitumor effects when used in combination with docetaxel (Figure 24.7) [22] and gemcitabine (Table 24.3). It can be anticipated that the additive or in certain settings synergistic antitumor effects recorded with these combinations were results of the distinct mechanisms of action when combining cytostatics with immunotherapy. It should be emphasized though, that inhibition of immune suppression by docetaxel and gemcitabine has been

FIGURE 24.7 Survival of syngeneic mice with B16-EpCAM lung tumors treated with C215Fab-SEA i.v. (A–C), (A) IFN-α i.p., *Source*: Reproduced from Reference 21, with permission from Elsevier. (B) anti-CTLA-4 mAb i.p., *Source*: Reproduced from Reference 74, with permission from Wolters Kluwer Health. (C) docetaxel i.p., *Source*: Reprinted from Reference 22, with permission from Elsevier, or the combinations and tumor growth in SCID mice with human tumors growing i.p. treated according to (D–F). (D) SCID mice with human renal cell carcinoma Caki-2 growing i.p. were treated with one or two cycles of activated human T lymphocytes (i.p.) day 5 or days 5 and 26 and 50 μg of naptumomab estafenatox (nap) (i.v.) days 5–8 or 5–8 and 26–29 or were treated with 200 μg of sunitinib (sun) (p.o.) days 1–25 and days 30–54, one cycle of activated human T lymphocytes (i.p.) day 26 and 5 μg of nap (i.v.) days 26–29, or the combination. (E) SCID mice with the human colon carcinoma Colo205 growing i.p. were treated with a series of 100 μg bevacizumab (bev) (i.p.) injections, one cycle of activated human T lymphocytes (i.p.) day 5 and 50 μg of nap (i.v.) days 5–8, or the combination. (F) SCID mice with the human NSCLC Calu-1 growing i.p. were treated with 200 μg of docetaxel (doc) (i.p.) at day 7, one cycle of activated human T lymphocytes (i.p.) day 2 and 50 μg of nap (i.v.) days 2–6, or the combination.

TABLE 24.3 Therapy of Tumor Metastases in the B16-EpCAM Tumor Model with C215Fab-SEA in Combination with Gemcitabine

Treatment	Mean Number of Lung Metastases Day 28 (n=8)	Mean Number of Lung Metastases Day 40 (n=8)
Control[a]	117	All animals dead
C215-SEA (10 μg/mouse i.v. days 3–6)	23	21
Gemcitabine (2.4 mg/mouse i.p. days 8, 11, 14, 17)	141	All animals dead
C215Fab-SEA + gemcitabine	11[b]	10[c]

[a]Nontreated tumor inoculated animals.
[b]$p < 0.001$, comparison to gemcitabine alone.
[c]$p < 0.05$, comparison to C215FabSEA alone.

shown [62,72,73] and could potentially have played a positive role inducing TTS effect enhancement. As was shown in the clinical COMBO study with naptumomab estafenatox and in preclinical models, docetaxel reduced production of potentially neutralizing anti-SAg antibodies when given in certain timely combinations with TTS [22].

Preclinical results indicate that T lymphocyte reactivation with TTS may be enhanced by interference with immune suppression and anergy. In attempts to enhance T lymphocyte antitumor effects, we have investigated combinations of TTS with IL-2 [28], IFN-α [21], and the mAbs recognizing CD25, CTLA-4, and Gr-1 [74]. These immune modulators have been shown to enhance immune function or inhibit immune suppression, and in timely combinations with C215Fab-SEA in the B16-EpCAM tumor model, they all show synergistic antitumor activity (Figure 24.7). IL-2 and IFN-α are approved for the treatment of multiple cancers and exerts immunomodulatory effects, which make them appropriate therapeutics to combine with TTS therapy. IL-2 and IFN-α enhanced and sustained CD8+ T-cell activation induced by treatment with the TTS C215Fab-SEA in the B16-EpCAM model, as reflected by increased and prolonged CTL activity as well as by augmented serum IFN-γ levels. C215Fab-SEA synergized with IL-2 and IFN-α in inhibiting tumor growth of pulmonary metastases of EpCAM-expressing B16 mouse melanoma cells as compared to monotherapy. In a long-term survival experiment using the same tumor model, the prolonged median survival time was dramatically increased with the combination of C215Fab-SEA and IFN-α as compared to monotherapy (Figure 24.7). Hence, the combination treatment provoked synergistic antitumor effects as measured by the number of lung tumors and markedly prolonged survival and the enhanced therapeutic efficacy correlated with a striking and sustained increase of CD8- and perforin-expressing tumor-infiltrating cells.

MAbs recognizing CD25, CTLA-4, and Gr-1 were used alone and in timely distinct combinations with C215Fab-SEA in the B16-EpCAM tumor model. CD25 and CTLA-4 are expressed by regulatory T (T_{reg}) cells (also expressing CD4 and Foxp3), and Gr-1 is expressed by granulocytic and monocytic myeloid-derived suppressor cells (MDSCs). Treatment with depleting anti-CD25 or anti-Gr-1 mAbs before the C215Fab-SEA cycle resulted in synergistic antitumor effects similar to the effects recorded with the blocking anti-CTLA-4 mAb when given during the C215Fab-SEA cycle (Figure 24.7). Within 48 h after C215Fab-SEA treatment, EpCAM expressing tumors were infiltrated and targeted by $CD4^+$ and $CD8^+$ T cells. In parallel, $CD3^+CD4^+Foxp3^+$ T_{reg} cells were expanded and infiltrated the tumors as potential modulators of the powerful SAg induced Th1 response. The TTS induced antitumor activity was dampened and short-term reactivation was inhibited possibly due to the suppressive effects by the T_{reg} cells. The expansion of T_{reg} cells was inhibited by anti-CTLA-4 treatment, and the CTL activity was enhanced and easily reactivated by a second TTS cycle. The combination of C215Fab-SEA with anti-CTLA-4 for treatment of B16-EpCAM tumors growing in the lungs of syngeneic mice showed classic synergy as compared to the monotherapies (Figure 24.7). $CD11b^+Gr\text{-}1^+Ly6C^{int}Ly6G^{high}$ and $CD11b^+Gr\text{-}1^+Ly6C^{high}Ly6G^-$ MDSCs increased in numbers in the spleen and in the vicinity of lung tumors in tumor bearing mice at 72–96 h after initiating TTS treatment. To interfere with these potentially suppressive cells, we used a depleting anti-Gr-1 mAb. The combination of C215Fab-SEA with anti-Gr-1 for treatment of B16-EpCAM tumors growing in the lungs of C57Bl/6 mice showed synergy as compared to the monotherapies. The antitumor synergy was seen when the anti-Gr-1 mAb was given the day before start of TTS treatment but not when given the day after last TTS injection. Furthermore, when an anti-Ly6G mAb was given in a similar setting, no additive effects were recorded indicating that the $CD11b^+Gr\text{-}1^+Ly6C^{high}Ly6G^-$ monocytic MDSCs represented the important cell type to target for therapy enhancement. Therefore, antitumor effects by TTS therapy can be greatly enhanced by interfering with T_{reg} cells and MDSC in tumor models suggesting similar beneficial effects in human clinical settings.

Furthermore and to evaluate naptumomab estafenatox in models for combination treatments with anti-angiogenic and cytostaic drugs, SCID mice were used. The compatibility of, and additive antitumor effects by the combination of naptumomab estafenatox with bevacizumab, docetaxel, and sunitinib were established in SCID mouse models with human tumors representing colon carcinomas, NSCLC, and RCC (Figure 24.7). The results indicate that multicycle TTS monotherapy is beneficial as compared to one cycle treatment; the combinations of TTS with bevacizumab, docetaxel, or sunitinib should provide clinical benefits as compared to the monotherapies.

The advantages as documented in the vast series of preclinical experiments with TTS fusion proteins in combination with different anticancer and immunomodulatory drugs are very promising. Two of the combination schedules have so far also been tested in patients, naptumomab estafenatox in combination with docetaxel for treatment of NSCLC (concluded Phase I) and in combination with IFN-α for treatment of RCC (ongoing Phase II/III).

24.7 CONCLUSIONS

TTS fusion proteins have shown proof of principle in preclinical immune pharmacology and antitumor models as well as in clinical trials. Selective immune activation and targeting of T lymphocytes to tumors by TTS therapeutics have been documented. Immune mediated inhibition of tumor growth in the mouse and in patients has been recorded with different TTS entities including the optimized TTS naptumomab estafenatox. Clinical Phase I and Phase IIa data show promising results with RCC and NSCLC patients compatible with present frontline treatments [75–78]. Naptumomab estafenatox has shown a favorable safety profile as compared to many other anticancer drugs and can therefore, in addition to monotherapy, be used in combination with other anticancer treatments. To further increase the therapeutic efficiency of the TTS fusion proteins, treatment combinations to inhibit anti-SAg antibody production with, for example, docetaxel and modulation of immune suppressive activity by anti-CTLA-4 antibody treatment, for instance, have been successfully evaluated. Naptumomab estafenatox is presently in pivotal Phase II/III clinical trials add-on to IFN-α for treatment of RCC.

REFERENCES

1. Dohlsten M, Hedlund G, Akerblom E, Lando PA, Kalland T. (1991) Monoclonal antibody-targeted superantigens: a different class of anti-tumor agents. *Proc. Natl. Acad. Sci. USA* 88, 9287–9291.
2. Dohlsten M, Abrahmsen L, Bjork P, Lando PA, Hedlund G, Forsberg G, et al. (1994) Monoclonal antibody-superantigen fusion proteins: tumor-specific agents for T-cell-based tumor therapy. *Proc. Natl. Acad. Sci. USA* 91, 8945–8949.
3. Dohlsten M, Hansson J, Ohlsson L, Litton M, Kalland T. (1995) Antibody-targeted superantigens are potent inducers of tumor-infiltrating T lymphocytes in vivo. *Proc. Natl. Acad. Sci. USA* 92, 9791–9795.
4. Carlsson R, Fischer H, Sjogren HO. (1988) Binding of staphylococcal enterotoxin A to accessory cells is a requirement for its ability to activate human T cells. *J. Immunol.* 140, 2484–2488.
5. Fischer H, Dohlsten M, Lindvall M, Sjogren HO, Carlsson R. (1989) Binding of staphylococcal enterotoxin A to HLA-DR on B cell lines. *J. Immunol.* 142, 3151–3157.
6. Petersson K, Forsberg G, Walse B. (2004) Interplay between superantigens and immunoreceptors. *Scand. J. Immunol.* 59, 345–355.
7. Choi YW, Kotzin B, Herron L, Callahan J, Marrack P, Kappler J. (1989) Interaction of Staphylococcus aureus toxin "superantigens" with human T cells. *Proc. Natl. Acad. Sci. USA* 86, 8941–8945.
8. Hedlund G, Dohlsten M, Herrmann T, Buell G, Lando PA, Segren S, et al. (1991) A recombinant C-terminal fragment of staphylococcal enterotoxin A binds to human MHC class II products but does not activate T cells. *J. Immunol.* 147, 4082–4085.
9. Kappler J, Kotzin B, Herron L, Gelfand EW, Bigler RD, Boylston A, et al. (1989) V beta-specific stimulation of human T cells by staphylococcal toxins. *Science* 244, 811–813.
10. Schad EM, Zaitseva I, Zaitsev VN, Dohlsten M, Kalland T, Schlievert PM, et al. (1995) Crystal structure of the superantigen staphylococcal enterotoxin type A. *EMBO J.* 14, 3292–3301.
11. Hedlund G, Dohlsten M, Lando PA, Kalland T. (1990) Staphylococcal enterotoxins direct and trigger CTL killing of autologous HLA-DR+ mononuclear leukocytes and freshly prepared leukemia cells. *Cell Immunol.* 129, 426–434.
12. Hedlund G, Dohlsten M, Petersson C, Kalland T. (1993) Superantigen-based tumor therapy: *in vivo* activation of cytotoxic T cells. *Cancer Immunol. Immunother.* 36, 89–93.
13. Dohlsten M, Lando PA, Hedlund G, Trowsdale J, Kalland T. (1990) Targeting of human cytotoxic T lymphocytes to MHC class II-expressing cells by staphylococcal enterotoxins. *Immunology* 71, 96–100.
14. Dohlsten M, Hedlund G, Kalland T. (1991) Staphylococcal-enterotoxin-dependent cell-mediated cytotoxicity. *Immunol. Today* 12, 147–150.
15. Akerblom E, Dohlsten M, Bryno C, Mastej M, Steringer I, Hedlund G, et al. (1993) Preparation and characterization of conjugates of monoclonal antibodies and staphylococcal enterotoxin A using a new hydrophilic cross-linker. *Bioconjug. Chem.* 4, 455–466.
16. Kalland T, Dohlsten M, Lind P, Sundstedt A, Abrahmsen L, Hedlund G, et al. (1993) Monoclonal antibodies and superantigens: a novel therapeutic approach. *Med. Oncol. Tumor Pharm.* 10, 37–47.
17. Forsberg G, Forsgren M, Jaki M, Norin M, Sterky C, Enhorning A, et al. (1997) Identification of framework residues in a secreted recombinant antibody fragment that control production level and localization in *Escherichia coli*. *J. Biol. Chem.* 272, 12430–12436.

18. Forsberg G, Skartved NJ, Wallen-Ohman M, Nyhlen HC, Behm K, Hedlund G, et al. (2010) Naptumomab estafenatox, an engineered antibody-superantigen fusion protein with low toxicity and reduced antigenicity. *J. Immunother.* 33, 492–499.

19. Erlandsson E, Andersson K, Cavallin A, Nilsson A, Larsson-Lorek U, Niss U, et al. (2003) Identification of the antigenic epitopes in staphylococcal enterotoxins A and E and design of a superantigen for human cancer therapy. *J. Mol. Biol.* 333, 893–905.

20. Bjork P, Jonsson U, Svedberg H, Larsson K, Lind P, Dillner J, et al. (1993) Isolation, partial characterization, and molecular cloning of a human colon adenocarcinoma cell-surface glycoprotein recognized by the C215 mouse monoclonal antibody. *J. Biol. Chem.* 268, 24232–24241.

21. Sundstedt A, Celander M, Hedlund G. (2008) Combining tumor-targeted superantigens with interferon-alpha results in synergistic anti-tumor effects. *Int. Immunopharmacol.* 8, 442–452.

22. Sundstedt A, Celander M, Ohman MW, Forsberg G, Hedlund G. (2009) Immunotherapy with tumor-targeted superantigens (TTS) in combination with docetaxel results in synergistic anti-tumor effects. *Int. Immunopharmacol.* 9, 1063–1070.

23. Litton MJ, Dohlsten M, Hansson J, Rosendahl A, Ohlsson L, Kalland T, et al. (1997) Tumor therapy with an antibody-targeted superantigen generates a dichotomy between local and systemic immune responses. *Am. J. Pathol.* 150, 1607–1618.

24. Forsberg G, Ohlsson L, Brodin T, Bjork P, Lando PA, Shaw D, et al. (2001) Therapy of human non-small-cell lung carcinoma using antibody targeting of a modified superantigen. *Br. J. Cancer* 85, 129–136.

25. Lando PA, Dohlsten M, Ohlsson L, Kalland T. (1995) Tumor-reactive superantigens suppress tumor growth in humanized SCID mice. *Int. J. Cancer* 62, 466–471.

26. Litton MJ, Dohlsten M, Lando PA, Kalland T, Ohlsson L, Andersson J, et al. (1996) Antibody-targeted superantigen therapy induces tumor-infiltrating lymphocytes, excessive cytokine production, and apoptosis in human colon carcinoma. *Eur. J. Immunol.* 26, 1–9.

27. Rosendahl A, Hansson J, Sundstedt A, Kalland T, Dohlsten M. (1996) Immune response during tumor therapy with antibody-superantigen fusion proteins. *Int. J. Cancer* 68, 109–113.

28. Rosendahl A, Kristensson K, Carlsson M, Skartved NJ, Riesbeck K, Sogaard M, et al. (1999) Long-term survival and complete cures of B16 melanoma-carrying animals after therapy with tumor-targeted IL-2 and SEA. *Int. J. Cancer* 81, 156–163.

29. Rosendahl A, Kristensson K, Hansson J, Ohlsson L, Kalland T, Dohlsten M. (1998) Repeated treatment with antibody-targeted superantigens strongly inhibits tumor growth. *Int. J. Cancer* 76, 274–283.

30. Rosendahl A, Kristensson K, Hansson J, Riesbeck K, Kalland T, Dohlsten M. (1998) Perforin and IFN-gamma are involved in the antitumor effects of antibody-targeted superantigens. *J. Immunol.* 160, 5309–5313.

31. Litton MJ, Dohlsten M, Rosendahl A, Ohlsson L, Sogaard M, Andersson J, et al. (1999) The distinct role of CD4+ and CD8+ T-cells during the anti-tumour effects of targeted superantigens. *Br. J. Cancer* 81, 359–366.

32. Giantonio BJ, Alpaugh RK, Schultz J, McAleer C, Newton DW, Shannon B, et al. (1997) Superantigen-based immunotherapy: a Phase I trial of PNU-214565, a monoclonal antibody-staphylococcal enterotoxin A recombinant fusion protein, in advanced pancreatic and colorectal cancer. *J. Clin. Oncol.* 15, 1994–2007.

33. Alpaugh RK, Schultz J, McAleer C, Giantonio BJ, Persson R, Burnite M, et al. (1998) Superantigen-targeted therapy: Phase I escalating repeat dose trial of the fusion protein PNU-214565 in patients with advanced gastrointestinal malignancies. *Clin. Cancer Res.* 4, 1903–1914.

34. Lando PA, Dohlsten M, Hedlund G, Akerblom E, Kalland T. (1993) T cell killing of human colon carcinomas by monoclonal-antibody-targeted superantigens. *Cancer Immunol. Immunother.* 36, 223–228.

35. Ouyang Q, Vilien M, Juhl BR, Larsen LG, Binder V. (1987) CEA and carbohydrate antigens in normal and neoplastic colon mucosa. An immunohistochemical study. *Acta Pathol. Microbiol. Immunol. Scand. A* 95, 177–183.

36. Haglund C, Lindgren J, Roberts PJ, Kuusela P, Nordling S. (1989) Tissue expression of the tumour associated antigen CA242 in benign and malignant pancreatic lesions. A comparison with CA 50 and CA 19-9. *Br. J. Cancer* 60, 845–851.

37. Levy M, Visokai V, Lipska L, Topolcan O. (2008) Tumor markers in staging and prognosis of colorectal carcinoma. *Neoplasma* 55, 138–142.

38. Shaw DM, Connolly NB, Patel PM, Kilany S, Hedlund G, Nordle O, et al. (2007) A Phase II study of a 5T4 oncofoetal antigen tumour-targeted superantigen (ABR-214936) therapy in patients with advanced renal cell carcinoma. *Br. J. Cancer* 96, 567–574.

39. Borghaei H, Alpaugh K, Hedlund G, Forsberg G, Langer C, Rogatko A, et al. (2009) Phase I dose escalation, pharmacokinetic and pharmacodynamic study of naptumomab estafenatox alone in patients with advanced cancer and with docetaxel in patients with advanced non-small-cell lung cancer. *J. Clin. Oncol.* 27, 4116–4123.

40. Tordsson JM, Ohlsson LG, Abrahmsen LB, Karlstrom PJ, Lando PA, Brodin TN. (2000) Phage-selected primate antibodies fused to superantigens for immunotherapy of malignant melanoma. *Cancer Immunol. Immunother.* 48, 691–702.

41. Tordsson J, Lavasani S, Ohlsson L, Karlstrom P, Svedberg H, Abrahmsen L, et al. (2000) A3–a novel colon and pancreatic cancer reactive antibody from a primate phage library selected using intact tumour cells. *Int. J. Cancer* 87, 559–568.

42. Riesbeck K, Billstrom A, Tordsson J, Brodin T, Kristensson K, Dohlsten M. (1998) Endothelial cells expressing an inflammatory phenotype are lysed by superantigen-targeted cytotoxic T cells. *Clin. Diagn. Lab. Immunol.* 5, 675–682.

43. Ragnarsson L, Stromberg T, Wijdenes J, Totterman TH, Weigelt C. (2001) Multiple myeloma cells are killed by syndecan-1-directed superantigen-activated T cells. *Cancer Immunol. Immunother.* 50, 382–390.

44. Gidlof C, Carlson B, Dohlsten M, Totterman TH. (1998) Antibody-directed superantigen-mediated T-cell killing of myeloid leukaemic cell line cells. *Eur. J. Haematol.* 60, 233–239.

45. Gidlof C, Dohlsten M, Lando P, Kalland T, Sundstrom C, Totterman TH. (1997) A superantigen-antibody fusion protein

for T-cell immunotherapy of human B-lineage malignancies. *Blood* 89, 2089–2097.

46. Hole N, Stern PL. (1988) A 72 kD trophoblast glycoprotein defined by a monoclonal antibody. *Br. J. Cancer* 57, 239–246.
47. Hole N, Stern PL. (1990) Isolation and characterization of 5T4, a tumour-associated antigen. *Int. J. Cancer* 45, 179–184.
48. Southgate TD, McGinn OJ, Castro FV, Rutkowski AJ, Al-Muftah M, Marinov G, et al. (2010) CXCR4 mediated chemotaxis is regulated by 5T4 oncofetal glycoprotein in mouse embryonic cells. *PLoS One* 5, e9982.
49. Starzynska T, Marsh PJ, Schofield PF, Roberts SA, Myers KA, Stern PL. (1994) Prognostic significance of 5T4 oncofetal antigen expression in colorectal carcinoma. *Br. J. Cancer* 69, 899–902.
50. Southall PJ, Boxer GM, Bagshawe KD, Hole N, Bromley M, Stern PL. (1990) Immunohistological distribution of 5T4 antigen in normal and malignant tissues. *Br. J. Cancer* 61, 89–95.
51. Jones H, Roberts G, Hole N, McDicken IW, Stern P. (1990) Investigation of expression of 5T4 antigen in cervical cancer. *Br. J. Cancer* 61, 96–100.
52. Naganuma H, Kono K, Mori Y, Takayoshi S, Stern PL, Tasaka K, et al. (2002) Oncofetal antigen 5T4 expression as a prognostic factor in patients with gastric cancer. *Anticancer Res.* 22, 1033–1038.
53. Starzynska T, Rahi V, Stern PL. (1992) The expression of 5T4 antigen in colorectal and gastric carcinoma. *Br. J. Cancer* 66, 867–869.
54. Starzynska T, Wiechowska-Kozlowska A, Marlicz K, Bromley M, Roberts SA, Lawniczak M, et al. (1998) 5T4 oncofetal antigen in gastric carcinoma and its clinical significance. *Eur. J. Gastroenterol. Hepatol.* 10, 479–484.
55. Wrigley E, McGown AT, Rennison J, Swindell R, Crowther D, Starzynska T, et al. (1995) 5T4 oncofetal antigen expression in ovarian carcinoma. *Int. J. Gynecol. Cancer* 5, 269–274.
56. Connor ME, Stern PL. (1990) Loss of MHC class-I expression in cervical carcinomas. *Int. J. Cancer* 46, 1029–1034.
57. Griffiths RW, Gilham DE, Dangoor A, Ramani V, Clarke NW, Stern PL, et al. (2005) Expression of the 5T4 oncofoetal antigen in renal cell carcinoma: a potential target for T-cell-based immunotherapy. *Br. J. Cancer* 93, 670–677.
58. Cheng JD, Babb JS, Langer C, Aamdal S, Robert F, Engelhardt LR, et al. (2004) Individualized patient dosing in Phase I clinical trials: the role of escalation with overdose control in PNU-214936. *J. Clin. Oncol.* 22, 602–609.
59. Motzer RJ. (2003) Prognostic factors and clinical trials of new agents in patients with metastatic renal cell carcinoma. *Crit. Rev. Oncol. Hematol.* 46(Suppl), S33–S39.
60. Motzer RJ, Bacik J, Schwartz LH, Reuter V, Russo P, Marion S, et al. (2004) Prognostic factors for survival in previously treated patients with metastatic renal cell carcinoma. *J. Clin. Oncol.* 22, 454–463.
61. Robinson MK, Alpaugh RK, Borghaei H. (2010) Naptumomab estafenatox: a new immunoconjugate. *Expert Opin. Biol. Ther.* 10, 273–279.
62. Kodumudi KN, Woan K, Gilvary DL, Sahakian E, Wei S, Djeu JY. (2010) A novel chemoimmunomodulating property of docetaxel: suppression of myeloid-derived suppressor cells in tumor bearers. *Clin. Cancer Res.* 16, 4583–4594.
63. Montero A, Fossella F, Hortobagyi G, Valero V. (2005) Docetaxel for treatment of solid tumours: a systematic review of clinical data. *Lancet Oncol.* 6, 229–239.
64. Toschi L, Finocchiaro G, Bartolini S, Gioia V, Cappuzzo F. (2005) Role of gemcitabine in cancer therapy. *Future Oncol.* 1, 7–17.
65. Chow LQ, Eckhardt SG. (2007) Sunitinib: from rational design to clinical efficacy. *J. Clin. Oncol.* 25, 884–896.
66. Jenab-Wolcott J, Giantonio BJ. (2009) Bevacizumab: current indications and future development for management of solid tumors. *Expert Opin. Biol. Ther.* 9, 507–517.
67. Schadendorf D, Algarra SM, Bastholt L, Cinat G, Dreno B, Eggermont AM, et al. (2009) Immunotherapy of distant metastatic disease. *Ann. Oncol.* 20(Suppl 6), vi41–vi50.
68. McDermott DF. (2009) Immunotherapy of metastatic renal cell carcinoma. *Cancer* 115, 2298–2305.
69. Rech AJ, Vonderheide RH. (2009) Clinical use of anti-CD25 antibody daclizumab to enhance immune responses to tumor antigen vaccination by targeting regulatory T cells. *Ann. N Y Acad. Sci.* 1174, 99–106.
70. Sarnaik AA, Weber JS. (2009) Recent advances using anti-CTLA-4 for the treatment of melanoma. *Cancer J.* 15, 169–173.
71. Peranzoni E, Zilio S, Marigo I, Dolcetti L, Zanovello P, Mandruzzato S, et al. (2010) Myeloid-derived suppressor cell heterogeneity and subset definition. *Curr. Opin. Immunol.* 22, 238–244.
72. Sinha P, Clements VK, Bunt SK, Albelda SM, Ostrand-Rosenberg S. (2007) Cross-talk between myeloid-derived suppressor cells and macrophages subverts tumor immunity toward a type 2 response. *J. Immunol.* 179, 977–983.
73. Suzuki E, Kapoor V, Jassar AS, Kaiser LR, Albelda SM. (2005) Gemcitabine selectively eliminates splenic Gr-1+/CD11b+ myeloid suppressor cells in tumor-bearing animals and enhances antitumor immune activity. *Clin. Cancer Res.* 11, 6713–6721.
74. Sundstedt A, Celander M, Eriksson H, Törngren M, Hedlund G. (2012) Monotherapeutically nonactive CTLA-4 blockade results in greatly enhanced antitumor effects when combined with tumor-targeted superantigens in a B16 melanoma model. *J Immunother*. 35(4): 344–353.
75. Rini BI. (2009) Metastatic renal cell carcinoma: many treatment options, one patient. *J. Clin. Oncol.* 27, 3225–3234.
76. Di Lorenzo G, Autorino R, Sternberg CN. (2009) Metastatic renal cell carcinoma: recent advances in the targeted therapy era. *Eur. Urol.* 56, 959–971.
77. Azzoli CG, Baker S, Jr, Temin S, Pao W, Aliff T, Brahmer J, et al. (2009) American society of clinical oncology clinical practice guideline update on chemotherapy for stage IV non-small-cell lung cancer. *J. Clin. Oncol.* 27, 6251–6266.
78. Langer CJ, Besse B, Gualberto A, Brambilla E, Soria JC. (2010) The evolving role of histology in the management of advanced non-small-cell lung cancer. *J. Clin. Oncol.*

IIC TARGETING: FUSION PROTEINS ADDRESSING SPECIFIC CELLS, ORGANS, AND TISSUES

25

FUSION PROTEINS WITH A TARGETING FUNCTION

Stefan R. Schmidt

Rentschler Biotechnologie GmbH, Laupheim, Germany

25.1 INTRODUCTION

This part deals with the many approaches to selectively target specific organs, cells, or even intracellular localizations with fusion proteins.

This requires a high specificity to epitopes that are unique for the respective targets. Over the last decades, systematic investigations identified a long list of cell-specific surface antigens. This is of primary interest for targeted therapies against cancer, where it is critical to only address malignant cells while sparing normal cells. Cancer cells frequently overexpress growth factor receptors or specific tumor-associated antigens [1]. In some cases, even intracellular proteins such as heat shock proteins appear aberrantly on the surface of cancer cells, and can be reached by antibodies [2]. Intracellular targets sometimes become accessible because of an increase of abnormally permeable and degenerating cells in necrotic regions of tumors. These epitopes are highly selective cancer targets since they are not accessible in normal tissues [3]. However, targeted therapy is nowadays also focusing on the processes that lead to malignant transformation, proliferation, survival, and angiogenesis [4]. In addition, intracellular mislocalization of proteins can cause disease and generate targets for therapeutics [5]. Other molecules that can be found on cell membranes are receptors, channels, or adhesion proteins. Recognition motifs for cellular receptors include protein sequences or oligosaccharide patterns. In the case of the extracellular matrix, glycoproteins and collagen fibrils build a network to which the receptors such as integrins can bind. Both receptors and matrix molecules play a role in human disease and represent targets for intervention [6].

Components of therapeutic fusion proteins addressing surface molecules can be the natural ligands, peptide domains, antibodies, or other designer molecules based on novel scaffolds [7]. Targeted drug delivery is a highly relevant strategy, particularly for large therapeutic molecules such as proteins to separate them from regions in the body where they are quickly degraded or because systemic distribution might never reach a sufficiently efficacious dose.

25.2 TARGETING ORGANS

Targeting to certain organs is quite useful from a pharmacological perspective. The desirable accumulation of drugs in specific tissues on one hand increases the local concentration while on the other hand reduces systemic exposure that could be harmful. The challenge is to find highly specific surface molecules and simultaneously to identify ways to overcome organ boundaries.

25.2.1 Brain

Targeting the brain is more complicated than reaching any other organ, since the brain is separated from normal blood flow by the blood–brain barrier (BBB). This barrier is formed by brain capillary endothelial cells. These cells are fused together by tight junctions, making them impermeable [8].

Fusion Protein Technologies for Biopharmaceuticals: Applications and Challenges, First Edition. Edited by Stefan R. Schmidt.

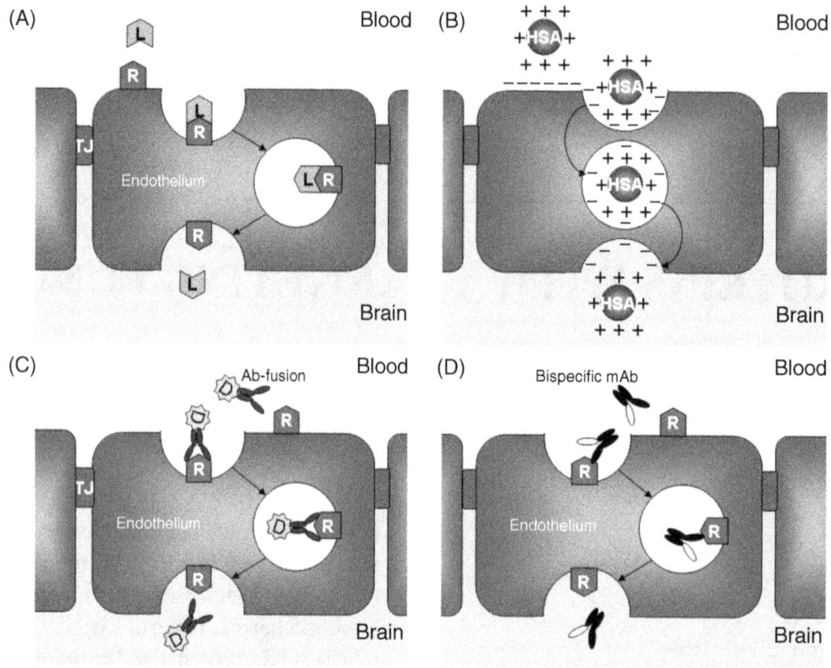

FIGURE 25.1 Transport of large molecules through the blood–brain barrier (BBB). The BBB is formed of endothelial cells that are connected through tight junctions (TJ). (A) Large molecules that are the ligands (L) of specific receptors (R) at the luminal side of the BBB are transported via receptor-mediated transcytosis (RMT) to the apical side. Typical ligands are insulin, transferring, or fusion proteins. (B) Other large cationic molecules such as human serum albumin (HSA) can enter the brain by adsorptive-mediated transcytosis. This principle relies on the negative charge of endothelial cells at the luminal side. (C) In the molecular Trojan horse (MTH) approach, antibodies that bind to receptors of the RMT pathway are utilized. Fusion of drug (D) molecules to the antibody facilitates a passage into the brain. (D) Similarly, MTH can be performed with bispecific antibodies as well. One arm docks to the receptor, the other arm binds a brain-specific target.

Only small lipid soluble molecules are able to pass by lipid-mediated free diffusion. Other molecules depend on transporter proteins or receptor mediated transcytosis (RMT). The receptors for insulin (IR) and transferrin (TfR) are highly expressed at the BBB, thus being interesting targets for transporting molecules to the brain [9]. Current strategies for BBB transport are summarized in Figure 25.1.

Initially, transferrin (Tf) fusion proteins were designed, which should cross the BBB via the TfR. The first described construct consisted of Tf fused to an antibody and successfully passed over the BBB. However, the concentration of this fusion protein in the brain reached only $^1/_3$ of an anti-TfR antibody [10]. Later, Tf connected by helical linkers to nerve growth factor (NGF) as payload was tested in cell culture, but never *in vivo* [11].

Other members of the transferrin family are lactoferrin (Lf) and melanoferrin (Mf). Both can cross the BBB as well and have been used as carrier molecule. Despite the sixfold shorter plasma half-life of Lf than Tf, the brain uptake was much better for Lf when tested in rats [12]. On the other hand, Mf has 10–15 times higher brain transport rates than the other iron-binding proteins Tf and Lf. After an initial accumulation of Mf in liver and kidney, the level of Mf in brain begins to rise to a maximum uptake at 6 h post intravenous (i.v.) injection. Overall, the enrichment of Mf in brain is better than many other carrier systems [13]. Although no fusion proteins utilizing Mf or Lf have been described so far, both have a good potential as transporter. Right now they served mainly as large molecule carrier to shuttle conjugated small molecules across the BBB. This approach might be superior to direct fusion of therapeutic proteins or peptides since more drug molecules can simultaneously be transported by a single carrier.

Owing to the excellent uptake of the antibody against TfR, a wide range of fusion partners was evaluated. Simultaneously, antibodies against the IR were generated. Comparing the BBB delivery of TfR- and IR-directed antibodies, the IR antibody was ninefold better, and therefore selected for all further therapeutic development [14]. Utilizing an antibody as molecular Trojan horse (MTH) combines the advantage of long half-life and stabilization of labile neurotrophic factors with the potential of dimerization. So far, molecules from the class of neurotrophic factors, lysosomal enzymes, and antibodies have been fused to the MTH antibodies [15]. One particularly interesting approach is the removal of Aβ amyloid fibrils through a specific single-chain antibody fragment (scFv) connected to the IR antibody. This fusion protein reaches the brain through IR-mediated import. In the brain, the scFv binds to the amyloid plaque, disintegrating the fibrils. Aβ molecules captured by the antibody are then exported via a neonatal Fc receptor (FcRn) mechanism. This asymmetric reverse transcytosis of antibody molecules reduces the Aβ concentration in the brain significantly [16].

Although antibodies against the TfR have proven to be less efficient for RMT, a new approach was launched quite recently. Here the affinity of the antibody against the TfR was modulated, and it was demonstrated that lower affinity anti-TfR antibody variants reach higher brain concentrations. The underlying mechanism is that high affinity antibodies remain bound to the TfR at the BBB, whereas low affinity ones are released and can penetrate the brain. The selected antibody was configured as a bispecifc molecule whose other arm binds and inhibits β-secretase (BACE1). This enzyme digests the amyloid protein into the Aβ peptides causing Alzheimer's disease (AD). The new molecule could be a novel therapeutic for AD [17].

Another mechanism for BBB transport involves the low-density lipoprotein receptor-related protein (LRP). In a transcytosis process, LRP shuttles other proteins containing a Kunitz protease inhibitor (KPI) domain across the BBB. Initially, this concept was validated by demonstrating the transfer of aprotinin, which contains a KPI, to the brain. The transport was 10 times more efficient than transferrin uptake into the brain. By a bioinformatic approach, further natural peptide sequences were identified sharing homology to the C-terminal region of aprotinin. Some were further optimized by site-directed mutagenesis and had up to sevenfold better efficacy in transcytosis than the original aprotinin [18]. The involvement of LRP suggests a receptor-mediated transport mechanism through the low-density lipoprotein (LDL) receptor. So far, no fusion proteins using these peptide sequences have been reported, but this seems to be a logical next step.

Recently, an Apolipoprotein A1 and interferon α (ApoA1-IFNα) fusion protein was described that is capable to cross the BBB in a saturable manner involving a receptor [19]. Therefore, ApoA1 might prove to be a valuable addition to the arsenal of BBB crossing protein motifs.

Besides the RMT over the BBB, other strategies to target the brain rely on viral or bacterial proteins that are involved in infections of the central nervous system. For example, the neurotropic virus coat glycoprotein of rabies virus can be split into peptides. One such 43 amino acid containing peptide (RDP) is sufficient for brain targeting. The mechanism utilizes retrograde axonal transport via the peripheral nerves into the brain. To demonstrate the ability of brain targeting, a fusion protein comprising β-galactosidase (β-Gal) and the RDP peptide was generated. After 1 h after i.v. injection, enzymatic activity could be detected in the hypocampus of mice [20]. A shorter peptide from rabies virus coat glycoprotein with only 39 residues was tested with β-Gal, luciferase, and brain-derived neurotrophic factor (BDNF) in a similar setting. The uptake into neuronal cells was very fast and the BDNF fusion protein protected nerves in a stroke animal model [21].

Another class of neuronal targeting proteins can be derived from tetanus toxin C (TTC). The full length Tetanus neurotoxin has a high specificity for the synapses of motor neurons. Similarly, to RDP, the Tetanus toxin is carried to the central nervous system by retrograde axonal transport. TTC represents the carboxy-terminal domain of the post-translationally processed heavy chain. TTC is nontoxic and the key domain required for internalization of the toxin. A wide range of large molecules has been fused to TTC and injected into animals [22]. Several studies were undertaken with a glial-derived neurotrophic factor (GDNF)-TTC fusion. GDNF-TTC was transported fast to the central nervous system of mice after intramuscular injection [23]. A similar GDNF-containing construct increased survival of mice being symptomatic for amyotrophic lateral sclerosis (ALS) by 9 days when administered intramuscularly [24].

For the sake of completion, it should also be mentioned that cell-penetrating peptides such as TAT are also able to cross BBB. This was demonstrated with a number of fusion partners, the earliest being β-Gal [25]. A more recent example is GDNF that was used to protect mice from ischemic stroke [26]. Further details on CPP can be seen in Section 25.3.2 and Chapter 26.

25.2.2 Liver

The liver as the major organ for metabolism is involved in a number of diseases and is the target for a wide range of drugs. However, so far, only limited success could be reported for targeted delivery of large biological therapeutics to the liver. Frequently, specific receptors of glycoproteins such as the asialoglycoprotein receptor (ASGP-R) are used as a means to address this organ [27]. Other approaches utilize cyclic peptides containing an Arg-Gly-Asp (RGD) motif that binds to the collagen type VI receptor on hepatocytes [28]. The highly upregulated platelet-derived growth factor β receptor (PDGFβR) in cystic fibrosis represents another molecule for targeted intervention. Cyclic peptides with a high affinity to PDGFβR were coupled to albumin and interferon-γ. These conjugates were specifically taken up by hepatic stellate cells (HSC) in an acute liver injury mouse model and improved vital functions in liver [29]. Despite the increasing understanding of binding motifs, none of these strategies has led to the design of liver-specific fusion proteins yet.

Interferon-α (IFN-α) that is applied to treat chronic viral hepatitis is one of the therapeutic proteins that would benefit from liver targeting. Recently, a new molecule was described that utilizes the presence of scavenger receptor class B type I (SR-BI) on liver cells. This receptor binds high-density lipoproteins (HDLs) with their main component apolipoprotein A-I (ApoA-I). Combining ApoA-I with IFN-α in a fusion protein solves two issues simultaneously. It provides liver targeting and prolongs the half-life of the relatively small IFN-α. The circulation of this bifunctional protein is shorter than an albumin fusion protein, but ApoA-I-IFN-α is more potent and less toxic than IFN-α alone [30]. A more detailed description of this concept can be found in Chapter 29.

25.2.3 Lung

The lung as organ with direct contact to the environment could be a preferable area for targeted therapies. In the past years, inhalable insulin was in the public focus [31]. Additionally, studies of pulmonary uptake of somatropin became available [32]. But both molecules are rather small, and not as complicated to deliver as larger proteins.

Similar to transport across the BBB, the TfR has also been evaluated for the epithelial transcytosis of large biomolecules in the lung. In a first attempt, granulocyte colony-stimulating factor (G-CSF) was conjugated to Tf, and a threefold higher apical-to-basolateral transcytosis was demonstrated in primary cultured rat alveolar epithelial cell monolayers for the conjugate, compared to G-CSF alone [33].

The best-studied approach for pulmonary delivery of proteins relies on the presence of the FcRn in lung epithelia. This receptor can also be found in the kidney and intestine. There it is responsible for the uptake of maternal immunoglobulin from the milk and the transfer to the bloodstream. However, the receptor is still present in adult humans, too. On the basis of this mechanism, the FcRn can be used as entry point for pulmonary delivery. The FcRn is expressed ubiquitously in alveolar epithelia cells, but is higher expressed in the upper and central airways, the bronchi. Most of the Fc-fusion proteins that so far have been tested for inhaled administration were monomers with one of the heavy chains without a therapeutic protein attached to it. Monomeric erythropoietin (EPO) Fc fusion had a twofold better affinity to the FcRn and a 30% longer half-life than the dimeric variant. Compared to the original EPO, monomeric EPO-Fc achieved a much higher systemic concentration when administered at equimolar concentrations in primates [34]. The better distribution of the monomer might be an effect of smaller size and better transport, less steric hindrance, and lower tendency to adhere to mucosal surfaces owing to only one glycosylated EPO molecule. In a Phase I clinical study, an EPO-Fc dimer reached 70% bioavailability in the central lung regions [35]. Recently, an interferon-β (IFN-β) monomeric Fc fusion was tested for inhalation in primates. The monomer was fourfold more active than the corresponding dimer, but still 2.3-fold less potent than the original IFN-β. However, the presence of the Fc-part improved the half-life threefold [36]. Besides IFN-β, IFN-α also was successfully tested as monomeric Fc-fusion protein for pulmonary delivery [37]. Monomeric Fc-fusion proteins are discussed in Chapter 7.

In some instances, lung epithelia might not be accessible from the airway. For example, in cystic fibrosis (CF) a thick mucus layer covers the lung surface. In this case, delivering therapeutic proteins is almost impossible by inhalation. Therefore, a novel strategy was designed that reaches the lung epithelia through the polymeric immunoglobulin receptor (pIgR). This receptor is present on the basolateral surface of epithelial cells. There it internalizes and transfers polymeric antibodies (IgA and IgM) to the apical surface of the cell where the antibodies are released to the lumen. This transfer is essentially irreversible because the secretory component (SC) to which the polymeric Igs are attached to is cleaved from the receptor [38].

Many CF patients suffer from bacterial infections, to which the body reacts with the release of neutrophil elastase (NE), a protease that destroys *Pseudomonas*, but also elastin in the airways. Usually, NE is inhibited by α_1-antitrypsin (A1AT), but the natural secretion is insufficient to counteract the massive NE release in CF. Therefore, A1AT was fused to a single-chain Fv antibody directed against the SC of the pIgR. In a human tracheal xenograft model, which resembles human lung epithelia on nude mice, it could be demonstrated that the SC-specifc Fv-A1AT fusion protein was successfully transferred from the blood stream to the apical lumen of lung epithelia. It was found that the fusion protein was sevenfold more enriched in the lumen than A1AT without the receptor-specific Fv. In general, the transfer efficiency was more dependent on the expression level of pIgR than on the concentration of the fusion protein [39]. Immunoglobulin-based transport in the lung is summarized in Figure 25.2.

The upper airways and the nasal epithelia as entry point for viral infections have been utilized as target for a novel concept to protect from influenza. Influenza A and B viruses bind with their hemagglutinin-neuraminidase to surface sialic acids. The prototypic *N*-acetylneuraminic acid (Neu5Ac) is usually connected to the penultimate galactose residues either by α(2,3)- or α(2,6)-linkages. Particularly, the Neu5Ac-α(2,6)-Gal that can be found on respiratory epithelia is the preferred receptor for human viruses. It has been demonstrated that the absence of this specific glycosylation drastically reduces the infection efficiency and protects from influenza viruses. Therefore, a fusion protein was designed that contained an exoglycosidase connected to a cell surface anchoring sequence. As broad-spectrum glycosidase, the catalytic domain of a sialidase from *Actinomyces viscosus* was selected, whereas the targeting domain was derived from the human protein amphiregulin. This 20 amino acid long

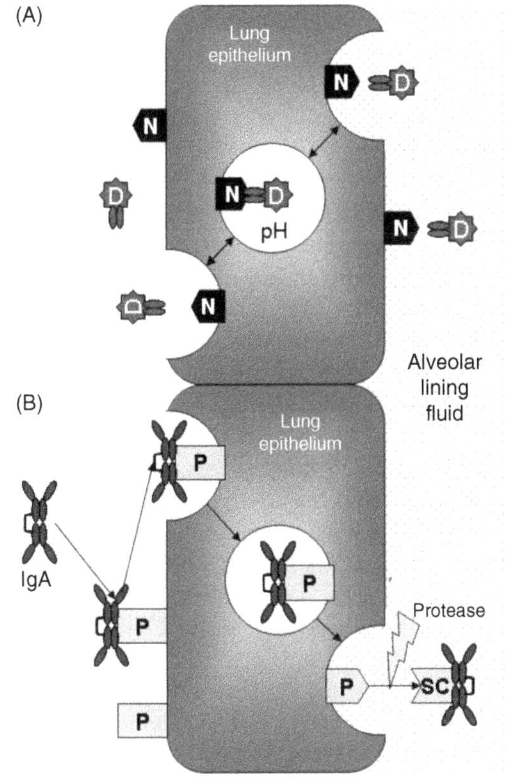

FIGURE 25.2 Transport to and from the lung. The lung is an entry point for therapeutic proteins as well as a target. (A) The neonatal (N) Fc receptor binds antibodies, Fc-fusion proteins (D), or albumin containing formulations. This transcytosis can transfer proteins bidirectionally and relies on pH-dependent binding to N. (B) In cases where pulmonary delivery is impossible, transport to the lung can be facilitated through the polymeric immunoglobulin receptor (P). It transfers polymeric antibodies such as IgA from the basal to the apical side. This passage is essentially irreversible because the secretory component (SC) to which IgA is attached to is cleaved from the receptor.

domain has a high affinity to heparin and glycosaminoglycans that can be found on surface of the respiratory epithelial [40]. Although the enzyme is not of human origin, a low level of immunogenicity can be expected, because *A. viscosus* is a normal inhabitant of human oral and gastrointestinal epithelia. The concept of this drug avoids the generation of resistances because it alters the structure of the human receptor and does not directly address a viral protein. This fusion protein, also called Fludase®, has shown both safety and efficacy in animal studies [41,42]. Currently, Phase II clinical trials are ongoing[1].

[1] http://clinicaltrials.gov/ct2/show/NCT01037205

25.2.4 Bone

Bones are highly specialized parts of the body that constitute the endoskeleton. Therefore, it is very relevant to develop specific drug delivery techniques to treat bone-related diseases, such as hypophosphatasia (HPP). In HPP, a critical enzyme, the tissue-nonspecific isozyme of alkaline phosphatase (TNALP) is nonfunctional, which leads to extracellular accumulation of phosphate containing substrates that cause rickets or osteomalacia. A fusion protein was designed consisting of a truncated TNALP without its hydrophobic C-terminus connected to an Fc-part and a deca-aspartate sequence (D10) to target mineralizing tissue. This construct could rescue mice with a TNALP deletion from all skeletal and dental disease or epilepsy. Furthermore, a positive correlation between dosis and survival was demonstrated [43]. The bone-targeting sequence of aspartate repeats was first identified when analyzing several noncollagenous bone proteins, which contained all repetitive acidic amino acid repeats of aspartate or glutamate that could bind to hydroxyapatite (HA) which is the main constituent of hard tissue. Fluorescein conjugated to hexameric aspartate had a bone half-life of 14 days in contrast to 1 h for fluorescein alone [44]. Several acidic oligopeptide-tagged drugs have been tested in the meantime for their ability of bone targeting [45]. More details on the alkaline phosphatase fusion protein are explained in Chapter 28.

25.2.5 Thrombus Targeting

In complex organisms, not only organs, tissues, and cells exist, but also aggregates of large molecules can be found. Some of these aggregates fulfill a natural function such as blood clotting, which is an essential functionality that allows survival of injuries that otherwise would lead to lethal blood losses. However, vascular blood clots cause serious diseases such as myocardial infarction, pulmonary embolism, or deep vein thrombosis. Treatment of these various forms of thrombi is always accompanied by a hemorrhagic risk. This dilemma between maximal thrombolysis and minimal bleeding risk has led to the development of novel fusion proteins. They combine the fibrinolytic potential of staphylokinase (SAK) with the anticoagulant activity of hirudin (HIR) and recognition sequences that target coagulation factors.

SAK binds to plasminogen; this complex then converts plasminogen to plasmin, which dissolves fibrin clots. Treatment with SAK should always be combined with an anticoagulant to prevent the formation of secondary clots. One suitable agent is hirudin that blocks thrombin, thus prohibiting the conversion of fibrinogen to fibrin. However, HIR can cause systemic bleeding. Based on the experience that blocking the N-terminus of HIR reduces activity and rapid degradation of SAK as C-terminal fusion partner the novel construct-contained SAK at the N-terminus linked by a factor Xa (FXa) sensitive spacer to HIR. This fusion protein combines three features: presence of active FXa in proximity to a thrombus, attenuation of HIR activity by an N-terminal extension, and therefore low systemic anticoagulant activity [46]. A variant of this molecule was equipped with a thrombin-sensitive spacer replacing the FXa recognition domain. This fusion protein benefits from the high concentration of thrombin at blood clots and the ability of free HIR to inhibit thrombin. So, the release of HIR is efficiently regulated by the amount of thrombin present. This fusion protein demonstrated an improved thrombolytic effect and a lower bleeding risk in animal studies [47].

25.3 INTRACELLULAR DELIVERY

The vast majority of currently approved and marketed antibodies is directed against extracellular targets. This follows the natural task of humoral response to neutralize infectious agents or foreign objects such as bacteria and viruses by capturing them. Binding of antibodies to surface antigens labels the cells for later destruction by the immune system. Owing to their large size of ~150 kDa antibodies would usually not be able to move across the cell membrane. However, it would be attractive to utilize the high specificity of antibodies also to address intracellular targets. In many cases, they are internalized through receptor-mediated endocytosis. It is not always obvious which type of receptor supports internalization; therefore, a functional screen was established to identify both targets and antibodies that allow intracellular delivery. For this purpose, a Fab library was created that links the antibody fragment to the intracellularly active Pseudomonas exotoxin A. Only Fab fragments binding to internalizing epitopes were toxic to the cells. This technique was called fusogenic and served to find new epitope–antibody pairs that might be useful to treat cancer and allow the characterization of cellular entry points [48].

25.3.1 Enzyme Replacement Therapy

A special case for targeted therapy is enzyme replacement. In a number of rare diseases, a single enzyme is dysfunctional and immediate cure can only be achieved by supplying cells with the required enzyme from external sources. However to accomplish that, the respective enzyme has to be transferred into the cell to be localized in organelles such as the lysosome. This type of enzymatic diseases is therefore termed lysosomal storage disease (LSD). The uptake of the enzymes is often mediated through the mannose/*N*-acetylglucosamine (Man/GlcNAc) receptor. Unfortunately, heterologous protein expression in mammalian systems does not lead to that glycosylation of the enzyme, which is required to be efficiently taken up by the defective cells. Therefore, a complicated process was established to artificially modify the glycosylation of glucocerebrosidase (GCD), the enzyme lacking in Gaucher's disease, to obtain terminal mannose residues to facilitate enzyme internalization. Alternatively, high mannose glycosylation could be achieved by changing to a plant system as expression host and forcing the trafficking of the recombinant enzyme to the storage vacuole where the mannosylation is generated. This requires the fusion of GCD with a C-terminal targeting and retention signal encoding the sequence DLLVDTM [49]. It could be demonstrated that macrophages from patients with Gaucher's disease efficiently internalized this GCD fusion protein and maintained its functionality in the cells. Currently, this protein has completed Phase III trials and awaits marketing approval [50].

An alternative entry point for enzyme replacement therapy (ERT) is the low-density lipoprotein receptor (LDLR) family. Bound ligands of these receptors are delivered directly to the lysosome. During the expression, LDLRs bind the receptor-associated protein (RAP) to prevent any association with their normal ligands during their passage through the secretory pathway toward the cell surface. The high affinity of RAP and the absence of RAP in the plasma can be exploited as a tool to mediate specific protein uptake via the LDLR to the lysosome. For this purpose, RAP was fused to the N-terminus of α-L-iduronidase (IDU) and acid α-glucosidase (GAA), the enzymes missing in Hurler/Scheie syndrome and in Pompe syndrome, respectively. For the production of these constructs, mutant Chinese hamster ovary (CHO) cell lines had to be used to prevent association of RAP to endogenous LDLR already during the expression. Both RAP-GAA and IDU fusions were 70-fold and 43-fold better internalized by primary fibroblasts from Pompe and Hurler disease patients than the enzymes without RAP fusion. Both fusion constructs had a high stability in serum, but were processed in the lysosome, which resulted in cleavage from the RAP domain. The much higher endocytosis capacity of the LDLR system is certainly an advantage compared to mannose receptors. Owing to the trans-epithelial trafficking ability of some LDLR, the RAP fusion system could even enable ERT over the BBB [51].

Despite the success with mannose 6-phosphate (Man6-P) glycosylated enzymes, this approach suffers sometimes from manufacturing issues or rapid clearance from the bloodstream by cross-reactive receptors. Therefore, another approach for ERT was established, the glycosylation-independent lysosomal targeting (GILT). It is based on the ability of insulin-like growth factor 2 (IGF-2) to bind to the bifunctional, IGF-2 cation-independent Man6-P receptor. This is particularly useful if cells must be targeted that lack the mannose receptor. Further advantages are that IGF-2 binds to a different site on the receptor than Man6-P and that the entry point to the endocytic pathway for lysosomal targeting is identical to Man6-P glycosylated proteins. IGF-2 had to be truncated to the amino acids 8–67 to avoid binding to other receptors and was fused to the C-terminal end of β-glucuronidase (GUS). This fusion protein was effective in clearing pathological lysosomal storage in a mouse model of mucopolysaccharidosis type VII. Interestingly, the GILT variant of GUS could also reach osteoblasts that could not be treated with the original enzyme [52]. Fusion protein based strategies for LSD are summarized in Figure 25.3.

A receptor-independent enzyme delivery to treat LSD utilizes the protein transduction domain of the human immunodeficiency virus (HIV) transacting activator of transcription (TAT) protein. GUS connected to the 11 amino acids of TAT at the C-terminus is internalized by adsorptive endocytosis in a receptor and glycosylation independent manner. The underlying mechanism is dependent on a positive net charge of the molecule that can adhere to negatively proteoglycans at the cell surface. Overall, the GUS-TAT fusion was less efficiently uptaken by this transport than via the Man6-P receptor, but was more efficiently delivered to the kidney [53]. The independence of Man6-P receptors could also be demonstrated with glucocerebrosidase-TAT fusions that were delivered to fibroblasts lacking the Man6-P receptors [54]. TAT as fusion partner was also successfully employed to restore lipoamide dehydrogenase (LAD; also known as E3) deficiency. TAT connected to the amino terminus of LAD was delivered to the mitochondria of E3-deficient mice in liver, heart, and brain [55].

25.3.2 Cell Penetrating Peptides

The aforementioned TAT peptide belongs to the family of cell-penetrating peptides (CPP). A number of these peptides have been used as targeting modules of fusion proteins. One example is the EPO-TAT fusion for brain delivery.

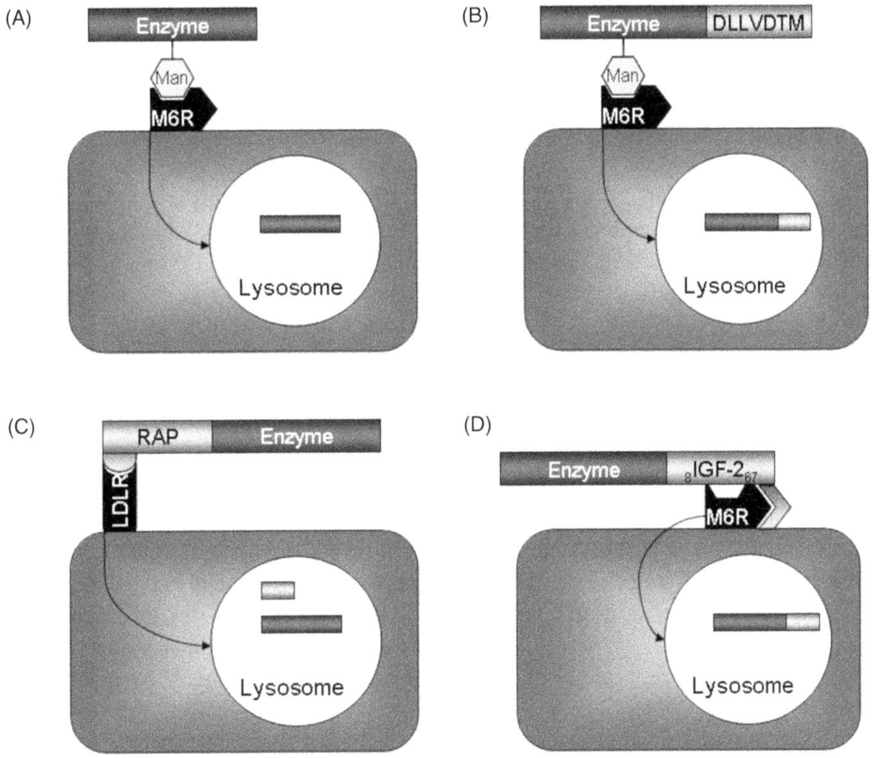

FIGURE 25.3 Enzyme replacement strategies for lysosomal targeting. Many lysosomal storage diseases can be treated by enzyme replacement strategies. (A) Originally lysosomal enzymes contain a specific glycosylation signature (Man) that mediates their transfer to the lysosome via mannose receptors (M6R). (B) Enzymes can be artificially mannosylated by attaching a short peptide sequence to their C-terminus. When expressed in plants, proteins with this tag are translocated to the storage vacuole where they are mannosylated. (C) Instead of the mannose receptor, the low-density lipoprotein (LDLR) receptor can be utilized as well. Ligands of LDLR are transferred to the lysosome. The receptor-associated protein (RAP) binds with high affinity to LDLR but is cleaved off from its payload in the lysosome. (D) The cation-independent mannose-6-phosphate receptor does also recognize of insulin-like growth factor 2 (IGF-2). A truncated form of IGF-2 fused to the C-terminal of an enzyme targets the fusion protein to the lysosome. This is called glycosylation-independent lysosomal targeting.

Intravenous administration of EPO-TAT resulted in a sixfold higher cerebrospinal fluid (CSF) concentration than EPO without the TAT peptide, indicating an improved BBB crossing [56]. A more universal concept is the combination of CPP with antibodies to reach intracellular targets. In a first attempt, TAT was fused to the B domain of staphylococcal protein A, which can bind the Fc fragment of IgGs. The noncovalent complex of the fusion protein and a fluorescently labeled, attached antibody could penetrate mouse fibroblast cells as detected by confocal microscopy [57].

Many other CPPs have been identified and characterized since the first observation that TAT could translocate to the nucleus [58]. But all share common features; CPPs are typically less than 30 amino acids long and have a high positive net charge. The simplest form is polyarginine (Arg_8). Besides polycationic peptides, amphipatic variants can also facilitate cell penetration. The transmembrane transport works with both: covalent linkage to the payload and noncovalent complexes. One general downside of CPP is the lack of specificity for cell types that results from the receptor independent uptake. Instead of receptors, the interaction between CPPs and the cell takes place via cell surface proteoglycans. Clustering of these surface molecules triggers the remodeling of the actin network and activates small GTPases. After this event, the CPPs can enter the cell either through changed membrane fluidity or through perturbation, clathrin, or caveolin-dependent endocytosis or macropinocytosis. The major limitation after passage into the cell is the escape from the endosome [59].

The use of CPP which have also been called protein transduction domains (PTD) has of course included antibody fragments as well. Antibodies fused to PTD can be termed transbody because they are able to translocate across cell membranes. An alternative to transbodies are intrabodies that rely on direct expression in a cell after transfection with recombinant DNA. However, there a safety concerns that restrict this form of gene therapy. Cell permeable antibodies however would certainly be safer, but suffer from a limited half-life [60].

CPPs are the carrier of choice to transfer proteins into the cell to correct an aberrant cellular status caused by the absence or malfunction of a protein. A good example is the introduction of the tumor suppressor p53 into cancer cells to restore normal function. The TAT-p53 fusion protein increased life span of mice suffering from terminal peritoneal carcinomatosis sixfold and even created disease free animals [61]. Further applications and examples of CPPs can be found in Chapter 26.

25.3.3 Antibody-Mediated Transduction

Cancer therapy that utilizes the introduction of tumor suppressor proteins can also be initiated by antibody-mediated transduction (AMT). This concept is based on the observation that certain autoantibodies can enter cells. One of these antibodies, called 3E10 from mice with lupus nephritis, is directed against DNA but not cytotoxic. Initially, the full murine antibody was conjugated to catalase. After incubation with the conjugate, the enzyme reached the cytoplasm and nuclei of cells and remained active. To further improve cellular uptake, one aspartate residue in the V_H region was exchanged to asparagine. Additionally, the antibody was miniaturized to a single chain fragment that maintained the full cell penetration ability [62]. The intracellular route of 3E10 was unclear at the beginning. But then it was found out that the nuclear transport of this autoantibody is dependent on the presence of the equilibrative nucleoside transporter ENT2. This finding confirmed that here the nucleoside salvage pathway is used for protein transport [63].

Owing to the unique nucleus targeting, the scFv variant of 3E10 was also tested in fusion to a p53-derived peptide. Since the elimination of cancer cells with a carboxy-terminal p53 fragment was not satisfactory, in later studies, full-length p53 was used. In the majority of tested cancer cells, the killing efficiency reached 90% [64]. The same fusion protein significantly reduced liver metastases in a colon cancer model in mice *in vivo* [65]. A further example is the 3E10-mediated transfer of nuclear transcription factor, FOXP3, into cancer cells. The fusion protein induced apoptosis and diminished tumor burden in a Balb/c mice colon cancer model [66].

Besides its applications in cancer therapy, the 3E10 scFv was also used to transfer microdystrophin to myoblasts to treat Duchenne muscular dystrophy [67] or heat shock protein 70 (Hsp70) to protect neurons from hydrogen peroxide toxicity [68].

25.3.4 Endosome Escape and Cytosol Translocation

The most frequently used strategy for internalization relies on receptor-mediated uptake by endocytosis. However, the bound ligands end up in the endosome that later fuse with the lysosome with the risk of degradation by low pH or proteases. Therefore, mechanisms are needed that overcome endosomal entrapment of internalized molecules. Natural proteins that are able to reach the cytosol after internalization are the different bacterial toxins. Figure 25.4 gives a schematic overview of the best-studied toxins containing a translocation domain.

The Pseudomonas exotoxin A (PE) consists as many other toxins from microorganisms of three domains with different functions. Before entering the cell, the C-terminal lysine is removed by a protease to expose the sequence REDL at the C-terminus. Domain I binds to the α2-macroglobulin receptor/low-density LRP which is internalized via coated pits. Proteolytic cleavage by furin and the reduction of disulfides separates the N-terminal binding domain I from the C-terminal rest. Passage to the cytosol is enabled by retrograde endoplasmic transport through the trans-Golgi

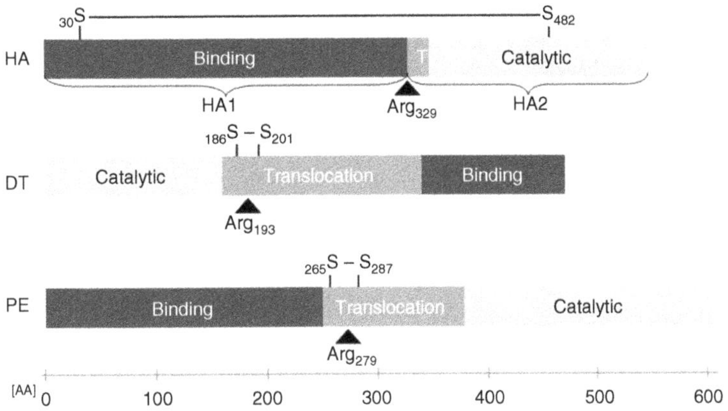

FIGURE 25.4 natural translocation domains for endosomal escape. Many toxins contain translocation domains to facilitate endosomal escape to execute their toxic activity in the cytosol. All of the three domain toxins are cleaved to separate the binding from the catalytic domain. After cleavage, both are connected by only a disulfide bridge that is usually reduced to separate the subunits. The hemagglutinin (HA) of the influenza virus contains a very short translocation domain. Diphtheria toxin (DT) and Pseudomonas exotoxin a (PE) have translocation domains beyond 100 amino acids [AA]. The triangle shows the position of the proteolytic split.

network involving the KDEL receptor that binds the REDL sequence of the toxin. The domain II mediates the translocation from the endoplasmic reticulum to the cytosol via low pH induced tryptophane membrane anchor. In the cytosol, the catalytic domain III rapidly inactivates elongation factor 2 (EF2) by ADP ribosylation [69].

The immunogenicity risk of the PE translocation domain II can be reduced by minimizing it to only 10 amino acids (residues 273–282) while still maintaining its functionality. This was demonstrated with a trimodular fusion protein containing a single-chain antibody against human epidermal growth factor receptor-2 (HER2), the minimal translocation domain of PE, and the BH3-interacting domain death agonist (BID) to trigger apoptosis. This molecule efficiently eliminated cancer HER2 positive cancer cells in an animal model [70]. Truncated translocation domains of PE have been used frequently to transfer apoptosis inducing molecules into cells [71].

The diphtheria toxin (DT) is split into two fragments that are still connected through a disulfide bridge. The amino terminal fragment contains the catalytic domain, and the carboxy-terminal polypeptide consists of the translocation and receptor-binding domain. Attachment to the heparin binding epidermal growth factor-like precursor (hb-EGF) on the cell surface triggers internalization via clathrin coated vesicles, which are then converted into early endosomes. Upon acidification of the endosomal lumen, the transmembrane domain unfolds and inserts a pore into the endosome membrane. The reduction of the disulfide bridges between the two toxin fragments releases the catalytic domain into the cytosol [72]. Despite its efficient escape from the endosome, the translocation domain of DT has so far not been used separately as building block for novel fusion proteins addressing intracellular targets. Until now, all therapeutic applications of DT focus on its use as immunotoxin, utilizing the catalytic domain for ADP ribosylation. Further references can be seen in Part IIb.

Other mechanisms to escape from the endosome can be derived from viruses. They contain fusogenic peptides that undergo a conformational change during the lowering of the pH in the endosomes. For instance, the first 23 N-terminal amino acids of the hemagglutinin subunit 2 (HA2) of the influenza virus form a α-helix that can insert into the endosome membrane under acidic conditions. A fusion consisting of this HA2 derived peptide, p53, and polyarginine as CPP induced apoptosis fivefold more efficient than the fusion protein without HA2 [73].

On the basis of the observation of pH induced conformational conversion, an artificial amphiphilic peptide was designed. Owing to its repetitive nature of the four amino acids Glu-Ala-Leu-Ala, it was termed GALA peptide. It consists of 30 residues and changes from a random coil to α-helix when the pH is lowered from 7 to 5. The helix inserts into the endosomal membrane and creates a pore through which molecules can passage to reach the cytosol [74]. So far, this peptide has only been used as additive but not as component of a fusion protein. In Table 25.1 sequences of translocation domains are collected.

25.4 ORAL DELIVERY

The majority of small molecule drugs can easily traverse membranes or barriers; hence they are orally administered.

TABLE 25.1 Endosomal Escape Mechanisms

Protein	Mechanism	Sequence	Length [aa]
Hemagglutinin (HA2)	Fusion	GLFGAIAGFIEGGWTGMIDG	20
Diphtheria toxin	Fusion	INLDWDVIRD-----QVVHNSYNRP	177
Pseudomonas exotoxin A	Pore	GGSLAALTAH-----TGNDEAGAAS	112
Melittin	Pore	GIGAVLKVLTTGLPALISWIKRKRQQ	26
GALA peptide	Fusion	WEAALAEALAEALAEHLAEALAEALEALAA	30

The opposite is true for large, protein-based therapeutics that reach their targets by invasive procedures such as subcutaneous or intravenous injections. A major limitation is the sensitivity of proteins to proteolytic degradation during the passage through the digestive tract. Besides proteolysis, low absorption and bioavailability are other major obstacles that cause dosing problems [75]. Encapsulation would represent an obvious solution. Alternative strategies focus on transferrin as carrier molecule relying on receptor-mediated transcytosis because TfRs are highly expressed in the gastrointestinal tract, and Tf is resistant to trypsin and chymotrypsin. The first payload was insulin, conjugated to Tf. Rats receiving this molecule orally displayed a slow but prolonged hypoglycaemic effect [76]. Later, G-CSF was fused to the amino terminus of Tf. Replacing chemical conjugation by genetic fusion generated homogeneous proteins that are more suitable for therapy. Initially, a very short dipeptide linker consisting of leucine and glutamate was used. The bioactivity of the G-CSF-Tf fusion protein was 90% less than for G-CSF alone, whereas the half-life was much improved. Interestingly, only the fusion protein caused a measurable effect in mice when administered orally. This effect lasted twice as long as subcutaneous injection [77]. The efficacy of this G-CSF-Tf fusion protein could be further increased by optimizing the length and sequence of the spacer connecting both molecules. The best variant with a 10-fold higher activity contained a long helical linker of 50 AA [78].

The same helical linker was used to generate an orally administrable human growth hormone (hGH) transferrin fusion protein. As expected, hGH-Tf showed a weight gain effect in rats, but the effect was lower for the oral administration than for the subcutaneous injection [79].

The intestine epithelia also contain the FcRn that remains present in adult humans. The FcRn fulfills its primary role in the uptake of maternal immunoglobulin present in milk. As already described earlier, binding of the CH2 and CH3 domains of the Fc part mediates transcytosis of the globulin from the intestine to the bloodstream. Initially, polarized T84 cells were used as model system to verify FcRn-mediated transcytosis of hGH-Fc fusions. A molecule containing a hinge region between hGH and Fc was better uptaken than a similar variant with only a short flexible linker. It could also be demonstrated that a dimeric structure of the Fc is required for optimal binding to the FcRn [80]. Previously, the ability of FcRn-induced intestinal transcytosis of large, orally administered molecules was proven with follicle stimulating hormone (FSH) fused to Fc. Newborn rats absorbed FSH-Fc after oral dosing. Two types of molecules were tested: one connected the β- and α-domain of FSH via a linker and another the hinge region with the heavy IgG1 chain in a single polypeptide, ultimately forming a homodimer with two FSH units. The other variant contained either the α- or β-domain of FSH fused to a heavy chain, generating a heterodimer with only one FSH molecule per fusion protein. At an age of 10 days, rats received daily a single oral dose of the different fusion proteins for 14 days. Both molecules achieved a half-life of more than 60 h, but the heterodimer was significantly more potent [81].

25.5 CONCLUSIONS AND FUTURE PERSPECTIVES

Much progress in the field of drug targeting has been made in the last decades. Owing to the advancement of antibodies, now a broad selection of binding entities is available to address extracellular targets. The reduction of antibody specificity to small binding fragments is now an essential part of many fusion protein constructs. In addition, the understanding of disease-related changes of surface molecules and organ-specific receptors has contributed a lot to the refinement of targeting techniques. Certainly, the field of targeted therapy that addresses cancer is the most advanced but probably also the most dispersed. Cancer seems to be the primary testing ground for fusion proteins because it combines the need for targeting with aim of toxicity, two activities that are nicely covered by bifunctional molecules.

But it is still very difficult to transfer large protein molecules into cells or across membranes such as the BBB. Some of the entry points are well understood now and will be exploited extensively in the future. However, more work is needed to establish alternatives and to identify more tissue- and organ-specific surface markers. Oral delivery that would allow a more patient friendly administration of therapeutic protein is still in its infancy. But maybe other concepts such as encapsulation instead of fusion protein approaches can cover that gap.

Fusion proteins will continue to expand the toolbox by allowing almost unlimited possibilities to combine targeting entities with effector modules. The strategies presented in this part give a broad overview on the state of the art in protein-based targeting and should inspire the reader to design novel therapeutic molecules.

REFERENCES

1. Allgayer H, Fulda S. (2008) An introduction to molecular targeted therapy of cancer. *Adv. Med. Sci.* 53(2), 130–138.
2. Weidle UH, Maisel D, Klostermann S, Schiller C, Weiss EH. (2011) Intracellular proteins displayed on the surface of tumor cells as targets for therapeutic intervention with antibody-related agents. *Cancer Genomics Proteomics.* 8(2), 49–63.
3. Sharifi J, Khawli LA, Hu P, King S, Epstein AL. (2001) Characterization of a phage display-derived human monoclonal antibody (NHS76) counterpart to chimeric TNT-1 directed against necrotic regions of solid tumors. *Hybrid. Hybridomics* 20(5–6), 305–312.
4. Yim KL, Cunningham D. (2011) Targeted drug therapies and cancer. *Recent Results Cancer Res.* 185, 159–171.
5. Hung M-C, Link W. (2011) Protein localization in disease and therapy. *J. Cell. Sci.* 124(Pt 20), 3381–3392.
6. Heino J, Käpylä J. (2009) Cellular receptors of extracellular matrix molecules. *Curr. Pharm. Des.* 15(12), 1309–1317.
7. Gebauer M, Skerra A. (2009) Engineered protein scaffolds as next-generation antibody therapeutics. *Curr. Opin. Chem. Biol.* 13(3), 245–255.
8. Brightman MW. (1977) Morphology of blood–brain interfaces. *Exp. Eye Res.* 25(Suppl), 1–25.
9. Pardridge WM. (1986) Receptor-mediated peptide transport through the blood-brain barrier. *Endocr. Rev.* 7(3), 314–330.
10. Shin SU, Friden P, Moran M, Olson T, Kang YS, Pardridge WM, et al. (1995) Transferrin-antibody fusion proteins are effective in brain targeting. *Proc. Natl. Acad. Sci. USA* 92(7), 2820–2824.
11. Park E, Starzyk RM, McGrath JP, Lee T, George J, Schutz AJ, et al. (1998) Production and characterization of fusion proteins containing transferrin and nerve growth factor. *J. Drug. Target.* 6(1), 53–64.
12. Ji B, Maeda J, Higuchi M, Inoue K, Akita H, Harashima H, et al. (2006) Pharmacokinetics and brain uptake of lactoferrin in rats. *Life Sci.* 78(8), 851–855.
13. Karkan D, Pfeifer C, Vitalis TZ, Arthur G, Ujiie M, Chen Q, et al. (2008) A unique carrier for delivery of therapeutic compounds beyond the blood-brain barrier. *PLoS ONE* 3(6), e2469.
14. Pardridge WM. (1997) Drug delivery to the brain. *J. Cereb. Blood Flow Metab.* 17(7), 713–731.
15. Pardridge WM. (2008) Re-engineering biopharmaceuticals for delivery to brain with molecular Trojan horses. *Bioconjug. Chem.* 19(7), 1327–1338.
16. Boado RJ, Zhang Y, Zhang Y, Xia C-F, Pardridge WM. (2007) Fusion antibody for Alzheimer's disease with bidirectional transport across the blood-brain barrier and abeta fibril disaggregation. *Bioconjug. Chem.* 18(2), 447–455.
17. Yu YJ, Zhang Y, Kenrick M, Hoyte K, Luk W, Lu Y, et al. (2011) Boosting brain uptake of a therapeutic antibody by reducing its affinity for a transcytosis target. *Sci. Transl. Med.* 3 (84), 84ra44.
18. Demeule M, Régina A, Ché C, Poirier J, Nguyen T, Gabathuler R, et al. (2008) Identification and design of peptides as a new drug delivery system for the brain. *J. Pharmacol. Exp. Ther.* 324(3), 1064–1072.
19. Fioravanti J, Medina-Echeverz J, Ardaiz N, Gomar C, Parra-Guillén ZP, Prieto J, et al. (2012) The fusion protein of IFN-α and apolipoprotein A-I crosses the blood–brain barrier by a saturable transport mechanism. *J. Immunol.* 188(8), 3988–3992.
20. Xiang L, Zhou R, Fu A, Xu X, Huang Y, Hu C. (2011) Targeted delivery of large fusion protein into hippocampal neurons by systemic administration. *J. Drug Target.* 19(8), 632–636.
21. Fu A, Wang Y, Zhan L, Zhou R. (2012) Targeted delivery of proteins into the central nervous system mediated by rabies virus glycoprotein-derived peptide. *Pharm. Res.* Available at http://www.ncbi.nlm.nih.gov/pubmed/22231987. Retrieved 2012 February 29.
22. Toivonen JM, Oliván S, Osta R. (2010) Tetanus toxin C-fragment: the courier and the cure? *Toxins (Basel)* 2(11), 2622–2644.
23. Li J, Chian R-J, Ay I, Kashi BB, Celia SA, Tamrazian E, et al. (2009) Insect GDNF: TTC fusion protein improves delivery of GDNF to mouse CNS. *Biochem. Biophys. Res. Commun.* 390(3), 947–951.
24. Ciriza J, Moreno-Igoa M, Calvo AC, Yague G, Palacio J, Miana-Mena FJ, et al. (2008) A genetic fusion GDNF-C fragment of tetanus toxin prolongs survival in a symptomatic mouse ALS model. *Restor. Neurol. Neurosci.* 26(6), 459–465.
25. Schwarze SR, Ho A, Vocero-Akbani A, Dowdy SF. (1999) *In vivo* protein transduction: delivery of a biologically active protein into the mouse. *Science* 285(5433), 1569–1572.
26. Kilic E, Kilic U, Hermann DM. (2005) TAT-GDNF in neurodegeneration and ischemic stroke. *CNS Drug Rev.* 11(4), 369–378.
27. Wu J, Nantz MH, Zern MA. (2002) Targeting hepatocytes for drug and gene delivery: emerging novel approaches and applications. *Front. Biosci.* 7, d717–d725.
28. Du S-L, Pan H, Lu W-Y, Wang J, Wu J, Wang J-Y. (2007) Cyclic Arg-Gly-Asp peptide-labeled liposomes for targeting drug therapy of hepatic fibrosis in rats. *J. Pharmacol. Exp. Ther.* 322(2), 560–568.
29. Bansal R, Prakash J, de Ruijter M, Beljaars L, Poelstra K. (2011) Peptide-modified albumin carrier explored as a novel strategy for a cell-specific delivery of interferon gamma to treat liver fibrosis. *Mol. Pharm.* 8(5), 1899–1909.
30. Fioravanti J, González I, Medina-Echeverz J, Larrea E, Ardaiz N, González-Aseguinolaza G, et al. (2011) Anchoring interferon alpha to apolipoprotein A-I reduces hematological toxicity while enhancing immunostimulatory properties.

Hepatology Available at http://www.ncbi.nlm.nih.gov/pubmed/21425312. Retrieved 2011 May 27.

31. Heinemann L. (2011) New ways of insulin delivery. *Int. J. Clin. Pract. Suppl.* (170), 31–46.
32. Nelson HS, Busse WW, Sanger M, Cutler G, Ellwanger C, Chipman JJ. (2009) Short-term safety of somatropin inhalation powder in adults with mild to moderate asthma. *Allergy Asthma Proc.* 30(3), 325–332.
33. Widera A, Kim K-JJ, Crandall ED, Shen W-C. (2003) Transcytosis of GCSF-transferrin across rat alveolar epithelial cell monolayers. *Pharm. Res.* 20(8), 1231–1238.
34. Bitonti AJ, Dumont JA, Low SC, Peters RT, Kropp KE, Palombella VJ, et al. (2004) Pulmonary delivery of an erythropoietin Fc fusion protein in non-human primates through an immunoglobulin transport pathway. *Proc. Natl. Acad. Sci. USA* 101(26), 9763–9768.
35. Dumont JA, Bitonti AJ, Clark D, Evans S, Pickford M, Newman SP. (2005) Delivery of an erythropoietin-Fc fusion protein by inhalation in humans through an immunoglobulin transport pathway. *J. Aerosol Med.* 18(3), 294–303.
36. Vallee S, Rakhe S, Reidy T, Walker S, Lu Q, Sakorafas P, et al. (2012) Pulmonary administration of interferon Beta-1a-fc fusion protein in non-human primates using an immunoglobulin transport pathway. *J. Interferon Cytokine Res.* 32 (4), 178–184.
37. Dumont JA, Low SC, Peters RT, Bitonti AJ. (2006) Monomeric Fc fusions: impact on pharmacokinetic and biological activity of protein therapeutics. *BioDrugs* 20(3), 151–160.
38. Kim K-J, Malik AB. (2003) Protein transport across the lung epithelial barrier. *Am. J. Physiol. Lung Cell Mol. Physiol.* 284 (2), L247–259.
39. Ferkol T, Cohn LA, Phillips TE, Smith A, Davis PB. (2003) Targeted delivery of antiprotease to the epithelial surface of human tracheal xenografts. *Am. J. Respir. Crit. Care Med.* 167 (10), 1374–1379.
40. Malakhov MP, Aschenbrenner LM, Smee DF, Wandersee MK, Sidwell RW, Gubareva LV, et al. (2006) Sialidase fusion protein as a novel broad-spectrum inhibitor of influenza virus infection. *Antimicrob. Agents Chemother.* 50(4), 1470–1479.
41. Hedlund M, Aschenbrenner LM, Jensen K, Larson JL, Fang F. (2010) Sialidase-based anti-influenza virus therapy protects against secondary pneumococcal infection. *J. Infect. Dis* 201 (7), 1007–1015.
42. Larson JL, Kang S-K, Choi BI, Hedlund M, Aschenbrenner LM, Cecil B, et al. (2011) A safety evaluation of DAS181, a sialidase fusion protein, in rodents. *Toxicol. Sci.* 122(2), 567–578.
43. Yadav MC, Lemire I, Leonard P, Boileau G, Blond L, Beliveau M, et al. (2011) Dose response of bone-targeted enzyme replacement for murine hypophosphatasia. *Bone* 49(2), 250–256.
44. Kasugai S, Fujisawa R, Waki Y, Miyamoto K, Ohya K. (2000) Selective drug delivery system to bone: small peptide (Asp)6 conjugation. *J. Bone Miner. Res.* 15(5), 936–943.
45. Takahashi-Nishioka T, Yokogawa K, Tomatsu S, Nomura M, Kobayashi S, Miyamoto K-I. (2008) Targeted drug delivery to bone: pharmacokinetic and pharmacological properties of acidic oligopeptide-tagged drugs. *Curr. Drug Discov. Technol.* 5(1), 39–48.
46. Shi B, Yu A, Liu Y, Li J, Jin J, Dong C, et al. (2007) Locally activity-released bifunctional fusion protein enhances antithrombosis and alleviates bleeding risk. *J. Thromb. Thrombolysis* 24(3), 283–292.
47. Wang X, Zhang G, Wang L, Zhang Q, Qin Y, Wu C, et al. (2009) A fusion protein with improved thrombolytic effect and low bleeding risk. *Thromb. Haemost.* 102(6), 1194–1203.
48. Cizeau J, Torres MGP, Cowling SG, Stibbard S, Premsukh A, Entwistle J, et al. (2011) Fusogenics: a recombinant immunotoxin-based screening platform to select internalizing tumor-specific antibody fragments. *J. Biomol. Screen.* 16(1), 90–100.
49. Shaaltiel Y, Bartfeld D, Hashmueli S, Baum G, Brill-Almon E, Galili G, et al. (2007) Production of glucocerebrosidase with terminal mannose glycans for enzyme replacement therapy of Gaucher's disease using a plant cell system. *Plant Biotechnol. J.* 5(5), 579–590.
50. Aviezer D, Brill-Almon E, Shaaltiel Y, Hashmueli S, Bartfeld D, Mizrachi S, et al. (2009) A plant-derived recombinant human glucocerebrosidase enzyme–a preclinical and phase I investigation. *PLoS ONE* 4(3), e4792.
51. Prince WS, McCormick LM, Wendt DJ, Fitzpatrick PA, Schwartz KL, Aguilera AI, et al. (2004) Lipoprotein receptor binding, cellular uptake, and lysosomal delivery of fusions between the receptor-associated protein (RAP) and alpha-L-iduronidase or acid alpha-glucosidase. *J. Biol. Chem.* 279(33), 35037–35046.
52. LeBowitz JH, Grubb JH, Maga JA, Schmiel DH, Vogler C, Sly WS. (2004) Glycosylation-independent targeting enhances enzyme delivery to lysosomes and decreases storage in mucopolysaccharidosis type VII mice. *Proc. Natl. Acad. Sci. USA* 101(9), 3083–3088.
53. Orii KO, Grubb JH, Vogler C, Levy B, Tan Y, Markova K, et al. (2005) Defining the pathway for Tat-mediated delivery of beta-glucuronidase in cultured cells and MPS VII mice. *Mol. Ther.* 12(2), 345–352.
54. Lee KO, Luu N, Kaneski CR, Schiffmann R, Brady RO, Murray GJ. (2005) Improved intracellular delivery of glucocerebrosidase mediated by the HIV-1 TAT protein transduction domain. *Biochem. Biophys. Res. Commun.* 337(2), 701–707.
55. Rapoport M, Salman L, Sabag O, Patel MS, Lorberboum-Galski H. (2011) Successful TAT-mediated enzyme replacement therapy in a mouse model of mitochondrial E3 deficiency. *J. Mol. Med.* 89(2), 161–170.
56. Zhang F, Xing J, Liou AK-F, Wang S, Gan Y, Luo Y, et al. (2010) Enhanced Delivery of Erythropoietin Across the Blood-Brain Barrier for Neuroprotection against Ischemic Neuronal Injury. *Transl. Stroke Res.* 1(2), 113–121.
57. Mie M, Takahashi F, Funabashi H, Yanagida Y, Aizawa M, Kobatake E. (2003) Intracellular delivery of antibodies using TAT fusion protein A. *Biochem. Biophys. Res. Commun.* 310(3), 730–734.
58. Frankel AD, Pabo CO. (1988) Cellular uptake of the tat protein from human immunodeficiency virus. *Cell* 55(6), 1189–1193.

59. Heitz F, Morris MC, Divita G. (2009) Twenty years of cell-penetrating peptides: from molecular mechanisms to therapeutics. *Br. J. Pharmacol.* 157(2), 195–206.

60. Heng BC, Cao T. (2005) Making cell-permeable antibodies (Transbody) through fusion of protein transduction domains (PTD) with single chain variable fragment (scFv) antibodies: potential advantages over antibodies expressed within the intracellular environment (Intrabody). *Med. Hypotheses* 64(6), 1105–1108.

61. Snyder EL, Meade BR, Saenz CC, Dowdy SF. (2004) Treatment of terminal peritoneal carcinomatosis by a transducible p53-activating peptide. *PLoS Biol.* 2(2), E36.

62. Weisbart RH, Baldwin R, Huh B, Zack DJ, Nishimura R. (2000) Novel protein transfection of primary rat cortical neurons using an antibody that penetrates living cells. *J. Immunol.* 164(11), 6020–6026.

63. Hansen JE, Tse C-M, Chan G, Heinze ER, Nishimura RN, Weisbart RH. (2007) Intranuclear protein transduction through a nucleoside salvage pathway. *J. Biol. Chem.* 282(29), 20790–20793.

64. Weisbart RH, Hansen JE, Chan G, Wakelin R, Chang SS, Heinze E, et al. (2004) Antibody-mediated transduction of p53 selectively kills cancer cells. *Int. J. Oncol.* 25(6), 1867–1873.

65. Hansen JE, Fischer LK, Chan G, Chang SS, Baldwin SW, Aragon RJ, et al. (2007) Antibody-mediated p53 protein therapy prevents liver metastasis *in vivo*. *Cancer Res.* 67(4), 1769–1774.

66. Heinze E, Baldwin S, Chan G, Hansen J, Song J, Clements D, et al. (2009) Antibody-mediated FOXP3 protein therapy induces apoptosis in cancer cells *in vitro* and inhibits metastasis *in vivo*. *Int. J. Oncol.* 35(1), 167–173.

67. Weisbart RH, Hansen JE, Nishimura RN, Chan G, Wakelin R, Chang SS, et al. (2005) An intracellular delivery vehicle for protein transduction of micro-dystrophin. *J. Drug Target.* 13(2), 81–87.

68. Hansen JE, Sohn W, Kim C, Chang SS, Huang NC, Santos DG, et al. (2006) Antibody-mediated Hsp70 protein therapy. *Brain Res.* 1088(1), 187–196.

69. Weldon JE, Pastan I. (2011) A guide to taming a toxin—recombinant immunotoxins constructed from Pseudomonas exotoxin A for the treatment of cancer. *FEBS J.* 278(23), 4683–4700.

70. Qiu X-C, Xu Y-M, Wang F, Fan Q-Y, Wang L-F, Ma B-A, et al. (2008) Single-chain antibody/activated BID chimeric protein effectively suppresses HER2-positive tumor growth. *Mol. Cancer Ther.* 7(7), 1890–1899.

71. Zhang L, Zhao J, Wang T, Yu C-J, Jia L-T, Duan Y-Y, et al. (2008) HER2-targeting recombinant protein with truncated pseudomonas exotoxin A translocation domain efficiently kills breast cancer cells. *Cancer Biol. Ther.* 7(8), 1226–1231.

72. Murphy JR. (2011) Mechanism of diphtheria toxin catalytic domain delivery to the eukaryotic cell cytosol and the cellular factors that directly participate in the process. *Toxins (Basel)* 3(3), 294–308.

73. Michiue H, Tomizawa K, Wei F-Y, Matsushita M, Lu Y-F, Ichikawa T, et al. (2005) The NH2 terminus of influenza virus hemagglutinin-2 subunit peptides enhances the antitumor potency of polyarginine-mediated p53 protein transduction. *J. Biol. Chem.* 280(9), 8285–8289.

74. Li W, Nicol F, Szoka FC, Jr. (2004) GALA: a designed synthetic pH-responsive amphipathic peptide with applications in drug and gene delivery. *Adv. Drug Deliv. Rev.* 56(7), 967–985.

75. Shen W-C. (2003) Oral peptide and protein delivery: unfulfilled promises? *DDT* 8(14), 607–608.

76. Xia CQ, Wang J, Shen WC. (2000) Hypoglycemic effect of insulin-transferrin conjugate in streptozotocin-induced diabetic rats. *J. Pharmacol. Exp. Ther.* 295(2), 594–600.

77. Bai Y, Ann DK, Shen W-C. (2005) Recombinant granulocyte colony-stimulating factor-transferrin fusion protein as an oral myelopoietic agent. *Proc. Natl. Acad. Sci. USA* 102(20), 7292–7296.

78. Bai Y, Shen W-C. (2006) Improving the oral efficacy of recombinant granulocyte colony-stimulating factor and transferrin fusion protein by spacer optimization. *Pharm. Res.* 23(9), 2116–2121.

79. Amet N, Wang W, Shen W-C. (2010) Human growth hormone-transferrin fusion protein for oral delivery in hypophysectomized rats. *J. Control Release* 141(2), 177–182.

80. Lee CH, Woo JH, Cho KK, Kang SH, Kang SK, Choi YJ. (2007) Expression and characterization of human growth hormone-Fc fusion proteins for transcytosis induction. *Biotechnol. Appl. Biochem.* 46(Pt 4), 211–217.

81. Low SC, Nunes SL, Bitonti AJ, Dumont JA. (2005) Oral and pulmonary delivery of FSH-Fc fusion proteins via neonatal Fc receptor-mediated transcytosis. *Hum. Reprod.* 20(7), 1805–1813.

26

CELL-PENETRATING PEPTIDE FUSION PROTEINS

Andrés Muñoz-Alarcón, Henrik Helmfors, Kristin Karlsson, and Ülo Langel

Department of Neurochemistry, Stockholm University, Stockholm, Sweden

26.1 INTRODUCTION

Protein interactions regulate many of the fundamental cellular processes such as gene expression and cellular signaling. Consequently, molecules with protein modulating properties provide an attractive source of potential therapeutic agents. Although small molecules remain the dominating therapeutic paradigm, development of traditional small molecule pharmaceuticals relies heavily on chance discovery of molecules that possess both cell-permeable properties and the ability to modulate biological processes. Many such small molecules contain insufficient biochemical information needed to obtain high specificity while avoiding undesirable and possibly toxic side effects. On the other hand, the structures of macromolecules, such as proteins and peptides, are better able to contain the information required to obtain therapeutic agents showing superior biological activity, specificity, and low toxicity. However, full-length proteins and larger peptides are difficult to produce and most often inherently cell impermeable. Regardless, the promising aspects of macromolecular therapeutics have prompted the development of a wide range of delivery vectors capable of overcoming the challenge posed by the cellular membranes that otherwise prohibit intracellular delivery.

In 1994, the first member of a new class of peptides—cell-penetrating peptides (CPPs)—with membrane penetrating ability was discovered [1]. CPPs constitute a heterogeneous group of relatively short cationic and/or amphipathic peptides, up to 40 amino acids in length. They possess the ability to gain entry to cells independent of membrane receptors [2] and they exhibit no cell-type specificity [3]. Furthermore, they possess a low cellular toxicity and the capacity to promote intracellular delivery of covalently or noncovalently conjugated bioactive cargoes. Several CPPs have been shown to enable delivery of biologically active proteins into cells when synthesized as a recombinant fusion protein or covalently conjugated to full-length proteins [4]. Subsequently, CPPs have been employed as vectors for intracellular delivery of a variety of molecules that have difficulties in entering cells by themselves [5,6]. Although peptides derived from protein interaction sites generate more extensive contacts with proteins than small hydrophobic molecules and have been successfully delivered using CPPs, peptides are often only capable of partially mimicking the activity of full-length proteins. As a result, proteins, albeit harder to deliver, are in many ways more attractive as potential drugs. The concept of intracellular delivery of CPP-fusion proteins with biological activity is commonly referred to as protein transduction.

26.2 TYPICAL APPLICATIONS AND INDICATIONS

In order to address the challenges in chemotherapy, imaging, diagnostics and mechanistic chemical biology novel strategies, and agents that enable or enhance the passage of drugs

Fusion Protein Technologies for Biopharmaceuticals: Applications and Challenges, First Edition. Edited by Stefan R. Schmidt.

TABLE 26.1 Selection of Proteins Delivered by CPPs

Protein	CPP	Application	References
Astrocytic phosphoprotein PEA-15[a]	Tat	Prevention of apoptosis	[17]
Avidin	Penetratin, pVEC, Tat, Transportan	Protein transduction efficiency of four cell-penetrating peptides	[18]
B-cell lymphoma-extra large (Bcl-xL)[a]	Tat	Protection against ischemic brain injury and neuronal apoptosis	[19]
Calcineurin[a]	Tat	Transduction of TAT fusion proteins into osteoclasts and osteoblasts	[20]
Caspase-3[a]	Tat	Targeted apoptosis initiation	[21]
Cell division control protein 42 (CDC42)[a]	Tat	Rho protein transduction, osteoblast effect	[22–24]
Cre recombinase[a]	MTS, Tat penetratin,	Manipulation of gene structure and function	[25]
Cu, Zn superoxide dismutase (SOD)[a]	Tat	Transduction of Cu, Zn-superoxide dismutase into human chondrocytes	[26]
Cyclin-dependent kinase 2 (Cdk2)[a]	Tat	Retinoblastoma tumor suppressor protein regulation	[27,28]
Domain III of Pseudomonas exotoxin A (PE)[b]	Tat	Protein delivery	[29]
Double-stranded RNA-binding domain (DRBD)[a]	Tat	Enhancement of CPP-mediated RNA delivery	[30]
Transcription factor (E2F)[a]	Tat	Apoptosis activation	[31]
Enhanced green fluorescent protein(eGFP)[a]	R8, Tat, VP22	Peptide-mediated therapy for pancreatic cancer. Intracellular delivery to muscle cells. Cellular uptake of histidine-tagged proteins	[32–34]
Exonuclease III (ExoIII)[a]	Tat	Transduction and targeted delivery into mitochondria of breast cancer cells	[12]
Flp recombinase[a]	VP22	Intracellular delivery of functional Flp recombinase	[35]
Forkhead box P3 (Foxp3)[a]	HHph-1	Therapeutic effect in autoimmune and allergic disease models	[16]
Fragment antigen-binding (antibody fragment) (FAB)[b,c]	REV, Tat	Enhancing uptake of FAB fragments	[36,37]
Transcription factor (GATA4)[a]	VP22	Therapeutic effect on ischemic cardiomyopathy	[38]
Green fluorescent protein(GFP)[a]	POD, R9, Tat, Transportan, VP22	Primary brain cell delivery is dependent on glycosaminoglycan expression. POD mediates delivery of recombinant proteins to retina, cornea, and skin. Fluorescent protein delivery via arginine-rich peptides to plant cells. Cellular translocation of proteins by transportan	[39–43]
Heat shock protein 70 (Hsp-70)[a]	Tat	Preventing neuronal cell death in *in vitro*/vivo models of Parkinson's disease. Neuroprotection of neuronal precursor cells after focal cerebral ischemia in mice	[44,45]
Hemagglutinin[a]	Tat	Transduction of TAT fusion proteins into osteoclasts and osteoblasts	[20]
Herpes simplex virus type 1 thymidine kinase (HSV1-TK)[a]	Tat, VP22	Intracellular delivery of proteins	[46]
Horse radish peroxidase (HRP)[b]	Tat	Protein delivery	[29]
Inhibitor of Kappa B (IκBα)[a]	Tat	Diminished infiltration of leukocytes into the site of inflammation	[47]
Insulin	K9, R9, Tat	Increased pulmonary absorption of insulin	[48]
Nuclear receptor corepressor fragments (N-CoR)[a]	Tat	Unblockage of differentiation in leukemia cells	[49]
Ovalbumin[a]	TLM	Enhancement of protein vaccination	[50]
p16[a]	Tat	Cancer therapy	[51]
p27[a]	Tat	Manipulation of cell migration	[52]

TABLE 26.1 *(Continued)*

Protein	CPP	Application	References
p53[a,d]	VP22	Cancer therapy	[53–55]
Peroxiredoxin 5 and 6 (PRDX5&6)[a]	Tat	Protection against high-glucose-induced cytotoxicity in retinal pericytes	[56]
Rac[a]	Tat	Treatment of inflammatory diseases and cancer	[22–24]
Red fluorescent protein (RFP)[a]	R9, Tat	Fluorescent protein delivery via arginine-rich peptides to plant cells	[41]
Rho[a]	Tat	Treatment of inflammatory diseases and cancer	[22–24]
RNAse A[b]	Tat	Protein transduction of active enzyme	[29]
Single chain variable fragment (scFV)[a]	Penetratin, Tat	Intracellular delivery of scFV fragments targeting various proteins	[13,57,58]
Suppressor of cytokine signaling 3 (SOCS3)[a]	MTM	Suppression of apoptosis and inflammation	[15]
T7 RNA polymerase[a]	VP22	Delivery of active enzyme	[35]
β-galactosidase[a]	Penetratin, Tat, VP22	Protein delivery	[29,43]

[a]Recombinant expression, C-terminal.
[b]Maleimide–cysteine bond. Illustrated in Figure 26.1.
[c]Disulfide bond (Cys–Cys).
[d]Recombinant expression, N-terminal.

or probes across biological barriers are required. CPPs have already contributed greatly to improve drug delivery *in vivo* to treat diseases such as cancer, ischemia, inflammation, and autoimmune diseases [7].

Protein transduction is an emerging technology with the potential to overcome many of the limitations of small molecule pharmaceuticals and recombinant genetic approaches, the latter often suffering from difficulties achieving sustained and appropriately regulated expression. Currently, proteins exceeding 100 kDa have successfully been delivered into cells in culture and mammalian model systems. The sheer size of macromolecules possible to transport across the cell membrane makes CPP-mediated delivery very promising, potentially enabling delivery of a large repertoire of proteins.

A solution to the problem of tissue nonspecificity of CPPs is the development of so-called activatable CPPs. Constructs have been described that are "activated" either outside or inside the cells [8–10].

Cell-penetrating fusion proteins have been used for a large variety of applications and the list of successfully delivered proteins is ever expanding. To date, several groups have reported successful delivery of CPP-fusion proteins *in vitro* [11–14] and a growing number of transducible proteins covering a wide range of sizes and functional classes have been delivered *in vivo* [15,16]. Table 26.1 presents an extensive selection of proteins that have been successfully delivered into cells for various applications as well as the CPP that mediated delivery.

Many of the fusion proteins presented have applications associated with cancer therapeutics, which can range from proapoptotic and anti-proliferative agents to tumor imaging and migration inhibitors. Other common applications include immunosuppression, apoptosis protection, diagnostics, and vaccines.

26.3 TECHNOLOGICAL ASPECTS

26.3.1 Creating CPP-Fusion Proteins

Production of CPP-fusion proteins is fairly straightforward and relies mostly on traditional means of cloning, creating an expression vector, transfection to and subsequent expression in *Escherichia coli*, resulting in either a N- or C-terminal attachment of the peptide, N-terminal attachment being the most common. Here, we give three examples of different vectors expressing different proteins in fusion with CPPs. Two of the examples utilize the TAT peptide and the last example is a fusion between the VP22 peptide and enhanced green fluorescent protein (eGFP). In this section, a brief general description of the methods used for purification of the fusion proteins and for synthesis of peptides is included.

We also give an example of expressed protein ligation as an alternative method for production of CPP-fusion proteins, and describe how this method may have advantages over the much more common cloning method. In addition, there is a short section about the utilization of cysteine to make disulfide bonds between peptides and proteins. A selection of common methods for creating CPP-fusion peptides are listed in Table 26.1, and illustrated in Figure 26.1.

26.3.2 Recombinant Fusion Protein Expression Systems

The most common method by far, as can be seen in Table 26.1, for generating fusion proteins is by recombinant

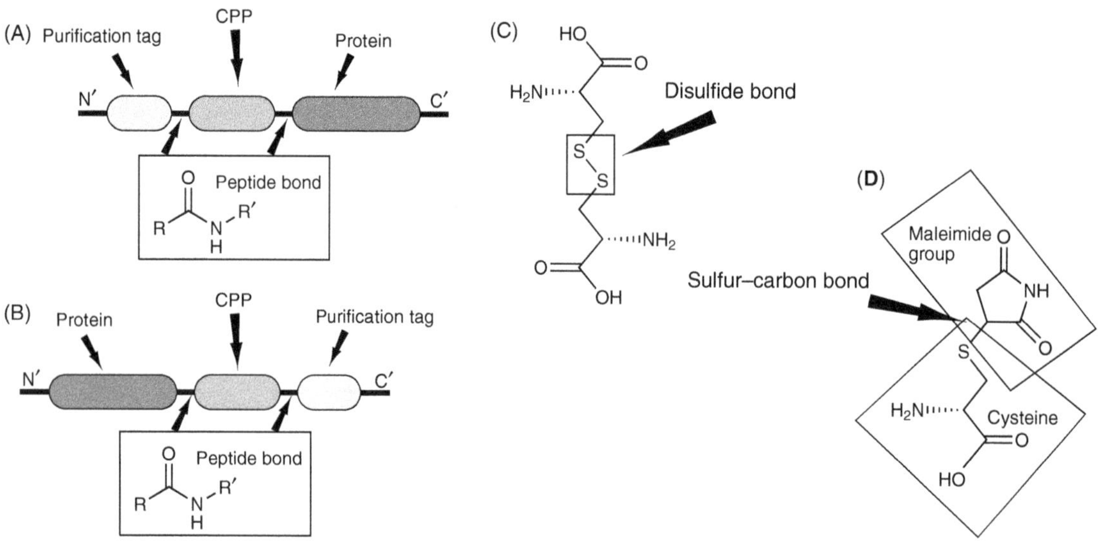

FIGURE 26.1 Methods for producing cell-penetrating fusion proteins. (A and B) Schematic drawing of the segment of DNA that is inserted to a vector for expression of the most common construct for CPP-fusion proteins, in the resulting fusion protein the CPP is either flanked by two glycine residues to allow for free rotation or there are no additional amino acids between the CPP and the protein. (C and D) Illustration of the two available methods for using cysteines to create covalent linkages between CPPs and proteins.

expression in *E. coli.* A vector containing the gene for a protein and either N- or C-terminal DNA encoding for a CPP is generated by inserting cDNA for the protein into an existing vector such as the pTAT-HA vector [59]. This vector contains an N-terminal hemagglutinin (HA)-tag and a hexahistidine (His-6)-tag followed by the TAT peptide. The two tags allow for standardized purification utilizing a tandem strategy for affinity-based chromatography. This vector was first described by Nagahara [52]. The vector has since then been extensively used for creating fusions of TAT and various proteins in for example [4,20–22,24,31,33,41,44,45,56,59–77].

Another example, in this case, novel vector, created for a specific application, is a vector expressing a VP22-eGFP fusion protein [32]. This vector was produced using a commercially available kit (Invitrogen's Voyager Protein Production Kit), in accordance with the manufacturers instructions.

A third example is the fusion between B-cell lymphoma-extra large (Bcl-xl) and TAT [19] where the authors started by isolating the Bcl-xl by PCR from a rat brain cDNA library, subcloned into the pSPORT1 vector (Invitrogen, San Diego, CA), amplified, purified, and verified it (with sequencing analysis). They then reconstructed the ET-30 expression plasmid to allow the generation of hemagglutinin (HA) tagged Bcl-xl-TAT fusion protein that contained the TAT-peptide sequence. The final cDNA (which replaced the enterokinase cloning site in the ET-30 plasmid) encodes peptide sequences in the following order: His6-tag, thrombin cleavage site (SGLVPRGS), TAT protein transduction domain (PTD), HA-tag, and Bcl-xl, and the resulting vector was designated as pET-PTD-HA-Bcl-xl. Expression was carried out in *E. coli* and the purification was performed using a Ni-NTA superflow agarose column. The His6-tag was removed by thrombin digestion and a second round through the Ni-NTA column removed the His6-tag.

On a side note, there have been indications that it may make a difference on which terminus (N or C) of the protein the CPP is fused. In a study where VP22 was fused both N- and C-terminally with p53, the N-terminally fused construct showed no activity, whereas the C-terminal showed a sixfold increase [55]. This is something that perhaps should be considered and evaluated for each protein and peptide combination.

26.3.3 Disulfide Bond

A method that employs the disulfide bond between cysteines or a carbon–sulfur bond between a carbon and a sulfur, that is, between a cysteine and a maleimide group, can also be used to produce CPP-fusion proteins. For example, a Fab antibody fragment labeled with Iodine 125 and three different CPPs (REV, TAT, and ANP) have been conjugated by synthesizing the peptides with an additional cysteine and then allowing these peptides to react with the radio-labeled antibody fragment treated with N-(6-maleimidocaproyloxy) succinimide ester (EMCS) [78]. The functional groups are illustrated in Figure 26.1.

26.3.4 Biotin-Avidin Fusion Model

Säälik et al. have shown a method for assessing the potential for protein delivery without the need to go through all the steps needed to clone, express, and purify using prokaryotic expression methods. They have a method that is based on the use of a biotin-tagged peptide, synthesized through conventional solid-phase peptide synthesis (SPPS), where one of the amino acids in the peptide sequence is tagged with biotin, thus creating a model for a peptide protein fusion. Delivery is limited to just two proteins, avidin and streptavidin, which bind very strongly to biotin (with a dis-association constant of about 10^{-14} mol/L). This method is perhaps most beneficial if one wishes to assess and compare the ability of different CPPs to deliver tightly bound (mimicking fused) protein cargo. The technical details for this strategy can be found in the paper [18].

26.3.5 Expressed Protein Ligation

Yu et al. have described a method called expressed protein ligation [79] for preparation of fusion proteins. Fusion proteins expressed in *E. coli* can be contaminated by endotoxins; this contamination can induce a multitude of unwanted physiological responses ranging from fever to death. The endotoxin problem is worsened by the fact that endotoxins are negatively charged and the TAT peptide and oligoarginine peptides are positively charged, thus making the separation of endotoxins from fusion proteins of TAT and other arginine-rich peptides potentially difficult. The peptides CG-R12 [NH2-Cys-Gly-$(Arg)_{12}$-amide] and CG-dR12 [NH2-Cys-Gly-D$(Arg)_{12}$-amide] were synthesized using Fmoc solid-phase peptide synthesis using a Rink amide resin using the standard protocol of a Shimadzu PSSM-8 peptide synthesizer. For the preparation of FITC-GC-R12 and FITC-R12, a γ-aminoburyl moiety was employed as a connecting linker between N-terminal FITC and the arginine peptides. A vector with eGFP, a mini-intein, and chitin binding domain (CBD) was prepared using a commercial system (IMPACT, New England Biolabs, Ipswich, MA). The intein-CBD domain allows for purification using a chitin bead column followed by on-column cleavage to generate a thioester of eGFP. This product can then be covalently linked with the cysteine in the CG-R12 peptide. The method has the additional benefit over His6-tagged fusion proteins that the purification tag is removed before the chemical ligation.

26.3.6 Purification

Recombinantly expressed proteins, fusion proteins, and synthetically produced peptides need to be purified. Recombinant proteins exist only in a mixture together with the endogenous proteins of the host organism. Peptides that are synthesized manually or using a synthesizer are not pure due to the fact that the chemical yield of every reaction that adds an additional amino acid to the peptide chain is not 100%.

26.3.7 Purification Tag

A histidine tag is a hexahistidine (six consecutive histidines) motif that is usually placed at the C- or N-terminus of a recombinant protein. It is used for purification of proteins expressed recombinantly in *E. coli*. The least cumbersome way to add a histidine tag is to utilize a vector that already has the coding sequence right after the start codon or directly before the stop codon. There are commercially available vectors from many manufacturers.

The hemagglutinin tag is another purification tag with the amino acid sequence YPYDVPDYA that works in a similar fashion as the hexahistidine tag.

Histidine- and hemagglutinin-tags are not the only tags employed to purify recombinant proteins. An example of another tag that has had widespread use is the FLAG-tag, it has recently been employed for CPP-fusion proteins [81]. The FLAG-tag is a polypeptide tag (DYKDDDDK) that can be used in much the same way as the His6-tag; in addition, it has the added benefit that there are antibodies available for the tag, which makes it possible to use FLAG-tagged proteins in conjunction with antibody-based assays.

26.3.8 Affinity Chromatography

The use of affinity for purification of proteins has been a method used since at least the end of the 1960s [82]. Here a protein's affinity for a substrate or an inhibitor is utilized. The substrate is attached to a cross-linked porous polymer contained in a column. The proteins that do not interact with the material in the column will pass through immediately, whereas the proteins that have affinity for the substrate that is bound in the column will be retarded. Elution is achieved by changing the pH or ion strength or by adding a competitive inhibitor in solution.

The use of histidine-tagged proteins expressed recombinantly in *E.coli* simplifies the process of finding a suitable substrate for the protein that is to be purified. The histidine tag has a high affinity for metal ions and can be purified using metal-chelate affinity chromatography [83]. The protein of interest binds to the column via interactions with the metal ions and can then be eluted using a solution of EDTA, ammonium chloride, imidazole, histidine, glycine, or high levels of competing metal ions, the most common being imidazole. Affinity chromatography using histidine-tagged proteins usually yields relatively pure products. Commercial kits for purification of histidine-tagged proteins are available from a number of different manufacturers. Affinity purification using a histidine tag in combination with a hemagglutinin tag, which utilizes a similar affinity-based interaction for retardation in a column, is a common method for purification of CPP-fusion proteins and has been utilized several times recently [40,45,84].

26.3.9 Solid Phase Peptide Synthesis and High Performance Liquid Chromatography

Solid-phase peptide synthesis (SPPS), first described by Merrifield [80], is the standard method for synthesizing CPPs that are not expressed recombinantly together with their fusion partner protein in bacteria. High performance liquid chromatography (HPLC) is the standard method for purification of peptides synthesized using SPPS [1].

26.4 SUCCESSES AND FAILURES IN PRECLINICAL AND CLINICAL RESEARCH

There are roughly four major therapeutic areas that have generated a great deal of research, namely cancer, neuroprotection, cardioprotection, and inflammation/infection.

26.4.1 Preclinical Research

Currently used anticancer therapeutics are unspecific and affect normal cells as well as cancer cells, which give side effects. In order to circumvent this, much effort is put into enhancing the specificity toward cancer cells in different ways. The use of macromolecules has been of interest since tumor cells often accumulate them due to leaky vasculature; tumor cells also have an increased temperature compared to normal cells. The use of an elastin-like polypeptide (ELP) fused to TAT were modified to aggregate in cells that had an elevated temperature and tested in four different cancer cell lines with encouraging result in specificity [85]. A specific inhibition of cell migration and proliferation of cancer cells was also seen in animal models of ovarian cancer [86]. Recently, the CPP Bac-7 was fused to ELP and a cell-cycle inhibitor, p21, to produce Bac-ELP1-p21, which was reported to be more specific and specifically accumulated in ovarian cancer cell lines with a temperature above 39°C and aggregated. No toxic effects were seen in cells at physiological temperature [87].

A different therapeutic approach to specifically target cancer cells is by activating the tumor suppressor protein, p53, which have attracted attention since it is mutated in a very high percentage of several different types of cancers [53,55,88]. Previous studies have used adenovirus-mediated transduction of p53 into cancer cells resulted in cell-cycle arrest and apoptosis and clinical studies have been performed [55,89]. The VP22-p53 fusion protein was shown to transduce between cells and induce apoptosis in a p53 negative cancer cell line in 1998 [53] but since then, the therapeutic utility of the VP22-p53 fusion protein have been under discussion since the intercellular trafficking properties of VP22 seems to vary with different cargoes [88,90,91].

Using the immunsystem is also one possible therapeutic approach to cancer treatment. Dendric cells present antigens on their surface for T cells in the immune system and by fusing a T-cell peptide or protein to dendritic cells were reported to specifically enhance the immune response 3- to 10-fold when compared to free peptide/protein [74,92].

The possibility of a novel cancer treatment that can circumvent the difficulties with pharmaceutical resistance among certain cancers is coadministration of chemotherapy and immunotherapy, the anticancer drug temozolomide and TAT-survivin-pulsed dendritic cells gave an increased survival rate in animal models at lower concentrations than temozolomide and TAT-survivin-pulsed dendritic cells alone [93]. One well known approach in coadministration is the herpes simplex virus type I thymidine kinase, HSV-TK, that enhances the sensitivity of tumor cells for the drug ganciclovir but the clinical relevance is limited due to low transduction efficiency [94]. The fusion of TAT to HSV-TK showed an increased cytotoxicity in 3 out of 12 human tumor cell lines [94].

In autoimmune diseases, such as arthritis and diabetes, the immune system targets substances and tissues normally present in the body. In arthritis, interleukin 1 (IL-1) plays a key role by activating the immune response but also by controlling the production of reactive oxygen species (ROS) that mediate degradation of collagen [26]. The cellular defense against ROS is, among others, superoxide dismutase (SOD), a protein that regulates the redox state in the cell and are known to cause inflammatory response [26,95,96]. Human cartilage contains extracellular SOD and patients with osteoarthritis have as much as a fourfold less SOD [97]. A topical formulation of TAT-SOD fusion protein has been administrated into murine ears in an inflammation model [95]. The TAT-SOD fusion protein increased the concentration of SOD and reduced the proinflammatory response created by administrating 12-O-tetradecanoylphorbol-13-acetate, TPA, both *in vivo* and *in vitro* [95]. This is promising for a possible new therapeutic treatment for arthritis.

NF-κB is an important mediator in inflammation by regulating proinflammatory cytokines and its activity is regulated by IκBα [47,98]. A nondegradable mutant of IκBα, named srIκBα, fused to TAT efficiently transduced and inhibited NF-κB activation [98]. This could be of interest for treatment of different inflammatory diseases.

Essential for immunology self-tolerance are a specific type of T cells called regulatory T cells or T_{reg} cells [16]. These T_{reg} cells are dependent of the transcription factor Foxp3, which have been identified as a possible therapeutic target for autoimmune diseases, such as diabetes and arthritis, and also asthma. Clinical applications have been limited due to several problems in isolating T_{reg} cells and the fact that the use of virus-mediated gene delivery gave very low transfection efficacy [16]. A human-derived CPP fused to

Foxp3, HHph-1-Foxp3, fusion protein inhibited cytokine production and T-cell activation that resulted in less tissue inflammation in an animal model after both local and systemic administration [16]. The fusion protein seemed to be specific for T-cell functions and biological effect was seen 4 days after the termination of treatment [16]. One possible therapeutic approach for diabetes is transplantation by viable islet cells. There are several limitations such as several steps of organ procurement, loss of surviving signals, and limitations of nutrients and oxygen that affect the transplantable islet viability and yield [99]. Using two anti-apoptotic proteins fused to TAT, namely TAT-BH4 and TAT-Bcl-xl, protected islets from staurosporine-induced apoptosis [99]. Treatment of islets with either TAT-HA or TAT-Bcl-xl fusion proteins showed 20–40% less dead cells, respectively, after 48 h compared to controls [99]. Diabetes needs to be carefully monitored and the lifelong treatment of injections and/or diet.

One of the major causes of hepatocellular carcinoma, cirrhosis, and chronic hepatitis is infection by the hepatitis B virus (HBV) [100]. In chronic HBV infection, the HBV nucleocapsid or core antigen, called HBcAg, is the only antigen that generates a prominent immune response [100,101]. There is a distinct difference in HBcAg-specific immune response between the patients with chronic HBV and the patient that can clear the virus. The specific response from cytotoxic T cells to HBcAg was strongly reduced or even below detection in patients with chronic HBV [100]. HBcAg-TAT fusion protein could effectively penetrate dendric cells and induce HBcAg-specific cytotoxic T lymphocytes in animal models [100]. A HBV core protein fused to VP22 together with a dominant negative mutant significantly inhibited the HBV replication [102].

The anti-apoptotic protein Bcl-xl fused to TAT was effectively transduced into neurons when intraocularly injected in animal models [61]. The survival rate of retinal ganglion cells was two times higher in TAT-Bcl-xl treated mice compared to control 14 days postsurgery [61]. Intravenous administration after ischemic insult resulted in an efficacious transduction of the brain tissue and reduced brain injury *in vivo* [103]. TAT-Bcl-xl effect on postischemia was examined in animal models where induction of cerebral ischemia was performed followed by intravenous injection of fusion protein [45]. Behavioral studies using the rota rod, water maze, and tight rope test were performed and the TAT-Bcl-xl treated mice. At day 4 after the induced ischemia, the mice were sacrificed and the extent of brain injury was assessed using terminal deoxynucleotidyl transferase dUTP nick end labeling (TUNEL) and infarct size analysis. The effects of TAT-Bcl-xl on hippocampal endogenous neurogenesis were examined after induced ischemia in mice [63]. TAT-Bcl-xl treated mice had increased survival of ischemia-induced newborn immature and mature neuronal cells.

Another protein of interest for treatment of ischemic injury is glial cell-derived neurotrophic factor (GDNF), which promotes neuronal survival. TAT-GDNF fusion protein was reported to significantly reduce brain injury when administered intravenously after ischemic insult in animal models [103].

Nogo-A extracellular peptide residue (NEP1-40) promotes axonal regrowth and recovery after CNS injury but the use is limited by the inability to cross the blood–brain barrier [104]. Importantly, TAT-NEP1-40 fusion protein is able to cross the blood–brain barrier and inhibit neuronal apoptosis after induced ischemia in animal models [104].

One therapeutic approach to treat osteoporosis is by inhibiting the NF-κB proteins that are involved in chronic inflammatory diseases and essential for osteoclast formation by the IκBα protein [47,60,98]. The TAT-fused IκBα protein efficiently entered osteoclasts and blocked bone resorption in a nontoxic fashion [60]. TAT fused to a protein such as calcineurin Aα or a small peptide such as hemagglutinin gave a very quick transduction into both osteoclasts and osteoblasts in primary cell culture and activity could even be detected for up to 5 days, which could be a therapeutic approach for treating osteoporosis [20]. The TAT-calcineurin Aα fusion protein was active and regulated osteoblast differentiation and maintaining of the protein levels could be seen in 50% of the cells up to 5 days after a single administration [20].

Antibodies have been reported to efficiently and specifically internalize cells through a TAT fusion protein in a nontoxic manner [105]. The use of CPP linked to antibodies for intracellular targets have a limited half-life and do not cause any permanent genetic alteration which otherwise was the major concern of recombinant DNA technology for clinical purposes and detection [106].

26.4.2 CPPs Used in Clinical Trials

The therapeutic interest in CPPs has so far resulted in a few clinical trials, which are summarized in the Table 26.2.

Several therapeutics against cancer used CPP as transduction technique. P28 is a novel CPP, developed by CDG therapeutics, from the protein azurin excreted by *Pseudomonas aeruginosa* [107]. P28 forms an intracellular complex with the tumor suppressor protein p53, which inhibits proliferation and induce apoptosis and it is tested with the intention to treat solid tumors, currently in a phase I study.

SN38 is an effective topoisomerase I inhibitor proven to be efficient antitumor agent, but the low solubility limits the clinical use [108]. DTS-108 is a water-soluble version of the parent drug SN38 linked to a highly charged Vectocell peptide [108]. DTS-108 has shown to have lower toxicity in preclinical animal studies.

Cancer cells can differentiate in hypoxic environment. The hypoxia inducible factor (HIF) contributes to cell

TABLE 26.2 Substances in Clinical Trials

Name	CPP and Cargo	Targeted Disease	Phase	References
KAI-9803	TAT-PKC δ	Myocardial infarction and cerebral ischemia	Phase IIb	[115,122,123]
RT001	A lysine rich domain in between two TAT— botulinum toxin type A	Wrinkles Excessive sweating	Phase IIb Phase I	[124]
AVI-5126	[(R-AHx-R)4]—NEUGENE antisense	Cardiovascular retenosis and cardiovascular disease	Phase II	[125]
AZX100	YARA[a]-phosphorylated HSP20	Dermal scarring	Phase II	[118,119,126]
AM-111	TAT-c-Jun N-terminal kinase (JNK) inhibitors	Hearing loss	Phase IIb	[127–129]
XG-102	TAT-c-Jun N-terminal kinase (JNK) inhibitors	Stroke	Phase I	[130–132]
PSORBAN®	Polyarginine (R8)-CsA	Psoriasis	Phase II discontinued	[117]
P28	p28[b] forms a complex with p53 inside the cell	Advanced cancer	Phase I	[122] [107]
DTS-108	Vectocell- SN38 a topoisomerase I inhibitor	Colorectal cancers	Phase I	[108,133]
POP33	PTD-ODD-caspase-3	Pancreatic cancer	Will start clinical studies shortly	[111]
	CPP-DRBD-siRNA	Leukaemia and glioblastoma but attractive for several diseases	Preclinical	[112]
	Truncated TAT-p53	Cancer	Will start clinical studies shortly	[113]
AVI-5038	(RXRRBR)2 - PMO	Duchenne muscular dystrophy	Aborted just before clinical studies	[120,121,125]

[a]YARA has the amino acid sequence YARAAARQARA.
[b]p28 is a CPP derived from the protein azurin excreted by *Pseudomonas aeruginosa.*

survival in hypoxic environment and is in turn degraded by oxygen-dependent degradation (ODD). Control of HIF-1α is of interest for the treatment of cancers. TAT-ODD-caspase-3 fusion protein, abbreviated TOP3, was shown to reduced tumor masses when injected into tumor bearing mice and specifically targeted hypoxic cells that were HIF-1α active but experienced no effect on normoxic cells [109,110]. The combination of both TOP3 and irradiation gave a significant long-term suppression of tumor growth [110]. Efforts to find a new CPP with an even higher transduction activity than TAT resulted in a prodrug called POP33. POP33 has displayed effective anticancer properties, suppressing malignant progression of pancreatic tumor cells in animal models by selectively inducing apoptosis via increase of caspase 3 activity [111]. Clinical studies are reported to start shortly. Short interfering RNA, siRNA, have been shown to have a gene silencing effect but the clinical use have been limited due to poor internalization. A new technology to deliver siRNA has been presented by Traversa therapeutics by fusing it to a double stranded RNA-binding domain fused with multiple CPPs, and preclinical studies are currently performed [112].

Justensen et al. (2007) presents another therapeutic method for the treatment of cancer. Here a 11 amino acid sequence of the TAT peptide is fused to the tumor suppressor protein p53 with the intention to treat cancer [113]. The preclinical studies have shown promising results and clinical studies are reported to start shortly.

CPPs have shown encouraging protective effects in different ischemia models for both neuroprotection and cardioprotection, which has been an unmet therapeutic need. Cardiovascular protective effects due to an inactivation of protein kinase C, PKC, have been seen in animal models [114]. KAI pharmaceuticals have a substance named KAI-9803, which is a TAT-conjugate with protein kinase C δ peptide inhibitor developed for the treatment of acute myocardial infarction [115]. KAI-9803 significantly reduced infarct size and the protection ability against reperfusion injury was maintained for up to 1 week after intravenous injection (kaipharma.com).

Using inhibitors for c-Jun N-terminal kinase (JNK), which is responsible for stress stimuli and involved in apoptosis, Xigen and Auris Medical have collaborated and present two different compounds, named XG-102 and AM-111, in clinical studies for the treatment of stroke and hearing loss, respectively.

AVI-5126 is a polyarginine and aminohexanoic-based CPP developed by AVI Biopharma that targets the c-myc gene. This antisense technology is currently in a Phase II clinical trial for treatment of cardiovascular diseases.

Cyclosporine A (CsA) is an immunosuppressive agent that can be used orally to treat acute inflammatory skin conditions. Systematically, administration of CsA showed signs of toxicity and trials with topical administration have proven ineffective due to very low skin penetration [116]. Psorban® is a topical formulation for treatment of psoriasis developed by CellGate, where CsA is linked to TAT that has been proven effective in Phase II clinical studies. (For review see [117]). Alopecia areata is an autoimmune disease that causes hair loss [116]. It was speculated that formulations that are effective for psoriasis most likely would be effective in alopecia areata as well.

Excessive sweating is a condition that can affect social life and is treated by injections of botulinum toxin (Botox). The method is time consuming, requiring as many as 120 injections to treat the palm of the hands. A possible new therapeutic method has been presented by Revance Therapeutics. RT001 is a topical gel where a botulinum toxin type A is fused to a CPP for treatment of excessive sweating, but it has also been effective in treating wrinkles around the eyes, so called lateral canthal lines or crow's feet wrinkles.[1]

AZX100 is a phosphorylated analog of the heat shock protein 20, HSPA20, attached to the CPP named YARA developed by Capstone[2] [118,119]. It is currently in a Phase II clinical trial for inhibition of dermal scarring.

AVI-5038 is a polyarginine and aminohexanoic, (RXRRBR)2, based CPP conjugated to a PMO that has been developed by AVI Biopharma for treatment of Duchenne Muscular Dystrophy.[3] Clinical studies were planed to start in February 2010, but signs of toxicity were seen in doses at 15 mg/kg and the clinical studies were discontinued [120,121].

26.5 ALTERNATIVES/VARIANTS OF THIS APPROACH

The ability of CPP to deliver cargo is not limited to proteins and several reports have proven the ability of using CPPs to greatly enhance cell internalization for cargoes such as nucleic acids, imaging agents, nanoparticles, liposomes, small pharmaceutical molecules, DNA, and peptides.

In order to use CPPs as pharmaceuticals, modifications are needed to enhance specificity, and one interesting approach is activatable cell-penetrating peptides (ACPP). ACPPs are polycationic cell-penetrating peptides covalently bound to a polyanionic domain, which inhibits cell internalization until the covalently binding is cleaved [10,134]. This reported method is designed toward cancer therapy where the covalent binding is cleaved by metalloproteases, which are predominantly found at higher concentrations in cancer cells [10,134,135]. ACPPs linked to nanoparticles together with the dyes Cy5 and/or gadolinium for imaging gave a high uptake in tumor cells that was between 4- and 15-folds higher than ACCPs [135]. The results indicate that this technique could be used in fluorescence-guided surgery, since studies have shown that tumors as small as 200 μm in diameter can be detected in animal studies and a significantly lower residual cancer cells [9,135]. The plasma half-life was proven to be around 9 h, a major improvement from the free ACPP, which have a plasma half-life of <10 min [135]. ACPPs have shown to be effective in cancer detection, cancer imaging, and cancer treatment where normal cells are mostly unaffected, in contrast to presently available cancer treatment [9].

Substance P is a neuropeptide that has been modified by Kossiakoff and colleagues to selectively and efficiently transduce cells expressing the neurokinin-1 receptors (NK1R) [136]. The NK1R is overexpressed in certain cancers, specifically brain tumors, and the tumor specificity could be useful in cancer treatment [136]. The substance P modified peptide conjugated to a synthetic antibody fragment specifically internalized and accumulated in tumors, which could be useful in imaging and cancer diagnosis [136].

26.6 CONCLUSIONS AND FUTURE PERSPECTIVES

CPPs have enormous therapeutic potential [137]. Intracellular delivery mediated by CPPs need not be restricted to proteins alone. In addition to protein transduction, CPPs provide efficient means for delivery of various macromolecules. For example, delivery of cancer chemotherapeutic drugs and larger DNA constructs could be enhanced. Although many of the studies performed on CPP-fusion proteins aim at studying biological mechanisms *in vitro* or designing therapeutic applications for use *in vivo*, *ex vivo* applications could also prove interesting. Stem cell expansion through growth stimulation via intracellular delivery of proteins such as SV40 large T antigen has been used as an example of one such potential application [138,139].

It is not certain that the best way to introduce proteins to cells using CPPs it to covalently couple them together. Another method that has shown promising results in other applications such as siRNA delivery is the coincubation strategy. This is a method that does not require the use of prokaryotic expression systems. Instead, this strategy utilizes interactions based on the positive charge of CPPs and negative charges in for example siRNA [140,141]. This method has also been employed for delivery of proteins, where the formation of stable complexes relied on hydrophobic interactions between the CPP Pep-1 and two different proteins GFP and β-Gal; a transfection rate for both the

[1] http://www.revance.com, htpp://clinicalTrials.gov NCT00888914
[2] http://www.capstonethx.com
[3] http://www. aviobio.com

proteins of >80% was observed, indicating that this could also be a path to protein delivery [142].

Improved understanding of CPPs and the uptake mechanisms involved in CPP-mediated delivery will greatly facilitate advances in both drug delivery and future therapeutic design. To date, the internalization mechanism of CPP remains a topic of much debate, though it appears that different CPPs use distinct cellular translocation pathways that in turn can be influenced by parameters such as cell type, cargo, and concentration [143,144]. The exact mechanisms of CPP-mediated transduction across the cell membrane remain largely unknown though several studies [145,146] suggest arginine content distribution can play a significant role in transduction efficiency. Furthermore, cell surface heparan sulfate appears to mediate peptide uptake in certain cases [147], suggesting a possibility to exploit cell surface molecules to increase protein transduction efficiency. Because CPPs, and especially CPP-cargo conjugates, have been shown to internalize largely via endocytotic mechanisms, endosomal escape is now discussed as a possible route for efficient CPP-cargo delivery [11,148].

Tools such as CPP sequence prediction models [149] for improving rational design of CPP may aid in the production of next generation of CPPs that have improved uptake, specificity, and reduced toxicity. Another means to improve CPP-fusion protein therapeutics is to modify the protein fused to the CPP enhancing desired biological properties.

Protein transduction has progressed significantly and the number of preclinical and clinical trials has increased markedly. Studies show encouraging results where the drugs appear to be safe and specific. Owings to a short plasma half-life and rapid distribution in tissue, the risk of generating immune responses is considered to be slim [115]. However, potential issues with specificity, toxicity [91], stability, and immunogenicity [103] impede the use of CPPs as pharmaceutical components. Nevertheless, the growing numbers of reports of successful *in vivo* transduction and promising clinical trials are both exciting and encouraging.

ACKNOWLEDGMENTS

This work was supported by grants from the Swedish Research Council (VR-NT), the Knut and Alice Wallenberg Foundation, the Swedish Governmental Agency for Innovation Systems (VINNOVA-SAMBIO 2006), and the Swedish Center for Biomembrane Research, Stockholm.

REFERENCES

1. Derossi D, Joliot A, Chassaing G, Prochiantz A. (1994) The third helix of the Antennapedia homeodomain translocates through biological membranes. *J. Biol. Chem.* 269, 10444–10450.
2. Derossi D, Calvet S, Trembleau A, Brunissen A, Chassaing G, Prochiantz A. (1996) Cell internalization of the third helix of the Antennapedia homeodomain is receptor-independent. *J. Biol. Chem.* 271, 18188–18193.
3. Kerkis A, Hayashi M, Yamane T, Kerkis I. (2006) Properties of cell penetrating peptides (CPPs). *IUBMB Life* 58, 7–13.
4. Schwarze S, Ho A, Vocero-Akbani, A, Dowdy S. (1999) *In vivo* protein transduction: delivery of a biologically active protein into the mouse. *Science* 285, 1569–1572.
5. Mäe M, Langel Ü. (2006) Cell-penetrating peptides as vectors for peptide, protein and oligonucleotide delivery. *Curr. Opin. Pharmacol.* 6, 509–514.
6. Futaki S. (2006) Oligoarginine vectors for intracellular delivery: design and cellular-uptake mechanisms. *Biopolymers* 84, 241–249.
7. Snyder EL, Dowdy S. (2005) Recent advances in the use of protein transduction domains for the delivery of peptides, proteins and nucleic acids *in vivo*. *Expert Opin. Drug Deliv.* 2, 43–51.
8. Jiang T, Olson E, Nguyen Q, Roy M, Jennings P, Tsien R. (2004) Tumor imaging by means of proteolytic activation of cell-penetrating peptides. *Proc. Natl. Acad. Sci. USA* 101, 17867–17872.
9. Nguyen QT, Olson E, Aguilera T, Jiang T, Scadeng M, Ellies L, et al. (2010) Surgery with molecular fluorescence imaging using activatable cell-penetrating peptides decreases residual cancer and improves survival. *Proc. Natl. Acad. Sci. USA* 107, 4317–4322.
10. Olson E, Aguilera T, Jiang T, Ellies L, Nguyen Q, Wong E, et al. (2009) *In vivo* characterization of activatable cell penetrating peptides for targeting protease activity in cancer. *Integrative Biology* 1, 382–393.
11. Wadia J, Stan R, Dowdy S. (2004) Transducible TAT-HA fusogenic peptide enhances escape of TAT-fusion proteins after lipid raft macropinocytosis. *Nat. Med.* 10, 310–315.
12. Shokolenko IN, Alexeyev MF, Ledoux SP, Wilson, GL. (2005) TAT-mediated protein transduction and targeted delivery of fusion proteins into mitochondria of breast cancer cells. *DNA Repair* 4, 511–518.
13. Theisen DM, Pongratz C, Wiegmann K, Rivero F, Krut O, Kronke M. (2006) Targeting of HIV-1 Tat traffic and function by transduction-competent single chain antibodies. *Vaccine* 24, 3127–3136.
14. Cao L, Si J, Wang W, Zhao X, Yuan X, Zhu H, et al. (2006) Intracellular localization and sustained prodrug cell killing activity of TAT-HSVTK fusion protein in hepatocelullar carcinoma cells. *Mol. Cells* 21, 104–111.
15. Jo D, Liu D, Yao S, Collins R, Hawiger J. (2005) Intracellular protein therapy with SOCS3 inhibits inflammation and apoptosis. *Nat. Med.* 11, 892–898.
16. Choi J, Shin J, Sohn M, Harding M, Park J, Tobiasova Z, et al. (2010) Cell-permeable Foxp3 protein alleviates autoimmune disease associated with inflammatory bowel disease and allergic airway inflammation. *Proc. Natl. Acad. Sci. USA* 107, 18575–18580.

17. Embury J, Klein D, Pileggi A, Ribeiro M, Jayaraman S, Molano R, et al. (2001) Proteins linked to a protein transduction domain efficiently transduce pancreatic islets. *Diabetes* 50, 1706–1713.
18. Säälik P, Elmquist A, Hansen M, Padari K, Saar K, Viht K, et al. (2004) Protein cargo delivery properties of cell-penetrating peptides. A comparative study. *Bioconjug. Chem.* 15, 1246–1253.
19. Cao G, Pei W, Ge H, Liang Q, Luo Y, Sharp F, et al. (2002) *In vivo* delivery of a Bcl-xL fusion protein containing the TAT protein transduction domain protects against ischemic brain injury and neuronal apoptosis. *J. Neurosci.* 22, 5423–5431.
20. Dolgilevich S, Zaidi N, Song JB, Abe E, Moonga BS, Sun L. (2002) Transduction of TAT fusion proteins into osteoclasts and osteoblasts. *Biochem. Biophys. Res. Commun.* 299, 505–509.
21. Vocero-Akbani AM, Heyden N, Lissy N, Ratner L, Dowdy S. (1999) Killing HIV-infected cells by transduction with an HIV protease-activated caspase-3 protein. *Nat. Med.* 5, 29–33.
22. Chellaiah M, Soga N, Swanson S, Mcallister S, Alvarez U, Wang D, et al. (2000) Rho-A is critical for osteoclast podosome organization, motility, and bone resorption. *J. Biol. Chem.* 275, 11993.
23. Hall D, Cui J, Bates M, Stout B, Koenderman L, Coffer P, et al. (2001) Transduction of a dominant-negative H-Ras into human eosinophils attenuates extracellular signal-regulated kinase activation and interleukin-5-mediated cell viability. *Blood* 98, 2014–2021.
24. Soga N, Namba N, Mcallister S, Cornelius L, Teitelbaum S, Dowdy S, et al. (2001) Rho family GTPases regulate VEGF-stimulated endothelial cell motility. *Exp. Cell Res.* 269, 73–87.
25. Jo D, Nashabi A, Doxsee C, Lin Q, Unutmaz D, Chen J, et al. (2001) Epigenetic regulation of gene structure and function with a cell-permeable Cre recombinase. *Nat. Biotech.* 19, 929–933.
26. Kim H, Kim D, Park J, Choi S. (2006) Transduction of Cu, Zn-superoxide dismutase mediated by an HIV-1 Tat protein basic domain into human chondrocytes. *Arthritis Res. Ther.* 8, R96.
27. Ezhevsky S, Ho A, Becker-Hapak M. Davis P, Dowdy S. (2001) Differential regulation of retinoblastoma tumor suppressor protein by G(1) cyclin-dependent kinase complexes *in vivo*. *Mol. Cell. Biol.* 21, 4773–4784.
28. Klekotka P, Santoro S, Ho A, Dowdy S, Zutter MM. (2001) Mammary epithelial cell-cycle progression via the alpha(2) beta(1) integrin: unique and synergistic roles of the alpha(2) cytoplasmic domain. *Am. J. Pathol.* 159, 983–992.
29. Fawell S, Seery J, Daikh Y, Moore C, Chen L, Pepinsky B, et al. (1994) Tat-mediated delivery of heterologous proteins into cells. *Proc. Natl. Acad. Sci. USA* 91, 664–668.
30. Eguchi A, Meade B, Chang Y, Fredrickson C, Willert K, Puri N, et al. (2009) Efficient siRNA delivery into primary cells by a peptide transduction domain-dsRNA binding domain fusion protein. *Nat. Biotech.* 27, 567–571.
31. Lissy NA, Davis P, Irwin M, Kaelin W, Dowdy S. (2000) A common E2F-1 and p73 pathway mediates cell death induced by TCR activation. *Nature* 407, 642–645.
32. Boenicke L, Chu K, Pauls R, Tams C, Kruse M, Kurdow R, et al. (2003) Efficient dose-dependent and time-dependent protein transduction of pancreatic carcinoma cells *in vitro* and *in vivo* using purified VP22-EGFP fusion protein. *J. Mol. Med.* 81, 205–213.
33. Caron N, Torrente Y, Camirand G, Bujold M, Chapdelaine P, Leriche K, et al. (2001) Intracellular delivery of a Tat-eGFP fusion protein into muscle cells. *Molecular Therapy* 3, 310–318.
34. Futaki S, Niwa M, Nakase I, Tadokoro A, Zhang Y, Nagaoka M, et al. (2004) Arginine carrier peptide bearing Ni(II) chelator to promote cellular uptake of histidine-tagged proteins. *Bioconjug. Chem.* 15, 475–481.
35. Bennett R, Dalby B, Guy P. (2002) Protein delivery using VP22. *Nat. Biotech.* 20(1), 20.
36. Anderson D, Nichols E, Manger R, Woodle D, Barry M, Fritzberg A. (1993) Tumor cell retention of antibody Fab fragments is enhanced by an attached HIV TAT protein-derived peptide. *Biochem. Biophys. Res. Commun.* 194, 876–884.
37. Kameyama S, Okada R, Kikuchi T, Omura T, Nakase I, Takeuchi T, et al. (2006) Distribution of immunoglobulin Fab fragment conjugated with HIV-1 REV peptide following intravenous administration in rats. *Mol. Pharmacol.* 3, 174–180.
38. Bian J, Popovic Z, Benejam C, Kiedrowski M, Rodriguez L, Penn M. (2007) Effect of cell-based intercellular delivery of transcription factor GATA4 on ischemic cardiomyopathy. *Circ. Res.* 100, 1626–1633.
39. Johnson L, Cashman S, Read S, Kumar-Singh R. (2010) Cell penetrating peptide POD mediates delivery of recombinant proteins to retina, cornea and skin. *Vision Res.* 50, 686–697.
40. Simon M, Gao S, Kang W, Banta S, Morrison BR. (2009) TAT-mediated intracellular protein delivery to primary brain cells is dependent on glycosaminoglycan expression. *Biotechnol. Bioeng.* 104, 10–19.
41. Chang M, Chou J, Lee H. (2005) Cellular internalization of fluorescent proteins via arginine-rich intracellular delivery peptide in plant cells. *Plant Cell Physiol.* 46, 482–488.
42. Pooga M, Kut C, Kihlmark M, Hällbrink M, Fernaeus S, Raid R, et al. (2001) Cellular translocation of proteins by transportan. *FASEB J.* 15, 1451–1453.
43. Elliott G, O'Hare P. (1997) Intercellular trafficking and protein delivery by a herpesvirus structural protein. *Cell* 88, 223–233.
44. Nagel F, Falkenburger B, Tonges L, Kowsky S, Poppelmeyer C, Schulz J, et al. (2008) Tat-Hsp70 protects dopaminergic neurons in midbrain cultures and in the substantia nigra in models of Parkinson's disease. *J. Neurochem.* 105, 853–864.
45. Doeppner T, Nagel F, Dietz G, Weise J, Tonges L, Schwarting S, et al. (2009) TAT-Hsp70-mediated neuroprotection and increased survival of neuronal precursor cells after focal

cerebral ischemia in mice. *J. cereb. blood flow metab.* 29, 1187–1196.

46. Dilber MS, Phelan A, Aints A, Mohamed A, Elliott G, Smith C, et al. (1999) Intercellular delivery of thymidine kinase prodrug activating enzyme by the herpes simplex virus protein, VP22. *Gene Ther.* 6, 12–21.
47. Blackwell N, Sembi P, Newson J, Lawrence T, Gilroy D, Kabouridis P. (2004) Reduced infiltration and increased apoptosis of leukocytes at sites of inflammation by systemic administration of a membrane-permeable IkappaBalpha repressor. *Arthritis Rheum.* 50, 2675–2684.
48. Patel EN, Wang J, Kim K, Borok Z, Crandall ED, Shen W. (2009) Conjugation with cationic cell-penetrating peptide increases pulmonary absorption of insulin. *Mol. Pharm.* 6, 492–503.
49. Racanicchi S, Maccherani C, Liberatore C, Billi M, Gelmetti V, Panigada M, et al. (2005) Targeting fusion protein/corepressor contact restores differentiation response in leukemia cells. *EMBO J.* 24, 1232–1242.
50. Bleifuss E, Kammertoens T, Hutloff A, Quarcoo D, Dorner M, Straub P, et al. (2006) The translocation motif of hepatitis B virus improves protein vaccination. *Cell. Mol. Life Sci.* 63, 627–635.
51. Ezhevsky S, Nagahara H, Vocero-Akbani AM, Gius D, Wei M, Dowdy S. (1997) Hypo-phosphorylation of the retinoblastoma protein (pRb) by cyclin D:Cdk4/6 complexes results in active pRb. *Proc. Natl. Acad. Sci. USA* 94, 10699–10704.
52. Nagahara H, Vocero-Akbani AM, Snyder E, Ho A, Latham D, Lissy N, et al. (1998) Transduction of full-length TAT fusion proteins into mammalian cells: TAT-p27Kip1 induces cell migration. *Nat. Med.* 4, 1449–1452.
53. Phelan A, Elliott G, O'Hare P. (1998) Intercellular delivery of functional p53 by the herpesvirus protein VP22. *Nat. Biotech.* 16, 440–443.
54. Wills K, Atencio I, Avanzini J, Neuteboom S, Phelan A, Philopena J, et al. (2001) Intratumoral spread and increased efficacy of a p53-VP22 fusion protein expressed by a recombinant adenovirus. *J. Virol.* 75, 8733–8741.
55. Zender L, Kock R, Eckhard M, Frericks B, Gosling T, Gebhardt T, et al. (2002) Gene therapy by intrahepatic and intratumoral trafficking of p53-VP22 induces regression of liver tumors. *Gastroenterology* 123, 608–618.
56. Kubo E, Singh DP, Fatma N, Akagi Y. (2009) TAT-mediated peroxiredoxin 5 and 6 protein transduction protects against high-glucose-induced cytotoxicity in retinal pericytes. *Life Sci.* 84, 857–864.
57. Avignolo C, Bagnasco L, Biasotti B, Melchiori A, Tomati V, Bauer I, et al. (2008) Internalization via Antennapedia protein transduction domain of an scFv antibody toward c-Myc protein. *FASEB J.* 22, 1237–1245.
58. Shin I, Edl J, Biswas S, Lin P, Mernaugh R, Arteaga C. (2005) Proapoptotic activity of cell-permeable anti-Akt single-chain antibodies. *Cancer Res.* 65, 2815–2824.
59. Becker-Hapak M, Mcallister S, Dowdy S. (2001) TAT-mediated protein transduction into mammalian cells. *Methods* 24, 247–256.
60. Abu-Amer Y, Dowdy SF, Ross F, Clohisy J, Teitelbaum S. (2001) TAT fusion proteins containing tyrosine 42-deleted IkappaBalpha arrest osteoclastogenesis. *J. Biol. Chem.* 276, 30499–30503.
61. Dietz GP, Kilic E, Bähr M. (2002) Inhibition of neuronal apoptosis *in vitro* and *in vivo* using TAT-mediated protein transduction. *Mol. Cell. Neurosci.* 21, 29–37.
62. Doeppner T, Dietz G, El Aanbouri M, Gerber J, Witte O, Bähr M, et al. (2009) TAT-Bcl-x(L) improves survival of neuronal precursor cells in the lesioned striatum after focal cerebral ischemia. *Neurobiol. Dis.* 34, 87–94.
63. Doeppner T, Dietz G, Weise J, Bähr M. (2010) Protection of hippocampal neurogenesis by TAT-Bcl-x(L) after cerebral ischemia in mice. *Exp. Neurol.* 223, 548–556.
64. Doeppner T, El Aanbouri M, Dietz G, Weise J, Schwarting S, Bähr M. (2010) Transplantation of TAT-Bcl-xL-transduced neural precursor cells: long-term neuroprotection after stroke. *Neurobiol. Dis.* 40, 265–276.
65. Guelen L, Paterson H, Gaken J, Meyers M, Farzaneh F, Tavassoli M. (2004) TAT-apoptin is efficiently delivered and induces apoptosis in cancer cells. *Oncogene* 23, 1153–1165.
66. Gustafsson A, Sayen M, Williams S, Crow M, Gottlieb R. (2002) TAT protein transduction into isolated perfused hearts: TAT-apoptosis repressor with caspase recruitment domain is cardioprotective. *Circulation* 106, 735–739.
67. Hudecz F, Banoczi Z, Csik G. (2005) Medium-sized peptides as built in carriers for biologically active compounds. *Med. Res. Rev.* 25, 679–736.
68. Kilic E, Dietz G, Hermann D, Bähr M. (2002) Intravenous TAT-Bcl-Xl is protective after middle cerebral artery occlusion in mice. *Ann. Neurol.* 52, 617–622.
69. Kilic U, Kilic E, Dietz G, Bähr M. (2003) Intravenous TAT-GDNF is protective after focal cerebral ischemia in mice. *Stroke* 34, 1304–1310.
70. Ribeiro M, Klein D, Pileggi A, Molano RD, Fraker C, Ricordi C, et al. (2003) Heme oxygenase-1 fused to a TAT peptide transduces and protects pancreatic beta-cells. *Biochem. Biophys. Res. Commun.* 305, 876–881.
71. Schwarze S, Dowdy S. (2000) *In vivo* protein transduction: intracellular delivery of biologically active proteins, compounds and DNA. *TrendsPharmacol. Sci.* 21, 45–48.
72. Souktani R, Pons S, Guegan C, Bouhidel O, Bruneval P, Zini R, et al. (2009) Cardioprotection against myocardial infarction with PTD-BIR3/RING, a XIAP mimicking protein. *J. Mol. Cell. Cardiol.* 46, 713–718.
73. Tinega A, Pelle R, Kang'A, S. Gicheru M, Taracha E, Nene V, et al. (2009) Fusion of a cell penetrating peptide from HIV-1 TAT to the Theileria parva antigen Tp2 enhances the stimulation of bovine CD8+ T cell responses. *Vet. Immunol. Immunopathol.* 130, 107–113.
74. Viehl C, Tanaka Y, Chen T, Frey D, Tran A, Fleming T, et al. (2005) Tat mammaglobin fusion protein transduced dendritic cells stimulate mammaglobin-specific CD4 and CD8 T cells. *Breast Cancer Res. Treat.* 91, 271–278.
75. Vitali R, Cesi V, Tanno B, Ferrari-Amorotti G, Dominici C, Calabretta B, et al. (2008) Activation of p53-dependent

responses in tumor cells treated with a PARC-interacting peptide. *Biochem. Biophys. Res. Commun.* 368, 350–356.

76. Wadia J, Dowdy S. (2005) Transmembrane delivery of protein and peptide drugs by TAT-mediated transduction in the treatment of cancer. *Adv. Drug Deliv. Rev.* 57, 579–596.
77. Wheeler D, Dunsmore K, Wong H. (2003) Intracellular delivery of HSP70 using HIV-1 Tat protein transduction domain. *Biochem. Biophys. Res. Commun.* 301, 54–59.
78. Kameyama S, Horie M, Kikuchi T, Omura T, Tadokoro A, Takeuchi T, et al. (2007) Acid wash in determining cellular uptake of fab/cell-permeating peptide conjugates. *Biopolymers* 88, 98–107.
79. Yu HH, Nakase I, Pujals S, Hirose H, Tanaka G, Katayama S, et al. (2010) Expressed protein ligation for the preparation of fusion proteins with cell penetrating peptides for endotoxin removal and intracellular delivery. *Biochim. Biophys. Acta.* 1798, 2249–2257.
80. Merrifield RB. (1963) Solid phase peptide synthesis. I. The synthesis of a tetrapeptide'. *J. Am. Chem. Soc.* 85, 2149–2154.
81. Planel S, Salomon A, Jalinot P, Feige J, Cherradi N. (2010) A novel concept in antiangiogenic and antitumoral therapy: multitarget destabilization of short-lived mRNAs by the zinc finger protein ZFP36L1. *Oncogene.* 29(45), 5989–6003.
82. Cuatrecasas P, Wilchek M, Anfinsen C. (1968) Selective enzyme purification by affinity chromatography. *Proc. Natl. Acad. Sci. USA* 61, 636–643.
83. Hochuli E, Bannwarth W, Döbeli H, Gentz R, Stüber D. (1988) Genetic approach to facilitate purification of recombinant proteins with a novel metal chelate adsorbent. *Nat. Biotech.* 6, 1321–1325.
84. Gao S, Simon M, Morrison BR, Banta S. (2009) Bifunctional chimeric fusion proteins engineered for DNA delivery: optimization of the protein to DNA ratio. *Biochim. Biophys. Acta.* 1790, 198–207.
85. Massodi I, Raucher D. (2007) A thermally responsive Tat-elastin-like polypeptide fusion protein induces membrane leakage, apoptosis, and cell death in human breast cancer cells. *J. Drug Target.* 15, 611–622.
86. Massodi I, Bidwell GR, Davis A, Tausend A, Credit K, Flessner M, et al. (2009) Inhibition of ovarian cancer cell metastasis by a fusion polypeptide Tat-ELP. *Clin. Exp. Metastasis* 26, 251–260.
87. Massodi I, Moktan S, Rawat A, Bidwell GR, Raucher D. (2010) Inhibition of ovarian cancer cell proliferation by a cell cycle inhibitory peptide fused to a thermally responsive polypeptide carrier. *Int. J. Cancer* 126, 533–544.
88. Zender L, Kuhnel F, Kock R, Manns M, Kubicka S. (2002) VP22-mediated intercellular transport of p53 in hepatoma cells *in vitro* and *in vivo*. *Cancer Gene Therapy* 9, 489–496.
89. Harris MP, Sutjipto S, Wills K, Hancock W, Cornell D, Johnson D, et al. (1996) Adenovirus-mediated p53 gene transfer inhibits growth of human tumor cells expressing mutant p53 protein. *Cancer Gene Therapy* 3, 121–130.
90. Fang B, Xu B, Koch P, Roth J. (1998) Intercellular trafficking of VP22-GFP fusion proteins is not observed in cultured mammalian cells. *Gene Therapy* 5, 1420–1424.
91. Zavaglia D, Lin E, Guidetti M, Pluquet O, Hainaut P, Favrot M, et al. (2005) Poor intercellular transport and absence of enhanced antiproliferative activity after non-viral gene transfer of VP22-P53 or P53-VP22 fusions into p53 null cell lines *in vitro* or *in vivo*. *J. Gene Med.* 7, 936–944.
92. Wang H, Fu T, Wang G, Zeng G, Perry-Lalley DM, Yang J, et al. (2002) Induction of CD4(+) T cell-dependent antitumor immunity by TAT-mediated tumor antigen delivery into dendritic cells. *J. Clin. Invest.* 109, 1463–1470.
93. Kim C, Woo S, Park J, Kim H, Park M, Park S, et al. (2007) Enhanced antitumour immunity by combined use of temozolomide and TAT-survivin pulsed dendritic cells in a murine glioma. *Immunology* 122, 615–622.
94. Rautsi O, Lehmusvaara S, Ketola A, Maatta A, Wahlfors J, Pellinen R. (2008) Characterization of HIV-1 TAT peptide as an enhancer of HSV-TK/GCV cancer gene therapy. *Cancer Gene Therapy* 15, 303–314.
95. Song H, Lee J, Ju S, Yoo K, Won M, Kwon H, et al. (2008) Topical transduction of superoxide dismutase mediated by HIV-1 Tat protein transduction domain ameliorates 12-O-tetradecanoylphorbol-13-acetate (TPA)-induced inflammation in mice. *Biochem. Pharmacol.* 75, 1348–1357.
96. Trouba K, Hamadeh H, Amin R, Germolec D. (2002) Oxidative stress and its role in skin disease. *Antioxid. Redox Signal.* 4, 665–673.
97. Regan E, Flannelly J, Bowler R, Tran K, Nicks M, Carbone B, et al. (2005) Extracellular superoxide dismutase and oxidant damage in osteoarthritis. *Arthritis Rheum.* 52, 3479–3491.
98. Kabouridis P, Hasan M, Newson J, Gilroy D, Lawrence T. (2002) Inhibition of NF-kappa B activity by a membrane-transducing mutant of I kappa B alpha. *J. Immunol.* 169, 2587–2593.
99. Klein D, Ribeiro M, Mendoza V, Jayaraman S, Kenyon N, Pileggi A, et al. (2004) Delivery of Bcl-XL or its BH4 domain by protein transduction inhibits apoptosis in human islets. *Biochem. Biophys. Res. Commun.* 323, 473–478.
100. Chen X, Lai J, Pan Q, Tang Z, Yu Y, Zang G. (2010) The delivery of HBcAg via Tat-PTD enhances specific immune response and inhibits Hepatitis B virus replication in transgenic mice. *Vaccine* 28, 3913–3919.
101. Wang S, Han Q, Zhang N, Chen J, Liu Z, Zhang G, et al. (2010) HBcAg18-27 epitope fused to HIV-Tat 49-57 adjuvanted with CpG ODN induces immunotherapeutic effects in transgenic mice. *Immunol. Lett.* 127, 143–149.
102. Yi J, Gong W, Wang L, Ling R, Chen J, Yun J. (2005) VP22 fusion protein-based dominant negative mutant can inhibit hepatitis B virus replication. *World J. Gastroenterol.* 11, 6429–6432.
103. Kilic E, Kilic U, Hermann D. (2006) TAT fusion proteins against ischemic stroke: current status and future perspectives. *Front. Biosci.* 11, 1716–1721.
104. Gou X, Wang Q, Yang Q, Xu L, Xiong L. (2010) TAT-NEP1-40 as a novel therapeutic candidate for axonal regeneration and functional recovery after stroke. *J. Drug Target.* 19(2), 86–95.

105. Mie M, Takahashi F, Funabashi H, Yanagida Y, Aizawa M, Kobatake E. (2003) Intracellular delivery of antibodies using TAT fusion protein A. *Biochem. Biophys. Res. Commun.* 310, 730–734.
106. Heng B, Cao T. (2005) Making cell-permeable antibodies (Transbody) through fusion of protein transduction domains (PTD) with single chain variable fragment (scFv) antibodies: potential advantages over antibodies expressed within the intracellular environment (Intrabody). *Med. Hypotheses* 64, 1105–1108.
107. CDG-Therapeutics. P28. Available at http://www.cdgti.com/. Accessed 2010 Dec 20.
108. Meyer-Losic F, Nicolazzi C, Quinonero J, Ribes F, Michel M, Dubois V, et al. (2008) DTS-108, a novel peptidic prodrug of SN38: *in vivo* efficacy and toxicokinetic studies. *Clin. Cancer Res.* 14, 2145–2153.
109. Harada H, Hiraoka M, Kizaka-Kondoh S. (2002) Antitumor effect of TAT-oxygen-dependent degradation-caspase-3 fusion protein specifically stabilized and activated in hypoxic tumor cells. *Cancer Res.* 62, 2013–2018.
110. Harada H, Kizaka-Kondoh S, Li G, Itasaka S, Shibuya K, Inoue M, et al. (2007) Significance of HIF-1-active cells in angiogenesis and radioresistance. *Oncogene* 26, 7508–7516.
111. Kizaka-Kondoh S, Itasaka S, Zeng L, Tanaka S, Zhao T, Takahashi Y, et al. (2009) Selective killing of hypoxia-inducible factor-1-active cells improves survival in a mouse model of invasive and metastatic pancreatic cancer. *Clin. Cancer Res.* 15, 3433–3441.
112. Traversa-Therapeutics. CPP-DRBD siRNA. Available at http://www.traversathera.com/. Accessed 2010 Dec 20.
113. Justesen S, Buus S, Claesson M, Pedersen A. (2007) Addition of TAT protein transduction domain and GrpE to human p53 provides soluble fusion proteins that can be transduced into dendritic cells and elicit p53-specific T-cell responses in HLA-A∗0201 transgenic mice. *Immunology* 122, 326–334.
114. Inagaki K, Begley R, Ikeno F, Mochly-Rosen D. (2005) Cardioprotection by epsilon-protein kinase C activation from ischemia: continuous delivery and antiarrhythmic effect of an epsilon-protein kinase C-activating peptide. *Circulation* 111, 44–50.
115. Chen L, Harrison S. (2007) Cell-penetrating peptides in drug development: enabling intracellular targets. *Biochem. Soc. Trans.* 35, 821–825.
116. Price V. (2003) Therapy of alopecia areata: on the cusp and in the future. *J. Investig. Dermatol. Symp. Proc.* 8, 207–211.
117. Delcroix M, Riley LW. (2010) Cell-penetrating peptides for antiviral drug development. *Pharmaceuticals* 3(3), 448–470.
118. Lopes LB, Brophy CM, Furnish E, Flynn CR, Sparks O, Komalavilas P, et al. (2005) Comparative study of the skin penetration of protein transduction domains and a conjugated peptide. *Pharm. Res.* 22, 750–757.
119. Lopes LB, Furnish E, Komalavilas P, Flynn C, Ashby P, Hansen A, et al. (2009) Cell permeant peptide analogues of the small heat shock protein, HSP20, reduce TGF-beta1-induced CTGF expression in keloid fibroblasts. *J. Invest. Dermatol.* 129, 590–598.
120. Hassane FS, Saleh AF, Abes R, Gait MJ, Lebleu B. (2010) Cell penetrating peptides: overview and applications to the delivery of oligonucleotides. *Cell. Mol. Life Sci.* 67, 715–726.
121. Moulton H, Moulton J. (2010) Morpholinos and their peptide conjugates: therapeutic promise and challenge for Duchenne muscular dystrophy. *Biochim. Biophys. Acta.* 1798, 2296–2303.
122. Järver P, Langel Ü, Langel Ü. (2004) The use of cell-penetrating peptides as a tool for gene regulation. *Drug Discov. Today* 9, 395–402.
123. KAI-Pharmaceuticals. KAI-9803. Available at http://www.kaipharma.com/. Accessed 2010 Dec 20.
124. Revance-Therapeutics. RT001. Available at http://www.revance.com/. Accessed 2010 Dec 20.
125. Avi-BioPharma. AVI-5126. Available at http://www.avibio.com/. Accessed 2010 Dec 20.
126. Capstone-Therapeutics. AZX100. Available at http://www.capstonethx.com/. Accessed 2010 Dec 20.
127. Xigen. AM-111. Available at http://www.xigen.ch/. Accessed 2010 Dec 20.
128. Auris-Medical. AM-111. Available at http://www.aurismedical.com/. Accessed 2010 Dec 20.
129. Beckham J, Goody R, Clarke P, Bonny C, Tyler K. (2007) Novel strategy for treatment of viral central nervous system infection by using a cell-permeating inhibitor of c-Jun N-terminal kinase. *J. Virol.* 81, 6984–6992.
130. Eshraghi A, Wang J, Adil E, He J, Zine A, Bublik M, et al. (2007) Blocking c-Jun-N-terminal kinase signaling can prevent hearing loss induced by both electrode insertion trauma and neomycin ototoxicity. *Hear. Res.* 226, 168–177.
131. Liu J, Zhao Y, Patzer A, Staak N, Boehm R, Deuschl G, et al. (2010) The c-Jun N-terminal kinase (JNK) inhibitor XG-102 enhances the neuroprotection of hyperbaric oxygen after cerebral ischaemia in adult rats. *Neuropathol. Appl. Neurobiol.* 36, 211–224.
132. Wang J, Van De Water TR, Bonny C, De Ribaupierre F, Puel JL, Zine A. (2003) A peptide inhibitor of c-Jun N-terminal kinase protects against both aminoglycoside and acoustic trauma-induced auditory hair cell death and hearing loss. *J. Neurosci.* 23, 8596–8607.
133. Drais-Pharmaceuticals. DTS-108. Available at http://www.draispharma.com/. Accessed 2010 Dec 20.
134. Aguilera TA, Olson ES, Timmers MM, Jiang T, Tsien RY. (2009) Systemic *in vivo* distribution of activatable cell penetrating peptides is superior to that of cell penetrating peptides. *Integrative Biology* 1, 371–381.
135. Olson E, Jiang T, Aguilera T, Nguyen Q, Ellies L, Scadeng M, et al. (2010) Activatable cell penetrating peptides linked to nanoparticles as dual probes for *in vivo* fluorescence and MR imaging of proteases. *Proc. Natl. Acad. Sci. USA* 107, 4311–4316.
136. Rizk S, Luchniak A, Uysal S, Brawley C, Rock R, Kossiakoff A. (2009) An engineered substance P variant for receptor-mediated delivery of synthetic antibodies into tumor cells. *Proc. Natl. Acad. Sci. USA* 106, 11011–11015.

137. Foerg C, Merkle H. (2008) On the biomedical promise of cell penetrating peptides: limits versus prospects. *J.Pharm. Sci.* 97, 144–162.

138. Li L, Schlag PM, Blankenstein T. (1997) Transient expression of SV 40 large T antigen by Cre/LoxP-mediated site-specific deletion in primary human tumor cells. *Human Gene Therapy* 8, 1695–1700.

139. Ford K, Souberbielle B, Darling D, Farzaneh F. (2001) Protein transduction: an alternative to genetic intervention? *Gene Therapy* 8, 1–4.

140. Gros E, Deshayes S, Morris M, Aldrian-Herrada G, Depollier J, Heitz F, et al. (2006) A non-covalent peptide-based strategy for protein and peptide nucleic acid transduction. *Biochim. Biophys. Acta.* 1758, 384–393.

141. Mäe M, El Andaloussi S, Lundin P, Oskolkov N, Johansson H, Guterstam P, et al. (2009) A stearylated CPP for delivery of splice correcting oligonucleotides using a non-covalent co-incubation strategy. *J. Control. Release* 134, 221–227.

142. Morris M, Depollier J, Mery J, Heitz F, Divita G. (2001) A peptide carrier for the delivery of biologically active proteins into mammalian cells. *Nat. Biotech.* 19, 1173–1176.

143. El-Andaloussi S, Holm T, Langel Ü. (2005) Cell-penetrating peptides: mechanisms and applications. *Curr. Pharm. Des.* 11, 3597–3611.

144. Trehin R, Merkle HP. (2004) Chances and pitfalls of cell penetrating peptides for cellular drug delivery. *Eur. J. Pharm. Biopharm.* 58, 209–223.

145. Futaki S, Suzuki T, Ohashi W, Yagami T, Tanaka S, Ueda K, et al. (2001) Arginine-rich peptides. An abundant source of membrane-permeable peptides having potential as carriers for intracellular protein delivery. *J. Biol. Chem.* 276, 5836–5840.

146. Ho A, Schwarze S, Mermelstein S, Waksman G, Dowdy S. (2001) Synthetic protein transduction domains: enhanced transduction potential *in vitro* and *in vivo*. *Cancer Res.* 61, 474–477.

147. Tyagi M, Rusnati M, Presta M, Giacca M. (2001) Internalization of HIV-1 tat requires cell surface heparan sulfate proteoglycans. *J. Biol. Chem.* 276, 3254–3261.

148. Magzoub M, Pramanik A, Gräslund A. (2005) Modeling the endosomal escape of cell-penetrating peptides: transmembrane pH gradient driven translocation across phospholipid bilayers. *Biochemistry* 44, 14890–14897.

149. Hällbrink M, Kilk K, Elmquist A, Lundberg P, Lindgren M, Jiang Y, et al. (2005) Prediction of cell-penetrating peptides. *Int. J. Pept. Res. Ther.* 11, 249–259.

27

CELL-SPECIFIC TARGETING OF FUSION PROTEINS THROUGH HEPARIN BINDING

JIAJING WANG, ZHENZHONG MA, AND JEFFREY A. LOEB

Department of Neurology, Center for Molecular Medicine and Genetics, Wayne State University, Detroit, MI, USA

27.1 WHY TARGET HEPARAN-SULFATE PROTEOGLYCANS WITH FUSION PROTEINS?

The current revolution in functional genomic technologies has led to the identification of novel genes that have the potential to become powerful new drugs to treat human diseases from cancer to neurodegenerative disorders. Those drugs are often called biopharmaceuticals because they consist of recombinant proteins and/or nucleic acids and can be used to initiate or block-specific biological processes to treat human diseases. One of the foremost obstacles is getting these powerful biopharmaceuticals to diseased tissues specifically. This is critical not only to achieve sufficient levels of drug for efficacy but also to minimize unwanted side effects in other tissues [1,2]. However, many new biopharmaceutical drugs have failed in clinical trials despite extensive preclinical testing both in culture and in animal models. In many cases, this is due to the lack of specific targeting resulting in unwanted toxicity and/or failed efficacy. An important example comes from neurotrophic factors that have been used to treat neurodegenerative disorders, such as brain-derived neurotrophic factor (BDNF; Regeneron) and glial cell line-derived neurotrophic factor (GDNF; Amgen), that failed in clinical trials in part because of poor targeting [3].

One of the ways that nature has learned to target growth factors and viruses to specific cells is through specific interactions with cell-surface molecules many of which reside in the extracellular matrix (ECM). The ECM provides cells with not only scaffold and support but also appears to be critical for important cellular and tissue functions including differentiation, proliferation, migration, and chemotaxis [4]. The ECM is composed of heterogeneous mixtures of various key proteins, such as collagens and proteoglycans. Proteoglycans produce some of the most diverse set of structures that give each cell, or portions of that cell, a unique cell-surface signature. Proteoglycans are large transmembrane or extracellular secreted proteins with a majority of their molecular composition consisting of covalently attached glycosaminoglycan (GAG) side chains. GAGs can be classified as keratan sulfate (KS), chondroitin sulfate

Fusion Protein Technologies for Biopharmaceuticals: Applications and Challenges, First Edition. Edited by Stefan R. Schmidt.

(CS), dermatan sulfate (DS), heparan sulfate (HS), and hyaluronan based on their sugar compositions. Except hyaluronan, the other GAGs display modifications of sugar molecules, consisting mostly of sulfated hydroxyl groups and epimerization of sugar structures [5,6]. Compared to KS, CS, and DS, HS contains the greatest number of different modifications. Consequently, heparan sulfate proteoglycans (HSPGs) have been suggested to be one of the most information-condensed biological molecules in nature [7]. In fact, HSPGs are highly specific "targets" for a wide variety of proteins, growth factors, and viruses as shown schematically in Figure 27.1.

In this chapter, the biochemical and regulatory mechanisms on how HSPGs provide tissue-specific targeting as well as how fusion proteins that bind to HSPGs can take advantage of this natural targeting system will be described as a rationale for developing targeted therapeutics for human diseases.

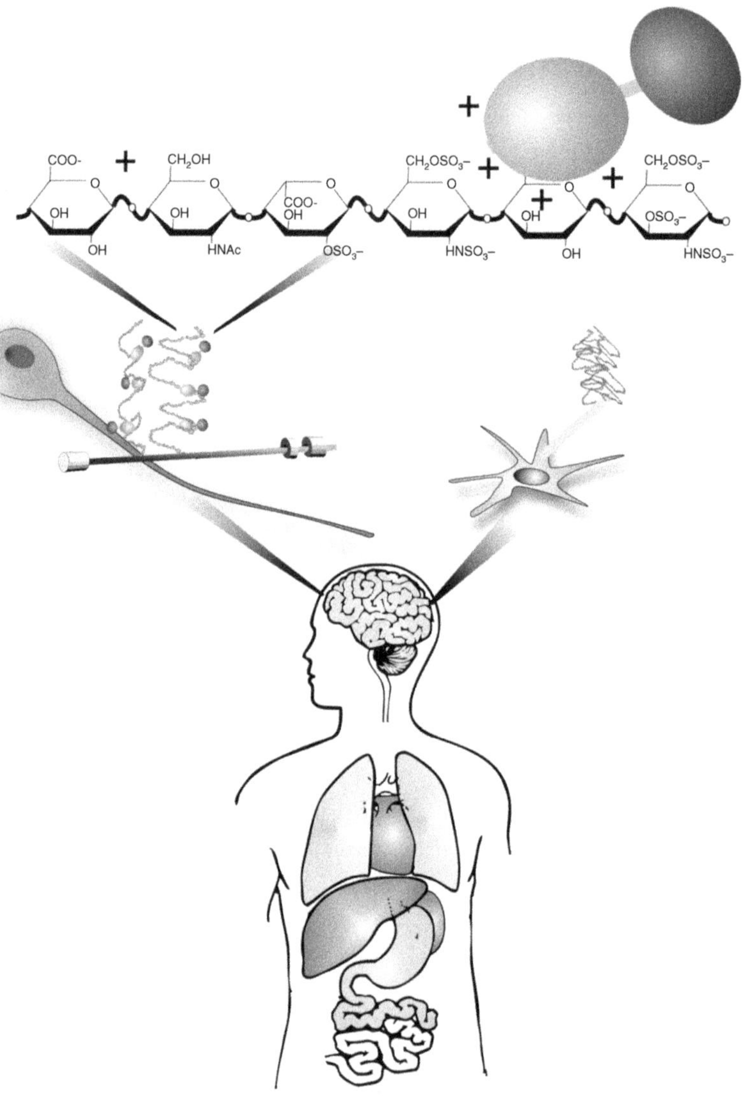

FIGURE 27.1 Heparan-sulfate proteoglycans in body and organs. Heparan-sulfate proteoglycans (HSPGs) consist of long chains of negatively charged sulfated sugars called glycosaminoglycans (GAGs) that are covalently attached to either secreted or membrane-bound proteins on the cell surface of almost all cells in the body. Their diversity in structure allows selective binding and concentration of proteins, such as growth factors and viral proteins through highly specific, positively charged heparin-binding domains (shown in red) to target specific cell types located in specific tissues. Once targeted to the cell surface, other portions of the protein are used to activate receptors or to enter the cell, in the case of viruses.

27.2 HEPARAN SULFATE STRUCTURE AND BIOSYNTHESIS CREATE DIVERSITY AND A TEMPLATE FOR TARGETING SPECIFICITY

To be a reliable biopharmaceutical target, HSPGs need to provide sufficient diversity and specificity for distinct biological contexts. Heparin, the well-known anticoagulant drug, is actually a processed and highly sulfated form of HS. While heparin is found mostly in connective-tissue type mast cells, HSPGs are found ubiquitously and abundantly on virtually every cell surface and in the ECM of most tissues. The diversity of HSPGs results first from a large number of core proteins that serve to distribute them to different regions of the extracellular space and second from the enormous variety of negatively charged, sulfated HS structures (See Figure 27.1 for structural schematics).

Most cells bear multiple types of proteoglycan core proteins. HSPGs can be generally categorized into three major classes based on the core protein primary structure [8–10]. Matrix HSPGs are secreted into ECM and include perlecan, agrin, and collagen XVIII. Syndecans are type I transmembrane HSPGs that carry the HS chains on their extracellular domains near the N-terminus. Most vertebrates have four syndecan genes, whereas all invertebrates have only one. Glypicans are the third class of HSPGs that are also anchored on the cell surface, but through a glycophosphatidylinositol linkage. Vertebrates typically have six glypican genes, whereas invertebrates only have two. The extracellular domains of betaglycan and a splice variant of CD44 (v3) are also subject to HS attachment. Each core protein often has several attachment sites capable of bearing multiple GAG chains. For example, the glypicans typically contain up to three HS side chains. Additionally, there are a number of proteoglycans that have more than one type of GAG chain, creating diversity of structure even within the same protein backbone. For example, neuropilin-1 is a proteoglycan that has combinations of HS, CS, and DS [11].

Besides the large number of core proteins available for HS decoration, HS itself is one of the most complex biomolecules. After a core protein is translated, many membrane-bound glycotransferases and modifying enzymes in the endoplasmic reticulum and the Golgi apparatus function in a specific and sequential manner to assemble HS chains [12]. HS synthesis initiates with the transfer of xylose to specific serine residues in a consensus sequence of serine-glycine with one or more flanking acidic residues. Attachment of two D-galctose (Gal) residues and a glucuronic acid (GlcA) completes the formation of the core-protein linkage tetrasaccharide. The formation of the linkage tetrasaccharide is identical in CS, DS, and HS. HS chains elongate by consecutive addition of alternating glucuronic acid (GlcA) and N-acetylglucosamine (GlcNAc) onto the linkage tetrasaccharide. This is catalyzed by enzymes that belong to the exostosis family of proteins (EXT) [13]. As the chain polymerizes, it undergoes a series of modifying enzymatic reactions that generate further heterogeneity of HS structure as well as give HSPGs their strongly negative electrostatic charge. These modifications include GlcNAc N-deacetylation/N-sulfation catalyzed by N-deacetylase/N-sulfotransferase (NDST) family of enzymes, epimerization of glucosamine to iduronic acid, 2-O-sulfation, and sulfation of GlcNAc by 6-O-sulfotransferases (6OST) or 3-O-sulfotranserases (3OST). The six potential modifications on every disaccharide lead to a great level of sequence diversity. Twenty-three unique disaccharides out of as many as 48 different disaccharides that can occur theoretically have thus far been identified from vertebrates [12]. The net result of this biosynthetic process is wide diversity of HS chains with domains of contiguous disaccharides containing N-sulfation, O-sulfation, and iduronic acid interspersed by variable tracts of unmodified domains. Not surprisingly, the binding sites for HS-interacting ligands, such as growth factors, occur in highly modified segments or at the borders between modified and unmodified HS segments [14].

The complexity and diversity of sugar structures does not end with biosynthesis. Further processing occurs once the HSPG reaches the plasma membrane, often as a means to carry out a biological process. Specific sulfate groups can be removed from the HS chains by plasma membrane-bound endosulfatases. For example, Sulf1 and Sulf2 are extracellular HSPG-specific sulfatases that catalyze the removal of sulfate from the C-6 position of glucosamine [15,16]. The GPI anchors of glypicans are subject to cleavage to shed the HSPGs from the cell surface [17]. Core proteins can also undergo proteolytic cleavage to be released into the ECM. For example, Syndecan-1 and -4 are expressed on epithelial cells in the lung [18]. Upon inflammation, they are released into the bronchoalveolar fluid mediated by the disintegrin-like metalloproteinase ADAM17. Extracellular heparanase is an endo-β-D-glucuronidase that cleaves HS side chains at certain sites [19]. They are strongly implicated in structural remodeling of the ECM as a consequence of HS cleavage. They can regulate growth factor and/or chemokines signaling by selectively mobilizing HS-bound ligands [20].

In summary, the various core proteins and essentially limitless combinations of distinct sugar compositions and modifications that occur during biosynthesis and postassembly processing contribute to an enormous structural diversity of HSPGs. The key to this structural heterogeneity is to use it in specific biological contexts and in segregated ways that are closely linked to the proteins that interact with the various HS structures. Examples of how these are used in nature and how one can harness this diversity for drug targeting are discussed later in this chapter.

27.3 TISSUE-SPECIFIC EXPRESSION OF HSPGs AND THE ENZYMES THAT MODIFY THEM

In order for HSPGs to be useful for protein targeting, they need to be expressed in a cell- and tissue-type-specific manner that is regulated by distinct physiological and/or pathological signals. For example, three of the four syndecans are distributed in tissue-specific patterns [10,21]. Syndecan-1 is common to many epithelia, developing mesenchyme, and some leukocytes. Syndecan-2 is enriched not only in many mesenchymal cell types, but in developing neural tissues as well. Syndecan-3 is mostly present in neural tissue and in the musculoskeletal system. In contrast, syndecan-4 is widely distributed across the organism. Similarly, the HSPG agrin is present on most cell types, but only neurons produce an alternatively spliced isoform, Z+ agrin, that produces an 8–19 amino acid domain that increase its acetylcholine receptor clustering activity dramatically and also encodes a heparin-binding domain (HBD) that acts synergistically with another growth factor called neuregulin1 [22]. Specific tissue distribution can also be regulated by environmental stress. Syndecan-2 is expressed in amoeboid microglial cells of the corpus callosum of the brain in neonatal rats during development and increases significantly upon a hypoxic exposure [23].

However, simply expressing a given core protein, does not necessarily mean that the structure of the HS chain is the same. The size, structure, and ligand affinity of the HS chains on a single proteoglycan species may differ significantly when expressed by different cell types [24]. For example, syndecan-1 from different cell types shows varied lengths of the HS chains as well as highly reproducible differences in the amount of certain sugar modifications that include the number of highly sulfated domains. Moreover, while knocking out the syndecan-1 gene in mice produces enhanced leukocyte adhesion under inflammatory conditions and delayed skin and cornea wound healing, it does not affect the cell-specific composition of HS in the affected tissue [25]. This finding suggests that much of the cell-specificity of HS structures depends mostly on the regulation and expression of the biosynthetic enzymes that determine the specific HS structures. Consistently, enzymes that polymerize and modify HS chains are expressed in a tissue-specific and developmental stage-dependent manner to produce distinct HS sequences. For example, NDSTs are the first enzymes engaged in the modification of the sugar backbone and are believed to create substrates for further enzymatic activity along the polymer [26]. There are four isozymes in the vertebrate family of NDSTs. Although they share the same overall structure and possess a high degree of protein identity, the expression profiles are different. NDST1 and NDST2 are expressed abundantly and ubiquitously in all tissues, whereas NDST2 and NDST4 are restricted to adult brain and fetal tissues. Both the type of enzyme expressed and the level of the expression determine the extent of N-sulfation of HS.

Diversity of cell-specific HS expression has been shown in an elegant study that generated HS saccharide-specific monoclonal antibodies [27]. Each of the monoclonal antibodies generated against a specific HS structure showed a spatially restricted pattern of expression, demonstrating that certain HS structures are tissue-specific. Since commonly occurring saccharide structures are not likely to elicit an immune response in the host, a phage display technique, which does not rely on the host immune response, was used for the generation of these antibodies that revealed the segregation of distinct HS structures in specific cellular locations.

The expression and activity of enzymes that process either the core proteins or the HS of HSPGs are often closely coupled to biological processes. Under normal conditions, growth factors and other proteins bound to HSPGs are kept in an inactive, sequestered state. Following a pathological condition, such as tissue injury or microbial infection, metalloproteinases are activated that cleave the ectodomains of the core proteins, resulting in their shedding from the cell surface as well as the release of their bound factors that can then become functionally active [18]. Enzymes that specifically degrade HS, such as heparanases, have long been correlated with tumor metastasis in various cancers [19]. Cleavage of HS leads to enhanced cell mobility as a consequence of HS remodeling in ECM of epithelial and endothelial cells.

Thus, in addition to their immense diversity in structure, the biologically driven synthesis and degradation of HSPGs can result in spatially restricted signaling that is consistent and reproducible. It is not surprising that both growth factors as well as viruses have evolved to fully exploit these properties of HSPGs and therefore, it seems a natural next step to exploit these natural "barcodes" for enhanced drug efficacy and targeted drug delivery.

27.4 HEPARIN-BINDING PROTEINS AND GROWTH FACTORS

Many cells are constantly poised to respond to a large variety of growth and differentiation factors through cell-surface receptors. In maintaining biological homeostasis, however, it is important that these potent signaling factors are only presented to such a cell at the right place and time. One way that this can be achieved is through the localized release and selective binding of ligands to the ECM through highly specific interactions with HSPGs. While these are generally fairly low-affinity interactions (micromolar to high nanomolar range), they serve to concentrate the ligand near its high-affinity cell-surface receptor as well as protect it from proteolysis [28]. Not surprisingly, protein–HS interactions are highly electrostatic through positively charged amino acid sequences that bind to the strongly anionic HS; and occurs with a diverse repertoire of proteins including growth factors, cytokines, chemokines, cell-surface receptors, matrix components, enzymes, and viruses (for review see [29]). As many of these HS-binding proteins are characterized and purified by affinity chromatography on columns of heparin, they are often collectively referred to as heparin-binding proteins, even though the endogenous binding partners are HSPGs.

While not used for targeting, one of the best-studied examples of a heparin-binding protein is antithrombin [30]. Heparin is used therapeutically to bind to antithrombin and results in the inactivation of thrombin, Factor Xa, and other members of the coagulation cascade. High-affinity binding requires a specific pentasaccharide within heparin that interacts with positively charged amino acids on the protein. The subsequent conformational change accelerates its binding and inhibition of the proteases involved in coagulation.

Perhaps, the most extensively studied heparin-binding growth factor is fibroblast growth factor (FGF). FGFs are a large family of monomeric growth factors that function in cell growth, angiogenesis, wound healing, and embryonic development. Heparin-binding forms of FGF bind HS not only for tissue targeting but also to enhance their signaling through high-affinity membrane receptors [31]. A tissue-binding assay has been developed that shows that specific forms of FGF bind selectively to some tissue and cells and not others based on their HS composition [32]. In addition to localizing FGF in various tissue locations, interactions with HSPGs significantly enhance signal transduction by lowering the concentration of FGF required for signaling initiation and extending the response duration. There is evidence to suggest that HS further activates signaling by forming a bridging complex between FGF and its receptor, that also has a heparin-binding site, to produce a ternary complex that promotes FGF signaling [33].

In addition to targeting and enhancing receptor activation, proteins bound to HS are protected from proteolytic degradation. HS has been shown to protect cytokines from endogenous proteolytic cleavage during the inflammatory process so that they can continue to carry out their important roles [34]. Interferon-γ (IFN-γ) is a cytokine that is critical for innate and adaptive immunity against viral and intracellular bacterial infections. It binds with high affinity to HS and heparin molecules through its carboxyl-terminal domain. Unbound IFN-gamma is cleaved rapidly at the carboxyl-terminal side, resulting in removal of at least 18 amino acids and inactivation of the cytokine. When bound to heparin, the plasma clearance of IFN-gamma is decreased greatly and the extent of its carboxyl-terminal domain degradation is limited to less than 10 amino acids, giving rise to sixfold increases of the cytokine activity. This resistance to proteolysis and resultant longer half-life of heparin-binding proteins is an attractive feature that could be incorporated into engineered biopharmaceuticals that bind heparin.

In addition to using heparin-binding proteins as receptor agonists, nature has also used heparin-binding capability to produce targeted antagonists. An example comes from the Wnt signaling pathway [35]. Wnt proteins are a large family of structurally related secreted heparin-binding glycoproteins that mediate fundamental biological processes such as cell polarity and proliferation, tissue patterning, and tumorigenesis. The Wnt signaling is initiated by the binding of Wnt proteins to a membrane receptor complex consisting of the Frizzled (Fz) family and members of the low-density lipoprotein receptor-related family. HS augments the action of secreted Wnt antagonists, some of which are also heparin-binding proteins. An example is the secreted frizzled-related protein-1 (sFRP-1) that belongs to a class of extracellular antagonists that modulate Wnt signaling pathways by preventing ligand–receptor interactions between the Wnt and the Frizzled membrane receptor. Heparin can specifically enhance recombinant sFRP-1 accumulation in a cell type-specific manner that requires O-sulfation in heparin and FGF signaling. Moreover, it seems that heparin also directly stabilizes the sFRP-1 protein *in vitro*.

27.5 VIRUSES TARGET CELLS THROUGH HEPARIN BINDING

It is not surprising that if nature has created a system for tissue-specific targeting of proteins using HSPGs that viruses have evolved to take full advantage of this system. In fact, a number of viruses use HSPGs as their specific targets for viral entry that lead to tissue-specific infectivity [36]. Herpes simplex virus type 1 (HSV-1) binds to cells through interactions of viral glycoprotein D with cell-surface HS chains consisting 3-O-sulfation (3-OS) generated by 3-O sulfotransferase (3OST). These interactions promote some selectivity in targeting HSV-1 infections to specific cell populations such as the dorsal root ganglion cells where the virus can lay dormant for long periods. There are six isoforms of 3OST in humans [37,38]. 3OST-2, -3, and -4 have been reported to make cells susceptible to HSV-1 entry and produce two disaccharides, one of which is the tetrasulfated disaccharide (Di-tetraS) unit [39]. While the 3OST-3 transcripts are highly expressed in spleen and liver, the 3OST-2 and -4 are expressed specifically in brain. Consistently, Di-tetraS is relatively highly expressed in spleen and liver. Although the interaction does not suffice for viral entry, bound viruses can then interact with other host determinants for internalization.

Another family of viruses that use HSPGs for cell recognition are the adeno-associated virus type 2 (AAV-2). Mutagenesis experiments of AAV-2 capsid proteins showed that a group of basic amino acids contribute to heparin binding and that heparin binding influences the infectivity [40]. Importantly, the capsid of the AAV-2 vector, which is being used for human gene therapy of congenital muscle disorders, has been shown to be taken up into human dendritic cells producing activation of capsid-specific T cells. This uptake is thought to be due to heparin binding and has the potential to produce toxicity in human clinical trials [41,42]. Consistently, engineered or natural variants of AAV lacking heparin-binding capability were less likely to activate T cells directed against the capsid and did not seem to compromise gene transfer.

In summary, the use of heparin-binding protein motifs as a source of tissue-specific targeting is widely used in nature from growth and differentiation factors to viruses. The "barcode" provided by HSPGs in the ECM is often used as low-affinity targeting system that often requires a second, more highly specific receptor–ligand interaction to exert its biological effects. As will be described later, we are currently developing such a system of tissue-specific targeting for biopharmaceutical development to enhance tissue-specific delivery and produce sustained efficacy using the HBD from the neuregulin 1 growth factor.

27.6 DISSECTING HEPARIN-BINDING PROTEIN DOMAINS FOR TISSUE-SPECIFIC TARGETING

Given the tremendous diversity, yet specificity, of HS structures in different tissues that are critical for normal development and function [43–45], it is equally important that there be a similar diversity in specificity of heparin-binding proteins. This is in fact the case with a wide variety of peptide structures that show selectivity toward specific HS structures. Understanding how HS specificity occurs in these naturally occurring heparin-binding proteins will not only provide clues to study the biological roles of HSPGs, but can also generate novel methods for the development of tissue-specific targeting of biopharmaceuticals.

One method that has been used to identify the HBDs within proteins that bind HSPGs is site-directed mutagenesis of specific amino acids. Site-directed mutagenesis was used to identify key positively charged amino acid residues required for the vaccinia viral envelope protein A27 to interact with both heparin and HSPGs for cell-surface attachment [46]. Using site-directed mutagenesis and solution NMR, a "KKPE" amino acid segment with a turn-like conformation was identified as a key structural region that mediates specific heparin binding. In addition, spontaneous mutants have been found reveal specific structural features required for heparin binding. The elbow knee synostosis (Eks) mouse shows elbow and knee joint synosotsis, and premature fusion of cranial sutures [47]. A missense mutation in the FGF9 gene was identified in the Eks mouse was identified that alters the binding of FGF9 to HSPGs through preventing homodimerization of the FGF9 protein. Since monomeric FGF9 binds to heparin with lower affinity than dimeric FGF9, the lack of dimerization results in increased diffusion of the altered FGF9 protein through developing tissues leading to ectopic FGF9 signaling, and repression of joint and suture development.

An alternative method used to define HBDs is to generate synthetic peptides replicating putative HBDs from a given protein and study these peptides for their heparin-binding properties. This has been done for the FGF receptor [48]. A specific sequence in one of the immunoglobulin-like loops in the extracellular domain of the FGF receptor tyrosine kinase was identified to interact with heparin, independent of FGF. In fact, a synthetic peptide of this sequence inhibited both heparin as well as FGF binding to the FGF receptor. Consistently, lysine residues within this sequence were required for this interaction and an antibody against this domain was also an antagonist of FGF-stimulated cell growth. Another example where peptides have been used to map heparin-binding portions is from the cell fusion glycoprotein of human respiratory syncytial virus (RSV-F) [49]. Synthetic overlapping peptides derived from the F-protein sequence were assayed for heparin–agarose affinity resulting in the identification of 15 peptides representing eight linear HBDs.

Although many studies on HS–protein interactions have been reported, a common consensus sequence for protein–HS binding has yet been established [50]. However, there is a general principle that ionic interactions between positively charged amino acids and negatively charged sulfate groups from HS are critical for protein–HS binding. There is also some data from crystal structural studies of heparin–protein complexes to suggest that arginine residues contribute more than lysine residues, despite their identical charges [51]. While the affinity and specificity of the binding is mainly determined by the secondary and/or tertiary spacing of these basic residues, in some cases binding between multiple proteins to HSPGs might require the formation of multi-protein complexes, where HS stabilizes the interactions.

27.7 FUSION PROTEINS INCORPORATING HBDs

Given that proteins that bind heparin are targeted to selective tissues, have increased activity, and are resistant to proteolysis, incorporating these features into biopharmaceuticals could be an effective way to improve therapeutics. In addition to our work that will be described below using the neuregulin 1's HBD, two other efforts to generate fusion proteins that bind heparin have been undertaken. Vascular endothelial growth factor A (VEGF-A) is a vascular permeability factor and a potent inducer of vascular leakage that can interact with HSPGs [52]. VEGF-E from the genome of parapox orf virus (OV) shares about 25% amino acid identity to mammalian VEGF-A, but does not have heparin-binding capability. Recombinant proteins were generated VEGF receptor agonist that stimulates proliferation and sprout formation of macro- and microvascular endothelial cells. At the same time, the fusion protein was a more potent competitor of VEGF-A than the wild-type VEGF-E in binding to primary human endothelial cells.

Insulin-like growth factor-1 (IGF-1) is a small and highly diffusible growth factor well known as an important mediator of cell growth and differentiation [53]. However, systemic administration of IGF-1 has significant unwanted side effects in tissues other than those of interest [54]. To try to solve this problem, a fusion protein consisting of the IGF-1 polypeptide and the HBD of heparin-binding epidermal growth factor-like growth factor (HB-EGF) was developed [55]. This fusion protein bound selectively to heparin as well as cell surfaces, without showing any reduction in biological activity. Moreover, the fusion protein was selectively retained by cartilage explants enriched with HSPGs, suggesting an improved and localized delivery of the growth factor to where it was needed.

27.8 THE NEUREGULIN 1 GROWTH FACTOR HAS A UNIQUE AND HIGHLY SPECIFIC HBD

The neuregulin 1 (NRG1) gene produces a large family of alternatively spliced growth and differentiation factors with a wide range of important functions in nervous system and heart development as well as in human diseases ranging from breast cancer to schizophrenia [56–62]. Although NRG1 has many isoforms generated through alternative splicing, all forms share a highly conserved and tightly folded epidermal growth factor (EGF) -like domain, which binds EGF receptor family members HER3 (erbB3) and HER4 (erbB4), which can heterodimerize with HER2 (erbB2) to transduce signals through an intracellular tyrosine kinase domain [57,60]. The receptor activation initiates several downstream signaling pathways including the mitogen-activated protein (MAP) kinase and phosphatidylinositol 3-kinase (PI3K) and mediates diverse effects on cellular processes including survival, proliferation, and differentiation [63]. A key feature of the secreted forms of NRG1 is a spatially separated heparin-binding domain (HBD) that is unique from other HBDs because of its disulfide-linked C2 immunoglobulin domain structure that allows it to be capable of maintaining its structural specificity for high-affinity heparin binding [64]. This C2 loop consists of a long stretch of alternating positively charged amino acids and is separated from the EGF-like domain by a glycosylated spacer. The net result is a highly specific HBD that recognizes specific sulfate groups on HS and results in precise tissue localization during development [65–68].

Recent studies from our group have shown that the interaction of NRG1's HBD with HSPGs leads to targeted deposition of the factor in specific regions of the developing nervous system [67]. *In vitro* studies suggest that the HBD facilitates the accumulation of NRG1 at HSPG-rich synaptic basal lamina and provides a sustained source of ligand required for sustained erbB receptor activation, important for the induction of acetylcholine receptor gene expression [1,65,67,69]. Structural heparin-binding studies using various desulfated heparins showed the importance as well as specificity of these sulfate groups by comparing the effects of fully sulfated heparin to either completely desulfated or partially desulfated heparin using a gel mobility shift assay [68]. Both the completely desulfated heparin and the De-N-sulfated sugars were unable to bind NRG1, suggesting a critical importance of N-sulfation for NRG1-heparin binding (Figure 27.2a). Removing the 2-O- and 6-O-sulfate groups, however, also reduced NRG1-heparin binding, but to a lesser extent (Figure 27.2b). Moreover, N-sulfate groups were needed for NRG1 to accumulate on the cell surface and activate its erbB receptors. Thus, NRG1 has a unique, structurally distinct, and highly tissue-specific HBD that is critical for its tissue localization during development as well as for optimal signal transduction.

27.9 USING NEUREGULIN'S HBD TO GENERATE A TARGETED NEUREGULIN ANTAGONIST

Because of the established role of NRG1 and its family of EGF receptors to promote the growth of breast cancer or other types of solid tumors as well as its potential role in neurological and psychiatric diseases, augmenting or blocking NRG1-erbB signaling has become an attractive therapeutic strategy [1,70–73]. A number of efforts have been made to block NRG signaling, including the FDA approved humanized monoclonal antibody drug against the erbB2 receptor called Herceptin [74]. It has become a clinically effective adjuvant therapy for a subgroup of breast cancer patients [75,76] but has to be used at extremely high concentrations that have been linked to cardiac side effects and only works on a subgroup of breast cancer patients that overexpress HER2 [77,78]. A major obstacle of blocking NRG1 signaling in both cancer and developmental studies comes from its ability to become highly concentrated in the ECM where it can produce sustained HER activation required for downstream gene activations [1,65]. All these studies provide a rationale to develop novel effective drugs with better targeting specificity that block NRG1 in breast cancers or other tumors that proliferate in response to NRG1 signaling.

To solve this important problem, we generated a novel fusion protein called HBD-S-H4 that fuses the human NRG1's HBD to the soluble ectodomain of human

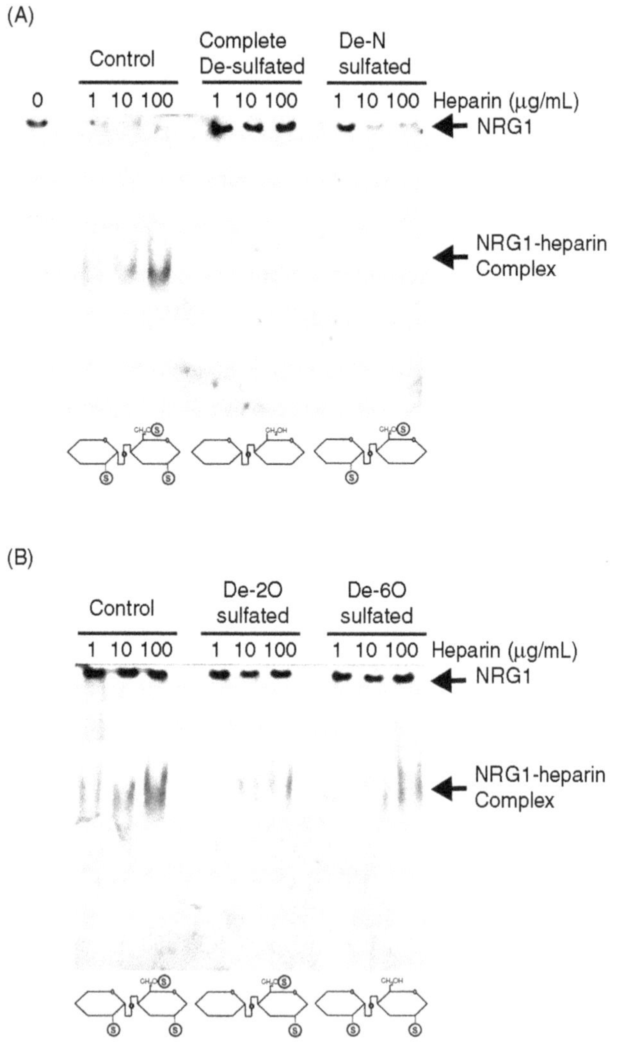

FIGURE 27.2 N-Sulfation is more important for heparin-NRG1 binding than 2-O- and 6-O- sulfation. Parallel gel shift assays were performed using the following: (A) Fully sulfated heparin, completely desulfated heparin, and De-N-sulfated heparin or (B) Fully sulfated heparin, De-2-O-sulfatd heparin, and De-6-O-sulfated heparin at 0, 1, 10, and 100 μg/mL. Completely desulfated and De-N-sulfated heparin were not able to bind and shift NRG1, whereas De-2-O- and De-6-O- sulfated heparins shifted NRG1 better than De-N-sulfated heparin but less than fully sulfated heparin. Schematic representations of each of the modified heparins are shown at the bottom of each gel with sulfate group marked with an S. *Source*: This research was originally published in J. Biol. Chem. Reference 68.

epidermal growth factor receptor 4 (HER4/H4). The HER4 ectodomain has high affinity for NRG1's EGF-like domain and can thus be used as a dominant negative NRG1 antagonist. By fusing HER4 to NRG1's HBD, the antagonist should be targeted to the same HS rich cell surfaces that bind NRG1 and therefore effectively disrupt NRG1 signaling [79].

As a first step, we generated a series of fusion proteins to ask what the optimal arrangement is between the HBD fused to H4 for both heparin binding and NRG1 antagonism (Figure 27.3). The constructs we tested placed the HBD either N-terminal or C-terminal to H4, and then compared these to H4 expressed alone. Figure 27.3a shows a heparin column-binding assay that uses increasing concentrations of NaCl to elute bound fusion proteins as a means to screen for heparin-binding affinity. While both N- and C-terminal fusion proteins could bind to the column, the N-terminal HBD fusion bound heparin only when it was separated from the H4 domain by a spacer region. The spacer domain used was the endogenous, glycosylated spacer domain from NRG1 that normally separates the HBD from NRG1's EGF-like domain. In fact, this fusion protein (HBD-S-H4) was not only optimal for heparin binding but also the most potent fusion protein for blocking NRG1 activity (Figure 27.3b). The spacer domain may be required to allow an optimal protein conformation necessary for heparin binding or to prevent steric inhibition between the other two domains. In comparison, the H4 construct without the HBD, showed little effect blocking receptor phosphorylation. Once concentrated in the ECM, heparin binding enables the antagonist to have long-lasting effects that cannot be removed even after vigorous washing steps to remove any proteins not bound to cells (Figure 27.3c). All these findings further demonstrate that, even when removed from its native protein, NRG1's isolated HBD retains high heparin-binding affinity when fused to another protein and can be effectively used to generate heparin-binding fusion proteins with enhanced and sustained activity.

Further experiments provided additional evidence of the importance of HS binding for the HBD-S-H4 fusion protein. Using wild-type cells and mutant cells that lack the ability to synthesize HS, we found that HBD-S-H4 adheres only to those cells that produce HS in the ECM, whereas the H4 control protein (without the HBD) adheres to neither (Figure 27.4a, b). Furthermore, heparinase treatment of wild-type cells (that selectively degrades cell-surface HS) significantly decreases HBD-S-H4 binding. Finally, a heparin-coated plate was used to determine that the Kd of HBD-S-H4 was 60 nM (Figure 27.4c). All these results demonstrate that the increased potency of HBD-S-H4 fusion protein is in part due to its ability to interact with specific HS in the matrix and concentrate on cell surfaces of cells that express HSPGs.

27.10 TISSUE TARGETING AND THERAPEUTIC EFFICACY OF A HEPARIN-TARGETED NRG1 ANTAGONIST FUSION PROTEIN

Given the retained high-affinity heparin binding of the HBD-S-H4 fusion, we next asked whether the *in vivo* targeting

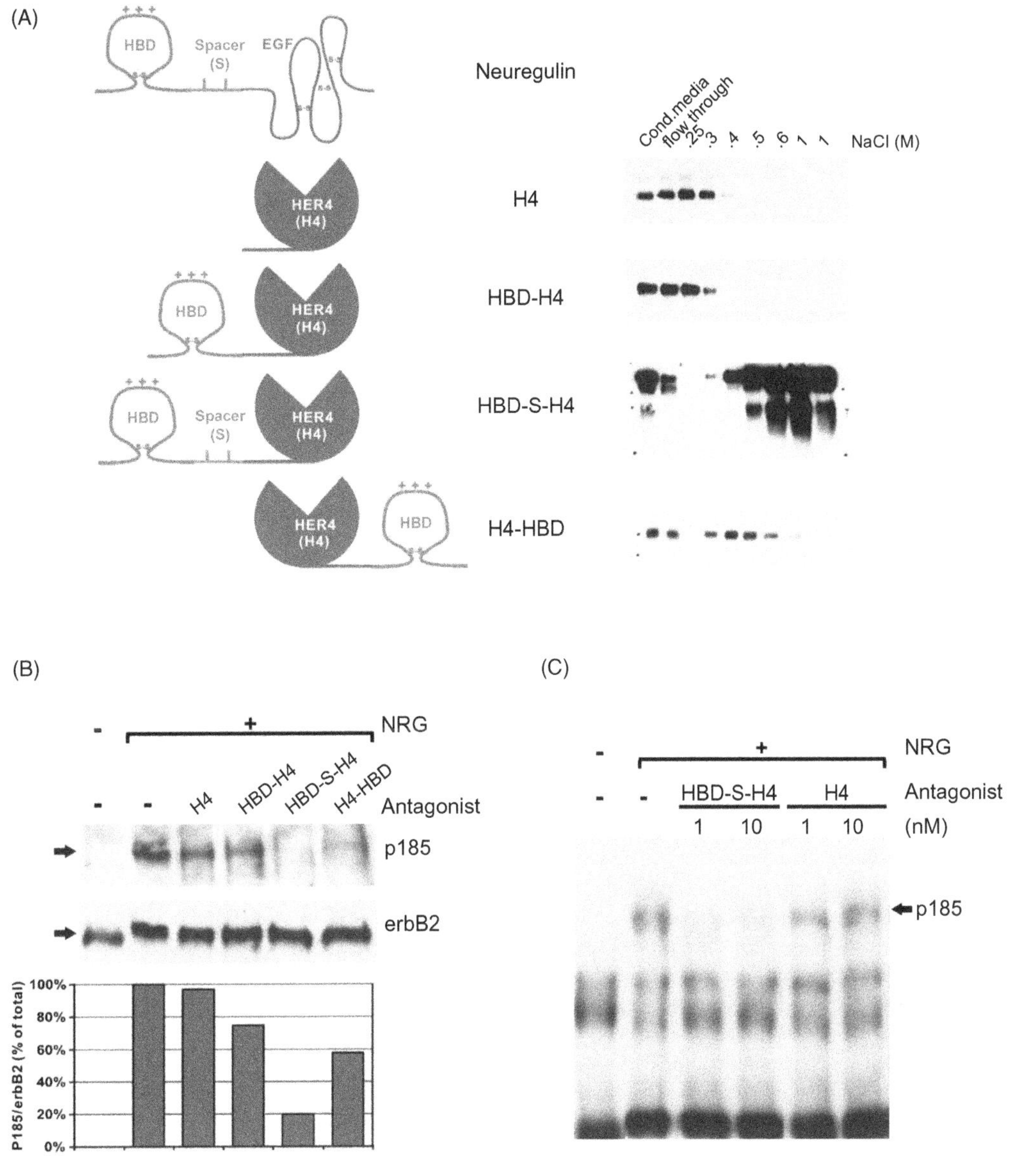

FIGURE 27.3 Neuregulin's heparin-binding domain retains heparin-binding ability and potentiates an HER4 antagonist. (A) Schematic diagram of secreted NRG and four antagonist constructs. The heparin-binding domain (HBD) was fused N-terminal or C-terminal to extracellular-binding domain of HER4 receptor (H4) with or without a spacer domain (S). H4 alone was used as a control. The heparin column flow through and successive elutions with increasing concentrations of NaCl were measured by immunoblotting. The HBD-S-H4 protein had the highest affinity for heparin. (B) Comparable amounts of each fusion protein were premixed with 50 pM NRG and applied to rat L6 muscle cells to determine the effect of each protein on the phosphorylation of erbB receptors (p185) normalized to erbB2 levels (bottom gel). Quantitation of phosphorylated erbB protein to total erbB2 protein reveals that HBD-S-H4 is the most potent antagonist at blocking NRG signaling with about 80% reduction of erbB activation. (C) L6 cells were treated with either purified HBD-S-H4 or H4 as a control at 1 and 10 nM for 1 h. The cells were then washed to remove unbound fusion protein and then challenged with 75 pM NRG. While H4 alone had no residual effects, HBD-S-H4 had sustained effects completely blocking NRG-induced erbB receptor phosphorylation. *Source*: This research was originally published in J. Biol. Chem. Reference 79.

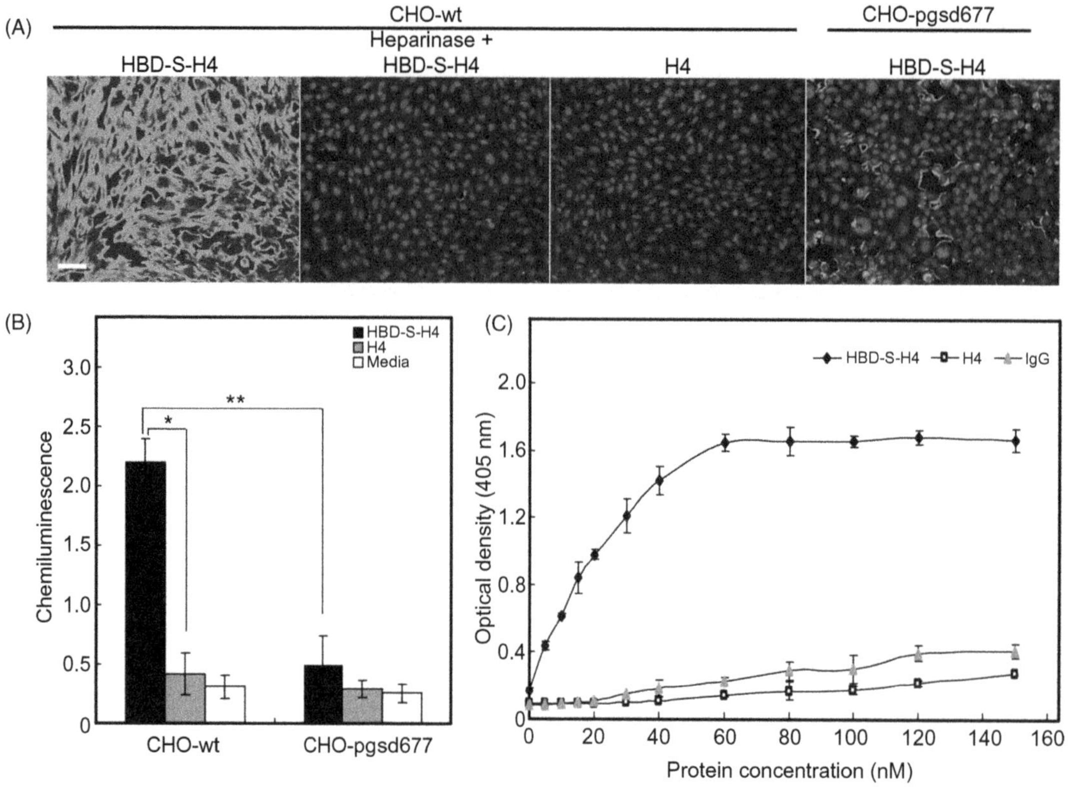

FIGURE 27.4 NRG1's heparin-binding domain targets HBD-S-H4 to cell surfaces through heparan sulfate interactions. (A) CHO-wt and CHO-pgsd677 (mutant) cells, with deficient heparan sulfate synthesis, were incubated with biotinylated HBD-S-H4 or H4 (50 nM) with or without heparinase treatment. Compared to the other conditions, HBD-S-H4 treated CHO-wt cells had the highest green fluorescence intensity that appeared to concentrate in ECM between cells. Scale bar is 50 μm. (B) A binding assay of protein constructs to CHO cells shows the enhanced binding ability of HBD-S-H4 to wild-type CHO cells (CHO-wt), but not to mutant cells. The H4 construct did not bind to these cells. * and ** indicate significant differences at $P < 0.001$ and <0.005, respectively. (C) Increasing concentrations of biotin-conjugated HBD-S-H4, H4, or IgG were added to heparin coated 96-well plates to produce a saturation-binding curve. The amount of HBD-S-H4 bound to heparin increased until saturation of binding sites. A calculated dissociation constant (Kd) is 14.7 nM. *Source*: This research was originally published in J. Biol. Chem. Reference 79. (See the color version of this figure in Color Plates section.)

specificity that leads to NRG1 deposition on specific tissues during chick embryonic development is preserved with the HBD-S-H4 antagonist. Within the developing spinal cord, NRG1 is highly expressed in and released by motor and sensory neurons where upon it adheres to specific tissues and cell types [60,67,80,81]. The tissue distribution of exogenously added HBD-S-H4 was compared to NRG1 tissue distribution as well as the H4 construct in developing embryos. Protein distribution of HBD-S-H4 was determined using a specific antibody that only recognizes the human HBD domain of NRG1 and does not cross-react with endogenous chicken NRG1 [82] and compared to a chicken NRG1 specific antibody staining (Figure 27.5). HBD-S-H4 became concentrated in the same regions of the spinal cord and peripheral nerve as endogenous NRG1. In addition, the fusion protein also accumulated outside the nervous system in regions that do not normally express NRG1, most likely because of high concentrations of HSPGs in these regions as well. To demonstrate that HBD-S-H4 is bound through ionic interactions to HSPGs, we showed that we could remove it from histological sections by high salt treatment [67,79]. In fact, a double-labeled region of the ventral nerve root shows that the same regions of peripheral nerve where chicken NRG1 is concentrated is also enriched with high levels of HBD-S-H4 (Figure 27.5b).

Tissue-specific targeting of HBD-S-H4 to the exact same regions where the NRG1 agonist is targeted in HS-rich cell surfaces thus places the NRG1 antagonist fusion protein on an even footing with the natural agonist. This novel, entirely "humanized" fusion protein could therefore be a powerful

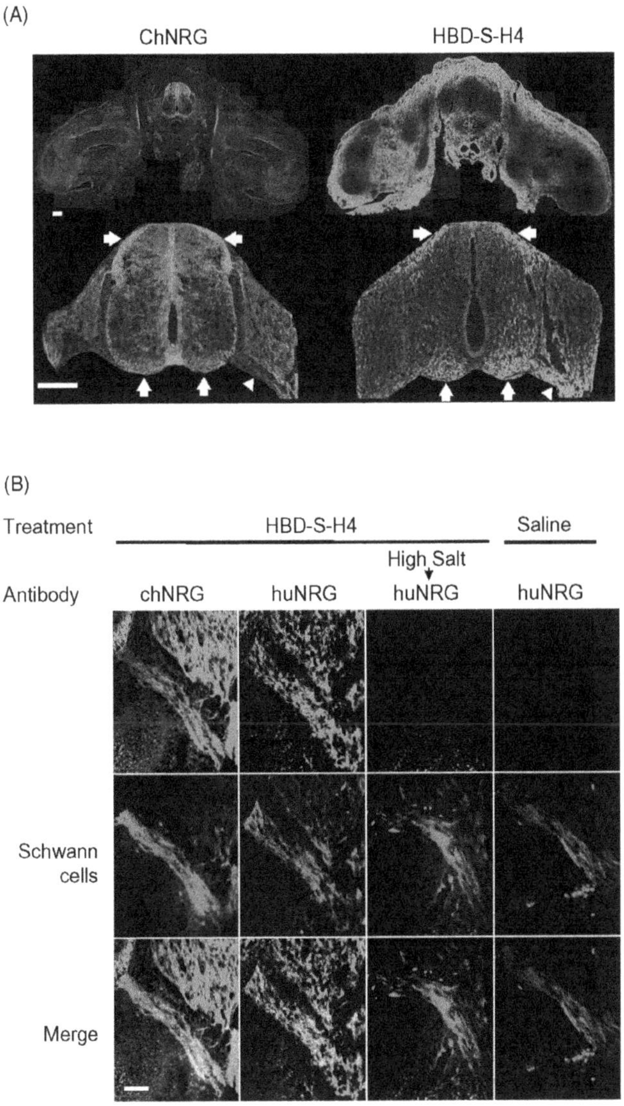

FIGURE 27.5 HBD-S-H4 targets to the same regions of the developing nervous system where NRG accumulates. (A) 20 μg of HBD-S-H4 was added to the chorioallantoic membrane of embryonic chicken embryos. Tissue sections through the spinal cord 2 days later at embryonic day 7 show a comparison between endogenous chicken NRG expression (green, left) and HBD-S-H4 distribution (green, right) at low power (top). Higher power images (bottom panel) focusing on the spinal cord show that HBD-S-H4 adhered to the same regions as endogenous NRG along axonal tracts in the spinal cord (arrows) and along the ventral nerve root (arrowhead). Sections were counterstained for nuclei with DAPI (blue). Scale bars are 200 μm. (B) Both endogenous chicken NRG (chNRG, green) and HBD-S-H4 (huNRG, green) were concentrated in the ventral nerve root identified by a Schwann cell marker (red). High salt (1M NaCl for 1.5 h) treatment on sections of HBD-S-H4 removed the signal in the ventral root. Control treatments with saline did not reveal any immunoreactivity with the same antibody in the ventral nerve root. Scale bar is 50 μm. *Source*: This research was originally published in J. Biol. Chem. Reference 79. (See the color version of this figure in Color Plates section.)

targeted antagonist that could be used in disease states where NRG1 signaling is deleterious as well as a useful research tool to study NRG1's multiple critical functions during development [79]. We and others have shown that NRG1 signaling is critical for Schwann cell survival during development. Using the HBD-S-H4 antagonist, we show a marked increase in Schwann cell apoptosis precisely in the regions where HBD-S-H4 becomes concentrated (as shown in Figure 27.5). We also found that Schwann cell differentiation from precursors to more mature stages was inhibited in both motor and sensory axons with HBD-S-H4, suggesting NRG1 is a critical survival and differentiation factor required for Schwann cell normal development (data not shown).

Since the fusion protein successfully targets NRG1 signaling *in vivo* during development, we next asked whether this completely humanized fusion protein could be useful therapeutically in disease models. We found that as some human breast epithelial cells become more malignant, they develop autocrine NRG1 signaling that promotes proliferation [1,2]. When the antagonist HBD-S-H4 was applied to a highly malignant breast cancer cell line, the fusion protein significantly decreased both autocrine phospohorylation as well as cell proliferation rate similar to Herceptin [79] (Figure 27.6). *In vivo*, HBD-S-H4 has been used to block NRG1 signaling in a rat model of chronic pain [83,84]. NRG1 promotes microglial proliferation and chemotaxis that directly contributes to the development of neuropathic pain after sciatic nerve injury. Intrathecal injection of HBD-S-H4 in these rats significantly reduced microglial cell proliferation and significantly reduced pain-related behaviors including mechanical pain hypersensitivity and cold allodynia.

27.11 CONCLUSIONS AND FUTURE PERSPECTIVES

HSPGs give tissue surfaces a unique molecular signature due to their diversity in both core protein and GAG structures. This diversity is regulated by highly specific enzymes that regulate their synthesis and modifications through post-translational processing. This result means for a given cell-surface region to become receptive to specific growth and differentiation factors in a spatially and temporally regulated manner. In addition to growth factors, cytokines, and chemokines, virus have also evolved to use this natural delivery system with the emergence of a number of structurally distinct HBDs.

We and others have taken advantage of this natural-targeting system using a strategy of generating fusion proteins with naturally occurring or engineered HBDs to guide biopharmaceuticals to achieve local delivery and to minimize systematic side effects at the same time. NRG1's HBD

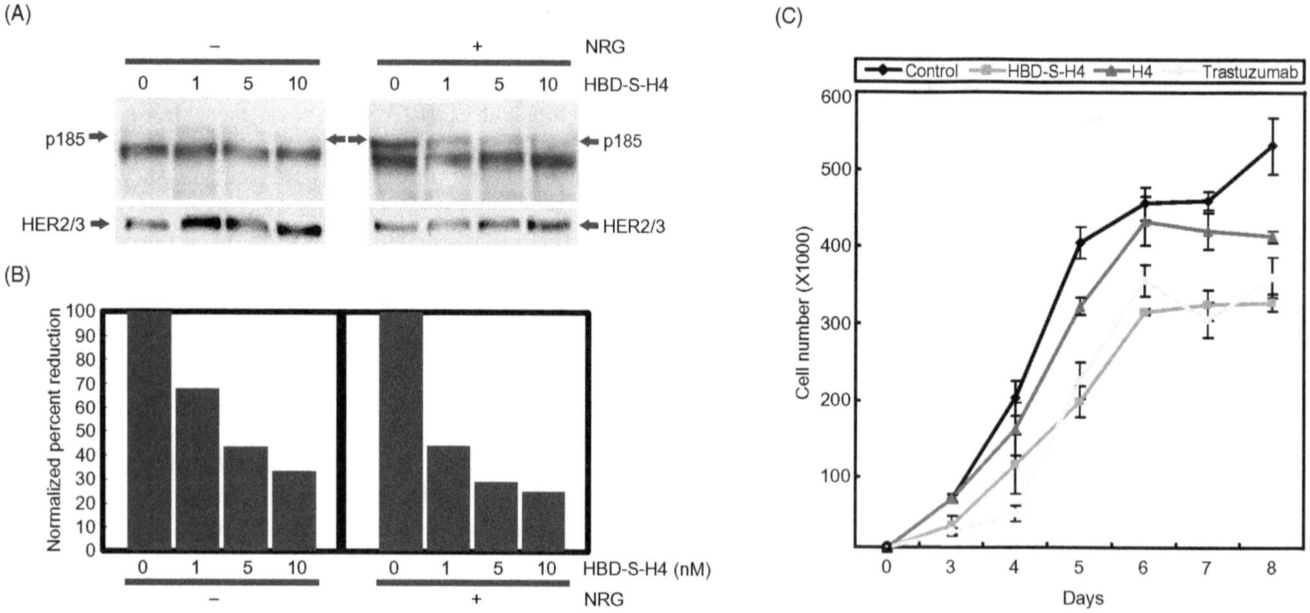

FIGURE 27.6 HBD-S-H4 blocks the growth of malignant human breast cancer MCF10CA1 cells through disruption of autocrine and paracrine NRG signaling. (A) Increasing concentrations of HBD-S-H4 were added to MCF-10CA1 breast cancer cells for 45 min without or with 75 pM recombinant NRG. The top panel shows that HBD-S-H4 blocked both autocrine (without added NRG) and paracrine (with NRG) HER2/3 receptor phosphorylation (p185). The lower panel shows the same Western blot reprobed with HER2 and HER3 antibodies. (B) Quantitation of the p185 signal normalized to HER2/3 levels is shown. (C) Growth curves of MCF10CA1 cells were compared without or with HBD-S-H4, H4, or trastuzumab (Herceptin) treatment. HBD-S-H4 (1 nM) markedly reduced the proliferation rate of these cancer cells compared to an equal molar amount of H4, and was comparable to treatment with trastuzumab. *Source*: This research was originally published in J. Biol. Chem. Reference 79.

is unique in that it is an easily detachable, stand-alone C2 immunoglobulin domain that can retain its targeting specificity both *in vitro* and *in vivo* when fused to another polypeptide. Early data suggest that this completely humanized fusion protein may be promising as an anti-NRG1 biopharmaceutical in diseases ranging from chronic pain to cancer. It serves a good example on how it is possible to improve therapeutics by targeting biopharmaceuticals to where they are needed with minimal toxicity to other tissues and thus resolves one of the most important obstacles in the development of biological therapeutics. It is tempting to further speculate that subtle modifications in this HBD or other potential HBD sequences could modify its interactions with HS sufficiently to change tissue-specific targeting specificity that could lead to targeting vectors for tissues other than where NRG1 goes.

REFERENCES

1. Li Q, Ahmed S, Loeb JA. (2004) Development of an autocrine neuregulin signaling loop with malignant transformation of human breast epithelial cells. *Cancer Res.* 64(19), 7078–7085.
2. Yao B, Rakhade SN, Li Q, Ahmed S, Krauss R, Draghici S, et al. (2004) Accuracy of cDNA microarray methods to detect small gene expression changes induced by neuregulin on breast epithelial cells. *BMC Bioinformatics* 599.
3. Thoenen H, Sendtner M. (2002) Neurotrophins: from enthusiastic expectations through sobering experiences to rational therapeutic approaches. *Nat. Neurosci.* 5(Suppl), 1046–1050.
4. Aszodi A, Legate KR, Nakchbandi I, Fassler R. (2006) What mouse mutants teach us about extracellular matrix function. *Annu. Rev. Cell Dev. Biol.* 22, 591–621.
5. Bulow HE, Hobert O. (2006) The molecular diversity of glycosaminoglycans shapes animal development. *Annu. Rev. Cell Dev. Biol.* 22, 375–407.
6. Toole BP. (2004) Hyaluronan: from extracellular glue to pericellular cue. *Nat. Rev. Cancer* 4(7), 528–539.
7. Turnbull J, Powell A, Guimond S. (2001) Heparan sulfate: decoding a dynamic multifunctional cell regulator. *Trends Cell Biol.* 11(2), 75–82.
8. Iozzo RV. (1998) Matrix proteoglycans: from molecular design to cellular function. *Annu. Rev. Biochem.* 67, 609–652.
9. Kramer KL, Yost HJ. (2003) Heparan sulfate core proteins in cell-cell signaling. *Annu. Rev. Genet.* 37, 461–484.

10. Couchman JR. (2010) Transmembrane signaling proteoglycans. *Annu. Rev. Cell Dev. Biol.* 26, 89–114.
11. Shintani Y, Takashima S, Asano Y, Kato H, Liao Y, Yamazaki S, et al. (2006) Glycosaminoglycan modification of neuropilin-1 modulates VEGFR2 signaling. *EMBO J.* 25(13), 3045–3055.
12. Esko JD, Selleck SB. (2002) Order out of chaos: assembly of ligand binding sites in heparan sulfate. *Annu. Rev. Biochem.* 71, 435–471.
13. Nadanaka S, Kitagawa H. (2008) Heparan sulphate biosynthesis and disease. *J. Biochem.* 144(1), 7–14.
14. Lindahl U, Li JP. (2009) Interactions between heparan sulfate and proteins-design and functional implications. *Int. Rev. Cell Mol. Biol.* 276, 105–159.
15. Morimoto-Tomita M, Uchimura K, Werb Z, Hemmerich S, Rosen SD. (2002) Cloning and characterization of two extracellular heparin-degrading endosulfatases in mice and humans. *J. Biol. Chem.* 277(51), 49175–49185.
16. Dhoot GK, Gustafsson MK, Ai X, Sun W, Standiford DM, Emerson CP., Jr. (2001) Regulation of Wnt signaling and embryo patterning by an extracellular sulfatase. *Science* 293 (5535), 1663–1666.
17. Sharom FJ, Radeva G. (2004) GPI-anchored protein cleavage in the regulation of transmembrane signals. *Subcell Biochem.* 37, 285–315.
18. Pruessmeyer J, Martin C, Hess FM, Schwarz N, Schmidt S, Kogel T, et al. (2010) A disintegrin and metalloproteinase 17 (ADAM17) mediates inflammation-induced shedding of syndecan-1 and -4 by lung epithelial cells. *J. Biol. Chem.* 285(1), 555–564.
19. Barash U, Cohen-Kaplan V, Dowek I, Sanderson RD, Ilan N, Vlodavsky I. (2010) Proteoglycans in health and disease: new concepts for heparanase function in tumor progression and metastasis. *FEBS J.* 277(19), 3890–3903.
20. Uchimura K, Morimoto-Tomita M, Bistrup A, Li J, Lyon M, Gallagher J, et al. (2006) HSulf-2, an extracellular endoglucosamine-6-sulfatase, selectively mobilizes heparin-bound growth factors and chemokines: effects on VEGF, FGF-1, and SDF-1. *BMC Biochem.* 72.
21. Xian X, Gopal S, Couchman JR. (2010) Syndecans as receptors and organizers of the extracellular matrix. *Cell Tissue Res.* 339(1), 31–46.
22. Li Q, Esper RM, Loeb JA. (2004) Synergistic effects of neuregulin and agrin on muscle acetylcholine receptor expression. *Mol. Cell Neurosci.* 26(4), 558–569.
23. Kaur C, Sivakumar V, Yip GW, Ling EA. (2009) Expression of syndecan-2 in the amoeboid microglial cells and its involvement in inflammation in the hypoxic developing brain. *Glia* 57(3), 336–349.
24. Kato M, Wang H, Bernfield M, Gallagher JT, Turnbull JE. (1994) Cell surface syndecan-1 on distinct cell types differs in fine structure and ligand binding of its heparan sulfate chains. *J. Biol. Chem* 269(29), 18881–18890.
25. Ledin J, Staatz W, Li JP, Gotte M, Selleck S, Kjellen L, et al. (2004) Heparan sulfate structure in mice with genetically modified heparan sulfate production. *J. Biol. Chem.* 279(41), 42732–42741.
26. Kjellen L. (2003) Glucosaminyl N-deacetylase/N-sulphotransferases in heparan sulphate biosynthesis and biology. *Biochem. Soc. Trans.* 31(2), 340–342.
27. van den Born J, Salmivirta K, Henttinen T, Ostman N, Ishimaru T, Miyaura S, et al. (2005) Novel heparan sulfate structures revealed by monoclonal antibodies. *J. Biol. Chem.* 280(21), 20516–20523.
28. Sadir R, Imberty A, Baleux F, Lortat-Jacob H. (2004) Heparan sulfate/heparin oligosaccharides protect stromal cell-derived factor-1 (SDF-1)/CXCL12 against proteolysis induced by CD26/dipeptidyl peptidase IV. *J. Biol. Chem.* 279(42), 43854–43860.
29. Bernfield M, Gotte M, Park PW, Reizes O, Fitzgerald ML, Lincecum J, et al. (1999) Functions of cell surface heparan sulfate proteoglycans. *Annu. Rev. Biochem.* 68, 729–777.
30. Bourin MC, Lindahl U. (1993). Glycosaminoglycans and the regulation of blood coagulation. *Biochem. J.* 289(Pt 2), 313–330.
31. Mohammadi M, Olsen SK, Ibrahimi OA. (2005) Structural basis for fibroblast growth factor receptor activation. *Cytokine Growth Factor Rev.* 16(2), 107–137.
32. Friedl A, Filla M, Rapraeger AC. (2001) Tissue-specific binding by FGF and FGF receptors to endogenous heparan sulfates. *Methods Mol. Biol.* 171, 535–546.
33. Forsten-Williams K, Chua CC, Nugent MA. (2005) The kinetics of FGF-2 binding to heparan sulfate proteoglycans and MAP kinase signaling. *J. Theor. Biol.* 233(4), 483–499.
34. Lortat-Jacob H, Baltzer F, Grimaud JA. (1996) Heparin decreases the blood clearance of interferon-gamma and increases its activity by limiting the processing of its carboxyl-terminal sequence. *J. Biol. Chem.* 271(27), 16139–16143.
35. Zhong X, Desilva T, Lin L, Bodine P, Bhat RA, Presman E, et al. (2007) Regulation of secreted Frizzled-related protein-1 by heparin. *J. Biol. Chem.* 282(28), 20523–20533.
36. Rostand KS, Esko JD. (1997) Microbial adherence to and invasion through proteoglycans. *Infect. Immun.* 65(1), 1–8.
37. Shworak NW, Liu J, Petros LM, Zhang L, Kobayashi M, Copeland NG, et al. (1999) Multiple isoforms of heparan sulfate D-glucosaminyl 3-O-sulfotransferase. Isolation, characterization, and expression of human cdnas and identification of distinct genomic loci. *J. Biol. Chem.* 274(8), 5170–5184.
38. Sinha BK. (1975) Renal amyloidosis in the buffalo (a case report). *Indian J. Pathol. Microbiol.* 18(2), 110–111.
39. Mochizuki H, Yoshida K, Shibata Y, Kimata K. (2008) Tetrasulfated disaccharide unit in heparan sulfate: enzymatic formation and tissue distribution. *J. Biol. Chem.* 283(45), 31237–31245.
40. Kern A, Schmidt K, Leder C, Muller OJ, Wobus CE, Bettinger K, et al. (2003) Identification of a heparin-binding motif on adeno-associated virus type 2 capsids. *J. Virol.* 77(20), 11072–11081.
41. Vandenberghe LH, Wang L, Somanathan S, Zhi Y, Figueredo J, Calcedo R, et al. (2006) Heparin binding directs activation of T cells against adeno-associated virus serotype 2 capsid. *Nat. Med.* 12(8), 967–971.

42. Manno CS, Pierce GF, Arruda VR, Glader B, Ragni M, Rasko JJ, et al. (2006) Successful transduction of liver in hemophilia by AAV-Factor IX and limitations imposed by the host immune response. *Nat. Med.* 12(3), 342–347.
43. Fuster MM, Esko JD. (2005). The sweet and sour of cancer: glycans as novel therapeutic targets. *Nat. Rev. Cancer* 5(7), 526–542.
44. Gorsi B, Stringer SE. (2007) Tinkering with heparan sulfate sulfation to steer development. *Trends Cell Biol.* 17(4), 173–177.
45. Bishop JR, Schuksz M, Esko JD. (2007) Heparan sulphate proteoglycans fine-tune mammalian physiology. *Nature* 446 (7139), 1030–1037.
46. Shih PC, Yang MS, Lin SC, Ho Y, Hsiao JC, Wang DR, et al. (2009) A turn-like structure "KKPE" segment mediates the specific binding of viral protein A27 to heparin and heparan sulfate on cell surfaces. *J. Biol. Chem.* 284(52), 36535–36546.
47. Harada M, Murakami H, Okawa A, Okimoto N, Hiraoka S, Nakahara T, et al. (2009) FGF9 monomer-dimer equilibrium regulates extracellular matrix affinity and tissue diffusion. *Nat. Genet.* 41(3), 289–298.
48. Kan M, Wang F, Xu J, Crabb JW, Hou J, McKeehan WL. (1993) An essential heparin-binding domain in the fibroblast growth factor receptor kinase. *Science* 259(5103), 1918–1921.
49. Crim RL, Audet SA, Feldman SA, Mostowski HS, Beeler JA. (2007). Identification of linear heparin-binding peptides derived from human respiratory syncytial virus fusion glycoprotein that inhibit infectivity. *J. Virol.* 81(1), 261–271.
50. Hileman RE, Fromm JR, Weiler JM, Linhardt RJ. (1998) Glycosaminoglycan-protein interactions: definition of consensus sites in glycosaminoglycan binding proteins. *Bioessays* 20(2), 156–167.
51. Fromm JR, Hileman RE, Caldwell EE, Weiler JM, Linhardt RJ. (1995) Differences in the interaction of heparin with arginine and lysine and the importance of these basic amino acids in the binding of heparin to acidic fibroblast growth factor. *Arch. Biochem. Biophys.* 323(2), 279–287.
52. Heil M, Mitnacht-Krauss R, Issbrucker K, van den Heuvel J, Dehio C, Schaper W, et al. (2003) An engineered heparin-binding form of VEGF-E (hbVEGF-E). Biological effects in vitro and mobilization of precursor cells. *Angiogenesis* 6(3), 201–211.
53. Delafontaine P, Song YH, Li Y. (2004) Expression, regulation, and function of IGF-1, IGF-1R, and IGF-1 binding proteins in blood vessels. *Arterioscler. Thromb. Vasc. Biol.* 24(3), 435–444.
54. Jabri N, Schalch DS, Schwartz SL, Fischer JS, Kipnes MS, Radnik BJ, et al. (1994) Adverse effects of recombinant human insulin-like growth factor I in obese insulin-resistant type II diabetic patients. *Diabetes* 43(3), 369–374.
55. Tokunou T, Miller R, Patwari P, Davis ME, Segers VF, Grodzinsky AJ, et al. (2008) Engineering insulin-like growth factor-1 for local delivery. *FASEB J.* 22(6), 1886–1893.
56. Esper RM, Pankonin MS, Loeb JA. (2006) Neuregulins: versatile growth and differentiation factors in nervous system development and human disease. *Brain Res. Rev.* 51(2), 161–175.
57. Buonanno A, Fischbach GD. (2001) Neuregulin and ErbB receptor signaling pathways in the nervous system. *Curr. Opin. Neurobiol.* 11(3), 287–296.
58. Lupu R, Cardillo M, Cho C, Harris L, Hijazi M, Perez C, et al. (1996). The significance of heregulin in breast cancer tumor progression and drug resistance. *Breast Cancer Res. Treat.* 38 (1), 57–66.
59. Mei L, Xiong WC. (2008) Neuregulin 1 in neural development, synaptic plasticity and schizophrenia. *Nat. Rev. Neurosci.* 9(6), 437–452.
60. Falls DL. (2003) Neuregulins: functions, forms, and signaling strategies. *Exp. Cell Res.* 284(1), 14–30.
61. Gilmour LM, Macleod KG, McCaig A, Sewell JM, Gullick WJ, Smyth JF, et al. (2002) Neuregulin expression, function, and signaling in human ovarian cancer cells. *Clin. Cancer Res.* 8(12), 3933–3942.
62. Fischbach GD, Rosen KM. (1997) ARIA: a neuromuscular junction neuregulin. *Annu. Rev. Neurosci.* 20, 429–458.
63. Yarden Y, Sliwkowski MX. (2001) Untangling the ErbB signalling network. *Nat. Rev. Mol. Cell. Biol.* 2(2), 127–137.
64. Loeb JA. (2003) Neuregulin: an activity-dependent synaptic modulator at the neuromuscular junction. *J. Neurocytol.* 32(5–8), 649–664.
65. Li Q, Loeb JA. (2001) Neuregulin-heparan-sulfate proteoglycan interactions produce sustained erbB receptor activation required for the induction of acetylcholine receptors in muscle. *J. Biol. Chem.* 276(41), 38068–38075.
66. Loeb J, Fischbach G. (1995) ARIA can be released from extracellular Matrix through cleavage of a heparin-binding domain. *J. Cell Biol.* 130(1), 127–135.
67. Loeb JA, Khurana TS, Robbins JT, Yee AG, Fischbach GD. (1999) Expression patterns of transmembrane and released forms of neuregulin during spinal cord and neuromuscular synapse development. *Development* 126(4), 781–791.
68. Pankonin MS, Gallagher JT, Loeb JA. (2005) Specific structural features of heparan sulfate proteoglycans potentiate Neuregulin-1 signaling. *J. Biol. Chem.* 280(1), 383–388.
69. Meier T, Masciulli F, Moore C, Schoumacher F, Eppenberger U, Denzer AJ, et al. (1998) Agrin can mediate acetylcholine receptor gene expression in muscle by aggregation of muscle-derived neuregulins. *J. Cell Biol.* 141(3), 715–726.
70. Montero JC, Rodriguez-Barrueco R, Ocana A, Diaz-Rodriguez E, Esparis-Ogando A, Pandiella A. (2008) Neuregulins and cancer. *Clin. Cancer Res.* 14(11), 3237–3241.
71. Slamon DJ, Clark GM, Wong SG, Levin WJ, Ullrich A, McGuire WL. (1987) Human breast cancer: correlation of relapse and survival with amplification of the HER-2/neu oncogene. *Science* 235(4785), 177–182.
72. Mendelsohn J, Baselga J. (2000) The EGF receptor family as targets for cancer therapy. *Oncogene* 19(56), 6550–6565.
73. Slamon DJ, Godolphin W, Jones LA, Holt JA, Wong SG, Keith DE, et al. (1989) Studies of the HER-2/neu proto-oncogene in human breast and ovarian cancer. *Science* 244(4905), 707–712.

74. Carter P, Presta L, Gorman CM, Ridgway JB, Henner D, Wong WL, et al. (1992) Humanization of an anti-p185HER2 antibody for human cancer therapy. *Proc. Natl. Acad. Sci. USA* 89 (10), 4285–4289.
75. Slamon DJ, Leyland-Jones B, Shak S, Fuchs H, Paton V, Bajamonde A, et al. (2001) Use of chemotherapy plus a monoclonal antibody against HER2 for metastatic breast cancer that overexpresses HER2. *N. Engl. J. Med.* 344(11), 783–792.
76. Vogel CL, Cobleigh MA, Tripathy D, Gutheil JC, Harris LN, Fehrenbacher L, et al. (2002) Efficacy and safety of trastuzumab as a single agent in first-line treatment of HER2-overexpressing metastatic breast cancer. *J. Clin. Oncol.* 20(3), 719–726.
77. Krauss WC, Park JW, Kirpotin DB, Hong K, Benz CC. (2000) Emerging antibody-based HER2 (ErbB-2/neu) therapeutics. *Breast Dis.* 11, 113–124.
78. Untch M, Himsl I, Kahlert S, Lueck HJ, Eidtmann H, Du Bois A, et al. (2004) Anthracycline and trastuzumab in breast cancer treatment. *Oncology (Williston Park)* 18(14 Suppl 14), 59–64.
79. Ma Z, Li Q, An H, Pankonin MS, Wang J, Loeb JA. (2009) Targeting human epidermal growth factor receptor signaling with the neuregulin's heparin-binding domain. *J. Biol. Chem.* 284(46), 32108–32115.
80. Nave KA, Salzer JL. (2006) Axonal regulation of myelination by neuregulin 1. *Curr. Opin. Neurobiol.* 16(5), 492–500.
81. Jessen KR, Mirsky R. (2005) The origin and development of glial cells in peripheral nerves. *Nat. Rev. Neurosci.* 6, 671–681.
82. Pankonin MS, Sohi J, Kamholz J, Loeb JA. (2009) Differential distribution of neuregulin in human brain and spinal fluid. *Brain Res.* 1258, 1–11.
83. Calvo M, Zhu N, Tsantoulas C, Ma Z, Grist J, Loeb JA, et al. (2010) Neuregulin-ErbB signaling promotes microglial proliferation and chemotaxis contributing to microgliosis and pain after peripheral nerve injury. *J. Neurosci.* 30(15), 5437–5450.
84. Calvo M, Zhu N, Grist J, Ma Z, Loeb JA, Bennett DL. (2011) Following nerve injury neuregulin-1 drives microglial proliferation and neuropathic pain via the MEK/ERK pathway. *Glia.* 59(4), 554–568.

28

BONE-TARGETED ALKALINE PHOSPHATASE

José Luis Millán

Sanford Children's Health Research Center, Sanford-Burnham Medical Research Institute, La Jolla, CA, USA

28.1 DETAILED DESCRIPTION OF THE CONCEPT

Bone-targeted alkaline phosphatase (a.k.a. sALP-Fc-D_{10}, ENB-0040 or asfotase alfa) is a protein therapeutic designed to prevent and treat the clinical manifestations of hypophosphatasia (HPP), a rare inherited form of rickets and osteomalacia caused by inactivating mutations in the *ALPL* gene, which encodes the tissue-nonspecific isozyme of alkaline phosphatase (TNAP).

Hypophosphatasia is a highly variable disease. The severity and type of inheritance depend on the nature of the *ALPL* mutation and to date, 239 missense or splice site mutations having been catalogued.[1] The prevalence of the most severe forms of the disease is ~1:100,000, although it is markedly higher in a small Canadian Mennonite population [1,2]. Inheritance can be autosomal recessive or dominant, and penetrance is variable, resulting in wide clinical expressivity. HPP differs from other forms of rickets and osteomalacia in that, that serum levels of calcium and phosphorus are generally normal or even elevated [3]. The natural substrates of TNAP that accumulate endogenously in HPP include inorganic pyrophosphate (PP_i) [4], an inhibitor of mineralization [5], and pyridoxal 5′-phosphate (PLP) [6,7], the major circulating form of vitamin B6 [8]. Rickets or osteomalacia develops in HPP because high extracellular levels of PP_i block hydroxyapatite crystal growth. Consequently, severely affected patients can acquire hypercalcemia and hyperphosphatemia [3]. High PP_i concentrations also lead to cementum, dentin, and alveolar bone phenotypes in HPP [9]. Diminished hydrolysis of PLP by TNAP explains the pyridoxine-responsive seizures observed in some severely affected HPP patients [7,10] and invariably noted in untreated $Akp2^{-/-}$ (TNAP-knockout) mice [11–14], a murine model that faithfully mimics infantile HPP.

HPP severity is inversely related to age of onset [3]. In its most extreme form (the perinatal variety) HPP causes death in utero or in early infancy. In these cases, there is profound hypomineralization of the skeleton, frequently with foreshortened and deformed limbs. Death generally results from respiratory insufficiency due to pulmonary hypoplasia and biomechanical compromise due to rachitic disease of the chest. Radiographic examination reveals hypomineralization, which in some cases may appear as almost total absence of bony structures. Infants diagnosed with HPP generally appear normal at birth but present with signs and symptoms within the first 6 months of postnatal life, most often failure to thrive or respiratory failure. Other clinical features may include functional craniosynostosis, increased intracranial pressure and papilledema, as well as the occurrence of nontraumatic fractures. Radiographic examination reveals skeletal hypomineralization and rickets. Hypercalcemia and hypercalciuria are not uncommon and nephrocalcinosis may occur. Mortality, usually due to pulmonary complications caused by rib

[1] http://www.sesep.uvsq.fr/03_hypo_mutations.php

Fusion Protein Technologies for Biopharmaceuticals: Applications and Challenges, First Edition. Edited by Stefan R. Schmidt.

fractures and rachitic deformity of the rib cage, has been reported to be as high as 50% [15]. If neonatal seizures (generally pyridoxine-responsive) occur, they are a grave prognostic symptom: a recent review of the literature found that all reported HPP infants with neonatal seizures had a fatal outcome by 18 months of age [10]. HPP in children is often diagnosed when there is premature exfoliation of deciduous teeth, defined as loss before the age of 5 years. Tooth loss is due to an absence of mineralized cementum, which is required for the periodontal ligaments to tether the tooth root to the surrounding bone. Rickets and its associated bony deformities are common, resulting in short stature in some patients. Walking is frequently delayed and may be characterized by a peculiar waddling gait. Nontraumatic fractures are not uncommon. HPP may also present later in life, although upon questioning adults may report a history of early tooth loss or rickets during childhood. Osteomalacia is the hallmark of HPP in adults and often presents as bone pain in the feet resulting from recurrent, poorly healing stress fractures. X-ray examination may reveal the presence of pseudofractures, osteopenia, and chondrocalcinosis. Odontohypophosphatasia, in which pathology is limited to the teeth, is the mildest form of the disease.

There is no established medical therapy for HPP [3]. Case reports of attempted enzyme replacement therapy (ERT) using intravenous (IV) infusions of ALP-rich plasma from Paget's bone disease patients, purified human liver ALP, or purified human placental ALP indicate failure to rescue affected infants [16–19]. These outcomes suggest that ALP activity must be increased not merely in the circulation but in the HPP skeleton itself. This hypothesis has been supported by favorable outcomes seen in two girls with infantile HPP following transplant of mesenchyme-derived cells, which likely contained small numbers of TNAP-expressing osteoblasts and chondrocytes that became distributed throughout the skeleton [20,21].

The drug ENB-0040, developed by Enobia Pharma Inc. (Montréal, Canada), consists of the TNAP polypeptide sequence devoid of its GPI anchoring domain (for a complete description of the TNAP structure see reference [22]), which was replaced by a human Fc region to allow rapid chromatographic purification followed by a deca-aspartate acid (D_{10}) sequence to confer bone-targeting properties due to its high affinity of D_{10} for bone mineral. The modeled 3D structure of bone-targeted TNAP is shown in Figure 28.1, along with *in vitro* evidence of enhanced binding to hydroxyapatite and *in vivo* support of the drug's ability to home to the mineralizing skeleton.

28.2 TECHNICAL ASPECTS

Bone-targeted TNAP was initially designated sALP-Fc-D_{10} [23,24] due to its chimeric composition: namely, removal of the C-terminal hydrophobic sequence encoding the GPI-anchor to create a soluble secreted enzyme (sALP); extension of the coding sequence of each TNAP subunit with the human IgG Fc region (γ1 form) to allow purification using Protein A chromatography (-Fc); and addition of a mineral-binding deca-aspartate (D_{10}) sequence [25–27] to the C-terminus of each Fc fragment. A subsequent publication [28] has used the name ENB-0040 for this reagent, while asfotase alfa is the name approved by the International Nonproprietary Name (INN) Program managed by the World Health Organization. We will continue to use sALP-FcD_{10} and ENB-0040 throughout this chapter.

The cDNA encoding the fusion protein was inserted into the pIRES vector (Clontech, San Diego, CA) in the first multiple cloning site located upstream of the IRES using NheI and BamHI endonuclease restriction sites. The dihydrofolate reductase (DHFR) gene was inserted into the second multiple cloning site downstream of the IRES using SmaI and XbaI endonuclease restriction sites. The resulting vector was transfected into Chinese hamster ovary (CHO-DG44) cells lacking both DHFR gene alleles [29] using the Lipofectamine transfection kit (Invitrogen, San Diego, CA). The purification procedure of sALP-FcD_{10} was described in detail [23] and is only briefly summarized here. The sALP-FcD_{10} concentration in spent media was ~3.5 mg/L, as assessed by TNAP enzymatic activity. Culture supernatants were concentrated, dialyzed against PBS and loaded onto Protein A-Sepharose columns (Hi-Trap 5 mL, GE Health Care) equilibrated with PBS. Bound proteins were eluted with 100 mM citrate buffer, pH 4.0. Collected fractions were immediately adjusted to pH 7.5 with 1 M Tris, pH 9.0. Fractions containing most of the eluted material were dialyzed against 150 mM NaCl, 25 mM sodium phosphate buffer, pH 7.4, containing 0.1 mM $MgCl_2$, 20 μM $ZnCl_2$, and filtered through a 0.22 μm (Millipore, Millex-GP) membrane under sterile conditions. The overall yield of the purification procedure was 50%, with a purity surpassing 95% as assessed by Sypro ruby-stained SDS-PAGE. Purified sALP-FcD_{10} preparations were stored at 4°C and remained stable for several months. This enzyme preparation (research grade) was used in two preclinical studies [23,24]. The purification of cGMP clinical grade ENB-0040 (Asfostase alfa) is much more elaborate, with additional chromatographic and filtration steps [30].

Recombinant ENB-0040 is a soluble glycoprotein of 726 amino acids, which after purification on Protein-A Sepharose and analyzed in reducing SDS-PAGE conditions migrates as a broad band with an apparent molecular mass of ~90,000 Da. ENB-0040 catalytic activity was comparable to that of HPLC-purified glycosylation isoforms of human TNAP using either the artificial colorimetric substrate *p*-nitrophenylphosphate at pH 9.8 or the physiological substrates inorganic pyrophosphate and pyridoxal-5′-phosphate at pH 7.4 [31]. The affinity of purified ENB-0040

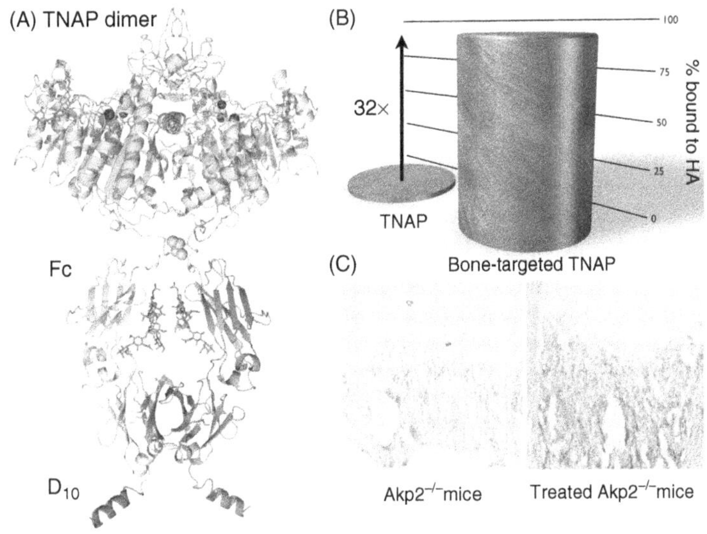

FIGURE 28.1 (A) Three-dimensional modeling of ENB-0040, constructed by Dr. Boguslaw Stec, Sanford-Burnham Medical Research Institute, La Jolla, CA. The sequence of the whole construct was analyzed. The sequences of fragments corresponding to particular molecular elements constituting the ENB-0040 molecule were submitted to the FFAS03, the molecular similarity searches against the structures present in the PDB (1ZED for alkaline phosphatase and 1HZH and 1IGZ for the antibody moiety). As a result, the closest sequence analogs of individual fragments, that is, the alkaline phosphatase, the Fc antibody fragment were selected. The full sequence models were created using the model building tools present in FFAS03 (Modeller) [68]. Program was used to create these molecular fragments. Subsequently, the models were positioned against each other in Coot (the crystallographic modeling software) in order to make required necessary connections to create the whole molecule. The linkers as well as the C-terminal Asp-rich regions were modeled by inserting the missing sequences in the extended conformation for the linker and helical conformation for the poly-Asp tail. Finally, the structure with optimized double disulfide bridge was energy minimized in Sybyl. The model shows rigid AP and Fc modules connected by a highly flexible linker reminiscent of the neck area of the mature antibody. Two disulfides bonds are marked by spheres. The terminal poly-Asp region is exposed on the opposite site of the AP module. The whole structure is dimeric that conforms to the preferred oligomeric state of the AP as well as the Fc region of the antibody. The figures and final visualization was performed in Pymol. (B) The affinity of the purified ENB-0040 for hydroxyapatite mineral was compared to that of soluble TNAP purified from bovine kidney. Total activity was the sum of enzymatic activity recovered in both free and bound fractions and was found to be 84% and 96% of the spiked enzymatic activity for the bovine and ENB-0040 forms of enzyme, respectively. ENB-0040 bound 32-fold more efficiently to reconstituted hydroxyapatite (HA) than did unmodified bovine kidney TNAP. Furthermore, most recombinant ENB-0040 utilized in the assay was accounted for by summing enzymatic activity recovered in bound and nonbound fractions, indicating that ENB-0040 binding to mineral does not significantly alter its enzymatic activity . (C) Histochemical staining for ALP activity in long bones of an ENB-0040-treated *Akp2*$^{-/-}$ mouse compared with an age-matched untreated *Akp2*$^{-/-}$ mouse. (Data taken from reference [23]).

protein for hydroxyapatite mineral was contrasted to that of soluble TNAP derived from bovine kidney. ENB-0040 binds 32-fold more efficiently to reconstituted hydroxyapatite than does bovine kidney TNAP (Figure 28.1B). Furthermore, most of the recombinant ENB-0040 protein introduced in the assay could be accounted for by summing enzymatic activity recovered in both the bound and nonbound fractions [23]. This analysis indicated that ENB-0040 binding to mineral does not significantly alter its enzymatic activity.

The ENB-0040 product is a sterile, colorless, aqueous solution with no visible particulates and is available at either 40 mg/mL or 100 mg/mL drug in 25 mM sodium phosphate buffer, pH 7.4, and 150 mM sodium chloride, packaged in a 2 mL glass vial. The product is dispensed in an extractable volume of 0.5 mL, without additional formulation. Investigational supplies for clinical use should be stored refrigerated from 2 to 8°C. ENB-0040 is compatible with infusion bags at 19–23°C for up to 2 h.

While the Fc region was incorporated into ENB-0040 primarily to facilitate purification, Fc is known to considerably increase the half-life of short lived proteins, as discussed at length elsewhere in this book, thus allowing us to use lower doses and less frequent injections. The pharmacokinetics (PK) and tissue distribution properties of ENB-0040 were determined in studies of adult and newborn mice comparing different administration routes [23]. Initially, using a single IV injection of 5 mg/kg ENB-0040 into adult wild-type (WT) mice, the authors found the circulating half-life to be 34 h, with prolonged retention of radioactive iodine-labeled ENB-0040 in bone, with as much as 1 μg/g of bone (wet) weight. Skeletal levels of bone-targeted test material are apparently highly stable, as no significant decrease in radiolabeled ENB-0040 was observed during the course of the experiment. Conversely, no sustained accumulation of sALP-FcD_{10} was observed in muscle, as radiolabeled enzyme levels in this tissue decreased in parallel with ENB-0040 enzymatic activity in blood. Figure 28.1C shows a histochemical staining for ALP activity in untreated and treated $Akp2^{-/-}$ bone sections revealing the successful homing of ENB-0040 to bone tissue. Because $Akp2^{-/-}$ mice die between days 12 and 16, and IV injection is very difficult in such small animals, PK analysis of ENB-0040 was planned using intraperitoneal (IP) and subcutaneous (SC) administration in newborn WT mice. However, IP injection proved unreliable due to the high intra-abdominal pressure seen in young animals that led to unpredictable losses of drug solution through the injection site. Since SC injections proved reproducible, SC injection route was used to generate PK data used to predict circulating levels of ENB-0040 achieved after repeated daily doses. Circulating ENB-0040 reached steady state serum concentrations oscillating between C_{min} and C_{max} values of 26.4 and 36.6 μg/mL, respectively, after 5–6 daily doses at 10 mg/kg. Prediction validity was tested using 5 daily SC injections of 10 mg/kg ENB-0040. Circulating ALP activity measured 24 h after the last injection (C_{min}) was in good agreement with predicted concentrations. In WT mice, serum TNAP levels measured in the same conditions were found to be 0.58 μg/mL. Thus, the authors calculated that the SC injection regimen would achieve steady-state circulating ENB-0040 concentrations ~50 times higher than normal WT TNAP concentrations. In clinical attempts to use ERT by injections of Paget's bone disease plasma, purified liver ALP, or purified placental ALP, only an eightfold elevation in serum ALP activity had been achieved [16,17,19].

28.3 APPLICATIONS AND INDICATIONS

Preclinical trials using daily SC injections of newborn $Akp2^{-/-}$ mice with purified ENB-0040 blocked disease manifestations in this model of infantile HPP, including early mortality, rickets, epileptic seizures, lack of an acellular cementum layer in teeth, as well as elevated concentrations of inorganic pyrophosphate, calcium, and pyridoxal-5-phosphate (PLP) in plasma. As described earlier in this chapter, disease severity, which is inversely proportional to age at onset, varies from a generally fatal perinatal form to an adult form characterized principally by nontraumatic fractures and bone pain. ENB-0040 is currently under investigation for treatment of life-threatening (perinatal and infantile HPP) childhood and adult HPP. As reviewed by Drake and Koshla [32] in an Editorial that accompanied the first report of successful treatment of HPP mice [23], this bone-targeted form of TNAP shows great promise for treatment of human HPP.

As with any protein, the formation of antibodies is a potential concern. Anti-ENB-0040 antibodies were detected in some TNAP knockout mice before and after ENB-0040 treatment, but antibodies were not associated with clinical symptoms or lessening of enzyme effect over time.

Because ENB-0040 contains the human IgG1 Fc region, a possible interaction with various immune cells was considered. The likelihood of such an interaction is low, since monkeys have a homologous receptor for the Fc region of human IgG1, and no clinical signs of adverse immunological reactions were seen even at very high doses of ENB-0040 in primate toxicology studies.

One potential concern regarding ENB-0040 treatment is that it could induce hypocalcemia or hypophosphatemia as bone mineralization is stimulated. This potential outcome was examined in the mouse HPP model, and no evidence of either biochemical abnormality was detected, either acutely or with chronic ENB-0040 administration.

Another potential concern related to ENB-0040 treatment in HPP is vitamin B_6 toxicity and a resulting peripheral neuropathy [33]. The primary circulating form of Vitamin B_6 is PLP, which must be dephosphorylated to pyridoxal (PL) to enter cells. In animal studies, ENB-0040 treatment has not resulted in excessively high PL levels and there has been no evidence of peripheral toxicity, even at very high ENB-0040 doses.

The possibility that ENB-0040 could induce ectopic calcification has been investigated in animal safety studies, as we have shown that TNAP expression and a fibrillar scaffold is sufficient to promote mineralization of any extracellular matrix [34] and, in particular, that TNAP upregulation in vascular smooth muscle cells causes medial vascular calcification [35,36]. Histological evaluation of organs and blood vessels in ENB-0040 treated mice and rats failed to demonstrate any such pathology. Monkey studies indicated microscopic evidence of reversible, dose-dependent calcium deposition at injection sites.

28.4 PRECLINICAL AND CLINICAL RESEARCH

Akp2$^{-/-}$ mice, which were created via homologous recombination by insertion of the Neo cassette into exon 6 of the mouse gene encoding TNAP (*Akp2*) [12], show no detectable TNAP mRNA or protein. Phenotypically, *Akp2*$^{-/-}$ knockout mice closely mimic infantile HPP [14]. Like HPP patients, *Akp2*$^{-/-}$ mice exhibit global deficiency in TNAP activity, endogenous accumulation of the ALP substrates PP_i and PLP, and postnatally acquire impaired mineralization of skeletal matrix leading to rickets or osteomalacia. *Akp2*$^{-/-}$ mice manifest stunted growth, and develop radiographically and histologically apparent rickets together with epileptic seizures and apnea, and die between postnatal days 10 and 12 [11,12,14].

The first disease efficacy study utilized SC ENB-0040 injections of newborn *Akp2*$^{-/-}$ mice daily for 15 days at 1 mg/kg per dose [23]. μCT analysis of animals treated at this dosage showed no attenuation of skeletal disease in the calvarium, and the proximal tibial growth plate (physis) showed excessive widening of the hypertrophic zone, consistent with early rickets in both ENB-0040- and vehicle-injected *Akp2*$^{-/-}$ animals. Increasing the daily dose to 2 mg/kg improved the general appearance, body weight and tail length of treated *Akp2*$^{-/-}$ mice, which also showed a normal growth rate. Next, *Akp2*$^{-/-}$ mice underwent 15 days of daily SC injections at 8.2 mg/kg ENB-0040. Treated animals had greater body weight than vehicle-treated mice and had plasma PP_i concentrations in the normal range. In this short-term experiment, ENB-0040 treatment minimized hypomineralization in the feet and reduced the number of mice exhibiting severely dysmorphic rib cages. Similarly, the hind limbs appeared healthy in all treated animals [23]. Long-term (52 day) survival and complete prevention of skeletal defects and epilepsy were obtained with daily SC injections of 8.2 mg/kg ENB-0040, an outcome accompanied by normal plasma PL concentrations and unremarkable calcium concentrations.

ENB-0040 treatment also prevented hypomineralization of alveolar bone and dentin [23], and acellular cementum, which is typically missing or reduced in *Akp2*$^{-/-}$ mice, formed normally [24]. Microcomputed tomography revealed a consistent reduction of mineralization in both alveolar bone and root dentin (and root analog dentin in the incisor) in the control (vehicle alone injected) *Akp2*$^{-/-}$ mice (Figure 28.2, left panels)—a feature not present in the ENB-0040-treated *Akp2*$^{-/-}$ mice, which showed complete mineralization of alveolar bone and dentin. Immunohistochemical staining for osteopontin, revealed absence of acellular cementum along the root surface in untreated *Akp2*$^{-/-}$ mice, while ENB-0040-treated *Akp2*$^{-/-}$ mice had an unremarkable acellular cementum layer comparable with that seen in WT mice (Figure 28.2, right panels) [24].

A subsequent preclinical study defined the dose/response relationship between increasing amounts of cGMP (current good manufacturing practice)-produced ENB-0040 using bolus SC injections and the therapeutic response after 43 days, in anticipation of clinical trials [28]. Endpoints were survival, body weight, bone length of the tibiae and femora, and bone mineralization defects, as assessed using radiographs of the feet, rib cage, and lower limbs. To date, these parameters have served as accurate indicators of correction of the HPP phenotype in this animal model. Also, radiographic manifestations of HPP are readily observed in affected infants and are thus an important endpoint in preclinical proof-of-concept studies to extrapolate from mice to humans. In addition, we used μCT (Figure 28.3) and histomorphometric analysis to evaluate improvement in *Akp2*$^{-/-}$ bone mineralization status for representative age groups undergoing treatment. We documented a clear relationship between daily ENB-0040 dose and the percentage of mice with normal bony structures of the foot, rib cage, and lower limbs. We focused on establishing an effective dose in 80% of the mice (ED_{80}). In mice, that dose was ~3.2 mg/kg/day for feet, ~2.8 mg/kg/day for rib cage, and ~2.9 mg/kg/day for lower limbs. These ED_{80} doses are consistent with our previous experience showing enhanced skeletal integrity following short term, 15-day treatment of *Akp2*$^{-/-}$ animals using 2 mg/kg/day of ENB-0040 [23].

Studies using ENB-0040 produced under cGMP conditions confirm and extend our previous data regarding prevention of HPP in Akp2$^{-/-}$ mice by administration of the bone-targeted form of TNAP from birth [23] and helped to set the stage for clinical trials now underway for HPP patients.[2] The ED_{80} along with enzyme concentrations measured in animal efficacy studies and ENB-0040 PK data were used to estimate the minimum effective concentrations for *Akp2*$^{-/-}$ mice. Initial dosing of ENB-0040 for clinical trials was then based on: (1) dose-response data [28] in the *Akp2*$^{-/-}$ mouse model of infantile HPP, (2) efficacy observed in treated *Akp2*$^{-/-}$ mice that achieved serum ALP activity in the range of 650–1000 U/L; (3) the no-observed-adverse-effect level (NOAEL) in more sensitive species (rats and monkeys) established in 1-month IV toxicology studies and 1-month IV/SC bridging and tolerability studies in rats; (4) a safety factor of 10 applied to the NOAEL; and (5) results from a multicenter, open-label, dose-escalating study of the safety, tolerability, and pharmacology of ENB-0040 in six adults with HPP who received one dose of 3 mg/kg IV and then either 1 or 2 mg/kg ENB-0040 SC once weekly for 3 weeks.

An investigational new drug (IND) application to use ENB-0040 to treat HPP was filed in June of 2008. To date, two clinical studies have been completed, three are ongoing, and two additional studies are being initiated.[2] Preliminary clinical data have been presented only as abstracts at

[2] http:// www.clinicaltrials.gov

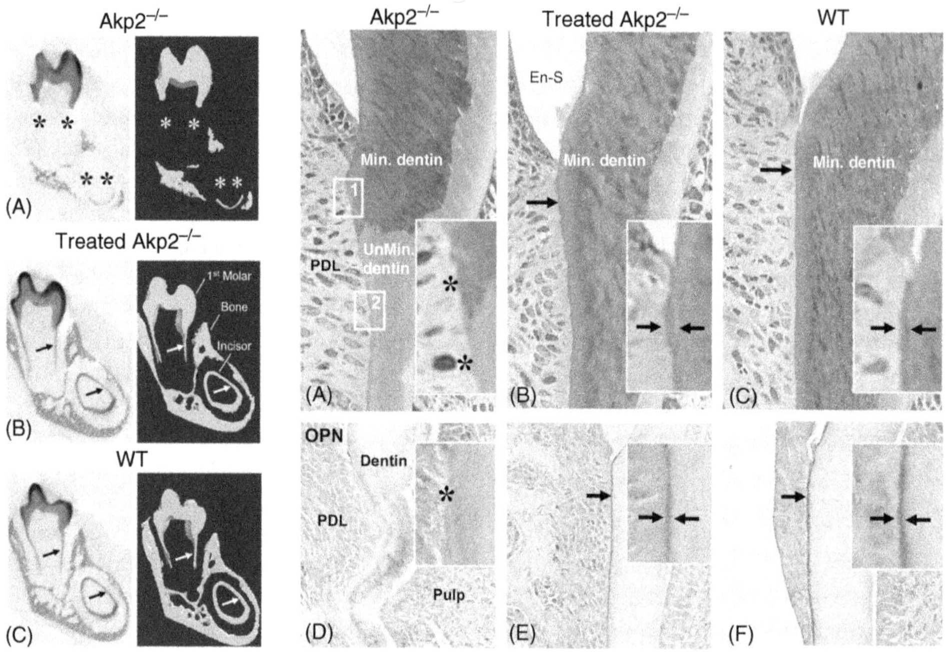

FIGURE 28.2 Left panels: Targeting TNAP to mineral restores bone and tooth mineralization defects seen in TNAP-deficient (Akp2$^{-/-}$) mice, as assessed by microcomputed tomography. Coronal views of mandibles from 16-day-old mice (vehicle-injected Akp2$^{-/-}$, ENB-0040-treated Akp2$^{-/-}$, and wild-type [WT]) as a single X-ray slice (images on the left), or as reconstructions of multiple slices (images on the right). (A) Vehicle-injected Akp2$^{-/-}$ mice show extensive regions of unmineralized root (and root analog) dentin (asterisks), as well as surrounding hypomineralized alveolar bone. (B) ENB-0040-treated Akp2$^{-/-}$ mice show complete mineralization of all incisor and molar tooth tissues and alveolar bone, which are indistinguishable from WT tissues (C). Arrows indicate mineralized dentin and acellular cementum combined defects. Right panels: Presence of acellular cementum in ENB-0040-treated Akp2$^{-/-}$ mice visualized by histology and immunohistochemistry with anti-osteopontin antibodies. (A–C) Plastic coronal sections of the first molar near the cemento-enamel junction, showing acellular cementum along the root surface in ENB-0040-treated Akp2$^{-/-}$ mice. Tissues appear comparable to those seen in WT mice (arrows and insets). This acellular cementum layer is absent (asterisks) in vehicle-injected Akp2$^{-/-}$ mice. Rectangles 1 and 2 in (a) indicate sites adjacent to mineralized (Min.) and unmineralized (UnMin.) dentin, respectively. Unmineralized dentin is commonly seen in TNAP-deficient mice. Electron micrographs of boxed areas 1 and 2 are shown in Fig. 4. (D–F) Immunohistochemical localization of osteopontin (red, arrows, and inset), as a marker of acellular cementum, confirms histologic observations in corresponding panels a–c showing acellular cementum following ENB-0040 treatment (Akp2$^{-/-}$-treated) but absence of a discrete immunostained layer (asterisk in inset) in vehicle-treated Akp2$^{-/-}$ mice. PDL, periodontal ligament; En-S, enamel space after decalcification. Magnification bars equal 100 μm. OPN, osteopontin. *Source*: The final definitive version of this figure has been published in Journal of Dental Research, Reference 24. 90(4): 470–476, January 2011, by SAGE publication, Inc, All rights reserved ©.

scientific meetings, but appear to indicate remarkable improvement of the skeletal condition in HPP infants and children treated with ENB-0040 [37–46].

28.5 ALTERNATIVES/VARIANTS OF THIS APPROACH

To date, HPP treatment has been essentially symptomatic [3]. For example, infants with hypercalcemia may require restriction of dietary calcium or administration of calciuretics, or patients with craniosynostosis may need surgical intervention. Likewise, fractures are treated by prolonged casting or stabilization with orthopedic hardware, and dental hygiene is carefully monitored in individuals with dental manifestations of the disease. Earlier attempts at more definitive HPP treatment have met with limited success. Use of a bisphosphonate (a synthetic analog of pyrophosphate) in one infant reportedly had no discernible effect on the skeleton, and the infant experienced progressive

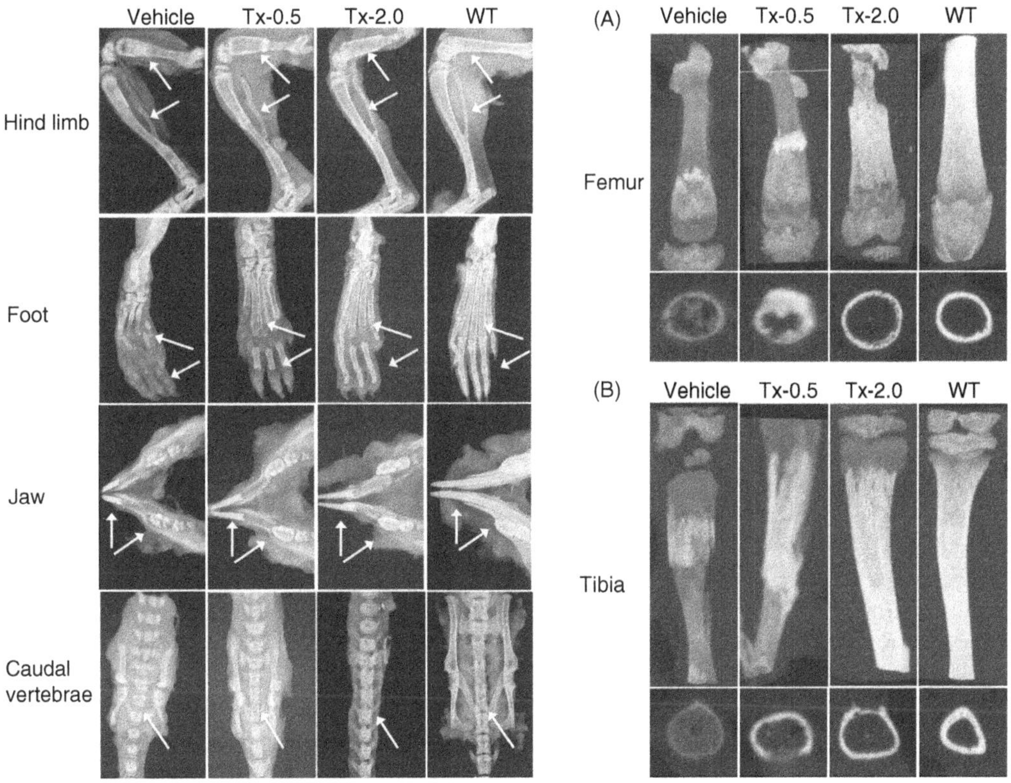

FIGURE 28.3 Left panels: Representative radiographs of hind limb, foot, jaw bones, and caudal vertebrae specimens from 22-day-old $Akp2^{-/-}$ mice treated with vehicle, 0.5 mg/kg/day ENB-0040 (Tx-0.5), or 2.0 mg/kg/day ENB-0040 (Tx-2.0), and untreated WT mice (radiographic magnification 5×). Arrows indicate improved mineralization of tibia, femur, metatarsals, finger bones, incisors and molars, fracture healing in the tibia and femur, and reduced spaces between adjacent spinal vertebrae in treated versus untreated mice. Right panels: μCT images of (A) femora and (B) tibiae of 22-day-old $Akp2^{-/-}$ mice treated with vehicle, Tx-0.5, or Tx-2.0 compared with untreated WT mice. Images clearly show improved tissue mineral density and callus formation at fracture sites of drug-treated mice. Transaxial views are shown at the bottom. *Source*: Reproduced from Reference 28, Bone 49(2): 250–256, August 2011, with permission of the International Bone and Mineral Society.

disease with death at 14 months of age [47]. Bone marrow cell transplantation has been attempted in two severely affected infants with some clinical improvement, but significant morbidity remains [20,21].

Anabolic treatment with parathyroid hormone 1–34, teriparatide apparently benefitted one adult patient [48]; however, others have reported even an increase in osteomalacia with this treatment [49]. Furthermore, prolonged use of this agent is not recommended nor approved for the treatment of rickets in children because its use is associated with increased incidence of osteosarcoma in rats [50]. Some patents have been filed relevant to an anti-sclerostin agent developed primarily to treat osteoporosis as a means of promoting osteoblastic activity and consequently stimulating TNAP activity in HPP patients [51]. My laboratory is also developing small molecule modulators of TNAP activity that could be used to upregulate residual TNAP activity seen in some HPP patients [52]. However, all of these anabolic treatments, including teriparatide, anti-sclerostin, and TNAP activators, rely on the existence of a sufficiently mild *ALPL* mutation such that the enzyme retains sufficient catalytic activity amenable to upregulation. Such a circumstance may apply to a subset of HPP patients, presumably those with milder, adult forms of the disease or odonto HPP, but will likely not be effective in more severe forms of the disease.

As described earlier in this chapter, ERT with either alkaline phosphatase-rich sera from patients with Paget's disease or purified liver or placental ALP was not found to be clinically beneficial [16–19]. This outcome could be due to the short half-life of the administered protein in patients or lack of homing of the administered enzyme to the site of mineralization, namely, the immediate environment of chondrocyte- and osteoblast-derived matrix vesicles where TNAP functions to control PP_i levels and promote extravesicular mineralization [53–55]. Both of those issues have been resolved with the design of bone-targeted alkaline phosphatase, ENB-0040, which has a half-life of 34 h in plasma of adult mice and more than 300 h in bone tissue. Nevertheless, currently optimized treatment requires repeated administration of large amounts of enzyme for long-term correction. In preclinical experiments, mice received daily injections of ENB-0040 and in the initial clinical trials, HPP patients received SC injections of bone-targeted TNAP three times a week [37–46].

We have recently shown that a single injection of a lentiviral vector harboring TNAP-D_{10}, that is, soluble TNAP linked directly to D_{10}, resulted in sustained ALP expression and phenotypic correction in *Akp2*$^{-/-}$ neonatal mice [56], including prevention of epileptic seizures and skeletal defects and preservation of a normal life-span. In another recent study, we demonstrated that a single IV injection of neonatal *Akp2*$^{-/-}$ mice with an adeno-associated viral vector expressing TNAP-D_{10} was as effective as the lentiviral vector treatment [57]. In both of these viral approaches, the Fc region was not included in the construct because neither purification of the enzyme nor an increase in plasma half-life was required. Indeed, these viral delivery systems utilize either the liver (lentiviral vector) or skeletal muscle (adeno-associated viral vector) as factories of bone-targeted TNAP for the life of the animal. These delivery systems show promise as an alternative means of delivery of bone-targeted TNAP that would decrease treatment frequency and greatly reduce treatment cost.

Lentiviral-mediated gene transfer has proven effective for long-term expression of transgenes in nondividing cells. Thus, we used an HIV-1-based lentiviral vector for our study [58]. Although HIV-1 pathogenicity is almost negligible in the current modified version of the lentiviral vector, insertional mutagenesis remains a major concern when using integrating vectors [59]. To minimize the possibility of proto-oncogene activation, our self-inactivating lentiviral vector contains the insulator element from the chicken β-globin locus [60]. Nevertheless, the oncogenic potential of integrated lentiviral vectors in differentiated tissues requires further examination in a long-term, follow-up study. Integrating lentiviral vectors have been used for stable gene transfer into hematopoietic cells in ex-vivo protocols [61,62], but their application for *in vivo* gene transfer is limited [63], and there are no clinical protocols for systemic lentivirus injection. In contrast, AAV vectors are suitable for *in vivo* gene delivery and have been widely used for enzyme replacement therapies for genetic diseases, such as lysosomal storage disorders and hemophilia [64–67]. Our data suggest that systemic delivery of an AAV vector is a promising and realistic approach to treat HPP. Furthermore, using the AAV8 vector, we confirmed that TNAP-D_{10} had a higher affinity for bone than TNAP lacking the D_{10} moiety both *in vitro* and *in vivo* [57]. However, we also demonstrated that TNAP lacking D_{10} could be effective to treat HPP mice, indicating that sustained TNAP production may be the most critical component of the treatment, more so than bone-targeting. This issue merits further investigation, as it is important not only for drug design but also to elucidate the pathophysiology of HPP and other mineralization disorders. We must clearly determine whether it is local or systemic concentrations of PP_i that influence skeletal mineralization, as well as medial vascular calcification, a disease caused by upregulated TNAP levels in vascular smooth muscle cells [35,36].

To conclude, while viral delivery represents a powerful research tool, viral vector safety remains a concern and gene therapy remains a distant promise. In contrast, ERT has a very good track record of safety and efficacy and is readily approved by regulators.

28.6 CHALLENGES IN PRODUCTION AND DEVELOPMENT

While there are many advantages to combining distinct functional domains into a single fusion protein, as discussed extensively in this book, there are also unique challenges related to fusion protein production and purification. Issues related to ENB-0040 purification are amply discussed in an article by Dr. Thomas P. Loisel, Group Leader at Enobia Pharma Inc [30]. Dr. Loisel indicated that ENB-0040 did not fit well into a platform process, even though it possessed three potential affinity sites and each target site had a platform purification unit operation. One site was the antibody Fc region for Protein A binding, but the product was too sensitive to low pH for typical Protein A elution conditions (a $\text{pH} < 4$ typically for elution and <3.7 to inactivate viruses). Therefore, a high pH elution process ($\text{pH} = 11$) was developed as an alternative. The other two affinity sites (catalytic site and polyaspartate sequence) proved to be unusable, as did ion exchange chromatography. The final process developed for ENB-0040 purification added hydrophobic interaction chromatography and a mixed-mode resin with hydrophobic and anion exchange characteristics along with solvent and detergent treatment to replace low pH virus inactivation [30].

28.7 CONCLUSIONS AND FUTURE PERSPECTIVES

To date, there is compelling evidence that ENB-0040 treatment is an efficacious, well-tolerated ERT in HPP mice and preliminary information indicates a similar effect in infants and juveniles with moderate to severe HPP. The success of ERT for HPP using bone-targeted alkaline phosphatase is unprecedented. This approach has gone from concept design to the clinic (bench to beside) in 3 years. I attribute this to four factors: (1) focused commitment of Enobia Pharma to ENB-0040, the only drug in the pipeline; (2) the availability of a well-characterized HPP animal model developed in our laboratory and our deep understanding of HPP pathophysiology in mice; (3) superb clinical expertise by Dr. Michael P. Whyte, the world's expert on HPP; and (4) last but not least, the dedicated efforts of advocacy groups[3,4] that helped inform patients, their families and physicians about progress made in research and therapeutic fronts and also assisted in identifying HPP patients suitable for inclusion in clinical trials.

ACKNOWLEDGMENTS

I thank Dr. Philippe Crine, Ph.D., Vice President and Chief Scientific Officer of Enobia Pharma Inc. for generously sharing data and company reports with me and for his comments on this chapter.

REFERENCES

1. Fraser D. (1957) Hypophosphatasia. *Am. J. Med.* 22, 730–746.
2. Chodirker BN, Evans JA, Seargeant LE, Cheang MS, Greenberg CR. (1990) Hyperphosphatemia in infantile hypophosphatasia: implications for carrier diagnosis and screening. *Am. J. Hum. Genet.* 46, 280–285.
3. Whyte MP. (2001) Hypophosphatasia. In: Scriver CR, Beaudet AL, Sly WS, Valle D (Eds), *The Metabolic and Molecular Bases of Inherited Disease*. 8th ed. McGraw-Hill, New York, pp. 5313–5329.
4. Russell RG. (1965) Excretion of inorganic pyrophosphate in hypophosphatasia. *The Lancet* 2, 461–464.
5. Fleisch H, Russell RG, Straumann F. (1966) Effect of pyrophosphate on hydroxyapatite and its implications in calcium homeostasis. *Nature* 212, 901–903.
6. Whyte MP, Mahuren JD, Vrabel LA, Coburn SP. (1985) Markedly increased circulating pyridoxal-5′-phosphate levels in hypophosphatasia. Alkaline phosphatase acts in vitamin B6 metabolism. *J. Clin. Invest.* 76, 752–756.
7. Whyte MP, Mahuren JD, Fedde KN, Cole FS, McCabe ER, Coburn SP. (1988) Perinatal hypophosphatasia: tissue levels of vitamin B6 are unremarkable despite markedly increased circulating concentrations of pyridoxal-5′-phosphate. Evidence for an ectoenzyme role for tissue-nonspecific alkaline phosphatase. *J. Clin. Invest.* 81, 1234–1239.
8. Coburn SP, Lewis DN, Fink WJ, Mahuren JD, Schaltenbrand WE, Costill DL. (1988) Human vitamin B-6 pools estimated through muscle biopsies. *Am. J. Clin. Nutr.* 48, 291–294.
9. Van den Bos T, Handoko G, Niehof A, Ryan LM, Coburn SP, Whyte MP, (2005) Cementum and dentin in hypophosphatasia. *J. Dent. Res.* 84, 1021–1025.
10. Baumgartner-Sigl S, Haberlandt E, Mumm S, Scholl-Burgi S, Sergi C, Ryan L, (2007) Pyridoxine-responsive seizures as the first symptom of infantile hypophosphatasia caused by two novel missense mutations (c.677T>C, p.M226T; c.1112C>T, p.T371I) of the tissue-nonspecific alkaline phosphatase gene. *Bone* 40, 1655–1661.
11. Waymire KG, Mahuren JD, Jaje JM, Guilarte TR, Coburn SP, MacGregor GR. (1995) Mice lacking tissue non-specific alkaline phosphatase die from seizures due to defective metabolism of vitamin B-6. *Nat. Genet.* 11, 45–51.
12. Narisawa S, Fröhlander N, Millán JL. (1997) Inactivation of two mouse alkaline phosphatase genes and establishment of a model of infantile hypophosphatasia. *Dev. Dyn.* 208, 432–446.
13. Narisawa S, Wennberg C, Millán JL. (2001) Abnormal vitamin B6 metabolism in alkaline phosphatase knock-out mice causes multiple abnormalities, but not the impaired bone mineralization. *J. Pathol.* 193, 125–133.
14. Fedde KN, Blair L, Silverstein J, Coburn SP, Ryan LM, Weinstein RS, (1999) Alkaline phosphatase knock-out mice recapitulate the metabolic and skeletal defects of infantile hypophosphatasia. *J. Bone Miner. Res.* 14, 2015–2026.
15. Caswell AM, Whyte MP, Russell RG. (1991) Hypophosphatasia and the extracellular metabolism of inorganic pyrophosphate: clinical and laboratory aspects. *Crit. Rev. Clin. Lab. Sci.* 28, 175–232.
16. Whyte MP, Valdes RJr, Ryan LM, McAlister WH. (1982) Infantile hypophosphatasia: enzyme replacement therapy by intravenous infusion of alkaline phosphatase-rich plasma from patients with Paget's bone disease. *J. Pediatr.* 101, 379–386.
17. Whyte MP, McAlister WH, Patton LS, Magill HL, Fallon MD, Lorentz WB, Jr, et al. (1984) Enzyme replacement therapy for infantile hypophosphatasia attempted by intravenous infusions of alkaline phosphatase-rich Paget plasma: results in three additional patients. *J. Pediatr.* 105, 926–933.
18. Weninger M, Stinson RA, Plenk H, Jr, Bock P, Pollak A. (1989) Biochemical and morphological effects of human hepatic alkaline phosphatase in a neonate with hypophosphatasia. *Acta Paediatr. Scand.* 360(Suppl), 154–160.
19. Whyte MP, Habib D, Coburn SP, Tecklenburg F, Ryan L, Fedde KN, (1992) Failure of hyperphosphatasemia by intravenous infusion of purified placental alkaline phosphatase (ALP) to correct severe hypophosphatasia: evidence against a role for circulating ALP in skeletal mineralization. *J. Bone. Miner. Res.* 7, S155.
20. Whyte MP, Kurtzberg J, McAlister WH, Mumm S, Podgornik MN, Coburn SP, et al. (2003) Marrow cell transplantation for infantile hypophosphatasia. *J. Bone. Miner. Res.* 18, 624–636.
21. Cahill RA, Wenkert D, Perlman SA, Steele A, Coburn SP, McAlister WH, et al. (2007) Infantile hypophosphatasia: transplantation therapy trial using bone fragments and cultured osteoblasts. *J Clin. Endocrinol. Metabol.* 92, 2923–2930.

[3] http://www.hypophosphatasie.com

[4] http://www.softbones.org

22. Millán JL. (2006) *Mammalian alkaline phosphatases, from biology to applications in medicine and biotechnology*. Wiley-VCH Verlag GmbH & Co., Weinheim, Germany.
23. Millán JL, Narisawa S, Lemire I, Loisel TP, Boileau G, Leonard P, et al. (2008) Enzyme replacement therapy for murine hypophosphatasia. *J. Bone Miner. Res.* 23, 777–787.
24. McKee MD, Nakano Y, Masica DL, Gray JJ, Lemire I, Heft R, et al. (2011) Enzyme replacement therapy prevents dental defects in a model of hypophosphatasia. *J. Dent. Res.* 90, 470–476.
25. Nishioka T, Tomatsu S, Gutierrez MA, Miyamoto K, Trandafirescu GG, Lopez PL, et al. (2006) Enhancement of drug delivery to bone: characterization of human tissue-nonspecific alkaline phosphatase tagged with an acidic oligopeptide. *Mol. Genet. Metab.* 88, 244–255.
26. Kasugai S, Fujisawa R, Waki Y, Miyamoto K, Ohya K. (2000) Selective drug delivery system to bone: small peptide (Asp)6 conjugation. *J. Bone Miner. Res.* 15, 936–943.
27. Yokogawa K, Miya K, Sekido T, Higashi Y, Nomura M, Fujisawa R, et al. (2001) Selective delivery of estradiol to bone by aspartic acid oligopeptide and its effects on ovariectomized mice. *Endocrinology* 142, 1228–1233.
28. Yadav MC, Lemire I, Leonard P, Boileau G, Blond L, Beliveau M, et al. (2011) Dose response of bone-targeted enzyme replacement for murine hypophosphatasia. *Bone* 49, 250–256.
29. Urlaub G, Kas E, Carothers AM, Chasin LA. (1983) Deletion of the diploid dihydrofolate reductase locus from cultured mammalian cells. *Cell* 33, 405–412.
30. Rea DW, Ultee ME, Chen SX, Loisel TP. (2008) Solutions for purification of Fc-fusion proteins *BioPharm Int.* March 20–25
31. Linder CH, Narisawa S, Millán JL, Magnusson P. (2009) Glycosylation differences contribute to distinct catalytic properties among bone alkaline phosphatase isoforms. *Bone* 45, 987–993.
32. Drake MT, Khosla S. (2008) Bone-targeted replacement therapy for hypophosphatasia. *J. Bone Miner. Res.* 23, 775–776.
33. Collmann H, Mornet E, Gattenlohner S, Beck C, Girschick H. (2009) Neurosurgical aspects of childhood hypophosphatasia. *Child. Nerv. Syst.* 25, 217–223.
34. Murshed M, Harmey D, Millán JL, McKee MD, Karsenty G. (2005) Unique coexpression in osteoblasts of broadly expressed genes accounts for the spatial restriction of ECM mineralization to bone. *Genes Dev.* 19, 1093–1104.
35. Narisawa S, Harmey D, Yadav MC, O'Neill WC, Hoylaerts MF, Millán JL. (2007) Novel inhibitors of alkaline phosphatase suppress vascular smooth muscle cell calcification. *J. Bone Miner. Res.* 22, 1700–1710.
36. Lomashvili KA, Garg P, Narisawa S, Millán JL, O'Neill WC. (2008) Upregulation of alkaline phosphatase and pyrophosphate hydrolysis: potential mechanism for uremic vascular calcification. *Kidney Int.* 73, 1024–1030.
37. Whyte MP, Landy H, Edgar T, Weber T, Kishnani P, Wenkert D, et al. (2009) ENB-0040, a bone-targeted human recombinant tissue nonspecific alkaline phosphatase (TNSALP) fusion protein, in the treatment of infants and adults with hypophosphatasia. *Bone* 45(Suppl 2), S57.
38. Greenberg CR, Mhanni A, Catte D, Kishnani P, Weber T, McGinn M, (2009) Interim results of adults and infants with hypophosphatasia treated with bone-targeted human recombinant alkaline phosphatase ENB-0040. *Mol. Genet. Metab.* 98, 1–16, Abstract 121.
39. Whyte MP, Greenber CR, Edgar T, Van Sickle BJ, Hamdam M, Salman N, et al. (2009) *Hypophosphatasia: treatment of life-threatening disease using bone-targeted human recombinant tissue-nonspecific alkaline phosphatase.* Annual meeting of the American Society of Human Genetics, Friday October 23, Abstract 138.
40. Greenberg CR, Craig S, McGinn M, Simmons J, Russell W. (2010) *Life-threatening hypophosphatasia in infants and young children: results of long-term treatment with ENB-0040, a bone-targeted, enzyme replacement-therapy (ERT), and an algorithm for patient management.* American Society of Human Genetics, 60th November 06, Abstract 13.
41. Whyte MP, Greenberg CR, Wenkert D, McAlister WH, Madson KL, Reeves AL, (2010) *Hypophosphatasia in children: enzyme replacement therapy using bone-targeted, tissue-nonspecific alkaline phosphatase.* Annual meeting of the American Society of Human Genetics, 60th, November 06, Abs 3070/W.
42. Whyte M, Greenberg CR, Wenkert D, McAlister WH, Madson KL, Reeves AL, (2010) Hypophosphatasia: enzyme replacement therapy for children using bone-treated, tissue-nonspecific alkaline phosphatase. *J. Bone Miner. Res.* 25, (Suppl 1) Abstract 1016.
43. Whyte MP, Greenberg CR, Wenkert D, McAlister WH, Madson KL, Reeves AL, (2010) Hypophosphatasia (HPP): enzyme replacement therapy (EzRT) for children using bone-targeted, tissue nonspecific alkaline phosphatase. *Ann. Endocr. Rev.* 31, (3 Suppl 1) Abstract OR11-4.
44. Greenberg CR, Landy H, Taylor J, Simmons J, Salman N, Bober M, et al. (2011) *Hypophosphatasia: one year of treatment of life-threatening disease using bone-targeted human recombinant tissue-nonspecific alkaline phosphatase.* Annual Clinical Genetics Meeting, Abstract 4.
45. Greenberg CR, Whyte MP, Wenkert D, McAlister WH, Madson KL, Reeves AL, et al. (2011) *Hypophosphatasia: a six-month controlled study in affected children using bone-targeted, tissue nonspecific alkaline phosphatase. Annual Clinical Genetics Meeting, Abstract 259.*
46. Bishop N, Greenberg C, Craig S, McGinn M, Simmons J, Russell W, et al. (2011) Life-threatening hypophosphatasia (HPP): results of up to two years bone-targeted enzyme replacement therapy (ERT) in infants and young children. *Bone* 48(Suppl 2), S82.
47. Deeb AA, Bruce SN, Morris AA, Cheetham TD. (2000) Infantile hypophosphatasia: disappointing results of treatment, *Acta Paediatr.* 89, 730–733.
48. Whyte MP, Mumm S, Deal C. (2007) Adult hypophosphatasia treated with teriparatide. *J.Clin. Endocrinol. Metabol.* 92, 1203–1208.

49. Gagnon C, Sims NA, Mumm S, McAuley SA, Jung C, Poulton IJ, J.Clin. Endocrinol. Metabol. (2010) Lack of sustained response to teriparatide in a patient with adult hypophosphatasia. *J.Clin. Endocrinol. Metabol.* 95, 1007–1012.

50. Vahle JL, Long GG, Sandusky G, Westmore M, Ma YL, Sato M. (2004) Bone neoplasms in F344 rats given teriparatide [rhPTH(1-34)] are dependent on duration of treatment and dose. *Toxicol. Pathol.* 32, 426–438.

51. Modulators of sclerostin binding partners for treating bone related disorders. Patent no. WO 2008/092894 A1, published Aug 7, 2008.

52. Chung TDY, Sergienko E, Millán JL. (2010) Assay format as a critical success factor for identification of novel inhibitor chemotypes of tissue-nonspecific alkaline phosphatase from high-throughput screening. *Molecules* 15, 3010–3037.

53. Hessle L, Johnsson KA, Anderson HC, Narisawa S, Sali A, et al. (2002) Tissue-nonspecific alkaline phosphatase and plasma cell membrane glycoprotein-1 are central antagonistic regulators of bone mineralization. *Proc. Natl. Acad. Sci. USA.* 99, 9445–9449.

54. Harmey D, Hessle L, Narisawa S, Johnson KA, Terkeltaub R, Millán JL. (2004) Concerted regulation of inorganic pyrophosphate and osteopontin by Akp2, Enpp1, and Ank: an integrated model of the pathogenesis of mineralization disorders. *Am. J. Pathol.* 164, 1199–1209.

55. Anderson HC, Harmey D, Camacho NP, Garimella R, Sipe JB, Tague S, et al. (2005) Sustained osteomalacia of long bones despite major improvement in other hypophosphatasia-related mineral deficits in tissue nonspecific alkaline phosphatase/nucleotide pyrophosphatase phosphodiesterase 1 double-deficient mice. *Am. J. Pathol.* 66, 1711–1720.

56. Yamamoto S, Orimo H, Matsumoto T, Iijima O, Narisawa S, Maeda T, et al. (2011) Prolonged survival and phenotypic correction of Akp2(−/−) hypophosphatasia mice by lentiviral gene therapy. *J. Bone Miner. Res.* 26, 135–142.

57. Matsumoto T, Yamamoto S, Iijima O, Orimo H, Miyake K, Miyake N, et al. (2011) Rescue of severe infantile hypophosphatasia Mice by AAV-mediated sustained expression of soluble alkaline phosphatase. *Hum. Gene Ther.* 22, 1355–1364.

58. Hanawa H, Yamamoto M, Zhao H, Shimada T, Persons DA. (2009) Optimized lentiviral vector design improves titer and transgene expression of vectors containing the chicken beta-globin locus HS4 insulator element. *Mol. Ther.* 17, 667–674.

59. Nienhuis AW, Dunbar CE, Sorrentino BP. (2006) Genotoxicity of retroviral integration in hematopoietic cells. *Mol. Ther.* 13, 1031–1049.

60. Hanawa H, Hargrove PW, Kepes S, Srivastava DK, Nienhuis AW, Persons DA. (2004) Extended beta-globin locus control region elements promote consistent therapeutic expression of a gamma-globin lentiviral vector in murine beta-thalassemia. *Blood* 104, 2281–2290.

61. Cartier N, Hacein-Bey-Abina S, Bartholomae CC, Veres G, Schmidt M, Kutschera I, et al. (2009) Hematopoietic stem cell gene therapy with a lentiviral vector in X-linked adrenoleukodystrophy. *Science* 326, 818–823.

62. Kaiser J. (2009) Gene therapy. Beta-thalassemia treatment succeeds, with a caveat. *Science* 326, 1468–1469.

63. Jarraya B, Boulet S, Ralph GS, Jan C, Bonvento G, Azzouz M, et al. (2009) Dopamine gene therapy for Parkinson's disease in a nonhuman primate without associated dyskinesia. *Sci. Transl. Med.* 1(2), 2ra4.

64. Manno CS, Pierce GF, Arruda VR, Glader B, Ragni M, Rasko JJ, et al. (2006) Successful transduction of liver in hemophilia by AAV-Factor IX and limitations imposed by the host immune response. *Nat. Med.* 12, 342–347.

65. Passini MA, Bu J, Fidler JA, Ziegler RJ, Foley JW, Dodge JC, et al. (2007) Combination brain and systemic injections of AAV provide maximal functional and survival benefits in the Niemann-Pick mouse. *Proc. Natl. Acad. Sci. USA.* 104, 9505–9510.

66. Kurai T, Hisayasu S, Kitagawa R, Migita M, Suzuki H, Hirai Y, et al. (2007) AAV1 mediated co-expression of formylglycine-generating enzyme and arylsulfatase a efficiently corrects sulfatide storage in a mouse model of metachromatic leukodystrophy. *Mol. Ther.* 15, 38–43.

67. Ogawa K, Hirai Y, Ishizaki M, Takahashi H, Hanawa H, Fukunaga Y, et al. (2009) Long-term inhibition of glycosphingolipid accumulation in Fabry model mice by a single systemic injection of AAV1 vector in the neonatal period. *Mol. Genet. Metab.* 96, 91–96.

68. Jaroszewski L, Rychlewski L, Li Z, Li W, Godzik A (2005) FFAS03: a server for profile-profile sequence alignments. *Nucleic Acids Res.* 33, W284–W288.

29

TARGETING INTERFERON-α TO THE LIVER: APOLIPOPROTEIN A-I AS A SCAFFOLD FOR PROTEIN DELIVERY

Jessica Fioravanti, Jesús Prieto, and Pedro Berraondo

Division of Hepatology and Gene Therapy, Center for Applied Medical Research, University of Navarra, Pamplona; Liver Unit, University Clinic, CIBER-EHD, Pamplona, Navarra, Spain

29.1 DETAILED DESCRIPTION OF THE CONCEPT

29.1.1 Interferons

Interferons (IFN) were discovered by Lindemann and Isaacs in 1957, who detected their ability to induce interference by heat-inactivated influenza virus with the growth of live virus in fresh pieces of chorioallantoic membrane [1]. IFNs are potent biologically active cytokines synthesized and secreted by virtually all eukaryotic cells. Three classes of IFNs have been identified, which are classified according to the receptor complex they signal through (Figure 29.1). The type II class of IFN comprises the single interferon-γ (IFN-γ) gene product that binds the interferon-γ receptor (IFNGR) complex, and mediates broad immune responses to pathogens other than viruses. The number of functional genes identified that encode type I IFNs has subsequently grown: 17 genes have now been described in humans. All are intronless and cluster on chromosome 9 [2]. There are 13 IFN-αs, whereas there is only one type of IFN-β, IFN-ω, IFN-ε, or IFN-κ. Among mammals, the number of type I IFN genes is variable; some have unique types (for example, IFN-δ occurs only in pigs and IFN-τ only in ruminants) and others are devoid of a particular type (for example, IFN-ω in mice). Consistent with the universal biological definition of IFNs (proteins able to induce relatively species-specific antiviral effects), all type I IFNs, which are mostly nonglycosylated proteins of 165–200 amino acids, share homologies that range from 30% to 85% within a species. Essentially, all have relatively high specific potencies (10^7–10^9 antiviral units per milligram protein). The type III IFN family with three subtypes of IFN-λ, which are coproduced with IFN-β, activate the same main signaling pathway as type I IFNs but have evolved a completely different receptor structure [3].

Production of IFNs, both *in vitro* and *in vivo*, is transient and requires stimulation. In the course of the discovery of IFNs, after identification of live or heat-inactivated influenza viruses as inducers, other microbial products, including those of bacteria, protozoa, and RNA and DNA viruses, were recognized also to induce IFN [4]. It was also shown that microbial nucleic acids, lipids, polysaccharides, or proteins trigger synthesis of IFNs through activation of toll-like receptors (TLRs). The discovery of double-stranded (ds) ribonucleic acids (RNAs), both natural and synthetic, as a potent IFN inducer [5], led to the simplistic perspective that viruses induce IFNs by the production of dsRNA; in reality, it is only one of the viral gene products responsible for induction. Nonetheless, the study of cellular responses that lead to

Fusion Protein Technologies for Biopharmaceuticals: Applications and Challenges, First Edition. Edited by Stefan R. Schmidt.

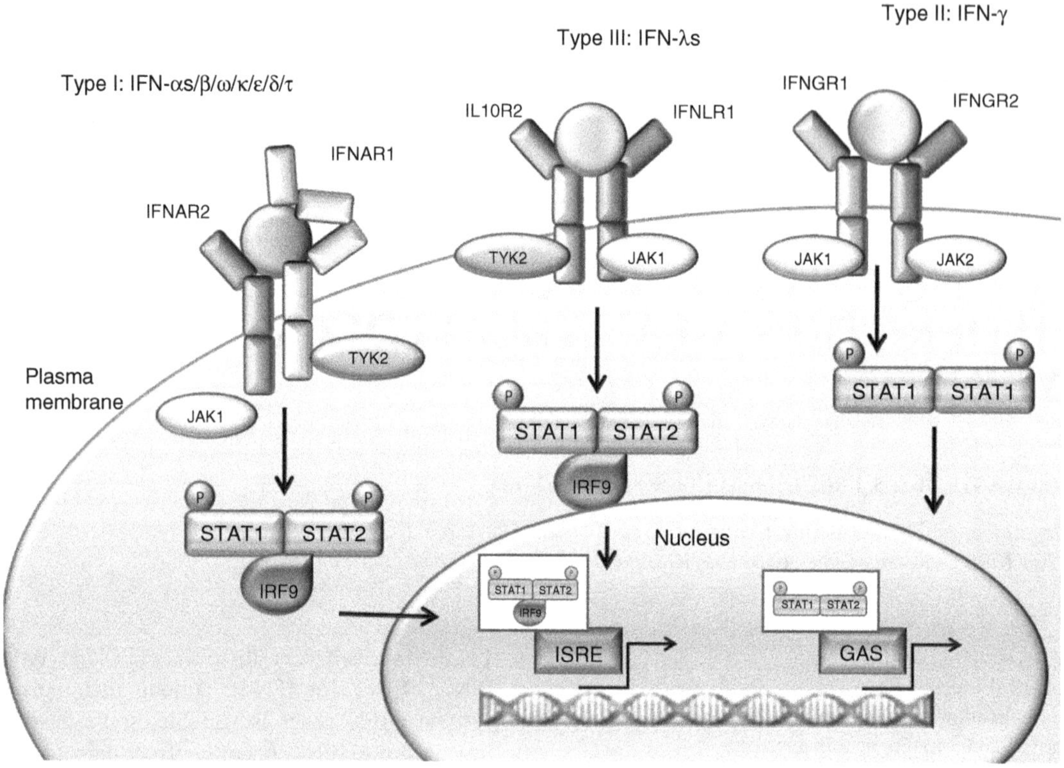

FIGURE 29.1 The IFN signaling pathways. The type I interferons (IFNs) interact with the IFN receptor 1 (IFNAR1) and IFNAR2. The type II IFN-γ with IFN-γ receptor 1 and 2 (IFNGR1 and IFNGR2), and type III IFN-λs with IFN-λ receptor 1 (IFNLR1) and interleukin 10 receptor 2 (IL10R2). All these receptors are associated with two JAK family kinases: either JAK1 and TYK2 for type I and III IFNs or JAK1 and JAK2 for type II IFN. These kinases phosphorylate STAT1 and STAT2, which after dimerization end promoting gene transcription, complexed or not to IRF9. GAS, IFN-γ-activated site; IRF9, IFN regulatory factor 9; ISRE, IFN-stimulated response element; STAT1/2, signal transducers and activators of transcription 1/2; and P, phosphate.

induction was headed by the analysis of dsRNA-mediated signaling pathways (reviewed by Borden et al. [6]).

Because of their pleiotropic effects, these proteins have therapeutic potential for a wide range of disorders. IFN-α, discovered initially for its antiviral action, is also a key mediator of the innate and adaptive immune response. It combatcs viruses directly by inhibiting viral replication (displaying its antiviral properties), or by playing immunoregulatory roles, that help to shape the innate and adaptive immune responses acting directly or indirectly on NK cells, T cells, B cells, and dendritic cells (DCs) [7]. IFN-α also plays a key role in the regulation of cell growth and differentiation, exerting antiproliferative [8] and/or proapoptotic effects [9].

29.1.2 Stabilization of IFN-α

IFN-α is indicated for the treatment of certain viral infections, including hepatitis B and C, and cancers. But the short circulating half-life of unmodified IFN-α makes frequent dosing necessary. To improve the pharmacokinetics of IFN-α and decrease dosing frequency, several stabilizing strategies have been developed, such as association with chemical species such as 2-sulfo-9-fluorenylmethoxycarbonyl [10], recombinant long-acting fusion protein of human albumin-IFN-α2b [11], or other approaches such as linking the protein to larger structures such as polyethylene glycol (PEG). Two peginterferons are currently available for treatment of viral hepatitis: peginterferon-α-2a (Pegasys®) [12] and peginterferon-α-2b (PegIntron®) [13]. They differ in the IFN-α that is modified, and in the size and nature of the pegylation. IFN-α-2a and IFN-α-2b are recombinant molecules differing by a single amino acid. In contrast, the covalent PEG is significantly different, resulting in differences in pharmacokinetic and pharmacodynamic properties, with an impact on their clinical use and performance [14]. Pegylation results in a longer serum half-life than standard IFN-α and maintenance of therapeutic concentrations throughout the dosing interval, allowing for once-weekly administration. This conjugation decreases the renal clearance of the IFN-α and, therefore, significantly increases its half-life from about 5 h to almost 90 h, which in turn allows a reduction in the required frequency of treatments [15,16]. PEG-IFN-α preparations have replaced the use of standard IFN-α for the treatment of hepatitis B and C, because of greater convenience and efficacy. When compared with oral regimens, PEG-IFN-α offers several advantages such as lack of development of drug resistance and greater rates of HBeAg seroconversion. Unfortunately, the prolonged peginterferon therapy necessary to control chronic hepatitis C virus (HCV) or B virus (HBV) infections was often associated with serious side effects such as fatigue, fever, and myalgias. These are symptoms of many acute virus infections, possibly because such effects are due to the induction of IFNs by the infecting agents. Usually, these symptoms respond to treatment with nonsteroidal anti-inflammatory agents [17]. In some patients, treatment with IFNs has also resulted in psychiatric problems such as depression, anxiety, and excessive irritability, which may require treatment with psychoactive pharmaceuticals. More severe toxicities, such as cytopenias and autoimmune disorders, also have been reported in patients treated with IFNs [18].

Another effort to stabilize IFN-α was made by Osborn et al. [11], when they developed the technology of fusion to human albumin. This is the most prevalent serum protein in the circulatory system, with an unusually long half-life (19 days). They not only fused human albumin to IFN-α (alb-IFNα), but also to a wide range of proteins to treat human diseases, such as growth hormone [19], insulin [20], or other small bioactive peptide hormones, such as the β-natriuretic peptide [21]. This albumin-IFN-α treatment was developed by Human Genome Sciences (HGS) and Novartis. On October 5, 2010 both companies announced that they had stopped drug development in response to an FDA ruling.

The last strategy of IFN-α stabilization to reach clinical trial was Locteron®. Locteron® is a new formulation of recombinant IFN-α-2b in poly(ether-ester) microspheres (PolyActive, OctoPlus N.V., Leiden, the Netherlands) [22]. The recombinant IFN-α-2b component of Locteron® (BLX-883) is synthesized in a Lemna aquatic plant expression system. Phase I and II clinical studies were completed. The results of these clinical trials indicate that Locteron® can be dosed once every 2 weeks. The best advantage of this new formulation over the pegylated IFN is an improved tolerability profile with fewer, less severe, and shorter-lasting flu-like symptoms [23].

29.1.3 IFN-α Liver Targeting

The second major limitation of IFN-α as a drug is the lack of organ-specific activity. The main indication of IFN-α is the treatment of viral hepatitis B and C and the target organ is therefore the liver. However, the IFN receptor is expressed in virtually all the cells of the organism. IFN-α signaling in other cell types than hepatocytes may induce unwanted adverse effects.

The first attempts to target IFN to the liver were performed by Kasama et al. [24]. They produced asialo IFN-β by treating natural IFN-β with neuraminidase. They demonstrated that the desialylated glycoprotein was recognized and taken up by the liver via asialoglycoprotein receptor. However, the half-life in circulation was reduced and the biological activity compromised. This latter limitation was solved years later and the benefits of IFN liver targeting were demonstrated for the first time in a murine model of HBV. Asialo IFN not only increased the interferon-stimulated genes (ISGs) expression in the liver but also reduced the HBV virions in the sera of athymic nude mice transfected with HBV plasmid [25]. To target the nonglycosylated IFN-α at the asialoglycoprotein receptor, IFN-α was linked [26] or complexed [27] in pullulan. Pullulan is a linear, nonionic polysaccharide with a repeated unit of maltotriose condensed through α-1,6 linkage. This compound was able to enhance the ISGs expression in the liver mediated by IFN-α.

Yuan et al. [28] developed a new formulation for IFN-α that increased the short half-life and solved the lack of liver-specific affinity. This approach was based on the reversible lipidation of IFN-α by conjugating palmitoylcysteine to cysteinyl residues of IFN-α via disulfide bonds. This palmitoylated IFN-α behaved like a controlled release conjugation. Upon reduction of the disulfide bonds inside the body, unmodified IFN-α is slowly released to the circulation, increasing the half-life. In addition, reversible lipidization alters tissue distribution by an unknown mechanism. Palmitoylated IFN-α accumulates in the liver and exerts an enhanced activity.

Another strategy to increase the local concentration of IFN-α in the liver is to synthesize the cytokine in this organ. The IFN-α gene can be introduced into a gene therapy vector and this vector will infect the liver and deliver the genetic material into the hepatocytes. Finally, the heterologous gene will be transcribed and translated to produce the therapeutic protein. Proof of the principle of this approach was established using a helper-dependent adenovirus vector [29]. This vector predominantly infects the liver and lacks all viral coding sequences, minimizing the innate immune response and allowing long-term expression of the transgene. The IFN-α gene was under the control of an inducible promoter regulated by doxycycline. This promoter allowed the authors to regulate the level of IFN-α produced and to switch the production of the cytokine on and off.

Unfortunately, experiments performed in the best animal model of chronic hepatitis infection (the woodchucks infected with the woodchuck hepatitis virus) revealed several drawbacks of the long-term expression of IFN-α in the liver of infected individuals [30]. In this study, the IFN-α gene was introduced into the hepatocytes using an adenoassociated virus (AAV) type 2. This virus infects the liver and allows the long-term expression of the transgene in $<10\%$ of total hepatocytes. However, most of the woodchucks eliminated the transgene after 2 months and, therefore the antiviral efficacy of this intervention was limited. Nowadays, after several failures in clinical trials, it is clear that the inflammation limits the long-term expression of transgenes. So, long-term expression of an inflammatory cytokine is a challenge [31]. The only woodchuck with sustained expression of IFN-α exemplified other limitations of the IFN-α gene therapy. Three months after AAV infection, this animal had to be sacrificed because of the development of severe blood cytopenia and liver steatosis. Sustained liver expression of IFN-α proved lethal for the animal.

29.1.4 Apolipoprotein A-I as a Scaffold for Peptide and Protein Delivery

Lipids are insoluble and they need to be carried as lipoproteins to be transported in plasma. Lipoproteins consist of a hydrophobic nucleus of nonpolar lipids (triglycerides and esterified cholesterol), enveloped within a hydrophilic surface containing amphipathic lipids (phospholipids and nonesterified cholesterol) [32].

Lipoproteins are classified on the basis of the hydrated density, a function of the lipid/protein contents. This enables the lipoproteins to be separated by ultracentrifugation.

The main functions of lipoproteins are (i) transport of exogenous and endogenous triglycerides, (ii) transport and removal of cholesterol, and (iii) transport of hydrophobic molecules such as liposoluble vitamins. The protein components of the lipoproteins are formed by the apolipoproteins. These proteins are involved in the assembly of the lipoproteins and they act as coactivators of enzymes and as ligands of receptors for their cellular uptake.

Apolipoproteins are a heterogeneous family with different sequences and sizes. They are classified as Apo As (A-I, A-II, A-IV, and A-V), apo Bs (B-48 and B-100), apo Cs (C-I, C-II, and C-III), apo D, apo E, apo J, apo F, apo J, apo H, and apo M [33]. Apolipoprotein A-I is the major protein constituent of the high-density lipoprotein (HDL) particles. HDLs play a central role in reverse cholesterol transport, the process by which excess cholesterol in peripheral tissues is esterified in plasma and transported to the liver for excretion (Figure 29.2). This function of HDLs is believed to explain the strong inverse correlation between plasma HDL levels and coronary heart disease [34]. Apo A-I is the preferential acceptor of cell cholesterol [35], acting as cofactor for the lecithin cholesterol acyltransferase (LCAT) enzyme [36] and as a ligand for the putative HDL receptor: scavenger receptor class B type I (SR-BI) [37,38].

Apo A-I is synthesized both in the liver and in the intestine as pre-pro-apo A-I, and is then processed to the mature form of 243 residues [39]. Then, functional interactions between apo A-I and ABCA1 are necessary for the initial lipidation of apo A-I. Lipidated apo A-I proceeds to form discoidal HDL particles that can be converted to spherical particles by the action of LCAT. Discoidal and spherical HDL can interact functionally with SR-BI and these interactions lead to selective lipid uptake and net efflux

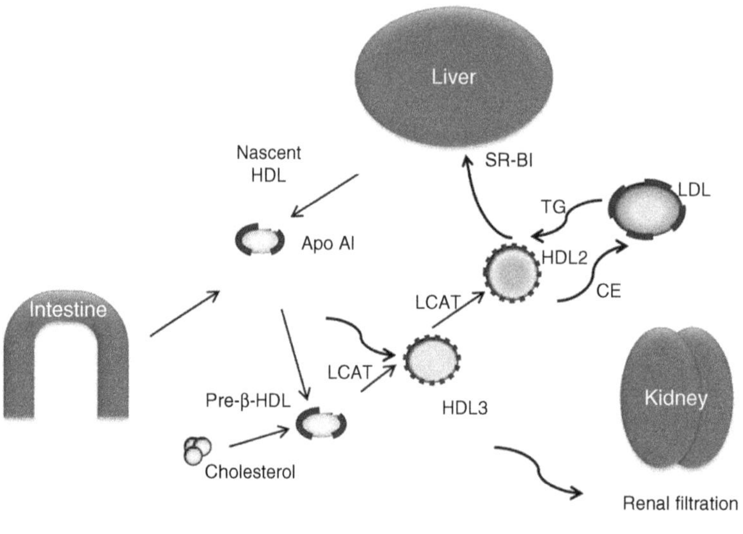

FIGURE 29.2 HDL metabolism. Apo A-I is synthesized both in liver and intestine. After lipidation steps, discoidal HDLs are converted to spherical HDLs by LCAT, and these HDLs can interact with SR-BI to promote cholesterol efflux. HDL, high-density lipoprotein; LDL, low-density lipoprotein; Apo, apolipoprotein; CE, cholesterol ester; TG, triglyceride; LCAT, lecithin cholesterol acyltransferase; SR-BI, scavenger receptor class B type I.

of cholesterol [40]. Finally, the HDL is taken up by the liver via SR-BI, where the protein can be recycled or catabolized.

Cholesterol-loaded HDLs therefore reach the liver by the circulating blood, where cholesterol is internalized by hepatocytes. It has also been demonstrated that after systemic administration of apo A-I, this exogenous apo A-I is accumulated in the liver [41]. Thus, drug-loaded HDLs have been used as vehicles for liver-targeted delivery [42,43]. Encapsulated drugs are hydrophobic or amphiphilic compounds incorporated into the hydrophobic nucleus of nonpolar lipids of HDLs [44], which can be taken up by SR-BI-mediated mechanisms [45–49]. In addition to the liver, many tumor cells overexpress SR-BI and therefore HDLs complexes have proven to be useful as drug targeting for cancer chemotherapy [50–55].

HDLs present three main advantages over other drug nanocarriers. First, the structure of HDLs, which consist of a hydrophobic nucleus enveloped within a hydrophilic surface, provides an ideal vehicle for transporting amphiphilic drugs, which is a challenge for other drug delivery systems [56]. Secondly, since they are human-derived nanoparticles, they are devoid of danger signals that activate the immune system [56]. Finally, the uptake into cells is mediated by SR-BI and, therefore, there is no need of additional targeting strategies [57]. There are two potential disadvantages of HDLs as drug carriers. The first limitation is the availability of a large amount of natural HDLs, and the safety issues because of the use of a human-derived material. This limitation is circumvented by the development of a form of HDL derived synthetically and termed reconstituted HDL (rHDL) [54,58]. These rHDLs are created by a general method that allows the generation of rHDL particles with a specified phospholipid and apolipoprotein content. These rHDLs are analogous to native nascent HDL, displaying all of its physiological functions. The diameters of these rHDL disks vary with the number of apo A-I molecules per particle (2, 3, or 4) and their conformation [58,59].

Secondly, this method is only suitable for hydrophobic compounds that can be housed inside artificial membranes in contact with the hydrophobic fraction of the phospholipids. This category of compound excludes most of the peptides and proteins of clinical interest. To extend the benefit of the HDL stabilization and liver targeting to therapeutic peptide and proteins, fusion protein of apolipoprotein A-I and the therapeutic protein can be generated. The proof of principle of this approach was performed with IFN-α [60].

29.1.5 Stabilization and Liver Targeting of IFN-α by Apolipoprotein A-I

Since apo A-I displays a long plasma half-life comparable to the half-life of PEG-IFN-α [61,62], and has a previously demonstrated liver-tropism [41], it was chosen as a stabilizing and liver-targeting moiety for an important cytokine in liver disease treatments: IFN-α. This strategy could combine the advantages of stabilized and liver-targeted formulations.

To confirm that the fusion protein exhibits long half-life and liver tropism, the mouse IFN-α gene was fused to the mouse apo A-I gene (IA) (Figure 29.3). A small linker of three amino acids (GAP) was used, although this is dispensable. The chimeric gene was cloned into an eukaryotic expression plasmid, and the plasmids coding IFN-α or the fusion protein (IA) were administered by hydrodynamic injection. The half-life of IA was 2.6-fold that of IFN-α, in agreement with the fact that apo A-I half-life is about 2–3 times that of IFN-α, and comparable to the half-life of pegylated IFN-α [10,11,62–65]. Moreover, the half-life of IA was only 1.6 times less than IFN-α bound to albumin (albumin-IFN-α), the formulation with the longest half-life reported to date for an IFN-α conjugate [66].

As a consequence of the superior half-life of IA and the liver tropism, hepatic expression of ISGs at day 3 following injection of a plasmid encoding IA or IFN-α was higher with the former treatment and even at day 3 after hydrodynamic injection of plasmids encoding IA or albumin-IFN-α, the expression of ISGs in the liver tended to be higher with IA treatment, although serum concentration of IA was half that of albumin-IFN-α at this time point.

Finally, these properties were formally demonstrated using recombinant IA (rIA). Recombinant IFN-α (rIFN-α) presented a sharp decay in mouse plasma levels, while the concentration of rIA decreased slowly. Moreover, rIA accumulated in the liver 2.5 h after i.v. administration, while rIFN-α was undetectable.

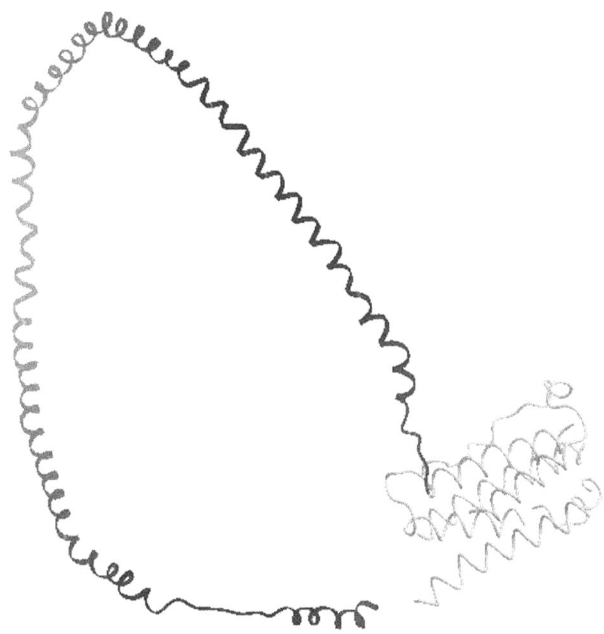

FIGURE 29.3 Molecular representation of the fusion protein of interferon-α (in gray) and apolipoprotein A-I (in black).

IA preserved the antiviral activity even when incorporated into HDLs. Purified HDLs containing IA promoted STAT-1, -2, and -3 phosphorylation in a manner comparable to the same antiviral units of recombinant IFN-α. *In vivo*, the enhanced stability of IA in circulation allows a greater antiviral activity reflected in an increase in the percentage of mice that survive a lethal challenge with encephalomyocarditis virus.

29.1.6 Unexpected Properties of the Fusion Protein of Interferon-α and Apolipoprotein A-I

A clear difference between IA and IFN-α was observed when cell viability and cytotoxicity were analyzed. It is noteworthy that IFN-α at the dose used for signaling experiments caused an increase in cell death, while the same antiviral units of HDL-IA caused no cytotoxic effect, cell viability being similar to that of control wells.

In keeping with the differential effects of IA and IFN-α on cell proliferation, those two molecules were found not to be comparable in their actions on the hematopoietic system. Thus, 3 days after plasmid injection, platelets and leukocytes were significantly higher in IA-treated mice than in IFN-α or albumin-IFN-α treated mice (Figure 29.4).

IA also differs from IFN-α and from albumin-IFN-α because of its immunomodulatory properties. *In vivo* killing assays against a Balb/c immunodominant β-galactosidase epitope were performed in mice, which 7 days previously had been given a hydrodynamic coinjection of a plasmid encoding LacZ together with pIFN, pIA, pALF, or pApo. These studies showed that the group treated with pIA exhibited a cytolytic activity significantly greater than the other three groups. Differences were also observed between pIFN and pIA in the induction of protective immunity in a murine model of vaccination against CT-26 colon cancer. Vaccination was performed by subcutaneous administration of peptide AH-1 (which corresponds to the tumor-immunodominant epitope) in combination with hydrodynamic injection of pApo, pIFN, or pIA. Ten days later, animals received a subcutaneous injection of CT-26 cells. The adjuvant treatment with pIA, but not with pIFN, was associated with significant protection against tumor growth as compared to control group given pApo.

IFN-α has been shown to induce activation-dependent cell death in lymphocytes [67,68]. Since data suggested that IA might be more effective than IFN-α in promoting the expansion of stimulated T cells, the lymphocyte proliferation and viability following T-cell stimulation with anti-CD3 and anti-CD28 in the presence of IFN-α or the same antiviral units of HDL-IA was analyzed. Flow cytometry analysis of stimulated lymphocytes showed that the number of large blast cells was markedly reduced in the presence of IFN-α, while values were similar to controls when HDL-IA was added to the culture. In these experiments, the lymphocyte proliferation was assessed by CFSE dilution and lymphocyte death by 7-AAD incorporation. Both cell proliferation and cell death were similar in control wells and in wells containing HDL-IA. In contrast, in the presence of IFN-α, lymphocyte proliferation was reduced and the number of nonviable lymphocytes was greatly increased. These data are consistent with the notion that preservation of viability of activated T cells may underlie the higher effectiveness of IA in boosting T-cell immunity.

As mentioned, IA is transported in plasma bound to HDLs, which act as nanocarriers for the fusion protein, enabling the interaction of IA with both SR-BI and the IFN-α receptor. After binding to SR-BI, HDLs mediate the uptake of cholesteryl esters and phospholipids from the cells and promote cytoprotective functions by mechanisms imperfectly understood involving multiple interactions between HDLs (lipid or protein moiety) and cell surface receptors [69]. SR-BI is expressed at low levels in a wide variety of cells, and at high levels in liver, adrenal glands, ovaries, testis, intestinal cells, phagocytes, and endothelial cells [38,70]. To investigate whether combined interaction with SR-BI and IFN-α receptor was important in mediating the superior immunostimulatory activity of IA as compared to IFN-α, an *in vivo* killing against β-galactosidase epitope was performed, following a vaccination with

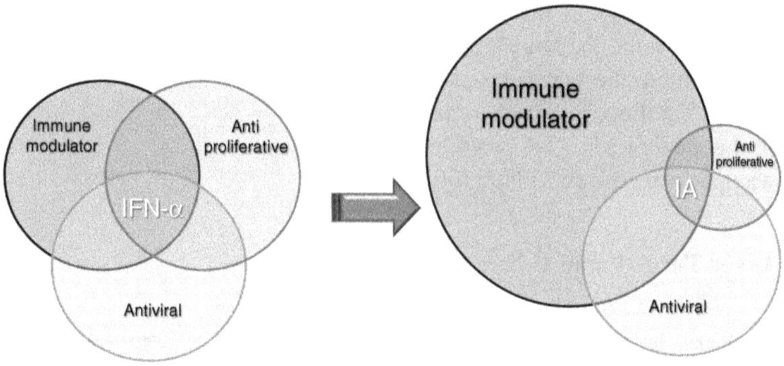

FIGURE 29.4 Scheme of the differential effects between interferon-α and the fusion molecule of interferon-α and apolipoprotein A-I.

hydrodynamic coinjection of a plasmid encoding LacZ together with pIFN or pIA or pApo in wild-type mice and in SR-BI$^{+/-}$ or SR-BI$^{-/-}$ animals. In agreement with previous data, pIA showed an eightfold increase in the specific lysis compared to that induced by pIFN in wild-type mice. However, this difference was reduced to twofold in SR-BI$^{+/-}$ animals and the cytotoxicity values were similar for IA and IFN-α in SR-BI null mice. Thus, IA requires interaction with SR-BI to display maximum adjuvant activity.

29.2 TECHNOLOGICAL ASPECTS

When the fusion protein is produced directly in the liver by gene therapy vectors, most of the protein is able to complex in HDLs. This high incorporation is probably because the protein is produced inside the transduced hepatocytes, following the natural apo A-I pathways, and therefore only 2–8% of apo A-I is found lipid-free in plasma, as it occurs physiologically. When administered intravenously, ~10% of the recombinant protein can also be detected in the HDL enriched fraction, and this percentage of incorporation is also observed when IA is administered by intratumoral injection of a viral vector expressing IA, and therefore delivered to plasma after being produced by tumor cells. An important question is, therefore, whether HDL incorporation of the fusion protein IA could be playing an essential role. All the important functions of IA (long half-life in circulation, liver targeting, reduced hematological toxicity, and enhanced immune stimulation) have been demonstrated both with the recombinant protein and with HDL-containing IA. In addition, to avoid the need for IA complexing to HDLs to be immunologically functional, an *in vivo* killing assay was performed in Apo A-I knockout mice (that lack HDLs) and in wild-type mice receiving rIA and ovalbumin protein as antigen. In both cases, rIA was able to act as an effective adjuvant increasing the cytotoxic T-cell-mediated specific lysis.

Therefore, the recombinant fusion protein can be used directly and the *in vitro* reconstitution of synthetic HDLs is not required. However, there is a lot of experience concerning the feasibility and security of this approach due to the clinical trials that have used reconstituted HDLs for the treatment of cardiovascular diseases [71–80]. Incubation of the recombinant fusion protein with phospholipids may lead to the formation of nanoparticles that resemble natural HDLs and which can be infused to patients.

Both components of the new fusion molecule have been produced and purified previously using different technologies. In spite of the amphiphilic character, autoaggregation, and degradation of apo A-I, a wide variety of heterologous protein production systems have been used successfully. They include bacteria [81,82], yeast [83], insect cells [84], hamster ovary cells [85,86], and more recently plant-based expression systems [87,88]. Regarding IFN-α, the recombinant protein produced in *Escherichia coli* was approved for clinical use since 1986. Most of the human IFN-α species are devoid of carbohydrates and therefore, bacteria are still the most widely used production system although many other systems have been tested [89–91]. The production of the recombinant fusion protein has been successfully performed both in *E. coli* and in yeast using standard technology.

29.3 TYPICAL APPLICATIONS AND INDICATIONS

IFN-α was first discovered for its antiviral activity [1] but its promising antitumor activity was soon appreciated [92]. The differential effects of IA versus IFN-α make this novel molecule a candidate, not only for the conventional IFN-α-based therapies, but also for new applications.

The major indication for using a liver-targeted IFN-α formulation such as IA is viral hepatitis. Nowadays, the treatment of HCV patients is based on combination of PEG-IFN and ribavirin plus/minus direct-acting antivirals [93–95]. The reduced toxicity of the new fusion protein could increase the response rate to these combinations and could perhaps rescue those patients, who failed to respond to conventional therapy. On the other hand, IA might be particularly useful in chronic HBV infection. In this infection, the viral genome remains in hepatocytes in the form of covalently closed circular (ccc) DNA maintaining transcriptional activity despite inhibition of replication with nucleoside/nucleotides analogs. Only an effective immune response generated against hepatocytes containing ccc HBV-DNA will be able to clear the infection. In this regard, the enhanced immunostimulatory activity of the new fusion protein can be crucial.

IFN-α also displays antitumor activity through its ability to halt cell proliferation and to induce apoptosis of tumor cells [96,97]. In addition, IFN-α may also block tumor progression by acting on stromal cells where it has been shown to inhibit angiogenesis by suppressing the proliferation and migration of endothelial cells [98,99]. Most importantly, IFN-α plays a central role in the stimulation of antitumor immune responses, being able to enhance the trafficking, expansion, and lytic activity of NK cells and T lymphocytes [100,101]. Although these properties make IFN-α very attractive for cancer therapy, clinical use of this cytokine is limited by its toxicity. An additional limitation is due to the fact that IFN-α plays a dual role in the biology of the T lymphocytes. On the one hand, it stimulates clonal expansion but, on the other hand, it sensitizes T cells to the antigen-induced cell death normally occurring at the end of an immune response [68]. In addition, IFN-α causes myelosuppression, which may result in severe leukopenia and thrombocytopenia precluding continuation of the therapy [102]. IA may be a good candidate to substitute IFN-α in its oncologic indications. Although IA does not exhibit direct antiproliferative activity, it can activate antitumor immunity in a very efficient manner. Moreover, its reduced hematological toxicity might enable prolonged treatment without significant peripheral cytopenia.

Finally, another indication for IA is in the field of therapeutic vaccines. These consist of the administration of an antigen plus appropriate adjuvants aiming at activating an adaptive immune response, which can eliminate a chronic infection or control the growth of a neoplasm. However, the limited success of therapeutic vaccines in their clinical applications makes the development of new adjuvants an important medical need. The fusion protein of IFN-α and apolipoprotein A-I represents a completely new type of adjuvant, which warrants clinical testing.

29.4 ALTERNATIVES AND VARIANTS OF THIS APPROACH

This approach has two potential alternatives: first, modification in the apo A-I moiety and secondly, modifications in the cytokine fused to apo A-I.

The apo A-I can be substituted by apo A-I mutants or by different apolipoproteins. A number of apo A-I mutants have been characterized [103]. While most of them are linked to pathological phenotypes, two variants, apo A-I Milano and apo A-I Paris, appear to exhibit some unique, favorable features in cardiovascular diseases [72,104,105]. Therefore, these mutants are an attractive alternative to wild-type apo A-I but the preservation of the properties of IA must be carefully studied.

Apo A-I can also be replaced by different apolipoproteins found in HDLs. This issue was studied using apo E or apo F. Plasmids encoding these fusion proteins were prepared, and then administered to mice by hydrodynamic injection. IFN-α fused to apo F was unstable and was not expressed. However, IFN-α fused to apo E was expressed and, similarly to IA, manifested liver targeting and little hematological toxicity but unlike IA did not exhibit differences with respect to IFN-α in half-life or immunostimulatory properties. Therefore, apo A-I seems to be the ideal candidate among the different apolipoproteins as a scaffold moiety. The stabilizing and liver-targeting properties are likely to be acquired by any therapeutic protein/peptide fused to apo A-I. Therapeutic proteins with a molecular weight lower than 50 kDa can benefit from this strategy since they usually display a short half-life in circulation. However, the newly acquired biological and pharmacodynamic properties of other therapeutic proteins when they are fused to apo A-I must be carefully evaluated.

29.5 CONCLUSIONS AND FUTURE PERSPECTIVES

The fusion protein of mouse IFN-α and mouse apolipoprotein A-I has demonstrated outstanding properties in the murine models. Moreover, these properties are being reproduced in woodchucks infected chronically with the woodchuck hepatitis virus. The translation of these properties to the clinical field remains the main challenge for the new fusion molecule. The first steps toward a Phase I clinical trial are being performed by HepaCyL Therapeutics. This trial will provide valuable information on the pharmacodynamic and toxicologic profile of IA in healthy volunteers. Subsequent Phase II trials in patients with chronic hepatitis C and chronic hepatitis B will inform us about the antiviral and immunostimulatory activities of IA compared to classical PEG-IFN. If preclinical data pertaining to the ability of IA to strongly activate cellular immunity while exhibiting reduced toxicity is confirmed in humans, this fusion molecule might be a key component in combination treatments aimed at fighting viral infections or neoplastic growth.

In conclusion, fusion of IFN-α to apolipoprotein A-I creates a molecule possessing IFN activity, prolonged half-life, and liver targeting. Moreover, this association confers novel properties on IA, which include diminished hematological toxicity and enhanced immunostimulatory activity. IA may constitute a promising alternative to recombinant IFN-α for a diversity of clinical applications, particularly for patients with low platelet and/or leukocyte counts.

REFERENCES

1. Isaacs A, Lindenmann J. (1957) Virus interference. I. The interferon. *Proc. R. Soc. Lond. B. Biol. Sci.* 147(927), 258–267.
2. Hardy MP, Owczarek CM, Jermiin LS, Ejdeback M, Hertzog PJ. (2004) Characterization of the type I interferon locus and identification of novel genes. *Genomics* 84(2), 331–345.
3. Uze G, Monneron D. (2007) IL-28 and IL-29: newcomers to the interferon family. *Biochimie* 89(6–7), 729–734.
4. Akira S, Uematsu S, Takeuchi O. (2006) Pathogen recognition and innate immunity. *Cell* 124(4), 783–801.
5. Field AK, Tytell AA, Lampson GP, Hilleman MR. (1967) Inducers of interferon and host resistance. II. Multistranded synthetic polynucleotide complexes. *Proc. Natl. Acad. Sci. U S A* 58(3), 1004–1010.
6. Borden EC, Sen GC, Uze G, Silverman RH, Ransohoff RM, Foster GR, et al. (2007) Interferons at age 50: past, current and future impact on biomedicine. *Nat. Rev. Drug Discov.* 6(12), 975–990.
7. Prchal M, Pilz A, Simma O, Lingnau K, von Gabain A, Strobl B, et al. (2009) Type I interferons as mediators of immune adjuvants for T- and B cell-dependent acquired immunity. *Vaccine* 27(Suppl 6), G17–20.

8. Kalie E, Jaitin DA, Abramovich R, Schreiber G. (2007) An interferon alpha 2 mutant optimized by phage display for IFNAR1 binding confers specifically enhanced antitumor activities. *J. Biol. Chem.* 282(15), 11602–11611.
9. Thyrell L, Arulampalam V, Hjortsberg L, Farnebo M, Grander D, Pokrovskaja Tamm K. (2007) Interferon alpha induces cell death through interference with interleukin 6 signaling and inhibition of STAT3 activity. *Exp. Cell Res.* 313(19), 4015–4024.
10. Shechter Y, Preciado-Patt L, Schreiber G, Fridkin M. (2001) Prolonging the half-life of human interferon-alpha 2 in circulation: Design, preparation, and analysis of (2-sulfo-9-fluorenylmethoxycarbonyl)7- interferon-alpha 2. *Proc. Natl. Acad. Sci. U S A* 98(3), 1212–1217.
11. Osborn BL, Olsen HS, Nardelli B, Murray JH, Zhou JX, Garcia A, et al. (2002) Pharmacokinetic and pharmacodynamic studies of a human serum albumin-interferon-alpha fusion protein in cynomolgus monkeys. *J. Pharmacol. Exp. Ther.* 303(2), 540–548.
12. Perry CM, Jarvis B. (2001) Peginterferon-alpha-2a (40 kD): a review of its use in the management of chronic hepatitis C. *Drugs* 61(15), 2263–2288.
13. Glue P, Fang JW, Rouzier-Panis R, Raffanel C, Sabo R, Gupta SK, et al. (2000) Pegylated interferon-alpha2b: pharmacokinetics, pharmacodynamics, safety, and preliminary efficacy data. Hepatitis C Intervention Therapy Group. *Clin. Pharmacol. Ther.* 68(5), 556–567.
14. Foster GR. (2004) Review article: pegylated interferons: chemical and clinical differences. *Aliment. Pharmacol. Ther.* 20(8), 825–830.
15. Hoofnagle JH. (2009) A step forward in therapy for hepatitis C. *N. Engl. J. Med.* 360(18), 1899–1901.
16. Fried MW, Shiffman ML, Reddy KR, Smith C, Marinos G, Goncales FL, Jr., et al. (2002) Peginterferon alfa-2a plus ribavirin for chronic hepatitis C virus infection. *N. Engl. J. Med.* 347(13), 975–982.
17. Sharma SD. (2010) Hepatitis C virus: molecular biology and current therapeutic options. *Indian J. Med. Res.* 131, 17–34.
18. Bacon BR, Shiffman ML, Mendes F, Ghalib R, Hassanein T, Morelli G, et al. (2009) Retreating chronic hepatitis C with daily interferon alfacon-1/ribavirin after nonresponse to pegylated interferon/ribavirin: DIRECT results. *Hepatology* 49(6), 1838–1846.
19. Osborn BL, Sekut L, Corcoran M, Poortman C, Sturm B, Chen G, et al. (2002) Albutropin: a growth hormone-albumin fusion with improved pharmacokinetics and pharmacodynamics in rats and monkeys. *Eur. J. Pharmacol.* 456(1–3), 149–158.
20. Duttaroy A, Kanakaraj P, Osborn BL, Schneider H, Pickeral OK, Chen C. et al. (2005) Development of a long-acting insulin analog using albumin fusion technology. *Diabetes* 54(1), 251–258.
21. Wang W, Ou Y, Shi Y. (2004) AlbuBNP, a recombinant B-type natriuretic peptide and human serum albumin fusion hormone, as a long-term therapy of congestive heart failure. *Pharm. Res.* 21(11), 2105–2111.
22. van Dijkhuizen-Radersma R, Hesseling SC, Kaim PE, de Groot K, Bezemer JM. (2002) Biocompatibility and degradation of poly(ether-ester) microspheres: *in vitro* and *in vivo* evaluation. *Biomaterials* 23(24), 4719–4729.
23. De Leede LG, Humphries JE, Bechet AC, Van Hoogdalem EJ, Verrijk R, Spencer DG. (2008) Novel controlled-release Lemna-derived IFN-alpha2b (Locteron): pharmacokinetics, pharmacodynamics, and tolerability in a Phase I clinical trial. *J. Interferon Cytokine Res.* 28(2), 113–122.
24. Kasama K, Utsumi J, Matsuo-Ogawa E, Nagahata T, Kagawa Y, Yamazaki S, et al. (1995) Pharmacokinetics and biologic activities of human native and asialointerferon-beta s. *J. Interferon Cytokine Res.* 15(5), 407–415.
25. Eto T, Takahashi H. (1999) Enhanced inhibition of hepatitis B virus production by asialoglycoprotein receptor-directed interferon. *Nat. Med.* 5(5), 577–581.
26. Xi K, Tabata Y, Uno K, Yoshimoto M, Kishida T, Sokawa Y, et al. (1996) Liver targeting of interferon through pullulan conjugation. *Pharm. Res.* 13(12), 1846–1850.
27. Tabata Y, Matsui Y, Uno K, Sokawa Y, Ikada Y. (1999) Simple mixing of IFN with a polysaccharide having high liver affinity enables IFN to target to the liver. *J. Interferon Cytokine Res.* 19(3), 287–292.
28. Yuan L, Wang J, Shen WC. (2008) Lipidization of human interferon-alpha: a new approach toward improving the delivery of protein drugs. *J. Control. Release.* 129(1), 11–17.
29. Aurisicchio L, Bujard H, Hillen W, Cortese R, Ciliberto G, La Monica N, et al. (2001) Regulated and prolonged expression of mIFN(alpha) in immunocompetent mice mediated by a helper-dependent adenovirus vector. *Gene Ther.* 8(24), 1817–1825.
30. Berraondo P, Ochoa L, Crettaz J, Rotellar F, Vales A, Martinez-Anso E, et al. (2005) IFN-alpha gene therapy for woodchuck hepatitis with adeno-associated virus: differences in duration of gene expression and antiviral activity using intraportal or intramuscular routes. *Mol. Ther.* 12(1), 68–76.
31. Breous E, Somanathan S, Bell P, Wilson JM. (2011) Inflammation causes loss of adeno-associated virus-mediated transgene expression in mouse liver. *Gastroenterology.* 141(1), 348–57.
32. Forti N, Diament J. (2006) High-density lipoproteins: metabolic, clinical, epidemiological and therapeutic intervention aspects. An update for clinicians. *Arq. Bras. Cardiol.* 87(5), 671–679.
33. Mahley RW, Innerarity TL, Rall SC, Jr., Weisgraber KH. (1984) Plasma lipoproteins: apolipoprotein structure and function. *J. Lipid Res.* 25(12), 1277–1294.
34. Sirtori CR, Calabresi L, Franceschini G. (1999) Recombinant apolipoproteins for the treatment of vascular diseases. *Atherosclerosis* 142(1), 29–40.
35. Rothblat GH, Mahlberg FH, Johnson WJ, Phillips MC. (1992) Apolipoproteins, membrane cholesterol domains, and the regulation of cholesterol efflux. *J. Lipid Res.* 33(8), 1091–1097.
36. Banka CL, Bonnet DJ, Black AS, Smith RS, Curtiss LK. (1991) Localization of an apolipoprotein A-I epitope critical for activation of lecithin-cholesterol acyltransferase. *J. Biol. Chem.* 266(35), 23886–23892.

37. Krieger M. (1999) Charting the fate of the "good cholesterol": identification and characterization of the high-density lipoprotein receptor SR-BI. *Annu. Rev. Biochem.* 68, 523–558.
38. Acton S, Rigotti A, Landschulz KT, Xu S, Hobbs HH, Krieger M. (1996) Identification of scavenger receptor SR-BI as a high density lipoprotein receptor. *Science* 271(5248), 518–520.
39. Brewer HB, Jr., Fairwell T, LaRue A, Ronan R, Houser A, Bronzert TJ. (1978) The amino acid sequence of human APOA-I, an apolipoprotein isolated from high density lipoproteins. *Biochem. Biophys. Res. Commun.* 80(3), 623–630.
40. Zannis VI, Chroni A, Krieger M. (2006) Role of apoA-I, ABCA1, LCAT, and SR-BI in the biogenesis of HDL. *J. Mol. Med.* 84(4), 276–294.
41. Kim SI, Shin D, Choi TH, Lee JC, Cheon GJ, Kim KY, et al. (2007) Systemic and specific delivery of small interfering RNAs to the liver mediated by apolipoprotein A-I. *Mol. Ther.* 15(6), 1145–1152.
42. Feng M, Cai Q, Shi X, Huang H, Zhou P, Guo X. (2008) Recombinant high-density lipoprotein complex as a targeting system of nosiheptide to liver cells. *J. Drug Target* 16(6), 502–508.
43. Feng M, Cai Q, Huang H, Zhou P. (2008) Liver targeting and anti-HBV activity of reconstituted HDL-acyclovir palmitate complex. *Eur. J. Pharm. Biopharm.* 68(3), 688–693.
44. Counsell RE, Pohland RC. (1982) Lipoproteins as potential site-specific delivery systems for diagnostic and therapeutic agents. *J. Med. Chem.* 25(10), 1115–1120.
45. Balazs Z, Panzenboeck U, Hammer A, Sovic A, Quehenberger O, Malle E, et al. (2004) Uptake and transport of high-density lipoprotein (HDL) and HDL-associated alpha-tocopherol by an *in vitro* blood-brain barrier model. *J. Neurochem.* 89(4), 939–950.
46. Connelly MA, Williams DL. (2004) Scavenger receptor BI: a scavenger receptor with a mission to transport high density lipoprotein lipids. *Curr. Opin. Lipidol.* 15(3), 287–295.
47. Krieger M. (2001) Scavenger receptor class B type I is a multiligand HDL receptor that influences diverse physiologic systems. *J. Clin. Invest.* 108(6), 793–797.
48. Reboul E, Klein A, Bietrix F, Gleize B, Malezet-Desmoulins C, Schneider M, et al. (2006) Scavenger receptor class B type I (SR-BI) is involved in vitamin E transport across the enterocyte. *J. Biol. Chem.* 281(8), 4739–4745.
49. Vishnyakova TG, Bocharov AV, Baranova IN, Chen Z, Remaley AT, Csako G, et al. (2003) Binding and internalization of lipopolysaccharide by Cla-1, a human orthologue of rodent scavenger receptor B1. *J. Biol. Chem.* 278(25), 22771–22780.
50. Cao WM, Murao K, Imachi H, Yu X, Abe H, Yamauchi A, et al. (2004) A mutant high-density lipoprotein receptor inhibits proliferation of human breast cancer cells. *Cancer Res.* 64(4), 1515–1521.
51. Lacko AG, Nair M, Paranjape S, Johnso S, McConathy WJ. (2002) High density lipoprotein complexes as delivery vehicles for anticancer drugs. *Anticancer Res.* 22(4), 2045–2049.
52. Lacko AG, Nair M, Paranjape S, Mooberry L, McConathy WJ. (2006) Trojan horse meets magic bullet to spawn a novel, highly effective drug delivery model. *Chemotherapy* 52(4), 171–173.
53. Lou B, Liao XL, Wu MP, Cheng PF, Yin CY, Fei Z. (2005) High-density lipoprotein as a potential carrier for delivery of a lipophilic antitumoral drug into hepatoma cells. *World J. Gastroenterol.* 11(7), 954–959.
54. McConathy WJ, Nair MP, Paranjape S, Mooberry L, Lacko AG. (2008) Evaluation of synthetic/reconstituted high-density lipoproteins as delivery vehicles for paclitaxel. *Anticancer Drugs* 19(2), 183–188.
55. Zhang X, Chen B. Recombinant high density lipoprotein reconstituted with apolipoprotein AI cysteine mutants as delivery vehicles for 10-hydroxycamptothecin. *Cancer Lett.* 298(1), 26–33.
56. Hamidi M, Foroozesh M, Zarrin A. (2006) Lipoproteins: from physiological roles to drug delivery potentials. *Crit. Rev. Ther. Drug Carrier Syst.* 23(6), 497–523.
57. Fidge NH. (1999) High density lipoprotein receptors, binding proteins, and ligands. *J. Lipid Res.* 40(2), 187–201.
58. Jonas A. (1986) Reconstitution of high-density lipoproteins. *Methods Enzymol.* 128, 553–582.
59. Wald JH, Krul ES, Jonas A. (1990) Structure of apolipoprotein A-I in three homogeneous, reconstituted high density lipoprotein particles. *J. Biol. Chem.* 265(32), 20037–20043.
60. Fioravanti J, Gonzalez I, Medina-Echeverz J, Larrea E, Ardaiz N, Gonzalez-Aseguinolaza G, et al. (2011) Anchoring interferon alpha to apolipoprotein A-I reduces hematological toxicity while enhancing immunostimulatory properties. *Hepatology* 53(6), 1864–1873.
61. Hovorka R, Nanjee MN, Cooke CJ, Miller IP, Olszewski WL, Miller NE. (2006) Mass kinetics of apolipoprotein A-I in interstitial fluid after administration of intravenous apolipoprotein A-I/lecithin discs in humans. *J. Lipid. Res.* 47(5), 975–981.
62. Nanjee MN, Crouse JR, King JM, Hovorka R, Rees SE, Carson ER, et al. (1996) Effects of intravenous infusion of lipid-free apo A-I in humans. *Arterioscler. Thromb. Vasc. Biol.* 16(9), 1203–1214.
63. Wang YS, Youngster S, Grace M, Bausch J, Bordens R, Wyss DF. (2002) Structural and biological characterization of pegylated recombinant interferon alpha-2b and its therapeutic implications. *Adv. Drug. Deliv. Rev.* 54(4), 547–570.
64. Pepinsky RB, LePage DJ, Gill A, Chakraborty A, Vaidyanathan S, Green M, et al. (2001) Improved pharmacokinetic properties of a polyethylene glycol-modified form of interferon-beta-1a with preserved *in vitro* bioactivity. *J. Pharmacol. Exp. Ther.* 297(3), 1059–1066.
65. Tape C, Kisilevsky R. (1990) Apolipoprotein A-I and apolipoprotein SAA half-lives during acute inflammation and amyloidogenesis. *Biochim. Biophys. Acta* 1043(3), 295–300.
66. Subramanian GM, Fiscella M, Lamouse-Smith A, Zeuzem S, McHutchison JG. (2007) Albinterferon alpha-2b: a genetic fusion protein for the treatment of chronic hepatitis C. *Nat. Biotechnol.* 25(12), 1411–1419.

67. Kaser A, Nagata S, Tilg H. (1999) Interferon alpha augments activation-induced T cell death by upregulation of Fas (CD95/APO-1) and Fas ligand expression. *Cytokine* 11(10), 736–743.

68. Dondi E, Roue G, Yuste VJ, Susin SA, Pellegrini S. (2004) A dual role of IFN-alpha in the balance between proliferation and death of human CD4+ T lymphocytes during primary response. *J. Immunol.* 173(6), 3740–3747.

69. Radojkovic C, Genoux A, Pons V, Combes G, de Jonge H, Champagne E, et al. (2009) Stimulation of cell surface F1-ATPase activity by apolipoprotein A-I inhibits endothelial cell apoptosis and promotes proliferation. *Arterioscler. Thromb. Vasc. Biol.* 29(7), 1125–1130.

70. Acton SL, Scherer PE, Lodish HF, Krieger M. (1994) Expression cloning of SR-BI, a CD36-related class B scavenger receptor. *J. Biol. Chem.* 269(33), 21003–21009.

71. Schonfeld G, Pfleger B, Roy R. (1975) Structure of human high density lipoprotein reassembled *in vitro*. Radioimmunoassay studies. *J. Biol. Chem.* 250(19), 7934–7950.

72. Nissen SE, Tsunoda T, Tuzcu EM, Schoenhagen P, Cooper CJ, Yasin M, et al. (2003) Effect of recombinant ApoA-I Milano on coronary atherosclerosis in patients with acute coronary syndromes: a randomized controlled trial. *JAMA* 290(17), 2292–2300.

73. Eriksson M, Carlson LA, Miettinen TA, Angelin B. (1999) Stimulation of fecal steroid excretion after infusion of recombinant proapolipoprotein A-I. Potential reverse cholesterol transport in humans. *Circulation* 100(6), 594–598.

74. Nicholls SJ, Tuzcu EM, Sipahi I, Schoenhagen P, Crowe T, Kapadia S, et al. (2006) Relationship between atheroma regression and change in lumen size after infusion of apolipoprotein A-I Milano. *J. Am. Coll. Cardiol.* 47(5), 992–997.

75. Spieker LE, Sudano I, Hurlimann D, Lerch PG, Lang MG, Binggeli C, et al. (2002) High-density lipoprotein restores endothelial function in hypercholesterolemic men. *Circulation* 105(12), 1399–1402.

76. Tardif JC, Gregoire J, L'Allier PL, Ibrahim R, Lesperance J, Heinonen TM, et al. (2007) Effects of reconstituted high-density lipoprotein infusions on coronary atherosclerosis: a randomized controlled trial. *JAMA* 297(15), 1675–1682.

77. Nanjee MN, Cooke CJ, Garvin R, Semeria F, Lewis G, Olszewski WL, et al. (2001) Intravenous apoA-I/lecithin discs increase pre-beta-HDL concentration in tissue fluid and stimulate reverse cholesterol transport in humans. *J. Lipid Res.* 42(10), 1586–1593.

78. Bisoendial RJ, Hovingh GK, Levels JH, Lerch PG, Andresen I, Hayden MR, et al. (2003) Restoration of endothelial function by increasing high-density lipoprotein in subjects with isolated low high-density lipoprotein. *Circulation* 107(23), 2944–2948.

79. Patel S, Drew BG, Nakhla S, Duffy SJ, Murphy AJ, Barter PJ, et al. (2009) Reconstituted high-density lipoprotein increases plasma high-density lipoprotein anti-inflammatory properties and cholesterol efflux capacity in patients with type 2 diabetes. *J. Am. Coll. Cardiol.* 53(11), 962–971.

80. Shaw JA, Bobik A, Murphy A, Kanellakis P, Blombery P, Mukhamedova N, et al. (2008) Infusion of reconstituted high-density lipoprotein leads to acute changes in human atherosclerotic plaque. *Circ. Res.* 103(10), 1084–1091.

81. Panagotopulos SE, Witting SR, Horace EM, Nicholas Maiorano J, Sean Davidson W. (2002) Bacterial expression and characterization of mature apolipoprotein A-I. *Protein Expr. Purif.* 25(2), 353–361.

82. Ryan RO, Forte TM, Oda MN. (2003) Optimized bacterial expression of human apolipoprotein A-I. *Protein Expr. Purif.* 27(1), 98–103.

83. Feng MQ, Cai QS, Song DX, Dong JB, Zhou P. (2006) High yield and secretion of recombinant human apolipoprotein AI in *Pichia pastoris*. *Protein Expr. Purif.* 46(2), 337–342.

84. Pyle LE, Fidge NH, Barton PA, Luong A, Sviridov D. (1997) Production of mature human apolipoprotein A-I in a baculovirus-insect cell system: propeptide is not essential for intracellular processing but may assist rapid secretion. *Anal. Biochem.* 253(2), 253–258.

85. Schmidt HH, Genschel J, Haas R, Buttner C, Manns MP. (1997) Expression and purification of recombinant human apolipoprotein A-I in Chinese hamster ovary cells. *Protein Expr. Purif.* 10(2), 226–236.

86. Mallory JB, Kushner PJ, Protter AA, Cofer CL, Appleby VL, Lau K, et al. (1987) Expression and characterization of human apolipoprotein A-I in Chinese hamster ovary cells. *J. Biol. Chem.* 262(9), 4241–4247.

87. Chiaiese P, Minutolo M, Arciello A, Guglielmini F, Piccoli R, Filippone E. (2011) Expression of human apolipoprotein A-I in Nicotiana tabacum. *Biotechnol. Lett.* 33(1), 159–165.

88. Nykiforuk CL, Shen Y, Murray EW, Boothe JG, Busseuil D, Rheaume E, et al. (2011) Expression and recovery of biologically active recombinant Apolipoprotein AI(Milano) from transgenic safflower (*Carthamus tinctorius*) seeds. *Plant Biotechnol. J.* 9(2), 250–263.

89. Loignon M, Perret S, Kelly J, Boulais D, Cass B, Bisson L, et al. (2008) Stable high volumetric production of glycosylated human recombinant IFNalpha2b in HEK293 cells. *BMC Biotechnol.* 8, 65.

90. Salunkhe S, Soorapaneni S, Prasad KS, Raiker VA, Padmanabhan S. (2011) Strategies to maximize expression of rightly processed human interferon alpha2b in *Pichia pastoris*. *Protein Expr. Purif.* 71(2), 139–146.

91. Ohya K, Matsumura T, Ohashi K, Onuma M, Sugimoto C. (2001) Expression of two subtypes of human IFN-alpha in transgenic potato plants. *J. Interferon Cytokine Res.* 21(8), 595–602.

92. Gresser I, Bourali C, Levy JP, Fontaine-Brouty-Boye D, Thomas MT. (1969) Increased survival in mice inoculated with tumor cells and treated with interferon preparations. *Proc. Natl. Acad. Sci. U S A* 63(1), 51–57.

93. Zeuzem S, Andreone P, Pol S, Lawitz E, Diago M, Roberts S, et al. (2011) Telaprevir for retreatment of HCV infection. *N. Engl. J. Med.* 364(25), 2417–2428.

94. Jacobson IM, McHutchison JG, Dusheiko G, Di Bisceglie AM, Reddy KR, Bzowej NH, et al. (2011) Telaprevir for

previously untreated chronic hepatitis C virus infection. *N. Engl. J. Med.* 364(25), 2405–2416.

95. McHutchison JG, Manns MP, Muir AJ, Terrault NA, Jacobson IM, Afdhal NH, et al. (2011) Telaprevir for previously treated chronic HCV infection. *N. Engl. J. Med.* 362(14), 1292–1303.
96. Sangfelt O, Erickson S, Castro J, Heiden T, Einhorn S, Grander D. (1997) Induction of apoptosis and inhibition of cell growth are independent responses to interferon-alpha in hematopoietic cell lines. *Cell Growth Differ.* 8(3), 343–352.
97. Thyrell L, Erickson S, Zhivotovsky B, Pokrovskaja K, Sangfelt O, Castro J, et al. (2002) Mechanisms of Interferon-alpha induced apoptosis in malignant cells. *Oncogene* 21(8), 1251–1262.
98. Sidky YA, Borden EC. (1987) Inhibition of angiogenesis by interferons: effects on tumor- and lymphocyte-induced vascular responses. *Cancer Res.* 47(19), 5155–5161.
99. Indraccolo S. (2010) Interferon-alpha as angiogenesis inhibitor: learning from tumor models. *Autoimmunity* 43(3), 244–247.
100. Swann JB, Hayakawa Y, Zerafa N, Sheehan KC, Scott B, Schreiber RD, et al. (2007) Type I IFN contributes to NK cell homeostasis, activation, and antitumor function. *J. Immunol.* 178(12), 7540–7549.
101. Curtsinger JM, Valenzuela JO, Agarwal P, Lins D, Mescher MF. (2005) Type I IFNs provide a third signal to CD8 T cells to stimulate clonal expansion and differentiation. *J. Immunol.* 174(8), 4465–4469.
102. Kirkwood JM, Bender C, Agarwala S, Tarhini A, Shipe-Spotloe J, Smelko B, et al. (2002) Mechanisms and management of toxicities associated with high-dose interferon alfa-2b therapy. *J. Clin. Oncol.* 20(17), 3703–3718.
103. Jonas A, von Eckardstein A, Kezdy KE, Steinmetz A, Assmann G. (1991) Structural and functional properties of reconstituted high density lipoprotein discs prepared with six apolipoprotein A-I variants. *J. Lipid Res.* 32(1), 97–106.
104. Bruckert E, von Eckardstein A, Funke H, Beucler I, Wiebusch H, Turpin G, et al. (1997) The replacement of arginine by cysteine at residue 151 in apolipoprotein A-I produces a phenotype similar to that of apolipoprotein A-IMilano. *Atherosclerosis* 128(1), 121–128.
105. Franceschini G, Sirtori CR, Capurso A, 2nd, Weisgraber KH, Mahley RW. (1980) A-IMilano apoprotein. Decreased high density lipoprotein cholesterol levels with significant lipoprotein modifications and without clinical atherosclerosis in an Italian family. *J. Clin. Invest.* 66(5), 892–900.

PART III

BEYOND THE TRIPLE T-PARADIGM

IIIA NOVEL CONCEPTS, NOVEL SCAFFOLDS

30

SIGNAL CONVERTER PROTEINS

MARK L. TYKOCINSKI

Department of Pathology, Anatomy and Cell Biology, Jefferson Medical College, Thomas Jefferson University, Philadelphia, PA, USA

30.1 INTRODUCTION

Fusion proteins that feature an ability to convert cell surface signals constitute a new protein therapeutic frontier. Such *signal converter proteins*, or SCP, modulate signals that are exchanged between cell surface proteins. A diverse set of multifunction signal converters have emerged in recent years, showcasing a spectrum of signal-converting functionalities. Uncannily, each new SCP seems to surprise, whether by its unexpectedly high efficacy, multiplicity of cellular targets, or mechanisms of action not anticipated from the known activities of the chimeric protein's component parts. The functional tapestry for signal converters is enriched by the very nature of their *in vivo* operative settings, since they effect their signal modulation within complex and dynamic cellular networks. Pinning diverse and inherently multifunctional SCP onto multicellular frameworks generates richness in matrices of cellular interaction and therapeutic outputs.

Over the past decade, the signal converter story has steadily evolved, with interesting twists and turns along the way. The catalog of SCP types has expanded, with the original straightforward paradigm—conversion of contact-dependent signaling between two interacting cells—giving way to more intricate paradigms. Redirection of signaling is one such newer paradigm, and it itself has spawned its own variants. Converted signals can be redirected not only to third-party cells, but also to receptors on the very same cell in an auto-signaling, loop-back mode. Remarkably, the same SCP can sometimes redirect signals in more than one direction, manifesting both intercellular- and auto-signaling capacities. Moreover, the array of cell types and diseases that can be targeted by SCP has continuously expanded. Early studies geared toward immune inhibition for autoimmune and inflammatory diseases set the stage for newer SCP designed for other purposes. These include immune activation in the context of immunization, and even entirely different non-immunological ends, for example, cancer cell eradication. This chapter details some of the evolving signal converter paradigms and their associated SCP categories, putting them within an historical context and mapping a future landscape with ever expanding SCP possibilities.

30.2 HISTORICAL ROOTS OF SIGNAL CONVERSION: ARTIFICIAL VETO CELL ENGINEERING AND PROTEIN PAINTING

The experimental path that led to SCP has its deeper roots in a seminal discovery made a number of years ago in the field

Fusion Protein Technologies for Biopharmaceuticals: Applications and Challenges, First Edition. Edited by Stefan R. Schmidt.

of immunoregulation. The pivotal insight, that immunoregulatory cells can be generated through gene or protein transfer, prompted the subsequent development of fusion proteins that can be used to accomplish this cellular engineering feat. SCP emerged as a preferred fusion protein option.

According to one of cellular immunology's central dogmas, antigen-presenting cells (APC), such as dendritic cells (DC), drive T-cell activation through contact-dependent intercellular signaling. Surface-associated major histocompatibility (MHC)·peptide antigen complexes and costimulators provide the classic APC-anchored first and second T-cell-activating signals, respectively. While the focus has traditionally been on the APC's role as a T-cell activator, early immunology literature also hinted at another APC functional dimension, casting it as a T-cell inhibitor.

30.2.1 Artificial Veto Cell Engineering

The view of an APC as a T-cell inhibitor arose from a provocative observation decades ago in transplantation immunology. Specifically, intravenous injection of allogeneic cells into an immunologically incompetent recipient induces suppression, rather than sensitization, of alloimmune responses. This finding formed the basis of two-step pretransplant transfusion strategies that entailed first reducing the pool of immunocompetent T cells, through irradiation and drugs, and thereafter inducing nonresponsiveness to specific alloantigens by exposing the regenerating T-cell repertoire to alloantigenic cells. Once alloantigen-presenting lymphoid cells were identified as the central drivers of this suppressive effect, the notion of the *veto cell* emerged—a cell inhibiting a second cell that recognizes it [1,2]. Cells capable of suppressing, or "vetoing," antigen-specific responders have been referred to by a variety of other names as well, "deletional APC" and "inhibitory APC" among them. Functionally, the veto effect has been invoked beyond the pretransplant transfusion phenomenon, to explain a variety of other immunological phenomena, such as self-tolerance and resistance to graft-versus-host disease.

While the veto cell, in its original incarnation, was a $CD8^+$ lymphoid cell that inhibits a class I alloreative cytotoxic T-cell precursor, this narrow definition gave way to a broader view of this cell category. The repertoire of cells capable of mediating veto grew, with the list encompassing a mélange of cell categories; some poorly defined, for example, lymphokine-activated killer cells, activated $CD4^+$ T cells, a subset of $CD8^+$ murine dendritic cells, and even a nonlymphoid cell type, thymic epithelial cells. Furthermore, approaches were developed for expanding veto cells *in vitro*, for example, by cloning T cells with lymphokines, or incubating naive lymph node cells with relatively high concentrations of IL-2. These *in vitro* stimulatory methods were directed toward eliciting and expanding naturally occurring cells with veto capacity. This beckoned the next critical advance—artificially engineering veto cells, using gene or protein transfer to introduce into APC additional molecules that trigger the inhibition or destruction of target T cells, in effect creating immunoregulatory APC Trojan horses that we named *artificial veto cells,* or AVC [3,4].

Given the centrality of contact-dependent signaling at the APC-T-cell nexus, leveraging this type of juxtacrine signaling for purposes of AVC engineering was a logical way to go. A compelling paradigm emerged: the generation of AVC, with the capacity to inhibit T cells in an antigen-specific manner, by altering the trans (intercellular) signaling properties of APC via neoexpression of T-cell inhibitory proteins on their surfaces. Our group coined the term "coinhibitor" to characterize such surface-associated, trans-signaling inhibitory proteins, drawing a parallel between them and their well-established counterparts, the costimulators [3]. As defined, costimulators and coinhibitors alike mediate APC-to-T-cell contact-dependent signaling, and operate alongside and coordinately with MHC·peptide antigen complexes to modulate antigen-specific T cells. In one case, there is T-cell activation and in the other inhibition. Thus, while the "veto" term served to conceptually link this artificially engineered cell to the natural veto cell of an earlier immunology literature, the "coinhibitor" term, which has since entered the immunology mainstream, emphasized the mechanistic parallel with costimulators.

AVC engineering via neoexpression of surface coinhibitors has proven to be a robust concept. A variety of cell surface coinhibitors have emerged, but somewhat paradoxically, the lymphoid surface molecule CD8, one of the classical cis-acting T-cell co-receptors, was the first molecule to be invoked as a trans-signaling coinhibitor. Specifically, while CD8 is primarily thought of as a lymphoid co-receptor essential for cytotoxic T-cell activation, this same surface molecule, when artificially neoexpressed on APC, proved capable of signaling outward, and in doing so, inhibiting antigen-specific responder T cells. The proposed mechanism for the latter is inhibitory reverse signaling through T-cell-resident MHC class I, CD8's cognate counter-receptor. This ties into a longstanding observation that certain agonist anti-MHC class I monoclonal antibodies can inhibit the proliferation of TCR-triggered T cells [5]. Our original identification of CD8 as a coinhibitor, along with our broader formulation of the CD8-engineered AVC concept [4,6–9], has been amply substantiated by other investigators [10–12]. The artificial veto cell has even been periodically rediscovered, sometimes recast under a new name [13–16].

When it comes to AVC engineering, CD8 has given way to more conventional coinhibitors. Since an apoptotic mechanism has been proposed for the CD8-to-MHC class I inhibitory pathway, it was reasonable to turn to other signaling axes more classically linked to T-cell apoptosis.

One especially attractive candidate was the Fas ligand (FasL; CD95L; CD178):Fas (CD95; TNFRSF6) signaling axis, which plays a major role in shaping the repertoires of both T and B cells and maintaining peripheral tolerance [17]. Various threads in the literature have pointed to veto-like function for FasL when naturally expressed on certain cells [18], and we [19] and others [13,14,18,20–22] have attested to the veto function of APC with enforced FasL expression, for example, in the induction of alloantigen-specific T-cell tolerance and clonal deletion *in vivo*.

Gene transfer technology was pivotal at the earliest stages of the AVC story. We established the original link between CD8 and juxtacrine coinhibition back in 1989 through antisense RNA-mediated inhibition of CD8 in T cells. This first application of antisense RNA technology to T cells enabled the demonstration that veto inhibition by natural, alloantigen-presenting cloned $CD8^+$ T cells requires the presence of CD8 on their surfaces [7]. Having established that CD8 on cloned human T veto cells accounts for their inhibitory activity, we proceeded to demonstrate that certain allo-APC can be converted into AVC by neoexpressing CD8 on their surfaces via sense gene transfer [8,23].

30.2.2 Protein Painting

While gene transfer offered the simplest experimental approach for neoexpressing coinhibitors on APC surfaces, protein transfer soon emerged as a more interesting option. In principle, protein transfer offers a number of advantages over gene transfer, in terms of controlling protein levels on cell surfaces, co-expressing combinations of proteins, limiting exogenous protein half-lives on modified cells, and accommodating other therapeutic dictates. The first protein transfer modalities applied to APC engineering were geared toward coating cell surfaces *ex vivo*, which we referred to as *protein painting*. Over time, the palette of *fusion protein paints* was expanded.

The first fusion protein paints were genetically engineered, glycosyl-phosphatidylinositol (GPI)-modified derivatives of transmembrane proteins. Because GPI proteins are amphiphilic, they remain soluble upon purification, even after depletion of solubilizing detergents within cell membrane extracts. In turn, this solubility feature makes it possible to combine purified, detergent-depleted GPI proteins with cells without concern for cell lysis, and when combined in this way, the fusion proteins spontaneously incorporate into the cell membranes. Thus, GPI proteins are inherently protein paints. We [24] and others [25] originally identified the GPI modification signal sequence within decay-accelerating factor (DAF; CD55), a natural GPI-anchored protein, and proceeded to demonstrate that appending the DAF signal sequence to the carboxyl-terminus of any target protein yields a GPI-modified variant of that protein. Thus, using this chimerization approach, any protein of interest can be produced as a GPI-modified, protein paint variant. A series of subsequent studies validated the trans-signaling functionality of artificial GPI proteins, once appended to cell surfaces [26–31].

Although elegant in conception, GPI protein paints posed challenges when it came to scaling up protein production since they necessitated purification from complex membrane lysates. To bypass this inherent limitation, we looked for alternative protein paints that could be produced more expeditiously as soluble recombinant proteins. In this vein, we devised a simple two-component protein transfer method, which uses protein·$Fc_{\gamma 1}$ derivatives conjugated to membrane-incorporating, chemically palmitated protein A. Significantly, both components of these protein·$Fc_{\gamma 1}$:palmitated protein A conjugates can be produced efficiently as soluble recombinant proteins.

The experimental utility of GPI protein and protein·$Fc_{\gamma 1}$:palmitated protein A conjugate paints was established in a series of immunobiological studies. For instance, painting of peptide antigen-loaded MHC class I·GPI complexes sensitized cellular targets to cytotoxic T-cell-mediated destruction [27]. Protein transfer of such MHC:peptide antigen complexes allows for optimization of APC antigenic displays, and in particular, titration of antigenic density, which is significant given the importance of optimizing epitope density on APC surfaces when seeking to stimulate naive T cells [32].

Moving from first to second APC-to-T-cell signal, costimulator·GPI derivatives were shown to retain costimulator function despite GPI modification and could be used to enhance tumor cell immunogenicity [28,33,34]. The ability to finely titrate costimulator·$Fc_{\gamma 1}$:palmitated protein A conjugates onto APC surfaces permitted the first demonstration of costimulator receptor activation thresholds [35], which mirrored the T-cell receptor activation threshold phenomenon that had already been firmly established in the literature. In turn, the deciphering of quantitative and synergistic features of costimulation, leveraging protein painting as an experimental tool, set the stage for the use of costimulator paints for active cancer immunotherapy. Specifically, our group demonstrated that effective cellular cancer vaccines can be generated by painting of tumor cells *in situ* with a combination of four costimulator·$Fc_{\gamma 1}$:palmitated protein A conjugates [36].

This substantiation of costimulator paints was paralleled by analogous proof-of-principle validation of coinhibitor paints. For CD8 qua inhibitor, this included our CD8·GPI [8], as well as a bispecific antibody, with one arm holding soluble CD8 and the other latching on to MHC class I resident on the AVC membrane [12]. Our laboratory further documented the feasibility of generating AVC by painting cells with FasL (CD95L) [19]. In the process, we exploited protein painting to shed light on how responding T cells quantitatively integrate opposing (costimulatory and coinhibitory) trans signals concurrently delivered by APC. Specifically, costimulator (either B7-1·$Fc_{\gamma1}$ or $Fc_{\gamma1}$·4-1BBL) and pro-apoptotic ($Fc_{\gamma1}$·FasL) Fc fusion proteins were quantitatively painted, in varying ratios, onto surrogate APC precoated with palmitated protein A. By this approach, we demonstrated that at a given level of TCR triggering, the quantitative balance between costimulator (B7-1 or 4-1BBL) and FasL dictates the magnitude of the proliferative T-cell response. Furthermore, when the costimulator density is kept constant, there is also a quantitative balance between TCR-directed and FasL signals.

Thus, AVC engineering and coinhibitor painting nucleated a compelling concept—fusion proteins can be used to functionally engineer immunoregulatory cells. The protein paints invoked in these earlier studies required bringing the paint into direct contact with the cells to be painted, whether *ex vivo* or in confined *in vivo* compartments (such as tumor beds). This beckoned the development of next-generation fusion protein paints that could be injected systemically and would home to their cellular targets *in vivo*.

30.3 TRANS SIGNAL CONVERTER PROTEINS

A novel fusion protein class, designated as *trans signal converter proteins* (TSCP), was developed with *in vivo* engineering of AVC in mind. The driving concept was to combine within a single fusion protein homing and effector elements, together geared toward converting an intercellular signal between two cells. By judiciously choosing the homing element, one can elicit additional functional effects that reinforce the efficacy of the effector element. This was in evidence in our paradigmatic AVC-generating TSCP, CTLA-4·FasL [37]. The CTLA-4 (CD152) component within this fusion protein serves a dual purpose, targeting the protein to APC surfaces bearing counter-receptors for this homing domain [B7-1 (CD80) and B7-2 (CD86)], and at the same time, passively blocking their costimulator function. The FasL (CD95L) component of this fusion protein sends an active inhibitory, pro-apoptotic trans (intercellular) signal to its cognate Fas receptor (CD95), which is upregulated on activated T cells. Thus, this first TSCP, with a costimulator receptor·coinhibitor (CoSR·CoI) configuration, functions as a signal converter, as it exchanges an active inhibitory (apoptosis-inducing) signal for a T-cell-activating costimulator one that is being passively blocked.

Interestingly, there are other functional advantages that inhere in CTLA-4·FasL, arising from known functional properties of the component proteins and their associated signaling axes. B7 (CD80, CD86) proteins bind to CD28 and CTLA-4 (CD152), thereby transmitting stimulatory and inhibitory signals to T cells, respectively [38]. Increased B7-triggering of the CD28 costimulator receptor interferes with Fas-mediated apoptosis [39], and conversely, passive blockade of this costimulator pathway using soluble CTLA-4·Ig increases cellular susceptibility to Fas-mediated apoptosis [21,40]. As another CTLA-4 derivative, CTLA-4·FasL shares this capacity to potentiate Fas-induced apoptosis through costimulator interference, but with an added twist—in this instance, both the costimulator blockade function, and the FasL signaling function that is being potentiated, are uniquely integrated into a single fusion protein.

Other CTLA-4·FasL functional attributes stem from its association with cell surfaces. This chimeric protein anchors FasL to cell surfaces, thereby tapping into the higher efficacy of surface-associated FasL [41,42]. Yet, as a two-faced, Janus-like entity, CTLA-4·FasL is at the same time anchoring CTLA-4 to activated T-cell surfaces. This in-and-of-itself may have some interesting side benefits. CTLA-4 on T cells has been ascribed a cell-extrinsic, antigen-independent mechanism of action whereby it can "catch" and trans-endocytose its B7 (B7-1 and -2) ligands away from antigen-presenting cell surfaces, thereby preempting CD28 costimulation [43,44]. The CTLA-4 anchored to Fas^+ cells via CTLA-4·FasL could well do the same. Furthermore, the CTLA-4 end of CTLA-4·FasL has yet another potential, that is, to reverse signal through its cognate B7 ligand and thereby induce B7-bearing cells to produce suppressive indoleamine 2,3-dioxygenase (IDO) [45].

In essence, CTLA-4·FasL, as well as other CoSR·CoI TSCP, can be thought of as *in vivo superpaints*, designed not just for coinhibitor painting *per se*, but also for their added complementary functions. An ability to generate AVC *in vivo*, their core function, is piggy-backed onto their inherent costimulator blocking capacity. Costimulator blockade has been explored in great depth, and has made it into the clinic. The B7-blocker CTLA-4·Ig has received the most attention, with its well-documented capacities to induce anergic hyporesponsiveness and prolong allograft survival [46,47]. However, CTLA-4·Ig has its limitations, including inability to reproducibly establish a permanent state of peripheral tolerance *in vivo* [48] and to prevent graft rejection in various transplantation settings [49]. By reaching beyond passive blockade and melding it with active inhibition, CTLA-4·FasL provides a route for overcoming CTLA-4·Ig's limitations. This is especially important where there are multiple costimulator pathways in play.

CTLA-4·FasL exhibits striking potency *in vitro*, when compared to CTLA-4·Ig or soluble FasL, alone or in combination [37,50–52]. Interestingly, CTLA-4·FasL not only deletes activated T cells, but also induces anergic proliferative hyporesponsiveness, as evidenced by its ability to inhibit secondary responses in mixed lymphocyte reactions [50]. This dual effect fits with other evidence in the literature pointing to both Fas-mediated deletion [15,20,21,53] and anergy [54]. CTLA-4·FasL-mediated inhibition of allogeneic responses was documented in the *in vivo* context as well, with evidence for antigen-induced cell death of alloreactive cells [51]. The demonstration of CTLA-4·FasL's *in vivo* efficacy has also been extended to animal models of autoimmune and alloimmune pathogenesis, including autoimmune diabetes [55] and cardiac and corneal transplantation [56,57].

30.4 EXPANDING TRANS SIGNAL CONVERSION OPTIONS: REDIRECTING SIGNALS

As defined, TSCP convert intercellular signals between two interacting cells. In looking to new functional SCP categories, we envisioned one that would bring third-party cells into the fold. Along these lines, we conceived of *trans signal redirecting proteins* (TSRP), which would redirect converted signals to third-party cells, outside of the original cellular dyad. While for TSCP, the cell being modulated bears receptors for both the blocked and converted signals, for TSRP, these two receptors reside on different cells. In effect, one is using a fusion protein to elicit "bystander veto." This phenomenon is exemplified by the composite grafting of allogeneic islets of Langerhans with syngeneic myoblasts expressing FasL (CD95L) to induce apoptosis in infiltrating mononuclear cells and protect the islet graft from immune rejection [58].

We designed Fn14·TRAIL as a paradigmatic immune-inhibitory TSRP. This fusion protein taps into two well-studied TNF superfamily signaling axes, TWEAK:Fn14 (CD266) and TRAIL (CD253):TRAIL-R (TRAIL-R1, CD261; TRAIL-R2, CD262; TRAIL-R3, CD263; TRAIL-R4, CD264), which are linked to inflammation and immunoregulation, respectively [59,60]. TWEAK (TNF-related weak inducer of apoptosis), a type II transmembrane protein, is expressed on a range of immune and nonimmune cell types [61], and is upregulated on activated monocyte/macrophages. TWEAK's counter-receptor, Fn14 (fibroblast growth factor-inducible 14 kDa protein), is also widely expressed on many cell types, with the notable exceptions of T and B cells. Fn14 on other immune cells (monocyte/macrophages, NK cells, DC) is upregulated by interferon (IFN)-γ, and though generally low in normal tissues, is overexpressed in the context of tissue injury.

Functionally, the TWEAK/Fn14 axis contributes to tissue repair and remodeling [62]. TWEAK induces several proinflammatory cytokines and adhesion molecules, yet suppresses production of cytokines (IFN-γ, IL-12) linked to the transition from innate to adaptive Th1 immunity. TWEAK promotes the proliferation of some cell types (astrocytes, endothelial cells, and certain human tumor cell lines) [59]. However, TWEAK:Fn14 signaling effects go beyond cytokine production and cell proliferation *per se*; encompassing a broader set of functions that contribute to autoimmune inflammatory disorders. TWEAK increases the permeability of the neurovascular unit [63,64], and its endogenous expression in the central nervous system is elevated during autoimmune encephalomyelitis and acute cerebral ischemia. Moreover, TWEAK has proangiogenic activity [65], which is of interest given the association between angiogenesis and autoimmune pathogenesis [66]. TWEAK exacerbates neurodegeneration and disease severity in experimental autoimmune encephalitis (EAE), a murine model for multiple sclerosis [67,68]. The induction of inhibitory anti-TWEAK or anti-Fn14 antibodies, via vaccination with the extracellular domain of either TWEAK or Fn14, ameliorates autoimmune encephalomyelitis manifestations in rodent models [69].

TRAIL (TNF-related apoptosis-inducing ligand; CD253) binds to a number of different cognate receptors of the TNF receptor superfamily, some activating and others decoy [70,71]. The activating receptors in humans are TRAIL-R1 (CD261), TRAIL-R2 (CD262), and osteoprotegrin, and in mice the sole activating receptor is DR5. Virtually all cells of the immune system (T/B/NK cells, DC, monocytes, granulocytes) upregulate surface TRAIL in response to IFN and other activation signals. TRAIL inhibits autoimmune manifestations in animal models of EAE [72–75] and autoimmune diabetes [76,77]. Evidence for TRAIL's capacity to inhibit EAE has come from experiments invoking TRAIL$^{-/-}$ knockout mice, soluble TRAIL receptor or neutralizing anti-TRAIL antibodies capable of blocking TRAIL function, and embryonic stem cell-derived dendritic cells co-expressing TRAIL and auto-antigen [73–75,78]. Interestingly, in multiple sclerosis patients, soluble TRAIL has emerged as a response marker for IFN-β therapy [79], with those most likely to respond to treatment showing early and sustained soluble TRAIL induction after therapy.

On the basis of this constellation of activities associated with the TWEAK:Fn14 and TRAIL:TRAIL-R signaling axes, we chose to bridge them within a chimeric Fn14·TRAIL protein [80]. The Fn14 component of Fn14·TRAIL binds to surface TWEAK at inflammatory sites, while the TRAIL ligand component triggers the inhibitory, pro-apoptotic TRAIL-R receptors on third-party (Fn14-negative) activated T cells in these locations. Fn14·TRAIL has a number of interesting features as follows:

- As a blocker of TWEAK, this fusion protein interferes with a variety of TWEAK proinflammatory activities, including triggering of cells of the innate immune system, blocking angiogenesis, and preventing TWEAK-mediated opening of the blood–brain barrier (in the context of autoimmune pathogenesis in the central nervous system). Presumably, key targets are TWEAK$^+$ professional APC (macrophages, DC) and other cells capable of antigen-presenting function (such as endothelial cells).

- As a signal converter, Fn14·TRAIL replaces a proinflammatory signal with one that knocks out bystander activated T cells.
- In bridging the TWEAK:Fn14 and TRAIL:TRAIL-R pathways, this chimeric protein interferes with both innate and adaptive immunity.
- Fn14·TRAIL tethers TRAIL to the cell surface, with surface-associated TRAIL known to be more functional than soluble TRAIL [81].

Thus Fn14·TRAIL, by virtue of the highly pleiotropic functions attributed to its component parts, has the potential to mediate a complex array of activities, which collectively provide a beneficial effect in the setting of autoimmunity. This was corroborated with our demonstration that Fn14·TRAIL prevents auto-antigen-induced EAE in mice, with an efficacy that is substantially greater than that for either of its component parts (as Fn14·Ig and soluble TRAIL derivatives) when used in isolation or in combination [80]. This study further suggested that Fn14·TRAIL can reverse the enhanced permeability of the blood–brain barrier in diseased animals, providing yet another explanation for the high therapeutic efficacy of this fusion protein in the EAE model.

30.5 FROM TRANS TO CIS SIGNAL CONVERSION: DRIVING AUTO-SIGNALING

For TSCP and TSRP, the receptor for the converted signal resides on second- and third-party cells, respectively. But what if the receptor for the converted signal is co-localized on the same cell that bears the surface signaling ligand that is being converted? This kernel of a thought seeded another category of SCP, designed to redirect converted signals back to the same cell. Such *cis loop-back proteins* (CLBP) generate auto-signaling loops, and by judiciously choosing the components of a given CLBP, one can take advantage of forward- and reverse-signaling features of the fusion protein's surface molecular targets to reinforce the efficacy of the converted signal. In this way, CLBP can be configured to double as both signal converting and dual co-signaling agents.

The concept of autocrine signaling has traditionally been framed in terms of a cell secreting signal molecules that can bind its own surface receptors. This type of signaling is presumably most effective when it occurs in the vicinity of neighboring cells of the same type since under these circumstances there are higher local concentration of the secreted signal. This engenders a community effect that encourages groups of identical cells to make the same developmental decisions and respond coordinately as a group. However, in the special case where autocrine signaling is driven by surface-associated ligands, the demand for cellular congregation is mitigated, as each cell carries its own triggers in immediate proximity to their cognate receptors. CLBP can thus be viewed as fusion proteins that serve to anchor ligands in proximity to their receptors and thereby foster cis auto-signaling, and in effect, community independence. There was a compelling logic to turn to FasL (CD95L, CD178) in the design of a paradigmatic CLBP. Within hours after their antigen receptors are engaged, T cells upregulate their expression of Fas (CD95) receptors, which acquire apoptosis-inducing functionality upon prolonged cell activation [82]. Importantly, antigen receptor engagement is also accompanied by transient neoexpression of FasL on activated T cells [83]. Thus, with concurrent expression of Fas and FasL by the same and other activated T cells during an immune response [84–86], a T cell might have only 3 or 4 days to perform its effector function before being targeted for destruction by suicide or fratricide/sororicide [85,87–89]. The appreciation of surface FasL as both an autocrine- and paracrine-acting agent instantly made it an interesting candidate for CLBP design.

Our first paradigmatic CLBP, CD40·FasL, was geared toward promoting autoinhibition of activated effector T cells [90]. The CD40 component binds to CD40L (CD154) neo-expressed on activated T cells, while the FasL component triggers the pro-apoptotic Fas receptor on these cells. The logic for pairing these particular proteins was several – fold:

- Selective targeting of activated T cells is made possible by choosing CLBP components that engage surface counter-receptors (CD40L and Fas) that are neoexpressed, and acquire full functionality, following T-cell activation [82].
- CD40·FasL can convert the outbound immune-activating signal emanating from CD40L into a pro-apoptotic FasL signal redirected back to the same cell.
- The two signaling pathways are functionally connected, in that reverse signaling through CD40L sensitizes activated T cells to Fas-mediated apoptosis [91]. Thus, CD40·FasL can function as a dual-triggering agent, by generating a CD40L reverse signal that potentiates the Fas-mediated pro-apoptotic signal and renders the cells more susceptible to activation-induced apoptosis.
- CD40·FasL tethers FasL to cell surfaces, thereby invoking surface-associated FasL's higher efficacy, as compared to soluble FasL [41,42].

This collection of mechanistic features exemplify the kind of elegant design that is possible when selecting SCP components, as one draws upon interlacing and reinforcing functionalities in the pairing process.

The functionality of CD40·FasL was documented experimentally, with strong evidence that a cis autoinhibitory "loop-back" mechanism of action is indeed operative [90]. The latter included the demonstration, through both soft agar clone formation and cell dilution analysis that CD40·FasL inhibition does not require intercellular contact. Moreover, the notion that cis loop-back signaling can be artificially elicited using fusion proteins derives support from another line of investigation. Our group has demonstrated that when costimulators are painted on to the surfaces of T cells, in the form of CoS·$Fc_{\gamma 1}$:palmitated protein A conjugates, they can trigger their cognate costimulator receptors on the same cells, thereby yielding auto-activating T cells [92]. Parenthetically, this finding also establishes that it is not just autoinhibition that can be artificially elicited with fusion proteins, but also auto-costimulation.

The feasibility of using fusion proteins to autotrigger receptors at cell surfaces raises an intriguing cis/trans *paradox*. Given that ligands and their cognate receptors are often naturally co-expressed on the same cell surfaces (LFA-3 and CD2 on T cells, as but one of many examples), why is there no constitutive receptor triggering, or alternatively, competitive blockade of incoming trans signals? And on the flip side, why do painted ligands, whether proteins paints or CLBP, elicit autotriggering so readily, while the natural ligand:receptor pairs seem to ignore each other? While autocrine signaling by soluble secreted ligands, such as growth factors, has been dealt with extensively in the literature, it is curious that this cis/trans paradox for membrane-associated ligands has been essentially ignored. As the catalog of CLBP grows, new insights into the determinants of, and constraints upon, autotriggering will likely be forthcoming.

30.6 MECHANISTIC DIVIDENDS OF CHIMERIZATION

While signal conversion is the initial design focus for each SCP, greater functional dimensionality is often revealed. Intriguingly, an SCP can function in ways that its component parts cannot replicate, alone or even in combination. The emergence of such chimerization-dependent properties has pointed to an important fusion protein principle—an SCP can be more than the sum of its parts.

CTLA-4·FasL is a case in point. As set forth earlier, CTLA-4·FasL, our paradigmatic TSCP, exhibited exceptionally high efficacy, when compared to its soluble CTLA-4 or FasL components, alone or in combination. This observation prompted deeper investigation into mechanisms of action unique to chimeric CTLA-4·FasL [52]. The absence of a CTLA-4·FasL effect on cell cycle entry and progression pointed away from the kind of classical anergy associated with CTLA-4·Ig. However, an intriguing mechanistic clue came from the further observation that CTLA-4·FasL elicits apoptosis as early as 24 h after T-cell activation, hinting that some kind of coordinate signaling might be capacitating the Fas receptor. The molecular underpinnings for this capacitation turned out to be the ability of CTLA-4·FasL, but not CTLA-4·Ig, agonistic anti-Fas antibody, or the two in combination, to abrogate the usual increase in expression of the anti-apoptotic protein, c-FLIP. This framed a model for CTLA-4·FasL action wherein there is coordinate triggering of a death receptor (Fas, CD95) and suppression of activation-driven induction of a pivotal anti-apoptotic protein (c-FLIP) that otherwise dampens the activity of that death receptor [93]. This coupled effect offers an explanation not just for CTLA-4·FasL's high efficacy, but also for its distinctive capacity to promote early T-cell apoptosis.

This CTLA-4·FasL example captures a recurring theme that has emerged from SCP studies to date. Chimerization provides functional dividends that cannot be recapitulated by simply delivering the fused components as a combination of separate molecular entities. New mechanisms and functions are emergent within fusion proteins. The growing list of chimerization dividends for SCP includes abilities to

- elicit new, sometimes unexpected, functionalities (as exemplified by CTLA-4·FasL's effect on c-FLIP)
- co-localize targeted molecules on cell membranes, sometimes generating macromolecular assemblies associated with enhanced function
- deliver ligands to preselected cell types and organ sites *in vivo*, thereby concentrating therapeutic agent at the right location (as exemplified by Fn14·TRAIL delivering TRAIL to inflamed TWEAK$^+$ endothelium)
- tether ligands to membranes, thereby eliciting higher order ligand function (as exemplified by the various FasL-containing SCP that yield membrane-anchored FasL)
- enhance functional avidity by enforcing multimerization (discussed later in this chapter).

The CTLA-4·FasL:c-FLIP connection represents a situation in which a unique SCP mechanism of action could not have been readily predicted based upon known functions of its components. In other instances, however, SCP mechanistic insights flow in a more straightforward manner from new information that becomes available on protein functional repertoires. CTLA-4·FasL is illustrative here again, in light of new insights into the CTLA-4 component's functional repertoire. For example, though largely studied as an inhibitory receptor on effector T cells, CTLA-4 has subsequently been linked to regulatory T cells. In particular, soluble CTLA-4 (in the form of CTLA-4·Ig) can indirectly drive the generation of CD4$^+$CD25^{hi} regulatory T cells, by reverse signaling through B7-2 on dendritic cells and thereby eliciting from them the potential to induce regulatory T cells [94]. This new functional dimension for one particular CTLA-4 derivative may well hold for others, including CTLA-4-containing SCP. If so, an SCP designed for deleting pathogenic T cells may prove to convey additional benefits in artificially promoting regulatory T-cell expansion, in this way further amplifying this chimeric protein's anti-inflammatory therapeutic potential.

The mechanistic picture for any given SCP is likely to evolve over time, not only as a consequence of new functionalities ascribed to its constituents, but also as a byproduct of a better understanding of structural and contextual determinants that dictate surface protein action and interaction. Seemingly well-delineated ligand-receptor pairs commonly give way to more complex protein interaction matrices, with ligands proving to have more than one receptor and vice versa. Many surface proteins exhibit functionally significant structural dynamism, with cell state-dependent binding and functional properties [95]. Different isoforms of a receptor can mediate opposing effects on a cell, even when engaging the same counter-receptor [96]. The interplay and interdependency between signaling systems creates an increasingly intricate web of interaction that extends the reach of any given SCP component signal and its outward functional ripple. Through these and other ways, the functional picture for each SCP invariably becomes more intricate.

30.7 TARGETING MULTIPLE DISEASES WITH INDIVIDUAL SIGNAL CONVERTERS

SCP mechanisms of action are diverse, and so are the disease applications. As each individual SCP is probed more fully, it seems to invariably reveal therapeutic potential beyond that which was originally planned in the original SCP design. Thus, SCP multifunctionality, as it unfolds, often serves to extend the therapeutic range of each to a variety of cellular targets and pathogenic processes.

Lymphoid-directed, FasL (CD95L)-containing SCP are a case example of this sort of therapeutic range extension. CTLA-4·FasL was first designed with autoimmunity and transplantation diseases in mind, and to this end, it was configured as a trans signal converter capable of substituting an APC-to-T-cell inhibitory signal (FasL/CD95L) in place of a costimulation signal (B7-1/CD80; B7-2/CD86). However, co-expression of the fusion protein's cognate counter-receptors, CD80/86 and CD95L, on hematopoietic malignancies of the B-cell lineage prompted an evaluation of its apoptosis-inducing potential against these transformed cell types. In this context, CTLA-4·FasL would be additionally functioning in a cis loop-back, pro-apoptotic autotriggering mode, that is, as a CLBP as opposed to a TSCP. CTLA-4·FasL's efficacy in this cancer setting was indeed substantiated by us [97]. Furthermore, the same held true for another FasL-containing SCP, CD40·FasL, which had also been developed at the outset as an inhibitor of activated autoimmune and alloimune T cells. In this case, it had been originally designed to function primarily via a cis loop-back mechanism that additionally invokes bidirectional signaling reinforcement. Its further therapeutic ability to eradicate transformed T cells expressing the cognate counter-receptors, CD40L (CD154) and Fas (CD95), draws mechanistically upon the same CLBP mode of activity [97]. Of note, in addition to eliciting transformed T-cell suicide, CD40·FasL can in parallel promote Fas-dependent T-T fratricide, in bridging Fas^+ target cells with neighboring $CD40L^+$ tumor cells, or even infiltrating, activated $CD40L^+$ T cells.

Fn14·TRAIL provides yet another example, in this case leading to a cancer therapy for solid tumors. Originally, this SCP was showcased as a TSRP, capable of interfering with a proinflammatory intercellular signal (TWEAK) and redirecting a converted inhibitory signal (TRAIL; CD253) that targets activated autoimmune T cells. However, neoexpression of TWEAK and its TRAIL counter-receptors on the surfaces of various solid tumor types, including those of breast, liver, pancreas, brain, and colon [98–102], has opened up another avenue for leveraging Fn14·TRAIL therapeutically. By deploying Fn14·TRAIL in a cancer context, one can draw upon its additional capacity to function as a CLBP to autotrigger apoptosis in molecularly defined tumor cells. In this way, cancer cells have come to share the limelight with activated T cells as prime targets for single multifaceted SCP, and in the process, a therapeutic agent for autoimmunity has morphed into an anticancer therapeutic for solid tumors.

Fn14·TRAIL's applicability to cancer therapy draws in yet other ways upon this SCP's mechanistic diversity. Various known TWEAK functions happen to be ones that are expected to promote tumor progression under certain circumstances. These functions relate to the promotion of proliferation, invasion, angiogenesis, and inflammation. As a TWEAK blocker, Fn14·TRAIL can thus exert anticancer effects in these pathogenic realms. From another perspective, Fn14·TRAIL can be viewed as a more potent soluble TRAIL derivative. Soluble TRAIL has been under active development as a cancer therapeutic capable of inducing tumor cell apoptosis [103], taking advantage of the fact that TRAIL receptors are expressed by, or are upregulated in, many cancer types. Fn14·TRAIL, by virtue of its cell-associating and multimerization characteristics, turns out to be far more effective than soluble TRAIL in promoting tumor cell apoptosis. In this context, it is mirroring monocyte/macrophages, which rapidly express TRAIL after activation and thereby acquire the ability to efficiently kill tumor cells [60,104].

Extrapolating from these examples, a new class of cancer therapeutics can be defined—fusion protein inducers of tumor cell auto-apoptosis. By drawing upon known inducers of tumor cell apoptosis as building blocks, including ones that are being actively developed for the clinic, design is streamlined. TRAIL-containing SCPs exemplify this nicely. In effect, this constitutes a new category of personalized onco-medicine, which looks to higher order tumor cell surface protein co-expression profiles: CTLA-4·FasL for $CD80/86^{+}CD95^{+}$ B-cell malignancies, CD40·FasL for $CD154^{+}CD95^{+}$ T-cell malignancies, Fn14·TRAIL for $TWEAK^{+}TRAIL\text{-}R^{+}$ solid tumors, and so on. Thus, kinase inhibitors and SCP can be thought of as complementary personalized onco-medicine therapeutics, attacking cancer cells from within and without, respectively.

30.8 STRUCTURAL CONSTRAINTS IN SCP DESIGN

Just as functional requirements have guided SCP design, so too have structural considerations. The type I (extracellular amino-terminus) versus type II (extracellular carboxyl-terminus) status of the parental membrane proteins constrains their placement within fusion proteins. When it came to GPI protein paints, the predecessors for SCP in the protein transfer developmental line, the focus was exclusively on type I membrane proteins, which were modified by appending a GPI modification signal sequence to their carboxyl termini. The Fc fusion protein paints that followed accommodated both type I and II proteins, albeit calling for alternative positioning of the Fc domain in each. That is, for type I and II Fc fusions, the Fc domain was positioned at the carboxyl or amino termini of the fusion proteins, respectively. This served to free up the business ends of the respective molecules and minimize chances that a covalently linked Fc appendage would interfere with their activity.

In transitioning to SCP, the most elegant approach was to generate type I·II fusions, an orientation that leaves both business ends of the chimeric protein free. In fact, all of the SCP set forth thus far have a type I·II structural configuration. By accommodating both type I and II proteins within SCP, there is the added advantage of encompassing at once members of the immunoglobulin (predominantly type I) and TNF (predominantly type II) superfamilies, which together dominate the cell membrane protein repertoire, within the immune system and beyond.

Interesting stoichiometries and fortuitous higher order multimer configurations arise when one couples trimeric TNF superfamily members to nontrimeric type I membrane proteins. This is exemplified by Fn14·TRAIL, where our three-dimensional model suggests that trimerization, enforced by trimeric TWEAK on the otherwise monomeric Fn14 component, yields a neo-Fn14 trimer, which can in turn stabilize the active, trimeric state of the fused TRAIL component [80]. In the case of CTLA-4·FasL, our modeling suggests that interplay of dimeric CTLA-4 with trimeric FasL allows for a hexameric configuration, representing at once a dimer of FasL trimers and a trimer of CTLA-4 dimers. This is functionally significant since hexamerization affords optimal functionality for FasL, and perhaps other TNF family members [41,105].

The TNF superfamily can be readily mined to generate other type I·II SCP of interest. However, this fusion formula unduly narrows the scope of potential SCP. In some instances, the desired SCP components happen to be both type II membrane proteins. To circumvent type II·II fusions in such cases, one can substitute derivatives from agonistic antibodies (by definition, type I) at the amino termini of the chimeric proteins, thereby allowing for a type I·II SCP. While such fusion proteins lack the simplicity of type I·II ligand·ligand SCP, they have proven functional in our hands. Production of antibody·TNF ligand fusions necessitates special steps to enrich for the active trimeric forms.

Where the SCP components are to come from two type I proteins, one can similarly consider antibody·ligand fusions. However, in this instance, type I·I ligand·ligand fusions may also be acceptable, given ample precedents for the functionality of such fusions, as evidenced by scFv proteins.

30.9 CODING SCP FUNCTIONAL REPERTOIRES

The multifunctionality of individual SCP invites new ways of codifying their respective functional repertoires. A *functional bar code* could be devised to concisely capture each SCP's documented functionalities. For instance, a coding schema might be structured around an "SCP | *fx*" format, wherein *fx* conveys the full set of known functional capabilities attributed to a particular SCP. A starting set of primary SCP *fx* designations might include

- t2c — converts trans signals between two cells
- tr — redirects converted trans signals to third-party cells
- t/bd — signals bidirectionally in trans
- c/ud — signals uni-directionally in cis
- c/bd — signals bidirectionally in cis
- neo/t — elicits neoexpression of a trans signaling ligand or receptor
- neo/c — elicits neoexpression of a cis signaling ligand or receptor

As an example, the coding schema for Fn14·TRAIL would be:

$$\text{Fn14} \cdot \text{TRAIL: SCP}|\text{tr}\{\text{tweak}^{+}\ \text{inflamm}\text{ø}\ \text{act Teff}\}+ \text{c/ud}\{\text{tumor} > \text{liver, pancreas, breast, brain, colon}\}$$

Decoded, this designation conveys that Fn14·TRAIL is an SCP that has two known functional modes: (1) a TSRP mode, in which it redirects trans signals from TWEAK-positive inflammatory cells to activated, effector T cells and (2) a CLBP mode, in which it directs loop-back signals, uni-directionally, on liver, pancreas, breast, brain, and colon cancer cells.

In similar manner, other paradigmatic SCP set forth earlier in this chapter would be characterized as follows:

CTLA-4 · FasL: SCP|t2c{apcø act Teff} + neo/c{dcø Treg} +c/ud{tumor > hemat}
CD40 · FasL: SCP|c/bd{act T eff; tumor > hemat}

For CTLA-4·FasL, this label conveys that it is an SCP with three well-defined functional modes: (1) a TSCP mode, in which it converts an intercellular signal exchanged between two cells, an APC and an activated effector T cell; (2) a neoexpression mode, in which it promotes neo-expression on DC surfaces of a regulatory T-cell stimulatory signal; and (3) a CLBP mode in which it directs loop-back signals, uni-directionally, on hematopoietic tumor cells. In the case of CD40·FasL, the designation specifies that there is in essence one well-characterized signaling mode—a bidirectional signaling CLBP mode—that plays out in two distinct cellular contexts, that is, activated, effector T cells and hematopoietic tumor cells.

Clearly, the functional bar code associated with a given SCP will evolve over time, often substantially, as new functional capabilities are uncovered. The inherent flexibility of such a coding system bypasses the reductionist trap when trying to pigeonhole individual SCP into monolithic functional categories. Since this nomenclature pegs functions to particular cellular contexts, it conveniently accommodates the functional heterogeneity that follows from the pleiotropic expression, across diverse cell types, of so many of the surface molecules targetable by SCP.

Embedded within this SCP code is directionality of signaling, which implicitly touches on a growing appreciation for Escher-like surface receptor-ligand functional duality. That is, many surface proteins initially labeled as "receptors" turn out to additionally act as signal-sending ligands, and reciprocally, many characterized at the outset as intercellular signaling ligands prove to be able to reverse signal in a signal-receiving mode. Among the continuously expanding compendium of molecular pairs associated with bidirectional signaling are CD40L (CD154):CD40 (TNFRSF5); OX40L (CD252; TNFSF4):OX40 (CD134); GITRL (TNFSF18):GITR (CD357; TNFRSF18), CD80/86:CTLA-4/CD28 and CD70:CD27 (TNFRSF7). Indeed, bidirectional signaling seems to be more a rule than an exception. The growing appreciation that "receiver" proteins can send and "sender" proteins can receive highlights the frequent arbitrariness of the "receptor" and "ligand" labels, which does the disservice of tunneling thinking when it comes to fusion protein design.

The term "moonlighting protein" has been applied to individual proteins displaying unrelated functional properties, which are of special interest when viewed from the systems biology vantage point [106]. While the discourse has largely centered on metabolic enzymes, along with signaling, structural, and nucleic acid binding proteins, the receptor-ligand duality of cell surface proteins is essentially just another form of moonlighting.

The fact that cell surface proteins moonlight, manifesting bidirectional signaling potential, has clear implications when predicting and deciphering SCP effects. It translates into an SCP's capacity, via either of their chimerized components, to at once serve as an intercellular signal blocker, a reverse-signaling ligand, and a competitive inhibitor of the receptor function. In turn, this array of functional potentialities, mediated by both ends of a single fusion protein, serves to amplify cellular reach and therapeutic efficacy.

This particular coding system does not purport to convey all the mechanisms of action associated with a given SCP, for example, CTLA-4-FasL's ability to prevent the upregulation of c-FLIP associated with T-cell activation. That is not to say that this or other functional coding systems could not be adapted to capture deeper mechanistic features that tie into signal converter functionalities and efficacy. Even for those SCP that have been studied most intensively to date, the mechanistic tree is almost certain to extend new branches. Functional richness of SCP will continue to emanate from both the broad tissue distribution and functional pleiotropism of their surface molecular targets.

30.10 EXPANDING THE CATALOG OF INHIBITORY SCP

The costimulator receptor·coinhibitor (CoSR·CoI) TSCP motif exemplified by CTLA-4·FasL, along with the more general inflammatory signal blocker·lymphoid inhibitor TSRP motif featured in Fn14·TRAIL, provide robust templates for generating additional T-cell-directed SCP inhibitors. In mixing and matching CoSR and CoI components, one can draw upon a diverse set of known costimulator receptors {for example, CD28, ICOS (CD278), OX40 (CD134), 4-1BB (CD137), GITR (TNFRSF18), and CD27; reviewed in References [107–110], and a more limited set of coinhibitor options [FasL (CD95L; CD178), TRAIL (CD253), PD-L1 (B7-H1; CD274), PDL2 (B7-DC; CD273), HVEM (TNFRSF14) [111–113]]. An expanded SCP palette, encompassing more chimeric CoSR·CoI permutations, will enable preferential targeting of selected T-cell subsets, corresponding to different lineage, activation, and differentiation states. While the cell surface molecular fingerprints of T-cell subsets are by no means discrete, they do offer some opportunity for configuring SCP with a certain degree of subset selectivity and leveraging differences in costimulator requirements and coinhibitor sensitivities. Furthermore, since the pivotal pathogenic T cells vary for different autoimmune diseases, and even different clinical phases of the same disease, a diverse collection of SCP, targeting a range of target cells, could ultimately provide the means for tailoring autoimmune therapies.

Another consideration is the respective affinities of fusion protein components for their respective counter-receptors. Binding affinities between costimulator receptors and their cognate ligands vary considerably, ranging over at least three log orders. While high affinity interaction should benefit *in vivo* homing, lower affinity interaction may offer a better therapeutic window for those costimulators that are promiscuously expressed, but are very highly expressed on pathogenic T cells. In the case of monoclonal antibodies, above a threshold affinity, other factors appear to dictate optimal *in vivo* targeting [114]. Moreover, for CoSR·CoI SCP, an "affinity ratio" (defined as the CosR:CoS affinity divided by the CoI:CoIR affinity) might govern *in vivo* localization and function.

Immune inhibitory SCP need not be confined to the CoSR·CoI motif. In place of the CoSR component, one might substitute elements that bind to other classes of surface molecules, for example, ligands (or antibody derivatives) that engage chemokine receptors (CR). There is substantial data available pegging chemokine receptor expression to homing and other functional properties of distinct lymphoid cell subsets [115], making them interesting subjects for CR·CoI SCP design.

Yet another inhibitory SCP class would bring together two coinhibitor elements, generating CoI·CoI fusions with the capacity to elicit dual-signaling autoinhibitory loops. The coinhibitor programmed cell death 1 ligand (PD-L1; B7-H1; CD274) [111,116] is of special interest for such fusions in light of its being a type I transmembrane protein. Elegant type I·II chimeric proteins can be developed by fusing PD-L1 to inhibitory TNF superfamily members, generating SCP such as PD-L1·FasL and PD-L1·TRAIL. These CoI·CoI SCP would target, in a cis loop-back mode, activated T cells co-expressing PD-L1's receptor, PD-1 [117], along with Fas or TRAIL-R receptors. Here again, the SCP would have other mechanistic possibilities. Interestingly, PD-L1 has significant affinity for B7-1 (CD80) [118]. This affords PD-L1-containing SCP the potential for trans signal conversion, immune modulation via reverse signaling through B7-1, and elicitation of immunosuppressive cytokine production by PD-1-triggered monocytes [119]. There might even be a fortuitous synergy between the trans signal conversion and cis loop-back mechanisms in this case, in light of the demonstration that costimulator blockade requires an intact PD-L/PD-1 signaling axis [120]. Moreover, the therapeutic reach of these SCP spans autoimmunity and cancer, in light of reported associations of PD-L1:PD-1 interaction with these disease entities [121–123]. Since the functional repertoire of PD-1's other ligand, PD-L2 (CD273), diverges from that of PD-L1, for example, with respect to T-cell subset orientation [124], substituting PD-L2 for PD-L1 within these CoI·CoI SCP would modify their functional profiles.

There are yet other ways to invoke SCP for the inhibition of immune responses. SCP that can amplify regulatory T-cell numbers and function would serve to indirectly inhibit effector T cells. To this end, one can select SCP components that are known to positively influence regulatory T-cell pathways. This points to yet another mechanistic angle for PDL1-containing SCP since PD-1 triggering not only inhibits effector T cells, but also stimulates regulatory T cells [125]. CTLA-4 (CD152) [126], ICOS (CD278) [127–129], and CD30 [130,131] also offer interesting options in this regard.

One can also broaden the inhibitory SCP landscape by reaching beyond lymphoid and cancer cell targets. Such SCP would conceptually fit into a larger therapeutic framework—the use of SCP for what might be termed *SCP-mediated cell deletion therapy*, where the goal is to selectively eliminate diverse pathogenic cells using death-inducing SCP. In the immunological realm, mast cells [132,133], natural killer T(NKT) cells [134–136], and natural killer (NK) cells of the female reproductive tract [137,138] are particularly interesting cellular targets, given the ever-growing spectrum of diseases linked to them. Beyond the immune system, there is also much room for creativity in the choice of pathogenic cell targets and associated SCP designs. Out-of-the-box thinking might even bring into the fold certain under-studied, high-prevalence conditions such as postsurgical adhesions. SCP cell deletion therapies could be devised to target the cells driving inappropriate fibrogenesis in these conditions. Other SCP-based cell deletion therapies might target virally infected cells, parenchymal cells driving inflammatory disorders, and many other types of pathogenic cells.

SCP could be invoked to indirectly sensitize target çells to other death-inducing agents. Along these lines, one intriguing approach would be to promote macrophage-mediated clearance of damaged target cells by interfering with macrophage-directed "don't eat me" signals, such as the one sent by surface CD47, a member of the immunoglobulin superfamily. Blocking anti-CD47 antibody enables macrophage-mediated phagocytosis of cancer cells, including acute myeloid leukemia cells [139] and bladder cancer cells [140]. A SIRPα·TRAIL SCP fusion, incorporating CD47's counter-receptor, SIRPα (CD172A), could serve to induce tumor cell apoptosis (via the TRAIL end), while simultaneously sensitizing them to macrophage-mediated apoptotic cell destruction and clearance (via the SIRPα end), in an auto-signaling mode. There is considerable reach for this SCP since CD47's role in modulating macrophage tumor surveillance likely extends to a range of tumor types [141]. Additional dividends for a SIRPα·TRAIL SCP would be promotion of tumor antigen representation by macrophage and pro-apoptotic signaling in trans to neighboring

TRAIL-sensitive cells. Furthermore, the discovery that thrombospondin-1 acts via CD47 to inhibit nitric oxide signaling throughout the vascular system [142] might even connect SIRPα-containing SCP to vascular physiology, and in particular, tumor vasculature. Going even further, complementary fusions could be contemplated that tap into the opposing "eat me" signaling system—surface calreticulin binding to CD91 (LDL receptor-related protein)—that sends a prophagocytic signal to macrophages and serves to activate them [143].

30.11 IMMUNE ACTIVATING SCP

Although the focus has been on inhibitory signaling, SCP can be readily developed for cellular activation. To transition from T-cell inhibition to activation, the most straightforward approach is to reformulate T-cell-directed TSCP from a CoSR·CoI to a coinhibitor receptor·costimulator (CoIR·CoS) configuration. To generate these reconfigured fusion proteins, one can draw upon the same menu of CoS:CoSR and CoI: CoIR molecular pairs. For instance, the inhibitory CoSR·CoI SCP CTLA-4·FasL can be refashioned into an activating CoIR·CoS SCP by again tapping into the FasL:Fas trans signaling pathway, this time incorporating Fas as a CoIR component.

In flipping from CoSR·CoI to CoIR·CoS, however, full mirror imaging is not always feasible. There may be binding asymmetries peculiar to given costimulator and coinhibitor pathways that preclude simplistic CoSR-to-CoS and CoI-to-CoIR reversals. Such asymmetries can arise, for example, when a single surface ligand has affinity for both a CoSR and a CoIR, a case in point being the two B7 proteins (CD80 and CD86), each of which has affinity for both the CoSR CD28 and the CoIR CTLA-4. Yet, even in such cases, the pathway can still often be invoked for SCP design by considering granular details pertaining to differential affinities, expression levels, dynamics of expression, and cellular distributions. The CD80/CD86 dyad represents just such a situation, where differences across these categories influence how interactions with the CD28 versus CTLA-4 receptors play out [144–146]. Even more complicated situations can generate asymmetries, such as when a single surface protein (such as HVEM, TNFRSF14) can function as both a CoI (as a DC surface protein binding to the inhibitory BTLA (CD272) and CD160 receptors on T cells) and a CoSR (as a T-cell surface protein receiving DC-derived LIGHT costimulator signals) [112,113]. These kinds of complexities demand significant experimental empiricism for validating net functional outputs.

While morphing CoSR·CoI into CoIR·CoS is the most direct approach for devising signal converters geared to T-cell activation, there are other immune-activating SCP options, some of which would function indirectly. SCP capable of downmodulating regulatory T cells is one of them. Since regulatory T cells put the brakes on effector T cells, the latter can be unleashed using SCP capable of downmodulating the generation and suppressive activity of regulatory T cells. Such regulatory T-cell-inhibiting SCP would be especially useful in the context of cancer immunotherapy, given that tumor-infiltrating regulatory T cells interfere with antitumor effector T-cell immunity within tumor beds [147–150].

In designing regulatory T-cell-directed SCP, two TNF superfamily ligand:receptor pairs stand out: GITRL:GITR and OX40L:OX40. Intriguingly, each of these signaling axes has been ascribed dual, complementary functional capacities, effector T-cell activation, and regulatory T-cell inhibition. That is, molecular agonists for GITR (glucocorticoid-induced TNF receptor; TNFRSF18) can coordinately costimulate and coinhibit effector and regulatory T cells, respectively [151–153]. The same holds true for OX40 receptor agonists [154,155]. Our group posited that an anti-GITR Ab·OX40L fusion protein, coupling an agonistic anti-GITR antibody derivative to OX40 ligand, might be especially effective in eliciting coordinate and synergistic effector T-cell-activating, regulatory T-cell-inhibiting effects. In addition to generating a bidirectional, cis loop-back auto-signaling loop between the GITR and OX40 receptors, dual-signaling anti-GITR Ab·OX40L also has the potential to function as a trans signal converter. That is, its binding to GITR on T cells would interfere with GITR's outward looking ability to trigger reverse signaling through GITR ligand on plasmacytoid dendritic cells, which otherwise elicits an IDO-dependent immune-inhibitory effect [156]. We have substantiated that anti-GITR Ab·OX40L does indeed have the ability to inhibit the generation and suppressive activity of $CD4^+CD25^+FoxP3^+$ regulatory T cells, while simultaneously costimulating effector T cells.

Another path to T-cell activation invokes fusion proteins with the capacity to promote the maturation and T-cell stimulatory capacity of DC. Here too, there are a variety of fusion protein options, exemplified by anti-CD40 Ab · OX40L, which couples an agonistic anti-CD40 antibody derivative to OX40 ligand. This chimeric protein features a number of potential mechanisms of action. Functioning as a trans bidirectional signaling agent, bridging DC and naïve T cells, it has the potential to promote DC maturation (via its agonistic anti-CD40 Ab end) and simultaneously costimulate effector T cells and inhibit regulatory T cells (via its OX40 ligand end). This chimeric protein also features trans signal conversion potential, in that it can competitively block CD40-driven, reverse signaling through CD40L-bearing monocytes, which otherwise induces a form of monocytic activation-induced cell death. Interestingly, anti-CD40 · Ab OX40L, like various other SCP that we originally configured with immune modulation in mind, has anticancer therapeutic possibilities as well. This follows from its inherent ability to bridge in trans $CD40^+$ tumor cells and tumor-infiltrating T cells, leading to tumor cell apoptosis, enhanced killing by tumor-directed cytotoxic T lymphocytes, and inhibition of local regulatory T cells. These functional properties are anticipated on the basis of findings that CD40 engagement on carcinomas and some B-cell malignancies results in direct antiproliferative and apoptotic effects, due to upregulation of Fas and TRAIL receptors [157]. As an aside, this also means that anti-CD40 Ab OX40L might also be expected to sensitize tumor cells to Fas- or TRAIL-containing pro-apoptotic SCP.

The constellation of signals amenable to conversion goes well beyond those that serve as activators, proliferators, differentiators, inhibitors, or inducers of cell death. Clearly, there is room for much creativity along these lines. Additional categories of signals that are targetable by SCP are, by example, addressins, adhesins, and cell surface-associated chemokines and their receptors.

30.12 EXPERIMENTAL TOOLS FOR SCREENING SCP CANDIDATES

Given the breadth of SCP possibilities, robust experimental approaches for vetting fusion protein candidates become essential. Protein production poses an experimental bottleneck. Since producing panels of candidate SCP proteins for functional screening is simply too cumbersome, gene transfer represents a preferable option. One experimental path, which we have deployed with success, starts with *in vivo* gene transfer as a gateway for achieving functional validation and garnering first-order efficacy information that enables early prioritization of SCP candidates.

By example, we have recently leveraged a robust transposon-based mammalian expression system for *in vivo* SCP evaluation [80]. This expression system enables systemic expression of the fusion protein candidates, with sustained serum levels over months.

As an alternative to transposon-based systemic expression of SCP, one can turn to local gene delivery for SCP functional screening. In this case, functional readouts that assess local SCP effects in a straightforward manner are essential. This is exemplified by our coupling of intradermal naked SCP-encoding DNA transfer with delayed type hypersensitivity suppression as a functional readout for local T-cell immunomodulation. This system has allowed us to screen the immune inhibitory potential of several new CosR·CoI TSCP candidates. This gene transfer-based, rapid functional screening approach pointed to OX40·TRAIL as a particularly interesting SCP candidate for memory T-cell inhibition.

Intradermal injection is but one of multiple local gene delivery options. Another is intrathecal gene delivery. We have developed the capability of administering SCP expression constructs intrathecally in mice, demonstrating prolonged SCP expression in the central nervous system by this approach. There may be substantial advantages in using intrathecal administration for screening SCP candidates relevant to autoimmune disorders of the central nervous system.

30.13 SCP FRONTIERS: MINING THE SURFACE PROTEIN INTERACTOME, REWIRING CELLULAR NETWORKS

The SCPs described in this chapter were developed in an eclectic manner and represent but a small sampling of those that might be generated. A more deliberate and informed approach toward expanding SCP libraries would be empowered by a systematic cataloging of surface proteins of therapeutic relevance. Indeed, the most comprehensive approach would draw upon a *cell surface protein interactome* (CSPI), which maps protein interaction networks. The Human Interactome Database constitutes one platform being developed to visualize and integrate information pertaining to protein interaction networks and disease associations for each protein in the human proteome [158,159]. A fully annotated CSPI, narrowing in on the subset of proteins relevant to cell contact-dependent interactions and providing cell distribution and protein function data for them, would substantially advance SCP design. In filtering the large matrix of potential SCP protein pairs that would emerge from such an exercise, one would conceptually navigate across several dimensions: interactive cell pairs or clusters linked to specific pathogenic processes; cell surface molecular pairs of special significance for the functional interplay among these pathogenic cell groupings; choice of SCP functionality most appropriate for targeting the cell surface molecules and cellular interactions of interest.

In thinking about SCP function *in vivo*, one per force enters a realm of higher order complexity, as one marries multifunctional SCP with multicellular networks. As discussed throughout this chapter, individual SCP commonly reveal multiple functionalities and mechanisms of action, even in the simplest of contexts where there are but one or two cell types in play. As one adds more cell types to a given interactive mix, the functional consequences ripple outward and are amplified. Furthermore, additional systemic complexity emerges when one considers the wide cellular distribution of some of the cognate counter-receptors that are targeted by SCP. This in effect can bring additional cellular networks into the fold. To decipher changes in the net output from SCP-perturbed cellular networks, more sophisticated readouts will be called for, invoking ever more multiplexed analytic tools, for example, cytometric fingerprinting [160] and high-dimensional microfluidic dynamic arrays [161].

Yet, despite the potential complexity of SCP:network overlays, they create new therapeutic opportunity and can even streamline interventions. By targeting critical nodes (with CLBP) or links (with TSCP or TSRP) within cellular networks, and leveraging the highly selective molecular targeting offered by SCP, there is the theoretical potential to impact network frameworks in an almost surgical way. Moreover, more distant scenarios envision the administration of SCP in combination to simultaneously target more than one node or link, generate new cellular bridges, and thereby functionally tune or even rewire whole cellular networks, in ways that often might not be achievable through the use of simpler, single function agents. SCP, by virtue of their two binding element constitution, can, paradoxically, provide for higher order cellular selectivity in effecting such network modulation.

These systems are not static. The cellular landscape *in vivo* is in flux, in terms of composition and location, as are the cell surface protein displays, conformations, and topologies of component cells. This dynamism of cells and proteins adds yet another layer of complexity to SCP mechanisms and therapeutic design. Looking into the future, one can imagine next-level interactome analyses, whether dealing with cellular webs or cell surface protein fingerprints, that more comprehensively capture repertoires of networked cellular interactions, as they are instantiated within specific tissue architectures and cellular trafficking flows. Scanning these dynamic cell and protein systems, advanced SCP designers might one day sift through and identify in an informed way those cells and proteins that are most targetable.

SCP can be viewed as part of a vanguard of next-generation, molecularly tailored medicines. In its earliest iteration, personalized medicine keyed in on singular molecular targets. In the realm of personalized onco-medicine, for example, those targets were predominantly kinases or isolated cell surface molecules. By contrast, next-generation individualized therapeutics could start to embrace cellular and molecular complexity. SCP provide one kind of therapeutic tool for such an embrace, as systems biology and fusion protein therapeutics are brought together in compelling new ways. Multifunctional SCP, when weaved into dynamic cellular and molecular networks, create a rich tapestry of therapeutic possibility.

REFERENCES

1. Fink PJ, Shimenkovitz RP, Bevan MJ. (1988) Veto cells. *Annu. Rev. Immunol.* 6, 115–137.
2. Rammensee H. (1989) Veto function *in vitro* and *in vivo*. *Intern. Rev. Immunol.* 4, 175–191.
3. Tykocinski ML, Kaplan DR. (1993) Prospects for anti-rejection therapies based upon CD8-dependent immunoregulation. *Kidney Int. Suppl.* 39, S120–3.
4. Tykocinski ML, Kaplan DR, Medof ME. (1996) Antigen-presenting cell engineering. The molecular toolbox. *Am. J. Pathol.* 148, 1–16.
5. Smith DM, Bluestone JA, Jeyarajah DR, Newberg MH, Engelhard VH, Thistlethwaite JR, et al. (1994) Inhibition of T cell activation by a monoclonal antibody reactive against the alpha3 domain of human MHC class I molecules. *J. Immunol.* 153, 1054–1067.
6. Hambor JE, Tykocinski ML, Kaplan DR. (1988) Functional consequences of anti-sense RNA-mediated inhibition of CD8 surface expression in a human T cell clone. *J. Exp. Med.* 168, 1237–45.
7. Kaplan DR, Hambor JE, Tykocinski ML. (1989) An immunoregulatory function for the CD8 molecule. *Proc. Natl. Acad. Sci. USA* 86, 8512–8515.
8. Hambor JE, Kaplan DR, Tykocinski ML. (1990) CD8 functions as an inhibitory ligand in mediating the immunoregulatory activity of $CD8^+$ cells. *J. Immunol.* 145, 1646–1652.
9. Tykocinski ML, Kaplan DR. (1992) A multifunctional perspective of the CD8 molecule. In: Herberman RB, Lotzova E (Eds), *NK Cell Mediated Cytotoxicity. Receptors, Signaling and Mechanisms*. CRC Press, Ann Arbor, pp. 393–408.
10. Sambhara SR, Miller RG. (1991) Programmed cell death of T cells signaled by the T cell receptor and the alpha3 domain of class I MHC. *Science* 252, 1424–1427.
11. Sambhara SR, Miller RG. (1994) Reduction of CTL antipeptide response mediated by $CD8^+$ cells whose class I MHC can bind the peptide. *J. Immunol.* 152, 1103–1109.
12. Qi Y, Berg R, Singleton MA, Debrick JE, Staerz UD. (1996) Hybrid antibody mediated veto of cytotoxic T lymphocyte responses. *J. Exp. Med.* 183, 1973–1980.
13. Matsue H, Matsue K, Walters M, Okumura K, Yagita H, Takashima A. (1999) Induction of antigen-specific immunosuppression by CD95L cDNA-transfected 'killer' dendritic cells. *Nature Med.* 5, 930–937.
14. Min WP, Gorczynski R, Huang XY, Kushida M, Kim P, Obataki M, et al. (2000) Dendritic cells genetically engineered to express Fas ligand induce donor-specific hyporesponsiveness and prolong allograft survival. *J. Immunol.* 164, 161–167.
15. Reich-Zeliger S, Zhao Y, Krauthgamer R, Bachar-Lustig E, Reisner Y. (2000) Anti-third party $CD8^+$ CTLs as potent veto cells: Coexpression of CD8 and FasL is a prerequisite. *Immunity* 13, 507–515.
16. Wu B, Wu JM, Miagkov A, Adams RN, Levitsky HI, Drachman DB. (2001) Specific immunotherapy by genetically engineered APCs: the "guided missile" strategy. *J. Immunol.* 166, 4773–4779.
17. Krueger A, Fas SC, Baumann S, Krammer PH. (2003) The role of CD95 in the regulation of peripheral T-cell apoptosis. *Immunol. Rev.* 193, 58–69.
18. Suss G, Shortman K. (1996) A subclass of dendritic cells kills CD4 T cells via Fas/Fas-ligand-induced apoptosis. *J. Exp. Med.* 183, 1789–1796.
19. Chen A, Zheng G, Tykocinski ML. (2003) Quantitative interplay between activating and pro-apoptotic signals dictates T cell responses. *Cell. Immunol.* 221, 128–137.

20. Zhang H, Su X, Liu D, Liu W, Yang P, Wang Z, et al. (1999) Induction of specific T cell tolerance by Fas ligand-expressing antigen-presenting cells. *J. Immunol.* 162, 1423–1430.
21. Lu L, Qian S, Hershberger PA, Rudert WA, Lynch DH, Thomson AW. (1997) Fas ligand (CD95L) and B7 expression on dendritic cells provide counter-regulatory signals for T cell survival and proliferation. *J. Immunol.* 158, 5676–5684.
22. George JF, Sweeney SD, Kirklin JK, Simpson EM, Goldstein DR, Thomas JM. (1998) An essential role for Fas ligand in transplantation tolerance induced by donor bone marrow. *Nat. Med.* 4, 333–335.
23. Hambor JE, Weber MC, Tykocinski ML, Kaplan DR. (1990) Regulation of allogeneic responses by expression of CD8 alpha chain on stimulator cells. *Int. Immunol.* 2, 879–883.
24. Tykocinski ML, Shu HK, Ayers DJ, Walter EI, Getty RR, Groger RK, et al. (1988) Glycolipid reanchoring of T-lymphocyte surface antigen CD8 using the 3' end sequence of decay-accelerating factor's mRNA. *Proc. Natl. Acad. Sci. USA* 85, 3555–3559.
25. Caras IW, Weddell GN, Davitz MA, Nussenzweig V, Martin JDW. (1987) Signal for attachment of a phopholipid membrane anchor in decay accelerating factor. *Science* 238, 1280–1283.
26. Weber MC, Groger RK, Tykocinski ML. (1994) A glycosylphosphatidylinositol-anchored cytokine can function as an artificial cellular adhesin. *Exp. Cell Res.* 145, 1646–1652.
27. Huang JH, Greenspan NS, Tykocinski ML. (1994) Alloantigenic recognition of artificial glycosyl phosphatidylinositol-anchored HLA-A2.1. *Mol. Immunol.* 31, 1017–1028.
28. Brunschwig EB, Fayen JD, Medof ME, Tykocinski ML. (1999) Protein transfer of glycosyl-phosphatidylinositol (GPI)-modified murine B7-1 and B7-2 costimulators. *J. Immunotherapy* 22, 390–400.
29. Geho DH, Fayen JD, Jackman RM, Moody BD, Porcelli SA, Tykocinski ML. (2000) Glycosyl-phosphastidylinositol reanchoring unmasks distinct antigen-presenting pathways for CD1b and CD1c. *J. Immunol.* 65, 1272–1277.
30. Medof ME, Nagarajan S, Tykocinski ML. (1996) Cell surface engineering with GPI-anchored proteins. *FASEB J.* 10, 574–586.
31. Fayen JD, Tykocinski ML, Medof ME. (2000) Glycerolphosphoinositide anchors for membrane-tethering proteins. *Methods Enzymol.* 327, 351–368.
32. De Bruijn MLH, Schumacher TNM, Nieland JD, Ploegh HL, Kast WM, Melief CJM. (1991) Peptide loading of empty major histocompatibility complex molecules on RMA-S cells allows the induction of primary cytotoxic T lymphocyte responses. *Eur. J. Immunol.* 21, 2963–2970.
33. McHugh RS, Ahmed SN, Wang YC, Sell KW, Selvaraj P. (1995) Construction, purification, and functional incorporation on tumor cells of glycolipid-anchored human B7-1 (CD80). *Proc. Natl. Acad. Sci. USA* 92, 8059–8063.
34. Brunschwig EB, Levine E, Trefzer U, Tykocinski ML. (1995) Glycosylphosphatidylinositol-modified murine B7-1 and B7-2 retain costimulator function. *J. Immunol.* 155, 5498–5505.
35. Chen A, Zheng G, Tykocinski ML. (2000) Hierarchical costimulator thresholds for distinct immune responses: application of a novel two-step fc fusion protein transfer method. *J. Immunol.* 164, 705–711.
36. Zheng G, Chen A, Sterner RE, Zhang PJ, Pan T, Kiyatkin N, et al. (2001) Induction of antitumor immunity via intratumoral tetra-costimulator protein transfer. *Cancer Res.* 61, 8127–8134.
37. Huang JH Tykocinski ML. (2001) CTLA-4-fas ligand functions as a trans signal converter protein in bridging antigen-presenting cells and T cells. *Int. Immunol.* 13, 529–539.
38. Rudd CE, Taylor A, Schneider H. (2009) CD28 and CTLA-4 coreceptor expression and signal transduction. *Immunol. Rev.* 229, 12–26.
39. Noel PJ, Boise LH, Green JM, Thompson CB. (1996) CD28 costimulation prevents cell death during primary T cell activation. *J. Immunol.* 157, 636–642.
40. Li Y, Li XC, Zheng XX, Wells AD, Turka LA, Strom TB. (1999) Blocking both signal 1 and signal 2 of T-cell activation prevents apoptosis of alloreactive T cells and induction of peripheral allograft tolerance. *Nat. Med.* 5, 1298–1302.
41. Holler N, Tardivel A, Kovacsovics-Bankowski M, Hertig S, Gaide O, Martinon F, et al. (2003) Two adjacent trimeric Fas ligands are required for Fas signaling and formation of a death-inducing signaling complex. *Mol. Cell. Biol.* 23, 1428–1440.
42. Schneider P, Tschopp J. (2000) Apoptosis induced by death receptors. *Pharm. Acta Helv.* 74, 281–286.
43. Schmidt EM, Wang CJ, Ryan GA, Clough LE, Qureshi OS, Goodall M, et al. (2009) Ctla-4 controls regulatory T cell peripheral homeostasis and is required for suppression of pancreatic islet autoimmunity. *J. Immunol.* 182, 274–282.
44. Friedline RH, Brown DS, Nguyen H, Kornfeld H, Lee J, Zhang Y, et al. (2009) $CD4^+$ regulatory T cells require CTLA-4 for the maintenance of systemic tolerance. *J. Exp. Med.* 206, 421–434.
45. Baban B, Hansen AM, Chandler PR, Manlapat A, Bingaman A, Kahler DJ, et al. (2005) A minor population of splenic dendritic cells expressing CD19 mediates IDO-dependent T cell suppression via type I IFN signaling following B7 ligation. *Int. Immunol.* 17, 909–919.
46. Judge TA, Wu Z, Zheng XG, Sharpe AH, Sayegh MH, Turka LA. (1999) The role of CD80, CD86, and CTLA4 in alloimmune responses and the induction of long-term allograft survival. *J. Immunol.* 162, 1947–1951.
47. Koenen HJ, Joosten I. (2000) Blockade of CD86 and CD40 induces alloantigen-specific immunoregulatory T cells that remain anergic even after reversal of hyporesponsiveness. *Blood* 95, 3153–3161.
48. Wallace PM, Rodgers JN, Leytze GM, Johnson JS, Linsley PS. (1995) Induction and reversal of long-lived specific unresponsiveness to a T-dependent antigen following CTLA4Ig treatment. *J. Immunol.* 154, 5885–5895.
49. Newell KA, He G, Guo Z, Kim O, Szot GL, Rulifson I, et al. (1999) Cutting edge: blockade of the CD28/B7 costimulatory pathway inhibits intestinal allograft rejection mediated by $CD4^+$ but not $CD8^+$ T cells. *J. Immunol.* 163, 2358–62.

50. Dranitzki-Elhalel M, Huang JH, Schmidt W, Rachmilewitz J, Tykocinski ML. (2003) CTLA-4-FasL induces alloantigen-specific hyporesponsiveness. *J. Immunol.* 170, 5842–5850.
51. Dranitzki-Elhalel M, Huang JH, Rachmilewitz J, Pappo O, Parnas M, Schmidt W, et al. (2006) CTLA-4-FasL inhibits allogeneic responses *in vivo*. *Cell. Immunol.* 239, 129–135.
52. Orbach A, Rachmilewitz J, Parnas M, Huang JH, Tykocinski ML, Dranitzki-Elhalel M. (2007) CTLA-4·FasL induces early apoptosis of activated T cells by interfering with anti-apoptotic signals. *J. Immunol.* 179, 7287–7294.
53. Zhang HG, Heike Y, Yang P, Wang Z, Wang X, Curiel DT, Zhou T, Mountz JD. (1998) Induction of specific T-cell tolerance by adenovirus-transfected, Fas ligand-producing antigen presenting cells. *Nature Biotechnol.* 16, 1045–1049.
54. Chai JG, Bartok I, Scott D, Dyson J, Lechler R. (1998) T:T antigen presentation by activated murine $CD8^+$ T cells induces anergy and apoptosis. *J. Immunol.* 160, 3655–3665.
55. Jin Y, Qu A, Wang GM, Hao J, Gao X, Xie S. (2004) Simultaneous stimulation of Fas-mediated apoptosis and blockade of costimulation prevent autoimmune diabetes in mice induced by multiple low-dose streptozotocin. *Gene Ther.* 11, 982–991.
56. Feng YG, Jin YZ, Zhang QY, Hao J, Wang GM, Xie SS. (2005) CTLA4-fas ligand gene transfer mediated by adenovirus induces long-time survival of murine cardiac allografts. *Transplant. Proc.* 37, 2379–2381.
57. Shi W, Chen M, Xie L. (2007) Prolongation of corneal allograft survival by CTLA4-FasL in a murine model. *Graefes Arch. Clin. Exp. Ophthalmol.* 245, 1691–1697.
58. Lau HT, Yu M, Fontana A, Stoeckert CJ, Jr. (1996) Prevention of islet allograft rejection with engineered myoblasts expressing FasL in mice. *Science* 273, 109–112.
59. Winkles JA. (2008) The TWEAK-Fn14 cytokine-receptor axis: discovery, biology and therapeutic targeting. *Nat. Rev. Drug Discov.* 7, 411–425.
60. Falschlehner C, Schaefer U, Walczak H. (2009) Following TRAIL's path in the immune system. *Immunology* 127, 145–154.
61. Maecker H, Varfolomeev E, Kischkel F, Lawrence D, LeBlanc H, Lee W, et al. (2005) TWEAK attenuates the transition from innate to adaptive immunity. *Cell* 123, 931–944.
62. Mittal A, Bhatnagar S, Kumar A, Lach-Trifilieff E, Wauters S, Li H, et al. (2010) The TWEAK-Fn14 system is a critical regulator of denervation-induced skeletal muscle atrophy in mice. *J. Cell Biol.* 188, 833–849.
63. Polavarapu R, Gongora MC, Winkles JA, Yepes M. (2005) Tumor necrosis factor-like weak inducer of apoptosis increases the permeability of the neurovascular unit through nuclear factor-kappa B pathway activation. *J. Neurosci.* 25, 10094–10100.
64. Zhang X, Winkles JA, Gongora MC, Polavarapu R, Michaelson JS, Hahm K, et al. (2006) TWEAK-Fn14 pathway inhibition protects the integrity of the neurovascular unit during cerebral ischemia. *J. Cereb. Blood Flow Metab.* 27, 534–544.
65. Jakubowski A, Browning B, Lukashev M, Sizing I, Thompson JS, Benjamin CD, et al. (2002) Dual role for TWEAK in angiogenic regulation. *J. Cell. Sci.* 115, 267–274.
66. Kirk SL, Karlik SJ. (2003) VEGF and vascular changes in chronic neuroinflammation. *J. Autoimmun.* 21, 353–363.
67. Desplat-Jego S, Varriale S, Creidy R, Terra R, Bernard D, Khrestchatisky M, et al. (2002) TWEAK is expressed by glial cells, induces astrocyte proliferation and increases EAE severity. *J. Neuroimmunol.* 133, 116–123.
68. Potrovita I, Zhang W, Burkly L, Hahm K, Lincecum J, Wang MZ, et al. (2004) Tumor necrosis factor-like weak inducer of apoptosis-induced neurodegeneration. *J. Neurosci.* 24, 8237–8244.
69. Mueller AM, Pedre X, Kleiter I, Hornberg M, Steinbrecher A, Giegerich G. (2005) Targeting fibroblast growth factor-inducible-14 signaling protects from chronic relapsing experimental autoimmune encephalomyelitis. *J. Neuroimmunol.* 159, 55–65.
70. Spierings DC, de Vries EG, Vellenga E, van den Heuvel FA, Koornstra JJ, Wesseling J, et al. (2004) Tissue distribution of the death ligand TRAIL and its receptors. *J. Histochem. Cytochem.* 52, 821–831.
71. Zauli G, Secchiero P. (2006) The role of the TRAIL/TRAIL receptors system in hematopoiesis and endothelial cell biology. *Cytokine Growth Factor Rev.* 17, 245–257.
72. Song K, Chen Y, Goke R, Wilmen A, Seidel C, Goke A, et al. (2000) Tumor necrosis factor-related apoptosis-inducing ligand (TRAIL) is an inhibitor of autoimmune inflammation and cell cycle progression. *J. Exp. Med.* 191, 1095–1104.
73. Hilliard B, Wilmen A, Seidel C, Liu TS, Goke R, Chen Y. (2001) Roles of TNF-related apoptosis-inducing ligand in experimental autoimmune encephalomyelitis. *J. Immunol.* 166, 1314–1319.
74. Hirata S, Senju S, Matsuyoshi H, Fukuma D, Uemura Y, Nishimura Y. (2005) Prevention of experimental autoimmune encephalomyelitis by transfer of embryonic stem cell-derived dendritic cells expressing myelin oligodendrocyte glycoprotein peptide along with TRAIL or programmed death-1 ligand. *J. Immunol.* 174, 1888–1897.
75. Cretney E, McQualter JL, Kayagaki N, Yagita H, Bernard CC, Grewal IS, et al. (2005) TNF-related apoptosis-inducing ligand (TRAIL)/Apo2L suppresses experimental autoimmune encephalomyelitis in mice. *Immunol. Cell Biol.* 83, 511–519.
76. Lamhamedi-Cherradi SE, Zheng S, Tisch RM, Chen YH. (2003) Critical roles of tumor necrosis factor-related apoptosis-inducing ligand in type 1 diabetes. *Diabetes* 52, 2274–2278.
77. Mi QS, Ly D, Lamhamedi-Cherradi SE, Salojin KV, Zhou L, Grattan M, et al. (2003) Blockade of tumor necrosis factor-related apoptosis-inducing ligand exacerbates type 1 diabetes in NOD mice. *Diabetes* 52, 1967–1975.
78. Lamhamedi-Cherradi SE, Zheng SJ, Maguschak KA, Peschon J, Chen YH. (2003) Defective thymocyte apoptosis

and accelerated autoimmune diseases in TRAIL-/- mice. *Nat. Immunol.* 4, 255–260.

79. Wandinger KP, Lunemann JD, Wengert O, Bellmann-Strobl J, Aktas O, Weber A, et al. (2003) TNF-related apoptosis inducing ligand (TRAIL) as a potential response marker for interferon-beta treatment in multiple sclerosis. *Lancet* 361, 2036–2043.
80. Razmara M, Hilliard B, Ziarani AK, Murali R, Yellayi S, Ghazanfar M, et al. (2009) Fn14-TRAIL, a chimeric intercellular signal exchanger, attenuates experimental autoimmune encephalomyelitis. *Am. J. Pathol.* 174, 460–474.
81. Wajant H, Moosmayer D, Wuest T, Bartke T, Gerlach E, Schonherr U, et al. (2001) Differential activation of TRAIL-R1 and -2 by soluble and membrane TRAIL allows selective surface antigen-directed activation of TRAIL-R2 by a soluble TRAIL derivative. *Oncogene* 20, 4101–4106.
82. Ehrenschwender M, Wajant H. (2009) The role of FasL and Fas in health and disease. *Adv. Exp. Med. Biol.* 647, 64–93.
83. Vignaux F, Vivier E, Malissen B, Depraetere V, Nagata S, Golstein P. (1995) TCR/CD3 coupling to Fas-based cytotoxicity. *J. Exp. Med.* 181, 781–786.
84. Suda T, Okazaki T, Naito Y, Yokota T, Arai N, Ozaki S, et al. (1995) Expression of the Fas ligand in cells of T cell lineage. *J. Immunol.* 154, 3806–3813.
85. Dhein J, Walczak H, Baumler C, Debatin KM, Krammer PH. (1995) Autocrine T-cell suicide mediated by APO-1 (Fas/CD95). *Nature* 373, 438–441.
86. Alderson MR, Tough TW, Davis-Smith T, Braddy S, Falk B, Schooley KA, et al. (1995) Fas ligand mediates activation-induced cell death in human T lymphocytes. *J. Exp. Med.* 181, 71–77.
87. Russell JH, Rush B, Weaver C, Wang R. (1993) Mature T cells of autoimmune lpr/lpr mice have a defect in antigen-stimulated suicide. *Proc. Natl. Acad. Sci. USA* 90, 4409–4413.
88. Ju ST, Panka DJ, Cui H, Ettinger R, el Khatib M, Sherr DH, et al. (1995) Fas (CD95)/FasL interactions required for programmed cell death after T-cell activation. *Nature* 373, 444–448.
89. Brunner T, Mogil RM, LaFace D, Yoo NJ, Mahboubi A, Echeverri F, et al. (1995) Cell-autonomous Fas (CD95)/Fas-ligand interaction mediates activation-induced apoptosis in T-cell hybridomas. *Nature* 373, 441–444.
90. Dranitzki-Elhalel M, Huang JH, Sasson M, Rachmilewitz J, Parnas M, Tykocinski ML. (2007) CD40·FasL inhibits human T cells: evidence for an auto-inhibitory loop-back mechanism. *Int. Immunol.* 19, 355–363.
91. Blair PJ, Riley JL, Harlan DM, Abe R, Tadaki DK, Hoffmann SC, et al. (2000) CD40 ligand (CD154) triggers a short-term CD4(+) T cell activation response that results in secretion of immunomodulatory cytokines and apoptosis. *J. Exp. Med.* 191, 651–660.
92. Zheng G, Chen A, Weber M, Tykocinski M. (2003) T cells painted with B7-1 and 4-1BB ligand auto-costimulate themselves. *FASEB-Experimental Biology 2003, Abstr 414.3.*
93. Lavrik IN, Krammer PH. (2012) Regulation of CD95/Fas signaling at the DISC. *Cell Death Differ.* 19, 36–41.
94. Razmara M, Hilliard B, Ziarani AK, Chen YH, Tykocinski ML. (2008) CTLA-4 x ig converts naive $CD4^{+}CD25$- T cells into $CD4^{+}CD25+$ regulatory T cells. *Int. Immunol.* 20, 471–483.
95. Rouleau M, Bernard A, Vernant JP, Charpentier B, Senik A. (1993) Apoptosis of $CD8^{+}$/CD57+ T cells is induced by some combinations of anti-CD2 mAb. *J. Immunol.* 151, 3547–3551.
96. Novak TJ, Farber D, Leitenberg D, Hong S, Johnson P, Bottomly K. (1994) Isoforms of the transmembrane tyrosine phosphatase CD45 differentially affect T cell recognition. *Immunity* 1, 109–119.
97. Orbach A, Rachmilewitz J, Shani N, Isenberg Y, Parnas M, Huang JH, et al. (2010) CD40·FasL and CTLA-4·FasL fusion proteins induce apoptosis in malignant cell lines by dual signaling. *Am. J. Pathol.* 177, 3159–3168.
98. Tran NL, McDonough WS, Donohue PJ, Winkles JA, Berens TJ, Ross KR, et al. (2003) The human Fn14 receptor gene is up-regulated in migrating glioma cells *in vitro* and overexpressed in advanced glial tumors. *Am. J. Pathol.* 162, 1313–1321.
99. Zhao H, Langerod A, Ji Y, Nowels KW, Nesland JM, Tibshirani R, et al. (2004) Different gene expression patterns in invasive lobular and ductal carcinomas of the breast. *Mol. Biol. Cell* 15, 2523–2536.
100. Ozawa F, Friess H, Kleeff J, Xu ZW, Zimmermann A, Sheikh MS, et al. (2001) Effects and expression of TRAIL and its apoptosis-promoting receptors in human pancreatic cancer. *Cancer Lett.* 163, 71–81.
101. Kawakita T, Shiraki K, Yamanaka Y, Yamaguchi Y, Saitou Y, Enokimura N, et al. (2005) Functional expression of TWEAK in human colonic adenocarcinoma cells. *Int. J. Oncol.* 26, 87–93.
102. Kawakita T, Shiraki K, Yamanaka Y, Yamaguchi Y, Saitou Y, Enokimura N, et al. (2004) Functional expression of TWEAK in human hepatocellular carcinoma: possible implication in cell proliferation and tumor angiogenesis. *Biochem. Biophys. Res. Commun.* 318, 726–733.
103. Wu GS. (2009) TRAIL as a target in anti-cancer therapy. *Cancer Lett.* 285, 1–5.
104. Washburn B, Weigand MA, Grosse-Wilde A, Janke M, Stahl H, Rieser E, et al. (2003) TNF-related apoptosis-inducing ligand mediates tumoricidal activity of human monocytes stimulated by newcastle disease virus. *J. Immunol.* 170, 1814–1821.
105. Shirley S, Morizot A, Micheau O. (2011) Regulating TRAIL receptor-induced cell death at the membrane: a deadly discussion. *Recent. Pat. Anticancer Drug Discov.* 6, 311–323.
106. Jeffery CJ. (2003) Moonlighting proteins: old proteins learning new tricks. *Trends Genet.* 19, 415–417.
107. Hurwitz AA, Kwon ED, van Elsas A. (2000) Costimulatory wars: the tumor menace. *Curr. Opin. Immunol.* 12, 589–596.

108. Coyle AJ, Gutierrez-Ramos J-C. (2001) The expanding B7 superfamily: increasing complexity in costimulatory signals regulating T cell function. *Nature Immunol.* 2, 203–209.

109. Watts TH. (2005) TNF/TNFR family members in costimulation of T cell responses. *Annu. Rev. Immunol.* 23, 23–68.

110. Li XC, Rothstein DM, Sayegh MH. (2009) Costimulatory pathways in transplantation: challenges and new developments. *Immunol. Rev.* 229, 271–293.

111. Okazaki T, Honjo T. (2007) PD-1 and PD-1 ligands: from discovery to clinical application. *Int. Immunol.* 19, 813–824.

112. Cai G, Freeman GJ. (2009) The CD160, BTLA, LIGHT-/HVEM pathway: a bidirectional switch regulating T-cell activation. *Immunol. Rev.* 229, 244–258.

113. del Rio ML, Lucas CL, Buhler L, Rayat G, Rodriguez-Barbosa JI. (2010) HVEM/LIGHT/BTLA/CD160 cosignaling pathways as targets for immune regulation. *J. Leukoc. Biol.* 87, 223–235.

114. Nielsen UB, Adams GP, Weiner LM, Marks JD. (2000) Targeting of bivalent anti-ErbB2 diabody antibody fragments to tumor cells is independent of the intrinsic antibody affinity. *Cancer Res.* 60, 6434–6440.

115. Lanzavecchia A. (2000) Understanding dendritic cell and T-lymphocyte traffic through the analysis of chemokine receptor expression. *Immunol. Rev.* 177, 134–140.

116. Chemnitz JM, Parry RV, Nichols KE, June CH, Riley JL. (2004) SHP-1 and SHP-2 associate with immunoreceptor tyrosine-based switch motif of programmed death 1 upon primary human T cell stimulation, but only receptor ligation prevents T cell activation. *J. Immunol.* 173, 945–954.

117. Keir ME, Francisco LM, Sharpe AH. (2007) PD-1 and its ligands in T-cell immunity. *Curr. Opin. Immunol.* 19, 309–314.

118. Butte MJ, Pena-Cruz V, Kim MJ, Freeman G.J, Sharpe AH. (2008) Interaction of human PD-L1 and B7-1. *Mol. Immunol.* 45, 3567–3572.

119. Said EA, Dupuy FP, Trautmann L, Zhang Y, Shi Y, El-Far M, et al. (2010) Programmed death-1-induced interleukin-10 production by monocytes impairs $CD4^+$ T cell activation during HIV infection. *Nat. Med.* 16, 452–459.

120. Tanaka K, Albin MJ, Yuan X, Yamaura K, Habicht A, Murayama T, et al. (2007) PDL1 is required for peripheral transplantation tolerance and protection from chronic allograft rejection. *J. Immunol.* 179, 5204–5210.

121. Ansari MJ, Salama AD, Chitnis T, Smith RN, Yagita H, Akiba H, et al. (2003) The programmed death-1 (PD-1) pathway regulates autoimmune diabetes in nonobese diabetic (NOD) mice. *J. Exp. Med.* 198, 63–69.

122. Mozaffarian N, Wiedeman AE, Stevens AM. (2008) Active systemic lupus erythematosus is associated with failure of antigen-presenting cells to express programmed death ligand-1. *Rheumatology (Oxford)* 47, 1335–1341.

123. Thompson RH, Gillett MD, Cheville JC, Lohse CM, Dong H, Webster WS, et al. (2004) Costimulatory B7-H1 in renal cell carcinoma patients: indicator of tumor aggressiveness and potential therapeutic target. *Proc. Natl. Acad. Sci. USA* 101, 17174–17179.

124. Habicht A, Kewalaramani R, Vu MD, Demirci G, Blazar BR, Sayegh MH, et al. (2007) Striking dichotomy of PD-L1 and PD-L2 pathways in regulating alloreactive CD4(+) and CD8 (+) T cells *in vivo*. *Am. J. Transplant.* 7, 2683–2692.

125. Amarnath S, Mangus CW, Wang JC, Wei F, He A, Kapoor V, et al. (2011) The PDL1-PD1 axis converts human TH1 cells into regulatory T cells. *Sci. Transl. Med.* 3(111) ra120.

126. Jain N, Nguyen H, Chambers C, Kang J. (2010) Dual function of CTLA-4 in regulatory T cells and conventional T cells to prevent multiorgan autoimmunity. *Proc. Natl. Acad. Sci. USA* 107, 1524–1528.

127. Akbari O, Freeman GJ, Meyer EH, Greenfield EA, Chang TT, Sharpe AH, et al. (2002) Antigen-specific regulatory T cells develop via the ICOS-ICOS-ligand pathway and inhibit allergen-induced airway hyperreactivity. *Nat. Med.* 8, 1024–1032.

128. Burmeister Y, Lischke T, Dahler AC, Mages HW, Lam KP, Coyle AJ, et al. (2008) ICOS controls the pool size of effector-memory and regulatory T cells. *J. Immunol.* 180, 774–782.

129. Herman AE, Freeman GJ, Mathis D, Benoist C. (2004) $CD4^+CD25+$ T regulatory cells dependent on ICOS promote regulation of effector cells in the prediabetic lesion. *J. Exp. Med.* 199, 1479–1489.

130. Dai Z, Li Q, Wang Y, Gao G, Diggs LS, Tellides G, et al. (2004) $CD4^+CD25+$ regulatory T cells suppress allograft rejection mediated by memory $CD8^+$ T cells via a CD30-dependent mechanism. *J. Clin. Invest.* 113, 310–317.

131. Zeiser R, Nguyen VH, Hou JZ, Beilhack A, Zambricki E, Buess M, et al. (2007) Early CD30 signaling is critical for adoptively transferred $CD4^+CD25+$ regulatory T cells in prevention of acute graft-versus-host disease. *Blood* 109, 2225–2233.

132. Galli SJ. (1997) Complexity and redundancy in the pathogenesis of asthma: reassessing the roles of mast cells and T cells. *J. Exp. Med.* 186, 343–347.

133. Brown JM, Wilson TM, Metcalfe DD. (2008) The mast cell and allergic diseases: role in pathogenesis and implications for therapy. *Clin. Exp. Allergy* 38, 4–18.

134. Umetsu DT, Dekruyff RH. (2010) Natural killer T cells are important in the pathogenesis of asthma: the many pathways to asthma. *J. Allergy Clin. Immunol.* 125, 975–979.

135. Braun NA, Covarrubias R, Major AS. (2010) Natural killer T cells and atherosclerosis: form and function meet pathogenesis. *J. Innate Immun.* 2, 316–324.

136. Gomez-Diaz RA, Aguilar MV, Meguro EN, Marquez RH, Magana EG, Martinez-Garcia MC, et al. (2011) The role of natural killer T (NKT) cells in the pathogenesis of type 1 diabetes. *Curr. Diabetes Rev.* 7, 278–283.

137. Kopcow HD, Karumanchi SA. (2007) Angiogenic factors and natural killer (NK) cells in the pathogenesis of preeclampsia. *J. Reprod. Immunol.* 76, 23–29.

138. Sikora J, Mielczarek-Palacz A, Kondera-Anasz Z. (2011) Role of natural killer cell activity in the pathogenesis of endometriosis. *Curr. Med. Chem.* 18, 200–208.

139. Majeti R, Chao MP, Alizadeh AA, Pang WW, Jaiswal S, Gibbs KD, Jr, et al. (2009) CD47 is an adverse prognostic

factor and therapeutic antibody target on human acute myeloid leukemia stem cells. *Cell* 138, 286–299.

140. Chan KS, Espinosa I, Chao M, Wong D, Ailles L, Diehn M, et al. (2009) Identification, molecular characterization, clinical prognosis, and therapeutic targeting of human bladder tumor-initiating cells. *Proc. Natl. Acad. Sci. USA* 106, 14016–14021.
141. Matozaki T, Murata Y, Okazawa H, Ohnishi H. (2009) Functions and molecular mechanisms of the CD47-SIRPalpha signalling pathway. *Trends Cell Biol.* 19, 72–80.
142. Isenberg JS, Roberts DD, Frazier WA. (2008) CD47: a new target in cardiovascular therapy. *Arterioscler. Thromb. Vasc. Biol.* 28, 615–621.
143. Martins I, Kepp O, Galluzzi L, Senovilla L, Schlemmer F, Adjemian S, et al. (2010) Surface-exposed calreticulin in the interaction between dying cells and phagocytes. *Ann. NY Acad. Sci.* 1209, 77–82.
144. Sansom DM. (2000) CD28, CTLA-4 and their ligands: Who does what and to whom? *Immunology*, 101, 169–177.
145. Vasu C, Wang A, Gorla SR, Kaithamana S, Prabhakar BS, Holterman MJ. (2003) CD80 and CD86 C domains play an important role in receptor binding and co-stimulatory properties. *Int. Immunol.* 15, 167–175.
146. Nolan A, Kobayashi H, Naveed B, Kelly A, Hoshino Y, Hoshino S, et al. (2009) Differential role for CD80 and CD86 in the regulation of the innate immune response in murine polymicrobial sepsis. *PLoS One* 4, e6600.
147. Miller AM, Lundberg K, Ozenci V, Banham AH, Hellstrom M, Egevad L, et al. (2006) $CD4^{+}$CD25high T cells are enriched in the tumor and peripheral blood of prostate cancer patients. *J. Immunol.* 177, 7398–7405.
148. Bates GJ, Fox SB, Han C, Leek RD, Garcia JF, Harris AL, et al. (2006) Quantification of regulatory T cells enables the identification of high-risk breast cancer patients and those at risk of late relapse. *J. Clin. Oncol.* 24, 5373–5380.
149. Gao Q, Qiu SJ, Fan J, Zhou J, Wang XY, Xiao YS, et al. (2007) Intratumoral balance of regulatory and cytotoxic T cells is associated with prognosis of hepatocellular carcinoma after resection. *J. Clin. Oncol.* 25, 2586–2593.
150. Oble DA, Loewe R, Yu P, Mihm MC, Jr. (2009) Focus on TILs: prognostic significance of tumor infiltrating lymphocytes in human melanoma. *Cancer. Immun.* 9, 3.
151. Shimizu J, Yamazaki S, Takahashi T, Ishida Y, Sakaguchi S. (2002) Stimulation of CD25(+)CD4(+) regulatory T cells through GITR breaks immunological self-tolerance. *Nat. Immunol.* 3, 135–142.
152. Ji HB, Liao G, Faubion WA, Abadia-Molina AC, Cozzo C, Laroux FS, et al. (2004) Cutting edge: the natural ligand for glucocorticoid-induced TNF receptor-related protein abrogates regulatory T cell suppression. *J. Immunol.* 172, 5823–5827.
153. Ko K, Yamazaki S, Nakamura K, Nishioka T, Hirota K, Yamaguchi T, et al. (2005) Treatment of advanced tumors with agonistic anti-GITR mAb and its effects on tumor-infiltrating Foxp3+CD25+$CD4^{+}$ regulatory T cells. *J. Exp. Med.* 202, 885–891.
154. Vu MD, Xiao X, Gao W, Degauque N, Chen M, Kroemer A, et al. (2007) OX40 costimulation turns off Foxp3+ tregs. *Blood* 110, 2501–2510.
155. Piconese S, Valzasina B, Colombo MP. (2008) OX40 triggering blocks suppression by regulatory T cells and facilitates tumor rejection. *J. Exp. Med.* 205, 825–839.
156. Puccetti P, Fallarino F. (2008) Generation of T cell regulatory activity by plasmacytoid dendritic cells and tryptophan catabolism. *Blood Cells Mol. Dis.* 40, 101–105.
157. Messmer D, Kipps TJ. (2005) CD154 gene therapy for human B-cell malignancies. *Ann. N Y Acad. Sci.* 1062, 51–60.
158. Sanchez C, Lachaize C, Janody F, Bellon B, Roder L, Euzenat J, et al. (1999) Grasping at molecular interactions and genetic networks in drosophila melanogaster using FlyNets, an internet database. *Nucleic Acids Res.* 27, 89–94.
159. Elefsinioti A, Sarac OS, Hegele A, Plake C, Hubner NC, Poser I, et al. (2011) Large-scale de novo prediction of physical protein-protein association. *Mol. Cell. Proteomics* 10, M111.010629.
160. Rogers WT, Moser AR, Holyst HA, Bantly A, Mohler ER, 3rd, Scangas G, et al. (2008) Cytometric fingerprinting: quantitative characterization of multivariate distributions. *Cytometry A.* 73, 430–441.
161. Spurgeon SL, Jones RC, Ramakrishnan R. (2008) High throughput gene expression measurement with real time PCR in a microfluidic dynamic array. *PLoS One* 3, e1662.

31

SOLUBLE T-CELL ANTIGEN RECEPTORS

PETER R. RHODE

Altor BioScience Corporation, Miramar, FL, USA

31.1 SOLUBLE T-CELL ANTIGEN RECEPTOR (STAR) FUSION TECHNOLOGY AND UTILITIES

Disease-targeting monoclonal antibodies have enjoyed enormous success in the clinic and marketplace. For >20 such drugs currently approved for clinical use, annual sales exceed \$30 billion. However, many valuable targets are inaccessible to monoclonal antibodies because they are not displayed on the surface of diseased cells. Through the antigen presentation pathway, these antibody-inaccessible intracellular proteins (as well as normally secreted and cell-surface proteins) are processed into small peptides by the diseased cells and displayed in the context of a major histocompatibility complex (MHC) molecule (also referred to as HLA molecules in humans). Such peptide antigen/MHC (pMHC) complexes are then recognized by antigen-specific T-cell receptor (TCR) molecules on T lymphocytes. The ability of TCRs to bind antigenic pMHC molecules on tumor or virus-infected cells highlights their potential as targeting agents for cancer and viral therapeutics. We have captured this potential with the development of Altor BioScience's STAR technology platform.

Native α/β TCRs are highly variable membrane-bound glycoproteins consisting of two polypeptide chains, each of which contains of one immunoglobulin (Ig)-like variable (V) domain, one Ig-like constant (C) domain, a hydrophobic transmembrane region, and a short cytoplasmic region [1]. Similar to the antibody variable domains, the V regions of TCR α and β chains contain short stretches of sequence where the variability between different TCRs is concentrated, and these form the hypervariable or complementarity-determining regions (CDRs). Three CDRs in the α chain are juxtaposed to three similar regions in the β chain to form the binding domain of the TCR that specifically recognizes its cognate antigenic pMHC complex [1].

Each mature α/β T cell usually expresses a single unique TCR on its surface generated by rearrangement of the TCR loci during T-cell development. However, there is tremendous TCR diversity within the T-cell repertoire based on the ability to create different TCRs through random recombination of V and joining (J) segments of the TCR α locus and V, J, and diversity (D) segments of the TCR β locus, random insertion and deletions of nucleotides at the junctions between these segments and subsequent functional pairing of different rearranged TCR α and β chains. It has been estimated that 10^{15} distinct TCRs are possible [2]. Additionally, the targets for TCR binding are equally diverse given that MHC class I and class II molecules are highly polygenic and polymorphic and each MHC molecule is capable of promiscuous binding to many different peptide sequences. Thus, the pMHC/TCR system has evolved as a

Fusion Protein Technologies for Biopharmaceuticals: Applications and Challenges, First Edition. Edited by Stefan R. Schmidt.

means to display and specifically recognize a wide range of peptide antigens derived from infection agents or malignant cells that threaten the survival of the host. Consequently, there is considerable interest in developing TCR-based agents to target this distinct class of antigens, particularly those derived from intracellular proteins, which constitute a majority of the disease-relevant antigens [3,4].

31.1.1 Three Domains Single-Chain TCR (scTCR) Format and scTCR-Fusions

Despite structural similarities with extracellular antibody domains, production of soluble recombinant TCRs has proven difficult. The main technical limitations are low expression yields and instability of individual α or β chains and unstable α/β pairing in the absence of anchoring transmembrane domains and/or interchain disulfide bonds. To overcome these issues, a single-chain approach, similar to that used with single-chain (sc) Fv antibody domains, has been explored where the TCR Vα and Vβ domains are covalently linked by a flexible peptide sequence [5–8]. While initial attempts to effectively generate soluble TCR Vα-Vβ single-chain molecules were not generally successful [6–8], subsequent structural studies showed that a prominent elbow region at the N-terminus of the Cβ domain makes numerous contacts with the Vβ domain, suggesting co-expression of these domains may promote for proper folding of soluble protein [9]. In contrast, less extensive interactions between the Cα and Vα domains are consistent with observations that the Cα domain was dispensable for expression of individual Vα chains or scTCR molecules [10–13]. Based on these finding, we and others have employed a three domain Vα-linker-Vβ-Cβ format to express scTCR molecules [5,12–16]. This three domains scTCR format has been successfully used to generate membrane-bound proteins [13] or soluble-secreted proteins from mammalian cells [5,14,15], insoluble/refolded proteins from bacterial cells [12], and coat protein fusions displayed on the surface of bacteriophage [16]. In each case, the scTCR retained the ability to specifically recognize the appropriate cognate pMHC complex. Additionally, inclusion of the invariant TCR Cβ domain provides a convenient means of purifying scTCR molecules via immunoaffinity chromatography using anti-murine or human TCR Cβ antibodies [5,14,15,17]. We have used this three-domain format to express soluble scTCRs with over 30 different pMHC specificities and hundreds of different unique scTCR variants confirming the broad utility of this approach for generating functional TCR molecules with different sequences.

Another limitation with TCR-based targeting is the fact that soluble TCRs *per se* do not possess any functional effector activity and binding of TCRs to pMHC complexes does not induce target cell death. As a result, we have developed a fusion protein approach to link the scTCR domain to molecules such as cytokines (STAR-Ck), immunoglobulin constant regions (STAR-Ig), and single-chain antibodies (STAR-Ab) (Figure 31.1). Such genetic fusions are made to the C-terminal TCR Cβ region of the scTCR which is distal to the TCR-binding domain and does not impact interaction between the scTCR and its cognate pMHC target. In each of the fusions, the scTCR component functions to guide and deliver the therapeutic agent or effector function to the disease site where the cytokine or antibody binding/effector domains stimulate and augment an immune response. Using a similar approach, we have also generated scTCR-conjugated imaging agents and therapeutic drugs. Current development is focused on improving the activities and lowering the systematic toxicities of FDA-approved agents such as interleukin-2 (IL-2) (Proleukin®), GM-CSF (Leukine®), and doxorubicin, and promising drug candidates, such as IL-15, for treatment of malignant

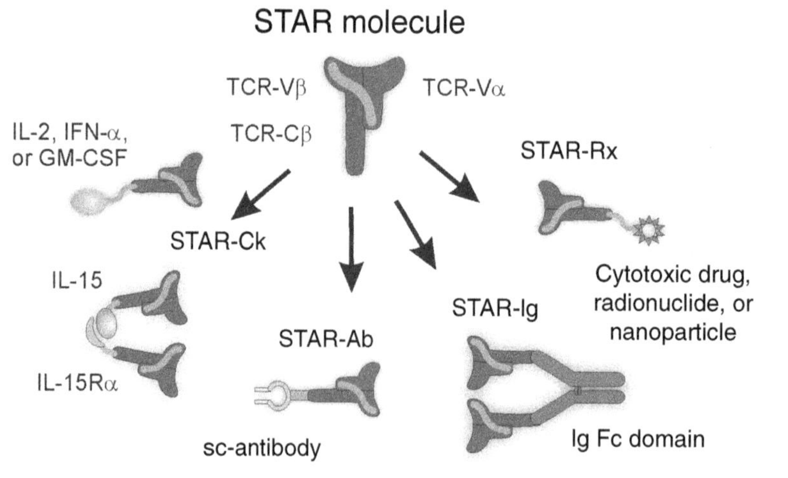

FIGURE 31.1 Diagram of STAR fusion proteins.

tumors, and IFN-α (Pegasys®) and small molecule viral inhibitors for chronic viral infections in humans. We are also testing the bispecific fusions linking the scTCR domains to anti-CD3 antibodies or IgG Fc effector domains as a means of redirecting T cell and natural killer (NK) cell responses, respectively, against tumors. As described in the following section, the fused cytokine and antibody domains retain their native functional activity and in many cases, exhibit improved pharmacokinetics *in vivo*. Thus, the scTCR-based fusions are being employed in a drug delivery system to enhance efficacy and lower the therapeutic dosage of currently approved drugs and to provide more convenient dosing regimens with decreased toxic side effects.

31.1.2 Mutagenesis and Multimerization for Improved Target Recognition

Naturally occurring TCRs generally interact with pMHC complexes at relatively low affinities, typically in the range of 1–100 μM [1]. This is due to the lack of somatic hypermutation and affinity maturation during T-cell development, the natural process of positive and negative selection of T-cells in the thymus, and the role of other membrane-bound co-receptors in stabilizing and augmenting TCR-mediated signaling. Several mutagenesis and selection approaches including phage, yeast, and mammalian cell display technologies have been employed to increase the affinity of the TCR to their targets, in some cases to picomolar affinities [18,19]. Generally, mutations in the TCR α and β CDR2 and CDR3 regions can provide variants with enhanced pMHC affinity either through direct contacts with the pMHC complex or stabilization of the TCR in a optimal binding conformation, though in some cases these variants also show a loss in peptide specificity suggesting increased interactions with the MHC component [20]. We and others have found that certain CDR3 mutations can enhance specific binding of the scTCR to its cognate pMHC complex and also improve interactions with epitope variants without a general loss of peptide specificity [21]. Such enhanced affinity TCRs have been found to useful in recognizing viral targets known to evolve during infection to avoid host T-cell immune responses. In the case of human immunodeficiency virus (HIV) antigen-specific TCR, we have isolated several different variants that bind most of the known escape mutations for immunodominant epitopes derived from vpr, p17, and p24 gag proteins. These reagents have the potential to broadly recognize HIV-infected cells carrying the HIV escape variants.

We have also employed a strategy for creating soluble scTCR as multimers with increase avidity to target pMHC molecules. In one example, STAR Multimer reagents have been created that consist of a soluble scTCR containing a biotin moiety. The biotinylated TCR molecules are multimerized following addition to streptavidin carrying detectable labels. The increased avidity of STAR Multimers results in an apparent higher pMHC binding affinity than that observed for the monomeric scTCR [15]. Thus, STAR Multimers are able to specifically and stably bind endogenous pMHC complexes under various experimental conditions [15]. We are exploring similar approaches using multiple copies of the scTCR to coat the surface of nanoparticles. These targeted nanoparticles containing therapeutic drugs have been found to be effective in reducing growth of tumors bearing the TCR-specific pMHC antigen in animal models.

Fusion of the scTCR domains to functional protein scaffold domains, such as the IgG Fc effector domains described earlier, have been used to create dimeric antibody-like molecules [14]. We have expanded this strategy to novel functional scaffolds including soluble dimerization domains of the IL-15/IL-15 receptor α (IL-15Rα) complex [22]. In each case, these multimeric scTCR fusions exhibit higher functional affinity to their cognate pMHC molecules than the monomeric form and can effectively target effector activity of the scaffold domains against cells expressing the pMHC. Additionally, we have found that scTCR fusions to other receptor binding proteins (i.e., cytokines or scAbs) can be used *in vitro* or *in vivo* to decorate receptor-positive cells with multiple scTCR binding domains. These cells are then able to specifically recognize the pMHC-positive target cells at the disease site in much the same manner as a native TCR-bearing T cell [23]. These findings, described in more detail in the following section, indicate that high-affinity interactions between monomeric TCR and pMHC proteins are not required for effective localization and retention of the dimeric/multimeric scTCRs or the scTCR-coated immune cells in target tissues carrying the cognate pMHC complexes [23]. Such results are consistent with recent theoretical models and experimental data of tumor uptake obtained using targeting molecules in different sizes and affinities [24].

31.2 EXPRESSION AND PURIFICATION OF RECOMBINANT STAR FUSION PROTEINS

STAR fusion proteins are produced in a soluble, active and fully glycosylated form by a recombinant Chinese hamster ovary (CHO) cell expression system. Mammalian cell post-translational modifications such as glycosylation are known to provide improved pharmacokinetics and lower immunogenicity of therapeutic proteins. In addition, glycosylation at particular sites in the TCR β chain may play a role in proper folding of the molecule [25]. Successful expression of TCRs of many different sequences clearly indicates that the "single-chain" format permits functional folding without the requirement of additional stabilizing modifications, denaturation/refolding steps or specialized production cells.

As indicated earlier, this format facilitates genetic fusions to other biologically active protein domains, allowing the generation of the STAR-Ck, STAR-Ig, and STAR-Ab reagents [5,14]. These advantages have allowed us to rapidly develop a broad platform of scTCR-based reagents directed against novel cancer and viral targets.

Generally, antigen-responsive human T-cell clones or pMHC tetramer-sorted T-cell lines obtained from patients with cancer or viral infections have served as the source to generate cDNAs encoding disease antigen-specific TCR α and β chains [15,23]. We have also isolated TCR genes from high-affinity T-cell clones obtained from peptide immunized HLA-transgenic mice [5]. Additionally, following the strategy of generating human antibodies in mice, we are creating transgenic mice containing unrearranged human TCR loci in the background of transgenic HLA alleles. Successful generation and characterization of such mice has recently been described by Li et al. [26]. This and our studies confirm that transgenic human TCR α and β loci can utilize mouse recombination enzymes and functionally rearrange during T-cell development to generate a diverse repertoire of human α/β TCRs [26]. Additionally, human α/β TCR loci transgenic mice carrying a HLA-A*0201 transgene could be immunized with human tumor peptide antigens to generate reactive T-cells with antigen-specific TCRs similar to those obtained from cancer patients [26]. Because of sequence divergence between human and mouse proteins, the human α/β TCR loci transgenic mice likely are not tolerized to most human tumor antigens allowing generation of T cells to many target antigens through standard immunization methods [26]. It has previously been shown that HLA transgenic mice lacking human CD4 or CD8 co-receptor transgenes can be used to generate TCRs, which are co-receptor independent and exhibit higher binding affinity toward the cognate peptide/HLA complexes [27].

Once isolated by standard PCR-based methods, the TCR α and β chain cDNA are constructed as scTCR genes in a mammalian expression vector followed by transfection into CHO cells. Short-term expression that permits assessment of the secreted fusion protein levels and pMHC-binding activity by standardized ELISAs using anti-TCR Cβ Abs and pMHC tetramer reagents. These simple methods allow rapid identification of functionally paired α/β chains and initial evaluation scTCR affinity and productivity. Suitable fusion proteins are then selected for larger scale production in stably transfected CHO cell lines in suspension culture conditions, a standard method for therapeutic protein manufacture. Using these methods, production of scTCR/IgG1 fusion proteins can reach the 0.5 g/L level, comparable to that observed for structurally similar antibody molecules. Further improvement in expression to the gram per liter level is expected to be highly feasible to support late phase clinical development or product commercialization.

The STAR fusion proteins are then purified via affinity chromatography using anti-TCR Cβ antibody or protein A (for scTCR/Ig fusion) coupled resin followed by an ion exchange chromatography step [5,14,15,17]. For GMP production of clinical product, robust viral inactivation and removal steps (i.e., low pH load, viral filtration) are also included in the purification process. The final product is formulated in a buffered saline solution and extensively characterized for composition, purity, and potency (i.e., TCR binding activity and effector domain function) using validated assays. In such a formulation, the STAR-Ck final vialed product (ALT-801) is stable for at least 4 years when stored at 5°C. As a result, the "industry-standard" upstream and downstream production methods developed for these fusion proteins are well suited to support late stage clinical development and potentially therapeutic product commercialization.

31.3 CLINICAL AND RESEARCH PRODUCT APPLICATIONS

As indicated earlier, we are utilizing the novel STAR platform technology to capture the precise, disease-targeting properties of TCRs in creating next-generation targeted therapeutic molecules for cancer and viral infection. Additionally, STAR reagents are being developed for diagnostic applications to identify disease-associated antigens derived from intracellular proteins as biomarkers [15] and for research application to better understand antigen presentation and T-cell immunology [28,29].

31.3.1 Cancer Therapeutics

Our strategy is to develop scTCR-based targeted therapeutics against broad-based tumor-associated antigen. Our lead anticancer agent, ALT-801, is a recombinant soluble scTCR/IL-2 cytokine fusion protein that recognizes cancer cells through p53-derived targets. Human p53 tumor suppressor protein is a particular interest for cancer target as it is mutated and overexpressed in up to 50% of all human cancers [30]. Normally, this protein that acts to arrest the proliferation of cells in response to various stresses, including DNA damage, hypoxia, and cell cycle aberrations [31]. When mutated, it loses its ability to suppress abnormal proliferation and accumulates within the cells [31]. As a result, p53 mutation and overexpression in cancer cells correlates with transformation events and tumor aggression and is associated with lower overall survival rates and resistance to chemotherapeutic intervention [32]. Since only low amounts of native p53 are detectable in normal tissue, the differential display of p53 in tumor versus normal tissues creates the opportunity to therapeutically target this protein. However, p53 is an intracellular protein and is not displayed on the cell surface, thus is not accessible to antibody-based agents. As with other intracellular proteins, p53 is processed and p53 peptides are presented at the cell surface in the context of HLA molecules. We and others have found that a peptide epitope (aa264-272) of p53 presented by HLA-A*0201 is displayed at high levels on the surface of different human tumor cells and tissues, whereas normal tissues do not present detectable levels of this complex [15,33,34]. Since this epitope is within a region of p53 that is rarely mutated, its cell-surface display serves as an attractive broad-based target for tumors that overexpress p53.

Cytotoxic T-lymphocytes (CTLs) that specifically recognize human p53 (aa264-272) peptide/HLA-A*0201 complex have been generated from HLA-A*0201/K^b-transgenic mice that were immunized with the p53 peptide [35]. The α/β TCR genes were isolated from a clonal line of these cells and constructed in a single-chain format genetically fused to human IL-2 to generate the soluble 264scTCR/IL-2 fusion protein [5]. The ALT-801 fusion protein is a chimeric form of 264scTCR/IL-2 wherein the mouse TCR β-constant domain of 264scTCR/IL-2 was replaced with the human counterpart [17]. This replacement is expected to significantly lower the immunogenicity of the TCR portion of the molecule in humans. As described in the following section, ALT-801 has been thoroughly characterized in preclinical efficacy and toxicology studies and in a Phase I human clinical study in patients with p53-overexpressing ($p53^+$) metastatic malignancies. This fusion protein is currently being evaluated in Phase II clinical studies in patients with melanoma and muscle invasive or metastatic urothelial cancer with addition indications in the planning stage. Using a similar strategy, we have linked the p53 peptide-specific scTCR domain to the human IgG1 domain to create the antibody-like 264scTCR/IgG1 and corresponding chimeric ALT-802 fusion proteins [14], the characterization of which is described in the following section.

We have also extended this approach to generate targeted fusion proteins with scTCR domains that recognize peptides specific to the melanoma differentiation antigens, gp100 and MART-1, and the cancer testis antigen, MAGE-A3, that are expressed at high frequency by human tumors including melanoma, nonsmall cell lung carcinoma (NSCLC), head and neck cancer, hepatocellular cancer, and multiple myeloma [36,37]. In each case, these scTCR domains are functionally active in recognizing their cognate peptide/HLA-A*0201 complexes presented on the cell surface. Overall, these fusion proteins have the potential to treat a broad range of cancers in HLA-A*0201-positive individuals, who represent ~50% of the U.S./European population. Expansion of this strategy to tumor peptide antigens displayed by other common HLA class I haplotypes, such as HLA-A24 in the Asian population, could be readily implemented based on current knowledge of tumor antigen presentation and tumor-specific T-cell responses [38].

31.3.2 Viral Therapeutics

Based on the success of our cancer-specific STAR program and due to the lack of effective preventive vaccines or viral therapies, we are taking a similar approach toward developing STAR molecules recognizing viral antigens from HIV, hepatitis C (HCV), and cytomegalovirus (CVM) antigens. These STAR molecules are being engineered into various formats as fusion proteins and drug or radioisotope conjugates to deliver antiviral therapies to diseased cells. To create the scTCR domains, we have established collaborations with leading viral immunologists who have provided Altor with virus-reactive human CTLs or TCR genes obtained from infected donors. From these reagents, we generated over 20 functional scTCRs reactive to different immunodominant viral peptide/HLA complexes. In some cases, these scTCRs have been further engineered though site-specific mutagenesis to generate variants with improved binding affinity to the cognate viral peptide antigen as well as known escape mutants that arise during the course of viral infection. Thus, our collection of antiviral STAR molecules provides a broad range of specific reactivities against infected cells. In the case of HIV, we have developed a panel of STAR molecules capable of recognizing epitopes that are restricted by different HLA molecules including those associated with disease protection (HLA-B*5701 and HLA-B*2705) or found at high frequency (i.e., HLA-A*0201 in the U.S./European populations and HLA-A24 in the Asian population). Through our collaborations, we have demonstrated that these fusion proteins are capable of recognizing viral peptide presented on HIV-infected cells and directing cytotoxicity via conjugated IgG Fc domains or radioisotopes against cells displaying HIV antigens. These agents could form the basis of a new treatment strategy of HIV, whereby the STAR fusion proteins target the reservoir of HIV-infected cells in patients currently on approved antiviral therapies. Potentially, such a strategy could result in the elimination of viral infection as opposed to suppression of viral replication with continued residual disease as observed with current treatments. A similar approach is being taken for HCV and CVM specific STAR fusion proteins where our preclinical testing is focused on a STAR-Ck fusion between the approved therapeutic cytokine IFN-α and a high-affinity scTCR domain specific recognizing a peptide derived from the HCV protein NS5 presented in the context of HLA-A*0201. Testing is underway to determine if this fusion protein capable of delivering IFN-α to HCV-infected cells, thereby improving its efficacy while reducing systemic side effects.

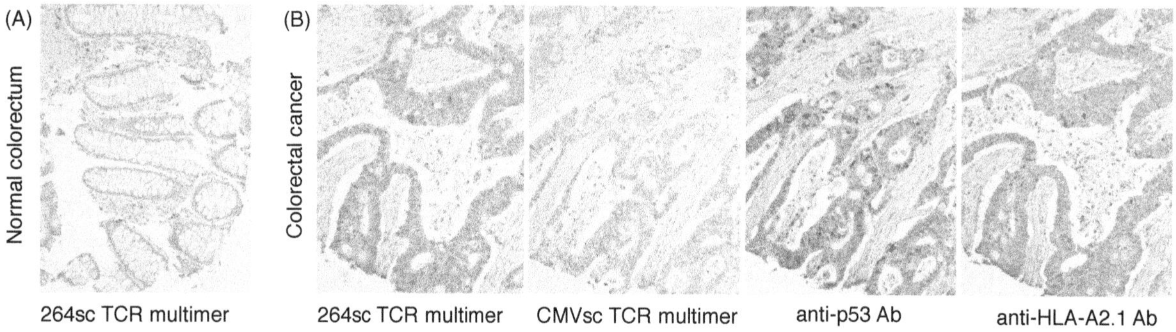

FIGURE 31.2 Detection of p53 (aa264-272)/HLA-A*0201 complexes on formalin-fixed paraffin-embedded tissue microarrays (TMAs) from normal or cancerous human tissues with STAR Multimer. (A) TMAs from normal colorectal or breast tissues were stained with HRP-conjugated 264scTCR multimer. (B) TMAs from cancerous colorectal and breast tissues stained with HRP-conjugated 264scTCR multimer, negative control HRP-conjugated CVMscTCR multimer, anti-HLA-A2 Ab, or anti-p53 Ab (magnification ×200). The results show differential staining of p53$^+$ tumor specimens compared to normal tissue by the p53-specific STAR Multimer. *Source*: Reproduced with the publisher's permission from Reference [15].

31.3.3 Diagnostic and Research Applications

In addition to their potential as targeted therapeutics, STAR molecules have utility as diagnostic or research reagents. As described earlier, we have found that multimeric forms of STAR molecules can be used to specifically detect endogenous peptide antigens displayed in the context of MHC molecules on antigen-presenting cells (APCs); cancer cells and tissues; and virally infected cells, including formalin-fixed, paraffin-embedded tissue, and microarray specimens. Additionally, we have developed simple and sensitive staining procedures using STAR reagents to differentiate human tumors from normal tissues (Figure 31.2) [15]. These methods are being employed in the clinical development of ALT-801 to identify patients whose tumor biopsies are positive for p53 (aa264-272)/HLA-A*0201 complexes. Thus, STAR reagents can be used to select cancer patients who would most likely to receive benefit from targeted therapies. This approach is commonly used with antibody-based diagnostic methods, such as the HercepTest™ for determination of HER2 protein overexpression in breast cancer patients prior to selection of Herceptin™ treatment.

We are also exploring use of STAR reagents as tools for viral vaccine development. Current methods to examine presentation of peptide antigens derived from HIV and other viruses are not fully quantitative and cannot evaluate events in real time or on an individual cell basis. As a result, there is no clear consensus as to which viral peptide antigens are actually presented to the immune system at different stages of infection. With our panel of viral peptide-specific STAR reagents and simple cytometry methods, we can directly quantitate peptide antigen presentation on HIV-infected cells. Efforts are underway to use these methods to fully evaluate the levels and timing of HIV antigen presentation in cells on a per cell basis during the course of HIV infection. The results of these studies in conjunction with evolving knowledge of CD8$^+$ T-cell responses and HIV escape from T-cell immunity should aid in the selection of the most relevant HIV targets for vaccine development. Successful application of this approach could also be used for other viruses and infectious agents to more fully understand the antigen presentation in regard to optimization of vaccine design.

We have found that STAR reagents are capable of detecting as little as 300 pMHC complexes per cell by flow cytometry and can distinguish peptide antigens that differ by a single amino acid [15]. Thus, STAR Multimers have the specificity and sensitivity to characterize peptide antigen presentation in different experimental settings in much the same way pMHC tetramers have been used to characterize antigen-specific T-cell populations. We have also found that these regents can be readily used to evaluate pMHC complexes on the surface of APCs and dendritic cells and could have utility in characterizing autologous cellular immunotherapies, such as the recently approved prostate cancer vaccine, Provenge®. Current methods to characterize the potency of such cell-based therapies have relied on T-cell response assays or surrogate assessment of APC activation [39]. With sensitivity comparable to T cell-based methods, STAR Multimer staining could provide a more direct and quantitative approach to optimizing and validating the production of APC-based vaccines.

Based on the unique binding capabilities, we have been offering STAR Multimers to the research community to expand the utility and application of these reagents in other areas of interest. For example, we have created a STAR Multimer reagent capable of detecting adeno-associated virus 2 (AAV) caspid peptide/HLA-B*0702 complexes on cells by flow cytometry and confocal microscopy methods [28]. Using this reagent, Pien et al. [28] demonstrated that low levels of AAV capsid-derived antigen are endogenously presented on human hepatocytes following AAV transduction. The STAR reagent could also be used to protect transduced hepatocytes from CTL-mediated cytolysis directed against cognate and cross-reactive epitopes derived from different AAV serotypes. These results provide insight as to how low levels of capsid antigen presentation and subsequent immune responses may limit the effectiveness of AAV-based gene therapies [28]. Additionally, STAR Multimers have been used to quantitate antigen presentation in addressing more fundamental questions on the relationships between antigen dose and events leading to T-cell activation [29]. Together with their potential in clinical diagnostic and vaccine development applications described earlier, these studies confirm that STAR reagents can be used in simple, direct, sensitive, and quantitative methods to advance the understanding of endogenous antigen presentation and related immune responses.

31.4 PRECLINICAL TESTING USING STAR FUSION PROTEINS

31.4.1 Preclinical Efficacy, Pharmacokinetic, and Toxicity Studies

ALT-801 and its original nonchimeric murine version, 264scTCR/IL-2, have been extensively characterized *in vitro* as described early [5,17]. Initial studies of 264scTCR/IL-2 showed that it retains the ability to specifically recognize the p53 (aa264-272) peptide displayed in the context of HLA-A*0201 on peptide-loaded cells as well as on melanoma cells and that the IL-2 portion of the molecule exhibits full biological activity [5]. The fusion protein also was capable of mediating conjugation of effector and target cells bearing the appropriate IL-2 receptor and p53 peptide/HLA-A*0201 complexes. The affinity (K_D) of the 264scTCR/IL-2 protein for p53 peptide/HLA-A*0201 complex was determined to be in the range of 10^{-6}–10^{-7} M [5]. Similar analysis carried out with ALT-801 indicated that it retained both the TCR binding activity and IL-2 biological activity of the 264scTCR/IL-2 fusion protein [17].

The antitumor activities of ALT-801 and 264scTCR/IL-2 fusion proteins have been evaluated in xenograft tumor models in immunocompromised mice [5,17,23]. Initially, we demonstrated that 264scTCR/IL-2 treatment could significantly reduce tumor burden in an experimental mouse model of lung metastasis with $p53^+$/HLA-A*0201^+ A375 human melanoma cells [5]. Studies were also conducted to directly compare the antitumor activity of the p53-specific scTCR/IL-2 fusions to recombinant human IL-2 in more clinically relevant tumor models [17,23]. Nude mice bearing primary tumors were injected intravenously (i.v.) with 264scTCR/IL-2 (from 0.15 to 1.6 mg/kg), an equivalent dose of IL-2, or PBS daily for 4 days, followed by treatment every other day for a total of nine doses. With this treatment regimen, 264scTCR/IL-2 inhibited the growth of subcutaneous tumors derived from the A375 melanoma cells in a dose-dependent manner and exhibited significantly better antitumor activity than IL-2 alone. Moreover, treatment with 264scTCR/IL-2 resulted in significant tumor growth retardation in mice bearing large A375 tumors (Figure 31.3A) [17]. The chimeric ALT-801 fusion protein was also shown to exhibit potent dose-dependent antitumor activity, equivalent to that of murine version, in this model (Figure 31.3B) [17]. In addition, the p53-specific 264scTCR/IL-2 fusion was significantly more effective than IL-2 in inhibiting the growth of tumors derived from $p53^+$/HLA-A*0201^+ PANC1 human pancreatic adenocarcinoma cells (Figure 31.3C) and $p53^+$/HLA-A*0201^+ MDA-MB-231 human mammary adenocarcinoma. However, tumor outgrowth is not inhibited in 264scTCR/IL-2 treated mice bearing HLA-A*0201-negative human tumors (Figure 31.3D), confirming that antigen targeting plays a role in the efficacy of 264scTCR/IL-2 against $p53^+$/HLA-$A2^+$ tumors [17]. It also demonstrates that ALT-801 has substantial antitumor activity against $p53^+$ tumors regardless of the tumor origin.

A major limitation with recombinant human IL-2 therapy is the thrice-daily i.v. dosing regimen required due to the molecule's short circulatory half-life. In contrast, evaluation of ALT-801 in a humanized HLA-A2.1/K^b-transgenic mouse strain indicated that the fusion protein has a fivefold longer terminal half-life (~2 h vs 0.4 h: IL-2) and is cleared >90-fold slower (0.5 mL/h vs 42.0 mL/h: IL-2) than IL-2 [23]. Results of these studies also indicate that the scTCR and IL-2 domains of ALT-801 are not cleaved *in vivo*, but instead remain stably associated throughout the *in vivo* metabolism and clearance of ALT-801. Thus, it is clear that the generation of the fusion between the IL-2 and the scTCR domain provides a significant improvement in the pharmacokinetic profile of the cytokine, consistent with the results observed with other cytokine fusion proteins [17,23,40].

To understand differential effects of tumor antigen targeting and enhanced pharmacokinetics on ALT-801 efficacy, A375 tumor bearing mice were treated with ALT-801 or a control scTCR/IL-2 fusion, MART1scTCR/IL-2, which does not recognize A375 tumor cells. The results shown in Figure 31.3E indicate that ALT-801 treatment continues to exhibit potent antitumor activity compared to PBS and that MART1scTCR/IL-2 provides intermediate but significant inhibition of A375 tumor growth in this model. In separate studies, MART1scTCR/IL-2 was found to display equivalent IL-2 bioactivity as ALT-801 and a serum half-life

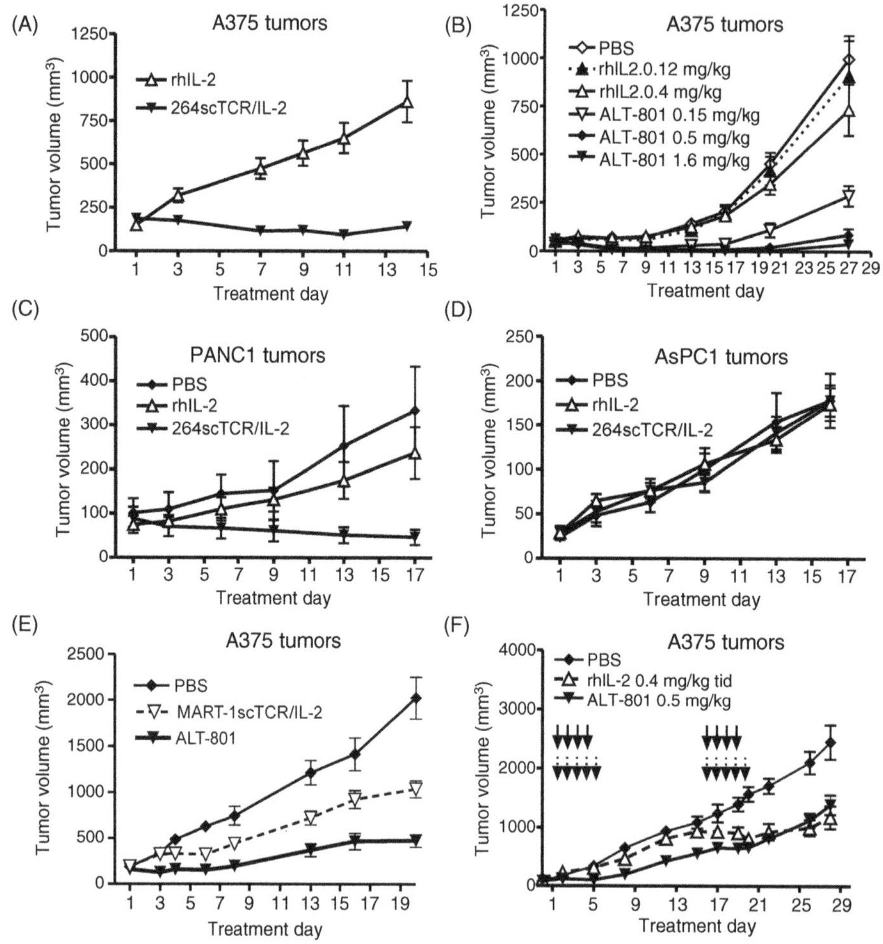

FIGURE 31.3 Effect of 264scTCR/IL-2 and ALT-801 on growth of subcutaneous human tumor xenografts in nude mice. (A) A375 melanoma tumor cells (1×10^6 cells) were injected subcutaneously into nude mice and tumors were allowed to grow to 200 mm^3. Starting on treatment day 1, mice were administered 264scTCR/IL-2 (1.6 mg/kg) or IL-2 (0.4 mg/kg, molar equivalent dose) for 4 days, followed by treatment every other day for a total of nine doses. (B) Subcutaneous A375 tumor-bearing nude mice were treated i.v. with ALT-801 (1.6 mg/kg, 0.5 mg/kg, or 0.15 mg/kg) or IL-2 (0.4 mg/kg or 0.12 mg/kg) as described in (A). (C and D) Nude mice were injected s.c. with p53$^+$/HLA-A2.1$^+$ PANC1 (C) or p53$^-$/HLA-A2.1$^-$ AsPC1 pancreatic tumor cells (D). Tumors were allowed to establish to ~25–100 mm^3 and mice were treated with molar equivalent doses of IL-2 or 264scTCR/IL-2, or the dose equivalent volume of PBS as described in (A). (E) Subcutaneous A375 tumor-bearing nude mice were treated i.v. with either 0.5 mg/kg ALT-801, 0.5 mg/kg control TCR/IL-2 fusion (MART-1scTCR/IL-2) or PBS with four daily injections, followed by a 10-day rest period and then four more daily injections. (F) Subcutaneous A375 tumor-bearing nude mice were treated with 0.5 mg/kg ALT-801 as described in (E) (dosing indicated by solid arrows) or with a 3.2-fold higher molar equivalent amount per dose (0.4 mg/kg) of IL-2 every 8 h for 5 days followed by a 10-day rest period and then every 8 h for five more daily doses (dosing indicated by dashed arrows). For each study, tumor volumes (5–8 mice per group) were measured at least twice a week and were plotted as mean ± SEM. *Source*: (A), (C), and (D) are reproduced with the publisher's permission from Reference [17]. (E) is reproduced with the publisher's permission from Reference [23].

of 3.2–3.6 h in HLA-A2 transgenic mice, about twice that of ALT-801 and 10-fold that of IL-2 [23]. Thus, these results suggest that increasing the biological half-life of IL-2 without specific tumor binding, as is the case for the MART1scTCR/IL-2 fusion, leads to better antitumor activity than is seen with IL-2 alone in these models. However, the combination of improved IL-2 serum half-life and tumor recognition activity of the TCR domain creates an even more potent antitumor agent as exemplified by the ALT-801 fusion protein.

Efficacy studies were also conducted to fully compare ALT-801 treatment with that of Proleukin-based thrice-daily high-dose IL-2 administration. A375 tumor-bearing nude mice treated with 0.5 mg/kg ALT-801 or PBS daily for four injections followed by a 10-day rest period and then four more daily injections or with 0.4 mg/kg IL-2 every 8 h for 5 days followed by a 9-day rest period and then every 8 h for five more daily injections. In this study, each single dose of IL-2 had 3.2 times the activity as the ALT-801 dose and the cumulative dose of IL-2 was 12-fold higher than that of ALT-801. As shown in Figure 31.3F, both ALT-801 and IL-2 treatment inhibited tumor growth to a similar degree following two dosing cycles. However, the IL-2 regimen also exhibited more toxicity based on clinical observations (hypoactivity, mottled skin, and hunched postures) and decreased body weight than was observed in ALT-801-treated mice (J. Wen, unpublished data). The potency of ALT-801 observed at a dose as low as 0.15 mg/kg dose in the A375 tumor model (Figure 31.3B) [17]. Given that ALT-801 was as effective as a 12-fold higher cumulative dose of IL-2 in reducing tumor burden without the overt signs of toxicity, this study provides additional support that ALT-801 could offer safer and more effective treatment for cancer than current high-dose IL-2 therapies.

The acute single-dose and multidose toxicity of ALT-801 was also extensively evaluation in HLA-A*0201/K^b transgenic mice. This strain was selected since the presence of HLA-A*0201 domain, for which the 264scTCR is restricted, may influence the pharmacokinetics and/or toxicity of this protein and should give a more relevant view of the parameters than other nonhuman animal models. At an single dose of up to 7.5 mg/kg, equivalent to over 50 times the single dose of recombinant human IL-2 approved for treatment of cancer patients, exhibited no drug related mortalities and no or minimal differences were observed between ALT-801 and PBS-treated groups in terms of gross pathology, hematology (including lymphocyte counts), and serum chemistry. The primary difference observed in the histopathology was a moderate increase in immune inflammation in the liver in the ALT-801 group compared with control animals. These results are expected based on lymphocyte infiltration previously observed in immune organs of recombinant IL-2-treated mice [41].

Various toxicology studies employing multidose ALT-801 treatment regimens in HLA-A*0201 transgenic mice also demonstrated that two cycles of four daily consecutive i.v. injections at up to 0.5 mg/kg of ALT-801 did not result in obvious toxicological effects. Additionally, no adverse clinical signs were observed at doses up to 1.85 mg/kg given nine times over a 2-week period, though dose- and schedule-dependent changes, including increases in normalized lymph node, lung, pancreas, and spleen weights, decreases in serum alkaline phosphatase and increases in serum AST/ALT levels, and changes in hematology and tissue histopathology, were observed in some ALT-801 treated mice compared to the control group. These ALT-801-induced changes return to control levels by 2 weeks after treatment. Overall, the changes observed following repeated ALT-801 administration in HLA-A2.1/K^b-transgenic mice appeared to be consistent with the reported biological activity of IL-2 in other mouse models [41] and no addition toxicity effects due to the scTCR domain were evident.

A GLP study was also conducted to determine the reactivity of ALT-801 to normal human tissues by immunohistochemical techniques. In a panel of 35 normal human tissue specimens from at least three unrelated donors, tissue-cross reactivity of ALT-801 was noted primarily in the bone marrow, T-cell regions of the spleen, tonsil, and lymph nodes. It has been well characterized that expression of human IL-2 receptor subunits gives rise to various mid- to high-affinity receptor forms on the surface of T and B cells, NK cells, and monocytes [42]. As ALT-801 functionally binds to the IL-2 receptor, the observed tissue cross-reactivity of immune organs was expected. There was no tissue cross-reactivity of ALT-801 with other tissues or cell types. These findings are consistent with the results of biodistribution studies reported for other IL-2 fusion proteins including the nontargeted albumin/IL-2 fusion [43]. Expression of HLA-A*0201 was also assessed on the panel of human tissues but no correlation between ALT-801 reactivity of these tissues and the HLA-A*0201 status of the tissues or donors was observed. Thus, p53-specific scTCR domain of ALT-801 does not appear to exhibit detectable reactivity to normal human tissues, consistent with previous studies of this TCR [15].

31.4.2 Antitumor Mechanism-of-Action Studies of ALT-801

Initial analysis of residual tumors from ALT-801-treated mice provided evidence that the antitumor activity of ALT-801 is likely mediated by the innate immune system via NK cell activation and tumor infiltration [17]. The effect of ALT-801 on immune cells was further investigated in A375 tumor-bearing nude mice over time during the course of treatment [23]. Daily administration of either ALT-801 or the control MART1scTCR/IL-2 fusion for 3 or more days resulted in an increase in IL-2 receptor (IL-2R) expression on immune cells, consistent with the known biological activities of IL-2. These results suggest that ALT-801 treatment can augment subsequent immune responsiveness (via IL-2R-positive cells) to the fusion protein. In addition, repeated administration of ALT-801 (1.6 mg/kg i.v. daily) in A375 tumor-bearing mice resulted in massive treatment-dependent infiltration of $CD45R^+$ cells into the tumors (Figure 31.4A). This effect was not observed following treatment with MART1scTCR/IL-2 or PBS, verifying the target specificity of this effect (Figure 31.4B). The tumor

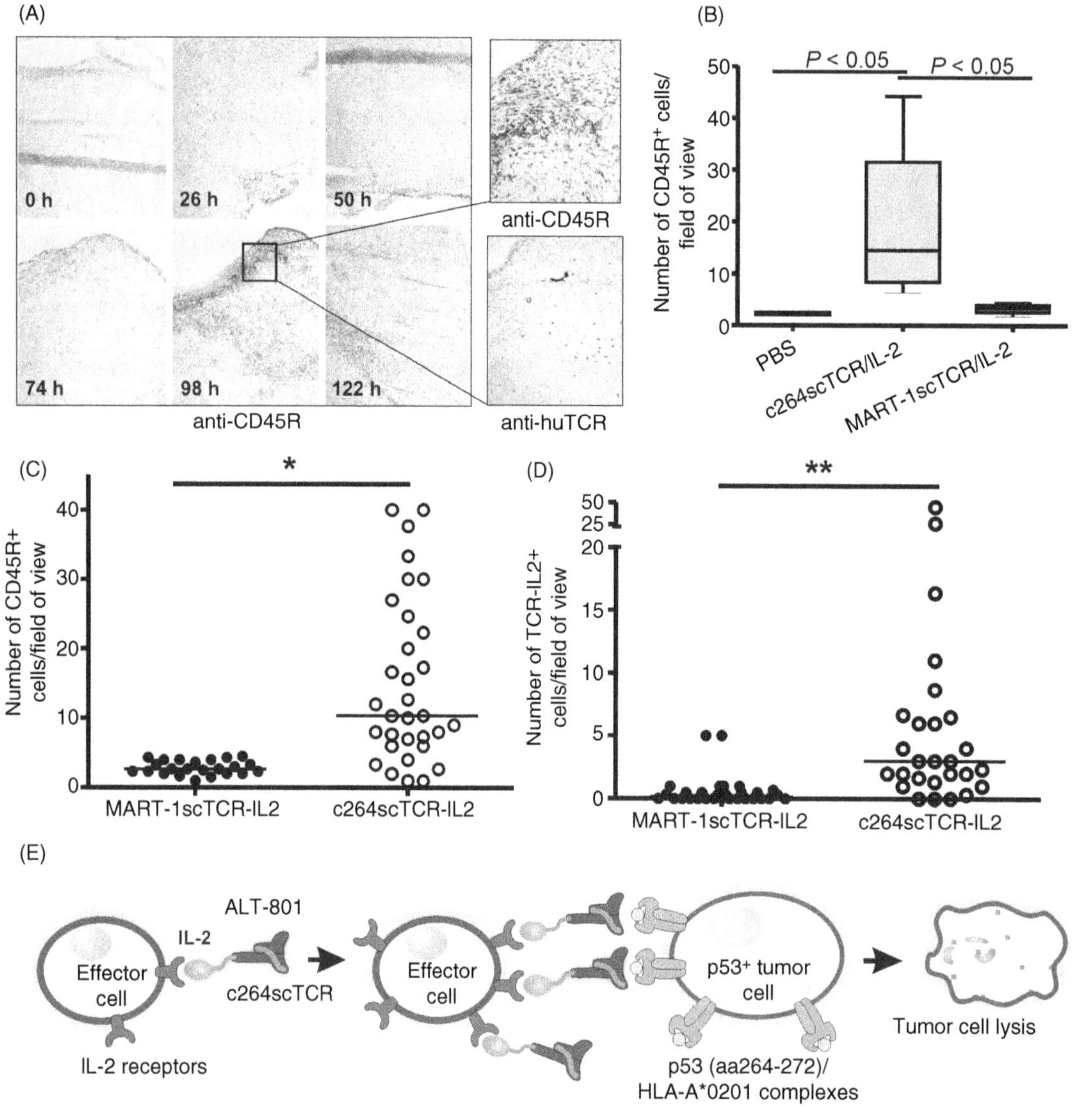

FIGURE 31.4 ALT-801-specific trafficking of immune cells to the tumor site. (A) A375 tumor bearing nude mice were treated i.v. with ALT-801 (1.6 mg/kg) at 0, 24, 48, 72, and 120 h. At the indicated times, tumors were collected and cryosections were stained with anti-CD45R and antihuman TCR mAb. Shown on the left, increasing amounts $CD45R^+$ cells (dark stained cells) were observed in the A375 tumors as treatment with ALT-801 continued for 3–5 days. Tumor cryosections from the 98 h sample also showed positive staining with the antihuman TCR antibody indicating cells bearing the ALT-801 fusion protein (see higher magnification on right). (B) Tumor samples were obtained 24 h post-treatment from A375 tumor-bearing nude mice injected daily for 4 days with ALT-801 (0.5 mg/kg), MART-1scTCR/IL-2 (0.5 mg/kg), or PBS. Tumor sections stained with anti-CD45R mAb and the positive cells were counted. The median number of $CD45R^+$ cells per field of view is presented in the graph. (C and D) Tumor samples were obtained 24 h post-treatment from A375 tumor-bearing nude mice injected with ALT-801- or MART-1scTCR/IL-2-coated splenocytes. Tumor sections stained with anti-CD45R (C) or anti-huTCR mAb (D). Symbols represent number of positively staining cells per field of view and bars represent median value. (E) Model of ALT-801 mechanism of action in binding to and activating IL-2 receptor-bearing immune effector cells and subsequently localizing these cells in the tumor site via pMHC specific interaction with tumor cells. *Source*: (A–D) are reproduced with the publisher's permission from Reference [23].

localized immune cells were not stained with B-cell marker-specific antibodies (i.e., anti-CD19 antibodies) indicating that they are primarily NK cells. At the point of peak NK cell infiltration, ALT-801 was also detected with anti-huTCR antibody in the tumor sections, suggesting that the fusion protein is associated with the immune effector cells. In addition, the time course of NK cell infiltration correlates well with the level of tumor growth inhibition observed in TCR/IL-2 efficacy studies using this xenograft tumor model [23].

Adoptive cell transfer experiments further verified the tumor targeting effects of ALT-801 [23]. *In vivo* activated splenocytes were isolated and incubated with ALT-801 or MART1scTCR/IL-2 fusion proteins prior to transfer into A375 tumor bearing nude mouse recipients. Significantly, more accumulation of $CD45R^+$ cells was observed in tumors from nude mice receiving splenocytes incubated with ALT-801 than in tumors from mice treated with splenocytes preincubated with the control MART1scTCR/IL-2 (Figure 31.4C, D). In addition, when the tumor sections were stained for either ALT-801 or MART1scTCR/IL-2 fusion proteins, significantly more cells bound with detectable levels of ALT-801 were observed infiltrating into the tumors.

Based on these studies [23], a model has been proposed in which ALT-801 first binds to the intermediate-affinity IL-2 receptor, which is constantly expressed on NK and other immune cells, and stimulates the cells to express high- and intermediate-affinity IL-2 receptors. Subsequently administered ALT-801 binds these receptors to form a stable complex and through the TCR domain retargets the NK cells to the p53/HLA-A*0201 antigen on the tumors. The activated NK cells then mediate their cytotoxic effects against the tumors (Figure 31.4E). The cytotoxic effects of the activated NK cells against tumor may not need the assistance of the TCR once they are inside the tumor. It has been established that NK cells could be stimulated by stress proteins produced by cancer cells through the NK cells activating receptors [44]. This "cell-mediated" mechanism of action differs from that of most therapeutic antibodies that rely on high-affinity interaction of soluble antibody with the tumor antigens. This model also predicts that a high-affinity TCR may not be necessary for efficacy if multiple copies of scTCRs bound to the surface of activated NK cells provide sufficient avidity to permit stable interactions with antigen-positive target cells.

31.4.3 Preclinical Efficacy Studies of Anticancer p53 Specific scTCRIgG1 Fusion Proteins

As indicated earlier, we have generated antibody-like fusion molecules by linking the scTCR domain to IgG constant domain, particularly the CH2-CH3 domains comprising the Fc dimerization fragment. We have demonstrated that the p53-specific 264scTCR/IgG1 fusion was capable of binding Fc receptors (FcR) on U937 human monocytic cells and p53 (aa 264-272)/HLA-A*0201 complexes on tumor cells [14]. This fusion protein was capable of promoting target cell and effector cell conjugation and mediating effector cell killing of target cells presenting p53 (aa 264-272)/HLA-A*0201 complexes by a mechanism similar to antibody-dependent cellular cytotoxicity (ADCC).

An experimental metastasis lung model was used to determine whether 264scTCR/IgG1 exhibits antitumor activity *in vivo* [14]. Female athymic nude mice were injected with the highly metastatic $p53^+$/HLA-A*0201^+ human melanoma subclone, A375-C15N, and tumors were allowed to establish for 3 days. Mice were then treated with varying doses of 264scTCR/IgG1, anti-STX mAb (an isotype control mAb), or vehicle. Forty-two days after tumor cell injection, lung nodules were counted. As shown in Figure 31.5A, 264scTCR/IgG1 reduced lung metastasis in a dose-dependent manner reaching maximum effect at a dose of 1 mg/kg, whereas the isotype control failed to reduce metastasis. These data indicate that 264scTCR/IgG1 exhibits targeted activity *in vitro* and potent anti-metastatic activity *in vivo*, confirming that replacement of the antibody recognition domains with that of a scTCR can provide an antibody-like molecule with novel tumor antigen specific functionality [14].

Using a similar strategy to the development of ALT-801, we have created a chimeric version of the 264scTCR/IgG1 fusion protein (ALT-802). Additionally, to assess the role of IgG effector function in this context, we have generated a variant of this protein (ALT-802LALA) that contains two mutations in the Fc domain known to reduce FcR-binding activity [45]. As expected, the ALT-802 fusion protein was capable of binding FcR on U937 cells and directing human immune effector cell cytolytic activity against p53 (aa 264-272)/HLA-A*0201-positive target cells, whereas the ALT-802LALA fusion exhibited decreased FcR binding and little or no ADCC-like effector activity. The antitumor activity of ALT-802 was examined in an NSCLC xenograft model using a derivative of the $p53^+$ human lung adenocarcinoma A549 cell line [46] that had been transfected to express HLA-A2*0201 (A2-A549-L1). By 10–15 days after i.v. injection into nude mice, the A2-A549-L1 cells form lung nodules that closely resemble the clinical features of human NSCLC. When mice were treated after lung tumor establishment (10-day post-tumor cell injection) with ALT-802 (3–10 mg/kg), ALT-802LALA (3 mg/kg), or PBS (equivalent volume) i.v. twice-weekly for 3–4 weeks, ALT-802 administration reduced the number and size of the pulmonary tumor nodules compared to PBS treatment (Figure 31.5B) (J. Wen, unpublished data). ALT-802LALA fusion protein did not exhibit this antitumor effect consistent with its lack of ADCC-like activity. Mice

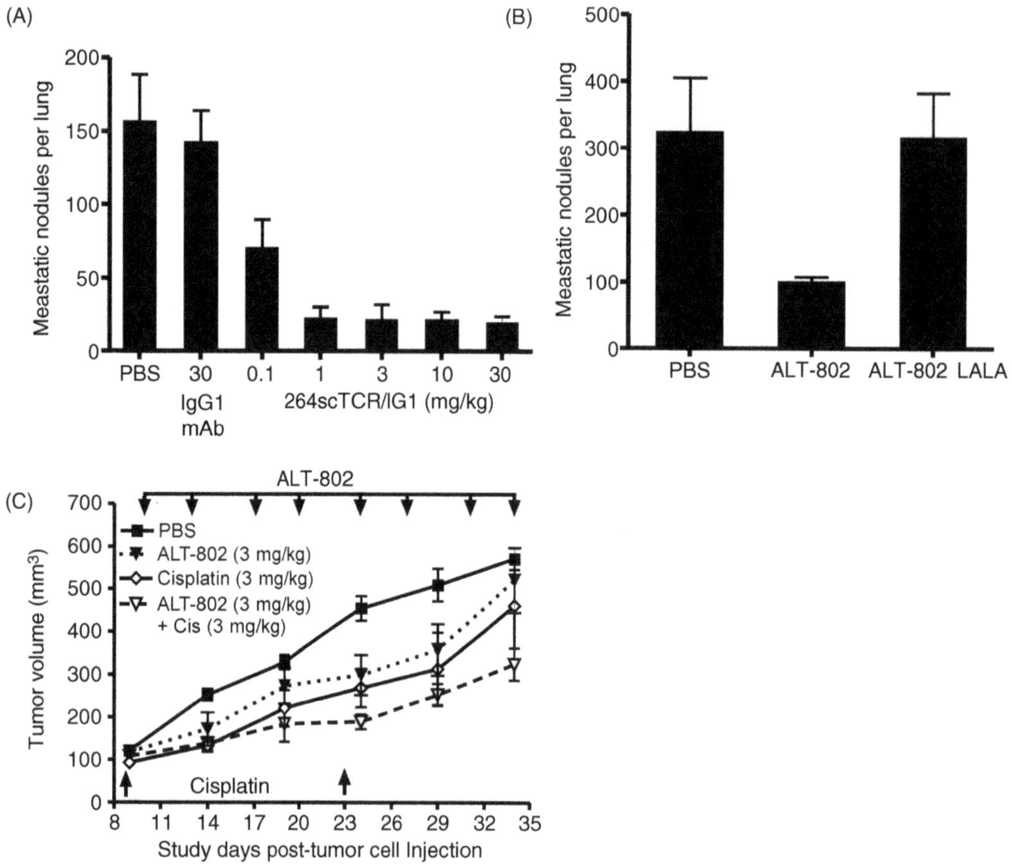

FIGURE 31.5 Antitumor activity of 264scTCR/IgG1 and ALT-802 in xenograft tumor models. (A) Female athymic nude mice were injected i.v. with $p53^+$ HLA-A*0201^+ A375-C15N human melanoma tumor cells and treated with PBS, 30 mg/kg anti-STX2 mAb (irrelevant IgG1 isotype control), or 264scTCR/IgG1 at doses ranging from 0.1 to 30 mg/kg as described previously [14]. Forty-two days after tumor cell injection, the lungs were removed, lung nodules were counted, and the mean number of lung nodules (±SEM) was plotted. *Source*: (A) is reproduced with publisher's permission from Reference 14 (B) Nude mice were injected i.v. with $p53^+$ HLA-A*0201^+ A549-A2 human nonsmall cell lung tumor cells on study day 0. The tumor-bearing mice were treated i.v. on study days 10, 13, 17, 20, 24, 27, and 33 with 3 mg/kg ALT-802 or ALT-802LALA or an equivalent volume of PBS. Thirty-four days after tumor cell injection, the lungs were removed, lung nodules were counted, and the mean number of lung nodules (±SEM) was plotted. (C) Nude mice bearing subcutaneous $p53^+$ HLA-A*0201^+ human melanoma A375-C15N tumor cells were treated with PBS, cisplatin, ALT-802, or cisplatin + ALT-802 as indicated in the figure legend. Tumor volumes were measured every other day and means ± SEM were plotted.

treated i.v. with up to eight doses at 10 mg/kg ALT-802 over a 4-week period did not exhibit any clinical signs of toxicity. Pharmacokinetic analysis has also been conducted indicating that ALT-802 has a half-life of ~38 h in transgenic HLA-A*0201 mice and 28 h in cynomolgus monkeys, which is similar to that of other IgG fusion proteins (J. Egan, unpublished data). Together these results indicate that ALT-802 exhibits potent Fc-mediated antitumor activity against lung tumors in our experimental metastatic model with potential use as a cancer therapeutic in humans.

31.4.4 Evaluation of Combination p53-Specific STAR Fusion Protein and Chemotherapy Regimens in Tumor Efficacy Models

Through development of novel cancer therapies over the past 20 years, it is apparent that combination regimens, including biological and chemotherapeutic agents, provide more effective outcomes than respective monotherapies [47,48]. These effects may be result from one agent sensitizing the tumor cells to the cytotoxic activity of another agent. For example, antibody-based approaches have been shown to be active in targeting tumor antigens induced by chemotherapy [49]. Since various cytotoxic chemotherapies, such as cisplatin, are know lead to cell cycle arrest and p53 induction by tumors [50], we were interested in evaluating the combination of p53-specific STAR fusion proteins and cisplatin in our animal efficacy models.

Initial studies confirmed the cisplatin induced expression of p53 (aa 264-272)/HLA-A*0201 complexes on the high aggressive human melanoma A375 subclone (A375C15NL1) and human NSCLC A2-A549-L1 cells following *in vitro* incubation. In the subcutaneous A375C15NL1 melanoma mouse xenograft model, monotherapy with ALT-802 (i.e., twice-weekly dosing for 4 weeks) or cisplatin (i.e., every 2 weeks) exhibits similar antitumor effects. However, combined therapy with ALT-802 and cisplatin exhibited significantly better antitumor activity than that of ALT-802 monotherapy (Figure 31.5C) (W. Xu, J. Wen, unpublished data). Similarly, the combination of ALT-801 treatment (i.e., three doses per weekly cycle, a 10-day rest, and another three doses per weekly cycle) with a single-dose cisplatin showed enhanced efficacy in this model compared to ALT-801 or cisplatin treatment alone. The synergistic effects of cisplatin and fusion protein treatment are possibly due to cisplatin-mediated increases in p53 peptide antigen display on the tumors and subsequent enhanced tumor targeting of the p53-specific scTCR domain. Since cisplatin is a component of standard-of-care therapy for advanced bladder cancer, NSCLC and several other cancers and been used in biochemotherapy regimens with high-dose IL-2 for treatment of melanoma [48], it is of interest to further evaluate the beneficial effects of ALT-801 and ALT-802 in combination with cisplatin in the clinical setting.

31.5 CLINICAL DEVELOPMENT OF ALT-801

31.5.1 Monotherapy Clinical Study with ALT-801

With the favorable preclinical testing results, the clinical objective for ALT-801 is to determine if this fusion protein has potential as a targeted immunotherapeutic with better antitumor potency, a lower toxicity profile and more convenient dosing schedule compared to that of human recombinant IL-2. It is also of particular interest to expand clinical utility of IL-2-based immunotherapy beyond its two FDA-approved indications, metastatic melanoma and renal cell carcinoma. Toward these aims, we have completed an open-label, dose-escalating Phase I/IIa clinical trial (Clinicaltrials.gov: NCT00496860) in which 26 patients with various metastatic, refractory malignancies were enrolled and infused with at least one and as many as eight doses of ALT-801 per treatment course [51]. Results indicate that ALT-801 is well tolerated at the maximum-tolerated dose (MTD) level exhibiting the general spectrum of side effects of high-dose IL-2 therapy but with a much lesser degree of severity and less grade III/IV effects on liver and kidney function. ALT-801 treatment resulted in immune stimulation with increased serum interferon-γ (IFN-γ) but not TNF-α ~4–8 h postdose and elevated NK cell activation following the initial dosing cycle. The serum half-life of ALT-801 was about 4 h consistent with the animal pharmacokinetic data, indicating that the half-life of fusion protein is considerably longer than the <15 min distribution and 85 min elimination half-lives reported for recombinant human IL-2 [52]. Furthermore, the maximum serum ALT-801 concentrations observed were consistent with the ALT-801 dose, in contrast to the mere 30% serum recovery observed in subjects receiving IL-2 therapy. ALT-801 treatment exhibits evidence of dose-dependent clinical benefit to cancer patients with 40% of the patients having stable disease, several with tumor shrinkage and one patient with a complete response (CR). The patient with the CR has been disease-free for 2 years postdosing without other anticancer therapy. This suggests that ALT-801 can provide a durable response in patients with metastatic malignances and mirrors that of high-dose IL-2 treatment but without the severe toxicities associated with Proleukin therapy. In addition, stable disease was observed in various indications (neuroendocrine, prostate, head and neck cancer) that are not typically sensitive to IL-2 treatment, suggesting tumor targeting with the scTCR domain can broaden the therapeutic utility of the cytokine. These data are also consistent with our observations in the preclinical studies described earlier.

31.5.2 ALT-801/Chemotherapy Phase II Studies

With the encouraging Phase I results and the demonstrated synergy between ALT-801 and cisplatin in mouse tumor efficacy studies, Phase Ib/II studies are underway for an ALT-801/cisplatin treatment regimen in patients with metastatic melanoma (Clinicaltrials.gov: NCT01029873) and an ALT-801/cisplatin/gemcitabine treatment regimen in patients with muscle invasive or metastatic bladder cancer (Clinicaltrials.gov: NCT01326871). We are particularly interested in advanced/metastatic bladder cancer since no new drugs for this indication have been approved in the past 20 years and current antibody-based cancer therapies have yet to demonstrate significant improvement over standard-of-care chemotherapy [53,54]. Notably, mutation and subsequent tumor cell overexpression of p53 may represent an important yet unexploited target for advanced/metastatic bladder cancer [32,55]. Recent analysis of over 3400 bladder cancer patients revealed a highly significant correlation between detectable p53 overexpression in tumors and both tumor grade/stage and tumor progression and mortality in patients with advanced bladder cancer [55]. In our ongoing clinical studies, patient is selected based on serologic HLA-A2 testing followed by assessment of tumor biopsy specimens for immunohistochemically identifiable p53 (aa 264-272)/HLA-A*0201 complex. Multicycle ALT-801/chemotherapy treatment is administered under an outpatient setting with the initial objective of determining the safety profile and tolerability during a dose escalation stage of the studies. Upon establishment of the MTD regimen, patient enrollment will be expanded based on the two-stage study design to further evaluate the safety of the ALT-801/chemotherapy regimens and assess tumor objective response and clinical benefit rates and patient survival rates.

Phase II trials are also planned that will further evaluate the efficacy and safety of ALT-801 treatment in an expanded population of patients with prostate and head and neck cancers as ALT-801 monotherapy showed some evidence of clinical benefit in these indications in the completed Phase I trial. The safety profile and efficacy results of ALT-801/chemotherapy being defined in the current Phase Ib/II studies will be used in designing these and future controlled studies of ALT-801-based treatment regimens for cancer indications. In addition, successful demonstration of clinical activity of the scTCR-mediated tumor-targeting approach will allow clinical advancement of ALT-802 for NSCLC and other tumor antigen-specific STAR fusion proteins (i.e., anti-MAGE-A3, gp100, or MART-1) currently in research and preclinical studies in our laboratory.

31.6 ALTERNATIVES/VARIANTS OF THIS APPROACH

31.6.1 Development of Novel TCR-Targeted IL-15 Superagonist Fusions

The success of utilizing scTCR domains in targeted immunotherapeutic approaches has prompted us to explore fusions with other cytokines and effector molecules including TNF-α, GM-CSF, and IL-15. Particularly, there has been considerable interest recently in the potential of the immunostimulatory cytokine IL-15 to supplant IL-2 as a therapeutic agent for cancer. This is due to the distinct and contrasting roles of IL-15 and IL-2 in the life cycle of effector lymphocytes [56]. Both IL-2 and IL-15 are capable of stimulating immune cells via interactions with shared signaling receptor components, IL-2Rβ and IL-2RγC, but unique α chain subunits (IL-2Rα and IL-15Rα) are required for high-affinity binding of each cytokine to the IL-2Rβγ complex. The effectiveness of IL-2 as an anticancer therapeutic has been questioned based on its pivotal role in the maintenance of $CD4^+$ $CD25^+$ T-regulatory cells, promoting CTL differentiation at the expense of memory cell formation, and supporting activation-induced cell death (AICD) that leads to the elimination of stimulated T cells and induction of T-cell tolerance [56]. In contrast, IL-15 inhibits IL-2-induced AICD and supports long-lasting $CD8^+$ T-cell memory and effector responses [56], suggesting that IL-15 may have significant advantages over IL-2 for the treatment of cancer [57]. Indeed, a recent NCI review listed IL-15 as the most promising product candidate among 12 immunotherapy drugs that could potentially cure cancer [58].

We have generated and characterized a number of different soluble scTCR fusion proteins to IL-15 and to novel IL-15 super agonists and antagonists. We have shown that scTCR/IL-15 fusion molecules containing an N72D mutation in the IL-15 domain exhibit a three-to-fourfold increase in IL-15 biological activity, whereas those with an IL-15 D8N mutation exhibit little or no activity [59]. While IL-15 superagonist-based fusion proteins could serve as targeted immunotherapeutics for cancer and infectious diseases, an IL-15 antagonist capable of inhibiting IL-15 responsive cells at the disease site may have therapeutic potential in treating allograft rejection and inflammatory autoimmune diseases, particularly if memory $CD8^+$ T cells play a role in disease pathology [56]. A nontargeted IL-15 mutant/Fcγ2a antagonist protein has already been shown to be effective at inhibiting islet and cardiac allograft rejection and preventing development and progression of arthritis in experimental animal models [60,61].

IL-15 can be effectively transpresented to IL-2Rβγ-bearing immune cells by the IL-15Rα chain on APCs or by soluble IL-15Rα domains [56,62,63]. As a result, we have generated scTCR/IL-15Rα fusion proteins capable of forming functional soluble multimeric complexes with the scTCR/IL-15 fusions (Figure 31.1). These fusion protein complexes have several advantages as targeted immunotherapeutics by providing improved and more biologically relevant functional activity and an increased serum half-life compared to IL-15 alone as well as multivalent-targeting activity directed to the TCR-specific tumor antigen [22,59]. In fact, we have found that the IL-15 and soluble IL-15Rα domains could be used as a protein scaffold much like the antibody constant domains to create various novel molecules, including bivalent complexes containing the same scTCR domains, bispecific complexes containing different scTCRs, scAb, and other binding domains, and heterodimeric complexes of TCR α and β chains [22]. As a further refinement, we have also generated scTCR-targeted IL-15/IL-15Rα complexes comprising IgG Fc domains. Such complexes are capable of binding and activating both IL-15RβγC-bearing and Fc receptor-bearing human immune cells and mediating both cytokine-induced and ADCC activity against target cells. We have found that administration of mice or cynomologous monkeys with tetrameric scTCR/IL-15: scTCR/IL-15Rα/Fc complexes resulted in significant induction of NK and $CD8^+$ T-cell proliferation and immune responsiveness. The p53-specific scTCR/IL-15:scTCR/IL-15Rα/Fc complexes were also capable of mediating antitumor activity against subcutaneous $p53^+$ human tumors in the A375 melanoma xenograft model described earlier. Together these results indicate that multivalent scTCR targeted IL-15 molecules could represent highly potent and novel immunotherapeutics for treating cancer and viral infections.

31.6.2 Cell-Based Therapies

Adoptive T-cell therapy employing autologous engineered T cells generated with gene transfer techniques using tumor antigen-specific TCR genes is currently a focus of intense interest in the field of cancer treatment and has been the subject of a number of recent reviews [64,65]. Generally, these approaches employ introduction of α/β TCR genes into the patient's peripheral blood lymphocytes via viral expression vector systems. Following expansion and *in vitro* functional characterization, the genetically modified autologous lymphocytes are adoptively transferred into lymphodeleted patients, who are subsequently treated IL-2 to promoter expansion of the transferred cells. Strategies to improve stable expression, functional pairing, and signaling of the transferred TCR genes as well as affinity improvement for antigen specific recognition by the engineered lymphocytes are currently being investigated. In early stage clinical trials, adoptive transfer of autologous T-cells expressing recombinant TCRs specific to MART-1, gp100, NY-ESO-1, and carcinoembryonic antigen (CEA) has been reported to provide objective antitumor response rates of 12–66% in patients with metastatic melanoma, colorectal cancer, and synovial cell sarcoma [36,66–68]. Additionally, destruction of normal melanocytes in the skin, eye, and ear following MART-1- and gp-100-specific T-cell transfer and colitis following CEA-specific T-cell transfer suggest that these therapies can exhibit potent immune cell-mediated reactivity against normal tissues expressing the target antigen [36,68].

We are exploring similar T cell based therapeutic strategies employing our portfolio of scTCR domains linked to lymphocyte signaling domains as novel chimeric receptors reactive to tumor and viral antigen targets. Additionally, we have been interested in evaluating therapies combining soluble STAR fusion proteins with adoptively transferred lymphocytes. In preclinical studies, we have shown in that ALT-801 can stably bind to IL-2R-bearing lymphocytes. Following adoptive transfer into $p53^+$ tumor-bearing mice, the ALT-801-treated lymphocytes preferentially accumulate at the tumors site, suggesting tumor targeting via cell-surface bound ALT-801 (Figure 31.4C) [23]. Based on these results, we have initiated a Phase I clinical trial that will employ ALT-801 in combination with NK cell-based donor lymphocyte infusion therapy in patients with acute myeloid leukemia (AML). It has previously been reported that adoptive transfer of haploidentical NK cells together with IL-2 therapy following nonmyeloablative immunosuppressive conditioning resulted in complete hematologic remission in 5 of 19 patients with poor-prognosis AML [69]. The current study is designed to evaluate the feasibility of using ALT-801 as a substitute for IL-2 for the activation of haploidentical NK cells prior to adoptive transfer and as a targeted immunostimulatory therapy post-cell transfer. The outcome of this study may provide an assessment of the safety and antitumor efficacy of ALT-801-targeted cellular therapy compared to nontargeted approaches. Similar strategies using adoptively transferred autologous T cells in combination with STAR-cytokine or scTCR/anti-CD3 scAb fusion protein are also being investigated and may provide more practical clinical applications for treating cancer or virally infected patients than genetically engineered lymphocyte approaches.

31.7 CHALLENGES

STAR fusion-base therapies represent personalized medicine in the eligible patients who express the appropriate HLA allele and display the peptide antigen on the diseased tissue recognized by the scTCR domain. As a result, the patient population likely to benefit from such treatments represents a subset of the total population afflicted with the disease. Consequently, selection of broad-based disease-relevant pMHC targets is an important consideration while employing this strategy. For our lead anticancer immunotherapeutic, we are targeting tumor cells presenting a peptide epitope (aa264-272) of p53 in the context of HLA-A*0201, the most common MHC class I allele. The aa264-272 region of p53 is rarely mutated in cancer cells, its cell-surface display likely will serve as a more broadly presented antigenic target for tumors with abnormal p53 expression than peptide epitopes bearing specific p53 mutations, the most frequent of which accounts for only 5% of possible p53 variants [70]. As indicated earlier, p53 overexpression is commonly observed in ~50% of tumors of different origins [30]. Consistent with these findings, we have found that 40% of the patients with different metastatic malignancies screened in our Phase I/II clinical studies ($n > 200$) were HLA-A2 positive and had tumors that presented the p53 (aa264-272)/HLA-A*0201 complex [51]. These values compare favorably to the 22% tumor positivity rate of HER-2 gene amplification/protein overexpression observed in breast cancer patients as a screening criterium for anti-HER-2 mAb-based therapies [71].

Another important consideration for TCR-based therapies is the level of antigenic pMHC complexes displayed on the diseased cells. Antigen presentation is a complex process that is influenced by the level of endogenous or exogenous protein present, appropriate degradation of these proteins by the proteasome machinery, the ability of the degraded peptides to bind the groove of the particular MHC molecules and the successful transport and stability of the pMHC complex on the cell surface. Although we have found that the p53 (aa264-272)/HLA-A*0201 complex to be displayed at up to 2000 complexes per cell on the surface of different human tumor cells [15], other researchers using pMHC-specific reagents or T-cell lines have shown only moderate to very low expression of tumor cell presentation of HLA-A*0201-bound peptides derived from well-characterized tumor-associated antigens including gp100, MART-1/melan-A, NY-ESO-1/LAGE-1, and telomerase reverse transcriptase [72–74]. For example, NY-ESO-1-positive tumor cells were found to present only 10-50 NY-ESO-1 (aa157-165)/HLA-A*0201 complexes per cell, a level which may be insufficient for a targeted protein-based therapeutic approach [73]. Additionally, intracellular expression levels of the tumor-associated antigens may not directly correlate with the level of the derived peptide/HLA-A*0201 complexes displayed on the cell surface or immune cell reactivity to these complexes [72,75]. These findings clearly indicate that detail characterization of the target pMHC levels and efficacy analysis against different antigen-positive target cells are required during the preclinical assessment of any given TCR-based therapy. Moreover, concurrent development of companion diagnostic methods to identify subjects with the appropriate disease-related pMHC presentation is likely an important component in the clinical evaluation of such an approach.

31.8 CONCLUSIONS AND FUTURE PERSPECTIVES

ALT-801, a p53-specific scTCR-IL-2 fusion protein, represents the first-in-class soluble TCR-based protein therapeutic to enter clinical studies. As described in this review, rigorous preclinical studies established the potent-targeted antitumor activity, mechanism-of-action and favorable pharmacokinetic and safety profiles of this molecule. Initial clinical results of ALT-801 in patents with metastatic malignancies are encouraging. However, further evaluation in Phase II/III clinical studies is required to establish proof-of-concept of this approach in patients. We are conducting clinical studies of ALT-801 in different cancer indications, where p53 overexpression in thought to play a role in disease progression, and in combination with different chemotherapy regimens, which may augment antitumor activity of ALT-801. Positive results establishing ALT-801 efficacy in such studies will facilitate clinical development of other p53-specific scTCR fusion molecules, including ALT-802, in cancer indications and establish the STAR technology platform as a next-generation targeted immunotherapeutic approach for treatment of patients with cancer and viral infections. As indicated, we are currently developing a portfolio of STAR fusion proteins reactive to tumor and viral antigens as follow-up clinical candidates to the p53-specific molecules.

Additionally, we are evaluating the targeted activity of novel immune effector domain fusion proteins against tumor and virally infected cells in various preclinical models. Of these proteins, the IL-15:IL-15Rα-Ig Fc complex equipped with scTCRs or scFv antibody domains shows great potential as a targeted immunotherapeutic fusion capable of promoting cytokine-induced NK cell and $CD8^+$ T-cell responses as well as ADCC and complement activation mediated by the Fc domain. Moreover, IL-15 has been shown to augment ADCC activity of NK cells, suggesting synergy between these effector domains [76]. Such multivalent antigen-specific fusion proteins are actively being investigated in our laboratory to assess whether this approach is capable of further improving the therapeutic activity of clinically proven antibody molecules.

ACKNOWLEDGMENTS

The work to develop the STAR technology was supported in part by grants from the National Institute of Health (CA088615, CA105816, AI056782, CA097550, CA117211, A1083130, CA139810, and AI093218), the Food and Drug Administration (FDR03452) and the Bill & Melinda Gates Foundation. The author thanks Dr. Hing C. Wong for helpful suggestions and critical review of the manuscript and Drs. Jack Egan, Jinghai Wen, and Wenxin Xu of Altor BioScience for allowing inclusion of unpublished data.

REFERENCES

1. van der Merwe PA, Davis SJ. (2003) Molecular interactions mediating T cell antigen recognition. *Annu. Rev. Immunol.* 21, 659–684.
2. Davis MM, Bjorkman PJ. (1988) T-cell antigen receptor genes and T-cell recognition. *Nature* 334, 395–402.
3. Admon A, Barnea E, Ziv T. (2003) Tumor antigens and proteomics from the point of view of the major histocompatibility complex peptides. *Mol. Cell. Proteomics* 2, 388–398.
4. Casiano CA, Mediavilla-Varela M, Tan EM. (2006) Tumor-associated antigen arrays for the serological diagnosis of cancer. *Mol. Cell. Proteomics* 5, 1745–1759.

5. Card KF, Price-Schiavi SA, Liu B, Thomson E, Nieves E, Belmont H, et al. (2004) A soluble single-chain T-cell receptor IL-2 fusion protein retains MHC-restricted peptide specificity and IL-2 bioactivity. *Cancer Immunol. Immunother.* 53, 345–357.
6. Novotny J, Ganju RK, Smiley ST, Hussey RE, Luther MA, Recny MA, et al. (1991) A soluble, single-chain T-cell receptor fragment endowed with antigen- combining properties. *Proc. Natl. Acad. Sci. USA* 88, 8646–8650.
7. Hoo WF, Lacy MJ, Denzin LK, Voss EWJr, Hardman KD, Kranz DM. (1992) Characterization of a single-chain T-cell receptor expressed in *Escherichia coli. Proc. Natl. Acad. Sci. USA* 89, 4759–4763.
8. Wulfing C, Pluckthun A. (1994) Correctly folded T-cell receptor fragments in the periplasm of *Escherichia coli*. Influence of folding catalysts. *J. Mol. Biol.* 242, 655–669.
9. Bentley GA, Boulot G, Karjalainen K, Mariuzza RA. (1995) Crystal structure of the beta chain of a T cell antigen receptor. *Science* 267, 1984–1987.
10. Garcia KC, Degano M, Stanfield RL, Brunmark A, Jackson MR, Peterson PA, et al. (1996) An alphabeta T cell receptor structure at 2.5 A and its orientation in the TCR-MHC complex. *Science* 274, 209–219.
11. Ward ES. (1992) Secretion of T cell receptor fragments from recombinant *Escherichia coli* cells. *J. Mol. Biol.* 224, 885–890.
12. Plaksin D, Polakova K, McPhie P, Margulies DH. (1997) A three-domain T cell receptor is biologically active and specifically stains cell surface MHC/peptide complexes. *J. Immunol.* 158, 2218–2227.
13. Chung S, Wucherpfennig KW, Friedman SM, Hafler DA, Strominger JL. (1994) Functional three-domain single-chain T-cell receptors. *Proc. Natl. Acad. Sci. USA* 91, 12654–12658.
14. Mosquera LA, Card KF, Price-Schiavi SA, Belmont HJ, Liu B, Builes J, et al. (2005) *In vitro* and *in vivo* characterization of a novel antibody-like single-chain TCR human IgG1 fusion protein. *J. Immunol.* 174, 4381–4388.
15. Zhu X, Belmont HJ, Price-Schiavi S, Liu B, Lee HI, Fernandez M, et al. (2006) Visualization of p53 (264-272)/HLA-A*0201 complexes naturally presented on tumor cell surface by a multimeric soluble single-chain T cell receptor. *J. Immunol.* 176, 3223–3232.
16. Weidanz JA, Card KF, Edwards A, Perlstein E, Wong HC. (1998) Display of functional alphabeta single-chain T-cell receptor molecules on the surface of bacteriophage. *J. Immunol. Methods* 221, 59–76.
17. Belmont HJ, Price-Schiavi S, Liu B, Card KF, Lee HI, Han KP, et al. (2006) Potent antitumor activity of a tumor-specific soluble TCR/IL-2 fusion protein. *Clin. Immunol.* 121, 29–39.
18. Li Y, Moysey R, Molloy PE, Vuidepot AL, Mahon T, Baston E, et al. (2005) Directed evolution of human T-cell receptors with picomolar affinities by phage display. *Nat. Biotechnol.* 23, 349–354.
19. Richman SA, Kranz DM. (2007) Display, engineering, and applications of antigen-specific T cell receptors. *Biom. Eng.* 24, 361–373.
20. Holler PD, Chlewicki LK, Kranz DM. (2003) TCRs with high affinity for foreign pMHC show self-reactivity. *Nat. Immunol.* 4, 55–62.
21. Donermeyer DL, Weber KS, Kranz DM, Allen PM. (2006) The study of high-affinity TCRs reveals duality in T cell recognition of antigen: specificity and degeneracy. *J. Immunol.* 177, 6911–6919.
22. Wong RL, Liu B, Zhu X, You L, Kong L, Han KP, et al. (2011) Interleukin-15:Interleukin-15 receptor {alpha} scaffold for creation of multivalent targeted immune molecules. *Protein Eng. Des. Sel.* 24, 373–383.
23. Wen J, Zhu X, Liu B, You L, Kong L, Lee HI, et al. (2008) Targeting activity of a TCR/IL-2 fusion protein against established tumors. *Cancer Immunol. Immunother.* 57, 1781–1794.
24. Schmidt MM, Wittrup KD. (2009) A modeling analysis of the effects of molecular size and binding affinity on tumor targeting. *Mol. Cancer Ther.* 8, 2861–2871.
25. Kuball J, Hauptrock B, Malina V, Antunes E, Voss RH, Wolfl M, et al. (2009) Increasing functional avidity of TCR-redirected T cells by removing defined N-glycosylation sites in the TCR constant domain. *J. Exp. Med.* 206, 463–475.
26. Li LP, Lampert JC, Chen X, Leitao C, Popovic J, Muller W, et al. (2010) Transgenic mice with a diverse human T cell antigen receptor repertoire. *Nat. Med.* 16, 1029–1034.
27. Sherman LA, Hesse SV, Irwin MJ, La Face D, Peterson P. (1992) Selecting T cell receptors with high affinity for self-MHC by decreasing the contribution of CD8. *Science* 258, 815–818.
28. Pien GC, Basner-Tschakarjan E, Hui DJ, Mentlik AN, Finn JD, Hasbrouck NC, et al. (2009) Capsid antigen presentation flags human hepatocytes for destruction after transduction by adeno-associated viral vectors. *J. Clin. Invest.* 119, 1688–1695.
29. Henrickson SE, Mempel TR, Mazo IB, Liu B, Artyomov MN, Zheng H, et al. (2008) T cell sensing of antigen dose governs interactive behavior with dendritic cells and sets a threshold for T cell activation. *Nat. Immunol.* 9, 282–291.
30. Vogelstein B, Lane D, Levine AJ. (2000) Surfing the p53 network. *Nature* 408, 307–310.
31. Vousden KH, Prives C. (2005) P53 and prognosis: new insights and further complexity. *Cell* 120, 7–10.
32. Strano S, Dell'Orso S, Di Agostino S, Fontemaggi G, Sacchi A, Blandino G. (2007) Mutant p53: an oncogenic transcription factor. *Oncogene* 26, 2212–2219.
33. Kuball J, Schmitz FW, Voss RH, Ferreira EA, Engel R, Guillaume P, et al. (2005) Cooperation of human tumor-reactive CD4+ and CD8+ T cells after redirection of their specificity by a high-affinity p53A2.1-specific TCR. *Immunity* 22, 117–129.
34. Cohen CJ, Zheng Z, Bray R, Zhao Y, Sherman LA, Rosenberg SA, et al. (2005) Recognition of fresh human tumor by human peripheral blood lymphocytes transduced with a bicistronic retroviral vector encoding a murine anti-p53 TCR. *J. Immunol.* 175, 5799–5808.
35. Theobald M, Biggs J, Dittmer D, Levine AJ, Sherman LA. (1995) Targeting p53 as a general tumor antigen. *Proc. Natl. Acad. Sci. USA* 92, 11993–11997.

36. Johnson LA, Morgan RA, Dudley ME, Cassard L, Yang JC, Hughes MS, et al. (2009) Gene therapy with human and mouse T-cell receptors mediates cancer regression and targets normal tissues expressing cognate antigen. *Blood* 114, 535–546.

37. Caballero OL, Chen Y-T. (2009) Cancer/testis (CT) antigens: Potential targets for immunotherapy. *Cancer Sci.* 100, 2014–2021.

38. Sakakura K, Chikamatsu K, Furuya N, Appella E, Whiteside TL, DeLeo AB. (2007) Toward the development of multi-epitope p53 cancer vaccines: An *in vitro* assessment of CD8+ T cell responses to HLA class I-restricted wild-type sequence p53 peptides. *Clin. Immunol.* 125, 43–51.

39. Sheikh NA, Jones LA. (2008) CD54 is a surrogate marker of antigen presenting cell activation. *Cancer Immunol. Immunother.* 57, 1381–1390.

40. Melder RJ, Osborn BL, Riccobene T, Kanakaraj P, Wei P, Chen G, et al. (2005) Pharmacokinetics and *in vitro* and *in vivo* antitumor response of an interleukin-2-human serum albumin fusion protein in mice. *Cancer Immunol. Immunother.* 54, 535–547.

41. Anderson TD, Hayes TJ, Powers GD, Gately MK, Tudor R, Rushton A. (1993) Comparative toxicity and pathology associated with administration of recombinant IL-2 to animals. *Int. Rev. Exp. Pathol.* 34(Pt A), 57–77.

42. Minami Y, Kono T, Miyazaki T, Taniguchi T. (1993) The IL-2 receptor complex: its structure, function, and target genes. *Annu. Rev. Immunol.* 11, 245–268.

43. Yao Z, Dai W, Perry J, Brechbiel MW, Sung C. (2004) Effect of albumin fusion on the biodistribution of interleukin-2. *Cancer Immunol. Immunother.* 53, 404–410.

44. Hayakawa Y, Smyth MJ. (2006) NKG2D and cytotoxic effector function in tumor immune surveillance. *Semin. Immunol.* 18, 176–185.

45. Hessell AJ, Hangartner L, Hunter M, Havenith CE, Beurskens FJ, Bakker JM, et al. (2007) Fc receptor but not complement binding is important in antibody protection against HIV. *Nature* 449, 101–104.

46. Kang Y, Omura M, Suzuki A, Oka T, Nakagami Y, Cheng C, et al. (2006) Development of an orthotopic transplantation model in nude mice that simulates the clinical features of human lung cancer. *Cancer Sci.* 97, 996–1001.

47. Bracci L, Proietti E, Belardelli F. (2007) IFN-gamma and novel strategies of combination therapy for cancer. *Ann. NY Acad. Sci.* 1112, 256–268.

48. O'Day SJ, Atkins MB, Boasberg P, Wang H-J, Thompson JA, Anderson CM, et al. (2009) Phase II multicenter trial of maintenance biotherapy after induction concurrent biochemotherapy for patients with metastatic melanoma. *J. Clin. Oncol.* 27, 6207–6212.

49. Rubinfeld B, Upadhyay A, Clark SL, Fong SE, Smith V, Koeppen H, et al. (2006) Identification and immunotherapeutic targeting of antigens induced by chemotherapy. *Nat. Biotechnol.* 24, 205–209.

50. Wang D, Lippard SJ. (2005) Cellular processing of platinum anticancer drugs. *Nat. Rev. Drug Discov.* 4, 307–320.

51. Fishman MN, Thompson JA. Pennock GK, Gonzalez R, Diez LM, Daud AI, et al. (2011) Phase I trial of ALT-801, an interleukin-2/T-cell receptor fusion protein targeting p53 (aa264-272)/HLA-A*0201 complex, in patients with advanced malignancies. *Clin. Cancer Res.* 17, 7765–7775.

52. Konrad MW, Hemstreet G, Hersh EM, Mansell PW, Mertelsmann R, Kolitz JE, et al. (1990) Pharmacokinetics of recombinant interleukin 2 in humans. *Cancer Res.* 50, 2009–2017.

53. Hussain MH, MacVicar GR, Petrylak DP, Dunn RL, Vaishampayan U, Lara PNJr, et al. (2007) Trastuzumab, paclitaxel, carboplatin, and gemcitabine in advanced human epidermal growth factor receptor-2/neu-positive urothelial carcinoma: results of a multicenter Phase II National Cancer Institute trial. *J. Clin. Oncol.* 25, 2218–2224.

54. Hahn NM, Stadler WM, Zon RT, Waterhouse DM, Picus J, Nattam SR, et al. (2009) A multicenter Phase II study of cisplatin (C), gemcitabine (G), and bevacizumab (B) as first-line chemotherapy for metastatic urothelial carcinoma (UC): Hoosier Oncology Group GU-0475. *J. Clin. Oncol.* 27, 5018.

55. Goebell PJ, Groshen SG, Schmitz-Drager BJ. (2010) p53 immunohistochemistry in bladder cancer—a new approach to an old question. *Urol. Oncol.* 28, 377–388.

56. Waldmann TA. (2006) The biology of interleukin-2 and interleukin-15: implications for cancer therapy and vaccine design. *Nat. Rev. Immunol.* 6, 595–601.

57. Fehniger TA, Cooper MA, Caligiuri MA. (2002) Interleukin-2 and interleukin-15: immunotherapy for cancer. *Cytokine Growth Factor Rev.* 13, 169–183.

58. Cheever MA. (2008) Twelve immunotherapy drugs that could cure cancers. *Immunol. Rev.* 222, 357–368.

59. Zhu X, Marcus WD, Xu W, Lee HI, Han K, Egan JO, et al. (2009) Novel human interleukin-15 agonists. *J. Immunol.* 183, 3598–3607.

60. Ferrari-Lacraz S, Zanelli E, Neuberg M, Donskoy E, Kim YS, Zheng XX, et al. (2004) Targeting IL-15 receptor-bearing cells with an antagonist mutant IL-15/Fc protein prevents disease development and progression in murine collagen-induced arthritis. *J. Immunol.* 173, 5818–5826.

61. Zheng XX, Gao W, Donskoy E, Neuberg M, Ruediger M, Strom TB, et al. (2006) An antagonist mutant IL-15/Fc promotes transplant tolerance. *Transplantation* 81, 109–116.

62. Stoklasek TA, Schluns KS, Lefrancois L. (2006) Combined IL-15/IL-15Ralpha immunotherapy maximizes IL-15 activity *in vivo*. *J. Immunol.* 177, 6072–6080.

63. Rubinstein MP, Kovar M, Purton JF, Cho JH, Boyman O, Surh CD, et al. (2006) Converting IL-15 to a superagonist by binding to soluble IL-15R{alpha}. *Proc. Natl. Acad. Sci. USA* 103, 9166–9171.

64. Morgan RA, Dudley ME, Rosenberg SA. (2010) Adoptive cell therapy: genetic modification to redirect effector cell specificity. *Cancer J.* 16, 336–341.

65. Thomas S, Stauss HJ, Morris EC. (2010) Molecular immunology lessons from therapeutic T-cell receptor gene transfer. *Immunology* 129, 170–177.

66. Morgan RA, Dudley ME, Wunderlich JR, Hughes MS, Yang JC, Sherry RM, et al. (2006) Cancer regression in patients after

transfer of genetically engineered lymphocytes. *Science* 314, 126–129.

67. Robbins PF, Morgan RA, Feldman SA, Yang JC, Sherry RM, Dudley ME, et al. (2011) Tumor regression in patients with metastatic synovial cell sarcoma and melanoma using genetically engineered lymphocytes reactive with NY-ESO-1. *J. Clin. Oncol.* 29, 917–924.
68. Parkhurst MR, Yang JC, Langan RC, Dudley ME, Nathan DA, Feldman SA, et al. (2011) T cells targeting carcinoembryonic antigen can mediate regression of metastatic colorectal cancer but induce severe transient colitis. *Mol. Ther.* 19, 620–626.
69. Miller JS, Soignier Y, Panoskaltsis-Mortari A, McNearney SA, Yun GH, Fautsch SK, et al. (2005) Successful adoptive transfer and *in vivo* expansion of human haploidentical NK cells in patients with cancer. *Blood* 105, 3051–3057.
70. Soussi T, Beroud C. (2003) Significance of TP53 mutations in human cancer: a critical analysis of mutations at CpG dinucleotides. *Hum. Mutat.* 21, 192–200.
71. Ross JS, Slodkowska EA, Symmans WF, Pusztai L, Ravdin PM, Hortobagyi GN. (2009) The HER-2 receptor and breast cancer: ten years of targeted Anti-HER-2 therapy and personalized medicine. *Oncologist* 14, 320–368.
72. Michaeli Y, Denkberg G, Sinik K, Lantzy L, Chih-Sheng C, Beauverd C, et al. (2009) Expression hierarchy of T cell epitopes from melanoma differentiation antigens: unexpected high level presentation of tyrosinase-HLA-A2 Complexes revealed by peptide-specific, MHC-restricted, TCR-like antibodies. *J. Immunol.* 182, 6328–6341.
73. Purbhoo MA, Sutton DH, Brewer JE, Mullings RE, Hill ME, Mahon TM, et al. (2006) Quantifying and imaging NY-ESO-1/LAGE-1-derived epitopes on tumor cells using high affinity T cell receptors. *J. Immunol.* 176, 7308–7316.
74. Purbhoo MA, Li Y, Sutton DH, Brewer JE, Gostick E, Bossi G, et al. (2007) The HLA A*0201-restricted hTERT(540-548) peptide is not detected on tumor cells by a CTL clone or a high-affinity T-cell receptor. *Mol. Cancer Ther.* 6, 2081–2091.
75. Weidanz JA, Nguyen T, Woodburn T, Neethling FA, Chiriva-Internati M, Hildebrand WH, et al. (2006) Levels of specific peptide-HLA class I complex predicts tumor cell susceptibility to CTL killing. *J. Immunol.* 177, 5088–5097.
76. Carson WE, Giri JG, Lindemann MJ, Linett ML, Ahdieh M, Paxton R, Anderson D, et al. (1994) Interleukin (IL) 15 is a novel cytokine that activates human natural killer cells via components of the IL-2 receptor. *J. Exp. Med.* 180, 1395–1403.

32

HIGH-AFFINITY MONOCLONAL T-CELL RECEPTOR (mTCR) FUSIONS

NIKOLAI M. LISSIN, NAMIR J. HASSAN, AND BENT K. JAKOBSEN

Immunocore Ltd, Abingdon, UK

32.1 INTRODUCTION: THE T CELL RECEPTOR (TCR) AS A TARGETING MOLECULE

Monoclonal antibodies, notwithstanding the extremely broad range of uses for which they can be applied, cannot be employed to target intracellular proteins (Figure 32.1). For cancer, in particular, this constitutes a severe limitation because secreted or cell surface-bound proteins represent only a fraction, probably 10–15%, of malignant state-associated antigens. T-cell receptors (TCRs) are antibody-like molecules comprising either the alpha/beta (αβ) or gamma/delta (γδ) heterodimers, expressed on the surface of T cells. While the functions of the γδ receptor are still debated, the αβ type of TCR, which is expressed on the vast majority of T cells, recognizes antigenic peptides derived from cellular proteins and presented on the cell surface as peptide-major histocompatibility complex (MHC) complexes (in humans also called peptide-human leukocyte antigen [HLA]). Similarly to antibodies, the TCR chains are encoded in the germ line by an array of rearranging genes, resulting in the expression of hugely diverse αβ combinations, each one unique to a particular T-cell clone. Selection in the thymus, however, limits the repertoire, such that the estimated number of naturally occurring T-cell variants is $\sim 2.5 \times 10^7$ per individual [1,2]. The T cells that are released into the peripheral lymphoid system have very low affinities for their relevant peptide-HLA antigens, this being a necessary feature to avoid autoimmune reactions. However, the low antigen affinities of naturally occurring T cells lead to immune escape by cancer cells, which achieve a selective advantage by presenting reduced peptide-MHC antigen levels [3]. Moreover, TCRs specific for cancer antigens are generally of particularly low affinity because the majority of tumor-associated peptide antigens (TAPAs) are self-antigens that are either over-represented or, in some cases, mutated. The close relationship to self-antigens, which are subject to negative T-cell selection in the thymus, means that only T cells with very low affinity TCRs are available to recognize most TAPAs. This stands in contrast to viral peptide antigens, which are generally quite distinct from self-peptide antigens and therefore can attract T cells with higher affinity TCRs. Significantly, antiviral T-cell responses are far more efficacious than anticancer ones, and at least some of this difference is associated with unequal TCR affinities for the two classes of antigen.

T-cell repertoires are governed (restricted) by each individual's particular types of MHC molecules (class I and II, with their corresponding subclasses) in the context of which antigenic peptides are presented; a restriction which will equally apply in the use of TCR-based therapeutics. MHC class I present 8–11 (typically 9) amino acid residue-long peptides, which are surveyed by CD8+ cytotoxic T cells. Longer peptides (typically 15–24 amino acid residues long)

Fusion Protein Technologies for Biopharmaceuticals: Applications and Challenges, First Edition. Edited by Stefan R. Schmidt.

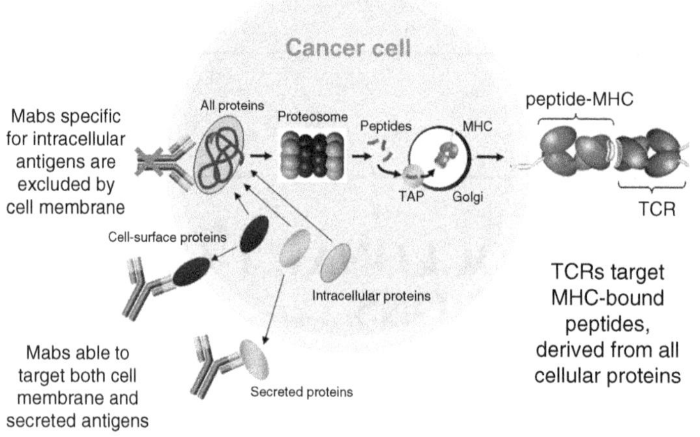

FIGURE 32.1 TCRs as potential targeting molecules. The T-cell receptors (TCRs) bind to the antigenic peptides derived from all proteins within a living cell and presented on the cell surface in the context of the major histocompatibility complex (peptide-MHC). This gives TCRs an advantage over antibodies, as the latter are unable to target intracellular proteins. However, to fulfill their roles as targeting molecules, TCRs must be re-engineered as soluble proteins, with the affinities toward their peptide-MHC ligands vastly increased.

are presented by MHC class II on specialized antigen presenting cells and recognized by CD4+ helper T cells. TCRs derived from MHC class I-restricted T cells are the main focus of interest for therapeutics because MHC class I antigens are presented on virtually all cells of the body and present target antigens relevant for both cancers and viral infections. A few examples of tumor-associated antigens are given in Table 32.1. An important

TABLE 32.1 Examples of Tumor-Associated Antigens (TAA)

Tumor-Associated Antigen (TAA)	Main Cancer Indications
Cancer testis antigens (CTA)	
MAGE-A3 (melanoma antigen-A3)	Melanoma, myeloma, head and neck, non-small cell lung cancer (NSCLC), esophageal, ovarian, and others
NY-ESO-1 (New York esophageal squamous cell carcinoma 1)	Myeloma, melanoma, NSCLC, esophageal, bladder, ovarian, and others
Differentiation antigens	
Gp100 (glycoprotein 100)	Melanoma
Melan-A/Mart-1 (melanoma antigen recognized by T cells 1)	Melanoma
NY-BR-1 (Ankyrin repeat domain-containing protein 30A)	Breast
PAP (prostatic acid phosphatase)	Prostate
Overexpressed antigens	
Survivin	Leukemia, lymphomas, esophageal, breast, colon, NSCLC and liver
hTERT (human telomerase reverse transcriptase)	Leukemia and other common cancers
WT-1 (Wilm's tumor-1)	Leukemia and some solid cancers
P450 CYP1B1 (cytochrome p450 1B1)	Many solid cancers including breast, ovarian, colorectal, and prostate
Rhamm	Many solid and hematopoietic cancers
Mutated antigens	
BCR-ABL	Chronic myeloid leukemia and acute myeloid leukemia
P53	Head and neck
Oncofetal antigens	
AFP (alpha-fetoprotein)	Liver
OFA-iLRP (oncofetal antigen)	Many solid and hematopoietic cancers

aspect to note is that the potential applicability of a TCR to a patient population is defined by the prevalence of both the presenting MHC allele and of the antigenic protein from which the peptide antigen is derived. MHC, or HLA, types differ considerably in their overall prevalence and geographical distribution. HLA-A2, for example, is found in ~50% of the European and North American populations and ~30% of Asian populations. The equivalent prevalence of HLA-A24 in the same populations is 18 and 65%, respectively.[1] While HLA type limits the proportion of patients that can be targeted with a specific TCR, the almost universal choice of, particularly intracellular, antigens means that peptide-MHC antigens can offer better patient coverage than other cell-surface antigens such as those that can be targeted by antibodies. Her2-neu as an antibody target in breast cancer, for example, is present in ~22% of the patient population. [2] Some intracellular antigens, in contrast, are present at very high frequencies in certain tumor types. Gp100, a protein expressed in virtually 100% of malignant melanomas, is one such example [4]. A TCR recognizing gp100 presented by HLA-A2 can therefore target ~50% of patients in Europe and North America. IMCgp100, a high-affinity gp100 specific TCR fused to an anti-CD3 scFv, has entered clinical trials in metastatic melanoma both the United Kingdom and the United States in 2010.

32.2 ENGINEERED HIGH-AFFINITY MONOCLONAL TCRs (mTCR)

From a protein engineering perspective, an immediate advantage is that, in contrast to the antigens bound by antibodies, peptide-MHC complexes are easily produced by a standard *in vitro* refolding procedure. The peptide of interest is synthesized, combined with the corresponding MHC type polypeptide and β2m and purified after refolding [5]. TCRs themselves differ from antibodies in two important characteristics. First, TCRs do not naturally exist as soluble, noncell-bound molecules. Second, the affinities of naturally occurring TCRs to their corresponding peptide-HLA ligands fall into a narrow range of about 1–300 μM [6–9], which is too low for them to be used in any therapeutic or diagnostic applications. Structurally, the extracellular part of a TCR resembles an antibody Fab fragment, with its similarly sized alpha and beta chains, each comprising a membrane-distal N-terminal variable domain and a membrane-proximal constant domain, with an intrachain disulfide bond stabilizing each of the four domains. An interchain disulfide bond, located near the cell surface in the constant domains, links the two chains together (Figure 32.2A). Many strategies have been tried to produce a soluble TCR framework. An scFv-like single chain design, offering the advantage of a single gene expression, might appear to be the simplest approach and single chain versions of some TCRs, using various designs [10], have been produced. However, the strategy is not generally applicable and the stability of conformation between the two variable domains is uncertain. For achieving maximal structural stability and integrity, as well as for avoiding immunogenicity, an authentic heterodimer comprising the two chains, truncated downstream of the native interchain disulfide, provides a more dependable solution. Such designs have been tried, but found unworkable, due to aggregation of chains, probably caused by incorrect disulfide bond formation in the process of protein folding. An alternative design, in which fos and jun leucine zippers were fused C-terminally to the truncated chains [11], provided good stability of the heterodimers and was successfully applied to all tested TCRs (Immunocore, unpublished data). For therapeutic applications, however, such a protein design would be too unstable. Instead, with the aim of generating a simple, generally applicable and highly stable TCR framework, several locations within the constant domains were investigated by molecular modeling for the introduction of a new interchain disulfide link [12]. Several suitable locations were identified, the theoretically optimal one being between TRAC threonine 48 and TRBC serine 57, which were therefore mutated to cysteines. Expressed separately in *Escherichia coli*, the chains were refolded *in vitro*, resulting in a highly stable, fully human soluble protein, which became known as the monoclonal TCR (mTCR). A large number of individual TCRs of varying peptide specificity, as well as HLA restriction, have been successfully produced as mTCRs, establishing this as the optimal generic soluble TCR format for crystallographic studies as well as for potential therapeutic uses.

The crystal structure model of a tumor-associated NY-ESO-1 peptide-specific 1G4 mTCR [12] docking onto its cognate peptide-HLA-A2 is shown in Figure 32.2B. The TCR's ligand-binding surfaces, in an analogous fashion to those of antibodies, are formed by six complementary determining regions (CDRs), which are hyper-variable loops protruding from the membrane-distal end of the variable domains. Counting from the N-terminus of each chain, these are called CDR α1, α2, α3 and β1, β2, β3, respectively. While the amino acid sequences of CDRs 1 and 2 are determined in the germ line and thus are fixed in each type of variable domain (of which there are 54 α and 67 β sequences in humans), the CDR3s are formed as a result of gene recombination, and consequently unique for each T-cell clone. Despite the significant variations in the docking mode observed in crystal structures of TCR/peptide-HLA complexes [12–18], one characteristic feature is that the CDR3s of both chains only make

[1] http://www.allelefrequencies.net/hla.

[2] Stakeholder Insight, Datamonitor 2010.

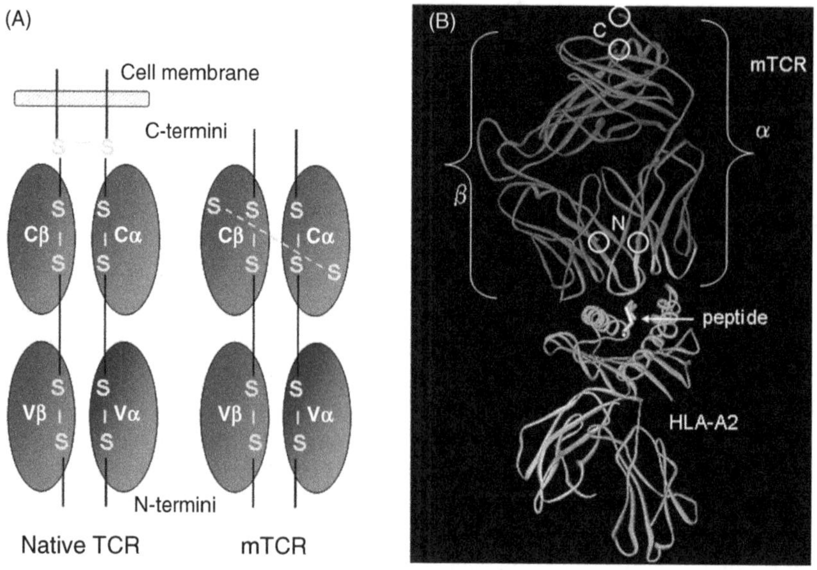

FIGURE 32.2 The monoclonal TCR (mTCR). (A) The design of the mTCR framework. Natural αβ TCRs (left) are disulfide-linked heterodimers, with each chain comprising the N-terminal variable domain and the constant domain, followed by a transmembrane region and a short C-terminal intracellular signal-transducing tail. In the mTCR (right), the chains are truncated, and the natural interchain disulfide bond replaced by an engineered one, linking the two chains within their constant domains, producing an exceptionally stable and fully human soluble TCR molecule. (B) Crystal structure-based model of 1G4 mTCR bound to NY-ESO peptide-MHC. The ligand, as with all MHC class I members, consists of the heavy chain (shown in green) normally anchored via its C-terminus to the cell surface, with the nine aa-long antigenic peptide (grey tube) buried within the groove formed between the two helices of the apical α3 domain, as well as the noncovalently associated invariant β2m (shown in yellow). The TCR, a heterodimer of an α chain (shown in red) and a β chain (blue), contacts the antigenic surface through its CDRs (highlighted by light cyan and light magenta colors). The centrally positioned CDR3 loops of both chains predominantly contact the peptide, whereas the CDR1 and the CDR2 loops (appearing the farthest from the groove) contribute mostly to the contacts with the relatively conserved HLA a3 domain. Note the location of the N- and C-termini of both TCR chains (encircled), which can potentially be fusion points for the attachment of the effector function moieties. (See the color version of this figure in Color Plates section.)

contact with the peptide, whereas CDR1s and CDR2s are interacting predominantly with the relatively conserved surfaces of the HLA component. Thus, CDR3s must be of principal importance in determining peptide specificity, while the affinity towards the HLA allele is mostly governed by the remaining CDR loops, in most cases mainly the CDR2s.

The low affinity of wild-type TCRs originates from the selection process, which all T cells undergo during maturation in the thymus, and reflects the necessity to maintain a delicate balance between recognition of foreign antigens and tolerance of self-antigens. Importantly, it has been shown that no inherent genetic or structural limitations exist on TCR affinity [19], encouraging the application of molecular display technologies in a similar manner as they are used for selecting antibody variants with enhanced affinities [20,21]. Yeast display was the first technique successfully used to produce TCR molecules with affinities increased by up to 100-fold [19,22]. More recently, phage display, which has significant methodological advantages over yeast display techniques, has been adapted for the selection of affinity matured mTCRs [12,23–25]. This technology does not require mutation of the TCR framework, can be used to screen libraries of 10^{10} variants, does not require clonotypic antibodies and, importantly, can be applied to any mTCR.

The approach currently used for the production of affinity matured mTCRs is summarized in Figure 32.3. Briefly, genetic constructs are made in which the mTCR β chain is fused to phage gene III coat protein, while the α

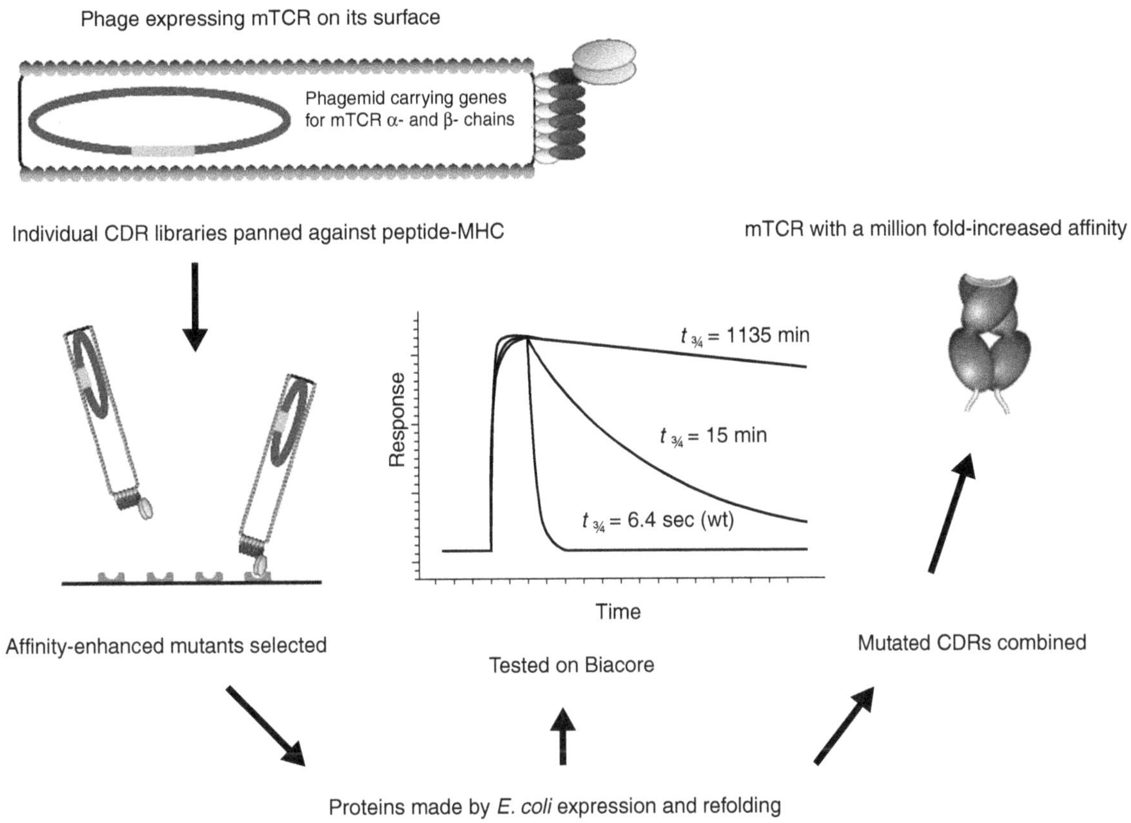

FIGURE 32.3 Affinity maturation of mTCRs with phage display. Mutations are being introduced into the CDR regions of the α and β chains. Expressed on phage, the individual CDR mutants are panned against the specific peptide-MHC. This typically increases the affinity from micromolar to nanomolar range. At the next stage, the optimum CDR sequences are brought together in mTCR format, the chains expressed in *E. coli*, refolded in desired combinations and the affinity is tested on the Biacore. This is an iterative process and enables the enhancement of the affinity into the picomolar range.

chain is encoded as a free (nonfused) chain. The resulting phage particles display the fused TCR β chains paired with the secreted α chains, disulfide-linked to each other. PCR mutagenesis is used to introduce variability to the targeted regions (one CDR at a time), and the libraries are panned on the relevant immobilized peptide-MHC. Creation of second-generation phage TCR libraries, and repeated selections against decreasing concentrations of antigen, result in isolation of TCRs with further incremental improvements in affinity. At this stage, a number of selected mutants are made as soluble mTCRs through *E. coli* expression and *in vitro* refolding, and their affinities measured using a surface plasmon resonance microchip-based detector (Biacore). mTCRs featuring mutations in just one or two of their CDRs generally exhibit affinities in the nanomolar range, the average increase being ~1000-fold. Ultimately, selections of the most promising mutations in several CDRs are combined, usually generating mTCR variants exhibiting affinities in the picomolar range. Thus, the phage display selection process typically produces affinity enhancements of a million fold compared to the original native TCR. Examples of mTCRs affinity matured by this approach are shown in Table 32.2. The high-affinity TCRs can be labeled, for example, with a fluorescent molecule, and used to visualize or quantitate the specific antigen on cells (Figure 32.5A). To our knowledge, these high-affinity mTCRs constitute the first tool that has enabled direct detection of cell-surface peptide-MHC, an application that has been used in vaccine validation and which could potentially be used for future diagnostic purposes [26,27]. No detectable levels of nonspecific binding to irrelevant peptides or antigen negative cells have been observed with any of the soluble mTCRs listed in Table 32.2.

TABLE 32.2 Examples of mTCR Affinity Maturation

Specificity	k_a	k_d	K_D	$T_{1/2}$
NY-ESO	4.0×10^4	1.3×10^{-1}	32 μM	7 s
	6.6×10^5	1.7×10^{-5}	26 pM	19 h
MAGE A3	ND	ND	281 μM	ND
	3.4×10^5	2.2×10^{-5}	58 pM	9 h
Telomerase	2.8×10^3	1.1×10^{-1}	38 μM	3 s
	7.5×10^4	4.9×10^{-6}	65 pM	39 h
gp100	ND	ND	26 μM	ND
	5.3×10^5	5.8×10^{-6}	11 pM	33 h
WT1	ND	ND	96 μM	ND
	9.9×10^5	1.6×10^{-5}	16 pM	12 h

For each specificity, wild type: top, affinity enhanced: bottom (ND: not determined due to kinetics too fast for measurements).

32.3 mTCR-BASED FUSION PROTEINS FOR THERAPEUTIC APPLICATIONS

Unlike antibodies, mTCRs do not encompass an Fc fragment or any other effector function and targeted cell killing, therefore, requires conjugation or fusion to a suitable cytotoxic mediator. Because any particular tumor-associated, peptide-MHC antigen is presented at very low copy numbers per cell, typically in the range 10–150, the effector function must have very high cytotoxic potential. This rules out some of the conjugates used for therapeutic antibodies, for example, radioactive labeling, which requires several thousand targets per cell to achieve a cytotoxic effect. Bearing this difference in mind, the strategies for choosing an effector function to be linked to mTCRs can follow similar paths to those explored with antibodies. The two main modes of action that can be employed are (a) a direct mechanism delivering a payload, for example, a cytotoxin, to the cancer cell surface, or (b) an indirect mechanism whereby immune activity is redirected against the cancer cell.

The first approach involves conjugating or fusing a cell toxin molecule to the mTCR. As no amplification of the killing mechanism can be achieved and the antigen target number per cell is very low, the toxin needs to be of very high potency, even in a situation where high mTCR affinity enables almost saturating binding to the antigen. While a single molecule of the *Pseudomonas* exotoxin A-derived PE38 is, in theory, sufficient to kill a eukaryotic cell, this is not achievable in practice, probably because the toxin requires internalization and mTCRs, when bound to antigen, largely remain on the cell surface (unpublished). PE38-fused mTCRs were found to be capable of killing peptide-pulsed target cells (which can present nonphysiologically high numbers of antigen per cell), but were not potent enough for efficiently destroying naturally presenting tumor cells (unpublished).

With the indirect approach, the protein fused to the mTCR must convey a potent activation signal to the immune system, for example, T cells, in the vicinity of the target cells. Such a protein moiety could be a cytokine, a receptor, a bacterial superantigen, or an antibody fragment. Many such possibilities have already been tested on antibody fusions. With this approach also, the potency of the fused effector function is the key property. The cytokine interleukin-2 (IL-2), fused to a tumor-specific, high-affinity NY-ESO mTCR, showed some promise in stimulating CD8+ cytotoxic T lymphocytes (CTLs) to exhibit activity against NY-ESO peptide-pulsed targets but, considering the toxicity of IL-2 versus the efficacy that could be achieved, this option was not considered attractive as a potential therapeutic (unpublished).

An alternative way of addressing the potency problem would be through increasing the payload, that is, the number of effector function moieties conjugated to each mTCR molecule. However, the opportunities for generating direct genetic fusions, which are suitable for larger scale manufacturing purposes, are limited. Although four potential fusion points exist on mTCRs, counting the N- and C-termini of the αβ heterodimer (Figure 32.2), only one is typically used, due to aggravating expression and folding problems. Tandem effector function molecules [28], while in theory attractive, in many cases have limited applications in practice for the same reasons. The future here may lie with nongenetic conjugates, involving newly developed protein nanoparticles as well as dendritic polymers and possibly other advanced delivery systems.

32.4 IMMUNE-MOBILIZING MONOCLONAL TCRs AGAINST CANCER (ImmTAC)

Probably, the most potent cytotoxic antibody-based fusions reported to date are the so-called BiTEs, which comprise a targeting antibody fragment linked with a T-cell-activating anti-CD3 scFv antibody [29–31]. A series of fusion proteins, consisting in each case of a high-affinity cancer-specific mTCR and an anti-CD3 scFv, have been produced and optimized. The TCR α and TCR β-scFv chains comprising these fusion molecules, called ImmTACs, are expressed separately at extremely high levels in *E. coli.* The ImmTAC is then generated by mixing and refolding the two chains, which recovers a yield in excess of 10% and produces protein that is homogeneous and, by release assays, 100% active.

Functionally, the mTCR part of the fusion binds to an antigenic peptide-MHC on the target cell, while the other end of the molecule, by engaging the ε-chain of the CD3 glycoprotein, is capable of activating any CD8+ CTL, independently of the intrinsic antigen specificity of the latter (Figure 32.4). Several of these CTL-activating agents have been comprehensively tested in cellular assays with promising results. Figure 32.5 shows the properties of one

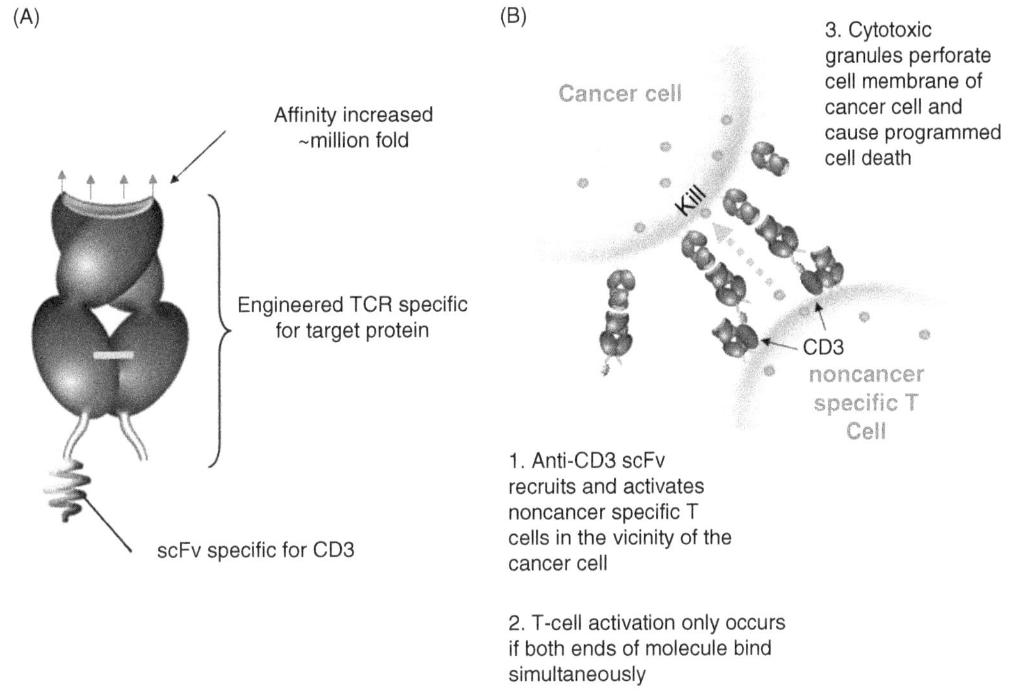

FIGURE 32.4 Redirection of T cells against cancer with ImmTAC. (A) An ImmTAC is a fusion protein comprising a picomolar affinity mTCR, specific for a chosen cancer-related peptide-MHC, and an anti-CD3 scFv. (B) The mTCR component binds very strongly to the ligand on a target cell, while the anti-CD3 moiety activates a proximal T cell independently of its original specificity. The effect of this redirection is the killing of the engaged cancer cell.

such fusion, ImmTAC-NY-ESO, for which the antigen is strongly associated with several cancers (Table 32.1). The use of biotinylated mTCRs is an effective tool for confirming epitope presentation as low as 10–50 epitopes per cell, on both cell lines and fresh tumor biopsies [26]. While the algorithms designed to predict peptide-MHC presentation provide valuable clues, high-affinity mTCRs can confirm the actual processing and presentation of these epitopes *in vivo*. In Figure 32.5A, NY-ESO-1 presentation was quantified on the IM9 cancer cell line and found to be in the range of 2–13 epitopes per cell. Despite this low level of antigen presentation, ImmTAC-NY-ESO was potent and highly specific in activating CD8+ CTLs mixed with IM9 cells (Figure 32.5B, left). The interferon-γ (IFN-γ) release assay (Figure 32.5B, left) illustrates that, while the ImmTAC was showing efficacy at concentrations as low as 10pM, the stimulation was specific because it was dramatically reduced when the corresponding nonfused mTCR was added as a competitor (middle line in Figure 32.5B, left in Figure 32.5B). No CTL activation was observed with an irrelevant cell line lacking the expression of NY-ESO antigen but containing the correct HLA-type, A2 (bottom line in figure 32.5B, left in Figure 32.5B). Importantly, this illustrates that the ImmTAC only activates when a multivalent contact is established between the target cell and the T cell (top line in figure 32.5B, left); this feature is crucial for avoiding systemic toxicity when applied as a therapeutic. Results from cell killing assays (Figure 32.5B, right) illustrate that the ImmTAC-activated CTLs were not only activated but also capable of efficiently destroying the target cells.

To investigate the importance of mTCR antigen-binding affinity for potency, a number of ImmTACs based on NY-ESO mTCRs, whose affinities varied significantly, were produced (Figure 32.5C, top). All these variants activated, in a dose dependent-manner, unstimulated purified CD8+ T cells in the presence of IM9 targets (Figure 32.5C, bottom). As anticipated, however, the ImmTACs based on the mTCRs of higher affinities were progressively more potent, achieving a profound effect at concentrations as low as 10pM in the case of the highest affinity. This highlights the importance of affinity maturation, and demonstrates the capability of mTCRs as targeting molecules. Down-regulation of MHC expression is a well-documented mechanism by which cancer cells evade immune surveillance [24,32]. Thus, to be able to eradicate tumors, an immunotherapy needs to be effective in the presence of low levels of peptide-MHC. The A375 cancer cell line, which expresses MAGE A3, exhibits very low levels of HLA-A1 (also positive by genotype, Figure 32.6 top) MHC I (Figure

(A)

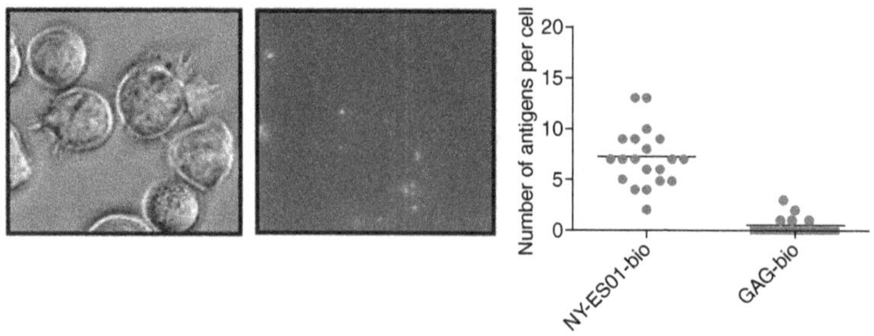

(B)

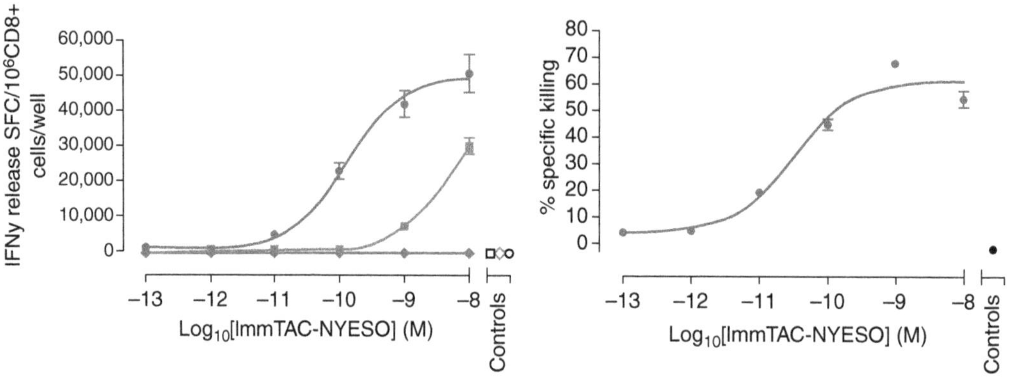

(C)

Clone ID	K_D (nM)	$T_{1/2}$ (s)
NYE-1	215	201
NYE-2	65.1	1040
NYE-3	2.9	4470
NYE-4	6	3590
NYE-5	1.1	16300

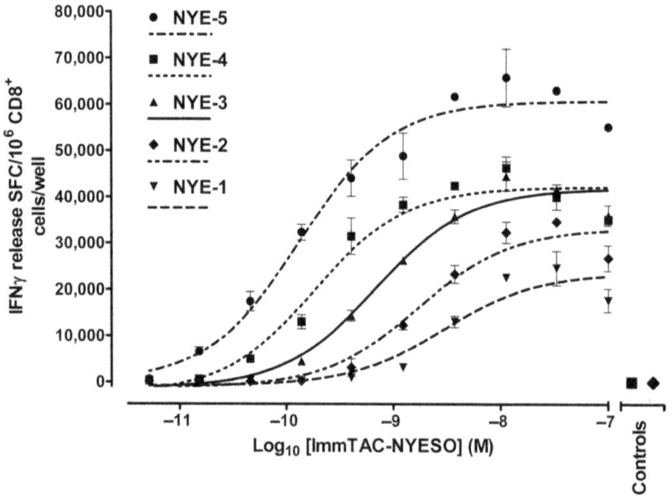

FIGURE 32.5 Targeting NY-ESO-1 presenting cancer cells with ImmTAC. (A) NY-ESO-1 peptide: MHC epitopes on the surface of cancer cells were determined using the high-affinity NY-ESO-1 mTCR (see Table 32.2). NY-ESO-1-positive IM9 cells were stained with biotinylated high-affinity NY-ESO mTCR and streptavidin-PE, and the individual epitopes observed as individual spots (middle panel), were counted using phase-contrast microscopy. The graph shows that the number of NY-ESO-1 epitopes presented were in the range 2–13 per cell. A high-affinity TCR

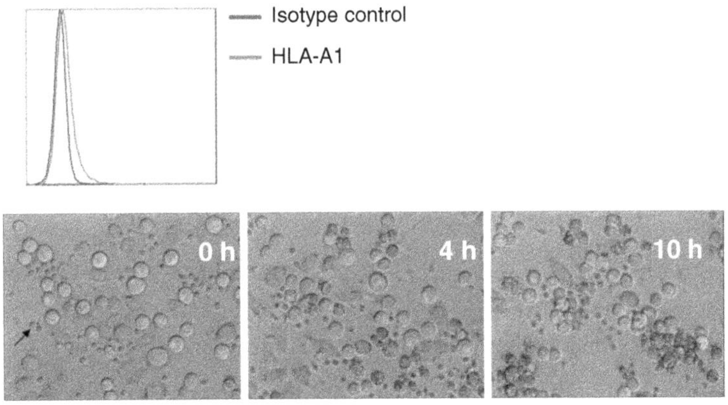

FIGURE 32.6 ImmTAC-MAGE displays high potency despite low ligand densities. Top: The expression of MHC class I on A375 cells. Flow cytometry shows only a low level of HLA-A1 expression (red), against an isotype control (blue). Bottom: Visualization of cell killing by CD8+ T cells in the presence of ImmTAC-MAGE. Time-lapse fluorescence microscopy over a period of 10 h shows progressive lysis of the MAGE-A3 positive A375 tumor cells (large red-stained cells) by the CD8+ T cells (small unstained cells, indicated by the black arrow) in the presence of ImmTAC-MAGE at 50 pM concentration. While the population of the A375 cells disappears entirely by the end of the observation period, the control MAGE-A3 negative cells (Colo205, large green-cells) remain intact. (No effect was registered in the presence of an irrelevant ImmTAC—data not shown.) (See the color version of this figure in Color Plates section.)

32.6 , top). Yet in the presence of 50pM ImmTAC-MAGE, the cancer cells were targeted and killed by purified CD8+ CTLs over the course of 10 h (Figure 32.6, bottom), while the HLA-A1^{+} MAGE-A3 negative cells (Colo205), remained unaffected.

32.5 CONCLUSIONS AND FUTURE PERSPECTIVES

T-cell receptors (TCRs) recognize cell surface-presented antigens; each consisting of an endogenously processed peptide carried in the context of a major histocompatibility complex molecule (peptide-MHC). This path of antigen presentation allows TCRs to detect the presence of intracellular proteins, which are largely inaccessible to antibodies.

The mTCR and ImmTAC technologies, which have their roots in research carried out at Oxford University and have been developed by Immunocore Ltd, have demonstrated that it is possible to target cancer cells with precious few antigens per cell. These types of cells will normally escape immune detection. However, ImmTACs redirect T cells with good efficacy when presented with as little as 10 targets per cell.

While naturally selected TCRs exhibit only micromolar affinities toward their peptide-MHC ligands, mTCRs have

FIGURE 32.5 ***(continued)*** specific for the HIV gag epitope was used as a negative control. (B) Biological activity of ImmTAC-NY-ESO-1. Purified CD8+ T cells were incubated overnight with IM9 cells and increasing concentrations of ImmTAC-NY-ESO-1. Dose-dependent IFN-γ release (demonstrating T-cell activation, left) and IM9 specific lysis (right) was observed. While significant levels of T-cell activation were observed in both assays at NY-ESO-ImmTAC concentrations as low as 10pM (top line), the addition of free (non-CD3-fused) high-affinity NY-ESO mTCR abrogated the effect (middle line, left), demonstrating the specificity of the interaction. A control cell line expressing HLA-A2 and lacking NY-ESO-1 expression was unaffected by the addition of the full titration range of ImmTAC (bottom line, left) or the top concentration of 10nM protein (closed circle, right). Further controls include CD8+ T cells with 10nM ImmTAC-NY-ESO-1 (open square, left), CD8+ T cells with IM9 cells (open diamond, left) and IM9 cells with 10nM ImmTAC-NY-ESO-1 (open circle, left). (C) Affinity-driven potency of ImmTAC-NY-ESO. ImmTACs were made with several NY-ESO mTCRs, ranging in their peptide: MHC-binding affinities from 1.1 to 215 nM (top). The IFN-γ release assays (bottom), shows that enhancing mTCR affinity enhances cellular potency as demonstrated by increased T-cell activation.

been engineered to exhibit picomolar levels without compromising specificity. These high-affinity mTCRs have been fused to a range of effector function polypeptides including protein toxins, cytokines, or single chain antibodies (scFv) to enable highly selective targeted killing of malignant cells.

Furthermore, TCRs can be developed against peptides derived from almost any intracellular antigen. Therefore, this strategy could open up a new therapeutic platform, addressing, by a very wide margin, the largest class of specific targets that the biology of cancer cells can afford. Cancer vaccines have attempted to use this class of targets but are limited by activating only naturally selected T cells that have low or, in many cases, no sensitivity against the low copy number of antigens presented on cancer cells. Therefore, ImmTACs, by means of their exquisite antigen sensitivity and high potency, may offer the first opportunity to address the huge spectrum of cancer relevant targets presented as peptide-MHC complexes.

A gp100-targeting ImmTAC, IMCgp100, has entered Phase I clinical trials in October 2010 for metastatic melanoma in both the United States and the United Kingdom. More cancer-specific ImmTACs are being developed. Additionally, mTCRs are being assessed for therapeutic applications against viral diseases and autoimmune diseases, such as type I diabetes. Experimentation has shown that the mTCR platform can be used to generate fusion constructs with a wide spectrum of effector moieties so that specific requirements can be addressed in different cancers, viral, and autoimmune diseases.

ACKNOWLEDGMENTS

We thank Karen Pulford for help and suggestions in preparing figures and chapter and our colleagues at Immunocore for help and comments.

REFERENCES

1. Arstila TP, Casrouge A, Baron V, Even J, Kanellopoulos J, Kourilsky P. (1999) A direct estimate of the human alphabeta T cell receptor diversity. *Science* 286(5441), 958–961.
2. Robins HS, Campregher PV, Srivastava SK, Wacher A, Turtle CJ, Kahsai O, et al. (2009) Comprehensive assessment of T-cell receptor beta-chain diversity in alphabeta T cells. *Blood* 114(19), 4099–4107.
3. Dunn GP, Old LJ, Schreiber RD. (2004) The three Es of cancer immunoediting. *Annu. Rev. Immunol.* 22, 329–360.
4. Luyten GP, van der Spek CW, Brand I, Sintnicolaas K, de Waard-Siebinga I, Jager MJ, et al. (1998) Expression of MAGE, gp100 and tyrosinase genes in uveal melanoma cell lines. *Melanoma Res.* 8(1), 11–16.
5. Garboczi DN, Hung DT, Wiley DC. (1992) HLA-A2-peptide complexes: refolding and crystallization of molecules expressed in *Escherichia coli* and complexed with single antigenic peptides. *Proc. Natl. Acad. Sci. U. S. A.* 89(8), 3429–3433.
6. Davis MM, Boniface JJ, Reich Z, Lyons D, Hampl J, Arden B, et al. (1998) Ligand recognition by alpha beta T cell receptors. *Annu. Rev. Immunol.* 16, 523–544.
7. Garcia KC, Degano M, Speir JA, Wilson IA. (1999) Emerging principles for T cell receptor recognition of antigen in cellular immunity. *Rev. Immunogenet.* 1(1), 75–90.
8. Cole DK, Pumphrey NJ, Boulter JM, Sami M, Bell JI, Gostick E, et al. (2007) Human TCR-binding affinity is governed by MHC class restriction. *J. Immunol.* 178(9), 5727–5734.
9. van der Merwe PA, Davis SJ. (2003) Molecular interactions mediating T cell antigen recognition. *Annu. Rev. Immunol.* 21, 659–684.
10. Schlueter CJ, Schodin BA, Tetin SY, Kranz DM. (1996) Specificity and binding properties of a single-chain T cell receptor. *J. Mol. Biol.* 256(5), 859–869.
11. Willcox BE, Gao GF, Wyer JR, O'Callaghan CA, Boulter JM, Jones EY, et al. (1999) Production of soluble alphabeta T-cell receptor heterodimers suitable for biophysical analysis of ligand binding. *Protein Sci.* 8(11), 2418–2423.
12. Boulter JM, Glick M, Todorov PT, Baston E, Sami M, Rizkallah P, et al. (2003) Stable, soluble T-cell receptor molecules for crystallization and therapeutics. *Protein Eng.* 16(9), 707–711.
13. Cole DK, Rizkallah PJ, Gao F, Watson NI, Boulter JM, Bell JI, et al. (2006) Crystal structure of HLA-A∗2402 complexed with a telomerase peptide. *Eur. J. Immunol.* 36(1), 170–179.
14. Sami M, Rizkallah PJ, Dunn S, Molloy P, Moysey R, Vuidepot A, et al. (2007) Crystal structures of high affinity human T-cell receptors bound to peptide major histocompatibility complex reveal native diagonal binding geometry. *Protein Eng. Des. Sel.* 20(8), 397–403.
15. Willcox BE, Gao GF, Wyer JR, Ladbury JE, Bell JI, Jakobsen BK, et al. (1999) TCR binding to peptide-MHC stabilizes a flexible recognition interface. *Immunity* 10(3), 357–365.
16. Kersh GJ, Allen PM. (1996) Structural basis for T cell recognition of altered peptide ligands: a single T cell receptor can productively recognize a large continuum of related ligands. *J. Exp. Med.* 184(4), 1259–1268.
17. Kersh GJ, Allen PM. (1996) Essential flexibility in the T-cell recognition of antigen. *Nature* 380(6574), 495–498.
18. Wilson IA, Garcia KC. (1997) T-cell receptor structure and TCR complexes. *Curr. Opin. Struct. Biol.* 7(6), 839–848.
19. Holler PD, Holman PO, Shusta EV, O'Herrin S, Wittrup KD, Kranz DM. (2000) In vitro evolution of a T cell receptor with high affinity for peptide/MHC. *Proc. Natl. Acad. Sci. USA* 97 (10), 5387–5392.
20. Hoogenboom HR. (2005) Selecting and screening recombinant antibody libraries. *Nat. Biotechnol.* 23(9), 1105–1116.
21. Rothe A, Nathanielsz A, Hosse RJ, Oberhäuser F, Strandmann EP, Engert A, et al. (2007) Selection of human anti-CD28 scFvs from a T-NHL related scFv library using ribosome display. *J. Biotechnol.* 130(4), 448–454.

22. Shusta EV, Holler PD, Kieke MC, Kranz DM, Wittrup KD. (2000) Directed evolution of a stable scaffold for T-cell receptor engineering. *Nat. Biotechnol.* 18(7), 754–759.
23. Boulter JM, Jakobsen BK. (2005) Stable, soluble, high-affinity, engineered T cell receptors: novel antibody-like proteins for specific targeting of peptide antigens. *Clin. Exp. Immunol.* 142(3), 454–460.
24. Dunn SM, Rizkallah PJ, Baston E, Mahon T, Cameron B, Moysey R, et al. (2006) Directed evolution of human T cell receptor CDR2 residues by phage display dramatically enhances affinity for cognate peptide-MHC without increasing apparent cross-reactivity. *Protein Sci.* 15(4), 710–721.
25. Li Y, Moysey R, Molloy PE, Vuidepot A-L, Mahon T, Baston E, et al. (2005) Directed evolution of human T-cell receptors with picomolar affinities by phage display. *Nat. Biotechnol.* 23(3), 349–354.
26. Purbhoo MA, Li Y, Sutton DH, Brewer JE, Gostick E, Bossi G, et al. (2007) The HLA A∗0201-restricted hTERT(540-548) peptide is not detected on tumor cells by a CTL clone or a high-affinity T-cell receptor. *Mol. Cancer Ther.* 6(7), 2081–2091.
27. Purbhoo MA, Sutton DH, Brewer JE, Mullings RE, Hill ME, Mahon TM, et al. (2006) Quantifying and imaging NY-ESO-1/LAGE-1-derived epitopes on tumor cells using high affinity T cell receptors. *J. Immunol.* 176(12), 7308–7316.
28. Zhong G, Zhang S, Li Y, Liu X, Gao R, Miao Q, et al. (2010) A tandem scFv-based fusion protein and its enediyne-energized analogue show intensified therapeutic efficacy against lung carcinoma xenograft in athymic mice. *Cancer Lett.* 295(1), 124–133.
29. Baeuerle PA, Reinhardt C. (2009) Bispecific T-cell engaging antibodies for cancer therapy. *Cancer Res.* 69(12), 4941–4944.
30. Dreier T, Lorenczewski G, Brandl C, Hoffmann P, Syring U, Hanakam F, et al. (2002) Extremely potent, rapid and costimulation-independent cytotoxic T-cell response against lymphoma cells catalyzed by a single-chain bispecific antibody. *Int. J. Cancer* 100(6), 690–697.
31. Nagorsen D, Bargou R, Ruttinger D, Kufer P, Baeuerle PA, Zugmaier G. (2009) Immunotherapy of lymphoma and leukemia with T-cell engaging BiTE antibody blinatumomab. *Leuk. Lymphoma* 50(6), 886–891.
32. Cabrera CM, Jiménez P, Cabrera T, Esparza C, Ruiz-Cabello F, Garrido F. (2003) Total loss of MHC class I in colorectal tumors can be explained by two molecular pathways: beta2-microglobulin inactivation in MSI-positive tumors and LMP7/TAP2 downregulation in MSI-negative tumors. *Tissue Antigens* 61(3), 211–219.

33

AMEDIPLASE

Stefano Evangelista and Stefano Manzini

Menarini Ricerche SpA, Preclinical Development, Florence, Italy

33.1 INTRODUCTION

Acute myocardial infarction (AMI) is the leading cause of cardiovascular mortality in the Western countries [1]. In the United States, the incidence of new myocardial infarction is high; accounting as 25.4% of the total annual deaths [2]. More than 250,000 people die within 1 h from the onset of symptoms and before ever reaching a hospital. The rapid restoration of coronary patency is the main goal of critical care therapy in subjects with AMI [3]. Thrombolytics represent the main pharmacological tool for rapid induction of coronary recanalization [4].

Amediplase is the result of a project aimed to identify and develop a recombinant plasminogen activator (PA) with outstanding fibrinolytic activity, high level of fibrin specificity, prolonged half-life, and no immunogenicity to be used as a single bolus agent in AMI.

All fibrinolytic agents act, directly or indirectly, as PAs, but they can be subdivided according to the fibrin dependency of their action in nonfibrin-specific and fibrin-specific agents. Nonfibrin-specific agents are streptokinase (SK), urokinase (u-PA, also known as two-chain urokinase, tcu-PA), and anistreplase (APSAC). On the other hand, fibrin-specific agents (i.e., agents whose proteolytic activity is greatly enhanced by fibrin binding) are alteplase (rt-PA), saruplase (unglycosylated human rscu-PA), reteplase, and tenecteplase (TNK-t-PA) differing for tPA length, glycosylation sites, and some crucial amino acids. See their schematic structure in Figure 33.1.

Even if thrombolytic efficacy of all these drugs has been established; each agent has different advantages and disadvantages in terms of time-to-reperfusion, reocclusion rate, systemic fibrinolysis, immunogenicity, and so on [5].

Different approaches have been explored to improve thrombolytic treatment, including the construction of mutants and variants of t-PA and u-PA and of chimeric PAs. t-PA is a single chain molecule that can be converted to a two-chain structure by a specific plasmin cleavage at Arg^{275}-Ile^{276} bond. The NH_2-terminal heavy chain (A chain) comprises four structural/functional domains: a finger region (F), a region similar to human EGF (E); and two kringles (K_1 and K_2). These domains are autonomous entities that regulate plasma clearance (E and F), lysine binding (K_1 and K_2), fibrin binding (F and K_2), while the protease activity resides in the COOH-terminal chain (B chain) [6]. Pro-urokinase, or single-chain u-PA (scu-PA) is a single chain glycoprotein that, at variance with t-PA, is enzymatically active only after limited cleavage (by plasmin or kallikrein) at the Lys^{158}-Ile^{159} bond to give two-chain u-PA (tcu-PA). The catalytic center is located in the COOH-terminal part of scu-PA. Although it has no affinity for fibrin, scu-PA preferentially cleaves fibrin-bound plasminogen, and the sites responsible for the formation of the functional complex appear also to be located in the COOH terminus of scu-PA [7].

Amediplase (K_2tu-PA, product code MEN 9036) was constructed as a hybrid PA comprising just the kringle 2

Fusion Protein Technologies for Biopharmaceuticals: Applications and Challenges, First Edition. Edited by Stefan R. Schmidt.

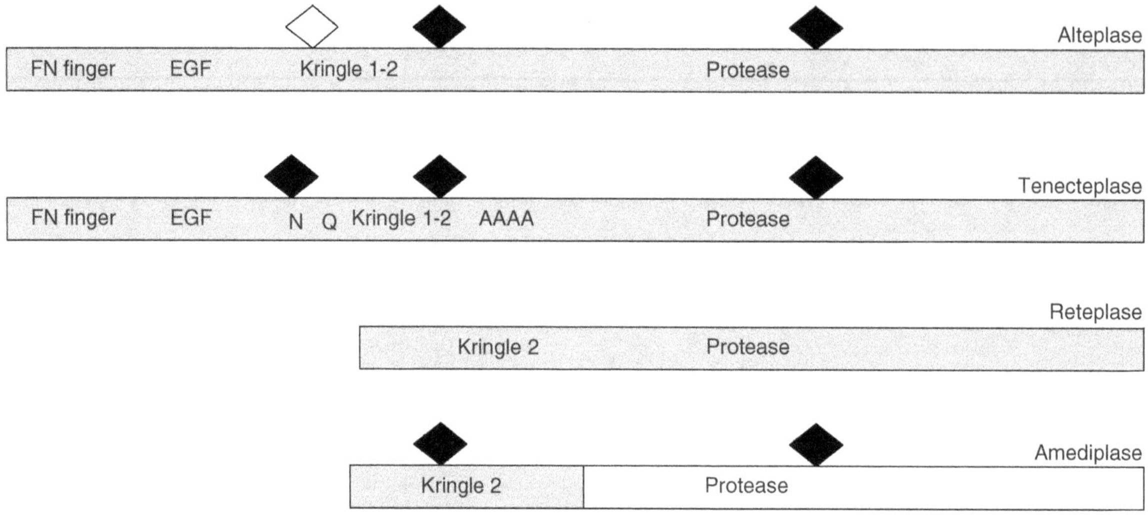

FIGURE 33.1 Schematic composition of the most recent thrombolytics. In dark grey the tPA portion, in light grey the sc-UPA portion. Dark diamond = N glycosylation site, light diamond = O-glycosilation site.

of t-PA (to maintain fibrin affinity, while losing domains responsible for fast plasma clearance) and the protease domain of scu-PA, fused at the plasmin cleavage site between Arg^{275} of t-PA and Ile^{159} of scu-PA (to form the Arg^{103}-Ile^{104} bond in the chimera). The advantage of this approach is that it eliminates from scu-PA two sites of possible cleavage leading to fibrin-aspecific or inactive u-PA forms (Glu^{143}-Leu^{144} and Arg^{156}-Phe^{157}) [8].

This rationale design of the molecule should result in a thrombolytic agent with relevant fibrinolytic potency (due to the scuPa protease domain), remarkable safety (due to the fibrin specificity linked to the kringle 2 of t-PA), prolonged plasma half-life allowing single bolus administration (resistance to metabolic degradation), and no immunogenicity (humanized protein).

33.2 SOURCE, PHYSICO-CHEMICAL PROPERTIES AND FORMULATION

33.2.1 Production Cell Line

The product amediplase was initially derived from the tissue culture supernatant of a permanent Chinese hamster ovary cell line (CHO CGI-1), which was obtained by deliberate genetic manipulation at Ciba Geigy Ltd, Basel, Switzerland. The parent cell line DUKXBI (ATCC CRL 9096) was supplied to Ciba Geigy by Professor L. Chasin [9]. Ciba Geigy modified the parent cell line by recombinant DNA technology to obtain the producer cell line CHO CGI-1, which produced and secreted CGP 42935 in the culture supernatant.

Following studies performed at Ciba Geigy, aimed to improving the stability and productivity of the original cell line CHO CGI-1, led to a modified expression vector and to the new cell line, renamed CGI-2. The expression vector is a pBR322-based plasmid containing a dihydrofolate reductase selection marker in addition to the hybrid PA gene and appropriate promoters, splicing signals, and so on required for effective expression.

CGI-2 cell was then transferred to A. Menarini S.r.l., Florence, Italy, and Menarini Biotech S.r.l., Pomezia, Italy for the production of amediplase (code MEN9036) for further preclinical and clinical experimentations.

33.2.2 Drug Substance

Amediplase (Figure 33.2) is a hybrid PA consisting of the kringle 2 domain of t-PA and the protease domain of u-PA. Amediplase contains 356 amino acids, 18 of which are cysteine residues forming 9 disulphide bonds. It has potentially two glycosylation sites, one in each domain, whereas the glycosylation site in the kringle domain is not or only lightly occupied, the other site is fully glycosylated. Glycosylation sites are occupied by N-linked complex oligosaccharide structures. No O-linked carbohydrates have been detected on amediplase. Treatment of amediplase with plasmin results in a single peptide bond cleavage between residues Arg^{103} and Ile^{104}, leading to the formation of two subunits, A and B, that are still connected through a disulphide bridge. While amediplase is a zymogen, the double chain form is the active form.

Production of amediplase drug substance is carried out at Menarini Biotech (Pomezia, Italy), a company of the Menarini Group. Currently, the production (in GMP compliance) has been scaled-up to an industrial level in 1500-L fermentors. A cycle consists of a semicontinuous fermentation in

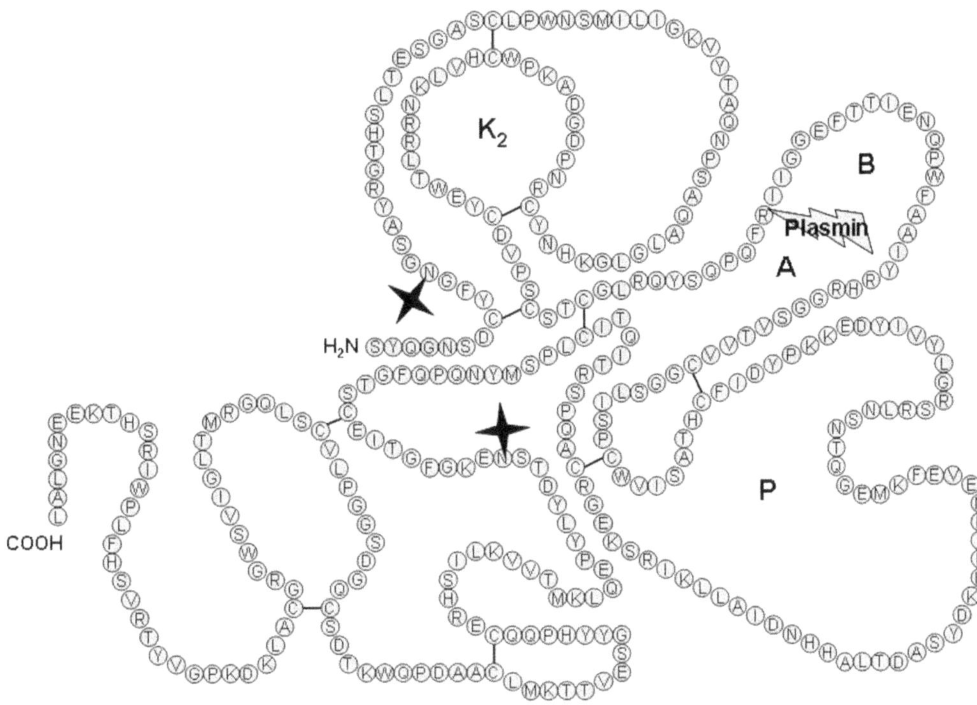

FIGURE 33.2 Primary sequence of amediplase. The plasmin cleavage site is indicated and splits amediplase into an A chain and a B chain. The diamonds denote positions of N-linked oligosaccharide; K2, kringle-2 domain; P, serine protease domain.

which culture supernatant is removed periodically, retaining the cells in the fermentor. Amediplase is recovered and purified from the supernatant "harvest." Each cycle yields up to 12 of such "harvests."

Currently, the biotechnological process is characterized by two fermentation steps following initial expansion of the seed in stationary flasks and spinners. The purification consists of four chromatographic steps, low pH viral inactivation, ultrafiltration/diafiltration followed by a final filtration step including a viral removal. Transfected CHO cells grow in a nutrient medium containing no antibiotics.

Amediplase drug substance is stable at −20°C for at least 24 months and at 4°C for at least 4 weeks. The quality of the amediplase drug substance in terms of assay, potency, and purity (in particular the absence of endotoxins, contaminants, and adventitious agents as bacteria, mycoplasma, and viruses) complies with the EEC requirements for biotechnological products.

33.2.3 Drug Product

Amediplase is available as lyophilized enzyme in vials containing 100 mg of active principle. Additional excipients are present in the lyophilized product: sucrose, succinic acid, and sodium hydroxide. Sucrose serves as a bulking agent and cake stabilizer and it is present at a concentration fivefold higher than amediplase by weight. The others excipients, succinic acid and sodium hydroxide, are present to buffer and to adjust the formulation at a pH compatible with amediplase stability.

One vial amediplase 100 mg (powder for injection) contains

Amediplase	100.00 mg
Sucrose	500.00 mg
Succinic acid	8.86 mg
Sodium hydroxide	qs to pH 4.0[a]
Nitrogen	qs[b]

[a]theoretical value 1.35 mg/vial.
[b]used for pre-venting.

For reconstitution of the lyophilized product, 10 mL water for injections (WfI) is prescribed.

The stability has been tested using a bracketing approach. A 50-mg as well as a 120-mg amediplase containing formulation was investigated. The stability data support a shelf life of 24 months for both formulations indicating a comparable shelf life for the 100-mg strength. However, a new stability study according to ICH has been started for the 100-mg formulation. The final product should be stored not above 25°C. After reconstitution, the drug product has to be administered immediately.

33.3 PRECLINICAL STUDIES

33.3.1 *In Vitro* Biological Properties

The recombinant hybrid retains some important properties of the parent molecules t-PA and scu-PA. Like t-PA, it requires fibrin to fully express PA activity thus granting a fibrin-specific thrombolytic activity. Like scu-PA, it behaves as a zymogen, being insensitive to plasma protease inhibitors, including the fast acting PAI-1, and requiring conversion to an active (two-chain) form for full expression of enzymatic activity. The latter property may further increase the fibrin specificity by limiting the activation of the molecule to the thrombus surface where plasmin is generated and protected from inhibition [10].

33.3.1.1 Plasminogen Activator Activity The specific activity of amediplase, measured by amidolytic assay and fibrin plate is 2.1×10^{-6} and 2.5×10^{6} t-PA equivalent units per milligram, respectively. On a molar basis, the hybrid is twofold more active than t-PA. PA activity of amediplase is reduced by more than 98% in the absence of CNBr-digested fibrinogen, indicating that, like t-PA, the hybrid requires fibrin for the expression of its activity [10].

33.3.1.2 In Vitro Thrombolytic Activity This property was studied using a ^{125}I-labeled (human) blood clot (125 μL) immersed in autologous plasma (1 mL). Amediplase, at a concentration of 500 ng/mL, induces clot lysis at a rate comparable to that of a similar concentration of t-PA whereas, at lower concentrations, it is less active than the reference enzyme. Dose–response curves indicate that the extent of lysis by amediplase is not linear with the enzyme concentration. This behavior resembles that of scu-PA [10].

33.3.1.3 Fibrin Specificity In plasma, in the absence of fibrin, amediplase induces little activation of fibrinolysis as indicated by the fact that significant plasminogen and α2-antiplasmin consumption is observed only at high concentrations (≥2 μg/mL). In this respect, the hybrid is well comparable to t-PA [10].

33.3.1.4 Inhibition by Plasma Protease Inhibitors Like scu-PA, the hybrid is insensitive to inhibition. Indeed, incubation in plasma for 24 h at 37°C, at a concentration of 500 ng/mL, is not associated with a noticeable inhibition of the activity (<10%). t-PA, under the same experimental conditions, is effectively inhibited, with a $t_{1/2}$ of about 90 min. Resistance to inhibition was also demonstrated using purified recombinant type 1 plasminogen activator inhibitor (rPAI-1) [10]. Amediplase has also been shown to be more resistant than t-PA [11] or scu-PA [12] to the antifibrinolytic activity of TAFI (thrombin activable fibrinolysis inhibitor). At therapeutic concentrations both amediplase and tenecteplase are marginally affected by TAFI [12].

33.3.1.5 Binding to Fibrin When the binding to forming fibrin was studied, a 40% binding of the hybrid was obtained at a starting fibrinogen concentration ≥3.2 mg/mL. This figure is significantly lower than that obtained with t-PA, which is almost totally (>80%) found in association with fibrin at a fibrinogen concentration as low as 0.2 mg/mL [10].

33.3.1.6 Clot Penetration Amediplase binding to fibrin clot is specific being fully inhibited by ε-aminocaproic. Plasma clot penetration was studied during pressure-driven fluid permeation. Amediplase, at variance to alteplase, was able to enter the clot without significant hindrance. This observation was also confirmed in diffusion-driven clot penetration study using confocal microscopy. Alteplase was detected on or close to the clot surface, while amediplase was also detected deep inside the clot [13].

In a further set of experiments (Figure 33.3), it has been shown that, independently from the size of the human clot, amediplase possesses, at therapeutic concentrations, a faster lysis time than tenececteplase or, even more, than scu-PA [12].

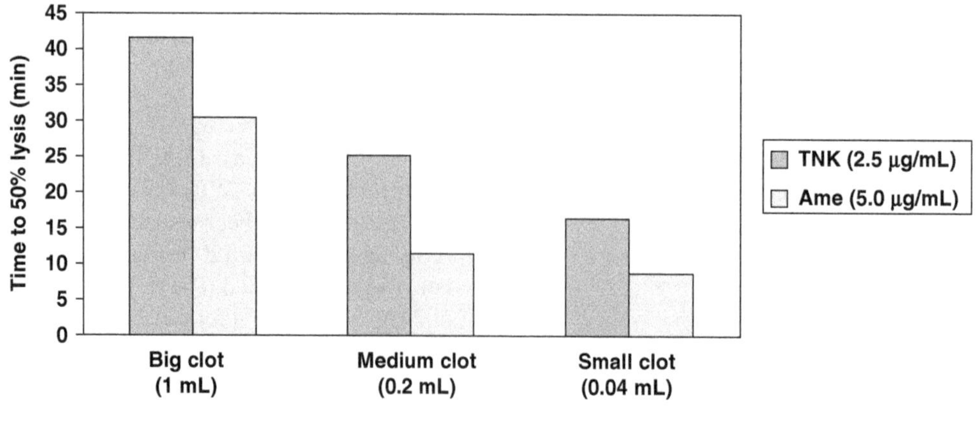

FIGURE 33.3 Comparison of *in vitro* lytic effect of amediplase (Ame) and tenecteplase (TNK) with clots of different size. See Reference [12].

33.3.2 Pharmacodynamics: *In Vivo* Studies

33.3.2.1 ***Thrombolytic Activity in Rabbits*** The thrombolytic activity of amediplase as compared to recombinant t-PA (rt-PA) was assessed by the well-known method of the lysis of standard-sized preformed ^{125}I-fibrinogen labeled thrombi produced in the external jugular veins of rabbits. For both drugs, the doses were 0.4, 0.8, and 1.2 mg/kg infused over 3 h, with saline as control.

Thrombolysis, measured as the percent reduction in thrombus weight, was 11% for saline, 38%, 66%, 89% for the three doses of amediplase, respectively; and 32%, 49%, 68% for the three doses of rt-PA. The thrombolytic activity of amediplase was significantly higher than that of rt-PA at 0.8 and 1.2 mg/kg ($p < 0.01$). At the highest dose (1.2 mg/kg), both amediplase and rt-PA produced a significant reduction of fibrinogen, α-2 antiplasmin, and plasminogen [14].

33.3.2.2 ***Thrombolytic and Haemorrhagic Effects in Rabbits*** The aim of this study was to evaluate the effects of bolus doses of rt-PA and amediplase on (1) lysis of radioactive thrombi (fibrinolysis); (2) accretion of new fibrin on preexisting thrombi during fibrinolysis (thrombus growth); (3) thrombolysis, the resultant of the two previous processes; and (4) blood loss. A jugular vein thrombosis model and an ear-bleeding model were adopted in rabbits. The doses of the two test drugs were 0.2, 0.4, and 0.8 mg/kg; saline was used as control. Fibrinolysis produced by saline was 11%, that produced by the three doses of amediplase was 39%, 57%, 83%, respectively and that produced by rt-PA was 35%, 54%, 78%, respectively. Amediplase was more effective than rt-PA at all doses in preventing the accretion of new fibrin on the thrombi and in producing thrombolysis. The two lower doses of amediplase and rt-PA did not produce systemic proteolysis and bleeding. The highest dose of amediplase produced a more intense systemic proteolysis and bleeding than the highest dose of rt-PA [15].

33.3.2.3 ***Fibrinolytic Activity of Amediplase in Rats*** The fibrinolytic activity of amediplase has been studied in a model of arterial thrombosis induced by FeCl3 in rat carotid artery. Amediplase was administered as a single bolus administration via the jugular vein at dosages of 1, 3, or 6 mg/kg. A dose-dependent fibrinolytic effect was observed. At the highest dose, there were 87.5% of animals fully reperfused, and this was achieved between 60 to 85 min following the bolus administration and lasted up to 105 min [16]. In this experimental model, thrombolytic activity and time to carotid patency were similar for amediplase (6 mg/kg) or tenececteplase (6 mg/kg) both administered as a bolus [17]. When bigger thrombus was obtained by increasing the time of occlusion, percentage of animals achieving complete reperfusion was slightly higher with amediplase than with tenecteplase [17].

33.3.2.4 **In Vivo** ***Thrombolytic Activity of Amediplase in Dogs*** The purpose of this investigation was to evaluate the *in vivo* thrombolytic action of amediplase in comparison to rt-PA in a model of coronary thrombosis in the dog. Thrombosis was produced through an anodal current applied to the intima of the circumflex coronary artery. Thrombolysis occurred in seven out of nine animals administered rt-PA at 25 mg total dose (1 mg/kg as a bolus, the balance over 1 h) and in all five animals given amediplase as a single intravenous bolus of 0.6 mg/kg. Mean time to thrombolysis after amediplase was 19.6 min compared to 36.3 min after rt-PA and coronary blood flow upon thrombolysis was greater and more protracted after amediplase than after rt-PA (Table 33.1). Both drugs decreased plasma fibrinogen. In conclusion, one amediplase dose injected quickly produced more rapid thrombolysis than the double dose of rt-PA infused over a longer period [18].

33.3.3 Pharmacokinetics

33.3.3.1 ***Pharmacokinetics of Amediplase in Monkeys*** Monkeys were chosen for kinetic studies because they are considered closer to humans than other mammals for haemostasis. Six cynomolgus monkeys were used. Amediplase was administered as a bolus at the dose of 0.5 mg/kg. Results show that kinetics of amediplase in monkeys are significantly slower than those of t-PA: the amediplase-α and amediplase-β half-lives were 6 and 47 min, respectively, as opposed to the reported values of <2.5 and 12–20 min for t-PA [19].

33.3.3.2 ***Pharmacokinetics of Amediplase in Rats*** Amediplase was administered as i.v. bolus of 0.3, 1, 3, or 6 mg/kg to male and female rats. The plasma concentration of amediplase decreased biphasically (Figure 33.4). Very high, dose-related, area under curves (AUCs) were measured. In the rat, the $t_{1/2}$ α was between 4 and 8 min and the $t_{1/2}$ β between 36 and 91 min, thus longer than literature values of alteplase. The volume of distribution values, V_c (31–49 mL/kg), V_{ss} (38–93 mL/kg), and V_{area} (342–439 ml/kg), strongly suggest that amediplase is mainly distributed in the central compartment (blood). The systemic clearance was in the range 2.18–7.67 mL/min/kg, suggesting a moderate hepatic metabolism [20].

TABLE 33.1 Thrombolytic Activity of Amediplase in Dogs

	n	Amediplase	*n*	rt-PA
Time to achieve reperfusion (min)	5	19.6 ± 4.3	7	36.3 ± 6.3
Time to achieve reocclusion (min)	4	124.3 ± 18.6	7	64.4 ± 20.2

[18]See Reference [18].

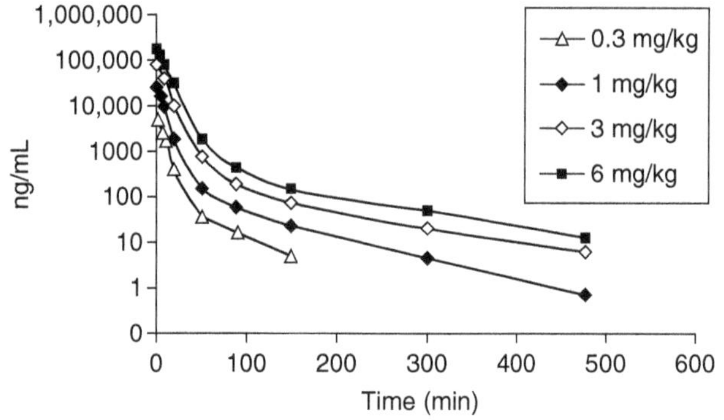

FIGURE 33.4 Semilogarithmic graphical representation of amediplase plasma concentration versus time following single intravenous administrations at different doses in rats. Data is expressed as mean ($n = 8$, 4 males + 4 females).

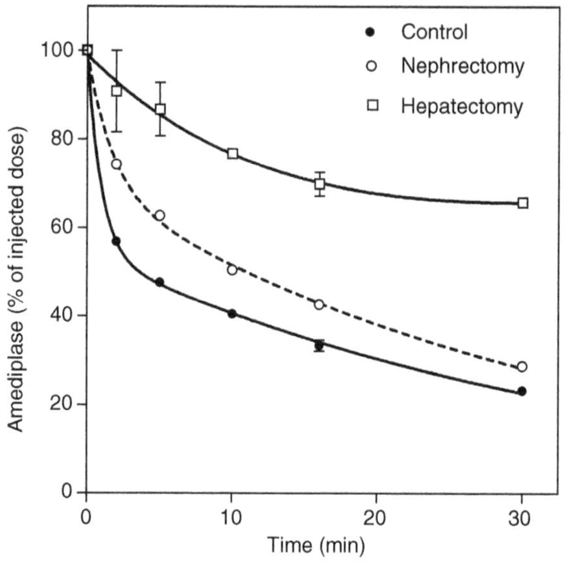

FIGURE 33.5 ^{125}I-Amediplase was injected intravenously in control rats (•), nephrectomized rats (○), or hepatectomized rats (□). At the indicated time point, plasma samples were taken and the radioactivity in the plasma was counted. Data are expressed as percentage of the injected dose (mean ± standard deviation, $n = 4$).

33.3.3.3 The **In Vivo** *Fate of Amediplase in Rats* The *in vivo* fate of ^{125}I-amediplase has been studied in the rats by intravenous administration of doses ranging from a trace dose (50 μg/kg body weight) to 1.0 mg/kg body weight. Amediplase is relatively rapidly cleared from plasma, mainly through hepatic route. Its kinetic profile is favorable as compared to alteplase. At the used doses, liver uptake is not saturated and is around 16–18%, as compared to 80% for alteplase. Parenchyma liver cells and Kupffer cells are the major contributors to hepatic uptake. Plasma clearance is substantially reduced in hepatectomized rats. Amediplase plasma clearance is marginally reduced in nephrectomized rats indicating a minor role of kidney in its clearance (Figure 33.5). Tissue distribution of amediplase, apart from liver and kidneys, is limited to muscle, bones, and skin [21].

33.3.4 Toxicology

Acute and subchronic toxicity studies in rat, dog, and monkeys have not revealed any unexpected toxicity. Acute toxicity studies in rats showed that the no-observed-adverse effect level (NOAEL) is 10 mg/kg. The acute NOAEL in dogs was 1.0 mg/kg and exaggerated pharmacological effects were observed at 3 and 10 mg/kg. Subacute toxicity in rats at 12.5, 25, and 50 mg/kg repeated for 14 days showed that 12.5 mg/kg/day is the NOAEL and toxic and/or pharmacological effects were reported at the higher doses. Subacute toxicity in dogs at 1 and 2 mg/kg repeated for at least 9 days showed that 2 mg/kg can be considered the NOAEL and longer treatment time were prevented by immunological response against a human protein. Subacute toxicity in monkeys at 0.5, 1, 2, and 4 mg/kg repeated for 14 days showed that 0.5 mg/kg/day is apparently the NOAEL and pharmacological effects were reported at the higher doses. Amediplase clinical formulation showed adequate local tolerance following single administration via intravenous, paravenous, or intra-arterial route in the rabbit. Amediplase is devoid of mutagenic properties. In the rat, amediplase at 12.5, 25, and 50 mg/kg from day 6 to day 15 postcoitus did not show teratogenic or embryofetal toxic effect, while being maternotoxic at the highest dose. In the rabbit, at 1, 3, and 10 mg/Kg from day 6 to day 18 postcoitus amediplase produced (similarly to tenececteplase) embryotoxic effects due to exaggerated pharmacology. In a battery of eight safety pharmacology tests (CNS, cardiovascular, gastrointestinal), amediplase did not reveal any unexpected toxic effects.

33.4 HUMAN STUDIES

33.4.1 1K2 Study: A Phase I Single Ascending Dose Study

The clinical development of amediplase started with the first administration to man in a Phase I single ascending dose study (coded 1K2) [22].

Twenty healthy male volunteers completed a two-session, double blind, placebo-controlled, crossover study, to evaluate the tolerability, pharmacokinetics, and pharmacodynamics of amediplase. Patients received placebo and one out of the following intravenous dose of amediplase, each separated by a 3-day wash out period: 0.1 mg/kg ($n = 2$), 0.25 mg/kg ($n = 6$), 0.5 mg/kg ($n = 6$) over 10 min and 0.25 mg/kg ($n = 3$) and 0.5 mg/kg ($n = 3$) over a 5–10-min period. The treatment was well tolerated. No clinically

TABLE 33.2 1K2 Study: Mean Pharmacokinetic Parameters at 0.5 mg/kg

	0.5 mg/kg in 10 min ($n=5$)	0.5 mg/kg in 5 min ($n=3$)
AUC (μg × min /mL)	93.5	163
$t_{1/2}$ α (min)	8.4	12.6
$t_{1/2}$ β (min)	36.7	69.3
Cl (mL/min/kg)	5.3	3.1
Vc (mL/kg)	78.9	61.5

significant changes were observed in blood pressure, pulse rate, serial 12-lead ECG recordings, laboratory biochemistry, and haematology. No serious adverse events were reported.

The pharmacokinetic results showed that the peak levels of amediplase in plasma were reached 1 min post infusion (with the only exception of the 0.5 mg/kg dose administered over 5 min in which peak levels were obtained at 2–3 min) and were directly proportional to the dose administered. The disappearance of amediplase from the circulation followed a biphasic kinetic, with a $t_{1/2}$ for the fast (α) phase ranging between 7 and 12 min, and $t_{1/2}$ for the terminal (β) phase ranging between 38 and 77 min (Table 33.2). Analysis of the fibrinolysis and coagulation parameters showed that antiplasmin/plasmin complex and D–dimer increased in a dose-dependent way (0.1–0.25 mg/kg), while a modest dose-dependent decrease was seen for plasminogen and α-2 antiplasmin; fibrinogen and prothrombin fragments 1+2 were within the normal range.

No antibody levels to amediplase were detected 15 days after drug administration. From the 1K2 study could be concluded:

- Amediplase, administered as a 5–10 min intravenous infusion up to 0.5 mg/kg, was well tolerated and did not induce immunogenicity.
- Amediplase was cleared from human plasma with $t_{1/2}\alpha$ of about 12 min, a value markedly superior to that reported for alteplase (4–5 min).
- Amediplase proved to be an efficient PA, as demonstrated by the increase of antiplasmin/plasmin complex and D–dimer starting from low doses (0.1–0.25 mg/kg).
- Amediplase did not affect circulating fibrinogen in a clinically significant manner in the studied population of healthy subjects, thus showing a good fibrin-specific activity.

33.4.2 2K2 Study: A Phase II, Multicenter, Open-Label, Dose-Finding, Pilot Study

This Phase II, open label, dose finding study (coded 2K2) has been carried out to evaluate the efficacy and safety of amediplase in patients with suspected AMI and further treated with heparin and aspirin [23,24].

Efficacy was evaluated by measuring infarct-related coronary artery patency (TIMI flow grade), at coronary angiogram performed 90 min after an i.v. bolus of amediplase. The minimum effective dose was defined as that achieving at least 50% TIMI flow grade 3. The safety profile and the 90-min angiographic data of amediplase (the latter evaluated both at the single institution and by a centralized expert panel) were reviewed by a safety steering committee. The pharmacokinetics and potential immunogenicity of amediplase and the effect on the hemostatic parameters were also assessed in a subgroup of patients.

One hundred forty-five patients received one out of single ascending doses of amediplase (20, 30, 50, 70, 80, or 90 mg). Data analysis was performed in 141 evaluable patients (i.e., patients treated with amediplase and with angiographic evaluation) following two approaches: the first one considering the administered amediplase dose as fixed dose (mg), the second one considering the administered amediplase dose as body weight-adjusted dose (mg/kg).

TIMI results by dose group are summarized in Table 33.3 . On the basis of the % TIMI 3 results, the effective doses were identified in the range 70–90 mg. When results were analyzed as amediplase doses adjusted by patient's body weight, TIMI 3 ≥ 50% was consistently achieved at 1.0–1.2 mg/kg (Table 33.4). The pharmacokinetics of amediplase was evaluated in 18 patients at 50 mg ($n=4$), 70 mg ($n=5$), 80 mg ($n=5$), and 90 mg ($n=4$).

Results showed a dose proportionality of C_{max} and AUC, with a $t_{1/2}$ α of about 12 min and a limited V_c. Median (and

TABLE 33.3 2K2 study: TIMI Results Grouped by Amediplase Dose Expressed as Fixed Dose

Dose (mg)	Total Number of Treated Patients	Number of Patients and % TIMI 0	Number of Patients and % TIMI 1	Number of Patients and % TIMI 2	Number of Patients and % TIMI 3
20	7	5 (71.4%)	1 (14.0 %)	0	1 (14.2%)
30	8	4 (50%)	0	4 (50%)	0
50	39	10 (25.6%)	3 (7.6%)	11 (28.2%)	15 (38.4%)
70	21	4 (19%)	3 (14.29%)	2 (9.5%)	11 (52.3%)
80	39	8 (20.5%)	3 (7.6%)	10 (3.9%)	18 (46.1%)
90	27	7 (25.9%)	1 (2.7%)	6 (18.5%)	13 (48.1%)

TABLE 33.4 2K2 Study: TIMI 3 Results Grouped by Amediplase Dose Expressed as Body Weight-Adjusted Doses

Dose (mg/kg)	Total Number of Treated Patients	Number of Patients with TIMI 3	% TIMI 3
0.2	6	1	17
0.3	3	0	0
0.4	5	0	0
0.5	9	2	22
0.6	13	3	23
0.7	19	8	42
0.8	23	13	57
0.9	10	4	40
1.0	21	11	52
1.1	11	6	55
1.2	10	6	60
1.3	10	4	40
1.4	1	0	0

range) values for relevant pharmacokinetic parameters are summarized in Table 33.5. Hemostatic parameters results referring to the 80-mg dose of amediplase are summarized in Table 33.6.

Overall, amediplase showed satisfactory fibrin specificity, with a moderate and transient consumption of fibrinogen and α-2-antiplasmin. The decrease in plasminogen and the increase of plasmin-α2-antiplasmin complexes and D-dimer reflected the fibrinolytic activity of the compound. PAI-1 levels appeared not to be affected by amediplase administration.

A modest increase of thrombin–antithrombin III complexes was restricted to the first hour, with no increase in prothrombin fragments 1+2, reflecting a transient and limited procoagulant effect of amediplase.

No anti-amediplase antibodies were detected in blood samples from the three patients analyzed. From the 2K2 study could be concluded:

- The effective dose range of amediplase achieving ~50% TIMI 3 was in the range 70–90 mg; when adjusted by patient's body weight, the effective dose range corresponded to 1.0–1.2 mg/kg.
- Amediplase appeared to be a safe treatment as judged by the low incidence of adverse events (including serious class-related adverse events).
- The pharmacokinetic profile of amediplase and in particular its relatively long half-life supports the bolus intravenous administration.
- The effects on coagulation parameters confirmed the fibrin specificity and the low procoagulant effect of amediplase.

33.4.3 3K2 Study: A Double-Blind, Randomized, Parallel Group Study to Compare the Efficacy and Safety of Two Doses of Amediplase

This Phase II, open label study (coded 3K2) has been carried out to compare the efficacy and safety of two different doses of amediplase (1.0 mg/kg vs 1.2 mg/kg), both given as an i.v. bolus injection, in patients with suspected AMI and further treated with heparin and aspirin [25,26].

Patients were randomly assigned, in a 1:1 ratio, to receive 1.0 mg/kg (Group A) or 1.2 mg/kg (Group B) of amediplase. The efficacy was evaluated by assessing the infarct-related

TABLE 33.5 2K2 Study: Median (Range) Amediplase Pharmacokinetic Parameters by Dose

	Dose (mg)			
	50 ($n=4$)	70 ($n=5$)	80 ($n=5$)	90 ($n=4$)
C_{max} (μg/mL)	9.7 (6.9–12.1)	11.1 (7.5–15.1)	14.2 (6.8–34.4)	20.6 (10.0–33.6)
AUC 0-∞ (μg × min/mL)	222 (152–296)	327 (150–334)	391 (152–987)	455 (356–607)
$t_{1/2}$ α (min)	8.6 (8.1–10.0)	12.0 (8.0–22)	12.0 (9.0–12.0)	13 (9–17)
$t_{1/2}$ β (min)	92 (45–125)	85 (79–140)	126 (84–253)	126 (93–195)
CL (mL/min)	236 (169–328)	213 (209–467)	204 (81–526)	203 (148–252)
V_c (L)	3.2 (2.4–3.8)	4.7 (4.0–7.7)	2.7 (1.5–10)	3.9 (2.2–6.4)
V_{area} (L/kg)	0.6 (0.3–0.7)	0.4 (0.3–0.7)	0.7 (0.1–1.5)	0.6 (0.4–0.8)

TABLE 33.6 2K2 Study: Mean Coagulation Parameters at 80 mg Dose

Time	*N*	PAI-1 (ng/mL)	PAP (μg/L)	D-Dimer (μg/L)	Fibrinogen (mg/dL)	Plasminogen (%)	α2-Antiplasmin (%)	TAT (μg/L)	F 1+2 (nM)
Predose	9	259.7 ± 39.3	4464.4 ± 2771.4	25.2 ± 11.5	341 ± 32.8	82.5 ± 7.4	95.5 ± 4.4	7.6 ± 2.9	2.4 ± 0.8
1 h	7	258.6 ± 57.8	23933.1 ± 1133.7	308 ± 54.6	189.4 ± 30.8	36.7 ± 5.8	32 ± 5.2	22.4 ± 9.3	3.3 ± 1.1
2 h	6	311.6 ± 55.9	22139.3 ± 1908.3	306.2 ± 72.2	243.3 ± 40.5	60.3 ± 8.8	41 ± 6.2	9.4 ± 2.4	2.5 ± 1.0
24 h	8	346.8 ± 45.3	1387.7 ± 401.4	25.9 ± 9.8	334 ± 31.2	69.1 ± 3.8	74.5 ± 2.0	11 ± 4.3	1.7 ± 0.4

coronary artery patency (TIMI flow grade 3) at coronary angiography performed 60 min after amediplase administration. Efficacy was also evaluated as corrected TIMI frame count at 60 min, resolution of ST elevation at 180 min after amediplase administration, 30-day mortality and incidence of in-hospital major nonfatal cardiovascular events. Safety was evaluated by incidence and severity of adverse events (including intracranial hemorrhages, major bleedings, and minor bleedings) vital signs, laboratory tests, and ECG recording. The pharmacokinetics and potential immunogenicity of amediplase and the effect on the coagulation system were also assessed in a subgroup of patients.

A total of 240 patients (120 in each treatment group) were randomized (ITT population) and all of them received the study drug (safety population). For the per-protocol (PP) population, 108 patients were included in Group A and 104 patients in Group B. Results of the 60-min angiograms by amediplase dose are summarized in Table 33.7.

A TIMI flow grade 3 was achieved in more than half of patients (53% at 1 mg/kg and 58% at 1.2 mg/kg). No significant differences were found between the two dose levels regarding the incidence of different grades of coronary artery patency. The mean corrected TIMI frame count at 1 mg/kg (50.01 ± 37.95) and at 1.2 mg/kg (46.38 ± 36.92) were comparable. At 3 h after amediplase administration, there was a greater proportion of patients with complete (≥70%) ST elevation resolution in Group A (58%) than in Group B (36%) (Table 33.8).

TABLE 33.7 3K2 Study: Results of the 60-min Coronary Angiogram by Amediplase Dose

	Number of patients (% ITT)	
	Amediplase 1.0 mg/kg (N = 120)	Amediplase 1.2 mg/kg (N = 120)
Is Infarct-related artery identifiable?[a]		
Yes	116 (97%)	118 (98%)
No	3 (3%)	2 (2%)
If yes, type		
Right coronary	60 (52%)	56 (47%)
Left anterior descending	41 (35%)	49 (42%)
Left circumflex	11 (9%)	11 (9%)
Left main	1 (1%)	1 (1%)
Other	3 (3%)	1 (1%)
TIMI flow grade achieved[b]		
Grade 0	21 (18%)	24 (20%)
Grade 1	9 (8%)	5 (4%)
Grade 2	24 (21%)	20 (17%)
Grade 3	62 (53%)	69 (58%)

[a]One patient died before undergoing coronary angiogram.
[b]TIMI flow grading as performed by a centralized expert panel; in one patient, only the local TIMI flow evaluation at performing institution was available.

TABLE 33.8 3K2 Study: Resolution of ST Elevation at 3 h by Amediplase Dose

	Amediplase 1.0 mg/kg (N = 120)	Amediplase 1.2 mg/kg (N = 120)
Sum of ST elevations at baseline		
N	45	57
Mean ± SD	1.48 ± 0.85	1.45 ± 0.81
Range	0.37–4.12	0.32–3.58
Sum of ST elevations at 3 h postdose		
N	45	53
Mean ± SD	0.46 ± 0.36	0.62 ± 0.48
Range	0.00–1.64	0.02–2.09
Schroeder's 3-component definition		
Complete (≥70%) ST elevation resolution	26 (58%)	19 (36%)
Partial (30–69%) ST elevation resolution	7 (16%)	11 (21%)
No (≤29%) ST elevation resolution	12 (27%)	23 (43%)

There was no difference between the two amediplase dose groups regarding 30-day mortality: 5 patients (4%) in Group A and 6 patients (5%) in Group B. No differences were also found with regard to nonfatal cardiovascular events and parameters of the coagulation system.

Pharmacokinetic, haemostasis, and immunogenicity parameters were also investigated in a subgroup of patients. The pharmacokinetic parameters (median, range) are summarized in Table 33.9. There were no differences between doses for all the considered parameters (except for AUC, as expected) [27].

Haemostasis parameters are summarized in Table 33.10. A marked decrease in plasminogen and increase in plasmin-α2-antiplasmin complexes confirmed the high fibrinolytic activity of amediplase. The fibrin specificity of amediplase is suggested by the marked increase in D-dimer and by the parallel consumption of α-2-antiplasmin. PAI-1

TABLE 33.9 3K2 Study: Median (Range) of Pharmacokinetic Parameters by Amediplase Dose

	Amediplase 1.0 mg/kg (N = 13)	Amediplase 1.2 mg/kg (N = 9)
Co (μg/mL)	20.5 (11.7–47.5)	22.1 (15.4–47.1)
AUC (μg/mL × min)	408 (194–825)	542 (212–935)
$t_{1/2}$ α (min)	11.5 (8.5–21.1)	14.1 (7.1–21.6)
$t_{1/2}$ β (min)	100 (55–581)	125 (73–824)
CL (mL/min × kg)	2.5 (1.2–5.1)	2.2 (1.3–5.7)
V_c (mL/kg)	48.8 (21.0–85.5)	54.2 (25.5–77.8)
V_{area} (mL/kg)	400 (173–2053)	401 (151–2376)

TABLE 33.10 3K2 Study: Haemostasis Parameters

	Amediplase 1.0 mg/kg				Amediplase 1.2 mg/kg			
	N = 21				*N* = 21			
	Predose	+1 h	+2 h	+24 h	Predose	+1 h	+2 h	+24 h
Prothrombin fragments, 1+2 (nM)								
N	18	18	18	17	17	18	17	16
Mean ± SD	3.6 ± 1.80	5.2 ± 1.96	5.5 ± 2.20	4.0 ± 1.79	4.1 ± 2.03	5.8 ± 2.19	5.9 ± 2.09	4.6 ± 1.89
Range	1.4–7.8	2.5–9.3	3.1–9.4	1.3–6.9	0.7–7.9	2.8–9.5	2.6–8.6	2.4–8.6
Plasmin α-2 antiplasmin complexes (μg/L)								
N	21	20	20	19	19	20	19	19
Mean ± SD	988 ± 706	5544 ± 2874	6230 ± 2411	1518 ± 997	1667 ± 1608	5577 ± 2295	5918 ± 2126	2547 ± 1043
Range	128–2548	1238–11541	2221–11034	136–3843	57–7787	1945–10340	3308–9908	563–3849
PAI-1 (ng/mL)								
N	20	20	20	19	17	18	17	17
Mean ± SD	204 ± 96	188 ± 89	227 ± 102	213 ± 94	200 ± 87	179 ± 61	188 ± 83	169 ± 94
Range	86–412	35–326	34–433	62–390	94–355	36–254	37–353	44–414
Fibrinogen (mg/dL)								
N	21	19	20	20	18	16	13	17
Mean ± SD	344 ± 99	210 ± 64	211 ± 66	274 ± 96	349 ± 79	189 ± 77	190 ± 79	220 ± 111
Range	199–563	68–289	65–384	114–499	218–473	67–302	79–348	86–411
D-dimer (μg/L)								
N	20	20	20	19	19	20	19	19
Mean ± SD	33 ± 45	166 ± 119	240 ± 126	48 ± 59	54 ± 82	248 ± 150	314 ± 135	162 ± 186
Range	7–162	1–390	38–475	4–243	3–316	5–489	54–527	2–539
Plasminogen (%)								
N	21	19	19	19	18	14	14	18
Mean ± SD	95 ± 23	47 ± 18	50 ± 17	69 ± 13	89 ± 28	33 ± 14	33 ± 19	58 ± 18
Range	55–138	18–91	18–79	50–98	40–139	17–59	15–74	32–96
α-2-antiplasmin (%)								
N	19	16	17	18	18	10	13	18
Mean ± SD	97 ± 118	37 ± 18	38 ± 16	68 ± 14	81 ± 27	28 ± 8	29 ± 17	62 ± 15
Range	53–121	17–88	18–84	38–90	17–110	19–40	16–78	20–81
a-PTT (s)								
N	17	12	9	19	14	6	5	15
Mean ± SD	25 ± 4	41 ± 15	43 ± 14	29 ± 8	32 ± 14	47 ± 12	45 ± 10	29 ± 6
Range	20–37	22–67	26–66	20–52	18–61	36–66	34–56	19–44

and a-PTT levels did not appreciably change after amediplase administration at both doses. All immunogenicity tests, performed in 34 patients, for IgG and IgM were negative.

Overall, 427 adverse events (AEs) were reported in 156 patients. AEs that were probably or highly probably related to amediplase accounted for 60 events in 42 patients. The most frequently reported AEs, during the in-hospitalization period were hemorrhage (12 events in 12 patients [10%] at 1 mg/kg and 17 events in 14 patients [12%] at 1.2 mg/kg), and hematuria (6 events in 6 patients [5%] at 1 mg/kg and 8 events in 8 patients [7%] at 1.2 mg/kg). The greatest proportion of bleeding events was minor in nature (78 events in 61 patients), with a similar distribution between dose groups (36 events in 30 patients [25%] at 1 mg/kg and 42 events in 31 patients [26%] at 1.2 mg/kg). Two events of intracranial hemorrhage were reported, one as a minor bleeding event at 1 mg/kg, and one as a major bleeding event at 1.2 mg/kg. Overall, the greatest proportion of AEs reported was mild in severity (287 events, 34% of the safety population). There were no remarkable differences between the two dose groups regarding the incidence of AEs. From the 3K2 study could be concluded:

- There was no difference in the incidence of coronary artery patency (primary efficacy parameter) between the body weight-adjusted dose of 1.0 and 1.2 mg/kg amediplase, as shown by the percentage of patients with TIMI flow grade 3 at angiography 60 min after amediplase administration.

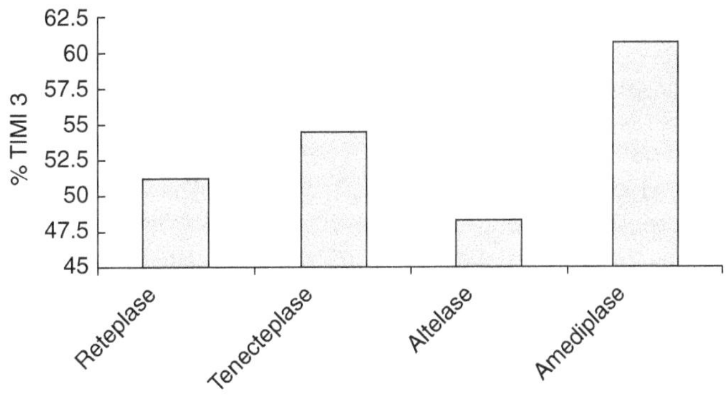

FIGURE 33.6 Effect of different thrombolytic on % TIMI 3 in similar Phase II angiographic studies.

- No significant differences between the treatment groups were observed for the secondary efficacy parameters: corrected TIMI frame count 60 min after amediplase administration; resolution of ST elevations at 180 min (3 h postdose); 30-day mortality; incidence of in-hospital major nonfatal cardiovascular events.
- The safety profiles of the body weight-adjusted doses of 1.0 and 1.2 mg/kg amediplase observed during this study did not reveal any remarkable differences between the two dose groups.

33.5 HISTORICAL COMPARISON WITH OTHER THROMBOLYTICS

Amediplase was compared to the reteplase, tenecteplase, and alteplase, for % TIMI 3 in similar angiographic trials [25,26,28,29]. As shown in Figure 33.6, at 60 min from administration of thrombolytic, it had the best performance on this parameter. This comparison has some obvious limitation because of the different conditions used in the trials and further and more complete studies shall necessary to better analyze the differences found.

33.6 CONCLUSIONS AND FUTURE PERSPECTIVES

Amediplase is the outcome of a project aimed to identify and develop a newer thrombolytic agent characterized by potent fibrinolytic activity, enhanced selectivity for the thrombus (fibrin-specificity), prolonged plasma half-life, absence of immunogenicity, and also a low manufacturing cost. In this regard, amediplase is considered a third-generation thrombolytic with advantages over the previous ones such as streptokinase, urokinase, and alteplase in terms of efficacy and handling.

Amediplase has confirmed in preclinical and clinical studies a promising pharmacodynamic and safety profile. Amediplase is endowed with a remarkable thrombolytic efficacy, has a relatively long-lasting plasmatic kinetic allowing a single bolus administration, is devoid of immunogenic properties in humans and may have a potentially lower incidence and severity of minor and major bleeding. Amediplase should therefore be considered as elective drugs for early thrombolysis (even in a prehospital setting) with dosing tailored to each single patient, and hence with a reduced risk of severe bleeding as side effects. The need of this kind of drug is still high irrespective of the large use of mechanical methods (angioplastic, stent) for rapid coronary reperfusion due to the fact that the hospitalization is not available everywhere, particularly in not Western countries.

Other potential indications are pulmonary embolism, deep venous thrombosis, ischemic stroke, and maintenance of patency of extracorporeal shunts where thrombolytics have been found active.

ACKNOWLEDGMENT

The authors would like to thank Mrs. Elena Oberto for her secretarial assistance.

REFERENCES

1. "WHO Disease and injury country estimates". World Health Organization. (2009) Available at http://www.who.int/healthinfo/global_burden_disease/estimates_country/en/index.html. Retrieved 2009 November 11.
2. Xu J, Kochanek, Murphy SL, Tejada-Vera B. (2010) Deaths: final data for 2007. *Nation. Vital Statist. Rep.* 58, 1–135.
3. Van de Werf F, Bax J, Betriu A, Bolmstrom-Lundqvist C, Crea F, Falk V, et al. (2008) Management of acute myocardial infarction: in patients with persistent ST-segment elevation. *Eur. Heart J.* 29, 2909–2945.
4. Menon V, Harrington RA, Hochman JS, Cannon CP, Goodman SD, Wilcox RG, et al. (2004) Thrombolysis and adjunctive therapy in acute myocardial infarction. *Chest* 126, 549S–575S.

5. Verstraete M. (2000) Third-generation thrombolytic drugs. *Am. J. Med.* 109, 52–58.
6. VanZonnefeld AJ, Verman H, Pannekoek H. (1986) Autonomous functions of structural domains on human tissue-type plasminogen activator. *Proc. Natl. Acad. Sci. USA* 83, 4670–4674.
7. Lijnen HR, Stump DC, Collen D. (1996) Single-chain urokinase-type plasminogen activator (prourokinase). In: Messerly FH (Ed), *Cardiovascular Drug Therapy*. 2nd Ed. WB Saunders, Philadelphia, pp. 1578–1587.
8. Nelles L, Lijnen HR, Collen D, Holmes WE. (1987) Characterization of a fusion protein consisting of amino acids 1 to 263 of tissue-type plasminogen activator and amino acids 144 to 411 of urokinase-type plasminogen activator. *J. Biol. Chem.* 262, 10855–10862.
9. Urlaub, G, Chasin LA. (1980) Isolation of Chinese hamster cell mutants deficient in dihydrofolate reductase activity. *Proc. Natl. Acad. Sci. USA* 77, 4216–4220.
10. Colucci M, Cavallo LG, Agnelli G, Mele A, Bürgi R, Heim J, et al. (1993) Properties of chimeric (tissue-type/urokinase-type) plasminogen activators obtained by fusion at the plasmin cleavage site. *Thromb. Haemost.* 69, 466–472.
11. Guimaraes AHC, Rijken DC. (2004) Thrombin activable fibrinolysis inhibitor (TAFI) affects fibrinolysis in a plasminogen activator concentration-dependent manner. *Thromb. Haemost.* 91, 473–479.
12. Guimaraes AHC, Barrett-Bergshoeff MM, Criscuoli M, Evangelista S, Rijken DC. (2006) Fibrinolytic efficacy of Amediplase, Tenecteplase and scu-PA in different plasma clot lysis models. *Thromb. Haemost.* 96, 325–330.
13. Rijken DC, Barrett-Bergshoeff MM, Jie AFH, Criscuoli M, Sakharov DV. (2004) Clot penetration and fibrin binding of Amediplase, achimeric plasminogen activator (K2-tu-PA). *Thromb. Haemost.* 91, 52–60.
14. Agnelli G, Pascucci C, Colucci M, Nenci GG, Mele A, Bürgi R, et al. (1992) Thrombolytic activity of two chimeric recombinant plasminogen activators (FK_2tu-PA and K_2tu-PA) in rabbits. *Thromb. Haemost.* 68, 331–335.
15. Agnelli G, Pascucci C, Nenci GG, Mele A, Bürgi R, Heim J. (1993) Thrombolytic and haemorrhagic effects of bolus doses of tissue-type plasminogen activator and a hybrid plasminogen activator with prolonged plasma half-life (Amediplase: CGP 42935). *Thromb. Haemost.* 70, 294–300.
16. Report on file at Menarini Ricerche SpA. Evaluation of fibrinolytic activity in a model of arterial thrombosis induced by ferric chloride in the rat following single intravenous administration. June 2001. CERB study n° 20010026PGR.
17. Report on file at Menarini Ricerche SpA. Thrombolytic activity of Amediplase and tenecteplase on FeCl3-induced arterial thrombosis in the rat. Menarini Ricerche. 2005. Report MRPO-2005-PHA004.
18. Lucchesi BR, Bürgi R, Heim J. (1991) The *in vivo* thrombolytic activity of CGP 42935, a hybrid plasminogen activator and fibrinolytic agent. *Coron. Artery Dis.* 2, 247–258.
19. Report on file at Menarini Ricerche SpA. Pharmacokinetics in Monkeys of a chimeric plasminogen activator (K_2tu-PA) linking the kringle2 domain of t-PA to the catalytic domain of scu-PA. April 1991.
20. Report on file at Menarini Ricerche SpA. Pharmacokinetics of Amediplase Following Single Intravenous Administration in Rats. October 2004. Menarini Ricerche Report MR0016.
21. Report on file at Menarini Ricerche SpA. The *in vivo* fate of the K_2TU-PA (Amediplase) in Rats. January 2004. Leiden/Amsterdam CDR Report.
22. Report on file at Menarini Ricerche SpA, Study 1K2. Evaluation of the safety, tolerance and pharmacokinetics of single sequentially rising doses of the plasminogen activator K2tu-PA in healthy volunteers. February 1995.
23. Vermeer F, Oldroyd K, Pohl J, Giannelli S, Charbonnier B. (2001) Safety and angiography data of Amediplase, a new fibrin specific thrombolytic agent, given as a single bolus to patients with acute myocardial infarction: the 2k2 dose finding trial. *Circulation* 104(17, Suppl. 2), abstract 2545.
24. Report on file at Menarini Ricerche SpA Study 2K2. A phase II, multicentre, open label, dose finding, pilot study of K2tu-PA in patients with suspected acute myocardial infarction treated with heparin and aspirin. July 2001.
25. Charbonnier B, Pluta W, De Ferrari G, Ceddia F, Capriati A, Van de Werf F. (2001) Evaluation of two weight adjusted single bolus doses of Amediplase to patients with acute myocardial infarction: the 3k2 trial. *Circulation* 104(17, Suppl. 2), abstract 2546.
26. Report on file at Menarini Ricerche SpA. Study 3K2. A double blind, randomised, parallel group study to compare the efficacy and safety of two doses of Amediplase given as a single i.v. bolus injection in patients with suspected acute myocardial infarction treated with heparin and aspirin. June 2003.
27. Report on file at Menarini Ricerche SpA. Menarini Ricerche Study n° 3K2", Menarini Ricerche Report MR0035. Pharmacokinetics of Amediplase in plasma of patients participating in the clinical trial "A double blind randomised, parallel group study to compare the efficacy and the safety of 2 doses of Amediplase given as a single i.v. bolus injection in patients with suspected acute myocardial infarction (AMI) treated with Heparin and Aspirin. May 2003.
28. Bode C, Smalling RW, Berg G, Burnett C, Lorch G, Kalbfleisch JM, et al. (1996) For the RAPID II investigators. Randomized comparison of coronary thrombolysis achieved with double-bolus reteplase (recombinant plasminogen activator) and front-loaded, accelerated alteplase (recombinant tissue plasminogen activator) in patients with acute myocardial infarction. *Circulation* 94, 891–898.
29. Cannon CP, Gibson CM, McCabe CH, Adgey AAJ, Schweiger MJ, Sequeira RF, et al. (1998) For the thrombolysis in myocardial infarction (TIMI) 10B investigators. TNK-tissue plasminogen activator compared with front-loaded alteplase in acute myocardial infarction results of the TIMI 10B trial. *Circulation* 98, 2805–2814.

34

BREAKING NEW THERAPEUTIC GROUNDS: FUSION PROTEINS OF DARPINS AND OTHER NONANTIBODY BINDING PROTEINS

Hans Kaspar Binz

Molecular Partners AG, Schlieren, Switzerland

34.1 INTRODUCTION

Monospecific antagonistic monoclonal antibodies and agonistic biologics are among the commercially most successful drugs. This success is in contrast to increasing evidence that several of these proteins exhibit limited therapeutic efficacy. Redundant signaling pathways and other escape mechanisms may lead to such limits. One way to achieve efficacy beyond the one of currently approved drugs is to combine these simple mechanisms of action at will, bringing together a multitude of functionalities in one drug. While fusion proteins are an ideal platform for such a task, antibody-based fusion approaches show clear technical limitations due to the molecular properties of the immunoglobulin fold and production issues. DARPins (designed ankyrin repeat proteins) are highly stable and more favorably behaved alternatives to antibodies, which can bind target molecules with high affinity and specificity. DARPins are currently in clinical development for various therapeutic applications. DARPins and other alternatives to antibodies clearly overcome the technical limitations of antibodies and enable fast research in a vast field of novel therapeutic innovations. This acceleration of research will lead to a more rapid identification of relevant novel therapeutic approaches and interesting new drugs, ultimately to the benefit of patients lacking appropriate treatment possibilities.

34.2 NOVEL SCAFFOLDS—ALTERNATIVES TO ANTIBODIES

34.2.1 Success and Limitations of Antibodies

The successful use of monoclonal antibodies for therapeutic applications has revolutionized modern medicine. Drugs with predefined target molecule-binding properties can be designed at will and due to a high degree of therapeutic validation; there is little toxicology risk on the molecule's side, minimizing side effects to mostly mechanism-related toxicology. Consequently, all major pharmaceutical companies run therapeutic antibody platforms for drug development. Currently, two main therapeutic approaches are used in the monoclonal antibody field: One antibody application is to target soluble cytokines removing them from the blood. Another successful application is to use antibodies specifically binding targets on tumor cells and recruiting effector functions to the site of binding. The success of both approaches is exemplified by several blockbuster antibody drugs. Recently, it became evident that the commercial success of monoclonal antibodies is only partly reflected in therapeutic efficacy. For example, bevacizumab currently struggles to deliver clinical benefit in all of its approved indications [1]. A next generation drug thus should build on

Fusion Protein Technologies for Biopharmaceuticals: Applications and Challenges, First Edition. Edited by Stefan R. Schmidt.

the success of monoclonal antibodies, yet increase the therapeutic efficacy.

In principle, an antibody is a natural example of a simple but innovative fusion protein: Binding moieties (Fv or Fab fragments) are combined with an effector and half-life extension function (Fc fragment). This, of course, inspired many antibody derivative approaches. Specificities were trimmed, valency was adapted, fragments were removed, and effector functions were reduced, extended, or adapted. In other cases, additional effector functions such as small molecule toxins or interleukins were added to antibodies. Many of these successful applications are presented in other chapters of this book.

With engineering antibodies, a new era of therapeutic protein engineering was started with a great prospect for novel therapeutics, ultimately to the benefit of patients. Considering that this research has been started about three decades ago, surprisingly few of these approaches have reached the market or marketability. This can be attributed to several limitations imposed by antibodies.

One of the limitations concerns the composition of immunoglobulin domains and their production. Antibodies are complex molecules that rely on disulfide bonds and glycosylation, requiring production in mammalian cells. To be able to better work with antibodies, for example, in selection procedures or for protein production, strong efforts were put in dismantling the antibody into single pieces, such as scFv or Fab fragments [2]. While very respectable progress was made, the main limitations of the immunoglobulin domain typically remain and a streamlined research process is only possible with large investments.

Another limitation is the conceptual limitation imposed by antibodies to our view on current drug development. The success of monoclonal antibodies has limited our thinking of what is possible with biologics. Limited tissue penetration and i.v. injection are just two of the aspects that often are not touched upon when developing new drugs.

To overcome such limitations and enable completely novel therapeutic approaches, two things must happen: (i) The research process must be fast-forward and as simple as possible allowing for rapid generation of novel proof-of-concept biologics. This can be achieved using antibody fragments if a considerable investment is made, or using novel approaches, that is alternatives to antibodies, that allow simple and straightforward generation of novel therapeutic candidates. (ii) We must think of novel therapeutic concepts, not limited by the limitations of antibodies. Biologics can not only be administered by i.v./s.c. but also by other routes. Biologics can be made to penetrate tissue efficiently. Any fusion protein is possible. Any combination of binding specificities and functionalities is feasible. If we think along these lines, we will come up with novel therapeutic approaches with clear benefit for the patients. This chapter highlights recent activities in the field of alternatives to antibodies with a focus on DARPins and fusion proteins.

34.2.2 Natural Binding Proteins

A look at natural human binding proteins reveals interesting insights. Next to immunoglobulin domains, nature has evolved a diverse set of other folds to be powerful protein–protein interaction mediators such as ankyrin repeat proteins [3,4], leucine-rich repeat proteins [5,6], fibronectins [7], as well as the SH2, SH3, and PDZ domains [8]. The prominence of repeat proteins is underlined by the fact that jawless vertebrates [9] and plants use this protein class as binding proteins in their adaptive immune systems, just like higher vertebrates use antibodies. The affinities of natural binding proteins to their target proteins cover a broad range. For example, the placental ribonuclease inhibitor quasi-covalently inhibits ribonuclease ($K_I = 5.9 \times 10^{-14}$ M) [10]. Ankyrin repeat proteins interact with their target proteins with high affinities in the picomolar range [11]. Other proteins show interactions that are more in the nanomolar to micromolar range, depending on the requirements to the binders [8,12]. While many binding proteins show specific binding, nature's strength is also in generating nonspecific binding molecules that can fulfill several tasks [13].

Many binding proteins are typically specified for binding certain types of target molecules. For example, SH2, SH3, and PDZ domain proteins are specialized in binding peptides [8]. Lipocalins are specialized in binding small chemicals [14]. Fibronectins are typically involved in extracellular matrix protein–protein interactions to connect cells [7]. Immunoglobulins and repeat proteins are different with respect to their target selectivity. They show more diverse target types, ranging from nonpeptides to peptides and proteins. For example, ankyrin repeat proteins were shown to be able to bind RNA [15], peptides [16], and proteins [11].

The chemical compositions and molecular structure of natural binding proteins are as diverse as one could expect from such a broad set of molecules. Some proteins such as the fibronectins and ankyrin repeat proteins do not contain disulphide bonds or free cysteines [4,7]. Other binding proteins contain disulfide bonds for stabilization similar as seen in immunoglobulin domains. Fibronectin has an immunoglobulin-like fold with loops similar to CDRs of antibodies [7]. Lipocalins have a beta-barrel structure and also displays loops [14]. Ankyrin repeat protein use a conceptually different approach to binding targets in that they consist of repeated units of a turn-helix-loop-helix-loop motif, forming an elongated protein domain with a large target-interaction surface exposed on rigid secondary structure elements (Figure 34.1) [4]. Size-wise, the ankyrin repeat domains of natural binding molecules are very flexible, ranging from two to more than thirty repeats [17], while typical ankyrin repeat proteins consist of four to six repeats,

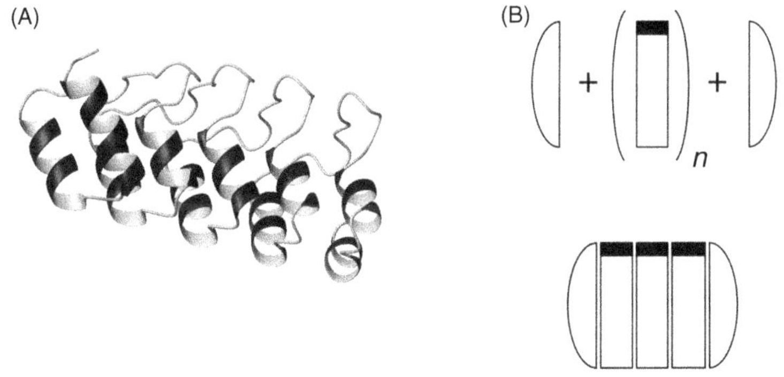

FIGURE 34.1 DARPins (designed ankyrin repeat proteins) [11]. (A) Structure of a DARPin shown in ribbon representation (PDB-ID: 1MJ0). The molecule comprises several repeats that stack to form one elongated domain. One repeat consists of a beta-turn, followed by a helix-loop-helix motif and a loop connecting to the next following beta turn. Special hydrophilic terminal repeats (capping repeats) shield the continuous hydrophobic core formed by the stacked central repeat modules. The molecule has the shape of a hand, where the palm of the hand is typically involved in the target interaction, and forms a large potential-interaction surface. The N term is at the left, the C term at the right side of the picture. (B) Schematic concept of generating DARPin libraries starting from capping repeats and varying numbers of a consensus-designed repeat modules. Randomized areas are indicated by black bars. Below, a schematic drawing of an N3C molecule is shown (three repeat modules flanked by an N- and a C-terminal capping repeat), exposing a large potential target-interaction surface. An example of such an N3C DARPin is shown in (A).

corresponding to about a tenth of an IgG immunoglobulin [4]. The flexibility of the repeat number is similarly seen in other repeat protein proteins [6]. Nonrepeat protein target-binding proteins typically have a fixed molecular size in the range of 6–35 kDa depending on the fold [8].

The stability of natural binding proteins is often trimmed to the biological requirements and can be diverse. For example, some ankyrin repeat proteins are extremely stable, comparable to designed ankyrin repeat proteins [18], while others, which require higher turnover due to their biological function, show low stability and fast turnover [19]. For most natural binding proteins analyzed, simple protein engineering allows increasing the thermodynamic stability, indicating that the different folds are intrinsically stable [8].

Overall, nature has evolved a large number of binding proteins that can serve as a source for alternatives to antibodies.

34.2.3 Tailoring Nonantibody Binding Proteins

Over the past 20 years, many of the naturally occurring binding proteins have been tested to serve as alternatives to antibodies. The list comprises proteins such as repeat proteins (see below) [20], fibronectin [7], protein Z [21], lipocalins [14], SH2, SH3, PDZ, γ-crystallin and others (reviewed, for example, in Binz et al. [8]). Typically, generating alternatives to antibodies involves (i) the choice of a scaffold (in this context scaffold refers to the protein which is used as a matrix for the following), (ii) randomization of one or several regions of the scaffold by library methods (typically based on knowledge of the sites of interaction of the scaffold), and (iii) selection of binding variants by state-of-the-art selection technologies such as ribosome display or phage display, and the characterization of the selected binders.

Already the first step is crucial. To be able to efficiently work with a scaffold, the protein should be expressable in a simple host at high levels, allowing for simple production and purification of variants. The scaffold should have a sufficient buffer of stability allowing randomization of the scaffold. In other words, the parental, nonrandomized scaffold should certainly exhibit a high degree of stability. Also, the protein should have a low potential to be immunogenic.

The next step classically consists of analyzing the natural binding site of the scaffold using literature or own experimental data, designing a randomization scheme, and generating a library of variants by DNA randomization methods. The better the understanding of the natural target-interaction of the scaffold is, the easier the design gets. The availability of 3D structures of the scaffold in complex with a target molecule is a great plus. In any case, the randomization scheme also has to be adapted to the subsequent selection/design process.

Typically, the DNA library of a scaffold is then transformed to the ribosome-display [22], phage-display [23], or yeast-display [24] format, and binding proteins are generated using these methods. Depending on the scaffold, the design, and the selection methods, high affinities, and high specificities can be generated. The outcome of this process and the analysis of the resulting binding molecules can help to refine a library design, and thus improve the process of generating binding molecules. Generating binding biologics from scratch using computer methods is currently not possible, as these methods are limited by the high degree of variability of biologics. Certain scaffolds, such as ankyrin repeat proteins, show binding sites on very rigid secondary structures. This should reduce the number of variables dramatically and actually might be the most attractive target for computer-designed binding proteins [25].

The following paragraph describes the process of tailoring nonantibody binding molecules using DARPins as a successful example to illustrate how to combine database knowledge, structural knowledge, published know-how, experimental tests, protein engineering, and library methods to yield one of the best platform technologies for the generation of binding molecules.

34.2.4 DARPins—Designed Ankyrin Repeat Proteins

For various reasons, repeat proteins are attractive scaffolds for the generation of alternatives to antibodies [20]. They are among the most frequent natural protein–protein interaction mediators across all phyla and are found in human at high frequency. Nature uses them in a similar fashion as antibodies, applying them in innate and adaptive immune systems, and involving them to inhibiting other proteins. Ankyrin repeat proteins are one prominent type of repeat proteins. They are involved in various cell-regulation mechanisms such as transcription factor inhibitors or cell signaling pathways [26] and are also important structural elements for cell integrity [27]. Interestingly, they also act as key molecules for information management, for example, in our hearing system [28]. An important feature of ankyrin repeat proteins is their occurrence as natural multivalent or multispecific proteins and very often, ankyrin repeat proteins are combined with other effector molecules combining different functionalities to a diverse set of possibilities [29]. In that respect, ankyrin repeat proteins are an expanded variant of antibodies. Importantly, they interact with their targets with high affinities and specificities, just as antibodies.

Using a consensus-design approach, we have generated designed ankyrin repeat proteins (DARPins) of varying repeat numbers (Figure 34.1) [4]. The design process involved sequence database analyses, structural analyses, computational design, as well as the use of experimental data [4]. Important features of the DARPins are the capping repeats and the consensus-designed repeat module. The design process had an important impact on the properties of DARPins. They can be expressed in soluble form in the cytoplasm of *Escherichia coli* at high levels, allowing for straightforward purification. The ankyrin repeat fold in combination with consensus design permitted the generation of highly stable molecules [18]. With increasing repeat numbers, the molecules get more and more stable, and very often, a four-repeat protein has a midpoint of thermal denaturation close to 100°C, and the analysis of larger constructs requires the addition of chemical denaturants to determine a midpoint of thermal unfolding. This high stability could be explained in part by a highly regular repeat fold, where the repeats stack perfectly to form an elongated repeat domain with a continuous and compact hydrophobic core in combination with a regular pattern of H-bonds [18]. A further important feature of DARPins is the absence cysteines in the fold, enabling specific introduction of cysteines at designated positions at will, allowing for easy chemical modification of the protein, including coupling of toxins or PEG-like molecules [30]. Also, DARPins are not glycosylated, which clearly facilitates production.

The analysis of the target interactions of natural ankyrin repeat proteins permitted the generation of libraries of DARPins [4]. We generated libraries of four or five repeats, with theoretical diversities of 5.2×10^{15} and 3.8×10^{23}, respectively. Using ribosome display or phage display, sampling between 10^{10} and 10^{12} variants, we could select single digit nanomolar to midpicomolar affinity binders to various targets just by doing regular selection rounds [11]. When increasing the selection pressure by, for example, off-rate selection, we could, still without any affinity maturation or additional randomization, select binders in the low picomolar range. Importantly, the selected DARPins exhibited high thermal stabilities and showed favorable expression properties equivalent to the nonselected DARPins, indicating that the DARPin scaffold is able to sustain a large number of modifications [11,31]. The regular ankyrin repeat fold is maintained in all selected DARPins whose structure has been determined so far [32].

The versatility of the DARPin platform has been proven at various instances. DARPins have been used as co-crystallizing agents [32], where their properties of high expression yield, ease of purification and rigid structure come into play very favorably. As diagnostic tools, the high specificity and affinity of DARPins is ideal, allowing for very sensitive assays [33]. DARPins have also been selected to act as intracellular kinase inhibitors [34], drug transporter (membrane protein) inhibitors [35], or enzyme inhibitors [36], whereby it became obvious that DARPins can be transformed to various types of inhibitors depending on the selection pressure, including competitive or allosteric inhibitors. Importantly, DARPins have all the favorable properties you would expect for a therapeutic biologic agent. Like for antibodies, DARPins can also be used to target tumor antigens. For example, high-affinity DARPins have been generated against the tumor targets EGFR [37] and ErbB2 [31]. By mutating a picomolar-affinity ErbB2-binding DARPins to variants of varying affinity and of varying size, we could show that for very small molecules such as single-domain DARPins, tumor accumulation and tumor-to-blood ratios were all the better, the higher the affinity was [30]. This was less pronounced for larger constructs. These results showed that in contrast to antibody fragments, the DARPins can efficiently be used for tumor targeting, showing enhanced tumor penetration and allowing for very high tumor loading. These are ideal prerequisites to generating powerful-targeted drugs.

Another application, which is straightforward for DARPins, is the generation of antagonist drug candidates. Indeed, with a powerful best-in-class candidate VEGF-antagonistic DARPin for the treatment of ophthalmic neovascularization diseases, Molecular Partners is currently conducting clinical trials in different indications in ophthalmology.[1]

Overall, DARPins are versatile therapeutic biologics candidates with clearly favorable molecular properties, having a wealth of therapeutic opportunities to venture [38,39].

34.3 NEW THERAPEUTIC CONCEPTS WITH NONANTIBODY BINDING PROTEINS

34.3.1 Natural Fusion Protein Examples

The immunoglobulin fold is often found naturally in combination with other domains or effector molecules. Antibodies are just one of several examples of immunoglobulin fold fusion proteins. They are naturally bivalent binding molecules combined with an effector function. The variations in domain numbers and domain compositions of human antibodies and shark or llama antibodies indicate that the fold is rather flexible. This is further underlined by the diversity of proteins that contain varying numbers of immunoglobulin domains.

A similar flexibility is observed for repeat proteins. Ankyrin repeat proteins, for example, are found naturally to exist in multivalent and multispecific formats in a large number of genes [29]. Typically, ankyrin repeat proteins are also found as natural fusion proteins in combination with various effector domains. Given this flexibility, they can fulfill the diverse set of functions that have been reported in the literature. Similarly, leucine-rich repeat proteins also occur as fusion proteins to effector moieties. Since protein A and fibronectin domains are also found in natural fusion proteins, nature appears to give us a hint that many natural folds are very flexible regarding fusion proteins.

Looking at the naturally occurring combinations of binding domains with effector domains reveals the flexibility of nature and indicates basically any fusion protein to be feasible. The ticket to novel therapeutics, however, rather should be sought in looking at what nature has not combined. Such nonexplored combinations could, for example, be equipping a hormone with a bispecific targeting domain and with a moiety for half-life extension, or combining several receptor antagonists to a multispecific superantagonist. These are only two simple possibilities of many more possibilities. To exploit these possibilities, a versatile, robust, and easy-to-handle fold is ideal. DARPins fully fit these requirements and are thus perfectly suited for this task [38,39].

34.3.2 Fusion Proteins with Nonantibody Binding Proteins

The research field of alternatives to antibodies is comparably new. It is thus clear that only a limited amount of data has been published especially for fusion proteins. The following provides a brief overview on the present status of published examples on fusion proteins with alternatives to antibodies. The field is currently very active, and as the technologies around alternatives to antibodies mature, more and more applications are expected to be published in the near future.

A strength of DARPins and other scaffolds is the fact that they can easily be fused head-to-tail to form bidomain or multidomain proteins, without compromising their biophysical or expression properties to a great extent. Several such examples have been published and are explained in the following (Figure 34.2).

34.3.2.1 Bivalent and Bispecific Molecules If two identical binding proteins are combined in one binding moiety, the resulting binder will exhibit a higher apparent affinity to the target, as rebinding or parallel binding of two targets may occur. In case of a homo-dimeric target, bivalency will lead to double binding and thus higher affinity. On the other hand, there are various reasons to combine two different binding domains in one binding moiety. For example, two different targets could be bound at the same time using a classical bispecific molecule. Taking a biparatopic-binding molecule, one target could be bound at two different epitopes, and thus higher affinity could be achieved or two different activities of a molecule could be blocked simultaneously, which could be interesting for, for example, molecules that interact with receptors via distinct interfaces.

Both bispecific and bivalent DARPins made of single-domain DARPins targeting different epitopes on the α chain of the high-affinity receptor for human immunoglobulin E (FcεRIα) were generated [40]. Importantly, the multidomain DARPins did not exhibit altered biophysical behavior compared to the single-domain DARPins. By screening the individual combinations of DARPins favorable combinations could be identified, leading to inhibitors which blocked IgE-mediated signaling (measured as inhibition of mediator release) to completion, showing equivalent results as omalizumab [40].

With single-domain DARPins, we can achieve affinities in the low picomolar range without the need for affinity

[1] http://www.molecularpartners.com

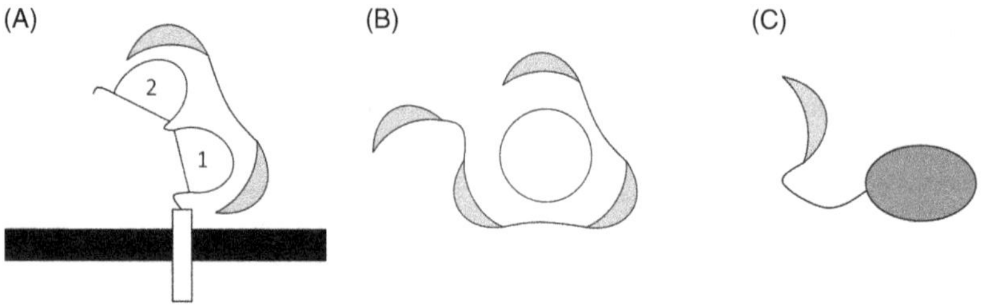

FIGURE 34.2 Schematic examples of DARPin fusion proteins. DARPins are represented as gray sickles. (A) Bispecific DARPins targeting the α chain of the membrane-bound high-affinity receptor for human immunoglobulin E (FcεRIα) in two different binding sites (Eggel et al. [40]). (B) Multidomain construct binding the knob of adenovirus as well as Her2, generated to retarget adenovirus to Her2-expressing cells. The knob is represented as white circle (Dreier et al. [44]). (C) DARPin-ETA fusion protein for targeted tumor therapy harnessing the cytotoxic activity of ETA. ETA is represented as dark gray oval (Martin-Killias et al. [47]).

maturation. We were thus curious to see, what would happen if one connected two high-affinity DARPins (90 pM affinity to the ErbB2 receptor [31]) in a row. Indeed, the construct, which again showed unchanged biophysical properties compared to the single-domain DARPin, exhibited an apparent quasi-covalent affinity [30]. The same DARPin could also easily be combined with other DARPins with no change of production properties. For example G3 was combined with a nonbinding DARPin. In this construct G3 had the same properties as single-domain G3, meaning that the fusion partner did not influence the activity of G3 [30].

Bispecific or bivalent approaches have also been published for protein Z, fibronectin, and lipocalin (see Table 34.1). Two lipocalins were fused head-to-tail resulting in a construct that could bind fluorescein and digoxigenin at the same time [41]. Similarly, bivalent protein Z variants binding Her2 or bispecific protein Z variants binding Her2 and EGFR were reported [42]. Recently, a fibronectin-derivative construct binding bispecifically to EGFR and IGF-1R has been reported to act as potent anticancer therapeutic [43]. Such molecules clearly can bring targeted therapy to a next level. It will be interesting to see what will happen with such molecules in a clinical context. Efficacy is likely to improve, yet, toxicity might like-wise also increase in parallel. With nonantibody binding proteins, it is easy to generate many combinations and rapidly achieve product candidates that can go to clinical testing, either validating or invalidating a certain targeting combination.

34.3.2.2 Multispecific Constructs Going beyond two-domain fusions is certainly more interesting. This is easily feasible with DARPins and was exemplified recently by a multidomain DARPin consisting of four repeat domains used to re-target adenovirus [44]. Binding DARPins against the knob of the virus were generated and then different valencies were assessed. A stoichiometry of 1 DARPin per knob subunit, corresponding to a trivalent DARPin, was found to be ideal, attaching the DARPin construct tightly to the virus and forming a stable complex. This trivalent DARPin was linked to a fourth DARPin targeting ErbB2. Importantly, the quadruple DARPin construct was stable on the virus and could efficiently be used to specifically guide

TABLE 34.1 Multispecific and Multidomain Molecules Involving Nonantibody Binding Proteins

Target(s)	Scaffold Involved	Number of Domains	Particularities	Reference
Her2	DARPin	2	Quasi-covalent binding by bivalency	[30]
FcεRIα	DARPin	2	Biparatopic receptor antagonist	[40]
Ad-fiber knob and Her2	DARPin	4	Trivalent knob binder and Her2 targeting	[44]
Her2	Affibody	2	Bivalent	[48]
EGFR	Affibody	2	Bivalent	[49]
Her2 and EGFR	Affibody	2	Bispecific	[42]
Her2 and HSA	Affibody	2	Half-life extended Her2 binder	[50]
TNF-α	Affibody	2–3	Bispecific or trispecific TNF inhibitors	[51]
Fluorescein and digoxigenin	Lipocalin	2	Bivalent	[41]
EGFR and IGF-1R	Fibronectin	2	Bivalent	[43]
IL-6 and IgG	Avimer	4	Half-life extended (IgG-binding) trivalent IL-6 antagonist	[52]

the virus to cells expressing the target molecule. This concept is even more interesting, as the ErbB2-targetting DARPin could in principle be exchanged by any other targeting DARPin, thus generating targeted adenovirus at will. Importantly, the constructs exhibited no substantial decrease in expression properties and also the general biophysical behavior was comparable to the one of a single-domain DARPin [44]. A projecting calculation in that report estimated the amount of a 1-L shake-flask culture to be sufficient to fully supply a midscale Phase I gene therapy trial with this system.

Multispecific binding proteins are certainly very promising beyond bispecific or individual binding proteins. Similar to this, a multitude of functionalities can be combined. Half-life engineering domains can be combined with two or three targeting domains and one or several disease pathways can be tackled simultaneously.

34.3.2.3 Critical Advantage of Multispecific Nonantibody Binding Proteins Bispecific and multispecific approaches are especially interesting as they may open new therapeutic modalities. This has very well been recognized by antibody companies and strong efforts are put in this area of research (see also other chapters in this book). The critical advantage of nonantibody binding molecules over antibodies is their flexibility in genetic engineering, ease in production, and high protein stability, making research fast-forward. Such approaches are more difficult to tackle with antibody platforms, as they might require elaborate and extremely streamlined cloning and production processes to work efficiently. Furthermore, protein properties of the final products will be very different, and especially protein stability might hamper the use of the resulting products in the case of antibody constructs. It is thus to be expected that multispecific nonantibody binding protein do have a critical advantage and will allow to rapidly assess the possibilities in multitargeting, both regarding efficacy and toxicology.

34.3.2.4 Combining Scaffolds with Other Effector Proteins Instead of combining two or more targeting moieties, the combination of a targeting moiety with innovative effector moieties is alluring. For antibodies, many different possibilities of fusion proteins have been explored to generate novel therapeutics approaches. Fusions to interleukins, interferons, cytokines, *Pseudomonas* exotoxin A (ETA) [45], RNase, and other have been reported [46]. Often these combinations were used for tumor therapy and good preclinical results have been achieved. Importantly, several of these constructs are indeed in clinical development.

In case of DARPins, ETA-fusions have been generated, which could efficiently be used for the treatment of cancer in tumor xenograft models in mouse, similarly to what has been shown with antibodies [45,47]. Importantly, the proteins could be expressed in soluble form in the cytoplasm of *E. coli* allowing for simple production and purification, enabling fast evaluation of this novel combination. Similarly, we have generated cytokine and interleukin fusions to DARPins. The favorable DARPin properties enabled high-level expression (Binz H.K., unpublished results). For example, TNF-α-fusions could be expressed in soluble form at levels of more than 200-mg/mL shake-flask culture with an ordinary *E. coli* XL1-blue strain in LB/Glucose medium. Purification was thus straightforward. In case of interleukin-2 (IL-2), the expression level was even higher and the protein was expressed quantitatively in inclusion bodies, which enabled to reach high purity of fully active protein in a single refolding step. The fusion proteins are typically fully active independent of the presence of the DARPin. Thus again, the platform of a nonantibody binding protein is offering a rapid way to assess the therapeutic possibilities of effector fusion proteins.

34.4 SCAFFOLD-FUSION PROTEINS BEYOND ANTIBODY POSSIBILITIES

With the limited background of antibody research, we are just about to explore the vast possibilities of fusion proteins to scaffolds. Currently, the scaffold-fusion protein approaches are very close to the antibody approaches. It is just a question of time until novel approaches will appear. One important advantage of scaffolds such as the DARPins is their flexibility regarding multispecificity and formatting, allowing basically generating any type of fusion protein with comparably little effort. Similar to this, novel combinations of several antagonists or targeting molecules with several agonists might become feasible, which were either not feasible with immunoglobulin domains, or which were only achievable with major investments of both time and money in production optimization. Here, an important step will be to not only look at cancer treatment but also apply the multifunctionality to other disease types.

Another important advantage of scaffolds such as DARPins is their robustness and chemical flexibility. This should facilitate chemical modification in a directed and much more controlled way than was possible for immunoglobulins. Depending on the fusion protein, one could thus add chemical modifications on top to create multifunctional molecules. On the other hand, stable scaffold-fusion proteins could be tested for novel routes of administration or penetration of, for example, the blood–brain barrier. The flexibility of the scaffolds should allow testing more approaches than could be tested with antibodies, increasing the chances for success dramatically.

Alternatives to antibodies as such will undoubtedly lead to a number of innovative therapeutics, enabling improved treatments of diseases or tackling unmet clinical needs.

Scaffold-fusion protein research has just started but already promising approaches are being tested in preclinical and clinical settings. The flexibility of the scaffolds will enable novel approaches, and, more importantly, lead to drugs with novel combined modes of action. This should ultimately result in more efficacious drugs and thus higher patient comfort than what is currently possible with mono-targeting antibodies. Considering the number of fusion-protein possibilities that are enabled using scaffolds, the concern will not be that innovative therapeutics are generated, but that the right functional combinations are chosen. The time and money that can be saved in the production of scaffold-fusion proteins should thus at least in part be spent in exploiting as many as possible approaches to enabling pinpointing the crucial novel therapeutic applications.

ACKNOWLEDGMENTS

I thank the many hands helping developing the DARPin technology to what it is today: A powerful and clinically validated technology to generate novel therapeutics. Patrik Forrer is acknowledged for advice to the present chapter.

REFERENCES

1. Ratner M. (2011) FDA panel votes to pull Avastin in breast cancer, again. *Nat. Biotechnol.* 29, 676.
2. Holliger P, Hudson PJ. (2005) Engineered antibody fragments and the rise of single domains. *Nat. Biotechnol.* 23, 1126–1136.
3. Sedgwick SG, Smerdon SJ. (1999) The ankyrin repeat: a diversity of interactions on a common structural framework. *Trends Biochem. Sci.* 24, 311–316.
4. Binz HK, Stumpp MT, Forrer P, Amstutz P, Plückthun A. (2003) Designing repeat proteins: well-expressed, soluble and stable proteins from combinatorial libraries of consensus ankyrin repeat proteins. *J. Mol. Biol.* 332, 489–503.
5. Kobe B, Deisenhofer J. (1994) The leucine-rich repeat: a versatile binding motif. *Trends Biochem. Sci.* 19, 415–421.
6. Stumpp MT, Forrer P, Binz HK, Plückthun A. (2003) Designing repeat proteins: modular leucine-rich repeat protein libraries based on the mammalian ribonuclease inhibitor family. *J. Mol. Biol.* 332, 471–487.
7. Koide A, Bailey CW, Huang X, Koide S. (1998) The fibronectin type III domain as a scaffold for novel binding proteins. *J. Mol. Biol.* 284, 1141–1151.
8. Binz HK, Amstutz P, Plückthun A. (2005) Engineering novel binding proteins from nonimmunoglobulin domains. *Nat. Biotechnol.* 23, 1257–1268.
9. Pancer Z, Mariuzza RA. (2008) The oldest antibodies newly discovered. *Nat. Biotechnol.* 26, 402–403.
10. Kobe B, Deisenhofer J. (1995) A structural basis of the interactions between leucine-rich repeats and protein ligands. *Nature* 374, 183–186.
11. Binz HK, Amstutz P, Kohl A, Stumpp MT, Briand C, Forrer P, et al. (2004) High-affinity binders selected from designed ankyrin repeat protein libraries. *Nat. Biotechnol.* 22, 575–582.
12. Ernst A, Gfeller D, Kan Z, Seshagiri S, Kim PM, Bader GD, et al. (2010) Coevolution of PDZ domain-ligand interactions analyzed by high-throughput phage display and deep sequencing. *Mol. Biosyst.* 6, 1782–1790.
13. Stirnimann CU, Petsalaki E, Russell RB, Muller CW. (2010) WD40 proteins propel cellular networks. *Trends Biochem. Sci.* 35, 565–574.
14. Beste G, Schmidt FS, Stibora T, Skerra A. (1999) Small antibody-like proteins with prescribed ligand specificities derived from the lipocalin fold. *Proc. Natl. Acad. Sci. USA* 96, 1898–1903.
15. Tanaka N, Nakanishi M, Kusakabe Y, Goto Y, Kitade Y, Nakamura KT. (2004) Structural basis for recognition of 2′,5′-linked oligoadenylates by human ribonuclease L. *EMBO J.* 23, 3929–3938.
16. Michaely P, Tomchick DR, Machius M, Anderson RG. (2002) Crystal structure of a 12 ANK repeat stack from human ankyrinR. *EMBO J.* 21, 6387–6396.
17. Mosavi LK, Cammett TJ, Desrosiers DC, Peng ZY. (2004) The ankyrin repeat as molecular architecture for protein recognition. *Protein Sci.* 13, 1435–1448.
18. Kohl A, Binz HK, Forrer P, Stumpp MT, Pluckthun A, Grutter MG. (2003) Designed to be stable: crystal structure of a consensus ankyrin repeat protein. *Proc. Natl. Acad. Sci. USA* 100, 1700–1705.
19. Yuan C, Li J, Selby TL, Byeon IJ, Tsai MD. (1999) Tumor suppressor INK4: comparisons of conformational properties between p16(INK4A) and p18(INK4C). *J. Mol. Biol.* 294, 201–211.
20. Forrer P. (2003) A novel strategy to design binding molecules harnessing the modular nature of repeat proteins. *FEBS Lett.* 539, 2–6.
21. Nord K, Gunneriusson E, Ringdahl J, Stahl S, Uhlen M, Nygren PA. (1997) Binding proteins selected from combinatorial libraries of an alpha-helical bacterial receptor domain. *Nat. Biotechnol.* 15, 772–777.
22. Hanes J, Plückthun A. (1997) *In vitro* selection and evolution of functional proteins by using ribosome display. *Proc. Natl. Acad. Sci. USA* 94, 4937–4942.
23. Steiner D, Forrer P, Stumpp MT, Pluckthun A. (2006) Signal sequences directing cotranslational translocation expand the range of proteins amenable to phage display. *Nat. Biotechnol.* 24, 823–831.
24. Boder ET, Wittrup KD. (1997) Yeast surface display for screening combinatorial polypeptide libraries. *Nat. Biotechnol.* 15, 553–557.
25. Karanicolas J, Corn JE, Chen I, Joachimiak LA, Dym O, Peck SH, et al. (2011) A de novo protein binding pair by computational design and directed evolution. *Mol. Cell* 42, 250–260.
26. Russo AA, Tong L, Lee JO, Jeffrey PD, Pavletich NP. (1998) Structural basis for inhibition of the cyclin-dependent kinase Cdk6 by the tumour suppressor p16INK4a. *Nature* 395, 237–243.

27. Lux SE, John KM, Bennett V. (1990) Analysis of cDNA for human erythrocyte ankyrin indicates a repeated structure with homology to tissue-differentiation and cell-cycle control proteins. *Nature* 344, 36–42.

28. Sotomayor M, Corey DP, Schulten K. (2005) In search of the hair-cell gating spring elastic properties of ankyrin and cadherin repeats. *Structure* 13, 669–682.

29. Letunic I, Doerks T, Bork P. (2009) SMART 6: recent updates and new developments. *Nucleic Acids Res.* 37, D229–232.

30. Zahnd C, Kawe M, Stumpp MT, de Pasquale C, Tamaskovic R, Nagy-Davidescu G, et al. (2010) Efficient tumor targeting with high-affinity designed ankyrin repeat proteins: effects of affinity and molecular size. *Cancer Res.* 70, 1595–1605.

31. Zahnd C, Wyler E, Schwenk JM, Steiner D, Lawrence MC, McKern NM, et al. (2007) A designed ankyrin repeat protein evolved to picomolar affinity to Her2. *J. Mol. Biol.* 369, 1015–1028.

32. Sennhauser G, Grutter MG. (2008) Chaperone-assisted crystallography with DARPins. *Structure* 16, 1443–1453.

33. Theurillat JP, Dreier B, Nagy-Davidescu G, Seifert B, Behnke S, Zurrer-Hardi U, et al. (2010) Designed ankyrin repeat proteins: a novel tool for testing epidermal growth factor receptor 2 expression in breast cancer. *Mod. Pathol.* 23, 1289–1297.

34. Amstutz P, Binz HK, Parizek P, Stumpp MT, Kohl A, Grutter MG, et al. (2005) Intracellular kinase inhibitors selected from combinatorial libraries of designed ankyrin repeat proteins. *J. Biol. Chem.* 280, 24715–24722.

35. Sennhauser G, Amstutz P, Briand C, Storchenegger O, Grutter MG. (2007) Drug export pathway of multidrug exporter AcrB revealed by DARPin inhibitors. *PLoS Biol.* 5, e7.

36. Schweizer A, Roschitzki-Voser H, Amstutz P, Briand C, Gulotti-Georgieva M, Prenosil E, et al. (2007) Inhibition of caspase-2 by a designed ankyrin repeat protein: specificity, structure, and inhibition mechanism. *Structure* 15, 625–636.

37. Steiner D, Forrer P, Pluckthun A. (2008) Efficient selection of DARPins with sub-nanomolar affinities using SRP phage display. *J. Mol. Biol.* 382, 1211–1227.

38. Stumpp MT, Amstutz P. (2007) DARPins: a true alternative to antibodies. *Curr. Opin. Drug Discov. Dev.* 10, 153–159.

39. Stumpp MT, Binz HK, Amstutz P. (2008) DARPins: a new generation of protein therapeutics. *Drug Discov. Today* 13, 695–701.

40. Eggel A, Baumann MJ, Amstutz P, Stadler BM, Vogel M. (2009) DARPins as bispecific receptor antagonists analyzed for immunoglobulin E receptor blockage. *J. Mol. Biol.* 393, 598–607.

41. Schlehuber S, Skerra A. (2001) Duocalins: Engineered Ligand-Binding Proteins with Dual Specificity Derived from the Lipocalin Fold. *Biol. Chem.* 382, 1335–1342.

42. Friedman M, Lindstrom S, Ekerljung L, Andersson-Svahn H, Carlsson J, Brismar H, et al. (2009) Engineering and characterization of a bispecific HER2 x EGFR-binding affibody molecule. *Biotechnol. Appl. Biochem.* 54, 121–131.

43. Emanuel SL, Engle LJ, Chao G, Zhu RR, Cao C, Lin Z, et al. (2011) A fibronectin scaffold approach to bispecific inhibitors of epidermal growth factor receptor and insulin-like growth factor-I receptor. *mAbs* 3, 38–48.

44. Dreier B, Mikheeva G, Belousova N, Parizek P, Boczek E, Jelesarov I, et al. (2011) Her2-specific multivalent adapters confer designed tropism to adenovirus for gene targeting. *J. Mol. Biol.* 405, 410–426.

45. Di Paolo C, Willuda J, Kubetzko S, Lauffer I, Tschudi D, Waibel R, et al. (2003) A recombinant immunotoxin derived from a humanized epithelial cell adhesion molecule-specific single-chain antibody fragment has potent and selective antitumor activity. *Clin. Cancer Res.* 9, 2837–2848.

46. Kaspar M, Zardi L, Neri D. (2006) Fibronectin as target for tumor therapy. *Int. J. Cancer* 118, 1331–1339.

47. Martin-Killias P, Stefan N, Rothschild S, Pluckthun A, Zangemeister-Wittke U. (2011) A novel fusion toxin derived from an EpCAM-specific designed ankyrin repeat protein has potent antitumor activity. *Clin. Cancer Res.* 17, 100–110.

48. Steffen AC, Wikman M, Tolmachev V, Adams GP, Nilsson FY, Stahl S, et al. (2005) *In vitro* characterization of a bivalent anti-HER-2 affibody with potential for radionuclide-based diagnostics. *Cancer Biother. Radiopharm.* 20, 239–248.

49. Tolmachev V, Friedman M, Sandstrom M, Eriksson TL, Rosik D, Hodik M, et al. (2009) Affibody molecules for epidermal growth factor receptor targeting *in vivo*: aspects of dimerization and labeling chemistry. *J. Nucl. Med.* 50, 274–283.

50. Tolmachev V, Wallberg H, Andersson K, Wennborg A, Lundqvist H, Orlova A. (2009) The influence of Bz-DOTA and CHX-A″-DTPA on the biodistribution of ABD-fused anti-HER2 Affibody molecules: implications for (114m)In-mediated targeting therapy. *Eur. J. Nucl. Med. Mol. Imaging* 36, 1460–1468.

51. Jonsson A, Wallberg H, Herne N, Stahl S, Frejd FY. (2009) Generation of tumour-necrosis-factor-alpha-specific affibody molecules capable of blocking receptor binding *in vitro*. *Biotechnol. Appl. Biochem.* 54, 93–103.

52. Silverman J, Liu Q, Lu Q, Bakker A, To W, Duguay A, et al. (2005). Multivalent avimer proteins evolved by exon shuffling of a family of human receptor domains. *Nat. Biotechnol.* 23, 1556–1561.

35

RESURGENCE OF BISPECIFIC ANTIBODIES

Patrick A. Baeuerle and Tobias Raum

Amgen Research (Munich) GmbH, Munich, Germany

35.1 A BRIEF HISTORY OF BISPECIFIC ANTIBODIES

Bispecific antibodies have a history of at least 25 years. With the simultaneous availability in the 1980s of monoclonal antibody (mAb), hybrid-hybridoma (or quadroma), and recombinant DNA technologies, it became possible to generate antibodies containing two different sets of variable domains. In a brief time period between 1983 and 1987, the most essential technologies and biomedical applications were already elaborated, which have been influential until today.

A first example of a bispecific mAb, which for a diagnostic purpose is binding to somatostatin and horse radish peroxidase, was described by Milstein and Cuello [1]. They fused two hydridoma lines into one hybrid-hybridoma line secreting various combinations of the heavy and light chains of the two mAbs, including the desired bispecific variant. In the mid-1990s, Lindhofer refined this technology by fusing hydridoma lines of rat and mouse origin. Secreted antibodies no longer had mixed heavy and light chains but formed one defined rat IgG2b/mouse IgG2a bispecific antibody species because of a preferential species-restricted heavy/light chain pairing [2]. Catumaxomab (Removab®) is a bispecific antibody using this technology which, in 2009, was the first bispecific antibody to obtain market approval [3].

In parallel to the quadroma technology, Brennan et al. [4] succeeded in 1985 in chemically cross-linking antibody fragments in high yield and free of monovalent species. Recently, chemical cross-linking of antibody fragments still is a viable technology for making bispecific antibodies. For instance, it was the basis for generating MDX-210, a HER2/CD64 (FcγRI)-bispecific $F(ab)_2$ antibody, which has been broadly explored in clinical trials [5].

Staerz et al. [6] were first to demonstrate redirected lysis of cancer cells by T cells with the help of a bispecific antibody cross-linking the two different cell types. Corvalan and Smith [7] produced an antibody bispecific for carcinoembryonic antigen (CEA) and vinca alkaloids, which can be considered an early prototype of today's popular antibody drug conjugate (ADC) format. Since then, we have seen a spade of different bispecific antibodies, which have been described in numerous review articles (for example, [8–10]).

Frequently, new bispecific antibodies are constructed for the purpose of redirected cancer cell lysis [10]. However, one should not forget that regular human IgG1 or IgG3 antibodies can function as nature's bispecific antibodies by tethering immune and target cells. Their bifunctionality resides in binding arms that attach to the target antigen, and in the CH2 domain of the Fcγ fragment, which can bind to Fcγ receptors on cytotoxic immune cells. Another apparently natural form of bispecific antibody results from Fab-arm exchange among antibodies of the human IgG4 class

Fusion Protein Technologies for Biopharmaceuticals: Applications and Challenges, First Edition. Edited by Stefan R. Schmidt.

[11]. Natural bispecific (polyreactive) antibodies have been reported to have superior protective activity against human immunodeficiency virus (HIV) [12].

For this review, we have grouped 37 bispecific antibody formats into seven distinct classes and present examples of each class in a separate figure. For each antibody construct, the bispecifically binding immunoglobulin (Ig) domains are shown in green and red colors. Particular structural features are highlighted and labeled, and all constructs are drawn at the same scale such that their relative sizes can be directly compared. The development stages of featured antibody development candidates represent a snapshot at the end of 2010. On the way forward, some candidates will disappear while others may newly appear in pipeline charts of companies. In the following sections, seven classes are described and their opportunities, specific issues, and challenges discussed.

35.2 ASYMMETRIC IgG-LIKE BISPECIFIC ANTIBODIES

The first bispecific antibody format is shown in Figure 35.1, where the two halves of antibodies are different, or asymmetric. If two distinct hybridoma lines are fused, the antibodies secreted by the resulting quadroma cells will be a mixture of up to 10 different species of which only one corresponds to the desired bispecific antibody. Seven species will have mispaired heavy and light chains and two correspond to the initial mAbs. This situation requires an extensive purification effort and can result in low yields of the correct bispecific antibody species, which is why researchers tried to find ways of selectively pairing the two different heavy chains.

One early solution resulted from the fusion of murine and rat hybridoma lines [2]. The preferential species-restricted heavy as well as light chain pairing of rat IgG2b/mouse IgG2a bispecific antibody species is leading to the correct bispecific antibody species with high yield (Figure 35.1A). The German Biotech Company Trion Pharma in collaboration with Fresenius Biotech has established a pipeline of bispecific trifunctional antibodies for cancer therapy based on their "Triomab" format. These antibodies are designed to bind with one arm to cancer cells expressing frequent tumor-associated antigens—such as EpCAM, HER2, CD20, or GD2—with their second arm to the invariant CD3 epsilon signaling chain of the T-cell receptor (TCR) on T cells, and with their "third arm," that is, the Fcγ part of rat IgG2b and mouse IgG2a, to antigen-presenting cells. It is believed that this trifunctional binding between cancer, T, and antigen-presenting cells

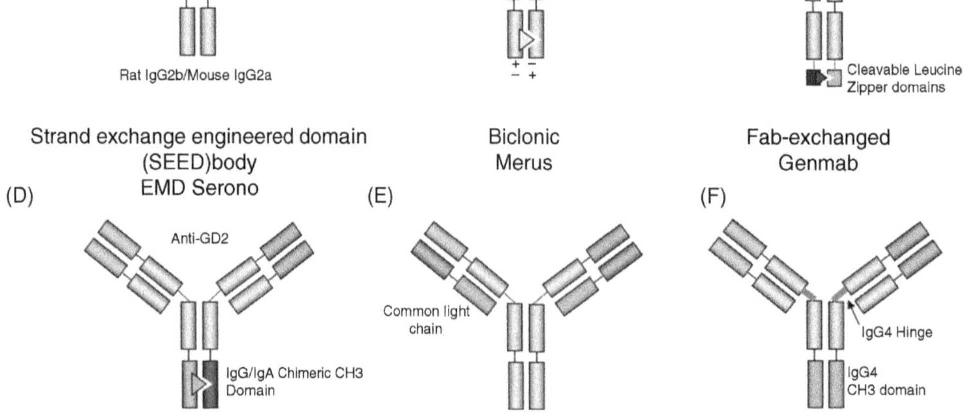

FIGURE 35.1 Asymmetric bispecific antibody formats. Red and green colors highlight variable antibody domains of distinct specificity. Gray and yellow label properly heterodimerized heavy and light chains derived from two different mAbs. Disclosed binding specificities and particular structural features are labeled. Figure 35.1b depicts three closely related but different formats. (See the color version of this figure in Color Plates section.)

provides for a high antitumor activity and is adding a vaccination effect to the redirecting lytic activities of the antibodies [13]. The EpCAM/CD3/Fcγ R-trifunctional Triomab catumaxomab (Removab®) has gained market approval in Europe in 2009 for the treatment of ascites resulting from EpCAM-expressing cancers [14]. The HER2/CD3-trifunctional antibody ertumaxomab is in formal Phase II clinical testing after having shown promising Phase I results [15], and the CD20/CD3- and GD2/CD3 trifunctional antibodies have been in pilot Phase I clinical studies. A major drawback of the Triomab approach is the high immunogenicity of rat and mouse antibody sequences, which trigger a neutralizing antibody response in nearly every patient, thereby pre-empting a retreatment [16]. With a serum half-life of 2.1 days and infrequent i.p. infusions, rather short exposures are achieved that may limit the time for efficient T-cell engagement. Nevertheless, high biological activity has been reported in clinical trials with distinct Triomabs.

Other groups have introduced structural elements for heterodimerization into the heavy chains of human, humanized, or human/murine chimeric origin with the promise of generating bispecific antibodies with much reduced immunogenicity. A very subtle, structure-guided approach took advantage of single amino acid exchanges in the CH3 domains creating bulky hydrophobic amino acid residues in one heavy chain and matching hydrophobic pockets in the other heavy chain (Figure 35.1B). This technology is referred to as "knobs-into-holes" [17]. Genentech/Roche published on a bispecific knobs-into-holes IgG that can cross-link the inhibitory Fcγ receptor CD32B with the high affinity FcεRI on mast cells for the treatment of asthma [18]. Researchers from Amgen have found another solution for heterodimer formation of heavy chains. They introduced by recombinant technology charged amino residues in the CH3 domain that interact with matching residues of the opposite charge [19].

Genentech/Roche presented in January 2010 another approach for pairing heavy chains called LUZ-Y antibodies [20]. This technology uses C-terminally fused leucine zipper domains derived from a heterodimerized transcription factor, which are later cleaved off the purified bispecific antibody with a protease (Figure 35.1C). Researchers from EMD Serono have likewise focused on the CH3 domain for heteromeric pairing of heavy chains. They developed a "strand exchange engineered domain" (SEED) as basis for their SEEDbodies [21] (Figure 35.1D).

A remaining issue of these three novel approaches is a potential mispairing of light chains with heterodimerized heavy chains. In theory, the properly paired bispecific antibody species will represent only 25% of possible light chain/heavy chain combinations. This will decrease yield and increase the complexity of purification. There are intense efforts ongoing to solve this issue. One possible solution has been developed by researchers from Roche [22]. Through crossover of heavy and light sequences within a knobs-into-holes bispecific antibody they crafted so-called CrossMAbs, which have properly paired variable domains. A CrossMAb, which is bispecific for VEGF-A and angiopoetin-2, has been characterized as an antiangiogenic drug candidate.

Another potential solution to mispairing of light chain has been worked out by the companies Merus and Genmab. Merus has developed so-called single V_L mAbs, which use only one light chain species to serve a large repertoire of human heavy chains [23]. Single V_L mAbs come from phage libraries or, in the future, will be generated by transgenic mice harboring the human vH gene repertoire and one human light chain gene. Evidently, the affinity and binding specificity of such antibodies solely resides within the heavy chain V_H, but seem to have uncomprimized activity. If two single V_L mAbs are simultaneously expressed in a producer line, a bispecific antibody called "biclonic" is formed (Figure 35.1E). By engineering the heavy chains, heterodimerization of biclonics can be forced to yield a high proportion of the desired bispecific variant.

On the basis of the natural Fab arm exchange observed for the IgG4 class [11], a new bispecific format has been developed by Genmab [24]. The Fab arm exchange by IgG4 structurally relies on both the IgG4 hinge region and critical amino acid residues in the IgG4 CH3 domain, predominantly Lys 409. By conferring these structural elements to other IgG subtypes, a Fab arm exchange leading to a bispecific antibody species can be forced *in vitro*. How this reaction is made irreversible has not yet been disclosed.

35.3 SYMMETRIC IgG-LIKE BISPECIFIC ANTIBODIES

A number of bispecific antibody formats have recently emerged that no longer rely on heteromeric pairing of heavy chains and hence result in a symmetric bispecific antibody. One such format has been developed by the company Domantis (now GSK) and is called "dual-targeting" (DT) Ig. It is combining two so called single-domain antibodies (dAbs) [25], which have been engineered from isolated heavy and light chain variable domains. A single-domain antibody is using only three complimentarity determining regions (CDRs) for antigen binding and no longer requires pairing with the respective other variable domain for achieving stability, solubility, or high binding specificity and affinity. If two distinct single-domain antibodies are replacing within an IgG molecule the otherwise cooperatively antigen-binding V_H/V_L domains, dual specificity is obtained (Figure 35.2A). A dual cytokine neutralizing antibody bispecific for IL-4 and IL-13, and a bispecific anti-VEGF/anti-EGFR mAb are in early development [26].

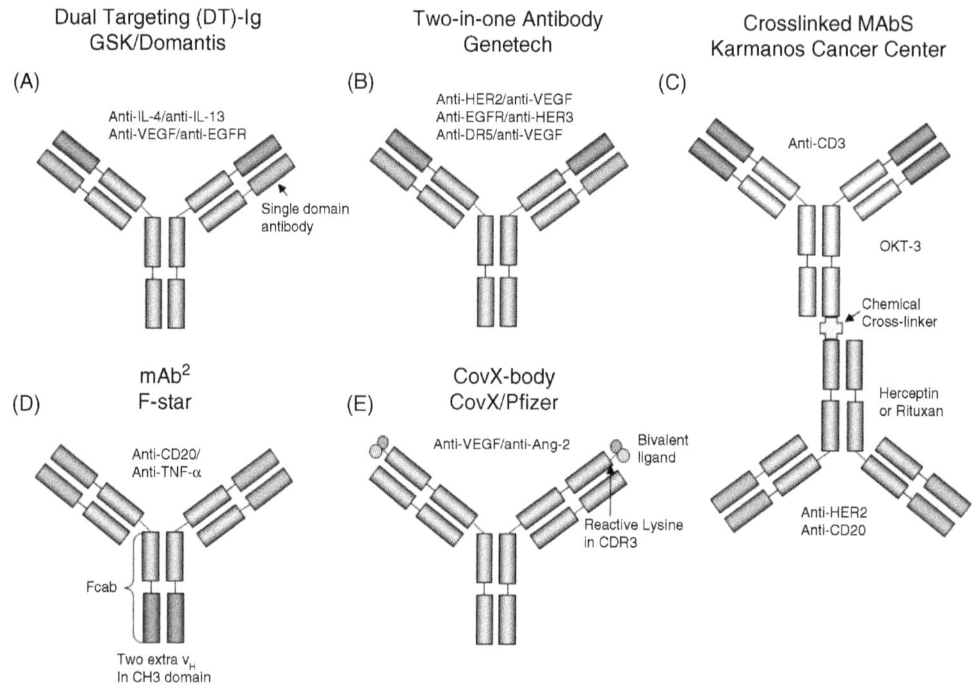

FIGURE 35.2 Symmetric bispecific antibody formats. Red and green colors highlight variable antibody domains of distinct specificity. Disclosed binding specificities and particular structural features are labeled. (See the color version of this figure in Color Plates section.)

On the basis of the observation that binding of the mAb trastuzumab (Herceptin®) to the HER2 antigen is primarily mediated by the three CDRs of the heavy chain variable domain, researchers at Genentech engineered the CDRs of the light chain variable domain to recognize a second antigen using phage display technology [27]. This approach has been successful and led to antibodies that in their variable heavy and light chain domains are bispecific for HER2 and VEGF, EGFR and HER3, or DR5 and VEGF, respectively [28]. Genentech's "two-in-one" antibodies are very similar in structure to GSK's DT IgG using single-domain antibodies (compare Figure 35.2A and B).

A rather simplistic approach to bispecific antibody generation has been taken by Lum and Davol (Figure 35.2C) [29]. Commercial mAb OKT3 (Orthoclone®) is chemically cross-linked at a 1:1 ratio with other commercial antibodies, such as trastuzmab (Herceptin®) or rituximab (Rituxan®). The resulting antibody conjugates are used for connecting cytotoxic CD3-expressing T cells with breast cancer cells overexpressing HER2 or lymphoma cells expressing CD20. In preclinical studies, this approach showed robust T-cell activation and redirected lysis [30]. In clinical practice, T cells isolated from patients are "armed" *ex vivo* with the bispecific antibody conjugate followed by their infusion, a procedure eventually avoiding systemic cytokine release reactions otherwise induced by the OKT-3 moiety of the conjugates. A Phase I study in 19 metastatic breast cancer patients with an OKT-3/trastuzumab conjugate has resulted in one partial response and 10 disease stabilizations [31].

The Biotech Company F-star has developed another solution for introduction of a second binding specificity into monoclonal IgG1 antibodies by engineering a second binding specificity into the C-terminal CH3 domain of the heavy chains (Figure 35.2D). The CH3 domain has loop structures that allow introduction of CDR diversity analogous to V_H and V_L domains by using phage display technology. The resulting bispecific antibodies with an additional binding specificity in the CH3 domains are referred to as "mAb^2 antibodies," and the isolated Fcγ part is called an "Fcab." The CH3 domain of mAb^2 antibodies and Fcabs essentially contains two identical single-domain antibodies that will bind their antigen by using only two CDR-like loops. The company has several preclinical programs. One is a bispecific anti-CD20/anti-TNF-α antibody in preclinical stage for treatment of rheumatoid arthritis [32]. F-star will collaborate with Boehringer Ingelheim on the development of up to seven mAb^2 or Fcab antibodies.

A unique approach to bispecific antibodies has been developed at the Scripps Research Institute and commercialized by CovX (now Pfizer). It is based on a catalytic antibody with a reactive lysine residue in its binding site. Dual binding specificity can be introduced to the antibody by covalently linking small bispecific molecules to the lysine residues [33]. An antiangiogenic bispecific "CovX-body" called CVX-241 (Pfizer) with dual specificity for VEGF and angiopoetin-2 is in Phase I clinical testing for treatment of cancer.

35.4 IgG-LIKE BISPECIFIC ANTIBODIES WITH FUSED ANTIBODY FRAGMENTS

A number of large biopharma companies have approached the development of bispecific antibodies by fusing extra variable domains to mAbs (Figure 35.3). For instance, Abbott is featuring its dual variable domain (DVD) bispecific antibody format [34]. The DVD format is extending the normal variable domains at the N-termini of heavy and light chains by two more variable domains and contains a total of four binding sites per antibody (Figure 35.3A). At least four early-stage programs have been presented in 2010 [35]. Three bispecific antibodies are neutralizing inflammatory mediators by DT of IL-1α and IL-1β, IL-12, and IL-18, or, of TNF-α and prostaglandin E2. A fourth bispecific antibody with dual specificity for CD3 and CD20 was designed to redirect T cells for lysis of lymphoma cells.

Researchers at Eli Lilly have fused a second variable domain to the C-termini of heavy chains (Figure 35.3B), creating a total of three binding sites per antibody molecule. In 2010, two early-stage programs were presented where bispecific antibodies target surface receptor combinations EGFR and IGF-1R, EGFR and VEGFR2, or TRAIL-R and LTRβ, respectively [36]. With the acquisition of Imclone, Eli Lilly gained access to the Di-diabody format for making bispecific antibodies [37].

Both ZymoGenetics (now BMS) and Biogen Idec have been fusing single-chain antibodies (scFvs) to the C-termini of mAb heavy chains (Figure 35.3D). ZymoGenetics has published a stable bispecific anti-IL-17A/anti-CD23 [38] and an anti-PDGFRβ/anti-VEGF-A antibody for treatment of inflammatory diseases. Biogen Idec is calling their approach "HERCULES." Their researchers published on generation and characteristics of a bispecific antibody where anti-LTRβ scFvs were fused either N- or C-terminally to the heavy chain of an anti-TRAIL-R antibody [39]. Roche also bought into this format, which they call "TvAb" for tetravalent bispecific antibody. One TvAb referred to as TAvi6 has disulfide bond-stabilized scFvs specific for angiopoetin-2 fused to the C-termini of anti-VEGF mAb bevacizumab (Avastin®) heavy chains [40]. Preclinical data showed superior antitumor activity in xenograft studies over bevacizumab.

Researchers at MedImmune/Astra Zeneca have been fusing scFvs also to the N-terminal position of IgG (Figure 35.3C). This allows design of not only bispecific but also of trispecific antibodies. A number of bispecific antibody candidates are in early development that were

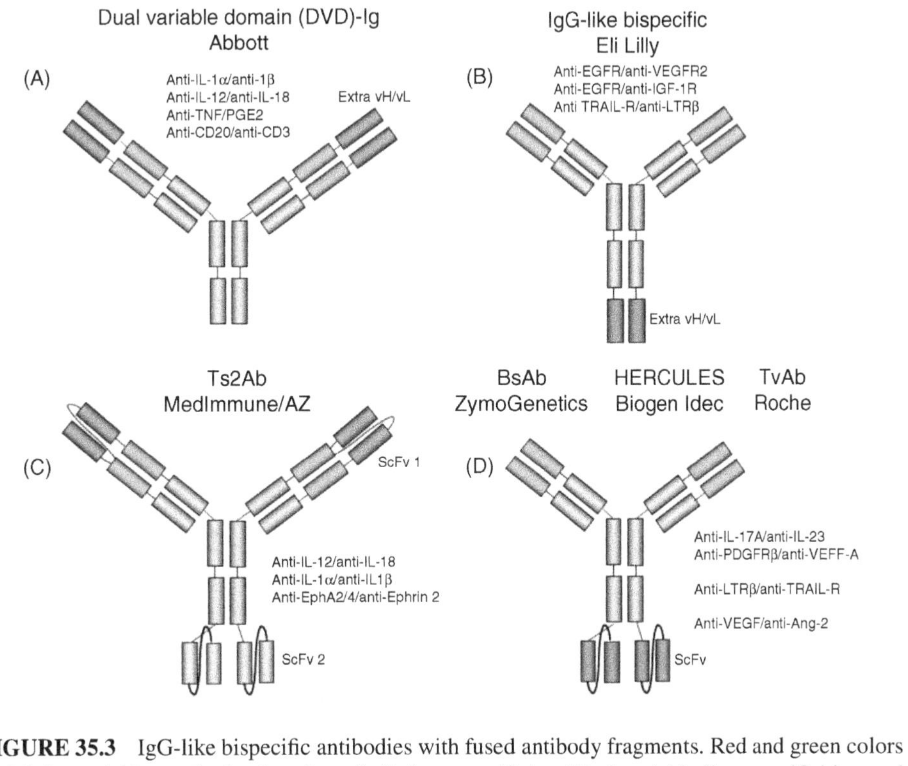

FIGURE 35.3 IgG-like bispecific antibodies with fused antibody fragments. Red and green colors highlight variable antibody domains of distinct specificity. Disclosed binding specificities and particular structural features are labeled. Single-chain antibodies (scFvs) show a peptide linker between variable domains. (See the color version of this figure in Color Plates section.)

designed to neutralize the proinflammatory cytokines IL-12 and IL-18, or IL-1α and IL-1β. A trispecific antibody has been mentioned, which binds the two tyrosine kinase receptors EphA2 and EphA4 as well as one of their ligands, Ephrin 2 [41].

All companies fusing extra domains to mAbs make a substantial effort to investigate and demonstrate the equivalency of their bispecific antibody constructs to regular mAbs with respect to productivity in Chinese hamster ovary (CHO) cells, stability, pharmacokinetics, aggregation behavior, and unencumbered binding affinity and specificity. It is expected that the performance of particular bispecific antibodies will mostly depend on the stability and properties of the fused scFv or V_H/V_L elements rather than on the bispecific format. It is obvious that all formats in Figure 35.3 exceed the size of a conventional mAb of 145–160 kDa by increments of 25 kDa. This may impact their capability to well penetrate into target tissue.

35.5 BISPECIFIC CONSTRUCTS BASED ON THE Fcγ FRAGMENT

While all bispecific antibody formats shown in Figures 35.1–35.3 still contain CH1/kappa or lambda domains, Figure 35.4 features bispecific antibodies that have been reduced to an IgG's Fcγ part as the core element for attachment of either single-chain antibodies or diabodies. Academic laboratories have published on bispecific antibodies attaching several different scFvs to the N- and C-termini of Fcγ part (Figure 35.4A) [42,43]. The company Trubion (now Emergent BioSolutions) is featuring a bispecific antibody format called "SCORPION" with a total of four scFvs attached to N- and C-termini of the Fcγ part (Figure 35.4B). A SCORPION construct with specificity for CD79B and HR (MHC II) has been presented [44]. A similar construct has been described by researchers from ZymoGenetics. They constructed an empowered anti-angiogenic antibody with dual specificity for VEGF-A and receptor tyrosine kinase PDGFRβ for treatment of age-related macular degeneration [45] and a further construct targeting CD19 and CD37.

MacroGenics is using an Fcγ part for attaching disulfide bond-stabilized diabodies they call DART for "dual affinity retargeting technology" (Figure 35.4C) [46]. Similar to SCORPION antibody constructs, Fc-DARTs can bivalently bind their target antigen, which will lead to high avidity binding. Another DART format is replacing V_H/V_L domains in IgG with stabilized diabodies.

The Chinese Government's National Engineering Research Center of Antibody Medicine has presented a bispecific antibody format consisting of two tandem single-chain antibodies $(scFv)_2$ N-terminally fused to a Fcγ domain

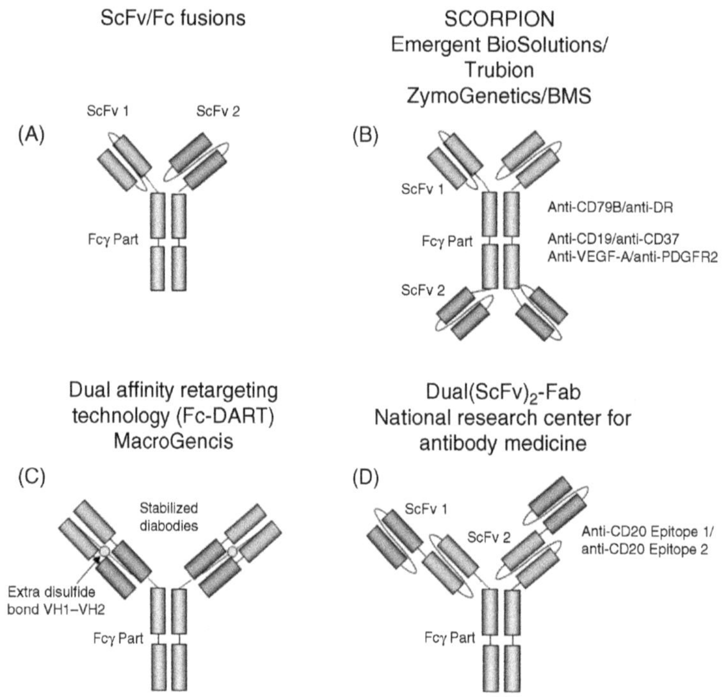

FIGURE 35.4 Bispecific constructs based on the Fcγ part of IgG. Red and green colors highlight variable antibody domains of distinct specificity. The Fcγ part is shown in grey. Disclosed binding specificities and particular structural features are labeled. Single-chain antibodies (scFvs) show a peptide linker between variable domains. (See the color version of this figure in Color Plates section.)

(Figure 35.4D) [47]. The antibody construct is recognizing two different epitopes of the CD20 antigen, binding to which combine the different functional properties of anti-CD20 mAbs rituximab (Rituxan®) and of atumomab (Azerra®). In xenograft studies, bispecific antibodies with dual anti-CD20 epitope specificity exceeded by far the antitumor activity and survival impact of the individual mAbs. Hence, dual epitope targeting by bispecific antibodies appears as a promising new approach to leverage established target antigens.

The Fcγ domain is responsible for the long serum half-life of IgG in the order of 2–3 weeks. This is conferred by binding of the Fcγ part to FcRn, the neonatal Fcγ receptor, on endothelial cells followed by receptor-mediated endocytosis, pH-controlled recycling, and final release of IgG back to the endothelial cell surface [48]. The Fcγ part is also responsible for mediating immune effector mechanisms such as antibody-dependent cellular cytotoxicity (ADCC), complement-dependent cytotoxicity (CDC), phagocytosis, and opsonisation, all of which rely on binding of the IgG's Fcγ part to Fcγ receptors I (CD64), II (CD32), or III (CD16) [49]. Finally, by its specific binding to *Staphylococcus aureus* proteins A and G, the Fcγ part can be employed for affinity purification of IgG and all Fcγ-containing bispecific constructs.

All bispecific antibodies shown in Figures 35.1–35.4 take advantage of the features of an Fcγ part. It must be pointed out that its use notoriously adds a third specificity to bispecific antibodies, or in other words, make bispecific antibodies in fact "trispecific." While this certainly is helpful for increasing serum half-life of an antibody construct through FcRn retention, the presence of a Fcγ part should be carefully considered and evaluated when constructing, for instance, T-cell-engaging antibodies. This is because the Fcγ part can turn all Fcγ receptor-bearing cells, including endothelial cells, into a target for redirected lysis by T cells through their surface binding of the T-cell-engaging antibody. Targeting of Fcγ receptor-bearing cells will occur even in the absence of the desired target cell population. The presence of an Fcγ part may explain the low systemic tolerability of T-cell-engaging catumaxomab of only 5 μg single dose in cancer patients [50].

Point mutations in the Fcγ part allow modulation of the interaction with Fcγ receptors leading to Fcγ variants of increased half-life or enhanced ADCC [51]. The smaller size of Fcγ fusions compared to the IgG format is thought to mediate better tumor penetration [52]. Improved tumor penetration is also expected for most of the bispecific constructs shown in following, many of which are even smaller than Fcγ fusion proteins.

35.6 BISPECIFIC CONSTRUCTS BASED ON Fab FRAGMENTS

A number of bispecific antibody formats solely use Fab fragments for binding to target antigens (Figure 35.5).

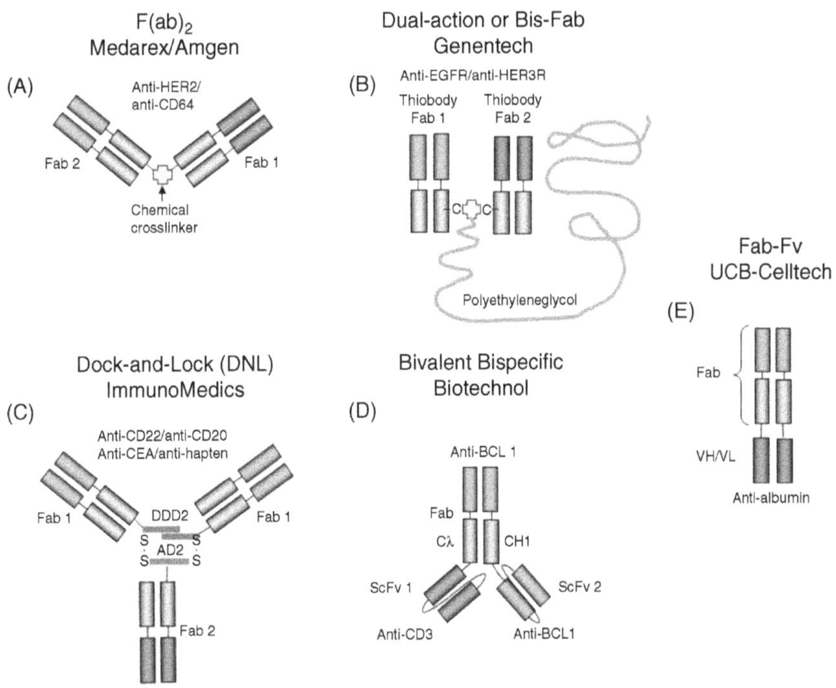

FIGURE 35.5 Bispecific constructs based on Fab fragments. Red and green colors highlight variable antibody domains of distinct specificity. The CH1/Cκ or λ domains are shown in grey. Disclosed binding specificities and particular structural features are labeled. Single-chain antibodies (scFvs) show a peptide linker between variable domains. (See the color version of this figure in Color Plates section.)

Chemical cross-linking or dimerization domains have been employed for joining two different Fab fragments, or, their recombinant fusion with scFvs. A bispecific construct called MDX-210 is based on two Fab fragments chemically cross-linked at the natural hinge region of IgG. The antibody construct has been in clinical development since the early 1990s [5]. It targets the tumor-asscociated tyrosine kinase receptor HER2 with one arm, and the high-affinity Fcγ receptor I (CD64) on macrophages with the second arm (Figure 35.5A). MDX-210 was designed to bind CD64 in a way not impacted by excess serum IgG and to cause redirected lysis of cancer cells by CD64-expressing cytotoxic macrophages.

Genentech/Roche revived chemical cross-linking for development of bispecific antibodies called "dual action Fab" (DAF) or "bis-Fab" (Figure 35.5B) [53]. In order to obtain a site-specific linkage, free cysteine residues were introduced in an accessible position on the surface of Fab fragments. The separately in *E. coli* produced and purified "thiobodies" are sequentially reacted with a chemical cross-linker to form the desired bispecific variant at high yield. By use of a special cross-linker containing polyethylene glycol, a means is given to prolong the serum half-life of the otherwise short-lived bispecific constructs. A bispecific antibody in this format is being developed for dual EGFR and HER3 receptor inhibition.

ImmunoMedics has worked out a very unique approach of connecting various Fab fragments with each other or with full IgG molecules [54,55]. For their "dock-and-lock" (DNL) platform, small dimerization domains derived from protein kinase A and its scaffold protein AKAP are employed. The C-terminally fused AD2 sequence is used to combine two Fab fragments of the same specificity (Figure 35.5C). This domain is then recognized by a DDD2 sequence, which is C-terminally fused to a second kind of Fab fragment. The resulting AD2/DDD2 homo- and heterodimerization complex ultimately combines three Fab fragments and is covalently stabilized by disulfide bridges. ImmunoMedics has published bispecific variants that combine CD20 and CD22-specific Fab fragments [54], and DNL antibodies for pre-targeting cancer cells for radioimmunotherapy [55].

The Portuguese Biotech Company Biotecnol has developed a bivalent bispecific antibody for T-cell engagement where two scFv are C-terminally fused to a Fab fragment. This format goes back to one developed by Schoonjans et al. [56]. Because the Fab fragment represents a heterodimerization domain, fusion of different scFvs to C-termini of CH1 and Cκ/λ domains will lead to homogenous bi- or tri-specific constructs. In the example shown in Figure 35.5D, the Fab-scFv construct recognizes CD3 on T cells and the BCL1 antigen on lymphoma cells. Both the Fab fragment and one of the two-fused scFvs are specific for BCL1 providing for bivalent target binding to lymphoma cells. A format called "heterominibody" has been developed by Micromet [57]. It just uses the heterodimerizing CH1/Cκ domain of the Fab fragment for fusion of scFvs and proinflammatory cytokines.

UCB-Celltech presented a novel bispecific format with extended serum half-life called "Fab-Fv" [58]. As shown in Figure 35.5E, a variable vH/vL (Fv) domain is C-terminally fused to a Fab fragment. By binding to serum albumin, the Fv domain confers a long serum half-life to the fusion protein.

Fab fragments provide attractive features as building blocks for bispecific antibodies. They already contain one specific binding site; they represent a heterodimerization element and allow fusion of additional V_H/V_L domains or single-chain antibodies to their N- or C-termini.

35.7 BISPECIFIC CONSTRUCTS BASED ON DIABODIES OR SINGLE-CHAIN ANTIBODIES

Since decades, bispecific antibodies are being developed that use diabodies or scFvs as sole structural basis [59]. Initial issues with the stability of small antibody fragments [60] can today be overcome by screening for fragments with high thermal stability, which typically translates into attractive biophysical features. Both scFvs and diabodies result from recombinantly fusing V_H and V_L domains with non-immunogenic peptide linker sequences [61,62]. If the peptide linker is short (between 5 and 9 amino acid residues) a diabody will result, which is composed of two polypeptide chains. For making a bispecific diabody, V_H and V_L domains from two different antibodies need to be fused with each other. If the linker between V_H and V_L domains is greater than 12 residues in length (usually between 15 and 20 residues), a scFv will result, which can bind to antigen as a single polypeptide chain. For construction of a bispecific antibody, at least two scFvs need to be fused in tandem by the help of a third peptide linker. Examples of bispecific antibody formats based on scFvs and diabodies are shown in Figure 35.6.

Bispecific T-cell engagers (BiTE), as developed by Amgen, consist of tandemly arranged scFvs (Figure 35.6A) that have been designed and optimized for redirected lysis of target cells with the help of cytotoxic T cells [63,64]. In its newest version, a primate cross-reactive, anti-CD3 human scFv is fused with a second human or humanized scFv specific for a tumor-associated antigen. Figure 35.6a lists several target antigens where the corresponding BiTE antibodies have been characterized in clinical and nonclinical studies (reviewed in Reference [64]). Properties of BiTE antibodies include (i) monovalent binding to CD3 and target antigen, (ii) activation of unstimulated T cells in a strictly target cell-dependent manner, (iii) support of serial lysis by activated T cells, (iv) extreme potency of redirected lysis at sub-picomolar concentrations, and (v) an

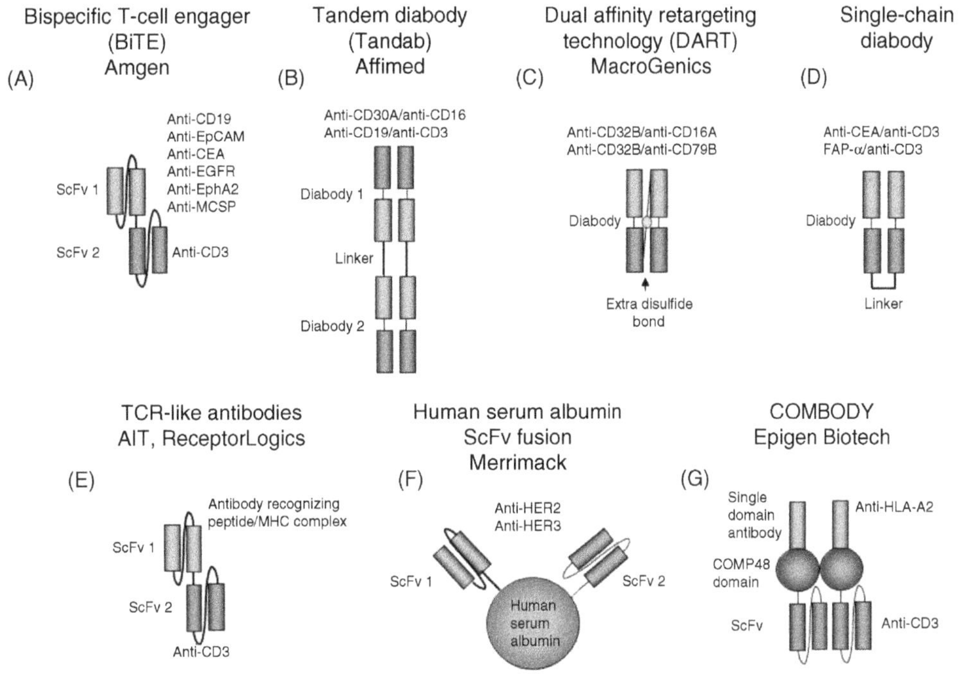

FIGURE 35.6 Bispecific antibody constructs based on diabodies or single-chain antibodies. Red and green colors highlight variable antibody domains of distinct specificity. Disclosed binding specificities and particular structural features are labeled. Single-chain antibodies (scFvs) show a peptide linker between variable domains. (See the color version of this figure in Color Plates section.)

induction of effector T-cell proliferation. It appears that for efficient and safe T-cell engagement BiTE antibodies all of these properties have to be combined. Prolonged infusion of the rather short-lived BiTE antibodies to patients allows for a sustained T-cell engagement against cancer cells as well as for an exquisite control of safety parameters by swift clearance of drug on discontinued infusion. Currently, three BiTE antibodies are in clinical development. The CD19/CD3-bispecific BiTE antibody blinatumomab is in a pivotal study in patients with acute lymphocytic leukemia (ALL). It has shown outstanding efficacy in patients with ALL [65], and non-Hodgkin lymphoma (NHL) [66]. An EpCAM/CD3-bispecific BiTE antibody called MT110 is in a Phase I study in cancer patients with EpCAM-expressing solid tumors, and a CEA-CD3-bispecific BiTE, developed in collaboration with MedImmune/Astra Zeneca, has just commenced a Phase I study in a similar cancer patient population. Three other BiTE antibodies are in preclinical development together with Sanofi-aventis, Boehringer Ingelheim, and Bayer Schering Pharma.

Fey and colleagues have developed a number of tandem scFv-based bi- and trispecific constructs for engagement of immune cells. Two are designed to redirect natural killer (NK) cells via CD16 to either CD123-expressing acute myeloid leukemia (AML) cells [67] or to both CD19-expressing leukemia and lymphoma, and to CD33-expressing AML cells [68]. The latter construct is called a "triplebody" and covalently aligns three scFvs to achieve targeting of two kinds of malignant cells. A third class of tandem scFv constructs is unique in that it specifically engages neutrophils via Fc alpha receptor I (CD89) against lymphoma cells expressing HLA class II molecules [69].

Tandem diabodies called "Tandabs" are developed by Affimed and use extra linkers to join two diabodies (Figure 35.6B) with the goal of increasing binding affinities and, eventually, potency and serum half-life [70]. Two Tandabs have been advanced into formal drug development. One is for engaging T cells via CD3 for lysis of CD19-expressing lymphoma and leukemia cells [71], and one for engaging NK cells via Fcγ receptor IIIa (CD16A) to CD30-expressing Hodgkin lymphoma cells [72]. The CD30/CD16-bispecific Tandab, called AMF13, has just commenced a Phase I clinical trial in Hodgkin's lymphoma patients, while the CD19/CD3-bispecific TandAb AMF11 is in formal preclinical development. One of their hallmarks is bivalent binding to effector as well as target cells. Bivalent target binding may promote potency through higher affinity binding but must be evaluated with great care. For instance, bivalent binding to CD3 on T cells—as also done by the anti-CD3 mAb OKT-3—may after initial, nonconditional T-cell activation cause down modulation of the TCR with impediment of further efficacy. On the other hand, bivalent binding to target antigens may down-modulate their surface expression with an additional negative impact on efficacy.

Although Tandabs have a size in excess of 100 kDa, the serum half-life of the CD19/CD3 bispecific Tandab in mice was surprisingly short with a t1/2β of 2.8 h [73]. In animal models, the CD19/CD3-bispecific Tandab required costimulation of human T cells with anti-CD28 antibody [71]. This may indicate that this bispecific format did not allow for single-agent activation of unstimulated T cells *in vivo*.

Another kind of diabody-based bispecific antibody constructs is based on MacroGenic's DART proteins (Figure 35.6C) of which at least two are in formal preclinical development. One is designed to engage NK cells via binding to their Fcγ receptor CD16A for lysis of CD32B-expressing lymphoma cells [74]. The other is cross-linking the B cell receptor subunit CD79B with the inhibitory Fcγ receptor CD32B on the surface of B cells for the treatment of rheumatoid arthritis [75]. In order to stabilize the diabodies used in DART constructs, extra cysteine residues for making an additional disulfide bond have been introduced. This technology will allow use of otherwise instable diabodies and prevent their aggregation. Diabodies with high thermodynamic stability can be also be obtained by other methods [76]. MacroGenics is developing bispecific DART antibody constructs together with Pfizer and Boehringer Ingelheim, and is about to commence a clinical study with their CD32B/CD16A-bispecific DART.

By introducing a peptide linker, Kontermann and colleagues have developed an alternative to disulfide-based diabody stabilization as used by DART by introduction of a peptide linker fusing the two chains of a diabody to form a "single-chain diabody" (Figure 35.6D). T cell-engaging single-chain diabodies with specificity for carcinoma embryonal antigen (CEA) or fibroblast-activating protein alpha and CD3 on T cells have shown high activity in cell-based assays, and human serum albumin (HSA) binding versions of these proteins a much prolonged serum half-life in mice [77].

Bispecific tandem scFv-based constructs for T-cell engagement are being developed by two more companies (Figure 35.6E). Applied Immune Technologies (AIT) and ReceptorLogics are using single-chain antibodies derived from "T-cell receptor mimic (TCRm) mAbs" for targeting tumor cells expressing a certain MHC class I antigen/peptide antigen complex [78]. With their TCR-like specificity, these antibody constructs can only bind target cells expressing the proper MHC class I/peptide complex and will thereby face the same limitations as the format shown in Figure 35.7C, which is discussed.

By fusing scFvs to both N- and C-terminus of HSA, a bispecific antibody was engineered by Merrimack that confers the long serum half-life of HSA to the fusion protein (Figure 35.6F) [79]. Similar to IgG, HSA is retained by the FcRn system [48]. A dual receptor-inhibiting antibody construct of Merrimack's called MM-111 is bispecific for HER2 and HER3 and commencing Phase II clinical development for the treatment of breast cancer patients. Several more multispecific antibodies in this format are in preclinical development.

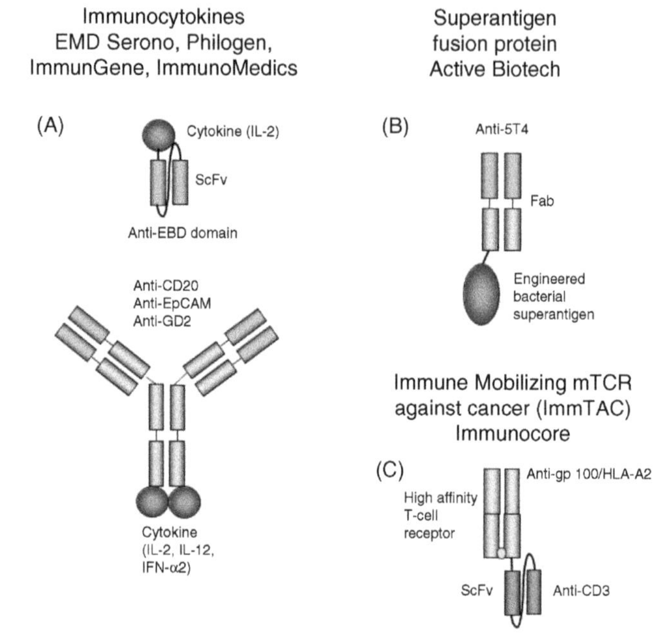

FIGURE 35.7 Bifunctional fusions of antibodies or fragments with unrelated proteins. Disclosed binding specificities and particular structural features are labeled. Single-chain antibodies (scFvs) show a peptide linker between variable domains. (See the color version of this figure in Color Plates section.)

Epigen Biotech has crafted a multivalent T-cell-engaging bispecific construct by fusing a single domain antibody specific for HLA-A2 with a scFv specific for CD3 using a coiled–coiled peptide domain from human cartilage matrix protein (COMP) 48 domain (Figure 35.6G) [80]. This construct is called a "COMBODY."

35.8 BIFUNCTIONAL FUSIONS OF ANTIBODIES OR FRAGMENTS WITH OTHER PROTEINS

While Figures 35.1–35.6 have described bispecific antibodies entirely based on immunoglobulin sequences (with exception of examples in Figures 35.6F and 35.6G), we will, for the sake of completeness, also discuss a selection of bifunctional proteins that combine immunoglobulin sequences with those of unrelated proteins (Figure 35.7). Because the second, immunoglobulin-unrelated protein is likewise binding to a protein target—be it a cytokine receptor or the TCR—such fusion proteins should also be considered bispecific antibodies.

A number of so-called immunocytokines are in preclinical and clinical development by several companies including EMD Serono, ImmunoMedics, Philogen, Bayer

Schering Pharma, and ImmunGene. Cytokines can be fused to an antibody fragment or to a complete IgG molecule (Figure 35.7A). The fused cytokines will bind their high-affinity receptors on immune cells and this way transiently connect immune cells to those cells targeted by the antibody or fragment. The same time, cytokines stimulate the cytokine receptor to elicit an immune effector function. Frequently selected cytokines for this approach are IL-2, IL-12, and IFN-α2. A concern of immunocytokines could be that their cytokine receptor binding may be of much higher affinity than cancer target binding. This would trigger receptors on effector cells before bridging of effector and target cells occurs. Another concern is that cytokines by their combination with antibodies become long-lived and thereby more toxic than short-lived cytokines. A number of immunocytokines are in clinical development. An anti-GD2 IgG/IL-2 immunocytokine has shown high clinical activity in juvenile neuroblastoma with 5 complete responses among 23 patients [81]. Intriguingly, this antibody has only a very short serum half-life of only 4 h. Disease stabilization has been observed in cancer patients in a Phase II trial treated with a single-chain anti-fibronectin EBD domain antibody fused to IL-2 [82].

Two fusion proteins shown in Figure 35.7 have been specifically developed for T-cell engagement. One from Active Biotech is a fusion between an anti-5T4-specific Fab fragment and an engineered bacterial superantigen, called naptumomab estafenatox or ANYARA® (Figure 35.7B) [83]. It can connect cancer cells expressing the fetal oncoprotein 5T4 to cytotoxic T cells by specific binding of the superantigen part to the TCR vβ-chain. Point mutations have much reduced the second specificity of the superantigen for MHC class molecules. An issue of the format still is the high immunogenicity of the bacterial superantigen. Pre-existing antibodies against the streptococcal antigen in patients require that doses have to be individually adjusted to match the level of anti-superantigen antibody titers [84]. Naptumomab estafenatox has completed several clinical studies showing signs of antitumor activity. A Phase III trial in renal cell cancer patients in combination with IFN-α treatment versus IFN-α alone is expecting readout in the first half of 2011.

Another T-cell-engaging format is called "Immune Mobilising mTCR Against Cancer" or "ImmTAC." The bispecific construct is a fusion between an anti-CD3-specific scFv and a high-affinity, disulfide bond-stabilized TCR [85]. ImmTACs are being developed by Immunocore and designed to engage T cells for lysis of cancer cells. Their lead product recognizes melanoma cells expressing a gp100-derived peptide antigen by HLA-A2. Other ImmTACs specific for other peptide antigens are under preclinical development. The gp100 peptide-specific ImmTAC has just commenced clinical Phase III testing in the United Kingdom.

One issue of this format and the one featured in Figure 35.6E is that a particular HLA/peptide complex will be present on cancer cells only at very low copy number. High affinity binding by TCR-like scFvs or single-chain TCRs in combination with high drug concentrations will be required for efficient lysis. Another problem arises if cancer cells become selected in patients for loss of MHC class I, β2 microglobulin or TAP transporter expression, or for altered proteasome activity. Such "immune-escaped" cancer cells can no longer be recognized by bispecific antibody constructs with TCR-like specificity. Finally, their activity will be HLA-restricted thereby limiting the patient population, and no relevant animal species will be available for preclinical toxicity assessment.

35.9 BISPECIFIC ANTIBODIES FOR VARIOUS FUNCTIONS: HOW TO SELECT THE RIGHT FORMAT?

This review demonstrates that numerous antibody formats are currently available for making bispecific antibodies. However, not every format will support a desired function and performance. For neutralizing two soluble ligands, for example, cytokines, an IgG-like bispecific antibody appears most desirable. This should neutralize ligands with high affinity, be of low immunogenicity and otherwise share properties with established ligand-neutralizing commercial antibodies such as anti-TNF human IgG1 adalimumab (Humira®) or anti-VEGF humanized IgG1 bevacizumab (Avastin®). Theoretically, all antibodies formats shown in Figures 35.1–35.4 qualify for this task. They all retain the Fcγ part mediating a long-serum half-life and clearance of antibody/ligand complexes via the reticulo-endothelial system. Bispecific antibody formats may however generically differ in target tissue penetration, distribution volume, compatibility with subcutaneous administration, clearance kinetics, aggregation behavior, protease resistance, productivity, immunogenicity, and other parameters. It is hard to predict which format will suit a neutralizing function best also, because biophysical properties of antibodies will largely depend on variable domains, which are likely to dominate the characteristic of any fusion protein.

For dual receptor blockade, other formats may be optimal than for ligand neutralization. Because anti-receptor antibodies bind to cell surfaces, it must be decided whether the antibody should confer as a third activity ADCC, which may worsen the safety profile but, on the other hand, add cytotoxicity to the dual receptor-blocking mode of action. In case immune effector mechanisms are not desired, the human IgG4 format, an ADCC/CDC-inactivated Fcγ1 mutant IgG, or a fusion of antibody fragments to HSA can be used. Good target tissue penetration, high potency of receptor blockade, and long serum half-life appear favorable. Certain formats or epitopes may allow simultaneous receptor binding by the antibody, while others only allow binding of the two receptors in a mutually exclusive manner. It is believed that formats at or below 100 kDa are more suitable for target tissue penetration than IgG-like fusion protein of 150–250 kDa [52]. A smaller size may translate into a higher biological activity when it comes to receptor binding on target tissue, but may not matter for neutralizing ligands in the blood stream or interstitial fluid.

For engagement of immune effector cells by bispecific antibodies, it will greatly matter which kind of immune cells shall be linked to target cells for mediating redirected lysis. For engagement of NK cells, presence of a Fcγ1 part is a frequently selected option because of the positive clinical experience with mostly ADCC-mediating antibodies rituximab and CAMPATH-1. Point mutations in the CH2 domain or manipulations avoiding the addition of fucose residues to the complex carbohydrate moiety in the CH2 domain have been shown to greatly enhance ADCC [51,86]. Moreover, Fcγ parts can engage immune cells other than NK cells, such as FcγR I (CD64)-expressing macrophages, or FcγR II (CD32)-expressing granulocytes. The use of anti-CD16A or anti-CD64 antibodies for cytotoxic immune cell engagement has been preferred in several instances over Fcγ1 domains, as done by particular Tandabs and DARTs, and by MDX-210. Arguments include that this way the inhibitory FcγR IIB (CD32B) will not be stimulated and that excess IgG in serum will not compete for drug binding to the Fcγ receptor.

For engagement of T cells, different design premises of bispecific antibodies may apply as for engagement of NK and other immune cells. Owing to the superior cytotoxic potential of T cells, T-cell engagement by bispecific antibodies holds the promise of treating cancer by completely eradicating target-expressing cancer cells. Owing to an enormously potent signal-amplifying cascade, T cells are highly responsive to minute stimulation of their TCR complex. In fact, single digit numbers of MHC/peptide complexes can fully activate the lytic potential of effector memory T cells [87, 90]. T cells do not bear Fcγ receptors and can therefore not be directly engaged by Fcγ parts of regular IgG antibodies. Therefore, most T cell-engaging bispecifics target an invariant sequence of the TCR complex.

The flip side is that by potential binding to every T cell, such bispecific antibodies can broadly activate the circulating T-cell compartment, which can cause serious cytokine-related adverse events. These were observed on first infusion of anti-CD28 IgG4 antibody TGN1412 to healthy volunteers [87], and are naturally triggered upon infection with bacteria secreting "superantigens," which can cross-link the TCR β-chain with MHC class II molecules [88]. Likewise, the bivalently CD3-binding murine mAb OKT-3 (Orthoclone®) can upon first infusion cause a cytokine release syndrome, which is followed by TCR down-modulation, T-cell anergy, and mutual redirected T-cell lysis (fratericide). The latter three reactions are clinically exploited for the prevention of acute transplant rejection. T-cell engagement and activation by bispecific antibodies has thus to be subtle, well controlled, and ideally not directly cross-linking the TCR complex. T-cell activation should rather occur in a highly conditional manner, only as long as target cells are present, and not trigger significant TCR down-modulation and T-cell anergy. This may require a delicate tuning of the anti-TCR binding arm for optimal binding affinity, epitope recognition and conformational impact on the bound TCR complex. Likewise, the binding affinity and epitope recognition of the second arm of the bispecific antibody for target cells must be optimal and well balanced with that for T-cell binding.

T cells engaged by BiTE antibodies can eliminate both dividing and non-dividing cells bearing the target antigen. This is in contrast to ADCs carrying an antiproliferative drug payload, which will predominantly eliminate dividing cells. Hence, great care has to be taken when selecting target antigens for T-cell-engaging antibodies.

Fcγ domains—as present in numerous bispecific antibody formats—bind to FcRn on endothelium and to other Fcγ receptors as found on most immune cells. As a consequence, these functionally "trispecific" antibodies can tether T cells to a plethora of Fcγ receptor-positive cells and eventually trigger this way T-cell activation and redirected lysis of Fcγ receptor-expressing normal cells. This potential issue of trispecific antibody catumaxomab, which may explain its low systemic tolerability, has obviously been managed by both its intraperitoneal administration and intrapatient dose escalation starting from very low microgram doses [3].

A notorious challenge with BiTE antibodies is the lack of appropriate animal models. This relates to the very limited repertoire of available and suitable TCR-binding antibodies. While xenograft studies in immunodeficient mice can still use human T cells as effector cells, T cells of rodents or macaques, as are typically used for preclinical safety studies, will not be recognized by bispecific antibodies. Unless, a species-reactive surrogate antibody is generated, the consequences of T-cell activation and expression of a particular target antigen on normal tissues can therefore not be assessed prior to Phase I clinical testing in patients. As a consequence, only minute doses delineated *in vitro* by the "minimum-anticipated biological effect level" (MABEL) approach can be used to define start doses for first-in-man studies, which may then require long-lasting escalation periods to reach active dose levels. Recently, a human/primate cross-reactive anti-CD3 human scFv antibody has been developed as a basis for constructing fully human/primate cross-reactive BiTE antibodies. Exemplary anti-EGFR BiTE antibodies were shown to cause the expected target-dependent adverse events in a macaque model [89]. The opportunity to explore potential target-specific adverse events in a relevant primate model early in preclinical development using the final BiTE antibody drug candidate will greatly facilitate future selection and development of T-cell-engaging antibodies.

REFERENCES

1. Milstein C, Cuello AC. (1983) Hybrid hybridomas and their use in immunohistochemistry. *Nature* 305, 537–540.
2. Lindhofer H, Mocikat R, Steipe B, Thierfelder S. (1995) Preferential species-restricted heavy/light chain pairing in rat/mouse quadromas. Implications for a single-step purification of bispecific antibodies. *J. Immunol.* 155, 219–225.
3. Heiss MM, Murawa P, Koralewski P, Kutarska E, Kolesnik OO, Ivanchenko VV, et al. (2010) The trifunctional antibody catumaxomab for the treatment of malignant ascites due to epithelial cancer: results of a prospective randomized Phase II/III trial. *Int. J. Cancer* 127, 2209–2222.
4. Brennan M, Davison PF, Paulus H. (1985) Preparation of bispecific antibodies by chemical recombination of monoclonal immunoglobulin G1 fragments. *Science* 229, 81–83.
5. Valone FH, Kaufman PA, Guyre PM, Lewis LD, Memoli V, Deo Y, et al. (1995) Phase Ia/Ib trial of bispecific antibody MDX-210 in patients with advanced breast or ovarian cancer that overexpress the proto-oncogene HER-2/neu. *J. Clin. Oncol.* 13, 2281–2292.
6. Staerz UD, Kanagawa O, Bevan MJ. (1985) Hybrid antibodies can target sites for attack by T cells. *Nature* 314, 628–631.
7. Corvalan JR, Smith W. (1987) Construction and characterization of a hybrid-hybrid monoclonal antibody recognising both carcinoembryonic antigen (CEA) and vinca alkaloids. *Cancer Immunol. Immunother.* 24, 127–132.
8. Carter P. (2001) Bispecific human IgG by design. *J. Immunol. Methods* 248, 7–15.
9. Kufer P, Lutterbuese R, Baeuerle PA. (2004) A revival of bispecific antibodies. *Trends Biotechnol.* 22, 238–244.
10. Müller D, Kontermann RE. (2007) Recombinant bispecific antibodies for cellular cancer immunotherapy. *Curr. Opin. Mol. Ther.* 9, 319–326.
11. Van der Neut Kolfschoten M, Schuurman J, Losen M, Bleeker WK, Martinez-Martinez P, Vermeulen E, et al. (2007) Anti-inflammatory activity of human IgG4 antibodies by dynamic Fab arm exchange. *Science* 317, 1554–1557.
12. Mouquet H, Scheid JF, Zoller MJ, Krogsgaard M, Ott RG, Shukair S, et al. (2010) Polyreactivity increases the apparent affinity of anti-HIV antibodies by heteroligation. *Nature* 467, 591–595.
13. Ströhlein MA, Siegel R, Jäger M, Lindhofer H, Jauch KW, Heiss MM. (2009) Induction of anti-tumor immunity by trifunctional antibodies in patients with peritoneal carcinomatosis. *J. Exp. Clin. Cancer Res.* 14, 18.
14. Seimetz D, Lindhofer H, Bokemeyer C. (2010) Development and approval of the trifunctional antibody catumaxomab (anti-EpCAM x anti-CD3) as a targeted cancer immunotherapy. *Cancer Treat. Rev.* 36, 458–467.
15. Kiewe P, Hasmüller S, Klahert S, Heinrigs M, Rack B, Marmé A, et al. (2008) *Clin. Cancer Res.* 12, 3085–3091.
16. Ruf P, Kluge M, Jäger M, Burges A, Volovat C, Heiss MM, et al. (2010) Pharmacokinetics, immunogenicity and bioactivity of the therapeutic antibody catumaxomab intraperitoneally administered to cancer patients. *Br. J. Clin. Pharmacol.* 69, 617–625.
17. Ridgway JB, Presta LG, Carter P. (1996) "Knobs-in-holes" engineering of antibody CH3 domains for heavy chain heterodimerization. *Protein Eng.* 9, 617–621.
18. Jackman J, Chen Y, Huang A, Moffat B, Scheer JM, Leong SR, et al. (2010) Development of a two-part strategy to identify a therapeutic human bispecific antibody that inhibits IgE receptor signalling. *J. Biol. Chem.* 285, 20850–20859.
19. Gunasekaran K, Pentony M, Shen M, garrett L, Forte C, Woodward A, et al. (2010) Enhancing antibody Fc heterodimer formation through electrostatic steering effects: applications to bispecific molecules and monovalent IgG. *J. Biol. Chem.* 285, 19637–19646.
20. Wranik B. (2010) Keystone Conference on Advances in Biopharmaceuticals, Keystone, CO, 2010 Jan 10.
21. Davis JH, Aperlo C, Li Y, Kurosawa E, Lan Y, Lo KM, et al. (2010) SEEDbodies: fusion proteins based on strand-exchange engineered domain (SEED) CH3 heterodimers in an Fc analogue platform for asymmetric binders or immunofusions and bispecific antibodies. *Protein Eng. Des. Sel.* 23, 195–202.
22. Klein C. (2010) Antibody Engineering Conference, San Diego, CA, 2010 Dec 2010.
23. Throsby M. (2010) Third Annual Antibody Engineering and Therapeutics Asia, Shanghai, China, 2010 Nov 11.
24. Schuurman J. (2010) Third Annual Antibody Engineering and Therapeutics Asia, Shanghai, China, 2010 Nov 11.
25. Ward ES, Güssow D, Griffiths AD, Jones PT, Winter G. (1989) Binding activities of a repertoire of single immunoglobulin variable domains secreted from *Escherichia coli*. *Nature* 341, 544–546.
26. Hamblin P. (2010) Sixth Annual PEGSummit, Boston, MA, 2010 May 20–21.
27. Bostrom J, Yu SF, Kan D, Appleton BA, Lee CV, Billeci K, et al. (2009) Variants of the antibody herceptin that interact with HER2 and VEGF at the antigen binding site. *Science* 323, 1610–1614.
28. Fuh G. (2010) 25th Annual Meeting of International Society for Biological Therapy of Cancer, Washington, 2010 Oct 2.
29. Lum LG, Davol PA. (2005) Retargeting T cells and immune effector cells with bispecific antibodies. *Cancer Chemother. Biol. Response Modif.* 22, 273–291.
30. Grabert RC, Cousens LP, Smith JA, Olson S, Gall J, Young WB, et al. (2006) Human T cells armed with Her2/neu bispecific antibodies divide, are cytotoxic, and secrete cytokines with repeated stimulation. *Clin. Cancer Res.* 12, 569–576.
31. Lum L. (2010) Sixth Annual PEGSummit, Boston, MA, 2010 May 20–21.
32. Fitzgerald K. (2010) Sixth Annual PEGSummit, Boston, MA, 2010 May 20.

33. Gavrilyuk JI, Wuellner U, Salahuddin S, Goswami RK, Sinha SC, Barbas CF, 3rd. (2009) An efficient chemical approach to bispecific antibodies and antibodies of high valency. *Bioorg. Med. Chem. Lett.* 19, 3716–3720.
34. Wu C, Ying H, Grinnell C, Bryant S, Miller R, Clabbers A, et al. (2007) Simultaneous targeting of multiple disease mediators by a dual-variable-domain Immunoglobulin. *Nat. Biotech.* 25, 1290–1297.
35. Ghayur T. (2010) Drug Discovery and Development Week, Burlingame, CA, 2010 August 3.
36. Ludwig D. (2010) Drug Discovery and Development Week, Burlingame, CA, 2010 Aug 3.
37. Lu D, Jimenez X, Zhang H, Atkins A, Brennan L, Balderes P, Bohlen P, et al. (2003) Di-diabody: a tetravalent bispecific antibody molecule by design. *J. Immunol. Methods* 279, 219–232.
38. Mabry R, Lewis KE, Moore M, McKernan PA, Bukowski TR, Bontadelli K, et al. (2010) Engineering of stable bispecific antibodies targeting IL-17A and IL-23. *Protein Eng. Des. Sel.* 23, 115–127.
39. Michaelson JS, Dermarest SJ, Miller B, Amatucci A, Snyder WB, Wu X, et al. (2009) Anti-tumor activity af stability-engineered IgG-like bispecific antibodies targeting TRAIL-R2 and LT-betaR. *MAbs* 1, 128–141.
40. Scheuer W. (2010) The NCI-EORTC-AACR Meeting in Berlin, Germany, 2010 Nov 19.
41. Dinasi N. (2010) Sixth Annual PEGSummit, Boston MA, 2010 May 20–21.
42. Natsume A, Wakitani M, Yamane-Ohnuki N, Shoji-Hosaka E, Niwa R, Uchida K, et al. (2006) Fucose removal from complex-type oligosaccharide enhances the antibody-dependent cellular cytotoxicity of single-gene-encoded bispecific antibodies linked to the antibody constant domain. *J. Biochem.* 140, 359–368.
43. Lu H, Shi M, Wang M, Xie Z, Hu M, Yu M, et al. (2008) *In vitro* and *in vivo* antitumor effect of a trivalent bispecific antibody targeting ErbB2 and CD16. *Cancer Biol. Ther.* 7, 1744–1750.
44. Blankenship JW, Strobel S, Tan P, Grosmaire L, Simon A, Clark C, et al. (2009) Anti-CD79B x anti-DR SCORPION molecule: A single chain, bispecific immunotherapeutic with potent *in vitro* activity against B-cell lymphoma. Poster at 100th Annual Meeting of the American Association for Cancer Research, Abstract No. 5465.
45. Mabry R, Gilbertson DG, Frank A, Vu T, Ardourel D, Ostrander C, et al. (2010) A dual-targeting PDGFRbeta/VEGF-A molecule assembled from stable antibody fragments demonstrates anti-angiogenic activity *in vitro* and *in vivo*. *MAbs* 2, 20–34.
46. Johnson S. (2010) Drug Discovery and Development Week, Burlingame, CA, 2010 Aug 2–4.
47. Guo J. (2010) Third Annual Antibody Engineering and Therapeutics Asia, Shanghai, China, 2010 Nov 11.
48. Roopenian DC, Akilesh S. (2007) FcRn: the neonatal Fc receptor comes of age. *Nat. Rev. Immunol.* 7, 715–725.
49. Nimmerjahn F, Ravetch JV. (2008) Fcgamma receptors as regulators of immune responses. *Nat. Rev. Immunol.* 8, 34–47.
50. Sebastian M, Passlick B, Friccius-Quecke H, Jäger M, Lindhofer H, Kanniess F, et al. (2007) Treatment of non-small cell lung cancer oatients with the trifunctional monoclonal antibody catumaxomab (anti-EpCAM x anti-CD3): a Phase I study. *Cancer Immunol. Immunother.* 56, 1637–1644.
51. Lazar GA, Dang W, Karki S, Vafa O, Peng JS, Hyun L, et al. (2006) Engineered antibody Fc variants with enhanced effector function. *Proc. Natl. Acad. Sci. USA* 103, 4005–4010.
52. Beckman RA, Weiner LM, Davis HM. (2007) Antibody constructs in cancer therapy: protein engineering strategies to improve exposure in solid tumors. *Cancer* 109, 170–179.
53. Sheer J. (2010) The Sixth Annual PEGSummit in Boston, MA, 2010 May 20–21.
54. Rossi EA, Goldenberg DM, Cardillo TM, Stein R, Chang CH. (2009) Hexavalent bispecific antibodies represent a new class of anticancer therapeutics: 1. Properties of anti-CD20/CD22 antibodies in lymphoma. *Blood* 113, 6161–6171.
55. Sharkey RM, Rossi EA, McBride WJ, Chang CH, Goldenberg DM. (2010) Recombinant bispecific monoclonal antibodies prepared by the dock-and-lock strategy for pretargeted radioimmunotherapy. *Semin. Nucl. Med.* 40, 190–203.
56. Schoonjans R, Willems A, Schoonooghe S, Fiers W, Grooten S, Mertens N. (2000) Fab chains as an efficient heterodimerization domain scaffold for the production of recombinant bispecific and trispecific antibody derivatives. *J. Immunol.* 165, 7050–7057.
57. Schanzer JM, Fichtner I, Baeuerle PA, Kufer P. (2006) Anti-tumor activity of a dual cytokine/single-chain antibody fusion protein for simultaneous delivery of GM-CSF and IL-2 to Ep-CAM expressing tumor cells. *J. Immunother.* 29, 477–488.
58. Humphreys D. (2010) Antibody Engineering Conference, San Diego, CA, 2010 Dec 8.
59. Nelson AL. (2010) Antibody fragments: hope and hype. *MAbs* 2, 77–83.
60. Mabry R, Snavely M. (2010) Therapeutic bispecific antibodies: the selection of stable single-chain fragments to overcome engineering obstacles. *IDrugs* 13, 543–549.
61. Huston JS, Mudgett-Hunter M, Tai MS, McCartney J, Warren F, Haber E, et al. (1991) Protein engineering of single-chain Fv analogs and fusion proteins. *Methods Enzymol.* 203, 46–88.
62. Todorovska A, Roovers RC, Dolezal, O, Kortt, AA, Hoogenboom, HR, Hudson, PJ. (2001) Design and application of diabodies, triabodies and tetrabodies for cancer targeting. *J. Immunol. Methods* 248, 47–66.
63. Baeuerle PA, Kufer P, Bargou R. (2009) BiTE: teaching antibodies to engage T-cells for cancer therapy. *Curr. Opin. Mol. Ther.* 11, 22–30.
64. Baeuerle PA, Reinhardt C. (2009) Bispecific T-cell engaging antibodies for cancer therapy. *Cancer Res.* 69, 4941–4944.
65. Handgretinger R, Zugmaier G, Henze G, Kreyenberg H, Lang P, von Stackelberg A. (2010) Complete remission after blinatumomab-induced donor T-cell activation in three pediatric patients with post-transplant relapsed acute lymphoblastic leukemia. *Leukemia*, October 14 [Epub ahead of print].

66. Bargou R, Leo E, Zugmaier G, Klinger M, Goebeler M, Knop S, et al. (2008) Tumor regression in cancer patients by very low doses of a T cell-engaging antibody. *Science* 321, 974–977.

67. Stein C, Kellner C, Kügler M, Reiff N, Mentz K, Schwenkert M, et al. (2010) Novel conjugates of single-chain Fv antibody fragments specific for stem cell antigen CD123 mediate potent death of acute myeloid leukaemia cells. *Br. J. Haematol.* 148, 879–889.

68. Schubert I, Kellner C, Stein C, Kügler M, Schwenkert M, Saul D, et al. (2011) A single-chain triplebody with specificity for CD19 and CD33 mediates effective lysis of mixed lineage leukemia cells by dual targeting. MAbs, 1, 3 [Epub ahead of print].

69. Guettinger Y, Barbin K, Peipp M, Bruenke J, Dechant M, Horner H, et al. (2010) A recombinant bispecific single-chain fragmentr variable specific for HLA class II and Fc alpha RI (CD89) recruits polymorphonuclear neutrophils for efficient lysis of malignant B lymphoid cells. *J. Immunol.* 184, 1210–1217.

70. Le Gall F, Reusch U, Little M, Kipriyanov SM. (2004) Effect of linker sequences between the antibody variable domains on the formation, stability, and biological activity of a bispecific tandem diabody. *Protein Eng. Des. Sel.* 17, 357–366.

71. reusch U, Le Gall F, Hensel M, Moledenhauer G, Ho AD, Little M, et al. (2004) Effect of tetravalent bispecific CD19xCD3 recombinant antibody construct and CD28 costimulation on lysis of malignant B cells from patients with chronic lymphocytic leukemia by autologous T cells. *Int. J. Cancer* 112, 509–518.

72. Arndt MA, Krauss J, Kipriyanov SM, Pfreundschuh M, Little M. (1999) A bispecific diabody that mediates natural killer cell cytotoxicity against xenotransplanted human Hodgkin's tumors. *Blood* 94, 2562–2568.

73. Cochlovius B, Kipiriyanov SM, Stassar MJ, Schuhmacher J, Bener A, Moldenhauer G, et al. (2000) Cure of Burkitt's lymphoma in severe combined immunodeficiency mice by T cells, tetravalent CD3 × CD19 tandem diabody, and CD28 costimulation. *Cancer Res.* 60, 4336–4341.

74. Johnson S, Burke S, Huang L, Gorlatov S, Li H, Wang H, et al. (2010) Effector cell recruitment with novel Fv-based dual-affinity retargeting proteins leads to potent tumor cytolysis and *in vivo* B-cell depletion. *J. Mol. Biol.* 399, 436–449.

75. Veri MC, Burke S, Huang L, Li H, Gorlatov S, Tuaillon N, et al. (2010) Therapeutic control of B cell activation via recruitment of Fcgamma receptor IIb (CD32B) inhibitory function with a novel bispecific antibody scaffold. *Arthritis Rheum.* 62, 1933–1943.

76. Villani ME, Di Carli M, Donini M, Traversini G, Lico C, Franconi R, et al. (2008) Validation of a stable recombinant antibodies repertoire for the direct selection of functional intracellular reagents. *J. Immunol. Methods* 329, 11–20.

77. Hopp J, Hornig N, Zettlitz KA, Schwarz A, Fuss N, Müller D, et al. (2010) The effects of affinity and valency of an albumin-binding domain (ABD) on the half-life of a single-chain diabody-ABD fusion protein. *Protein Eng. Des. Sel.* 23, 827–834.

78. Noy R, Eppel M, Haus-Cohen M, Klechevsky E, Mekler O, Michaeli Y, et al. (2005) T-cell receptor-like antibodies: novel reagents for clinical cancer immunology and immunotherapy. *Expert Rev. Anticancer Ther.* 5, 523–536.

79. Drummond D. (2010) Drug Discovery and Development Week, Burlingame, CA, 2010 August 2–4.

80. Zhu X, Wang L, Liu R, Flutter B, Li S, Ding J, et al. (2010) COMBODY: one-domain antibody multimer with improved avidity. *Immunol. Cell Biol.* 88, 667–675.

81. Shusterman SS, London, WB, Gillies SD, Hank, JA, Voss, SD, Seeger, RC, et al. (2010) Antitumor activity of Hu14.18-IL2 in patients with relapsed/refractory neuroblastoma: a children's oncology group (COG) Phase II study. *J. Clin. Oncol.* October 4 [Epub ahead of print].

82. Johannsen M, Spitaleri G, Curigliano G, Roigas J, Weikert S, Kempkensteffen C, et al. (2010) *Eur. J. Cancer* 46, 2926–2935.

83. Robinson MK, Alpaugh RK, Borghaei H. (2010) Naptumomab estefenatox: a new immunoconjugate. *Expert Opin. Biol. Ther.* 10, 273–279.

84. Cheng JD, Babb JS, langer C, Aamdal S, Robert F, Engelhardt LR, et al. (2004) Individualized patient dosing in Phase I clinical trials: the role of escalation with overdose control in PNU-214936. *J. Clin. Oncol.* 22, 602–609.

85. Liddy N, Molloy PE, Bennett AD, Boulter JM, Jakobsen BK, Li Y. (2010) Production of a soluble disulfide-linked TCR in the cytoplasm of *Escherichia coli* trxB got mutants. *Mol. Biotechnol.* 45, 140–149.

86. Okazaki A, Shoji-Hosaka E, Nakamura K, Wakitani M, Uchida K, kakita S, et al. (2004) Fucose depletion from human IgG1 oligosaccharide enhances binding enthalpy and association rate between IgG1 and FcgammaRIIIa. *J. Mol. Biol.* 336, 1239–1249.

87. Bugelski PJ, Achuthanadam R, Capocascale RJ, Treacy G, Bouman-Thio E. (2009) Monoclonal antibody-induced cytokine-release syndrome. *Expert Rev. Clin. Immunol.* 5, 499–521.

88. Dohlsten M, Hedlund G, Kalland T. (1991) Staphylococcal-enterotoxin-dependent cell-mediated cytotoxicity. *Immunol. Today* 12, 147–150.

89. Lutterbuese R, Raum T, Kischel R, Hoffmann P, Mangold S, Rattel B, et al. (2010) T cell-engaging BiTE antibodies specific for EGFR potently eliminate KRAS- and BRAF-mutated colorectal cancer cells. *Proc. Natl. Acad. Sci. USA* 107, 12605–12610.

90. Purbhoo MA, Irvine DJ, Huppa JB, Davis MM. (2004) T cell killing does not require the formation of a stable mature immunological synapse. *Nat. Immunol.* 5, 524–530.

36

NOVEL APPLICATIONS OF BISPECIFIC DART® PROTEINS

Syd Johnson, Bhaswati Barat, Hua W. Li, Ralph F. Alderson, Paul A. Moore, and Ezio Bonvini

MacroGenics, Inc., Rockville, MD, USA

36.1 INTRODUCTION

Bispecific antibodies, in the simplest sense, combine two different naturally or synthetically developed antigen-binding domains in a nonnatural manner in order to create a single molecule that can bind two different targets or two distinct epitopes on the same target. Such recombinant proteins can provide novel solutions for treating various diseases.

The most fully developed use of bispecific antibodies is for the redirected killing of tumors cells by immune effector cells. Early examples that entered clinical trials coupled a tumor-targeting antibody arm with an arm recognizing either CD3 on T cells [1], CD16 on natural killer (NK) cells or macrophages, [2,3] or CD64 on macrophages [4]. These early bispecific antibodies were generally derived by fusing two hybridomas forming a "quadroma" cell line that expressed the heavy (H) and light (L) chain of each individual monoclonal antibody (mAb) [5]. The desired bispecific molecule then could be purified from the multiple HL × HL combinations that were produced. Another nonrecombinant method chemically conjugated the Fab′ portion of one mAb to the Fab′ of a second mAb by coupling the fragments at their hinge region to obtain a bispecific F(ab′)2 molecule [6,7].

The dual targeting concept in oncology has also been explored for regulating cell processes and increasing target specificity. Bispecific antibodies have been developed that bind to two different targets on the same cell to simultaneously modulate two different signaling pathways (e.g., EGFR and IGF-1R [8] or two distinct receptors for the same ligand family (e.g., VEGFRI and VEGFR2) [9] or to enhance the specificity and avidity for targeted tumor cells (e.g., HER2 and HER3) [9,10]. For these approaches, effector cell recruitment and function may or may not be desirable, depending on whether the antibody exerts direct inhibitory effects, whether an Fc-mediated engagement further contribute to this inhibitory activity, or effector functions such as antibody-dependent cellular cytotoxicity (ADCC) are beneficial.

There are two general strategies for generating recombinant bispecific antibodies: fragment-based or immunoglobulin-like (Ig-like). Fragment-based strategies often employ one or more single-chain variable domains (scFv), [11,12] where a 12-16 amino acid linker tethers each respective V_L-V_H (or V_H-V_L) pair and a second linker links two scFv units on the same molecule. An alternate arrangement referred to as a diabody [13], uses shorter linkers between V_L and V_H (or vice-versa) such that the scFv domain cannot be formed but instead the interaction with a second chain creates a dimeric protein with two variable domains. In the latter case, the Fv from two respective mAbs A and B (V_L-A + V_H-B and V_L-B +

Fusion Protein Technologies for Biopharmaceuticals: Applications and Challenges, First Edition. Edited by Stefan R. Schmidt.

V_H-A) can either be expressed as separate chains (diabody) or combined in a single polypeptide that folds back on itself (single-chain diabody) [12,14]. With appropriate linker selection, one can also drive the latter to a dimeric tetravalent structure (tandem diabody) [15,16].

Bispecific Ig-like structures, that are monovalent for each antigen-binding domain, can also be constructed recombinantly in a number of ways. Several recombinant strategies that promote heterodimer formation and deter homodimerization of the two Fc-bearing chains have been developed by modifying interface sequences in CH3 of the respective chains. These include (1) a knobs-into-holes strategy that creates a protuberance on one strand by replacing small amino acid side chains with larger one, and a structurally complementary cavity on the other strand by replacing large side chains with smaller ones; (2) strand-exchange engineered domain (SEED) of CH3 composed of alternating segments of human IgA and IgG sequences; and (3) reversing the electrostatic interactions on the respective CH3 sequences [17–19] by swapping acidic for basic amino acids and vice-versa. However, it remains difficult to preferentially assemble different antigen-binding domains on opposite sides of the molecule in a homogeneous manner. This can be achieved by selecting two antibodies that use a common light chain; by expressing a scFv on one chain and a normal Fab on the other; or by a "cross-over" strategy in the Fab region of one arm. One "cross-over" approach is to replace the Cκ of one antibody with CH1 on the light chain and likewise the CH1 with the Cκ on the heavy chain [20].

Ig-like structures that are bivalent for each antigen can simply involve a scFv appended to the C-terminus of an IgG, but also can be made by fusing a second set of variable regions onto the existing variable regions (V_L-B onto V_L-A and V_H-B onto V_H-A) of the respective IgG light and heavy chains with flexible intervening linker sequences, yielding a dual-variable domain Ig (DVD-Ig) [21].

Bivalent bispecific antibodies can also be generated by linking two independent target-binding peptides to a scaffold antibody through chemical conjugation. One such bispecific antibody, designed to block angiogenesis through combined inhibition of VEGF and Ang2 is currently undergoing clinical evaluation [22].

36.2 DART® PROTEINS

Fragment-based and Ig-like strategies often both rely on noncovalently linked scFv or diabody structures to maintain the integrity of the binding unit. As the association between the V_L and V_H is quite variable, and there is always equilibrium between the associated and nonassociated forms, the stability of these molecules in solution and under conditions of use have been highly unpredictable. This is probably the major reason that bispecific antibodies have not, as a class, moved further in the clinic and product development.

The novel bispecific antibody platform called dual-affinity retargeting (DART®) is based on covalently linking Fv regions of two distinct antibodies. DART® proteins arrange the V_H and V_L domains specific for different antigens in a bispecific diabody arrangement with the order V_L(A)-V_H(B) and V_L(B)-V_H(A). The association of the two chains is stabilized through covalent carboxy-terminal disulfide linkage. DART® proteins have been focused mainly on two areas: (1) recruiting inhibitory receptors to activating receptors on the same cell to negatively modulate cell activation [23] and (2) recruiting effector cells to tumor cell targets for redirected cytolysis of the tumor cells [24,25].

DART® proteins are easily expressed in mammalian systems at high levels. During secretion, the two chains associate and become disulfide bonded in an intracellular compartment to form the heterodimeric bispecific unit. Preferential formation of the heterodimeric form can be promoted by the introduction of native Ig sequences upstream from the hinge of the Cκ and CH1 regions [24] or by the use of oppositely charged coil sequences on the respective chains [26]. These "E-coil" and "K-coil" sequences strongly favor heterodimer over homodimer formation, greatly simplifying downstream processing by minimizing the formation of unwanted homodimeric species.

For downstream processing of DART® proteins, both protein A and nonprotein A based schemes have been successfully developed. The former requires that one or both of the Fv sequences in the DART® be able to bind to protein A through the use of human V_H3 family sequence(s) (or an antibody sequence from another species, determined empirically to bind protein A) [27]. While not absolutely required, protein A binding is preferable in that the downstream steps can be adapted from established antibody purification processes that are standard in the industry. Nonprotein A based purification of DART® proteins for research use can be accomplished using an affinity matrix composed of one of the target antigens immobilized to Sepharose, or using a specific antibody that recognizes the associated E-coil/K-coil structure.

The remainder of this chapter focuses on the application of DART® technology for oncology and autoimmune diseases and introduces a Universal DART® (U-DART) platform; a technique for rapid screening of monoclonal antibodies to prioritize them for subsequent development into DART® molecules.

36.3 APPLICATION OF DART® TO CROSS-LINK INHIBITORY AND ACTIVATING RECEPTORS

Bispecific antibody-based approaches to autoimmune disease control have largely aimed at dual blockade of potential disease mediators. For instance, simultaneous blockade of the proinflammatory cytokines interleukin 1α and 1β (IL-1α, IL-1β) in a mouse collagen-induced arthritis (CIA) model of rheumatoid arthritis yielded positive results using a bispecific tetravalent IgG approach [21]. An alternative to dual mediator blockade is interference with signaling pathways by engaging negative regulatory loops. One such approach employs DART® molecules to interfere with B-cell activation by inducing inhibitory signaling via the Fc receptor, FcγRIIB (CD32B) [23].

The physiological role of inhibitory receptors such as CD32B is to dampen or raise the threshold for cellular activation. For example, CD32B engagement downmodulates activation initiated through the B-cell receptor complex (BCR) on B cells; through FcξRI on mast cells; and through FcRI, FcRIIA, or FcRIIIA on macrophages. In each case, the antigen itself serves as a scaffold for the formation of an immune complex linking the stimulatory and inhibitory receptors. When this occurs, the receptor tyrosine-based inhibitory motif (ITIM) of CD32B is phosphorylated by kinase(s) associated with the stimulatory receptor, initiating the inhibitory cascade, and abrogating activation and downstream effects. We used this well-established paradigm to demonstrate the applicability of the DART® platform to coligate the inhibitory and stimulatory receptors in the absence of antigen on human B cells.

CD32B is a component of an inhibitory receptor loop triggered when coligated with the BCR complex. A DART® protein designed to modulate B-cell function was constructed by pairing an Fv specific for CD32B with an Fv specific for the Ig-β component (CD79B) of the human BCR complex [23]. Two different DART® proteins that recognize different epitopes on CD32B and have the same specificity for CD79B were expressed in Chinese hamster ovary (CHO) cells and purified to homogeneity using antigen-affinity and size-exclusion chromatography (SEC). Bispecific enzyme-linked immunosorbent assays (ELISA) and surface plasmon resonance (SPR) analyses demonstrated that both of the arms could bind to both target antigens simultaneously and retained the affinity of the parent Fab fragments.

As reported [23], both DART® proteins specific for CD32B and CD79B demonstrated preferential simultaneous binding to both the activating and inhibitory receptors on the surface of the same B cells as evidenced by fluorescence-activated cell sorting (FACS) studies. The coligation of receptors decreased BCR-induced calcium mobilization and phosphorylation of selected signaling molecules, which in turn inhibited both B-cell proliferation and IgG secretion *in vitro*. Consistent with this mechanism of action, a DART® protein specific for the mouse orthologs of CD32B and CD79B reduced disease severity in a mouse model of collagen-induced inflammatory arthritis. This novel approach to B-cell function modulation forms the basis for translating inhibitory signaling into therapeutic tools for treating autoimmune and inflammatory diseases.

36.4 APPLICATION OF BISPECIFIC ANTIBODIES IN ONCOLOGY

The development of bispecific antibodies for recruiting immune effector cells to kill tumor cells has long been explored as a potential therapeutic agent. However, inefficient production and poor stability have prevented wide use. An early example is the use of humanized Fab anti-CD64 × humanized Fab anti-EGFR to engage cytotoxic immune cells other than T cells. This bispecific antibody was intended for CD64-directed immunotherapy in cancer patients, but failed in clinical trials [28]. Greater success has been achieved using bispecific antibodies to CD16 on NK and macrophages and CD3 on T cells [29–31]. More recently, blinatumomab (MT103), a bispecific T-cell engager (BiTE®) specific for CD19 and CD3, has successfully been used to treat patients with non-Hodgkin's lymphoma by redirecting cytotoxic T cells to kill tumor cells [32].

36.4.1 CD16-Based DART® Proteins

We recently reported on the application of DART® proteins for redirected killing, whereby an Fv specific for human CD16 (FcγRIII) on effector cells (e.g., NK cells) was coupled to an Fv specific for human CD32B, used here as a target on B cells [24]. DART® proteins were produced at high levels in mammalian cells and retained the binding activity of the respective parental Fv domains. Simultaneous binding of both antigens was evident from dual-affinity ELISA and SPR analysis. These proteins showed good stability upon extended storage of about 9 months in phosphate-buffered saline at 4°C and also in human serum at 37°C for up to 8 weeks.

Functionally, the DART® proteins demonstrated extremely potent, dose-dependent cytotoxicity in retargeting human peripheral blood mononuclear cells (PBMC) against CD32B-positive B-lymphoma cell lines. When cultured with human PBMC these proteins also mediated autologous B-cell depletion, measured as the residual fraction of CD20-positive B cells. The potency of these DART® molecules in ADCC assays was ~100-fold greater than that of the corresponding IgG bearing an Fc that was optimized for enhanced CD16 binding.

In vivo, the DART® proteins effectively depleted B cells in mice transgenic for both CD16A on the effector cells and CD32B on the B-cell targets. No depletion was observed in either hCD16A or hCD32B single transgenic animals, indicating that the presence of both receptors was required for the observed effect. Furthermore, DART® proteins showed potent *in vivo* antitumor activity in a mouse xenograft model using Raji cells, a human Burkitt's lymphoma B-cell line expressing CD32B antigen.

36.4.2 CD3/TCR-Based DART® Proteins

Given the potent activity of bispecific proteins that recruit T cells as effector cells, we developed DART® proteins combining either anti-CD3ε or anti-TCRβ (T-cell receptor β) with anti-CD19 as a B-cell targeting specificity [25]. Both of these "T-DART®" proteins exhibited extremely potent redirected T-cell mediated lysis of B-cell targets, either in an autologous setting or using B-cell tumor lines as targets in a cytotoxic T-lymphocyte assay. Moreover, the potency of these agents was greater than that achieved by a dual-scFv bispecific protein derived from identical anti-CD3 and anti-CD19 Fv sequences and differing only in the bispecific assembly strategy utilized. The CD19 × TCR T-DART® protein was further shown to be active *in vivo* in a human PBMC-reconstituted xenograft mouse model against Raji (lymphoblastoid) cell targets.

We have also constructed several T-DART® proteins against solid tumor targets including HER2. The HER2 × TCR DART® was made by combining the 4D5 Fv (the parent of trastuzumab, a mAb that interferes with the HER2/neu receptor) with the humanized anti-TCR Fv. The DART® protein was purified to homogeneity using affinity and SEC (Figure 36.1A). Both chains of DART®

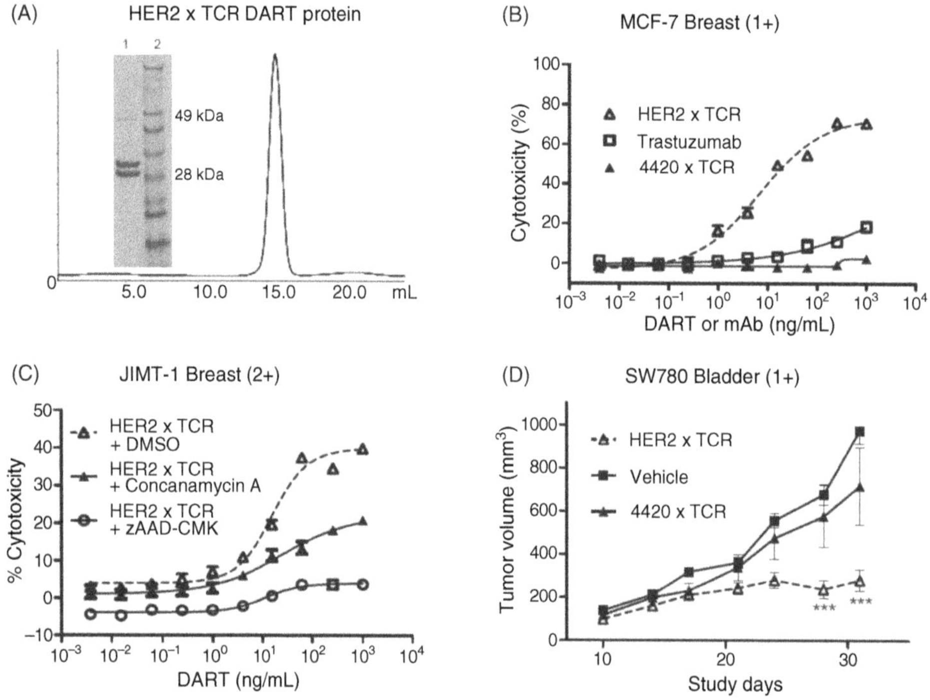

FIGURE 36.1 Anti-HER2 T-DART® for redirected T-cell killing against tumor cells. (A) SDS-PAGE analysis of purified DART® protein under reducing conditions. Lane 1, HER2 × TCR DART® and lane 2, molecular mass markers. Graph: Size exclusion chromatography performed on a Superdex 200HR 10/30 column demonstrates that under native conditions, a typical DART® protein forms a single peak with a retention time similar to that expected for an antibody Fab fragment. (B) Cytotoxicity of DART® protein against HER2 positive breast cancer cell lines. Antibody-dependent cellular cytotoxicity (ADCC) assays were performed using MCF-7 breast cancer cells with HercepTest score of 1+ cancer cell lines as target cells incubated with PBMC as effector cells at an effector-to-target ratio of 30:1. Assays were performed in triplicate at each concentration with HER2 × TCR DART®, trastuzumab and 4420 × TCR DART. Cell-mediated cytotoxicity was quantified by measuring lactate dehydrogenase (LDH) released into the culture medium. Curves were fit using nonlinear regression analysis. Error bars represent SD. (C) The DART® protein functions through the Granzyme B/Perforin pathway. ADCC assays were performed using JIMT-1 breast cancer cells with HercepTest score of 2+ cancer cell lines as target cells incubated with PBMC at an effector-to-target ratio of 30:1. Cytotoxicity of HER2 × TCR DART® measured by LDH release was evaluated in the presence of either DMSO, or Z-AAD-CMK (20 nM) or concanamycin A (1 mM). Error bars represent SD. (D) *In vivo* antitumor activity of DART® against HER2 positive (HercepTest score 1+) bladder cancer xenografts. NOD-SCID mice ($N = 8$ per treatment group) were implanted subcutaneously with a 1:1 ratio of 2.5×10^6 SW780 cells and human PBMC (5×10^6) that had been activated overnight with IL-2 (10 μg/mL) and mixed with matrigel. Mice were then treated with vehicle or the indicated DART® molecules (15 μg/animal) i.v. on days 0–4. Tumor volume was measured externally at each time point. Error bars represent SD. $^{***}P < 0.001$.

migrated around 29 kDa, similar to that expected for an antibody Fab fragment.

This HER2 × TCR DART® protein potently redirected cytotoxicity against HER2 positive target cells. Moreover, it retained high level of activity even against targets that expressed low surface HER2 density, with a HercepTest score as low as 1+ (Figure 36.1B, C), that were ineffectively lysed by trastuzumab. Addition of the granzyme B inhibitor, zAAD-CMK, or the perforin inhibitor, concanamycin A, blocked the cell-killing activity, demonstrating that this DART® functions through the granzyme B/perforin pathway (Figure 36.1C). This DART® protein also effectively limited establishment of HER2+ tumor xenografts in human PBMC-reconstituted NOD/SCID mice (Figure 36.1D).

36.5 U-DART CONCEPT FOR SCREENING DART® CANDIDATE TARGETS AND MABS

Given the applicability of the redirected killing approach to multiple cancer targets, a rapid screening method was needed in order to be able to evaluate antibodies "off the shelf" for use in the bispecific DART® format before embarking on more extensive lead generation, humanization, and engineering efforts.

For this purpose, we constructed a DART® protein whereby one arm targets a relevant molecule on the desired effector cell population (e.g., CD3, TCR, CD16) and the other arm is an Fv specific for fluorescein (derived from the 4-4-20 mAb). These proteins are termed "Universal DART®" or U-DART. In a U-DART-based screening process, the candidate antibody is simply conjugated with fluorescein isothiocyanate (FITC) and the fluoresceinated mAb then combined with the U-DART and evaluated for its intended biological effect. Hundreds of mAbs to known or unknown targets can be screened rapidly using this approach. The screen can serve multiple purposes, including prioritizing mAbs against known targets for conversion to T-DART® and clinical development as well as screening of mAbs against unknown targets for selection for use in target identification. This screening platform works with both monoclonal and polyclonal antibodies from commercial and proprietary sources.

As an example of this approach, a panel of 34 fluoresceinated mAbs was assessed for their ability to redirect T-cell lysis of tumor cells using a 4-4-20 × TCR U-DART (Figure 36.2A). These 34 murine mAbs were all reactive against a proprietary target and derived from numerous different hybridoma campaigns in which mice were immunized with human fetal or tumor-derived stem-like cell lines. The purified mAbs were conjugated with fluorescein (FL) to a similar degree and then combined with the 4-4-20 × TCR U-DART at a fixed concentration of each and used in a cytotoxic T-cell lymphocyte assay using A498 renal cell carcinoma as target cells and human PBMC as effector cells. The result of this screen, shown in Figure 36.2A, indicated that many of the fluoresceinated mAbs were able to redirect lysis of A498 cells in the presence of TCR U-DART, suggesting that the target was amenable for the development of T-DART® candidates. Furthermore, this screening process provided a preliminary ranking of relative potency. No activation or killing by the U-DART with effector cells was observed in the absence of a fluoresceinated partner mAb. In contrast, the mouse mAbs, being mostly IgG1, showed little to no ability to mediate ADCC. A few IgG2a mAbs showed limited activity that was Fc domain dependent (hence, relied on FcR-dependent effector mechanisms) and thus did not reflect T-cell mediated killing, the mechanism of action desired for T-DART® development. U-DART studies were subsequently expanded to other tumor cell lines expressing the target of interest.

Together with immunohistochemistry on normal and tumor tissue specimens, flow cytometry, epitope binning, and affinity analyses, the U-DART screens helped in selecting a subset of eight mAbs for further analysis. U-DART dose titration studies revealed that mAb #2 and mAb #6 consistently performed the best both in terms of EC_{50} and maximum lysis (Figure 36.2B). Consistent with the U-DART screen, a T-DART® candidate constructed with a humanized version of mAb #6 potently mediated lysis of target cells (Figure 36.2C). Thus, the U-DART screen contributes a relevant functional evaluation of large mAb panels at an early stage, assisting, together with other parameters, in identifying leads for further development.

A similar mAb screening process was also developed using 4-4-20 × CD16 U-DART for the functional screen of Fc-receptor (e.g., NK-cell) mediated redirected killing of tumor cells, and can be used to evaluate candidate mAbs for use in either a CD16-based DART® or alternatively an IgG antibody product with an Fc enhanced for CD16A binding.

36.6 U-DART CONCEPT FOR APPLICATIONS IN AUTOIMMUNE AND INFLAMMATORY DISEASE

As outlined earlier, one approach to control inflammatory diseases is based on the hypothesis that activation and inhibitory receptors on the surface of the same cell can be cross-linked using bispecific DART® proteins in order to dampen the responsiveness of that cell to activation and the amplification of downstream effects, such as cytokine releases, antibody production, or histamine release, depending on the cell type and application. The CD32B × CD79B DART® protein capable of cross-linking the inhibitory CD32B receptor and the activating CD79B component of

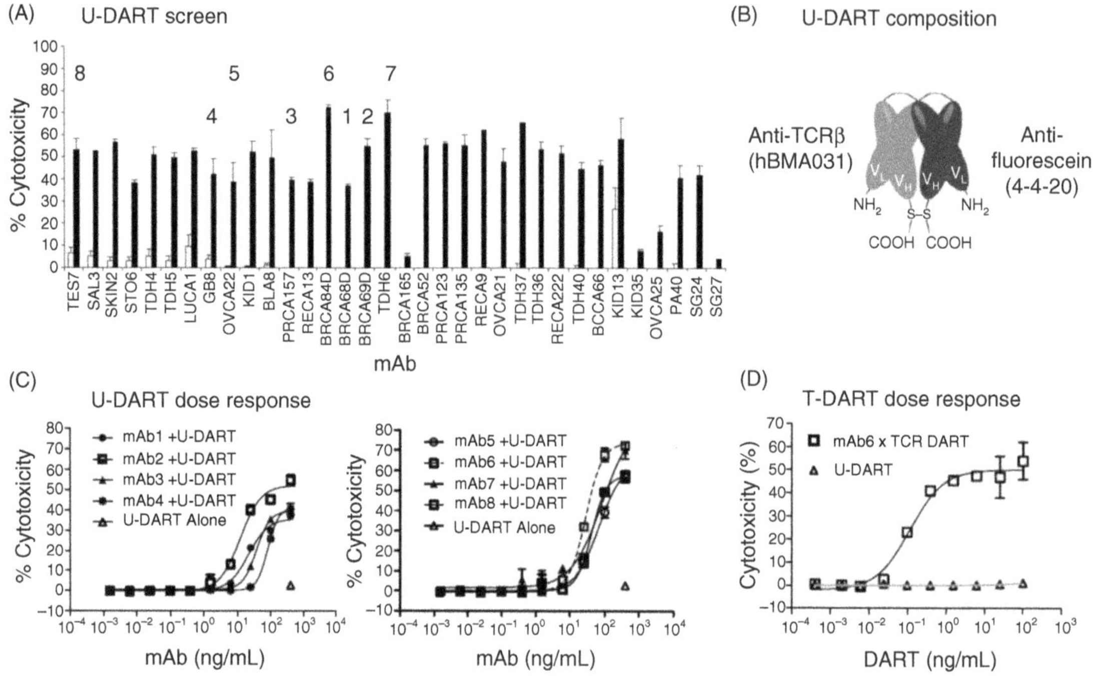

FIGURE 36.2 Application of U-DART® to screen DART® lead candidate among mAb panel. (A) U-DART® screening of 34 mAbs against A498 renal cancer target cells incubated with resting human PBMC effector cells (effector to target ratio = 25:1). Each fluorescein labeled mAb (100 ng/mL) was mixed with (solid bar) and without (open bar) an antifluorescein (4420) × TCR DART® protein (200 ng/mL) and added to the same 96-well culture plate. Cell-mediated cytotoxicity was quantified by measuring lactate dehydrogenase release. Error bars represent SD. mAbs 1–8 were selected for further screening. (B) Schematic representation of U-DART® molecule. The 4420 × TCR DART® consists of two covalently linked polypeptide chains that assemble to generate two specificities, one against TCRβ and the other against fluorescein. (C) U-DART® dose response against renal cancer cells. A498 renal cancer target cells were incubated with resting human PBMC effector cells at an effector to target ratio of 20:1. Each fluorescein labeled antibody (400 ng/mL) was mixed with 4420 × TCR DART® (800 ng/mL) and followed by 1:4 serial dilution and individual assays were done in duplicate. Cytotoxicity was quantified by measuring lactate dehydrogenase (LDH) release. Dose response curve of each antibody/U-DART® mixture was processed with GraphPad Prism software. (D) T-cell mediated cytotoxicity of mAb6 × TCR DART®. ADCC assays were performed using A498 cells with PBMC at an effector to target ratio of 20:1. Individual assays were done in duplicate and with a titration of antibodies: mAb6 × TCR DART® and U-DART®. Cytotoxicity was quantified by measuring LDH release.

the human BCR complex is an example of such approach aimed at inhibiting B-cell activation [23]. We termed such bispecific DARTs "immunomodulatory DARTs" or I-DART® (Figure 36.3).

A U-DART format was developed for screening of I-DART® candidates using an approach similar to that employed for redirected effector cell killing. In this U-DART® format, either component arm can be fixed (i.e., the CD32B or the CD79B arms of a B-cell targeted I-DART), and fluoresceinated mAb(s) representing candidates for the other arm can be evaluated. A test of this functional screening format was carried out using the CD32B and CD79B specificities in reciprocal U-DART formats. In the first format, the anti-CD79B Fv was used as the fixed arm (4420 × hCB3.1; Figure 36.4A—left panel) in a U-DART that can be combined with a fluoresceinated anti-CD32B mAb (CD32B mAb-FITC) to support bispecific engagement and downregulation of B cells. Conversely, the anti-CD32B Fv can be kept fixed resulting in a U-DART (4420 × h2B6; Figure 36.4A—center panel) that can be combined with a fluoresceinated anti-BCR mAb to downregulate B-cell activation. The U-DART proteins were expressed in eukaryotic cells (293-H or CHO-S), affinity purified and further purified by SEC. Simultaneous binding of both arms of 4420 × hCB3.1 U-DART were demonstrated by bispecific ELISA (Figure 36.4B). In this case, a soluble single-chain form of the human CD79A/B ectodomain was used to capture the DART® and detection was accomplished

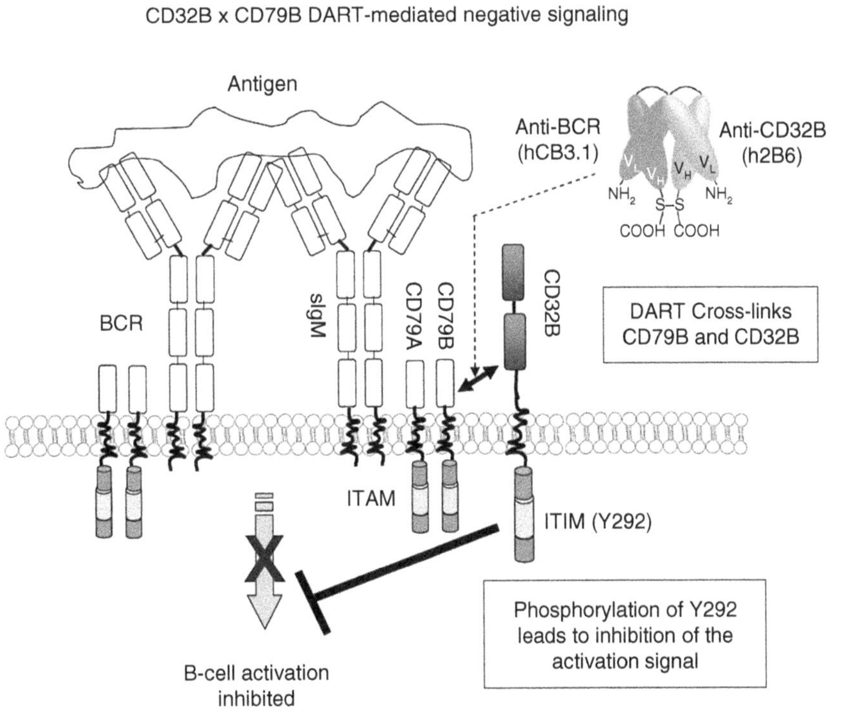

FIGURE 36.3 CD32B × CD79B DART® mediated negative signaling. In B-lymphocytes, recognition of an antigen by the B-cell receptor (BCR) induces a signal that can direct clonal expansion. CD79B of the BCR contains an immunoreceptor tyrosine-based activation motif (ITAM) in the cytoplasmic domain. Recruitment of the inhibitory receptor FcγRII (CD32B) to the BCR signaling complex by antigen-bound antibodies results in phosphorylation of tyrosine residue (Y292) of the immunoreceptor tyrosine-based inhibitory motif (ITIM) of CD32B and activation of the inhibitory pathway. Following this paradigm, the CD32B × CD79B DART® inhibits B-cell activation by directly cross-linking CD32B to the CD79B (Igβ) subunit of the BCR.

using FL-labeled alkaline phosphatase. 4420-hCB3.1 DART® was readily detected under these conditions, while the negative control, 4420 × 2B6 U-DART, was not captured and gave no signal.

Activation of human B cells isolated from PBMC by negative selection was initiated using anti-μ mAb cross-linked with F(ab)'2 of goat anti mouse IgG or using goat anti-μ polyclonal antibody. Activation was monitored using incorporation of 3H-thymidine, calcium flux, and IgG secretion. The results show that U-DART molecules in combination with the appropriate fluoresceinated partners were able to effectively inhibit proliferation in a dose-dependent manner (Figure 36.4C, D). Specifically, inclusion of the 4420 × hCB3.1 DART® in combination with fluoresceinated CD32B mAb resulted in an ~70% decrease in the amount of 3H-thymidine incorporation (Figure 36.4C). The aglycosylated N297Q form of the antibody was utilized in order to avoid FcRn engagement of antibody. Neither an irrelevant fluoresceinated mAb nor nonfluoresceinated CD32 mAb inhibited B-cell proliferation, indicating that specific binding of a fluoresceinated mAb at the site of activation was required for DART® inhibition. Interestingly, the 4420 × hCB3.1 U-DART also mediated B-cell inhibition when combined with a fluoresceinated anti-CD22 mAb demonstrating that alternative inhibitory receptors expressed on human B cells can be utilized as long as they are appropriately recruited to the activation complex. CD22 has multiple ITIMs and has been shown to regulate B-cell activation; its biological role, however, is less well understood than that of CD32B. Similar inhibition of B-cell proliferation was observed when the 4420 × 2B6 U-DART was used in combination with hCB3.1-agly-FITC (Figure 36.4D). The level of B-cell inhibition mediated by U-DART® approach was also consistent with the level mediated by the CD32BxCD79B I-DART® molecule itself supporting the predictive value of the U-DART approach (Figure 36.4D—upper panels). Recapitulation of the U-DART® activity in the I-DART® format was also reflected in the comparable inhibition of IgG secretion mediated by the combination of 4420 × 2B6 U-DART and

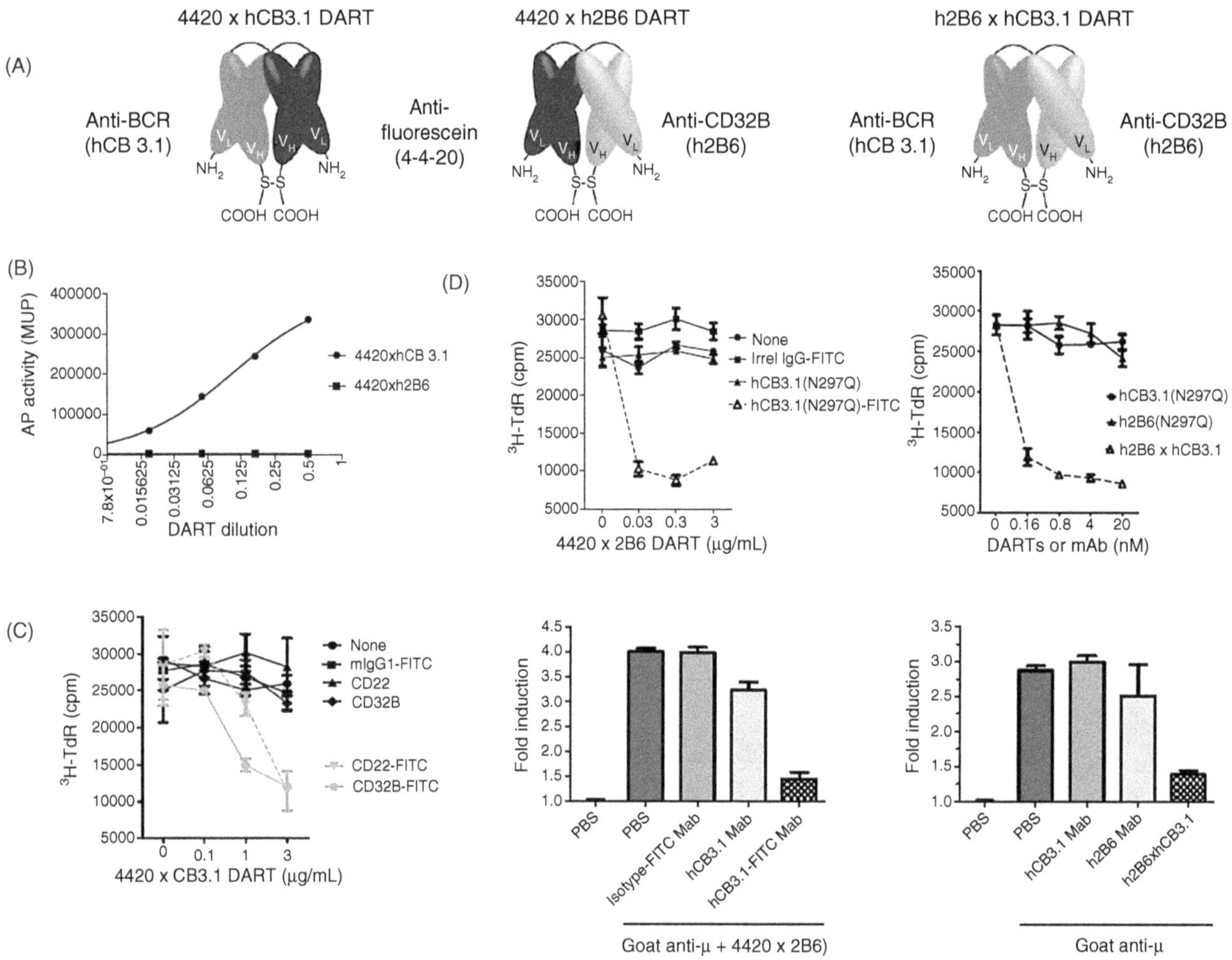

FIGURE 36.4 CD32B × CD79b cross-linking by dual affinity retargeting molecules activates the inhibitory pathway in human B cells. (A) Schematic representation of DART® molecules. (B) Dual-affinity ELISA assay demonstrates that for typical DART, both arms can simultaneously engage their respective targets. ELISA plates were coated with soluble human CD79a/b ectodomain followed by 4420 × hCB3.1 or 4420 × h2B6 U-DART® protein. (C) B-cell proliferation in the presence of 4420 × hCB3.1 U-DART. Purified human B cells activated with polyclonal F(ab)′2 fragment of goat anti-human IgM antibody (50 μg/mL) were treated with varying concentrations of 4420 × hCB3.1 DART® in the presence of a fixed concentration (3 μg/mL) of fluorescein-labeled anti-CD32B (2B6), anti-CD22 (HIB22), and isotype control mAbs or with unlabeled anti-CD32B and anti-CD22 mAbs. Following 48 h of culture, cell proliferation was assessed by incorporation of radiolabeled thymidine (1 μCi/well) added during the last 16 h of culture. (D) B-cell proliferation and Ig secretion mediated by 4420 × hCB3.1 U-DART® compared to h2B6 × hCB3.1 DART. Upper panels: activated human B cells were treated with varying concentration of 4420 × 2B6 U-DART® in combination with indicated mAbs (left panel) or to h2B6 × chCB3.1 DART, chCB3.1(N297Q) and h2B6(N297Q) mAbs alone (right panel) with cell proliferation assessed by thymidine incorporation following 48 h culture. Lower panels: activated human B cells were treated with 20 nM of 4420 × 2B6 U-DART in the presence of anti-CD79B Mab, anti-CD79b-FITC Mab, or an irrelevant FITC-labeled antibody of the same isotype (left panel) or to 20 nM of h2B6 × chCB3.1 DART®, anti-human CD79b mAb, or anti-human CD32B mAb (right panel). Supernatants were tested in ELISA using plates coated with polyclonal anti-human Ig Ab to capture secreted immunoglobulins by activated B cells. Bound human immunoglobulins were detected using a polyclonal anti-human κ-chain HRP-conjugated antibody.

chCB3.1-FITC mAb with that mediated by the h2B6 × hCB3.1 I-DART® (Figure 36.4D—lower panels).

36.6.1 Inhibition of Basophil Degranulation

To further demonstrate the versatility of the 4420 × CD32B U-DART, we also evaluated the ability of this DART® to inhibit granule release from basophils. In this case, we used RBL cells transfected with either wild-type human CD32B (RBL/CD32B) or a version mutated within the inhibitory ITIM motif (RBL/CD32B-Y292A). RBL/CD32B cells were activated using chimeric anti-DNP IgE and DNP-BSA or DNP-BSA-FITC together with DART. Significant, dose-dependent inhibitory activity was seen only when 4420 × CD32B U-DART and DNP-BSA-FITC were used (Figure 36.5), but not with DNP-BSA (nonfluoresceinated) (data not shown). As expected no inhibition was observed in RBL/CD32B-Y292A cells (Figure 36.5), although expression of the CD32B on the mutant cell line was similar to that of the wild-type transfectant (data not shown). In terms of specific signaling, increased phosphorylation of the CD32B ITIM was observed in those conditions that led to inhibition of hexaminidase release, consistent with its role in initiating the inhibitory cascade (Figure 36.5A).

In conclusion, the use of the antifluorescein Fv from the mAb 4-4-20 as one partner in several of the DART® molecules, in combination with novel or commercially available fluoresceinated mAbs, allows for the flexibility to interrogate these different modes of inhibition. The anti-hapten Fv was linked either to the Fv specific for the inhibitory receptor CD32B or to the Fv specific for the activating receptor, in this case CD79B. Such a DART®

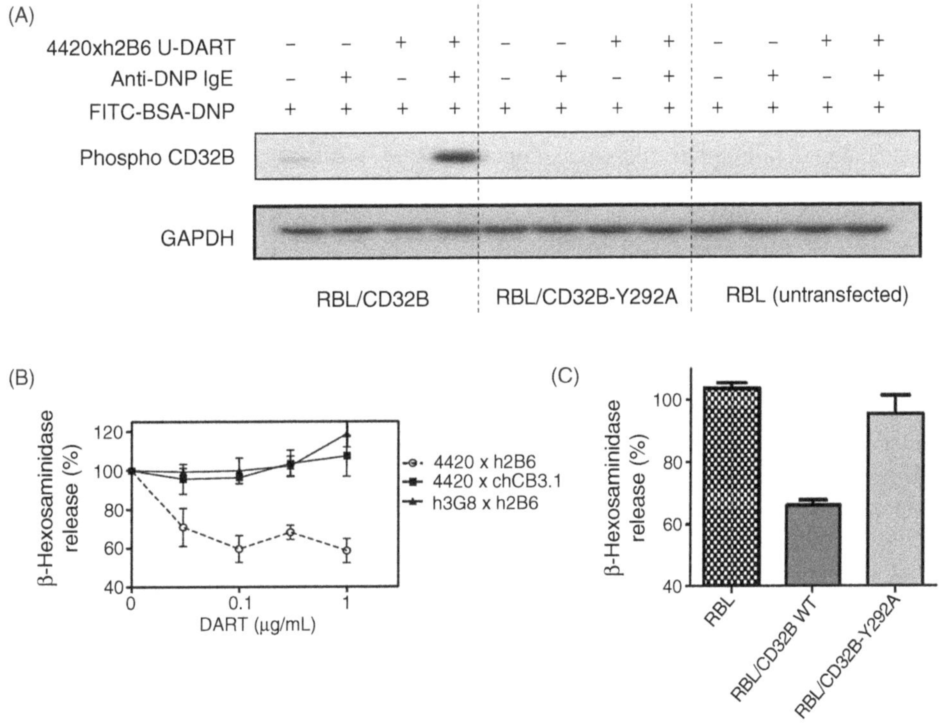

FIGURE 36.5 Receptor coligation activates the inhibitory signaling pathway. Rat basophilic leukemia RBL-2H3 (RBL) cells, RBL/CD32B (RBL cells stably expressing human CD32B) cells, and RBL/CD32B/Y292A (RBL cells stably expressing a nonsignaling form of human CD32B) were cultured in RPMI complete media. (A) Western blot analysis of whole cell lysate (1×10^6 cells) obtained from different types of purified RBL cells that were activated alone or in combination with anti-DNP IgE, FITC-BSA-DNP (0.5 μg/mL), and 4420 × 2B6 U-DART® (1 μg/mL). The blots were probed with anti-pCD32B. Equivalent loading was confirmed by stripping and reprobing with anti-GAPDH. (B) 2×10^4 cells/well were grown for overnight in 96-well plates in the presence of anti-DNP IgE. The following day all RBL/CD32B cells were activated with FITC-BSA-DNP in the presence of either 4420 × 2B6 U-DART (○) or 4420 × chCB3.1 DART® (■) or CD16 × CD32B DART® (▲). Finally supernatants were transferred to microtiter plates and β-hexosaminidase release was measured. (C) DART® mediated inhibition of β-hexoaminidase release. After overnight treatment with anti-DNP IgE, different RBL cell lines were activated with FITC-BSA-DNP in the presence of 1 μg/mL of 4420 × 2B6 U-DART.

when combined with the appropriate fluoresceinated mAb was equivalent in its *in vitro* biological activity to a DART® directly linking CD32B and CD79B.

36.7 CONCLUSIONS AND FUTURE PERSPECTIVES

DART® proteins are highly potent, stable, and manufacturable fragment-based bispecific antibodies. They are monovalent for each antigen; this is advantageous for redirected killing, which requires activation of the effector cell only upon engagement of the target cell. The DART® size of ~50 kDa, similar to a Fab fragment, is close to the threshold for renal clearance, and thus the half-life of these proteins should be in the same order of magnitude. It may be desirable in certain indications to extend the half-life so that at least weekly dosing can be contemplated in a clinical setting. This can be accomplished by fusion to an albumin-binding domain that has been deimmunized by removing potential T-cell helper epitopes, while retaining binding to serum albumin. Albumin binds to FcRn on a site separate from that of IgG, nonetheless resulting in similar recycling and reduced renal clearance.

In an alternate approach, the DART® structure can be combined in an Ig-like bispecific molecule, such that one or two of the DART® bispecific units is assembled with an Fc. This approach results in both increasing the molecular size and, more importantly, allowing binding to FcRn with ensuing recycling and extended half-life. In the case where two binding sites for each target are produced, complete retention of antigen binding has been observed compared to the parent mAbs. This structure has utility in applications where receptor agonism may be advantageous or is not a concern. In cases where agonism is a potential concern, structures monovalent for each antigen but containing an Fc have been expressed in several alternative configurations.

Just as the multiple applications of bispecific antibodies are now being realized through more robust configurations such as DART®, new possibilities both in research and for clinical application can be conceived. While challenges remain to their clinical use in a binary or "pretargeting" mode, the U-DART system has considerable utility in selecting the appropriate candidate antibodies for making the bispecific clinical product, be it a T-DART®, I-DART®, or Fc-engineered IgG.

REFERENCES

1. Seimetz D, Lindhofer H, Bokemeyer C. (2010) Development and approval of the trifunctional antibody catumaxomab (anti-EpCAM x anti-CD3) as a targeted cancer immunotherapy. *Cancer Treat. Rev.* 36(6), 458–467.
2. Weiner LM, Clark JI, Davey M, Li WS, Garcia de Palazzo I, Ring DB, et al. (1995) Phase I trial of 2B1, a bispecific monoclonal antibody targeting c-erbB-2 and Fc gamma RIII. *Cancer Res.* 55(20), 4586–4593.
3. Weiner LM, Alpaugh RK, Amoroso AR, Adams GP, Ring DB, Barth MW. (1996) Human neutrophil interactions of a bispecific monoclonal antibody targeting tumor and human Fc gamma RIII. *Cancer Immunol. Immunother.* 42(3), 141–150.
4. Pullarkat V, Deo Y, Link J, Spears L, Marty V, Curnow R, et al. (1999) A Phase I study of a HER2/neu bispecific antibody with granulocyte-colony-stimulating factor in patients with metastatic breast cancer that overexpresses HER2/neu. *Cancer Immunol. Immunother.* 48(1), 9–21.
5. Lanzavecchia A, Scheidegger D. (1987) The use of hybrid hybridomas to target human cytotoxic T lymphocytes. *Eur. J. Immunol.* 17(1), 105–111.
6. Tutt A, Stevenson GT, Glennie MJ. (1991) Trispecific F(ab')3 derivatives that use cooperative signaling via the TCR/CD3 complex and CD2 to activate and redirect resting cytotoxic T cells. *J. Immunol.* 147(1), 60–69.
7. Nitta T, Ishizawa A, Ito M, Sato K, Yagita H, Kumura K. (1990) [Induction of cytotoxicity from human lymphocytes coated with bispecific antibody against human glioma cells]. *No Shinkei Geka* 18(11), 1001–1006.
8. Lu D, Zhang H, Ludwig D, Persaud A, Jimenez X, Burtrum D, et al. (2004) Simultaneous blockade of both the epidermal growth factor receptor and the insulin-like growth factor receptor signaling pathways in cancer cells with a fully human recombinant bispecific antibody. *J. Biol. Chem.* 279(4), 2856–2865.
9. Lu D, Jimenez X, Zhang H, Wu Y, Bohlen P, Witte L, et al. (2001) Complete inhibition of vascular endothelial growth factor (VEGF) activities with a bifunctional diabody directed against both VEGF kinase receptors, fms-like tyrosine kinase receptor and kinase insert domain-containing receptor. *Cancer Res.* 61(19), 7002–7008.
10. Robinson MK, Hodge KM, Horak E, Sundberg AL, Russeva M, Shaller CC, et al. (2008) Targeting ErbB2 and ErbB3 with a bispecific single-chain Fv enhances targeting selectivity and induces a therapeutic effect *in vitro*. *Br. J. Cancer* 99 (9), 1415–1425.
11. Bird RE, Hardman KD, Jacobson JW, Johnson S, Kaufman BM, Lee SM, et al. (1988) Single-chain antigen-binding proteins. *Science* 242 (4877), 423–426.
12. Huston JS, Levinson D, Mudgett-Hunter M, Tai MS, Novotný J, Margolies MN, et al. (1988) Protein engineering of antibody binding sites: recovery of specific activity in an anti-digoxin single-chain Fv analogue produced in *Escherichia coli*. *Proc. Natl. Acad. Sci. USA* 85(16), 5879–5883.
13. Olafsen T, Cheung C-W, Yazaki PJ, Li L, Sundaresan G, Gambhir SS, et al. (2004) Covalent disulfide-linked anti-CEA diabody allows site-specific conjugation and radiolabeling for tumor targeting applications. *Protein Eng. Des. Sel.* 17(1), 21–27.
14. Holliger P, Prospero T, Winter G. (1993) "Diabodies": small bivalent and bispecific antibody fragments. *Proc. Natl. Acad. Sci. USA* 90(14), 6444–6448.

15. Arndt MA, Krauss J, Kipriyanov SM, Pfreundschuh M, Little M. (1999) A bispecific diabody that mediates natural killer cell cytotoxicity against xenotransplantated human Hodgkin's tumors. *Blood* 94(8), 2562–2568.
16. Cochlovius B, Kipriyanov SM, Stassar MJ, Schuhmacher J, Benner A, Moldenhauer G, et al. (2000) Cure of Burkitt's lymphoma in severe combined immunodeficiency mice by T cells, tetravalent CD3 × CD19 tandem diabody, and CD28 costimulation. *Cancer Res.* 60(16), 4336–4341.
17. Ridgway JB, Presta LG, Carter P. (1996) "Knobs-into-holes" engineering of antibody CH3 domains for heavy chain heterodimerization. *Protein Eng.* 9(7), 617–621.
18. Gunasekaran K, Pentony M, Shen M, Garrett L, Forte C, Woodward A, et al. (2010) Enhancing antibody Fc heterodimer formation through electrostatic steering effects: applications to bispecific molecules and monovalent IgG. *J. Biol. Chem.* 285(25), 19637–19646.
19. Muda M, Gross AW, Dawson JP, He C, Kurosawa E, Schweickhardt R, et al. (2011) Therapeutic assessment of SEED: a new engineered antibody platform designed to generate mono- and bispecific antibodies. *Protein Eng. Des. Sel.* 24(5), 447–454.
20. Schaefer W, Regula JT, Bähner M, Schanzer J, Croasdale R, Dürr H, et al. (2011) Immunoglobulin domain crossover as a generic approach for the production of bispecific IgG antibodies. *Proc. Natl. Acad. Sci. USA* 108(27), 11187–11192.
21. Wu C, Ying H, Grinnell C, Bryant S, Miller R, Clabbers A, et al. (2007) Simultaneous targeting of multiple disease mediators by a dual-variable-domain immunoglobulin. *Nat. Biotechnol.* 25(11), 1290–1297.
22. Doppalapudi VR, Huang J, Liu D, Jin P, Liu B, Li L, et al. (2010) Chemical generation of bispecific antibodies. *Proc. Natl. Acad. Sci. USA* 107(52), 22611–22616.
23. Veri M-C, Burke S, Huang L, Li H, Gorlatov S, Tuaillon N, et al. (2010) Therapeutic control of B cell activation via recruitment of Fcgamma receptor IIb (CD32B) inhibitory function with a novel bispecific antibody scaffold. *Arthritis Rheum.* 62(7), 1933–1943.
24. Johnson S, Burke S, Huang L, Gorlatov S, Li H, Wang W, et al. (2010) Effector cell recruitment with novel Fv-based dual-affinity re-targeting protein leads to potent tumor cytolysis and *in vivo* B-cell depletion. *J. Mol. Biol.* 399(3), 436–449.
25. Moore PA, Zhang W, Rainey GJ, Burke S, Li H, Huang L, et al. (2011) Application of dual affinity retargeting molecules to achieve optimal redirected T-cell killing of B-cell lymphoma. *Blood* 117(17), 4542–4551.
26. Litowski JR, Hodges RS. (2002) Designing heterodimeric two-stranded alpha-helical coiled-coils. Effects of hydrophobicity and alpha-helical propensity on protein folding, stability, and specificity. *J. Biol. Chem.* 277(40), 37272–37279.
27. Hakoda M, Kamatani N, Hayashimoto-Kurumada S, Silverman GJ, Yamanaka H, Terai C, et al. (1996) Differential binding avidities of human IgM for staphylococcal protein A derive from specific germ-line VH3 gene usage. *J. Immunol.* 157(7), 2976–2981.
28. Curnow RT. (1997) Clinical experience with CD64-directed immunotherapy. An overview *Cancer Immunol. Immunother.* 45(3–4), 210–215.
29. Müller D, Kontermann RE. (2007) Recombinant bispecific antibodies for cellular cancer immunotherapy. *Curr. Opin. Mol. Ther.* 9(4), 319–326.
30. Müller D, Kontermann RE. (2010) Bispecific antibodies for cancer immunotherapy: Current perspectives. *BioDrugs* 24(2), 89–98.
31. Baeuerle PA, Reinhardt C. (2009) Bispecific T-cell engaging antibodies for cancer therapy. *Cancer Res.* 69(12), 4941–4944.
32. Bargou R, Leo E, Zugmaier G, Klinger M, Goebeler M, Knop S, et al. (2008) Tumor regression in cancer patients by very low doses of a T cell-engaging antibody. *Science* 321(5891), 974–977.

37

STRAND EXCHANGE ENGINEERED DOMAIN (SEED): A NOVEL PLATFORM DESIGNED TO GENERATE MONO AND MULTISPECIFIC PROTEIN THERAPEUTICS

ALEC W. GROSS,[1] JESSICA P. DAWSON,[1] MARCO MUDA,[1,†] CHRISTIE KELTON,[1] SEAN D. MCKENNA,[1] AND BJÖRN HOCK[2]

[1]*EMD Serono Research Institute, Billerica, MA, USA*
[2]*Merck Serono, Protein Engineering and Antibody Technologies, Darmstadt, Germany*

37.1 INTRODUCTION

Monoclonal antibodies (mABs) have achieved outstanding commercial success as therapeutics in the last decade and of the top 10 best selling drugs in 2009, four belonged to this class [1]. On the other hand, the target space seems to be limited. In oncology for example, nine marketed products correspond to only six different targets (VEGF, EGFR, Her2, CD20, CD33, and CD52). For autoimmune and inflammatory diseases, four TNF-α-neutralizing mABs have been marketed (Infliximab, Adalimumab, Golimumab, and Certolizumab PEGol) [2]. The number of mAB launches validating new targets is limited to one or fewer per year demonstrating that new therapeutic target discovery is an elusive and costly business. Therefore, some next-generation mAB therapeutics will likely build on existing well validated targets by trying to exploit protein engineering technologies to improve antibody activity, for example, enhanced effector functions [3], introduce immunoconjugates [4] or tailored PK properties [5], or to achieve reduction of treatment costs.

Another strategy to enhance efficacy involves combining mABs with different modes of action (MOA): As an example, different targets in connected pathways such as EGFR and IGFR can be combined to produce synergistic MOAs [6] or to suppress resistance mechanisms as described for EGFR and cMET [7].

MAB combinations, although self-evident as a best in class strategy, are challenging from a development and marketing perspective: First, each single antibody as well as the combination likely require regulatory testing and approval from authorities such as FDA and EMEA leading to delays and a significant increase in development costs. Secondly, the interdependency of efficacy and pharmacokinetics (PK) needs to be carefully studied. In addition, combination of individual mABs leads to high costs. Based on these challenges bispecific antibody (bsABs) technologies become an attractive alternative because they offer a solution to combine synergistic MOAs in a single entity. In addition, bsABs open new MOAs as demonstrated by Micromet's BiTE technology [8].

Although the bsAB concept provides a striking solution for the combination of different MOA, production of those molecules turned out to be very challenging. Shortly after invention of the mAB technology [9], Milstein and Cuello [10,11] described the Quadroma technology for the

†Present Address: Merrimack Pharmaceuticals, Cambridge, MA, USA.

Fusion Protein Technologies for Biopharmaceuticals: Applications and Challenges, First Edition. Edited by Stefan R. Schmidt.

production of bsAB that is essentially a hybrid of two hybridomas. Although scientifically very elegant, it became apparent that production in industrial scale is not feasible because of the complexity of species produced [12]. Other early approaches included chemical crosslinking [13,14], tandem fusions of scFv or V-domains [15,16] or their fusion to the C-terminus of heavy chains [8]. Up to now, only one bsAB (Catumaxomab, [17]) has been approved in Europe for treatment of malignant ascites. Catumaxomab is a rat/mouse chimeric IgG2a displaying a high level of immunogenicity. Other bsABs in advanced clinical development such as Blinatumomab are based on tandem fusions of scFvs and therefore do not display mAB effector functions or favorable pharmacokinetic properties such as long serum half-life in humans.

A human heterodimeric Fc fusion retaining long serum half-life and mAB-like effector functions is the ideal vehicle for engineering therapeutic bsABs. Genentech's "knobs-into-holes" technology [18] offered a first technical solution by introducing different complementary mutations in the CH3 domain that favor heterodimerization.

Recently, Davis et al. [19] designed SEED (strand-exchange engineered domains), a novel bsAB platform based on the engineering of the CH3 domain. SEED is based on a rational design strategy that combines structurally related and conserved sequences of human IgG and IgA CH3 domains. By alternating sequences derived from both CH3 domains, two distinct SEED protein chains were created (designated AG and GA) creating asymmetric interfaces with modified β-sheet structures that preferentially heterodimerize (Figure 37.1). As CH3 domains drive the heterodimerization of immunoglobulin heavy chains, AG and GA chains expressed in the context of antibody heavy chains can be utilized to produce monovalent and monospecific or bivalent and bispecific antibody-like molecules.

In the development and design of the SEED molecule, special care was taken that the resulting bispecific molecule should bear the same biological properties as the parental immunoglobulin G (IgG) molecule. This includes long serum half-life, antibody-dependent cellular cytotoxicity (ADCC) and complement-dependent cytotoxicity (CDC), as well as excellent biophysical properties such as high protein homogeneity, protein A binding, and thermal stability. The task was made more complicated by the fact that IgA does not have all the same properties of IgG. Consequently, the final molecule was not simply IgG on one half and IgA on the other, but required some judicious addition of IgG sequences to both halves to ensure properties such as protein A and FcRn binding on each side of the protein. The resulting SEED molecule then required extensive testing in its earliest phases to guarantee that the reengineering did not disrupt any of the critical biological properties enumerated above. In addition, it was important to evaluate the manufacturability of this new molecule, including expression levels, ease of purification, thermal stability, and long-term storage potential. Only a SEED that passed all of these tests would be ready for incorporation into the therapeutic pipeline.

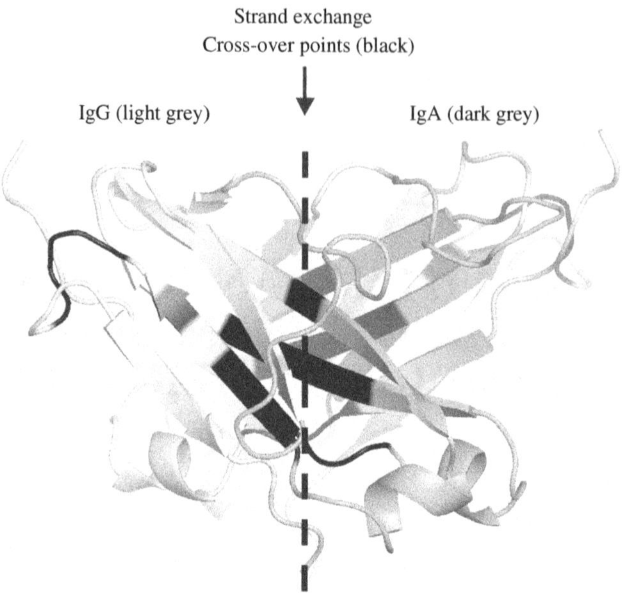

FIGURE 37.1 SEED design produces an asymmetric heterodimer protein scaffold. Diagram of the crystal structure determined at 2.5 Å of the AG/GA SEED heterodimer protein. The common axis of symmetry is denoted with a dashed line through the strand-exchange crossover segments (black). Sequence segments predominantly derived from IgA are shown on the right in dark grey and those derived from IgG on the left in light grey. The sequence on a single chain alternates between IgA and IgG such that in the folded tertiary structure one half of the strands and interface are derived from IgA (right side) and the other half from IgG (left side).

37.2 TECHNICAL ASPECTS

37.2.1 Workflow

For studies to assess the SEED design and SEED proteins, we established a standardized workflow (Figure 37.2A) to produce and characterize the purity and heterodimer content of SEED proteins before further downstream biophysical and biological experiments. For these assessments, SEED-based proteins were compared to IgG1 Fc-based proteins to test if SEED retains the desirable properties of therapeutic antibodies. Proteins were produced by transient transfection of 293 or Chinese hamster ovary (CHO) cells, followed by protein A-purification from culture medium and then preparative size-exclusion chromatography (SEC). Productivity of proteins was measured as the yield of protein A-purified protein obtained per millliter of cell culture medium. The

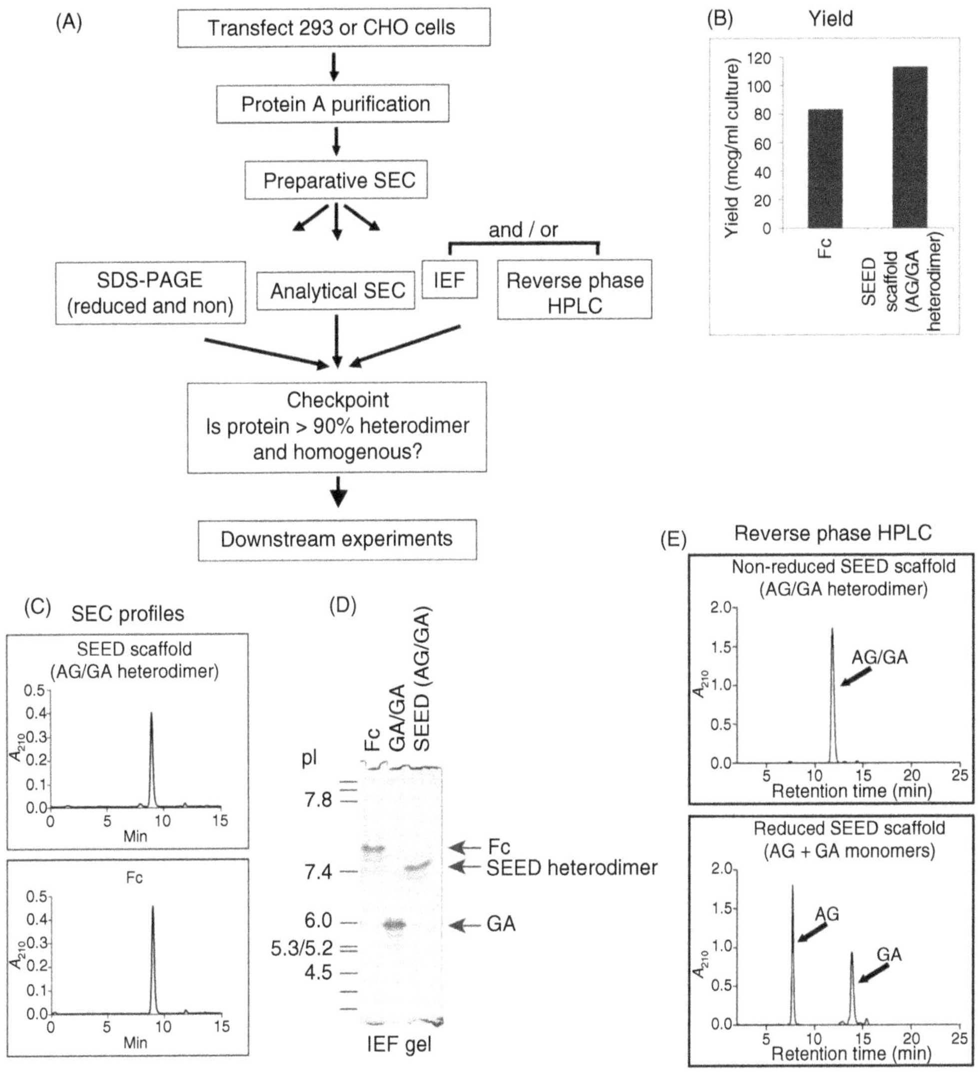

FIGURE 37.2 Expression, purification, and characterization of SEED proteins. (A) Standardized workflow to produce SEED proteins by transient transfection, purify from medium, and determine if the purified proteins pass the checkpoint for downstream experiments. SEC, size-exclusion chromatography; IEF, isoelectric focusing. (B) Example comparing yield after protein A purification of IgG1 Fc and SEED. Yield measured as microgram protein A-purified protein per milliliter cell culture medium. (C) Analytical SEC example comparing Fc and SEED scaffold proteins obtained after single-step protein A purification. (D) IEF gel analysis example showing distinct focusing of Fc, SEED heterodimer (AG/GA), and GA/GA homodimer. (E) Reverse phase HPLC example showing different elution times of SEED scaffold (AG/GA heterodimer) compared to individual AG and GA chains.

molecular weight and percent aggregate of the purified material was determined by SDS-PAGE (reduced and non-reduced) and analytical SEC. Since SEED proteins should be produced as heterodimers of the AG and GA SEED chains, which have the same size, additional methods were developed to measure AG/GA heterodimer proteins versus homodimer and individual chains, specifically isoelectric focusing (IEF) gel analysis and reverse phase HPLC analysis. The workflow is designed so after these characterizations there is a checkpoint that requires that the SEED proteins must be >90% heterodimer and homogenous before they can be used in further biophysical and biological experiments. Proteins were routinely obtained as >95% homogenous AG/GA heterodimers.

In Figure 37.2, representative data sets are shown to illustrate this workflow. For example, Fc- and SEED-based proteins are produced with similar yields (~20–100 μg/mL), and in this example the yields of both Fc and SEED scaffold

(AG/GA heterodimer) were comparable (~80–100 μg/mL) (Figure 37.2B). Analytical SEC profiles comparing Fc and SEED scaffold (AG/GA heterodimer) show that both types of proteins were homogenous and of the expected size (Figure 37.2C). Our methods to determine if the SEED protein is an AG/GA heterodimer are also illustrated in Figure 37.2. Owing to their different amino acid composition, the AG and GA SEED chains have different pI values and thus a protein that is a heterodimer of AG and GA chains focuses to a distinct position in IEF gel analysis (theoretical pI values: AG = 8.2; GA = 6.4; AG/GA = 7.3; and Fc = 7.5). The example in Figure 37.2D shows that the SEED heterodimer is clearly distinguished from Fc or a GA/GA homodimer. The elution profile of the SEED AG/GA heterodimer in reverse phase HPLC is also distinct form the elution of either the AG or GA SEED chains; the example in Figure 37.2E shows a nonreduced sample of SEED scaffold (AG/GA heterodimer) eluting at 11.8 min, which is clearly distinct from the individual AG and GA peaks (eluting at 7.7 and 13.9 min, respectively) seen when a sample of the SEED scaffold (AG/GA heterodimer) was first reduced with dithiothreitol into its separate chains.

Together, this workflow allowed us to determine that with transient transfection Fc- and SEED-based proteins are produced from cultured cells with a similar yield and with only minor fractions of higher weight aggregated material that can easily be removed with preparative SEC as needed. In contrast to Fc-based proteins, SEED-based proteins are produced by co-expression of two different protein chains, which in transient transfection is simply achieved by co-transfection of an AG and GA expression plasmid. Since the IEF gel and HPLC methods allow measurement of the percent heterodimer content of the proteins produced, we determined that with transfection of our standard optimal ratio of AG and GA plasmids an essentially pure AG/GA heterodimer protein is produced from cells.

37.2.2 Therapeutic Formats

To explore the use of the SEED heterodimer scaffold, we have produced a number of SEED-based proteins in different therapeutic formats (Figure 37.3). For example, SEED can be used with Fab domains to produce bivalent or monovalent Fab-SEED molecules (Figure 37.3A). This is accomplished by fusion of one or both SEED chains to the variable heavy chain (VH) and constant heavy chain (CH) domains of a Fab and co-expressing the SEED chains with the Fab light chain. SEED is also compatible with other binder domains; for example, we have fused scFv domains (Figure 37.3B) or VHH domains (Figure 37.3C) to one or both SEED chains to

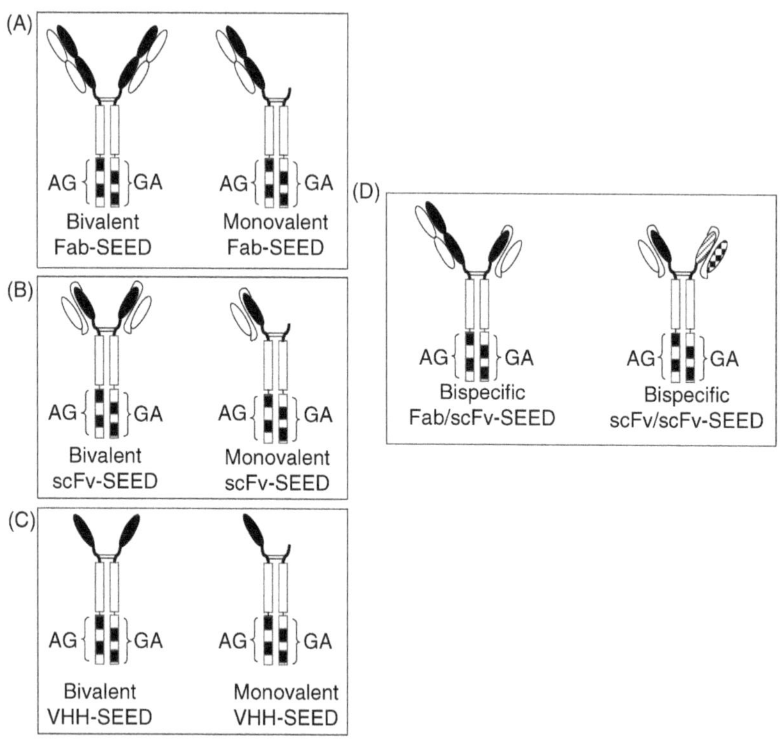

FIGURE 37.3 Versatile therapeutic formats of SEED proteins. SEED has been found to be compatible with different binder domains and in different combinations. Some examples: (A) Bivalent and monovalent Fab-SEED proteins. (B) Bivalent and monovalent ScFv-SEED proteins. (C) Bivalent and monovalent VHH-SEED proteins. (D) Bispecific Fab/ScFv-SEED and ScFV1/ScFV2-SEED.

produce bivalent and monovalent scFv-SEED and VHH-SEED proteins. Bispecific SEED-based proteins have also been produced, for example by fusing one SEED chain to a Fab's VH-CH domains and the other SEED chain to a scFv, or by fusing different scFv's to each of the SEED chains (Figure 37.3D). We have found that the SEED scaffold is compatible with fusion in different combinations to different types of binder domains, including Fab, scFv, and VHH domains, and that these domains still bind their targets.

37.2.3 Biophysical Properties

To assess the stability and biophysical properties of SEED proteins, we have performed a variety of assays with the SEED scaffold and also with the SEED-fusion proteins. Figure 37.4 shows examples of these types of physical analyses of SEED-based proteins. In our assays, we compared SEED- and Fc-based proteins to determine if SEED retains physical properties that are compatible with therapeutic use. For example Fc-based proteins are stable and tolerate a range of pH conditions; therefore changes in the pH environment could reveal if SEED-based proteins are significantly less stable than Fc-based proteins. Stability specifically in low pH is also a desirable therapeutic antibody property because of the low-pH viral inactivation steps required during production. Here, we show an example of Fc and SEED scaffold proteins exposed to different pH buffers for 30 min, after which the amount remaining of nonaggregated protein of the correct size was measured with analytical SEC and the area under the curve (AUC) of the SEC profile was plotted for each condition. SEED scaffold showed a similar stability to Fc in this range of pH conditions (Figure 37.4A).

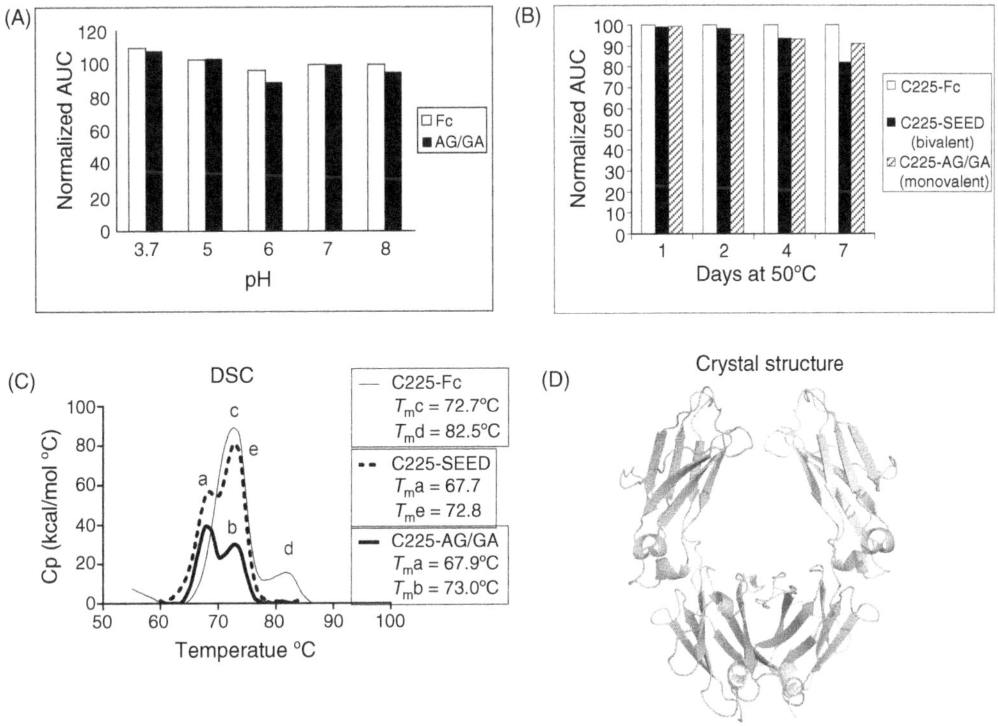

FIGURE 37.4 Physical testing of SEED molecules. (A) Stability at a different pH values. Example comparing Fc and SEED scaffold (AG/GA heterodimer) after being held at different pH values for 30 min. Amount of protein remaining after 30 min was measured by analytical SEC and the area under the curve (AUC) of the main protein peak was normalized to protein after control incubation. (B) Stability under temperature stress. Example compares stability of C225-Fc, C225-SEED (bivalent), and C225-AG/GA (monovalent) after proteins were held at 50°C for 1–7 days. Amount of protein remaining was measured by analytical SEC and AUC of main protein peak was normalized to control sample. (C) Differential scanning calorimetery (DSC) analysis. Example compares DSC profiles of C225-Fc, C225-SEED (bivalent), and C225-AG/GA (monovalent). T_m values shown are derived from the midpoint of the indicated peaks. Peaks: (A) CH2+ SEED domains; (B) single C225 Fab domain; (C) CH2+ two C225 Fab domains; (D) CH3 domains; and (E) two C225 Fab domains. (D) Physical analysis by crystal structure determination. Crystal structure of SEED scaffold (AG/GA heterodimer) was solved to 2.5 Å to determine the SEED design domain folding and interactions between the domains. (*Source*: Figure 37.4C reproduced from Reference [37] by permission of Oxford University Press.)

It is also important to test the SEED scaffold in therapeutic formats with fusion to other protein domains. To accomplish this we have compared the clinically validated antibody C225 (Fc-based) with C225-SEED fusion proteins in both bivalent and monovalent formats (formats illustrated in Figure 37.3A). For example, an accelerated thermal stability study was conducted by holding C225-Fc, C225-SEED (bivalent), or C225-AG/GA (monovalent) proteins at 50°C over a period of 7 days, then analytical SEC profiles of the postincubation samples were compared to a control held at 4°C to assess changes in aggregation or loss of protein due to precipitation. The amount of C225-Fc did not change over 7 days at 50°C, while a small fraction of C225-SEED and C225-AG/GA proteins were lost (Figure 37.4B). Differential scanning calorimetery (DSC) was also used as an alternate method to measure the stability of the proteins. DSC analysis showed that the SEED CH3 domain has a lower melting temperature than the CH3 domain of Fc. However, the melting temperature of the SEED CH3 domain is ~68°C, which is still compatible for use as a therapeutic protein, as neither protein productivity nor stability (tested by incubation in different pH or at 50°C) was greatly affected. An example comparing the DSC profiles of C225-Fc, C225-SEED (bivalent), and C225-AG/GA (monovalent) is shown in Figure 37.4C.

As a final example of our physical characterization of the SEED design, we show in Figure 37.4D the crystal structure of SEED, determined to 2.5 Å. This crystal structure confirms that the SEED design, as predicted, produces AG and GA SEED domains that each fold in a similar way to Fc CH3 domains. Furthermore, the crystal structure physically confirmed the specific interaction interface between AG and GA SEED domains.

In summary, our biophysical assays and structure determination confirm that SEED has physical properties compatible with therapeutic applications.

37.2.4 Biological Properties

In addition to biophysical properties, we have tested the biological properties of SEED-based proteins. An important part of the SEED design is that it should retain desirable biological properties of therapeutic antibodies, such as target binding, effector functions, and a long half-life *in vivo* (Figure 37.5). We again used the clinically validated antibody C225 to assess if SEED retains therapeutic antibody properties. The C225 antibody has been shown to bind to the EGF receptor extracellular domain and to inhibit EGF binding to its receptor. Figure 37.5A shows in a ligand competition assay that bivalent C225-SEED binds to EGFR and inhibits EGF binding in a similar way as C225-Fc, while monvalent C225-AG/GA and C225-GA/AG bind to EGFR but have a higher K_i in this assay.

An important feature of antibodies is their effector functions, specifically antibody-dependent ADCC and CDC. C225-Fc and C225-SEED (bivalent proteins) and monovalent C225-AG/GA and C225-GA/AG all had similar potency to induce ADCC against A431 target cells (Figure 37.5B).

CDC effector function was tested using a different antibody, hu14.18 Fab, which is known to induce CDC against M21 cells [20]. Bivalent hu14.18-Fc and hu14.18-SEED induced CDC against M21 target cells with similar efficacy (Figure 37.5C). The results from ADCC and CDC assays thereby demonstrated that SEED retains antibody-like effector functions.

Mouse PK studies (Figure 37.5d) were performed to test if SEED-based fusion proteins retain long *in vivo* half-life similar to antibodies. By comparing Fc and SEED scaffolds, and C225-Fc, C225-SEED, and C225-AG/GA proteins, we found that SEED-based proteins and Fc-based proteins had similar PK properties in mice, suggesting that engineering the SEED platform did not impair its *in vivo* half-life compared to IgG.

All together, our experiments (exemplified in Figure 37.5) showed that SEED retains the functional and biological properties of therapeutic antibodies while efficiently forming heterodimeric proteins.

37.3 POTENTIAL THERAPEUTIC APPLICATIONS

The evolution of biotherapeutics has been driven by both technological advances as well as by improved understanding of disease mechanisms. While the first effective protein-based drugs consisted of naturally derived molecules such as insulin and growth hormone purified from native sources, more recent developments in molecular biology have allowed the generation of engineered proteins suited to specific needs and purposes. In general, biotherapeutics have proven to be a very important drug class based on safety and the efficacy provided by their exquisite specificity. Antibodies in particular were identified early as having great potential in targeting disease-associated processes because of this characteristic. However, early antibodies generated in nonhuman hosts proved to be highly immunogenic, which greatly hindered their clinical utility. More recently, technologies allowing the humanization of nonhuman antibodies, together with the generation of new display technologies to generate fully human antibodies, have largely overcome this hurdle and, as a result, antibodies have rapidly become a major category of disease modifying therapeutics. While the list of clinically validated antibody targets continues to grow, it is also clear that in many cases the clinical efficacy observed with conventional antibodies is often much poorer than originally predicted. The reasons for this are complex and disease specific but may include the following: (1) multiple and often redundant signaling pathways involved in driving complex diseases are inadequately

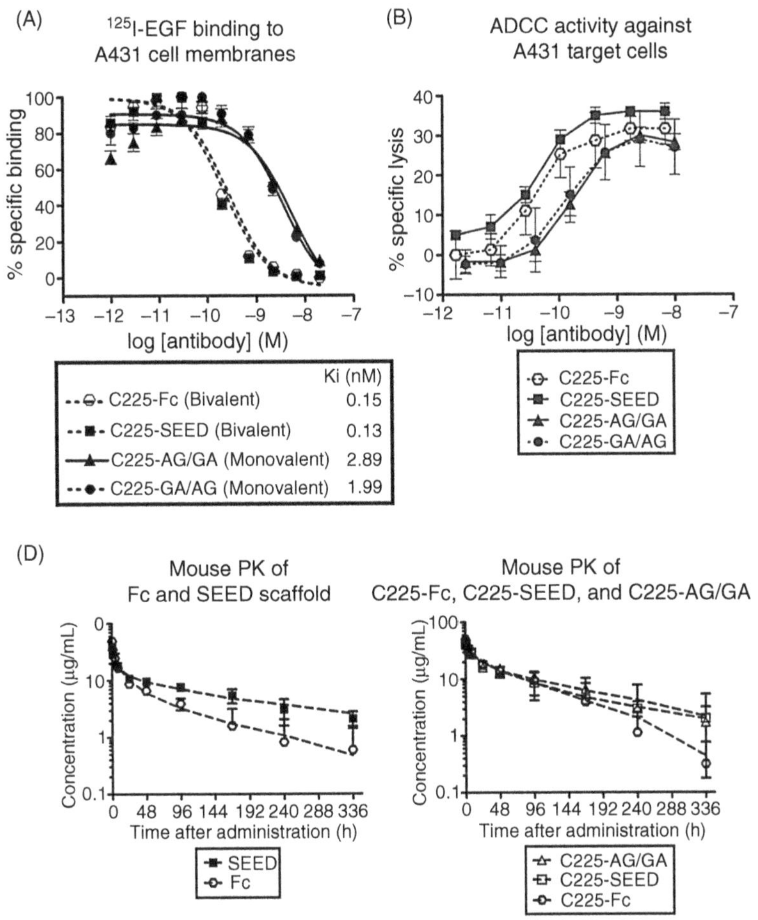

FIGURE 37.5 Biological function testing of SEED molecules. Biological properties of SEED proteins, C225-SEED (bivalent), C225-AG/GA (monovalent), C225-GA/AG (monovalent), were tested for *in vitro* cell binding and ADCC activity and for *in vivo* PK properties compared to the clinically validated antibody C225-Fc, and for CDC activity in comparison to hu14.18-Fc antibody. (A) Ligand competition assays to measure binding to EGFR. A431 cell membranes were incubated with ^{125}I-EGF (0.1 nM) and each antibody as indicated above. Incubations were performed in binding buffer for 1.5 h at 37°C. Plates were washed and bound radioactivity measured. Competition curves through the data points and K_i values were determined using GraphPad Prism software. (B) ADCC activity of human peripheral blood mononuclear cells against A431 target cells (100:1 effector:target ratio) was measured by standard ^{51}Cr-release assay and plotted as percent target cell specific lysis versus the concentration of each antibody as indicated above. Each data point is the mean ± SD of triplicates. (C) CDC activity induced against M21 target cells by the indicated proteins was measured by standard ^{51}Cr-release assay as described above. (D) Mouse PK study of Fc, SEED scaffold, C225-Fc, C225-SEED, and C225-AG/GA proteins. Plasma concentration profiles after single IV administration of 100 μg (~3.5 mg/kg) into nude mice. Each data point is the mean ± SD from three mice. (*Source*: Reproduced from Reference [37] by permission of Oxford University Press.)

addressed by modulating a single target, (2) antibody binding to a single epitope of even a key target protein may be insufficient by itself to modulate all of the biological activities of that protein, or (3) the bivalent nature of antibodies can lead to crosslinking and subsequent activation of target proteins. One potential solution to these obstacles is the creation of a highly flexible biotherapeutic platform that can be used to create antibody-like molecules in which the specificities and valencies can be tuned to meet a specific application.

Effectively treating cancer with single target-modifying drugs such as antibodies has been particularly difficult because of the high complexity of the oncogenic process that often involves mutations or other genetic abnormalities in parallel signaling pathways. In the case of glioblastoma multiforme it has been reported that tumor tissues, but not normal brain tissues, can be associated with activation of multiple kinase receptors, including EGFR, PDGFR-α, RET, and CSF-1R [7]. Based on these observations and supporting *in vitro* data, it has been proposed that such receptor tyrosine kinase coactivation is responsible for the poor clinical responses observed with monotherapies. Moreover, evidence is emerging to suggest that the treatment of cancer patients with a monotherapy may actually drive additional alterations in other pathways that ultimately lead to acquired drug resistance. In one example, selection of tumor cells resistant to the EGFR inhibitor cetuximab was associated with significant increases in phospho-ErbB2, phospho-ErbB3, and phospho-cMet [21]. It was noted that cMet in these resistant cells, but not in the parental cells, could be activated with EGF, suggesting that receptor tyrosine kinases may act in a cooperative manner to overcome drug-mediated inhibition. In a similar case, when a non-small cell lung carcinoma (NSCLC) cell line HCC827 was rendered resistant to the EGF receptor inhibitor gefitinib through prolonged low-dose exposure significantly heightened levels of phospho-ErbB3 and phospho-cMet were found. Notably, these cells were resistant to inhibitors of EGFR or cMet when used alone but were sensitive when a combination of inhibitors was used [22]. This result was further supported by an analysis of NSCLC patients' tumors. Those tumors that were originally sensitive to EGFR inhibitors but subsequently became resistant to these drugs had a high incidence of MET gene amplification. Other groups have found that combining inhibitors of IGF-1 receptors with EGF receptor antagonists could also result in enhancement of antiproliferative and proapoptotic activity in NSCLC cell lines [23]. In aggregate, numerous preclinical and clinical reports support the notion that the targeting of multiple pathways may be critical in affecting strong clinical responses in complex diseases such as cancer. As a result of such observations, testing combinations of drugs that target cooperative pathways is becoming increasingly common. Based on preclinical evidence demonstrating functional synergy, combinations of EGFR inhibitors and other targeted therapeutics such as mTOR inhibitors are progressing into clinical trials [24].

While the rationale for combining different targeted therapeutics continues to grow stronger, the regulatory process for the approval of investigational drug combinations, particularly where the individual drug candidates may have only marginal efficacy when used alone, remains difficult. Moreover, the relatively high price of biotherapeutics, including antibodies, can become prohibitively expensive when used as combinations. As a result, a number of groups have looked into the possibility of creating single molecules that can simultaneously inhibit multiple pathways. Within this new category of drug candidates are the antibody-like molecules engineered to have multiple binding specificities. With the recent approval of the Catumaxomab, a bispecific antibody derived from a mixed-specificity quadroma, and promising clinical results with the CD19/CD3 bispecific tandem binder, Blinatumomab, multispecific antibody-like molecules appear to have a very promising future. Although these particular molecules act through recruitment and engagement of acquired immune cells with one binding arm together with tumor cell targeting with the other arm, other publications have demonstrated enhanced preclinical activity in which the targets involved are members of parallel activation pathways. For example, as both osteopontin (OPN) and vascular endothelial growth factor (VEGF) play a role in tumor angiogenesis, a bispecific antibody was produced that inhibited the biological activities of both of these soluble factors [25]. This engineered antibody demonstrated much greater antiangiogenic activity *in vitro* and *in vivo* than either conventional parental antibody when tested individually. A similar observation was made by another group using a bispecific antibody targeting VEGF together with the receptor for PDGF-β [26]. In addition, Robinson et al. [27] created a bispecific antibody that bound both of the cancer-associated EGF receptor family members, ErbB2 and ErbB3. In this instance, the bispecific antibody achieved selective binding to cells that coexpressed both target proteins, presumably because of the increased avidity of this molecule for these cells rather than those that only expressed one of the target proteins. Owing to this selective targeting, it was hypothesized that such a molecule may have an improved safety profile versus monospecific antibodies.

One of the key technological hurdles in the development of clinical-grade bispecific antibodies is random pairing of heavy and light chains when chains comprising different specificities are expressed within the same cell. Maintaining the proper light and heavy chain association is necessary to retain the target specificity of the original antibodies. To ensure the correct pairing of heavy and light chains in a bispecific antibody, several strategies have been developed over the years. Creation of single chain variable fragments (scFv), formed by linking the heavy chain to the light chain with a flexible peptide linker, has proven to be a valuable method of generating a single binder with high heavy and light chain pairing fidelity [28]. While scFvs have reportedly suffered from stability and expression liabilities, improvements in the understanding of molecular interactions between the variable region domains has permitted this technology to advance to a clinically acceptable state, as exemplified by Blinatumomab currently in Phase II trials. Another solution to the chain-pairing problem takes advantage of the preferential association of mouse and rat light chains for their respective heavy chain counterparts [29]. Quadromas generated from a fusion of mouse and rat hybridomas have species-specific heavy and light chain pairing, which has allowed the production of bispecific antibodies. Although this technology has led to the generation of the clinically approved bispecific antibody Catumaxomab, its reliance on the use of nonhuman heavy and light chains increases the risk of immunogenicity that may ultimately limit the general utility of this approach in the clinic. An alternative approach to address the heavy and light chain pairing issue is the use of a single common light chain that can associate with different variable heavy chain domains [30]. This can be accomplished using antibody Fab fragment libraries containing broad heavy chain diversity but only a single light chain. While there is evidence that this approach is effective, it likely requires that the light chain contributes minimally to the overall binding affinity. The removal of the light chain entirely, though engineering of human heavy chains [31] or use of naturally occurring light chain-free camelid antibodies [32], eliminates the potential for random chain pairing. Although such an approach would negate any potential contribution to the overall affinity that light chains normally provide, some heavy chain only-based binders have been reported to attain affinities up to 100 pM [32].

All of the aforementioned binder technologies have been used as building blocks for the creation of multispecific antibodies through the genetic fusion to the C- and/or N-termini of conventional antibodies of differing specificity. For example, fusions of scFv moieties to the termini of conventional antibodies have been reported [33], as have fusions of additional VH and VL domains, creating dual variable domain immunoglogulins [34]. When fused to a conventional homodimeric Fc-based antibody, such molecules typically have four or more binding sites and a significantly larger size than traditional antibodies. The increased overall size of these molecules has the potential to impact production, stability, pharmacokinetics, tissue distribution, and tumor penetration [6]. To create a platform for bispecific antibodies possessing the biophysical characteristics, size, pharmacokinetic profile, and ease of manufacture as conventional antibodies, investigators have examined methods for selectively heterodimerizing heavy chains of immunoglobulins having different specificities. Such investigations have resulted in different solutions to preferentially drive this heavy chain heterodimerization. In the "knobs-into-holes" technology first described over a decade ago [18,30], heavy chains engineered with amino acid protuberances were coexpressed with heavy chains engineered to have corresponding concavities such that the maintenance of van der Waals contacts through high surface complementarity favored heterodimerization rather than homodimerization. In combination with a single common light chain, this technology proved successful in creating bivalent and bispecific antibodies [30]. Although this technology has been somewhat limited by the presence of homodimer contaminants during large-scale expression in mammalian cells, a recent report demonstrates that process improvements allow good production of bispecific antibodies in a bacterial expression system [35].

Another recent approach to preferentially drive heavy chain heterodimerization relies upon the introduction of oppositely charged amino acids in opposing heavy chains [36]. Results clearly demonstrate that this approach can be used to generate heterodimeric heavy chain pairs. Moreover, by fusing scFv domains with different specificities to each heavy chain, the resulting molecules could demonstrate true bispecific targeting and be readily expressed in mammalian cells.

The novel method for creating a heterodimerizing platform described within this chapter differs from the two technologies summarized above in that the SEEDs were designed by incorporating complementary amino acid sequence stretches from human IgA and IgG within the Fc CH3 domain of a normal antibody [19]. Results demonstrate the strong heterodimerizing propensity of the engineered heavy chains. Notably, the SEED-based antibodies retain the biophysical, biological, and pharmacokinetic properties of conventional antibodies [37]. The SEED platform has proven to be highly flexible when coupled with different binding domains, such as Fabs, scFv, and heavy chain only binders. Unlike either conventional antibodies or dual variable domain immunoglobulins, the heterodimerizing heavy chain-based platforms can be readily used to create monovalent antibodies [36–38]. Such molecules may have significant advantages over multivalent antibodies for antagonizing targets that can become activated upon crosslinking. One such target of significant therapeutic relevance is cMet. This target has proven difficult to functionally inhibit with conventional antibodies because of the propensity of bivalent binding to dimerize and activate the receptors, even when the ligand is completely blocked. However monovalent Fab fragments of these same activating antibodies can effectively antagonize ligand-mediated activation and lead to increased antitumor activity *in vivo* [39]. This observation was substantiated when the "knobs-into-holes"-based monovalent antibody MetMab, which includes a Fab as the single binding arm, was demonstrated to fully antagonize cMet activation, resulting in potent inhibition of c-Met dependent tumor cell growth in preclinical models [40]. Notably, MetMab has sufficient drug-like characteristics to be progressed into clinical trials. The SEED-based monovalent antibodies retain similar biophysical properties and expression/purification characteristics as conventional antibodies.

While monovalent antibodies generated from platforms such as knobs-into-holes and SEED can specifically bind targets with high affinity, the lack of a second arm may reduce overall avidity that, in certain circumstances, could impact efficacy. Use of a heterodimerizing platform provides the potential to create biparatopic molecules having two binding arms that recognize two nonoverlapping epitopes on a single target protein. By controlling the orientation and flexibility of the binding arms and careful selection of the target epitopes it may be possible to create high avidity protein clamps similar to the tandem scFv-based chelating recombinant antibodies (CRAbs) CRAbs have been demonstrated to achieve sub-nanomolar avidities through cooperative binding [41]. A CRAb-like biparatopic molecule based on the heterodimerizing SEED platform would be expected to having the avidity advantage of the CRAb together with the effector mechanisms and pharmacokinetic properties of normal antibodies. Alternatively, a biparatopic antibody oriented to favor interaction with different epitopes on neighboring cell-surface proteins has the potential to form a lattice of interconnected target proteins. Interestingly, such a phenomenon was observed using different conventional antibodies that bound distinct epitopes of EGF receptor [42]. In this case, a combination of two antibodies that bound to nonoverlapping epitopes of EGFR led to strong synergistic activation of complement-dependent target cell lysis. These results suggested that the lattice clustering induced by the antibodies was, at least in part, responsible for increased tumor cell killing. Although not formally demonstrated, it would be anticipated that a single bispecific antibody that bound to these same epitopes would likewise induce EGF receptor clustering and possibly increased cell lysis.

Other potential opportunities for SEED-based molecules include the enhancement of biologically active ligands, such as interferons and interleukins, through improved pharmacokinetics and ease of manufacture. Toward this end, a SEED-based molecule having a single IL-2 moiety fused to one chain of the molecule has been described that has prolonged pharmacokinetics when tested in mice [19]. Such an approach may have advantages in cases where two copies of an active ligand fused to a single molecule can interfere with protein production. Alternatively, the SEED platform may be amenable to generating cytokine traps created by fusing a pair of different receptor chains capable of creating a high affinity receptor complex to the two heavy chain arms. This approach may be less complex than approaches such as VEGF-Trap where multiple copies of receptor chains are fused to a homodimeric Fc scaffold [43].

37.4 FUTURE PERSPECTIVES

While preclinical activity has demonstrated the potential utility of the SEED platform, as with all engineered proteins, significant challenges can be expected in the future. For example, even when the interactive forces strongly favor heavy chain heterodimerization versus homodimerization, significant variation in expression rates of the different heavy chain genes may impact the final molecule. While gene dose can be more readily regulated in transiently transfected cells, the relative chain expression in stable transfection may be more difficult to control because of integration site, copy number, and translation efficiency variability. From a manufacturability stand-point, production complexities often increase in step with the number of different protein chains comprised in the product. Hence, molecules such as MetMab that include three or more distinct chains could be expected to have more challenges in production than a conventional antibody with two chains.

As found in other bispecific antibody platforms, nonFab binders often remain a weak link in the overall molecular stability. Significant effort is required to assess the biophysical attributes, such as melting temperature, of the binder arm components to select the ideal binders or improve suboptimal ones through stability engineering [44].

In principle, a single bispecific antibody that has two different specificities could have cost advantages over two conventional antibodies. However, as many companies have well established and economically optimized antibody manufacturing platforms, any additional costs seen with a bispecific antibody, whether due to expression or purification differences, may eliminate any potential cost advantage.

Finally, as with all nonnative proteins, there remains a risk of immunogenicity for engineered antibodies. Unlike

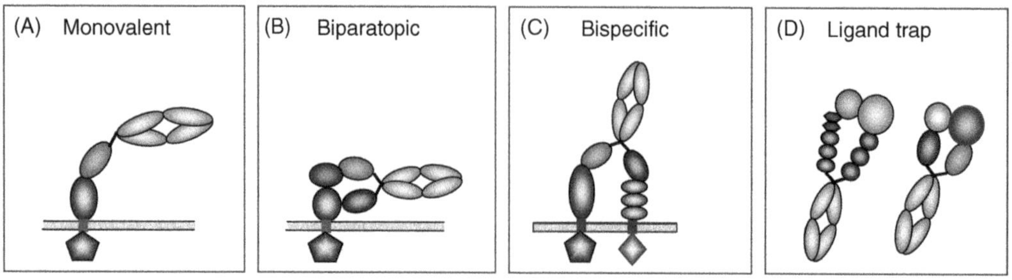

FIGURE 37.6 Potential applications for a heterodimeric Fc-based biotherapeutics platform. Applications include monovalent binding to reduce risk of agonism or receptor down-regulation (A), binding to two epitopes on same target (B), binding two distinct targets (C), and as a high affinity cytokine trap, as either a bispecific antibody or receptor chain fusion molecule (D).

the case for many recombinant ligand-based molecules such as erythropoietin, immunogenicity to therapeutic antibodies is rarely directly life threatening. However, reduced drug exposure due to development of antidrug antibodies can reduce efficacy and have a significant impact on the approvability and utility of such therapeutics. Unfortunately, no tool has yet been able to clearly predict immunogenicity in humans. As the key amino acid interactions that drive preferential heterodimerization in SEED reside largely at the interface of the IgG and IgA heavy chain residues within the CH3 domain core, the opportunity exists to alter the surface exposed residues to those most typical of an IgG in a surface veneering approach.

37.5 CONCLUSIONS

There are a growing number of engineered antibody platforms, ranging greatly in complexity that offer new opportunities to create biotherapeutic candidates to meet specific purposes. Within the catalog of options, the SEED platform offers certain advantages that enable significant flexibility. In particular, the heavy chain interface design strongly drives selective heterodimerization in a platform that can be highly expressed in conventional mammalian cell systems. Moreover, final products retain the critical drug-like properties of conventional antibodies, including expression, effector function, biophysical characteristics, and *in vivo* behavior. By addition of biologically active building blocks, whether they are binding arms, active ligands, or receptor components, a wide variety of mono or multifunctional therapeutic candidates can be constructed that can be designed to address many diverse clinical opportunities (Figure 37.6). The true potential of this platform awaits final validation in the clinic.

ACKNOWLEDGMENTS

We gratefully acknowledge the contributions, support, and expert advice from many scientists at EMD Serono, and in particular Chaomei He, Emmi Kurosawa, Rene Schweickhardt, Melanie Dugas, Maria Soloviev, Anna Bernhardt, David Fischer, John S. Wesolowski, Berend Neuteboom, Xuliang Jiang, Paul Towler, Kris Josephson, Aroop Sircar, Terry Brush, Nina Crowley, Angela Lim, Sue McKenzie, and Scott Lauder. In addition we thank Georg Feger, Steve Arkinstall, and Robert Campbell for their encouragement and support.

REFERENCES

1. Via MC. (2009) Monoclonal Antibodies: Pipeline Analysis and Competitive Assessment. Cambridge Healthtech Institute, Needham, MA, USA.
2. Licastro F, Chiappelli M, Ianni M, Porcellini E. (2009) Tumor necrosis factor-alpha antagonists: differential clinical effects by different biotechnological molecules. *Int. J. Immunopathol. Pharmacol.* 22, 567–572.
3. Strohl WR. (2009) Optimization of Fc-mediated effector functions of monoclonal antibodies. *Curr. Opin. Biotechnol.* 20, 685–691.
4. Teicher BA. (2009) Antibody-drug conjugate targets. *Curr. Cancer Drug Targets.* 9, 982–1004.
5. Presta LG. (2008) Molecular engineering and design of therapeutic antibodies. *Curr. Opin. Immunol.* 20, 460–470.
6. Lu D, Zhang H, Ludwig D, Persaud A, Jimenez X, Burtrum D, Balderes P, Liu M, Bohlen P, Witte L, Zhu Z. (2004) Simultaneous blockade of both the epidermal growth factor receptor and the insulin-like growth factor receptor signaling pathways in cancer cells with a fully human recombinant bispecific antibody. *J. Biol. Chem.* 279, 2856–2865.
7. Stommel JM, Kimmelman AC, Ying H, Nabioullin R, Ponugoti AH, Wiedemeyer R, Stegh AH, Bradner JE, Ligon KL, Brennan C, Chin L, DePinho RA. (2007) Coactivation of receptor tyrosine kinases affects the response of tumor cells to targeted therapies. *Science* 318, 287–290.
8. Baeuerle PA, Reinhardt C. (2009) Bispecific T-cell engaging antibodies for cancer therapy. *Cancer Res.* 69, 4941–4944.
9. Kohler G, Milstein C. (1975) Continuous cultures of fused cells secreting antibody of predefined specificity. *Nature* 256, 495–497.
10. Milstein C, Cuello AC. (1983) Hybrid hybridomas and their use in immunohistochemistry. *Nature* 305, 537–540.
11. Milstein C, Cuello AC. (1984) Hybrid hybridomas and the production of bi-specific monoclonal antibodies. *Immunol. Today* 5, 299–304.
12. Carter P. (2001) Bispecific human IgG by design. *J. Immunol. Methods* 248, 7–15.
13. Brennan M, Davison PF, Paulus H. (1985) Preparation of bispecific antibodies by chemical recombination of monoclonal immunoglobulin G1 fragments. *Science* 229, 81–83.
14. Glennie MJ, McBride HM, Worth AT, Stevenson GT. (1987) Preparation and performance of bispecific F(ab′ gamma)2 antibody containing thioether-linked Fab′ gamma fragments. *J. Immunol.* 139, 2367–2375.
15. Atwell JL, Pearce LA, Lah M, Gruen LC, Kortt AA, Hudson PJ. (1996) Design and expression of a stable bispecific scFv dimer with affinity for both glycophorin and N9 neuraminidase. *Mol. Immunol.* 33, 1301–1312.
16. George AJT, Huston JS. (1997) Bispecific antibody engineering. In: Zanetti M, Capra J. D. (Eds), *The Antibodies*, Vol. 4, Harwood Academic Publishers.
17. Seimetz D, Lindhofer H, Bokemeyer C. (2010) Development and approval of the trifunctional antibody catumaxomab (anti-EpCAM x anti-CD3) as a targeted cancer immunotherapy. *Cancer Treat. Rev.* 36, 458–467.
18. Ridgway JB, Presta LG, Carter P. (1996) Knobs-into-holes engineering of antibody CH3 domains for heavy chain heterodimerization. *Protein Eng.* 9, 617–621.

19. Davis JH, Aperlo C, Li Y, Kurosawa E, Lan Y, Lo KM, Huston JS. (2010) SEEDbodies: fusion proteins based on strand-exchange engineered domain (SEED) CH3 heterodimers in an Fc analogue platform for asymmetric binders or immunofusions and bispecific antibodies. *Protein Eng. Des. Sel.* 23, 195–202.
20. Mueller BM, Romerdahl CA, Gillies SD, Reisfeld RA. (1990) Enhancement of antibody-dependent cytotoxicity with a chimeric anti-GD2 antibody. *J. Immunol.* 144, 1382–1386.
21. Wheeler DL, Huang S, Kruser TJ, Nechrebecki MM, Armstrong EA, Benavente S, Gondi V, Hsu KT, Harari PM. (2008) Mechanisms of acquired resistance to cetuximab: role of HER (ErbB) family members. *Oncogene* 27, 3944–3956.
22. Engelman JA, Zejnullahu K, Mitsudomi T, Song Y, Hyland C, Park JO, Lindeman N, Gale CM, Zhao X, Christensen J, Kosaka T, Holmes AJ, Rogers AM, Cappuzzo F, Mok T, Lee C, Johnson BE, Cantley LC, Janne PA. (2007) MET amplification leads to gefitinib resistance in lung cancer by activating ERBB3 signaling. *Science* 316, 1039–1043.
23. Choi YJ, Rho JK, Jeon BS, Choi SJ, Park SC, Lee SS, Kim HR, Kim CH, Lee JC. (2010) Combined inhibition of IGFR enhances the effects of gefitinib in H1650: a lung cancer cell line with EGFR mutation and primary resistance to EGFR-TK inhibitors. *Cancer Chemother. Pharmacol.* 66, 381–388.
24. Adjei AA. (2006) Novel combinations based on epidermal growth factor receptor inhibition. *Clin. Cancer Res.* 12, 4446s–4450s.
25. Kou G, Shi J, Chen L, Zhang D, Hou S, Zhao L, Fang C, Zheng L, Zhang X, Liang P, Zhang X, Li B, Guo Y. (2010) A bispecific antibody effectively inhibits tumor growth and metastasis by simultaneous blocking vascular endothelial growth factor A and osteopontin. *Cancer Lett.* 299, 130–136.
26. Mabry R, Gilbertson DG, Frank A, Vu T, Ardourel D, Ostrander C, Stevens B, Julien S, Franke S, Meengs B, Brody J, Presnell S, Hamacher NB, Lantry M, Wolf A, Bukowski T, Rosler R, Yen C, Anderson-Haley M, Brasel K, Pan Q, Franklin H, Thompson P, Dodds M, Underwood S, Peterson S, Sivakumar PV, Snavely M. (2010) A dual-targeting PDGFRbeta/VEGF-A molecule assembled from stable antibody fragments demonstrates anti-angiogenic activity *in vitro* and *in vivo*. *MAbs*. 2, 20–34.
27. Robinson MK, Hodge KM, Horak E, Sundberg AL, Russeva M, Shaller CC, von MM, Shchaveleva I, Simmons HH, Marks JD, Adams GP. (2008) Targeting ErbB2 and ErbB3 with a bispecific single-chain Fv enhances targeting selectivity and induces a therapeutic effect *in vitro*. *Br. J. Cancer* 99, 1415–1425.
28. Huston JS, McCartney J, Tai MS, Mottola-Hartshorn C, Jin D, Warren F, Keck P, Oppermann H. (1993) Medical applications of single-chain antibodies. *Int. Rev. Immunol.* 10, 195–217.
29. Lindhofer H, Mocikat R, Steipe B, Thierfelder S. (1995) Preferential species-restricted heavy/light chain pairing in rat/mouse quadromas. Implications for a single-step purification of bispecific antibodies. *J. Immunol.* 155, 219–225.
30. Merchant AM, Zhu Z, Yuan JQ, Goddard A, Adams CW, Presta LG, Carter P. (1998) An efficient route to human bispecific IgG. *Nat. Biotechnol.* 16, 677–681.
31. Ward ES, Gussow D, Griffiths AD, Jones PT, Winter G. (1989) Binding activities of a repertoire of single immunoglobulin variable domains secreted from *Escherichia coli*. *Nature* 341, 544–546.
32. Harmsen MM, De Haard HJ. (2007) Properties, production, and applications of camelid single-domain antibody fragments. *Appl. Microbiol. Biotechnol.* 77, 13–22.
33. Dimasi N, Gao C, Fleming R, Woods RM, Yao XT, Shirinian L, Kiener PA, Wu H. (2009) The design and characterization of oligospecific antibodies for simultaneous targeting of multiple disease mediators. *J. Mol. Biol.* 393, 672–692.
34. Wu C, Ying H, Grinnell C, Bryant S, Miller R, Clabbers A, Bose S, McCarthy D, Zhu RR, Santora L, Davis-Taber R, Kunes Y, Fung E, Schwartz A, Sakorafas P, Gu J, Tarcsa E, Murtaza A, Ghayur T. (2007) Simultaneous targeting of multiple disease mediators by a dual-variable-domain immunoglobulin. *Nat. Biotechnol.* 25, 1290–1297.
35. Jackman J, Chen Y, Huang A, Moffat B, Scheer JM, Leong SR, Lee WP, Zhang J, Sharma N, Lu Y, Iyer S, Shields RL, Chiang N, Bauer MC, Wadley D, Roose-Girma M, Vandlen R, Yansura DG, Wu Y, Wu LC. (2010) Development of a two-part strategy to identify a therapeutic human bispecific antibody that inhibits IgE receptor signaling. *J. Biol. Chem.* 285, 20850–20859.
36. Gunasekaran K, Pentony M, Shen M, Garrett L, Forte C, Woodward A, Ng SB, Born T, Retter M, Manchulenko K, Sweet H, Foltz IN, Wittekind M, Yan W. (2010) Enhancing antibody Fc heterodimer formation through electrostatic steering effects: applications to bispecific molecules and monovalent IgG. *J. Biol. Chem.* 285, 19637–19646.
37. Muda M, Gross AW, Dawson JP, He C, Kurosawa E, Schweickhardt R, Dugas M, Soloviev M, Bernhardt A, Fischer D, Wesolowski JS, Kelton C, Neuteboom B, Hock B. (2011) Therapeutic assessment of SEED: a new engineered antibody platform designed to generate mono- and bispecific antibodies. *Protein Eng. Des. Sel.* 24, 447–454.
38. Jin H, Yang R, Zheng Z, Romero M, Ross J, Bou-Reslan H, Carano RA, Kasman I, Mai E, Young J, Zha J, Zhang Z, Ross S, Schwall R, Colbern G, Merchant M. (2008) MetMAb, the one-armed 5D5 anti-c-Met antibody, inhibits orthotopic pancreatic tumor growth and improves survival. *Cancer Res.* 68, 4360–4368.
39. Pacchiana G, Chiriaco C, Stella MC, Petronzelli F, De SR, Galluzzo M, Carminati P, Comoglio PM, Michieli P, Vigna E. (2010) Monovalency unleashes the full therapeutic potential of the DN-30 anti-Met antibody. *J. Biol. Chem.* 285, 36149–36157.
40. Martens T, Schmidt NO, Eckerich C, Fillbrandt R, Merchant M, Schwall R, Westphal M, Lamszus K. (2006) A novel one-armed anti-c-Met antibody inhibits glioblastoma growth *in vivo*. *Clin. Cancer Res.* 12, 6144–6152.
41. Neri D, Momo M, Prospero T, Winter G. (1995) High-affinity antigen binding by chelating recombinant antibodies (CRAbs). *J. Mol. Biol.* 246, 367–373.

42. Dechant M, Weisner W, Berger S, Peipp M, Beyer T, Schneider-Merck T, Lammerts van Bueren JJ, Bleeker WK, Parren PW, van de Winkel JG, Valerius T. (2008) Complement-dependent tumor cell lysis triggered by combinations of epidermal growth factor receptor antibodies. *Cancer Res.* 68, 4998–5003.

43. Holash J, Davis S, Papadopoulos N, Croll SD, Ho L, Russell M, Boland P, Leidich R, Hylton D, Burova E, Ioffe E, Huang T, Radziejewski C, Bailey K, Fandl JP, Daly T, Wiegand SJ, Yancopoulos GD, Rudge JS. (2002) VEGF-Trap: a VEGF blocker with potent antitumor effects. *Proc. Natl. Acad. Sci. USA* 99, 11393–11398.

44. Miller BR, Demarest SJ, Lugovskoy A, Huang F, Wu X, Snyder WB, Croner LJ, Wang N, Amatucci A, Michaelson JS, Glaser SM. (2010) Stability engineering of scFvs for the development of bispecific and multivalent antibodies. *Protein Eng. Des. Sel.* 23, 549–557.

38

CovX-BODIES

Abhijit Bhat, Olivier Laurent, and Rodney Lappe

CovX Research, Pfizer Worldwide R & D, San Diego, CA, USA

38.1 THE CovX-BODY CONCEPT

CovX-Body™ represents a new class of biotherapeutic agent created by the fusion of a relatively small molecular weight payload with an antibody, which serves as a carrier protein scaffold. The fused payload determines the targeting properties of the CovX-Body, whereas the antibody scaffold provides bivalency and long circulating half-life for the payload. The antibody scaffold prevents rapid renal clearance and also slows down the proteolytic and metabolic degradation of the fused pharmacophore and thereby provides an elegant solution to overcome the PK-PD problems of some peptides, small proteins, and traditional small molecules (Figure 38.1). The fusion reaction in a CovX-Body is mediated by a linker that is designed to react exclusively with a specific Lysine residue in the Fab region the carrier antibody. CovX-Bodies are one of the first examples of a class of site-specific antibody drug conjugates wherein the stoichiometry and the site of the drug attachment are precisely defined. CovX-Bodies are also unique in the respect that the scaffold antibody has no known targets *in vivo* and can be used across multiple therapeutic programs by changing the nature of the fused pharmacophore.

During the course of preclinical development of low molecular weight therapeutics such as peptides or small molecules, significant time and resources are devoted to improve the stability, metabolic liabilities, and pharmacokinetic profiles of the lead molecules. Often, medicinal chemistry programs are predominantly focused on improving the safety and pharmacokinetic properties of small molecules. This is often complicated by the fact that the chemical and structural modifications that result in increased potency or selectivity in small molecules also often result in poor pharmaceutical properties. Peptides have proven to be even more challenging for drug development as they suffer from rapid clearance, poor metabolic stability, and lack of oral bioavailability. Therapeutic antibodies, on the other hand, are enjoying growing success as biotherapeutic agents as they generally offer high selectivity and potency against their molecular targets along with prolonged circulating half lives [1]. The CovX-Body technology offers a way to confer antibody-like properties on small molecules and peptides and enables clinical development of molecules that may otherwise be compromised by their poor pharmaceutical properties. CovX-Bodies, like antibodies, work best against extracellular targets and are currently administered via intravenous or subcutaneous injections.

38.2 TECHNOLOGICAL ASPECTS

38.2.1 Scaffold Antibody

The origins of the scaffold antibody used to generate CovX-Bodies can be traced to the catalytic antibody work done at the Scripps Research Institute (La Jolla, CA) in the 1990s, where researchers pioneered a new process termed "reactive immunization" to generate antibody catalysts [2]. In general, the haptens used to generate catalytic antibodies up to that point were designed to mimic the geometric and electronic features of the proposed transition state of the reaction of interest.

Fusion Protein Technologies for Biopharmaceuticals: Applications and Challenges, First Edition. Edited by Stefan R. Schmidt.

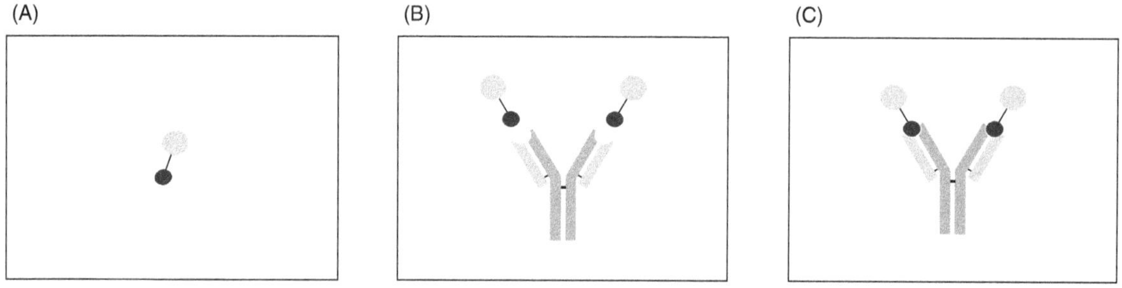

FIGURE 38.1 Generating a CovX-Body. (A) A payload functionalized with a specially designed linker is synthesized. The linker is designed to be recognized and covalently bind only to the active site of the antibody. (B) In a well-controlled reaction, the payload-linker is combined with CovX's proprietary antibody. (C) The antibody selectively recognizes and combines with the linker, to form a well-characterized, bivalent fusion molecule referred to as a CovX-Body. The fully assembled CovX-Body is isolated by simple purification.

Immunization with such haptens resulted in antibodies that catalyzed the target reactions by stabilizing the transition state intermediates. The Scripps researchers reasoned that by using reactive compounds as haptens they would be able to isolate antibodies wherein the interaction with the antigen would involve chemical reactivity in addition to nonpolar and electrostatic interactions. Antibodies identified from reactive immunization were expected to behave more like enzymes, wherein the active site amino acids participate in bond breaking and bond forming reactions. The strategy of reactive immunization was applied to identify antibody catalysts that would mimic class I aldolase enzymes. Class I aldolase enzymes have an active site Lysine residue whose ε-amino group reacts with a carbonyl group of the substrate to form a Schiff's base. The Schiff's base serves as an electron sink that facilitates the abstraction of H^+ from the C-α position resulting in the formation of an enamine, thus activating the substrate as an aldol donor (Figure 38.2). A simple phenethyl-1,3-diketone (**1**)

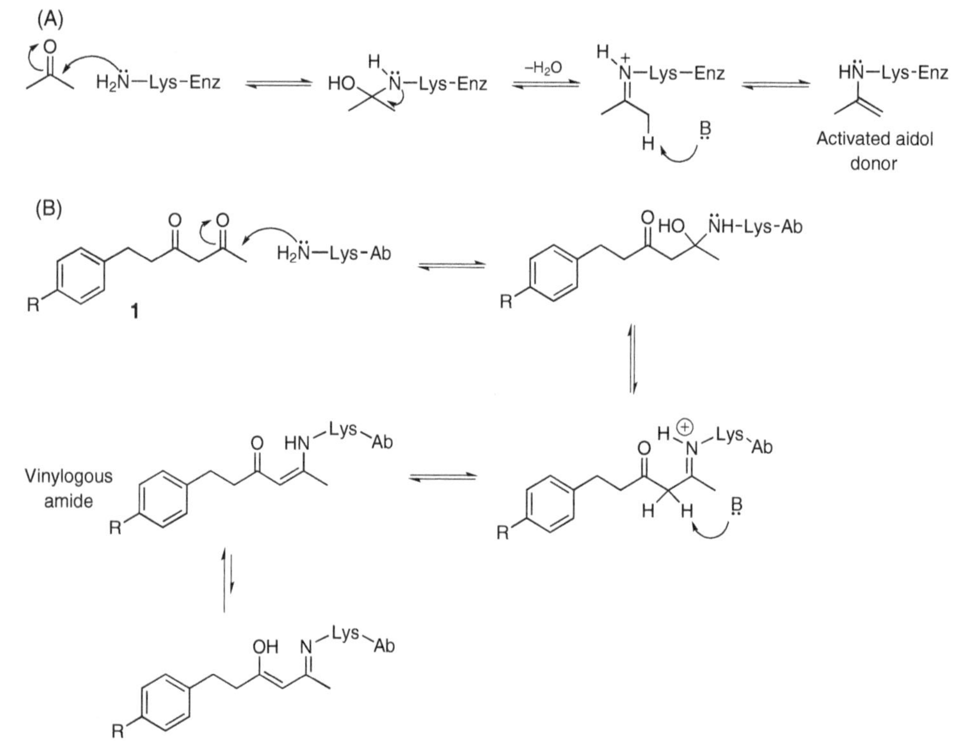

FIGURE 38.2 (A) Reaction mechanism of class I aldolase. (B) 1,3-diketone designed to trap lyinse ε-amino group in the active site of the catalytic antibody and facilitate the formation of enamine intermediate, analogous to the natural class I aldolase enzyme.

was used as a reactive hapten, designed to trap the lysine ε-amino group and facilitate the formation of an enamine intermediate. The diketone hapten, after coupling to keyhole limpet hemocyanin was used in the immunization of 129 G^{IX+} mice [3]. Twenty hybridomas producing antibodies that bound the 1,3-diketone were isolated from the mice. Two of these antibodies, m38C2 and m33F12, were shown to catalyze the synthetic and retrosynthetic aldol reactions, whereas the remaining 18 antibodies were found to be inactive as catalysts. Mechanistic studies of m38C2 and m33F12 revealed that the reaction with the hapten proceeded via the proposed enamine-vinylogous amide intermediate and these antibodies facilitated the aldol and retro-aldol reactions via a mechanism that mimics the natural class I aldolases. The catalytic activity of m38C2 and m33F12 can be attributed to an active site Lysine with a perturbed pKa. The structure of Fab' of m33F12 was determined at 2.15 Å resolution and in general was found to be identical to other known Fab structures [4]. The antigen binding pocket was found to be 11 Å deep and comprised of mostly nonpolar amino acid residues with a lysine residue, LysH93 at the base of the pocket. In most antibodies, H93 is an alanine next to a polar arginine H94; which in the case of m33F12 was found to be replaced by a nonpolar isoleucine residue. The structure revealed that no charged residues were close enough to form a salt-bridge or an H-bond with LysH93. It is hypothesized that this unusual hydrophobic microenvironment along with absence of an H-bond or a salt-bridge perturbs the pKa of ε-amino group of LysH93 to around 6. This enables LysH93 to function as a strong nucleophile at neutral pH when all other lysine side chains are protonated and non-nucleophilic. The structure of m38C2 was expected to be identical to m33F12 as all the residues that are in van der Waals contact with LysH93 are conserved in both the antibodies, except for m33F12 IleH94, which is a threonine in m38C2.

In the course of mechanistic studies of the aldolase antibodies m38C2 and m33F12 it became apparent that these antibodies readily react with 1,3-diketones to form a so-called reversible covalent bond via an enamine docking mechanism (Figure 38.3). Thus, any molecule that is functionalized with a 1,3-diketone linker will react with LysH93 of the aldolase antibody and be fused to the antibody via a reversible covalent bond. This effectively results in the catalytic, untargeted antibody being chemically programmed with the binding specificity of the molecule carrying the 1,3-diketone linker. This concept was demonstrated in 2003 by Rader et al. who synthesized a 1,3-diketone derivative of a potent RGD peptidomimetic small molecule that bound integrins $\alpha_v\beta_3$ and $\alpha_v\beta_5$. They showed that upon fusion with m38C2 the RGD peptidomimetic effectively mediated the cell surface binding of m38C2 to integrin expressing cells [5]. The antibody m38C2 was incapable of binding the integrin expressing cells in absence of a fused peptidomimetic. Through mouse experiments they were able to show that the integrin targeting complex of antibody m38C2 can be formed *in vivo* and this complex increases the circulating half-life of the integrin targeting small molecule and effectively inhibits tumor growth in animal models of Kaposi's sarcoma and colon cancer. The *in vivo* half-life of this prototype murine CovX-Body was projected to be around 2 h (based on the k_{off} rates for the complex of m38C2 and a 1,3-diketone linker, as determined by surface plasmon resonance). Remarkably, the authors found that the circulatory half life of the RGD targeting CovX-Body to be around 3 days. Additional successful examples of chemical programming of m38C2 with small molecule pharmacophores resulted in generation of highly potent and selective CovX-Bodies targeting the endothelin and CCR5 receptors (Figure 38.3) [6,7].

To develop CovX-Bodies for human therapeutic applications, the murine aldolase antibody m38C2 was humanized through grafting of the murine complementarity determining regions (CDRs) into a human antibody framework sequence. Human V_κ gene DPK9 and human J_κ gene JK4 were used as frameworks for the humanization of the κ light chain variable domain. Human V_H gene DP47 and human J_H gene JH4 were used as frameworks for the humanization of the heavy chain variable domain of m38C2. Related frameworks from the same light and heavy chain subgroups, V_κI and V_HIII, respectively, have been used in the humanization of widely used therapeutic antibodies, such as Herceptin® [8,9]. Combinations of light chains derived from germ line DPK9 with heavy chains derived from germ line DP47 are also frequently found in native human antibodies and have been used as templates for synthetic human antibody libraries [10,11]. All CDR residues, as well as defined framework residues in both light and heavy chain variable domains, were grafted from murine 38C2 onto the human framework. The selection of the grafted framework residues was based on the crystal structure of the murine antibody m33F12 described earlier. The grafted framework residues consist of five residues in the light chain and seven in the heavy chain and encompass residues that are likely to participate directly or indirectly in the catalytic activity of the aldolase antibody. These include the reactive LysH93, which is positioned in the framework region 3 (FR3) of the heavy chain. In addition to the CDR residues, a number of framework residues line the active site pocket. Among these, LeuL37, GlnL42, SerL43, ValL85, PheL87, ValH5, SerH40, GluH42, GlyH88, IleH89, and ThrH94 were grafted on the human framework. Six residues, SerH35, ValH37, TrpH47, TyrH95, TrpH103, and PheL98 which are conserved between murine monoclonal antibodies 33F12 and 38C2, are within a 5-Å radius of LysH93. These residues were conserved in the humanization process. The resulting humanized variable domains were grafted on human constant domains. The G1m(f) heavy chain and Km3 light chain allotypes were selected due to their high prevalence in pooled blood donor populations and

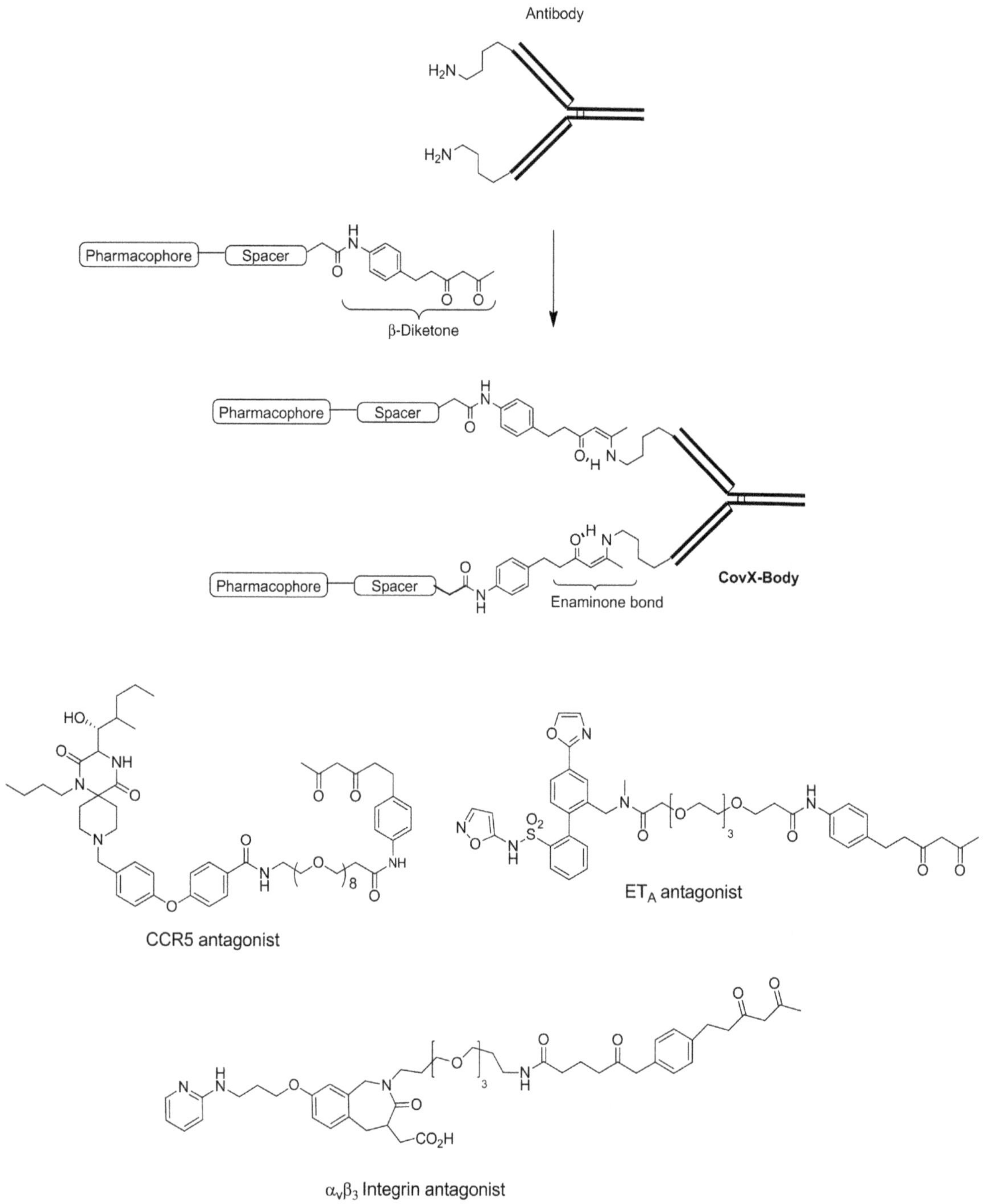

FIGURE 38.3 Formation of CovX-Body using 1,3-diketone based linker and structures of representative small molecule pharmacophores with 1,3-diketone linkers that were used to generate selective and potent CovX-Bodies targeting CCR5, endothelin A, and $\alpha_v\beta_3$ integrin receptors.

precedence in other commercial therapeutic antibodies, such as Synagis® and Erbitux®. The resulting humanized version of 38C2 referred to as CVX-2000, was stably transfected into Chinese hamster ovary cells, and serves as the common carrier scaffold for all CovX-Bodies in preclinical and clinical development. It is 98.2% human when compared to murine 38C2 antibody. Only 12 of the 667 light and heavy chain amino acids in CVX-2000 are retained from the murine sequence: seven in the heavy chain (out of 448), and five in the light chain (out of 219). On the basis of the 3D structure, all five of the light chain murine-sequence residues are buried (core) residues, and therefore unlikely to be of immunogenic potential. Four of the seven heavy chain retained murine residues are buried as well. Of the three predicted surface accessible murine-sequence heavy chain amino acids, one is buried deep in the hydrophobic binding pocket (GluH42).

The binding cleft of CVX-2000 was characterized functionally by fluorescence spectroscopy and structurally by X-ray crystallography. The unique reactivity and the apparent pKa of ~5–6 for the active site Lys H93 is maintained during the humanization process. The hydrophobicity of the binding cleft of CVX-2000 was investigated by the use of the hydrophobic probe molecule 1-anilinonaphthalene-8-sulfonic acid (ANS) [12]. ANS has very little fluorescence emission in aqueous solutions. However, when associated with a hydrophobic region of a protein, a substantial blue shift and increase in fluorescence with emission maximum around 460 nm is observed. When CVX-2000 is incubated alone in solution with ANS, a strong fluorescent emission is observed. The emission signal is significantly reduced and red-shifted after CVX-2000 has been conjugated to a diketone, presumably by displacement of ANS from the hydrophobic binding cleft. The resulting fluorescent emission spectrum of conjugated CVX-2000 is comparable to the control sample of polyclonal IgG incubated with ANS, as seen in Figure 38.4A. The apparent pKa of LysH93 of CVX-2000 was determined by measuring the binding activity of a model diketone substrate (2,4-pentanedione) as function of pH. The binding of 2,4-pentanedione was followed by measuring the UV absorbance signal at 316 nm that results from the vinylogous amide formed from the reaction of the 2,4-pentanedione with LysH93 in the active site of CVX-2000. Figure 38.4B shows the titration experiment of CVX-2000 with 2,4-pentanedione. This figure suggests an apparent pKa of approximately five for the two reactive LysH93s of CVX-2000 under equilibrium reaction conditions. This is substantially lower than the typical value for the lysine ε-amino group (~10.2) in an aqueous environment. At the standard fusion pH of 6.5, the Henderson–Hasselbach equation predicts a Lys ε-amino with a pKa of 5 to be ~150,000-fold more nucleophilic than a typical lysine side chain. Under those conditions over 99.9% of surface lysines are protonated, thus nonreactive. Therefore, of the 88 lysines present in the primary sequence of CVX-2000, the only lysines expected to be nucleophilic enough to fuse with the electrophilic linker would be the two LysH93 residues.

38.2.2 Linker Design

The early examples of CovX-Bodies targeting integrin, endothelin and CCR5 receptors utilized linkers based on the 1,3-diketone hapten used to raise 38C2. The advantage of this linker system was that the fusion reaction was, for all practical purposes, instantaneous and the fusion product resulted in the formation of a vinylogous amide in the active site that could be detected spectroscopically, which made it possible to easily follow the course of the fusion reaction and characterize the fusion products. However, a major concern with diketone linkers was that the fusion reactions between the antibody and the targeting-pharmacophore were mediated via a reversible, albeit, stable bond. The reversibility of the bond presented several challenges for the development of CovX-Bodies. For example, the strength of the interaction between the linker and the antibody, and thereby the degree of reversibility was expected to be influenced by the nature of the linker and the pharmacophore. This, in turn, introduced an added variable that needed to be studied and optimized for each class of fusion product. Additionally, the *in vitro* bioactivity assessment of CovX-Bodies was complicated by the fact that the observed activity could be a result of the payload released from the antibody rather

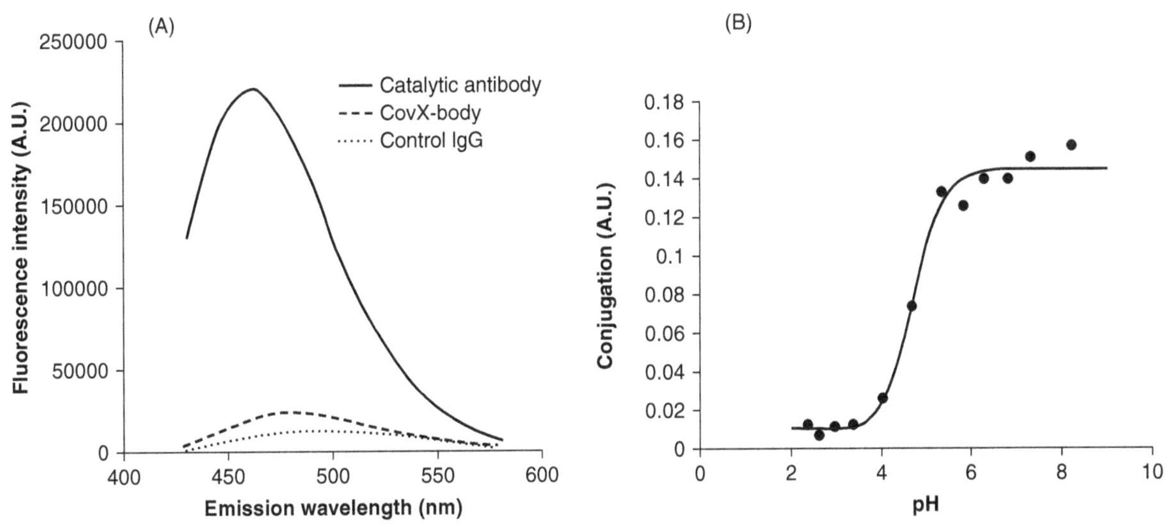

FIGURE 38.4 (A) Hydrophobicity of CVX-2000 binding cleft as determined by ANS probe. (B) Binding activity of CVX-2000 with pentanedione as a function of pH.

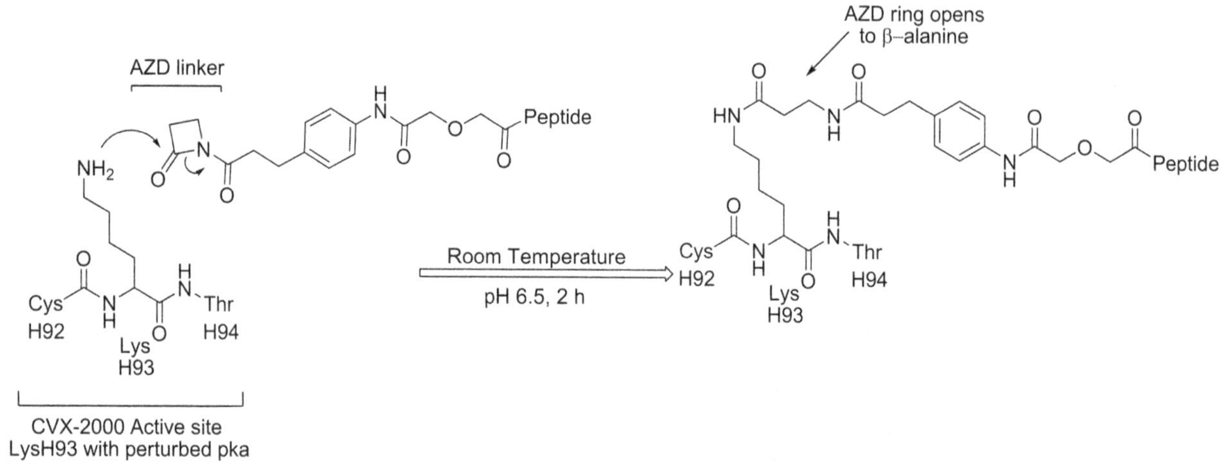

FIGURE 38.5 Irreversible fusion of AZD linker with LysH93 in the active site of CVX-2000.

than the CovX-Body itself. More importantly, the release of the payload due to the reversible enaminone bond could easily compromise the pharmacokinetic advantage provided by the carrier antibody scaffold.

The 1,3-diketone CovX-Bodies were found to be stable in *in vitro* assay systems and no detectable payload was released from these CovX-Bodies when subjected to dialysis, thus providing circumstantial proof that the *in vitro* activity was mediated by the CovX-Body rather than the free payloads. However, *in vivo* PK experiments with 1,3-diketone based CovX-Bodies revealed that the serum half-life of the fused pharmacophore was significantly lower than the serum half-life of the CVX-2000 scaffold, likely a result of the reversible 1,3-diketone linker system. To overcome this serious limitation of the 1,3-diketone system, an extensive screening campaign was undertaken to identify electrophilic linkers that would react selectively and irreversibly with LysH93. This led to the identification of a phenethyl acyl azetidinone (AZD) that forms a covalent conjugate with CVX-2000 (Figure 38.5). The fusion reaction with the AZD linker is an acylation of LysH93, where in the ε-amino group acts as a nucleophile to selectively react with the electrophilic azetidinone group. This results in the AZD ring opening and fusion of the linker with the antibody via a stable beta alanine peptide bond. Alkyl AZDs alone were found to be ineffective as selective inhibitors of CVX-2000, thereby highlighting the importance of the additional phenyl group in providing a recognition element and enabling an induced fit of the AZD linker in the active site of CVX-2000. The effect of the irreversible AZD versus the reversible 1,3-diketone linker system on the pharmacokinetic profile of the fused payload is shown in Figure 38.6.

Peptide mapping experiments were conducted to confirm the site of conjugation between CVX-2000 and AZD linkers.

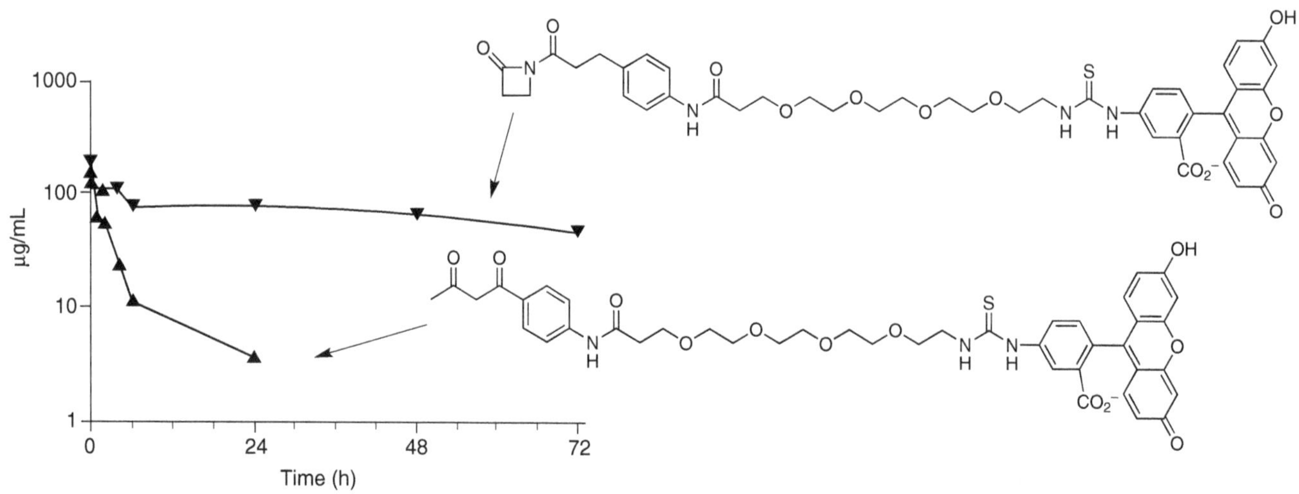

FIGURE 38.6 Comparison of the PK profile of PEGylated FITC dye when fused with CVX-2000 via the irreversible AZD linker or the reversible diketone linker.

CovX-Bodies were subjected to trypsin digestion and the resulting peptide fragments were analyzed by electrospray LC/MS/MS. The LC/MS/MS data was compared to the predicted fragmentation pattern and processed using Waters Protein Lynx Global server. Near complete identification of predicted peptide fragments was achieved along with confirmation that only the peptide fragment containing LysH93 was modified with the AZD linked peptide. A site directed mutant of CVX-2000 was expressed wherein the LysH93 was mutated to AlaH93. No fusion reaction was detected when LysH93Ala mutant of CVX-2000 was treated with a 3:1 stoichiometric excess of AZD-linked peptides, thereby confirming that LysH93 is required for the fusion reaction. Direct proof of the molecular specificity of the AZD conjugation to LysH93 was obtained when the crystal structure of the Fab fragment of CVX-2000 in complex with a peptide with an AZD linker was determined at 1.8 Å resolution. Consistent with peptide mapping results, each CVX-2000 Fab fragment was found to bind one AZD linker (Figure 38.7A). LysH93 was located at the bottom of a 12 Å long lipophilic tunnel, forming a covalent amide bond with the linker (Figure 38.7B) providing structural proof of the site of conjugation. This result perfectly agrees with the reaction mechanism of a nucleophilic attack of the LysH93 ε-amino group on the AZD carbonyl carbon, leading to ring opening and covalent conjugation via a β-alanine bond. As predicted, LysH93 was surrounded by nonpolar residues consistent with an environment contributing to the reduced pKa and increased reactivity for the LysH93 ε-amino group. TyrH95 and TyrL37, which occlude the binding site in the apo-crystal structure, adopt an open conformation in the covalent conjugate structure thereby enabling access to LysH93. The LysH93 ε-amino group also formed a 2.9 Å hydrogen bond with a water molecule that was stabilized by the hydroxyl groups of SerH106 and TyrL41. The side chains of TyrH95, PheH96, TyrH97, TrpH33, TyrL101, and TyrL37 lined the wall of the hydrophobic tunnel, consistent with the good stability of the covalently linked therapeutic conjugates.

The reaction between CVX-2000 and 1,3-diketone linkers is very rapid and often complete within a matter of minutes. Reaction of the ε-amino group of LysH93 with 1,3-diketone results in the formation of a vinylogous amide that can be detected spectroscopically. In contrast, the reaction between LysH93 and AZD linkers results in the formation of a β-alanine bond that lacks a characteristic spectroscopic signature. Resultantly, the kinetics of the fusion reaction of the AZD linkers are usually studied using a SEC/MS method. The SEC/MS method is capable of detecting and calculating the ratios of various species during the course of the conjugation reaction. At the start of the conjugation reaction, only the unconjugated CVX-2000 is detected; a few minutes after the onset of the fusion process, three major species are observed with masses consistent with, (1) unconjugated CVX-2000; (2) monovalent fusion product; and (3) divalent fusion product. The fusion reaction is deemed complete when only the divalent product is detected by SEC/MS. The reaction with AZD linker is slower than with the 1,3-diketone linkers but is usually complete within a few hours. The physicochemical properties of the payload being conjugated have an influence on the kinetics of the fusion reaction.

38.2.3 Choice of Payload

The range of payloads that have been fused with CVX-2000 to generate CovX-Bodies include small molecules, linear and cyclic peptides, cysteine knotted toxins, 20 kDa proteins, and

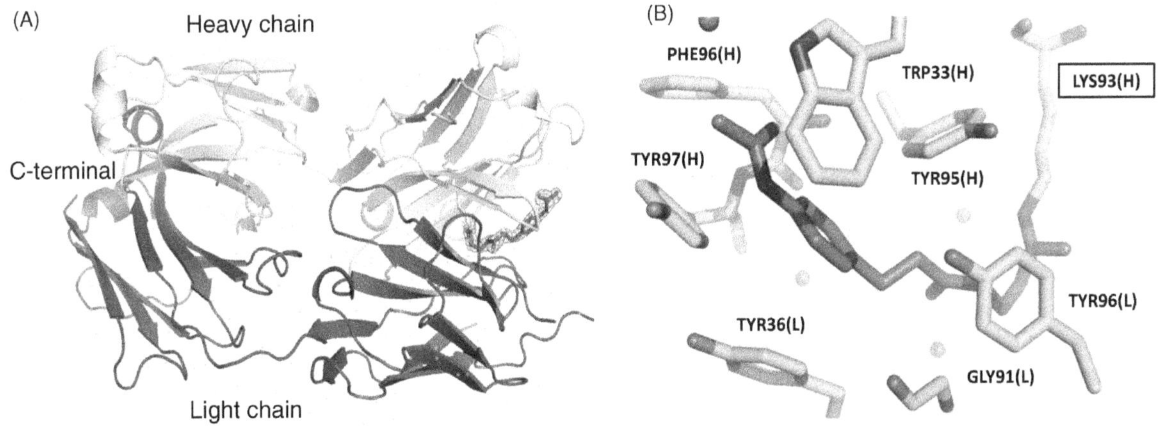

FIGURE 38.7 Crystal structure of CVX-2000 Fab in complex with AZD linker. (A) The AZD linker is shown in stick model with its omit (2Fo-Fc) electron density map contoured at 1.2 s level. Linker atoms outside of the hydrophobic tunnel were disordered and not observed in the crystal structure. (B) The covalent adduct between LysH93 and the AZD linker, as well as other key residues lining the hydrophobic tunnel.

oligonucleotides. However, most of the clinical stage applications of CovX-Bodies to date involve fusion with peptidic pharmacophores. Peptides hold tremendous therapeutic and commercial promise because they possess enormous structural diversity that are highly specific and potent, and generally have fewer safety issues relative to small molecules. Despite these specificity and potency advantages, peptides have been handicapped by their poor half-life, which is mostly a result of rapid renal clearance and proteolytic degradation by serum proteases. CovX-Body technology provides a means to markedly extend the half-life of peptide-based drugs without compromising their specificity and potency. This also presents opportunities to pursue and resurrect peptide leads that had previously been abandoned due to poor pharmacokinetic properties. There are precedents for the conjugation of peptides to scaffold proteins such as albumin, transferrin, and Fc fragments that have resulted in molecules that are presently undergoing clinical evaluation [13–16]. In contrast with peptide fusion constructs generated at the DNA level and expressed in cell cultures, the CovX-Body technology involves a chemical fusion approach that permits the use of peptides with non-natural structural features such as noncoded amino acids, peptoids, N- and C-terminus caps and stapled or otherwise constrained backbones. In addition, chemical fusion allows tremendous flexibility in modifying the linker position and composition for optimal efficacy, stability and physicochemical properties.

38.2.4 Manufacture, Process, Characterization

The scaffold antibody, CVX-2000 is expressed in Chinese hamster ovary cells and purified at kilogram scale for clinical use by a standard three-column process that includes protein-A, anion exchange, and cation exchange chromatographic steps. AZD tethered peptides are readily synthesized at process scales using standard Fmoc-solid phase peptide synthesis methods. The manufacturing process of CovX-Bodies is optimized for robustness and simplicity. AZD-peptides are solubilized shortly before use in either aqueous buffer or DMSO depending on their physical properties. They are then added directly to CVX-2000 in long-term storage buffer and left to react with the catalytic antibody, generally overnight at room temperature. Reaction progress between CVX-2000 and the AZD-peptide can be monitored using multiple orthogonal methods, such as reduction of CVX-2000 specific activity, displacement of bound ANS, reduced availability of CVX-2000 to form enaminone bonds with diketones, and presence of CovX-Body by LC/MS or by standard HPLC such as HIC.

The CovX-Bodies resulting from the reaction of the AZD-peptide and CVX-2000 are purified by tangential flow filtration to remove unreacted peptides and other low molecular weight process related impurities from the bulk drug product. When working with negatively charged peptides, that tend to remain associated with CovX-Bodies as a result of the Gibbs–Danan equilibrium, a charged membrane can be added to the purification process. In those situations, a cation exchanging filter has always been highly efficient at capturing unreacted peptides in a single pass while letting the CovX-Bodies flow through without any adsorption.

Drug Product release is performed using conventional assays classically used for the release of therapeutic monoclonal antibodies. Similarly, Drug Product sterility is ensured by classic isolation and filtration techniques in use with most parenteral products.

38.3 APPLICATIONS OF THE CovX-BODY TECHNOLOGY

38.3.1 CVX-060, a Selective Angiopoietin-2 Targeting CovX-Body

Angiopoietins are ligands of the endothelial receptor Tie-2 and play a crucial role in the tumor angiogenic switch. It is generally accepted that Angiopoietin-1 (Ang1) and Angiopoietin-2 (Ang2) have opposing effects on tumor angiogenesis, and that increased expression of Ang2 relative to Ang1 in tumors promotes angiogenesis in presence of other growth factors such as VEGF and PDGF [17]. The product concept for CVX-060 is based on the premise that selectively removing Ang2 from a tumor will restore the Ang2/Ang1 balance and inhibit angiogenesis and tumor growth [18]. A series of analogs of a selective Ang2 binding peptide were synthesized for fusion with CVX-2000, see Table 38.1. The effect of the tether attachment site on the properties of the resulting CovX-Body was evaluated by moving the tether position across the length of the peptide. This was done by a serial substitution of Lys at each position of the peptide and subsequently using the ε-amino group of the Lys side chain as a site for tethering the AZD linker. The resulting CovX-Bodies were tested in an *in vitro* Ang2-Tie2 competition ELISA assay. The pharmacokinetic profile of CovX-Bodies was studied in mice to identify molecules with the best combination of potency and pharmacokinetics. This "tether walk" exercise showed the dramatic effect of tether positioning on the potency and pharmacokinetic profile of CovX-Bodies. The molecules linked through either the N- or C-termini were active with IC50s of 1.8 and 0.2 nM, respectively. Tethering at certain internal residues (4, 6, 7, 10, and 13) resulted in complete loss of antagonist activity of CovX-Bodies, whereas tethering at positions 5, 14, and 17 resulted in significantly compromised potency. All the other internal sites of attachment resulted in active CovX-Bodies with potencies comparable to N- or C- terminal tethering. The pharmacokinetic profile of the CovX-Bodies showed a dependence on the tether site as molecules bearing a tether

TABLE 38.1 Influence of Tether Position the Bioactivity and PK Profile of Ang2 Targeting CovX-Bodies

Compound#	Sequence	IC_{50} (nM)	$T_{1/2}$ (h)
Peptide 1	QK(Ac)YQPLDELDK(Ac)TLYDQFMLQQG	98	0.1
1	K(0P)K(Ac)YQPLDELDK(Ac)TLYDQFMLQQG	1.8	24
2	QK(0P)YQPLDELDK(Ac)TLYDQFMLQQG	0.3	13
3	QK(Ac)K(0P)QPLDELDK(Ac)TLYDQFMLQQG	0.2	18
4	QK(Ac)YK(0P)PLDELDK(Ac)TLYDQFMLQQG	>1000	ND
5	QK(Ac)YQK(0P)LDELDK(Ac)TLYDQFMLQQG	44.2	ND
6	QK(Ac)YQPK(0P)DELDK(Ac)TLYDQFMLQQG	>1000	ND
7	QK(Ac)YQPLK(0P)ELDK(Ac)TLYDQFMLQQG	>1000	ND
8	QK(Ac)YQPLDK(0P)LDK(Ac)TLYDQFMLQQG	0.3	35
9	QK(Ac)YQPLDEK(0P)DK(Ac)TLYDQFMLQQG	2.0	72
10	QK(Ac)YQPLDELK(0P)K(Ac)TLYDQFMLQQG	>1000	ND
CVX-32	QK(Ac)YQPLDELDK(0P)TLYDQFMLQQG	0.3	74
12	QK(Ac)YQPLDELDK(Ac)K(0P)LYDQFMLQQG	0.1	56
13	QK(Ac)YQPLDELDK(Ac)TK(0P)YDQFMLQQG	>1000	ND
14	QK(Ac)YQPLDELDK(Ac)TLK(0P)DQFMLQQG	32.8	ND
15	QK(Ac)YQPLDELDK(Ac)TLYK(0P)QFMLQQG	0.2	66
16	QK(Ac)YQPLDELDK(Ac)TLYDK(0P)FMLQQG	0.3	94
17	QK(Ac)YQPLDELDK(Ac)TLYDQK(0P)MLQQG	19.5	ND
18	QK(Ac)YQPLDELDK(Ac)TLYDQFK(0P)LQQG	0.6	72
19	QK(Ac)YQPLDELDK(Ac)TLYDQFMK(0P)QQG	0.1	65
20	QK(Ac)YQPLDELDK(Ac)TLYDQFMLK(0P)QG	0.1	35
21	QK(Ac)YQPLDELDK(Ac)TLYDQFMLQK(0P)G	0.3	27
22	QK(Ac)YQPLDELDK(Ac)TLYDQFMLQQK(0P)	0.2	20
CVX-060	QK(Ac)YQPLDEK(Ac)DK(0P)TLYDQFMLQQG	0.5	110
CVX-060T	QK(Ac)YQPLDEK(Ac)DK(0P)TLYDQFMLQQG	342	ND
CVX-87	QK(Ac)YQPLDELDK(0P)TLFDQFMLQQG	0.2	61
Peptide 2	TNFMPMDDLEQRLYEQFILQQG	5.4	ND
CVX-37	(DFB)TNFMPMDDLEK(0P)RLYEQFILQQG	0.4	26
CVX-51	(0P)TNFMPMDDLEQRLYEQFILQQG	1.8	28

The *in vitro* potency of CovX-Bodies tethered at different residues, indicated by K(0P), in blocking Ang2–Tie2 interaction and their mouse pharmacokinetics. ND, not determined; 0P, no PEG unit between peptide and the AZD linker; K(Ac), acetylated Lysine; DFB, 1,5-difluorobenzoyl. Compounds 1–10, 12–22, CVX-32, CVX-060, CVX-87, CVX-37, and CVX-51 were all tested as CovX-Bodies, while peptides 1, 2, and CVX-060T (the pharmacophores of CVX-060) were also tested as controls. The IC_{50}s were average data from at least three independent preparations for each compound.

at internal sites showed better half-lives compared with the ones tethered near either termini. Thus, CovX-Body with an internal tether at position 11 was found to have optimal potency and pharmacokinetic profile. To improve its potency and pharmacokinetic profile, the peptide pharmacophore was subjected to a series of point mutations with natural and unnatural amino acids including several non-natural capping groups at the N-terminus. Thus, when Lys at position 9 was changed to acetyl Lys, the resulting CovX-Body, CVX-060 had a significantly improved half-life of 110 h in mice. As shown in Table 38.1, CovXing the lead peptide dramatically improves its potency and half-life, thus making it a viable development candidate. CVX-060 selectively binds Ang2 (Kd: 1.89×10^{-10} M; k_{off}: $5.69 \times 10^{-5}\,s^{-1}$) blocks the Ang2/Tie-2 interactions and does not interact with Ang1, Ang3, or Ang4. In staged Colo-205 xenograft models, the animals treated with CVX-060 (10 mg/kg, once a week), had a significant tumor growth inhibition (38–60%) coupled with reduction in intratumoral Ang2 protein levels (~50%) and reduction in microvessel density (34–48%). Additionally, the intratumoral levels of proangiogenic $Tie2^+$ $CD11b^+$ cells were reduced (by up to 85%) after treatment with CVX-060, demonstrating for the first time that selective Ang-2 targeting can block the recruitment of proangiogenic monocytes into tumors. CVX-060 demonstrated significant additive efficacy in xenograft models when combined with standard of care agents such as sunitinib, sorafenib, bevacizumab, irinotecan, and docetaxel. This demonstrates that CVX-060 is effective at reducing tumor angiogenesis and growth when administrated alone or in combination with standard of care agents.

The safety and efficacy of CVX-060 are currently being evaluated in patients with advanced solid tumors [19]. On the basis of the available datasets from 30 of 34 patients who were enrolled in an early Phase I study, CVX-060 was generally well tolerated in patients and a maximum tolerated dose (MTD) was not reached during trial. The preliminary pharmacokinetic analysis suggests dose proportional increases in area under the curve (AUC) and mean with mean ß half-life 5.5 days. The most common adverse events observed were proteinuria and fatigue. At all dose levels, patients showed a dramatic increase in total serum Ang2 levels following administration of CVX-060, thus providing a clinical proof of pharmacology and mechanism. Patients were assessed radiographically for response per RECIST criteria; of the 34 patients enrolled, 24 (71%) remained on study ≥8 weeks. In addition to the radiographic assessment, DCE-MRIs were conducted in selected patients to assess the effect of CVX-060 on tumour blood flow [20]. There were significant mean decreases in K_{trans} as measured by DCE-MRI at Day 4 ($p = 0.034$) and Day 28 ($p = 0.0025$) in 11 patients with data available at all 3 time points. A total of 10 out of 11 patients had decreases in K_{trans} with mean decreases of 12.9% (0.94–44%) at Day 4 and 25.6% (8–61%) at Day 28. The early clinical data and safety profile is encouraging and CVX-060 continues to be studied as a monotherapy and in combination with standards of care for the treatment of solid tumors.

38.3.2 CVX-241, an Example of a Bispecific Antibody

Most diseases have complex etiology and are driven by multiple mechanisms. It is therefore reasonable to expect that therapeutic agents that simultaneously engage multiple targets or pathways should offer better treatment outcomes. Several strategies for the creation of bispecific antibodies (BsAbs) have been proposed over the last two decades [21]. Unfortunately, most of the proposed BsAb formats suffer from lack of product homogeneity and challenging production problems [22]. Recent advances in antibody engineering techniques has enabled the creation of new recombinant formats like tandem single chain variable Fragment [23] (scFv), diabodies [24], tandem diabodies [25], two in one antibody [26], and dual variable domain antibodies [27] (DVD-Ig). These new formats solved some of the manufacturing issues but these scaffolds, with the exception of DVD-Ig and the two in one antibody, suffer from poor pharmacokinetics and therefore require frequent dosing or conjugation to other scaffolds [28] to improve half-life. Moreover, the complexity of generating new versions of these scaffolds slows down the cycle time for their optimization and lengthens the overall drug development time. CovX-Body technology offers a chemical solution for the creation of bi-specific antibodies referred to herein as bi-specific CovX-Bodies [29]. Bispecific CovX-Bodies are created by using a branched AZD linker that makes it possible to attach two different pharmacophores on each Fab arm as shown in Figure 38.8. This technology has been applied to the creation of CVX-241 a bispecific CovX-Body that targets both Ang2 and vascular endothelial growth factor (VEGF), two growth factors implicated in tumor angiogenesis.

VEGF plays a dominant role in tumor angiogenesis [30,31] and its inhibition has been implemented clinically, mostly in combination with chemotherapy, to treat cancers of the intestine, liver, brain, colon, kidney, breast, and lung. However, in a majority of cases, the benefit to patients from VEGF inhibition is modest at best. Ang2, as discussed earlier, is another major pro-angiogenic growth factor widely expressed in renal, skin, colon, breast, and many other cancers. Ang2 plays a complementary role to VEGF in the tumor angiogenesis [32]. The increased expression of both VEGF and Ang2 has also been shown to correlate with poor survival [33]. In human tumor, xenograft studies in mice, the physical combination of VEGF, and Ang2 antagonists showed enhanced efficacy relative to the use of either single agent [18]. Thus, simultaneous inhibition of both VEGF and Ang2 may provide better benefit to patients with cancers that express both VEGF and Ang2 than inhibiting either VEGF or Ang2 alone. Physical combination of VEGF and Ang2 antagonists is, however, greatly hindered by a number of factors including the high

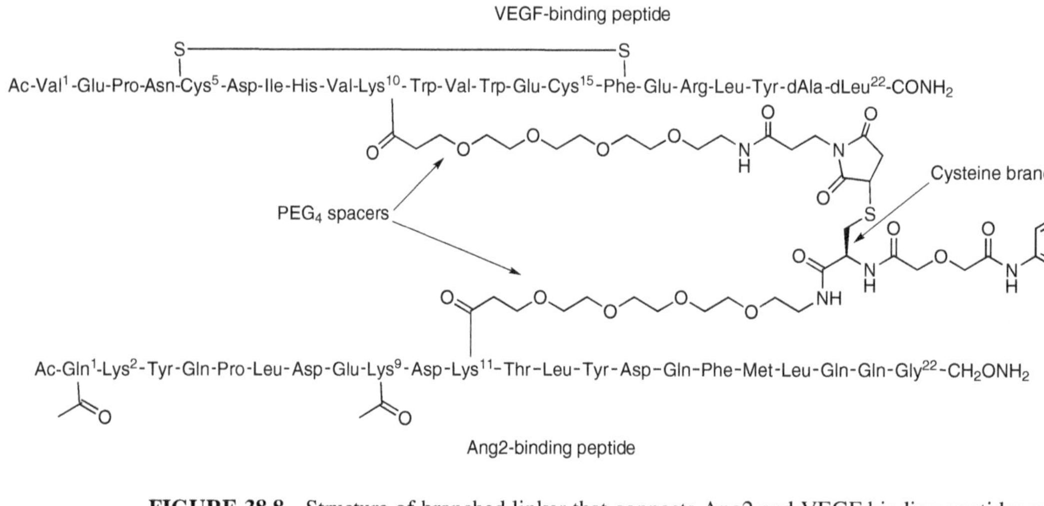

FIGURE 38.8 Structure of branched linker that connects Ang2 and VEGF binding peptides with AZD linker.

cost of each product, the requirement of multiple infusions and possible compliance and regulatory hurdles. CVX-241 is a novel, first-in-class bispecific antibody that neutralizes two important tumor angiogenesis growth factors *in vitro* as well as *in vivo* and shows preclinical anti-tumor efficacy equal to a combinations achieved by mixing distinct VEGF and Ang2 inhibiting agents that are not part of the same molecule.

CVX-241 was evaluated for its ability to inhibit the interaction between human VEGF and its receptor VEGFR2, as well as the interaction between human Ang2 and its receptor Tie-2 using competitive binding ELISAs. CVX-241 inhibited the binding of human rVEGF165 to human VEGFR2 with high potency, $IC_{50} = 0.9$ nM, comparable to the monofunctional VEGF binding CovX-Body. It also inhibited the binding of hAng2 to hTie-2 with similar potency to the Ang2 binding CovX-Body (CVX-060), $IC_{50} = 0.6$ nM. Additional *in vitro* binding experiments showed that CVX-241 was able to simultaneously bind both Ang2 and VEGF. Mouse PK studies with CVX-241 showed that the Ang2 binding activity had a half life of 91 h whereas the VEGF binding activity had a life of 92 h. In xenograft studies, CVX-241 showed an anti-tumor effect that was significantly greater and prolonged than either mono-specific CovX-Bodies alone and comparable with the physical combination of the two mono-specific CovX-Bodies. These results clearly demonstrated the superior therapeutic potential of the bispecific CovX-Body, CVX-241, over the corresponding monospecific CovX-Bodies and comparable efficacy to that of physical combination of two monospecific CovX-Bodies. To demonstrate that CVX-241 targets both Ang2 and VEGF *in vivo*, the effects of CVX-241 on tissue Ang2 levels and phosphorylated VEGFR2 (pVEGFR2) levels were assessed using immunofluorescence in Colo205 xenograft tumors. To measure Ang2 expression within the tumors, frozen sections of these tumors were stained with a FITC-labeled anti-Ang2 monoclonal antibody. In CVX-241-treated groups, Ang2 levels were significantly reduced by ~70% in comparison with the vehicle-treated group. To measure pVEGFR2 levels within the treated tumors, frozen sections of these tumor samples were also double-stained with a FITC-labeled VEGFR antibody and a rhodamine-labeled pVEGFR2 antibody, and the pVEGFR2 immunoreactivity was quantified. pVEGFR2/VEGFR2 was significantly reduced by CVX-241 treatment in a dose-dependent manner in comparison with the vehicle treated group. These data demonstrate that CVX-241 affects both Ang2 and VEGF pathways in the Colo205 xenograft model.

Peptide payload modification allows design of a bispecific CovX-Body with desired characteristics. As can be expected, there is a complex interplay between the composition of each peptide payload, the site of attachment of the tether to each peptide payload and the composition and length of the spacers. This versatility is a unique feature of the technology in that it facilitates critical optimization of the target binding affinity and pharmacokinetics of the payloads relative to each other. In this way, the pharmacokinetics and pharmacodynamics of the bi-specific CovX-Body can be optimally balanced in a controlled manner. Furthermore, replacing one or both of the peptide pharmacophores with a new peptide pharmacophore targeting a different protein can lead to a completely new bi-specific agent. For example, a novel bi-specific CovX-Body, targeting αvβ3 integrin and Ang2, was made by replacing the VEGF peptide of a bi-specific CovX-Body with a $\alpha_v\beta_3$ integrin binding peptide and tailoring the linker accordingly. By changing the peptide pharmacophores, this technology allows creation of multiple bi-specific products from one scaffold antibody.

38.3.3 CVX-343, Fusion of a Small Protein with CVX-2000

FGF21, a member of the FGF19 subfamily, plays an important role in regulating glucose and lipid homeostasis in animal models. Administration of FGF21 improves insulin sensitivity, reduces body weight, and reverses hepatic steatosis in diabetic rodents and monkeys. Due to its short half-life, daily injection of the protein is required for *in vivo* bioactivity. CVX-343 was created by covalently linking two FGF21 proteins to the Fab of CVX-2000 IgG1 mAb via an AZD linker [34]. Similar to the FGF21 protein, CVX-343 increased Glut1 mRNA expression in 3T3-L1 adipocytes with an EC_{50} of 3.4 nM. In *ob/ob* mice, a single SC administration of CVX-343 significantly reduced body weight gain, improved glucose tolerance, and lowered liver triglyceride levels in a dose-related manner compared with the vehicle control up to 6 days post dose, demonstrating the prolonged pharmacodynamics of this molecule *in vivo*. In diet-induced obese mice, once-weekly administration of 10 mg/kg CVX-343 for 2 weeks caused 7% weight loss without affecting food intake, significantly decreased glucose AUC by 17% in an oral glucose tolerance test, and lowered serum triglycerides and NEFA levels by 48 and 69%, respectively, in comparison with vehicle controls. In *db/db* mice, once-weekly administration of CVX-343 (10 mg/kg) improved glucose tolerance and increased pancreatic β-cell mass by 2.4-fold in the absence of a body weight effect. FGF21 CovX-Body CVX-343 shows comparable potency as the wild type FGF21 and prolonged efficacy across multiple preclinical models. This long-acting FGF21 molecule has an attractive therapeutic profile for type 2 diabetes and is an example of the application of CovX-Body technology beyond small molecules and peptides to therapeutically relevant small proteins.

REFERENCES

1. An Z. (2009) *Therapeutic Monoclonal Antibodies. From Bench to Clinic*. John Wiley and Sons, Inc., New Jersey, 711–762.
2. Wirsching P, Ashley JA, Lo CL, Janda KD, Lerner RA. (1995) Reactive immunization. *Science* 270, 1775–1782.

3. Wagner J, Lerner RA, Barbas CF, III. (1995) Efficient aldolase catalytic antibodies that use the enamine mechanism of natural enzymes. *Science* 270, 1797–1780.
4. Barbas CF. III, Heine A, Zhong G, Hoffman T, Gramatikova S, Björnested R, et al. (1997) Immune versus natural selection: antibody aldolases with enzymic rates but broader scope. *Science* 278, 2085–2092.
5. Rader C, Sinha SC, Popkov M, Lerner RA, Barbas CF. III. (2003) Chemically programmed monoclonal antibodies for cancer therapy: Adaptor immunotherapy based on a covalent antibody catalyst. *Proc. Natl. Acad. Sci.USA* 100(9), 5396–5400.
6. Doppalapudi VR, Tryder N, Li L, Aja T, Griffith D, Liao F, et al. (2007) Chemically programmed antibodies: endothelin receptor targeting CovX-Bodies. *Bioorg. Med. Chem. Lett.* 17, 501–506.
7. Liu D, Wood L, Wheeler W, Zhao J, Griffith D, Bradshaw C, et al. (2006) CCR5 Targeting CovX-Bodies. 30th National Medicinal Chemistry Meeting, June 25–29, University of Washington, Seattle.
8. Rader C, Turner JM, Heine A, Shabat D, Sinha S, Wilson IA, et al. (2003) A humanized aldolase antibody for selective chemotherapy and adaptor immunotherapy. *J. Mol. Biol.* 332, 889–899.
9. Carter P, Presta L, Gorman CM, Ridgway JB, Henner D, Wong WL, et al. (1992) Humanization of an anti-p185HER2 antibody for human cancer therapy. *Proc. Natl. Acad. Sci. USA* 89 (10), 5434–5441.
10. de Wildt RM, Hoet RM, van Venrooij WJ, Tomlinson IM, Winter G. (1999) Analysis of heavy and light chain pairings indicates that receptor editing shapes the human antibody repertoire. *J. Mol. Biol.* 285(3), 895–901.
11. de Wildt RM, Mundy CR, Gorick BD, Tomlinson IM. (2000) Antibody arrays for high-throughput screening of antibody-antigen. *Nat. Biotechnol.* 18(9), 989–994.
12. Kyte J, Doolittle RF. (1982) A simple method for displaying the hydropathic character of a protein. *J. Mol. Biol.* 157(1), 105–132.
13. Rosen CA, Haseltine WA. (2003) Albumin Fusion Proteins for Prolonged Shelf-Life of Therapeutic Proteins. PCT Int. Appl. WO 2003059934.
14. Leger R, Thibaudeau K, Robitaille M, Quraishi O, van Wyk P, Bousquet-Gagnon N, et al. (2004) Identification of CJC-1131-albumin bioconjugate as a stable and bioactive GLP-1(7–36) analog. *Bioorg. Med. Chem. Lett.* 14(17), 4395–4398.
15. Kim B, Zhou J, Martin B, Carlson OD, Maudsley S, Greig NH, et al. (2010) Transferrin fusion technology: A novel approach to prolonging biological half-life of insulinotroic peptides. *J. Pharmacol. Exp. Ther.* 334(3), 682–692.
16. Sutherland S. (2004) Peptibodies: The new cool technology. *Drug Discov. Today* 9(16), 683.
17. Huang H, Bhat A, Woodnutt G, Lappe R. (2010) Targeting the ANGPT-TIE2 pathway in malignancy. *Nat. Rev. Cancer* 10(8), 575–585.
18. Huang H, Lai J, Do J, Liu D, Li L, Del Rosario J, et al. (2011) Specifically targeting angiopoietin-2 inhibits angiogenesis, Tie2-expressing monocyte infiltration, and tumor growth. *Clin. Cancer Res.* 17(5), 1001–1011.
19. Rosen LS, Mendelson DS, Cohen RB, Gordon MS, Goldman JW, Bear IK, et al. (2010). First-in-Human Dose-Escalation Safety and PK Trial of a Novel Intravenous Humanized Monoclonal CovX Body Inhibiting Angiopoietin 2 [Abstract]. *J. Clin. Oncol.* 28:15s, Abstract 2524.
20. Cheng HM. (2007) Dynamic contrast enhanced magnetic resonance imaging in oncology drug development. *Curr. Clin. Pharmacol.* 2(2), 111–122.
21. Fischer N, Leger O. (2007) Bispecific antibodies: molecules that enable novel therapeutic strategies. *Pathobiology* 74, 3–14.
22. Hagemeyer CE, von Zur Muhlen C, von Elverfeldt D, Peter K. (2009) Single-chain antibodies as diagnostic tools and therapeutic agents. *Thromb. Haemost.* 101, 1012–1019.
23. Robinson MK, Hodge KM, Horak E, Sundberg ÅL, Russeva M, Shaller CC, et al. (2008) Targeting ErbB2 and ErbB3 with a bispecific single-chain Fv enhances targeting selectivity and induces a therapeutic effect *in vitro*. *Br. J. Cancer* 99, 1415–1425.
24. Hudson PJ, Kortt AA. (1999) High avidity scFv multimers, diabodies and triabodies. *J. Immunol. Methods* 231, 177–189.
25. Kipriyanov SM. (2009) Generation of bispecific and tandem diabodies. *Methods Mol. Biol.* 562, 177–193.
26. Bostrom J, Yu S, Kan D, Appleton BA, Lee CV, Billeci K, et al. (2009) Variants of the antibody herceptin that interact with HER2 and VEGF at the antigen binding site. *Science* 323, 1610–1614.
27. Wu C, Ying H, Grinnell C, Bryant S, Miller R, Clabbers A, et al. (2007) Simultaneous targeting of multiple disease mediators by a dual-variable-domain immunoglobulin. *Nat. Biotechnol.* 25, 1290–1297.
28. Constantinou A, Epenetos AA, Hreczuk-Hirst D, Jain S, Wright M, Chester KA, et al. (2009). Site-specific polysialylation of an antitumor single-chain Fv fragment. *Bioconjug. Chem.* 20(5), 924–931.
29. Doppalapudi VR, Huang J, Liu D, Jin P, Liu B, Li L, et al. (2010) Chemical generation of bispecific antibodies. *Proc. Natl. Acad. Sci. USA* 107(52), 22611–22616.
30. Carmeliet P. (2005) VEGF as a key mediator of angiogenesis in cancer. *Oncology* 69(Suppl. 3), 4–10.
31. Ferrara N, Gerber HP, LeCouter J. (2003) The biology of VEGF and its receptors. *Nat. Med.* 9, 669–676.
32. Moon WS, Rhyu KS, Kang MJ, Lee DG, Yu HC, Yeum JH, et al. (2003) Overexpression of VEGF and angiopoietin 2: a key to high vascularity of hepatocellular carcinoma? *Modern Pathol.* 16, 552–557.
33. Ochiumi T, Tanaka S, Oka S, Hiyama T, Ito M, Kitadai Y, et al. (2004) Clinical significance of angiopoietin-2 expression at the deepest invasive tumor site of advanced colorectal carcinoma. *Int. J. Oncol.* 24, 539–547.
34. Huang J, Rolzin P, Osothprarop T, Ishino T, Wiese J, Retting K, et al. (2011) CVX-343 A long acting FGF-21 CovX-Body demonstrates prolonged antidiabetic efficacy in diabetic mouse models. *71st American Diabetes Association Scientific Sessions,* June 23–28, San Diego CA.

39

MODULAR ANTIBODY ENGINEERING: ANTIGEN BINDING IMMUNOGLOBULIN Fc CH3 DOMAINS AS BUILDING BLOCKS FOR BISPECIFIC ANTIBODIES (mAb^2)

MAXIMILIAN WOISETSCHLÄGER,[1] FLORIAN RÜKER,[2] GEERT C. MUDDE,[1] GORDANA WOZNIAK-KNOPP,[2] ANTON BAUER,[1] AND GOTTFRIED HIMMLER[1]

[1]*f-star GmbH, Vienna, Austria*
[2]*Department of Biotechnology, Christian Doppler Laboratory for Antibody Engineering, University of Natural Resources and Life Sciences, Vienna, Austria*

39.1 INTRODUCTION

The immune system creates binding sites of high specificity and affinity in the variable domains of antibodies by generating sequence and consequently structural diversity in the complementarity determining regions (CDR) loops, which are located at the N-terminal ends of these domains. Sequence variations in the CDR loops of an antibody generally do not have a significant influence on the overall structure of the variable domain that carries them. This feature of variable domains is actually observed in a more general sense in domains belonging to the immunoglobulin-fold family, which are known to have a similar general shape in the core (β-barrel, and high structural variability in the loops. Furthermore, overall sequence similarity of immunoglobulin-fold domains is mainly below 25%, while their structural similarity is high, with a root-mean-square deviation of Cα atoms always below 3.9 Å [1].

We therefore set out to explore whether this inherent stability of the immunoglobulin fold allows loops of immunoglobulin domains other than the CDR loops to accommodate sequence variation without negatively impacting the overall structure and stability of the protein. As shown in Figure 39.1, the candidate loops of an immunoglobulin G1 (IgG1) for this kind of engineering are manifold, including the N- and C-terminal loops of the constant domains as well as C-terminal loops of the variable domains.

39.2 IMMUNOGLOBULIN Fc AS A SCAFFOLD

Successful proof-of-concept of engineering non-CDR loops has been obtained by randomizing the C-terminal loops of the third constant domain (CH3) of human IgG1 [2]. CH3 domains engineered with antigen-binding properties in the context of a complete Fc fragment (composed of the hinge, the CH2, and the CH3 domains) were able to bind the tumor-associated antigen HER2/neu with high affinity and specificity. Such antigen-binding Fc proteins are called Fcabs™ (Fc molecules with antigen-binding properties). In addition, such HER2/neu binding Fcabs, at a size of only ~50 kDa, also possess all attractive properties of complete antibodies such as their ability to trigger effector functions via binding to Fcγ receptors or C1q. Finally, owing to the functional presence of the FcRn-binding site, Fcabs also display the long *in vivo* half-life of antibodies.

Fusion Protein Technologies for Biopharmaceuticals: Applications and Challenges, First Edition. Edited by Stefan R. Schmidt.

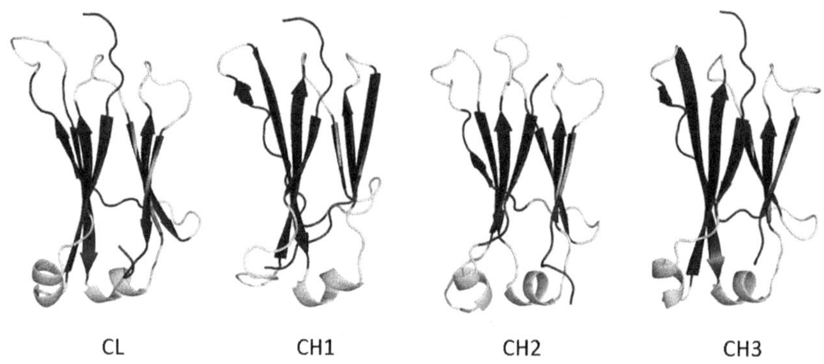

FIGURE 39.1 Ribbon presentation of the constant domains of an IgG1. Non-CDR loops are indicated in light gray and the beta sandwich core in dark gray. The structures are aligned such that the N-terminal ends are on the top and the C-terminal ends are on the bottom.

One attractive extension of the Fcab scaffold concept is the possibility of creating bispecific or multivalent antibodies. Modular replacement of the wild-type CH3 domain of an existing conventional antibody with a CH3 domain binding another or the same antigen is expected to yield antibody molecules, which structurally are minimally altered compared to the acceptor antibody. However, such proteins (which are referred to as mAb^2) are able to recognize antigen via their variable domains in a conventional fashion and via the engineered antigen-binding site located in the CH3 domain of the Fc. In analogy to the Fcab scaffold, mAb^2 proteins are expected to retain all functionalities of a normal antibody.

The following sections of this chapter describe the design and characterization of such bispecific mAb^2 proteins. The results of these experiments laid the foundation for the successful proof-of-concept of the mAb^2 approach.

39.3 DESIGN OF LIBRARIES BASED ON THE HUMAN IgG1 CH3 DOMAIN

For designing the Fcab libraries with mutations in the C-terminal loops of the CH3 domains, a number of criteria were applied in order to choose those residues in the sequence that

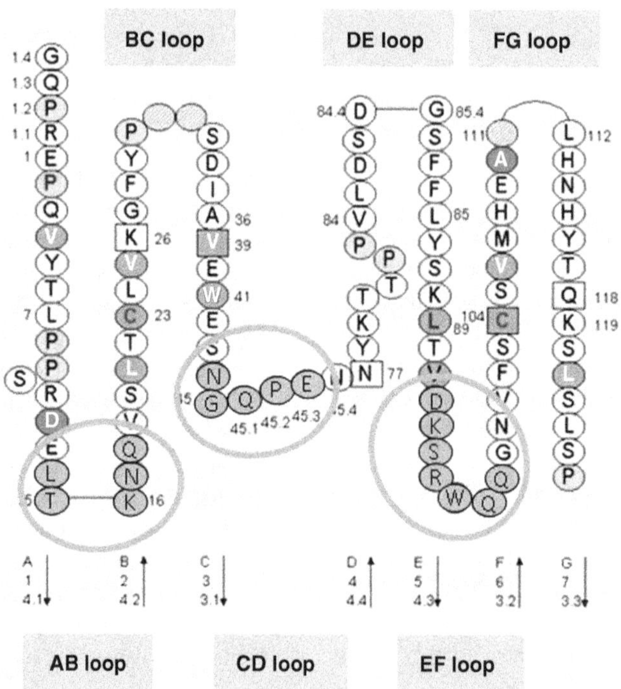

FIGURE 39.2 The fold of the CH3 domain of human IgG1 presented as string of pearls chain form. Residues that were randomized in the CH3 libraries are encircled and indicated in gray.

should be randomized to create the new binding surface. Solvent accessibility and relative structural independence of the residues located in the AB, CD, and EF loops were judged based on the crystal structure of an Fc fragment of human IgG1 (PDB code 1OQO). The degree of evolutionary conservation of the residues was evaluated using the ConSurf Server [3] and was taken as a further guide for selecting the residues to be mutated. Fcab libraries were designed essentially as described in Reference [2]. Together, the randomized positions in these libraries yield a coherent surface of ~700–1000 Å², which compares well with the surface areas buried in Fab–antigen interactions, which typically range from 600 to 900 Å² [4]. A schematic presentation of the secondary structure of the CH3 domain [5], highlighting the randomized positions, is given in Figure 39.2.

39.4 TNF-α-BINDING Fcab: SELECTION AND CHARACTERIZATION OF Fcab TNF353-2

For selection of an Fcab targeting the cytokine tumor necrosis factor α (TNF-α), a Fcab library was generated in the yeast expression vector pYD1 as described recently [2]. The resulting library expresses the α-agglutinin yeast adhesion receptor Aga-2 as a fusion protein with Fcabs. Upon transformation of the library into the *Saccharomyces cerevisiae* strain EBY100 (supplied transformed with the AGA1 gene), the Aga1 and the Aga2-Fcab fusion protein associate within the secretory pathway and upon translocation are displayed on the cell surface in an inducible fashion [6]. Yeast cells expressing Fcabs on the cell surface were incubated with biotinylated human TNF-α, and cells interacting with the cytokine were enriched by high-speed cell sorting essentially as described [2]. After five rounds of sorting, the best TNF-α binders were screened by titration of individual yeast clones with biotinylated antigen. One of these hits, Fcab TNF353-2, was chosen for further analysis. Fcab TNF353-2 was expressed in mammalian cells as soluble Fcab and purified by a single immuno-affinity chromatography step using a Protein A column. The purified Fcab eluted as a single peak from a size exclusion chromatography column at the expected volume indicating the absence of protein aggregates, fragments, or other contaminants. The purified Fcab bound specifically to immobilized TNF-α in an enzyme-linked immunosorbent assay (ELISA) with an EC_{50} in the single nano-molar range. This interaction appeared to be specific since no binding to irrelevant antigens (IgE, HER2/neu, or phycoerythrine) was detected (Figure 39.3).

For the generation of a prototype mAb², the CH3 domain of Fcab TNF353-2 was used to replace the CH3 domain of the anti-CD20 specific antibody rituximab. This antibody is used in the clinic for the treatment of blood malignancies [7]. The heavy chain gene of rituximab was modified to create a silent mutation leading to a unique XhoI restriction enzyme cleavage site at the border between the CH2 and CH3 domain in order to

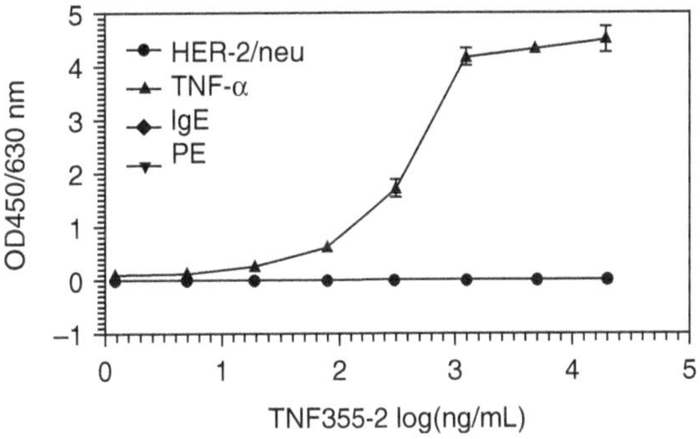

FIGURE 39.3 Antigen-specific binding of Fcab TNF353-2 to immobilized TNF-α·Bound Fcab was detected with antihuman CH2 domain antibodies. PE, phycoerythrine.

facilitate exchange for the CH3 domain of Fcab TNF353-2. HEK293 cells were transiently co-transfected with plasmids encoding the TNF-α binding heavy chain and the light chain of rituximab. After 5 days, supernatants were harvested and mAb² protein RX-TNF353-2 was purified by Protein A immuno-affinity chromatography. HPLC analysis showed that the purified mAb² eluted as single peak at the expected volume indistinguishable of the parental rituximab antibody.

To determine if the changes in the CH3 domain influenced binding to Fc receptors known to be important for immuno-effector functions (CD16a) and long *in vivo* half-life (FcRn), surface plasmon resonance measurements were carried out. At pH = 6, mAb² RX-TNF353-2 and its parent rituximab (RX) bound with very similar affinities to human FcRn. At pH = 7, both proteins completely lost their ability to interact with the receptor indicating that the TNF-α binding site in the mAb² did not interfere with the known pH-dependent FcRn association and dissociation phenotype of normal antibodies. Similar results were obtained when CD16a protein was allowed to bind to the antibodies captured on Protein A Biacore chips (Figure 39.4). These data suggested that the *in vivo* half-life and immuno-effector functions of the original rituximab antibody should be preserved in the mAb².

It is known that rituximab is able to initiate immuno-effector functions such as antibody-dependent cellular cytotoxicity (ADCC) and complement-dependent cytotoxicity (CDC). To determine the ADCC potency of the mAb² purified primary human B lymphocytes were opsonized with rituximab or the mAb² and then incubated with autologous natural killer (NK) cells as effector cells. After 4 h, the killed B-cells were enumerated by flow cytometry after staining dead cells with the fluorescent dye 7-amino-actinomycin D. The data demonstrated that both proteins mediated NK cell dependent killing of B cells in a dose-dependent manner and that the potencies of the mAb² and rituximab were very similar (Figure 39.5A). No dead B cells were measured in the absence of NK cells indicating that the mechanism of cell death was indeed ADCC (not shown). To measure CDC reactions, purified B cells were incubated

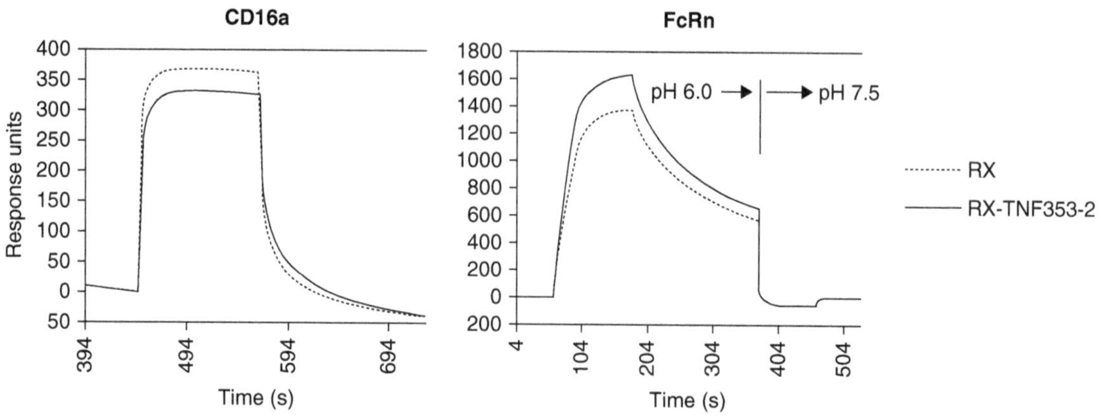

FIGURE 39.4 Binding of mAb^2 RX-TNF353-2 and rituximab (RX) to human CD16a and FcRn. Antibodies were captured with Biacore chips loaded with Protein A followed by addition of human CD16a. Biacore chips loaded with human FcRn were used to measure its interaction with the antibodies.

with antibodies in 100% autologous serum before killed B cells were quantified as above. Again, mAb^2 RX-TNF353-2 displayed the same CDC potency compared to unmodified rituximab (Figure 39.5B).

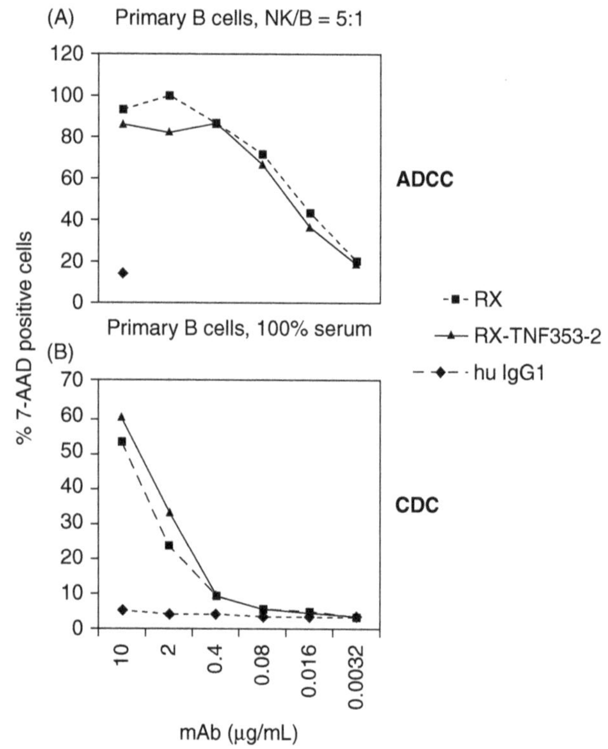

FIGURE 39.5 Immuno-effector functions of mAb^2 RX-TNF353-2 are unchanged compared to rituximab. (A) ADCC of opsonized primary human B cells incubated with autologous NK cells at a 5:1 ratio for 4 h. Dead cells were stained with 7-amino-actinomycin D (7-AAD) and the percentage of 7-AAD positive cells was measured by flow cytometry. (B) Opsonized human B cells were incubated in 100% autologous serum for 4 h. Dead cells were quantified by 7-AAD staining as stated earlier.

To assess if these data were specific for this particular mAb^2 or could be extended to other model types of mAb^2, the CH3 domain of Fcab TNF461-1 was used to replace the CH3 domain of the monoclonal antibody trastuzumab (HC). Trastuzumab is a monoclonal antibody recognizing the receptor HER2/neu and is approved for the treatment of HER2/neu overexpressing breast cancer patients [8]. Fcab TNF461-1 was selected for TNF-α binding by yeast display similarly as Fcab TNF353-2, but the composition of amino acids forming the binding site for the cytokine is quite different. The resulting mAb^2 HC-TNF461-1 and its parent trastuzumab were expressed in HEK 293 cells and purified to homogeneity as described earlier.

ADCC experiments were performed to assess the relative potencies of mAb^2 HC-TNF461-1 versus its parental antibody. The human lung cancer cell line Calu-3, which expresses HER2/neu on its surface, was used as target cells. Antibody opsonized Calu-3 cells were incubated with primary NK cells as effector cells for 4 h before dead cells were quantified. The data revealed that the mAb^2 was equipotent to trastuzumab to elicit ADCC killing of Calu-3 cells. Addition of excess of TNF-α did not alter the potency of HC-TNF461-1 to kill Calu-3 cells (Figure 39.6) demonstrating that TNF-α occupation of the binding site in the CH3 domain did not negatively influence NK-cell recruitment mediated by the CD16a binding site located in the CH2 domain. In the absence of NK cells, no dead Calu-3 cells were observed indicating that the mechanism of killing was due to ADCC (not shown). Collectively, these results demonstrated that the immuno-effector functionalities of two different mAb^2 proteins are unchanged compared to their parental monoclonal antibodies.

To determine if binding of mAb^2 RX-TNF353-2 to TNF-α was different compared to cytokine binding of the Fcab, ELISA experiments were performed. For both proteins, comparable EC_{50} values were calculated (EC_{50} TNF353-2:

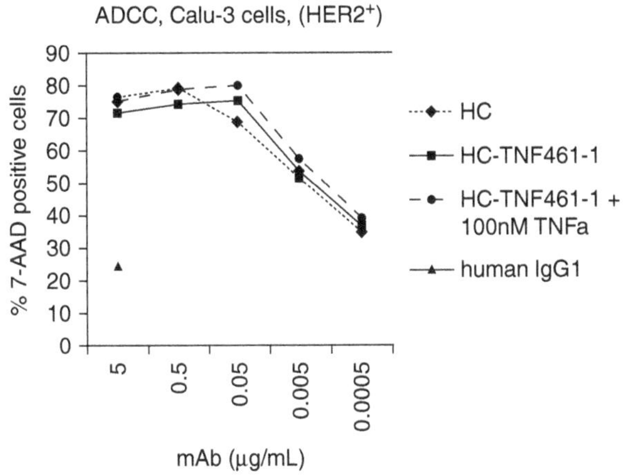

FIGURE 39.6 mAb^2 HC-TNF461-1 is equipotent to trastuzumab (HC) to kill Calu-3 cells by ADCC. Calu-3 cells were opsonized with increasing concentrations of antibodies and then incubated with human NK cells at a Calu-3:NK = 1:5 ratio for 4 h. Killed cells were detected with 7-AAD by flow cytometry.

17 nM, EC_{50} RX-TNF353-2: 23 nM) demonstrating the independence of the cytokine-binding site in the CH3 domain from its molecular context (Fcab or mAb^2) (Figure 39.7).

Finally, we tested if the mAb^2 was able to interact simultaneously with both antigens to prove its true bispecific functionality. For that purpose, a human CD20 positive B-cell line (CFB4.2) was incubated with increasing concentrations of rituximab or mAb^2. Afterwards, cell-bound proteins were detected either with an antihuman kappa chain specific monoclonal antibody coupled to a fluorescence dye or with biotinylated TNF-α followed by streptavidin coupled to a fluorescent dye. Incubation with the anti-light-chain-specific antibody revealed that both rituximab and RX-TNF353-2 could interact equally well with CD20 on the cell surface via their Fab arms indicating that the amino acid changes in the CH3 domain did not compromise binding to CD20. Importantly, incubation with TNF-α showed that only cells

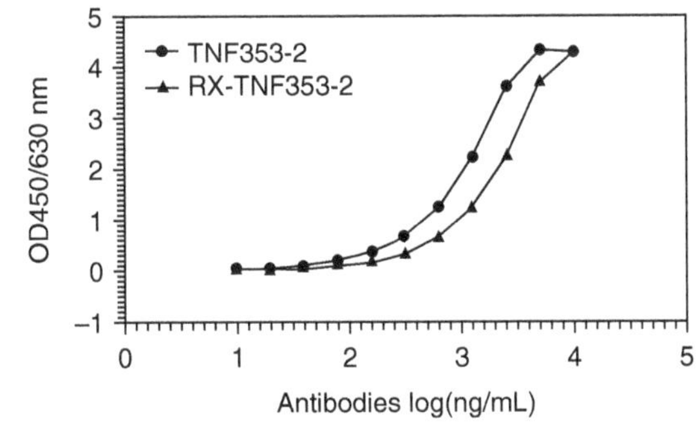

FIGURE 39.7 Comparable TNF-α binding of Fcab TNF353-2 and mAb^2 RX-TNF353-2. TNF-α was coated on ELISA plates and increasing concentrations of antibodies were added. Bound proteins were detected with antihuman CH2 domain antibodies.

which had previously been exposed to the mAb^2 could be detected (Figure 39.8), demonstrating that the mAb^2 could bind two different antigens at the same time.

The bispecific nature of mAb^2 HC-TNF461-1 could also be shown by a similar experimental setup. In this case, the mAb^2 or its parent trastuzumab were incubated with the human HER2/neu positive breast cancer cell line SKBR3. Similar to the experiment described earlier, biotinylated TNF-α specifically could detect the mAb^2 but not trastuzumab while both antibodies were detectable with similar sensitivities using antihuman light chain antibodies (not shown). These data demonstrated that both tested mAb^2 proteins were able to simultaneously bind antigen via their Fab arms and the binding site in the CH3 domain.

In summary, these data provided proof-of-principle for the concept that modular replacement of Fc CH3 domains of conventional monoclonal antibodies with antigen-binding CH3 domains is a viable strategy to generate bispecific antibodies (mAb^2). The affinities of mAb^2 proteins to their antigens are unchanged relative to the parental antibody or to the Fcab, which is the donor of the antigen-binding CH3 domain. In addition, immuno-effector functionalities such as ADCC or CDC are not compromised in the mAb^2 format. The functional ADCC data are in-line with Biacore results demonstrating that binding of the mAb^2 to the relevant Fc receptor CD16a is unchanged compared to the unmodified parent. Furthermore, the presence of the antigen-binding site in the CH3 domain does not interfere with binding to the FcRn receptor suggesting a long half-life of mAb^2 *in vivo*. In conclusion, mAb^2 proteins represent a novel class of bispecific proteins, which feature only minimal structural changes compared to conventional antibodies while retaining all their functional attributes.

Treatment of cancer patients with therapeutic monoclonal antibodies has been quite successful. It has become clear, however, that only a subpopulation of patients will benefit from such a therapy and that a significant percentage of treated patients will develop resistance to antibody therapy. In order to overcome these shortcomings, combination therapies have been successfully tried in preclinical animal models using two monoclonal antibodies targeting two different but interrelated tumor-relevant therapeutic targets. Such studies have demonstrated additive or even synergistic effects of the combination therapy versus the individual antibodies [9–11]. Consequently, clinical trials are now ongoing to translate these preclinical results into men. A combination therapy with two therapeutic antibodies could potentially be achieved with a single molecule such as a bispecific mAb^2 protein. Other attractive applications for mAb^2 molecules can be envisioned. For instance, improved efficacy and lower side effects may be expected by specifically targeting the mAb^2 to diseased tissue using one of the two specificities of the antibody while the second specificity supplies the therapeutic principle. Similarly, one could

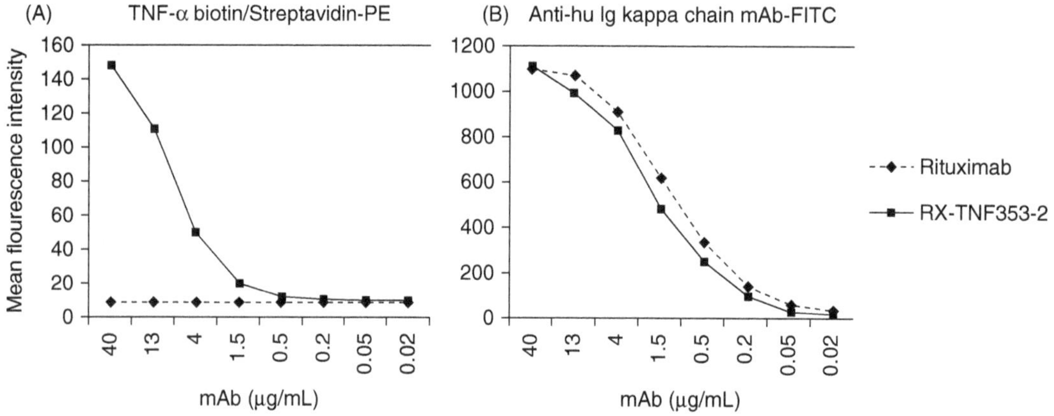

FIGURE 39.8 mAb^2 RX-TNF353-2 can interact simultaneously with two different antigens. The human CD20 positive B-cell line CFB4.2 was incubated with increasing concentrations of mAb^2 or rituximab. Cell-bound proteins were detected either with biotinylated TNF-α plus phycoerythrine (PE) conjugated streptavidin (A) or with fluorescein-iso-thiocyanate (FITC) conjugated antihuman kappa chain antibodies (B).

imagine the usefulness of a mAb^2 as antibody drug conjugate in which a toxic compound is specifically transported to the diseased tissue. Finally, mAb^2 proteins could be used to enhance immune responses or initiate alternative forms of immune responses by using one antigen specificity for the recruitment and activation of immune-effector cells, such as cytotoxic T cells [12]. Such a concept is currently being tested in cancer patients in the clinic.

39.5 CONCLUSIONS AND FUTURE PERSPECTIVES

During the last two decades, a number of different bispecific antibody or antibody-fragment formats have been developed (reviewed in [13,14]). They include bispecific IgGs and F(ab′) 2 fragments generated by variations of the hybridoma technology, by chemical cross-linking or by genetic engineering, as well as recombinant bispecific formats based on smaller fragments such as scFvs and single-domain antibodies. Furthermore, recombinant formats utilizing various fusion partners (HSA, Fc fragment, leucine zipper) to achieve bispecificity have been described. Among all these, heterodimeric IgG formats such as the knob-into-holes [15] or the SEEDbody [16] constructs are the ones that are most similar to normal, full-size IgG. Clinically, the most advanced full-size IgG platform is being developed by Trion Pharma, and is based on the ability of mouse IgG2a and rat IgG2b to heterodimerize, leading to monovalent bispecific mouse/rat IgG molecules [17]. However, immunogenicity of these nonhuman antibodies remains one of the major drawbacks of this platform as well as the demanding production process which is based on Quadroma technology. Among the fragment platforms, Micromet [18] and Merrimack have entered Phase I and Phase II studies with their bispecific scFv-based molecules. These compounds are biologically highly potent, yet their intrinsically short *in vivo* half-life limits their potential, similar to all other fragment-based platforms. These limitations can only be overcome by fusion of the antibody fragments with molecules that can provide the missing properties, such as Fc fragment of IgG, HSA, or HSA-binding moieties. Among the preclinical full-size IgG formats such as the knobs-into-holes IgG or the SEEDbody, their monovalent binding toward each of the recognized targets is considered to be a major difference when compared to a normal IgG antibody, which is always bivalent. Compared to these platforms, the mAb^2 format is the ideal solution, since it provides a normal IgG with its full functionality in terms of effector functions and *in vivo* half-life. Bivalent antigen binding via its Fab arms is fully retained, and bispecificity is achieved by engineering the C-terminal tip of the CH3 domain, which is otherwise not known to have any biological function in antibodies. Further validation of the mAb^2 concept *in vivo* with clinically relevant mAb^2 proteins is in progress.

ACKNOWLEDGMENTS

We thank Austria Wirtschaftsservice for funds provided under its “Preseed” program, the Austrian Research Promotion Agency (FFG) for funds provided under its “BRIDGE” program (project 813002 “Modular Antibody Design Based on Immunoglobulin Libraries”), and the Christian Doppler Forschungsgesellschaft. f-star provided funds as the industry partner in connection with the “BRIDGE” program and the Christian Doppler Laboratory for Antibody Engineering. We thank all colleagues from f-star GmbH for their valuable contributions to this work.

REFERENCES

1. Halaby DM, Poupon A, Mornon J. (1999) The immunoglobulin fold family: sequence analysis and 3D structure comparisons. *Protein Eng.* 12(7), 563–571.
2. Wozniak-Knopp G, Bartl S, Bauer A, Mostageer M, Woisetschlager M, Antes B, et al. (2010) Introducing antigen-binding sites in structural loops of immunoglobulin constant domains: Fc fragments with engineered HER2/neu-binding sites and antibody properties. *Protein Eng. Des. Sel.* 23(4), 289–297.
3. Landau M, Mayrose I, Rosenberg Y, Glaser F, Martz E, Pupko T, et al. (2005) ConSurf 2005: the projection of evolutionary conservation scores of residues on protein structures. *Nucleic Acids Res.* 33(Web Server issue), W299–302.
4. Coley AM, Gupta A, Murphy VJ, Bai T, Kim H, Foley M, et al. (2007) Structure of the malaria antigen AMA1 in complex with a growth-inhibitory antibody. *PLoS Pathog.* 3(9), 1308–1319.
5. Ruiz M, Lefranc MP. (2002). IMGT gene identification and Colliers de Perles of human immunoglobulins with known 3D structures. *Immunogenetics* 53(10–11), 857–883.
6. Boder ET, Wittrup KD. (1997) Yeast surface display for screening combinatorial polypeptide libraries. *Nat. Biotechnol.* 15(6), 553–557.
7. Coiffier B. (2007) Rituximab therapy in malignant lymphoma. *Oncogene* 26(25), 3603–3613.
8. Perez EA, Palmieri FM, Brock SM. (2009) Trastuzumab. *Cancer Treat. Res.* 151, 181–196.
9. Canon J, Bryant R, Roudier M, Osgood T, Jones J, Miller R, Coxon A, et al. (2010) Inhibition of RANKL increases the antitumor effect of the EGFR inhibitor panitumumab in a murine model of bone metastasis. *Bone* 46(6), 1613–1619.
10. Naumov GN, Nilsson MB, Cascone T, Briggs A, Straume O, Akslen LA, et al. (2009) Combined vascular endothelial growth factor receptor and epidermal growth factor receptor (EGFR) blockade inhibits tumor growth in xenograft models of EGFR inhibitor resistance. *Clin. Cancer Res.* 15(10), 3484–3494.
11. Pakkala S, Ramalingam SS. (2009) Combined inhibition of vascular endothelial growth factor and epidermal growth factor signaling in non-small-cell lung cancer therapy. *Clin. Lung Cancer* 10(Suppl. 1), S17–S23.
12. Herrmann I, Baeuerle PA, Friedrich M, Murr A, Filusch S, Ruttinger D, et al. (2010) Highly efficient elimination of colorectal tumor-initiating cells by an EpCAM/CD3-bispecific antibody engaging human T cells. *PLoS One* 5(10), e13474.
13. Kontermann RE. (2010) Alternative antibody formats. *Curr. Opin. Mol. Ther.* 12(2), 176–183.
14. Muller D, Kontermann RE. (2010) Bispecific antibodies for cancer immunotherapy: current perspectives. *BioDrugs* 24(2), 89–98.
15. Merchant AM, Zhu Z, Yuan JQ, Goddard A, Adams CW, Presta LG, et al. (1998) An efficient route to human bispecific IgG. *Nature Biotechnol.* 16(7), 677–681.
16. Davis JH, Aperlo C, Li Y, Kurosawa E, Lan Y, Lo KM, et al. (2010) SEEDbodies: fusion proteins based on strand-exchange engineered domain (SEED) CH3 heterodimers in an Fc analogue platform for asymmetric binders or immunofusions and bispecific antibodies. *Protein Eng. Des. Sel.* 23(4), 195–202.
17. Seimetz D, Lindhofer H, Bokemeyer C. (2010) Development and approval of the trifunctional antibody catumaxomab (anti-EpCAM x anti-CD3) as a targeted cancer immunotherapy. *Cancer Treat. Rev.* 36(6), 458–467.
18. Baeuerle PA, Kufer P, Bargou R. (2009) BiTE: Teaching antibodies to engage T-cells for cancer therapy. *Curr. Opin. Mol. Ther.* 11(1), 22–30.

40

DESIGNER FUSION MODULES FOR BUILDING MULTIFUNCTIONAL, MULTIVALENT ANTIBODIES, AND IMMUNOCONJUGATES: THE DOCK-AND-LOCK METHOD

EDMUND A. ROSSI,[1] DAVID M. GOLDENBERG,[2] AND CHIEN-HSING CHANG[3]

[1]*IBC Pharmaceuticals, Inc., Morris Plains, NJ, USA*
[2]*Garden State Cancer Center, Center for Molecular Medicine and Immunology, Morris Plains, NJ, USA*
[3]*Immunomedics, Inc., Morris Plains, NJ, USA*

40.1 INTRODUCTION

Advances in new technologies for both rational and combinatorial protein engineering have increased the variety and magnitude of potential molecules that may be designed and produced for biotechnological and biomedical applications. An essential factor for these accomplishments is that many natural proteins have their functional properties located in discrete domains, which can be manipulated as molecular modules and exploited as building blocks for devising artificial structures with multiple functions. Molecular modules derived from the binding domains of antibodies or nonantibodies, or those based on protein display scaffolds, currently have received the most attention, are described in several reviews [1–4]. Molecular modules that confer effector functions of proteins, for example, the Fc of an IgG or the catalytic domain of an enzyme, also constitute an important class of components in the architecture of designer proteins [5,6]. In addition, peptide motifs with the innate ability to self-associate are often built into molecular modules to facilitate the formation of dimeric, trimeric, or multimeric products composed of the same or different fusion polypeptides [7–11]. Nevertheless, we have recognized that innovative fusion proteins created by recombinant engineering with these molecular modules may be built into more complex structures to gain additional attributes that are highly desirable, yet not technically attainable, in the individual engineered construct. To date, such goals are commonly achieved with varied success by judicious application of conjugation chemistries. Well-known examples include pegylated cytokines to increase serum half-lives [12], biotinylated proteins to enable immobilization into microarrays [12,13], and intein-mediated assembly of protein-DNA chimeras to quantify specific molecules to which the protein binds [14].

Relatively new to the bioconjugate field is promising strategies for tethering two or more molecular modules of distinct functions into covalent or quasi-covalent assemblies following a binding event. We present here an overview of one such strategy known as Dock-and-Lock™ (DNL™) platform technology [15], which has the potential for making an unlimited number of bioactive molecules with multivalency, multifunctionality, and defined composition.

Fusion Protein Technologies for Biopharmaceuticals: Applications and Challenges, First Edition. Edited by Stefan R. Schmidt.

40.2 DDD/AD MODULES BASED ON PKA AND AKAP

The basis of the DNL method is the exploitation of the specific protein/protein interactions occurring in nature between the regulatory (R) subunits of cyclic adenosine monophosphate (cAMP)-dependent protein kinase (PKA) and the anchor domain (AD) of A-kinase anchor proteins (AKAPs) [16,17]. Two types of R subunits (RI and RII) are found in PKA and each has α and β isoforms. The R subunits have been isolated only as stable dimers with the dimerization domain shown to consist of the first 44 amino-terminal residues [18]. The AD of AKAPs for PKA is an amphipathic helix of 14–18 residues [19], which binds only to dimeric R subunits. For human RIIα, the AD binds to a hydrophobic surface formed by the 23 amino-terminal residues [20]. Thus, the dimerization domain and AKAP binding domain of human RIIα are located within the same N-terminal 44 amino acid sequence [18,21], and are referred to as the dimerization and docking domain (DDD).

For the DNL method, AD and DDD peptide sequences, which are modified with cysteine residues for covalent "locking" via disulfide bridges, are fused to a precursor protein (or other entity) to make DNL-modules, which are produced and stored independently (Figure 40.1A and C). The DDD-derivatized proteins (DDD-modules) spontaneously form stable homodimers; therefore, DDD-modules always comprise two copies of the precursor protein (Figure 40.1B). On demand, any DDD-module can be paired with any AD-module to generate a wide variety of stable conjugates comprising two and one copies of the DDD-module AD2-module precursors, respectively (Figure 40.1D).

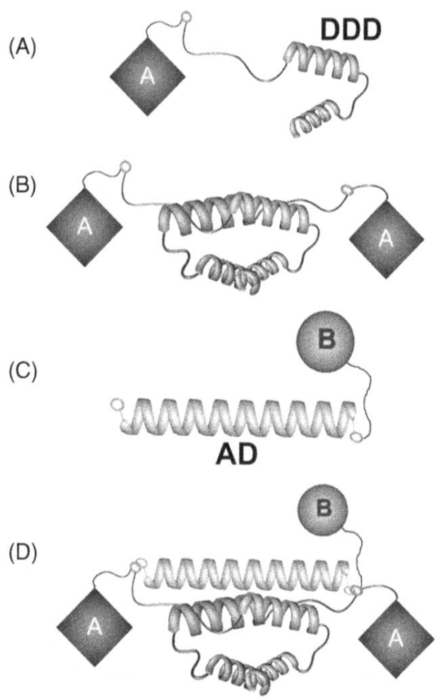

FIGURE 40.1 Basic features of the DNL method. (A) DDD-module of precursor A. (B) DDD-mediated dimer of precursor A. (C) AD-module of precursor B. (D) Stably tethered DNL conjugate comprising two copies of precursor A and one precursor B. Cysteine residues of DDD and AD, rings; "locking" disulfide bonds, interlocking rings. Modified with copyright permission from Ref. [22].

In our first study [15], we showed the DDD of human RIIα (designated DDD1) and the AD derived from AKAP-IS [23], a synthetic peptide optimized for RII-selective binding with a reported K_D of 4×10^{-10} M (designated AD1), to be an excellent pair of linker peptides for docking two entities into a noncovalent complex, which could be further locked into a stably tethered structure through the introduction of cysteine residues into both the DDD and AD at strategic positions to facilitate the formation of disulfide bonds. The cysteine-modified versions of AD1 and DDD1, designated AD2 (**C**GQIEYLAKQIVDNAIQQAG**C**) and DDD2 (**C**GHIQIPPGLTELLQGYTVEVLRQQPPDLVE-FAVEYFTRLREARA), respectively, are the preferred peptide linker pairing of the DNL method.

40.3 ADVANTAGES AND DISADVANTAGES OF THE DNL METHOD

40.3.1 Advantages of the DNL Method

Besides the unique feature that a DDD-derivatized precursor is always presented in two copies, additional advantages of the DNL method that together distinguish it from other site-specific conjugation methods [12–14,24–26] are summarized as follows:

DNL is modular. Each DDD- or AD-containing entity serves as a module and any DDD-module can be paired with any AD-module. Such modules can be produced independently, stored separately "on shelf," and combined "on demand." There is essentially no limit on the types of precursors that can be converted into a DDD- or AD-module, so long as the resulting modules do not interfere with the dimerization of DDD or the binding of DDD to AD.

DNL is versatile. Modules can be made recombinantly or synthetically. Recombinant modules, which may be produced in mammalian, microbial, or other expression systems, may include fusions of antibodies or antibody fragments, cytokines, enzymes, carrier proteins (e.g., human serum albumin and human transferrin), or a variety of natural or artificial non-antibody binding or scaffold proteins [1–4]. Furthermore, DDD or AD can be coupled to the amino-terminal or carboxyl terminal end or even positioned internally within the fusion protein, preferably with a spacer containing an appropriate length and composition of amino acid residues, provided that the binding activity of the DDD or AD and the desired activity of the polypeptide fusion partners are not compromised. Two or more AD peptides can be incorporated into AD modules to create a scaffold for attachment of multiple DDD modules.

Modules can also be made synthetically, to compose peptides, polyethylene glycol (PEG), dendrimers, nucleic acids, chelators with or without radioactive or non-radioactive metals, drugs, dyes, oligosaccharides, natural or synthetic polymeric substances, nanoparticles, fluorescent molecules, or quantum dots, depending on the intended applications [27,28].

DNL manufacturing is easy. The DNL method is basically a one-pot preparation and requires three simple steps to recover the product from the starting materials: (i) combine DDD- and AD-modules in stoichiometric amounts; (ii) add redox agents to facilitate the self-assembly of the DNL-conjugate; and (iii) purify by an appropriate affinity chromatography process. The DNL-modules can be purified and stored prior to their use in a DNL conjugation. However, purification of the modules is not necessary. DNL conjugation can be accomplished in mixtures of cell lysates and/or culture supernatant fluids containing the DNL-modules, with subsequent isolation of the DNL-conjugate by affinity chromatography. A single-step affinity purification process with commonly used affinity media, such as Protein A or immobilized metal, typically results in >95% purity of the DNL conjugate, which is sufficient for most preclinical applications. However, for manufacturing of clinical material, further processing steps, such as additional affinity chromatography, ion exchange chromatography, low pH treatment, and ultrafiltration, are required to ensure adequate removal of virus and other contaminants.

DNL results in quantitative yields of a homogeneous product with a defined composition and *in vivo* stability. The high-affinity binding between the DDD- and AD-modules results in nearly 100% conversion of each into the desired DNL product. The site-specific conjugation results in a preparation of defined and homogeneous molecular size and composition, for which the full activity of each module is usually preserved and *in vivo* integrity is sustained.

40.3.2 Disadvantages of the DNL Method

Although the modular nature of DNL is clearly an advantage for constructing many fusion proteins from a much smaller number of modules, and also allows the assembly of complex structures that would not be possible with conventional recombinant engineering, two (or possibly more) modules need to be synthesized in order to make a single product. Obviously, this results in increased production costs because of the establishment of two host cell lines and bioreactor production of each module.

Another potential complication, which we have yet to suffer from, is the potential for irreversible inactivation of precursor molecules during the mild redox conditions during DNL conjugation. The cysteine residues involved in DNL docking are exposed and readily reduced under mild conditions. Internal disulfides, which might be present in a precursor, are less affected under these conditions, and even if reduced, they are likely to reform on reoxidation. However, it is possible that some precursors might denature or aggregate during the redox reaction.

40.4 Fab-BASED MODULES

In our first study [15], the DNL method was applied to generate bispecific trivalent complexes comprising three Fab fragments by combining DDD-modules constructed from the Fab of hMN-14, a humanized monoclonal antibody (mAb) with specificity for the A3B3 domains of human carcinoembryonic antigen (CEACAM5) [29], with AD-modules constructed from the Fab of h679, a humanized mAb with specificity for the hapten histamine-succinyl-glycine (HSG) [30].

Three Fab-DDD-modules and two Fab-AD-modules were used to generate three different Tri-Fabs (TFs), each comprising two hMN-14 Fabs and one h679 Fab. Fab-based modules are purified by affinity chromatography using Protein L or KappaSelect affinity media. C_H1-DDD1-Fab-hMN-14 (Figure 40.2A) and C_H1-AD1-Fab-h679 (Figure 40.2D) are Fab-DDD- and Fab-AD-modules of the respective parent mAbs, fused at the carboxyl-terminal end of the F_d chain (C-terminal end of C_H1 domain) to DDD1 and AD1, respectively. On mixing of the two modules, the formation of a binary complex (Figure 40.2F) was readily demonstrated by SE-HPLC with the K_D determined by equilibrium gel filtration analysis [31] to be about 8 nM, which is presumably too weak of an affinity for *in vivo* applications. DDD1 was converted to DDD2 by incorporation of a cysteine residue at the amino-terminal end of the DDD peptide. Thus, the naturally dimeric DDD2-modules have two reactive cysteine residues. Two additional modules were generated for the hMN-14 Fab, *N*-DDD2-Fab-hMN-14 (Figure 40.2B) and C_H1-DDD2-Fab-hMN-14 (Figure 40.2C), where the DDD2 peptide was fused to the amino- or carboxyl-terminal end of the F_d chain, respectively. A cysteine residue was added to each end of AD1 to create AD2, which was included in the C_H1-AD2-Fab-h679 module (Figure 40.2E). Two stably tethered trivalent bispecific structures, referred to as TF1 (Figure 40.2G)

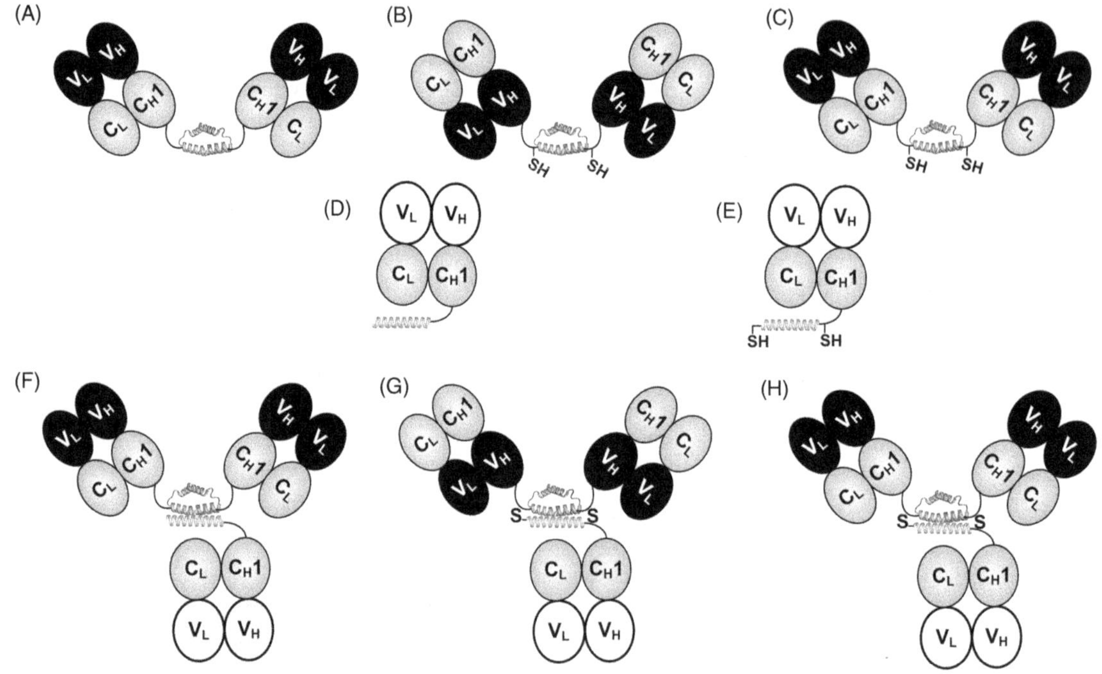

FIGURE 40.2 Fab-based DNL-modules and TFs. (A) C_H1-DDD1-Fab, a Fab-DDD-module without "locking" cysteine residues in the DDD, which is fused to the carboxyl-terminal end of the Fd chain. (B) *N*-DDD2-Fab, a Fab-DDD-module with added cysteine residues in the DDD, which is fused to the amino-terminal end of the Fd chain. (C) C_H1-DDD2-Fab, a Fab-DDD-module with cysteine residues in the DDD, which is fused to the carboxyl-terminal end of the Fd chain. (D) C_H1-AD1-Fab, a Fab-AD-module without "locking" cysteine residues in the AD, which is fused to the carboxyl-terminal end of the Fd chain. (E) C_H1-AD2-Fab, a Fab-AD-module with cysteine residues in the AD, which is fused to the carboxyl-terminal end of the Fd chain. (F) Binary, noncovalent complex of C_H1-DDD1-Fab-hMN-14 and C_H1-AD1-Fab-h679 (G) TF1, a covalent complex of N-DDD2-Fab-hMN-14 and C_H1-AD2-Fab-h679 (H) TF2, a covalent complex of C_H1-DDD2-Fab-hMN-14 and C_H1-AD2-Fab-h679. Variable (V, black or white) and constant (C, gray) domains of the heavy (H) and light (L) chains are represented as ovals. The DDD and AD peptides are shown as helices with the locations indicated for the reactive sulfhydryl groups (SH) of the engineered cysteine residues in AD2 and DDD2, and the disulfide bridges (S) that they form in TF1 and TF2. Modified with copyright permission from Ref. [22].

for the conjugate of *N*-DDD2-Fab-hMN-14 and C_H1-AD2-Fab-h679 and TF2 (Figure 40.2H) for the conjugate of C_H1-DDD2-Fab-hMN-14 and C_H1-AD2-Fab-h679, were produced in nearly quantitative yields and characterized extensively. TF1 and TF2 were isolated by affinity chromatography using the h679-binding hapten HSG, and were each resolved by size exclusion high performance liquid chromatography (SE-HPLC) as a single peak of the expected molecular size (~150 kDa). BIAcore demonstrated bispecific binding, which was confirmed by competition ELISA to be equivalent to hMN-14 IgG and h679 Fab, reflecting the retention of valency and binding affinity. Furthermore, TF1 and TF2 were stable for at least 7 days at 37°C in human or mouse serum. The superiority of TF2 as a pretargeting agent for diagnostic imaging has been demonstrated in numerous studies [32–35].

Since the generation of TF1 and TF2, the modular DNL method has allowed the rapid development of over 20 different trivalent bispecific Fab-based complexes, referred to as the TF series, by combinatorial pairing of monomeric Fab-AD2 and dimeric Fab-DDD2 modules [36–45]. The technology platform has also been expanded to generate a variety of conjugates, including multivalent, multifunctional antibodies, and immunocytokines, which are highlighted below.

40.5 IgG-AD2-MODULES

After establishing proof of principle with the TF series using Fab-based DDD2- and AD2-modules, we next developed AD2-modules for IgG, which allowed the synthesis of a variety of complex, IgG-based DNL conjugates. C_H3-AD2-IgG modules were produced recombinantly by appending, in frame, the coding sequence for

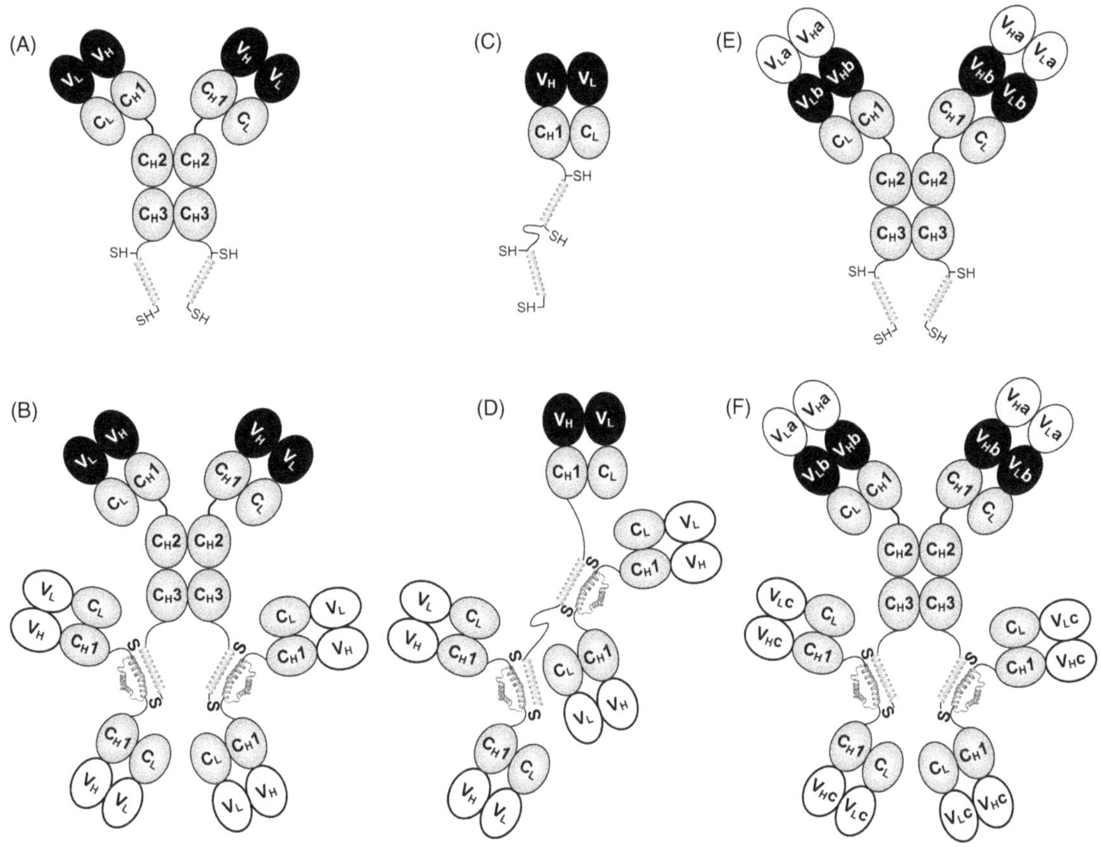

FIGURE 40.3 DNL modules for multivalent antibodies. (A) C_H3-AD2-IgG, an IgG-AD2 module having two AD2 peptides. (B) bsHexAb, comprising an IgG and two stabilized Fab dimers. (C) C_H1-$(AD2)_2$-Fab, a Fab module comprising tandem AD2 helices separated by a short peptide linker. (D) Bispecific antibody comprising five Fabs. (E) C_H3-AD2-DVD-IgG, a duel variable domain IgG-AD2 module, with binding specificities "a" and "b", and possessing two AD2 peptides. (F) A trispecific octavalent antibody comprising a dual variable domain IgG, having binding specificities "a" and "b", and two stabilized Fab dimers of binding specificity "c". Variable (V, black, white or light gray) and constant (C, dark gray) domains of the heavy (H) and light (L) chains are represented as ovals. The DDD2 and AD2 peptides are shown as helices with the locations indicated for the reactive sulfhydryl groups (SH) and disulfide bridges (S). Modified with copyright permission from Ref. [22].

AD2, preceded by a flexible peptide linker, to the 3′ end of the coding sequence of the C_H3 domain of IgG (Figure 40.3A). This method allowed the simple conversion of any existing IgG-expression plasmid vector into one for IgG-AD2 expression. Because IgG naturally forms a heterotetramer comprising two heavy and two light chains, IgG-AD2-modules possesses two AD2 peptides, having one on the carboxyl-terminal end of each heavy chain (Figure 40.3A). The IgG-AD2-modules are purified similar to IgG, using Protein A affinity chromatography. Each AD2 binds a dimeric X-DDD2 module, resulting in defined DNL structures comprising an IgG fused at its carboxyl terminus to four X groups, where X denotes an entity that could be a protein, peptide, polymer, drug, or other molecule.

40.6 HEXAVALENT ANTIBODIES

IgG-AD2-modules were first used to construct hexavalent antibodies (HexAbs), which were produced by combination of IgG-AD2 with the same Fab-DDD2 modules used in the TF series, highlighting the advantage of the modular nature of DNL. Initially, monospecific HexAbs were produced using IgG-AD2- and Fab-DDD2-modules derived from the same parental mAb. As an example, a HexAb was made for the humanized anti-CD20 mAb veltuzumab (a.k.a. hA20) [46], where C_H3-AD2-IgG-hA20 was combined with C_H1-DDD2-Fab-hA20 [47]. The DNL conjugation resulted in a homogeneous preparation of a conjugate comprising an F_c and six functional anti-CD20 Fabs, which each retain the binding affinity of veltuzumab. The construct, originally named Hex-hA20, has been designated 20-(20)-(20), using a standardized naming system where the first code indicates the IgG-AD2-module and the codes in parentheses indicate dimeric Fab-DDD2-modules. Compared to veltuzumab, 20-(20)-(20) exhibited threefold higher binding avidity by ELISA, and a threefold slower off-rate from live non-Hodgkin lymphoma (NHL) cells by flow cytometry, which demonstrated that all six binding arms can bind CD20 on cells. *In vitro*, 20-(20)-(20) inhibited proliferation of NHL cells at subnanomolar concentrations without the need for an additional crosslinking antibody. For 20-(20)-(20), some of the Fc-associated effector functions, including antibody-dependent cell-mediated cytotoxicity (ADCC), are retained, while others are apparently compromised. Unlike veltuzumab, 20-(20)-(20) did not effect complement-dependent cytotoxicity (CDC) *in vitro*. However, it is unclear if this is also the case *in vivo*, because recent experimental evidence suggests CDC mediated by similarly constructed HexAbs in whole blood. Even though 20-(20)-(20) is a nearly 2.5-fold larger molecule than veltuzumab, the circulating serum half-life is shorter for the former, suggesting that its interaction with the neonatal F_c receptor is altered. Although 20-(20)-(20) has reduced circulating half-life and CDC, it still demonstrated antitumor efficacy, which was comparable to veltuzumab at equivalent doses, in tumor-bearing mice. Bispecific hexavalent antibodies (bsHexAbs) are readily constructed using DNL by combining an IgG-AD2 module with a Fab-DDD2 module derived from a different parental mAb (Figure 40.3B). A large variety of bsHexAbs can be generated from a relatively small number of DNL modules. For example, 100 different bsHexAbs can be made combinatorially using 10 IgG-AD2 and 10 Fab-DDD2 modules.

The first bsHexAbs produced were derived from the humanized anti-CD22 mAb epratuzumab (a.k.a. hLL2) [48] and veltuzumab [49]. Combination of C_H3-AD2-IgG-hA20 with C_H1-DDD2-Fab-hLL2 produced 20-(22)-(22), which comprises veltuzumab with four Fabs of epratuzumab. A bsHexAb of the opposite configuration, 22-(20)-(20), which has four Fabs of veltuzumab fused to epratuzumab, was generated from C_H3-AD2-IgG-hLL2 and C_H1-DDD2-Fab-hA20. Characterization of the bsHexAbs demonstrated that the DNL conjugation resulted in highly purified covalent structures of the expected size and composition. Both 22-(20)-(20) and 20-(22)-(22) retain the binding properties of their parental Fab/IgGs with all six Fabs apparently capable of binding simultaneously. The bsHexAbs exhibited biological activities that were not observed using a mixture of the parental mAbs. Treatment of cells with the bsHexAbs, but not veltuzumab plus epratuzumab, resulted in translocation of both CD22 and CD20 into lipid rafts, induction of apoptosis and growth inhibition without second-antibody crosslinking, and homotypic adhesion. The bsHexAbs induced significant increases in the levels of phosphorylated p38 and PTEN, and also notable differences in signaling events from those incurred by crosslinking veltuzumab or rituximab with a secondary antibody [50]. Thus, the greatly enhanced direct toxicity of these bsHexAbs correlates with their ability to alter the basal expression of various intracellular proteins involved in regulating cell growth, survival, and apoptosis, with the net outcome leading to cell death. Indeed, the bsHexAbs killed lymphoma cells *in vitro* much more potently than the parental mAb mixture.

As observed previously for the monospecific HexAbs, both bsHexAbs exhibited ADCC, but not CDC (*in vitro*), and had shorter circulating half-lives than the parent mAbs. Intriguingly, 22-(20)-(20) and 20-(22)-(22) killed human lymphoma cells in preference to normal B cells in whole human blood (*ex vivo*), whereas the parental veltuzumab depleted malignant and normal B cells about equally. HexAbs produced with IgG-AD2 modules having the AD2 peptide fused at the carboxyl-terminal end of the light chain exhibited improved *in vivo* properties, including CDC and circulating half-lives that were similar to the parental mAbs, resulting in superior anti-tumor efficacy [51].

40.7 MORE ANTIBODY-BASED-MODULES AND MULTIVALENT ANTIBODIES

We have produced many bispecific TF and HexAb DNL conjugates, which are of 1×2 and 2×4 Fab formats, respectively. In addition, we have produced trispecific HexAbs ($2 \times 2 \times 2$) by combining two different Fab-DDD2-modules with an IgG-AD2-module of a third specificity (unpublished results). A wide variety of alternative formats of multispecific antibodies is attainable with the introduction of new types of DNL-modules. For example, we have produced a Fab-based module with tandem AD2 peptides fused to the carboxyl-terminal end of the F_d (Figure 40.3C). Combination with a Fab-DDD2 module resulted in a 1×4 bispecific conjugate comprising five Fabs, without an F_c (Figure 40.3D), demonstrating that multiple AD2s can be incorporated into a single module (unpublished data). Application of this concept to an IgG module (IgG-AD2-AD2) would allow the creation of a multispecific IgG having a total of 10 Fabs, via combination with existing Fab-DDD2 modules. AD2- and DDD2-modules could be generated for other types of antibody-based proteins/fragments including scFv, diabody, minibody, and dual variable domain IgG, providing a myriad of possible multispecific antibody formats. As a final example, we have generated a bispecific dual variable domain IgG-AD2 module (Figure 40.3E), which was combined with a Fab-DDD2 module to generate a $2 \times 2 \times 4$ trispecific octavalent antibody (Figure 40.3F, unpublished data).

40.8 NONANTIBODY-BASED DNL MODULES

DNL is not restricted to antibody-based constructs, as recombinant DDD2 or AD2 modules can be constructed for almost any protein. We have applied the DNL method to various cytokines, including interferon-α (IFN-α) [27,52,53], erythropoietin [54], and granulocyte colony-stimulating factor (G-CSF) [55], to generate various immunocytokines (Figure 40.4). The cytokine modules were produced in either *Escherichia coli* or mammalian cell culture with AD2 or DDD2 fused at their amino- or carboxyl-terminal ends (Figure 40.4A and B) [56]. We have utilized a DDD2-module of human protamine which, when paired with an IgG-AD2 module, can be used for targeted cellular delivery of nucleic acids, such as siRNA [57]. As an example of a module of a precursor that is both an enzyme and a toxin, a DDD2-module of ranpirnase (Rap, a cytotoxic ribonuclease from *Rana pipiens*) has been combined with various IgG-AD2 modules for targeted delivery of the toxin to tumor cells [58,59]. In addition to recombinant protein modules, nonprotein DNL-modules can be assembled using synthetic AD2 peptides coupled to a variety of synthetic precursor molecules [27,28].

40.9 IFN-α2b-DDD2 MODULE AND IMMUNOCYTOKINES

Our most utilized cytokine module, IFN-α2b-DDD2, which is produced with recombinant myeloma cell culture, has a carboxyl-terminal DDD appended to IFN-α2b (Figure 40.4A). We have combined IFN-α2b-DDD2 with various IgG-AD2-modules to create a panel of IgG-IFN-α for targeted delivery of IFN-α to various malignancies.

The prototype, 20-2b-2b (a.k.a. 20-2b), which is a conjugate of IFN-α2b-DDD2 and C_H3-AD2-IgG-hA20, comprises veltuzumab and four copies of IFN-α2b (Figure 40.4C) [52]. All of the IgG-IFN-α made by DNL, including 20-2b-2b, retain the binding specificity and affinity of their targeting mAb, and also exhibit IFN-α specific activity approaching that of recombinant IFN-α2b (rIFN-α2b) [52,53]. Cytokine fusion proteins made by traditional recombinant engineering or chemical conjugation often suffer extensive loss of biological activity. For example, where 20-2b-2b has similar activity to rIFN-α2b, traditional recombinant anti-CD20-IFN-α fusion proteins showed a 300-fold reduction in IFN-α activity [60]. We believe that high-level activity of the DNL constructs is maintained due to the site-specific conjugation, which is spatially removed from the functional domain.

Compared to veltuzumab, 20-2b-2b shows enhanced ADCC, but lacks CDC (*in vitro*), consistent with other constructs comprising C_H3-AD2-IgG-hA20. It inhibits *in vitro* proliferation of lymphoma cells and depletes them from whole human blood more potently than a combination of veltuzumab and IFN-α. Comparative pharmacokinetic (PK) analysis in mice demonstrated a longer circulating serum half-life for 20-2b-2b (23.4 h) compared to the commercial pegylated IFN-α drugs peginterferonalfa-2a (14.9 h) or peginterferonalfa-2b (9.3 h). Owing to high specific activity, extended PK and tumor targeting, 20-2b-2b demonstrated superior therapeutic efficacy compared to peginterferonalfa-2a, veltuzumab or nontargeting IgG-IFN-α

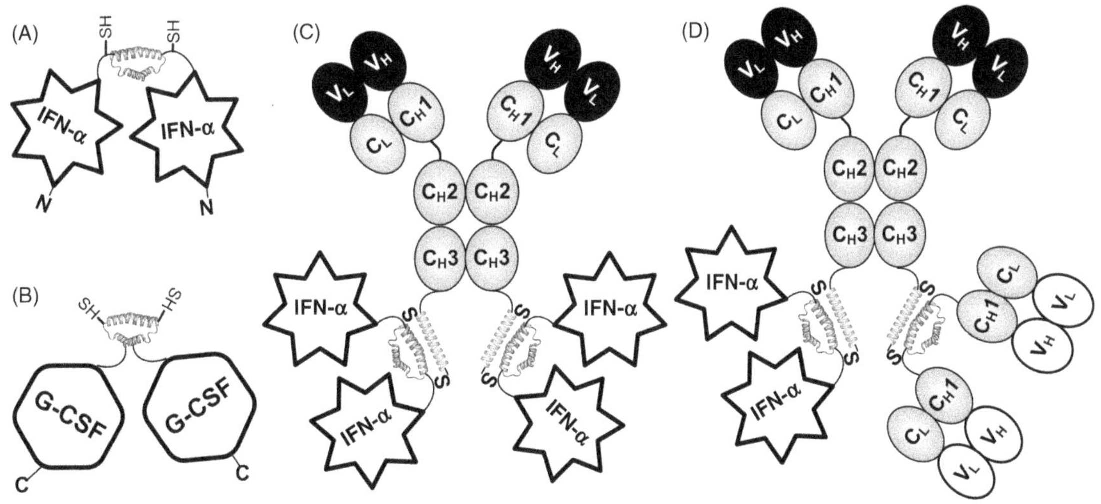

FIGURE 40.4 Cytokine modules and immunocytokines. (A) IFN-α-DDD2, a dimeric module comprising DDD2 fused to the amino-terminal end of IFN-α. (B) DDD2-G-CSF, a dimeric module comprising DDD2 fused to the carboxyl-terminal end of G-CSF. (C) IgG-2b-2b (e.g., 20-2b-2b), an immunocytokine comprising an IgG and four IFN-α groups. (D) IgG-$(Fab)_2$-2b (e.g., 20-C2-2b), a bispecific immunocytokine comprising an IgG, a stabilized Fab dimer and two IFN-α groups. Modified with copyright permission from Ref. [22].

in three human lymphoma xenograft models, even though mouse immune cells respond poorly to human IFN-α2b. Targeting IFN-α with an anti-CD20 mAb makes the immunocytokine more potent than either agent alone or in combination.

On the basis of encouraging preclinical results, 20-2b-2b, the prototype IgG-IFN-α, is now under development for CD20-targeted immunotherapy of NHL. CD20 is a preferred target for this disease, because it is expressed at high levels on the cell surface of many B-cell NHL, and its expression on normal cells is essentially limited to B cells. The potential benefits of therapy with 20-2b-2b are likely limited to NHL, hairy-cell leukemia, and possibly chronic lymphocytic leukemia (CLL) patients. Using the combination of the IFN-α2b-DDD2 module with C_H3-AD2-IgG-hL243, we have recently developed an IgG-IFN-α named C2-2b-2b, which has tetrameric IFN-α2b fused to an anti-HLA-DR mAb (hL243) [61]. HLA-DR is an attractive target because it is expressed on the cell surface of many hematopoietic malignancies [62]. The broad range and high-level expression of HLA-DR makes C2-2b-2b attractive for use in therapy of diverse malignancies. *In vitro*, C2-2b-2b inhibited 20 cell lines, including B-cell lymphoma (Burkitt, mantle cell and follicular), leukemia (hairy cell, acute myeloid leukemia (AML), acute lymphoblastic leukemia (ALL), and CLL), and myeloma. In most cases, this immunocytokine was more effective than CD20-targeted mAb-IFN-α or a mixture comprising the parental mAb and IFN-α. Responsiveness of each hematopoietic tumor cell line correlated with HLA-DR expression/density and sensitivity to IFN-α and hL243. C2-2b-2b induced more potent and longer-lasting IFN-α signaling compared to nontargeted IFN-α. *In vivo*, C2-2b-2b demonstrated superior efficacy compared to nontargeting mAb-IFN-α, peginterferonalfa-2a, or a combination of hL243 and IFN-α using human lymphoma and myeloma xenografts.

The modular nature of DNL enabled the production of the first bispecific immunocytokine, 20-C2-2b, which comprises two copies of IFN-α2b and a stabilized $F(ab)_2$of hL243 (humanized anti-HLA-DR) site-specifically linked to veltuzumab (Figure 40.4D) [53]. This was achieved by combining C_H3-AD2-IgG-hA20 with two DDD-modules, IFN-α2b-DDD2 and C_H1-DDD2-Fab-hL243. Owing to the random association of either DDD-module with the two AD2 groups, two side-products, 20-C2-C2 and 20-2b-2b are formed, in addition to 20-C2-2b, which was purified from the mixture by a multistep affinity chromatography process. The predicted biochemical and biological properties of 20-C2-2b were confirmed experimentally. It bound bispecifically to $CD20^+$/HLA-DR^+ cells. With two IFN-α2b groups, 20-2b-2b exhibited half of the level of IFN-α specific activity, compared to IgG-IFN-α with four IFN-α (e.g., 20-2b-2b). The bispecific IgG-IFN-α potently inhibited lymphoma and myeloma cell lines, and in some cases, more effectively than 20-2b-2b or C2-2b-2b [53].

IFN-α2b-DDD2 was also utilized in the development of a novel DNL-based PEGylation technology [27], which is beyond the focus of this review.

40.10 VARIATIONS ON THE DNL THEME

Several potential variations on the DNL method as described above are possible. In addition to the DDD sequence of human RIIα, other DDD sequences may be selected from human RIα, human RIβ, or human RIIβ. The DDD sequence of choice can be matched with a highly interactive AD sequence, which can be deduced from the literature [63] or determined experimentally.

40.10.1 Alternative Linker Modules

The X-type four-helix bundle dimerization motif that is a structural characteristic of the DDD [18,19] is found in other classes of proteins. For example, over 300 proteins that are involved in either signal transduction or transcriptional activation contain a module of 65–70 amino acids termed the sterile α motif (SAM) domain, which has a variation of the X-type four-helix bundle present on its dimerization interface. Such proteins include the S100 proteins (for example, S100B and calcyclin), and the hepatocyte nuclear factor (HNF) family of transcriptional factors (for example, HNF-1α and HNF-1β). In the case of S100B, this X-type four-helix bundle enables the binding of each dimer to two p53 peptides derived from the C-terminal regulatory domain (residues 367–388) with micromolar affinity [64]. Similarly, the N-terminal dimerization domain of HNF-1α (HNF-p1) was shown to associate with a dimer of DCoH (dimerization cofactor for HNF-1) via a dimer of HNF-p1 [65]. These naturally occurring systems may also be adapted as linker modules using the DNL approach. However, S100 proteins have biological activities, such as tumorigenesis, thus rendering them less desirable as alternative sources for the DDD.

40.10.2 Alternative Covalent Stabilization

In addition to the use of disulfide linkages for preventing the dissociation of the assembled components, other methods for enhancing the overall stability of the tethered structure may be employed. For example, various cross-linking agents or methods that are commercially available or used in research may be selected. A potentially useful agent is glutaraldehyde, which has been widely used for probing the structures of noncovalently associated multimeric proteins by cross-linking the constituting subunits to form stable conjugates. Also of interest are two chemical methods involving oxidative cross-linking of protein subunits. One is a proximity labeling technique that employs either hexahistidine-tagged [66], or

N-terminal glycine–glycine–histidine-tagged proteins [67]. These tags bind Ni^{2+} tightly and, when oxidized with a peracid, a Ni^{3+} species is produced to mediate a variety of oxidative reactions, including protein–protein cross-linking. Another technique, termed PICUP for photo-induced cross-linking of unmodified proteins, uses $[Ru(II)(bipy)_3]^{2+}$, ammonium persulfate, and visible light to induce protein–protein cross-linking [68]. A general concern, however, is that these approaches would be less specific and efficient than the DDD2/AD2 coupling systems.

40.11 CONCLUSIONS AND FUTURE PERSPECTIVE

Clinical evaluation has begun, or is planned, for several of the DNL-based fusion proteins described in this chapter. TF2 is currently under Phase I clinical investigation for imaging and therapy of colorectal and lung cancers (Clinicaltrial.gov nos. NCT01273402, NCT01221675, NCT00860860, and NCT00895323). Additional TFs are being manufactured for clinical use in pretargeted imaging and therapy of pancreatic (TF10) and prostate (TF12) cancers [36]. The immunocytokine 20-2b-2b is under development for CD20-targeted immunotherapy of lymphoma, and C2-2b-2b is under consideration as a therapeutic agent for a variety of hematopoietic cancers. Preliminary results suggest that some of the bsHexAbs, including 22-(20)-(20) and others, might be effective agents for the treatment of certain autoimmune diseases. Several additional multifunctional antibodies and immunocytokines have shown promise, but are at earlier stages in development.

It has been five years since the invention of the DNL method and it is still an evolving technology. We continue to produce new classes of DNL-modules, and with the introduction of each new module, the repertoire of potentially useful DNL conjugates increases exponentially. Thus far, much of our effort has focused on the creation of agents for cancer therapy via multivalent/multifunctional antibodies and antibody-targeted delivery of radionuclides, cytokines, or toxins. The indications and modalities for which the DNL method can be applied expand with the increasing library of DNL-modules and conjugates.

REFERENCES

1. Binz HK, Amstutz P, Pluckthun A. (2005) Engineering novel binding proteins from nonimmunoglobulin domains. *Nat. Biotechnol.* 23, 1257–1268.
2. Binz HK, Pluckthun A. (2005) Engineered proteins as specific binding reagents. *Curr. Opin. Biotechnol.* 16, 459–469.
3. Hey T, Fiedler E, Rudolph R, Fiedler M. (2005) Artificial, non-antibody binding proteins for pharmaceutical and industrial applications. *Trends Biotechnol.* 23, 514–522.
4. Hosse RJ, Rothe A, Power BE. (2006) A new generation of protein display scaffolds for molecular recognition. *Protein Sci.* 15, 14–27.
5. Lo KM, Sudo Y, Chen J, Li Y, Lan Y, Kong SM, et al. (1998) High level expression and secretion of Fc-X fusion proteins in mammalian cells. *Protein Eng.* 11, 495–500.
6. Pastan I. (2003) Immunotoxins containing Pseudomonas exotoxin A: a short history. *Cancer Immunol. Immunother.* 52, 338–341.
7. Kishore U, Strong P, Perdikoulis MV, Reid KB. (2001) A recombinant homotrimer, composed of the alpha helical neck region of human surfactant protein D and C1q B chain globular domain, is an inhibitor of the classical complement pathway. *J. Immunol.* 166, 559–565.
8. Kostelny SA, Cole MS, Tso JY. (1992) Formation of a bispecific antibody by the use of leucine zippers. *J. Immunol.* 148, 1547–1553.
9. Pack P, Muller K, Zahn R, Pluckthun A. (1995) Tetravalent miniantibodies with high avidity assembling in *Escherichia coli*. *J. Mol. Biol.* 246, 28–34.
10. Terskikh AV, Le Doussal JM, Crameri R, Fisch I, Mach JP, Kajava AV. (1997) "Peptabody": a new type of high avidity binding protein. *Proc. Natl. Acad. Sci. USA* 94, 1663–1668.
11. Willuda J, Kubetzko S, Waibel R, Schubiger PA, Zangemeister-Wittke U, Pluckthun A. (2001) Tumor targeting of mono-, di-, and tetravalent anti-p185(HER-2) miniantibodies multimerized by self-associating peptides. *J. Biol. Chem.* 276, 14385–14392.
12. Pepinsky RB, LePage DJ, Gill A, Chakraborty A, Vaidyanathan S, Green M, et al. (2001) Improved pharmacokinetic properties of a polyethylene glycol-modified form of interferon-beta-1a with preserved *in vitro* bioactivity. *J. Pharmacol. Exp. Ther.* 297, 1059–1066.
13. Tan LP, Lue RY, Chen GY, Yao SQ. (2004) Improving the intein-mediated, site-specific protein biotinylation strategies both *in vitro* and *in vivo*. *Bioorg. Med. Chem. Lett.* 14, 6067–6070.
14. Burbulis I, Yamaguchi K, Gordon A, Carlson R, Brent R. (2005) Using protein-DNA chimeras to detect and count small numbers of molecules. *Nat. Methods* 2, 31–37.
15. Rossi EA, Goldenberg DM, Cardillo TM, McBride WJ, Sharkey RM, Chang CH. (2006) Stably tethered multifunctional structures of defined composition made by the dock and lock method for use in cancer targeting. *Proc. Natl. Acad. Sci. USA* 103, 6841–6846.
16. Baillie GS, Scott JD, Houslay MD. (2005) Compartmentalisation of phosphodiesterases and protein kinase A: opposites attract. *FEBS Lett.* 579, 3264–3270.
17. Wong W, Scott JD. (2004) AKAP signalling complexes: focal points in space and time. *Nat. Rev. Mol. Cell Biol.* 5, 959–970.
18. Newlon MG, Roy M, Morikis D, Hausken ZE, Coghlan V, Scott JD, et al. (1999). The molecular basis for protein kinase A anchoring revealed by solution NMR. *Nat. Struct. Biol.* 6, 222–227.
19. Carr DW, Stofko-Hahn RE, Fraser ID, Bishop SM, Acott TS, Brennan RG, et al. (1991) Interaction of the regulatory subunit

(RII) of cAMP-dependent protein kinase with RII-anchoring proteins occurs through an amphipathic helix binding motif. *J. Biol. Chem.* 266, 14188–14192.

20. Colledge M, Scott JD. (1999) AKAPs: from structure to function. *Trends Cell Biol.* 9, 216–221.
21. Newlon MG, Roy M, Morikis D, Carr DW, Westphal R, Scott JD, et al. (2001) A novel mechanism of PKA anchoring revealed by solution structures of anchoring complexes. *EMBO J.* 20, 1651–1662.
22. Rossi EA, Goldenberg DM, Chang C-H. (2012) The dock-and-lock method combines recombinant engineering with site-specific covalent conjugation to generate multifunctional structures. *Bioconjug. Chem.* 23, 309–323.
23. Alto NM, Soderling SH, Hoshi N, Langeberg LK, Fayos R, Jennings PA, et al. (2003) Bioinformatic design of A-kinase anchoring protein-in silico: a potent and selective peptide antagonist of type II protein kinase A anchoring. *Proc. Natl. Acad. Sci. USA* 100, 4445–4450.
24. Backer MV, Patel V, Jehning BT, Backer JM. (2006) Self-assembled "dock and lock" system for linking payloads to targeting proteins. *Bioconjug. Chem.* 17, 912–919.
25. Deyev SM, Waibel R, Lebedenko EN, Schubiger AP, Pluckthun A. (2003) Design of multivalent complexes using the barnase*barstar module. *Nat. Biotechnol.* 21, 1486–1492.
26. Hodneland CD, Lee YS, Min DH, Mrksich M. (2002) Selective immobilization of proteins to self-assembled monolayers presenting active site-directed capture ligands. *Proc. Natl. Acad. Sci. USA* 99, 5048–5052.
27. Chang CH, Rossi EA, Cardillo TM, Nordstrom DL, McBride WJ, Goldenberg DM. (2009) A new method to produce monoPEGylated dimeric cytokines shown with human interferon-alpha2b. *Bioconjug. Chem.* 20, 1899–1907.
28. Wang Y, McBride B, Goldenberg D, Chang C-H. Novel antibody-dendrimer conjugayes efficiently complex plasmid DNA [abstract]. In: Proceedings of the 102th Meeting of the American Association of Cancer Research; April 2–6, 2011; Orlando, FL. Philadelphia, PA: AACR; 2011. Abst# 4445.
29. Blumenthal RD, Hansen HJ, Goldenberg DM. (2005) Inhibition of adhesion, invasion, and metastasis by antibodies targeting CEACAM6 (NCA-90) and CEACAM5 (Carcinoembryonic Antigen). *Cancer Res.* 65, 8809–8817.
30. Rossi EA, Sharkey RM, McBride W, Karacay H, Zeng L, Hansen HJ, et al. (2003) Development of new multivalent-bispecific agents for pretargeting tumor localization and therapy. *Clin. Cancer Res.* 9, 3886S–3896S.
31. Gegner JA, Dahlquist FW. (1991) Signal transduction in bacteria: CheW forms a reversible complex with the protein kinase CheA. *Proc. Natl. Acad. Sci. USA* 88, 750–754.
32. Schoffelen R, van der Graaf WT, Franssen G, Sharkey RM, Goldenberg DM, McBride WJ, et al. (2010) Pretargeted ^{177}Lu radioimmunotherapy of carcinoembryonic antigen-expressing human colonic tumors in mice. *J. Nucl. Med.* 51, 1780–1787.
33. Sharkey RM, Rossi EA, McBride WJ, Chang CH, Goldenberg DM. (2010) Recombinant bispecific monoclonal antibodies prepared by the dock-and-lock strategy for pretargeted radioimmunotherapy. *Semin. Nucl. Med.* 40, 190–203.
34. McBride WJ, Sharkey RM, Karacay H, D'Souza CA, Rossi EA, Laverman P, (2009) A novel method of ^{18}F radiolabeling for PET. *J. Nucl. Med.* 50, 991–998.
35. Sharkey RM, Karacay H, Vallabhajosula S, McBride WJ, Rossi EA, Chang CH, et al. (2008) Metastatic human colonic carcinoma: molecular imaging with pretargeted SPECT and PET in a mouse model. *Radiology* 246, 497–507.
36. Gold DV, Goldenberg DM, Karacay H, Rossi EA, Chang CH, Cardillo TM, et al. (2008) A novel bispecific, trivalent antibody construct for targeting pancreatic carcinoma. *Cancer Res.* 68, 4819–4826.
37. Goldenberg DM, Rossi EA, Sharkey RM, McBride WJ, Chang CH. (2008) Multifunctional antibodies by the Dock-and-Lock method for improved cancer imaging and therapy by pretargeting. *J. Nucl. Med.* 49, 158–163.
38. Karacay H, Sharkey RM, Gold DV, Ragland DR, McBride WJ, Rossi EA, et al. (2009) Pretargeted radioimmunotherapy of pancreatic cancer xenografts: TF10-90Y-IMP-288 alone and combined with gemcitabine. *J. Nucl. Med.* 50, 2008–2016.
39. Karacay H, Sharkey RM, McBride WJ, Rossi EA, Chang CH, Goldenberg DM. (2011) Optimization of hapten-peptide labeling for pretargeted immunoPET of bispecific antibody using generator-produced ^{68}Ga. *J. Nucl. Med.* 52, 555–559.
40. Schoffelen R, van der Graaf WT, Franssen G, Sharkey RM, Goldenberg DM, McBride WJ, et al. (2010) Pretargeted ^{177}Lu radioimmunotherapy of carcinoembryonic antigen-expressing human colonic tumors in mice. *J. Nucl. Med.* 51, 1780–1787.
41. Sharkey RM, Karacay H, McBride WJ, Rossi EA, Chang CH, Goldenberg DM. (2007) Bispecific antibody pretargeting of radionuclides for immuno single-photon emission computed tomography and immuno positron emission tomography molecular imaging: an update. *Clin. Cancer Res.* 13, 5577s–5585s.
42. Sharkey RM, Karacay H, Litwin S, Rossi EA, McBride WJ, Chang CH, et al. (2008) Improved therapeutic results by pretargeted radioimmunotherapy of non-Hodgkin's lymphoma with a new recombinant, trivalent, anti-CD20, bispecific antibody. *Cancer Res.* 68, 5282–5290.
43. Sharkey RM, Karacay H, Johnson CR, Litwin S, Rossi EA, McBride WJ, et al. (2009) Pretargeted versus directly targeted radioimmunotherapy combined with anti-CD20 antibody consolidation therapy of non-Hodgkin lymphoma. *J. Nucl. Med.* 50, 444–453.
44. Sharkey RM, Rossi EA, McBride WJ, Chang CH, Goldenberg DM. (2010) Recombinant bispecific monoclonal antibodies prepared by the dock-and-lock strategy for pretargeted radioimmunotherapy. *Semin. Nucl. Med.* 40, 190–203.
45. Sharkey RM, Rossi EA, Chang CH, Goldenberg DM. (2010) Improved cancer therapy and molecular imaging with multivalent, multispecific antibodies. *Cancer Biother. Radiopharm.* 25, 1–12.
46. Stein R, Qu Z, Chen S, Rosario A, Shi V, Hayes M, et al. (2004) Characterization of a new humanized anti-CD20 monoclonal antibody, IMMU-106, and its use in combination with the humanized anti-CD22 antibody, epratuzumab, for the therapy of non-Hodgkin's lymphoma. *Clin. Cancer Res.* 10, 2868–2878.

47. Rossi EA, Goldenberg DM, Cardillo TM, Stein R, Wang Y, Chang CH. (2008) Novel designs of multivalent anti-CD20 humanized antibodies as improved lymphoma therapeutics. *Cancer Res.* 68, 8384–8392.

48. Carnahan J, Stein R, Qu Z, Hess K, Cesano A, Hansen HJ, et al. (2007) Epratuzumab, a CD22-targeting recombinant humanized antibody with a different mode of action from rituximab. *Mol. Immunol.* 44, 1331–1341.

49. Rossi EA, Goldenberg DM, Cardillo TM, Stein R, Chang CH. (2009) Hexavalent bispecific antibodies represent a new class of anticancer therapeutics: 1. Properties of anti-CD20/CD22 antibodies in lymphoma. *Blood* 113, 6161–6171.

50. Gupta P, Goldenberg DM, Rossi EA, Chang CH. (2010) Multiple signaling pathways induced by hexavalent, monospecific, anti-CD20 and hexavalent, bispecific, anti-CD20/CD22 humanized antibodies correlate with enhanced toxicity to B-cell lymphomas and leukemias. *Blood* 116, 3258–3267.

51. Rossi EA, Chang C-H, Cardillo TM, Goldenberg DM. (2012) Optimization of multivalent bispecific antibodies and immunocytokines with improved in vivo properties. *Bioconjug. Chem.* [epub ahead of print] PMID 23116517.

52. Rossi EA, Goldenberg DM, Cardillo TM, Stein R, Chang CH. (2009) CD20-targeted tetrameric interferon-alpha, a novel and potent immunocytokine for the therapy of B-cell lymphomas. *Blood* 114, 3864–3871.

53. Rossi EA, Rossi DL, Stein R, Goldenberg DM, Chang CH. (2010) A bispecific antibody-IFNα2b immunocytokine targeting CD20 and HLA-DR is highly toxic to human lymphoma and multiple myeloma cells. *Cancer Res.* 70, 7600–7609.

54. Rossi D, Goldenberg D, Cardillo T, Rossi E, Chang C-H. Tetrameric erythropoietin anchored on IgG with improved biological activity [abstract]. In: Proceedings of the 100th Meeting of the American Association of Cancer Research; April 18–22, 2009; Denver, CO. Philadelphia, PA: AACR; 2009. Abst# 3255.

55. Wang Y, Loo M, Cardillo T, Goldenberg D, Chang C-H. Generation of a novel class of tetrameric granulocyte colony-stimulating factor with improved biological activity [abstract]. In: Proceedings of the 100th Meeting of the American Association of Cancer Research; April 18–22, 2009; Denver, CO. Philadelphia, PA: AACR; 2009. Abst# 3256.

56. Chang C-H, Rossi E, Loo M, Nordstrom D, Chan E, Cardillo T, et al. A modular method to prepare novel tetrameric cytokines, IFN, G-CSF, and EPO, with improved pharmacokinetics by the Dock-and-Lock (DNL) platform technology [abstract]. In: Proceedings of the 99th Meeting of the American Association of Cancer Research; April 12–16, 2008; San Diego, CA. Philadelphia, PA: AACR; 2008. Abst # 4906.

57. Gupta P, Goldenberg D, Chang C-H. Targeted delivery of siRNAs to diverse solid cancers by a novel immunoconjugate comprising tetrameric human protamine linked to an internalizing anti-TROP-2 humanized antibody [abstract]. In: Proceedings of the 101th Meeting of the American Association of Cancer Research; April 17–21, 2010; Washington, DC. Philadelphia, PA: AACR; 2010. Abst# LB-34.

58. Cardillo T, Rossi E, Arrojo R, Chan E, Goldenberg D, Chang C-H. A novel immunotoxin comprising quadruple RNase tethered to an internalizing anti-TROP-2 humanized MAb shows potent cytotoxicity against diverse solid tumors *in vitro* [abstract]. In: Proceedings of the 101th Meeting of the American Association of Cancer Research; April 17–21, 2010; Washington, DC. Philadelphia PA: AACR; 2010. Abst# 5346.

59. Rossi E, Chan E, Cardillo T, Trisal P, Goldenberg D, Chang C-H. Potent ribonuclease-based immunotoxins comprising quadruple ranpirnase (Rap) site-speceifically conjugated to B-cell lymphoma-targeting antibodies [abstract]. In: Proceedings of the 101th Meeting of the American Association of Cancer Research; April 17–21, 2010; Washington, DC. Philadelphia, PA: AACR; 2010. Abst# 5345.

60. Xuan C, Steward KK, Timmerman JM, Morrison SL. (2010) Targeted delivery of interferon-alpha via fusion to anti-CD20 results in potent antitumor activity against B-cell lymphoma. *Blood* 115, 2864–2871.

61. Rossi EA, Rossi DL, Cardillo TM, Stein R, Goldenberg DM, Chang CH. (2011) Preclinical studies on targeted delivery of multiple IFNα2b to HLA-DR in diverse hematologic cancers. *Blood* 118,1877–1884.

62. Stein R, Gupta P, Chen X, Cardillo TM, Furman RR, Chen S, et al. (2010) Therapy of B-cell malignancies by anti-HLA-DR humanized monoclonal antibody, IMMU-114, is mediated through hyper-activation of ERK and JNK MAP kinase signaling pathways. *Blood* 115, 5180–5190.

63. Burns-Hamuro LL, Ma Y, Kammerer S, Reineke U, Self C, Cook C, et al. (2003) Designing isoform-specific peptide disruptors of protein kinase A localization. *Proc. Natl. Acad. Sci. USA* 100, 4072–4077.

64. Rustandi RR, Drohat AC, Baldisseri DM, Wilder PT, Weber DJ. (1998) The Ca(2+)-dependent interaction of S100B(beta beta) with a peptide derived from p53. *Biochemistry* 37, 1951–1960.

65. Rose RB, Bayle JH, Endrizzi JA, Cronk JD, Crabtree GR, Alber T. (2000) Structural basis of dimerization, coactivator recognition and MODY3 mutations in HNF-1alpha. *Nat. Struct. Biol.* 7, 744–748.

66. Fancy DA, Melcher K, Johnston SA, Kodadek T. (1996) New chemistry for the study of multiprotein complexes: the six-histidine tag as a receptor for a protein crosslinking reagent. *Chem. Biol.* 3, 551–559.

67. Brown KC, Yu Z, Burlingame AL, Craik CS. (1998) Determining protein–protein interactions by oxidative cross-linking of a glycine–glycine–histidine fusion protein. *Biochemistry* 37, 4397–4406.

68. Fancy DA, Kodadek T. (1999) Chemistry for the analysis of protein–protein interactions: rapid and efficient cross-linking triggered by long wavelength light. *Proc. Natl. Acad. Sci. USA* 96, 6020–6024.

INDEX

Fusion Protein Technologies for Biopharmaceuticals: Applications and Challenges, First Edition. Edited by Stefan R. Schmidt.

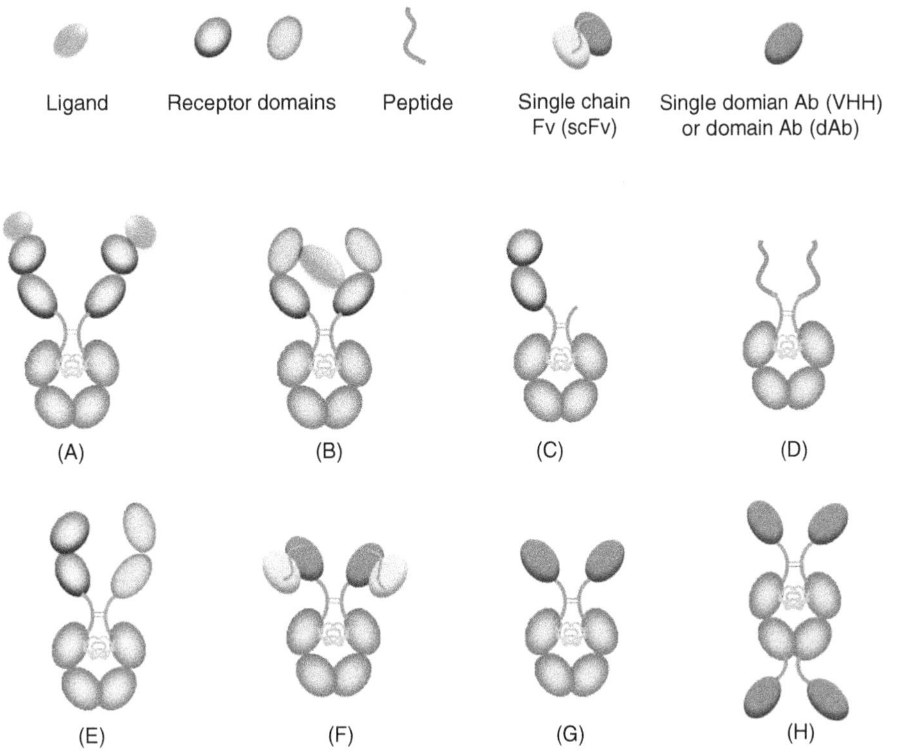

FIGURE 8.1 Schematic representation of selected Fc-fusion proteins. (A) Typical receptor-Fc-fusion protein in which ECD from the same receptor is fused to the Fc domain; (B) multiple ligand binding domains from different receptors are fused to the Fc domain; (C) monomeric-Fc fusion in which only one protein is fused to the Fc domain; (D) MIMETIBODY^{TM} platform; (E) two different Fc-fusion chains form a heterodimer Fc-fusion molecule; (F) single chain Fv (scFv) is fused to the Fc domain; (G) single domain Ab-Fc; (H) single-domain Abs are fused to both N- and C-termini of the Fc domain (Harbour Antibodies). *Source*: Reprinted from *Curr. Opin. Biotechnol.*, **20**(6), Huang C., Receptor-Fc fusion therapeutics, traps, and MIMETIBODY technology. pp. 692–699, Copyright (2009), with permission from Elsevier.

Fusion Protein Technologies for Biopharmaceuticals: Applications and Challenges, First Edition. Edited by Stefan R. Schmidt.

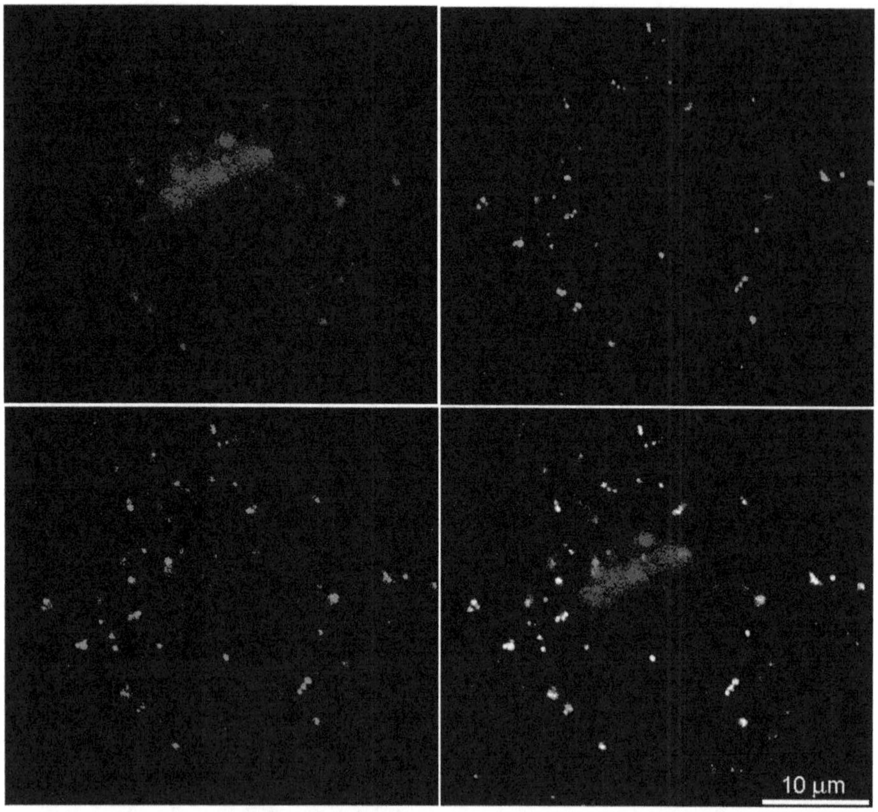

FIGURE 12.3 Molecular imaging of internalized SHG2210:mAb 35 complex in human muscle cells by confocal microscopy SHG2210 fusion protein and mAb 35 were directly labeled with Alexa fluor 546 and 488 dyes, respectively. Fusion protein and antibody were preincubated for 15 min to form complexes and then added to the human muscle cell line DB40 for 2 h on ice. Cells were then washed and switched to 37°C to initiate uptake of the complex. Uptake of complex was allowed to occur for 15 min, at which point cells were fixed and processed for immunofluorescence using antitransferrin receptor monoclonal antibody. Clockwise from top: anti-TF receptor (blue); mAb 35 (green); SHG2210 (red); merged image.

FIGURE 12.5 Molecular imaging of internalized SHG2210:mAb 35 complex in human muscle cells by confocal microscopy SHG2210 fusion protein and mAb 35 were directly labeled with Alexa fluor 546 and 488 dyes, respectively. Fusion protein and antibody were preincubated for 15 min to form complexes and then added to the human muscle cell line DB40 for 2 h on ice. Cells were then washed and switched to 37°C to initiate uptake of the complex. Uptake of complex was allowed to occur for 45 min, at which point cells were fixed and processed for immunofluorescence using anti-Lamp1 monoclonal antibody. Clockwise from top: mAb 35 (green); SHG2210 (red); Lamp-1 (blue); merged image.

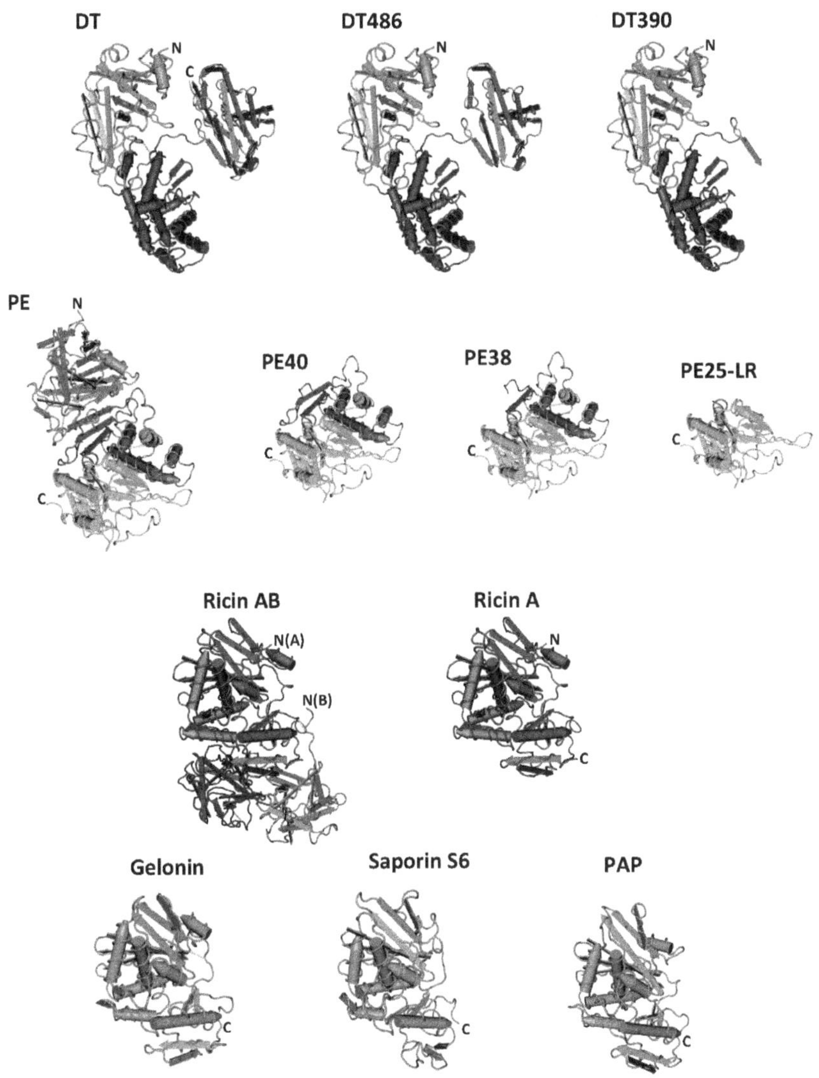

FIGURE 18.1 Backbone diagram of X-ray crystal structures of toxins used in immunotoxin preparation. N and C indicate N and C termini of proteins, respectively. PAP: pokeweed antiviral protein. *Diphtheria toxin*: catalytic domain (magenta), translocation domain (blue), binding domain (brown), DT (residues 1–535), DT486 (residues 1–486), DT390 (residues 1–390). *Pseudomonas exotoxin*: domain 1a (residues 1–252; magenta), domain 1b (residues 365–404; blue), domain II (residues 253–364; brown), domain III (405–613; green), PE (residues 1–613), PE40 (residues 253–613), PE38 (residues 253–364 and residues 381–613), and PE25-LR (residues 274–284 and residues 395–613). *Ricin*: chain A (brown) and chain B (magenta and blue). Ricin, gelonin, saporin, and PAP have a similar structure.

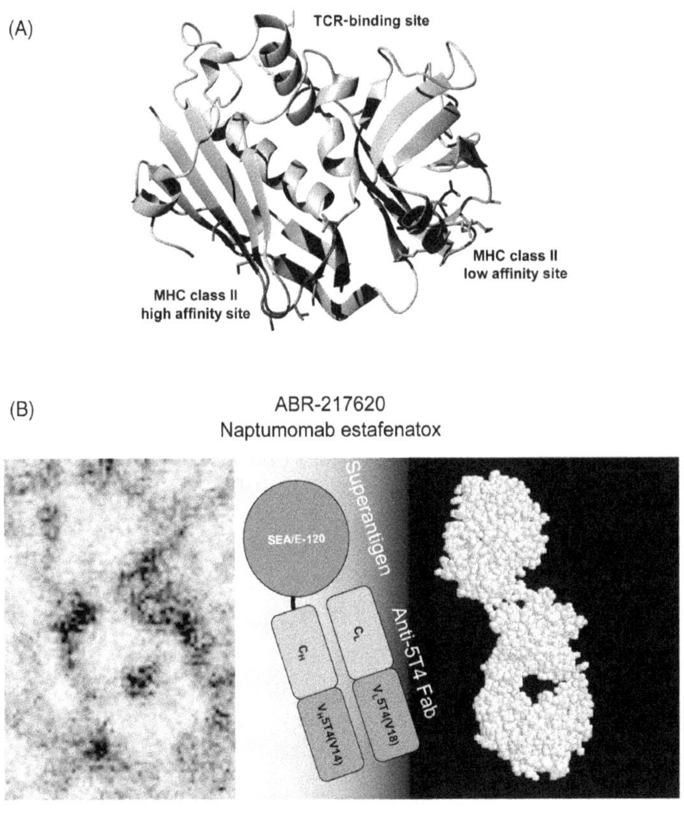

FIGURE 24.1 (A) Model of staphylococcal enterotoxin A (SEA) showing TCR (yellow), MHC class II high affinity (purple), and low affinity (green and red) binding sites. (B) Schematic picture of naptumomab estafenatox (middle), a three-dimensional model (right), and a scanning transmission electron microscopy (STEM) image of the 4×12 nm structure at $1{,}000{,}000 \times$ magnification (left). The STEM image is shown with the courtesy from Dr Tudor Arvinte.

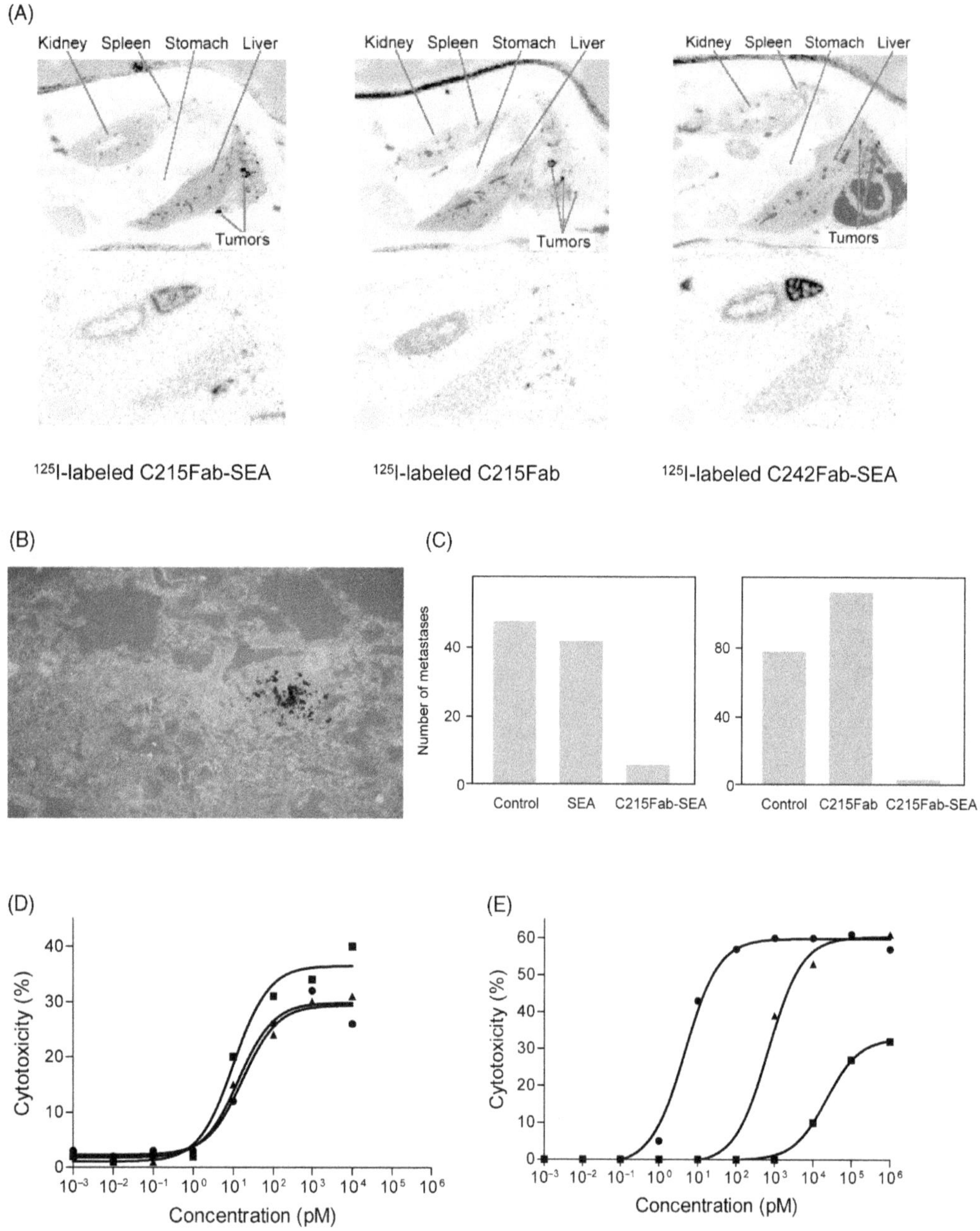

FIGURE 24.3 (A) Radioluminogram showing the localization of TTS/mAb molecules to the targeted tumors (B16 melanoma transfected with EpCAM antigen). The corresponding tissue section is shown above. (B) Tumor-selective recruitment of $CD8^+$ T cell (stained red) by C215Fab-SEA to a B16-EpCAM melanoma (black) in a syngeneic mouse lung. (C) Inhibition of growth of B16-EpCAM melanoma lung metastases after treatment with different TTS/mAb/SAg molecules. (D) SADCC against 5T4-expressing Caki-2 cells mediated by 4-h incubation with 5T4Fab-SEA (circle), anatumomab mafenatox (triangle), or naptumomab estafenatox (square) at the indicated concentrations. (E) SDCC against MHC class II expressing Raji cells mediated by 4-h incubation with 5T4Fab-SEA (circle), anatumomab mafenatox (triangle), or naptumomab estafenatox (square) at the indicated concentrations. *Source*: Permission from Wolters Kluwer Health.

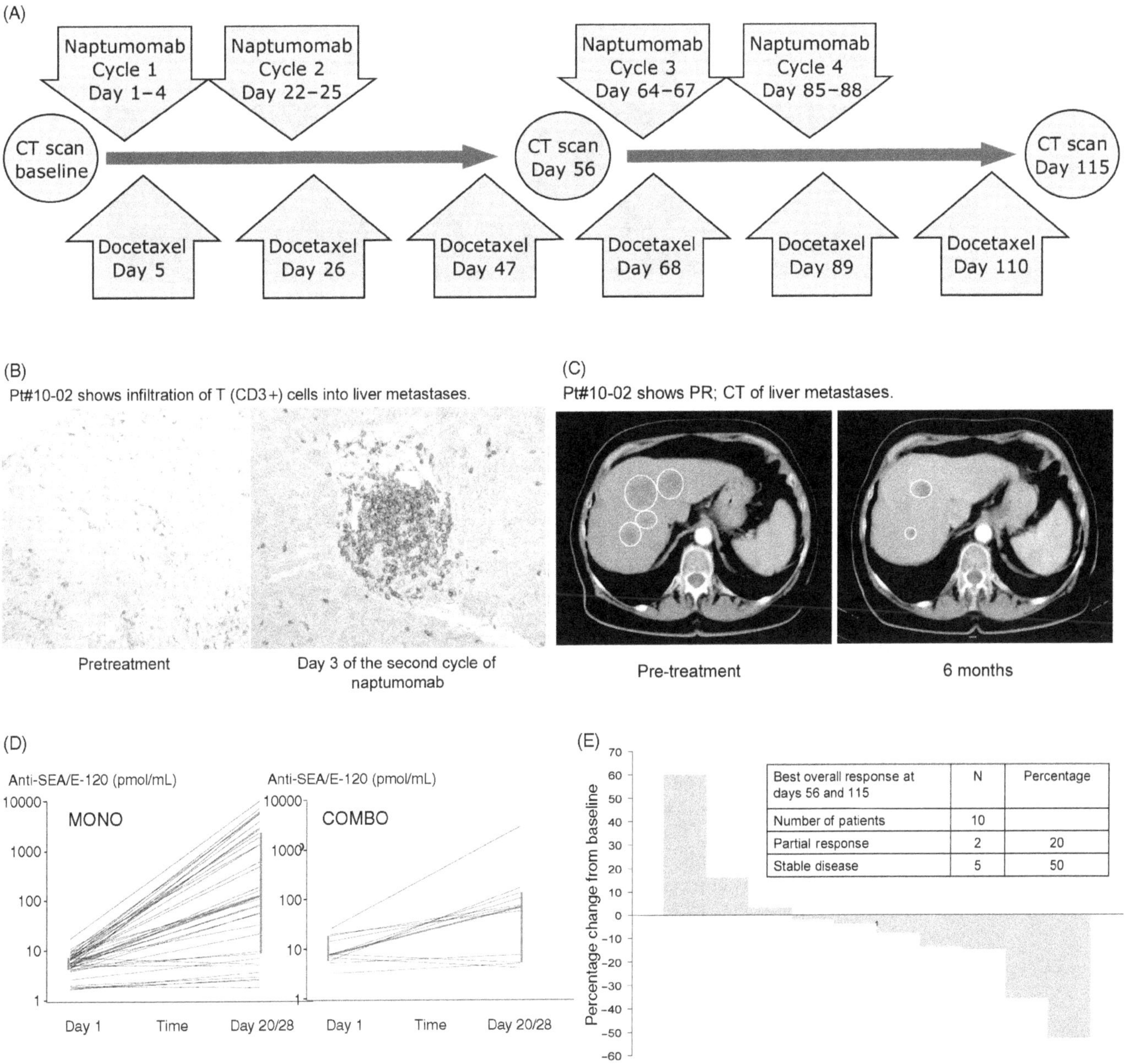

FIGURE 24.6 Naptumomab estafenatox was investigated in combination with docetaxel in a Phase I study with 13 NSCLC patients (COMBO). (A) Study treatment scheme. (B) Immunohistochemical staining (brown) for T lymphocytes (anti-CD3) of paraffin sections of liver metastasis from samples taken pretreatment and day 3 of the second cycle of 22 μg/kg naptumomab estafenatox treatment from a patient with a PR (pt#10-02). This patient had previously received three treatment regimes (including docetaxel), but had a PR after COMBO treatment. (C) CT scans pretreatment and 6 months after naptumomab estafenatox treatment start of pt#10-02. (D) Anti-SEA/E-120 antibody concentrations of individual patients at baseline (day 1) and after 3–4 weeks (Day 20/28); left: population MONO $n = 35$; right: population COMBO $n = 11$; red lines depict median ± quartiles. (E) Best overall tumor response according to RECIST criteria at days 56 and 115 of evaluable patients depicted as a waterfall plot. *Source*: Reprinted from Reference 39 with permission from American Society of Clinical Oncology.

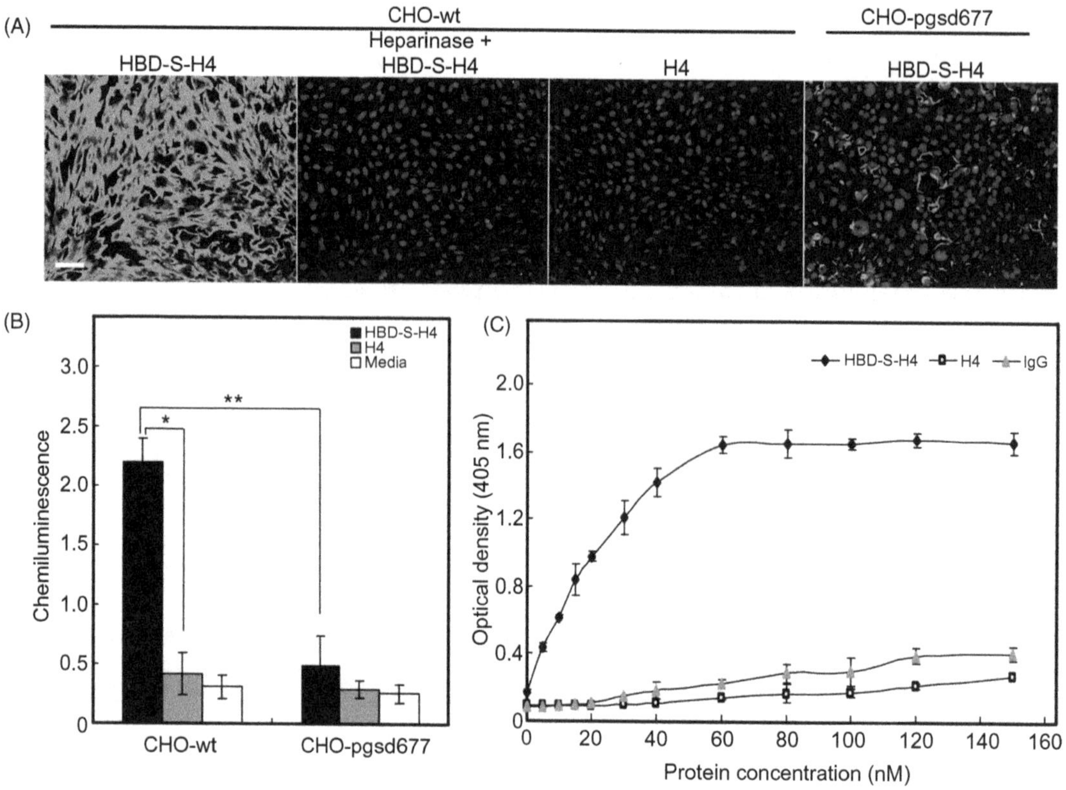

FIGURE 27.4 NRG1's heparin-binding domain targets HBD-S-H4 to cell surfaces through heparan sulfate interactions. (A) CHO-wt and CHO-pgsd677 (mutant) cells, with deficient heparan sulfate synthesis, were incubated with biotinylated HBD-S-H4 or H4 (50 nM) with or without heparinase treatment. Compared to the other conditions, HBD-S-H4 treated CHO-wt cells had the highest green fluorescence intensity that appeared to concentrate in ECM between cells. Scale bar is 50 μm. (B) A binding assay of protein constructs to CHO cells shows the enhanced binding ability of HBD-S-H4 to wild-type CHO cells (CHO-wt), but not to mutant cells. The H4 construct did not bind to these cells. * and ** indicate significant differences at $P < 0.001$ and <0.005, respectively. (C) Increasing concentrations of biotin-conjugated HBD-S-H4, H4, or IgG were added to heparin coated 96-well plates to produce a saturation-binding curve. The amount of HBD-S-H4 bound to heparin increased until saturation of binding sites. A calculated dissociation constant (Kd) is 14.7 nM. *Source*: This research was originally published in J. Biol. Chem. Reference 79.

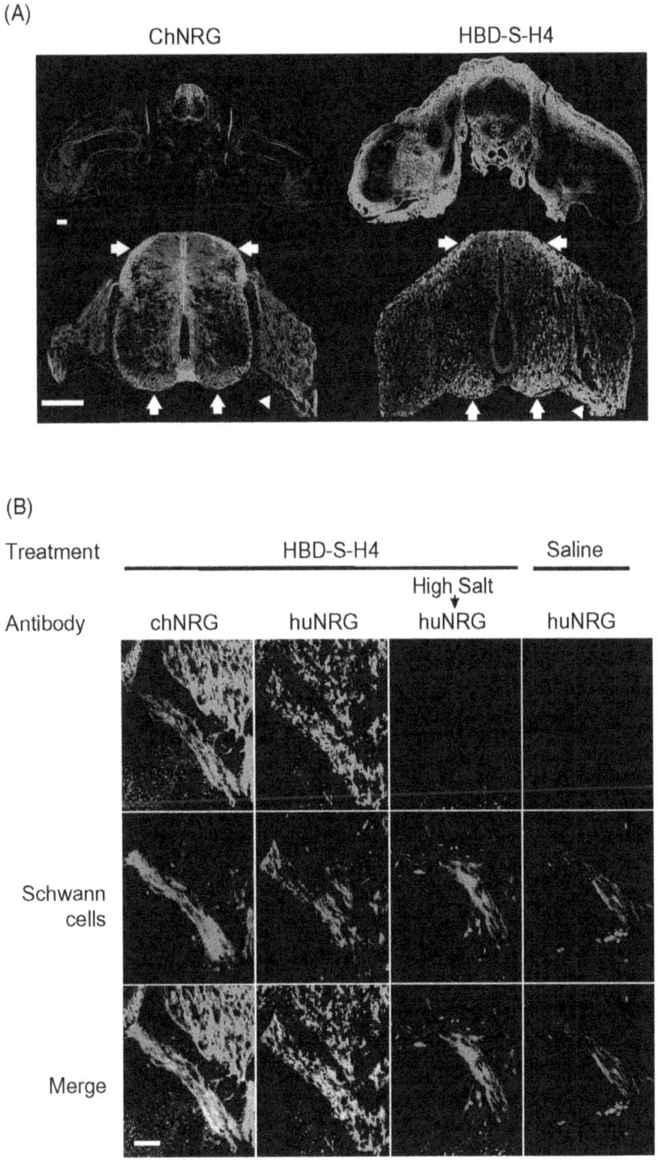

FIGURE 27.5 HBD-S-H4 targets to the same regions of the developing nervous system where NRG accumulates. (A) 20 μg of HBD-S-H4 was added to the chorioallantoic membrane of embryonic chicken embryos. Tissue sections through the spinal cord 2 days later at embryonic day 7 show a comparison between endogenous chicken NRG expression (green, left) and HBD-S-H4 distribution (green, right) at low power (top). Higher power images (bottom panel) focusing on the spinal cord show that HBD-S-H4 adhered to the same regions as endogenous NRG along axonal tracts in the spinal cord (arrows) and along the ventral nerve root (arrowhead). Sections were counterstained for nuclei with DAPI (blue). Scale bars are 200 μm. (B) Both endogenous chicken NRG (chNRG, green) and HBD-S-H4 (huNRG, green) were concentrated in the ventral nerve root identified by a Schwann cell marker (red). High salt (1M NaCl for 1.5 h) treatment on sections of HBD-S-H4 removed the signal in the ventral root. Control treatments with saline did not reveal any immunoreactivity with the same antibody in the ventral nerve root. Scale bar is 50 μm. *Source*: This research was originally published in J. Biol. Chem. Reference 79.

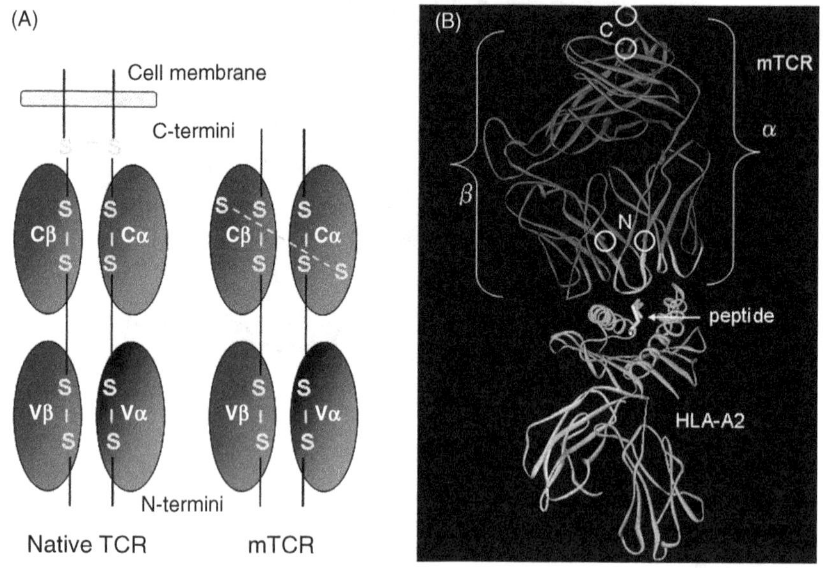

FIGURE 32.2 The monoclonal TCR (mTCR). (A) The design of the mTCR framework. Natural αβ TCRs (left) are disulfide-linked heterodimers, with each chain comprising the N-terminal variable domain and the constant domain, followed by a transmembrane region and a short C-terminal intracellular signal-transducing tail. In the mTCR (right), the chains are truncated, and the natural interchain disulfide bond replaced by an engineered one, linking the two chains within their constant domains, producing an exceptionally stable and fully human soluble TCR molecule. (*See text for full caption.*)

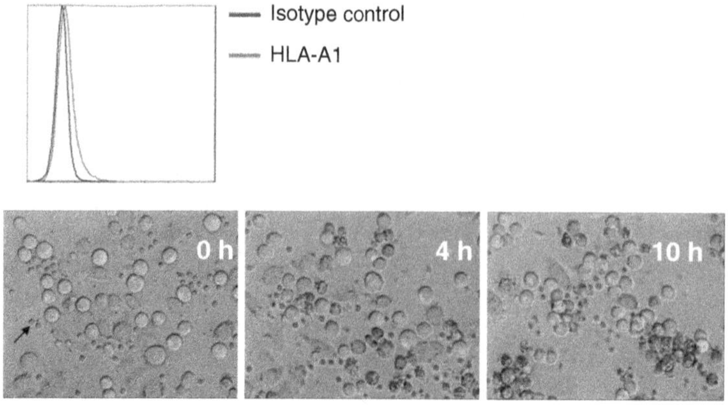

FIGURE 32.6 ImmTAC-MAGE displays high potency despite low ligand densities. Top: The expression of MHC class I on A375 cells. Flow cytometry shows only a low level of HLA-A1 expression (red), against an isotype control (blue). Bottom: Visualization of cell killing by CD8+ T cells in the presence of ImmTAC-MAGE. Time-lapse fluorescence microscopy over a period of 10 h shows progressive lysis of the MAGE-A3 positive A375 tumor cells (large red-stained cells) by the CD8+ T cells (small unstained cells, indicated by the black arrow) in the presence of ImmTAC-MAGE at 50 pM concentration. While the population of the A375 cells disappears entirely by the end of the observation period, the control MAGE-A3 negative cells (Colo205, large green-cells) remain intact. (No effect was registered in the presence of an irrelevant ImmTAC—data not shown.)

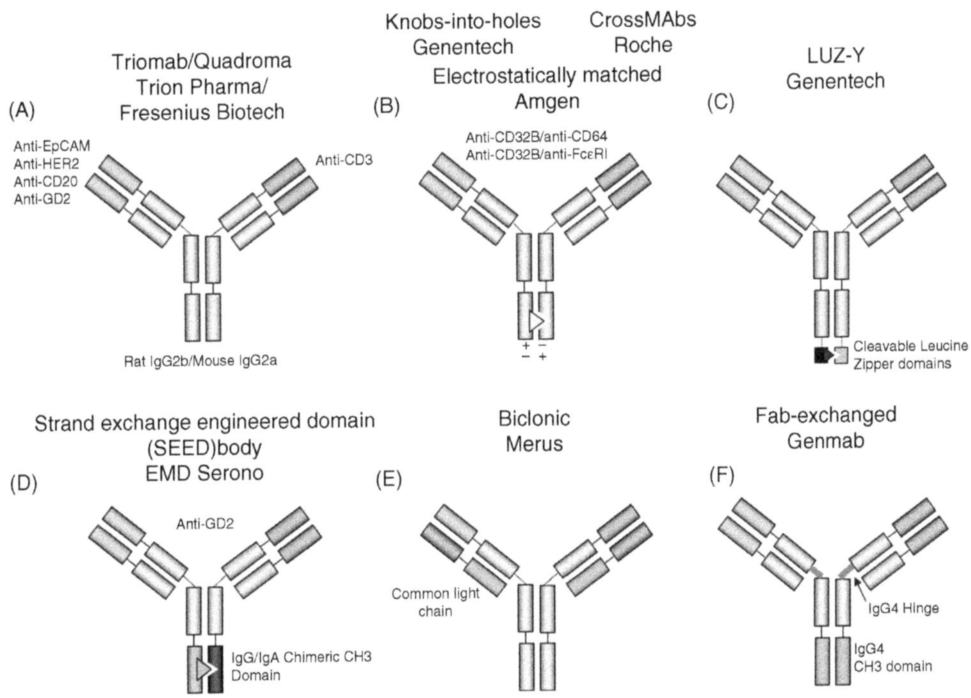

FIGURE 35.1 Asymmetric bispecific antibody formats. Red and green colors highlight variable antibody domains of distinct specificity. Gray and yellow label properly heterodimerized heavy and light chains derived from two different mAbs. Disclosed binding specificities and particular structural features are labeled. Figure 35.1b depicts three closely related but different formats.

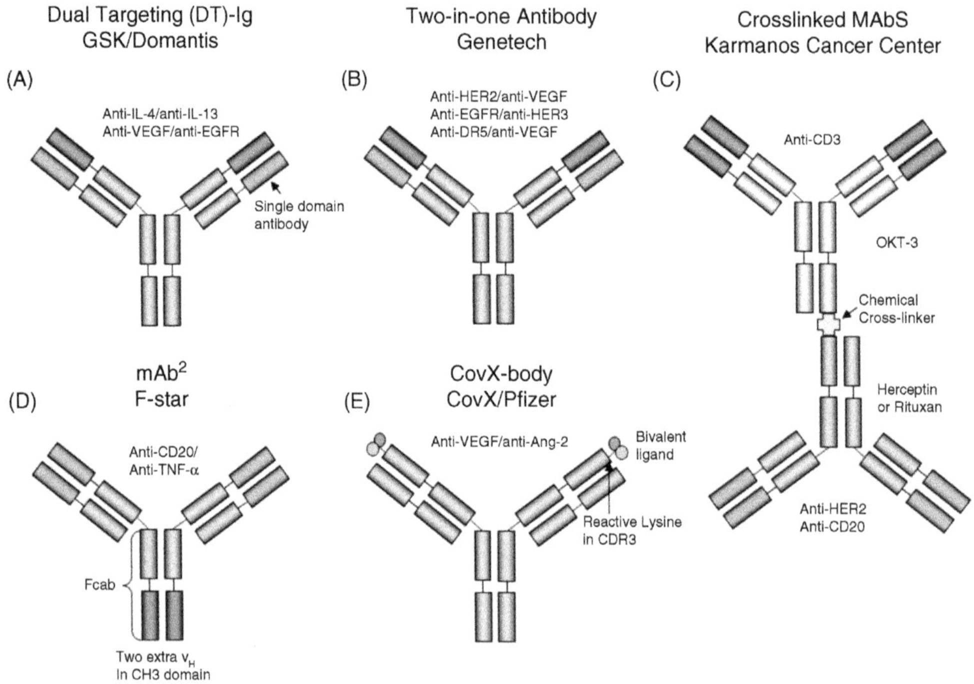

FIGURE 35.2 Symmetric bispecific antibody formats. Red and green colors highlight variable antibody domains of distinct specificity. Disclosed binding specificities and particular structural features are labeled.

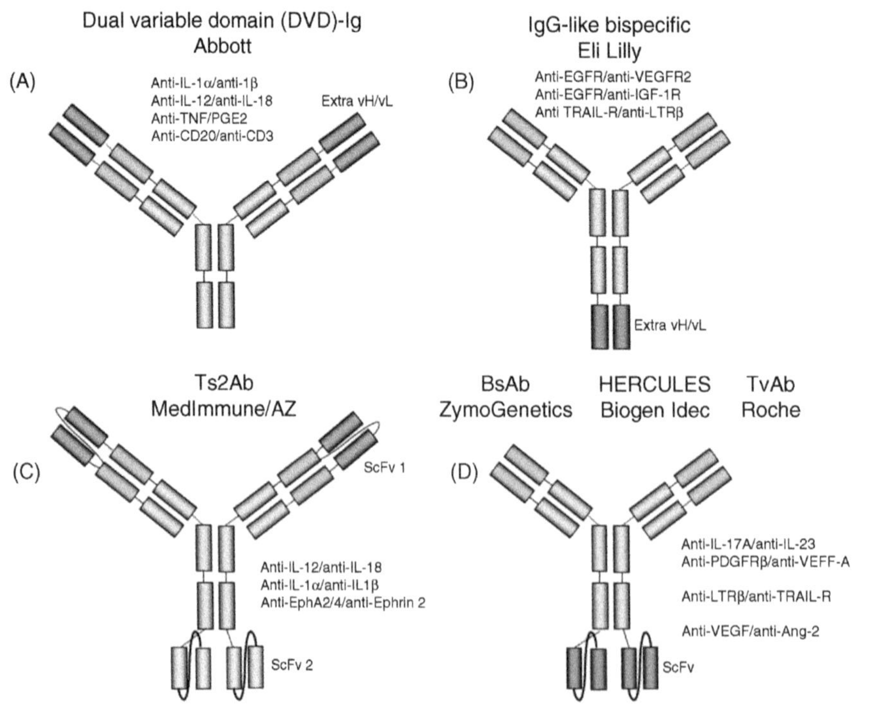

FIGURE 35.3 IgG-like bispecific antibodies with fused antibody fragments. Red and green colors highlight variable antibody domains of distinct specificity. Disclosed binding specificities and particular structural features are labeled. Single-chain antibodies (scFvs) show a peptide linker between variable domains.

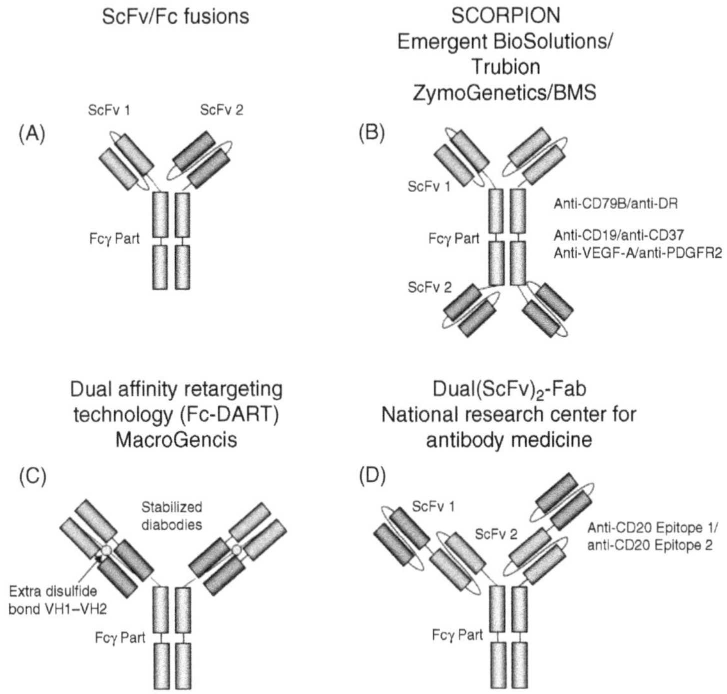

FIGURE 35.4 Bispecific constructs based on the Fcγ part of IgG. Red and green colors highlight variable antibody domains of distinct specificity. The Fcγ part is shown in grey. Disclosed binding specificities and particular structural features are labeled. Single-chain antibodies (scFvs) show a peptide linker between variable domains.

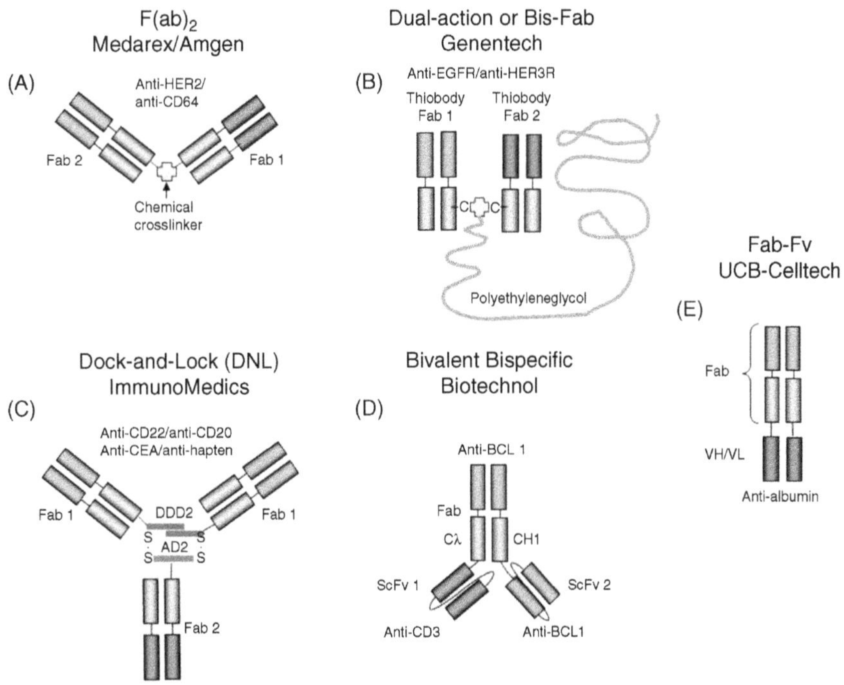

FIGURE 35.5 Bispecific constructs based on Fab fragments. Red and green colors highlight variable antibody domains of distinct specificity. The CH1/Cκ or λ domains are shown in grey. Disclosed binding specificities and particular structural features are labeled. Single-chain antibodies (scFvs) show a peptide linker between variable domains.

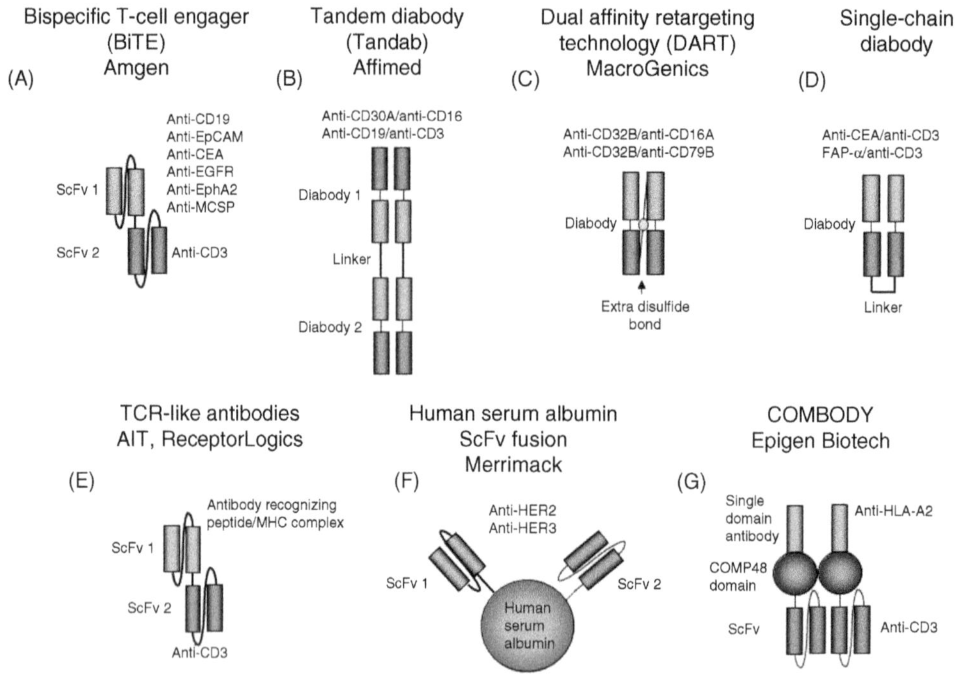

FIGURE 35.6 Bispecific antibody constructs based on diabodies or single-chain antibodies. Red and green colors highlight variable antibody domains of distinct specificity. Disclosed binding specificities and particular structural features are labeled. Single-chain antibodies (scFvs) show a peptide linker between variable domains.

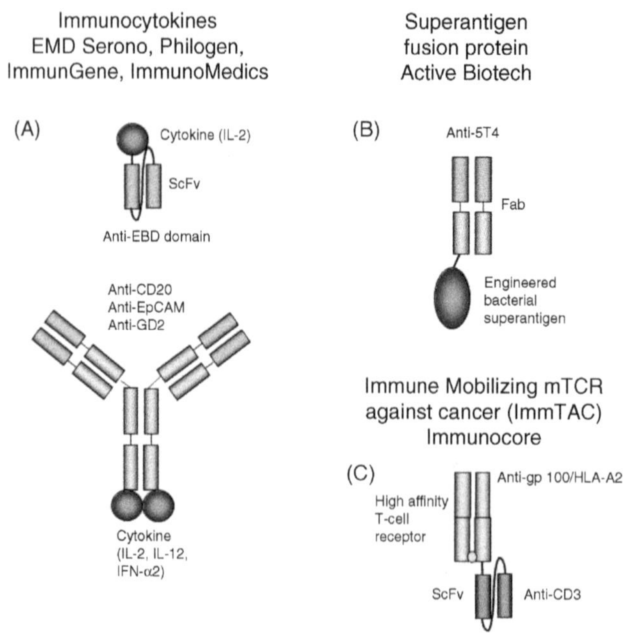

FIGURE 35.7 Bifunctional fusions of antibodies or fragments with unrelated proteins. Disclosed binding specificities and particular structural features are labeled. Single-chain antibodies (scFvs) show a peptide linker between variable domains.

www.ingramcontent.com/pod-product-compliance
Lightning Source LLC
LaVergne TN
LVHW062133200926
848484LV00007B/1810
9780470646274